D1609286

with Cat & Fetal Pig Dissections

LABORATORY MANUAL FOR

HUMAN ANATOMY & PHYSIOLOGY

FIFTH EDITION

CYNTHIA PRENTICE-CRAVER
Chemeketa Community College

TERRY R. MARTIN
Kishwaukee College

Mc Graw Hill

LABORATORY MANUAL FOR HUMAN ANATOMY & PHYSIOLOGY, FIFTH EDITION

This book is printed on acid-free paper.

1 2 3 4 5 6 7 8 9 LMN 27 26 25 24 23 22

ISBN 978-1-260-26520-0 (bound edition)
MHID 1-260-26520-X (bound edition)

Portfolio Manager: *Matthew Garcia*
Product Developer: *Krystal Faust*
Marketing Manager: *Valerie Kramer*
Content Project Managers: *Jeni McAtee, Tammy Juran*
Buyer: *Sandy Ludovissy*
Designer: *David W. Hash*
Content Licensing Specialist: *Lori Hancock*
Cover Image: *101cats/Getty Images*
Compositor: *MPS Limited*

mheducation.com/highered

CONTENTS

Surface Anatomy

Nervous System

General and Special Senses

Endocrine System

Cardiovascular System

Lymphatic System

Respiratory System

*These supplemental exercises are available in the eBook via Connect Anatomy & Physiology and also online for instructor distribution; see Instructor Resources via Connect Library tab.

PREFACE

InTOUCH | WITH Anatomy & Physiology Lab Courses

Author Cynthia Prentice-Craver's twenty-seven years of passion for and experience in teaching human anatomy and physiology, and her commitment to developing curriculum that stimulates student curiosity and enthusiasm, steered her cultivation of this fifth edition laboratory manual. Terry, the original author of this laboratory manual, had forty-year experience teaching anatomy and physiology courses, authored three different laboratory manuals, and was actively involved in the Human Anatomy and Physiology Society (HAPS) which drove his determination to create a laboratory manual with an innovative approach that would benefit students. Cynthia is honored to have had Terry Martin as her mentor and coauthor on previous editions, and she has worked to uphold the reputation and brand while developing this fifth edition laboratory manual. The *Laboratory Manual for Human Anatomy & Physiology* includes sixty-three laboratory exercises, three supplemental labs found online, six cat dissection labs, six fetal pig dissection labs, forty-six Virtual Lab lessons, and thirteen Physiology Interactive Lab Simulations (Ph.I.L.S) lessons. This laboratory manual is written to work well with any anatomy and physiology text.

Martin Lab Manual Series . . .

InTOUCH WITH Anatomy & Physiology Lab Courses

- Anatomy and Physiology REVEALED® icons are found in figure legends. These icons indicate that there is a direct link to APR available in the eBook provided with Connect® for this title.
- Incorporates **learning outcomes and assessments** to help students master important material.
- **Pre-Lab** assignments are printed in the lab manual. They will help students be more prepared for lab and save instructors time during lab.
- **Clear, concise** writing style facilitates more thorough understanding of lab exercises.
- **BIOPAC®** exercises use hardware and software for data acquisition, analysis, and recording.

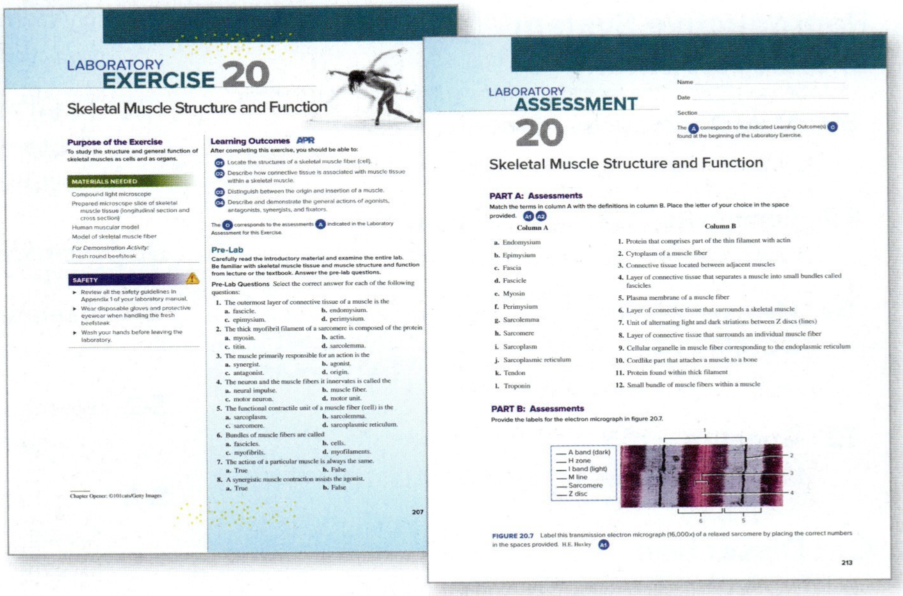

- **NEW! Virtual Labs,** with lessons available in Appendix 4, are simulations that correspond well to many of the laboratory activities in this 5e laboratory manual. The virtual lab lessons can be completed as pre-lab preparation, in-lab or on-line assignments, or post-lab review.
- **NEW!** Cadaver and anatomical models **(APR)** are incorporated throughout the lab manual. These anatomical model images correspond with typical models seen in a laboratory classroom and help students make an association with the cadaver.
- **Ph.I.L.S.** physiology lab simulations, with lessons available in Appendix 4, make otherwise difficult and expensive experiments a breeze through digital simulations.
- **Micrographs** incorporated throughout the lab aid students' visual understanding of difficult topics.
- **Instructor's Guide** is available online for quick and easy use by instructors.

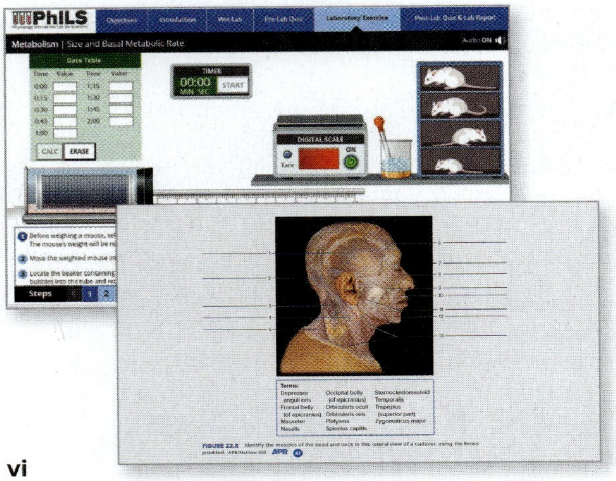

FEATURES OF THIS LABORATORY MANUAL

InTOUCH WITH Student Needs

- The procedures are clear, concise, and easy to follow. Relevant lists and summary tables present the contents efficiently. Histology micrographs and anatomical model and cadaver photos are incorporated in the appropriate locations within the associated labs.
- The pre-lab section includes quiz questions. It also directs the student to carefully read the introductory material and the entire lab to become familiar with its contents. If necessary, a textbook or lecture notes might be needed to supplement the concepts.
- **Terminologia Anatomica** is used as the source for universal terminology in this laboratory manual. Alternative names are included when a term is introduced for the first time.
- Laboratory assessments immediately follow each laboratory exercise.
- Histology photos (micrographs) are placed within the appropriate laboratory exercise.
- A section called "Study Skills for Anatomy and Physiology" is located in the front of this laboratory manual. This section was written by students enrolled in a Human Anatomy and Physiology course.
- Critical Thinking Activities and Assessments are incorporated within most of the laboratory exercises to enhance valuable critical thinking skills that students need throughout their lives.
- Cadaver images are incorporated with dissection labs.
- Images of anatomical models are included within certain laboratory exercises.

InTOUCH WITH Instructor Needs

- The instructor will find digital assets for use in creating customized lectures, visually enhanced tests and quizzes, and other printed support material.
- Some unique labs included are "Scientific Method and Measurements," "Chemistry of Life," "Fetal Skeleton," "Surface Anatomy," "Diabetic Physiology," "Metabolism," and "Genetics."
- The instructor's guide for *Laboratory Manual for Human Anatomy and Physiology, 5e* provides the answers to the pre-lab and assessment questions, as well as an estimated average time schedule.

(hand): 4FR/E+/Getty Images

- One or more laboratory exercise can be completed during a single laboratory session.

InTOUCH WITH Educational Needs

- Learning outcomes with icons **O** have matching assessments with icons **A** so students can be sure they have accomplished the laboratory exercise content. Outcomes and assessments include all levels of learning skills: remember, understand, apply, analyze, evaluate, and create.
- Assessment rubrics for entire laboratory assessments are included in Appendix 3.

InTOUCH WITH Technology

- **Anatomy & Physiology Revealed® 4.0**

  Detailed cadaver photographs blended together with a state-of-the-art layering technique provide a uniquely interactive dissection experience. Certain animations can be viewed to enhance student learning. Cat and fetal pig versions are also available.
- *Virtual Labs* Virtual Labs are included with Connect for this laboratory manual. There are forty-six Virtual Lab lessons located in Appendix 4, including a correlation guide.
- **PhILS** Physiology Interactive Lab Simulations (Ph.I.L.S.) is included with Connect for this laboratory manual. Thirteen lab simulation lessons are located in Appendix 4, including a correlation guide.
- **BIOPAC Systems, Inc.** (Registered to ISO 9001:2000) BIOPAC® exercises are included on four different body systems. BIOPAC® systems use hardware and software for data acquisition, analysis, and recording of information for an individual.
- **Practice ATLAS** Practice Atlas is included with Connect for this laboratory manual. It is an interactive tool that pairs images of anatomical models with cadaver photographs, allowing students to practice naming structures on both models and human bodies. New to this edition are additional labeling exercises that feature anatomical models and cadaver images from Practice Atlas.

GUIDED TOUR THROUGH AN EXERCISE

The laboratory exercises include a variety of special features that are designed to stimulate interest in the subject matter, to involve students in the learning process, and to guide them through the planned activities. These features include the following:

Purpose of the Exercise The purpose provides a statement about the intent of the exercise—that is, what will be accomplished.

Learning Outcomes The learning outcomes list what a student should be able to do after completing the exercise. Each learning outcome will have matching assessments indicated by the corresponding icon (A) in the laboratory assessment.

Materials Needed This section lists the laboratory materials that are required to complete the exercise and to perform the demonstrations and learning extensions.

Safety A list of laboratory safety guidelines is located in Appendix 1 of your laboratory manual. Each lab session that requires special safety guidelines has a safety section. Your instructor might require some modifications of these guidelines.

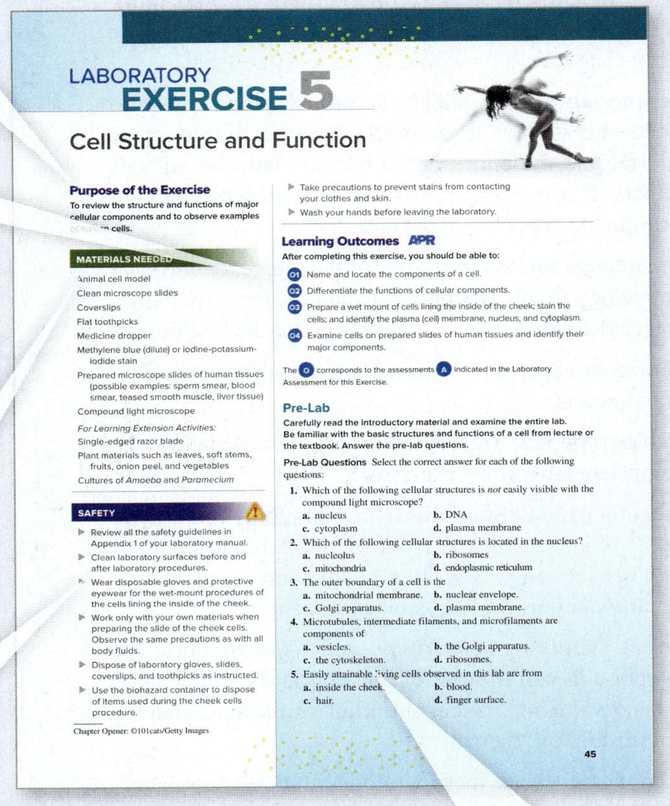

LABORATORY EXERCISE 5

Cell Structure and Function

Purpose of the Exercise

To review the structure and functions of major cellular components and to observe examples of human cells.

MATERIALS NEEDED

Animal cell model
Clean microscope slides
Coverslips
Flat toothpicks
Medicine dropper
Methylene blue (dilute) or iodine-potassium-iodide stain
Prepared microscope slides of human tissues (possible examples: sperm smear, blood smear, teased smooth muscle, liver tissue)
Compound light microscope

For Learning Extension Activities:
Single-edged razor blade
Plant materials such as leaves, soft stems, fruits, onion peel, and vegetables
Cultures of *Amoeba* and *Paramecium*

SAFETY

- Review all the safety guidelines in Appendix 1 of your laboratory manual.
- Clean laboratory surfaces before and after laboratory procedures.
- Wear disposable gloves and protective eyewear for the wet-mount procedures of the cells lining the inside of the cheek.
- Work only with your own materials when preparing the slide of the cheek cells. Observe the same precautions as with all body fluids.
- Dispose of laboratory gloves, slides, coverslips, and toothpicks as instructed.
- Use the biohazard container to dispose of items used during the cheek cells procedure.
- Take precautions to prevent stains from contacting your clothes and skin.
- Wash your hands before leaving the laboratory.

Chapter Opener: ©101cats/Getty Images

Learning Outcomes APR

After completing this exercise, you should be able to:

O1 Name and locate the components of a cell.
O2 Differentiate the functions of cellular components.
O3 Prepare a wet mount of cells lining the inside of the cheek; stain the cells; and identify the plasma (cell) membrane, nucleus, and cytoplasm.
O4 Examine cells on prepared slides of human tissues and identify their major components.

The (O) corresponds to the assessments (A) indicated in the Laboratory Assessment for this Exercise.

Pre-Lab

Carefully read the introductory material and examine the entire lab. Be familiar with the basic structures and functions of a cell from lecture or the textbook. Answer the pre-lab questions.

Pre-Lab Questions Select the correct answer for each of the following questions:

1. Which of the following cellular structures is *not* easily visible with the compound light microscope?
   a. nucleus b. DNA
   c. cytoplasm d. plasma membrane
2. Which of the following cellular structures is located in the nucleus?
   a. nucleolus b. ribosomes
   c. mitochondria d. endoplasmic reticulum
3. The outer boundary of a cell is the
   a. mitochondrial membrane. b. nuclear envelope.
   c. Golgi apparatus. d. plasma membrane.
4. Microtubules, intermediate filaments, and microfilaments are components of
   a. vesicles. b. the Golgi apparatus.
   c. the cytoskeleton. d. ribosomes.
5. Easily attainable living cells observed in this lab are from
   a. inside the cheek. b. blood.
   c. hair. d. finger surface.

45

Pre-Lab The pre-lab includes quiz questions and directs the student to carefully read introductory material and examine the entire laboratory contents after becoming familiar with the topics from a textbook or lecture. After successfully answering the pre-lab questions, the student is prepared to become involved in the laboratory exercise.

Introduction The introduction describes the subject of the exercise or the ideas that will be investigated. It includes all of the information needed to perform the laboratory exercise.

Procedure The procedure provides a set of detailed instructions for accomplishing the planned laboratory activities. Usually these instructions are presented in outline form so that a student can proceed efficiently through the exercise in stepwise fashion.

The procedures, often presented in parts, include a wide variety of laboratory activities and, from time to time, direct the student to complete various tasks in the laboratory assessments.

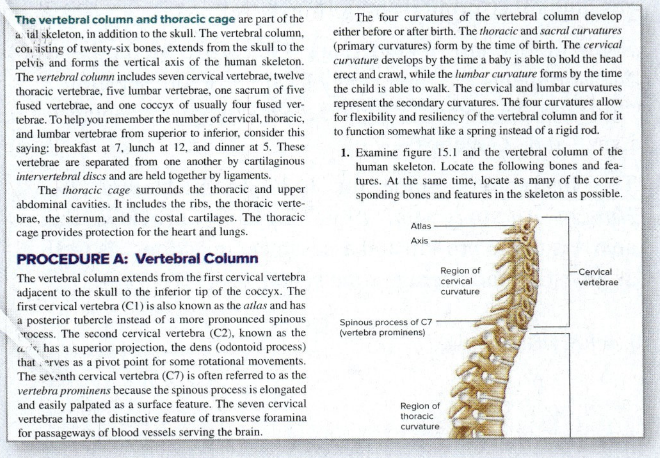

The vertebral column and thoracic cage are part of the axial skeleton, in addition to the skull. The vertebral column, consisting of twenty-six bones, extends from the skull to the pelvis and forms the vertical axis of the human skeleton. The *vertebral column* includes seven cervical vertebrae, twelve thoracic vertebrae, five lumbar vertebrae, one sacrum of five fused vertebrae, and one coccyx of usually four fused vertebrae. To help you remember the number of cervical, thoracic, and lumbar vertebrae from superior to inferior, consider this saying: breakfast at 7, lunch at 12, and dinner at 5. These vertebrae are separated from one another by cartilaginous *intervertebral discs* and are held together by ligaments.

The *thoracic cage* surrounds the thoracic and upper abdominal cavities. It includes the ribs, the thoracic vertebrae, the sternum, and the costal cartilages. The thoracic cage provides protection for the heart and lungs.

PROCEDURE A: Vertebral Column

The vertebral column extends from the first cervical vertebra adjacent to the skull to the inferior tip of the coccyx. The first cervical vertebra (C1) is also known as the *atlas* and has a posterior tubercle instead of a more pronounced spinous process. The second cervical vertebra (C2), known as the *axis*, has a superior projection, the dens (odontoid process) that serves as a pivot point for some rotational movements. The seventh cervical vertebra (C7) is often referred to as the *vertebra prominens* because the spinous process is elongated and easily palpated as a surface feature. The seven cervical vertebrae have the distinctive feature of transverse foramina for passageways of blood vessels serving the brain.

The four curvatures of the vertebral column develop either before or after birth. The *thoracic* and *sacral curvatures* (primary curvatures) form by the time of birth. The *cervical curvature* develops by the time a baby is able to hold the head erect and crawl, while the *lumbar curvature* forms by the time the child is able to walk. The cervical and lumbar curvatures represent the secondary curvatures. The four curvatures allow for flexibility and resiliency of the vertebral column and for it to function somewhat like a spring instead of a rigid rod.

1. Examine figure 15.1 and the vertebral column of the human skeleton. Locate the following bones and features. At the same time, locate as many of the corresponding bones and features in the skeleton as possible.

Atlas
Axis
Region of cervical curvature
Cervical vertebrae
Spinous process of C7 (vertebra prominens)
Region of thoracic curvature

Demonstration Activities Demonstration activities appear in separate boxes. They describe specimens, specialized laboratory equipment, or other materials of interest that an instructor may want to display to enrich the student's laboratory experience.

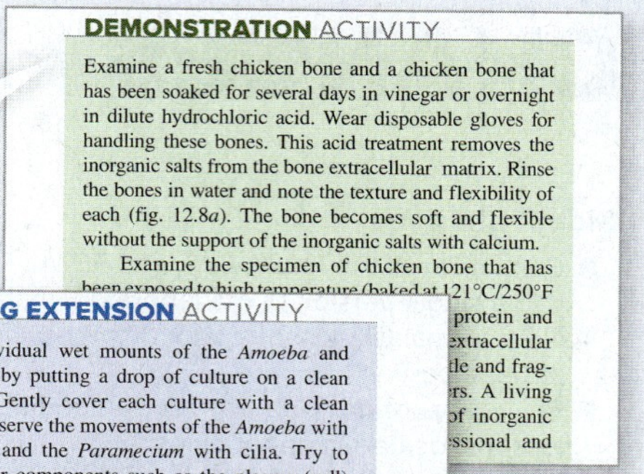
DEMONSTRATION ACTIVITY

Examine a fresh chicken bone and a chicken bone that has been soaked for several days in vinegar or overnight in dilute hydrochloric acid. Wear disposable gloves for handling these bones. This acid treatment removes the inorganic salts from the bone extracellular matrix. Rinse the bones in water and note the texture and flexibility of each (fig. 12.8*a*). The bone becomes soft and flexible without the support of the inorganic salts with calcium.

Examine the specimen of chicken bone that has

Learning Extension Activities Learning extension activities also appear in separate boxes. They encourage students to extend their laboratory experiences. Some of these activities are open-ended in that they suggest the student plan an investigation or experiment and carry it out after receiving approval from the laboratory instructor. Some of the figures are illustrated as line art or in grayscale. This will allow colored pencils to be used as a visual learning activity to distinguish various structures.

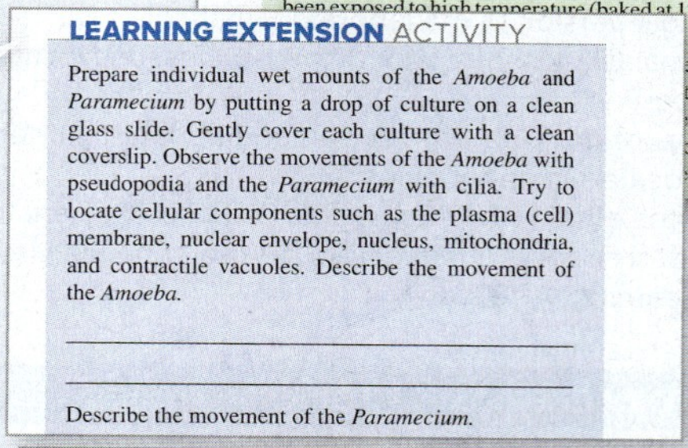
LEARNING EXTENSION ACTIVITY

Prepare individual wet mounts of the *Amoeba* and *Paramecium* by putting a drop of culture on a clean glass slide. Gently cover each culture with a clean coverslip. Observe the movements of the *Amoeba* with pseudopodia and the *Paramecium* with cilia. Try to locate cellular components such as the plasma (cell) membrane, nuclear envelope, nucleus, mitochondria, and contractile vacuoles. Describe the movement of the *Amoeba*.

Describe the movement of the *Paramecium*.

Histology Histology photos (micrographs) are placed within the appropriate exercise.

Illustrations Diagrams similar to those in a textbook often are used as aids for reviewing subject matter. Other illustrations provide visual instructions for performing steps in procedures or are used to identify parts of instruments or specimens. Micrographs are included to help students identify microscopic structures or to evaluate student understanding of tissues.

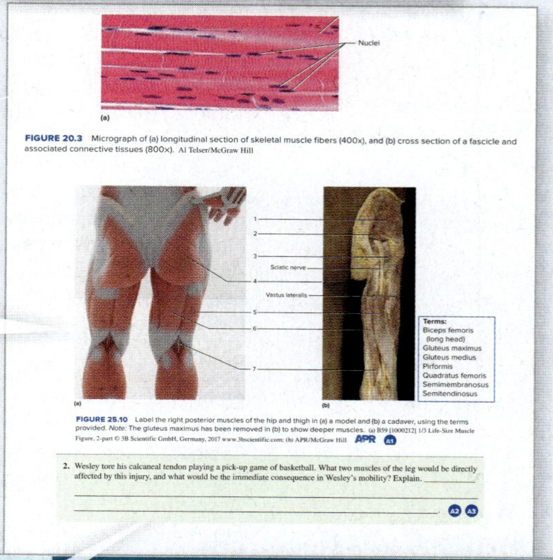

Laboratory Assessments A laboratory assessment form to be completed by the student immediately follows each exercise. These assessments include various types of review activities, spaces for sketches of microscopic specimens, tables for recording observations and experimental results, and questions dealing with the analysis of such data. Critical Thinking Assessments enhance higher-order thinking skills.

As a result of these activities, students will develop a better understanding of the structural and functional characteristics of their bodies and will increase their skills in gathering information by observation and experimentation. By completing all of the assessments, students will be able to determine if they were able to accomplish all of the learning outcomes.

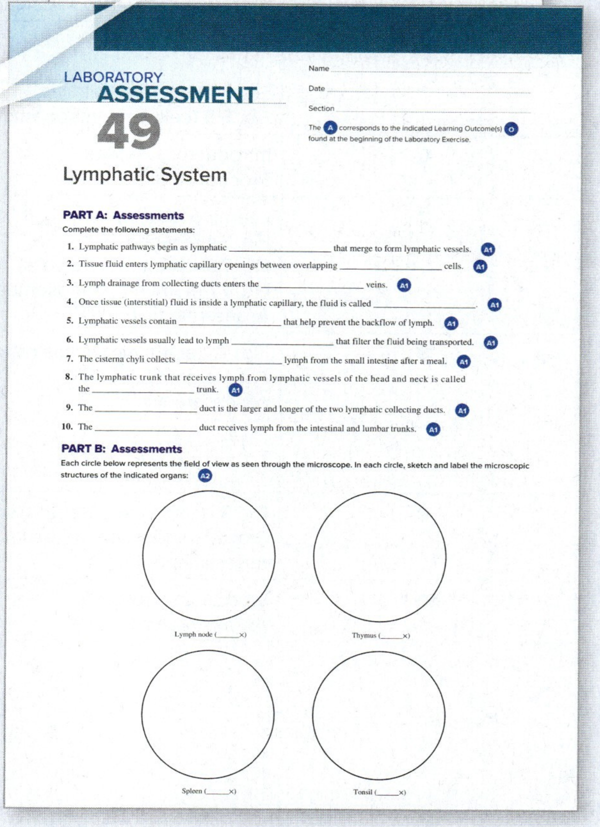
LABORATORY ASSESSMENT 49

Name
Date
Section

Lymphatic System

PART A: Assessments

Complete the following statements:

1. Lymphatic pathways begin as lymphatic ______ that merge to form lymphatic vessels.
2. Tissue fluid enters lymphatic capillary openings between overlapping ______ cells.
3. Lymph drainage from collecting ducts enters the ______ veins.
4. Once tissue (interstitial) fluid is inside a lymphatic capillary, the fluid is called ______.
5. Lymphatic vessels contain ______ that help prevent the backflow of lymph.
6. Lymphatic vessels usually lead to lymph ______ that filter the fluid being transported.
7. The cisterna chyli collects ______ lymph from the small intestine after a meal.
8. The lymphatic trunk that receives lymph from lymphatic vessels of the head and neck is called the ______ trunk.
9. The ______ duct is the larger and longer of the two lymphatic collecting ducts.
10. The ______ duct receives lymph from the intestinal and lumbar trunks.

PART B: Assessments

Each circle below represents the field of view as seen through the microscope. In each circle, sketch and label the microscopic structures of the indicated organs:

Lymph node (____x) Thymus (____x)

Spleen (____x) Tonsil (____x)

CHANGES TO THIS FIFTH EDITION

Global Changes

- Consolidated the MAIN, CAT, and FETAL PIG dissection versions into a single version of the laboratory manual.
- NEW Virtual Lab lessons added; forty-six lessons located in appendices.
- Ph.I.L.S. laboratory lessons revised and updated; thirteen lessons located in appendices.
- New photos added of micrographs, Practice Atlas anatomical models, and cadaver images from APR to enhance visuals for student learning. APR icons.
- Moved location of figure number and caption to below the illustration or image.
- Critical Thinking Assessment questions have added to their title the assessment Part in which they are located.

LABORATORY EXERCISE	TOPIC	CHANGE
2	Introductory material Procedure A (body cavities and membranes) Fig. 2.2*a* and 2.2*b* (thoracic serous membranes) Fig. 2.3 (abdominal serous membranes) Table 2.1 (directional terms meaning) Fig. 2.7 (planes) Fig. 2.9 (body surface regions) Fig. 2.11 (body cavities of head) Fig. 2.12 (serous membranes of heart) Fig. 2.13 (abdominal cavity serous membrane) Assessments: Part C Fig. 2.14 (planes) Fig. 2.15 (regions)	Revised components and improved depth Removed content from labels and added them to text Removed content from labels and added them to text Removed content from labels and added them to text Added terms Changed/revised figure Changed location of labels New figure Changed figure New figure Added questions Changed/revised figure New figure
3	Pre-Lab	Added question
4	Learning Outcomes Fig. 4.1 (microscope) Table 4.1 (microscope parts and their function) Fig. 4.5 (focus on objectives) Assessments: Part A Fig. 4.8 (microscope) Assessments: Part D Matching	Revised Learning Outcome 01 Changed/revised figure Revised table Cropped figure Added letter *e* directions to step 9 text Changed/revised figure Edited certain terms in Column A
5	Introductory material Fig. 5.4*b* (cell model) Fig. 5.5 (cellular components)	Provided examples of membranous and nonmembranous organelles New figure Changed/revised figure
6	Introductory material Procedure B (osmosis)	Revised part on filtration Revised components
7	Learning Outcomes Fig. 7.4 (anaphase) Fig. 7.6 (onion root tip cells) Fig. 7.7 (human chromosomes) Assessments: Part D	Revised Learning Outcome 01 Added (b) micrograph Replaced images; revised legend Replaced (b) image New CTA question
8	Fig. 8.1 (epithelial tissues histology) Assessments: Part B Assessments: Part C	Replaced (i) micrograph Revised questions; added question New CTA question Revised leader lines Revised field of view circles for drawings Added questions
9	Fig. 9.1 (connective tissues histology) Procedure (connective tissues) Assessments: Part B	Replaced (d) micrograph Expanded directions in steps 3 and 4 Added question
10	Introductory material Assessments: Part B	Revised components New CTA question

LABORATORY EXERCISE	TOPIC	CHANGE
11	Introductory material Fig. 11.2 (skin and hypodermis) Table 11.1 (layers of epidermis) Procedure (integumentary system) Fig. 11.6 (glands) Fig. 11.8 (strata of thick skin) Assessments: Part A	Revised components Added hair matrix label Added components Expanded directions in step 7 New figure with micrographs (a) and (b) New figure with model and micrograph images New CTA question
12	Introductory material Assessments: Part A Assessments: Part C	Improved depth of cells Added question New CTA question
13	Fig. 13.3 (major bones of skeleton) Assessments: Part A Assessments: Part C Assessments: Part D	Replaced illustrations Added question Added term and definition to number 1; new CTA question Renumbered 9-14
14	Learning Outcome 04 Procedure (skull) Fig. 14.9 (paranasal sinuses) Assessments: Part B Assessments: Part C Fig. 14.14 (paranasal sinuses) Fig. 14.15 (orbital bones)	Revised for clarity Added bone feature to number 3 Replaced illustrations Added bone feature Rearranged questions; added question to number 1; added a number 2 New figure with radiograph image New figure
15	Learning Outcomes 01 and 04 Procedure A (vertebral column) Fig. 15.8 (vertebral column)	Changed wording Modified components Replaced illustration with image; revised/added features to terms
16	Pre-Lab Introductory material Procedure A (pectoral girdle) Procedure B (upper limb) Fig. 16.9 (left shoulder radiograph) Assessments: Part D	Added question Revised components Revised and expanded components Added components; new CTA question Replaced radiograph image New CTA question
17	Introductory material Procedure A (pelvic girdle) Fig. 17.1 (bones of the pelvis) Procedure B (lower limb) Fig. 17.3 (femur and patella) Fig. 17.4 (tibia and fibula) Fig. 17.6 (male and female pelves) Fig. 17.7 (radiograph of pelvic region) Fig. 17.10 (features of coxal bone)	Expanded components Added components Added (a) anterolateral view; revised (b) anterosuperior view Added components Added label Added label New figure Added term to label Replaced illustrations with images; moved position of figure
19	Purpose of this Exercise Fig. 19.8 (knee joint) Assessments: Part E	Revised and expanded components New figure New CTA question
20	Pre-Lab Introductory material Fig. 20.3 (skeletal muscle) Procedure (skeletal muscle structure and function)	Added question Expanded components Added (a) longitudinal section micrograph Revised components
21	Introductory material and procedures	Updated branding and equipment terms
22	Pre-Lab Fig. 27.7 (anterior head and neck muscles)	Added question Added terms to (a); added new image of model as (b)
23	Pre-Lab Fig. 23.9 (posterior muscles) Fig. 23.11 (forearm muscles)	Added questions Modified legend; added new image of model as (b) Added new image of posterior muscles as (b)

LABORATORY EXERCISE	TOPIC	CHANGE
24	Pre-Lab Procedure B (muscles of the pelvic floor) Table 24.3 (pelvic floor) Fig. 24.5 (deep back) Assessments: Part B Fig. 24.7 (pelvic floor) Assessments: Part C	Added question Organized muscles Expanded components Added label; renumbered Improved clarity to numbers 2 and 9 New figure Added questions
25	Pre-Lab Introductory material Fig. 25.9 (anterior hip and thigh) Fig. 25.10 (posterior hip and thigh) Assessments: Part B Fig. 25.11 (lateral leg)	Added question Revised components Added new cadaver image as (b); added new terms box Added new image of model in (a) Added muscle and action Added new image of model in (b)
26	Fig. 26.2 (anterior head and neck) Fig. 26.3 (anterior upper torso) Fig. 26.4 (lateral upper shoulder and upper limb) Fig. 26.5 (anterior and posterior upper limb) Fig. 26.7 (plantar foot) Fig. 26.8 (lower limb) Table 26.1 (bony and soft tissue surface)	Replaced image of (a) Replaced image of (a); moved and removed some labels Replaced image Replaced images of (a) and (b) Replaced image Replaced images of (a), (b), and (c) Added three terms
27	Pre-Lab Fig. 27.8 Assessments: Part A Assessments: Part D	Added question Replaced image Added term and description New CTA question
28	Pre-Lab Introductory material Fig. 28.5 (spinal cord) Table 28.1 (nerve plexuses) Fig. 28.11 (lumbar and sacral plexuses) Fig. 28.15 (spinal cord cross-section model) Fig. 28.16 (nerves and plexi of arm and leg models) Assessments: Part D	Added question Revised components Added image of cadaver as (b) Removed and expanded components Added labels New figure New figure Added question
29	Pre-Lab	Added question
30	Pre-Lab Introductory material Procedure A: Meninges and brain Fig. 30.2 (cadaver brain) Fig. 30.12 (facial cranial nerve) Fig. 30.13 (median section brain) Fig. 30.14 (median section brain model)	Added question Expanded components Changed title; expanded components in steps 3, 4, and 5 Added image as (b) Replaced (b) with line art Added terms and renumbered New figure
31A	Introductory material and procedures	Updated branding and equipment terms
31B	Introductory material and procedures	Updated branding and equipment terms
32	Pre-Lab Fig. 32.5 (median section sheep brain) Assessments: Part A Fig. 32.8 (dorsal view sheep brain); Fig. 32.9 (midbrain and pineal gland); Fig. 32.10 (ventral view sheep brain)	Added question Revised labels Removed question; new CTA question New figures
34	Pre-Lab Assessments: Part C Assessments: Part D	Added question Added questions New CTA question
35	Introductory material Procedure A (structure and function of the eye) Fig. 35.12 (sagittal section of eye) Fig. 35.13 (eye model) Fig. 35.14 (sagittal section eye micrograph) Fig. 35.15 (tunics of eye micrograph) Assessments: Part B Fig. 35.16 (cow eye internal structures) and Fig. 35.17 (cow eye tunics)	Expanded components Expanded components in step 3 Added label New figures (a) and (b) Replaced image New figure New CTA question New figures

LABORATORY EXERCISE	TOPIC	CHANGE
36	Purpose of the Exercise Introductory material	Revised components Expanded components
37	Procedure A (structure and function of the ear) Fig. 37.3 (middle and inner ear) Fig. 37.10 (ear model) Assessments: Part C	Expanded components in step 2 Added label New figure New CTA question
38	Pre-Lab Assessments: Part A	Added question Added questions
39	Purpose of the Exercise Pre-Lab Table 39.1 (hypothalamic hormones) Table 39.2 Procedure (endocrine gland and histology) Fig. 39.8 (parathyroid gland) Fig. 39.3 (ovary) Assessments: Part B Fig. 39.16 (major endocrine organs) Assessments: Part C Assessments: Part D	Revised components Added questions New table Expanded components Revised components of pituitary gland (c), thyroid gland (b), parathyroid gland (b), adrenal gland (a), and ovary (b) Replaced micrograph Added antral label to tertiary Added question New figure Added questions New CTA question
40	Pre-Lab	Added question
41	Introductory material Table 41.1 (cellular components of blood) Table 41.2 (differential WBC count) Assessment: Part C	Revised components Expanded component Revised component New CTA question
42	Pre-Lab	Added questions
43	Pre-Lab Introductory material Fig. 43.3 (agglutinated RBCs) Fig. 43.5 (possible results of ABO test) Assessments: Part B Assessments: Part D Assessments: Part E	Added question Revised components Replaced micrograph Replaced image Added ABO blood testing results table Added Rh blood testing results table Moved Fig. 43.7 and results in separate assessment part; new CTA question
44	Pre-Lab Procedure A (the human heart) Procedure B (dissection of a sheep heart) Fig. 44.11 (frontal section of sheep heart) Fig. 44.13 (anterior heart region of cadaver) Fig. 44.14 (frontal section of heart model) Fig. 44.15 (exterior and interior features of heart model) Assessments: Part B	Revised question number 6 and added a question Revised and expanded components of steps 4 and 5 Revised and expanded components of steps 2, 9, and 10 Revised and added labels Revised and added labels Added label New figure New CTA questions
45	Pre-Lab Introductory material Table 45.1 (ECG components) Assessments: Part A and Part C Fig. 45.7 (heart chambers and conduction system)	Revised question numbers 3 and 5; added question Revised components Revised and expanded components Added questions Revised components
46	Materials Needed and Procedure A (setup) Introductory material	Updated branding and equipment terms Revised components
47	Learning Outcome 02 Pre-Lab Fig. 47.2 (neurovascular bundle) Fig. 47.9 (arteries chest and neck) Fig. 47.10 (arteries of shoulder and arm); fig. 47.15 (veins of shoulder and arm) Fig. 47.17 (veins of pelvis and lower limb) Assessments: Part C Assessments: Part D Fig. 47.20 (blood vessels of models)	Revised wording Revised question number 3; added questions Added new micrograph as (b) New figure Replaced illustrations Revised illustration Added question Added question; new CTA question New figure

LABORATORY EXERCISE	TOPIC	CHANGE
48	Pre-Lab Introductory material	Added questions Revised components
49	Learning Outcomes Pre-Lab Introductory material Fig. 49.9 (thymus) Assessments: Part A, Part C, and Part D	Revised Learning Outcomes 01 and 03 Added question Revised and added components Replaced micrograph Added questions
50	Fig. 50.6 (trachea and surrounding organs and tissues) Fig. 50.7 (tracheal wall) Fig. 50.11 (respiratory system) Fig. 50.12 (upper respiratory) Assessments: Part B	New figure Revised to form (a) and (b) micrographs Added image of model as (b) Replaced illustration with image of model as (a) and cadaver as (b) New CTA question
51	Introductory material Assessments: Part A Assessments: Part B	Revised components Added questions; new CTA question Revised test results table expected values
52	Materials Needed and Procedure A (setup)	Updated branding and equipment terms
53	Pre-Lab Assessments: Part A	Added question Added question
54	Pre-Lab Fig. 54.5 (esophagus) Procedure B (pharynx and esophagus) Fig. 54.6 (stomach features) Fig. 54.7 (stomach mucosa) Fig. 54.11 (microscopic liver) Procedure E (small and large intestines) Fig. 54.15 (small intestine histology) Fig. 54.16 (large intestine histology) Fig. 54.19 (abdominal digestive organs); Fig. 54.21 (villi of small intestine) Fig. 54.22 (large intestine)	Added question Replaced and added micrographs as (a) and (b) Revised and added components of step 4 Added label Added new micrograph of chief and parietal cells as (b) Added new micrograph of hepatic triad as (c) Revised components in step number 1; added components in step number 2 Added label in (a); added new micrograph of villus as (b) Added new micrograph of mucous gland as (b) New figures Added label
55	Pre-Lab	Added question
56	Pre-Lab Introductory material Assessments: Part C	Revised question number 1; added questions Revised and added components Revised questions 1, 3, and 4
57	Learning Outcomes Pre-Lab Introductory material Procedure A (kidney) Procedure C (ureter, urinary bladder, urethra) Fig. 57.8 (urinary bladder histology) Fig. 57.11 (parts of a nephron); Fig. 57.12 (blood vessels of kidney model); Fig. 57.13 (urinary organs and associated blood vessels model)	Revised Learning Outcomes 01, 02, and 05 Added question; revised question numbers 3 and 6 Revised and added components Revised and added components in step 2 Revised components Added new micrograph as (b) New figures
58	Pre-Lab Assessments: Part A	Added questions New CTA question
59	Pre-Lab Procedure A (male reproductive organs) Fig. 59.8 (male reproductive system)	Added question Revised components of step 2 Added new image of model as (b)
60	Introductory material Procedure B (microscopic anatomy) Fig. 60.7 (ovarian cortex histology) Fig. 60.10 (female reproductive system) Fig. 60.11 (female reproductive structures) Assessments: Part A Fig. 60.12 (ovary) Fig. 60.13 (ovary histology)	Revised and expanded components Revised and expanded components in steps 3 and 8 Added label Added new image of model as (b) Added labels; renumbering Added questions to number 2; new CTA question Revised illustration New figure

LABORATORY EXERCISE	TOPIC	CHANGE
61	Pre-Lab Introductory material Fig. 61.3 (sea urchin stages) Procedure C (human early development) Assessments: Part A Assessments: Part C	Revised question numbers 3 and 6; added questions Revised and expanded components Replaced blastula and gastrula images Revised components in step 2 Revised question numbers 7 and 12; added questions Revised question numbers 8 and 9; new CTA question
62	Purpose of the Exercise Learning Outcome Pre-Lab Introductory material Table 62.1 (examples of common human phenotypes); fig. 62.1 (representative genetic traits) Procedure C (genetics problems) Assessments: Part A Assessments: Part C	Revised component Revised Learning Outcome 03 Added question Revised and expanded components Removed tongue roller Added new steps 3 and 4 Revised test results table Added new question
63	Learning Outcomes Assessments: Part E Assessments: Part F Fig. 63.25 (ventral thorax muscles) Fig. 63.26 (dorsal muscles) Fig. 63.27 (medial hindlimb muscles) Fig. 63.28 (superficial hip and hindlimb muscles) Assessments: Part G	Revised Learning Outcome 01 Added questions in new Part Changed Part letter New figure New figure as (a) New figure New figure as (a) Changed Part letter
64	Fig. 64.14 (cardiovascular structures) Fig. 64.17 (cardiovascular structures) Fig. 64.18 (cadaver arteries and veins)	New figure as (a) Replaced figure Added term; renumbering
65	Fig. 65.3 (lower respiratory tract) Fig. 65.6 (lungs and respiratory structure)	Added labels New figure
66	Fig. 66.5 (digestive features)	Added label and renumbering in (a); replaced figure (b)
67	Fig. 67.7 (urinary features and associated blood vessels)	New figure
68	Fig. 68.7 (female reproductive structures) Fig. 68.8 (male reproductive structures)	Added new figure as (b) Added new figure as (b)
69	Learning Outcome Fig. 69.12 (head, neck, thorax, medial arm) Fig. 69.16 (hindlimb)	Revised Learning Outcome 01 New figure New figure
70	Fig. 70.4 (abdominal arteries and veins) Fig. 70.5 (trunk and neck veins)	Replaced figure as (b) Added new figure as (b)
71	Assessments: Part B Fig. 71.6 (lungs and respiratory tubes) Assessments: Part C	Added questions in new Part New figure Changed Part letter
72	Fig. 72.5 (salivary glands)	New figure
73	Assessments: Part B Fig. 73.3 (urinary features) Assessments: Part C	Added questions in new Part New figure Changed Part letter

ACKNOWLEDGMENTS

We value all of the support and encouragement from the staff at McGraw Hill Education, including Matthew Garcia, Krystal Faust, Valerie Kramer, Fran Simon, Jeni McAtee, Naveen Jhaldiyal, Tammy Juran, Lori Hancock, David Hash, and Michael Koot. We would like to give special recognition to Colin Wheatley for his insight, confidence, wisdom, warmth, and friendship, and to Jim Connely for his vision, instinct, passion, support, and leadership.

We are appreciative for the expertise of J and J Photography for numerous contributions. There have been valuable contributions from our students, who have supplied thoughtful suggestions and assisted in clarification of details.

Cynthia: I am immensely grateful to my extraordinary mentor and friend, Terry Martin, who is a person of genuine character. I would also like to thank my colleagues, students, and all instructors who provide support and feedback as they use this laboratory manual. For my supportive and encouraging sons—Forrest, Addison, Avery, Austin, and Aiden—I am incredibly appreciative. Finally, I am indebted to my husband and best friend, Bill Craver, for his patience, counsel, and enthusiasm throughout this labor of love, the fifth-edition laboratory manual and its digital content.

Terry: I am thankful to Dr. Norman Jenkins, Dr. David Louis, and Dr. Thomas Choice, retired presidents of Kishwaukee College, and Dr. Laurie Borowicz, president of Kishwaukee College, for their support, suggestions, and confidence in my endeavors. To my son Ross, an art instructor, I owe gratitude for his keen eye, creative suggestions, and creative cover illustrations of the second and third editions. I am appreciative to Sherrie Martin for advice, understanding, and devotion throughout the writing and revising.

Cynthia Prentice-Craver
Chemeketa Community College
4000 Lancaster Drive NE
Salem, OR 97309

Terry R. Martin
Kishwaukee College
21193 Malta Road
Malta, IL 60150

Reviewers

I would like to express my sincere gratitude to all reviewers of the laboratory manual who provided suggestions for its improvement. Their thoughtful comments and valuable suggestions are greatly appreciated. They include the following:

Ahmed Badr *California State University Los Angeles*
Janet Brodsky *Ivy Tech Community College*
Matthew M. Dougherty, D.O. *South College*
Edward Eivers *California State University Los Angeles*
Sam Furr-Rodgers *UNC Charlotte*
Kathryn Link *SUNY Alfred State College*
Dr. Bruce P. Maring *Daytona State College*
Kerrie McDaniel *Western Kentucky University*
Dr. Jennifer H. Miller PhD *Gateway Community and Technical College*
Dr. Albert Urazaev *Ivy Tech Community College*

ABOUT THE AUTHORS

William Craver

CYNTHIA PRENTICE-CRAVER, author of this fifth-edition laboratory manual, has been teaching anatomy and physiology at Chemeketa Community College for twenty-seven years. Her experience as a contributing author of the third edition, coauthor of the fourth edition, as well as her observations and engagement with students who use this laboratory manual, continually reinforce her excitement and passion for authoring. Her M.S. in Curriculum and Instruction, B.S. in Exercise Science, and extended undergraduate and graduate coursework in biological sciences have been instrumental in achieving the effective results in the online and on-campus courses she teaches. Teaching anatomy and physiology in these different formats has allowed Cynthia to explore and use different methods of content delivery that promote student involvement, learning, and confidence building. She is thrilled to be using the human cadaver lab at Chemeketa Community College in her teaching. Cynthia was a contributing author of the *Hole's Essentials Human Anatomy and Physiology, Fourteenth Edition* and coauthor with Charles Welsh of the *Hole's Human Anatomy and Physiology, Sixteenth Edition.* She is a member of the Human Anatomy and Physiology Society (HAPS) and the Textbook & Academic Authors Association (TAA). Her professional experiences include serving as program chair in the Life Sciences for eight years, serving on committees, and being a reviewer and advisor of textbooks and digital products. Beyond her professional pursuits, Cynthia's passions include reading and listening to books, attending exercise classes, hiking/walking outdoors, attending concerts, traveling, and spending time with her family and friends.

J and J Photography

This laboratory manual series was created by now-coauthor **TERRY R. MARTIN** of Kishwaukee College. Terry's teaching experience of over forty years, his interest in students and love for college instruction, and his innovative attitude and use of technology-based learning enhance the solid tradition of his other well-established laboratory manuals. Among Terry's awards are the Kishwaukee College Outstanding Educator, Kishwaukee College ICCTA Outstanding Educator Award, Kishwaukee College Faculty Board of Trustees Award of Excellence, Continued Excellence Award for Phi Theta Kappa Advisors, and John C. Roberts Community Service Award. Terry's professional memberships include the National Association of Biology Teachers (NABT), Human Anatomy and Physiology Society (HAPS), and The Nature Conservancy. Terry coauthored with Phillip Snider the *Laboratory Manual to Accompany Hole's Human Anatomy and Physiology,* Sixteenth Edition and the *Laboratory Manual to Accompany Hole's Essentials of Human Anatomy and Physiology,* Fourteenth Edition. Terry taught lecture and cadaver portions of EMT and paramedic classes. He has also been a faculty exchange member in Ireland. Terry locally supports historical preservation, natural areas, scouting, and scholarship. Through an established endowment to the Kishwaukee College Foundation, the "Terry & Sherrie Martin Health Careers Wing" was designated in 2014.

Laura Chiavini, Creative Communication Specialist, Kishwaukee College

TO THE STUDENT

The exercises in this laboratory manual will provide you with opportunities to observe various anatomical structures and to investigate certain physiological phenomena. Such experiences should help you relate specimens, models, microscope slides, and your body to what you have learned in the lecture and read about in the textbook.

Frequent variations exist in anatomical structures among humans. The illustrations in the laboratory manual represent normal (normal means the most common variation) anatomy. Variations from normal anatomy do not represent abnormal anatomy unless some function is impaired.

The following list of suggestions and study skills may make your laboratory activities more effective and profitable.

1. Prepare yourself before attending the laboratory session by reading the assigned exercise and reviewing the related sections of the textbook and lecture notes as indicated in the pre-lab section of the laboratory exercise. Answer the pre-lab questions. It is important to have some understanding of what will be done in the lab before you come to class.
2. Be on time. During the first few minutes of the laboratory meeting, the instructor often will provide verbal instructions. Make special note of any changes in materials to be used or procedures to be followed. Also listen carefully for information about special techniques to be used and precautions to be taken.
3. Keep your work area clean and your materials neatly arranged so that you can locate needed items. This will enable you to proceed efficiently and will reduce the chances of making mistakes.
4. Pay particular attention to the purpose of the exercise, which states what you are to accomplish in general terms, and to the learning outcomes, which list what you should be able to do as a result of the laboratory experience. Then, before you leave the class, review the outcomes and make sure that you can perform all of the assessments.
5. Precisely follow the directions in the procedure and proceed only when you understand them clearly. Do not improvise procedures unless you have the approval of the laboratory instructor. Ask questions if you do not understand exactly what you are supposed to do and why you are doing it.
6. Handle all laboratory materials with care. Some of the materials are fragile and expensive to replace. Whenever you have questions about the proper treatment of equipment, ask the instructor.
7. Treat all living specimens humanely and try to minimize any discomfort they might experience.
8. Although at times you might work with a laboratory partner or a small group, try to remain independent when you are making observations, drawing conclusions, and completing the activities in the laboratory reports.
9. Record your observations immediately after making them. In most cases, such data can be entered in spaces provided in the laboratory assessments.
10. Read the instructions for each section of the laboratory assessment before you begin to complete it. Think about the questions before you answer them. Your responses should be based on logical reasoning and phrased in clear and concise language.
11. At the end of each laboratory period, clean your work area and the instruments you have used. Return all materials to their proper places and dispose of wastes, including glassware or microscope slides that have become contaminated with human blood or body fluids, as directed by the laboratory instructor. Wash your hands thoroughly before leaving the laboratory.

Study Skills for Anatomy and Physiology

Students have found that certain study skills worked well for them while enrolled in Human Anatomy and Physiology. Although everyone has his or her learning style, there are techniques that work well for most students. Using some of the skills listed here can make your course more enjoyable and rewarding.

1. **Time management:** Prepare monthly, weekly, and daily schedules. Include dates of quizzes, exams, and projects on the calendar. On your daily schedule, budget several short study periods. Daily repetition alleviates cramming. Prioritize your tasks so that you still have time for work and leisure activities. Find an appropriate study atmosphere with minimum distractions.
2. **Note taking:** Look for the main ideas and briefly express them in your own words. Organize, edit, and review your notes soon after the lecture. Add textbook information to your notes as you reorganize them. Underline or highlight with different colors the important points, major headings, and key terms. Create concept maps and flow charts to show relationships and interconnections. Study your notes daily, as they provide sequential building blocks of the course content.
3. **Chunking:** Organize information into logical groups or categories. Study and master one chunk of information at

a time. For example, study the bones of the upper limb, lower limb, trunk, and head as separate study tasks.

4. **Mnemonic devices:** An *acrostic* is a combination of association and imagery to aid your memory. It is often in the form of a poem, rhyme, or jingle in which the first letter of each word corresponds to the first letters of the words you need to remember. **S**o **L**ong **T**op **P**art, **H**ere **C**omes **T**he **T**humb is an example of such a mnemonic device for remembering the eight carpals in a correct sequence. *Acronyms* are words formed by the first letters of the items to remember. *IPMAT* is an example of this type of mnemonic device to help you remember the phases of the cell cycle in the correct sequence. Try to create some of your own.
5. **Note cards/flash cards:** Make your own. Add labels and colors to enhance the material. Keep them with you; study them often and for short periods. Concentrate on a small number of cards at one time. Shuffle your cards and have someone quiz you on their content. As you become familiar with the material, you can set aside cards that don't require additional mastery.
6. **Recording and recitation:** An auditory learner can benefit by recording lectures and review sessions using a recording device. Many students listen to the taped sessions as they drive or just before going to bed. Reading your notes aloud can help also. Explain the material to anyone (even if there are no listeners) as if you are teaching. Talk about anatomy and physiology in everyday conversations.
7. **Study groups:** Small study groups that meet periodically to review course material and compare notes have helped and encouraged many students. However, keep the group on the task at hand. Work as a team and alternate leaders. This group often becomes a support group.

Practice sound study skills during your anatomy and physiology endeavor.

The Use of Animals in Biology Education*

The National Association of Biology Teachers (NABT) believes that the study of organisms, including nonhuman animals, is essential to the understanding of life on Earth. NABT recommends the prudent and responsible use of animals in the life science classroom. NABT believes that biology teachers should foster a respect for life. Biology teachers also should teach about the interrelationship and interdependency of all things.

Classroom experiences that involve nonhuman animals range from observation to dissection. NABT supports these experiences so long as they are conducted within the long-established guidelines of proper care and use of animals, as developed by the scientific and educational community.

As with any instructional activity, the use of nonhuman animals in the biology classroom must have sound educational objectives. Any use of animals, whether for observation or dissection, must convey substantive knowledge of biology. NABT believes that biology teachers are in the best position to make this determination for their students.

NABT acknowledges that no alternative can substitute for the actual experience of dissection or other use of animals and urges teachers to be aware of the limitations of alternatives. When the teacher determines that the most effective means to meet the objectives of the class do not require dissection, NABT accepts the use of alternatives to dissection, including models and the various forms of multimedia. The Association encourages teachers to be sensitive to substantive student objections to dissection and to consider providing appropriate lessons for those students where necessary.

To implement this policy, NABT endorses and adopts the "Principles and Guidelines for the Use of Animals in Precollege Education" of the Institute of Laboratory Animals Resources (National Research Council). Copies of the "Principles and Guidelines" may be obtained from the ILAR (2101 Constitution Avenue, NW, Washington, DC 20418; 202-334-2590).

*Adopted by the Board of Directors in October 1995. This policy supersedes and replaces all previous NABT statements regarding animals in biology education.

McGraw Hill connect®

Instructors: Student Success Starts with You

Tools to enhance your unique voice

Want to build your own course? No problem. Prefer to use an OLC-aligned, prebuilt course? Easy. Want to make changes throughout the semester? Sure. And you'll save time with Connect's auto grading too.

65%
Less Time Grading

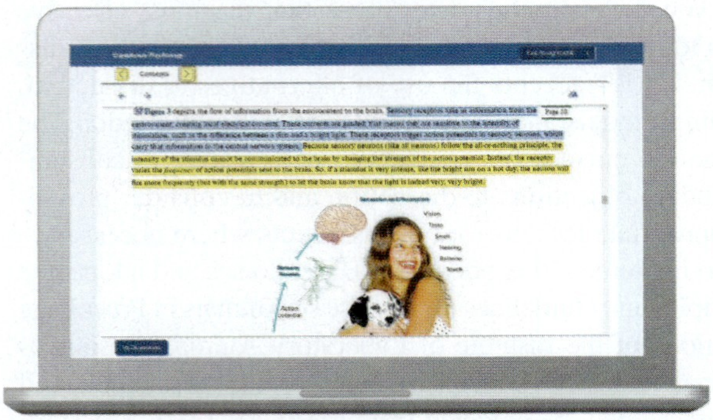

Laptop: McGraw Hill; Woman/dog: George Doyle/Getty Images

Study made personal

Incorporate adaptive study resources like SmartBook® 2.0 into your course and help your students be better prepared in less time. Learn more about the powerful personalized learning experience available in SmartBook 2.0 at **www.mheducation.com/highered/connect/smartbook**

Affordable solutions, added value

Make technology work for you with LMS integration for single sign-on access, mobile access to the digital textbook, and reports to quickly show you how each of your students is doing. And with our Inclusive Access program you can provide all these tools at a discount to your students. Ask your McGraw Hill representative for more information.

Padlock: Jobalou/Getty Images

Solutions for your challenges

A product isn't a solution. Real solutions are affordable, reliable, and come with training and ongoing support when you need it and how you want it. Visit **www.supportateverystep.com** for videos and resources both you and your students can use throughout the semester.

Checkmark: Jobalou/Getty Images

SUPPORT AT every step

Students: Get Learning that Fits You

Effective tools for efficient studying

Connect is designed to help you be more productive with simple, flexible, and intuitive tools that maximize your study time and meet your individual learning needs. Get learning that works for you with Connect.

Study anytime, anywhere

Download the free ReadAnywhere app and access your online eBook, SmartBook 2.0, or Adaptive Learning Assignments when it's convenient, even if you're offline. And since the app automatically syncs with your Connect account, all of your work is available every time you open it. Find out more at **www.mheducation.com/readanywhere**

"I really liked this app—it made it easy to study when you don't have your text-book in front of you."

- Jordan Cunningham,
Eastern Washington University

Calendar: owattaphotos/Getty Images

Everything you need in one place

Your Connect course has everything you need—whether reading on your digital eBook or completing assignments for class, Connect makes it easy to get your work done.

Learning for everyone

McGraw Hill works directly with Accessibility Services Departments and faculty to meet the learning needs of all students. Please contact your Accessibility Services Office and ask them to email accessibility@mheducation.com, or visit **www.mheducation.com/about/accessibility** for more information.

Top: Jenner Images/Getty Images, Left: Hero Images/Getty Images, Right: Hero Images/Getty Images

LABORATORY
EXERCISE 1

Scientific Method and Measurements

Purpose of the Exercise

To become familiar with the scientific method of investigation, learn how to formulate sound conclusions, and provide opportunities to use the metric system of measurements.

MATERIALS NEEDED

Meterstick
Calculator
Human skeleton

Learning Outcomes APR

After completing this exercise, you should be able to:

O1 Convert English measurements to the metric system, and vice versa.

O2 Calculate expected upper limb length and actual percentage of height from recorded upper limb lengths and heights.

O3 Apply the scientific method to test the validity of a hypothesis concerning the direct, linear relationship between human upper limb length and height.

O4 Design an experiment, formulate a hypothesis, and test it using the scientific method.

The O corresponds to the assessments A indicated in the Laboratory Assessment for this Exercise.

Pre-Lab

Carefully read the introductory material and examine the entire lab. Be familiar with the scientific method from lecture or the textbook. Answer the pre-lab questions.

Pre-Lab Questions Select the correct answer for each of the following questions:

1. To explain biological phenomena, scientists use a technique called
   a. the scientific method. **b.** the scientific law.
   c. conclusions. **d.** measurements.
2. Which of the following represents the correct sequence of the scientific method?
   a. analysis of data, conclusions, observations, experiment, hypothesis
   b. conclusions, experiment, hypothesis, analysis of data, observations
   c. observations, hypothesis, experiment, analysis of data, conclusions
   d. hypothesis, observations, experiment, analysis of data, conclusions
3. A hypothesis, verified continuously from experiments by many investigators, can become known as a
   a. control. **b.** variable.
   c. valid result. **d.** theory.
4. The most likely scientific unit for measuring the height of a person would be
   a. feet. **b.** centimeters.
   c. inches. **d.** kilometers.
5. Which of the following is *not* a unit of the metric system of measurements?
   a. centimeters **b.** liters
   c. inches **d.** millimeters

Chapter Opener: ©101cats/Getty Images

6. The variable that can be changed and is determined before the experiment starts is the
   a. dependent variable. **b.** hypothesis.
   c. independent variable. **d.** analysis.
7. The hypothesis is formulated from the results of the experiment.
   a. True **b.** False
8. A centimeter represents an example of a metric unit of length.
   a. True **b.** False

Scientific investigation involves a series of logical steps to arrive at explanations for various biological phenomena. It reflects a long history of asking questions and searching for knowledge. This technique, called the *scientific method,* is used in all disciplines of science. It allows scientists to draw logical and reliable conclusions.

The scientific method begins with making *observations* related to the topic under investigation. This step commonly involves the accumulation of previously acquired information and/or your observations of the phenomenon. These observations are used to formulate a tentative explanation known as the *hypothesis.* An important attribute of a hypothesis is that it must be testable. The testing of the proposed hypothesis involves designing and performing a carefully controlled *experiment* to obtain data that can be used to support, reject, or modify the hypothesis. During the experiment to test the proposed hypothesis, it is important to be able to examine only a single changeable factor, known as a *variable.* An *independent variable* is one that can be changed, but is determined before the experiment occurs; a *dependent variable* is determined from the results of the experiment.

An *analysis of data* is conducted using sufficient information collected during the experiment. Data analysis may include organization and presentation of data as tables, graphs, and drawings. From the interpretation of the data analysis, *conclusions* are drawn. (If the data do not support the hypothesis, you must reexamine the experimental design and the data, and if needed develop a new hypothesis.) The final presentation of the information is made from the conclusions. Results and conclusions are presented to the scientific community for evaluation through peer reviews, presentations at professional meetings, and published articles. If many investigators working independently can validate the hypothesis by arriving at the same conclusions, the explanation can become a *theory.* A theory serves as the explanation from a summary of known experiments and supporting evidence unless it is disproved by new information. The five components of the scientific method are summarized as

Observations
↓
Hypothesis
↓
Experiment
↓
Analysis of data
↓
Conclusions

Metric measurements are characteristic tools of scientific investigations. The English system of measurements is often used in the United States, so the investigator must make conversions from the English system to the metric system. Table 1.1 provides the conversion factors necessary to change from English to metric units.

PROCEDURE A: Using the Steps of the Scientific Method

This procedure represents a specific example of the order of the steps utilized in the scientific method. Each of the steps for this procedure will guide you through the proper sequence in an efficient pathway.

1. A correlation exists between the length of the upper and lower limbs and the height (stature) of an individual. For example, a person who has long upper limbs (the arm, forearm, and hand combined) tends to be tall. Make some visual observations of other people around you to observe a possible correlation.
2. From such observations, the following hypothesis can be formulated: The length of a person's upper limb is equal to 0.4 (40%) of the height of the person. To test this hypothesis, perform the following experiment.
3. Use a meterstick (fig. 1.1) to measure an upper limb length of ten subjects. Place the meterstick in the axilla (armpit) and record the length in centimeters to the end of the longest finger (fig. 1.2). Obtain the height of

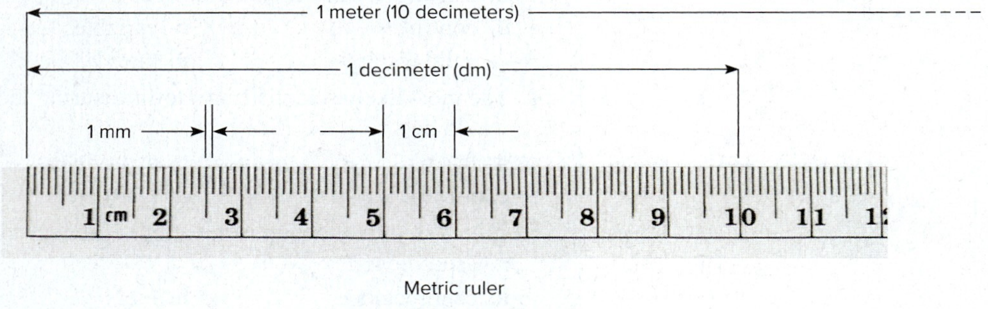

FIGURE 1.1 Metric ruler with metric lengths indicated. A meterstick length would be 100 cm (10 dm). (The image size is approximately to scale.)

Table 1.1 Metric Measurement System and Conversions

Measurement	Unit and Abbreviation	Metric Equivalent	Conversion Factor Metric to English (approximate)	Conversion Factor English to Metric (approximate)
Length	1 kilometer (km)	1,000 (10^3) m	1 km = 0.62 mile	1 mile = 1.61 km
	1 meter (m)	100 (10^2) cm 1,000 (10^3) mm	1 m = 1.1 yards = 3.3 feet = 39.4 inches	1 yard = 0.9 m 1 foot = 0.3 m
	1 decimeter (dm)	0.1 (10^{-1}) m	1 dm = 3.94 inches	1 inch = 0.25 dm
	1 centimeter (cm)	0.01 (10^{-2}) m	1 cm = 0.4 inches	1 foot = 30.5 cm 1 inch = 2.54 cm
	1 millimeter (mm)	0.001 (10^{-3}) m 0.1 (10^{-1}) cm	1 mm = 0.04 inches	
	1 micrometer (μm)	0.000001 (10^{-6}) m 0.001 (10^{-3}) mm		
Mass	1 metric ton (t)	1,000 (10^3) kg	1 t = 1.1 ton	1 ton = 0.91 t
	1 kilogram (kg)	1,000 (10^3) g	1 kg = 2.2 pounds	1 pound = 0.45 kg
	1 gram (g)	1,000 (10^3) mg	1 g = 0.04 ounce	1 pound = 454 g 1 ounce = 28.35 g
	1 milligram (mg)	0.001 (10^{-3}) g		
Volume (liquids and gases)	1 liter (L)	1,000 (10^3) mL	1 L = 1.06 quarts	1 gallon = 3.78 L 1 quart = 0.95 L
	1 milliliter (mL)	0.001 (10^{-3}) L 1 cubic centimeter (cc or cm^3)	1 mL = 0.03 fluid ounce 1 mL = 1/5 teaspoon 1 mL = 15–16 drops	1 quart = 946 mL 1 fluid ounce = 29.6 mL 1 teaspoon = 5 mL
Time	1 second (s)	1/60 minute	Same	Same
	1 millisecond (ms)	0.001 (10^{-3}) s	Same	Same
Temperature	Degrees Celsius (°C)		°F = 9/5 °C + 32	°C = 5/9 (°F − 32)

FIGURE 1.2 Measurement of upper limb length. ©J & J Photography

each person in centimeters by measuring them without shoes against a wall (fig. 1.3). The height of each person can also be calculated by multiplying each individual's height in inches by 2.54 to obtain his/her height in centimeters. Record all your measurements in Part A of Laboratory Assessment 1.

4. The data collected from all of the measurements can now be analyzed. The expected (predicted) correlation between upper limb length and height is determined using the following equation:

   Height × 0.4 = expected upper limb length

   The observed (actual) correlation to be used to test the hypothesis is determined by

   Length of upper limb/height = actual % of height

5. A graph is an excellent way to display a visual representation of the data. Plot the subjects' data in Part A of the laboratory assessment. Plot the upper limb length of each subject on the x-axis (independent variable) and the height of each person on the y-axis (dependent variable). A line is already located on the graph that represents a hypothetical relationship of 0.4 (40%) upper limb length compared to height. This is a graphic representation of the original hypothesis.
6. Compare the distribution of all of the points (actual height and upper limb length) that you placed on the graph with the distribution of the expected correlation represented by the hypothesis.
7. Complete Part A of the laboratory assessment.

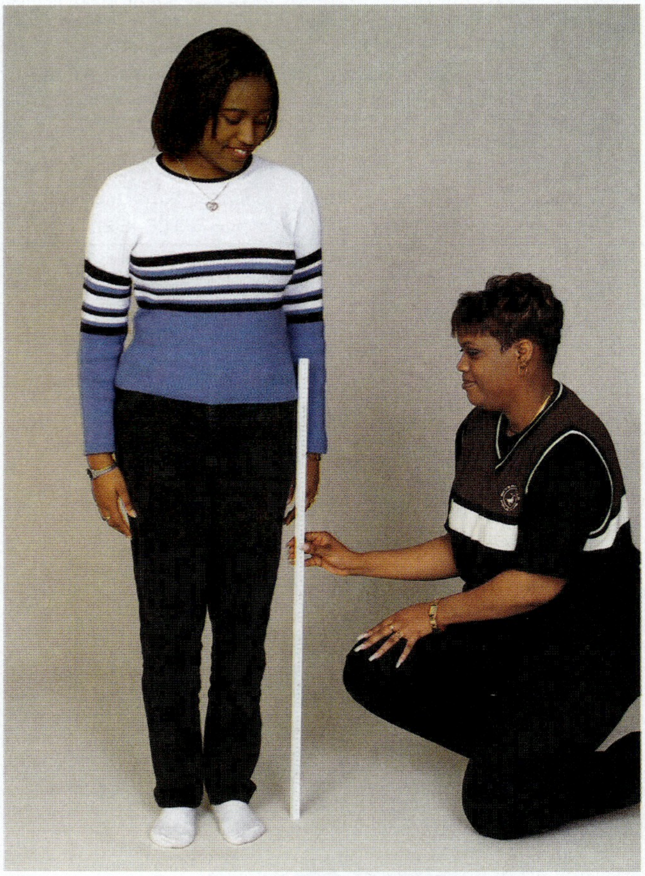

FIGURE 1.3 Measurement of height. ©J & J Photography

PROCEDURE B: Design an Experiment

You have completed the steps of the scientific method with guidance directions in Procedure A. This procedure will allow for less guidance and more flexibility using the scientific method.

CRITICAL THINKING ACTIVITY

You have probably concluded that there is some correlation of the length of body parts to height. Often, when a skeleton is found, it is not complete. It is occasionally feasible to use the length of a single bone to estimate the height of an individual. Observe human skeletons and locate the humerus bone in an upper limb or the femur bone in a lower limb. Use your observations to identify a mathematical relationship between the length of the humerus or femur and height. Formulate a hypothesis that can be tested. Make measurements, analyze data, and develop a conclusion from your experiment. Complete Part B of the laboratory assessment.

LABORATORY
ASSESSMENT 1

Name ____________________

Date ____________________

Section ____________________

The (A) corresponds to the indicated Learning Outcome(s) (O) found at the beginning of the Laboratory Exercise.

Scientific Method and Measurements

PART A: Assessments

1. Record measurements for the upper limb length and height of ten subjects. Use a calculator to determine the expected upper limb length and the actual percentage (as a decimal or a percentage) of the height for the ten subjects. Record your results in the following table. A2

Subject	Measured Upper Limb Length (cm)	Height* (cm)	Height × 0.4 = Expected Upper Limb Length (cm)	Actual % of Height = Measured Upper Limb Length (cm)/Height (cm)
1.				
2.				
3.				
4.				
5.				
6.				
7.				
8.				
9.				
10.				

*The height of each person can be calculated by multiplying each individual's height in inches by 2.54 to obtain his/her height in centimeters. A1

2. Plot the distribution of data (upper limb length and height) collected for the ten subjects on the following graph. The line located on the graph represents the *expected* 0.4 (40%) ratio of upper limb length to measured height (the original hypothesis). (The x-axis represents upper limb length, and the y-axis represents height.) Draw a line of *best fit* through the distribution of points of the plotted data of the ten subjects. Compare the two distributions (expected line and the distribution line drawn for the ten subjects). A3

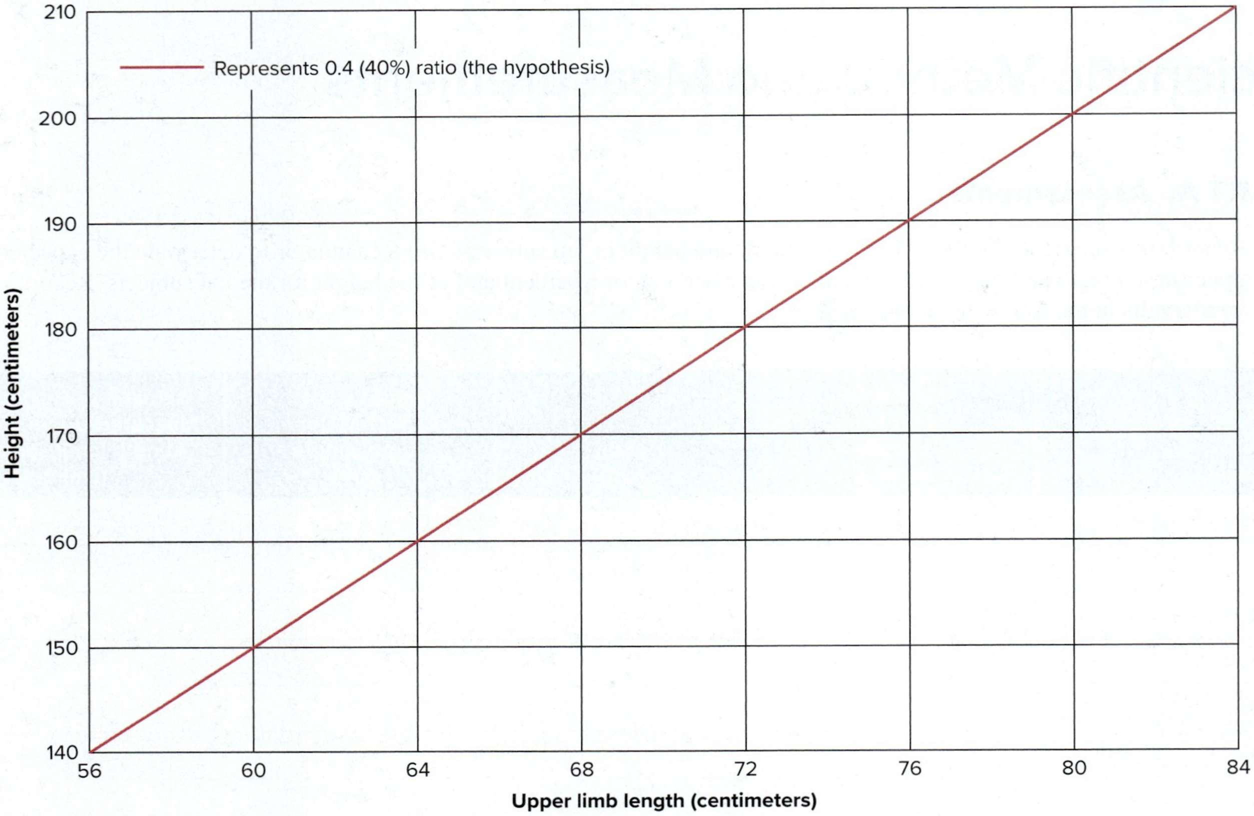

3. Does the distribution of the ten subjects' measured upper limb lengths support or reject the original hypothesis? ____________ Explain your answer. A3

PART B: Assessments

1. Describe your observations of a possible correlation between the humerus or femur length and height. **A4**

2. Write a hypothesis based on your observations. **A4**

3. Describe the design of the experiment that you devised to test your hypothesis. **A4**

4. Place your analysis of the data in this space in the form of a table and a graph. **A4**

   a. Table:

b. Graph:

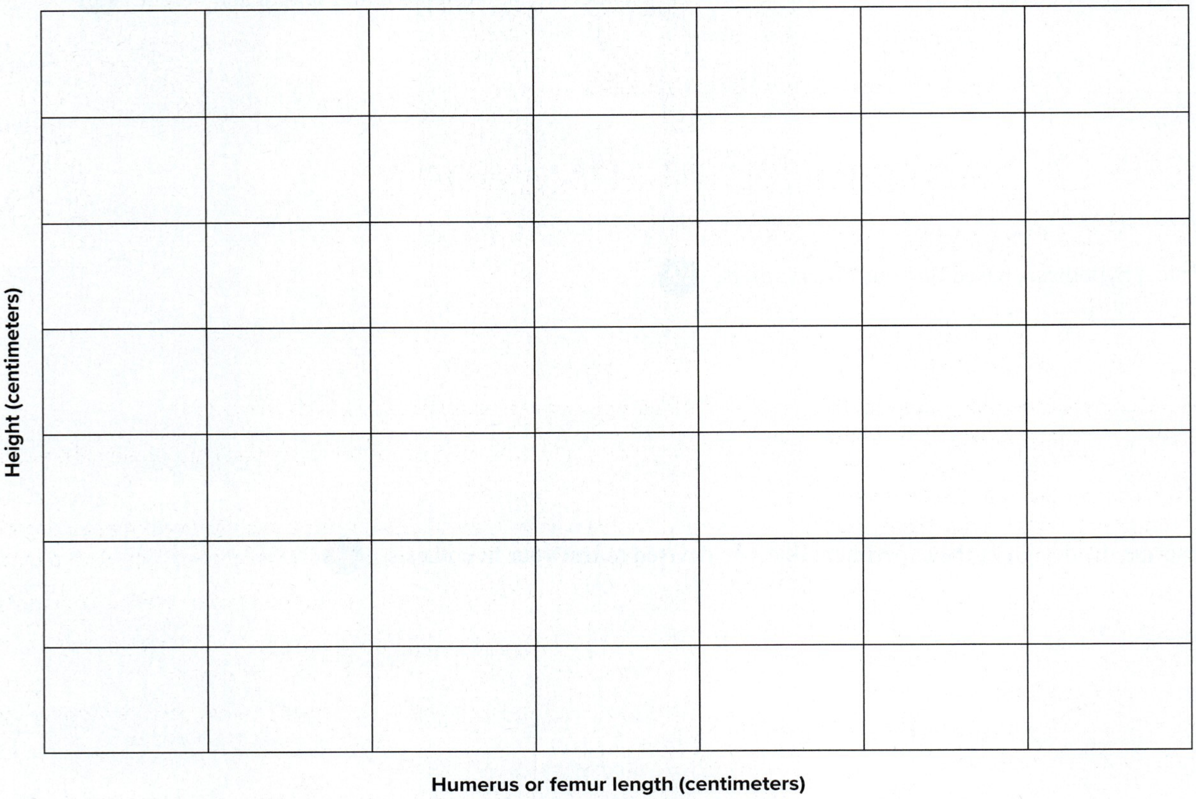

5. Based on an analysis of your data, what can you conclude? Did these conclusions confirm or refute your original hypothesis? A4

6. Discuss your results and conclusions with classmates. What common conclusion can the class formulate about the correlation between the humerus or femur length and height? A4

LABORATORY
EXERCISE 2

Body Organization, Membranes, and Terminology

Purpose of the Exercise

To review the organizational pattern of the human body, to review its organ systems and the organs included in each system, and to become acquainted with the terms used to describe the relative position of body parts, body sections, and body regions.

MATERIALS NEEDED

Dissectible human torso model (manikin)

Variety of specimens or models sectioned along various planes

Chapter Opener: ©101cats/Getty Images

Learning Outcomes APR

After completing this exercise, you should be able to:

O1 Locate and name the major body cavities and identify the membranes associated with each cavity.

O2 Associate the organs and functions included within each organ system and locate the organs in a dissectible human torso model.

O3 Select the terms used to describe the relative positions of body parts.

O4 Differentiate the terms used to identify body sections and identify the plane along which a particular specimen is cut.

O5 Label body regions and associate the terms used to identify body regions.

The O corresponds to the assessments A indicated in the Laboratory Assessment for this Exercise.

Pre-Lab

Carefully read the introductory material and examine the entire lab. Be familiar with body cavities, membranes, organ systems, and body regions from lecture or the textbook. Answer the pre-lab questions.

Pre-Lab Questions Select the correct answer for each of the following questions:

1. The basis for communication in anatomy and physiology assumes
  a. the person is lying down. **b.** relative positions.
  c. anatomical position. **d.** the person is sleeping.

2. Which of the following is *not* a body cavity?
  a. diaphragm **b.** thoracic
  c. cranial **d.** abdominopelvic

3. The pericardium is associated with the
  a. lung. **b.** intestine.
  c. liver. **d.** heart.

4. The _____ plane divides the body into left and right sides.
  a. frontal **b.** cranial
  c. sagittal **d.** transverse

5. The abdominopelvic cavity can be subdivided into
  a. pleural cavities. **b.** pericardial cavities.
  c. quadrants. **d.** vertebral canals.

6. The larynx is part of the _____ system.
  a. urinary **b.** respiratory
  c. lymphatic **d.** nervous

7. The epigastric region is a portion of the _____ cavity.
   a. pelvic b. pleural
   c. vertebral d. abdominal
8. In the posterior view, the cubital region is _____ to the carpal region.
   a. distal b. medial
   c. superficial d. proximal
9. The brachial surface region pertains to the wrist.
   a. True b. False
10. A frontal plane divides the body into anterior and posterior parts.
   a. True b. False

The major features of the organization of the human body include certain body cavities. A body cavity may contain an organ and/or a specific fluid. The *cranial cavity* contains the brain and the *vertebral canal (spinal cavity)* contains the spinal cord. These body cavities also contain cerebrospinal fluid (CSF). Two major cavities are found in the trunk of the body and are separated by the diaphragm, a skeletal muscle important in breathing. The *thoracic cavity* is located in the chest, superior to the diaphragm. This cavity is subdivided into a mediastinum containing primarily the heart, esophagus, and trachea, with the lungs located on either side of the mediastinum. The *abdominopelvic cavity* lies inferior to the diaphragm. This cavity is composed of an *abdominal cavity* and *pelvic cavity*. The entire abdominopelvic cavity is further subdivided into either nine regions or four quadrants. The large size of the abdominopelvic cavity, with its many visceral organs, warrants these further subdivisions into regions or quadrants for convenience and for accuracy in describing organ locations, injury sites, and pain locations.

The *pericardium*, *pleura*, and *peritoneum* are thin double-layered *serous membranes* that line the thoracic and abdominopelvic cavities. Each serous membrane has an outer parietal layer that forms the outer cavity wall which folds in on itself to form an inner visceral layer that covers the surface of an organ. Between each serous membrane is a narrow space called a cavity that is filled with serous fluid. This slippery fluid provides organs some freedom of movement without friction as they carry out their functions. In the thoracic cavity, the pericardium surrounds the heart. Between its parietal pericardium and visceral pericardium is the pericardial cavity that is filled with pericardial fluid. The thoracic cavity also includes a parietal pleura and visceral pleura surrounding each lung. Between the pleura layers is the pleural cavity that contains pleural fluid. Within the abdominopelvic cavity is the parietal peritoneum and visceral peritoneum with a peritoneal cavity containing peritoneal fluid. Several abdominopelvic organs, such as the kidneys and pancreas, are located just behind (are retroperitoneal to) the parietal peritoneum, thus lacking a mesentery (visceral peritoneum).

Some other body cavities include the orbital cavity of the eye, nasal cavity of the nose, oral cavity of the mouth, cavity of the middle ear, and synovial cavity of a movable joint such as the knee or elbow.

Although the human body functions as one entire unit, it is customary to divide the body into eleven body organ systems. In order to communicate effectively with each other about the body, scientists have devised anatomical terminology. Foremost in this task we use *anatomical position* as our basis for communication, including directional terms, body regions, and planes of the body. A person in anatomical position is standing erect, facing forward, with upper limbs at the sides and palms forward. This standard position allows us to describe relative positions of various body parts using such directional terms as left-right, anterior-posterior, proximal-distal, medial-lateral, and superior-inferior. Body regions include certain surface areas, portions of limbs, and portions of body cavities. In order to study internal structures, often the body is depicted as sectioned into a sagittal plane, frontal plane, or transverse plane.

PROCEDURE A: Body Cavities and Membranes

This procedure outlines the body cavities by location. Certain cavities contain thin layers of cells, called membranes, that line the cavity and cover the organs within the cavity. The figures included in Procedure A will allow you to locate and identify the associated membrane location and name appropriate for the cavity and the organs involved.

1. Study figures 2.1 to 2.5 to become familiar with body cavities and associated membranes.
2. Locate as many of the following features as you can on a dissectible human torso model (fig. 2.5):

body cavities

- cranial cavity—houses brain
- vertebral canal (spinal cavity)—houses spinal cord
- thoracic cavity
  - mediastinum—region between the lungs; includes pericardial cavity
  - pleural cavities (2)
- abdominopelvic cavity
  - abdominal cavity
  - pelvic cavity

diaphragm—separates thoracic and abdominopelvic cavities; functions in respiration

smaller cavities within the head

- oral cavity (mouth)
- nasal cavity with connected sinuses
- orbital cavity—houses eye and associated structures
- middle ear cavity (tympanic cavity)—air-filled and contains auditory ossicles

membranes and cavities

- pleural cavity—associated with lungs; contains pleural (serous) fluid
  - parietal pleura—outer layer; lines thoracic cavity wall
  - visceral pleura—inner layer; covers surface of lungs

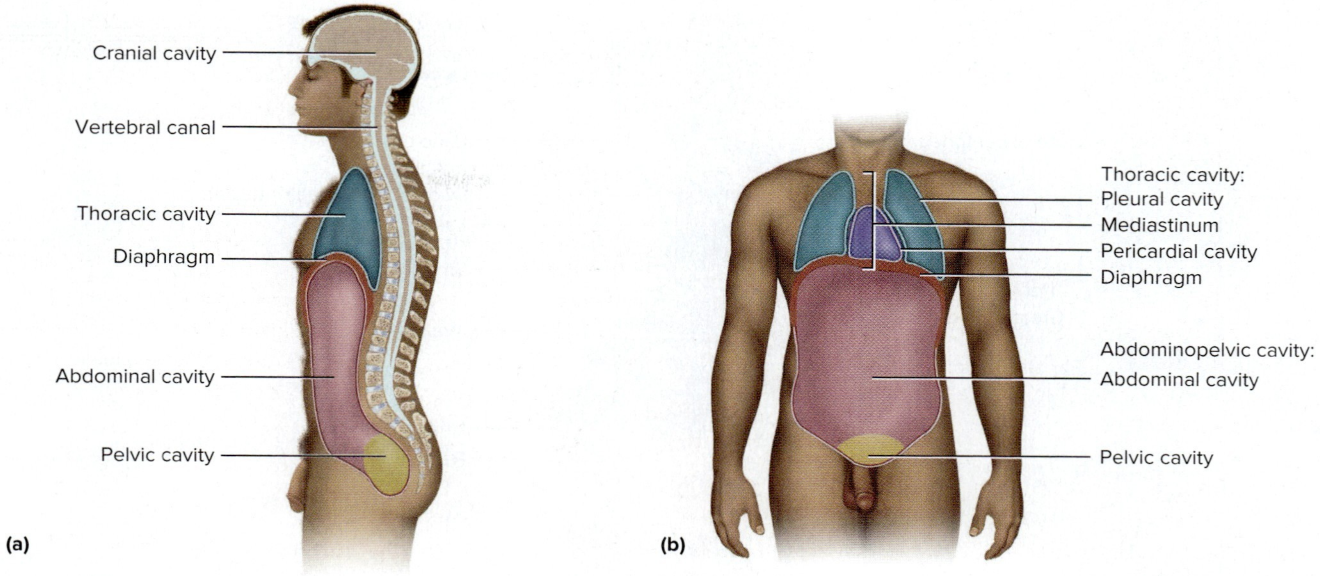

FIGURE 2.1 Major body cavities: (a) left lateral view; (b) anterior view. APR

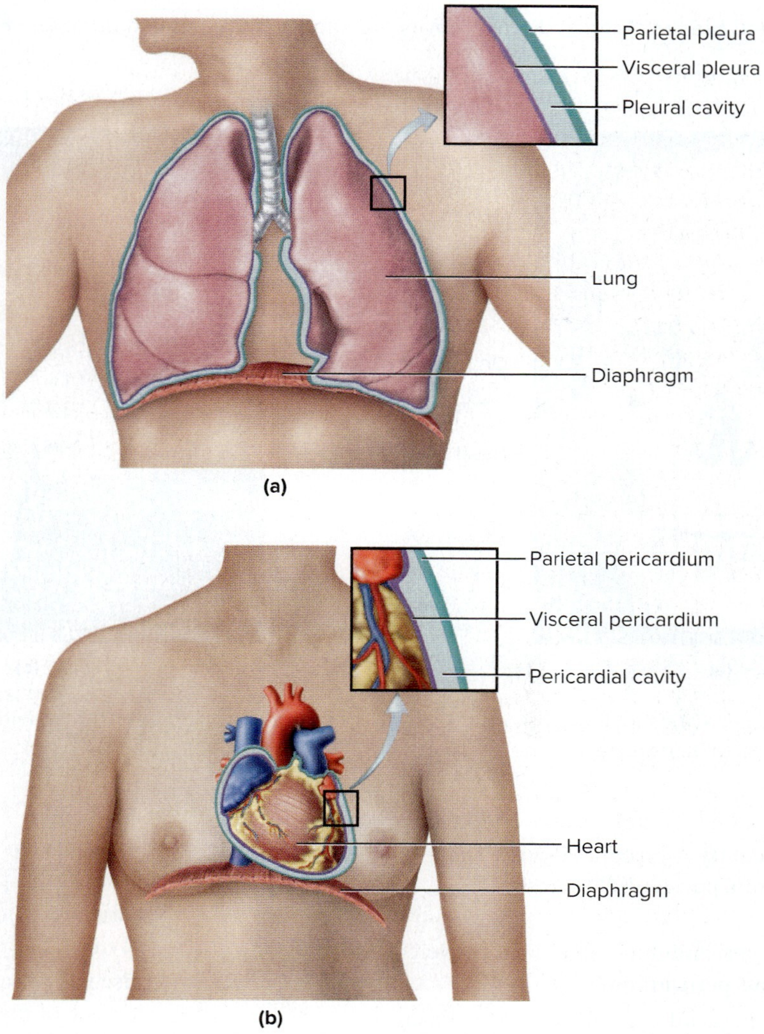

FIGURE 2.2 Thoracic serous membranes and cavities associated with (a) the lungs and (b) the heart. APR

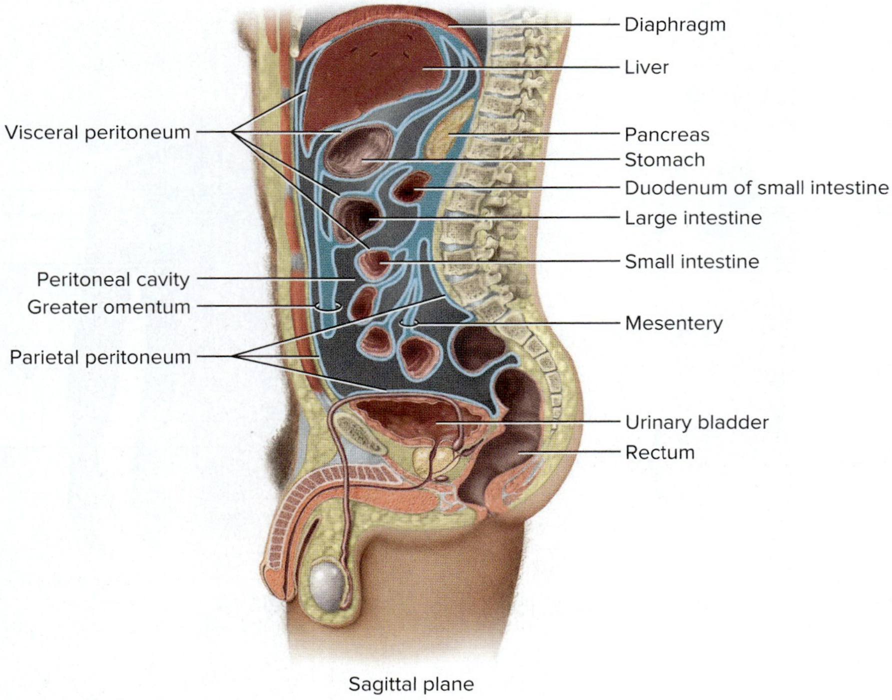

FIGURE 2.3 Serous membranes of the abdominal cavity are shown in this left lateral view. APR

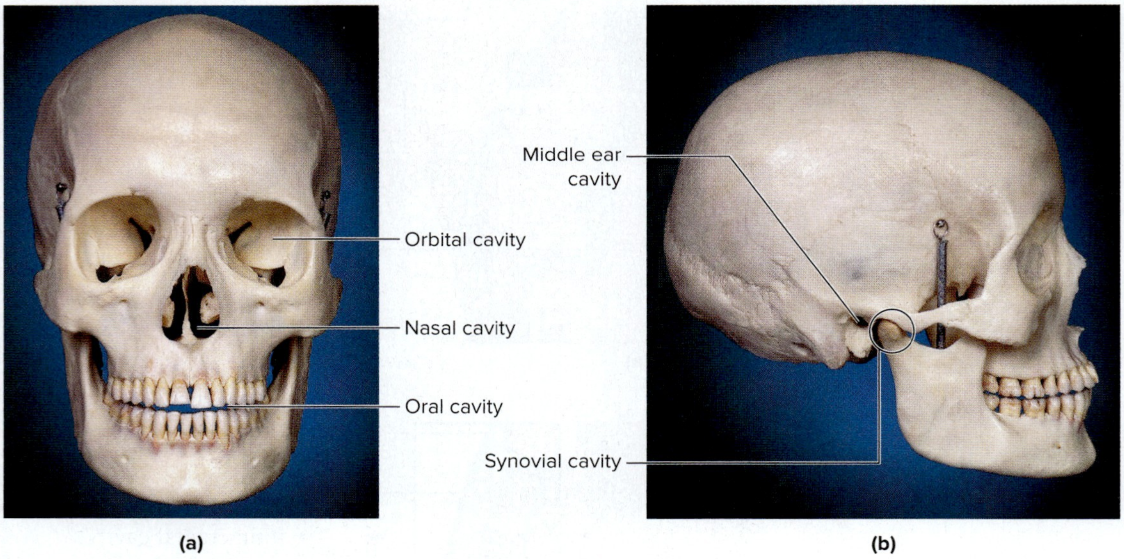

FIGURE 2.4 Other body cavities include the orbital, nasal, oral, middle ear, and synovial cavities, indicated in these images of the (a) anterior view and (b) lateral view of the skull.

- pericardial cavity—associated with heart; contains pericardial (serous) fluid
  - parietal pericardium—outer layer; covered by fibrous pericardium
  - visceral pericardium (epicardium)—inner layer; covers surface of heart
- peritoneal cavity—associated with abdominal organs; contains peritoneal (serous) fluid
  - parietal peritoneum—outer layer; lines cavity wall
  - visceral peritoneum—inner layer; covers surface of organs

3. Complete Part A of Laboratory Assessment 2.

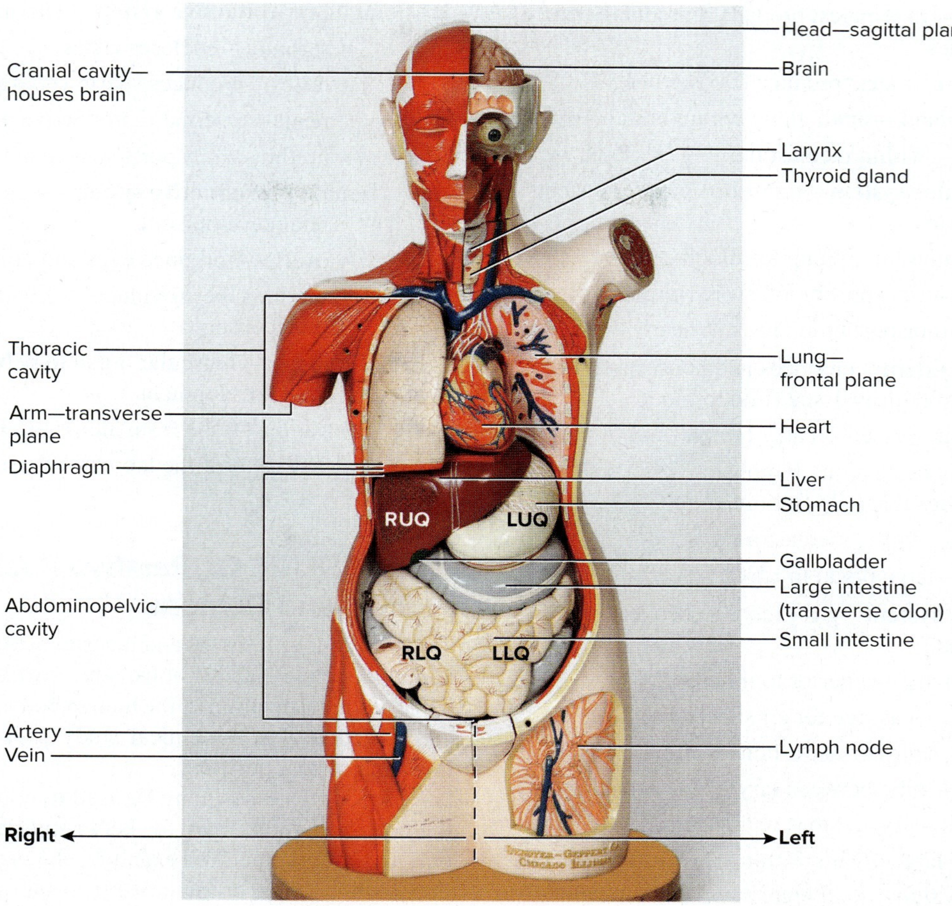

FIGURE 2.5 Dissectible human torso model with body cavities, abdominopelvic quadrants, body planes, and major organs indicated. The abdominopelvic quadrants include right upper quadrant (RUQ), left upper quadrant (LUQ), right lower quadrant (RLQ), and left lower quadrant (LLQ). ©J & J Photography

PROCEDURE B: Organ Systems

The eleven body systems are listed in this procedure, and a major function of each system is included. Representative major organs are listed within each of the systems. By using models, charts, and images to locate each major organ included within the system, you will have an early overview of each system of the human body. More detailed organs and functions are covered within each system during later laboratory exercises.

1. Use a dissectible human torso model (fig. 2.5) to locate the following systems and their major organs:

   integumentary system—protection
   - skin—composed of epidermis and dermis
   - accessory organs—such as hair and nails

   skeletal system—support and protection
   - bones—in head, torso, and limbs
   - ligaments—connect bones to bones

   muscular system—movement
   - skeletal muscles—allow voluntary movements
   - tendons—connect muscles to bones

   nervous system—detects changes; interprets sensory information; stimulates muscles and glands
   - brain—within cranial cavity
   - spinal cord—extends through vertebral canal
   - nerves—conduct impulses into and from brain and spinal cord

   endocrine system—secretes hormones
   - pituitary gland—attached to base of brain
   - thyroid gland—anterior neck; inferior to larynx
   - parathyroid glands—four glands; embedded on posterior thyroid gland
   - adrenal glands—located superior to kidneys
   - pancreas—most in left upper quadrant (LUQ); influence blood sugar

- ovaries—in females; produce reproductive hormones
- testes—in males; produce testosterone
- pineal gland—small gland within brain
- thymus—within mediastinum

cardiovascular system—transports gases, nutrients, and wastes
- heart—muscular pump for blood
- arteries—transport blood away from heart
- veins—transport blood back to heart

lymphatic system—produces and houses immune cells and returns tissue fluid to blood
- lymphatic vessels—carry lymph fluid
- lymph nodes—along lymphatic vessels; contain leukocytes that help fight infections
- thymus—within mediastinum
- spleen—large organ in LUQ

respiratory system—gas exchange between air and blood
- nasal cavity—superior to mouth cavity
- pharynx—passage for air superior to larynx
- larynx—anterior neck; houses vocal cords
- trachea—tube between larynx and bronchi
- bronchi—airway tubes within lungs
- lungs—large organs within thoracic cavity

digestive system—food breakdown and absorption
- mouth—contains tongue and teeth
- tongue—for food manipulation
- teeth—for biting and chewing food
- salivary glands—secrete saliva into mouth
- pharynx—passageway for food superior to esophagus
- esophagus—tube from pharynx to stomach
- stomach—between esophagus and small intestine; in LUQ
- liver—produces bile; in right upper quadrant (RUQ)
- gallbladder—stores bile; in RUQ
- pancreas—produces digestive enzymes; most in LUQ
- small intestine—tube from stomach to large intestine
- large intestine—tube from small intestine to anus

urinary system—removes liquids and wastes from blood
- kidneys—two large urine-forming organs in upper abdominopelvic cavity
- ureters—tubes from kidneys to urinary bladder
- urinary bladder—pelvic organ; stores urine
- urethra—tube from urinary bladder to external opening

male reproductive system—sperm production
- scrotum—encloses testes
- testes—produces sperm and hormones
- penis—external reproductive organ
- urethra—transports semen and urine

female reproductive system—egg production and fetal development
- ovaries—produce eggs and hormones
- uterine tubes (oviducts; fallopian tubes)—transport eggs
- uterus—muscular organ in pelvis; structure for fetal development
- vagina—tube from uterus to external opening

2. Complete Part B of the laboratory assessment.

PROCEDURE C: Relative Positions, Planes, and Regions

This procedure illustrates and incorporates planes (sections), abdominopelvic subdivisions, and surface regions using the anatomical position of the human body. Communications and directions for anatomical study assume the person is in an anatomical position.

Directional terms may be used to compare one part of the body to another part for better understanding of location and orientation. For example, the esophagus is posterior to the trachea, or the tarsal bones are distal to the tibia bone.

Anatomical planes (sections; cuts) of the entire body and individual organs allow a better understanding of the internal structure and function. Sectional anatomy is accomplished by cutting the body in standard ways to maximize the view of the particular organ. The anatomical planes include the *sagittal, frontal,* and *transverse* sections or cuts. Radiologic images by such techniques as a computed tomography (CT) scan or an magnetic resonance imaging (MRI) are examples of the advances and the importance of sectional anatomy for the diagnosis and treatment of medical situations.

The abdominopelvic area of the body is rather large and contains numerous viscera representing structural and functional parts of several body systems. People with abdominopelvic discomfort will often complain about it using terms like a "stomach ache." This is too general a term to be used for medical tests and procedures. Therefore, this abdominopelvic portion of the body is further subdivided into four *quadrants* and/or *nine regions.* Some medical disciplines prefer the use of quadrants, while others prefer the more specific use of regions.

The general surface of the entire body is subdivided into regions using anatomical terms appropriate for that particular portion of the body. This becomes helpful for descriptions of injuries or pain locations that are more specific than using broader general terms such as head, neck, and upper limb.

1. Observe the person standing in anatomical position (fig. 2.6). Anatomical terminology assumes the body is in anatomical position even though a person is often observed differently.
2. Study table 2.1 and figures 2.6 to 2.9 to become familiar with directional terms, anatomical planes, abdominopelvic quadrants and regions, and body surface regions.
3. Examine the sectioned specimens on the demonstration table and identify the plane along which each is cut.
4. Complete Parts C, D, E, and F of the laboratory assessment.

Table 2.1 Directional Terms and Their Meaning

Directional Term	Meaning
Left	Toward the left side of the body.
Right	Toward the right side of the body.
Anterior (ventral)	Toward the front or belly.
Posterior (dorsal)	Toward the back.
Superior (cephalad)	Toward the head or upward or above; pertains to the trunk anatomy.
Inferior (caudal)	Toward the tail or downward or below; pertains to the trunk anatomy.
Proximal	Toward the trunk or origin of a structure; pertains to the limb anatomy.
Distal	Further from the trunk or origin of a structure; pertains to the limb anatomy.
Medial	Toward the midline (median plane).
Lateral	Further from the midline (median plane).
Superficial	Toward the body surface.
Deep	Further away from the body surface.
Ipsilateral	On the same side of the body (right or left).
Contralateral	On opposite sides of the body (right and left).

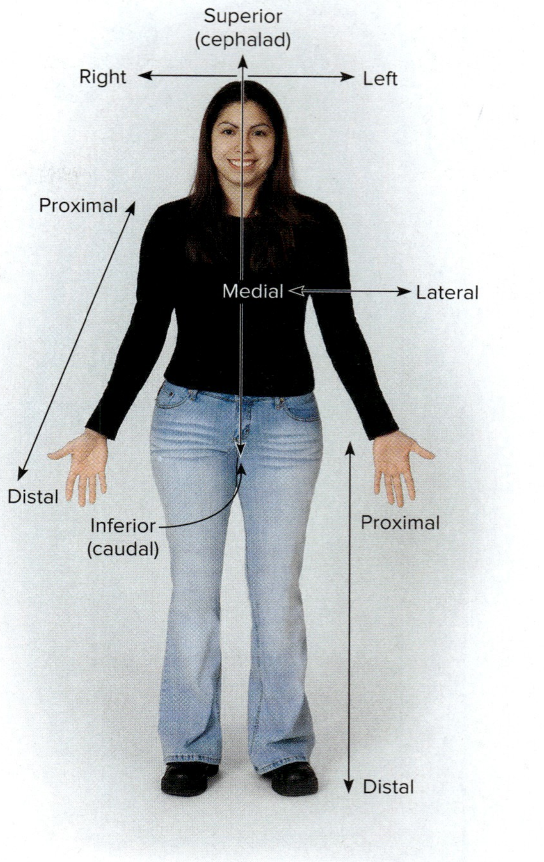

FIGURE 2.6 Anatomical position with directional terms indicated. The body is standing erect, face forward, with upper limbs at the sides and palms forward. When the palms are forward (supinated) the radius and ulna in the forearm are nearly parallel. This results in an anterior view of the body as shown. The relative positional terms are used to describe a body part's location in relation to other body parts. ©J & J Photography APR

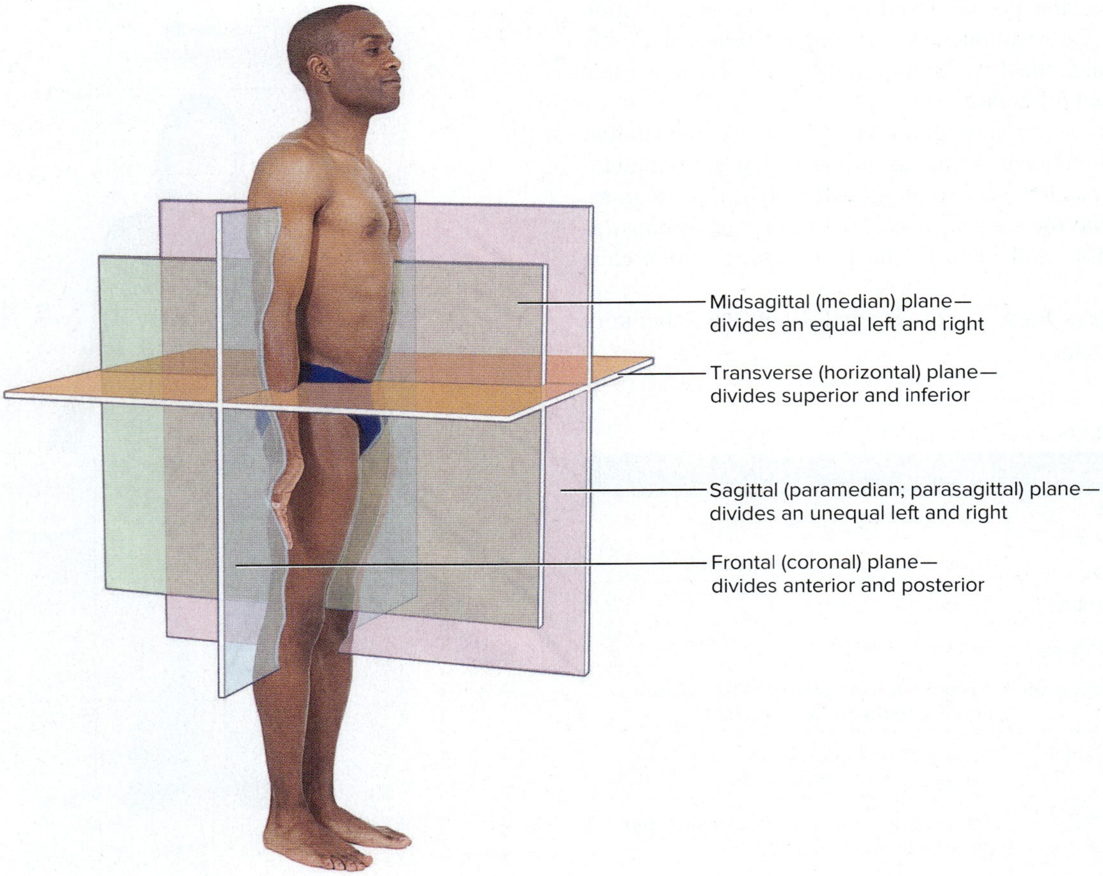

FIGURE 2.7 Anatomical planes (sections) of the body. Eric Wise/McGraw Hill APR

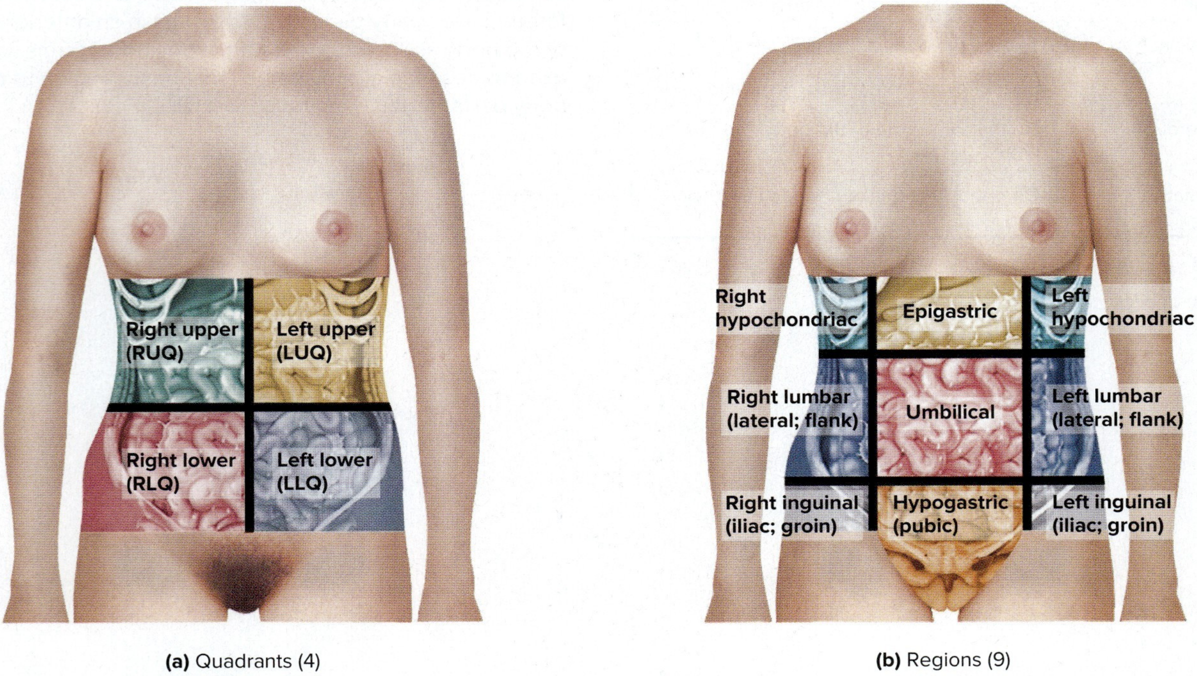

FIGURE 2.8 Abdominopelvic (a) quadrants and (b) regions. An area as large as the abdominopelvic cavity is subdivided into either four quadrants or nine regions for purposes of locating organs, injuries, or pain, or for performing medical procedures. APR

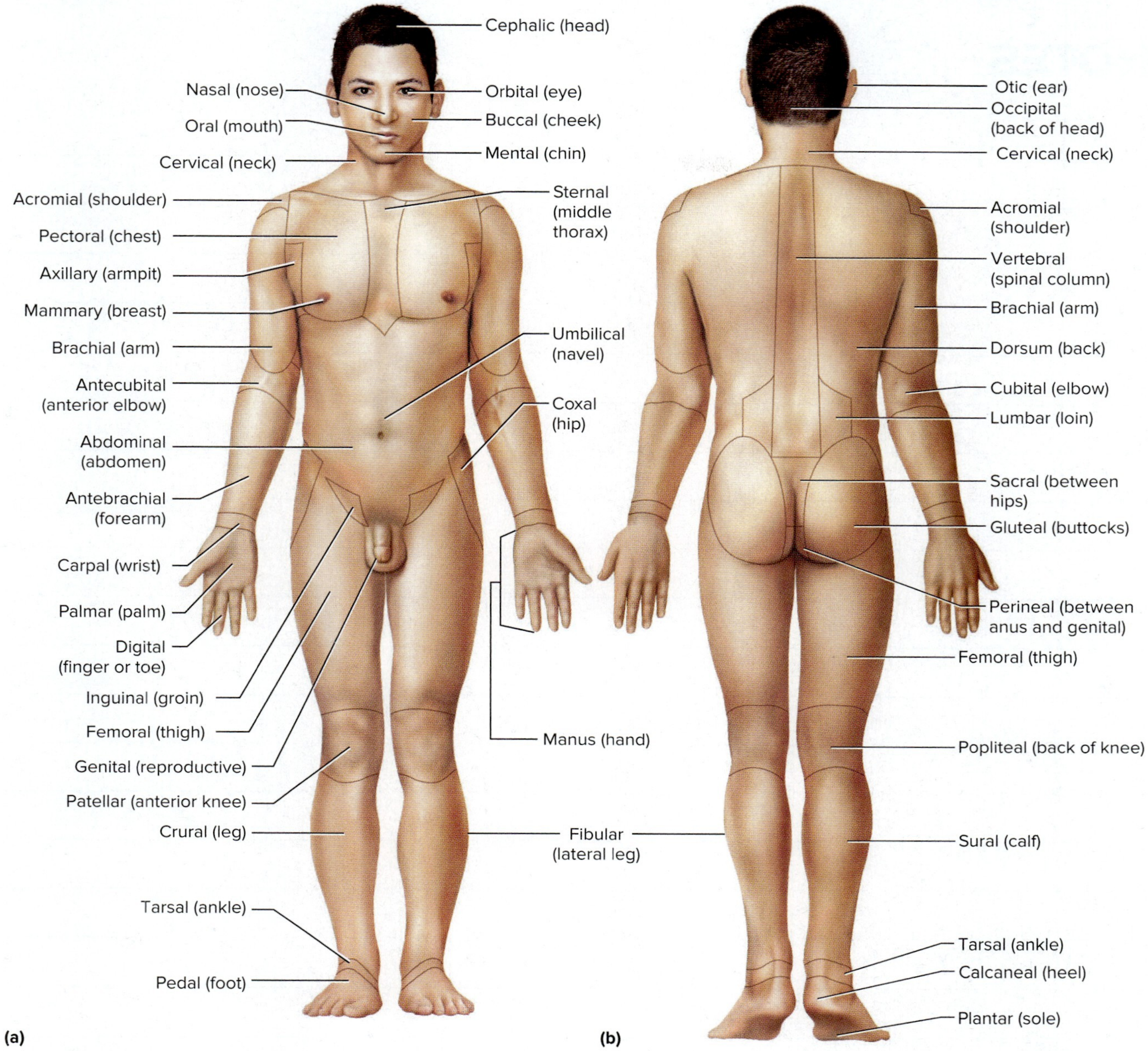

FIGURE 2.9 Diagrams of the body surface regions: (a) anterior view of regions; (b) posterior view of regions. APR

NOTES

LABORATORY
ASSESSMENT 2

Name ________________________________

Date ________________________________

Section ________________________________

The (A) corresponds to the indicated Learning Outcome(s) (O) found at the beginning of the Laboratory Exercise.

Body Organization, Membranes, and Terminology

PART A: Assessments

1. Match the body cavities in column A with the organs contained in the cavities in column B. Place the letter of your choice in the space provided. A1 A2

Column A	Column B
a. Abdominal cavity	______ 1. Liver
b. Cranial cavity	______ 2. Lungs
c. Pelvic cavity	______ 3. Spleen
d. Thoracic cavity	______ 4. Stomach
e. Vertebral canal (spinal cavity)	______ 5. Brain
	______ 6. Internal reproductive organs
	______ 7. Urinary bladder
	______ 8. Spinal cord
	______ 9. Heart
	______ 10. Small intestine

2. Label figures 2.10 to 2.13. A1 A2

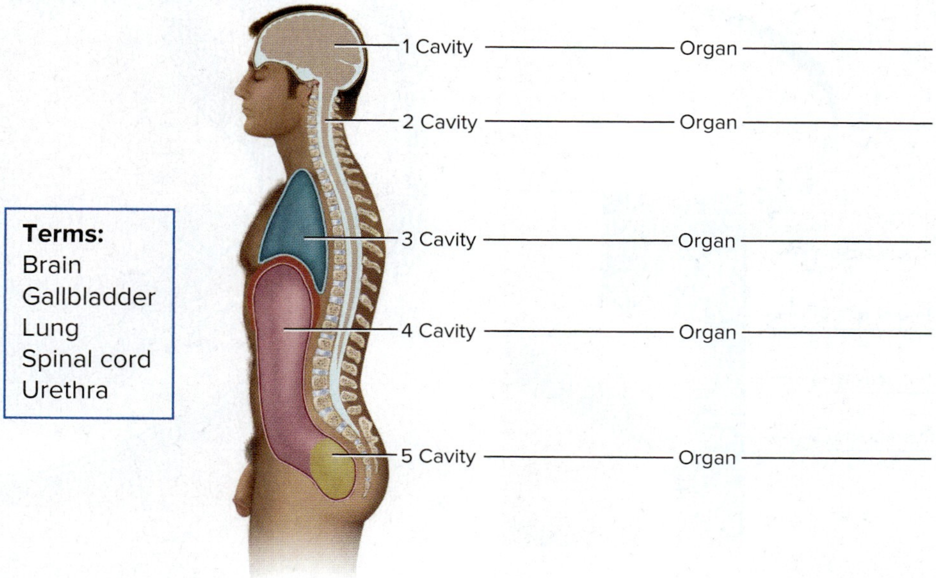

FIGURE 2.10 Label body cavities 1–5. Add a representative organ located in the cavity, using the terms provided.

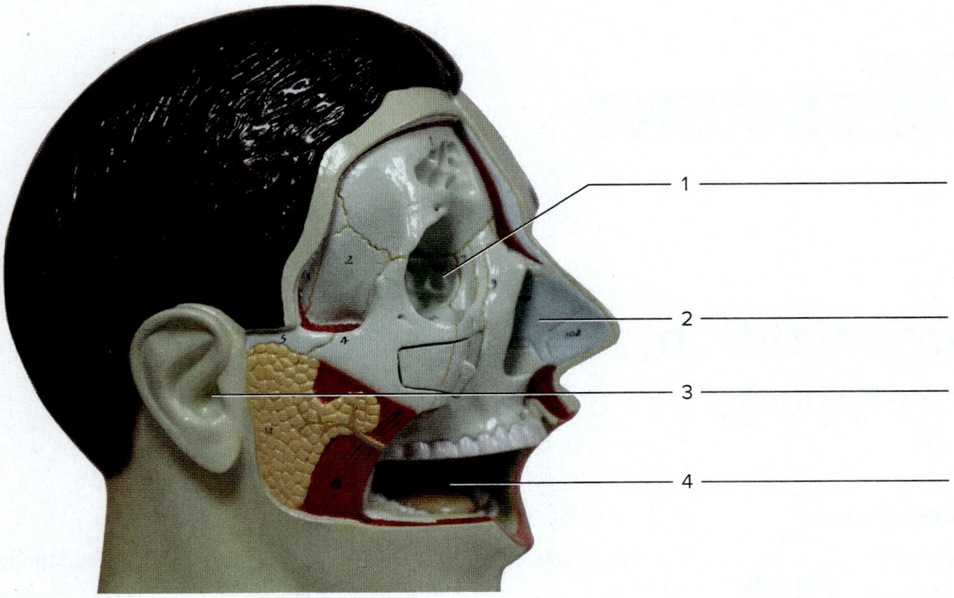

FIGURE 2.11 Label the body cavities of the head shown in this image of a model (lateral view). © 2017 Denoyer-Geppert Science Company

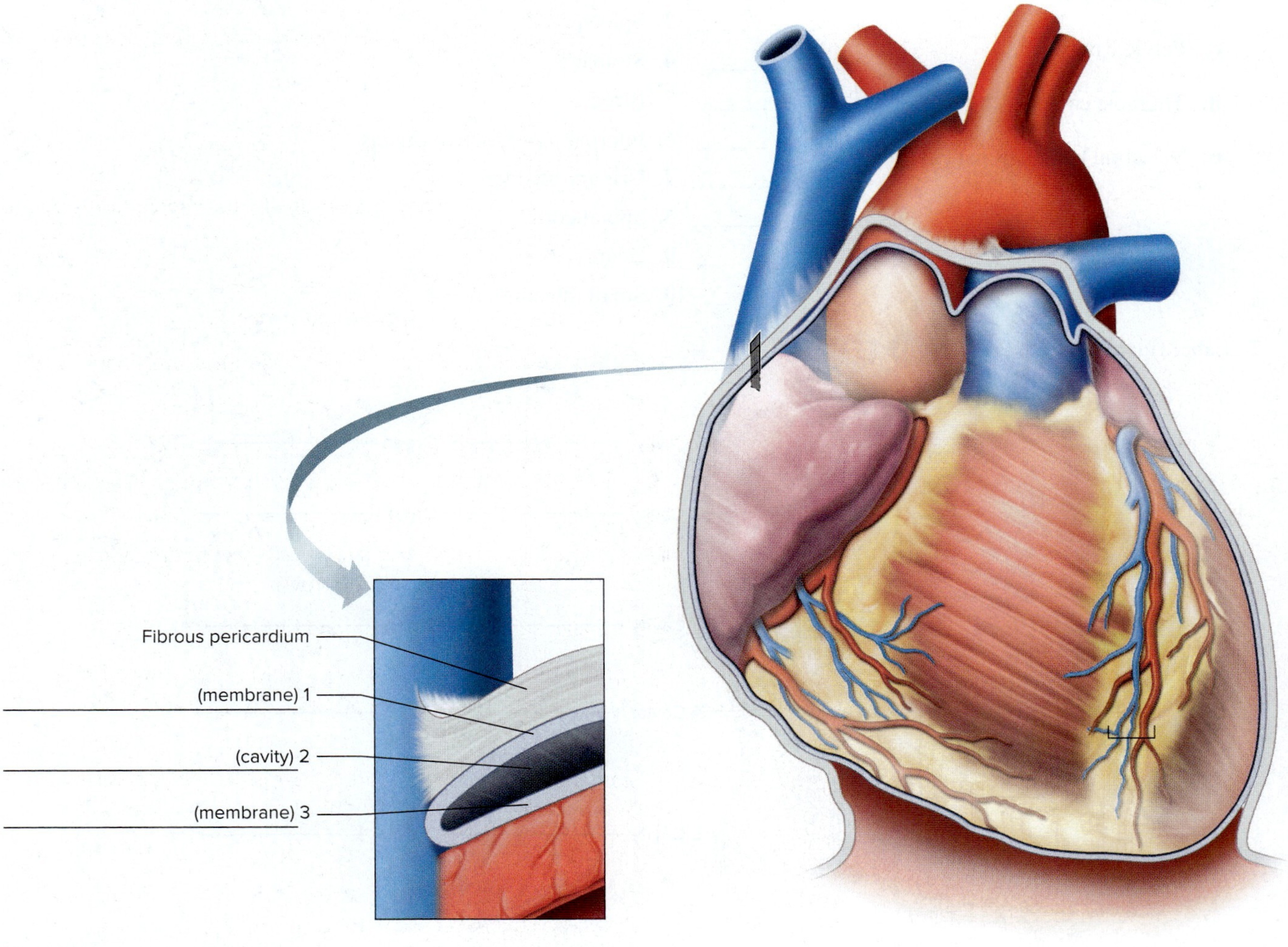

FIGURE 2.12 Label the specific serous membranes and cavity (1–3) shown in the frontal section of the heart.

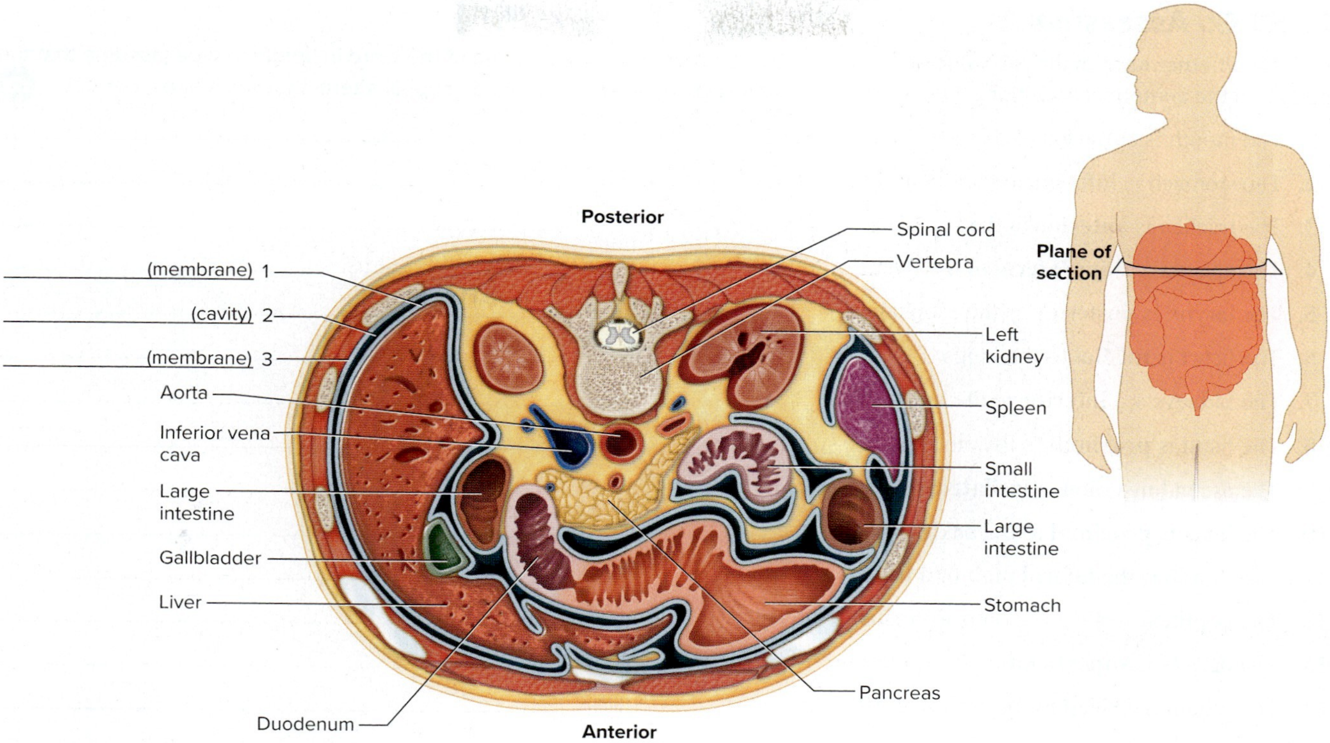

FIGURE 2.13 Label the specific serous membranes and cavity shown in this superior view of a transverse section of the abdominopelvic cavity.

PART B: Assessments

Match the organ systems in column A with the principal functions in column B. Place the letter of your choice in the space provided. A2

Column A	Column B
a. Cardiovascular system	______ 1. The main system that secretes hormones
b. Digestive system	______ 2. Provides an outer covering of the body for protection
c. Endocrine system	______ 3. Produces gametes (eggs and sperm) and offspring
d. Integumentary system	______ 4. Stimulates muscles to contract, interprets information from sensory organs, and functions in the integration and coordination of the body
e. Lymphatic system	______ 5. Provides a framework and support for soft tissues and produces blood cells in red marrow
f. Muscular system	______ 6. Exchanges gases between air and blood
g. Nervous system	______ 7. Transports excess fluid from tissues to blood and contains immune cells that help defend the body against infectious agents
h. Reproductive system	______ 8. Involves contractions that apply forces on body parts to cause movement, and creates most body heat
i. Respiratory system	______ 9. Removes liquid and wastes from blood and transports them to the outside of the body
j. Skeletal system	______ 10. Converts food molecules into forms that are absorbable
k. Urinary system	______ 11. Transports nutrients, wastes, and gases throughout the body

PART C: Assessments

Indicate whether each of the following sentences makes correct or incorrect usage of the word in boldface type (assume that the body is in the anatomical position). If the sentence is incorrect, in the space provided supply a term that will make it correct. A3

1. The mouth is **superior** to the nose. ________________
2. The stomach is **inferior** to the diaphragm. ________________
3. The trachea is **anterior** to the spinal cord. ________________
4. The appendix is **ipsilateral** to the liver. ________________
5. The larynx is **posterior** to the esophagus. ________________
6. The heart is **medial** to the lungs. ________________
7. The kidneys are **inferior** to the lungs. ________________
8. The hand is **proximal** to the elbow. ________________
9. The ascending colon is **ipsilateral** to the descending colon. ________________
10. The knee is **proximal** to the ankle. ________________
11. The thumb is the **lateral** digit of a hand. ________________
12. The popliteal surface region is **anterior** to the patellar region. ________________
13. The dermis is **superficial** to the epidermis. ________________
14. The medulla is **deep** to the cortex of the kidney. ________________

PART D: Assessments

Label figure 2.14.

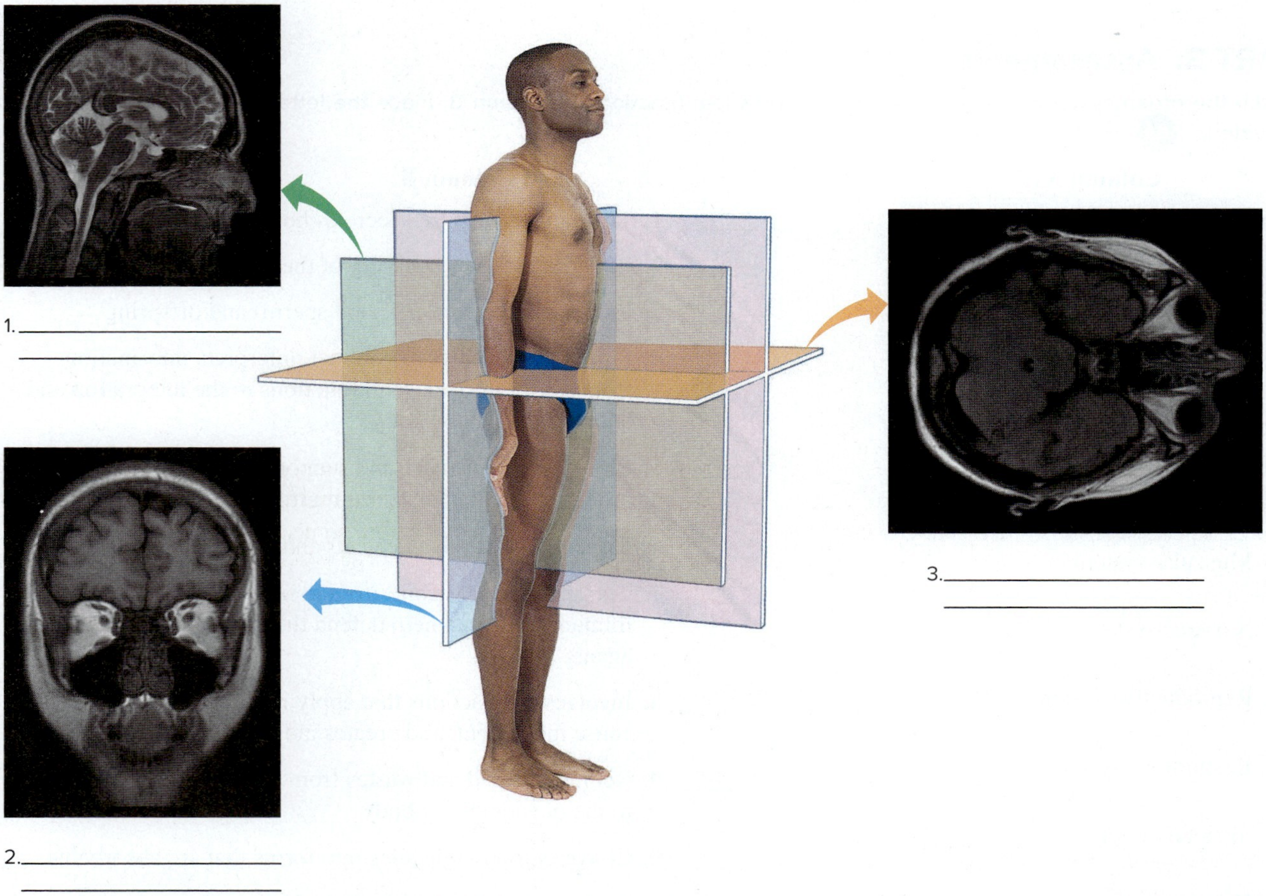

FIGURE 2.14 Label the planes of the sectioned (cut) body part. (1, 2, 3): kalewa/Shutterstock; (4): Eric Wise/McGraw Hill A4

PART E: Assessments

Label figure 2.15.

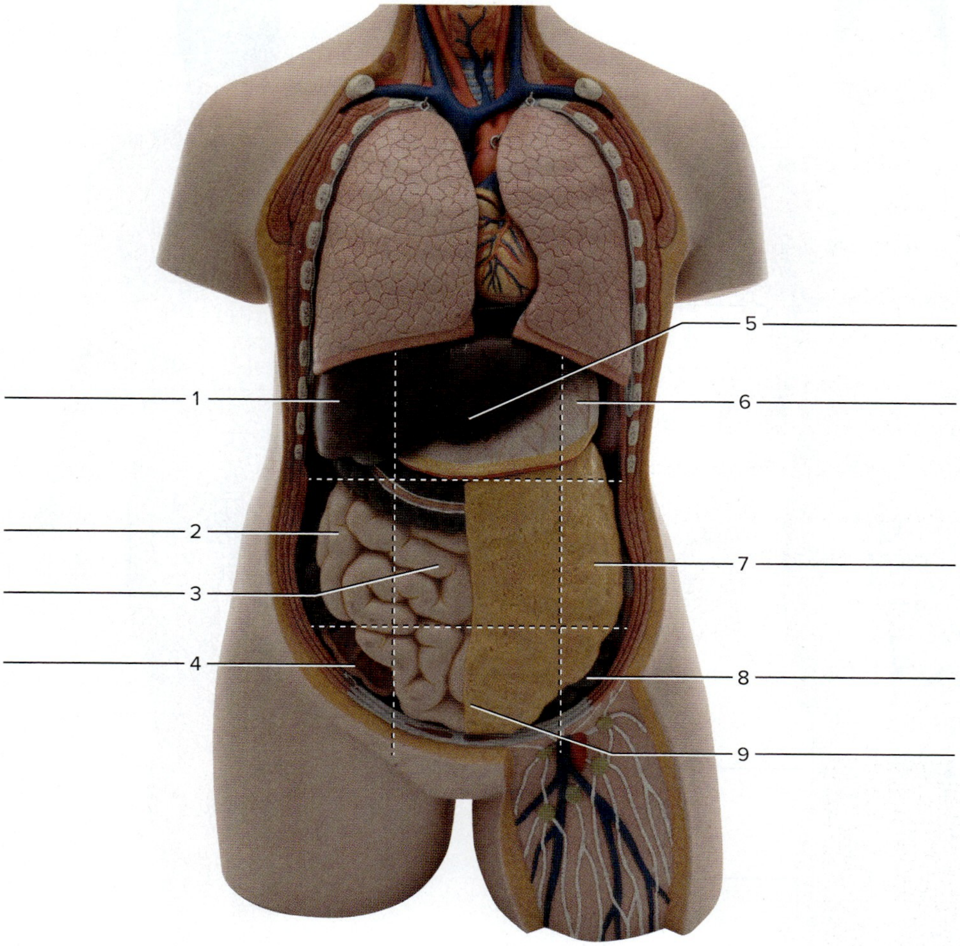

FIGURE 2.15 Label the nine abdominopelvic regions shown on the image of a torso model (anterior view). © 3B Scientific GmbH, Germany, 2017, www.3bscientific.com A5

CRITICAL THINKING ASSESSMENT—PART E

State the quadrant of the abdominopelvic cavity where the pain or sound would be located for each of the six conditions listed. In some cases, there may be more than one correct answer, and pain is sometimes referred to another region. This phenomenon, called *referred pain,* occurs when pain is interpreted as originating from some area other than the parts being stimulated. When referred pain is involved in the patient's interpretation of the pain location, the proper diagnosis of the ailment is more challenging. For the purpose of this exercise, assume the pain is interpreted as originating from the organ involved. A5

1. Stomach ulcer ______________________
2. Appendicitis ______________________
3. Bowel sounds ______________________
4. Gallbladder attack ______________________
5. Kidney stone in left ureter ______________________
6. Ruptured spleen ______________________

PART F: Assessments

Label figure 2.16.

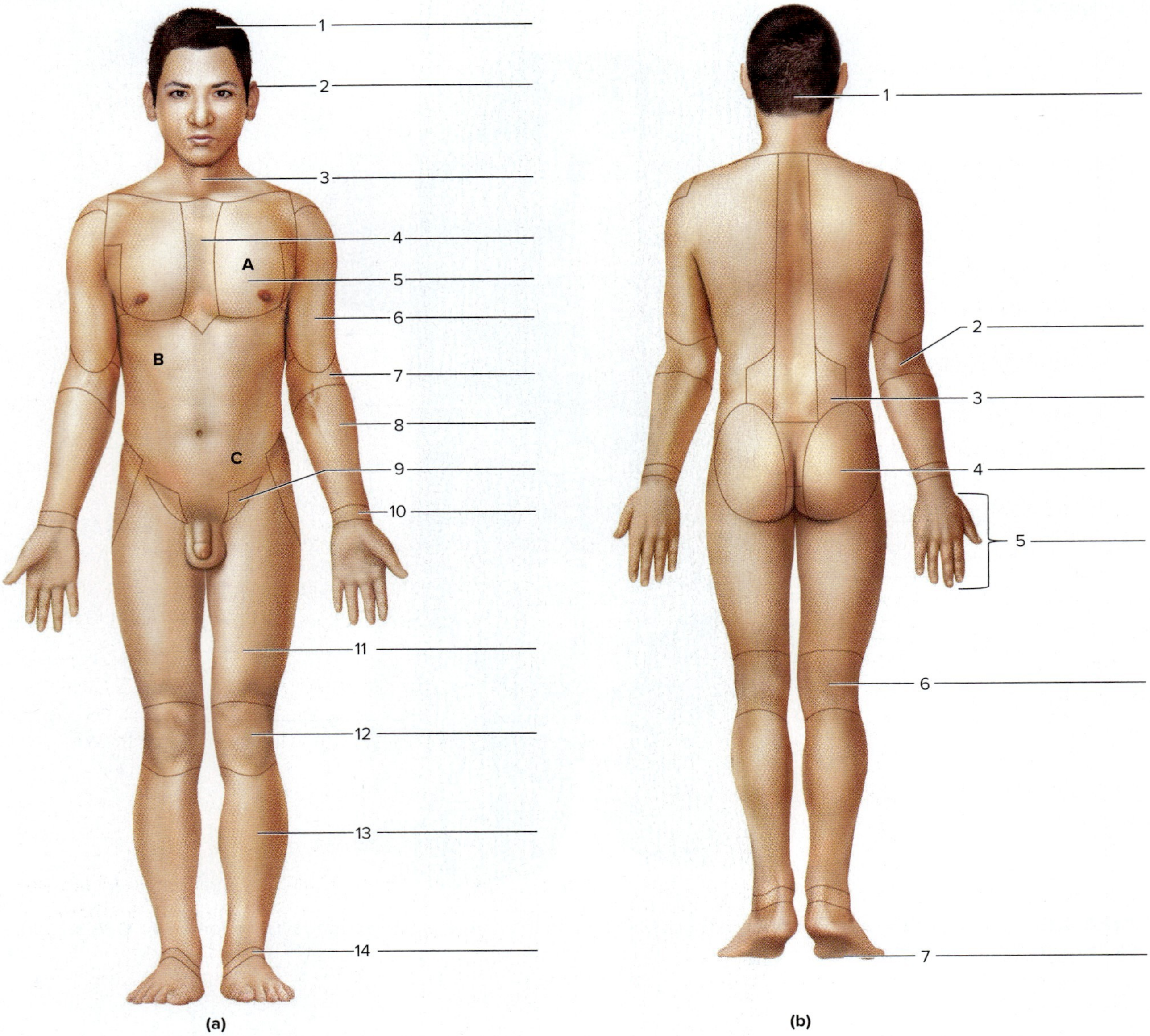

FIGURE 2.16 Label the indicated body surface regions: (a) anterior view; (b) posterior view. A5

CRITICAL THINKING ASSESSMENT—PART F

Examine the letters A, B, and C indicated on figure 2.16*a* that indicate a possible site of a deep puncture wound. Name the body cavity and the major internal organs that could be damaged from the injury. A1 A2

Injury Site	Body Cavity	Internal Organ
A		
B		
C		

LABORATORY
EXERCISE 3

Chemistry of Life

Purpose of the Exercise

To review pH and basic categories of organic compounds, and to differentiate between types of organic compounds.

MATERIALS NEEDED

For pH Tests:
- Chopped fresh red cabbage
- Beaker (250 mL)
- Distilled water
- Tap water
- Vinegar
- Baking soda
- Pipets for measuring
- Laboratory scoop for measuring
- Full-range pH test papers
- Seven assorted common liquids clearly labeled in closed bottles on a tray or in a tub
- Droppers labeled for each liquid

For Organic Tests:
- Test tubes
- Test-tube rack
- Test-tube clamps
- China marker
- Hot plate
- Beaker for hot water bath (500 mL)
- Pipets for measuring
- Benedict's solution
- Sudan IV dye
- Biuret reagent (or 10% NaOH and 1% $CuSO_4$)
- Iodine-potassium-iodide (IKI) solution
- Egg albumin
- 10% glucose solution
- Clear carbonated soft drink
- 10% starch solution
- Potatoes for potato water
- Distilled water
- Vegetable oil
- Brown paper
- Numbered unknown organic samples

SAFETY

- Review all safety guidelines in Appendix 1 of your laboratory manual.
- Clean laboratory surfaces before and after laboratory procedures using soap and water.
- Use extreme caution when working with chemicals.
- Wear protective eyewear.
- Wear disposable gloves while working with the chemicals.
- Take precautions to prevent chemicals from contacting your skin.
- Do not mix any of the chemicals together unless instructed to do so.
- Clean up any spills immediately and notify the instructor at once.
- Wash your hands before leaving the laboratory.

Chapter Opener: ©101cats/Getty Images

Learning Outcomes APR

After completing this exercise, you should be able to:

O1 Measure pH values of various substances through testing methods.

O2 Determine categories of organic compounds with basic colorimetric tests.

O3 Discover the organic composition of an unknown sample solution.

The O corresponds to the assessments A indicated in the Laboratory Assessment for this Exercise.

Pre-Lab

Carefully read the introductory material and examine the entire lab. Be familiar with pH and organic molecules from lecture or the textbook. Answer the pre-lab questions.

Pre-Lab Questions Select the correct answer for each of the following questions:

1. The most basic unit of matter is
   a. energy. **b.** an element.
   c. a molecule. **d.** a cell.
2. Which number would represent a neutral pH?
   a. 100 **b.** 0
   c. 7 **d.** 14
3. The building block of a protein is
   a. an amino acid. **b.** a monosaccharide.
   c. a lipid. **d.** starch.
4. The building block of a carbohydrate is
   a. glycerol. **b.** an amino acid.
   c. fatty acids. **d.** a monosaccharide.
5. The test to indicate the presence of starch uses which substance?
   a. IKI (iodine solution) **b.** Benedict's solution
   c. Sudan IV dye **d.** Biuret reagent

6. Using brown paper, a positive test for lipids results in a
   a. pinkish color change.
   b. yellow color change.
   c. translucent stain.
   d. blue to red color change.
7. Which of the following pH numbers would represent the strongest acid?
   a. 1
   b. 5
   c. 7
   d. 14
8. Which of the following solutions has the greatest H^+ concentration?
   a. Solution with pH 2
   b. Solution with pH 4
   c. Solution with pH 8
   d. Solution with pH 10
9. Red cabbage pigment can be used as a colorimetric indicator to determine pH value.
   a. True
   b. False

The complexities of the human body arise from the organization and interactions of chemicals. Organisms are made of matter, and the most basic unit of matter is the chemical element. The smallest unit of an element is an atom, and two or more of those can unite to form a molecule. All processes that occur within the body involve chemical reactions—interactions between atoms and molecules. We breathe to supply oxygen to our cells for energy. We eat and drink to bring chemicals into our bodies that our cells need. Water fills and bathes all of our cells and allows an amazing array of reactions to occur, all of which are designed to keep us alive. Chemistry forms the very basis of life and thus forms the foundation of anatomy and physiology.

The pH scale is from 0 to 14 with the midpoint of 7.0 (pure water) representing a *neutral* solution. If the solution has a pH lower than 7.0, it represents an *acidic* solution with more hydrogen ions (H^+) than hydroxide ions (OH^-). A *basic (alkaline)* solution has a pH higher than 7.0 with more OH^- ions than H^+ ions. The weaker acids and bases are closer to 7, with the strongest acids near 0 and the strongest bases near 14. A seemingly minor change in the pH actually represents a more significant change as there is a tenfold difference in hydrogen ion concentrations with each whole number represented on the pH scale. Our cells can only function within minimal pH fluctuations.

The *organic molecules* (biomolecules) that are tested in this laboratory exercise include *carbohydrates, lipids,* and *proteins.* Carbohydrates are used to supply energy to our cells. The polysaccharide starch is composed of simple sugar (monosaccharide) building blocks. Lipids include fats (triglycerides), phospholipids, and steroids that provide some cellular structure and energy for cells. The building blocks of most lipids include fatty acids and glycerol. Proteins compose important cellular structures, antibodies, and enzymes. Amino acids are the building blocks of proteins.

PROCEDURE A: The pH Scale

In this section, you will be able to determine the pH of various substances. The abbreviation pH is traditionally used to measure the power of H (hydrogen). The pH is a mathematical way to refer to the negative logarithm of the hydrogen ion (H^+) concentration in a solution. Each change of a number on the pH scale actually represents a tenfold increase or decrease of the hydrogen ion concentration.

The pH of various body fluids is very important in the homeostasis of a particular body system and the entire living organism. Blood, urine, saliva, and gastric secretions are examples of some body fluids with characteristic pH ranges for a healthy individual. Some body fluids like blood must fall within a very narrow pH range, while others like urine have a much wider normal range. In order to maintain normal ranges of the pH of body fluids, various buffering systems are involved to resist pH fluctuations. Some of these systems entail chemical buffers, such as the bicarbonate buffer system, and others consist of physiological buffers involving the kidneys and the lungs to help regulate the pH levels. The acid-base balance is important since our bodies are composed predominantly of water. Various acid-base relationships are covered within the discussion of each body system.

1. Study figure 3.1, which shows the pH scale. Note the range of the scale and the pH value that is considered neutral. Notice the pH of various foods and household items.

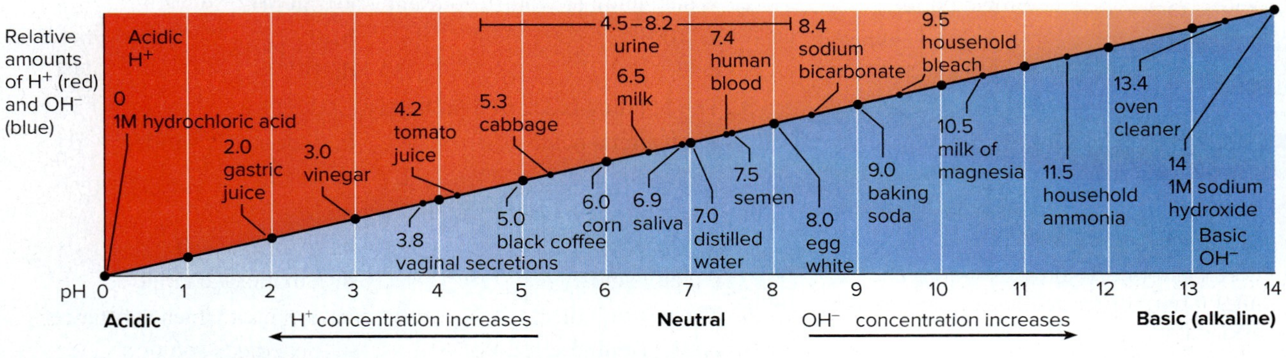

FIGURE 3.1 As the concentration of hydrogen ions (H^+) increases, a solution becomes more acidic and the pH value decreases. As the concentration of ions that combine with hydrogen ions (such as hydroxide ions) increases, a solution becomes more basic (alkaline) and the pH value increases. The pH values of some common substances are shown. APR

2. **Cabbage water tests.** Many tests can determine a pH value, but among the more interesting are colorimetric tests in which an indicator chemical changes color when it reacts. Many plant pigments, especially anthocyanins (which give plants color, from blue to red), can be used as colorimetric pH indicators. One that works well is the red pigment in red cabbage.
   a. Prepare cabbage water to be used as a general pH indicator. Fill a 250 mL beaker to the 100 mL level with chopped red cabbage. Add water to make 150 mL. Place the beaker on a hot plate and simmer the mixture until the pigments come out of the cabbage and the water turns deep purple. Allow the water to cool. (You may proceed to step 3 while you wait for this to finish.)
   b. Label three clean test tubes: one for water, one for vinegar, and one for baking soda.
   c. Place 2 mL of cabbage water into each test tube.
   d. To the first test tube, add 2 mL of distilled water and swirl the mixture. Note and record the color in Part A of Laboratory Assessment 3.
   e. Repeat this procedure for test tube 2, adding 2 mL of vinegar and swirling the mixture. Note and record the results in Part A of the laboratory assessment.
   f. Repeat this procedure for test tube 3, adding one laboratory scoop of baking soda. Swirl the mixture. Note and record the results in Part A of the laboratory assessment.
3. **Testing with pH paper.** Many commercial pH indicators are available. A very simple one to use is pH paper, which comes in small strips. Don gloves for this procedure so your skin secretions do not contaminate the paper, and to protect you from the chemicals you are testing.
   a. Test the pH of distilled water by dropping one or two drops of distilled water onto a strip of pH paper, and note the color. Compare the color to the color guide on the container. Note the results of the test as soon as color appears, as drying of the paper will change the results. Record the pH value in Part A of the laboratory assessment.
   b. Repeat this procedure for tap water. Record the results.
   c. Individually test the various common substances found on your lab table. Record your results.
   d. Complete Part A of the laboratory assessment.

PROCEDURE B: Organic Molecules

In this section, you will perform some simple tests to check for the presence of the following categories of organic molecules (biomolecules): protein, carbohydrate—sugar and starch, and lipid. Specific color changes will occur if the target compound is present. Please note the original color of the indicator being added so you can tell if the color really changed. For example, Biuret reagent and Benedict's solution are both initially blue, so an end color of blue would indicate no change. Use caution when working with these chemicals and follow all directions. Carefully label the test tubes and avoid contaminating any of your samples. Only the Benedict's test requires heating and a time delay for accurate results. Do not heat any other tubes. To prepare for the Benedict's test, fill a 500 mL beaker half full with water and place it on the hot plate. Turn the hot plate on and bring the water to a boil. To save time, start the water bath before doing the Biuret test.

1. **Biuret test for protein.** In the presence of protein, Biuret reagent reacts with peptide bonds and changes to violet or purple. A pinkish color indicates that shorter polypeptides are present. The color intensity is proportional to the number of peptide bonds; thus, the intensity reflects the length of polypeptides (amount of protein).
   a. Label six test tubes as follows: 1W, 1E, 1G, 1D, 1S, and 1P. The codes placed on the six test tubes used for the protein test include a number *1* referring to test 1 (biuret test for protein) followed by a letter. Letter *W* represents water (distilled); letter *E* represents egg (albumin); letter *G* represents glucose (10% solution); letter *D* represents drink (carbonated soft); letter *S* represents starch (solution); and letter *P* represents potato (juice). Similar codes are used for the tests for sugar (test 2) and starch (test 3) of Procedure B of this exercise.
   b. To each test tube, add 2 mL of one of the samples as follows:
   1W—2 mL distilled water
   1E—2 mL egg albumin
   1G—2 mL 10% glucose solution
   1D—2 mL carbonated soft drink
   1S—2 mL 10% starch solution
   1P—2 mL potato juice
   c. To each, add 2 mL of Biuret reagent, swirl the tube to mix it, and note the final color. [*Note:* If Biuret reagent is not available, add 2 mL of 10% NaOH (sodium hydroxide) and about ten drops of 1% $CuSO_4$ (copper sulfate).] **Be careful—NaOH is very caustic.**
   d. Record your results in Part B of the laboratory assessment.
   e. Mark a "+" on any tubes with positive results and retain them for comparison while testing your unknown sample. Put remaining test tubes aside so they will not be confused with future trials.
2. **Benedict's test for sugar (monosaccharides).** In the presence of sugar, Benedict's solution changes from its initial blue to a green, yellow, orange, or reddish color, depending on the amount of sugar present. Orange and red indicate greater amounts of sugar.
   a. Label six test tubes as follows: 2W, 2E, 2G, 2D, 2S, and 2P.
   b. To each test tube, add 2 mL of one of the samples as follows:
   2W—2 mL distilled water
   2E—2 mL egg albumin
   2G—2 mL 10% glucose solution

2D—2 mL carbonated soft drink
2S—2 mL 10% starch solution
2P—2 mL potato juice

c. To each, add 2 mL of Benedict's solution, swirl the tube to mix it, and place all tubes into the boiling water bath for 3 to 5 minutes.

d. Note the final color of each tube after heating and record the results in Part B of the laboratory assessment.

e. Mark a "+" on any tubes with positive results and retain them for comparison while testing your unknown sample. Put remaining test tubes aside so they will not be confused with future trials.

3. **Iodine test for starch.** In the presence of starch, iodine turns a dark purple or blue-black color. Starch is a long chain formed by many glucose units linked together side by side. This regular organization traps the iodine molecules and produces the dark color.

a. Label six test tubes as follows: 3W, 3E, 3G, 3D, 3S, and 3P.

b. To each test tube, add 2 mL of one of the samples as follows:

3W—2 mL distilled water
3E—2 mL egg albumin
3G—2 mL 10% glucose solution
3D—2 mL carbonated soft drink
3S—2 mL 10% starch solution
3P—2 mL potato juice

c. To each, add 0.5 mL of IKI (iodine solution) and swirl the tube to mix it.

d. Record the final color of each tube in Part B of the laboratory assessment.

e. Mark a "+" on any tubes with positive results and retain them for comparison while testing your unknown sample. Put remaining test tubes aside so they will not be confused with future trials.

4. **Tests for lipids.**

a. Label two separate areas of a piece of brown paper as "water" and "oil."

b. Place a drop of water on the area marked "water" and a drop of vegetable oil on the area marked "oil."

c. Let the spots dry several minutes, then record your observations in Part B of the laboratory assessment. Upon drying, oil leaves a translucent stain (grease spot) on brown paper; water does not leave such a spot.

d. In a test tube, add 2 mL of water and 2 mL of vegetable oil and observe. Shake the tube vigorously; then let it sit for 5 minutes and observe again. Record your observations in Part B of the laboratory assessment.

e. Sudan IV is a dye that is lipid-soluble but not water-soluble. If lipids are present, the Sudan IV will stain them a reddish color. Add a small amount of Sudan IV to the test tube that contains the oil and water. Swirl it; then let it sit for a few minutes. Record your observations in Part B of the laboratory assessment.

PROCEDURE C: Identifying Unknown Compounds

Now you will apply the information you gained with the tests in the previous section. You will retrieve an unknown sample that contains none, one, or any combination of the following types of organic compounds: protein, sugar, starch, or lipid. You will test your sample using each test from the previous section and record your results in Part C of the laboratory assessment.

1. Label four test tubes as follows: (1, 2, 3, and 4).
2. Add 2 mL of your unknown sample to each test tube.
3. **Test for protein.** Add 2 mL of Biuret reagent to tube 1. Swirl the tube and observe the color. Record your observations in Part C of the laboratory assessment.
4. **Test for sugar.** Add 2 mL of Benedict's solution to tube 2. Swirl the tube and place it in a boiling water bath for 3 to 5 minutes. Record your observations in Part C of the laboratory assessment.
5. **Test for starch.** Add several drops of iodine solution to tube 3. Swirl the tube and observe the color. Record your observations in Part C of the laboratory assessment.
6. **Test for lipid.** Add 2 mL of water to tube 4. Swirl the tube and note if there is any separation.
7. Add a small amount of Sudan IV to test tube 4, swirl the tube, and record your observations in Part C of the laboratory assessment.
8. Based on the results of these tests, determine if your unknown sample contains any organic compounds and, if so, what they are. Record and explain your identification in Part C of the laboratory assessment.

LABORATORY
ASSESSMENT 3

Name ____________________

Date ____________________

Section ____________________

The (A) corresponds to the indicated Learning Outcome(s) (O) found at the beginning of the Laboratory Exercise.

Chemistry of Life

PART A: Assessments

1. Results from cabbage water tests: A1

Substance	Cabbage Water	Distilled Water	Vinegar	Baking Soda
Color				
Acid, base, or neutral?				

2. Results from pH paper tests: A1

Substance Tested	Distilled Water	Tap Water	Sample 1: _____	Sample 2: _____	Sample 3: _____	Sample 4: _____	Sample 5: _____	Sample 6: _____	Sample 7: _____
pH Value									

3. Are the pH values the same for distilled water and tap water? ____________________

4. If not, what might explain this difference? ____________________

5. Draw the pH scale below, and indicate the following values: 0, 7 (neutral), and 14. Now label the scale with the sample names and indicate the location of the pH values for each substance you tested.

CRITICAL THINKING ASSESSMENT—PART A

A person's blood pH of 7.20 is considered acidosis, even though this pH is above 7 on the pH scale. Explain why. A1

____

____________________.

PART B: Assessments

1. **Biuret test results for protein.** Enter your results from the Biuret test for protein. A color change to purple indicates that protein is present; pink indicates that short polypeptides are present. A2

Tube	Contents	Color	Protein Present (+) or Absent (–)
1W	Distilled water		
1E	Egg albumin		
1G	Glucose solution		
1D	Soft drink		
1S	Starch solution		
1P	Potato juice		

2. **Benedict's test results for most sugars (monosaccharides).** Enter your results from the Benedict's test for sugars. A color change to green, yellow, orange, or red indicates that sugar is present. Note the color after the mixture has been heated for 3 to 5 minutes. A2

Tube	Contents	Color	Sugar Present (+) or Absent (–)
2W	Distilled water		
2E	Egg albumin		
2G	Glucose solution		
2D	Soft drink		
2S	Starch solution		
2P	Potato juice		

3. **Iodine test results for starch.** Enter your results from the iodine test for starch. A color change to dark blue or black indicates that starch is present. A2

Tube	Contents	Color	Sugar Present (+) or Absent (–)
3W	Distilled water		
3E	Egg albumin		
3G	Glucose solution		
3D	Soft drink		
3S	Starch solution		
3P	Potato juice		

4. **Lipid test results.** What did you observe when you allowed the drops of oil and water to dry on the brown paper? A2

____

What did you observe when you mixed the oil and water together? ______________________

____

What did you observe when you added the Sudan IV dye? A2 ______________________

CRITICAL THINKING ASSESSMENT—PART B

If a person were on a low-carbohydrate, high-protein diet, which of the six substances tested would the person want to increase? A2 ______________________

Explain your answer. ______________________

____

____

____

____

PART C: Assessments

1. What is the number of your sample? __________

2. Record the results of your tests on the unknown sample here: A3

Test Performed	Results
Biuret test for protein	
Benedict's test for sugars	
Iodine test for starch	
Water test for lipid	
Sudan IV test for lipid	

3. Based on these results, what organic compound(s) does your unknown sample contain? A3 ______________________

____

4. Explain your answer. A3 ______________________

____

NOTES

LABORATORY EXERCISE 4

Care and Use of the Microscope

Purpose of the Exercise

To become familiar with the major parts of a compound light microscope. Be able to use the microscope to observe small objects.

MATERIALS NEEDED

Compound light microscope
Lens paper
Microscope slides
Coverslips
Transparent plastic millimeter ruler
Prepared slide of letter *e*
Slide of three colored threads
Medicine dropper
Dissecting needle (needle probe)
Specimen examples for wet mounts
Methylene blue (dilute) or iodine-potassium-iodide stain

For Demonstration Activity:
Stereomicroscope (dissecting microscope)

SAFETY

- Review all the safety guidelines inside Appendix 1 of your laboratory manual.
- Clean laboratory surfaces before and after laboratory procedures.
- Wear disposable gloves for the wet mount procedures.
- Dispose of laboratory gloves, slides, coverslips, and specimens as instructed.
- Take precautions to prevent stains from contacting clothes and skin.
- Wash your hands before leaving the laboratory.

Chapter Opener: ©101cats/Getty Images

Learning Outcomes

After completing this exercise, you should be able to:

O1 Locate and identify the major parts of a compound light microscope and differentiate the functions of these parts.

O2 Calculate the total magnification produced by various combinations of eyepiece and objective lenses.

O3 Demonstrate proper use of the microscope to observe and measure small objects.

O4 Prepare simple wet mount microscope slides and sketch the objects you observed.

The O corresponds to the assessments A indicated in the Laboratory Assessment for this Exercise.

Pre-Lab

Carefully read the introductory material and examine the entire lab. Answer the pre-lab questions.

Pre-Lab Questions Select the correct answer for each of the following questions:

1. The human eye cannot perceive objects less than
a. 1 inch. **b.** 1 mm.
c. 0.1 mm. **d.** 1 cm.

2. The objective lenses of the compound light microscope are attached to the
a. stage. **b.** base.
c. body tube. **d.** rotating nosepiece.

3. Which objective lens provides the least total magnification?
a. scanning **b.** low power
c. high power **d.** oil immersion

4. The ________ increases or decreases the light intensity of the compound light microscope.
a. eyepiece **b.** stage
c. adjustment knob **d.** iris diaphragm

5. Basic lens cleaning is accomplished using
a. water. **b.** lens paper.
c. a paper towel. **d.** xylene.

6. When preparing a wet mount specimen for viewing, it should be covered with
a. clear paper. **b.** another glass slide.
c. a coverslip. **d.** transparent tape.

7. The total magnification achieved using a 10× objective lens with a 10× eyepiece lens is 20×.
   a. True **b.** False
8. A stereomicroscope is a good option to use for viewing rather large, opaque specimens.
   a. True **b.** False

The human eye cannot perceive objects less than 0.1 mm in diameter, so a microscope is an essential tool for the study of small structures such as cells. The microscope used for this purpose is the *compound light microscope.* It is called compound because it uses two sets of lenses: an eyepiece, or ocular lens, and an objective lens system. The eyepiece lens system magnifies, or compounds, the image reaching it after the image is magnified by the objective lens system. Such an instrument can magnify images of small objects up to about 1000×.

PROCEDURE A: Microscope Basics

This procedure includes the terminology for the parts of the compound light microscope, the purpose of the parts, and how to manipulate the components for suitable image viewing. The proper storage of the microscope, lens-cleaning techniques, and magnification determination are also included.

1. Familiarize yourself with the following list of rules for care of the microscope:
   a. Keep the microscope under its *dustcover* and in a cabinet when it is not being used.
   b. Handle the microscope with great care. It is an expensive and delicate instrument. To move it or carry it, hold it by its *arm* with one hand and support its *base* with the other hand (fig. 4.1).
   c. Always store the microscope with the lowest power objective in place. Always start with this objective when using the microscope.
   d. To clean the lenses, rub them gently with *lens paper* or a high-quality cotton swab. If the lenses need additional cleaning, follow the directions in the "Lens-Cleaning Technique" section that follows.
   e. If the microscope has a substage lamp, be sure the electric cord does not hang off the laboratory table where someone might trip over it. The bulb life can be extended if the lamp is cool before the microscope is moved.

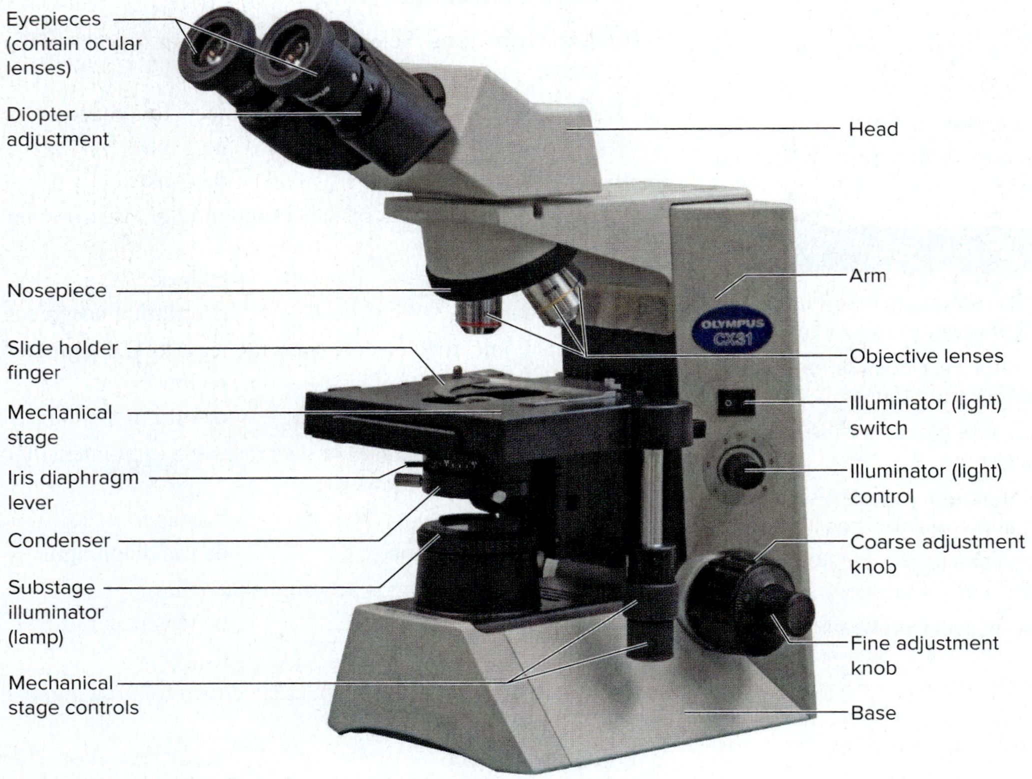

FIGURE 4.1 Major parts of a compound light microscope with a binocular (two eyepieces) head and a mechanical stage. The microscope you use may differ slightly from the one shown in this figure. James Redfearn/McGraw Hill

Table 4.1 Microscope Parts and Their Functions

Part	Function
Arm	On vertical side of the microscope, connects base and body; used to help carry the microscope
Base	Rests on flat surface and supports the microscope; used to help carry the microscope
Coarse adjustment knob	Used to get the image in focus under scanning or low-power objectives
Condenser	Concentrates and focuses the amount of light onto the microscope slide
Diopter adjustment	Adjusts focus of the eyepiece to compensate for vision differences between the two eyes
Eyepieces (ocular lenses)	Look through to make observations; magnifies the image 10×
Fine adjustment knob	Fine-tuning the focus of the image under high-power objective
Head (body)	Located between eyepiece and rotating nosepiece; provides support and path for light
Illuminator (light) control	Knob that regulates the brightness of light
Illuminator (light) switch	Turns on the light bulb
Iris diaphragm lever	Opens and closes the diaphragm, which adjusts the intensity of light entering the condenser; provides ability to control image contrast
Mechanical stage	Platform that holds the microscope slide
Mechanical stage controls	Knobs that move the microscope slide on the stage
Objective lenses	Scanning, low-power and high-power lenses that magnify the image of an object as it projects light upward into the head
Rotating nosepiece	Rotates objective lenses into position to achieve different magnifications
Slide holder finger; stage clips	Secures the microscope slide onto the mechanical stage
Substage illuminator	Lamp that allows light to enter the condenser, objective lens, and eyepiece

LENS-CLEANING TECHNIQUE

1. Moisten one end of a high-quality cotton swab (or lens-cleaning paper) with one drop of lens cleaner. Keep the other end dry.
2. Clean the optical surface with the wet end. Dry it with the other end, using a circular motion.
3. Use a hand aspirator to remove lingering dust particles.
4. Start with the scanning objective and work upward in magnification, using a new cotton swab or lens-cleaning paper for each objective.
5. When cleaning the eyepiece, do not open the lens unless it is absolutely necessary.
6. Use alcohol for difficult cleaning.

   f. Never drag the microscope across the laboratory table.
   g. Never remove parts of the microscope or try to disassemble the eyepiece or objective lenses.
   h. If your microscope is not functioning properly, report the problem to your laboratory instructor immediately.

2. Observe a compound light microscope and study figure 4.1 and table 4.1 to learn the names of its major parts and their functions. Your microscope might be equipped with a binocular head (body) with two eyepieces (fig. 4.2), or a monocular (one eyepiece) body. The lens system of a compound microscope includes three parts—the condenser, objective lens, and eyepiece (ocular lens).

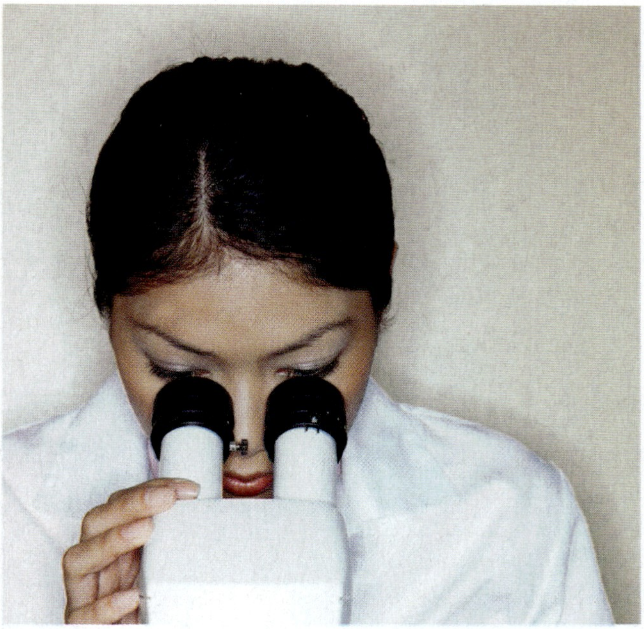

FIGURE 4.2 Scientist using a compound light microscope equipped with a binocular body and two eyepieces. Notice both eyes are open when viewing. George Doyle/Stockbyte/Getty Images

Light enters this system from a *substage illuminator (lamp)* and is concentrated and focused by a *condenser* onto a microscope slide (fig. 4.3). The condenser, which contains a set of lenses, usually is kept in its highest position possible.

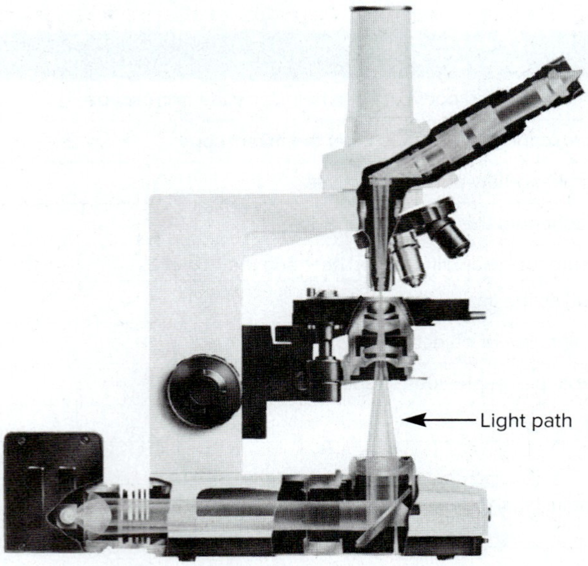

FIGURE 4.3 Microscope, showing the path of light through it. ©Nikon Instruments, Inc.

The *iris diaphragm,* located between the light source and the condenser, can be used to increase or decrease the intensity of the light entering the condenser and to control the contrast of the image. Locate the lever that operates the iris diaphragm beneath the stage and move it back and forth. Note how this movement causes the size of the opening in the diaphragm to change. (Some microscopes have a revolving plate called a disc diaphragm beneath the stage instead of an iris diaphragm. Disc diaphragms have different-sized holes to admit varying amounts of light.) Which way do you move the diaphragm to increase the light intensity (left or right)? ___________ Which way to decrease it (left or right)? ___________

After light passes through a specimen mounted on a microscope slide, it enters an *objective lens system.* This lens projects the light upward into the *head (body),* where it produces a magnified image of the object being viewed.

The *eyepiece (ocular) lens system* then magnifies this image to produce another image seen by the eye. Typically, the eyepiece lens magnifies the image ten times (10×). Look for the number in the metal of the eyepiece that indicates its power (fig. 4.4). What is the eyepiece power of your microscope? ___________

The objective lenses are mounted in a *rotating nosepiece* so that different magnifications can be achieved by rotating any one of several objective lenses into position above the specimen. Commonly, this set of lenses includes a scanning objective (4×), a low-power

(a) (b)

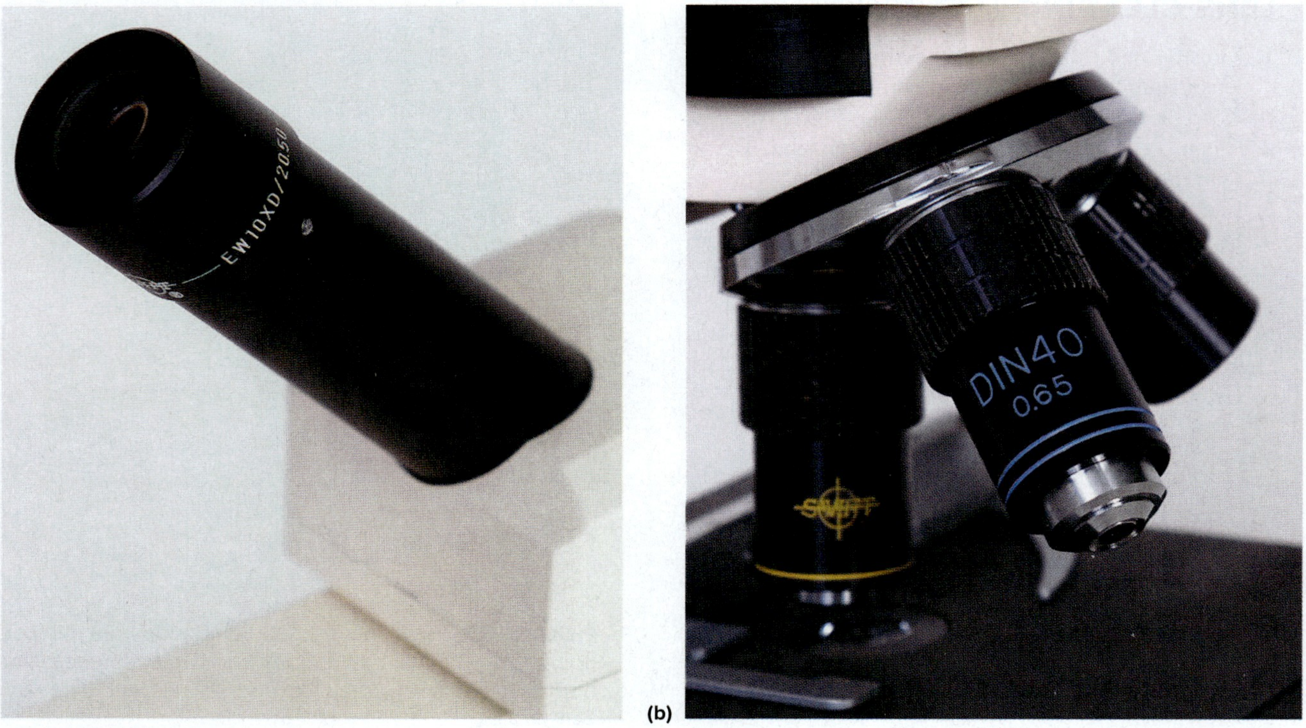

FIGURE 4.4 The powers of this (a) 10× monocular eyepiece and (b) 40× objective are marked in the metal. DIN is an international optical standard on quality optics. The 0.65 on the 40× objective is the numerical aperture, a measure of the light-gathering capabilities. (a, b): ©J & J Photography

Table 4.2 Microscope Lenses

Objective Lens Name	Common Objective Lens Magnification	Common Eyepiece Lens Magnification	Total Magnification
Scanning	4×	10×	40×
Low-power (LP)	10×	10×	100×
High-power (HP)	40×	10×	400×
Oil immersion	100×	10×	1,000×

Note: If you wish to observe an object under LP, HP, or oil immersion, locate and then center and focus the object first under scanning magnification.

objective (10×), and a high-power objective, also called a high-dry-power objective (about 40×). Sometimes an oil immersion objective (about 100×) is present. Look for the number printed on each objective that indicates its power. What are the objective lens powers of your microscope? ________________

To calculate the *total magnification* achieved when using a particular objective, multiply the power of the eyepiece by the power of the objective used. Thus, the 10× eyepiece and the 40× objective produce a total magnification of 10 × 40, or 400×. See a summary of microscope lenses in table 4.2.

3. Complete Part A of Laboratory Assessment 4.
4. Turn on the substage illuminator and look through the eyepiece. You will see a lighted, circular area called the *field of view.*

   You can measure the diameter of this field of view by focusing the lenses on the millimeter scale of a transparent plastic ruler. To do this, follow these steps:

   a. Place the ruler on the microscope stage in the spring clamp of a slide holder finger on a *mechanical stage* or under the *stage (slide) clips.* (*Note:* If your microscope is equipped with a mechanical stage, it may be necessary to use a short section cut from a transparent plastic ruler. The section should be several millimeters long and can be mounted on a microscope slide for viewing.)

   b. Center the millimeter scale in the beam of light coming up through the condenser, and rotate the scanning objective into position.

   c. While you watch from the side to prevent the lens from touching anything, raise the stage until the objective is as close to the ruler as possible, using the *coarse adjustment knob.* (*Note:* The adjustment knobs on some microscopes move the body and objectives downward and upward for focusing.)

   d. Look into the eyepiece and use the *fine adjustment knob* to raise the stage until the lines of the millimeter scale come into sharp focus.

   e. Adjust the light intensity by moving the *iris diaphragm lever* so that the field of view is brightly illuminated but comfortable to your eye. At the same time, take care not to overilluminate the field because transparent objects tend to disappear in bright light.

   f. Position the millimeter ruler so that its scale crosses the greatest diameter of the field of view. In addition, move the ruler so that one of the millimeter marks is against the edge of the field of view.

   g. In millimeters, measure the distance across the field of view.
5. Complete Part B of the laboratory assessment.
6. Most microscopes are designed to be *parfocal.* This means that when a specimen is in focus with a lower-power objective, it will be in focus (or nearly so) when a higher-power objective is rotated into position. Always center the specimen in the field of view using the *mechanical stage controls* before changing to higher objectives.

   Rotate the low-power objective into position, and then look at the millimeter scale of the transparent plastic ruler. If you need to move the low-power objective to sharpen the focus, use the fine adjustment knob.

   Adjust the iris diaphragm so that the field of view is properly illuminated. Once again, adjust the millimeter ruler so that the scale crosses the field of view through its greatest diameter, and position the ruler so that a millimeter mark is against one edge of the field. Measure the distance across the field of view in millimeters.
7. Rotate the high-power objective into position, while you watch from the side (fig. 4.5), and then observe the millimeter scale on the plastic ruler. *All focusing using*

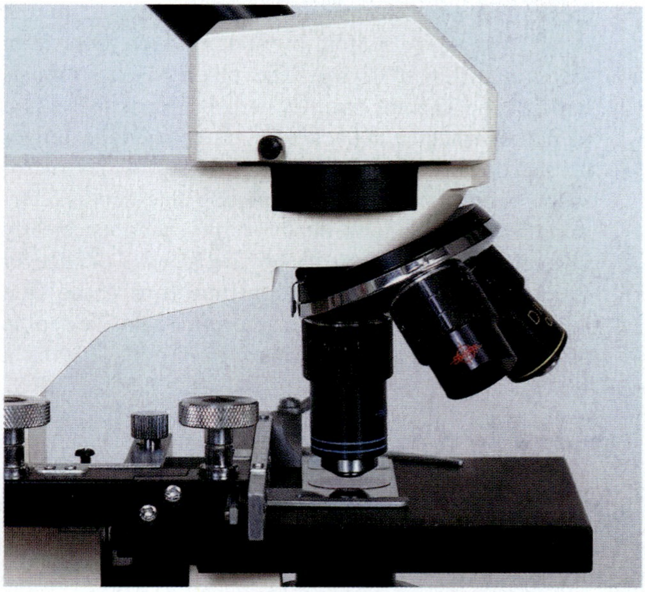

FIGURE 4.5 When you focus using a particular objective, you can prevent it from touching the specimen by watching from the side. ©J & J Photography

high-power magnification should be done only with the fine adjustment knob. If you use the coarse adjustment knob with the high-power objective, you can accidentally force the objective into the coverslip and break the slide. This is because the *working distance* (the distance from the objective lens to the slide on the stage) is much shorter when using higher magnifications.

Adjust the iris diaphragm for proper illumination. When using higher magnifications, more illumination usually will help you to view the objects more clearly. Measure the distance across the field of view in millimeters.

8. Locate the numeral 4 (or 9) on the plastic ruler and focus on it using the scanning objective. Note how the number appears in the field of view. Move the plastic ruler to the right and note which way the image moves. Slide the ruler away from you and again note how the image moves.
9. Observe a letter *e* slide. Directly view with the naked eye the orientation of the letter *e* on the slide. Then, place the letter *e* slide securely in position on the mechanical stage of the microscope and note the orientation of the letter *e* using the scanning, LP, and HP objectives. As you increase magnifications, note that the amount of the letter *e* shown is decreased. Move the slide to the left, then away from you, and note the direction in which the observed image moves. If a *pointer* is visible in your field of view, you can manipulate the pointer to a location (structure) within the field of view by moving the slide or rotating the eyepiece.
10. Examine the slide of the three colored threads using the low-power objective and then the high-power objective. Focus on the location where the three threads cross. Using the fine adjustment knob, determine the order from top to bottom by noting which color is in focus at different depths. The other colored threads will still be visible, but they will be blurred. Be sure to notice whether the stage or the body tube moves up and down with the adjustment knobs of the microscope being used before completing this depth determination. The vertical depth of the specimen clearly in focus is called the *depth of field (focus).* Whenever specimens are examined, continue to use the fine adjustment focusing knob to determine relative depths of structures clearly in focus within cells, giving a three-dimensional perspective. The depth of field is less at higher magnifications.
11. Complete Parts C and D of the laboratory assessment.

CRITICAL THINKING ACTIVITY

What was the sequence of the three colored threads from top to bottom? Explain how you came to that conclusion.

____

____

____

PROCEDURE B: Slide Preparation

This procedure includes techniques for making your own temporary slides. If your microscope has an oil immersion objective lens, the special procedures and care are described. For viewing thicker, opaque specimens, a stereomicroscope demonstration activity is included.

1. Prepare several temporary *wet mounts,* using any small, transparent objects of interest, and examine the specimens using the scanning objective, then the low-power objective, and then the high-power objective to observe their details. To prepare a wet mount, follow these steps (fig. 4.6):
   a. Wear proper safety equipment.
   b. Obtain a precleaned microscope slide and coverslip.
   c. Place a tiny, thin piece of the specimen you want to observe in the center of the slide, and use a medicine dropper to put a drop of water over it. Consult with your instructor if a drop of stain might enhance the image of any cellular structures of your specimen. If the specimen is solid, you might want to tease some of it apart with dissecting needles. In any case, the specimen must be thin enough that light can pass through it. Why is it necessary for the specimen to be so thin?

      ____

      ____

   d. Cover the specimen with a coverslip. Try to avoid trapping bubbles of air beneath the coverslip by slowly dragging the edge of the coverslip to the water or stain. When the edge of the coverslip makes contact with the water or stain, drop it from a 45-degree angle.
   e. If your microscope has an inclination joint, do not tilt the microscope while observing wet mounts because the fluid will flow.
   f. Place the slide under the stage (slide) clips or in the slide holder finger on a mechanical stage, and position the slide so that the specimen is centered in the light beam passing up through the condenser.
   g. Focus on the specimen using the scanning objective first. Next focus using the low-power objective, and then examine it with the high-power objective.
2. If an oil immersion objective is available, use it to examine the specimen. To use the oil immersion objective, follow these steps:
   a. Center the object you want to study under the high-power field of view.
   b. Rotate the high-power objective away from the microscope slide, place a small drop of immersion oil on the coverslip, and swing the oil immersion

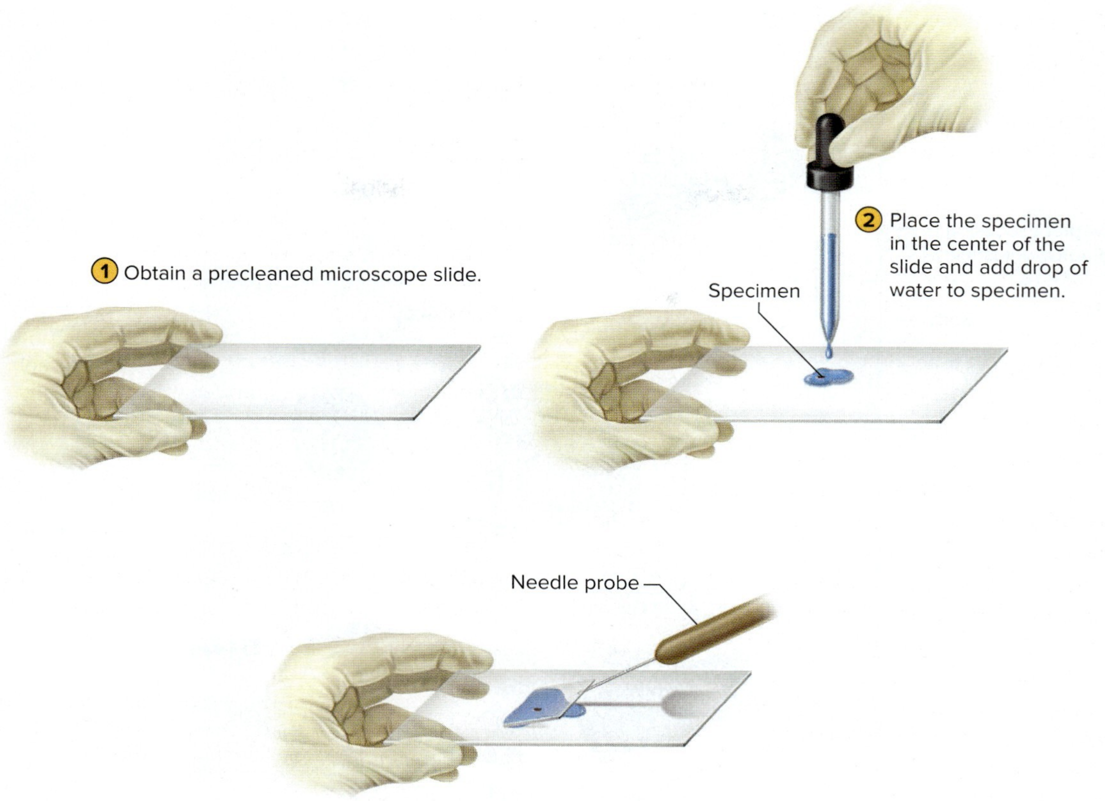

FIGURE 4.6 Steps in the preparation of a wet mount.

objective into position. Use the fine adjustment knob to achieve sharp focus.

c. You will need to open the iris diaphragm more fully for proper illumination. More light is needed because the oil immersion objective covers a very small lighted area of the microscope slide.

d. The oil immersion objective must be very close to the coverslip to achieve sharp focus, so care must be taken to avoid breaking the coverslip or damaging the objective lens. For this reason, never lower the objective when you are looking into the eyepiece. Instead, always raise the objective to achieve focus, or prevent the objective from touching the coverslip by watching the microscope slide and coverslip from the side if the objective needs to be lowered. *When using the oil immersion objective, only the fine adjustment knob should be used for focusing. Never switch back to one of the other three objective lenses while using oil immersion because this could damage the microscope.*

DEMONSTRATION ACTIVITY

A stereomicroscope (dissecting microscope) (fig. 4.7) is useful for observing the details of relatively large, opaque specimens. Although this type of microscope achieves less magnification than a compound light microscope, it has the advantage of producing a three-dimensional image rather than the flat, two-dimensional image of the compound light microscope. In addition, the image produced by the stereomicroscope is positioned in the same manner as the specimen, rather than being reversed and inverted, as it is by the compound light microscope.

Observe the stereomicroscope. The eyepieces can be pushed apart or together to fit the distance between your eyes. Focus the microscope on the end of your finger. Which way does the image move when you move your finger to the right? ______________________________ When you move it away from you? ______________________________

If the instrument has more than one objective, change the magnification to a higher power. Use the instrument to examine various small, opaque objects available in the laboratory.

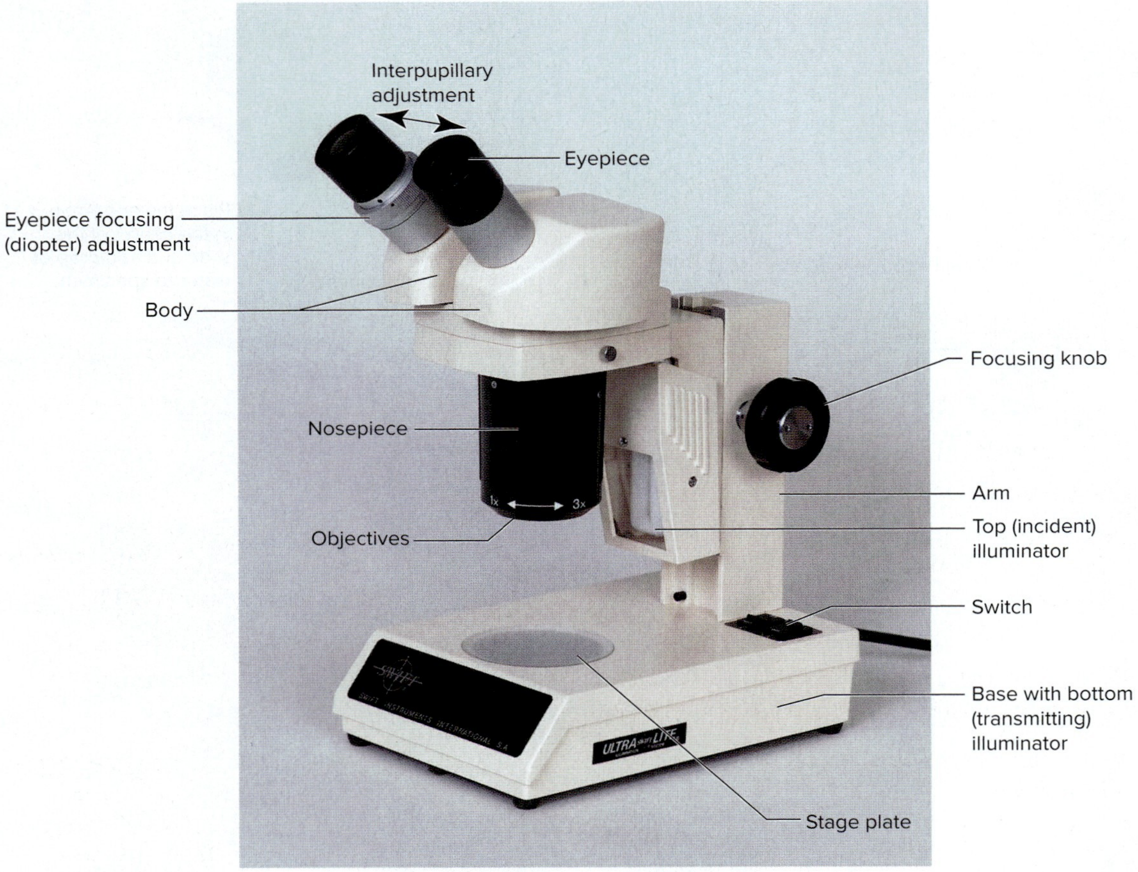

FIGURE 4.7 A stereomicroscope, also called a dissecting microscope. ©J & J Photography

3. When you have finished working with the microscope, remove the microscope slide from the stage and wipe any oil from the objective lens with lens paper or a high-quality cotton swab. Swing the scanning objective or the low-power objective into position. Wrap the electric cord around the base or arm of the microscope and replace the dustcover.
4. Complete Part E of the laboratory assessment.

Name ______________________________

Date ______________________________

Section ______________________________

The (A) corresponds to the indicated Learning Outcome(s) (O) found at the beginning of the Laboratory Exercise.

LABORATORY ASSESSMENT 4

Care and Use of the Microscope

PART A: Assessments

Revisit Procedure A, number 2; then complete the following:

1. What total magnification will be achieved if the 10× eyepiece and the 10× objective are used? (A2) ____________

2. What total magnification will be achieved if the 10× eyepiece and the 100× objective are used? (A2) ____________

PART B: Assessments

Revisit Procedure A, number 4; then complete the following:

1. Sketch the millimeter scale as it appears under the scanning objective magnification. (The circle represents the field of view through the microscope.)

2. In millimeters, what is the diameter of the scanning field of view? (A3) ____________

3. Microscopic objects often are measured in *micrometers*. A micrometer equals 1/1,000 of a millimeter and is symbolized by μm. In micrometers, what is the diameter of the scanning power field of view? (A3) ____________

4. If a circular object or specimen extends halfway across the scanning field, what is its diameter in millimeters? (A3) ____________

5. In micrometers, what is its diameter? (A3) ____________

PART C: Assessments

Complete the following:

1. Sketch the millimeter scale as it appears using the low-power objective. (A3)

2. What do you estimate the diameter of this field of view to be in millimeters? (A3) ____________

3. How does the diameter of the scanning power field of view compare with that of the low-power field? (A3)

____

4. Why is it more difficult to measure the diameter of the high-power field of view than that of the low-power field? (A3)

____

____

5. What change occurred in the intensity of the light in the field of view when you exchanged the low-power objective for the high-power objective? **A3** ______________________

6. Sketch the numeral 4 (or 9) or letter *e* as it appears through the scanning objective of the compound microscope. **A3**

7. What has the lens system done to the image of the numeral or letter? (Is it right side up, upside down, or what?) **A3** ______________________

8. When you moved the ruler to the right, which way did the image move? **A3** __________

9. When you moved the ruler away from you, which way did the image move? **A3** __________

PART D: Assessments

1. Label the microscope parts in figure 4.8. **A1**

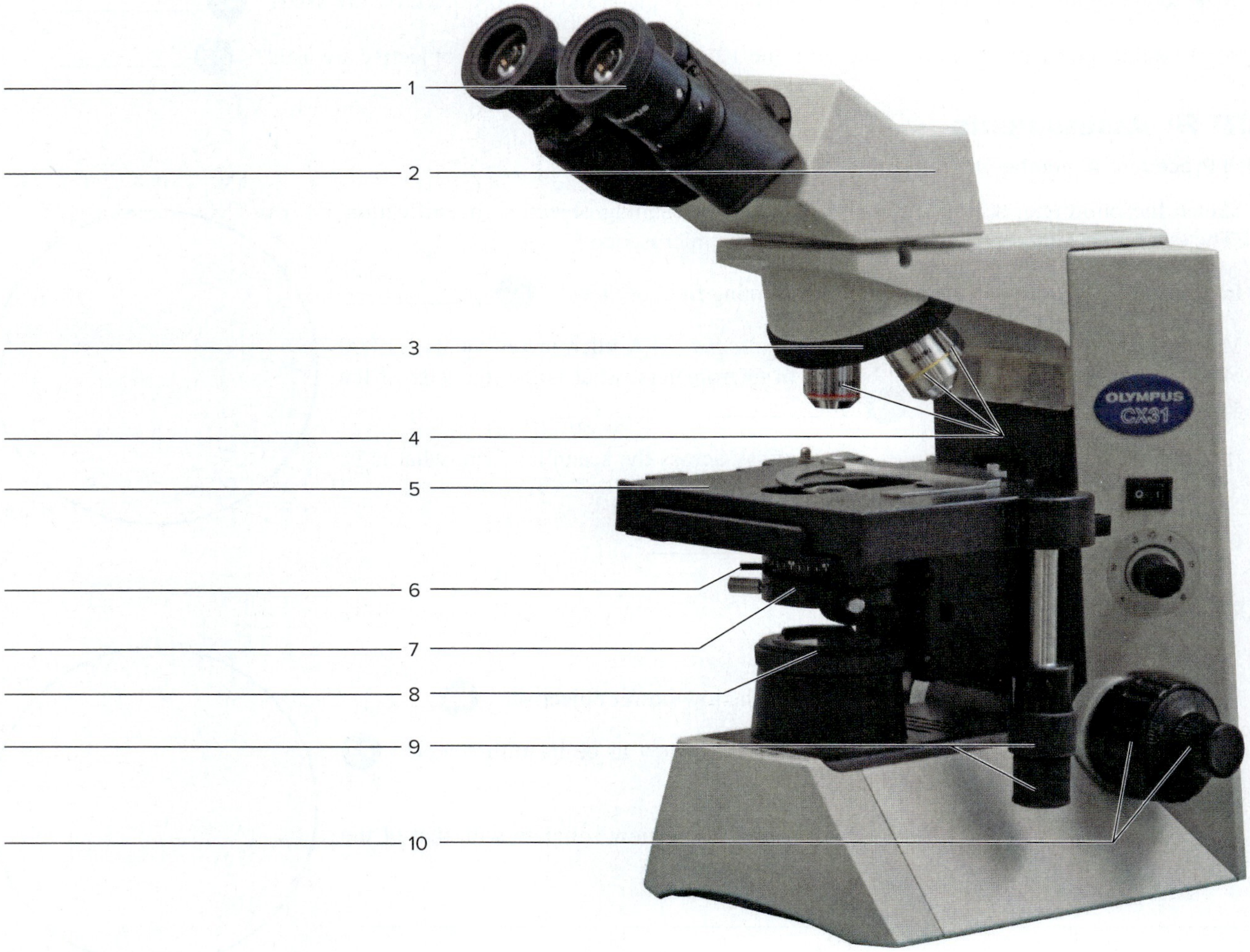

FIGURE 4.8 Identify the parts indicated on this compound light microscope. James Redfearn/McGraw Hill

2. Match the names of the microscope parts in column A with the descriptions in column B. Place the letter of your choice in the space provided. A1

Column A

- **a.** Adjustment knob (coarse)
- **b.** Arm
- **c.** Condenser
- **d.** Eyepiece (ocular lens)
- **e.** Field of view
- **f.** Iris diaphragm
- **g.** Mechanical stage
- **h.** Nosepiece
- **i.** Objective lens system
- **j.** Stage clip

Column B

_______ **1.** Increases or decreases the light intensity

_______ **2.** Platform that supports a microscope slide

_______ **3.** Concentrates light onto the specimen

_______ **4.** Causes stage (or objective lens) to move upward or downward

_______ **5.** After light passes through the specimen, it enters this lens system

_______ **6.** Holds a microscope slide in position

_______ **7.** Contains a lens (or two) at the top of the head (body)

_______ **8.** Serves as a handle for carrying the microscope

_______ **9.** Part to which the objective lenses are attached

_______ **10.** Circular area seen through the eyepiece

PART E: Assessments

Using the field-of-view circles provided, sketch the objects you observed using the microscope. For each sketch, include the name of the object, the magnification you used to observe it, and its estimated dimensions in millimeters and micrometers. A4

____

_______ ×

_______ mm

____

_______ ×

_______ mm

____

_______ ×

_______ mm

____

_______ ×

_______ mm

NOTES

LABORATORY
EXERCISE 5

Cell Structure and Function

Purpose of the Exercise

To review the structure and functions of major cellular components and to observe examples of human cells.

MATERIALS NEEDED

Animal cell model
Clean microscope slides
Coverslips
Flat toothpicks
Medicine dropper
Methylene blue (dilute) or iodine-potassium-iodide stain
Prepared microscope slides of human tissues (possible examples: sperm smear, blood smear, teased smooth muscle, liver tissue)
Compound light microscope

For Learning Extension Activities:
Single-edged razor blade
Plant materials such as leaves, soft stems, fruits, onion peel, and vegetables
Cultures of *Amoeba* and *Paramecium*

SAFETY

- Review all the safety guidelines in Appendix 1 of your laboratory manual.
- Clean laboratory surfaces before and after laboratory procedures.
- Wear disposable gloves and protective eyewear for the wet-mount procedures of the cells lining the inside of the cheek.
- Work only with your own materials when preparing the slide of the cheek cells. Observe the same precautions as with all body fluids.
- Dispose of laboratory gloves, slides, coverslips, and toothpicks as instructed.
- Use the biohazard container to dispose of items used during the cheek cells procedure.
- Take precautions to prevent stains from contacting your clothes and skin.
- Wash your hands before leaving the laboratory.

Learning Outcomes APR

After completing this exercise, you should be able to:

O1 Name and locate the components of a cell.
O2 Differentiate the functions of cellular components.
O3 Prepare a wet mount of cells lining the inside of the cheek; stain the cells; and identify the plasma (cell) membrane, nucleus, and cytoplasm.
O4 Examine cells on prepared slides of human tissues and identify their major components.

The O corresponds to the assessments A indicated in the Laboratory Assessment for this Exercise.

Pre-Lab

Carefully read the introductory material and examine the entire lab. Be familiar with the basic structures and functions of a cell from lecture or the textbook. Answer the pre-lab questions.

Pre-Lab Questions Select the correct answer for each of the following questions:

1. Which of the following cellular structures is *not* easily visible with the compound light microscope?
   a. nucleus **b.** DNA
   c. cytoplasm **d.** plasma membrane
2. Which of the following cellular structures is located in the nucleus?
   a. nucleolus **b.** ribosomes
   c. mitochondria **d.** endoplasmic reticulum
3. The outer boundary of a cell is the
   a. mitochondrial membrane. **b.** nuclear envelope.
   c. Golgi apparatus. **d.** plasma membrane.
4. Microtubules, intermediate filaments, and microfilaments are components of
   a. vesicles. **b.** the Golgi apparatus.
   c. the cytoskeleton. **d.** ribosomes.
5. Easily attainable living cells observed in this lab are from
   a. inside the cheek. **b.** blood.
   c. hair. **d.** finger surface.

6. A slide of human cheek cells can be stained to observe _____ under the microscope.
   a. individual membrane proteins
   b. Golgi apparatus
   c. the plasma membrane and nucleus
   d. mitochondria
7. Cellular energy is stored in
   a. ER. b. ATP. c. DNA. d. RNA.
8. The smooth ER possesses ribosomes.
   a. True b. False
9. The nuclear envelope contains nuclear pores.
   a. True b. False

Cells are the smallest units that possess the characteristics of life. All parts of the human body are composed of cells. Their arrangement and interactions result in the shape, organization, and construction of the body, and cells are responsible for carrying on its life processes. Using a compound light microscope and the proper stain, one can easily see the *plasma (cell) membrane*, the *cytoplasm*, and the *nucleus*. The cytoplasm is composed of a clear fluid, the *cytosol,* and numerous *cytoplasmic organelles* that are suspended in the cytosol.

The plasma membrane represents the selectively permeable boundary of a cell that controls what substances are allowed to enter or exit. It is primarily composed of phospholipids that form a bilayer with their nonpolar fatty acid tails facing one another in the center, and their polar phosphate heads facing the water located in extracellular fluid (ECF) outside the cell and in intracellular fluid (ICF), or cytosol, within the cell. The phospholipid bilayer is arranged this way because nonpolar molecules are hydrophobic (water-fearing) and essentially repel water because they are electrically neutral, whereas polar molecules are hydrophilic (water-loving) and attracted to water because they have a partial electrical charge. This amphipathic structure allows some smaller, nonpolar molecules, such as carbon dioxide and oxygen, to pass easily through the membrane. Charged or larger molecules rely on membrane proteins to assist in their passage, or they may not get through at all. The plasma membrane might possess extensions such as microvilli that increase the surface area, or cilia that assist in movement of substances, or flagella involved in cell motility.

Embedded between the phospholipids are transmembrane proteins that have a variety of functions, such as ion channels or pumps, enzymes, cell-adhesion molecules, and ligand receptors. Embedded cholesterol helps provide structural integrity to the plasma membrane. Glycolipids and glycoproteins contribute to the glycocalyx, the cell-identity marker.

Many cytoplasmic organelles have membrane boundaries similar to the plasma membrane and are referred to as membranous organelles. These organelles are involved in transportation, synthesis, or containment of substances within their membranes. Examples of membranous organelles include mitochondria, endoplasmic reticulum, and Golgi apparatus. Nonmembranous organelles, such as centrioles and ribosomes, lack a membrane.

Movement of substances across membranes can occur by passive processes, involving kinetic energy or hydrostatic pressure, or by active processes, using the cellular energy of adenosine triphosphate (ATP).

PROCEDURE: Cell Structure and Function APR

The boundary of the nucleus is a double-layered *nuclear envelope,* which has nuclear pores allowing the passage of genetic information. The nucleus contains fine strands of *chromatin* that consist of deoxyribonucleic acid (DNA) and structural and regulatory proteins. Chromatin condenses to form chromosomes during cell division. The nucleus is often referred to as the "control center" of the cell because it contains the genetic material and is required for cell division. The *nucleolus* within the nucleus is a nonmembranous structure composed of ribonucleic acid (RNA) and protein and is a site of ribosomal formation.

Cytoplasmic organelles carry out specialized metabolic functions. The *endoplasmic reticulum (ER)* has numerous canals that serve in transporting molecules such as proteins throughout the cytoplasm. *Ribosomes* synthesize proteins and are located free in the cytoplasm or on the surface of the endoplasmic reticulum *(rough ER).* If the ER lacks ribosomes on the surface, it is called *smooth ER,* and it is involved in lipid synthesis but does not serve as a region of protein synthesis. The *Golgi apparatus (complex),* composed of flattened membranous sacs, is the site for packaging glycoproteins for transport and secretion. *Mitochondria* (sing., *mitochondrion*) possess a double membrane and provide the main location for cellular energy production of ATP. Mitochondria are often referred to as the "powerhouses" of a cell because of the energy production. *Lysosomes* are membranous sacs that contain intracellular digestive enzymes for destroying debris and worn-out organelles. Lysosomes are nicknamed "suicide sacs" because they have the ability to destroy the entire cell. *Vesicles* are membranous sacs produced by the cell that contain substances for storage or transport, or form from pinching off pieces of the plasma membrane. A *cytoskeleton* contains microtubules, intermediate filaments, and microfilaments that support cellular structures within the cytoplasm and are involved in cellular movements.

1. Study figures 5.1 and 5.2 and table 5.1.
2. Observe an animal cell model and identify its major structures.
3. Complete Part A of Laboratory Assessment 5.
4. Prepare a wet mount of cells lining the inside of the cheek. To do this, follow these steps:
   a. Wear proper safety equipment.
   b. Gently scrape (force is not necessary and should be avoided) the inner lining of your cheek with the broad end of a flat toothpick.

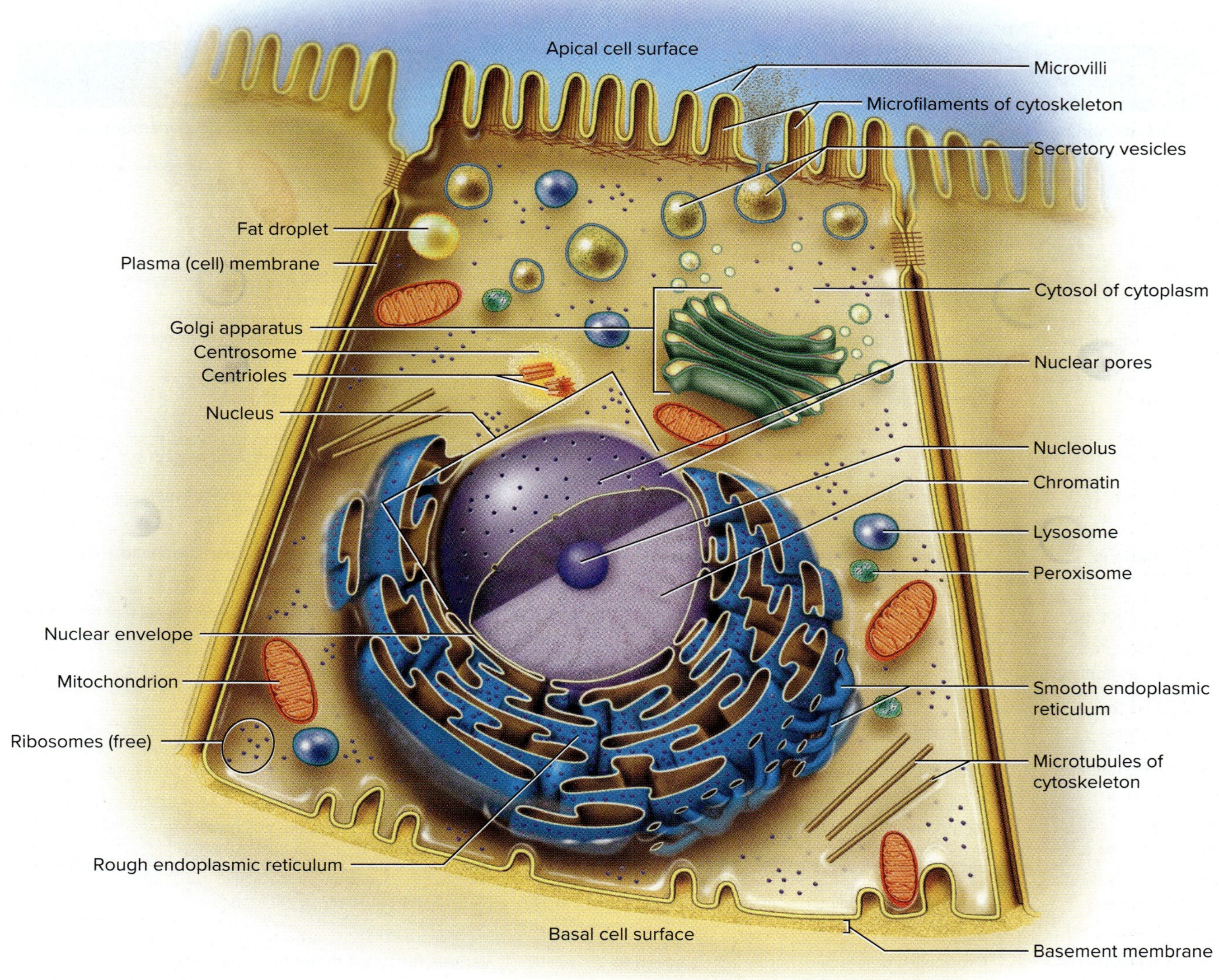

FIGURE 5.1 The structures of a composite cell. The structures are not drawn to scale. APR

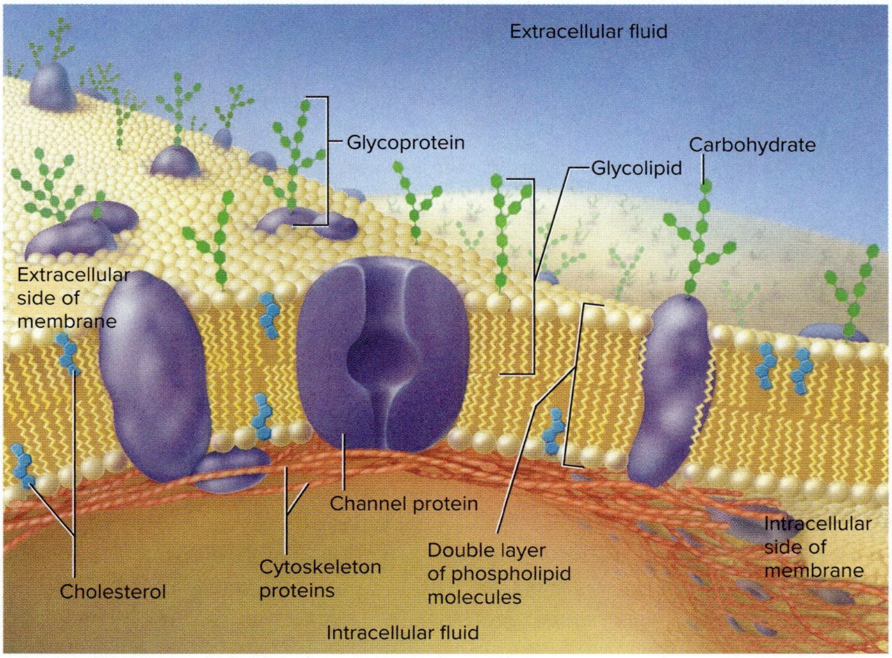

FIGURE 5.2 Structure of the plasma (cell) membrane. APR

Table 5.1 Cellular Structures and Functions

Cellular Component	Structure	Function
Plasma membrane	**Bilayer of phospholipids, glycolipids, and glycoproteins**	**Cell boundary and involved in various membrane transport mechanisms**
Microvilli	Numerous tiny folds containing microfilaments	Allow greater surface area for absorption
Cilia	Numerous short extensions containing microtubules	Move substances such as mucus
Flagellum	Long extension of sperm cell containing microtubules	Moves sperm
Nucleus	**Large structure**	**Center of cellular control**
Nuclear envelope	Double membrane boundary of nucleus with pores; continuous with rough ER	Passageway between nucleus and cytoplasm
Nucleolus	Nonmembranous body containing ribosomal RNA and protein	Location of ribosome formation
Chromatin	Strands of DNA and protein	DNA represents genetic control for protein synthesis
Cytoplasm	**Contents between plasma membrane and nucleus containing cytosol fluid and organelles**	**Location of numerous cellular processes**
Membranous organelles	**Organelles possessing a membranous covering**	**Various functions depending upon organelle**
Rough endoplasmic reticulum	Extensive membrane network with attached ribosomes	Site of protein synthesis with ribosomes; cellular transport
Smooth endoplasmic reticulum	Extensive membrane network lacking ribosomes	Site of lipid synthesis; cellular transport
Mitochondria	Possess a double membrane with folded inner membrane	Synthesis of ATP
Golgi apparatus (complex)	Series of flattened sacs often near nucleus	Packaging site of glycoproteins for transport and secretion
Vesicles	Membranous sacs containing various substances	Store or transport cellular substances
Lysosomes	Sacs containing digestive enzymes	Sites that destroy cellular debris and worn-out organelles
Peroxisomes	Small sacs containing oxidative enzymes; abundant in liver and kidney cells	Detoxify harmful cellular substances
Nonmembranous organelles	**Organelles that lack a membranous covering**	**Various functions depending upon organelle**
Ribosomes	Contain ribosomal RNA and protein; attached to rough ER or free in cytoplasm	Sites of protein synthesis
Centrosome	Area near nucleus containing a pair of centrioles	Formation center to produce and organize microtubules and spindle fibers for cell division
Centrioles	Cylindrical bodies of microtubules within the centrosome of the cytoplasm	Form mitotic spindle during cell division
Cytoskeleton: microtubules	Hollow cylinders of proteins in cilia, flagella, and centrioles	Provide support and motility within cell
Cytoskeleton: intermediate filaments	Various protein fibers	Provide internal cellular support for shape of cell
Cytoskeleton: microfilaments	Fine threads of a protein actin	Help maintain cell shape and change in shape; involved in some cell motility as muscle contraction

c. Stir the toothpick in a drop of water on a clean microscope slide and dispose of the toothpick as directed by your instructor.

d. Cover the drop with a coverslip.

e. Observe the cheek cells using the microscope at scanning objective, low-power objective, and then high-power objective. Compare your image with figure 5.3. To report what you observe, sketch a single cell in the field of view circle provided in Part B of the laboratory assessment.

5. Prepare a second wet mount of cheek cells, but this time, add a drop of dilute methylene blue or iodine-potassium-iodide stain to the cells. Cover the liquid with a coverslip and observe the cells with the microscope.

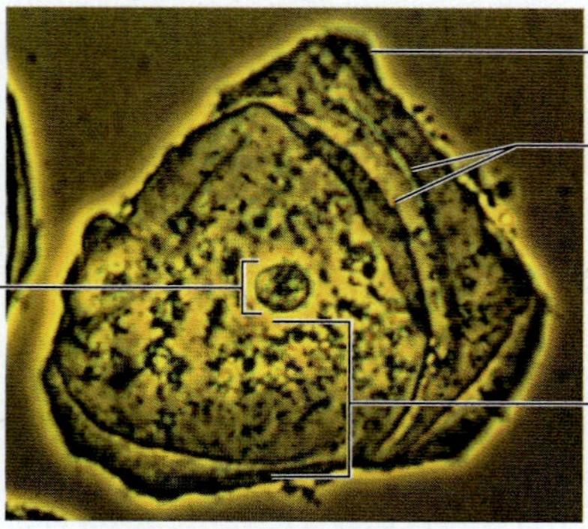

FIGURE 5.3 Iodine-stained cell from the inner cheek, as viewed through the compound light microscope using the high-power objective (400×). Ed Reschke

Add to your sketch any additional structures you observe in the stained cells.

6. Answer the questions in Part B of the laboratory assessment.
7. Observe all safety guidelines for disposal of materials used during this procedure.
8. Using the microscope, observe each of the prepared slides of human tissues at scanning objective, low-power objective, and then high-power objective. To report what you observe, sketch and label visible parts of a single cell of each type in the space provided in Part C of the laboratory assessment.
9. Complete Parts C and D of the laboratory assessment.

CRITICAL THINKING ACTIVITY

The cells lining the inside of the cheek are frequently removed for making observations of basic cell structure. The cells are from stratified squamous epithelium. Explain why these cells are used instead of outer body surface tissue.

____

____

____

LEARNING EXTENSION ACTIVITY

Investigate the microscopic structure of various plant materials. To do this, prepare tiny, thin slices of plant specimens using a single-edged razor blade. *(Take care not to injure yourself with the blade.)* Keep the slices in a container of water until you are ready to observe them. To observe a specimen, place it into a drop of water on a clean microscope slide and cover it with a coverslip. Use the microscope and view the specimen using scanning, low-power, and high-power magnifications. Observe near the edges where your section of tissue is most likely to be one cell thick. Add a drop of dilute methylene blue or iodine-potassium-iodide stain, and note if any additional structures become visible. How are the microscopic structures of the plant specimens similar to the human tissues you observed? _______________________

____

How are they different? _______________________

____

LEARNING EXTENSION ACTIVITY

Prepare individual wet mounts of the *Amoeba* and *Paramecium* by putting a drop of culture on a clean glass slide. Gently cover each culture with a clean coverslip. Observe the movements of the *Amoeba* with pseudopodia and the *Paramecium* with cilia. Try to locate cellular components such as the plasma (cell) membrane, nuclear envelope, nucleus, mitochondria, and contractile vacuoles. Describe the movement of the *Amoeba*.

____

____

Describe the movement of the *Paramecium.*

____

____

NOTES

LABORATORY
ASSESSMENT
5

Name ______________________________

Date ______________________________

Section ______________________________

The **A** corresponds to the indicated Learning Outcome(s) **O** found at the beginning of the Laboratory Exercise.

Cell Structure and Function

PART A: Assessments

1. Label the cellular structures in figure 5.4. **A1**

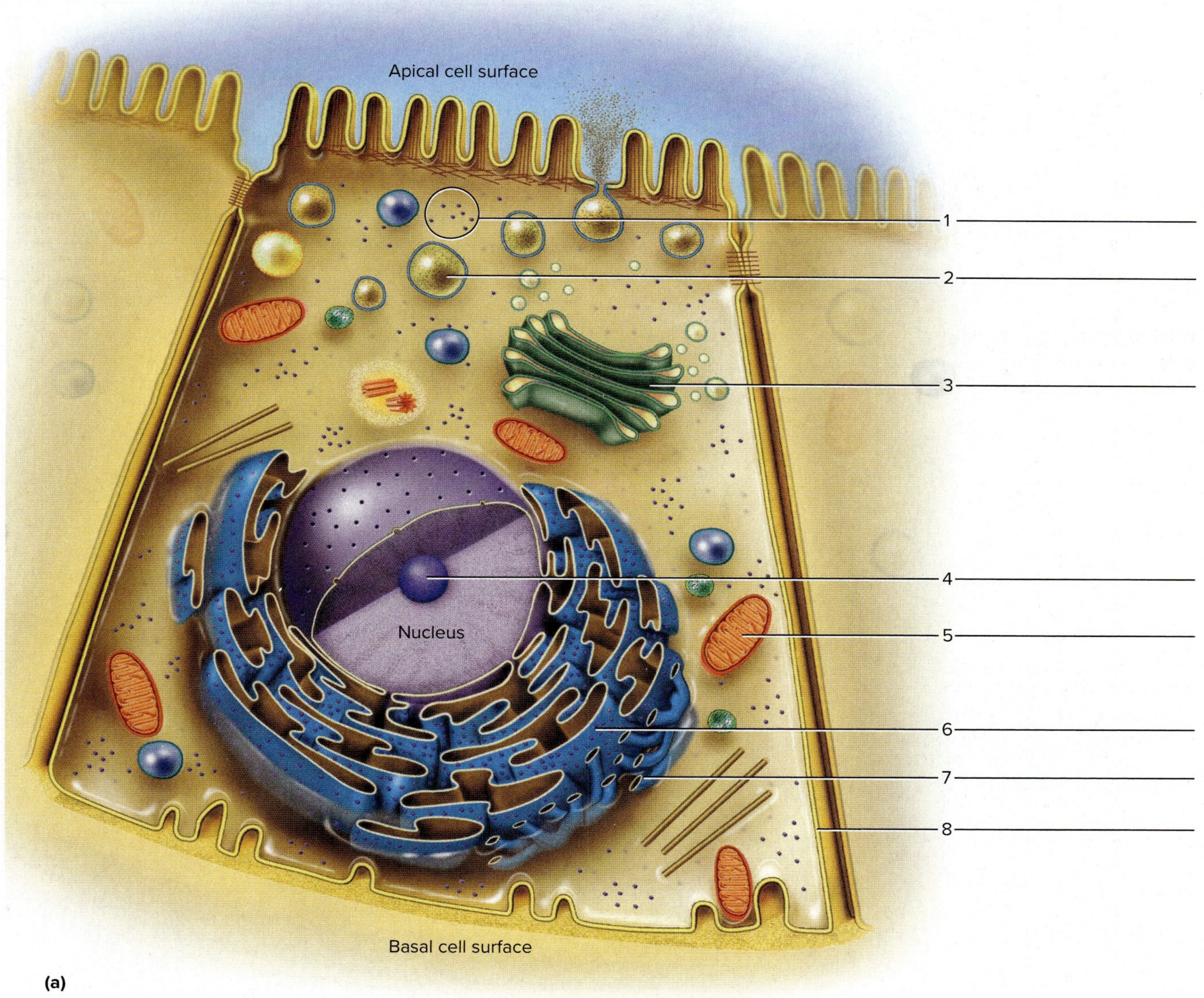

FIGURE 5.4 Label the indicated cellular structures of the (a) illustration of a composite cell and (b) an image of a cell model, using the terms provided [for (b) only].

LABORATORY
ASSESSMENT 5

Name ______________________________

Date ______________________________

Section ______________________________

The A corresponds to the indicated Learning Outcome(s) O found at the beginning of the Laboratory Exercise.

Cell Structure and Function

PART A: Assessments

1. Label the cellular structures in figure 5.4. A1

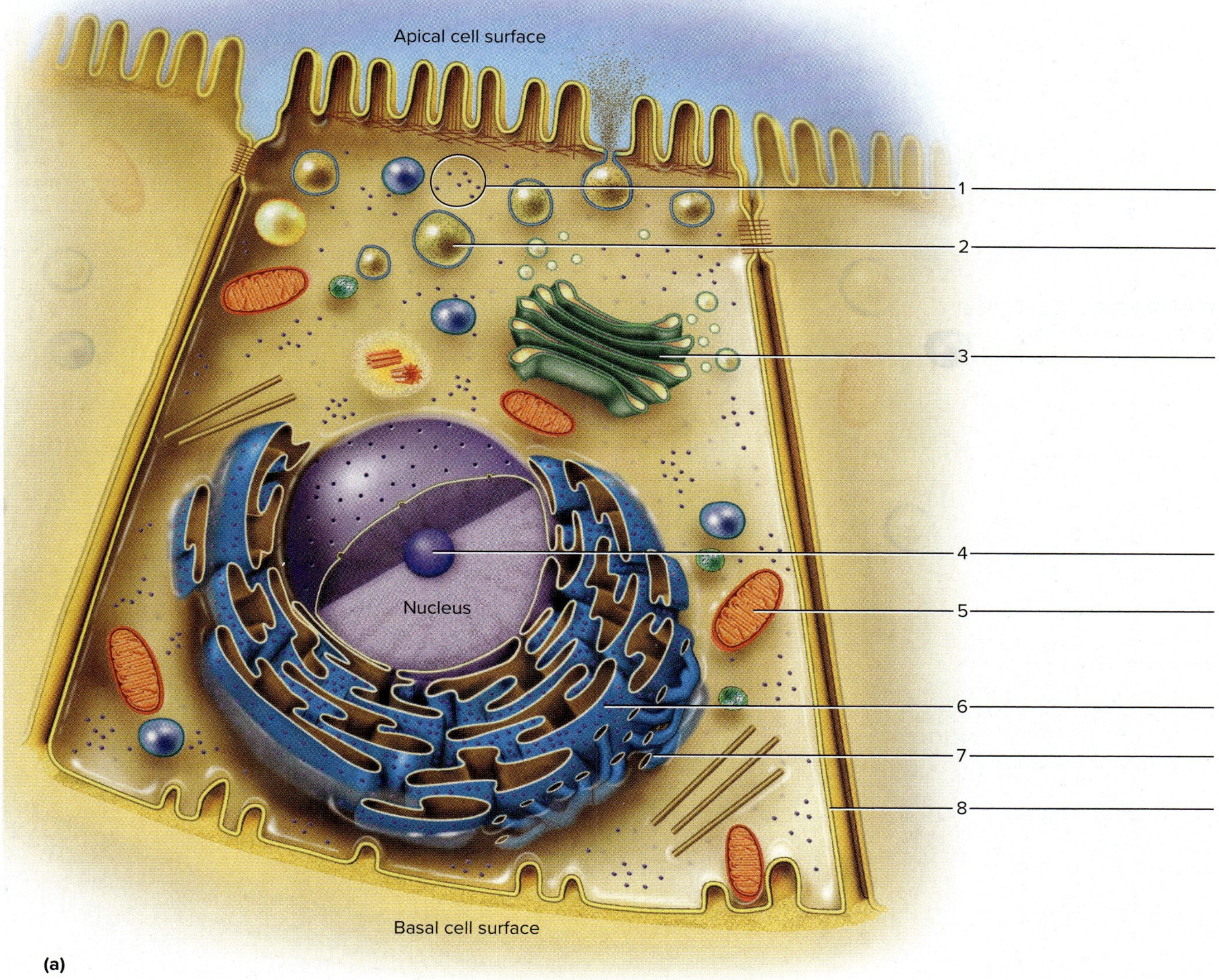

FIGURE 5.4 Label the indicated cellular structures of the (a) illustration of a composite cell and (b) an image of a cell model, using the terms provided [for (b) only].

space provided. A2

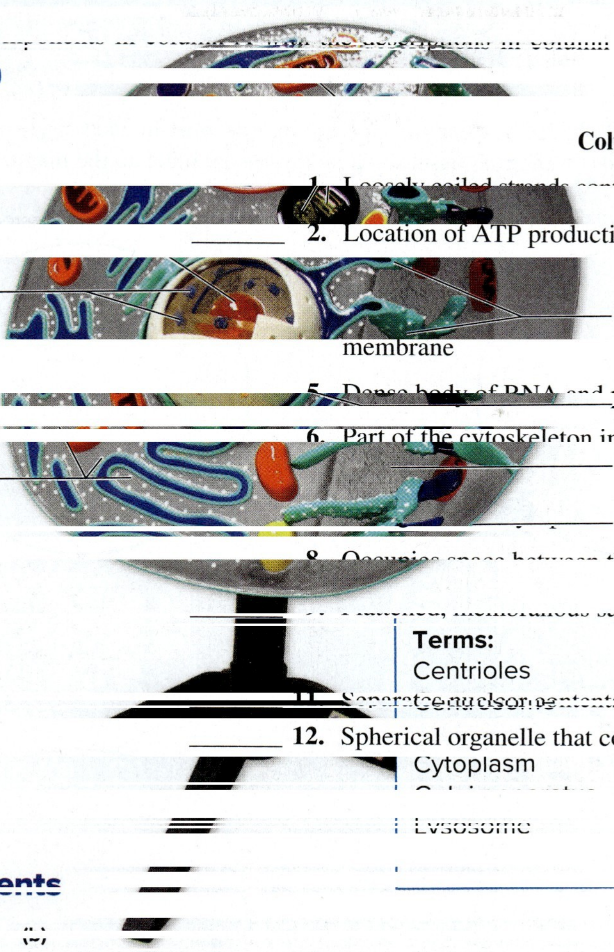

FIGURE 3.4 (Continued.)

Science Interactive Group

Column A		Column B
a. Chromatin	_______	1. Loosely coiled strands containing protein and DNA within the nucleus
b. Cytoplasm	_______	2. Location of ATP production for cellular energy
e. Lysosome		membrane
f. Microtubule	_______	5. Dense body of RNA and protein within the nucleus
	_______	6. Part of the cytoskeleton involved in cellular movement
i. Nucleolus	_______	8. Occupies space between the plasma membrane and the nucleus
	_______	9. Flattened, membranous sacs that modify and package a secretion
	_______	11. Separates nuclear contents from the cytoplasm
	_______	12. Spherical organelle that contains chromatin and the nucleolus

Terms:

Centrioles	Nuclear pore
Cytoplasm	Plasma membrane
Lysosome	Rough Endoplasmic Reticulum

PART B: Assessments

completed. (The circle represents the field of view through the

____

____

PART C: Assessments

Complete the following:

1. Each circle below represents the field of view as seen through the microscope. In each circle, sketch a single cell of each type you observed in the prepared slides of human tissues. Name the tissue, indicate the magnification used, and label the cellular components you recognize. **A4**

Tissue ______________________________

______________ ×

Tissue ______________________________

______________ ×

Tissue ______________________________

______________ ×

Tissue ______________________________

______________ ×

2. What do the various types of cells in these tissues have in common? ______________________________

____

3. What are the main differences you observed among these cells? ______________________________

____

____

4. Using a cell that you observed, explain how its structure helps the cell carry out its function.

____

____

PART D: Assessments

Electron micrographs represent extremely thin slices of cells. The micrograph in (a) represents an entire cell, while the micrograph in (b) represents a section of the cytoplasm of a cell. Compare the organelles shown in these micrographs with organelles of the animal cell model and figure 5.1.

Use the terms provided to identify the structures indicated by the leader lines and numbers in figure 5.5. A1

1. ______________________
2. ______________________
3. ______________________
4. ______________________
5. ______________________
6. ______________________
7. ______________________
8. ______________________
9. ______________________

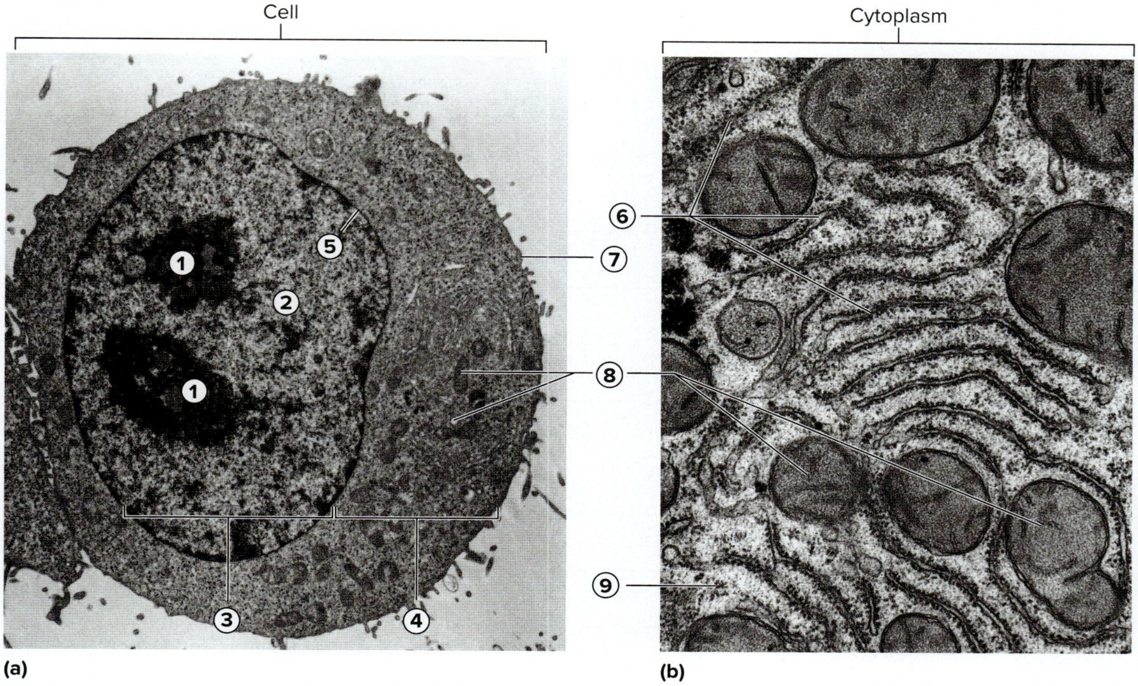

Terms:

Chromatin	Nucleoli
Cytoplasm	Nucleus
Mitochondria	Plasma membrane
Nuclear envelope	Ribosomes (free)
	Rough endoplasmic reticulum

FIGURE 5.5 Transmission electron micrographs of cellular components. Magnifications: (a) 16,000×; (b) 26,000×. Identify the numbered cellular structures, using the terms provided. A1
(a) David M. Phillips/Science Source; (b) Don W. Fawcett/Science Source

Answer the following question after observing the transmission electron micrographs in figure 5.5.

10. What cellular structures were visible in the transmission electron micrographs that were not apparent in the cells you observed using the compound light microscope? A1 ______________________

____

____

NOTES

LABORATORY
EXERCISE 6

Movements Through Membranes

Purpose of the Exercise

To become familiar with the physical processes by which substances move through membranes and to demonstrate those processes that are passive.

MATERIALS NEEDED

For Procedure A—Diffusion:

Petri dish
White paper
Forceps
Colored pencils
Potassium permanganate crystals
Millimeter ruler (thin and transparent)

For Procedure B—Osmosis:

Thistle tube
Molasses (or Karo dark corn syrup or 60% colored sucrose solution)
Ring stand and clamp
Beaker
Rubber band
Millimeter ruler
Selectively permeable (semipermeable) membrane (presoaked dialysis tubing of 1 5/16" or greater diameter)

For Procedure C—Hypertonic, Hypotonic, and Isotonic Solutions:

Test tubes
Marking pen
Test-tube rack
10 mL graduated cylinder
Medicine dropper
Uncoagulated animal blood
Distilled water
0.9% NaCl (aqueous solution)
3.0% NaCl (aqueous solution)
Clean microscope slides
Coverslips
Compound light microscope

For Procedure D—Filtration:

Glass funnel
Filter paper
Support stand and ring
Beaker
Powdered charcoal or ground black pepper
1% glucose (aqueous solution)
1% starch (aqueous solution)
Test tubes
10 mL graduated cylinder
Water bath (boiling water)
Benedict's solution
Iodine-potassium-iodide solution
Medicine dropper

For Alternative Activities:

Fresh chicken egg
Beaker
Laboratory balance
Spoon
Vinegar
Corn syrup (Karo)

SAFETY

- Review all safety guidelines in Appendix 1 of your laboratory manual.
- Clean laboratory surfaces before and after laboratory procedures.
- Wear disposable gloves and protective eyewear when handling chemicals and animal blood.
- Dispose of laboratory gloves and blood-contaminated items as instructed.
- Wash your hands before leaving the laboratory.

Learning Outcomes APR

After completing this exercise, you should be able to:

O1 Demonstrate the process of diffusion and identify examples of diffusion.

O2 Explain diffusion by preparing and interpreting a graph.

O3 Demonstrate the process of osmosis and identify examples of osmosis.

O4 Distinguish among hypertonic, hypotonic, and isotonic solutions and examine the effects of these solutions on animal cells.

O5 Demonstrate the process of filtration and identify examples of filtration.

The O corresponds to the assessments A indicated in the Laboratory Assessment for this Exercise.

Pre-Lab

Carefully read the introductory material and examine the entire lab. Be familiar with diffusion, osmosis, tonicity, and filtration from lecture or the textbook. Answer the pre-lab questions.

Pre-Lab Questions Select the correct answer for each of the following questions:

1. Which of the following membrane movements is *not* a passive process?
   a. osmosis **b.** diffusion
   c. active transport **d.** filtration

2. Osmosis is a membrane passage process pertaining to
   a. glucose solutions. b. hydrostatic pressure.
   c. saline solutions. d. water.
3. Which of the following represents an isotonic solution to human cells?
   a. 0.9% NaCl b. 3% NaCl
   c. 3% glucose d. 0.9% glucose
4. Filtration requires ________ for movements through membranes.
   a. hydrostatic pressure b. active transport
   c. pinocytosis d. phagocytosis
5. Which of the following does *not* influence the rate of diffusion?
   a. concentration gradient b. size of petri dish
   c. molecular size d. temperature
6. Dialysis tubing serves as a ________ for the osmosis experiment.
   a. cell wall b. permeable membrane
   c. selectively permeable membrane d. barrier membrane
7. Passive membrane passage requires ATP.
   a. True b. False
8. A cell will not swell or shrink when placed in an isotonic solution.
   a. True b. False

A plasma (cell) membrane functions as a gateway through which chemical substances and small particles may enter or leave a cell. Because of its bilayer of phospholipid molecules, the plasma membrane is *selectively permeable* in that it allows only certain substances to easily pass through it. These substances move through the membrane by *passive (physical) processes* such as diffusion, osmosis, and filtration, or by *active (physiological) processes* such as active transport, phagocytosis, or pinocytosis. Passive processes do not require the cellular energy of ATP, but occur due to molecular motion or hydrostatic pressure; active processes utilize ATP.

This laboratory exercise contains examples of passive membrane transport through either living plasma membranes or artificial membranes. *Diffusion* is the random motion of molecules from an area of higher concentration toward an area of lower concentration, thus moving down their concentration gradient. The rate of diffusion depends upon factors such as concentration gradient, molecular size, and temperature. For example, an increase in temperature would cause an increase in molecular motion, or kinetic energy, resulting in an increase in the rate of diffusion. Facilitated diffusion requires the help of carrier or channel proteins to passively transport certain substances through the plasma membrane, which is not represented in this laboratory exercise.

A special case of passive membrane passage, called *osmosis,* occurs when water molecules move through water channels (aquaporins) or between phospholipids of a selectively permeable membrane from an area of lower-solute concentration (more water) to an area of higher-solute concentration (less water). *Tonicity* refers to the osmotic pressure of a solution in relation to the intracellular fluid (ICF) of a cell. A solution is *isotonic* if the osmotic pressure is the same as the ICF of a cell, so water enters and exits the cell at the same rate. A 5% glucose solution and a 0.9% normal saline solution serve as isotonic solutions to human cells. A *hypertonic* solution has a higher concentration of solutes than the ICF of a cell; a *hypotonic* solution has a lower concentration of solutes than the ICF of a cell. The osmosis of water will occur out of a cell when it is immersed in a hypertonic solution, and the cell will shrink (crenate). Conversely, the osmosis of water will occur into a cell when it is immersed in a hypotonic solution, and the cell will swell, and may burst (lyse).

Filtration is the movement of water and solutes through a selectively permeable membrane by a physical pressure upon the membrane. In the body, blood pressure (blood hydrostatic pressure) creates a force that drives water and solutes (filtrate) through the plasma membrane. For example, blood pressure pushes fluids and small dissolved substances through capillary walls. Hypertension (high blood pressure) is a condition that would increase capillary filtration, as blood hydrostatic pressure is increased. This may lead to swelling (edema) in the tissues.

PROCEDURE A: Diffusion APR

Simple diffusion can occur in living organisms as well as in nonliving systems involving gases, liquids, and solids. In this procedure, diffusion will be examined in water without cells.

1. To demonstrate diffusion, refer to figure 6.1 as you follow these steps:
   a. Place a petri dish, half filled with water, on a piece of white paper that has a millimeter ruler positioned

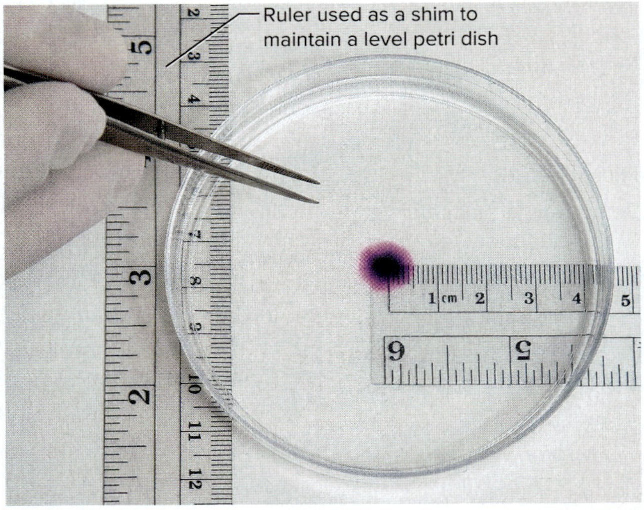

FIGURE 6.1 To demonstrate diffusion, place one crystal of potassium permanganate in the center of a petri dish containing water. Place the crystal near the millimeter ruler (positioned under the petri dish). ©J & J Photography

on the paper. Wait until the water surface is still. Allow approximately 3 minutes.

Note: The petri dish should remain level. A second millimeter ruler may be needed under the petri dish as a shim to obtain a level amount of the water inside the petri dish.

b. Using forceps, place one crystal of potassium permanganate near the center of the petri dish and near the millimeter ruler (fig. 6.1).

c. Measure the radius of the purple circle at 1-minute intervals for 10 minutes and record the results in Part A of Laboratory Assessment 6.

2. Complete Part A of the assessment.

LEARNING EXTENSION ACTIVITY

Repeat the demonstration of diffusion using a petri dish filled with ice-cold water and a second dish filled with very hot water. At the same moment, add a crystal of potassium permanganate to each dish and observe the circle as before. Record your findings in the data table in Part A of the laboratory assessment. What difference do you note in the rate of diffusion in the two dishes? How do you explain this difference? ______________

____

____

____

____

____

PROCEDURE B: Osmosis APR

For the osmosis experiment, an artificial membrane will serve as a model for an actual plasma membrane. A layer of dialysis tubing that has been soaked for 30 minutes can easily be cut open because it becomes pliable. This membrane model works as a selectively permeable membrane mimicking the plasma membrane of a cell.

When water moves across a selectively permeable membrane from the side with less solute concentration (more water) toward the side with higher solute concentration (less water), it is termed osmosis. Penetrating solute particles will move down their concentration gradients across a membrane by diffusion if the solute is able to pass through the membrane. However, if the solute particles are nonpenetrating because they are too large to pass across a selectively permeable membrane, only water will move across the selectively permeable membrane toward the higher concentration of nonpenetrating particles. This will change the volume of water on either side of the membrane, which influences the volume within cells or models used as simulated cells.

1. To demonstrate osmosis, refer to figure 6.2 as you follow these steps:

a. One person plugs the tube end of a thistle tube with a finger.

b. Another person then fills the bulb with molasses until it is about to overflow at the top of the bulb. Allow the molasses to enter the first centimeter of the stem, leaving the rest filled with trapped air.

c. Cover the bulb opening with a single-thickness piece of moist selectively permeable membrane.

d. Tightly secure the membrane in place with several wrappings of a rubber band.

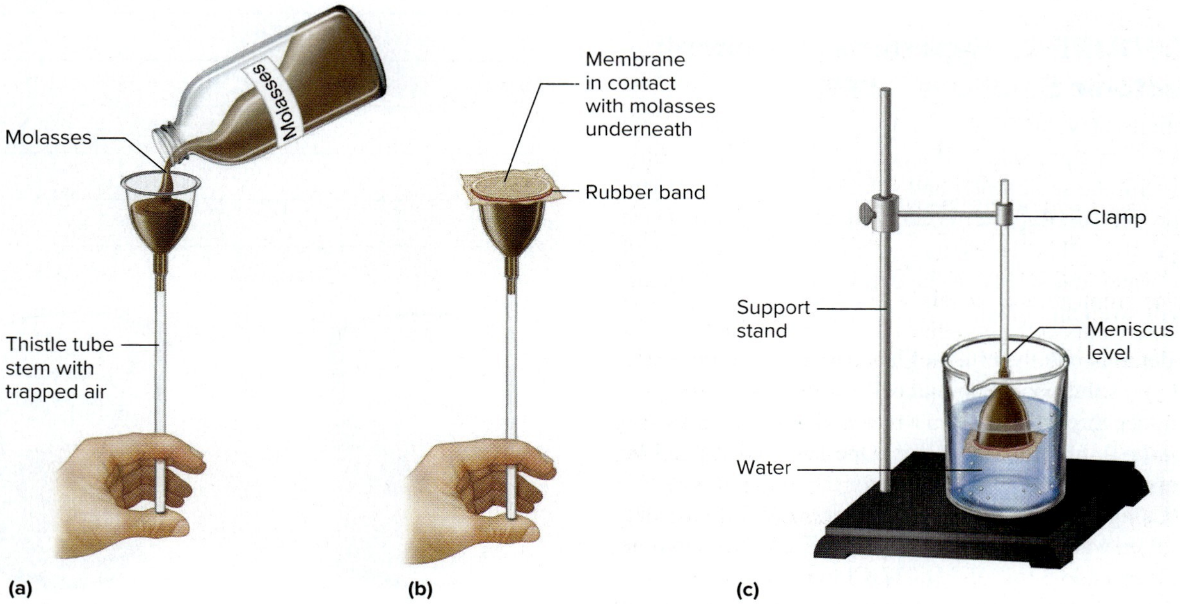

FIGURE 6.2 (a) Fill the bulb of the thistle tube with molasses; (b) tightly secure a piece of selectively permeable (semipermeable) membrane over the bulb opening; and (c) immerse the bulb in a beaker of water. *Note:* These procedures require the participation of two people.

e. Immerse the bulb end of the tube in a beaker of water. If leaks are noted, repeat the procedures.
f. Support the upright portion of the tube with a clamp on a support stand. Folded paper between the stem and the clamp will protect the thistle tube stem from breakage.
g. Mark the meniscus level of the molasses in the tube. *Note:* The best results will occur if the mark of the molasses is a short distance up the stem of the thistle tube when the experiment starts.
h. Measure the level changes after 10 minutes and 30 minutes and record the results in Part B of the laboratory assessment.

2. Complete Part B of the laboratory assessment.

ALTERNATE ACTIVITY

Eggshell membranes possess selectively permeable properties. To demonstrate osmosis using a natural membrane, soak a fresh chicken egg in vinegar for about 24 hours to remove the shell. Use a spoon to carefully handle the delicate egg. Place the egg in a hypertonic solution (corn syrup) for about 24 hours. Remove the egg, rinse it, and using a laboratory balance, weigh the egg to establish a baseline weight. Place the egg in a hypotonic solution (distilled water). Remove the egg and weigh it every 15 minutes for an elapsed time of 75 minutes. Explain any weight changes that are noted during this experiment.

____

____

PROCEDURE C: Hypertonic, Hypotonic, and Isotonic Solutions APR

Applications of volume changes to cells from osmosis are addressed in this procedure. If a cell is placed in an isotonic solution with the same water and solute concentration, there will be no net osmosis and the cell will retain its normal size and shape. If a cell is placed in a hypertonic solution, the cell will lose water and shrink; if the cell is in a hypotonic solution it will swell and burst.

1. To demonstrate the effect of hypertonic, hypotonic, and isotonic solutions on animal cells, follow these steps:
   a. Place three test tubes in a rack and mark them *tube 1, tube 2,* and *tube 3.* (*Note:* One set of tubes can be used to supply test samples for the entire class.)
   b. Using 10 mL graduated cylinders, add 3 mL of distilled water to tube 1; add 3 mL of 0.9% NaCl to tube 2; and add 3 mL of 3.0% NaCl to tube 3.
   c. Gently shake first, and then place three drops of fresh, uncoagulated animal blood into each of the tubes, and gently mix the blood with the solutions. Wait 5 minutes.
   d. Using three separate medicine droppers, remove a drop from each tube and place the drops on three separate microscope slides marked *1, 2,* and *3.*
   e. Cover the drops with coverslips and observe the blood cells, using the high-power objective of the microscope.
2. Complete Part C of the laboratory assessment.

ALTERNATE ACTIVITY

Various substitutes for blood can be used for Procedure C. Onion, cucumber, and cells lining the inside of the cheek represent three possible options.

ALTERNATE ACTIVITY

Like all vegetables, carrots and celery are composed of plant cells. These freshly cut vegetables tend to be crispier (and thus, more appealing to eat) if kept in a container of water and placed in the refrigerator. In this activity, you will demonstrate and explain how the tonicity of a solution affects the volume within cells. Use two equal-sized pieces of cut carrots. Tie a string around the circumference of each carrot so that there is no slack. Place one cut carrot in a beaker of 100 mL plain tap water, and place the second cut carrot in a beaker of 100 mL salt water (1 teaspoon salt dissolved in 100 mL water) (fig. 6.3). You may put them in the refrigerator (preferably) or at room temperature. After 48 to 72 hours, observe the tightness of each string. Using the concept of *osmosis*, explain the changes in the tightness of each string.

Plain water:______________________________

____

____

____

Salt water: ______________________________

____

____

____

Plain water   Salt water

FIGURE 6.3 Apparatus setup for determining osmosis of a cut carrot.

PROCEDURE D: Filtration

Filtration is another example of passive membrane passage. Unlike the previous experiments, hydrostatic pressure provides the mechanism for movements across a membrane instead of molecular motion. Water and solutes that pass through the membrane move simultaneously in the same direction. Solutes that are too large to cross the membrane remain on their original side of the membrane. Other than molecular size, filtration is a nonselective process, meaning that any molecule may be pushed across the membrane if pressure is great enough. Filtration occurs faster when hydrostatic pressure increases.

1. To demonstrate filtration, follow these steps:
   a. Place a glass funnel in the ring of a support stand over an empty beaker. Fold a piece of filter paper in half and then in half again. Open one thickness of the filter paper to form a cone. Wet the cone, and place it in the funnel. The filter paper is used to demonstrate how movement across membranes is limited by the size of the molecules, but it does not represent a working model of biological membranes.
   b. Prepare a mixture of 5 cc (approximately 1 teaspoon) powdered charcoal (or ground black pepper) and equal amounts of 1% glucose solution and 1% starch solution in a beaker. Pour some of the mixture into the funnel until it nearly reaches the top of the filter-paper cone. Care should be taken to prevent the mixture from spilling over the top of the filter paper. Collect the filtrate in the beaker below the funnel (fig. 6.4).
   c. Test some of the filtrate in the beaker for the presence of glucose. To do this, place 1 mL of filtrate in a clean test tube and add 1 mL of Benedict's solution. Place the test tube in a water bath of boiling water for 2 minutes and then allow the liquid to cool slowly. If the color of the solution changes to green, yellow, or red, glucose is present (fig. 6.5).
   d. Test some of the filtrate in the beaker for the presence of starch. To do this, place a few drops of filtrate in a test tube and add 1 drop of iodine-potassium-iodide solution. If the color of the solution changes to blue-black, starch is present.
   e. Observe any charcoal in the filtrate.
2. Complete Part D of the laboratory assessment.

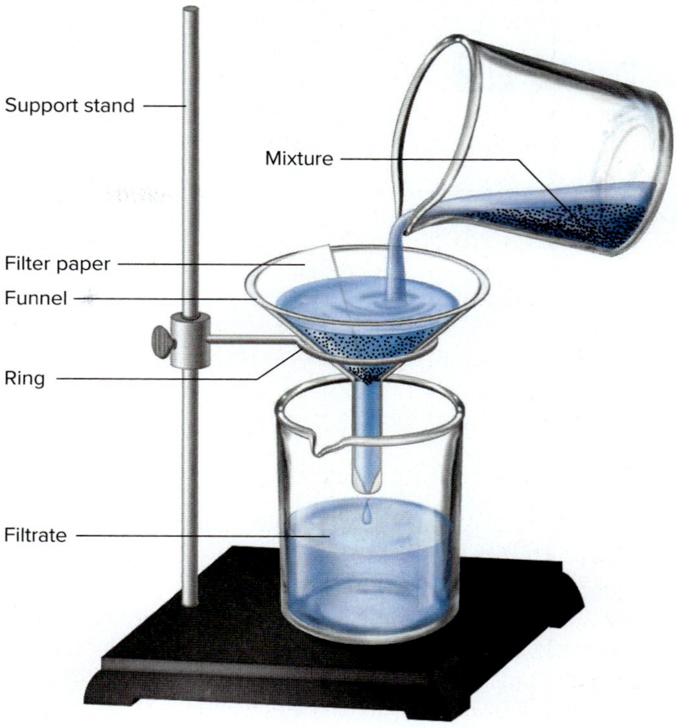

FIGURE 6.4 Apparatus used to illustrate filtration.

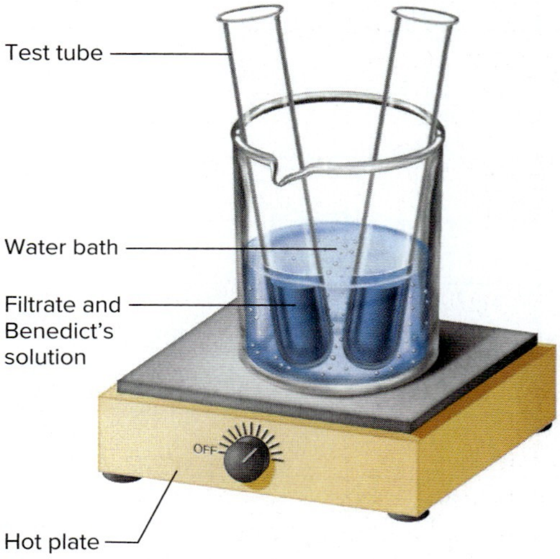

FIGURE 6.5 Heat the filtrate and Benedict's solution in a boiling water bath for 2 minutes.

NOTES

LABORATORY
ASSESSMENT 6

Name ____________________

Date ____________________

Section ____________________

The A corresponds to the indicated Learning Outcome(s) O found at the beginning of the Laboratory Exercise.

Movements Through Membranes

PART A: Assessments

Complete the following:

1. Enter data for changes in the movement of the potassium permanganate in the following table. If the Learning Extension Activity was performed, include those results in the table. A1

Elapsed Time	Radius of Purple Circle in Millimeters (mm) at Room Temperature	Learning Extension Activity: Radius of Purple Circle in Ice-Cold Water in mm	Learning Extension Activity: Radius of Purple Circle in Hot Water in mm
Initial			
1 minute			
2 minutes			
3 minutes			
4 minutes			
5 minutes			
6 minutes			
7 minutes			
8 minutes			
9 minutes			
10 minutes			

2. Prepare a graph that illustrates the diffusion distance of potassium permanganate in 10 minutes. Then, using a different colored pencil for ice-cold water and a third colored pencil for hot water, graph the data you obtained if you performed the Learning Extension Activity. A2

3. Interpret your graph. A2 ____________________

____

____

____

____

____

____

____

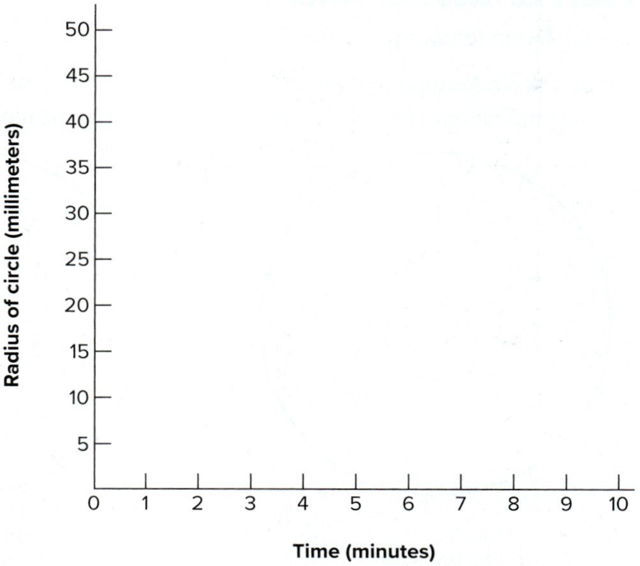

4. Using a drop of food coloring placed in water, explain the process of diffusion. **A1** ______________________________
____
____

CRITICAL THINKING ASSESSMENT—PART A

By answering yes or no, indicate which of the following provides an example of diffusion. **A1**

1. A perfume bottle is opened, and soon the odor can be smelled in all parts of the room. ____________
2. A tea bag is dropped into a cup of hot water, and, without being stirred, all of the liquid becomes the color of tea leaves. ____________
3. Water molecules move from a faucet through a garden hose when the faucet is turned on. ____________

PART B: Assessments

Complete the following:

1. What was the change in the level of molasses in 10 minutes? **A3** ____________ In 30 minutes? ____________
2. How do you explain this change? **A3** ______________________________
____
3. Using what you know about the process of osmosis, explain why celery and carrot pieces become more plump and crisp when placed overnight in a container of plain water. **A3** ______________________________
____

CRITICAL THINKING ASSESSMENT—PART B

By answering yes or no, indicate which of the following involves osmosis. **A3**

1. A fresh potato is peeled, weighed, and soaked in a strong salt solution. The next day, it is discovered that the potato has lost weight. ____________
2. Air molecules escape from a punctured tire as a result of high pressure inside. ____________
3. Plant seeds soaked in water swell and become several times as large as before soaking. ____________

PART C: Assessments

Complete the following:

1. In the circles representing the microscope field of view, sketch a few blood cells from each of the test tubes and indicate the magnification. Then, draw an arrow(s) indicating the movement of water (into the cell, out of the cell, or both equally). **A4**

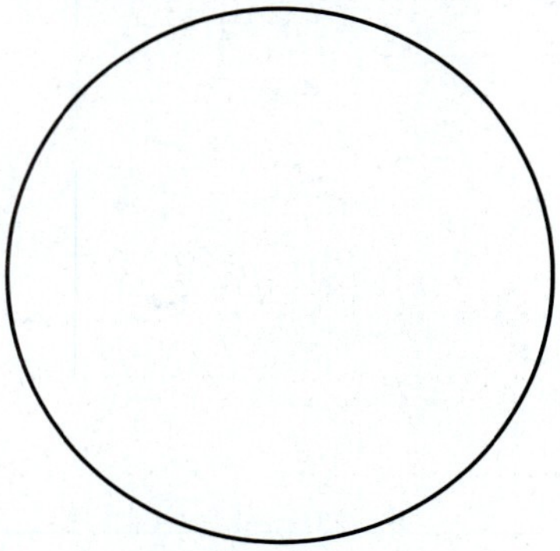

Tube 1
(distilled water)

__________×

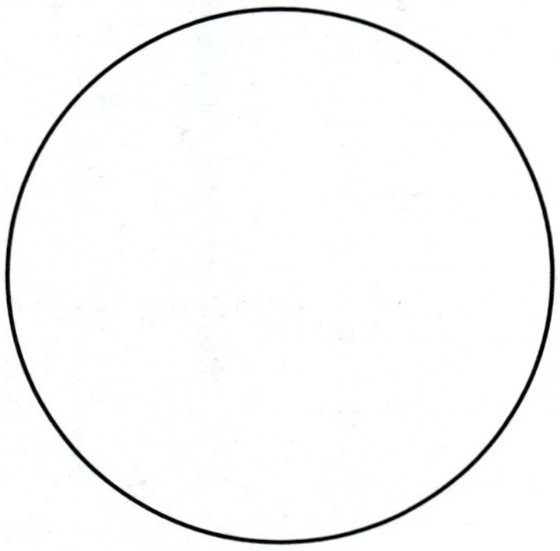

Tube 2
(0.9% NaCl)

__________×

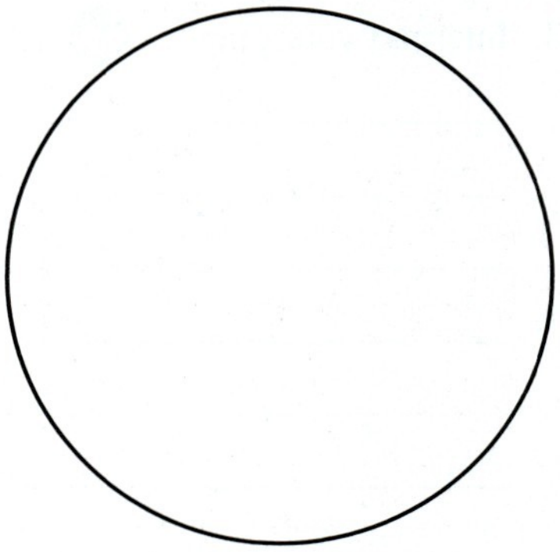

Tube 3
(3.0% NaCl)

__________×

2. Based on your results, which tube contained a solution hypertonic to the blood cells? ____________
Give the reason for your answer. **A4** ____

____

3. Which tube contained a solution hypotonic to the blood cells? ____________
Give the reason for your answer. **A4** ____

____

4. Which tube contained a solution isotonic to the blood cells? ____________
Give the reason for your answer. **A4** ____

____

5. Observe the RBCs shown in figure 6.6. Select the solutions in which the cells were placed to illustrate the effects of tonicity on cells. **A4**

Terms:
Hypertonic
Hypotonic
Isotonic

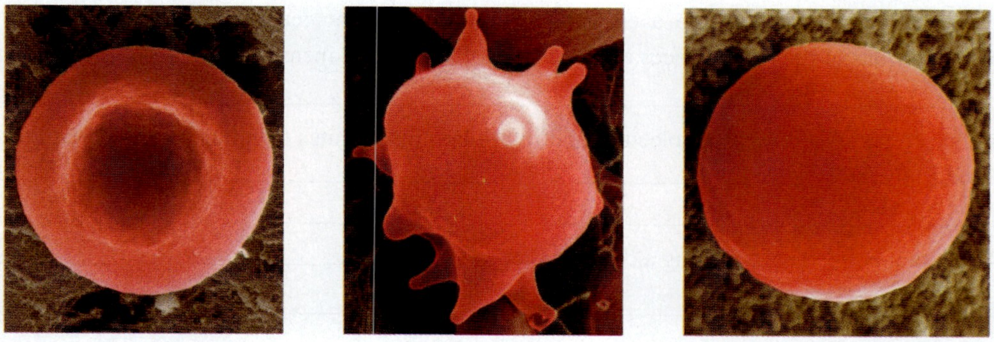

(a) ____________ (b) ____________ (c) ____________

FIGURE 6.6 Three blood cells placed in three different solutions: distilled water, 0.9% NaCl, and 3% NaCl. Select the solutions in which the cells were placed, using the terms provided. David M. Phillips/Science Source

CRITICAL THINKING ASSESSMENT—PART C

Explain the potential risk to a person who drinks unusually large quantities of plain water in a short period of time.
A3 **A4**

____

____

____

PART D: Assessments

Complete the following:

1. Which of the substances in the mixture you prepared passed through the filter paper into the filtrate? **A5** ____________

____

2. What evidence do you have for your answer to question 1? **A5** ________________________________

____

3. What force was responsible for the movement of substances through the filter paper? **A5** ____________

4. What substances did not pass through the filter paper? A5 ____

5. What factor prevented these substances from passing through? A5 ____

____

6. During exercise, skeletal muscles increase their metabolic activity, and capillary blood flow and pressure increase. Would this *increase* or *decrease* filtration in these capillaries? ____________ Explain how this benefits these active tissues. A5

____

____

CRITICAL THINKING ASSESSMENT—PART D

By answering yes or no, indicate which of the following involves filtration; then, provide your reasoning. A5

1. Oxygen molecules move into a cell and carbon dioxide molecules leave a cell because of differences in the concentrations of these substances on either side of the plasma membrane. ______________; ________________

   ____

2. Blood pressure forces water molecules and small dissolved solutes from the blood outward through the thin wall of a blood capillary. ________________________; ________________________________

   ____

3. Noninstant coffee is made using a coffeemaker. ________________; ______________________________

   ____

LABORATORY
EXERCISE 7

Cell Cycle

Purpose of the Exercise

To review the phases in the cell cycle and to observe cells in various phases of their life cycles.

MATERIALS NEEDED

Models of animal cells during mitosis

Compound light microscope

Microscope slides of whitefish mitosis (blastula)

For Learning Extension:

Microscope slides of allium root tip mitosis

For Demonstration Activity:

Microscope slide of human chromosomes from leukocytes in mitosis

Oil immersion objective

Learning Outcomes APR

After completing this exercise, you should be able to:

O1 Describe the cell cycle and locate structures that are a part of this process.

O2 Identify and sketch the phases (stages) in the life cycle of a particular cell.

O3 Arrange into a correct sequence a set of models or drawings of cells in various phases of their life cycles.

The O corresponds to the assessments A indicated in the Laboratory Assessment for this Exercise.

Pre-Lab

Carefully read the introductory material and examine the entire lab. Be familiar with the cell cycle from lecture or the textbook. Answer the pre-lab questions.

Pre-Lab Questions Select the correct answer for each of the following questions:

1. Which of the following is *not* part of interphase?
   a. cytokinesis **b.** first gap phase
   c. second gap phase **d.** synthesis phase

2. A cell in the stage of G-zero
   a. divides continuously. **b.** ceases cell division.
   c. exhibits cytokinesis. **d.** exists in a gap phase.

3. Which is the correct sequence of the M (mitotic) phase?
   a. telophase, prophase, metaphase, anaphase **b.** prophase, anaphase, metaphase, telophase
   c. prophase, telophase, metaphase, anaphase **d.** prophase, metaphase, anaphase, telophase

4. Which of the following events occurs in prophase?
   a. spindle disappears **b.** nuclear envelopes reassemble
   c. chromosomes condense **d.** sister chromatids separate

5. Which of the following events occurs during telophase?
   a. sister chromatids separate **b.** chromosomes align along equator
   c. cytokinesis **d.** spindle apparatus forms

6. Sister chromatids separate to opposite poles during
   a. telophase. **b.** anaphase.
   c. prophase. **d.** metaphase.

7. A human diploid cell has _____ chromosomes.
   a. 23 **b.** 26
   c. 46 **d.** 92

8. In animal cells, the indentation where the cells are pinching apart during telophase is called the _____.
   a. cell plate   b. cleavage furrow
   c. nuclear envelope   d. metaphase plate
9. A karyotype is used to count centrioles and spindle fibers.
   a. True   b. False
10. Daughter cells are formed from cell cycles.
   a. True   b. False

The cell cycle consists of the series of changes a cell undergoes from the time it is formed until it divides. Interphase and mitosis are phases of a cell cycle. It is through a form of development called *differentiation* that uncommitted cells become committed and are consequently able to carry out a particular function. This process occurs over many cell cycles. Typically, a newly formed diploid cell with 46 chromosomes grows to a certain size and then divides to form two new cells (*daughter cells*), each with 46 chromosomes. As cells replace injured or dying cells, or as growth is occurring, it is important that each new daughter cell has the same genetic information as the original parent cell, so that it can carry out identical functions.

Before the cell divides, it must synthesize biochemicals and other contents. This extensive period of preparation is called *interphase* and is divided into three phases. The S phase, when DNA synthesis occurs, is between two gap phases (G_1 and G_2), when cell growth occurs and cytoplasmic organelles duplicate. Interphase is followed by the *M (mitotic) phase,* which has two major portions: (1) division of the nucleus, called *mitosis,* and (2) division of the cell's cytoplasm, called *cytokinesis.* Mitosis includes four recognized phases: *prophase, metaphase, anaphase,* and *telophase.* Eventually, some specialized cells, such as skeletal muscle cells and most nerve cells, cease further cell division, but remain alive. A mature living cell that ceases cell division is in a state called G_0 (G-zero) for the remainder of its life span. This state can last from hours to decades.

A special type of cell division, called *meiosis,* occurs in the reproductive system to produce haploid gametes (egg and sperm) with 23 chromosomes. Meiosis is included in Laboratory Exercise 61.

PROCEDURE: Cell Cycle

A cell generally subsists for extended periods of time in interphase in which the metabolic processes within the nucleus and cytoplasm occur. Those processes were studied in Laboratory Exercise 5. The mitotic phase of cell division enables growth and repair of cells. As you examine cells in various phases of cell division in this lab, you will find that cells are most frequently found to be in interphase. The slides selected for your study possess abundant mitotic phases so that you can more efficiently locate some of the interesting cellular structures that can be distinguished during mitosis. Keep in mind that the cell cycle is referred to by phases, not steps, as one phase gradually becomes the next phase. You will probably observe some phases that are debatable as to which phase is represented, but this is common. The figures selected for this lab for comparison purposes represent phases that are close to the mid-portion of the phase.

1. Study the various phases of the cell's life cycle represented in figures 7.1 to 7.4.
2. Using figure 7.5 as a guide, observe the animal mitosis models and review the major events in a cell's life cycle represented by each of them. Be sure you can arrange these models in correct sequence if their positions are

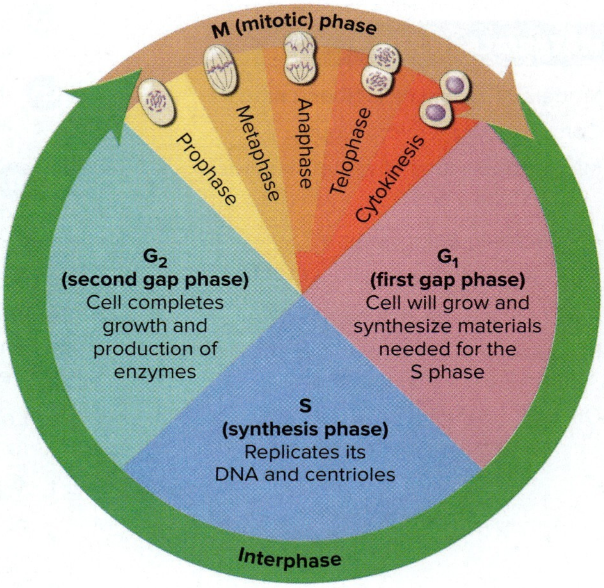

FIGURE 7.1 The cell cycle: interphase (G_1, S, and G_2) and M (mitotic) phase (mitosis and cytokinesis). APR

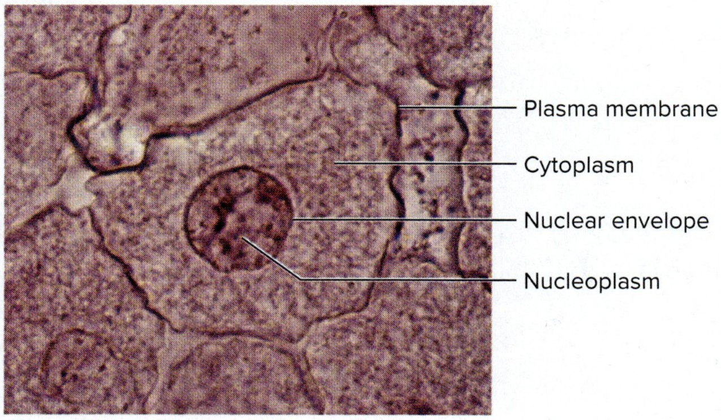

FIGURE 7.2 Whitefish blastula cell in interphase (400×). The nucleoplasm contains a fine network of chromatin.
Ed Reschke/Getty Images

changed. The acronym IPMAT can help you arrange the correct order of phases in the cell cycle. This includes interphase followed by the four phases of mitosis. Cytokinesis overlaps anaphase and telophase.

3. Complete Part A of Laboratory Assessment 7.
4. Obtain a slide of the whitefish mitosis (blastula).
   a. Examine the slide using the high-power objective of a microscope. The tissue on this slide was obtained from a developing embryo (blastula) of a fish, and many of the embryonic cells are undergoing mitosis. The chromosomes of these dividing cells are darkly stained.
   b. Search the tissue for cells in various phases of cell division. There are several sections on the slide. If you cannot locate different phases in one section, examine the cells of another section because the phases occur randomly on the slide. Frequently refer to figures 7.2, 7.3, and 7.5 as references for identifications.
   c. Each time you locate a cell in a different phase, sketch it in an appropriate circle in Part B of the laboratory assessment.
5. Complete Parts C and D of the laboratory assessment.

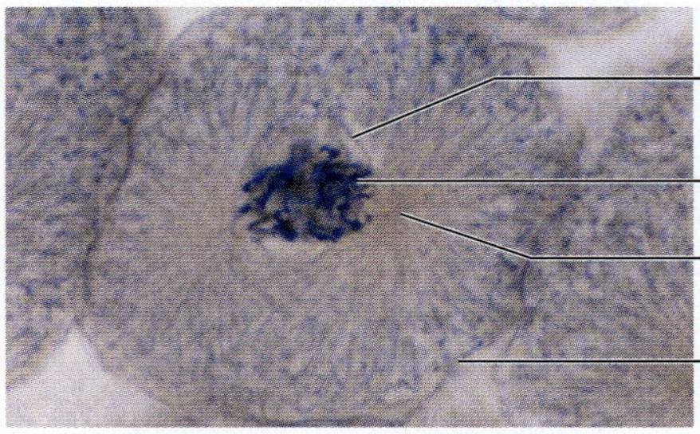

FIGURE 7.3 Whitefish blastula cell in prophase (400×). The chromosomes are more distinct, and the nuclear envelope is beginning to disintegrate. Ed Reschke/Getty Images

CRITICAL THINKING ACTIVITY

Which phase of the cell cycle was the most numerous in the blastula? ______________________

Explain your answer. ______________________

____

____

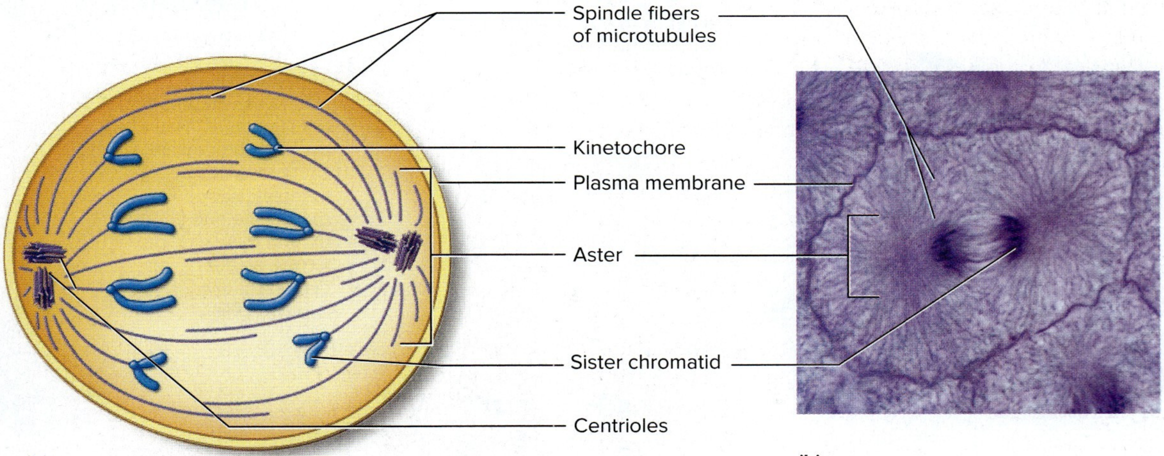

FIGURE 7.4 Structures in the dividing cell during anaphase. (a) Illustration and (b) micrograph of whitefish blastula. (b): Ed Reschke/Getty Images

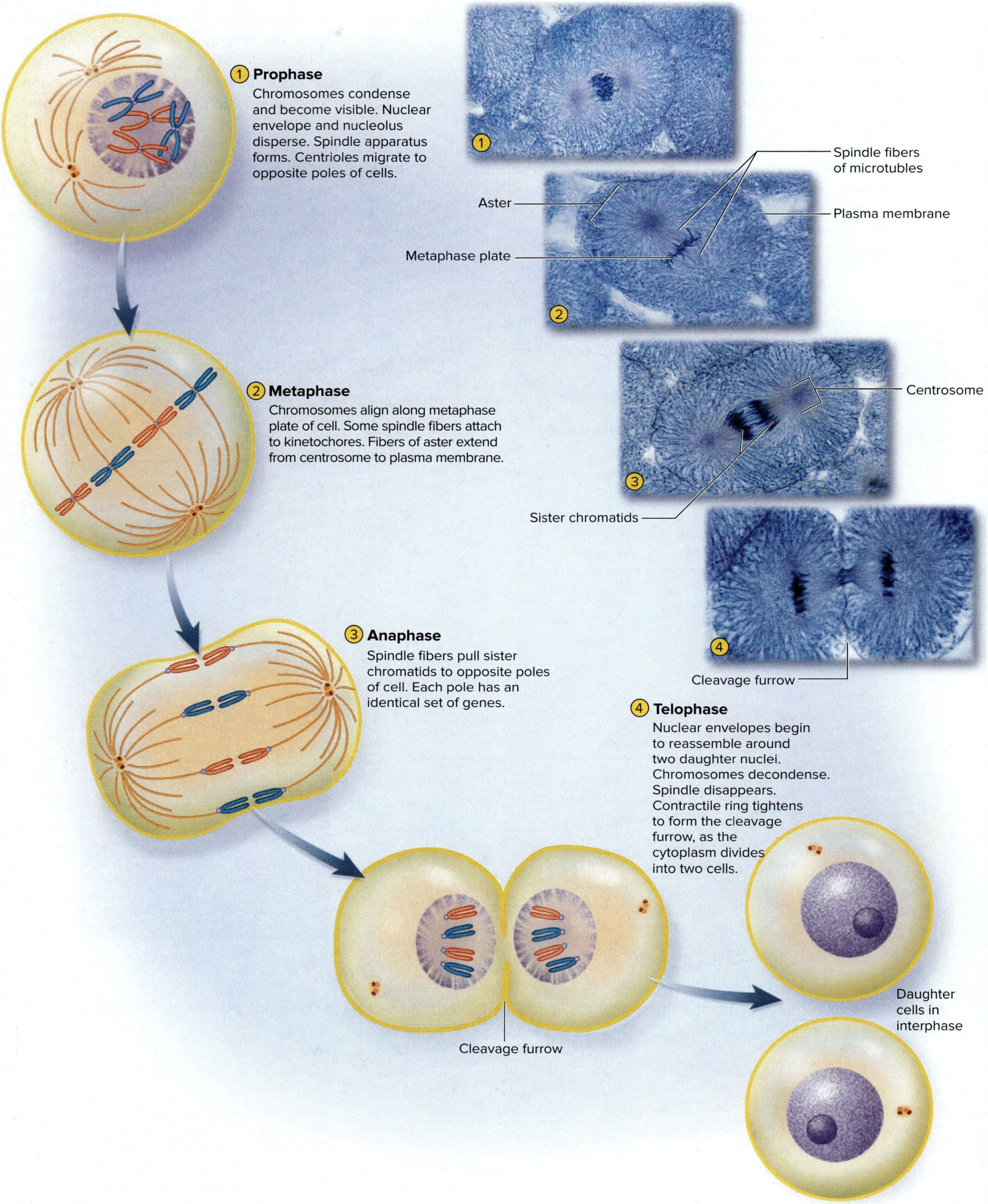

FIGURE 7.5 Mitosis drawings and photographs of the four phases and cytokinesis. (1, 2) Ed Reschke; (3, 4) Ed Reschke/Getty Images APR

LEARNING EXTENSION ACTIVITY

Identifying the phases of the cell cycle in a plant may improve understanding of an animal's cell cycle. Obtain an *Allium* onion root tip prepared microscope slide to compare the cell cycle of a plant to that of an animal (whitefish blastula). Examine the slide using the high-power objective of a microscope searching toward the root tip where there are more dividing cells. If you cannot locate all the different phases in one root tip, search another root tip on the slide. Instead of a cleavage furrow, a plant cell in telophase has a cell plate, which will eventually form the cell wall of each daughter cell. The cell wall is the outer boundary of plant cells, providing rigid structure and protection, and is not found in animal cells. Refer to figure 7.6 to help you identify the plant cell phases, and for comparison, refer to the whitefish blastula mitosis (fig. 7.5). Draw a line down the center of each circle (see below) that represents the field of view in Part B: Assessments and draw the plant (allium) cell on one side of the line and the animal (whitefish blastula) cell on the other side representing the same phase in the cell cycle.

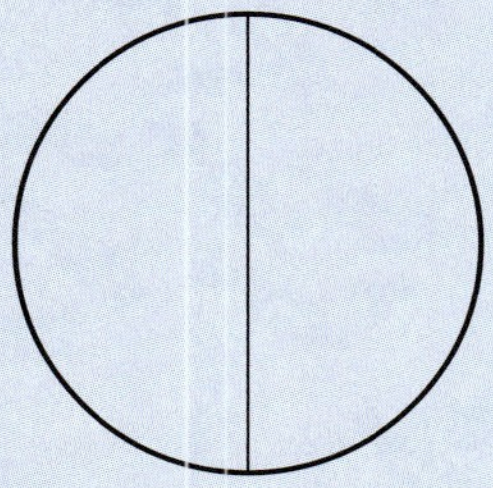

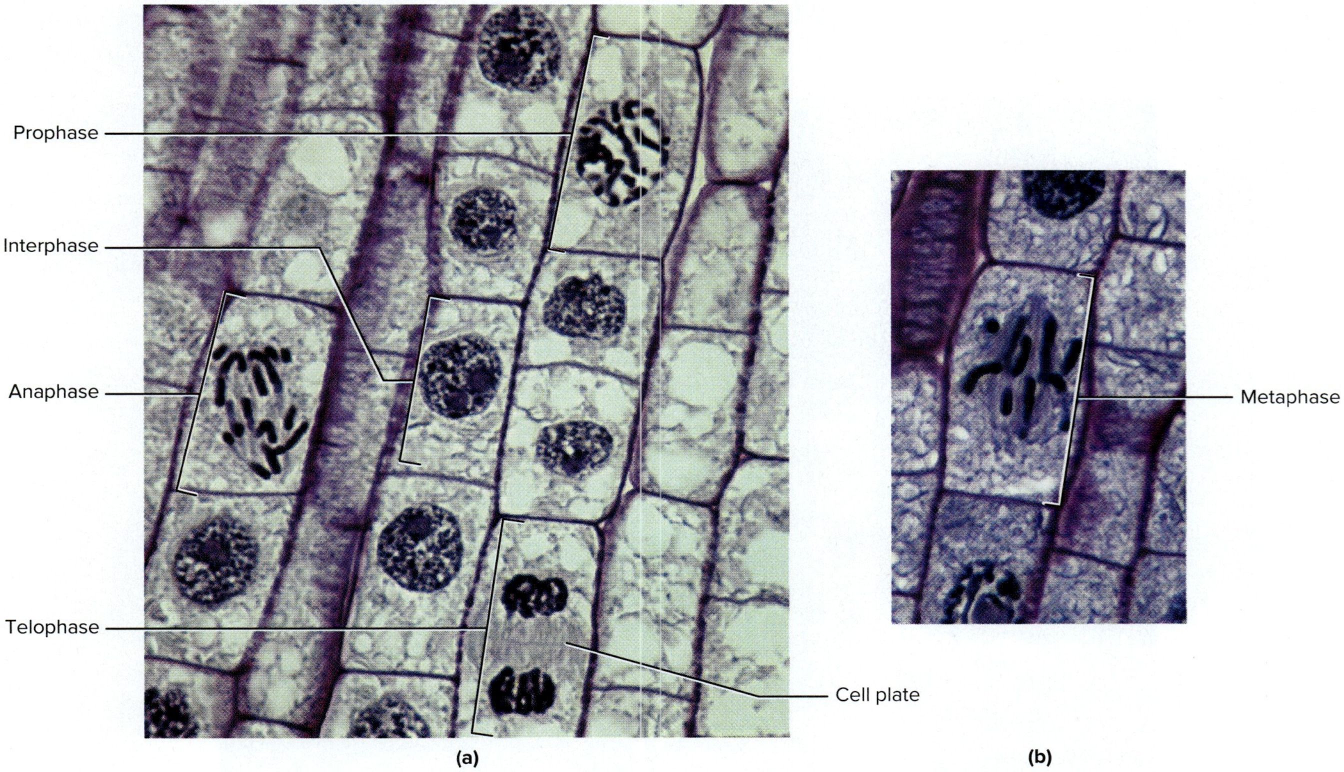

FIGURE 7.6 Allium onion root tip cells shown in various phases of their cell cycle (500×). In (a) the chromosomes are much more distinct in prophase than they are in interphase. The nucleus is swollen and the nuclear envelope is disintegrating in prophase. A nucleolus may be observable in interphase. Also shown is an onion cell in anaphase where sister chromatids have separated and are migrating to opposite poles of the cell, and an onion cell in telophase with a cell plate that is clearly seen. In (b) the chromosomes are aligned along the equator of the cell in metaphase. (a) Ed Reschke/Getty Images; (b) Ed Reschke/Stone/Getty Images

DEMONSTRATION ACTIVITY

Using the oil immersion objective of a microscope, see if you can locate some human chromosomes by examining a prepared slide of human chromosomes from leukocytes. The cells on this slide were cultured in a special medium and were stimulated to undergo mitosis. The mitotic process was arrested in metaphase by exposing the cells to a chemical called colchicine, and the cells were caused to swell osmotically. As a result of this treatment, the chromosomes were spread apart. A complement of human chromosomes should be visible when they are magnified about 1,000×. Each chromosome is double-stranded and consists of two sister chromatids joined by a common centromere (fig. 7.7).

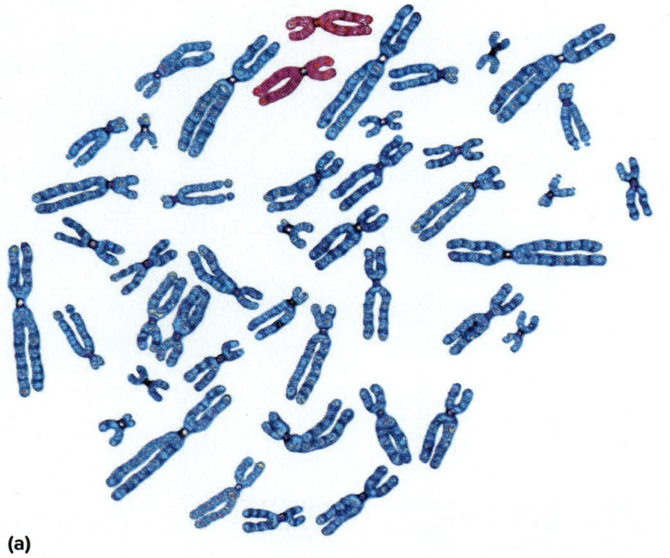

(a)

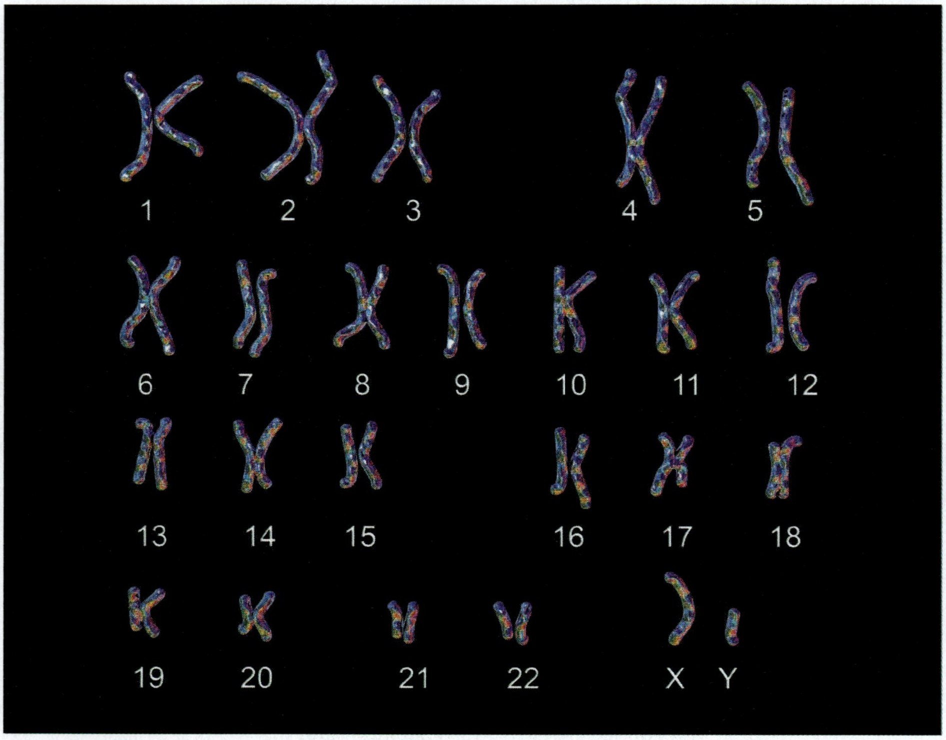

(b)

FIGURE 7.7 (a) A complement of human chromosomes of a female with two (duplicated) X chromosomes shown in red (1,000×). (b) A karyotype can be constructed by arranging the homologous chromosome pairs together in a chart. A karyotype can aid in the diagnosis of genetic conditions and abnormalities. The completed karyotype indicates a typical male with one (single-stranded) X and one Y chromosome. (a) James Cavallini/Science Source; (b) Kateryna Kon/Shutterstock APR

LABORATORY ASSESSMENT 7

Name ______________________________

Date ______________________________

Section ______________________________

The (A) corresponds to the indicated Learning Outcome(s) (O) found at the beginning of the Laboratory Exercise.

Cell Cycle

PART A: Assessments

Complete the table by listing the major events of each phase in the box. (A1)

	Phase	Major Events Occurring
Interphase	G_1, S, and G_2	
M (Mitotic) Phase	**Mitosis** Prophase	
	Metaphase	
	Anaphase	
	Telophase	
	Cytokinesis	

PART B: Assessments

Sketch an interphase cell and cells in different phases of mitosis to illustrate the whitefish cell's life cycle. Label the major cellular structures represented in the sketches and indicate cytokinesis locations. (The circles represent fields of view through the microscope.) A2

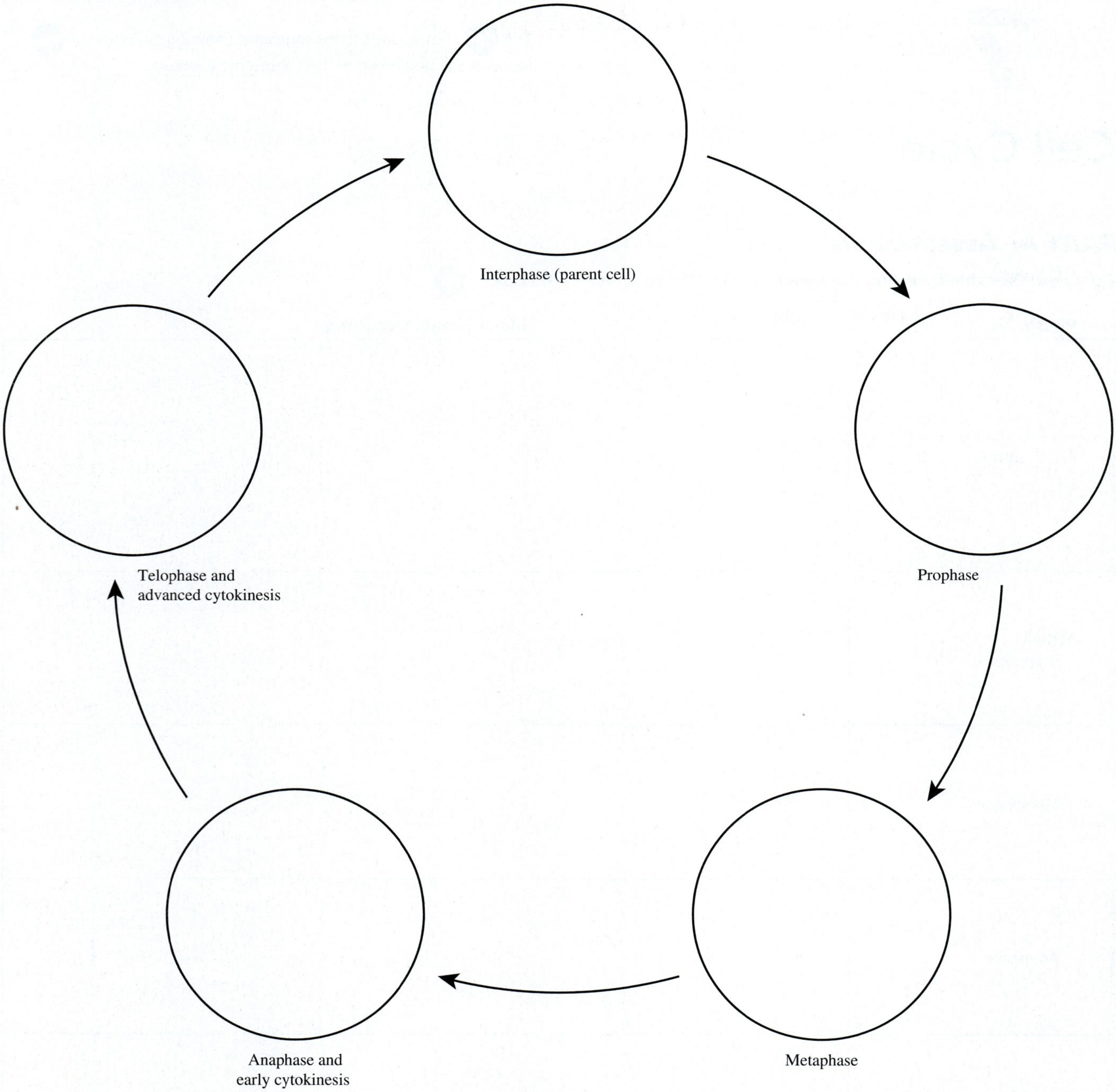

Magnification ___________________ ×

PART C: Assessments

Complete the following:

1. In what ways are the new cells (daughter cells), which result from a cell cycle, similar? A1 ________________________

____

2. Distinguish between mitosis and cytokinesis. A1 ______________________________

____

PART D: Assessments

1. Identify the mitotic phase represented by each of the micrographs in figure 7.8*a–d*. A3

a. ______________________ c. ______________________

b. ______________________ d. ______________________

2. Identify the structures indicated by numbers in figure 7.8 by placing the correct numbers in the spaces provided.

____ Aster
____ Centrosome
____ Chromatid
____ Cleavage furrow
____ Metaphase plate
____ Nuclear envelope
____ Plasma membrane
____ Spindle fiber

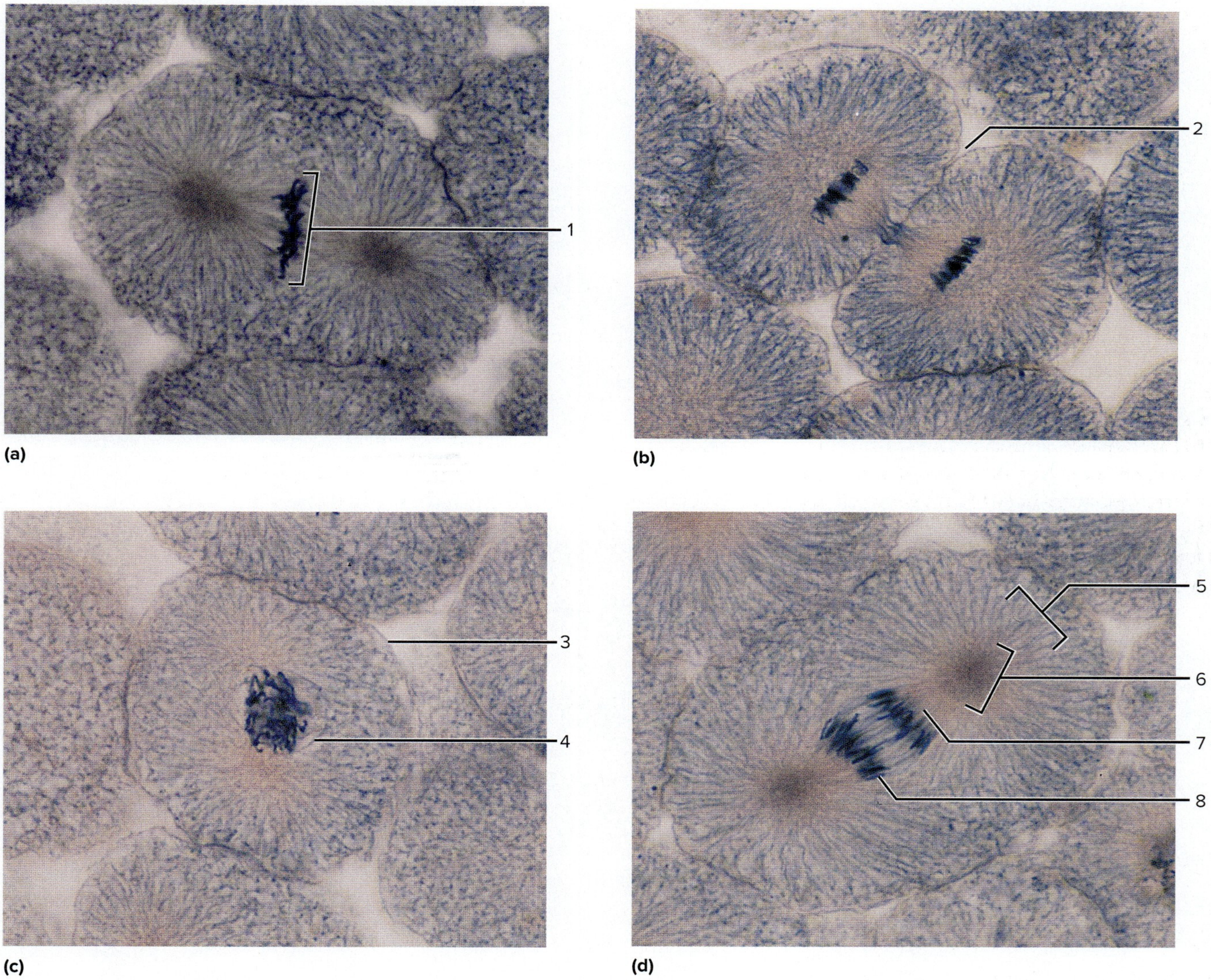

FIGURE 7.8 Identify the mitotic phase and structures of the cell in each of these micrographs of the whitefish blastula (400×). Place your answers in the spaces provided above. (a) Ed Reschke/Photolibrary/Stone/Getty Images; (b, c, d) Ed Reschke/Getty Images

CRITICAL THINKING ASSESSMENT—PART D

Predict what could happen if a cell were allowed to proceed in its cell cycle unregulated. A1

____

____

____

LABORATORY
EXERCISE 8

Epithelial Tissues

Purpose of the Exercise

To review the characteristics of epithelial tissues and to observe examples.

MATERIALS NEEDED

Compound light microscope

Prepared slides of the following epithelial tissues:

- Simple squamous epithelium (lung)
- Simple cuboidal epithelium (kidney)
- Simple columnar epithelium (small intestine)
- Pseudostratified (ciliated) columnar epithelium (trachea)
- Stratified squamous epithelium (nonkeratinized; esophagus)
- Stratified squamous epithelium (keratinized; epidermis)
- Stratified cuboidal epithelium (salivary gland)
- Stratified columnar epithelium (urethra)
- Transitional epithelium (urinary bladder)

For Learning Extension Activity:

Colored pencils

Chapter Opener: ©101cats/Getty Images

Learning Outcomes APR

After completing this exercise, you should be able to:

O1 Sketch and label the characteristics of epithelial tissues that you were able to observe.

O2 Differentiate the special characteristics of each type of epithelial tissue.

O3 Indicate a location and function of each type of epithelial tissue.

O4 Inventory the general characteristics of epithelial tissues that you were able to observe.

The O corresponds to the assessments A indicated in the Laboratory Assessment for this Exercise.

Pre-Lab

Carefully read the introductory material and examine the entire lab. Be familiar with epithelial tissues from lecture or the textbook. Answer the pre-lab questions.

Pre-Lab Questions Select the correct answer for each of the following questions:

1. The tissues that cover external and internal body surfaces are
a. epithelial. **b.** connective.
c. muscle. **d.** nervous.

2. The cells with a flat shape are named
a. stratified. **b.** cuboidal.
c. columnar. **d.** squamous.

3. The basal surface of epithelial tissue cells attach to the
a. cilia. **b.** basement membrane.
c. nucleus. **d.** cytoplasm.

4. If a structure cut has a circular appearance when observed, the sectional cut was a(n)
a. oblique section. **b.** cross section.
c. longitudinal section. **d.** lengthwise section.

5. Which of the following is *not* a function of epithelial tissues?
a. protection **b.** secretion
c. bind and support **d.** absorption

6. Which of the following epithelial tissues is more than one cell layer thick?
a. transitional **b.** simple squamous
c. pseudostratified columnar **d.** simple cuboidal columnar

7. Which of the following epithelial tissues does *not* have all cells resting on the basement membrane?
   a. stratified squamous
   b. pseudostratified columnar
   c. simple cuboidal
   d. simple squamous
8. The study of tissues is called histology.
   a. True   b. False
9. The basement membrane is composed of cilia.
   a. True   b. False

A tissue is composed of a layer or group of cells similar in size, shape, and function. The study of tissues is called *histology.* Within the human body, there are four major types of tissues: (1) *epithelial,* which cover the body's external and internal surfaces and compose most glands; (2) *connective,* which bind and support parts of the body; (3) *muscle,* which make movement possible; and (4) *nervous,* which conduct impulses from one part of the body to another and help to control and coordinate body activities.

Epithelial tissues are tightly packed single (simple) to multiple (stratified) layers composed of cells that provide protective barriers. The underside of this tissue contains an acellular basement membrane layer of adhesive cellular secretions and collagen through which the epithelial cells anchor to an underlying connective tissue. A unique type of simple epithelium, pseudostratified columnar, appears to be multiple cells thick. However, this is a false appearance because all cells bind to the basement membrane. Epithelial tissues have a free (apical) surface exposed to the outside or to an open space internally and a basal surface that attaches to the basement membrane. The cells of epithelial tissue closest to the basement membrane readily divide. Because epithelial tissues are avascular (lack blood vessels), the cells rely on the blood supply of underlying vascular connective tissues for nourishment and waste removal.

Many epithelial cell shapes, like squamous (flat), cuboidal, and columnar, exist and are used to name and identify the variations. The layers and their shape are often included as part of the epithelial tissue name; for example, "simple cuboidal epithelium" or "stratified squamous epithelium." The functions of epithelial tissues include protection, filtration, secretion, and absorption. Many of the prepared slides contain more than the tissue to be studied, therefore be certain that your view matches the correct tissue. Also be aware that stained colors of all tissues might vary.

The slides you observe might have representative tissue from more than one site in the body, but the basic tissue structure will be similar to those represented in this laboratory exercise. Simple squamous epithelium is located in the air sacs (alveoli) of the lungs and inner linings of the heart and blood vessels. Simple cuboidal epithelium is situated in kidney tubules, thyroid gland, liver, and ducts of salivary glands. A nonciliated type of simple columnar epithelium is located in the linings of the uterus, stomach, and intestines, and a ciliated type lines the uterine tubes. A ciliated type of pseudostratified columnar epithelium is positioned in the linings of the upper respiratory tubes. A keratinized type of stratified squamous epithelium is found in the epidermis of the skin, and a nonkeratinized type in the linings of the oral cavity, esophagus, vagina, and anal canal. Stratified cuboidal epithelium is represented in larger ducts of glands while stratified columnar epithelium lines the urethra of males. Transitional epithelium lines the urinary bladder, ureters, and part of the urethra. Various glands of the body are composed of primarily glandular epithelium. Studies of glands are found in Laboratory Exercises 11, 39, and 40.

PROCEDURE: Epithelial Tissues APR

This procedure represents the first of three laboratory exercises in a study of histology. So that you may clearly observe the details of tissue sections, the sectioned tissue has been stained, and the colors you observe will depend upon the particular stains used for the preparation. The slice of the histological section is usually only one or two cells thick, which will allow light to be transmitted through the entire section. Use the fine adjustment knob on the microscope to view the depth of the structures within the tissue section. Refer to the figures and the table in the laboratory exercise often as you examine each tissue.

1. Use the microscope to observe the prepared slides of types of epithelial tissues. As you observe each tissue, look for its special distinguishing features such as cell size, shape, and arrangement, and whether there are goblet cells, cilia, or microvilli present. Compare your prepared slides of epithelial tissues to the micrographs in figure 8.1 and the characteristics of each specific tissue in table 8.1 on p. 81.
2. As you observe the tissues in figure 8.1 and the prepared slides, note characteristics epithelial tissues have in common.
3. As you observe each type of epithelial tissue, prepare a labeled sketch of a representative portion of the tissue in Part A of Laboratory Assessment 8. Use a magnification close to what is shown in the lab manual of each tissue or specified by your instructor, and add any personal notes near your sketches that might assist your recognition of the tissues for any future assessments.
4. Test your ability to recognize each type of epithelial tissue. To do this, have a laboratory partner select one of the prepared slides, cover its label, and focus the microscope on the tissue. Then see if you can correctly identify the tissue.
5. Review the introductory material and table 8.1.
6. Complete Parts B and C of the laboratory assessment.

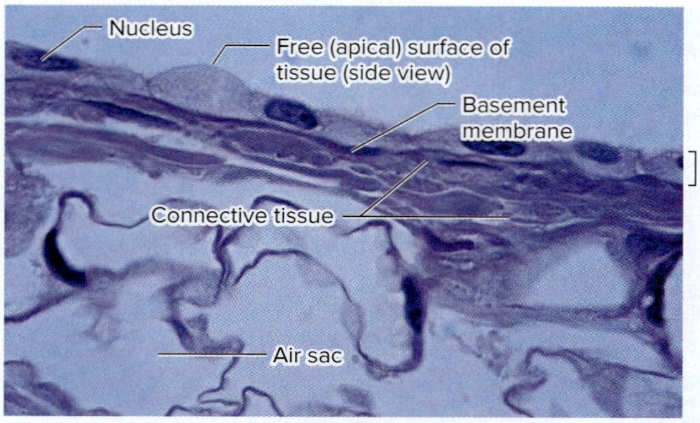

(a) Simple squamous epithelium (side view) (from lung) (250×) APR

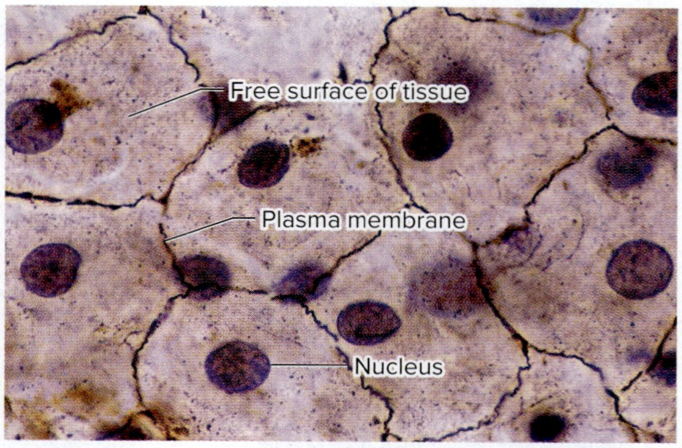

(b) Simple squamous epithelium (surface view) (250×)

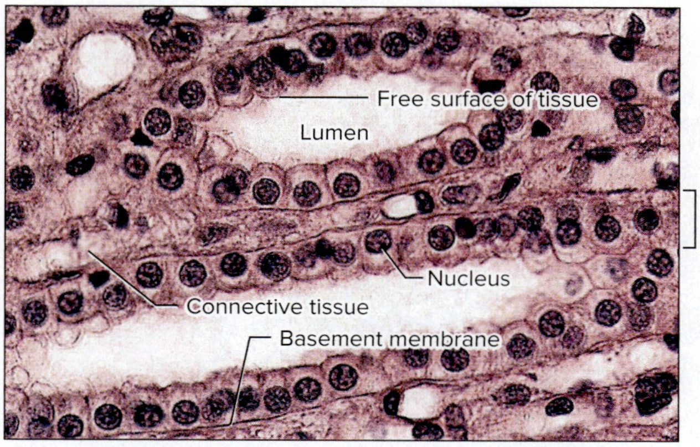

(c) Simple cuboidal epithelium (from kidney) (165×) APR

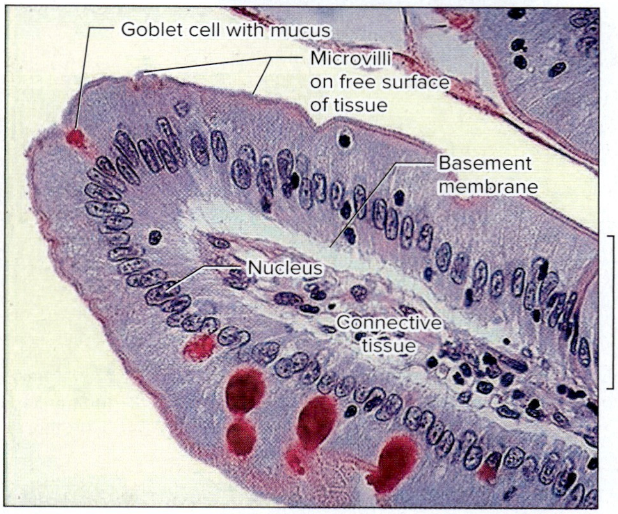

(d) Simple columnar epithelium (nonciliated) (from intestine) (400×) APR

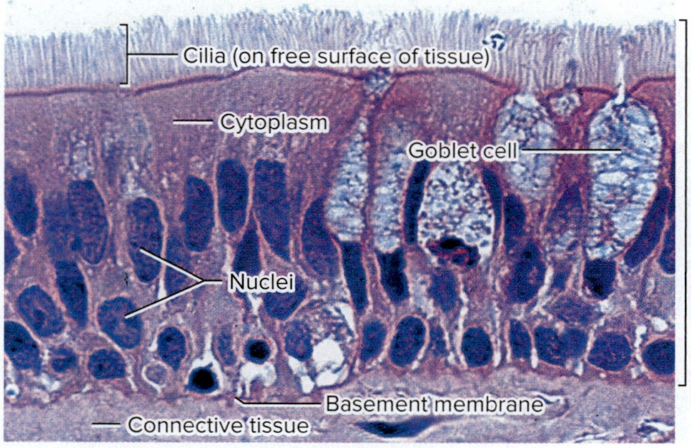

(e) Pseudostratified columnar epithelium with cilia (from trachea) (1,000×) APR

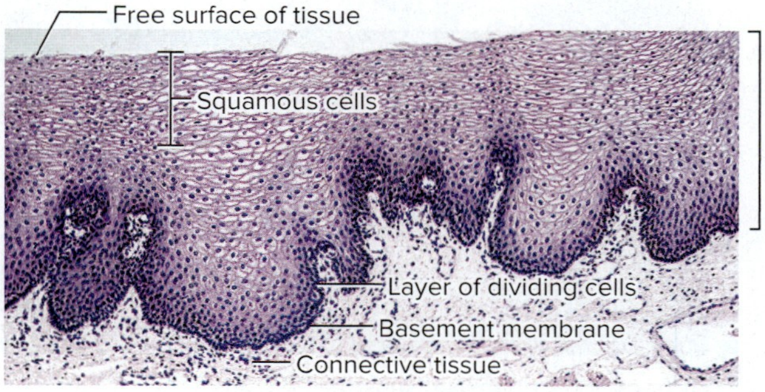

(f) Stratified squamous epithelium (nonkeratinized) (from esophagus) (100×) APR

FIGURE 8.1 Micrographs of epithelial tissues. *Note:* A bracket to the right of a micrograph indicates the tissue layer.
(a) Ed Reschke/Photolibrary/Stone/Getty Images; (b) Ed Reschke/Getty Images; (c, d) Victor P. Eroschenko; (e) Dennis Strete/McGraw Hill; (f) Al Telser/McGraw Hill

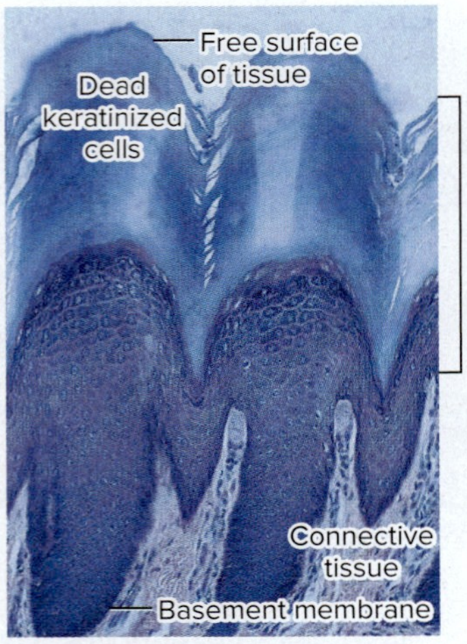

(g) Stratified squamous epithelium (keratinized) (from skin) (400×)

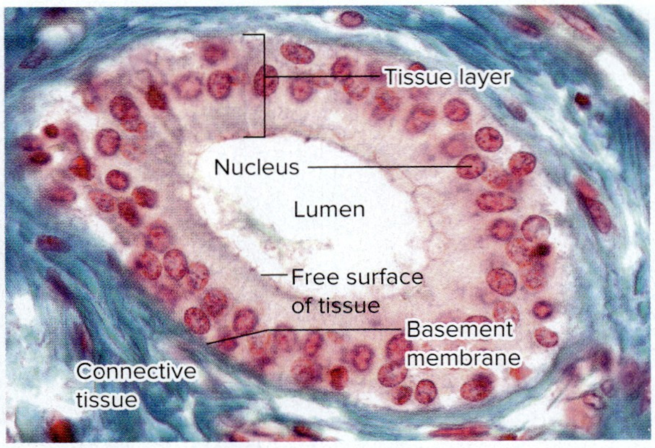

(h) Stratified cuboidal epithelium (from salivary gland) (600×)

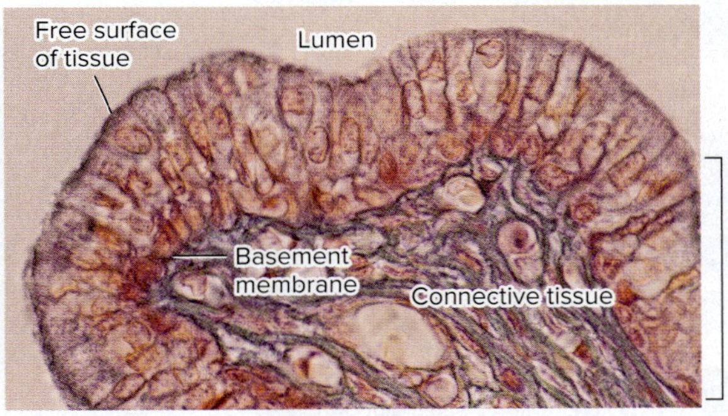

(i) Stratified columnar epithelium (from male urethra in penis) (250×)

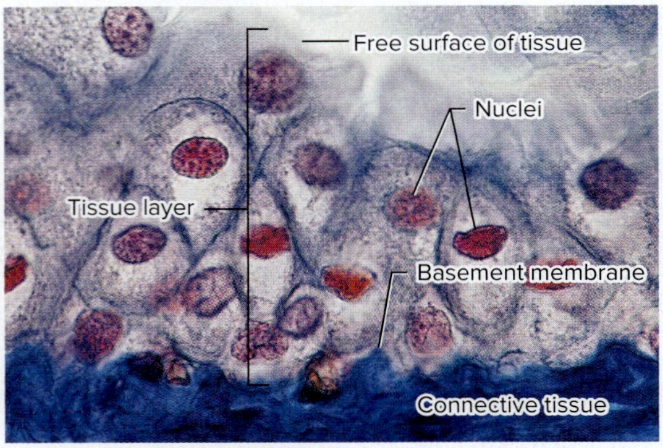

(j) Transitional epithelium (unstretched) (from urinary bladder) (675×)

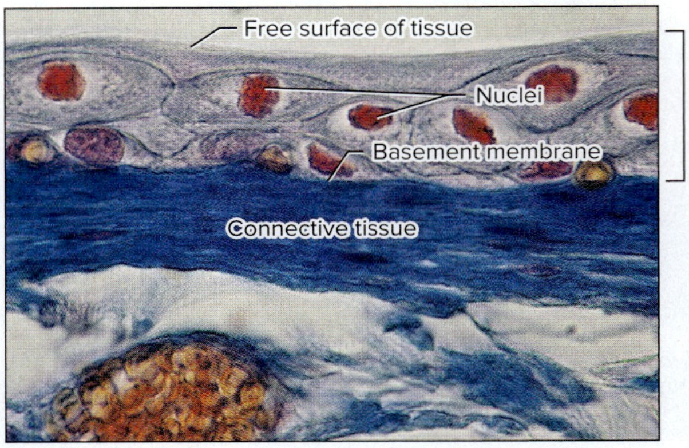

(k) Transitional epithelium (stretched) (from urinary bladder) (675×)

FIGURE 8.1 *Continued.* (g) Dennis Strete/McGraw Hill; (h) Al Telser/McGraw Hill; (i) Ed Reschke/Stone/Getty Images; (j, k) Ed Reschke

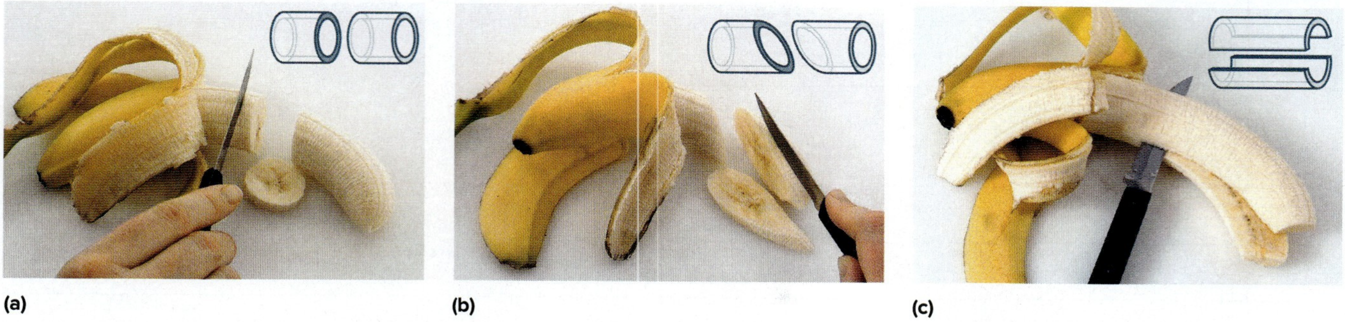

FIGURE 8.2 Three possible cuts of a banana: (a) cross section; (b) oblique section; (c) longitudinal section. Sections through an organ, as a body tube, frequently produce views similar to the cut banana. ©J & J Photography

LEARNING EXTENSION ACTIVITY

As you observe histology slides, be aware that tissues and some organs may have been sectioned in longitudinal (lengthwise cut), cross-sectional (cut across), or oblique (angular cut) ways. Observe figure 8.2 for how this can be demonstrated on cuts of a banana. The direction in which the tissue or organ was cut will result in a certain perspective when it is sectioned, just as the banana was cut in three different ways. Often a tissue slide has more than one tissue represented and has blood vessels or other structures cut in various ways. Locate an example of a longitudinal section, cross section, and oblique section of some structure on one of your tissue slides.

Table 8.1 Epithelial Tissues, Descriptions, Functions, and Representative Locations

Tissue Type	Descriptions	Functions	Representative Locations
Simple squamous epithelium	Single thin layer; flattened cells	Filtration; diffusion; osmosis; secretion	Air sacs (alveoli) of lungs; walls of capillaries; linings of blood vessels and ventral body cavity
Simple cuboidal epithelium	Single layer; cube-shaped cells	Secretion; absorption	Surface of ovaries; linings of kidney tubules; linings of ducts of certain glands
Simple columnar epithelium	Single layer; elongated narrow cells; some ciliated	Protection; secretion; absorption	Linings of uterus, stomach, gallbladder, and intestines
Pseudostratified columnar epithelium	Single layer; elongated cells; some cells do not reach free surface; often ciliated	Protection; secretion; movement of mucus and substances	Linings of respiratory passages
Stratified squamous epithelium (nonkeratinized)	Many layers; surface cells flattened and remain alive	Protection; resists abrasion	Linings of oral cavity, esophagus, vagina, and anal canal
Stratified squamous epithelium (keratinized)	Many layers; surface cells dead and keratinized	Protection; resists abrasion; retards water loss	Epidermis of skin
Stratified cuboidal epithelium	2 to 3 layers; cube-shaped cells	Protection; secretion	Linings of larger ducts of mammary glands, sweat glands, salivary glands, and pancreas
Stratified columnar epithelium	Superficial layer of elongated cells; basal layers of cube-shaped cells	Protection; secretion	Part of the male urethra; parts of the pharynx
Transitional epithelium	Many layers; cube-shaped and elongated cells; thinner layers when stretched	Distensibility; protection	Linings of urinary bladder and ureters and part of urethra
Glandular epithelium	Unicellular or multicellular	Secretion	Salivary glands; sweat glands; endocrine glands

NOTES

Name ______________________________

Date ______________________________

Section ______________________________

The (A) corresponds to the indicated Learning Outcome(s) (O) found at the beginning of the Laboratory Exercise.

LABORATORY ASSESSMENT 8

Epithelial Tissues

PART A: Assessments

Each circle below represents the field of view as seen through the microscope. In each circle, sketch a few cells of each type of epithelial tissue you observed that will clearly help you distinguish that tissue type. For each sketch, label the major characteristics and distinguishing features, indicate the magnification used, write an example of a location in the body, and provide a function. Alongside your sketch, add any personal notes that would assist your future ability to recognize the tissue. (A1) (A2) (A3)

Simple squamous epithelium (____×)
Location: ______________________________
Function: ______________________________

Simple cuboidal epithelium (____×)
Location: ______________________________
Function: ______________________________

Simple columnar epithelium (____×)
Location: ______________________________
Function: ______________________________

Pseudostratified columnar epithelium with cilia (____×)
Location: ______________________________
Function: ______________________________

Nonkeratinized stratified squamous epithelium (____×)
Location: ____________________________
Function: ____________________________

Keratinized stratified squamous epithelium (____×)
Location: ____________________________
Function: ____________________________

Stratified cuboidal epithelium (____×)
Location: ____________________________
Function: ____________________________

Stratified columnar epithelium (____×)
Location: ____________________________
Function: ____________________________

Transitional epithelium (____×)
Location: ____________________________
Function: ____________________________

__________________ epithelium (____×)
Location: ____________________________
Function: ____________________________

CRITICAL THINKING ASSESSMENT — PART A

As a result of your observations of epithelial tissues, which one(s) provide(s) the best protection? Explain your answer. A3 ______________________________

____

____

____

LEARNING EXTENSION ACTIVITY

Use colored pencils to differentiate various cellular structures in Part A. Select a different color for a nucleus, cytoplasm, plasma membrane, basement membrane, goblet cell, cilia, and microvilli whenever visible.

PART B: Assessments

Match the tissues in column A with the characteristics in column B. Place the letter of your choice in the space provided. (Some answers may be used more than once.) A2 A3

Column A

a. Simple columnar epithelium
b. Simple cuboidal epithelium
c. Simple squamous epithelium
d. Pseudostratified columnar epithelium
e. Stratified squamous epithelium
f. Stratified columnar epithelium
g. Stratified cuboidal epithelium
h. Transitional epithelium

Column B

______ **1.** Consists of several layers of varied-shaped cells, allowing an expandable lining

______ **2.** Commonly possesses cilia that move dust and mucus out of the respiratory airways

______ **3.** Single layer of flattened cells

______ **4.** Single row of elongated cells with nuclei located at different levels, and some cells don't reach the free surface

______ **5.** Forms walls of capillaries and air sacs of lungs

______ **6.** Elongated cells with elongated nuclei located near the basement membrane

______ **7.** Provides lining of urethra of males and parts of pharynx

______ **8.** Provides abrasion protection of skin epidermis and oral cavity

______ **9.** Forms inner lining of urinary bladder and ureters

______ **10.** Lines kidney tubules and ducts of salivary glands

______ **11.** Forms lining of stomach and intestines

______ **12.** Two or three layers of cube-shaped cells

______ **13.** Forms lining of oral cavity, anal canal, and vagina

______ **14.** Possesses microvilli in small intestine to increase surface area for better absorption

PART C: Assessments

As you examined each specific epithelial tissue, what were the general characteristics they possessed? List and describe those that you were able to observe. A4

____

____

____

____

____

____

____

CRITICAL THINKING ASSESSMENT — PART C

Which of these two epithelial tissues, simple cuboidal or stratified squamous, would you expect to heal more quickly if injured? Explain why. A4

____

____.

LABORATORY
EXERCISE 9

Connective Tissues

Purpose of the Exercise

To review the characteristics of connective tissues and to observe examples of the major types.

MATERIALS NEEDED

Compound light microscope

Prepared slides of the following:

- Areolar connective tissue
- Adipose tissue
- Reticular connective tissue
- Dense regular connective tissue
- Dense irregular connective tissue
- Elastic connective tissue
- Hyaline cartilage
- Fibrocartilage
- Elastic cartilage
- Bone (compact, ground, cross section)
- Blood (human smear)

For Learning Extension Activity:

Colored pencils

Chapter Opener: ©101cats/Getty Images

Learning Outcomes APR

After completing this exercise, you should be able to:

O1 Sketch and label the characteristics of connective tissues that you were able to observe.

O2 Differentiate the special characteristics of each of the major types of connective tissue.

O3 Indicate a location and function of each type of connective tissue.

O4 Inventory the general characteristics of connective tissues that you were able to observe.

The O corresponds to the assessments A indicated in the Laboratory Assessment for this Exercise.

Pre-Lab

Carefully read the introductory material and examine the entire lab. Be familiar with connective tissues from lecture or the textbook. Answer the pre-lab questions.

Pre-Lab Questions Select the correct answer for each of the following questions:

1. Which of the following is *not* a function of connective tissues?
   a. fill spaces **b.** provide support
   c. bind structures together **d.** body movement

2. The most abundant fibers of connective tissues are
   a. reticular. **b.** elastic.
   c. collagen. **d.** glycoprotein.

3. Which of the following is a loose connective tissue?
   a. fibrocartilage **b.** areolar
   c. spongy bone **d.** elastic connective

4. Which connective tissue has a liquid matrix?
   a. blood **b.** adipose
   c. reticular connective **d.** hyaline cartilage

5. The connective tissue that composes tendons and ligaments is
   a. dense regular connective. **b.** elastic connective.
   c. elastic cartilage. **d.** dense irregular connective.

6. The connective tissue _____ contains abundant fibers of collagen and cells in a solid matrix.
   a. fibrocartilage **b.** adipose
   c. bone **d.** areolar connective

7. Adipose, areolar, and reticular connective tissues are considered loose connective tissue types.
   a. True **b.** False

8. Blood and lymph are considered loose connective tissue types.
   a. True b. False
9. Cells secrete the fibers and a ground substance of a connective tissue.
   a. True b. False

Connective tissues, derived from embryonic tissue called mesenchyme, contain a variety of cell types and occur in all regions of the body. They bind structures together, provide support and protection, fill spaces, store fat, and transport blood cells.

Connective tissue cells are not as close together as epithelial cells and are often widely scattered in an abundance of a noncellular *extracellular matrix.* This matrix varies in quantity and form depending upon the specific tissue type and is like a filler material between the cells. The cells produce and secrete two components of the extracellular matrix known as the *ground substance* and *fibers.* The ground substance varies from a liquid to semisolid to solid, and has functions in support and as a medium for substances to move between cells and blood vessels. Fibers, the other component of the extracellular matrix, consist of fibrous proteins including *collagen, reticular,* and *elastic* fibers. Collagen fibers are thick threads, the most abundant of the three types, and they appear white in unstained tissues. Reticular fibers are fine threads of highly branched collagen with glycoprotein. Elastic fibers form complex networks of a highly branched protein called elastin, which is springlike and appears yellowish in unstained tissues. The fibers provide the binding properties within the tissue and between organs. You might compare connective tissue to making gelatin: the gelatin of various densities represents the ground substance, added fruit represents cells, and added strands such as vegetable sprouts represent the fibers.

Many of the prepared slides contain more than the tissue to be studied, so be certain that your view matches the correct tissue. Additional study of bone and blood will be found in Laboratory Exercises 12 and 41.

PROCEDURE: Connective Tissues APR

1. Use a microscope to observe the prepared slides of various connective tissues. As you observe each tissue, look for its special distinguishing features, such as extracellular matrix, cell types, and fiber types. Compare your prepared slides of connective tissues to the micrographs in figure 9.1 and the characteristics of each specific tissue in tables 9.1 and 9.2 and various cell types in table 9.3 on page 90.

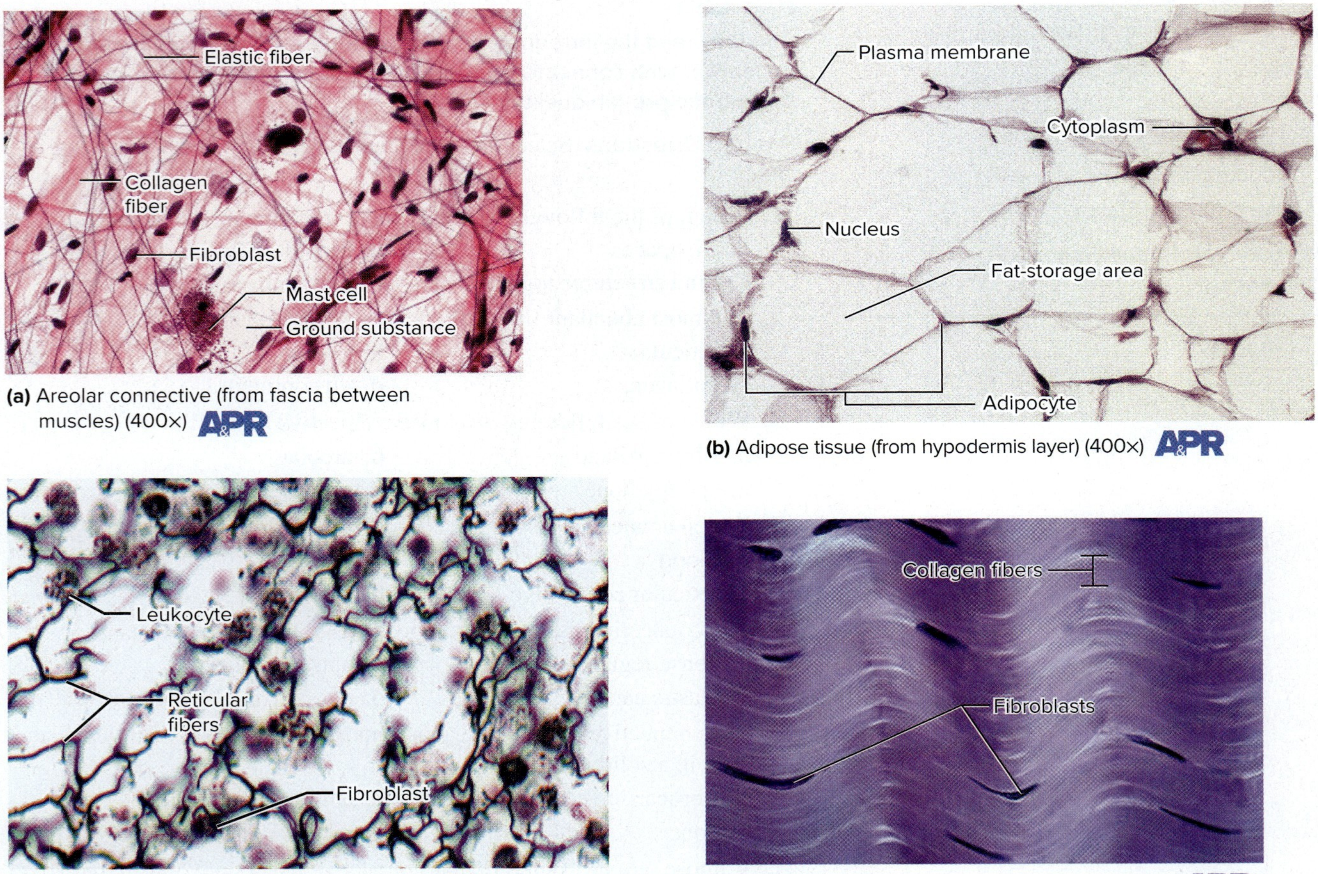

(a) Areolar connective (from fascia between muscles) (400×) APR

(b) Adipose tissue (from hypodermis layer) (400×) APR

(c) Reticular connective (from spleen) (400×) APR

(d) Dense regular connective (from tendon) (400×) APR

FIGURE 9.1 Micrographs of connective tissues. (a, b, c) Al Telser/McGraw Hill; (d) Ed Reschke/Getty Images

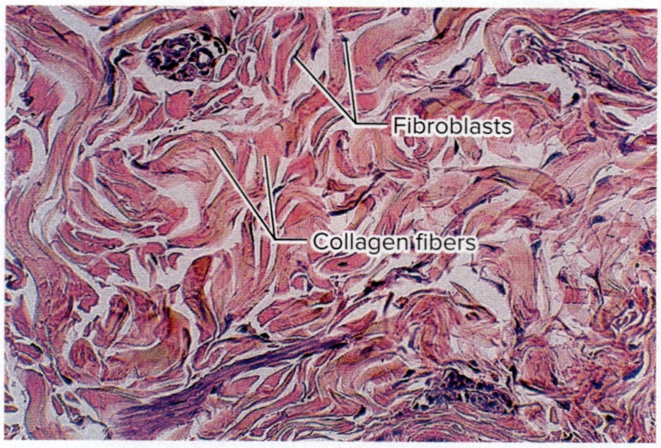

(e) Dense irregular connective (from dermis) (400×) APR

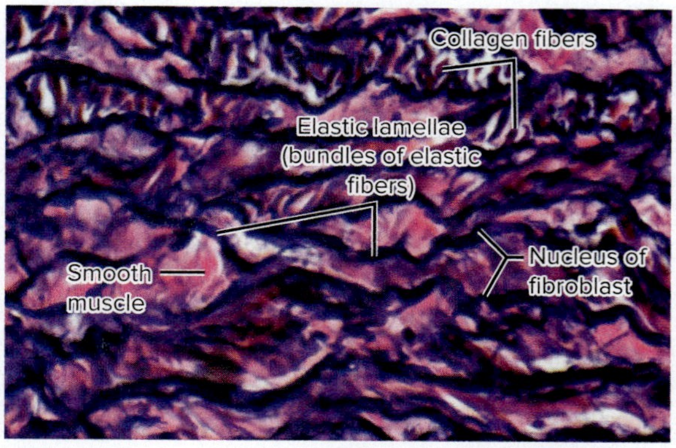

(f) Elastic connective tissue (from aorta) (400×) APR

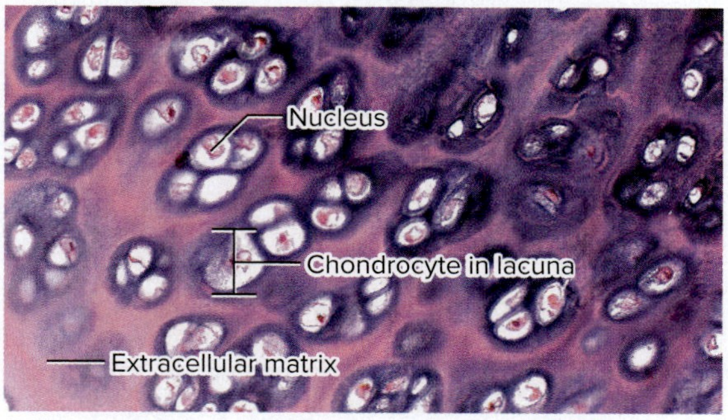

(g) Hyaline cartilage (from trachea) (200×) APR

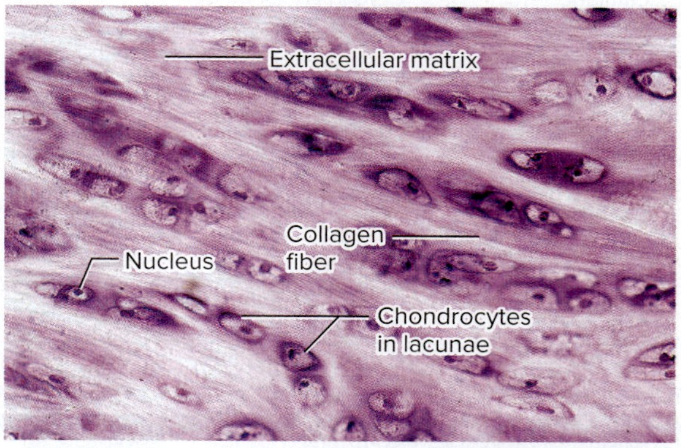

(h) Fibrocartilage (from intervertebral disc) (400×) APR

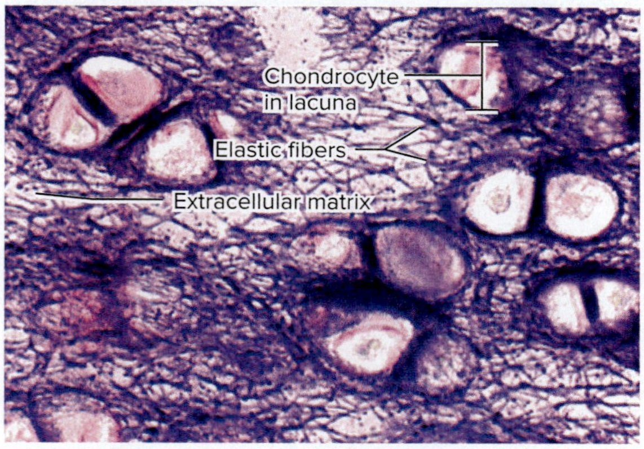

(i) Elastic cartilage (from ear) (400×) APR

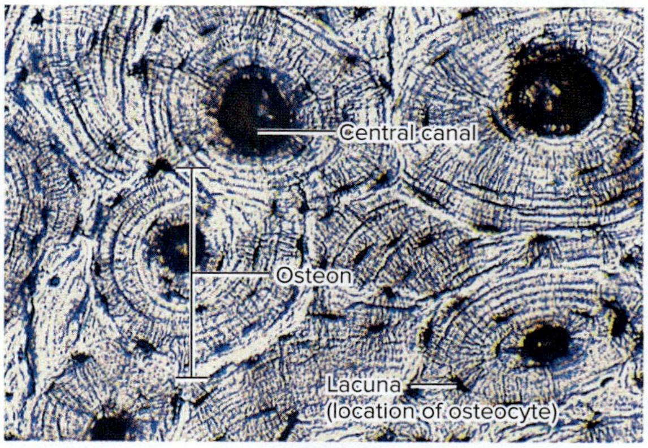

(j) Compact bone (200×) APR

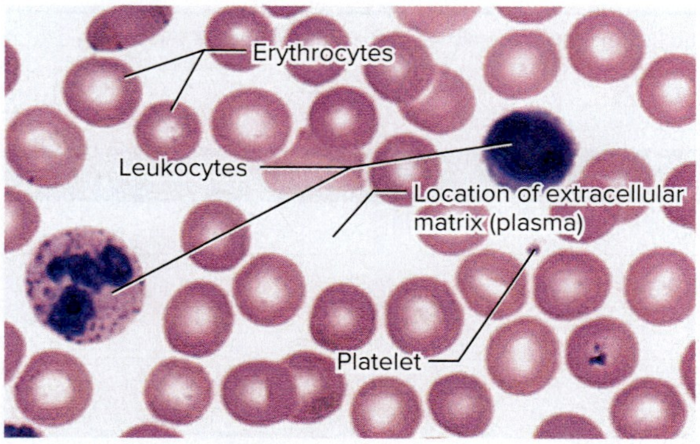

(k) Blood (400×) APR

FIGURE 9.1 *Continued.* (e, j) Dennis Strete/McGraw Hill; (f) Cynthia Prentice-Craver; (g, i, k) Al Telser/McGraw Hill; (h) Victor P. Eroschenko

2. As you observe the tissues in figure 9.1 and the prepared slides, note the characteristics connective tissues have in common.
3. As you observe each type of connective tissue, prepare a labeled sketch of a representative portion of the tissue in Part A of Laboratory Assessment 9. Use a magnification close to what is shown in the lab manual of each tissue or specified by your instructor, and add any personal notes near your sketches that might assist your recognition of the tissues for any future assessments.
4. Test your ability to recognize each type of connective tissue. To do this, have a laboratory partner select a slide, cover its label, and focus the microscope on this tissue. Then see if you correctly identify the tissue.
5. Complete Parts B and C of the laboratory assessment.

Table 9.1 Connective Tissues, Descriptions, and Functions

Tissue Type	Descriptions	Functions
Areolar connective	Cells in abundant fluid-gel matrix; loose arrangement of collagen and elastic fibers	Loosely binds organs; holds tissue fluids
Adipose	Cells in sparse fluid-gel matrix; closely packed cells	Protects; insulates; stores fat
Reticular connective	Cells in fluid-gel matrix; reticular fibers	Supportive framework of organs
Dense regular connective	Cells in fluid-gel matrix; parallel, wavy collagen fibers; limited vascularity	Tightly binds body parts
Dense irregular connective	Cells in fluid-gel matrix; random collagen fibers	Sustains tissue tension; durable
Elastic connective	Cells in fluid-gel matrix; collagen fibers densely packed; branched elastic fibers	Provides elastic quality
Hyaline cartilage	Cells in firm solid-gel matrix; fine network of collagen fibers; appears glassy; avascular	Supports; protects; provides framework
Fibrocartilage	Cells in firm solid-gel matrix; abundant collagen fibers; avascular	Supports; protects; absorbs shock
Elastic cartilage	Cells in firm solid-gel matrix; weblike elastic fibers; avascular	Supports; protects; provides flexible framework
Bone (osseous)	Cells in solid matrix; many collagen fibers; vascular	Supports; protects; provides framework; calcium storage
Blood	Cells and platelets in fluid matrix called plasma	Transports nutrients, wastes, and gases; defends against disease; clotting

Table 9.2 Types of Mature Connective Tissues and Representative Locations*

Loose Connective	Dense Connective	Cartilage	Bone	Liquid Connective
Areolar Connective Locations: around body organs; binds skin to deeper organs	*Dense Regular Connective* Locations: tendons; ligaments	*Hyaline Cartilage* Locations: nasal septum; larynx; costal cartilage; ends of long bones; fetal skeleton	*Compact Bone* Locations: bone shafts; beneath periosteum	*Blood* Locations: lumens of blood vessels; heart chambers
Adipose Locations: hypodermis (subcutaneous layer); around kidneys and heart; yellow bone marrow; breasts	*Dense Irregular Connective* Locations: dermis; heart valves; periosteum on bone	*Fibrocartilage* Locations: between vertebrae; between pubic bones; pads (meniscus) in knee	*Spongy (Cancellous) Bone* Locations: within ends of long bones; inside flat and irregular bones	*Lymph* Locations: lumens of lymphatic vessels
Reticular Connective Locations: spleen; thymus; lymph nodes; red bone marrow	*Elastic Connective* Locations: larger artery walls; vocal cords; some ligaments between vertebrae	*Elastic Cartilage* Locations: outer ear; epiglottis		

*This table represents a scheme to organize connective tissue relationships. Spongy bone and lymph tissues are not examined microscopically in this laboratory exercise.

Table 9.3 Cells Found in Various Connective Tissues, and Their Functions

Cells	Functions
Mast cell	Secretes histamine (a vasodilator) and heparin (an anticoagulant)
Fibroblast	Synthesizes extracellular matrix (protein fibers and ground substance)
Adipocyte	Stores fat, an energy reservoir
Chondroblast	Produces cartilage; called chondrocyte when isolated in lacunae
Osteoblast	Produces bone (osseous tissue); called osteocyte when isolated in lacunae
Leukocyte	White blood cell; fights off infection and pathogens
Erythrocyte	Red blood cell; transports gases (oxygen and carbon dioxide)
Platelet	Cell fragment; secretes chemicals to attract clotting proteins and to stimulate blood vessel growth; acts to reduce blood loss

LABORATORY

ASSESSMENT

9

Name ______________________________

Date ______________________________

Section ______________________________

The **A** corresponds to the indicated Learning Outcome(s) **O** found at the beginning of the Laboratory Exercise.

Connective Tissues

PART A: Assessments

Each circle below represents the field of view as seen through the microscope. In each circle, sketch a small section of each type of connective tissue you observed that will clearly help you distinguish that tissue type. For each sketch, label the major characteristics and distinguishing features, indicate the magnification used, write an example of a location in the body, and provide a function. Alongside your sketch, add any personal notes that would assist your future ability to recognize the tissue. **A1** **A2** **A3**

Areolar connective (_____×)
Location: ______________________________
Function: ______________________________

Adipose (_____×)
Location: ______________________________
Function: ______________________________

Reticular connective (_____×)
Location: ______________________________
Function: ______________________________

Dense regular connective (_____×)
Location: ______________________________
Function: ______________________________

Dense irregular connective (_____×)
Location: ____________________________
Function: ____________________________

Elastic connective (_____×)
Location: ____________________________
Function: ____________________________

Hyaline cartilage (_____×)
Location: ____________________________
Function: ____________________________

Fibrocartilage (_____×)
Location: ____________________________
Function: ____________________________

Elastic cartilage (_____×)
Location: ____________________________
Function: ____________________________

Compact bone (_____×)
Location: ____________________________
Function: ____________________________

Blood (_____×)
Location: ____________________________
Function: ____________________________

_______________ connective (_____×)
Location: ____________________________
Function: ____________________________

LEARNING EXTENSION ACTIVITY

Use colored pencils to differentiate various cellular structures in Part A. Select a different color for the cells, fibers, and ground substance whenever visible.

PART B: Assessments

Match the tissues in column A with the characteristics in column B. Place the letter of your choice in the space provided. (Some answers may be used more than once.) A2 A3

Column A	Column B
a. Adipose	______ **1.** Forms framework of outer ear
b. Areolar connective	______ **2.** Functions as heat insulator beneath skin
c. Blood	______ **3.** Contains large amounts of fluid and transports wastes and gases
d. Bone (compact)	______ **4.** Cells in a solid matrix arranged around central canal
e. Dense irregular connective	______ **5.** Binds skin to underlying organs
f. Dense regular connective	______ **6.** Main tissue of tendons and ligaments
g. Elastic cartilage	______ **7.** Forms the flexible part of the nasal septum and ends of long bones
h. Elastic connective	______ **8.** Pads between vertebrae that are shock absorbers
i. Fibrocartilage	______ **9.** Main tissue of dermis
j. Hyaline cartilage	______ **10.** Occurs in some ligament attachments between vertebrae and larger artery walls
k. Reticular connective	______ **11.** Forms supporting tissue in thymus and spleen
	______ **12.** Cells in a fluid-gel matrix with parallel collagen fibers
	______ **13.** Contains loose arrangement of elastic and collagen fibers
	______ **14.** Transports nutrients and defends against disease
	______ **15.** Matrix contains collagen fibers and calcium and phosphorus salts

PART C: Assessments

As you examined each specific connective tissue, what were the general characteristics they possessed? List and describe those that you were able to observe. A4

____

____

____

____

____

CRITICAL THINKING ASSESSMENTS—PART C

1. Abdominal impact injuries in the region of the left upper quadrant (LUQ) often involve the spleen. Explain the structural tissue characteristics within the spleen that make it so vulnerable to serious injury. A3 A4

____

____

____

____

____

2. Which of these two connective tissues, hyaline cartilage or bone (osseous), would you expect to heal more quickly if injured? Explain why. A4

____

____

LABORATORY
EXERCISE 10

Muscle and Nervous Tissues

Purpose of the Exercise

To observe muscle and nervous tissues and recognize the characteristics that make them unique.

MATERIALS NEEDED

Compound light microscope

Prepared slides of the following:

- Skeletal muscle tissue
- Smooth muscle tissue
- Cardiac muscle tissue
- Nervous tissue (spinal cord smear and/or cerebellum)

For Learning Extension Activity:

Colored pencils

Learning Outcomes APR

After completing this exercise, you should be able to:

O1 Sketch and label the characteristics of the different muscle and nervous tissues that you were able to observe.

O2 Differentiate the special characteristics of each type of muscle tissue and nervous tissue.

O3 Indicate an example location and function of each type of muscle tissue and nervous tissue.

The O corresponds to the assessments A indicated in the Laboratory Assessment for this Exercise.

Pre-Lab

Carefully read the introductory material and examine the entire lab. Be familiar with muscle tissues and nervous tissue from lecture or the textbook. Answer the pre-lab questions.

Pre-Lab Questions Select the correct answer for each of the following questions:

1. Which muscle tissue is under conscious control?
a. skeletal **b.** smooth
c. cardiac **d.** stomach

2. Which of the following organs lacks smooth muscle?
a. blood vessels **b.** stomach
c. iris **d.** heart

3. Which muscle tissue lacks striations?
a. cardiac **b.** skeletal
c. smooth **d.** voluntary

4. Which of the following conduct action potentials to other cells?
a. smooth muscle cells **b.** neurons
c. neuroglia **d.** skeletal muscle cells

5. Muscles of facial expression are
a. smooth muscles. **b.** skeletal muscles.
c. involuntary muscles. **d.** cardiac muscles.

6. Which muscle tissue is multinucleated?
a. skeletal **b.** smooth
c. stomach **d.** cardiac

7. Where is the nucleus located in a neuron?
a. dendrites **b.** cell body
c. cellular processes **d.** axon

8. Intercalated discs represent the junctions where cardiac muscle cells fit together.
a. True **b.** False

9. Both cardiac and skeletal muscle cells are under voluntary control.
a. True **b.** False

Chapter Opener: ©101cats/Getty Images

Muscle tissues are characterized by the presence of elongated cells, often referred to as muscle fibers, that can contract to create bodily movements. Many of our muscles are attached to the skeleton, but muscles are also components of many of our internal organs. During muscle contractions, considerable body heat is generated to help maintain our body temperature. Because more heat is generated than is needed to maintain our body temperature, much of the heat is dissipated from our bodies through the skin.

The three types of muscle tissues are *skeletal, smooth,* and *cardiac. Skeletal muscles* are usually under our conscious control and are considered voluntary. Although most skeletal muscles are attached to bones via tendons, the tongue, facial muscles, and voluntary sphincters do not attach directly to the bone. Functions of skeletal muscle include body movements, such as maintaining posture, breathing, speaking, and controlling waste eliminations, as well as helping to regulate blood sugar levels and heat production, and to provide protection. Skeletal muscles also aid the movement of lymph and venous blood during muscle contractions. Skeletal muscle cells are striated with an organized pattern of contractile proteins. These cells are multinucleated, providing the genetic instructions to synthesize proteins. *Smooth muscle* is under involuntary control and is located in many visceral organs, such as the urinary bladder, trachea, blood vessels, intestines, and uterus. This muscle type is also found in the iris of the eye and attached to hair follicles. Functions of smooth muscle include the motions of visceral organs (peristalsis), controlling pupil size, blood flow, airflow, and creating "goose bumps" when a person is too cold or frightened. Smooth muscle cells are not striated, and each cell has a fusiform shape with a nucleus located close to its center. *Cardiac muscle* is located only in the heart wall. It is under involuntary control and functions to pump blood. Cardiac muscle cells (cardiocytes) are striated and separated from one another by intercalated discs that contain gap junctions, which are avenues for electrical impulses, and desmosomes, which are mechanical connections that keep cells from pulling apart. Glycogen may be found surrounding the single nucleus of each cell.

Nervous tissues occur in the brain, spinal cord, and peripheral nerves. The tissue consists of two cell types: *neurons* and *neuroglia.* Neurons, also called nerve cells, contain a cell body (soma) with the nucleus and most of the cytoplasm, and cellular processes that extend from the cell body. Cellular processes include one to many dendrites and a single axon (nerve fiber). Neurons are considered excitable cells because they can generate electrical signals called action potentials (nerve impulses) that are conducted along the neuron to another neuron or a muscle or gland. Neuroglia (glial cells) of various types are more abundant than neurons; they cannot generate or conduct nerve impulses, but they have important supportive and protective functions for neurons. Additional study of nervous tissue will be found in Laboratory Exercise 27.

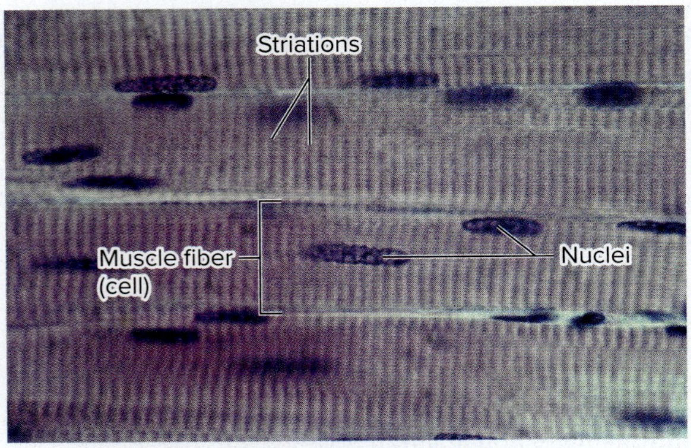

(a) Skeletal muscle (from leg) (100×) APR

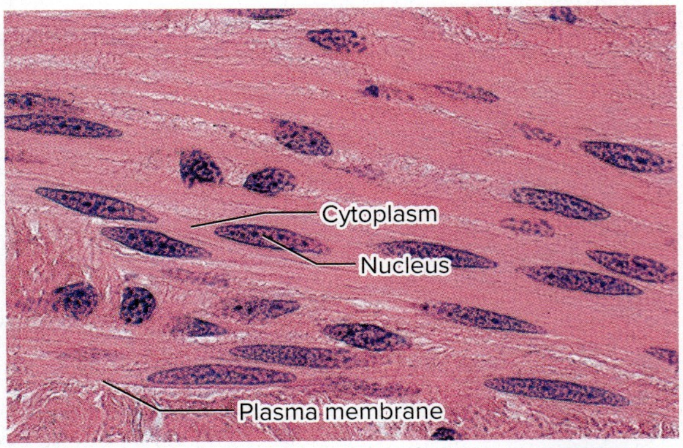

(b) Smooth muscle (from small intestine) (1,000×) APR

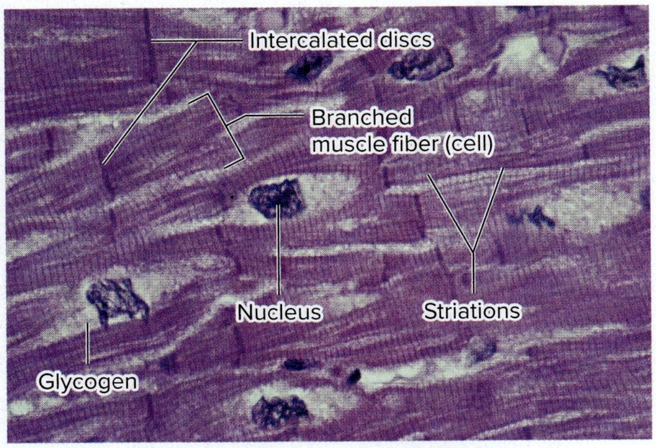

(c) Cardiac muscle (from heart) (250×) APR

FIGURE 10.1 Micrographs of muscle tissues.
(a) Ed Reschke/Photolibrary/Stone/Getty Images; (b) Dennis Strete/McGraw Hill; (c) Ed Reschke/Stone/Getty Images APR

PROCEDURE: Muscle and Nervous Tissues

1. Using the microscope, observe each of the types of muscle tissues on the prepared slides. Look for the special features of each type. Compare your prepared slides of muscle tissues to the micrographs in figure 10.1 and the characteristics of each specific tissue in tables 10.1 and 10.2.
2. As you observe each type of muscle tissue, prepare a labeled sketch of a representative portion of the tissue in Part A of Laboratory Assessment 10. Add any personal notes near your sketches that might assist your recognition of the tissues for future assessments.
3. Observe the prepared slide of nervous tissue and identify neurons (nerve cells), neuron cellular processes, and neuroglia (glial cells). Compare your prepared slide of nervous tissue to the micrograph in figure 10.2 and the characteristics in table 10.1.
4. Prepare a labeled sketch of nervous tissue along with any personal notes in Part A of the laboratory assessment.
5. Test your ability to recognize each of these muscle and nervous tissues by having your laboratory partner select a slide, cover its label, and focus the microscope on this tissue. Then see if you correctly identify the tissue.
6. Complete Part B of the laboratory assessment.

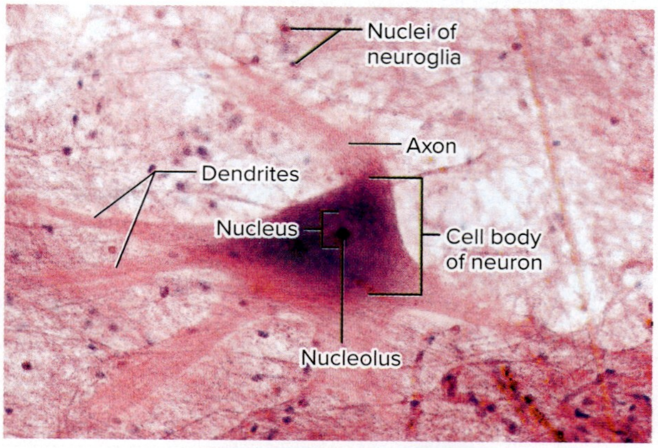

Nervous tissue (from spinal cord) (350×) APR

FIGURE 10.2 Micrograph of nervous tissue.
Alvin Telser/McGraw Hill APR

Table 10.1 Muscle and Nervous Tissues, Descriptions, Functions, and Representative Locations

Tissue Type	Descriptions	Functions	Representative Locations
Skeletal muscle	Long, threadlike cells; striated; many nuclei near plasma membrane	Voluntary movements of skeletal parts; facial expressions	Muscles usually attached to bones
Smooth muscle	Shorter fusiform (spindle-shaped) cells; single central nucleus	Involuntary movements of internal organs as peristalsis; control of blood vessel diameter	Walls of hollow internal organs
Cardiac muscle	Branched cells; striated; single nucleus (usually)	Heart contractions to pump blood; involuntary	Heart walls
Nervous tissue	Neurons with long cellular processes; neuroglia smaller and variable	Sensory reception and conduction of action potentials; neuroglia supportive	Brain, spinal cord, and peripheral nerves; glial cells in close proximity to neurons

Table 10.2 Muscle Tissue Characteristics

Characteristic	Skeletal Muscle	Smooth Muscle	Cardiac Muscle
Appearance of cells	Unbranched and relatively parallel	Short fusiform (spindle-shaped) cells tapering at ends	Branched and connected in complex networks
Striations (pattern of alternating dark and light bands across cells)	Present and obvious	Absent	Present but faint
Nucleus	Multinucleated	Uninucleated	Uninucleated (usually)
Intercalated discs (junction where cells fit together)	Absent	Absent	Present
Control	Voluntary (usually)	Involuntary	Involuntary

NOTES

Name ______________________________

Date ______________________________

Section ______________________________

The A corresponds to the indicated Learning Outcome(s) O found at the beginning of the Laboratory Exercise.

LABORATORY ASSESSMENT 10

Muscle and Nervous Tissues

PART A: Assessments

Each circle below represents the field of view as seen through the microscope. In each circle, sketch a few cells or fibers of each of the three types of muscle tissues and of nervous tissue that will clearly help you distinguish that tissue type. For each sketch, label the major characteristics and distinguishing features of the cells or fibers, indicate the magnification used, write an example of a location in the body, and provide a function. Alongside your sketch, add any personal notes that would assist your future ability to recognize the tissue. A1 A2 A3

Skeletal muscle tissue (_____×)
Location: ______________________________
Function: ______________________________

Smooth muscle tissue (_____×)
Location: ______________________________
Function: ______________________________

Cardiac muscle tissue (_____×)
Location: ______________________________
Function: ______________________________

Nervous tissue (_____×)
Location: ______________________________
Function: ______________________________

LEARNING EXTENSION ACTIVITY

In Part A, use colored pencils to differentiate various cellular structures.

PART B: Assessments

Match the tissues in column A with the characteristics in column B. Place the letter of your choice in the space provided. (Some answers may be used more than once.) A2 A3

Column A	Column B
a. Cardiac muscle	______ **1.** Coordinates, regulates, and integrates body functions
b. Nervous tissue	______ **2.** Contains intercalated discs
c. Skeletal muscle	______ **3.** Muscle that lacks striations
d. Smooth muscle	______ **4.** Striated and involuntary
	______ **5.** Striated and voluntary
	______ **6.** Contains neurons and neuroglia
	______ **7.** Muscle attached to bones
	______ **8.** Muscle that composes heart
	______ **9.** Moves food through the digestive tract
	______ **10.** Conducts impulses along cellular processes
	______ **11.** Muscle under conscious control
	______ **12.** Muscle of blood vessels and iris

CRITICAL THINKING ASSESSMENTS—PART B

1. The word *fiber* is used to describe a structure in three of the four primary (main) tissue types. In connective tissues, a *fiber* is a protein strand that is collagen, reticular, or elastic. The meaning of the word *fiber* is different in muscle tissue and in nervous tissue. Explain a *fiber* in each of these two tissue types. A2 ______________________________

____.

2. Deficient blood flow (ischemia) to the heart during a heart attack (myocardial infarction) may cause fibrosis (scarring) to occur in the injured area of the heart wall. A fibrotic scar consists of connective tissue produced by fibroblasts. Identify the pros and cons of fibrosis occurring in the injured cardiac muscle tissue. A2 ______________________________

____

____

____.

LABORATORY
EXERCISE 11

Integumentary System

Purpose of the Exercise

To observe the structures and tissues of the integumentary system and to review the functions of these parts.

MATERIALS NEEDED

Skin model
Hand magnifier or stereomicroscope
Forceps
Microscope slide and coverslip
Compound light microscope
Prepared microscope slide of human scalp or axilla
Prepared slide of dark (heavily pigmented) human skin
Prepared slide of thick skin (plantar or palmar)

For Learning Extension Activity:
Tattoo slide

Learning Outcomes APR

After completing this exercise, you should be able to:

O1 Locate and name the structures of the integumentary system.

O2 Describe the major functions of these structures.

O3 Distinguish the locations and tissues among epidermis, dermis, and the hypodermis.

O4 Identify and sketch the layers of the skin and associated structures observed on the prepared slide.

The O corresponds to the assessments A indicated in the Laboratory Assessment for this Exercise.

Pre-Lab

Carefully read the introductory material and examine the entire lab. Be familiar with skin layers and accessory structures of the skin from lecture or the textbook. Answer the pre-lab questions.

Pre-Lab Questions Select the correct answer for each of the following questions:

1. Which of the following is *not* a function of the integumentary system?
   a. protection **b.** excrete small amounts of waste
   c. movement **d.** aid in regulating body temperature
2. The two distinct skin layers are the
   a. epidermis and dermis. **b.** hypodermis and dermis.
   c. hypodermis and epidermis. **d.** dermis and hypodermis.
3. Apocrine sweat glands are located in _______ regions of the body.
   a. forehead **b.** axillary and genital
   c. palmar **d.** plantar
4. The hypodermis is composed of _______ tissues.
   a. adipose connective and stratified squamous epithelial
   b. areolar and dense irregular connective
   c. stratified squamous epithelial and adipose connective
   d. areolar and adipose connective
5. The _______ of the epidermis is only present in thick skin.
   a. stratum corneum **b.** stratum lucidum
   c. stratum spinosum **d.** stratum granulosum
6. Frequent cell division occurs in the _______ of the epidermis.
   a. stratum corneum **b.** stratum spinosum
   c. stratum granulosum **d.** stratum basale

Chapter Opener: ©101cats/Getty Images

7. The greatest concentration of melanin is in the dermis.
   a. True　　**b.** False
8. Thick skin of the palms and soles contains five strata of the epidermis.
   a. True　　**b.** False

The integumentary system includes the skin, hair, nails, and glands. These structures provide a protective covering for deeper tissues, aid in regulating body temperature, retard water loss, house sensory receptors, synthesize various glandular chemicals, and excrete small quantities of wastes.

The skin consists of two distinct layers. The superficial layer, the *epidermis,* consists of keratinized stratified squamous epithelium. The deepest layer, the *dermis,* consists of a superficial papillary region of areolar connective tissue and a thicker and deeper reticular region of dense irregular connective tissue. Beneath the dermis is the *hypodermis* (subcutaneous layer; superficial fascia) composed of primarily adipose connective tissue and some areolar connective tissue. The hypodermis is not considered a layer of the skin.

Accessory structures of the skin include *nails, hair follicles,* and *skin (cutaneous) glands.* The hair, which grows through a depression from the epidermis, possesses a *hair papilla* at the base of the hair. The papilla contains a network of capillaries that supply the nutrients for cell divisions required for hair growth within the *hair matrix* of the hair bulb. As the cells of the hair are forced toward the surface of the body, they become keratinized and pigmented and die. Attached to the follicle is the *arrector pili muscle* that can pull the hair to a more upright position, causing goose bumps when experiencing cold temperatures or fear.

Sebaceous glands secrete an oily sebum into hair follicles, which keep the hair and epidermal surface pliable and somewhat waterproof.

Sweat glands (*sudoriferous glands*) are distributed over most regions of the body and consist of two types of glands. The widespread *merocrine (eccrine) sweat glands* are most numerous on the palms, the soles of the feet, and the forehead. Their ducts open to the surface at a sweat pore. Their secretions increase during hot days, physical exercise, and stress; they serve a mini-excretory function and can help prevent our bodies from overheating. The *apocrine sweat glands* are most abundant in the axillary and genital regions. Apocrine sweat ducts open into the hair follicles and become active at puberty. Their secretions increase during stress and pain and have little influence on thermoregulation. Modified sweat glands of the integumentary system include *ceruminous glands* of the external auditory canal that secrete cerumen (ear wax) and *mammary glands* of the breasts that secrete milk.

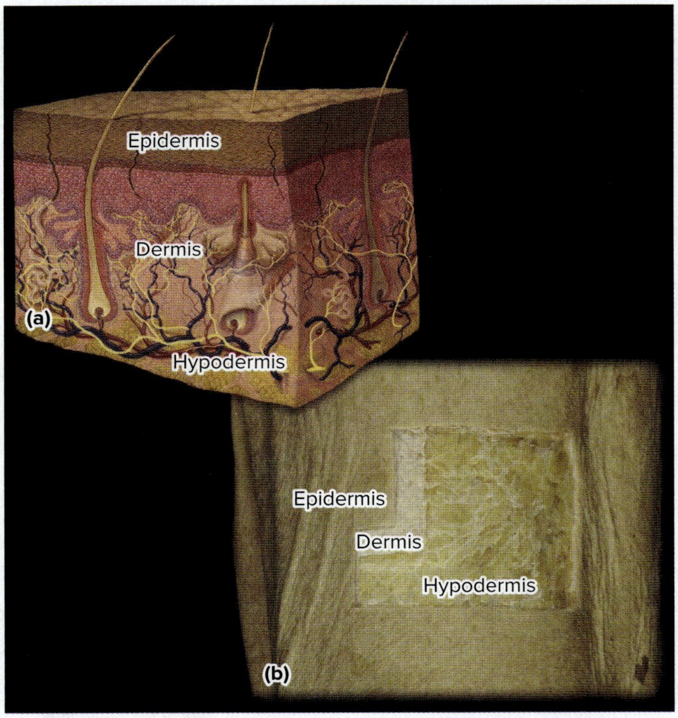

FIGURE 11.1 Layers of skin and deeper hypodermis indicated on (a) an illustration and (b) cadaver skin.
The University of Toledo/McGraw Hill APR

PROCEDURE: Integumentary System

In this procedure you will use a skin model and make comparisons to the figures in the lab manual to locate the layers and accessory structures of the skin. Hair structures will be observed from your own body with different magnifications and then compared to the additional detail observed when viewing prepared slides. Several micrographs of different magnifications are provided showing vertical sections of the skin layers and accessory structures. Use a combination of all of the micrographs as you observe the integument slides available in your laboratory.

1. Use figures 11.1 and 11.2 to locate as many of these structures as possible on a skin model.
2. Use figure 11.3 as a guide to locate the specific epidermal layers (strata) on a skin model. Note the locations and descriptions from table 11.1.
3. Use the hand magnifier or stereomicroscope and proceed as follows:
   a. Observe the skin, hair, and nails on your hand.
   b. Compare the type and distribution of hairs on the front and back of your forearm.

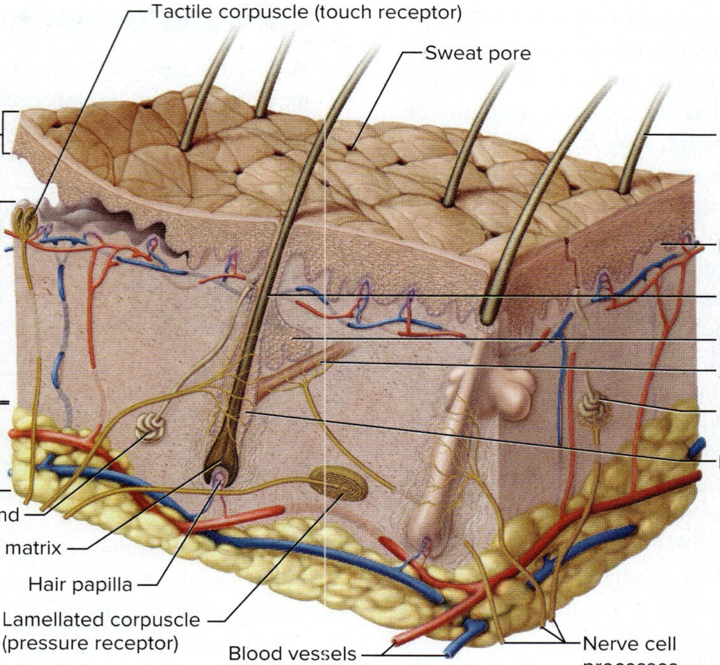

FIGURE 11.2 Vertical section of the skin and hypodermis (subcutaneous layer).

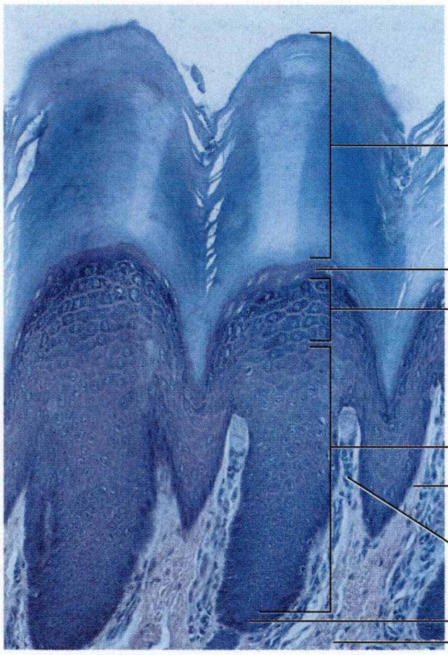

FIGURE 11.3 Epidermal layers in this section of thick skin from the fingertip (400×). Dennis Strete/McGraw Hill APR

Table 11.1 Layers of the Epidermis

Layer	Location	Descriptions
Stratum corneum	Most superficial layer	Many layers of keratinized, dead epithelial cells; appear scaly and flattened; resists water loss, absorption, and abrasion
Stratum lucidum (only present in thick skin)	Between stratum corneum and stratum granulosum on soles and palms of thick skin	Cells appear clear; nuclei, organelles, and plasma membranes no longer visible
Stratum granulosum	Beneath the stratum corneum (or stratum lucidum of thick skin)	Three to five layers of flattened granular cells; contain shrunken fibers of keratin and shriveled nuclei; cells secrete substances that create a waterproofing barrier that cuts off blood supply from below
Stratum spinosum	Beneath the stratum granulosum	Many layers of cells with centrally located, large, oval nuclei; develop fibers of keratin; deeper cells capable of dividing; cells becoming flattened in superficial portion
Stratum basale	Deepest layer	A single row of cuboidal or columnar cells; layer also includes melanocytes; frequent cell division; some cells become parts of more superficial layers

4. Study the structure of a fingernail in figure 11.4. Use a hand magnifier or a stereomicroscope to observe various features of a fingernail. The *nail plate* is composed of dead epithelial cells derived from the stratum corneum. The exposed *free edge, nail body,* and the *nail root* under the skin make up the nail plate. The pinkish hue of most of the nail body results from underlying blood within dermal blood vessels, whereas the crescent-shaped *lunula* area appears whitish because of a thicker region of the *nail bed* that lies beneath the nail plate that obscures the color of blood. The nail bed is composed of epithelial cells derived from the stratum basale. On the proximal end of the nail bed is the *nail matrix* where cell division occurs, resulting in nail growth. The narrow *cuticle* is a region of dead skin.
5. Use low-power magnification of the compound light microscope and proceed as follows:
   a. Pull out a single hair with forceps and mount it on a microscope slide under a coverslip.
   b. Observe the root and shaft of the hair and note the scalelike parts that make up the shaft.
6. Complete Parts A and B of Laboratory Assessment 11.

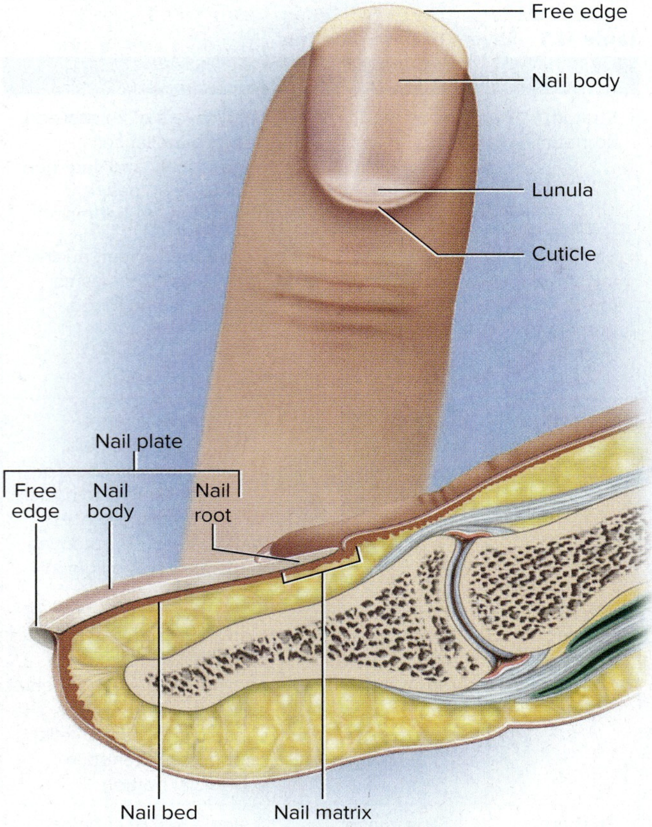

FIGURE 11.4 Fingernail structures as seen from a surface view and a sagittal section. APR

CRITICAL THINKING ACTIVITY

Explain the advantage of melanin granules being located in the deep layers of the epidermis, and not in the dermis or deeper hypodermis (fig. 11.5*c*).

____

____

____

____

____

____

7. As vertical sections of human skin are observed, remember that the lenses of the microscope invert and reverse images. It is important to orient the position of the epidermis, dermis, and hypodermis layers using scan magnification before continuing with additional observations. Compare all of your skin observations to the various micrographs in figure 11.5 and to the higher magnification micrographs of sebaceous and eccrine glands in figure 11.6. Use low-power magnification of the compound light microscope and proceed as follows:
   a. Observe the prepared slide of human scalp or axilla.
   b. Locate the epidermis, dermis, and hypodermis; a hair follicle; an arrector pili muscle; a sebaceous

LEARNING EXTENSION ACTIVITY

Observe a vertical section of human skin through a tattoo, using low-power magnification. Note the location of the dispersed ink granules within the upper portion of the dermis. From a thin vertical section of a tattoo, it is not possible to determine the figure or word of the entire tattoo as seen on the surface of the skin because only a small, thin segment of the image is in view at any one time. Compare this to the location of melanin granules found in dark (heavily pigmented) skin. Suggest reasons why a tattoo is permanent and a suntan is not.

____

____

____

____

____

____

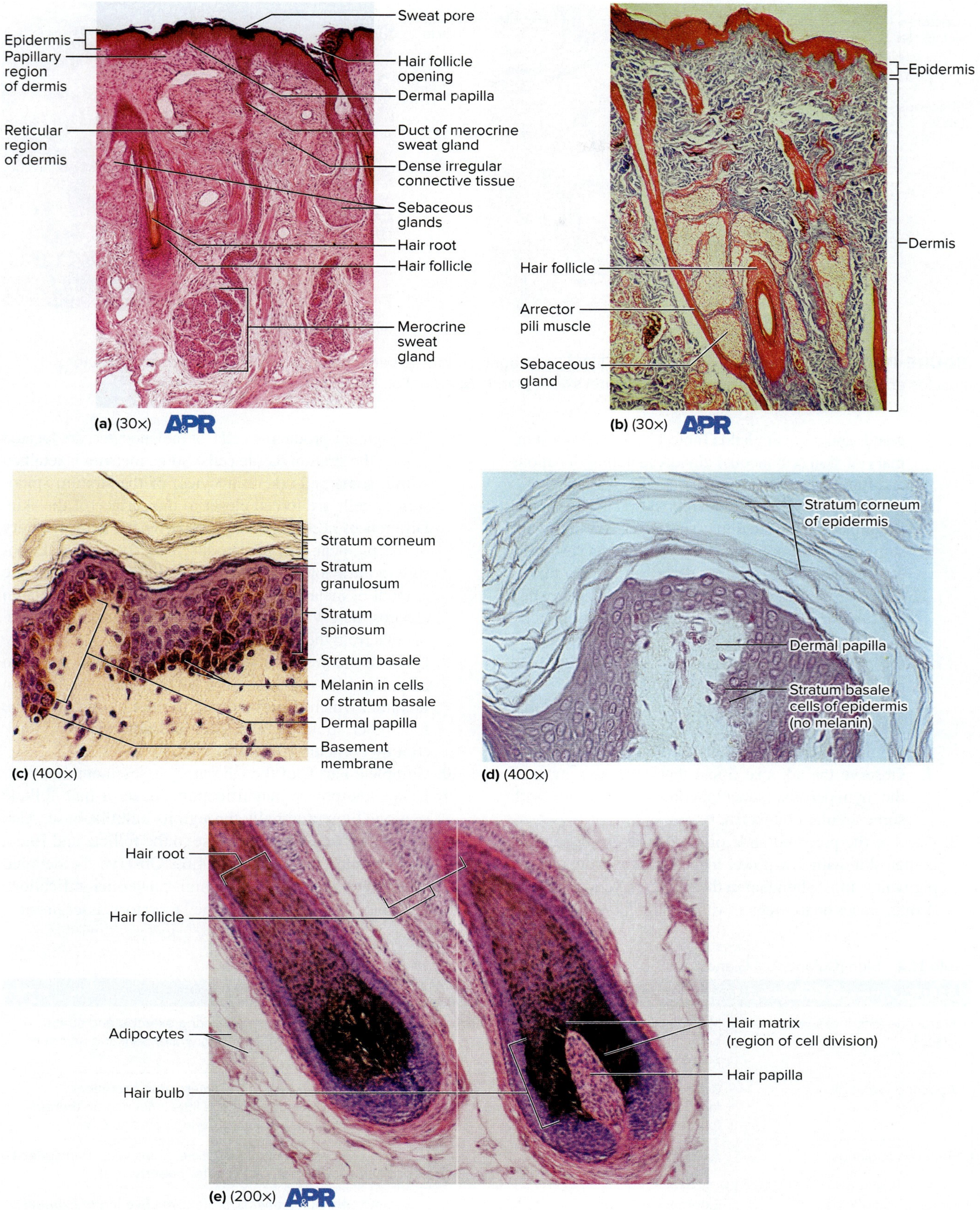

FIGURE 11.5 Features of human skin are indicated in these variously stained micrographs: (a) and (b) various structures of epidermis and dermis; (c) epidermis of dark skin; (d) epidermis of light skin; (e) base of two hair structures. (a) ©Per Kjeldsen; (b) Image Source/Getty Images; (c, d) Dennis Strete/McGraw Hill; (e) Cynthia Prentice-Craver

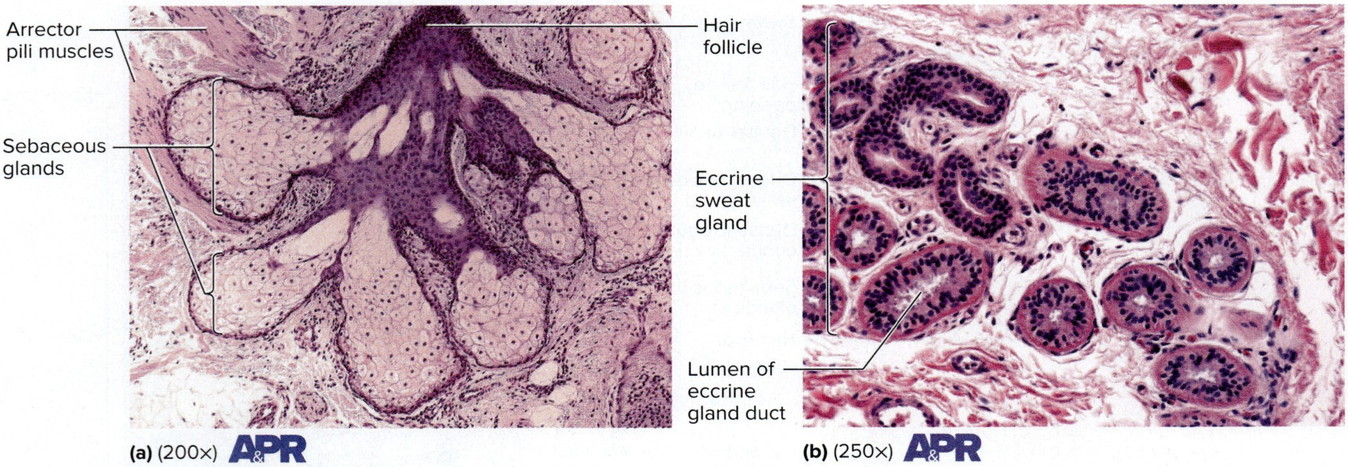

FIGURE 11.6 Some cutaneous glands shown in higher magnification: (a) sebaceous glands with associated arrector pili muscles and hair follicle; (b) an eccrine (merocrine) sweat gland. (a, b) Jose Luis Calvo/Shutterstock

gland; and a sweat gland. Table 11.2 includes a summary of skin (cutaneous) glands, including locations and functions.

c. Focus on the epidermis with high power and locate the stratum corneum, stratum granulosum, stratum spinosum, and stratum basale. Note how the shapes of the cells in these layers differ as described in table 11.1.
d. Observe the projections of the papillary layer called the *dermal papillae,* located in the dermis directly beneath the epidermis. Larger projections found in the palms of the hands and soles of the feet help produce unique fingerprints (and toeprints).
e. Observe the dense irregular connective tissue that makes up the bulk of the dermis (reticular region). Identify the collagen fibers of the matrix.
f. Observe the adipose tissue that composes most of the hypodermis (subcutaneous layer) along with some areolar connective tissue.

8. Observe the prepared slide of dark (heavily pigmented) human skin with low-power magnification. Note that the pigment is most abundant in the deepest layers of the epidermis. Focus on this region with the high-power objective. The pigment-producing cells, or melanocytes, are located among the stratum basale cells. Some melanin is retained within some cells (keratinocytes) of the stratum spinosum as cells are forced closer to the surface of the skin. Differences in skin color are primarily due to the quantity of the pigment *melanin* produced by these cells. Exposure to ultraviolet (UV) rays of sunlight can increase the amount of melanin produced, causing a suntan. Melanin absorbs the UV radiation, which helps protect the nuclei of cells from the harmful effects of the sun.
9. Observe the prepared slide of thick skin from the palm of a hand or the sole of a foot. Locate the stratum lucidum, which is present only in thick skin. Locate the four other strata of thick skin and note the very thick stratum corneum (fig. 11.3).
10. Complete Part C of the laboratory assessment.
11. Using low-power magnification, locate a hair follicle sectioned longitudinally through its bulblike base. Also locate a sebaceous gland close to the follicle and find a sweat gland (fig. 11.5 and 11.6). Observe the detailed structure of these parts with high-power magnification.
12. Complete Parts D and E of the laboratory assessment.

Table 11.2 Skin (Cutaneous) Glands

Skin (Cutaneous) Glands	Location	Functional Descriptions
Merocrine (eccrine) sweat glands	Most of body surface; ducts open onto body surface; abundant on forehead, palms, and soles of feet	Provide evaporative cooling during exercise and stress; they serve excretory function for water, salt, and nitrogen wastes
Apocrine sweat glands	Concentrations in axillary, genital areas; facial hair of males; ducts open into hair follicles	Secretions more viscous than merocrine secretions; more active after puberty; have little influence on thermoregulation; increased activity during pain and stress
Sebaceous glands	Associated with hair follicles	Secrete oily sebum into hair follicle; helps keep skin surface and hair pliable; more active after puberty
Ceruminous glands	External ear canal	Secrete cerumen (earwax); protective covering of external ear canal
Mammary glands	Within breasts	Produce breast milk during pregnancies; secrete milk during lactation into ducts that open on a nipple

LABORATORY
ASSESSMENT 11

Name ______________________

Date ______________________

Section ______________________

The (A) corresponds to the indicated Learning Outcome(s) (O) found at the beginning of the Laboratory Exercise.

Integumentary System

PART A: Assessments

1. Label the structures indicated in figures 11.7 and 11.8. (A1)

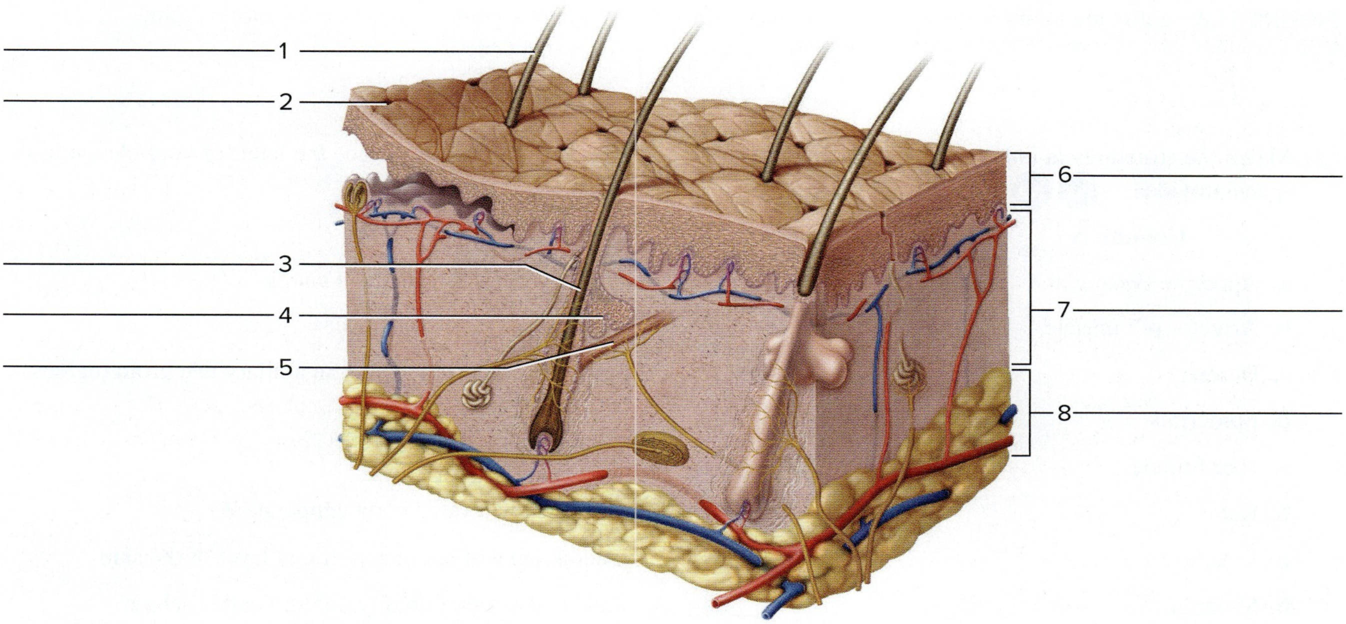

FIGURE 11.7 Label the features of the skin and layer beneath the skin.

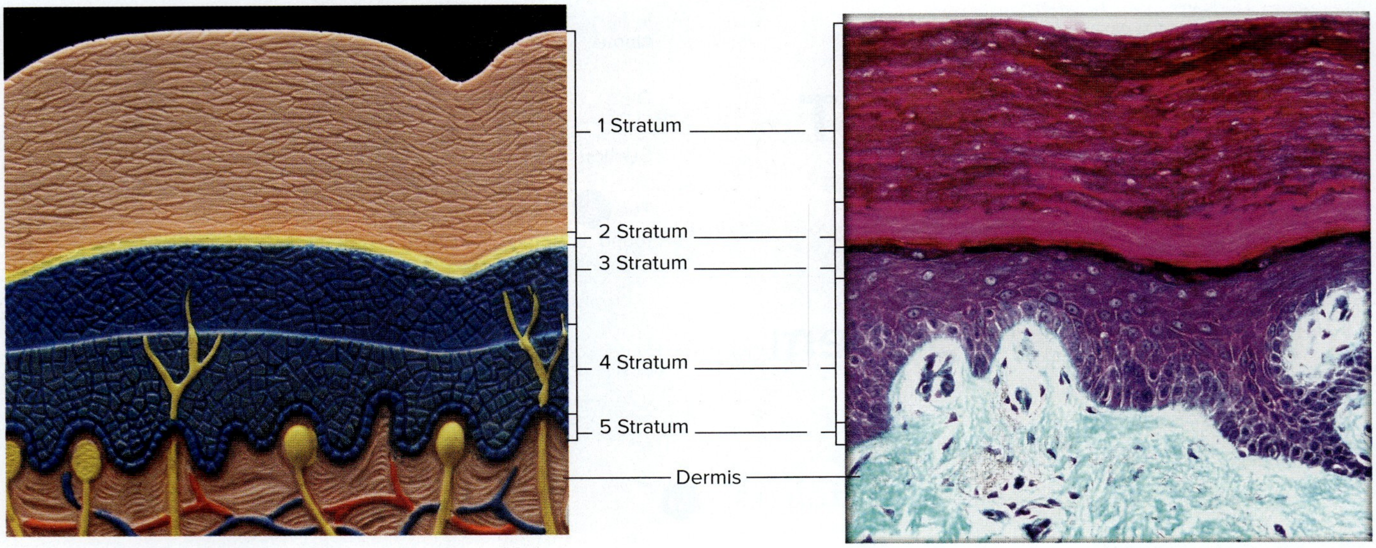

FIGURE 11.8 Label the strata of the epidermis of thick skin in the image of a model (left) and micrograph (right).

2. Match the structures in column A with the descriptions and functions in column B. Place the letter of your choice in the space provided. A1 A2

Column A	Column B
a. Apocrine sweat gland	______ **1.** Portion of nail body near the cuticle
b. Arrector pili muscle	______ **2.** Superficial layer of epidermis
c. Dermis	______ **3.** Become active after puberty in axillary and groin regions
d. Epidermis	______ **4.** Epidermal pigment
e. Hair follicle	______ **5.** Deepest layer of skin
f. Keratin	______ **6.** Responds to elevated body temperature
g. Lunula	______ **7.** General name of entire superficial layer of the skin
h. Melanin	______ **8.** Gland that secretes an oily mixture called sebum
i. Merocrine sweat gland	______ **9.** Tough protein of nails and hair
j. Sebaceous gland	______ **10.** Cell division and deepest layer of epidermis
k. Stratum basale	______ **11.** Tubelike part that contains the root of the hair
l. Stratum corneum	______ **12.** Causes hair to stand on end and goose bumps to appear

CRITICAL THINKING ASSESSMENT—PART A

As you spend a particularly hot, sunny afternoon swimming in the pool and talking with friends while lounging poolside, how is your integumentary system functioning to maintain homeostasis? A1 A2

____

____

____

PART B: Assessments

Complete the following:

1. How does the skin of your palm differ from that on the back (posterior) of your hand? A1 ____

2. Describe the differences you observed in the type and distribution of hair on the front (anterior) and back (posterior) of your forearm. A1 ____

3. Explain how a hair is formed. A2 ____

4. What portion of the nail plate is not visible on the body surface when using a hand magnifier? A1 ____________

PART C: Assessments

Complete the following:

1. Complete the following chart, and then circle the layer that is found *beneath* the skin. A3

Layer	Location	Tissue(s)	Distinguishing Characteristics
Epidermis			
Dermis			
Hypodermis			

2. How do the cells of stratum corneum and stratum basale differ? A3 ____

3. State the specific location of melanin observed in dark skin. A3 ____

4. Which is darker, the anterior forearm or the posterior forearm? ____________ Explain why. A3 ____

5. What special qualities, due to the presence of fibers, does the connective tissue of the dermis have? A3 ____

6. How is the structure of the thick skin of the palms of the hands and soles of the feet properly suited for its function? A4 ____

PART D: Assessments

Complete the following:

1. What part of the hair extends from the hair bulb to the hair shaft? A1 ________________

2. In which layer of skin are sebaceous glands found? A1 ________________

3. How are sebaceous glands associated with hair follicles, and what do they secrete? A1 ________________
____

4. The ducts of apocrine sweat glands open into ________________. A1

5. Which type of sweat gland is most important in maintaining normal body temperature? A1 ________________

PART E: Assessments

Sketch a vertical section of human skin, using the scanning objective. Label the skin layers and a hair follicle, a sebaceous gland, and a sweat gland. You may choose to use low-power and high-power objectives to add detail to your sketch, but be sure to draw a composite drawing to scale. A4

LABORATORY EXERCISE 12

Bone Structure and Classification

Purpose of the Exercise

To review the way bones are classified and to examine the structure and general function of a long bone and bone tissue.

MATERIALS NEEDED

Prepared microscope slide of ground compact bone

Human bone specimens, including long, short, flat, irregular, and sesamoid types

Human long bone, sectioned longitudinally

Fresh animal bones, sectioned longitudinally and transversely

Compound light microscope

Dissecting microscope

For Demonstration Activity:

Fresh chicken bones (radius and ulna from wings)

Vinegar or dilute hydrochloric acid

SAFETY

- Review all the safety guidelines in Appendix 1 of your laboratory manual.
- Wear disposable gloves and protective eyewear for handling fresh bones and for the demonstration of a bone soaked in vinegar or dilute hydrochloric acid.
- Wash your hands before leaving the laboratory.

Chapter Opener: ©101cats/Getty Images

Learning Outcomes APR

After completing this exercise, you should be able to:

O1 Classify bones into four groups based on their shapes and identify an example for each group.

O2 Locate the major structures of a long bone.

O3 Distinguish between compact and spongy bone.

O4 Differentiate the structural and functional characteristics of bones.

The O corresponds to the assessments A indicated in the Laboratory Assessment for this Exercise.

Pre-Lab

Carefully read the introductory material and the entire lab. Be familiar with bone structure and bone tissue from lecture or the textbook. Answer the pre-lab questions.

Pre-Lab Questions Select the correct answer for each of the following questions:

1. Which of the following tissues is *not* part of a bone as an organ?
   a. dense connective **b.** cartilage
   c. muscle **d.** blood

2. The organic matter of living bone includes
   a. calcium phosphate.
   b. collagen fibers.
   c. calcium carbonate.
   d. magnesium and fluoride.

3. The _____ is an example of a sesamoid bone.
   a. vertebra **b.** femur
   c. carpal **d.** patella

4. The epiphyseal plate represents the
   a. ends of the epiphyses.
   b. shaft between the epiphyses.
   c. growth zone of hyaline cartilage.
   d. membrane around a bone.

5. The central canal of a bone osteon contains
   a. blood vessels and nerves. **b.** osteocytes.
   c. red bone marrow. **d.** yellow bone marrow.

6. A _____ is an example of an irregular bone.
   a. femur **b.** carpal
   c. rib **d.** vertebra

7. A femur includes both compact and spongy bone tissues.
   a. True b. False
8. Chicken bones, with both organic and inorganic components, possess the quality of tensile strength.
   a. True b. False
9. Trabeculae are structural characteristics of compact bone.
   a. True b. False

A bone represents an organ of the skeletal system. As such, it is composed of a variety of tissues, including bone (osseous) tissue, cartilage, dense connective tissue, blood, and nervous tissue. Bones are not only alive but also multifunctional. They support and protect softer tissues, provide points of attachment for muscles, house blood-producing cells, help buffer the blood from extreme fluctuations in pH, and store inorganic salts.

The extracellular matrix of living bone is a combination of about one-third *organic matter* and two-thirds *inorganic matter.* The organic matter consists mostly of collagen fibers that provide flexibility and strength to withstand tension. The inorganic matter is mostly hydroxyapatite, complex salt crystals consisting of calcium phosphate. Lesser amounts of calcium carbonate and ions of magnesium, fluoride, and sodium become incorporated into the inorganic crystals.

Cells of bone consist of *osteogenic (stem) cells* that primarily differentiate into *osteoblasts* that help form bone, and when isolated in matrix, osteoblasts form *osteocytes* that are vital to bone homeostasis; *osteoclasts* dissolve bone to release calcium salts. Osteogenic cells and their progeny are derived from embryonic mesenchyme, while osteoclasts are descendants of bone marrow stem cells. Osteoblasts and osteoclasts are essential in the remodeling of bone, as they, respectively, deposit and release minerals in bone.

Bones are classified according to their shapes as *long, short, flat,* or *irregular.* Long bones are longer than they are wide and have expanded ends. Short bones are somewhat cube shaped, with similar lengths and widths. *Sesamoid bones,* a special type of short bone, are small and embedded within a tendon near joints where compression often occurs. The patella is a sesamoid bone that is included in the total skeletal number of 206. Any additional sesamoid bones that may develop in compression areas of the hand or foot are not considered among the 206 bones of the adult skeleton. Flat bones have wide surfaces, but they are sometimes curved, such as those of the cranium. Irregular bones have numerous shapes and often have articulations with more than one other bone. Although bones of the skeleton vary greatly in size and shape, they have much in common structurally and functionally.

CRITICAL THINKING ACTIVITY

Explain how bone cells (osteocytes) embedded in a solid ground substance obtain nutrients and eliminate wastes.

____

____

____

____

____

PROCEDURE: Bone Structure and Classification

During embryonic and fetal development, much of the supportive tissue is cartilage. Cartilage is retained in certain regions of the adult skeleton, such as on the articulating surfaces of movable joints, costal cartilage connects the ribs to the sternum, and intervertebral discs are between the vertebrae. Articular cartilage, composed of hyaline cartilage, is avascular and depends upon highly vascular bone tissue to obtain nourishment indirectly. It functions to provide a smooth, articulating surface that secretes a lubricating fluid to reduce friction and allow freer movement at a joint. Costal cartilages are also composed of hyaline cartilage, whereas the intervertebral discs are composed of fibrocartilage.

Bones contain compact and spongy bone components. The *compact bone* structure contains cylinder-shaped units called osteons. Each osteon contains a central canal that includes blood vessels and nerves in living bone and is encircled by concentric lamellae. The cells or osteocytes are located in concentric circles within almond-shaped cavities known as lacunae. The cellular processes of the osteocytes pass through microscopic canaliculi, which allow for the transport of nutrient and waste substances between cells and the central canal. The extracellular matrix occupies most of the area of an osteon.

Spongy bone does not possess the typical osteon units of compact bone. The osteocytes of spongy bone are located within a lattice of bony plates known as trabeculae. Canaliculi allow for diffusion of substances between the cells and the marrow that is positioned between the trabeculae.

1. Observe the individual bone specimens and arrange them into groups, according to the following shapes and examples (figs. 12.1 and 12.2).
   long—femur; humerus; phalanges
   short—carpals; tarsals; sesamoid bones
   flat—ribs; scapula; sternum; most cranial bones
   irregular—vertebra; some facial bones such as sphenoid
2. Become familiar with the structural features of a long bone indicated in figure 12.3.

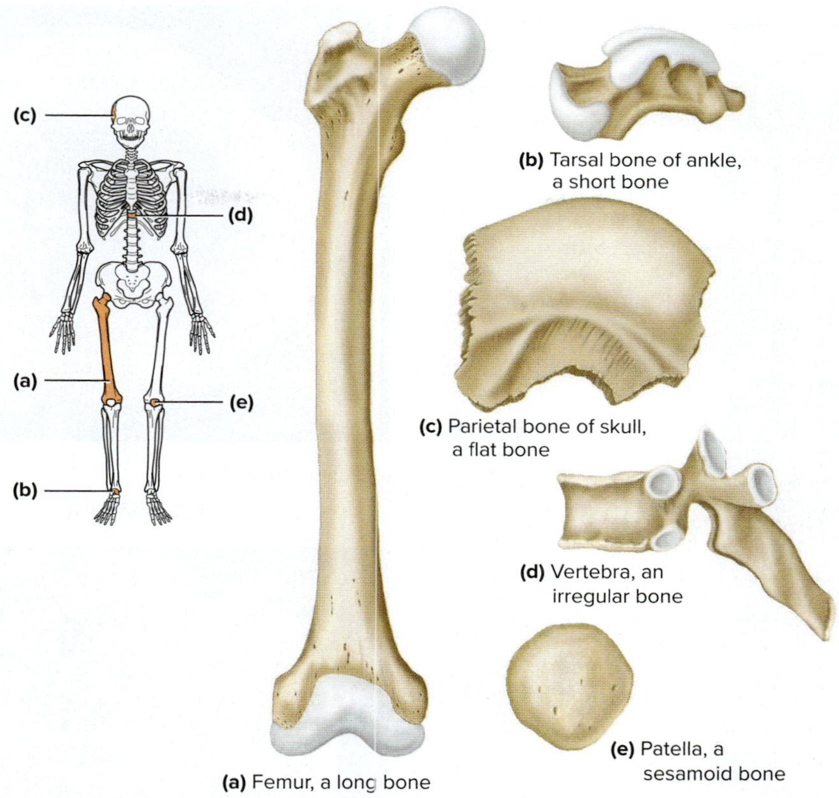

FIGURE 12.1 Representative examples of bones classified by shape with a skeleton location icon for the bones illustrated.

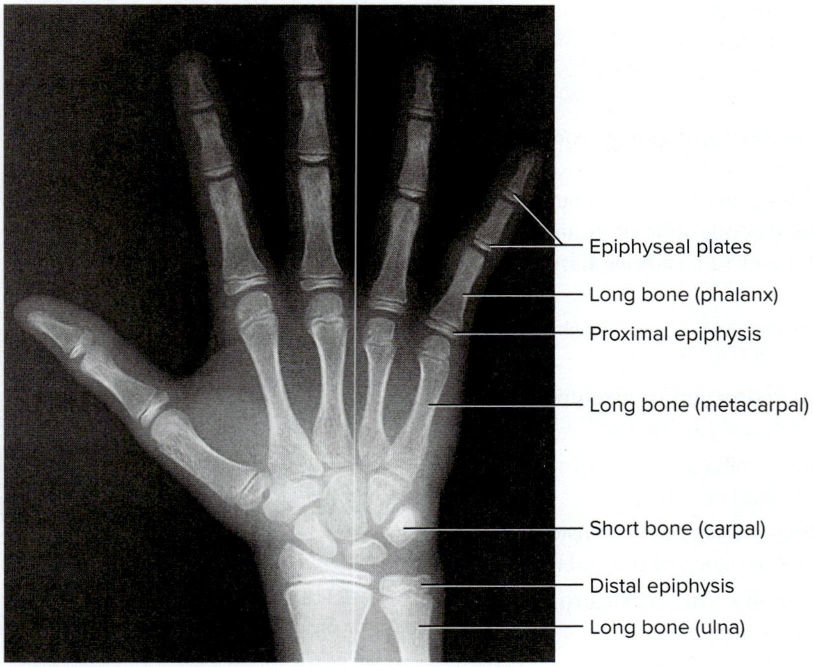

FIGURE 12.2 Radiograph (X ray) of a child's hand showing numerous epiphyseal plates. Only single epiphyseal plates develop in the hand and fingers. Utah Valley Regional Medical Center, Department of Radiology APR

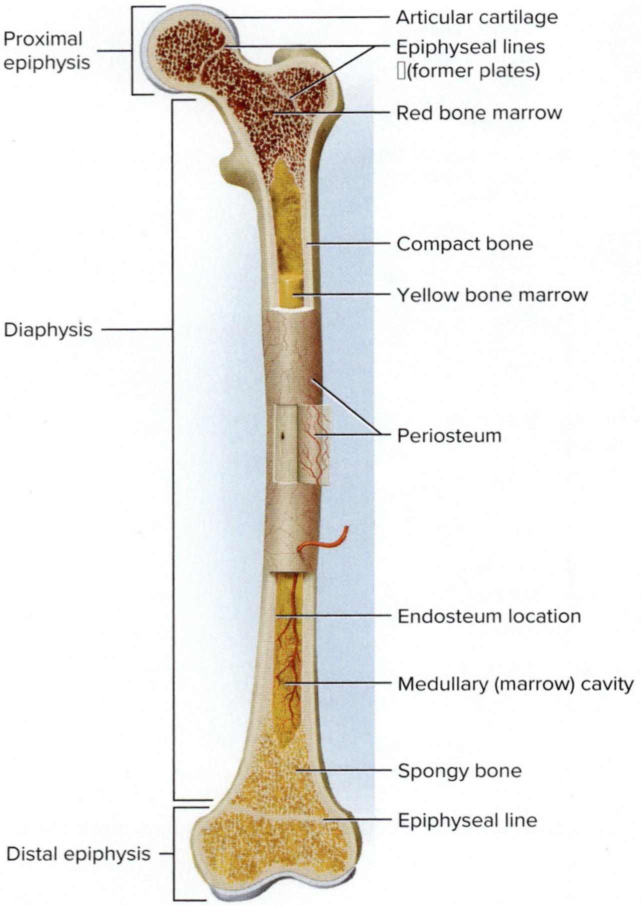

FIGURE 12.3 The major structures of a long bone (living femur) of an adult. Notice that the hemopoietic tissue is located in the proximal epiphysis of this long bone.

3. Note representative compact and spongy bone locations (fig. 12.4).
4. Examine the microscopic structure of bone tissue and observe a prepared microscope slide of ground compact bone. Use figures 12.5 and 12.6 of bone tissue to locate the following features:

   osteon (Haversian system)—cylinder-shaped unit of compact bone

   central canal (Haversian canal; osteonic canal)—contains blood vessels and nerves

   concentric lamella—concentric ring of matrix around central canal of osteon

   interstitial lamella—matrix between osteons composed of the remains of old osteons

   circumferential lamella—matrix that runs parallel to the surface of bone

   lacuna—small chamber for an osteocyte

   bone extracellular matrix—collagen and calcium phosphate

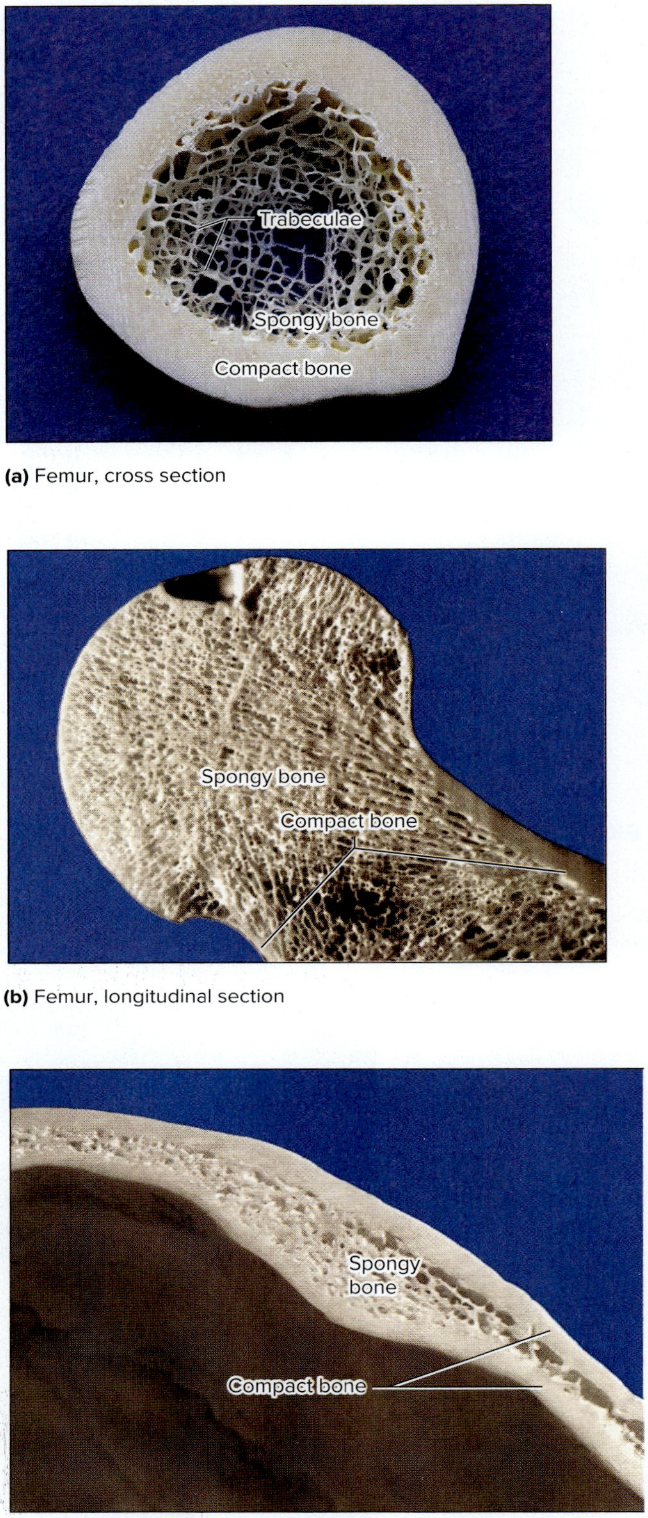

(a) Femur, cross section

(b) Femur, longitudinal section

(c) Skull bone, sectioned

FIGURE 12.4 Representative locations of compact and spongy bone. (a) Ed Reschke/Getty Images; (b, c) John W. Hole, Jr.

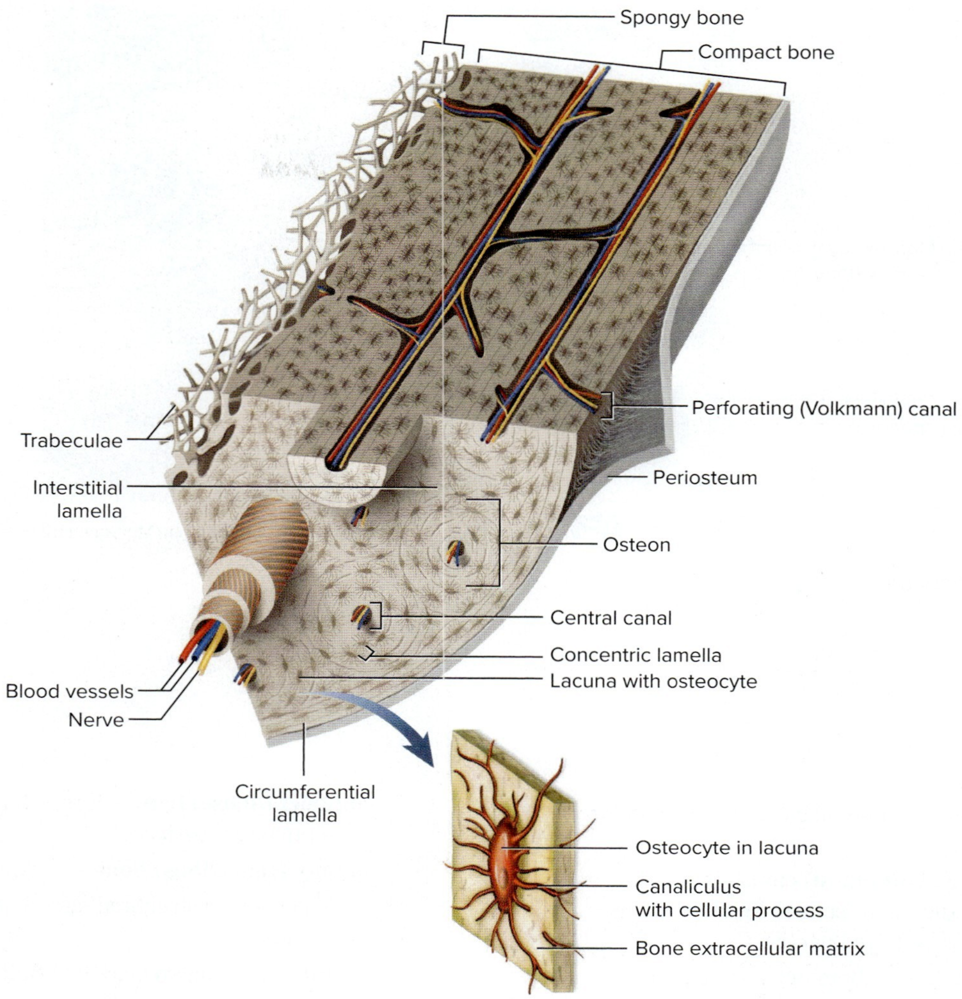

FIGURE 12.5 The structures associated with microscopic anatomy of a bone.

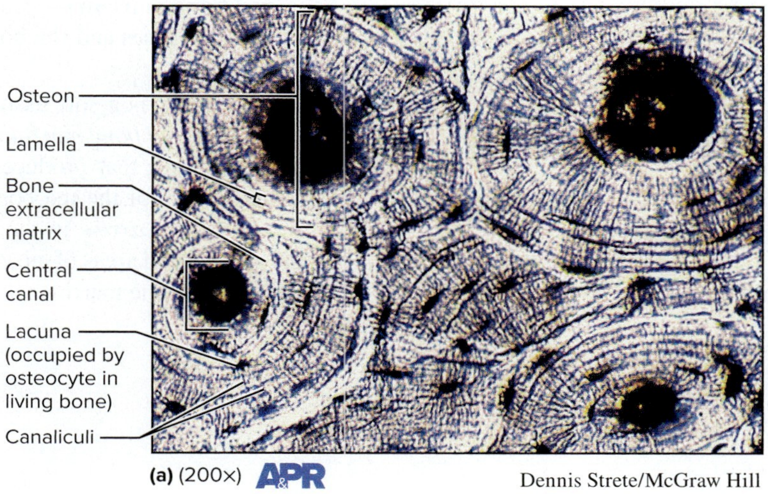

FIGURE 12.6 Micrographs of (a) ground compact bone tissue (200×), and (b) spongy bone of the metaphysis (100×).

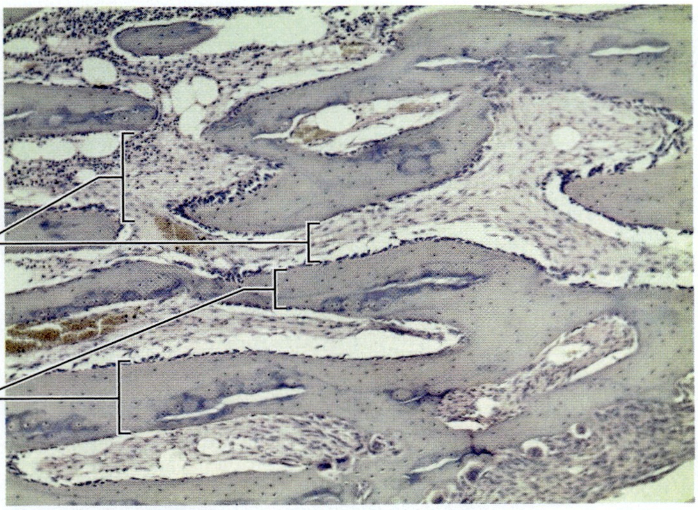

(b) (100×) APR

Stephen Sullivan/McGraw Hill

FIGURE 12.6 *Continued.*

- **canaliculus**—minute tube containing cellular process
- **perforating (Volkmann) canal**—runs perpendicular to central canal; contains blood vessels and nerves

5. Complete Part A of Laboratory Assessment 12.
6. Examine the sectioned long bones and hand X-ray image and locate the following (figs. 12.2–12.4):
   - **epiphysis**—enlarged ends
     - proximal—nearest limb attachment to torso
     - distal—farthest from limb attachment to torso
   - **epiphyseal plate**—growth zone of hyaline cartilage
   - **epiphyseal line**—site of original epiphyseal plate that has completely ossified
   - **articular cartilage**—on ends of epiphyses
   - **diaphysis**—shaft between epiphyses
   - **periosteum**—strong membrane around bone (except articular cartilage) of dense irregular connective tissue
   - **compact (dense) bone**—forms diaphysis and epiphyseal surfaces
   - **spongy (cancellous) bone**—within epiphyses
     - trabeculae—a structural lattice of plates in spongy bone
   - **medullary (marrow) cavity**—hollow chamber in diaphysis; contains bone marrow
   - **endosteum**—thin membrane of reticular connective tissue that lines the medullary cavity
   - **yellow bone marrow**—occupies medullary cavity and stores adipose tissue
   - **red bone marrow**—occupies spongy bone in some epiphyses and flat bones and produces blood cells
7. Bone marrow is a soft tissue located within spaces of bones. The *red bone marrow* is very prevalent in a child and is a tissue that produces blood cells *(hemopoietic tissue).* Much of the red bone marrow becomes a fatty *yellow bone marrow* in the adult. Figure 12.7 depicts the represented areas of an adult skeleton containing red and yellow bone marrow.

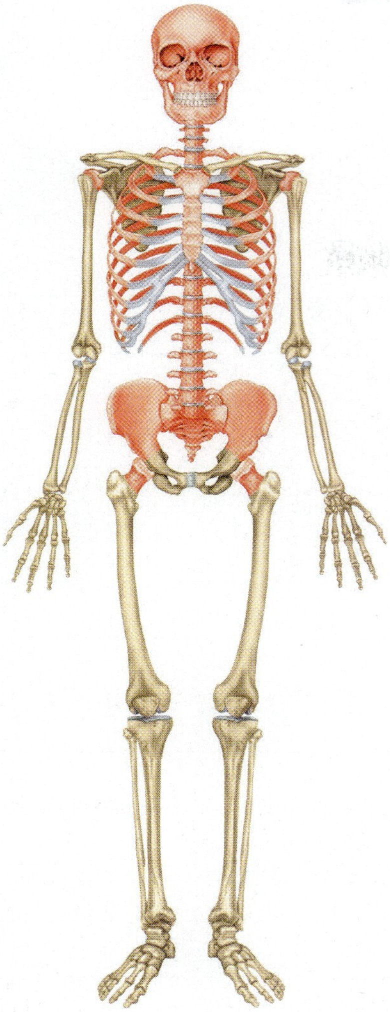

FIGURE 12.7 Locations of bones containing red and yellow bone marrow in an adult skeleton. Bones in red contain red bone marrow; most bones in yellow contain yellow bone marrow.

8. Use the dissecting microscope to observe the compact bone and spongy bone of the sectioned specimens. Also examine the marrow in the medullary cavity and the spaces within the spongy bone of the fresh specimen.
9. Complete Parts B and C of the laboratory assessment.

DEMONSTRATION ACTIVITY

Examine a fresh chicken bone and a chicken bone that has been soaked for several days in vinegar or overnight in dilute hydrochloric acid. Wear disposable gloves for handling these bones. This acid treatment removes the inorganic salts from the bone extracellular matrix. Rinse the bones in water and note the texture and flexibility of each (fig. 12.8*a*). The bone becomes soft and flexible without the support of the inorganic salts with calcium.

Examine the specimen of chicken bone that has been exposed to high temperature (baked at 121°C/250°F for 2 hours). This treatment removes the protein and other organic substances from the bone extracellular matrix (fig. 12.8*b*). The bone becomes brittle and fragile without the benefit of the collagen fibers. A living bone with a combination of the qualities of inorganic and organic substances possesses compressional and tensile strength.

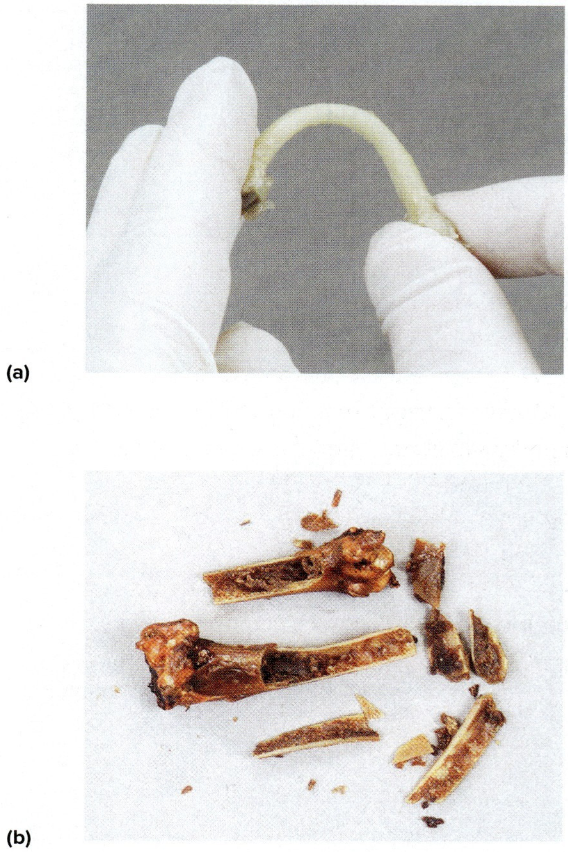

FIGURE 12.8 Results of fresh chicken bone demonstration: (a) soaked in vinegar; (b) baked in oven.

NOTES

LABORATORY
ASSESSMENT 12

Name ______________________

Date ______________________

Section ______________________

The (A) corresponds to the indicated Learning Outcome(s) (O) found at the beginning of the Laboratory Exercise.

Bone Structure and Classification

PART A: Assessments

Complete the following statements: (*Note:* Questions 1–7 pertain to bone classification by shape.)

1. A bone that has a wide surface is classified as a(an) ______________________ bone. A1
2. The bones of the wrist are examples of ______________________ bones. A1
3. The bone of the thigh is an example of a(an) ______________________ bone. A1
4. Vertebrae are examples of ______________________ bones. A1
5. The patella (kneecap) is an example of a special type of short bone called a(an) ______________________ bone. A1
6. The bones of the skull that form a protective covering for the brain are examples of ______________________ bones. A1
7. A bone that provides a rigid rod (lever) for skeletal muscle to pull upon during contraction is a(n) ______________________ bone. A1
8. Distinguish between the epiphysis and the diaphysis of a long bone. A2 ______________________
9. Describe where cartilage is found on the surface of a long bone. What function does cartilage serve in this location? A2 ______________________
10. Why doesn't the periosteum cover the articular cartilage of a long bone? A2 ______________________

PART B: Assessments

Complete the following:

1. Distinguish between the locations and tissues of the periosteum and those of the endosteum. A2 ______________________
2. What structural differences did you note between compact bone and spongy bone? A3 ______________________

3. How are these structural differences related to the locations and functions of these two types of bone? A3 ____________
____
____

4. From your observations, how does the marrow in the medullary cavity compare with the marrow in the spaces of the spongy bone? A4 ____________
____
____

5. The humerus is the proximal bone of an upper limb; the femur is the proximal bone of a lower limb. In the adult skeleton only certain portions of these long bones retain functional red bone marrow. Describe the specific regions of these bones that retain blood cell production sites. A4 ____________
____
____

PART C: Assessments

Identify the structures indicated in figures 12.9 to 12.12.

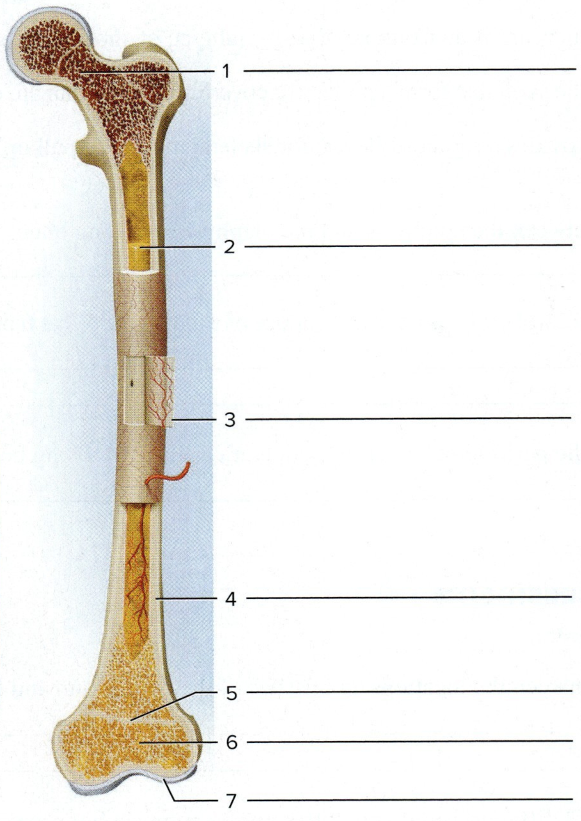

Terms:
Articular cartilage
Compact bone
Epiphyseal line
Periosteum
Red bone marrow
Spongy bone
Yellow bone marrow

FIGURE 12.9 Label the structures of this long bone, using the terms provided. A2 A3

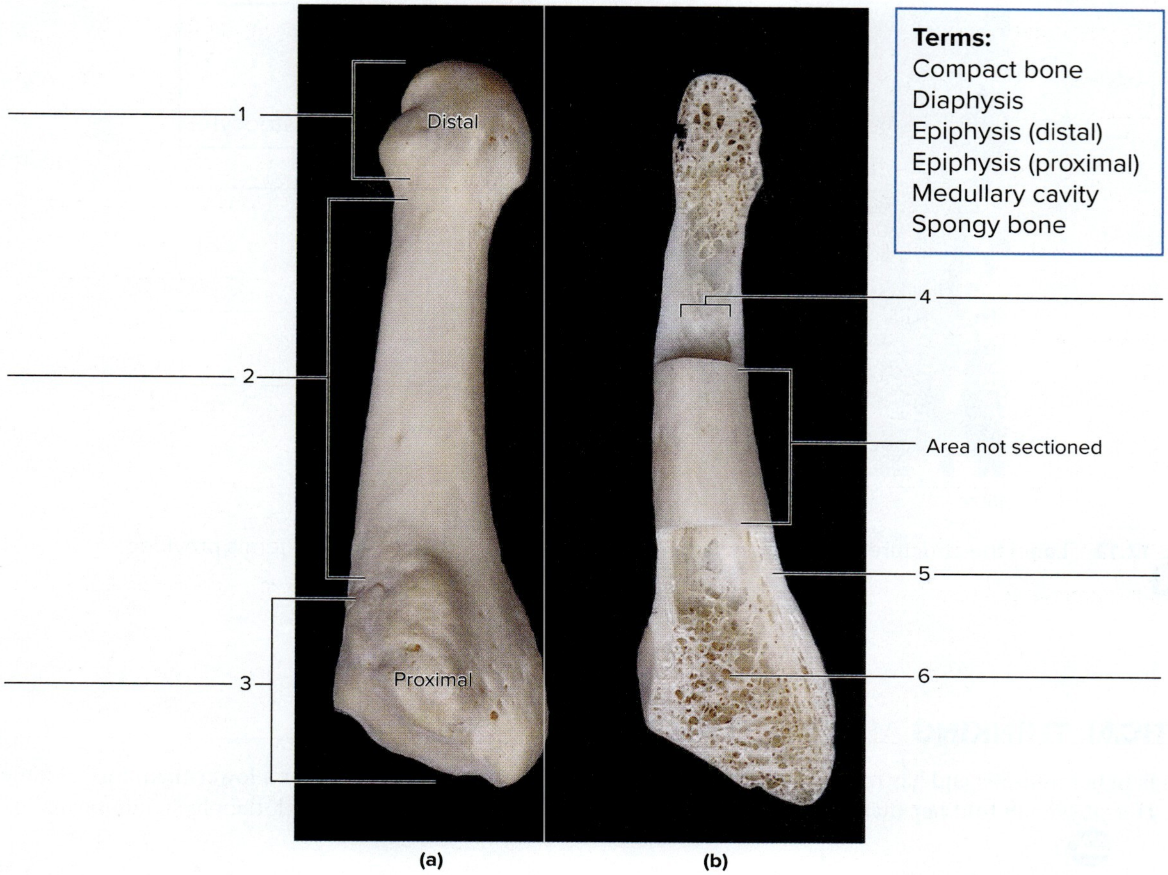

FIGURE 12.10 Identify the structures indicated in (a) the unsectioned long bone (fifth metatarsal) and (b) the partially sectioned long bone, using the terms provided. ©J & J Photography A2 A3

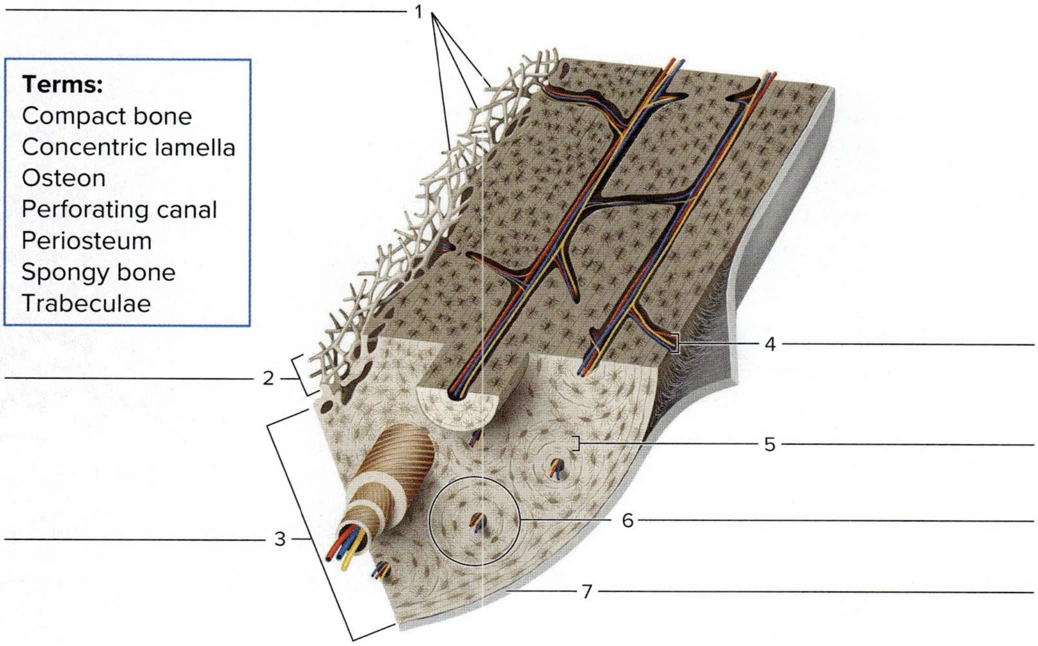

FIGURE 12.11 Label the structures associated with bone, using the terms provided. A3 A4

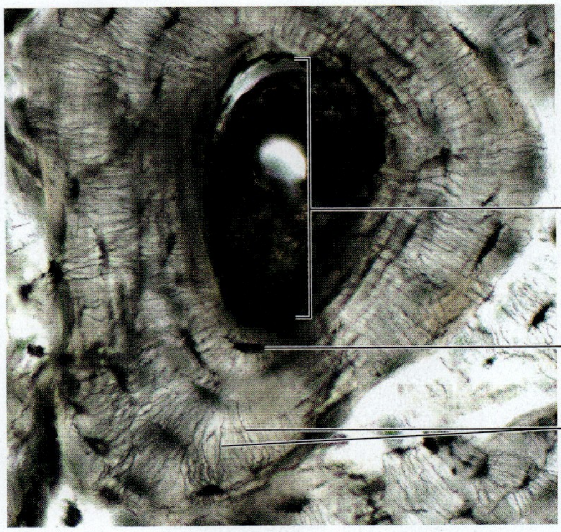

Terms:
Canaliculi
Central canal
Lacuna (with osteocyte)

1 ________________

2 ________________

3 ________________

160×

FIGURE 12.12 Label the structures of compact bone shown in the micrograph, using the terms provided. Al Telser/McGraw Hill A4

CRITICAL THINKING ASSESSMENT—PART C

Laura is in her mid-50s and has recently been diagnosed with osteopenia, a term used for a lower-than-normal bone mass. Her physician told her that this diagnosis is "an opportunity." What do you think the physician means? Explain. A4

____

____

____

LABORATORY EXERCISE 13

Organization of the Skeleton

Purpose of the Exercise

To explain the organization of the skeleton, introduce the major bones of the skeleton, and demonstrate the use of selected terms to describe bony structures.

MATERIALS NEEDED

Human skeleton, articulated
Human skeleton, disarticulated

For Demonstration Activity:
Radiographs (X rays) of skeletal structures
Stereomicroscope (dissecting microscope)

Learning Outcomes APR

After completing this exercise, you should be able to:

O1 Locate and label the major bones of the human skeleton.
O2 Distinguish between the axial skeleton and the appendicular skeleton.
O3 Associate the terms used to describe skeletal structures and locate examples of such structures on the human skeleton.

The O corresponds to the assessments A indicated in the Laboratory Assessment for this Exercise.

Pre-Lab

Carefully read the introductory material and examine the entire lab. Be familiar with the axial and appendicular skeleton from lecture or the textbook. Answer the pre-lab questions.

Pre-Lab Questions Select the correct answer for each of the following questions:

1. The vertebral column does *not* include a
a. rib. **b.** vertebra.
c. sacrum. **d.** coccyx.

2. The _____ bone is part of the pectoral girdle.
a. humerus **b.** sternum
c. rib **d.** scapula

3. A _____ is a very large projection on a bone.
a. fossa **b.** tubercle
c. trochanter **d.** facet

4. A _____ is a shallow depression on a bone.
a. head **b.** fossa
c. condyle **d.** tuberosity

5. Bones that might form in the skull, but are *not* considered in the total number, are
a. cartilaginous bones. **b.** sutural bones.
c. middle ear bones. **d.** sesamoid bones.

6. The _____ is a bone in the upper limb.
a. ulna **b.** scapula
c. clavicle **d.** fibula

7. A bone in the lower limb is the _____.
a. hip bone **b.** radius
c. humerus **d.** femur

8. A _____ is a depression type of bone feature (bone marking).
a. foramen **b.** crest
c. sulcus **d.** tuberosity

Chapter Opener: ©101cats/Getty Images

9. The bones of the skull, face, neck, and trunk make up the appendicular skeleton.

   a. True   **b.** False

The skeleton can be separated into two divisions: (1) the *axial skeleton*, which consists of the bones and cartilages of the head, neck, and trunk, and (2) the *appendicular skeleton*, which consists of the bones of the limbs and those that anchor the limbs to the axial skeleton. The bones that anchor the limbs include the pectoral and pelvic girdles.

Ossification is the formation of bone. This process occurs during fetal development and continues through childhood growth. Bone forms either from intramembranous origins or endochondral origins. During the growth of *intramembranous bones*, membrane-like connective tissue layers similar to the dermis develop in an area destined to become the flat bones of the skull. Eventually, bone-forming cells (osteoblasts) alter the membrane areas into bone tissue. In contrast, *endochondral bones* develop from hyaline cartilage and through the ossification process develop into most of the bones of the skeleton. In either type of ossification, both compact and spongy bones develop.

The number of bones in the adult skeleton is often reported to be 206. Males and females, although variations can exist, possess the same total bone number of 206. However, at birth the number of bones is closer to 275 as many ossification centers are still composed of cartilage, and some bones form during childhood. For example, each hip bone (coxal bone) includes an ilium, ischium, and pubis, and many long bones have three ossification centers separated by epiphyseal (growth) plates. The sternum, composed of a manubrium, body, and xiphoid process, becomes a single bone much later than when we reach our full height. Some people have additional bones not considered in the total number. *Sesamoid bones* other than the patellae may develop in the hand or the foot. Also, extra bones sometimes form in the skull within the sutures; these are called *sutural (wormian) bones.*

Special terminology is used to describe the features of a bone. The term used depends on whether the feature is a type of projection, articulation, depression, or opening. Many of these features can be noted when viewing radiographs. Some of the features can be *palpated* (touched) if they are located near the surface of the body.

PROCEDURE: Organization of the Skeleton

1. Study figure 13.1 and use it as a reference to examine the bones in the articulated human skeleton. As you locate the following bones, note the number of each in the skeleton. Palpate as many of the corresponding bones in your skeleton as possible.

   axial skeleton
   - skull
     - cranium (8)
     - face (14)
   - middle ear bone (6)
   - hyoid bone—supports the tongue (1)
   - vertebral column
     - vertebra (24)
     - sacrum (1)
     - coccyx (1)
   - thoracic cage
     - rib (24)
     - sternum (1)

   appendicular skeleton
   - pectoral girdle
     - scapula (2)
     - clavicle (2)
   - upper limbs
     - humerus (2)
     - radius (2)
     - ulna (2)
     - carpal (16)
     - metacarpal (10)
     - phalanx (28)
   - pelvic girdle
     - hip bone (coxal bone; os coxa; pelvic bone; innominate bone) (2)
   - lower limbs
     - femur (2)
     - tibia (2)
     - fibula (2)
     - patella (2)
     - tarsal (14)
     - metatarsal (10)
     - phalanx (28)

   Total 206 bones

2. Complete Part A of Laboratory Assessment 13.
3. Bone features (bone markings) can be grouped together in a category of *projections, articulations, depressions,* or *openings.* Within each category more specific examples occur. The bones illustrated in figure 13.2 represent specific examples of locations of specific features in the human body. Locate each of the following features on the example bone from a disarticulated skeleton, noting the size, shape, and location in the human skeleton.

   Projections: sites for tendon and ligament attachment

   crest—ridgelike

   epicondyle—superior to condyle

   line (linea)—slightly raised ridge

   process—prominent

   protuberance—outgrowth

   ramus—extension

spine—thornlike
trochanter—large; located on femur bone
tubercle—small knoblike
tuberosity—rough elevation

Articulations: where bones connect at a joint or articulate with each other
condyle—rounded process
facet—nearly flat
head—expanded end

Depressions: recessed areas in bones
alveolus—socket
fossa—shallow basin
fovea—tiny pit
notch—indentation on edge
sulcus—narrow groove

Openings: open spaces in bones
canal—tubular passage
fissure—slit
foramen—hole
meatus—tubelike opening
sinus—cavity

4. Complete Parts B, C, and D of the laboratory assessment.

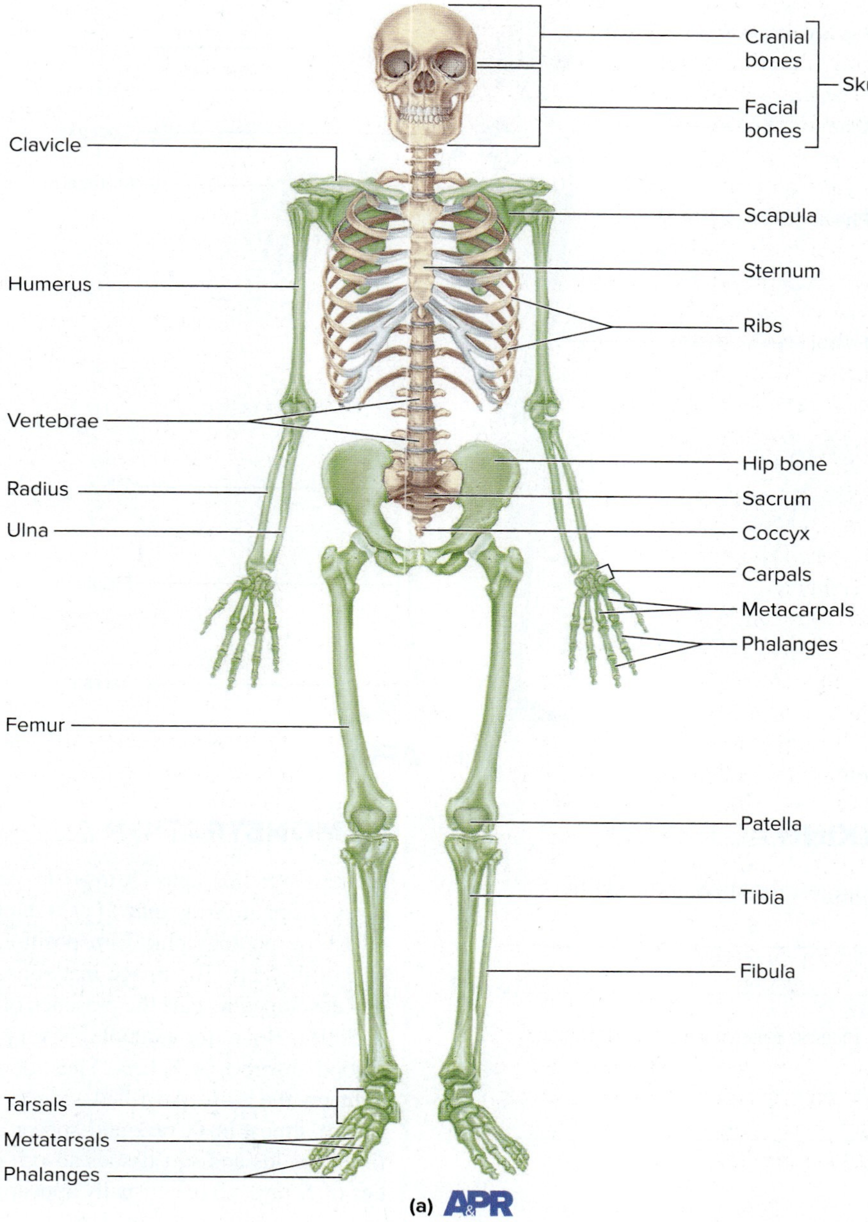

FIGURE 13.1 Major bones of the skeleton: (a) anterior view; (b) posterior view. The axial portion is tan, and the appendicular portion is green.

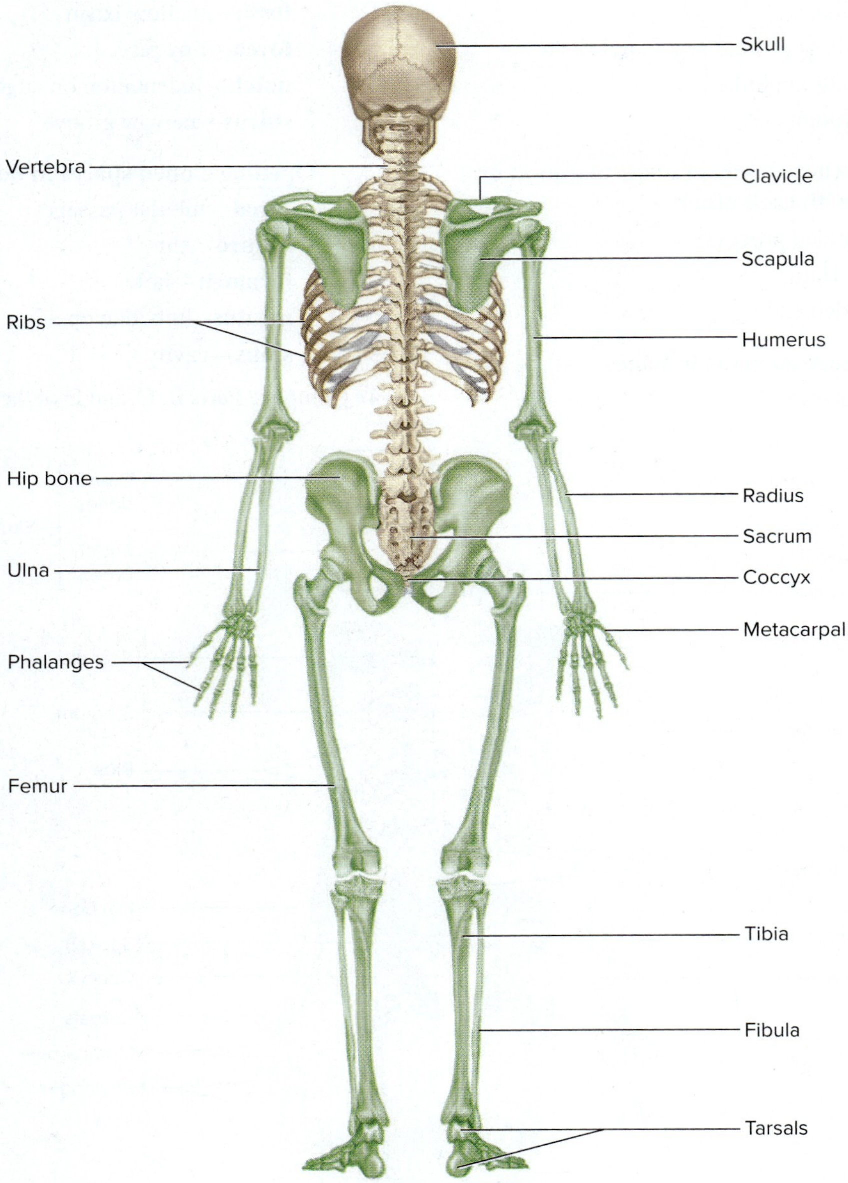

(b) APR

FIGURE 13.1 *Continued.*

CRITICAL THINKING ACTIVITY

Locate and name the largest foramen in the skull.

____

____

Locate and name the largest foramen in the skeleton.

____

____

DEMONSTRATION ACTIVITY

Images on radiographs (X rays) are produced by allowing X rays from an X-ray tube to pass through a body part and to expose photographic film positioned on the opposite side of the part. The image that appears on the film after it is developed reveals the presence of parts with different densities. Bone, for example, is very dense tissue and is a good absorber of X rays. Thus, bone generally appears light on the film. Air-filled spaces, on the other hand, absorb almost no X rays and appear as dark areas on the film. Liquids and soft tissues absorb intermediate quantities of X rays, so they usually appear in various shades of gray.

Examine the available radiographs of skeletal structures by holding each film in front of a light source. Identify as many of the bones and features as you can.

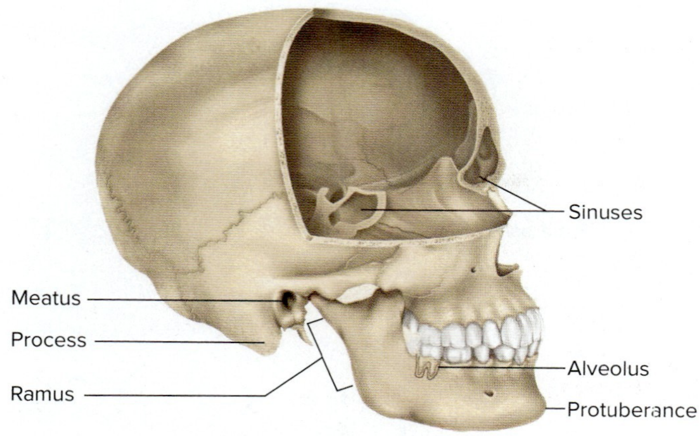

(a) Skull (lateral view)

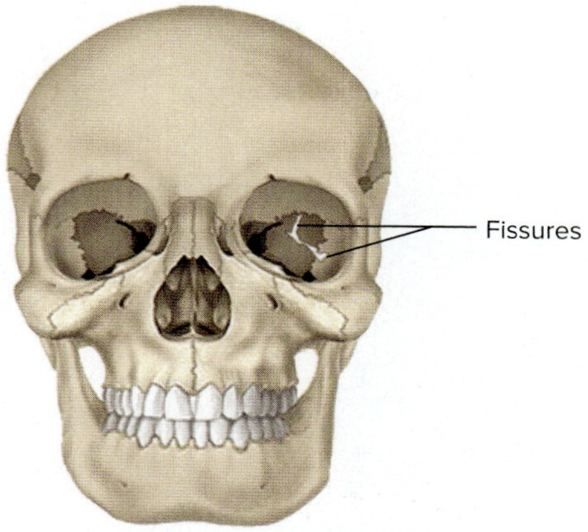

(b) Skull (anterior view) APR

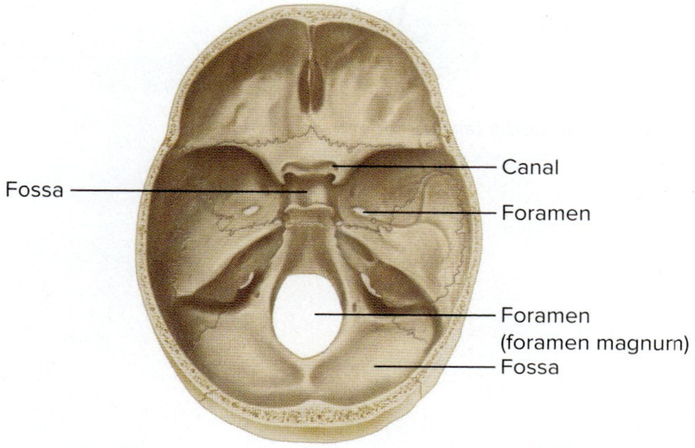

(c) Skull (cranial cavity, superior view) APR

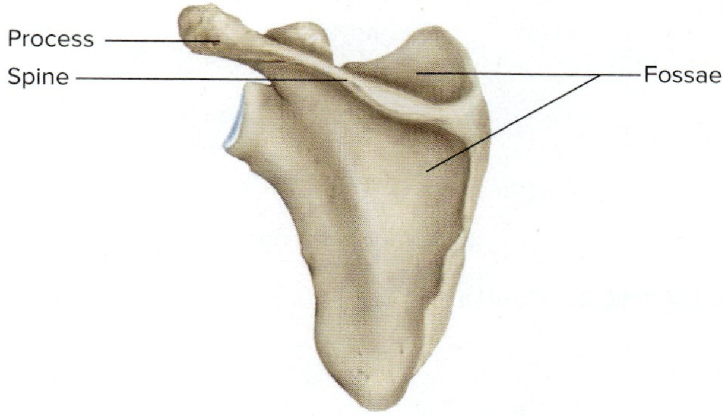

(d) Scapula (posterior view) APR

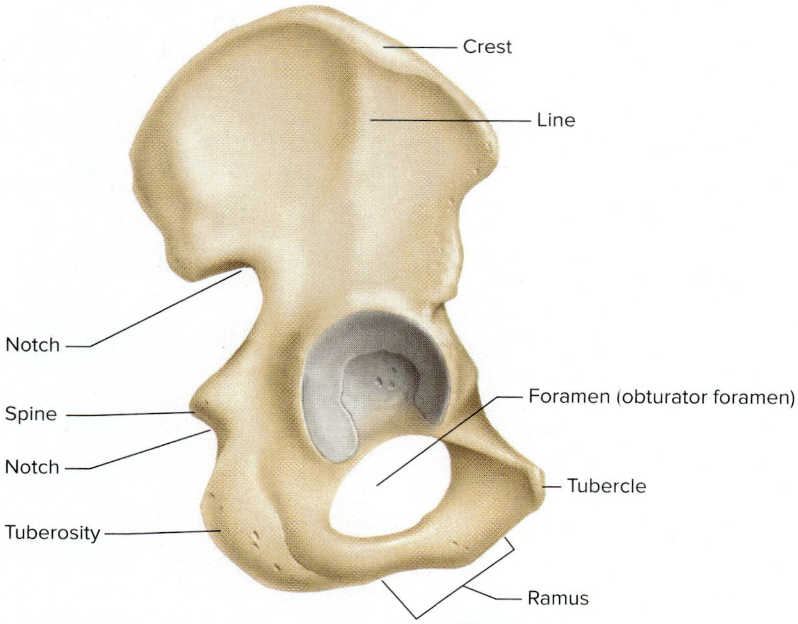

(e) Hip (coxal) bone (lateral view) APR

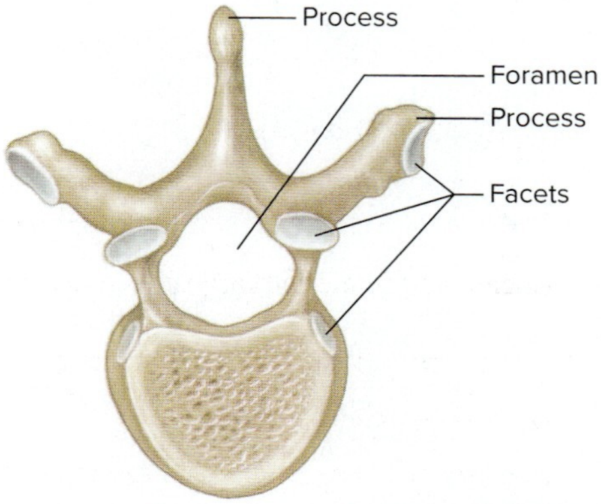

(f) Thoracic vertebra (superior view) APR

FIGURE 13.2 Representative examples of bone features (bone markings) on bones of the skeleton (a)–(h). The two largest foramina of the skeleton have complete labels.

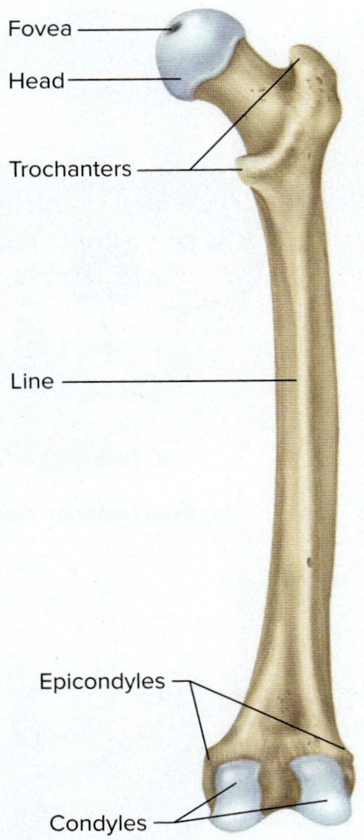

(g) Femur (posterior view) APR

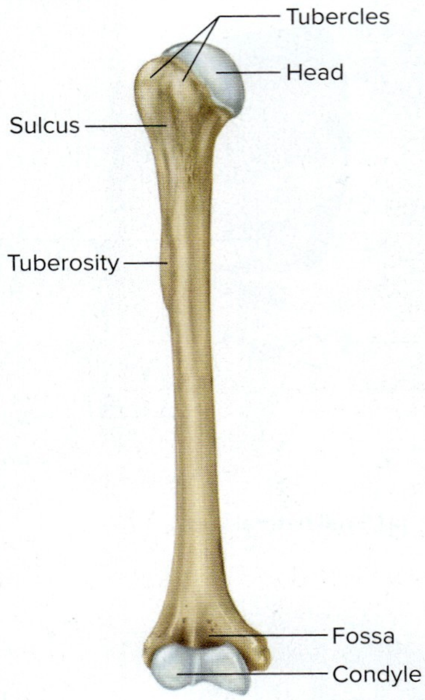

(h) Humerus (anterior view) APR

FIGURE 13.2 *Continued.*

LABORATORY
ASSESSMENT 13

Name ______________________________

Date ______________________________

Section ______________________________

The (A) corresponds to the indicated Learning Outcome(s) (O) found at the beginning of the Laboratory Exercise.

Organization of the Skeleton

PART A: Assessments

1. Label the bones indicated in figure 13.3.

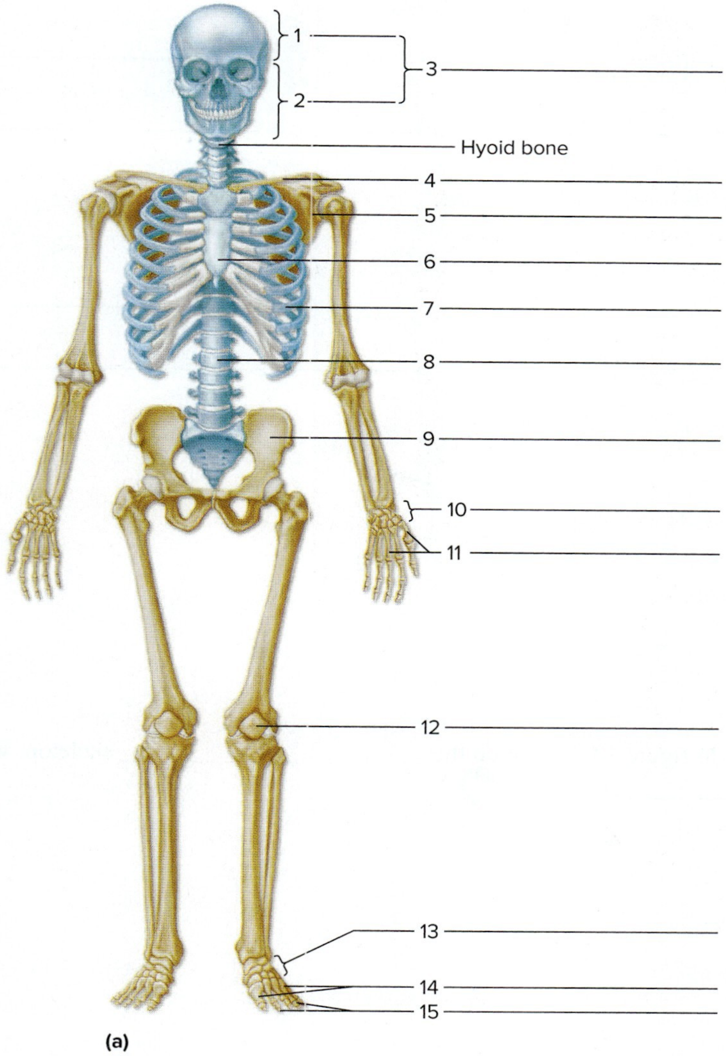

FIGURE 13.3 Label the major bones of the skeleton: (a) anterior view; (b) posterior view. A1

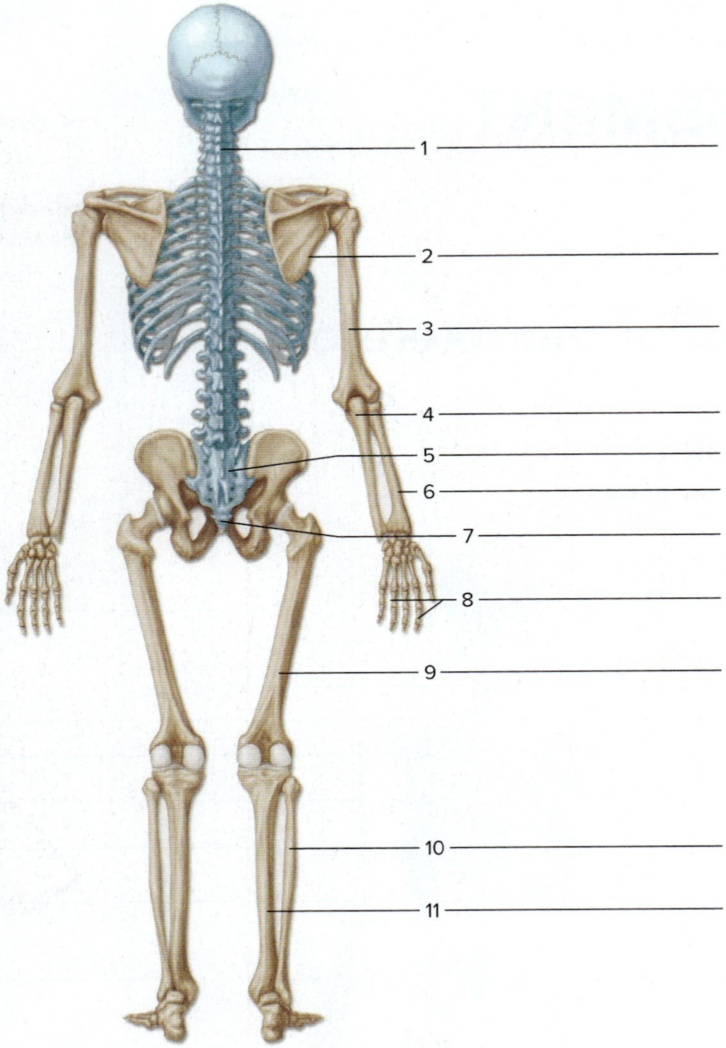

(b)

FIGURE 13.3 *Continued.*

2. The blue-shaded bones in figure 13.3 make up the ______________________ skeleton, while the other bones make up the ______________________ skeleton. A2

PART B: Assessments

1. Identify the bones indicated in figure 13.4.

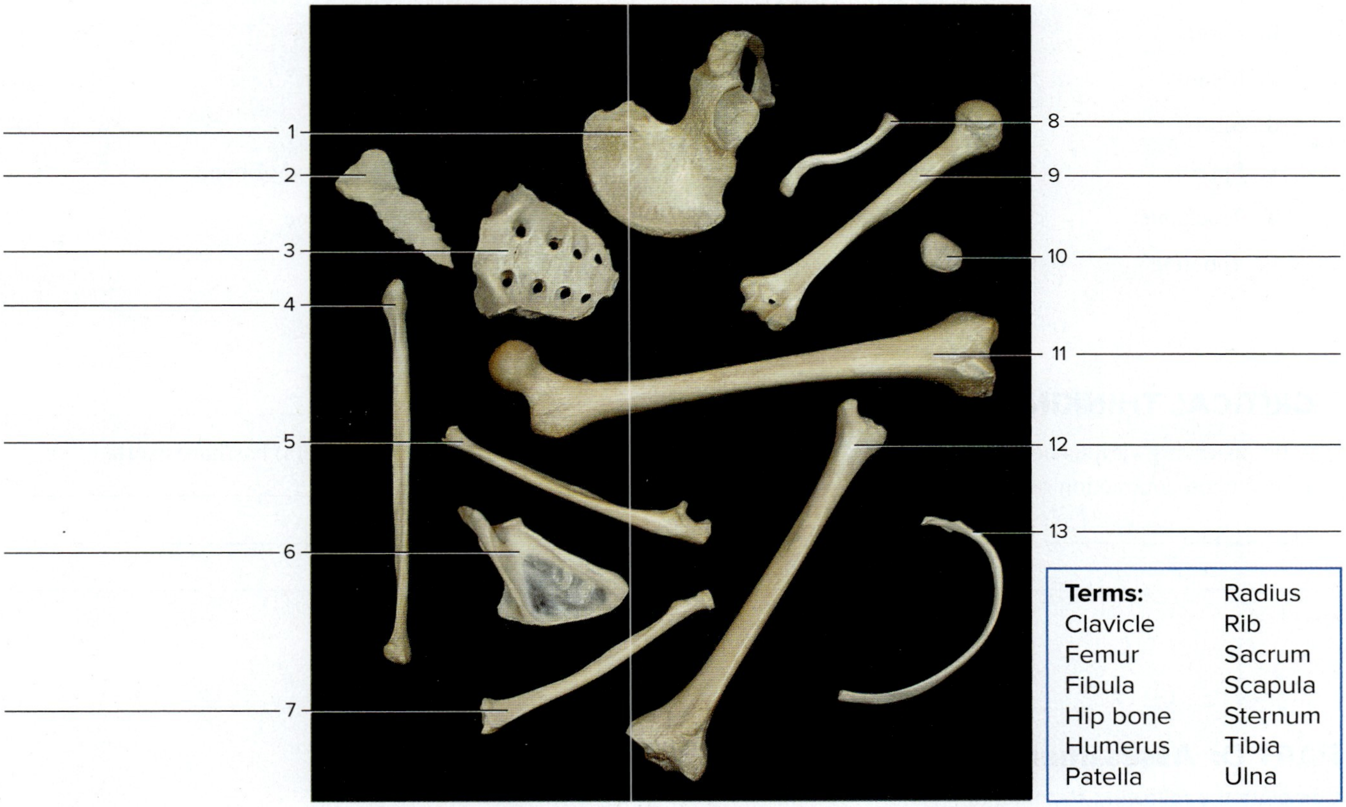

Terms:	Radius
Clavicle	Rib
Femur	Sacrum
Fibula	Scapula
Hip bone	Sternum
Humerus	Tibia
Patella	Ulna

FIGURE 13.4 Identify the bones in this random arrangement, using the terms provided. ©J & J Photography A1

2. List any of the bones shown in figure 13.4 that are included as part of the axial skeleton. A2 ______________________

____

PART C: Assessments

1. Match the terms in column A with the definitions in column B. Place the letter of your choice in the space provided. A3

Column A	Column B
a. Condyle	_____ **1.** Small, nearly flat articular surface
b. Crest	_____ **2.** Deep depression or shallow basin
c. Facet	_____ **3.** Rounded process
d. Foramen	_____ **4.** Rough, elevated projection
e. Fossa	_____ **5.** Opening or hole
f. Line	_____ **6.** Projection extension
g. Ramus	_____ **7.** Ridgelike projection
h. Tuberosity	_____ **8.** Slightly raised ridge

2. Match the terms in column A with the definitions in column B. Place the letter of your choice in the space provided. A3

Column A	Column B
a. Fovea	_______ 1. Tubelike opening
b. Head	_______ 2. Tiny pit or depression
c. Meatus	_______ 3. Small, knoblike projection
d. Sinus	_______ 4. Thornlike projection
e. Spine	_______ 5. Rounded enlargement at end of bone
f. Trochanter	_______ 6. Air-filled cavity within bone
g. Tubercle	_______ 7. Relatively large process

CRITICAL THINKING ASSESSMENT—PART C

After observing various bones of the body, what *specific* evidence suggests that skeletal muscle(s) has/have exerted a force upon contraction on the bone? A3 ____

____

____

PART D: Assessments

Complete the following statements:

1. The extra bones that sometimes develop between the flat bones of the skull are called ____________________ bones. A1
2. Small bones occurring in some tendons in the hand or foot are called ____________________ bones. A1
3. The cranium and facial bones compose the ____________________. A1
4. The ____________________ bone supports the tongue. A1
5. The ____________________ at the inferior end of the sacrum is composed of several fused vertebrae. A1
6. Most ribs are attached anteriorly to the ____________________. A1
7. The thoracic cage is composed of ____________________ pairs of ribs. A1
8. The scapulae and clavicles together form the ____________________ girdle. A1
9. Each wrist is composed of eight bones called ____________________. A1
10. The hip bones (coxal bones) are attached posteriorly to the ____________________. A1
11. The ____________________ bone covers the anterior surface of the knee. A1
12. The bones that articulate with the distal ends of the tibia and fibula are called ____________________. A1
13. All finger and toe bones are called ____________________. A1
14. Which of the following bones is *not* part of the appendicular skeleton: clavicle, femur, sternum, scapula? ____________________ A2

LABORATORY
EXERCISE 14

Skull

Purpose of the Exercise

To examine the structure of the human skull and to identify the bones and major features of the skull.

MATERIALS NEEDED

Human skull, articulated
Human skull, disarticulated
Human skull, sagittal section

Learning Outcomes APR

After completing this exercise, you should be able to:

O1 Locate and label the bones of the skull and their major features.
O2 Locate and label the major sutures of the cranium.
O3 Distinguish the skull bones that comprise the orbit of an eye.
O4 Locate and describe the sinuses of the skull and provide their functions.

The O corresponds to the assessments A indicated in the Laboratory Assessment for this Exercise.

Pre-Lab

Carefully read the introductory material and the entire lab. Be familiar with the skull from lecture or the textbook. Answer the pre-lab questions.

Pre-Lab Questions Select the correct answer for each of the following questions:

1. The _____ is a skull bone that is *not* interlocked along sutures.
a. mandible **b.** maxilla
c. temporal bone **d.** occipital bone

2. Sinuses include all of the following functions *except*
a. warming and humidifying the air during breathing.
b. providing resonance to our voices.
c. making the skull lighter.
d. attachment sites for muscles.

3. The _____ bone serves for tongue and larynx muscle attachments.
a. mandible **b.** maxilla
c. hyoid **d.** temporal

4. Which of the following is a facial bone?
a. sphenoid bone **b.** nasal bone
c. ethmoid bone **d.** frontal bone

5. The _____ suture is located between the two parietal bones.
a. squamous **b.** sagittal
c. coronal **d.** lambdoid

6. The two bones that make up the nasal septum are the _____
a. vomer and ethmoid **b.** sphenoid and maxilla
c. nasal and lacrimal **d.** zygomatic and temporal

7. The mastoid process, jugular foramen, and the external acoustic meatus are features of the _____ bone.
a. occipital **b.** maxillary
c. temporal **d.** ethmoid

Chapter Opener: ©101cats/Getty Images

8. A(n) _____ lightens the skull and acts to warm and humidify the air.
   a. nasal septum **b.** paranasal sinus
   c. auditory ossicle **d.** suture
9. Foramina, canals, and fissures serve as passageways for blood vessels and nerves.
   a. True **b.** False
10. The jugular foramen is larger than the foramen magnum.
   a. True **b.** False

A part of the axial skeleton, the human skull consists of twenty-two bones that, except for the lower jaw, are firmly interlocked along sutures. Eight of these immovable bones make up the *braincase,* or *cranium,* and thirteen more immovable bones and the mandible form the *facial skeleton.*

Bones located within the temporal bone are the *auditory ossicles* (*middle ear bones*). They include the *malleus, incus,* and *stapes* that are examined in Laboratory Exercise 37. The stapes is considered the smallest bone in the human body among the total number of 206. The hyoid bone is a nonarticulating bone that is suspended from the temporal bones and serves for tongue and larynx muscle attachments. A fractured hyoid bone is sometimes used as an indication of strangulation as the cause of death.

After intramembranous ossification of the flat bones of the skull is complete, a fibrous type of joint remains between them. These joints are considered immovable and are known as *sutures.* The suture between the two parietal bones is the *sagittal suture.* The *lambdoid suture* is between the occipital and parietal bones. The *coronal suture* is bordered by the frontal bone and the parietal bones. The *squamous suture* is primarily along the superior border of the temporal bone with the parietal bone.

Sinus cavities are found in the frontal bone, ethmoid bone, sphenoid bone, and both maxillary bones. Because they are associated with the nasal passages, they are called *paranasal sinuses.* These air-filled sinuses are lined with a mucous membrane that is continuous with the nasal passages. The sinuses function to lighten the skull, provide resonance to our voices, and assist in warming and humidifying the air during breathing. An infection of the large maxillary sinuses is often misinterpreted to be a toothache.

PROCEDURE: Skull APR

1. Examine human articulated, disarticulated, and sectioned skulls available in the laboratory. Identify the bones and features illustrated in figures 14.1 to 14.7 on the actual skulls. Note that the figures are in full color, with the bones labeled in **boldface** and feature labels not in boldface. It is best to work with a partner or in small groups.
2. Examine the **cranial bones** of the articulated human skull and the sectioned skull. Also observe the corresponding disarticulated bones. Locate the following

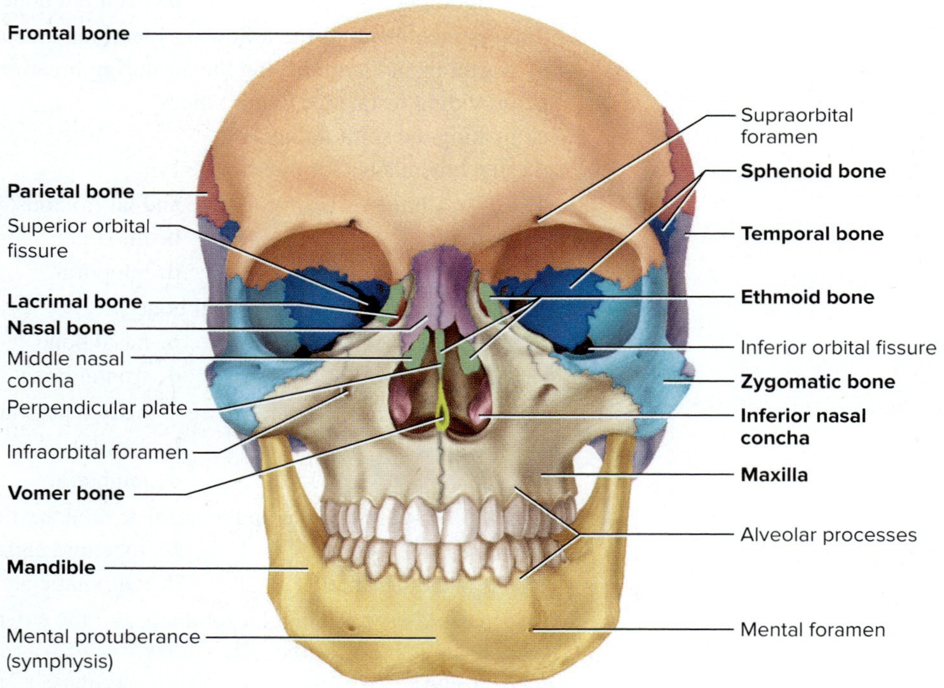

FIGURE 14.1 Bones and features of the skull, anterior view. (**Boldface** indicates a bone name; not boldface indicates a feature.) APR

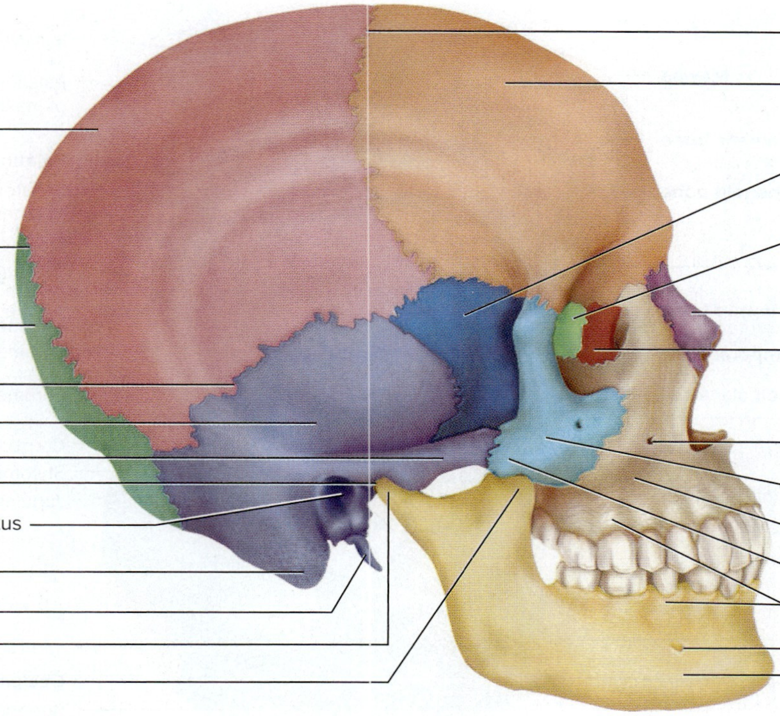

FIGURE 14.2 Bones and features of the skull, lateral view. (**Boldface** indicates a bone name; not boldface indicates a feature.) APR

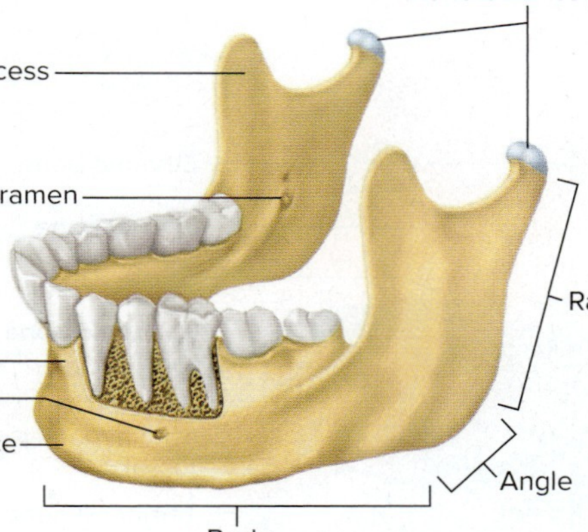

FIGURE 14.3 The mandible and its features. APR

bones and features in the laboratory specimens and, at the same time, palpate as many of these bones and features in your skull as possible.

frontal bone (1)

- supraorbital foramen (or notch)—transmits blood vessels and nerves
- frontal sinus—two present

parietal bone (2)

- sagittal suture—where parietal bones meet
- coronal suture—where parietal bones meet frontal bone

occipital bone (1)

- lambdoid suture—where parietal bones meet occipital bone
- external occipital protuberance—midline projection on posterior surface
- foramen magnum—where brainstem and spinal cord meet
- occipital condyle—articulates with first cervical vertebra

temporal bone (2)

- squamous suture—where parietal bones meet temporal bone
- external acoustic meatus—opening of outer ear
- mandibular fossa—articulates with mandible
- mastoid process—attachment site for muscles
- styloid process—long, pointed projection
- carotid canal—transmits internal carotid artery
- jugular foramen—transmits internal jugular vein

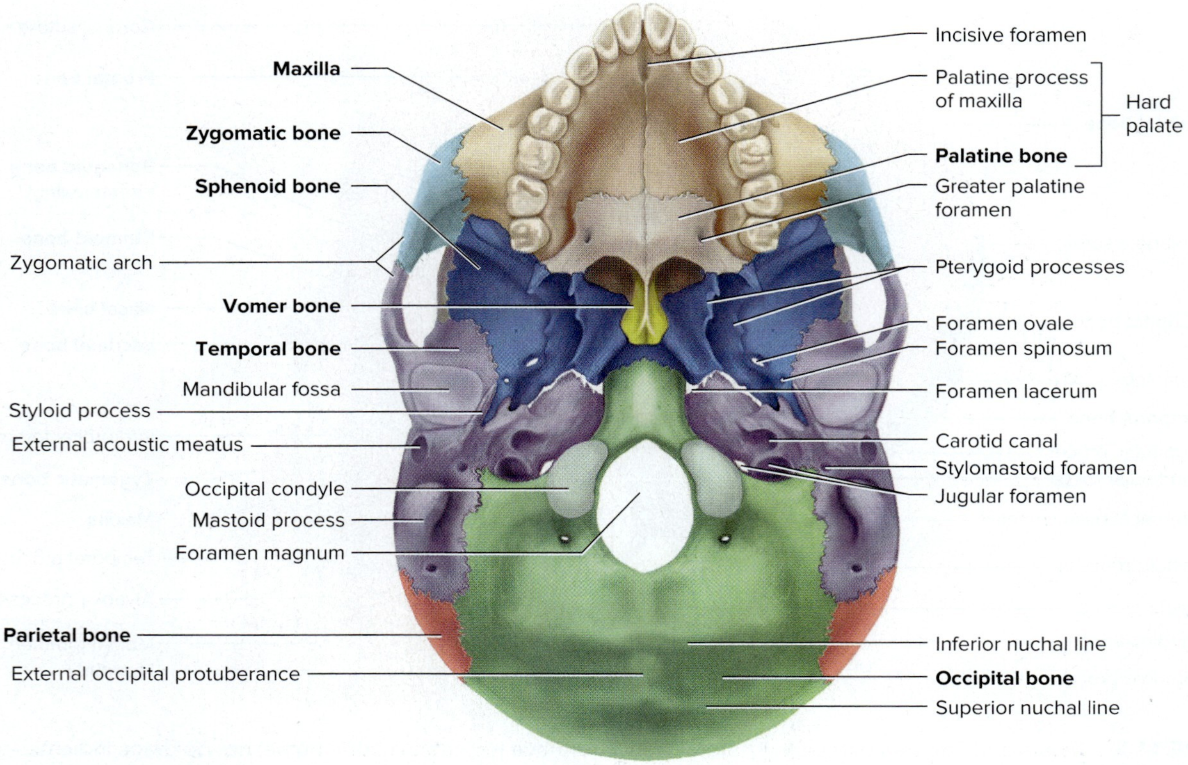

FIGURE 14.4 Bones and features of the skull, inferior view. (**Boldface** indicates a bone name; not boldface indicates a feature.) APR

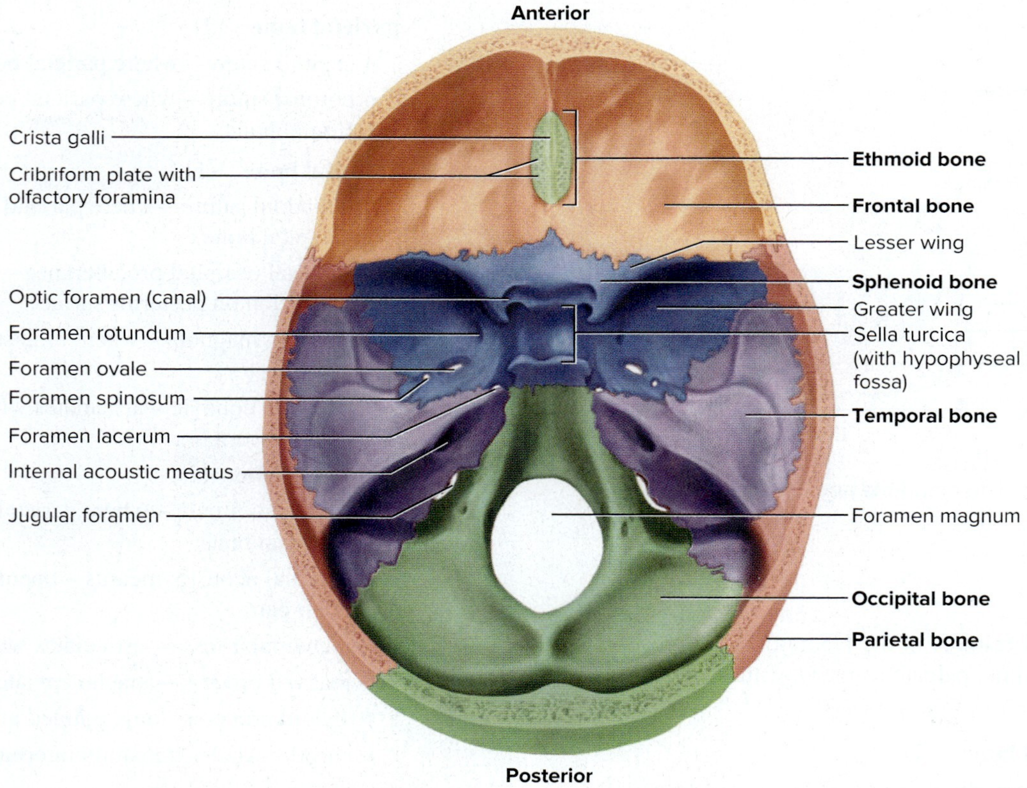

FIGURE 14.5 Bones and features of the floor of the cranial cavity, superior view. (**Boldface** indicates a bone name; not boldface indicates a feature.) APR

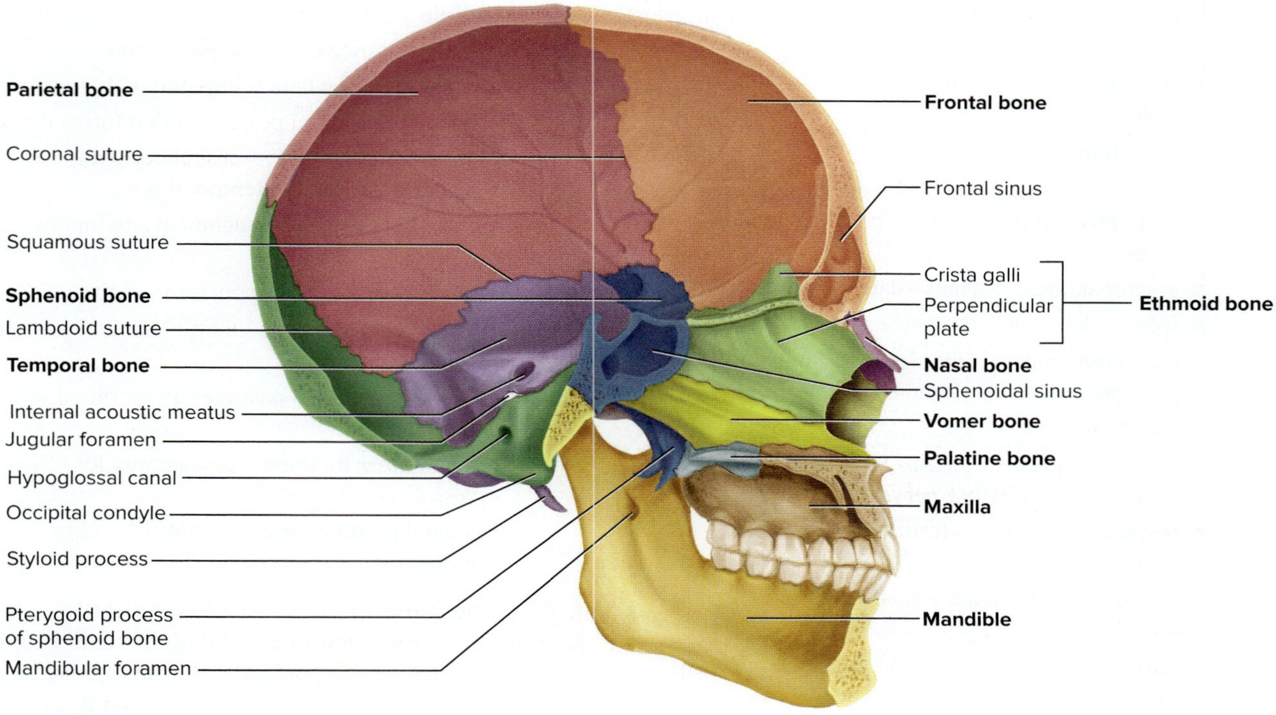

FIGURE 14.6 Bones and features of the sagittal section of the skull. (**Boldface** indicates a bone name; not boldface indicates a feature.) APR

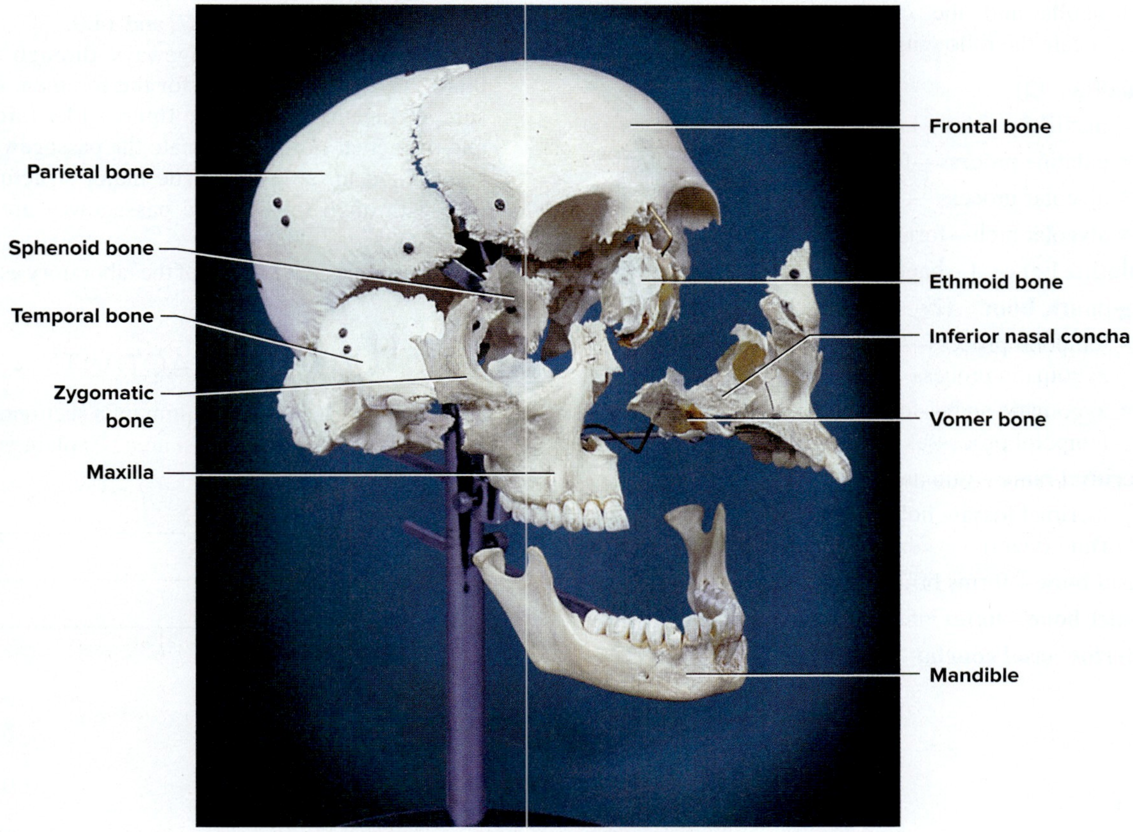

FIGURE 14.7 Bones of a disarticulated skull. ©J & J Photography APR

- internal acoustic meatus—opening for cranial nerves
- zygomatic process—forms much of zygomatic arch

sphenoid bone (1)

- sella turcica—saddle-shaped depression (hypophyseal fossa) where pituitary gland rests
- greater and lesser wings—lateral extensions
- sphenoidal sinus—two present
- pterygoid process—inferior extensions for chewing muscle attachments

ethmoid bone (1)

- cribriform plate—contains many olfactory foramina for olfactory nerves
- perpendicular plate—forms most of nasal septum
- superior nasal concha—small extensions in nasal cavity
- middle nasal concha—medium extensions in nasal cavity
- ethmoidal sinus—many air spaces (ethmoidal cells) present
- crista galli—attachment site of some brain membranes

3. Examine the **facial bones** of the articulated and sectioned skulls and the corresponding disarticulated bones. Locate the following:

maxilla (2)

- maxillary sinus—largest sinuses
- palatine process—forms anterior hard palate
- alveolar process—supports the teeth
- alveolar arch—formed by alveolar processes

palatine bone—L-shaped bones (2)

zygomatic bone (2)

- temporal process—extension that joins zygomatic process
- zygomatic arch—formed by zygomatic and temporal processes

lacrimal bone—thin and scalelike (2)

- lacrimal fossa—holds lacrimal sac that contains fluid (tears)

nasal bone—forms bridge of nose (2)

vomer bone—forms inferior part of nasal septum (1)

inferior nasal concha—largest of the conchae (2)

mandible (1)

- ramus—upward vertical projection
- angle—point where ramus and body meet
- body—horizontal projection that forms the chin
- mandibular condyle—articulates with mandibular fossa of temporal bone
- coronoid process—attachment site for chewing muscles
- alveolar process—supports the teeth
- mandibular foramen—admits blood vessels and nerves for teeth
- mental foramen—passageway for blood vessels and nerves
- mandibular foramen—passageway for blood vessels and nerves
- mental protuberance (symphysis)—chin projection

4. Complete Part A of Laboratory Assessment 14.
5. Study an anterior view of an orbit of the skull. Examine the bones that have a portion of the entire bone visible in an orbit (fig. 14.8). A disarticulated skull, as shown in figure 14.7, would be helpful to distinguish skull bones forming portions of an orbit. The orbit of an eye includes some cranial and facial bones.
6. Examine any skulls that are sectioned to expose the paranasal sinuses. Sinuses are located in the frontal, ethmoid, sphenoid, and maxillary bones. Compare the skulls with figures 14.6, 14.7, and 14.9.
7. Some of the major passageways through the skull bones have specific names for the foramen, canal, fissure, or meatus. Reexamine figures 14.1 through 14.6 and the actual skulls and locate the passageways of the skull listed in table 14.1. The major structures transmitted through the specific passageway are included in the table.
8. Complete Parts B, C, and D of the laboratory assessment.

CRITICAL THINKING ACTIVITY

Examine the inside of the cranium on a sectioned skull. What appears to be the weakest area? Explain your answer.

____

____

____

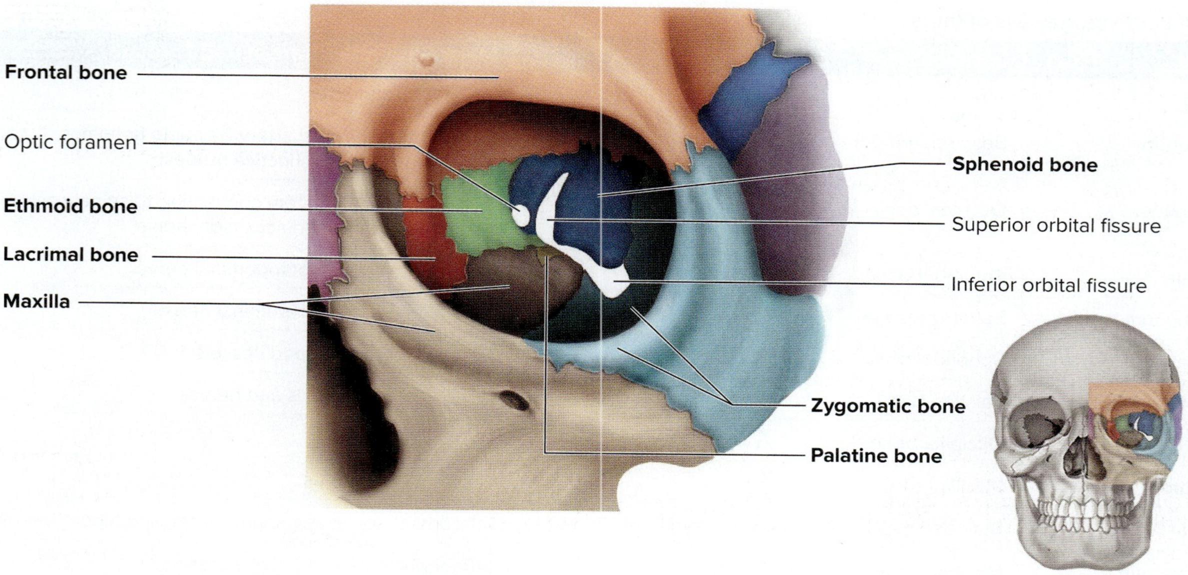

FIGURE 14.8 Anterior view of a left orbit. (**Boldface** indicates a bone name; not boldface indicates a feature.) APR

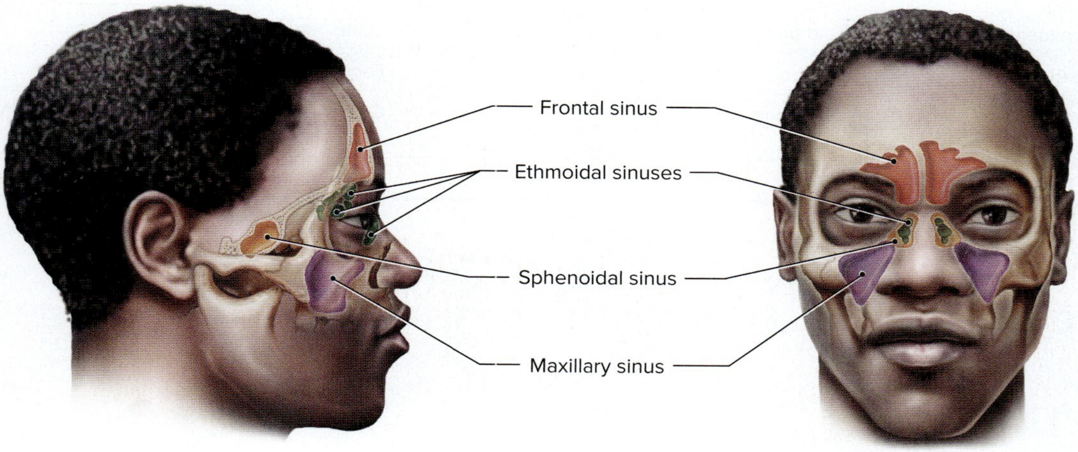

FIGURE 14.9 Paranasal sinuses are located in the frontal bone, ethmoid bone, sphenoid bone, and both maxillary bones. APR

Table 14.1 Major Passageways of the Skull

Passageway	Location Bone	Major Structures Transmitted Through
Carotid canal	Temporal bone	Internal carotid artery
Foramen lacerum	Between temporal, sphenoid, and occipital bones	Branch of pharyngeal artery (opening becomes mostly closed by connective tissues)
Foramen magnum	Occipital bone	Inferior part of brainstem connecting to spinal cord, vertebral arteries, and accessory nerve
Foramen ovale	Sphenoid bone	Mandibular division of trigeminal nerve
Foramen rotundum	Sphenoid bone	Maxillary division of trigeminal nerve
Foramen spinosum	Sphenoid bone	Middle meningeal blood vessels
Greater palatine foramen	Palatine bone	Palatine blood vessels and nerves
Hypoglossal canal	Occipital bone	Hypoglossal nerve
Incisive foramen	Maxilla	Nasopalatine nerves
Inferior orbital fissure	Between maxilla, sphenoid, and zygomatic bones	Infraorbital nerve, zygomatic nerve, and blood vessels
Infraorbital foramen	Maxilla	Infraorbital nerve and blood vessels
Internal acoustic meatus	Temporal bone	Branches of facial and vestibulocochlear nerves
Jugular foramen	Between temporal and occipital bones	Glossopharyngeal, vagus, and accessory nerves, and internal jugular vein
Mandibular foramen	Mandible	Inferior alveolar blood vessels and nerves
Mental foramen	Mandible	Mental nerve and blood vessels
Optic foramen (canal)	Sphenoid bone	Optic nerve and ophthalmic artery
Stylomastoid foramen	Temporal bone	Facial nerve and blood vessels
Superior orbital fissure	Sphenoid bone	Oculomotor, trochlear, and abducens nerves, and ophthalmic division of trigeminal nerve
Supraorbital foramen	Frontal bone	Supraorbital blood vessels and nerves

LABORATORY
ASSESSMENT 14

Name ____________________

Date ____________________

Section ____________________

The (A) corresponds to the indicated Learning Outcome(s) (O) found at the beginning of the Laboratory Exercise.

Skull

PART A: Assessments

Identify the numbered bones and features of the skulls indicated in figures 14.10 to 14.13.

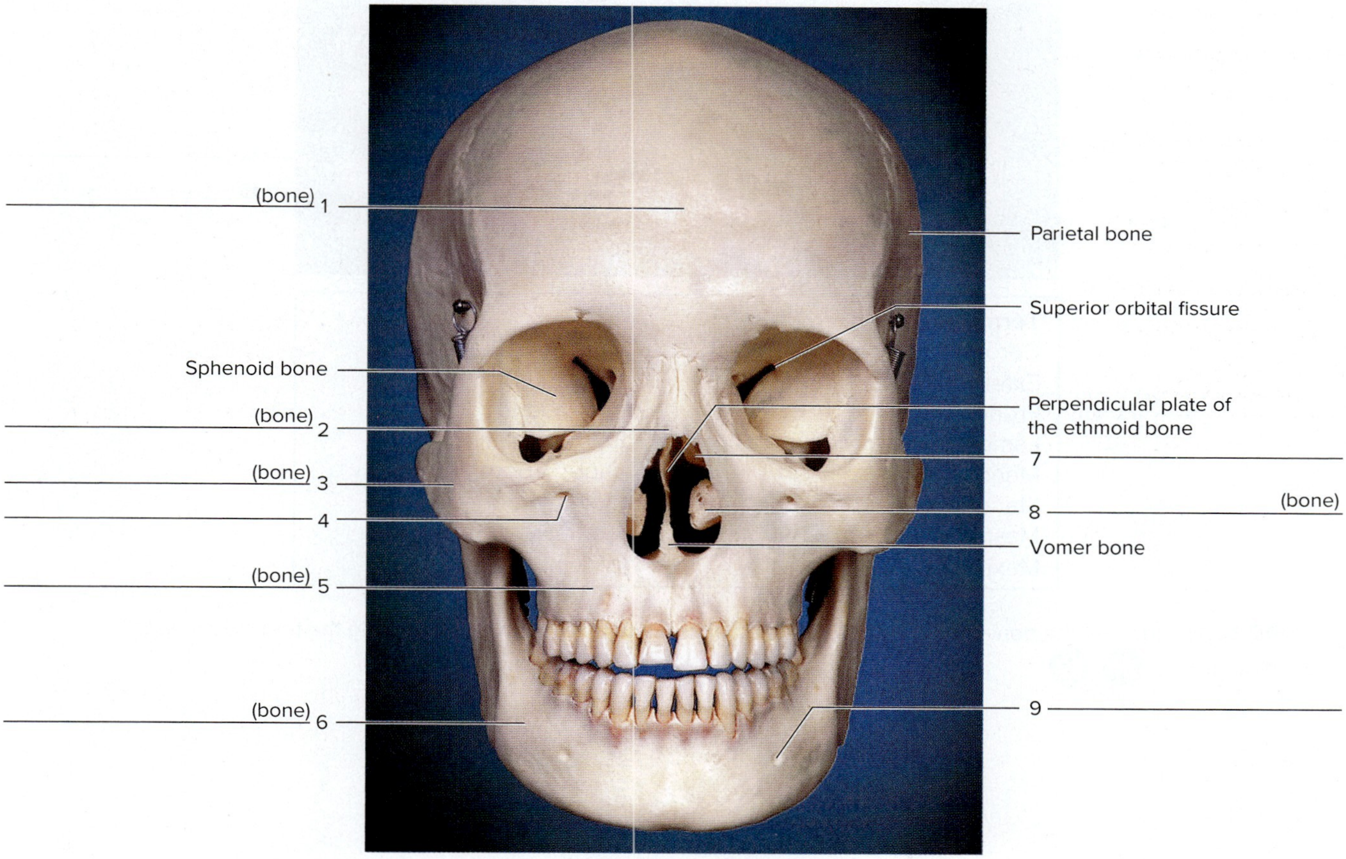

Terms:

Frontal bone	Mental foramen
Inferior nasal concha	Middle nasal concha (of ethmoid bone)
Infraorbital foramen	Nasal bone
Mandible	Zygomatic bone
Maxilla	

FIGURE 14.10 Identify the bones and features indicated on this anterior view of the skull, using the terms provided. (If the line lacks the word *bone*, label the particular feature.) ©J & J Photography (A1)

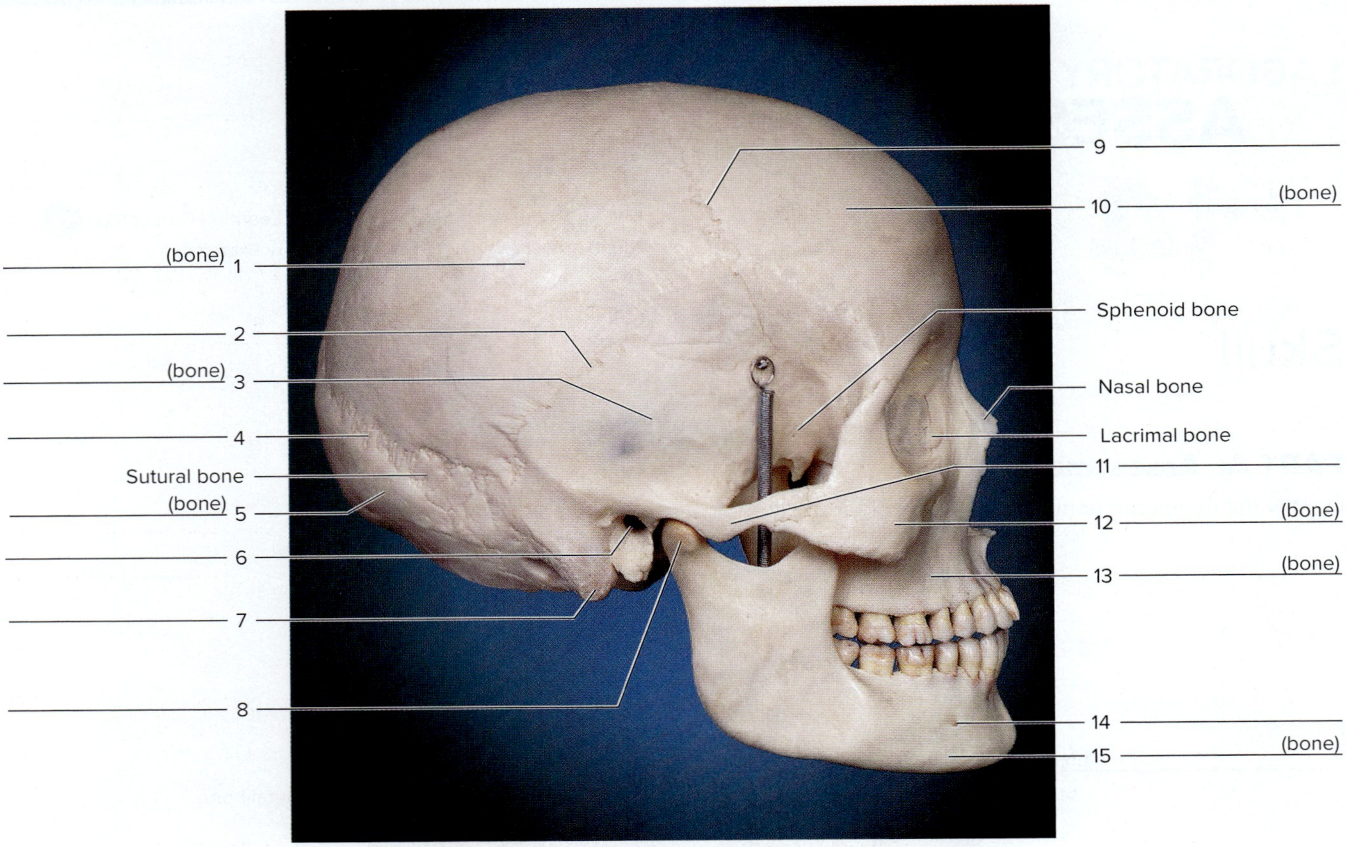

Terms:

Coronal suture
External acoustic meatus
Frontal bone
Lambdoid suture
Mandible
Mandibular condyle
Mastoid process
Maxilla
Mental foramen
Occipital bone
Parietal bone
Squamous suture
Temporal bone
Zygomatic bone
Zygomatic process (of temporal bone)

FIGURE 14.11 Identify the bones and features indicated on this lateral view of the skull, using the terms provided.

 A1 A2

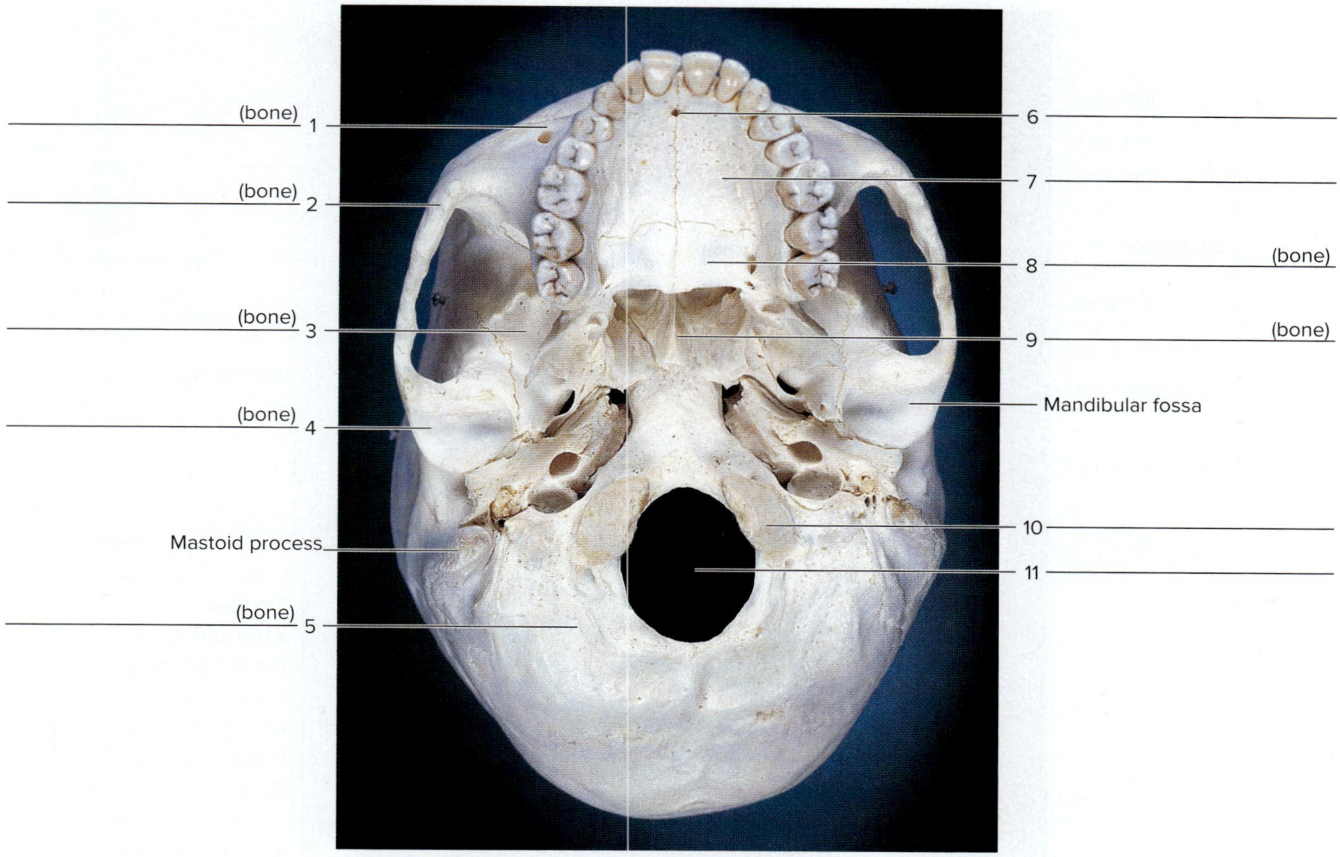

Terms:

Foramen magnum
Incisive foramen
Maxilla
Occipital bone
Occipital condyle
Palatine bone
Palatine process of maxilla
Sphenoid bone
Temporal bone
Vomer bone
Zygomatic bone

FIGURE 14.12 Identify the bones and features indicated on this inferior view of the skull, using the terms provided.

©J & J Photography A1

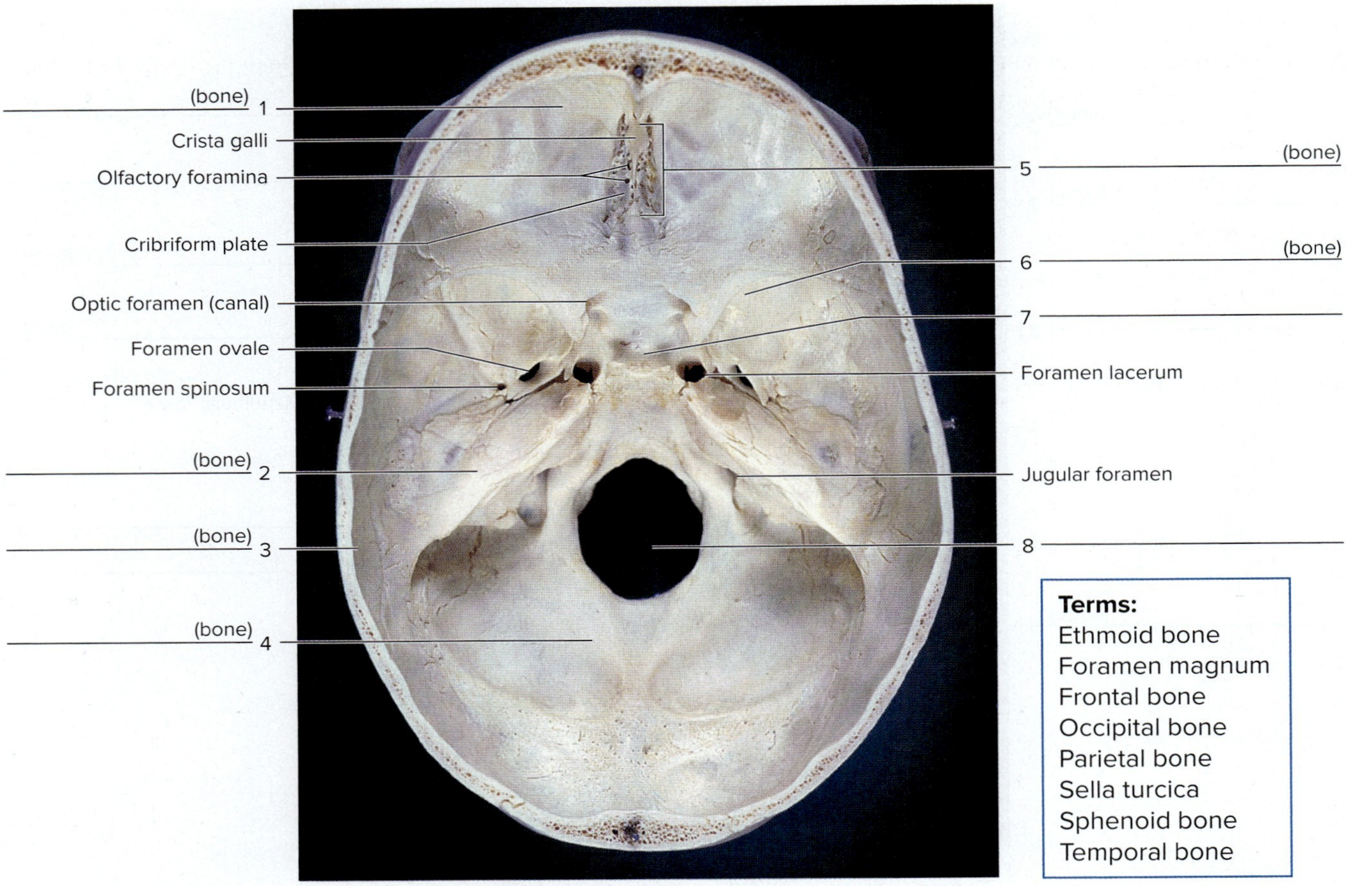

FIGURE 14.13 Identify the bones and features on this floor of the cranial cavity of a skull, using the terms provided.
©J & J Photography A1

CRITICAL THINKING ASSESSMENT—PART A

What bones and their features compose the temporomandibular joint (TMJ)? A1 ______________________________

____.

PART B: Assessments

Match the bones in column A with the features in column B. Place the letter of your choice in the space provided. (Some answers are used more than once.) A1

Column A

a. Ethmoid bone

b. Frontal bone

c. Occipital bone

d. Parietal bone

e. Sphenoid bone

f. Temporal bone

Column B

______ **1.** Forms sagittal, coronal, squamous, and lambdoid sutures

______ **2.** Cribriform plate

______ **3.** Crista galli

______ **4.** External acoustic meatus

______ **5.** Foramen magnum

______ **6.** Mandibular fossa

______ **7.** Mastoid process

______ **8.** Middle nasal concha

______ **9.** Occipital condyle

______ **10.** Perpendicular plate

______ **11.** Sella turcica

______ **12.** Styloid process

______ **13.** Supraorbital foramen

PART C: Assessments

1. Complete the following statements:

a. Name six cranial bones that are visible on a lateral view of a skull. A1 ________________________________

____

b. The ____________________ suture joins the frontal bone to the parietal bones. A2

c. The parietal bones are firmly interlocked along the midline by the ____________________ suture. A2

d. The ____________________ suture joins the parietal bones to the occipital bone. A2

e. The temporal bones are joined to the parietal bones along the ____________________ sutures. A2

f. Name the three cranial bones that contain sinuses. A4 ________________________________

____

g. Name a facial bone that contains a sinus. A4 ________________________________

h. Provide the functions of the paranasal sinuses. A4 ________________________________

____

2. Identify the numbered paranasal sinuses indicated in figure 14.14.

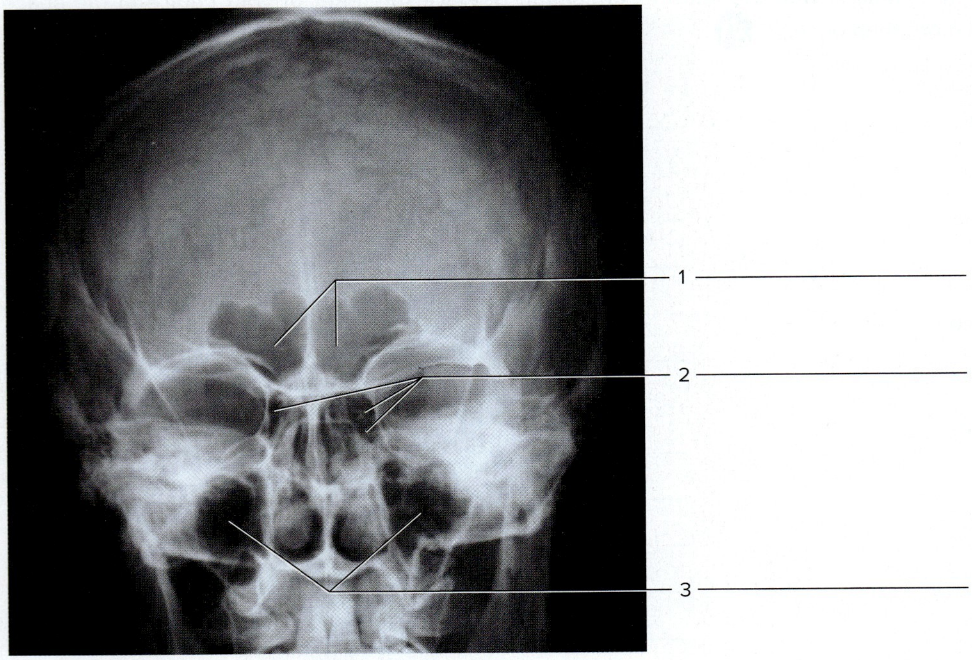

FIGURE 14.14 Identify the numbered paranasal sinuses shown in this X ray of the skull, anterior view. Dennis Strete A4

PART D: Assessments

1. Match the bones in column A with the characteristics in column B. Place the letter of your choice in the space provided. (Some answers are used more than once.) A1 A3 A4

Column A

a. Inferior nasal concha
b. Lacrimal bone
c. Mandible
d. Maxilla
e. Nasal bone
f. Palatine bone
g. Vomer bone
h. Zygomatic bone

Column B

______ 1. Forms bridge of nose
______ 2. Only movable bone in the facial skeleton
______ 3. Contains coronoid process
______ 4. Creates prominence of cheek inferior and lateral to the eye
______ 5. Contains sockets of upper teeth
______ 6. Forms inferior portion of nasal septum
______ 7. Forms anterior portion of zygomatic arch
______ 8. Scroll-shaped bone in nasal passage
______ 9. Forms anterior roof of mouth
______ 10. Contains mental foramen
______ 11. Forms posterior roof of mouth
______ 12. Small bone in medial wall of orbit
______ 13. Forms a very small portion of inferior orbit
______ 14. Contains large sinus inferior to orbit

2. Identify the numbered bones and features of the orbit of the eye indicated in figure 14.15.

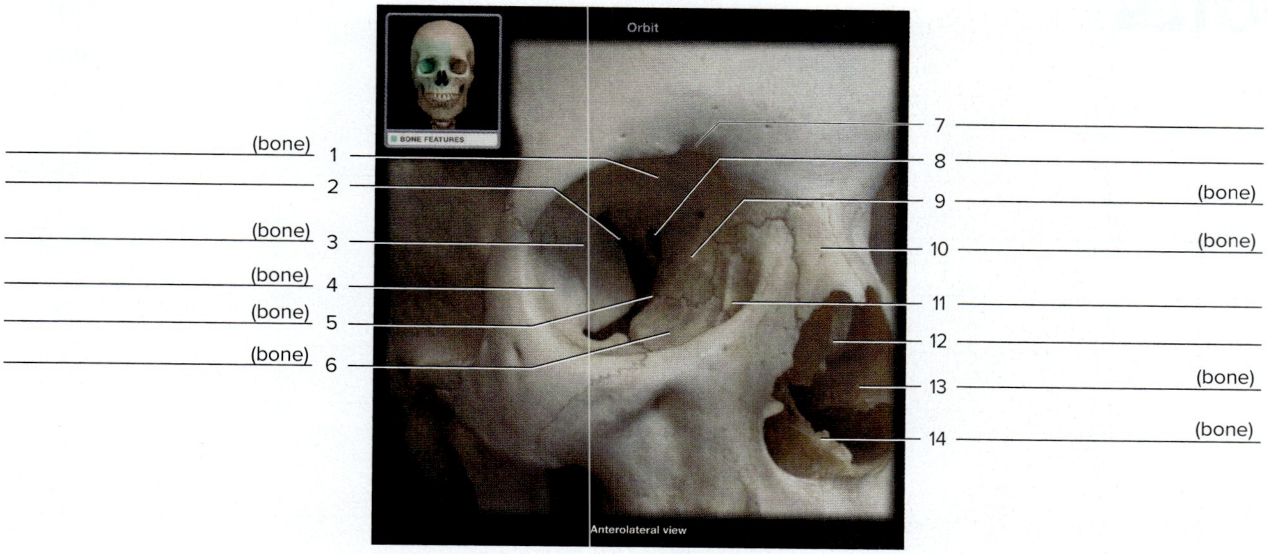

Terms:

Ethmoid	Palatine
Frontal	Perpendicular plate
Inferior nasal concha	Sphenoid
Lacrimal fossa	Superior orbital fissure
Maxilla	Superior orbital notch
Nasal	Vomer
Optic foramen/canal	Zygomatic

FIGURE 14.15 Identify the bones and features indicated in this anterolateral view of the orbit of the skull, using the terms provided. (If the line lacks the word *bone*, label the particular feature.) APR/McGraw Hill A3

CRITICAL THINKING ASSESSMENTS—PART D

1. What bones may be displaced in a person who has a deviated septum? Explain possible consequences of this condition. A1 A4 ________________

2. Assume that an orbit of a skull was damaged in an accident. Which bones of a skull might need to be considered for reconstruction in an orbital surgery? A3 ________________

NOTES

LABORATORY
EXERCISE 15

Vertebral Column and Thoracic Cage

Purpose of the Exercise

To examine the vertebral column and the thoracic cage of the human skeleton and to identify the bones and major features and functions of these parts.

MATERIALS NEEDED

Human skeleton, articulated

Samples of cervical, thoracic, and lumbar vertebrae

Human skeleton, disarticulated

Learning Outcomes APR

After completing this exercise, you should be able to:

O1 Identify the major features and functions of the vertebral column.

O2 Locate the features of a vertebra.

O3 Distinguish the cervical, thoracic, and lumbar vertebrae and the sacrum and coccyx.

O4 Identify the features and functions of the thoracic cage.

O5 Distinguish between true and false ribs.

The O corresponds to the assessments A indicated in the Laboratory Assessment for this Exercise.

Pre-Lab

Carefully read the introductory material and examine the entire lab. Be familiar with the vertebral column and the thoracic cage from lecture or the textbook. Answer the pre-lab questions.

Pre-Lab Questions Select the correct answer for each of the following questions:

1. The most superior bone of the vertebral column is the
   a. coccyx. **b.** vertebra prominens. **c.** axis. **d.** atlas.
2. The vertebral column possesses
   a. four curvatures. **b.** three curvatures.
   c. one curvature. **d.** no curvatures as it is straight.
3. Humans have _____ pairs of true ribs.
   a. two **b.** five **c.** seven **d.** twelve
4. The _____ ribs do *not* have costal cartilage attachments to the sternum.
   a. false **b.** floating **c.** true **d.** superior
5. Humans possess _____ cervical vertebrae.
   a. twenty-six **b.** twelve
   c. seven **d.** five
6. The superior end of the sacrum articulates with the
   a. coccyx. **b.** femur.
   c. twelfth thoracic vertebra. **d.** fifth lumbar vertebra.
7. The anterior (sternal) end of a rib articulates with a thoracic vertebra.
   a. True **b.** False
8. All cervical, thoracic, and lumbar vertebrae possess a vertebral foramen.
   a. True **b.** False
9. A feature of the second cervical vertebra is the dens.
   a. True **b.** False

Chapter Opener: ©101cats/Getty Images

The vertebral column and thoracic cage are part of the axial skeleton, in addition to the skull. The vertebral column, consisting of twenty-six bones, extends from the skull to the pelvis and forms the vertical axis of the human skeleton. The *vertebral column* includes seven cervical vertebrae, twelve thoracic vertebrae, five lumbar vertebrae, one sacrum of five fused vertebrae, and one coccyx of usually four fused vertebrae. To help you remember the number of cervical, thoracic, and lumbar vertebrae from superior to inferior, consider this saying: breakfast at 7, lunch at 12, and dinner at 5. These vertebrae are separated from one another by cartilaginous *intervertebral discs* and are held together by ligaments.

The *thoracic cage* surrounds the thoracic and upper abdominal cavities. It includes the ribs, the thoracic vertebrae, the sternum, and the costal cartilages. The thoracic cage provides protection for the heart and lungs.

PROCEDURE A: Vertebral Column

The vertebral column extends from the first cervical vertebra adjacent to the skull to the inferior tip of the coccyx. The first cervical vertebra (C1) is also known as the *atlas* and has a posterior tubercle instead of a more pronounced spinous process. The second cervical vertebra (C2), known as the *axis,* has a superior projection, the dens (odontoid process) that serves as a pivot point for some rotational movements. The seventh cervical vertebra (C7) is often referred to as the *vertebra prominens* because the spinous process is elongated and easily palpated as a surface feature. The seven cervical vertebrae have the distinctive feature of transverse foramina for passageways of blood vessels serving the brain.

The twelve thoracic vertebrae have facets for the articulation sites of the twelve pairs of ribs. Thoracic vertebrae are larger than the cervical vertebrae and have spinous processes that are rather long with an inferior angle.

The five lumbar vertebrae have the largest and thickest bodies, allowing better support and resistance to twisting of the trunk. The spinous processes of lumbar vertebrae are rather short and blunt.

The five sacral vertebrae of a child fuse into a single bone by the age of about 26. The posterior ridge known as the *medial sacral crest* represents fused spinous processes. The usual four coccyx vertebrae fuse into a single bone by the age of about 30.

CRITICAL THINKING ACTIVITY

Note the four curvatures of the vertebral column. What functional advantages exist with curvatures for skeletal structure instead of a straight vertebral column?

____

____

____

____

____

The four curvatures of the vertebral column develop either before or after birth. The *thoracic* and *sacral curvatures* (primary curvatures) form by the time of birth. The *cervical curvature* develops by the time a baby is able to hold the head erect and crawl, while the *lumbar curvature* forms by the time the child is able to walk. The cervical and lumbar curvatures represent the secondary curvatures. The four curvatures allow for flexibility and resiliency of the vertebral column and for it to function somewhat like a spring instead of a rigid rod.

1. Examine figure 15.1 and the vertebral column of the human skeleton. Locate the following bones and features. At the same time, locate as many of the corresponding bones and features in the skeleton as possible.

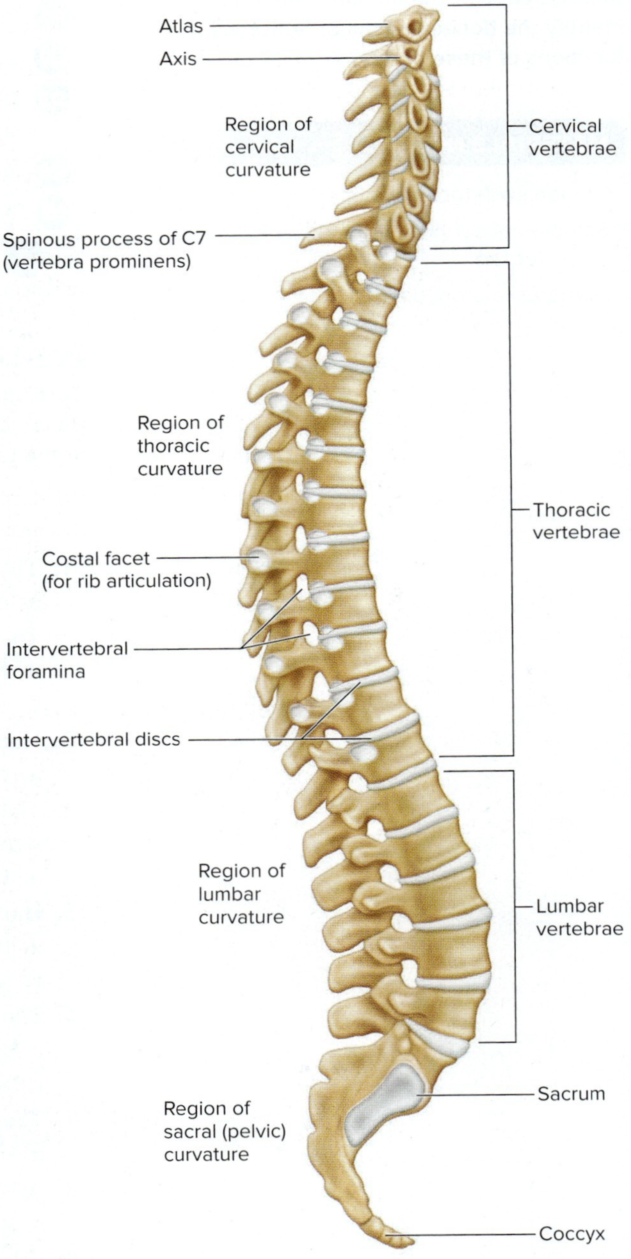

FIGURE 15.1 Bones and features of the vertebral column (right lateral view). APR

cervical vertebrae (7)
- atlas (C1)
- axis (C2)
- vertebra prominens (C7)

thoracic vertebrae (12)
lumbar vertebrae (5)
sacrum (1)
coccyx (1)
intervertebral discs—composed of fibrocartilage
vertebral canal—contains spinal cord
cervical curvature
thoracic curvature
lumbar curvature
sacral (pelvic) curvature
intervertebral foramina—passageway for spinal nerves; formed when superior and inferior vertebrae articulate

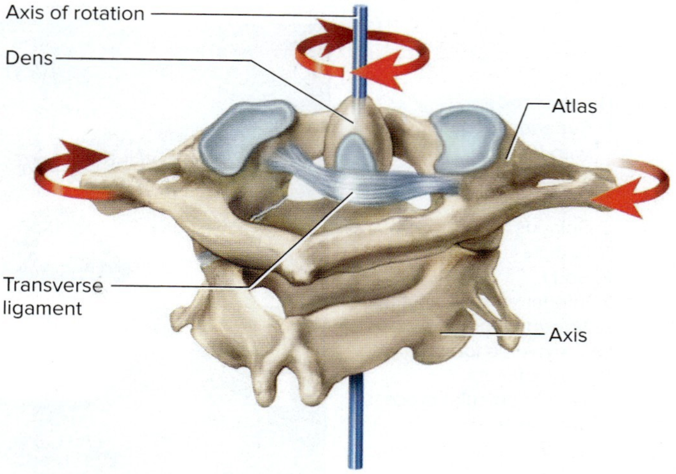

FIGURE 15.2 Articulation of atlas (C1) and axis (C2), showing the rotation of the atlas to turn the head from side to side. APR

Number code:
1. Superior articular facet
2. Vertebral foramen
3. Dens (odontoid process)
4. Facet that articulates with occipital condyle
5. Transverse process
6. Transverse foramen
7. Spinous process
8. Body
9. Posterior tubercle

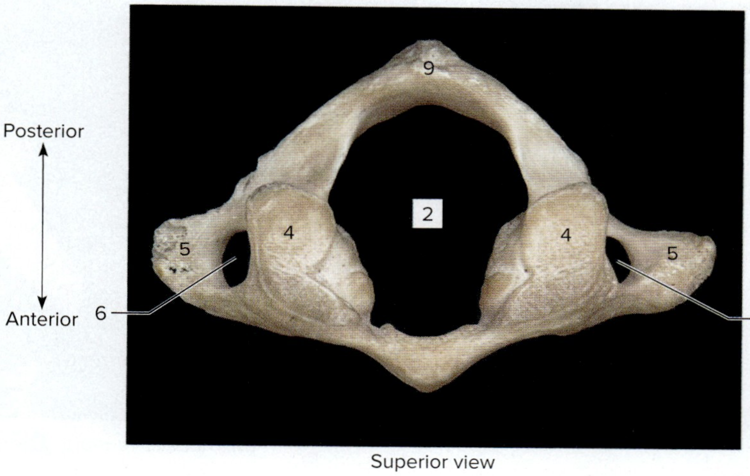

(a) Atlas (C1) APR

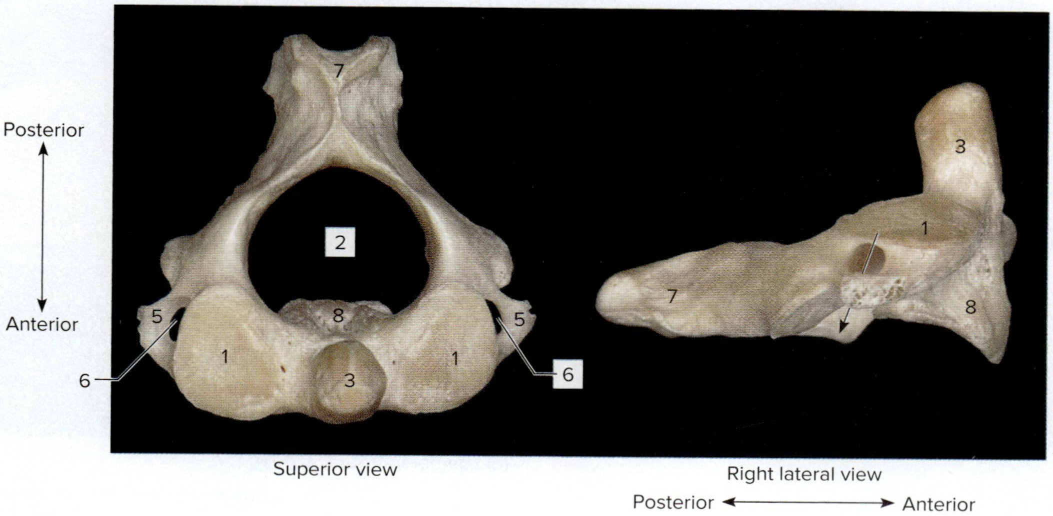

(b) Axis (C2) APR

FIGURE 15.3 The superior features of (a) the atlas, and the superior and right lateral features of (b) the axis. (The broken arrow indicates a transverse foramen.) ©J & J Photography

Number code:
1. Superior articular process with facet
2. Transverse process
3. Lamina
4. Spinous process
5. Pedicle
6. Body
7. Inferior vertebral notch
8. Vertebral foramen
9. Transverse foramen
10. Costal facets
11. Inferior articular process with facet

Superior views Right lateral views

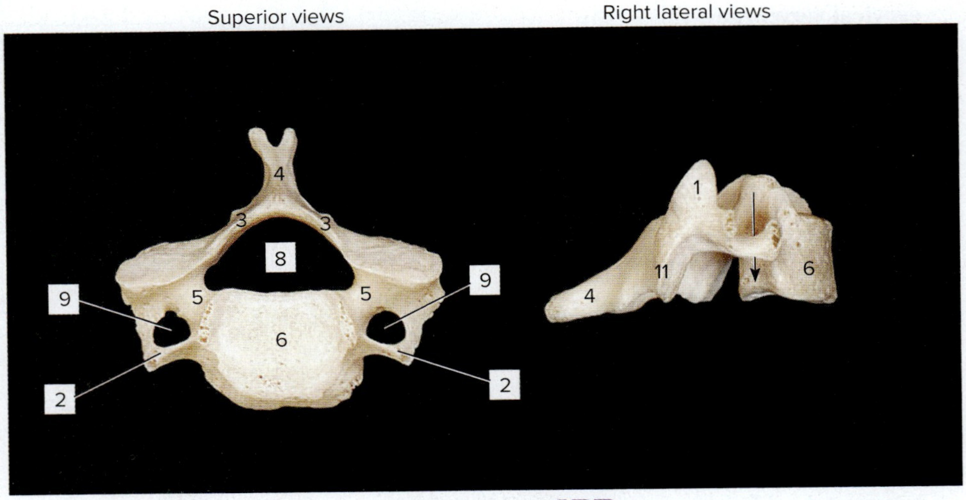

(a) Cervical vertebra APR

Posterior

Anterior

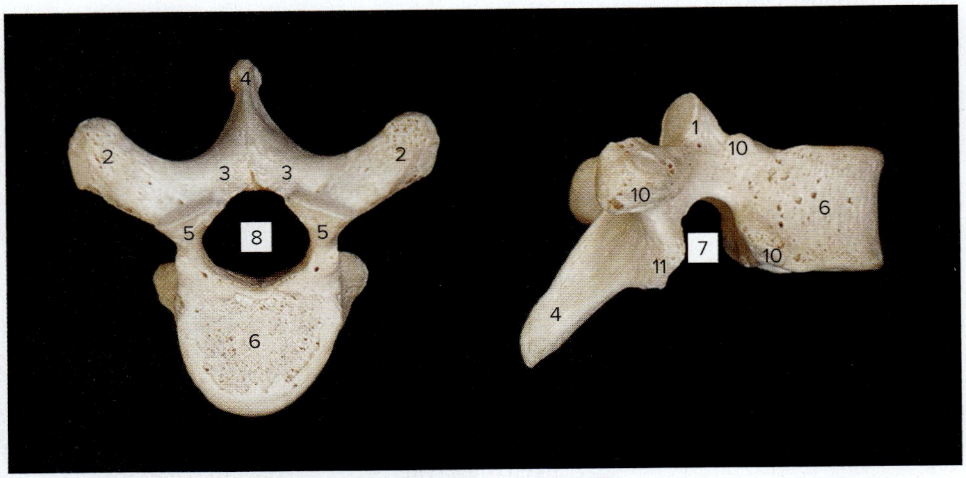

(b) Thoracic vertebra APR

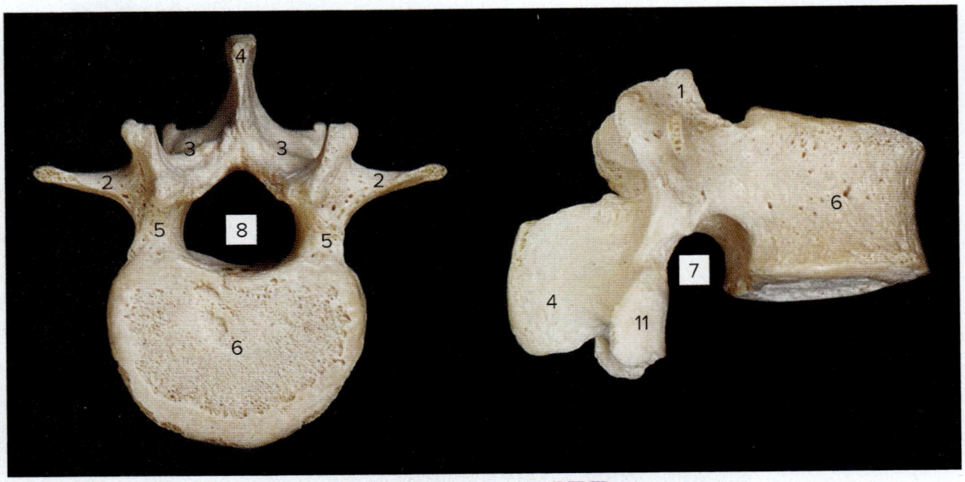

(c) Lumbar vertebra APR

Posterior ⟷ Anterior

FIGURE 15.4 The superior and right lateral features of the (a) cervical, (b) thoracic, and (c) lumbar vertebrae. (The broken arrow indicates a transverse foramen.) ©J & J Photography

2. Compare the available samples of cervical, thoracic, and lumbar vertebrae along with figures 15.2 to 15.4. Note differences in size and shapes and locate the following features:

   vertebral foramen—location of spinal cord

   body—largest part of vertebra; main support portion

   pedicles—form lateral area of vertebral foramen

   laminae—thin plates forming posterior area of vertebral foramen

   spinous process—posterior projection

   transverse processes—lateral projections

   facets—articulating surfaces

   superior articular processes—superior projections

   inferior articular processes—inferior projections

   inferior vertebral notch—space for nerve passage

   transverse foramina—only present on cervical vertebrae; passageway for blood vessels

   dens (odontoid process) of axis—superior process of C2 and is a pivot location at atlas

3. Examine the sacrum and coccyx along with figure 15.5. Locate the following features:

   sacrum—formed typically by five fused vertebrae

   - alae—pair of winglike expansions
   - superior articular process—superior projection with a facet articulation site
   - anterior sacral foramen—passageway for blood vessels and nerves
   - posterior sacral foramen—passageway for blood vessels and nerves
   - sacral promontory—anterior border of S1; important landmark for obstetricians
   - sacral canal—portion of vertebral canal
   - median sacral crest—area of fused spinous processes
   - sacral hiatus—inferior opening of vertebral canal
   - auricular surface—portion of sacrum that articulates with ilium to form sacroiliac joint

   coccyx—formed by three to five fused vertebrae

4. Complete Parts A and B of Laboratory Assessment 15.

PROCEDURE B: Thoracic Cage

The twelve thoracic vertebrae are associated with the twelve pairs of ribs. The superior seven pairs of ribs are connected directly to the sternum with costal cartilage and are known as *true ribs*. The five inferior pairs, known as *false ribs,* either connect indirectly to the sternum with costal cartilage or do not connect to the sternum. Pairs eleven and twelve are called the *floating ribs* because they only connect with the thoracic vertebrae and not with the sternum.

The manubrium, body, and xiphoid process represent three regions that eventually fuse into a single flat bone, the sternum. The heart is located mainly beneath the body portion of the sternum. Any chest compressions administered during cardiopulmonary resuscitation should be over the body area of the sternum, not the xiphoid process region. Chest compressions over the xiphoid process region can force the xiphoid process deep into the liver or the inferior portion of the heart and cause a fatal hemorrhage to occur.

The thoracic cage protects organs in the thoracic and upper abdominal cavities. The thoracic cage also supports the pectoral girdle and upper limbs, functions in breathing, and serves for various muscle attachments.

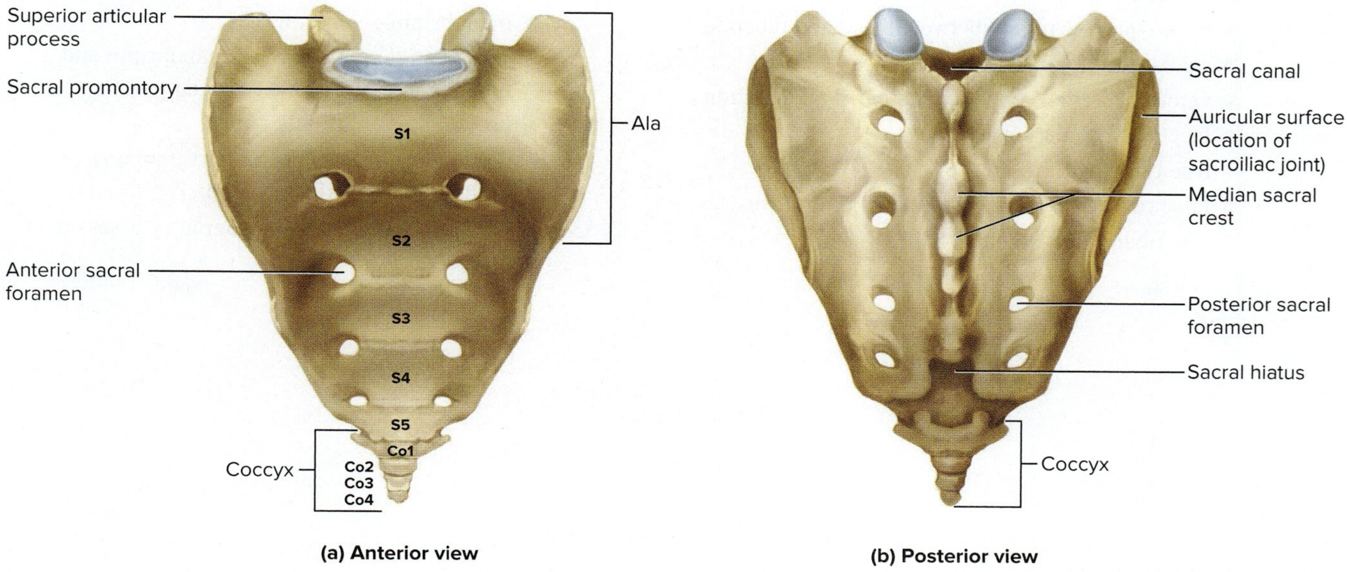

FIGURE 15.5 The sacrum and coccyx. APR

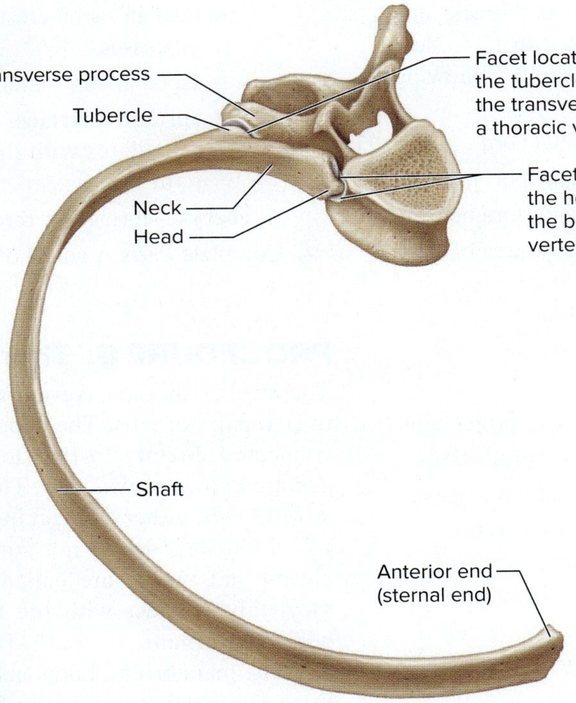

FIGURE 15.6 Superior view of a typical rib with the articulation sites with a thoracic vertebra.

1. Examine figures 15.6 and 15.7 and the thoracic cage of the human skeleton. Locate the following bones and features:

   rib
   - head—expanded end near thoracic vertebra
   - tubercle—projection near thoracic vertebra
   - neck—narrow region between head and tubercle
   - shaft—main portion
   - anterior (sternal) end—costal cartilage location
   - facets—articulation surfaces
   - true ribs—pairs 1–7
   - false ribs—pairs 8–12; includes floating ribs
     - floating ribs—pairs 11–12

   costal cartilages—hyaline cartilage

   sternum
   - jugular (suprasternal) notch—superior concave border of manubrium
   - clavicular notch—articulation site of clavicle
   - manubrium—superior part
   - sternal angle—junction of manubrium and body at level of second rib pair
   - body—largest, middle part
   - xiphoid process—inferior part; remains cartilaginous until adulthood

2. Complete Parts C and D of the laboratory assessment.

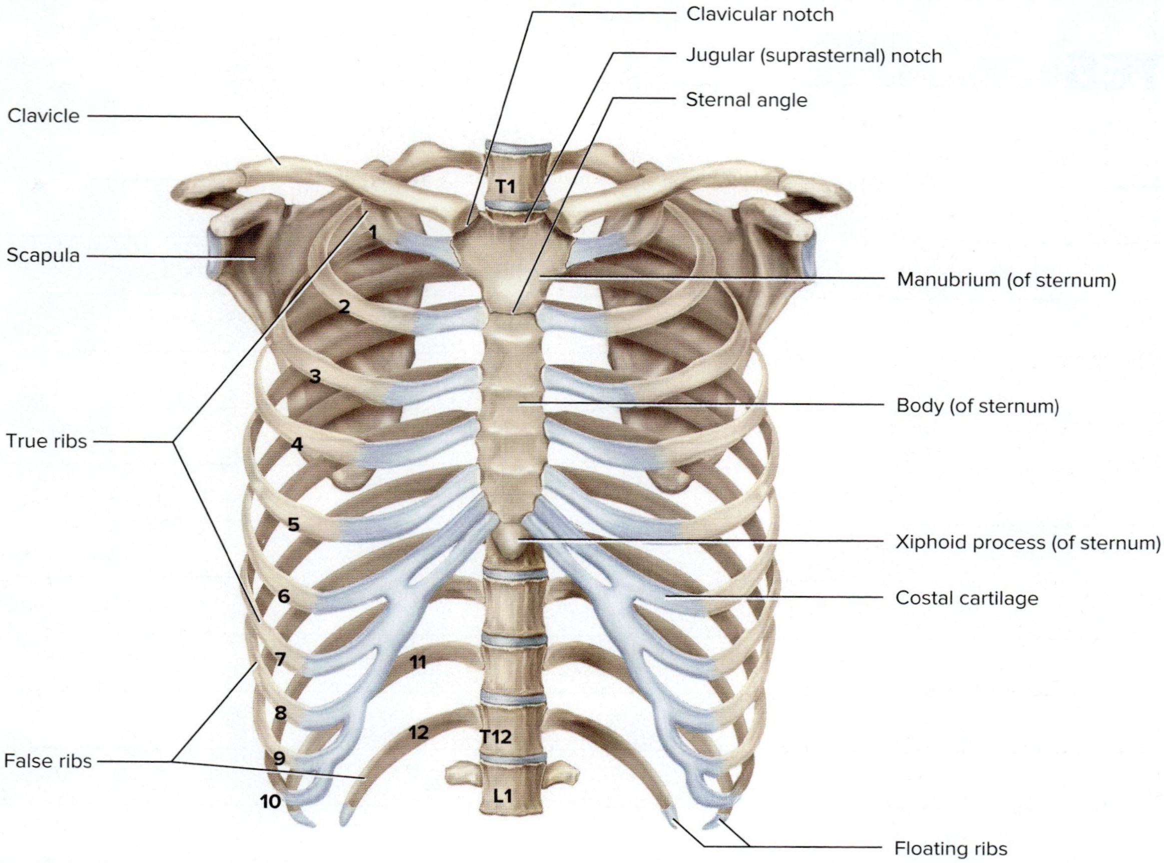

FIGURE 15.7 Bones and features of the thoracic cage (anterior view). APR

NOTES

LABORATORY
ASSESSMENT 15

Name ______________________________

Date ______________________________

Section ______________________________

The (A) corresponds to the indicated Learning Outcome(s) (O) found at the beginning of the Laboratory Exercise.

Vertebral Column and Thoracic Cage

PART A: Assessments

Complete the following statements:

1. The vertebral column encloses and protects the ______________________. A1
2. The vertebral column extends from the skull to the ______________________. A1
3. The seventh cervical vertebra is called the ______________________ and has an obvious spinous process surface feature that can be palpated. A2
4. The ______________________ of the vertebrae support the weight of the head and trunk. A2
5. The ______________________ separate adjacent vertebrae, and they soften the forces created by walking. A1
6. The intervertebral foramina provide passageways for ______________________. A1
7. Transverse foramina of ______________________ vertebrae serve as passageways for blood vessels leading to the brain. A3
8. The first vertebra also is called the ______________________. A3
9. When the head is moved from side to side, the first vertebra pivots around the ______________________ of the second vertebra. A2
10. The ______________________ vertebrae have the largest and strongest bodies. A3
11. The typical number of vertebrae that fuse in the adult to form the sacrum is ______________________. A1

PART B: Assessments

1. Based on your observations, compare typical cervical, thoracic, and lumbar vertebrae in relation to the characteristics indicated in the table. The table is partly completed. For your responses, consider characteristics such as size, shape, presence or absence, and unique features. A2 A3

Vertebra	Number	Size	Body	Spinous Process	Transverse Foramina
Cervical	7		Smallest	C2 through C6 are forked (bifid)	
Thoracic		Intermediate			
Lumbar					Absent

2. Identify the bones and features in figures 15.8 and 15.9.

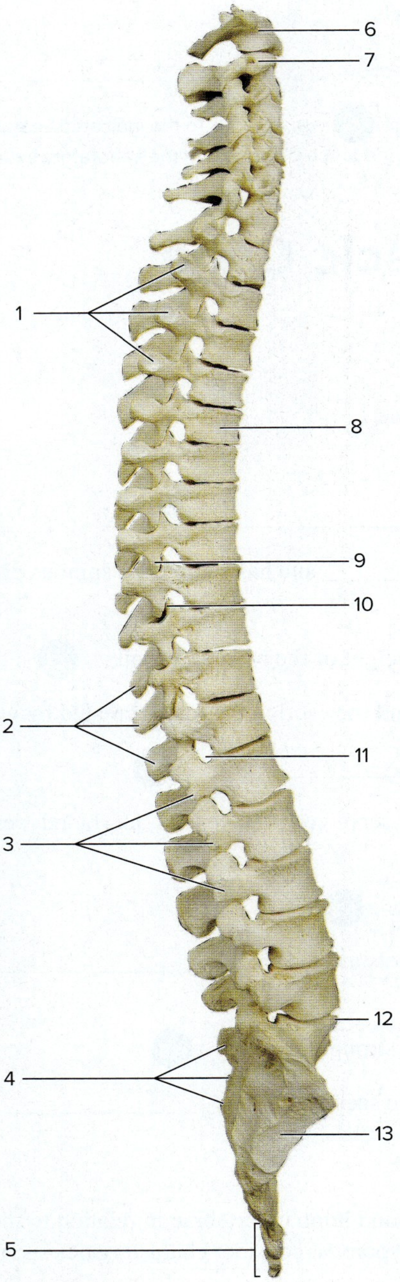

____	Atlas
____	Auricular surface of SI joint
____	Axis
____	Coccyx
____	Costal facets
____	Inferior articular process
____	Intervertebral foramen
____	Median sacral crest
____	Sacral promontory
____	Spinous processes
____	Superior articular process
____	Thoracic vertebra
____	Transverse processes

FIGURE 15.8 Label the bones and features of a lateral view of a vertebral column by placing the correct numbers in the spaces provided. A1 A2 A3

Terms:
Atlas
Axis
Body
Intervertebral disc
Spinous process
Transverse process

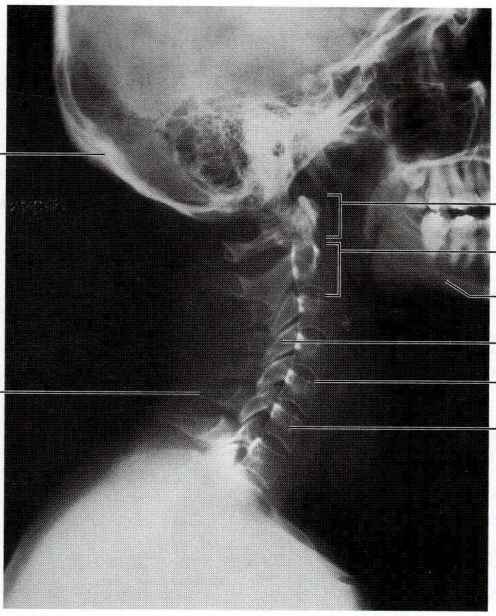

FIGURE 15.9 Identify the bones and features indicated in this radiograph of the neck (lateral view), using the terms provided. Stockbyte/Getty Images APR A1 A2

CRITICAL THINKING ASSESSMENT—PART B

An abnormal lateral curvature of the spine is called *scoliosis*, as shown in figure 15.10, which can be treated by wearing a brace and/or having surgery before the completion of skeletal growth. Predict possible consequences of this type of spinal curvature if left untreated. A1

____

____

____

____

____.

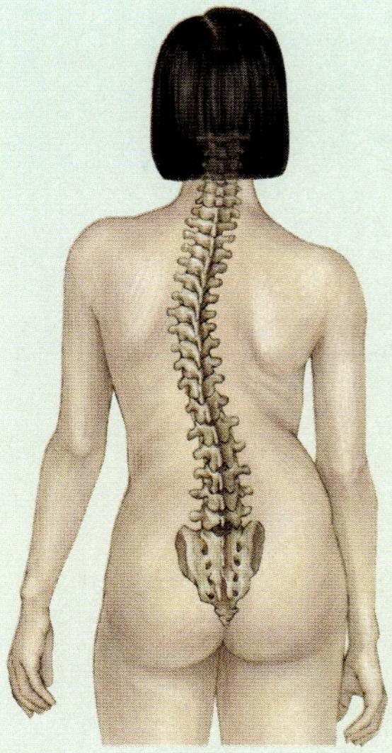

FIGURE 15.10 Scoliosis. APR

PART C: Assessments

Complete the following statements:

1. The manubrium, body, and xiphoid process form a bone called the ______________________. A4
2. The last two pairs of ribs that have no cartilaginous attachments to the sternum are sometimes called ________________ ribs. A5
3. There are ______________________ pairs of true ribs. A5
4. Costal cartilages are composed of ______________________ tissue. A4
5. The manubrium articulates with the ______________________ on its superior border. A4
6. List three general functions of the thoracic cage. A4 ____
____
7. The sternal angle indicates the location of the ______________________ pair of ribs. A4

PART D: Assessments

Identify the bones and features indicated in figure 15.11.

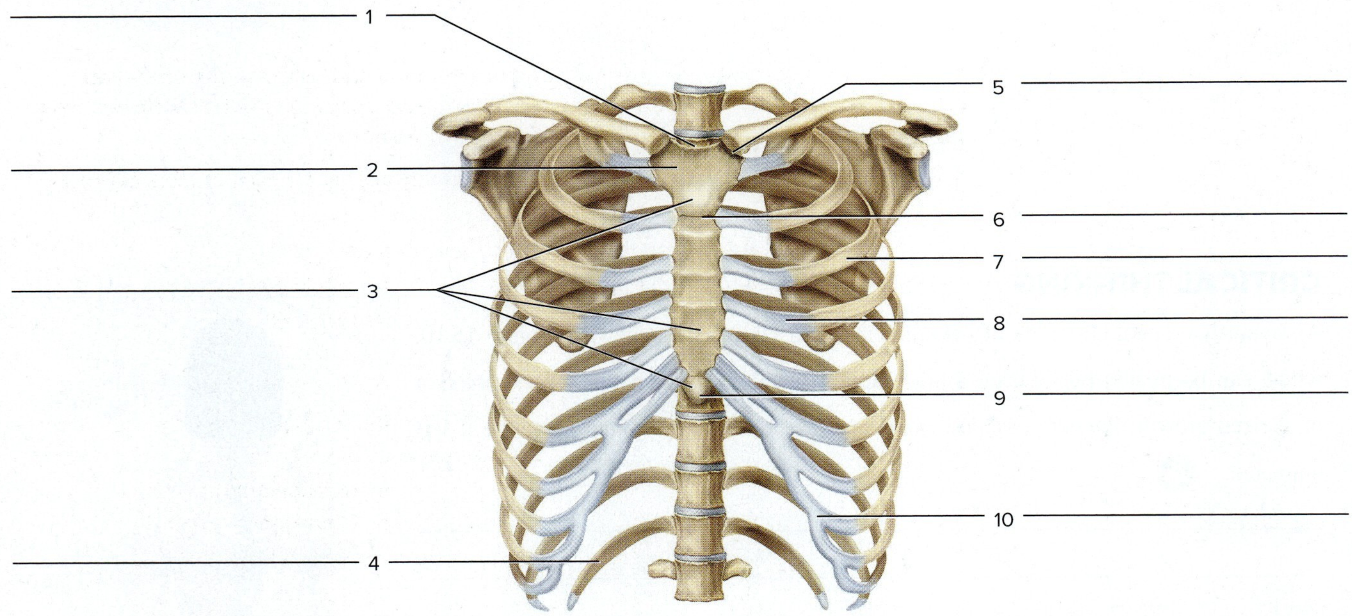

Terms:

Costal cartilage of false rib	Manubrium
Costal cartilage of true rib	Sternal angle
Clavicular notch	Sternum
Floating rib	True rib
Jugular notch	Xiphoid process

FIGURE 15.11 Label the bones and features of the thoracic cage, using the terms provided. A4 A5

LABORATORY EXERCISE 16

Pectoral Girdle and Upper Limb

Purpose of the Exercise

To examine the bones of the pectoral girdle and upper limb and to identify the major features of these bones.

MATERIALS NEEDED

Human skeleton, articulated
Human skeleton, disarticulated

Chapter Opener: ©101cats/Getty Images

Learning Outcomes APR

After completing this exercise, you should be able to:

O1 Locate and identify the bones of the pectoral girdle and their major features.

O2 Locate and identify the bones of the upper limb and their major features.

The O corresponds to the assessments A indicated in the Laboratory Assessment for this Exercise.

Pre-Lab

Carefully read the introductory material and examine the entire lab. Be familiar with the pectoral girdle and upper limb bones from lecture or the textbook. Answer the pre-lab questions.

Pre-Lab Questions Select the correct answer for each of the following questions:

1. The clavicle and the scapula form the
   a. pectoral girdle. **b.** pelvic girdle.
   c. upper limb. **d.** axial skeleton.
2. Anatomically, *arm* represents
   a. shoulder to fingers. **b.** shoulder to wrist.
   c. elbow to wrist. **d.** shoulder to elbow.
3. Which of the following is *not* part of the scapula?
   a. spine **b.** manubrium
   c. acromion **d.** supraspinous fossa
4. Which carpal is included in the proximal row?
   a. hamate **b.** capitate
   c. trapezium **d.** lunate
5. Which of the following is the most proximal part of the upper limb?
   a. styloid process of radius **b.** styloid process of ulna
   c. head of humerus **d.** medial epicondyle of humerus
6. Which of the following is the most distal feature of the humerus?
   a. anatomical neck **b.** deltoid tuberosity
   c. capitulum **d.** head
7. The lateral bone of the forearm is the radius.
   a. True **b.** False
8. The capitate is one of the eight carpals in a wrist.
   a. True **b.** False
9. The clavicle articulates with the sternum and the humerus.
   a. True **b.** False

The pectoral girdle (shoulder girdle) consists of two bones, a clavicle and a scapula, on each side of the body. The *pectoral girdle* represents an incomplete ring (girdle) of these bones. Posteriorly, the scapulae do not meet each other, but muscles extend from their medial borders to the vertebral column. Anteriorly, the two clavicles also do not meet as they articulate with the manubrium of the sternum. The pectoral girdle supports the upper limb and serves as the attachment point for various muscles that move the upper limb. This allows considerable flexibility of the shoulder. Relatively loose attachments of the pectoral girdle with the humerus allow a wide range of movements, but shoulder joint injuries are somewhat common.

Each upper limb includes a humerus in the arm, a radius and ulna in the forearm, and eight carpals, five metacarpals, and fourteen phalanges in the hand. (Anatomically, *arm* represents the region from shoulder to elbow, *forearm* is elbow to the wrist, and *hand* includes the wrist to the ends of digits.) These bones form the framework of the upper limb. They also function as parts of levers when the limbs are moved as muscles exert a force upon them.

PROCEDURE A: Pectoral Girdle

The clavicles and scapulae of the pectoral girdle provide attachments for muscles of the neck and trunk. The clavicle is not a straight bone, but rather it has two curves that make it slightly S-shaped. It can be palpated by moving the fingertips from the superior part of the sternum laterally. The clavicle serves as a brace bone to keep the upper limb to the side of the body. This is one of the bones of the body most commonly fractured because it is so close to the anterior surface and because the shoulder or upper limb is commonly used to break a fall. Fortunately, bones have tensile strength, and when a force is exerted upon the clavicle from the lateral side of the body, the bone will bend to some extent at both curves until the threshold is reached, causing a fracture to occur.

The scapula has many tendon attachment sites for muscles of the neck, trunk, and upper limb. Certain features of the posterior scapula can be easily palpated. Because the scapula does not connect directly to the axial skeleton and because it has so many muscle attachments, there is a great deal of flexibility for shoulder movements. Unfortunately, this much flexibility makes the humerus vulnerable to dislocation from the scapula. This type of injury can occur during gymnastics maneuvers or when suddenly pulling or swinging a child by an upper limb.

1. Examine figures 16.1 to 16.3 of the pectoral girdle. Locate the following features of the clavicle and the

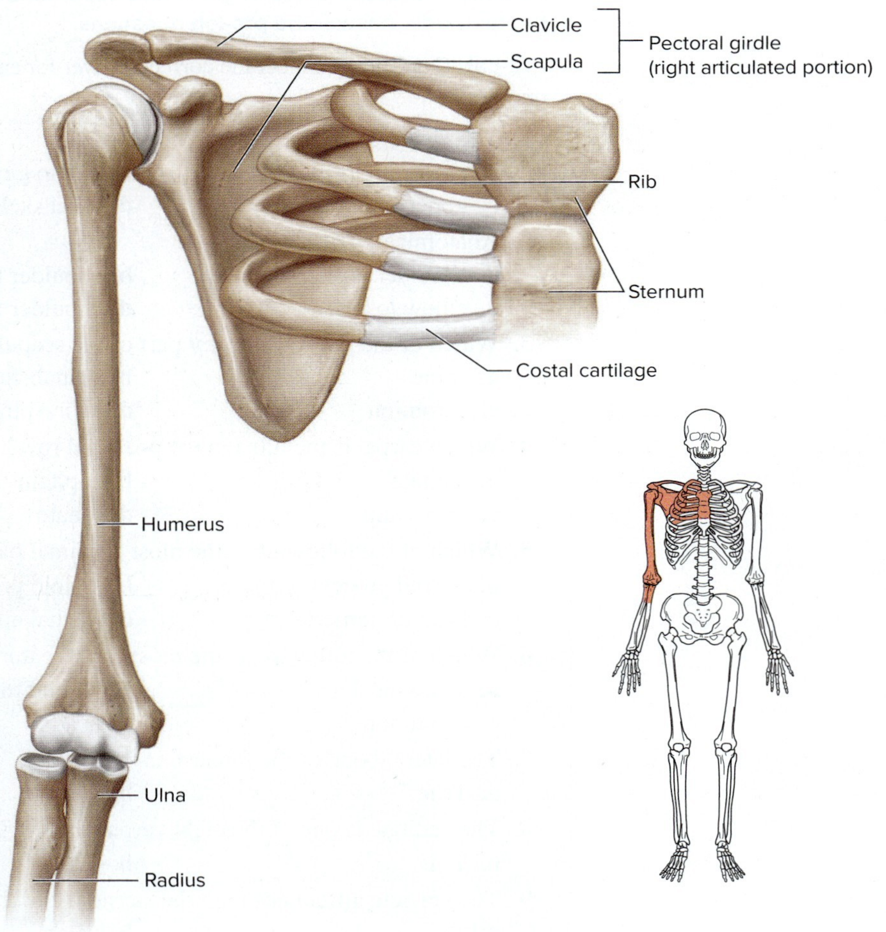

FIGURE 16.1 Bones and features of the right shoulder and upper limb (anterior view). APR

scapula. At the same time, locate as many of the corresponding surface bones and features of your own skeleton as possible.

clavicle

- sternal (medial) end—articulates with the manubrium of the sternum
- acromial (lateral) end—articulates with the acromion of the scapula

scapula

- spine—ridge
- acromion—lateral end of spine
- glenoid cavity—shallow socket; articulates with head of humerus
- coracoid process—beaklike projection
- borders
  - superior border
  - medial (vertebral) border
  - lateral (axillary) border

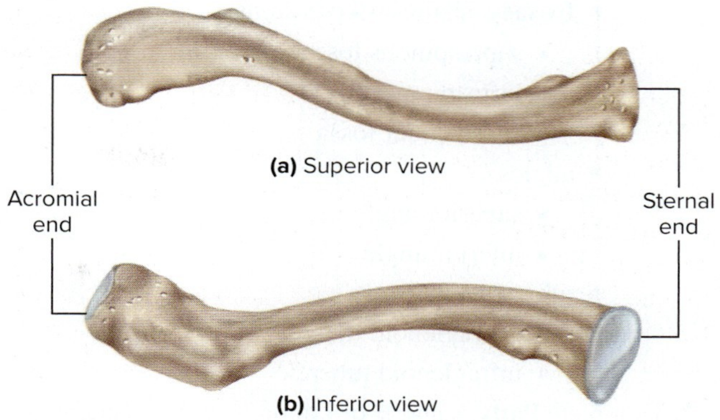

FIGURE 16.2 Right clavicle. The clavicle appears somewhat S-shaped when viewed from (a) the superior and (b) the inferior. APR

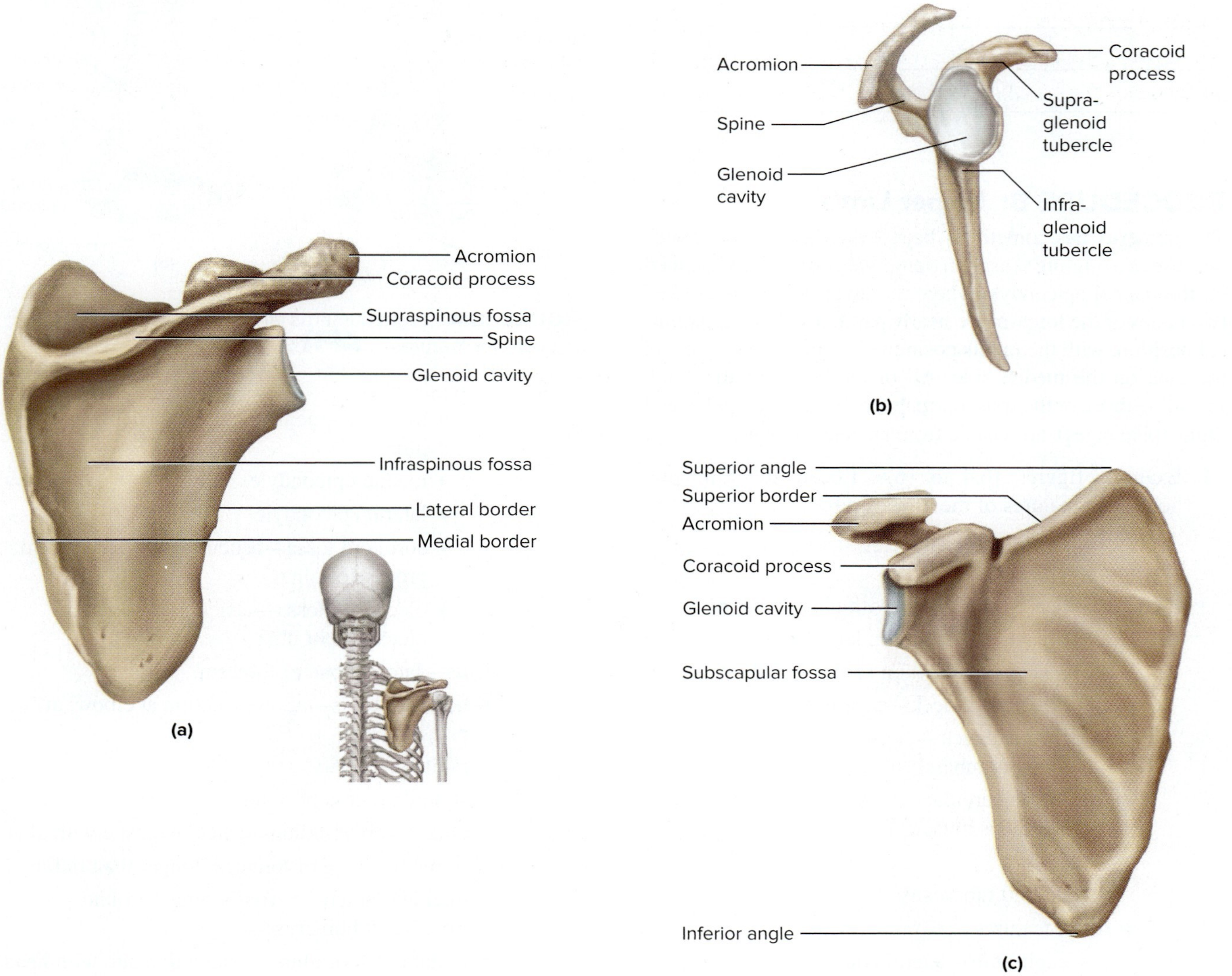

FIGURE 16.3 Right scapula (a) posterior surface, (b) lateral view, and (c) anterior surface. APR

- fossae—shallow depressions
  - supraspinous fossa
  - infraspinous fossa
  - subscapular fossa
- angles
  - superior angle
  - inferior angle
- tubercles
  - supraglenoid tubercle
  - infraglenoid tubercle

2. Complete Parts A and B of Laboratory Assessment 16.

CRITICAL THINKING ACTIVITY

Why is the clavicle a bone that can easily fracture?

____

____

____

____

____

____

PROCEDURE B: Upper Limb

The humerus has sometimes been called the "funny bone" because of a tingling sensation (temporary pain) if it is bumped on the medial epicondyle where the ulnar nerve passes. The two bones of the forearm are nearly parallel when in anatomical position, with the radius positioned on the lateral side and the ulna on the medial side. All of the bones of the hand, including those of the wrist (carpals), palm (metacarpals), and digits (phalanges), are visible from an anterior view.

1. Examine figures 16.4 to 16.6. Locate the following bones and features of the upper limb:

humerus

- proximal features
  - head—articulates with glenoid cavity
  - greater tubercle—on lateral side
  - lesser tubercle—on anterior side
  - anatomical neck—tapered region near head
  - surgical neck—common fracture site; distal to tubercles
  - intertubercular sulcus—furrow for tendon of biceps muscle
- shaft
  - deltoid tuberosity
- distal features
  - capitulum—lateral condyle; articulates with radius

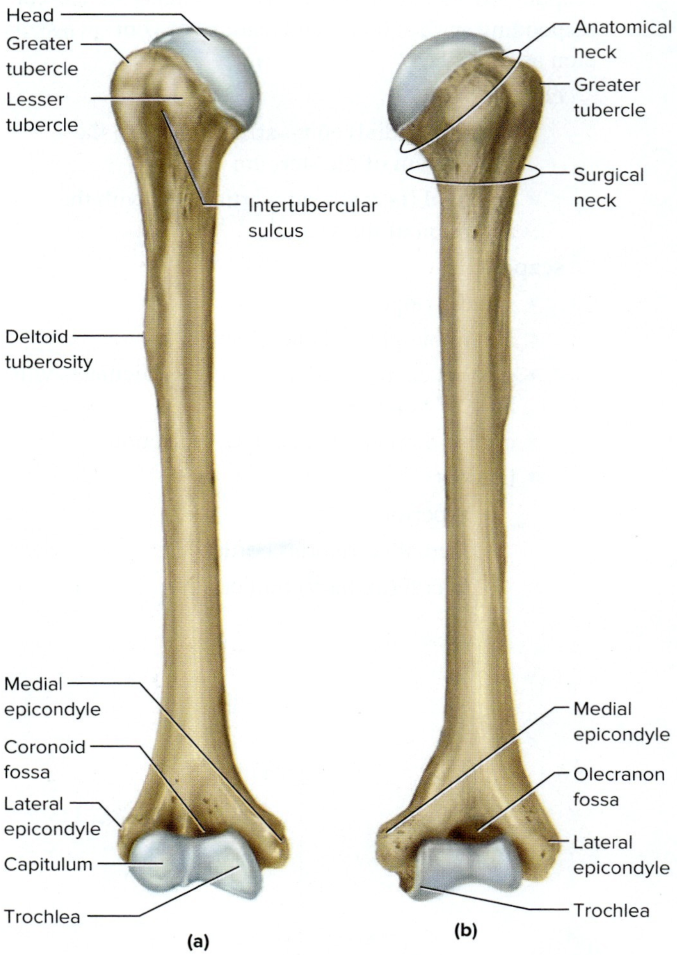

FIGURE 16.4 Right humerus (a) anterior features and (b) posterior features. APR

  - trochlea—medial condyle; articulates with ulna
  - medial epicondyle
  - lateral epicondyle
  - coronoid fossa—articulates with coronoid process of ulna
  - olecranon fossa—articulates with olecranon of ulna

radius—lateral bone of forearm

- head of radius—allows rotation at elbow; at proximal end
- radial tuberosity
- styloid process of radius
- ulnar notch of radius—articulation site with ulna

ulna—medial bone of forearm; longer than radius

- trochlear notch—partially wraps around trochlea of humerus
- radial notch of ulna—articulation site with head of radius

- olecranon (olecranon process)—bony point of posterior elbow
- coronoid process
- styloid process of ulna
- head of ulna—at distal end

carpal bones—positioned in two irregular rows
- proximal row (listed lateral to medial)
  - scaphoid
  - lunate
  - triquetrum
  - pisiform
- distal row (listed medial to lateral)
  - hamate
  - capitate
  - trapezoid
  - trapezium

The following mnemonic device will help you learn the eight carpals:

So **L**ong **T**op **P**art
Here **C**omes **T**he **T**humb

The first letter of each word corresponds to the first letter of a carpal. This device arranges the carpals in order for the proximal, transverse row of four bones from lateral to medial, followed by the distal, transverse row from medial to lateral, which ends nearest the thumb. This arrangement assumes the hand is in anatomical position.

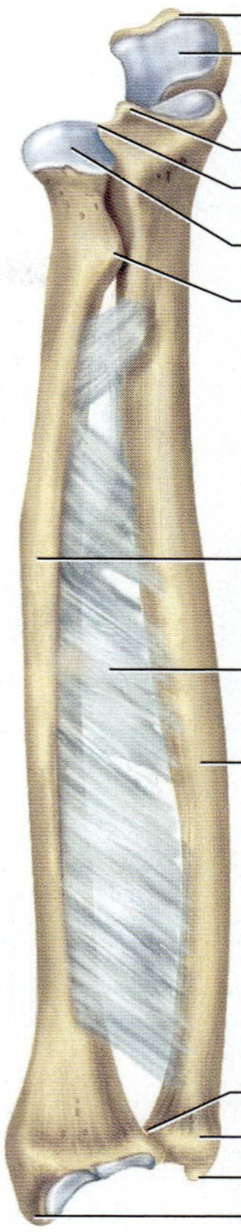

FIGURE 16.5 Anterior features of the right radius and ulna. APR

CRITICAL THINKING ACTIVITY

Create your own mnemonic for the eight carpal bones.

____

____

____

metacarpals (I–V)

phalanges—located in digits (fingers)
- proximal phalanx
- middle phalanx—not present in first digit
- distal phalanx

2. Complete Parts C, D, and E of the laboratory assessment.

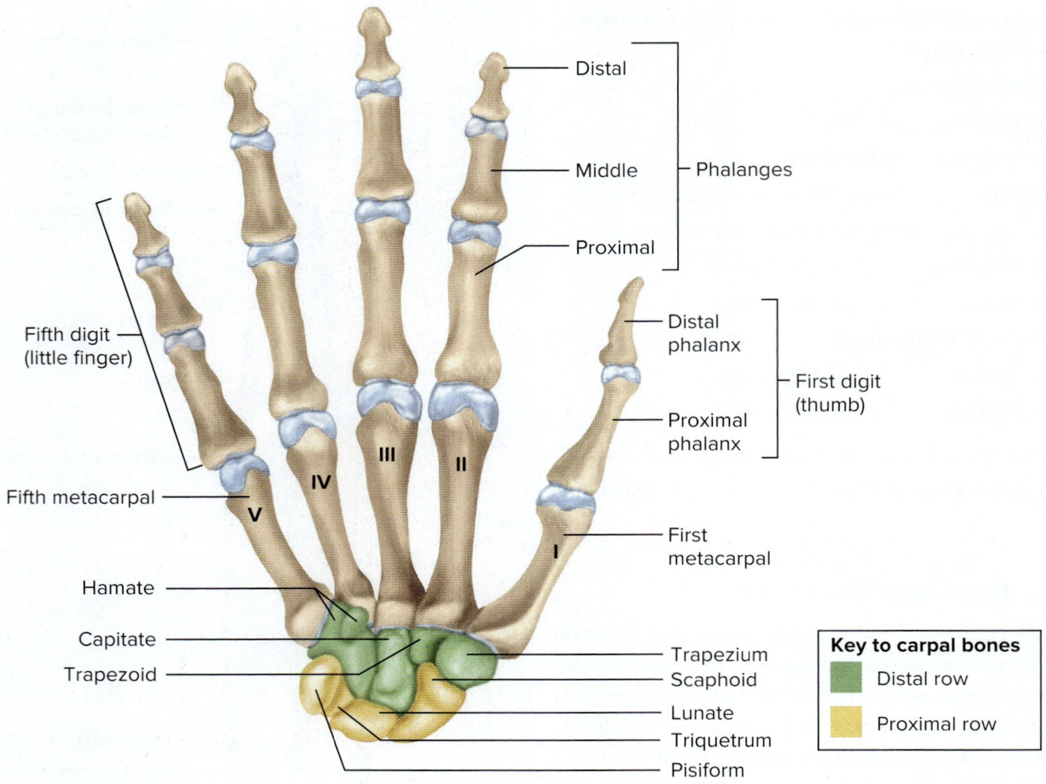

FIGURE 16.6 Anterior (palmar) view of the bones of the right hand. The proximal row of carpals are colored yellow and the distal row of carpals are colored green. APR

LABORATORY
ASSESSMENT 16

Name ________________________________

Date ________________________________

Section ________________________________

The A corresponds to the indicated Learning Outcome(s) O found at the beginning of the Laboratory Exercise.

Pectoral Girdle and Upper Limb

PART A: Assessments

Complete the following statements:

1. The pectoral girdle is an incomplete ring because it is open in the back between the ______________________. A1
2. The medial end of a clavicle articulates with the ______________________ of the sternum. A1
3. The lateral end of a clavicle articulates with the ______________________ process of the scapula. A1
4. The ______________________ is a bone that serves as a brace between the sternum and the scapula. A1
5. The ______________________ divides the posterior side of the scapula into unequal portions. A1
6. The lateral tip of the shoulder is the ______________________ of the scapula. A1
7. Near the lateral end of the scapula, the ______________________ process of the scapula curves anteriorly and inferiorly from the clavicle. A1
8. The glenoid cavity of the scapula articulates with the ______________________ of the humerus. A1

PART B: Assessments

Label the structures indicated in figure 16.7.

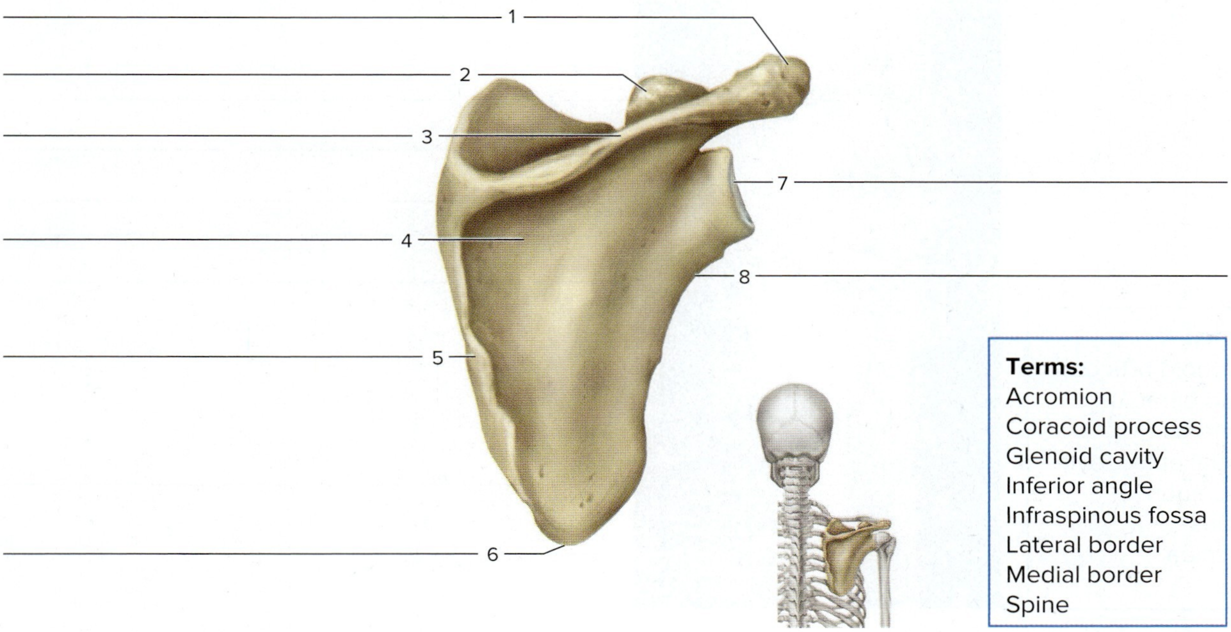

Terms:
Acromion
Coracoid process
Glenoid cavity
Inferior angle
Infraspinous fossa
Lateral border
Medial border
Spine

FIGURE 16.7 Label the posterior surface of the right scapula, using the terms provided. APR A1

PART C: Assessments

Match the bones in column A with the bones and features in column B. Place the letter of your choice in the space provided. (Some answers may be used more than once.) A2

Column A

a. Carpals
b. Humerus
c. Metacarpals
d. Phalanges
e. Radius
f. Ulna

Column B

______ **1.** Capitate
______ **2.** Coronoid fossa
______ **3.** Deltoid tuberosity
______ **4.** Greater tubercle
______ **5.** Five palmar bones
______ **6.** Fourteen bones in digits
______ **7.** Intertubercular sulcus
______ **8.** Lunate
______ **9.** Olecranon fossa
______ **10.** Radial tuberosity
______ **11.** Trapezium
______ **12.** Trochlear notch

PART D: Assessments

Identify the bones and features indicated in the radiographs of figures 16.8 to 16.10.

Terms:
Coronoid process
Head of radius
Humerus
Medial epicondyle
Olecranon
Radius
Trochlea
Ulna

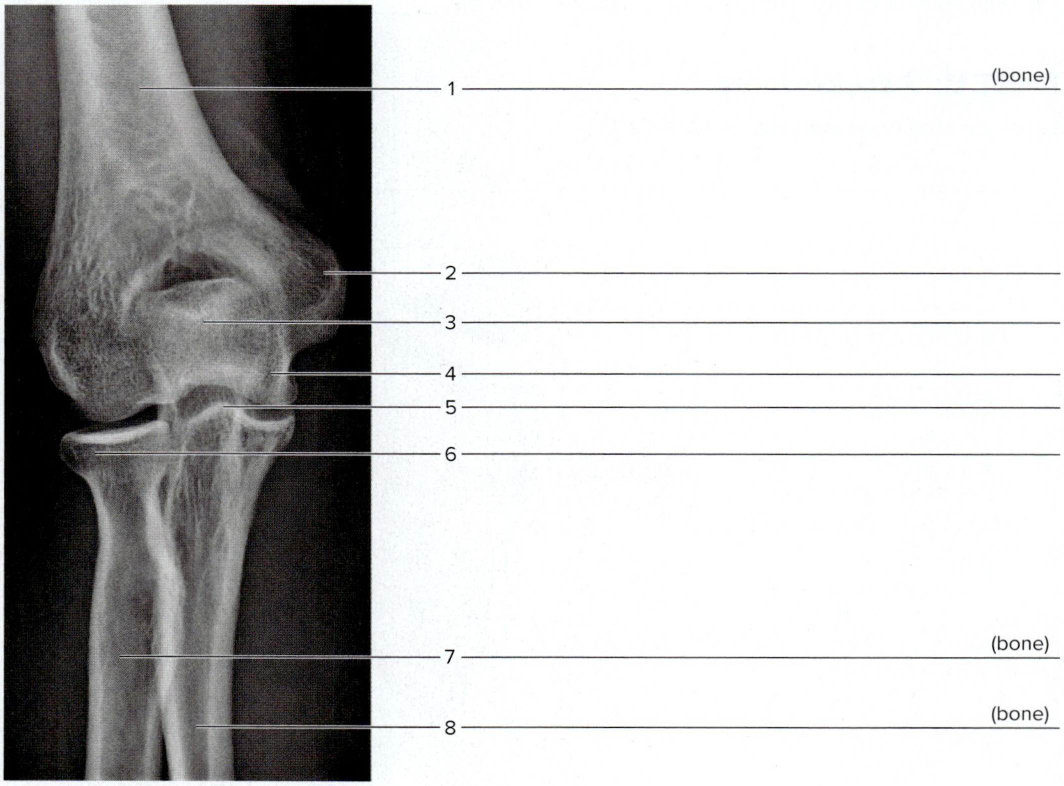

FIGURE 16.8 Identify the bones and features indicated on this radiograph of the right elbow (anterior view), using the terms provided. Dale Butler APR A2

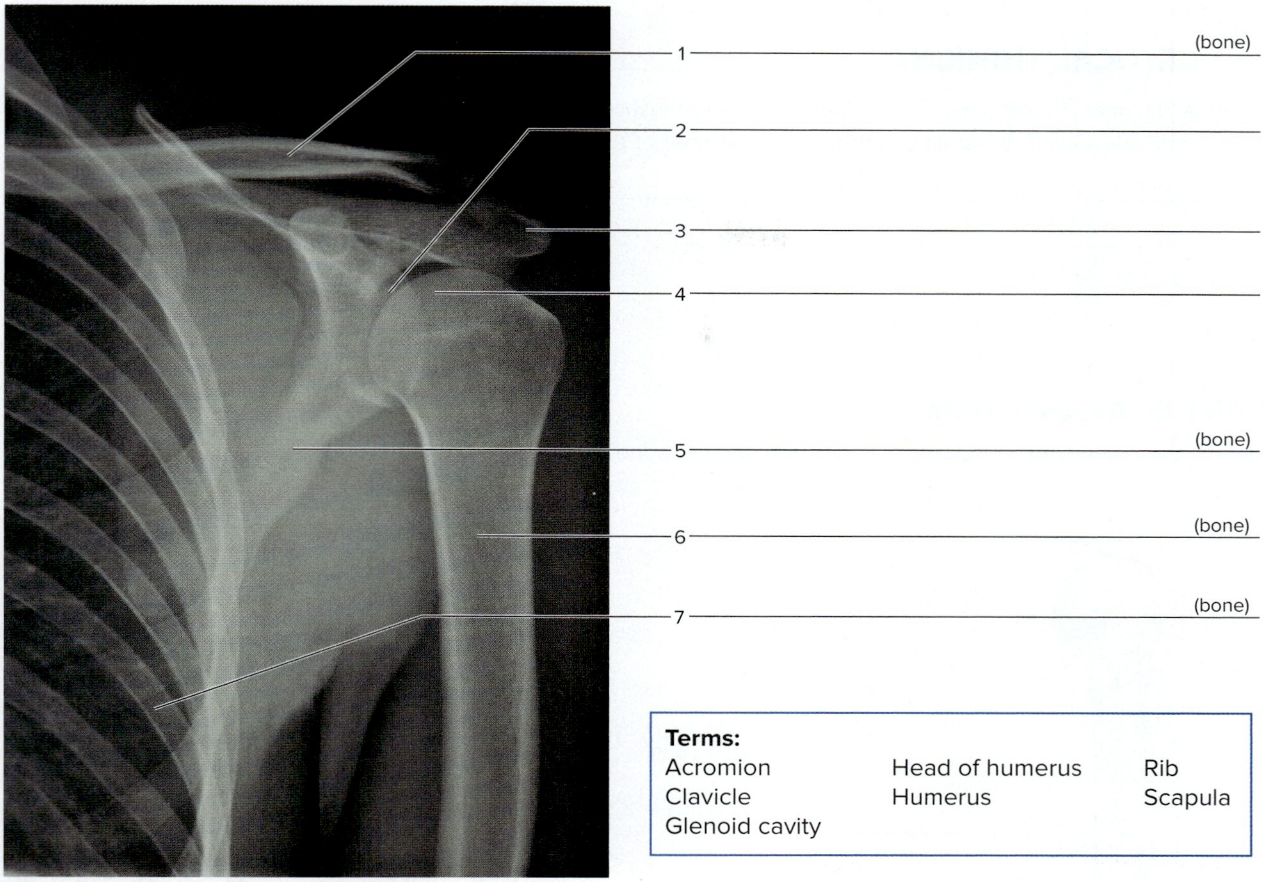

Terms:
Acromion
Clavicle
Glenoid cavity
Head of humerus
Humerus
Rib
Scapula

FIGURE 16.9 Identify the bones and features indicated on this radiograph of the anterior view of the left shoulder, using the terms provided. sanyanwuji/iStock/Getty Images APR A1 A2

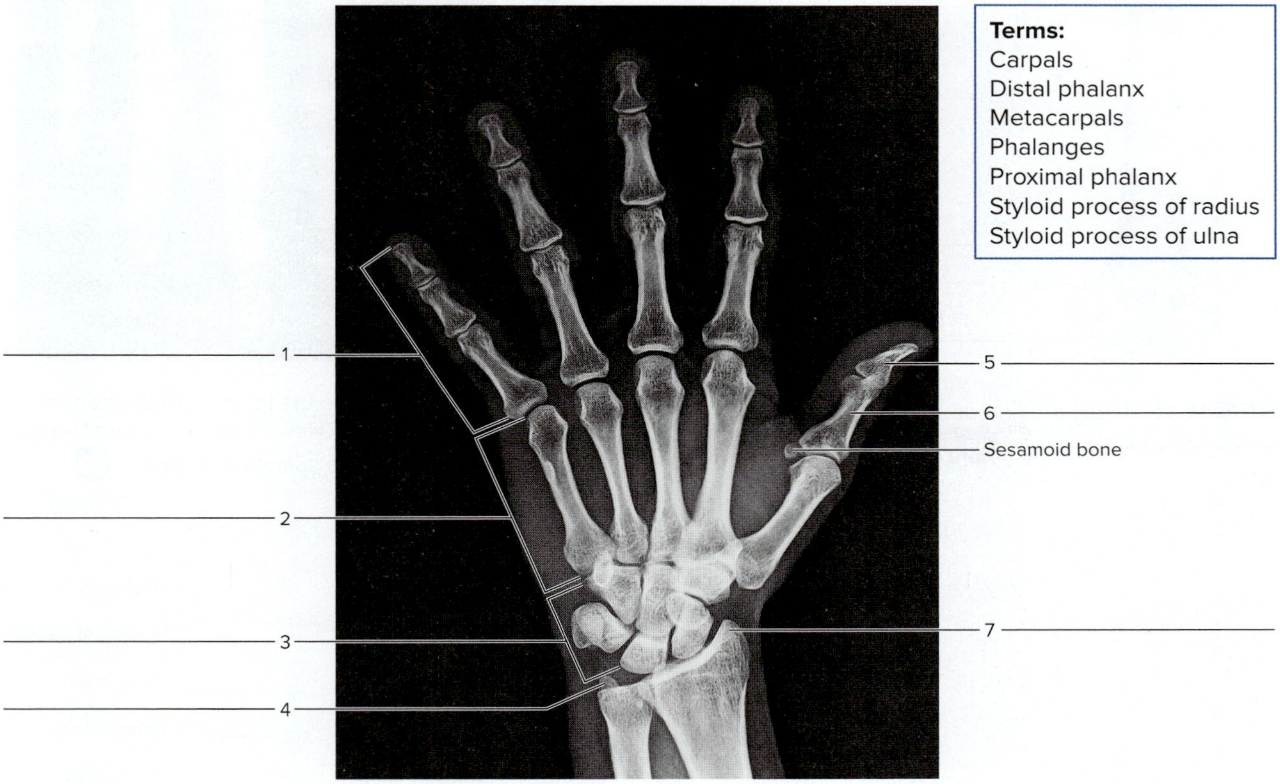

Terms:
Carpals
Distal phalanx
Metacarpals
Phalanges
Proximal phalanx
Styloid process of radius
Styloid process of ulna

FIGURE 16.10 Identify the bones indicated on this radiograph of the right hand (anterior view), using the terms provided. Dale Butler APR A2

CRITICAL THINKING ASSESSMENT—PART D

A six years old showing signs of puberty was prescribed by an endocrinologist to have a bone X ray taken of the child's left hand. What do you think is the purpose of this radiograph? A2

____

____

PART E: Assessments

Identify the features of a humerus in figure 16.11 and the bones of the hand in figure 16.12.

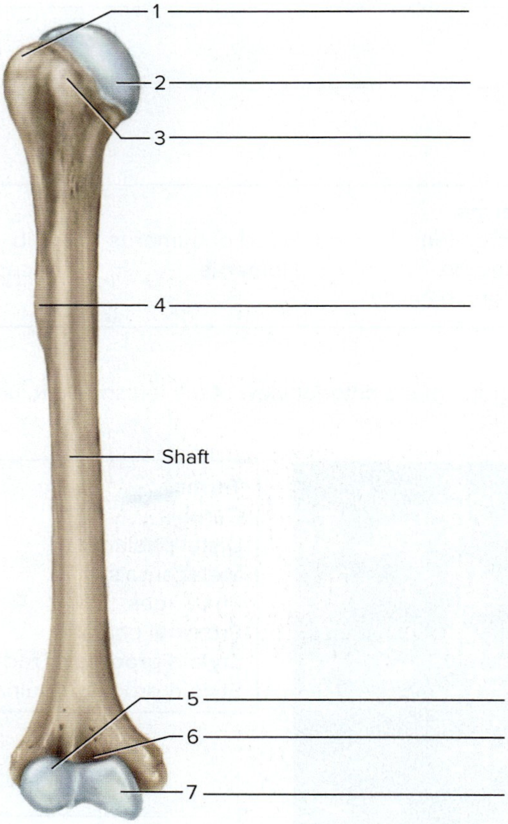

FIGURE 16.11 Label the anterior features of a right humerus. ©J & J Photography A2

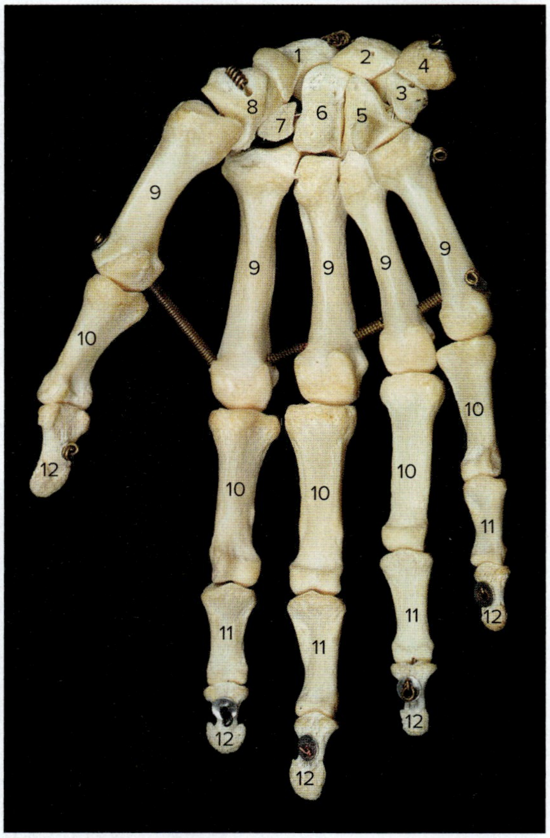

FIGURE 16.12 Complete the labeling of the bones numbered on this anterior view of the right hand by placing the correct numbers in the spaces provided. A2

______	Capitate	4	Pisiform
______	Distal phalanges	______	Proximal phalanges
______	Hamate	______	Scaphoid
______	Lunate	______	Trapezium
______	Metacarpals	______	Trapezoid
______	Middle phalanges	3	Triquetrum

LABORATORY
EXERCISE 17

Pelvic Girdle and Lower Limb

Purpose of the Exercise

To examine the bones of the pelvic girdle and lower limb and to identify the major features of these bones.

MATERIALS NEEDED

Human skeleton, articulated
Human skeleton, disarticulated
Male and female pelves

Chapter Opener: ©101cats/Getty Images

Learning Outcomes APR

After completing this exercise, you should be able to:

O1 Locate and identify the bones of the pelvic girdle and their major features.

O2 Differentiate between a male and female pelvis.

O3 Locate and identify the bones of the lower limb and their major features.

The O corresponds to the assessments A indicated in the Laboratory Assessment for this Exercise.

Pre-Lab

Carefully read the introductory material and examine the entire lab. Be familiar with the pelvic girdle and the lower limb bones from lecture or the textbook. Answer the pre-lab questions.

Pre-Lab Questions Select the correct answer for each of the following questions:

1. The two hip (coxal) bones articulate anteriorly at the
a. acetabulum. **b.** pubic arch.
c. sacroiliac joint. **d.** pubic symphysis.

2. Anatomically, *leg* refers to
a. the lower limb. **b.** hip to knee.
c. knee to ankle. **d.** hip to ankle.

3. The _____ is the largest portion of the hip (coxal) bone.
a. acetabulum **b.** ilium
c. ischium **d.** pubis

4. The _____ is the lateral bone in the leg.
a. tibia **b.** fibula
c. femur **d.** patella

5. Which of the following bones is *not* a tarsal bone?
a. metatarsal **b.** talus
c. calcaneus **d.** cuboid

6. The two bones that articulate to form the sacroiliac (SI) joint are the sacrum and the
a. femur. **b.** ilium.
c. ischium. **d.** pubis.

7. The ilium, ischium, and pubis are separate bones in a young child.
a. True **b.** False

8. Ischial spines, ischial tuberosities, and iliac crests are closer together in a pelvis of a female than in a pelvis of a male.
a. True **b.** False

9. Each digit of a foot has three phalanges.
a. True **b.** False

The pelvic girdle includes two hip (coxal) bones and the sacrum. Commonly called the *ossa coxae,* the hip bones articulate with each other anteriorly at the pubic symphysis. Posteriorly, the ilium of each hip bone articulates with the sacrum at a sacroiliac joint. The pelvic girdle provides support for the trunk of the body and provides attachments for the lower limbs. The bowl-shaped *pelvis* includes the pelvic girdle and all of its corresponding ligaments and muscles, providing support and protection of the viscera in the pelvic region of the abdominopelvic cavity. The *false (greater) pelvis* is located between the flare of the hip bones, and is separated from the *true (lesser) pelvis* below (inferiorly) by the pelvic brim. In the upper margin of the true pelvis is the *pelvic inlet*, a space enclosed by the pelvic brim. The *pelvic outlet* is a somewhat diamond-shaped space located in the lower margin of the true pelvis, with boundaries that include the coccyx, the ischial tuberosities, and the pubic symphysis. In females, the pelvic outlet is clinically important, because it must be large enough to successfully accommodate the fetal head during a vaginal delivery. Each acetabulum of a hip bone articulates with the head of the femur of a lower limb. The hip joint structures provide a more stable joint compared to a shoulder joint.

The bones of the lower limb form the framework of the thigh, leg, and foot. (Anatomically, *thigh* represents the region from hip to knee, *leg* is from knee to ankle, and *foot* includes the ankle to the end of the toes.) Each limb includes a femur in the thigh, a patella in the knee, a tibia and fibula in the leg, and seven tarsals, five metatarsals, and fourteen phalanges in the foot. These bones and the muscles attached to them provide weight-bearing support and locomotion and thus are considerably larger and possess more stable joints than those of an upper limb.

PROCEDURE A: Pelvic Girdle

The pelvic girdle supports the majority of the weight of the head, neck, and trunk. Each hip (coxal) bone originates in three separate ossification areas known as the ilium, ischium, and pubis. These are separate bones in a child but fuse into an individual hip bone. All three parts of the hip bone fuse within the acetabulum, and the ischium and pubis also fuse along the inferior portion of the obturator foramen. The acetabulum is a well-formed deep socket that articulates with the head of the femur. The obturator foramen is the largest foramen in the skeleton; it serves as a passageway for blood vessels and nerves between the pelvic cavity and the thigh.

Several of the bone features (bone markings) of the pelvis can be palpated because they are located near the surface of the body. The median sacral crest is near the middle of the posterior surface of the pelvis, somewhat superior to the coccyx. The ilium is the largest portion of the hip bone, and its iliac crest can be palpated along the anterior and lateral portions. By following the anterior portion of the iliac crest, the anterior superior iliac spine can be felt very close to the surface.

1. Examine figures 17.1 and 17.2.
2. Observe the bones of the pelvic girdle and locate the following:

   hip bone (coxal bone; os coxa; pelvic bone; innominate bone)

   - ilium
     - iliac crest
     - anterior superior iliac spine
     - anterior inferior iliac spine
     - posterior superior iliac spine
     - posterior inferior iliac spine
     - greater sciatic notch—passageway for sciatic nerve to lower limb
     - iliac fossa
     - auricular surface—portion of ilium that articulates with sacrum to form sacroiliac joint
   - ischium
     - ischial tuberosity—supports weight of body when seated
     - ischial spine
     - ischial ramus
     - lesser sciatic notch
   - pubis
     - pubic symphysis—cartilaginous joint between pubic bones
     - pubic tubercle
     - superior pubic ramus
     - inferior pubic ramus
     - pubic arch—between pubic bones of pelvis
   - acetabulum—formed by portions of ilium, ischium, and pubis
   - obturator foramen—formed by portions of ischium and pubis
   - pelvic inlet—space bound by pelvic brim
   - pelvic outlet—space inferior to pelvic inlet
3. Observe the male pelvis and the female pelvis. Use table 17.1 as a guide as comparisons are made between the pelves of males and females.
4. Complete Part A of Laboratory Assessment 17.

PROCEDURE B: Lower Limb

The femur of the lower limb is the longest and strongest bone of the skeleton. The neck of the femur has somewhat of a lateral angle and is the weakest part of the bone and a common fracture site, especially if the person has some degree of osteoporosis. This fracture site is usually called a "broken hip"; however, it is actually a broken femur in the region of the hip joint. Often this fracture is attributed to a fall, when

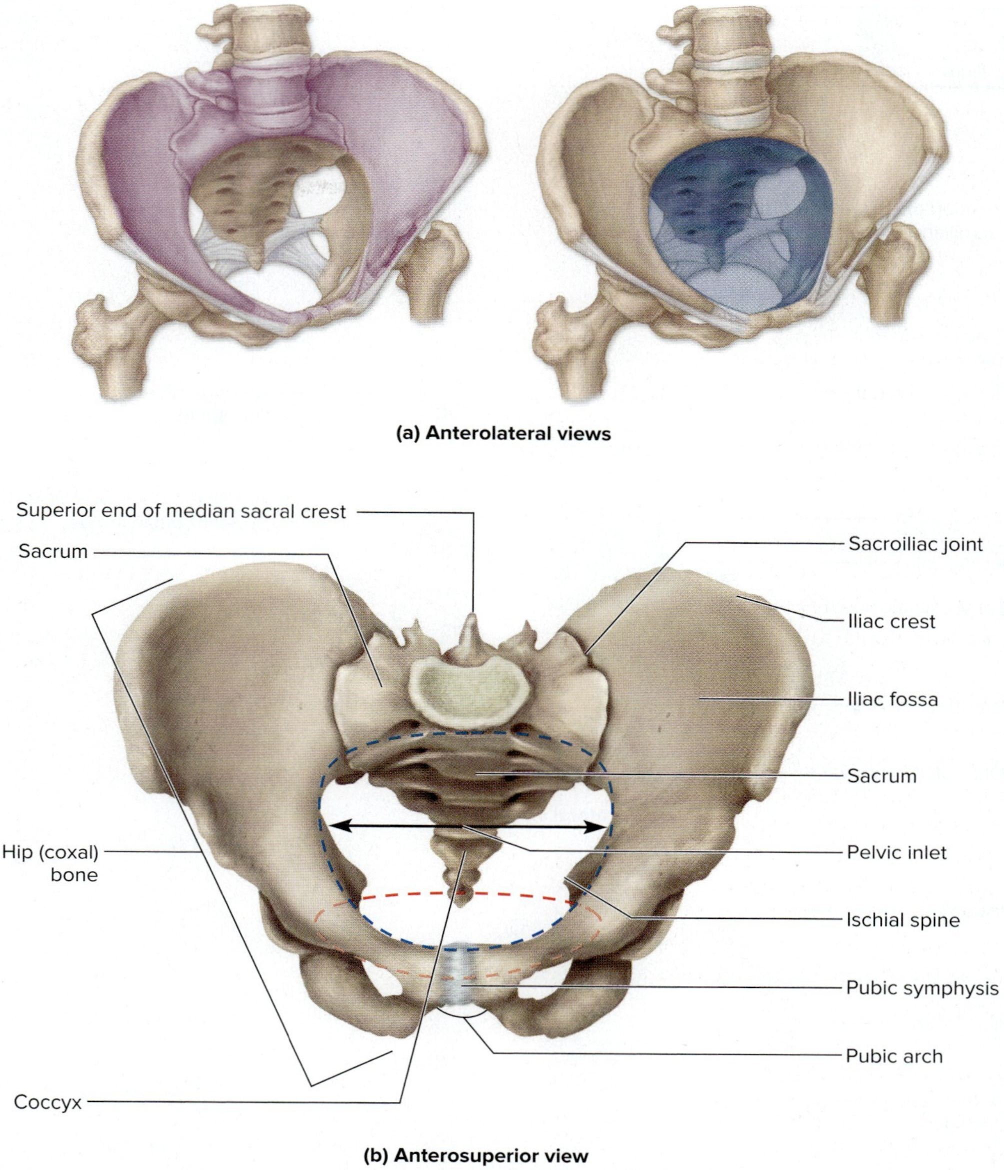

FIGURE 17.1 Bones of the pelvis. (a) Anterolateral view showing the false pelvis (on the left) and true pelvis (on the right); (b) anterosuperior view showing the pelvic brim (blue dashed lines) that encloses the pelvic inlet and pelvic outlet boundary (red dashed lines) of the true pelvis. APR

many times the fracture occurs first, followed by the fall. The greater trochanter can be palpated along the proximal and lateral region of the thigh about one hand length below the iliac crest. The medial and lateral epicondyles can be palpated near the knee.

Large muscles are positioned on the anterior and posterior surfaces as well as the lateral and medial surfaces of the femur and the hip joint region. The thigh can be pulled in any direction depending upon which muscle group is contracting at that time. The large anterior muscles of the thigh (quadriceps femoris) possess a common patellar tendon with an enclosed sesamoid bone, the patella. The muscle attachment continues distally to the tibial tuberosity.

1. Examine figures 17.3 to 17.5.
2. Observe the bones of the lower limb and locate each of the following:

    femur

    - proximal features
        - head
        - fovea capitis
        - neck
        - greater trochanter
        - lesser trochanter
    - shaft
        - gluteal tuberosity
        - linea aspera

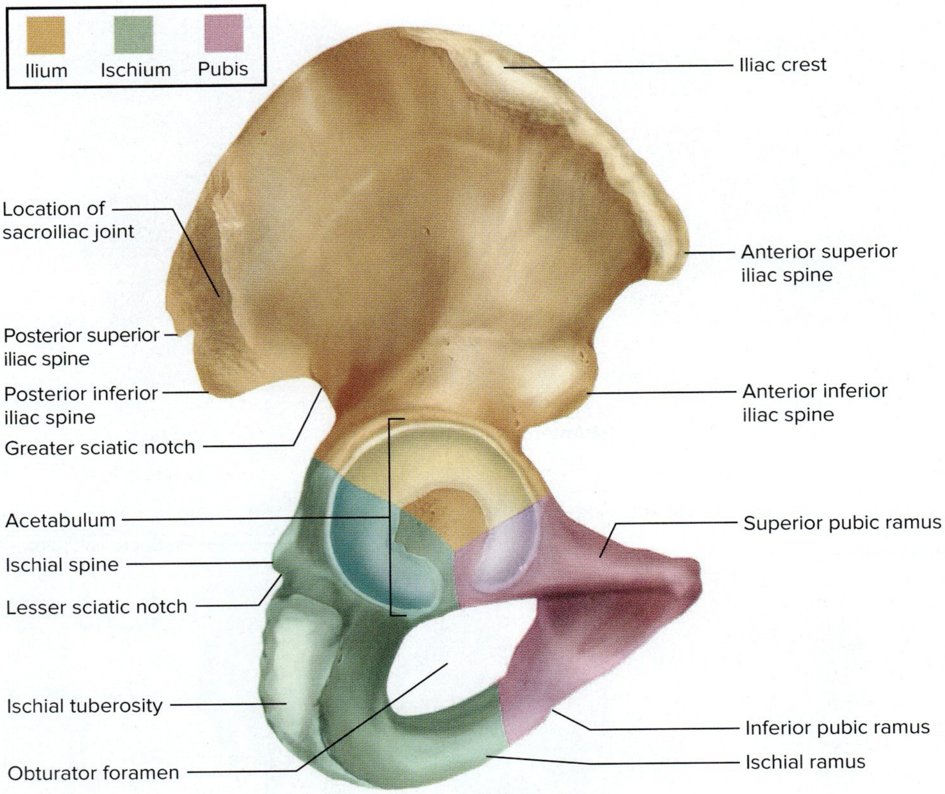

(a) Lateral view

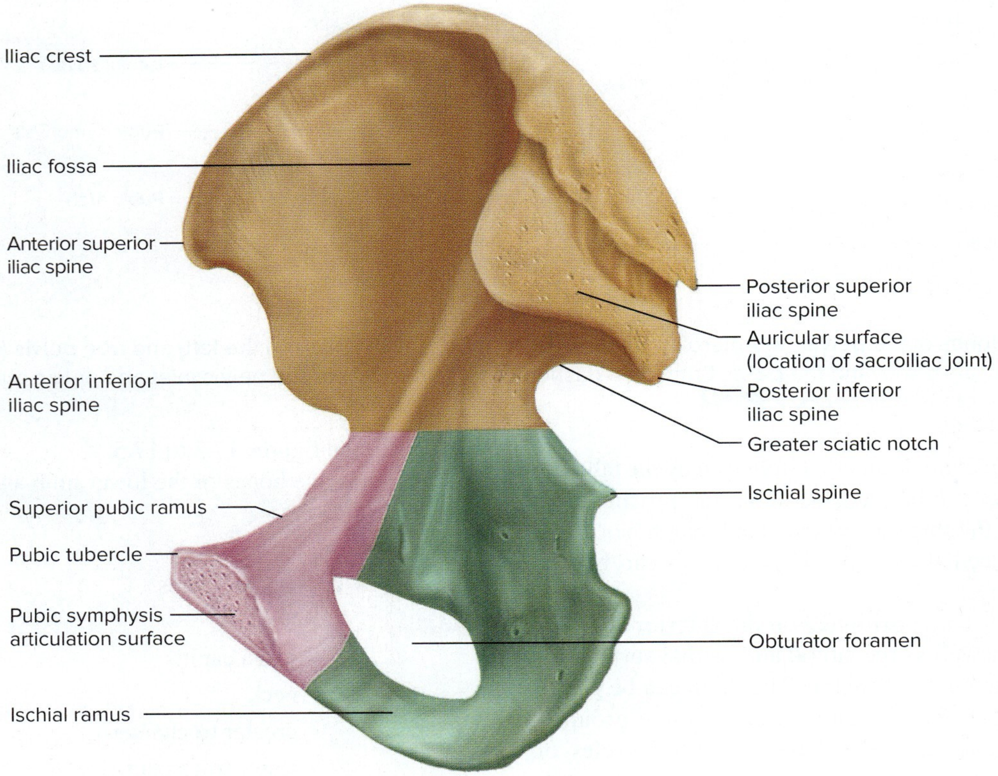

(b) Medial view

FIGURE 17.2 Right hip (coxal) bone: (a) lateral view; (b) medial view. The three colors used enable the viewing of fusion locations of the ilium, ischium, and pubis of the adult skeleton. APR

Table 17.1 Differences Between Male and Female Pelves

Structure of Comparison	Male Pelvis APR	Female Pelvis APR
General structure	Heavier, thicker bones and processes	Lighter, thinner bones and processes
Sacrum	Narrower and longer	Wider and shorter
Coccyx	Less movable	More movable
Pelvic outlet	Smaller	Larger
Greater sciatic notch	Narrower	Wider
Obturator foramen	Round	Triangular to oval
Acetabula	Larger; closer together	Smaller; farther apart
Pubic arch	Usually 90° or less; more V-shaped	Usually greater than 90°
Ischial spines	Longer; closer together	Shorter; farther apart
Ischial tuberosities	Rougher; closer together	Smoother; farther apart
Iliac crests	Less flared; closer together	More flared; farther apart

- distal features
  - lateral epicondyle
  - medial epicondyle
  - lateral condyle
  - medial condyle
  - patellar surface
- patella
  - apex

tibia—medial bone of leg
- medial condyle
- lateral condyle
- tibial tuberosity
- anterior border (crest; margin)
- medial malleolus
- intercondylar eminence

fibula—lateral bone of leg
- head
- lateral malleolus

tarsal bones
- talus
- calcaneus
- navicular
- cuboid
- lateral cuneiform
- intermediate (middle) cuneiform
- medial cuneiform

metatarsal bones

phalanges
- proximal phalanx
- middle phalanx—absent in first digit
- distal phalanx

3. Complete Parts B, C, and D of the laboratory assessment.

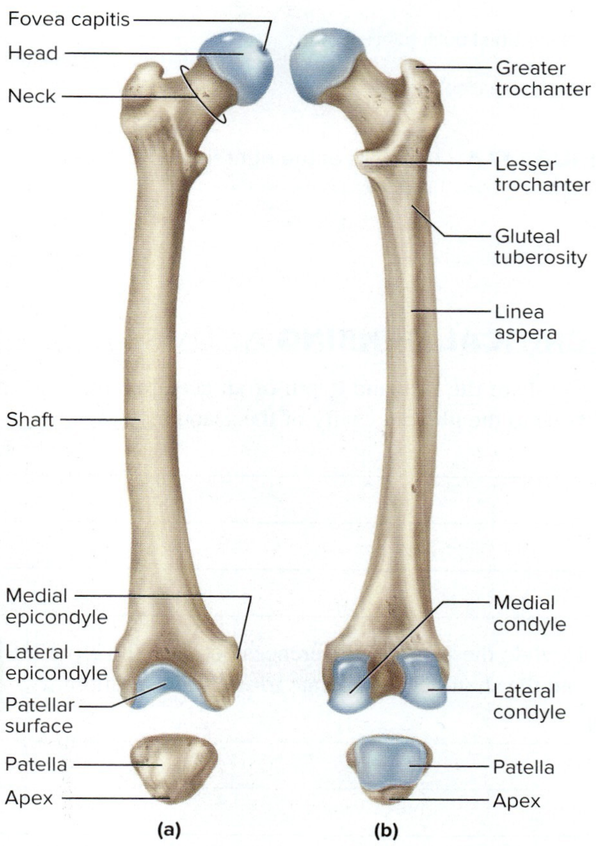

FIGURE 17.3 The features of (a) the anterior surface and (b) the posterior surface of the right femur. The anterior and posterior views of the patella are included in the figure. APR

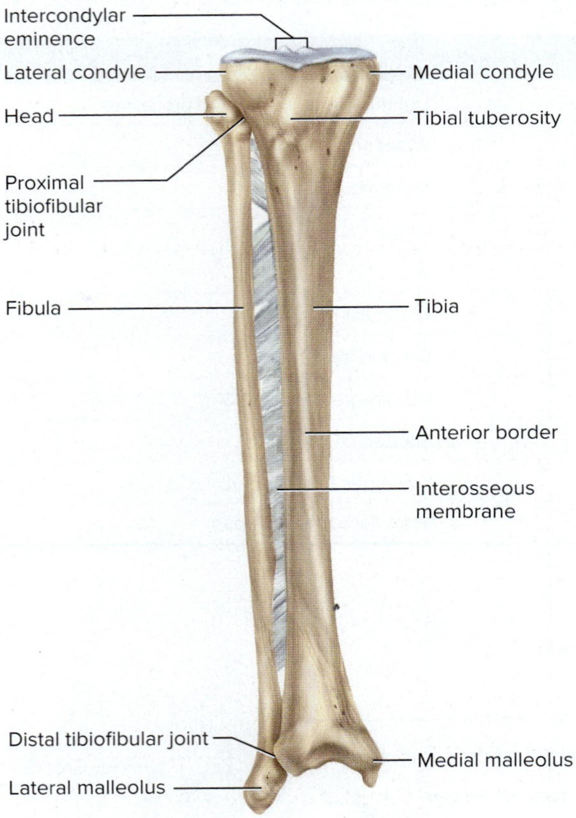

FIGURE 17.4 Features of the right tibia and fibula (anterior view). APR

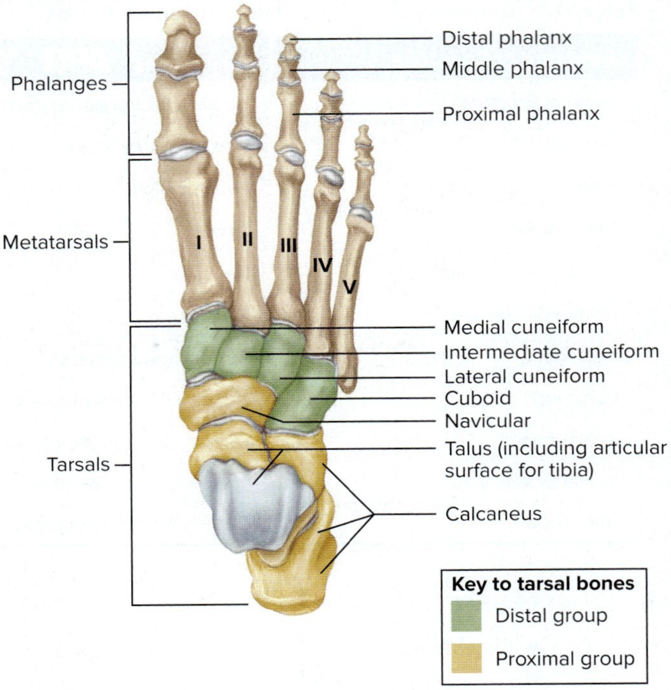

FIGURE 17.5 Superior surface view of right foot. The proximal group of tarsals are colored yellow and the distal group of tarsals are colored green. APR

CRITICAL THINKING ACTIVITY

Compare the size and depth of an acetabulum of a hip bone to the glenoid cavity of the scapula. ____________

____

____

____

____

How do these socket differences between the hip joint and the shoulder joint relate to strength and mobility?

____

____

____

____

____

CRITICAL THINKING ACTIVITY

As a review of the entire skeleton, use the disarticulated skeleton and arrange all of the bones in relative position to recreate their normal positions. Work with a partner or in a small group and place the bones on the surface of a laboratory table. Check your results with the articulated skeletons.

Name ______________________

Date ______________________

Section ______________________

The (A) corresponds to the indicated Learning Outcome(s) (O) found at the beginning of the Laboratory Exercise.

LABORATORY ASSESSMENT 17

Pelvic Girdle and Lower Limb

PART A: Assessments

Complete the following statements:

1. The pelvic girdle consists of two ______________________. A1
2. The head of the femur articulates with the ______________________ of the hip bone. A1
3. The ______________________ is the largest portion of the hip bone. A1
4. The distance between the ______________________ represents the shortest diameter of the pelvic outlet. A1
5. The pubic bones come together anteriorly to form a cartilaginous joint called the ______________________. A1
6. The ______________________ is the superior margin of the ilium that causes the prominence of the hip. A1
7. When a person sits, the ______________________ of the ischium supports the weight of the body. A1
8. The angle formed by the pubic bones below the pubic symphysis is called the ______________________. A1
9. The ______________________ is the largest foramen in the skeleton. A1
10. The ilium joins the sacrum at the ______________________ joint. A1

CRITICAL THINKING ASSESSMENT — PART A

Examine the male and female pelves shown below (fig. 17.6). Look for major differences between them. Note especially the flare of the iliac bones, the angle of the pubic arch, the distance between the ischial spines and ischial tuberosities, and the curve and width of the sacrum. In what ways are the differences you observed related to the function of the female pelvis as a birth canal? A2

____

____

____

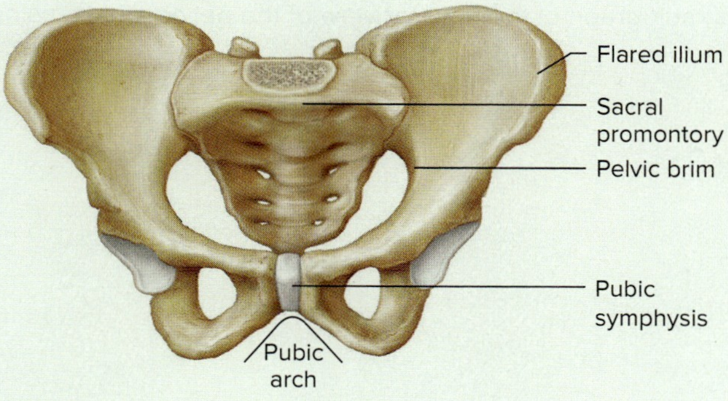

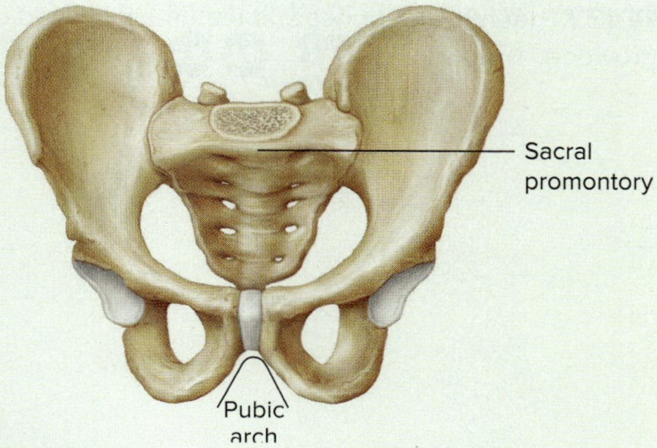

FIGURE 17.6 The features of the (a) female pelvis and (b) male pelvis.

PART B: Assessments

Match the bones in column A with the features in column B. Place the letter of your choice in the space provided. A3

Column A

a. Femur
b. Fibula
c. Metatarsals
d. Patella
e. Phalanges
f. Tarsals
g. Tibia

Column B

_____ 1. Middle phalanx
_____ 2. Lesser trochanter
_____ 3. Medial malleolus
_____ 4. Fovea capitis
_____ 5. Calcaneus
_____ 6. Lateral cuneiform
_____ 7. Tibial tuberosity
_____ 8. Talus
_____ 9. Linea aspera
_____ 10. Lateral malleolus
_____ 11. Sesamoid bone
_____ 12. Five bones that form the instep

PART C: Assessments

Identify the bones and features indicated in the radiographs of figures 17.7 to 17.9.

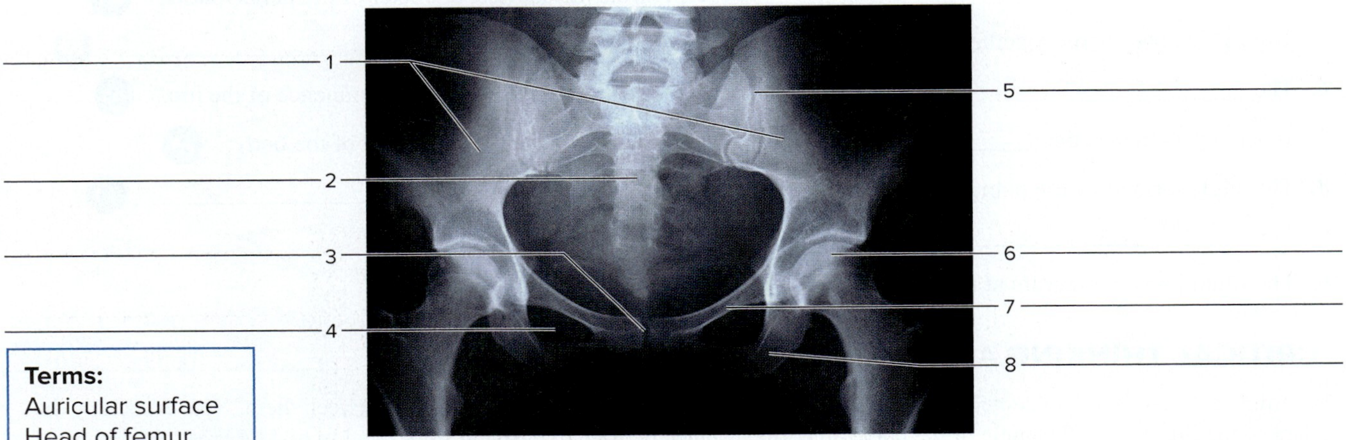

Terms:
Auricular surface
Head of femur
Ilium
Ischium
Obturator foramen
Pubic symphysis
Pubis
Sacrum

FIGURE 17.7 Identify the bones and features indicated on this radiograph of the anterior view of the pelvic region, using the terms provided. Dale Butler APR A1 A3

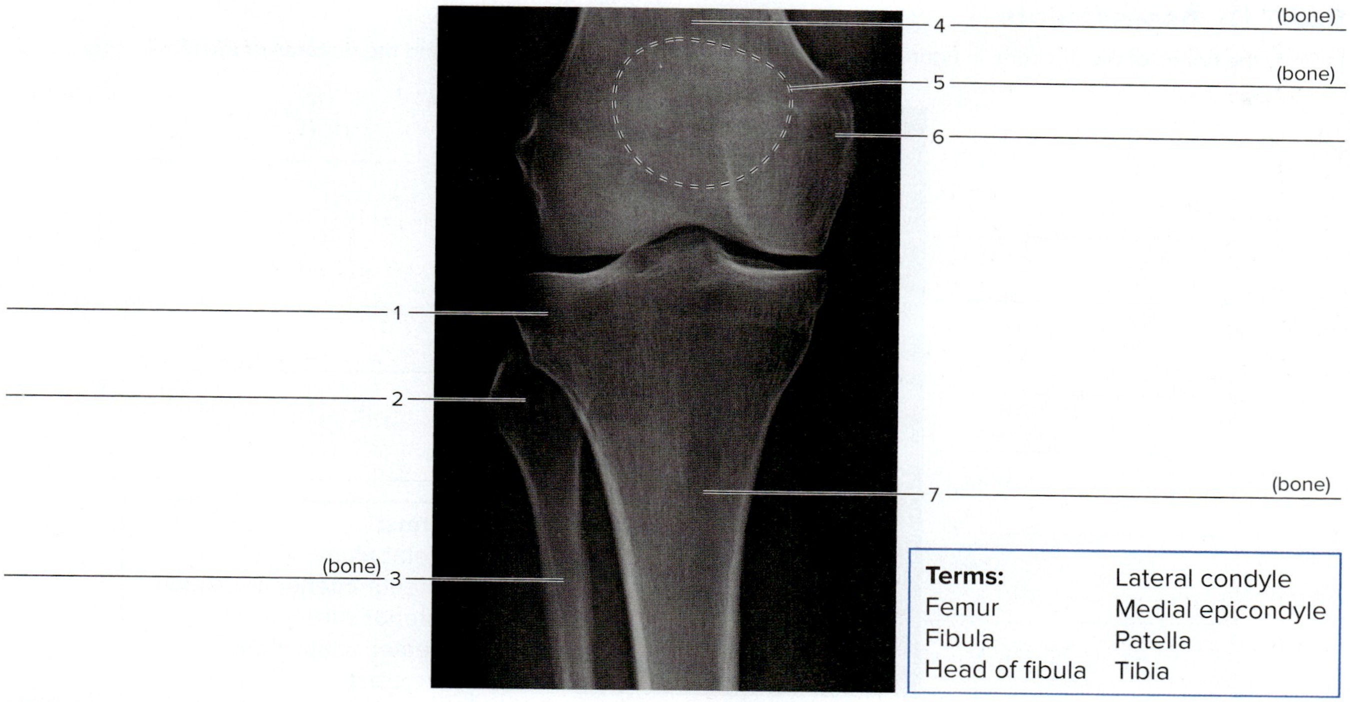

FIGURE 17.8 Identify the bones and features indicated in this radiograph of the right knee (anterior view), using the terms provided. Dale Butler APR A3

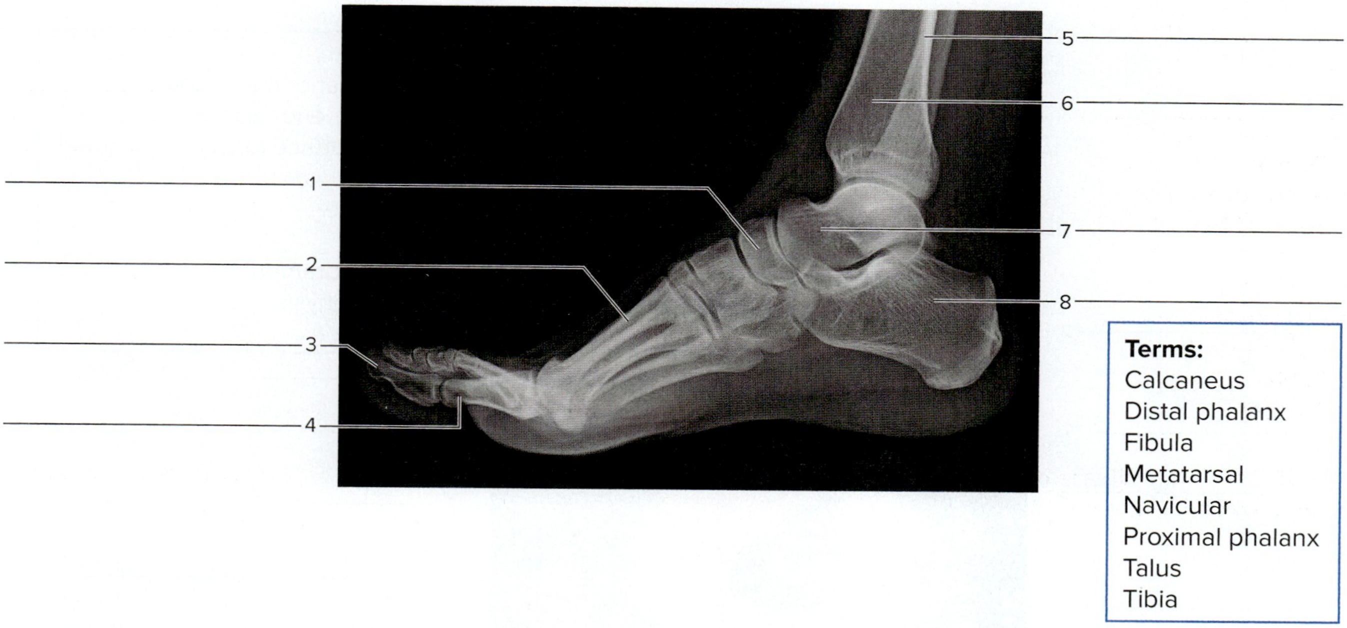

FIGURE 17.9 Identify the bones indicated in this radiograph of the right foot (medial view), using the terms provided. Dale Butler APR A3

PART D: Assessments

Identify the bones of the hip bone in figure 17.10, the features of a femur in figure 17.11, and the features of the foot in figure 17.12.

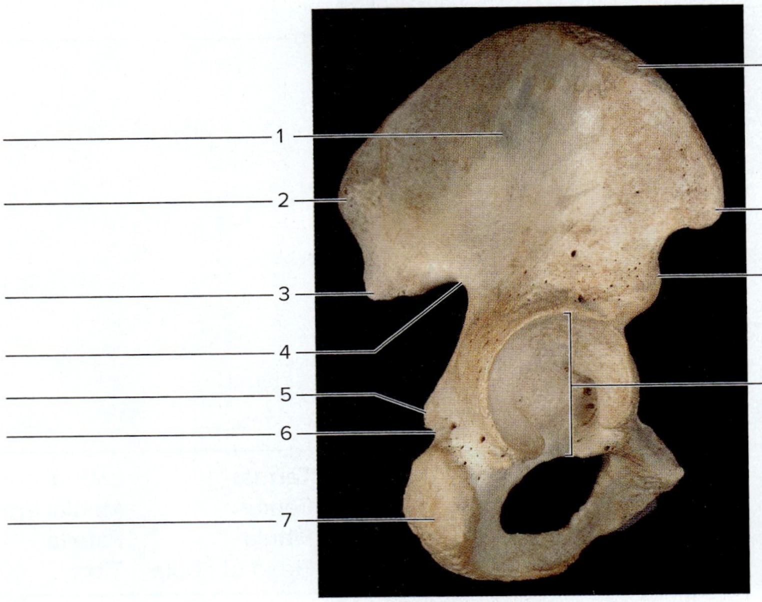

(a) Lateral view

Terms:
Acetabulum
Anterior inferior iliac spine
Anterior superior iliac spine
Greater sciatic notch
Iliac crest
Iliac fossa
Ischial spine
Ischial tuberosity
Lesser sciatic notch
Posterior inferior iliac spine
Posterior superior iliac spine

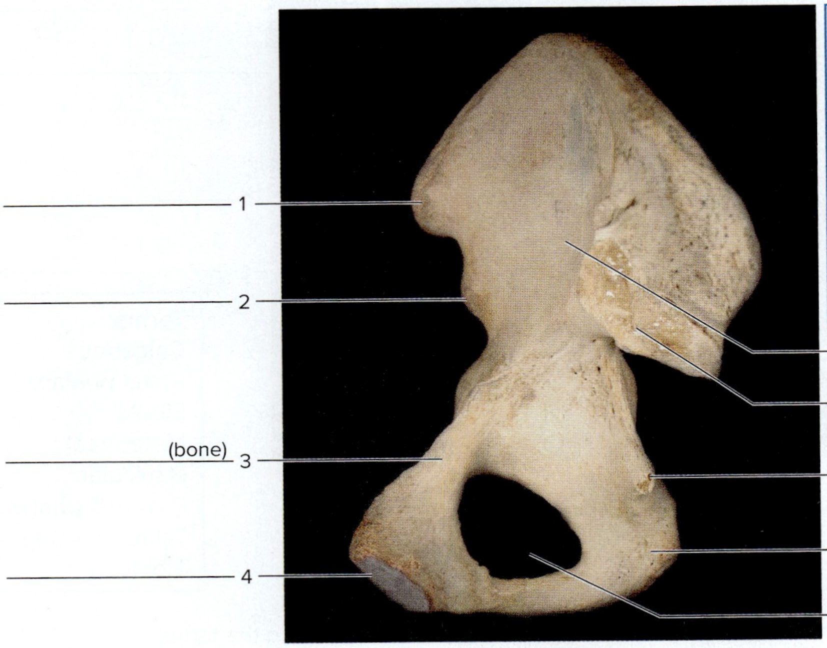

Terms:
Anterior inferior iliac spine
Anterior superior iliac spine
Auricular surface (of sacroiliac joint)
Ilium
Ischial spine
Ischium
Obturator foramen
Pubic symphysis
Pubis

(b) Medial view

FIGURE 17.10 Label the (a) lateral and (b) medial features of the right hip (coxal) bone, using the terms provided.

Dennis Strete A1

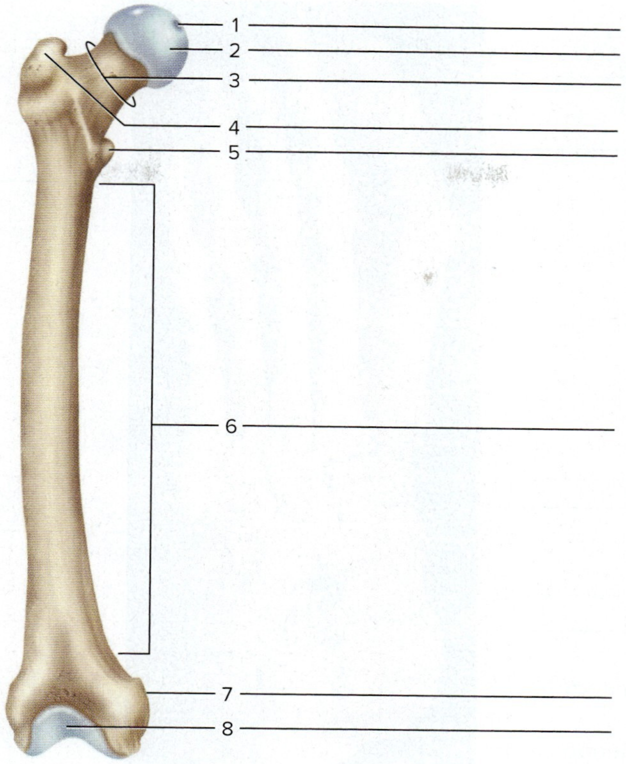

FIGURE 17.11 Label the anterior features of a right femur. A3

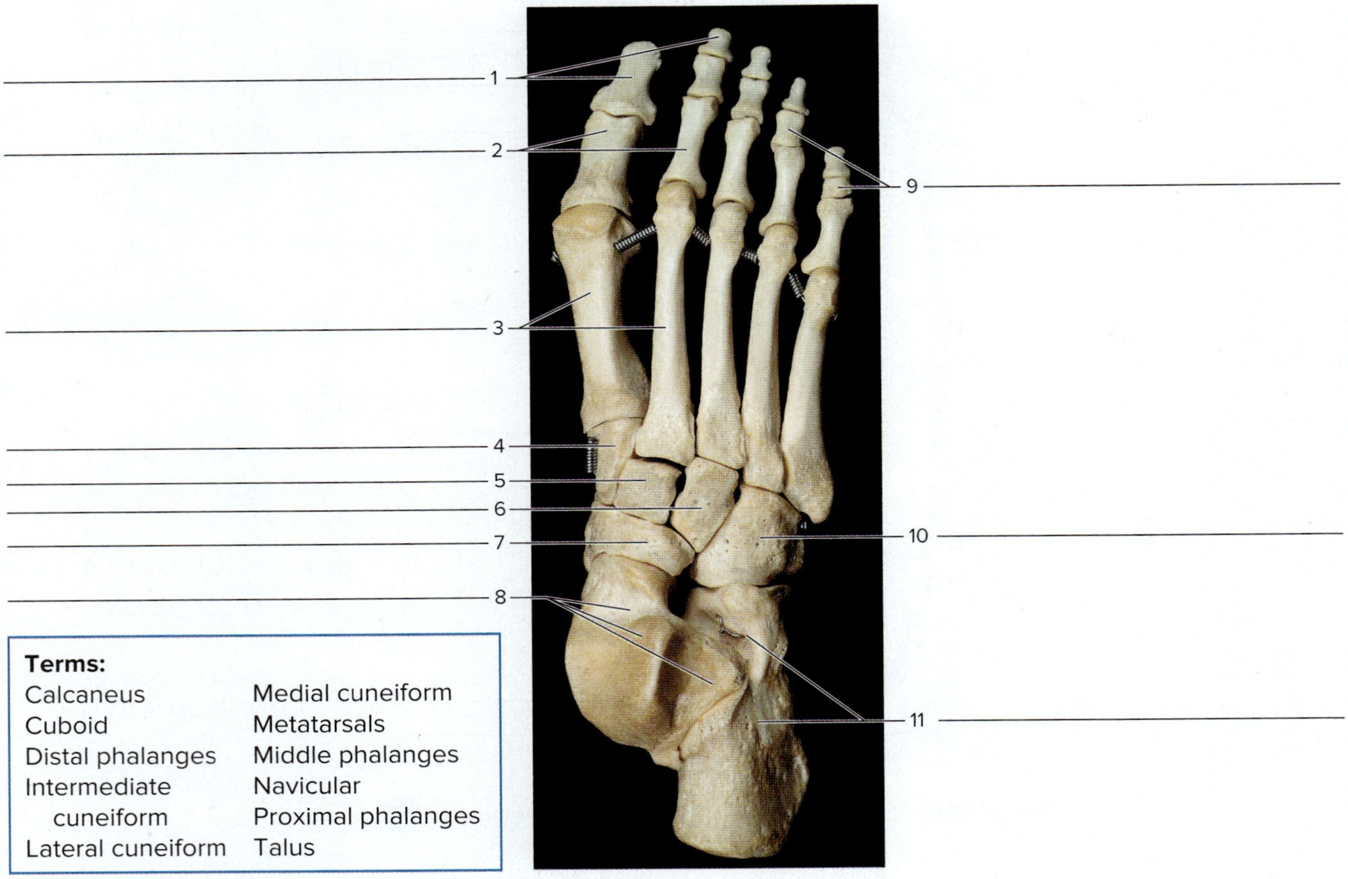

Terms:

Calcaneus	Medial cuneiform
Cuboid	Metatarsals
Distal phalanges	Middle phalanges
Intermediate cuneiform	Navicular
Lateral cuneiform	Proximal phalanges
	Talus

FIGURE 17.12 Identify the bones indicated on this superior view of the right foot, using the terms provided.

©J & J Photography **A3**

LABORATORY
EXERCISE 18

Fetal Skeleton

Purpose of the Exercise

To examine and make comparisons between fetal skeletons and adult skeletons.

MATERIALS NEEDED

Human skull
Human adult skeleton
Human fetal skeleton
Meterstick
Metric ruler

Note: If a fetal skeleton is not available, use the figures represented in this laboratory exercise with the metersticks included in the photographs.

Learning Outcomes

After completing this exercise, you should be able to:

O1 Locate and describe the function of the fontanels of a fetal skull.
O2 Measure and compare the size of the cranial bones and the facial bones between a fetal skull and the adult skull.
O3 Compare the fetal frontal and temporal bones to adult examples.
O4 Measure and compare the total height and the upper and lower limb lengths for a fetal skeleton and an adult skeleton.
O5 Locate and describe six fetal bones or body regions that are not completely ossified but become single bones or groups of bones in the adult skeleton.
O6 Estimate the gestational age of a fetus from measured lengths.
O7 Distinguish intramembranous ossification bones from endochondral ossification bones.

The O corresponds to the assessments A indicated in the Laboratory Assessment for this Exercise.

Pre-Lab

Carefully read the introductory material and examine the entire lab. Be familiar with bone development and a fetal skull from lecture or the textbook. Answer the pre-lab questions.

Pre-Lab Questions Select the correct answer for each of the following questions:

1. Development of the embryo and fetus progresses from
   a. medial toward lateral. **b.** cephalic toward caudal.
   c. anterior toward posterior. **d.** inferior toward superior.
2. The last fontanel to close is the
   a. posterior. **b.** sphenoidal.
   c. mastoid. **d.** anterior.
3. Fontanels function for all of the following *except*
   a. formation of the vertebrae.
   b. compression of the cranium during vaginal delivery.
   c. additional cranial skull growth after birth.
   d. additional brain development after birth.
4. Measurements of height after 20 weeks of development are from
   a. crown-to-heel. **b.** crown-to-rump.
   c. head-to-toes. **d.** sitting height.
5. The anterior fontanel involves all of the following bones *except* the
   a. left parietal bone. **b.** right parietal bone.
   c. occipital bone. **d.** frontal bone.

6. The occipital bone forms from intramembranous ossification.
   a. True b. False
7. The upper and lower limb bones develop from intramembranous ossification.
   a. True b. False
8. The skeleton of a newborn baby consists of fewer bones than 206.
   a. True b. False

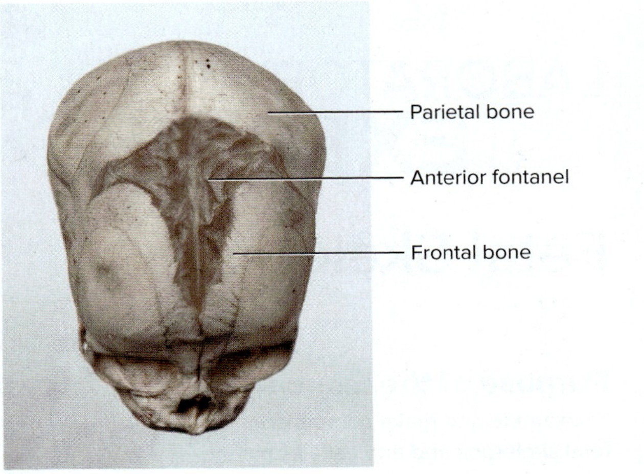

FIGURE 18.1 Superior view of the fetal skull.
©J & J Photography

A human adult skeleton consists of 206 bones. However, the skeleton of a newborn child may have nearly 275 bones due to the numerous *ossification centers* and *epiphyseal plates* present for many bones of the appendicular skeleton. Development of the embryo and the fetus progresses from cephalic toward caudal regions of the body. As a result, there is a noticeable difference between the proportions of the head, the entire body, and the upper and lower limbs between a fetus and an adult. Additionally, the bones most important for protection are among the early ossification sites.

The ossification process occurs during fetal development and continues through childhood growth. *Intramembranous ossification* originating from mesenchyme forms the flat bones of the cranium, some facial bones, and portions of the mandible and clavicles. *Endochondral ossification* originating from hyaline cartilage forms most of the bones of the skeleton.

PROCEDURE A: Skulls

1. Examine the fetal skull from three views (figs. 18.1–18.3). Areas of incomplete ossification of the flat bones of an infant's skull are called *fontanels* (fig. 18.1). Sometimes called "soft spots," fontanels allow for compression of the cranium during a vaginal delivery and for some additional brain development after birth. The posterior and lateral fontanels close during the first year after birth, but the anterior fontanel does not close until nearly two years of age. Locate the following fontanels:

   Mastoid fontanel (2)

   Sphenoidal fontanel (2)

   Posterior fontanel (1)

   Anterior fontanel (1)
2. Complete Part A of Laboratory Assessment 18.
3. Examine the anterior views of a fetal skull next to an adult skull (fig. 18.4). Use a meterstick and measure the height of the entire cranium (braincase) and the height of the face of the fetal skull (fig. 18.4*a*). The brain, which forms in early development, occupies the large cranial cavity. This gives a perspective of a large head. However, the small face is related to undersized maxillae, maxillary sinuses, mandible, lack of teeth eruptions, and nasal cavities. The facial portion of a fetal skull at birth represents about one-eighth of the entire skull,

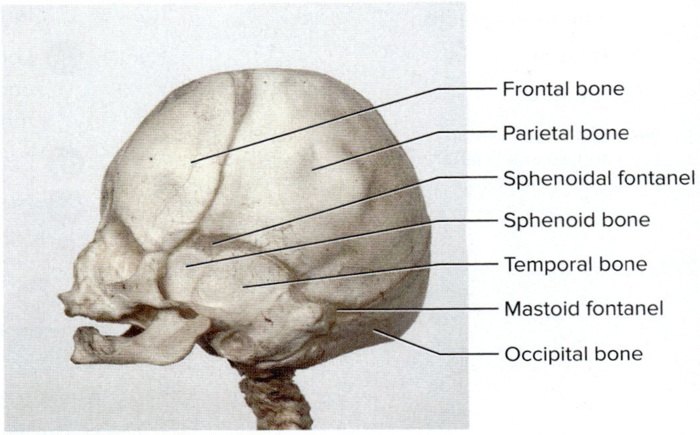

FIGURE 18.2 Lateral view of the fetal skull.
©J & J Photography

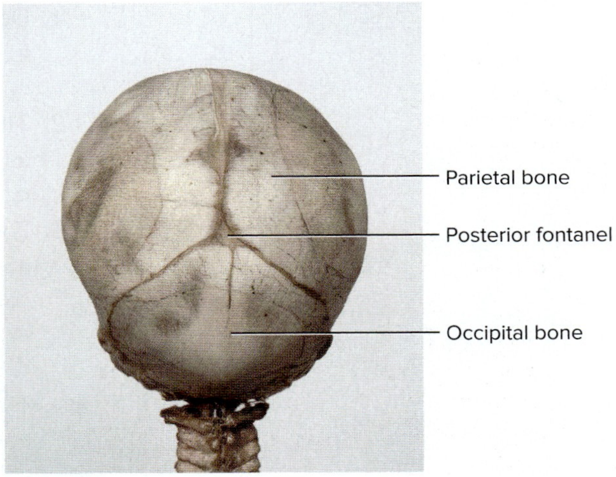

FIGURE 18.3 Posterior view of fetal skull.
©J & J Photography

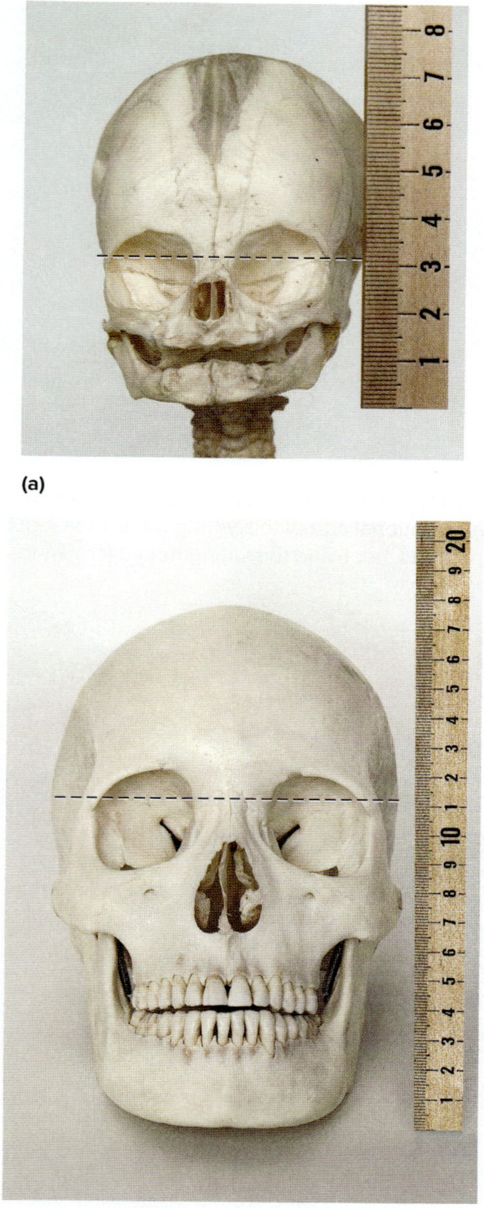

(a)

(b)

FIGURE 18.4 Anterior views of skulls next to metric scales: (a) fetal skull; (b) adult skull. The dotted lines indicate the division between the cranium and the face.
©J & J Photography

whereas the adult facial portion represents nearly one-half of the complete skull.

4. As growth of the fetus advances, the relative size of the cranium becomes smaller as the size of the face increases. Use a meterstick and measure the height of the cranium and face of an adult skull (fig. 18.4*b*).
5. Record your measurements in Part B of the laboratory assessment.
6. Make careful observations of the ossification of the frontal bone and the temporal bone.
7. Complete Part B of the laboratory assessment.

PROCEDURE B: Entire Skeleton

1. Observe and then measure the total height (crown-to-heel) of the entire fetal skeleton (fig. 18.5) and an adult skeleton. Compare the height of the fetal skeleton with the data in table 18.1. Observe and then measure

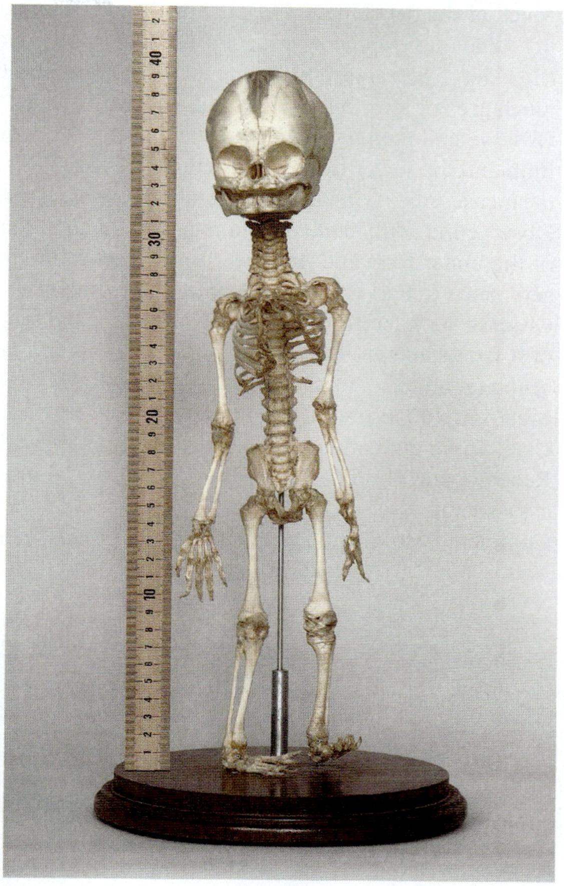

FIGURE 18.5 Anterior view of a fetal skeleton next to a metric scale. To estimate the gestational age of this fetus, measure the total height (crown-to-heel).
©J & J Photography

Table 18.1 Embryonic and Fetal Height (Length)

Gestational Age (Months)	Sitting Height (Crown-to-Rump)*	Total Height (Crown-to-Heel)*
1	0.6 cm	
2	3 cm	
3	7–9 cm	9–10 cm
4	12–14 cm	16–20 cm
5	18–19 cm	25–27 cm
6		30–32 cm
7		35–37 cm
8		40–45 cm
9		50–53 cm

*Measurements of height until about 20 weeks are more often from crown-to-rump; measurements after 20 weeks are more often from crown-to-heel.

the total height (crown-to-heel) and the sitting height (crown-to-rump) of the human fetus shown in figure 18.6. Figure 18.7 illustrates the proportionate changes of bodies during development from an embryo to the adult. Record the measurements in Part C of the laboratory assessment.

2. Much of the development of the upper limb occurs during fetal development, especially during the second trimester. The rapid development of the lower limb is much greater after birth (figs. 18.5 and 18.7).
3. Observe and then measure the total length of the upper limb and the lower limb of the fetus (fig. 18.5). Record the measurements in Part C of the laboratory assessment.
4. Observe the adult skeleton and measure the total length of the upper limb and the lower limb. Record the measurements in Part C of the laboratory assessment.
5. Observe very carefully a hip bone, sternum, and sacrum of a fetal skeleton. Record your observations in Part C of the laboratory assessment.
6. Make careful observations of the anterior knee, thoracic cage, wrist, and ankles of the fetal skeleton (fig. 18.5). Record your observations in Part C of the laboratory assessment.
7. Complete Parts C and D of the laboratory assessment.

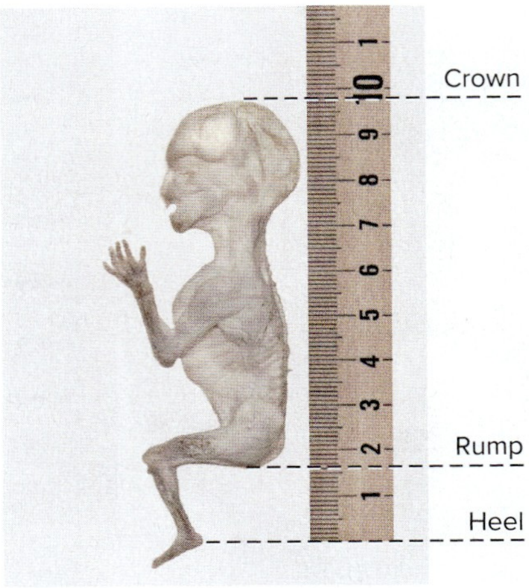

FIGURE 18.6 Human fetus next to a metric scale. To estimate the gestational age of this young fetus, measure the total height (crown-to-heel) and the sitting height (crown-to-rump).
©J & J Photography

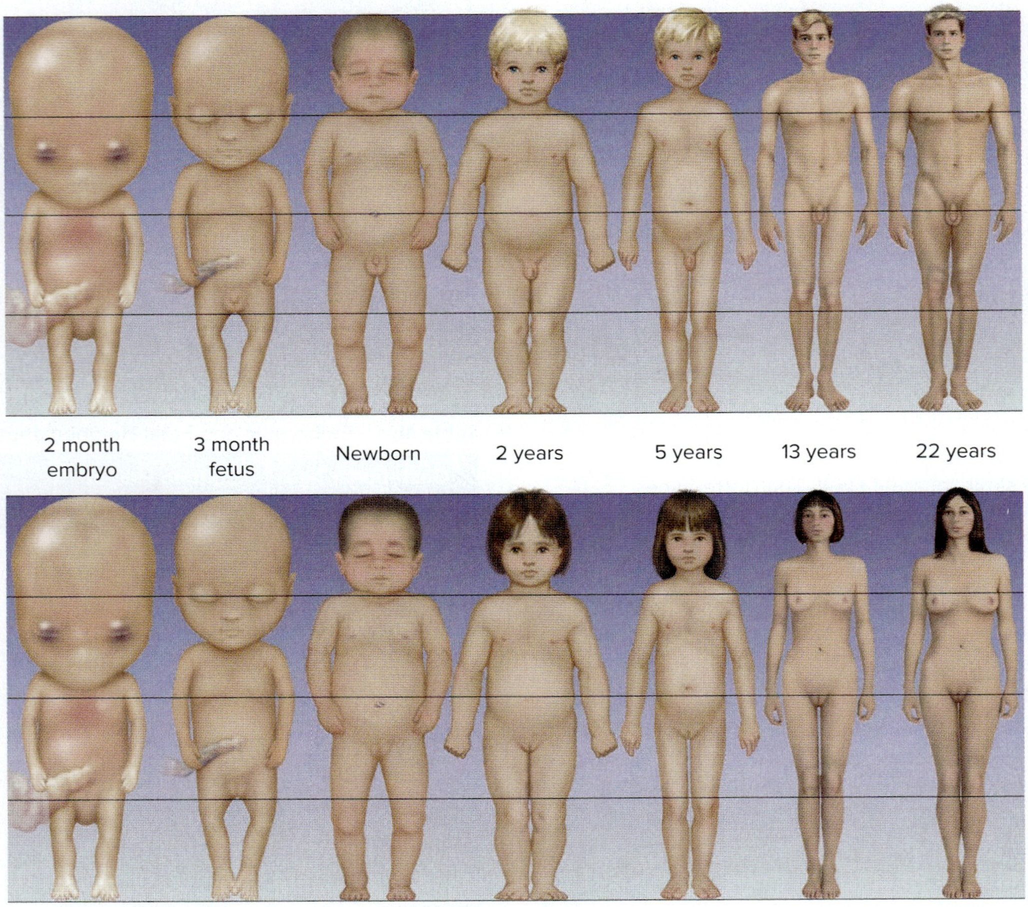

FIGURE 18.7 During development, proportions of the head, entire body, and upper and lower limbs change considerably.

LABORATORY
ASSESSMENT
18

Name ___________________________

Date ___________________________

Section ___________________________

The A corresponds to the indicated Learning Outcome(s) O found at the beginning of the Laboratory Exercise.

Fetal Skeleton

PART A: Assessments

Complete the following:

1. List all of the bones that form part of the border with the mastoid fontanel. A1

____

2. List all of the bones that form part of the border with the sphenoidal fontanel. A1

____

3. List all of the bones that form part of the border with the posterior fontanel. A1

____

4. List all of the bones that form part of the border with the anterior fontanel. A1

____

5. Describe the function of fontanels. A1

____

6. What is the final outcome of the fontanels? A1

____

PART B: Assessments

1. Enter the measurements of the fetal and adult skulls in the table. A2

Skull	Total Height of Skull (cm)	Cranial Height (cm)	Facial Height (cm)
Fetus			
Adult			

2. Compare the frontal bone of the fetal skull with that of the adult skull. Describe some unusual aspects of the fetal frontal bone as compared to the adult structure. A3

____

____

____

3. Compare the temporal bone of the fetal skull with that of the adult skull. Describe some unusual aspects of the fetal temporal bone as compared to the adult structure. A3

____

____

____

4. Summarize the measured and observed relationships between a fetal skull and an adult skull. A2

____

____

____

____

CRITICAL THINKING ASSESSMENT — PART B

Why can it be difficult to match the identifications of baby pictures with those of adults? A2

____

____

____

____

PART C: Assessments

1. Enter the measurements of the fetal and adult skeletons in the table. A4

Skeleton	Total Height (cm)	Upper Limb Length (cm)	Lower Limb Length (cm)
Fetus			
Adult			

2. Summarize the relationship of total body height to limb lengths for the fetal skeleton and the adult skeleton. A4

____

____

____

____

3. Locate and describe the discernible ossification of the following fetal bones and body regions and compare them with the adult skeleton. A5

   a. Hip bone:

   ____

   ____

b. Sternum:

c. Sacrum:

d. Anterior knee:

e. Thoracic cage:

f. Wrist (carpals) and ankle (tarsals):

4. Estimate the age of fetal development of the fetal skeleton in figure 18.5 and the one that is represented at your school. Estimate the age of the human fetus shown in figure 18.6. Use the data provided in table 18.1 for your determination of the estimates. A6

a. Fetal skeleton in figure 18.5: ________________

b. Fetal skeleton at your school: ________________

c. Fetus in figure 18.6: ________________

PART D: Assessments

Identify the fetal skeleton structures in figures 18.8 and 18.9.

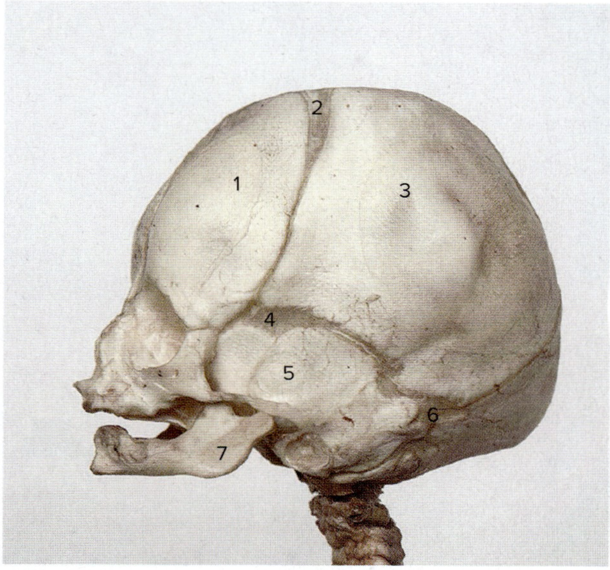

____ Anterior fontanel
____ Frontal bone
____ Mandible
____ Mastoid fontanel
____ Parietal bone
____ Sphenoidal fontanel
____ Temporal bone

FIGURE 18.8 Label the fetal bones and fontanels by placing the correct numbers in the spaces provided. ©J & J Photography A1 A5

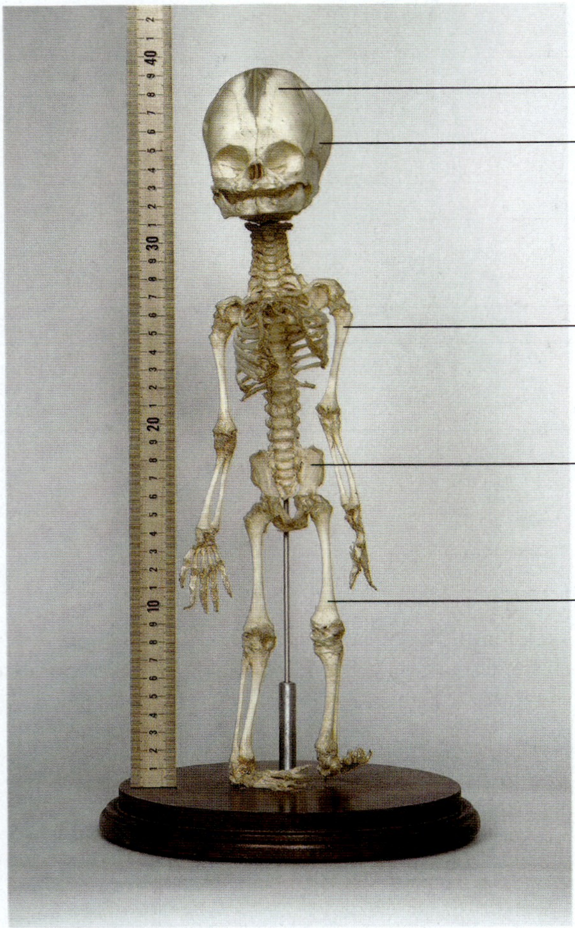

FIGURE 18.9 Distinguish the type of ossification formation by selecting the correct choice from the terms provided.

©J & J Photography A7

LABORATORY
EXERCISE 19

Joint Structure and Movements

Purpose of the Exercise

To examine examples of the three types of joints and their classification, to identify the major features of these joints, and to review the types of movements produced at synovial joints.

MATERIALS NEEDED

Human skull

Human skeleton, articulated

Models of synovial joints (shoulder, elbow, hip, and knee)

For Demonstration Activities:

Fresh animal joint (knee joint preferred)

Radiographs of major joints

SAFETY

- Review all the safety guidelines in Appendix 1 of your laboratory manual.
- Wear disposable gloves and protective eyewear when handling the fresh animal joint.
- Wash your hands before leaving the laboratory.

Learning Outcomes APR

After completing this exercise, you should be able to:

O1 Identify key features of fibrous, cartilaginous, and synovial joints.

O2 Distinguish functional characteristics of synarthroses, amphiarthroses, and diarthroses.

O3 Locate examples of each type of joint, structurally and functionally.

O4 Examine the structure and types of movements of the shoulder, elbow, hip, and knee joints.

O5 Identify the types of movements that occur at synovial joints.

The O corresponds to the assessments A indicated in the Laboratory Assessment for this Exercise.

Pre-Lab

Carefully read the introductory material and examine the entire lab. Be familiar with joint structures and movements from lecture or the textbook. Answer the pre-lab questions.

Pre-Lab Questions Select the correct answer for each of the following questions:

1. Fibrous joints are structural types containing
  a. cartilage. **b.** dense fibrous connective tissue.
  c. synovial membranes. **d.** synovial fluid.

2. Which synovial joint allows movements in all planes?
  a. saddle **b.** pivot
  c. ball-and-socket **d.** plane

3. Synarthrosis pertains to functional joints that are
  a. immovable. **b.** slightly movable.
  c. freely movable. **d.** synovial types.

4. Movement away from the midline is called
  a. rotation. **b.** adduction.
  c. extension. **d.** abduction.

5. Movements that combine flexion, abduction, extension, and adduction are called
  a. rotations. **b.** excursions.
  c. circumductions. **d.** hyperextensions.

6. The synovial membrane and the fibrous capsule combine and form the
  a. articular cartilage. **b.** periosteum.
  c. spongy bone. **d.** joint (articular) capsule.

Chapter Opener: ©101cats/Getty Images

7. Rotation occurs at a pivot type of synovial joint.
   a. True **b.** False
8. Dorsiflexion and plantar flexion are characteristic movements of the hand.
   a. True **b.** False

Joints are junctions between bones. Although they vary considerably in structure, they can be classified according to the type of tissue that binds the bones together. Thus, the three groups of structural joints can be identified as *fibrous joints, cartilaginous joints,* and *synovial joints.* Fibrous joints are filled with dense fibrous connective tissue, cartilaginous joints consist of two bones connected by hyaline cartilage or fibrocartilage, and synovial joints contain synovial fluid inside a joint cavity.

Joints can also be classified by the degree of functional movement allowed: *synarthroses* are immovable, *amphiarthroses* allow slight movement, and *diarthroses* allow free movement. Movements occurring at freely movable synovial joints are due to the contractions of skeletal muscles. In each case, the type of movement depends on the type of joint involved and the way in which the muscles are attached to the bones on either side of the joint.

PROCEDURE A: Types of Joints

Joints form where bones come together, or articulate with each other. Because there are variations in the types of connective tissues involved in the joint structure and variations in the degree of movement possible, two types of classifications are used. Structurally, joints are fibrous, cartilaginous, or synovial; functionally, joints are synarthrosis, amphiarthrosis, or diarthrosis. The joints accommodate various degrees of body movements. Some joints are only temporary; when complete ossification occurs, the entire joint becomes bone. Examples of these temporary joints include a right and left frontal bone becoming a single frontal bone, and the eventual complete ossification of the epiphyseal plates in the growth areas of the long bones.

1. Use table 19.1 and figure 19.1 as a reference and examine the human skull and articulated skeleton. Locate examples of the following types of structural and functional joints:

   fibrous joints
   - suture—synarthrosis
   - gomphosis—synarthrosis
   - syndesmosis—amphiarthrosis

Table 19.1 Classification of Joints

Structural Classification	Structural Features	Functional Classification and Movements	Location Examples
FIBROUS			
Suture	Dense fibrous connective tissue	Synarthrosis—immovable	Sutures of skull
Gomphosis	Periodontal ligament	Synarthrosis—immovable	Tooth sockets
Syndesmosis	Dense fibrous connective tissue	Amphiarthrosis—slightly movable	Tibiofibular joint
CARTILAGINOUS			
Synchondrosis	Hyaline cartilage	Synarthrosis—immovable	Epiphyseal plates
Symphysis	Fibrocartilage	Amphiarthrosis—slightly movable	Pubic symphysis; intervertebral discs
SYNOVIAL			
Plane (gliding)	All synovial joints contain articular cartilage, a synovial membrane, and a joint cavity filled with synovial fluid	Diarthrosis—uniaxial; sliding or twisting	Intercarpal; intertarsal; knee (between femur and patella)
Hinge		Diarthrosis—uniaxial; flexion and extension	Elbow; knee; interphalangeal
Pivot		Diarthrosis—uniaxial; rotation	Radioulnar at elbow; dens at atlas
Condylar (ellipsoid)		Diarthrosis—biaxial; flexion, extension, abduction, and adduction	Radiocarpal; metacarpophalangeal; knee (between femur and tibia when flexed)
Saddle		Diarthrosis—biaxial; variety of movements mainly in two planes	Base of thumb with trapezium
Ball-and-socket		Diarthrosis—multiaxial; movements in all planes	Hip; shoulder

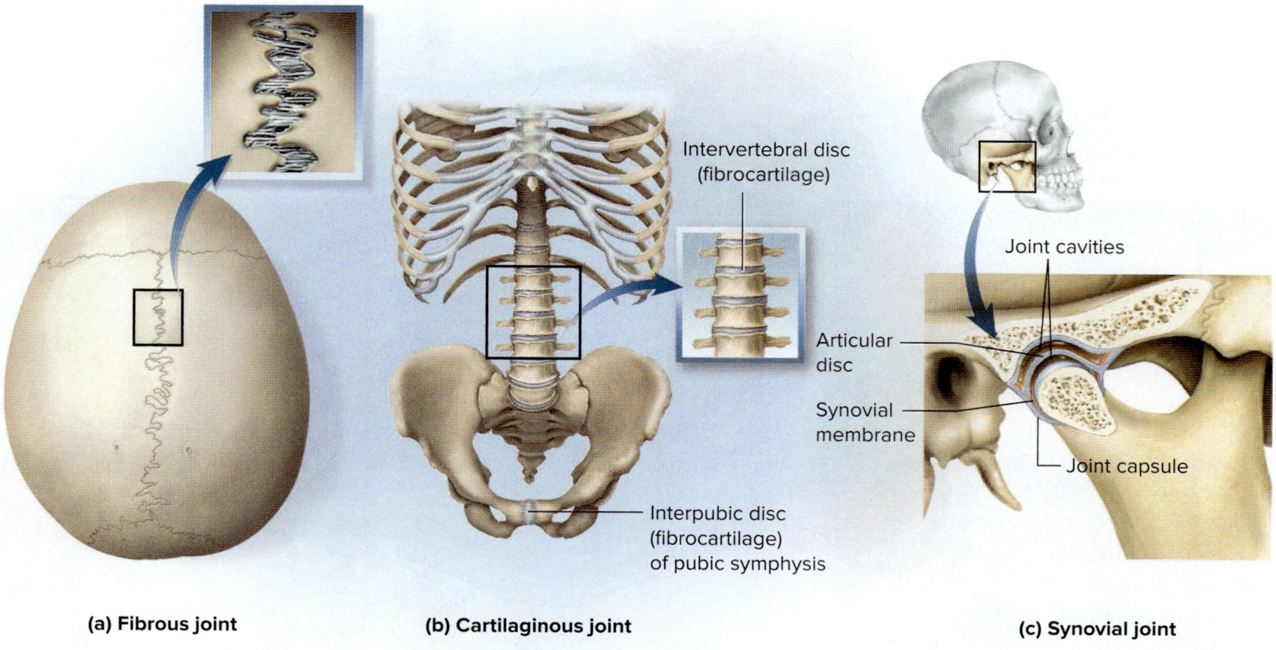

FIGURE 19.1 Examples of specific joint classifications. A type of fibrous joint called a suture (synarthrosis) is shown in (a); in (b) two examples of cartilaginous joints are shown, the interpubic disc at the pubic symphysis and intervertebral discs (both amphiarthrosis); and in (c) a synovial joint called the temporomandibular joint (diarthrosis) is shown.

cartilaginous joints

- synchondrosis—synarthrosis
- symphysis—amphiarthrosis

synovial joints—diarthrosis

2. Complete Part A of Laboratory Assessment 19.
3. Synovial, diarthrotic joints allow various degrees of freedom of movement. *Uniaxial* joints move in a single plane or axis; *biaxial* joints move in two planes or axes; *multiaxial* joints move in multiple planes or axes.
4. Using table 19.1 and figure 19.2 as references, locate examples of the following types of synovial joints in the skeleton. At the same time, examine the corresponding joints in the models and in your own body. Experiment with each joint to demonstrate the range of motion.

plane (gliding) joint	**hinge joint**
pivot joint	**condylar (ellipsoid) joint**
saddle joint	**ball-and-socket joint**

5. Complete Parts B and C of the laboratory assessment.

PROCEDURE B: Examples of Synovial Joints

Synovial joints have some characteristic structural features and functional characteristics. The structure of the synovial joint includes a joint (articular) capsule composed of two layers: an outer fibrous capsule and an inner synovial membrane. The fibrous capsule is composed of a dense connective tissue that is continuous with the periosteum of the bone, while the synovial membrane secretes a lubricating synovial fluid that fills the enclosed joint cavity. Articular (hyaline) cartilage covers the surface of the bone within the joint. Various ligaments and tendons also cross synovial joints.

In the knee, two crescent-shaped pieces of cartilage called menisci extend partway across the joint from the medial and lateral sides. The meniscus will help stabilize the knee joint, where a great amount of pressure occurs, and it also helps to absorb shock. Although the knee is basically a hinge joint, there are some condylar and plane components of the joint. The condylar component of the joint allows some limited rotation between the articulations of the rounded condyles of the femur and the tibia when the knee is partly in the flexed position. The plane joint between the patellar surface of the femur and the patella allows some sliding movements. The knee joint also consists of two cruciate ligaments that cross each other deep within the joint. The posterior cruciate ligament (PCL) helps prevent excessive flexion; the anterior cruciate ligament (ACL) helps prevent hyperextension of the knee joint. A torn ACL is a common type of knee injury.

1. Study the major features of a synovial joint in figure 19.3.
2. Examine models of the shoulder, elbow, hip, and knee joints.

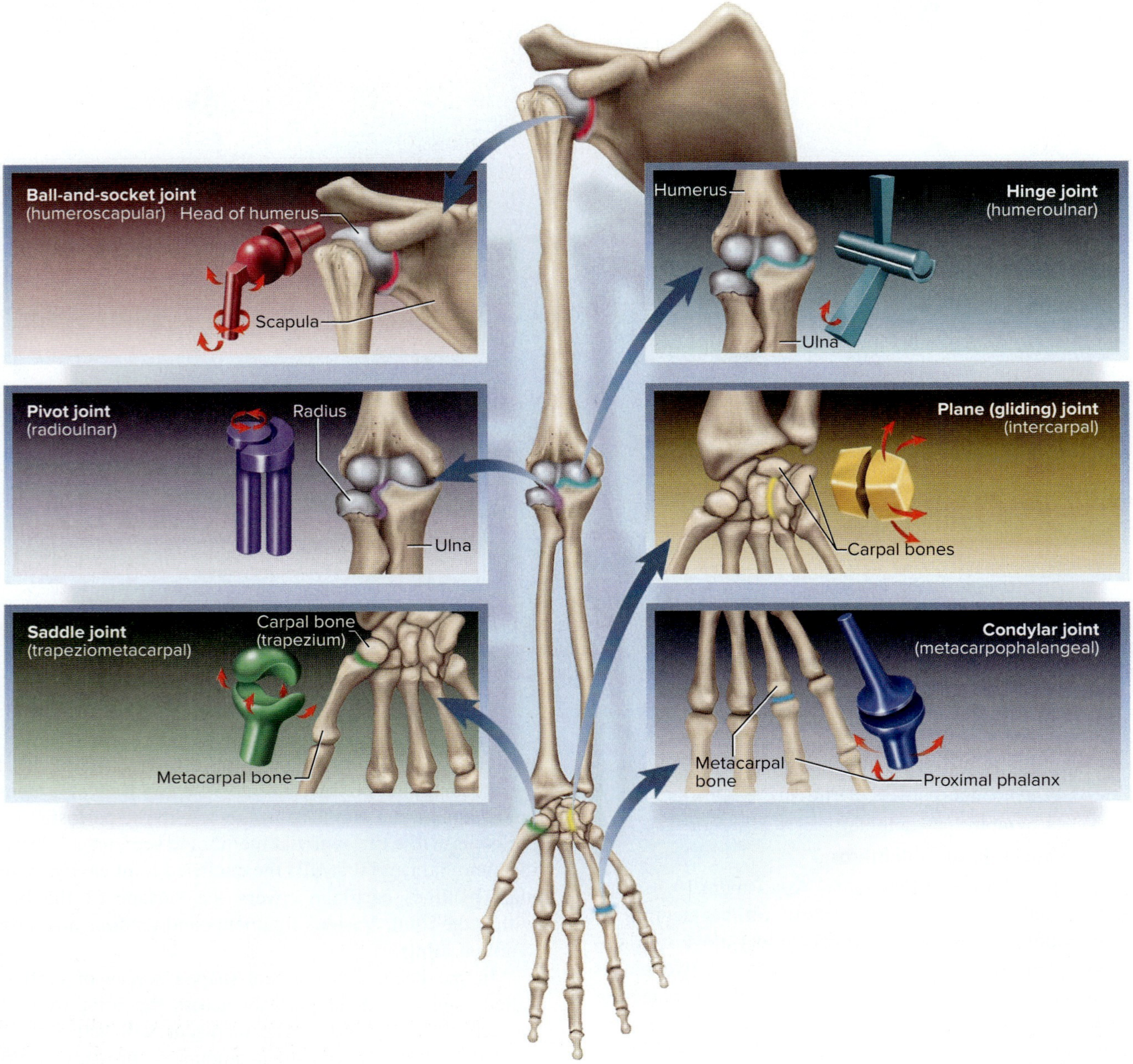

FIGURE 19.2 Examples of all types of synovial joints and the possible movements of each type of joint. APR

3. Locate the major knee joint structures of figure 19.4 that are visible on the knee joint model.
4. Complete Part D of the laboratory assessment.

DEMONSTRATION ACTIVITY

Examine a longitudinal section of a fresh synovial animal knee joint. Locate the dense connective tissue that forms the fibrous capsule and the hyaline cartilage that forms the articular cartilage on the ends of the bones. Locate the synovial membrane on the inner surface of the joint capsule. Does the joint have any semilunar cartilages (menisci)? ______________________________

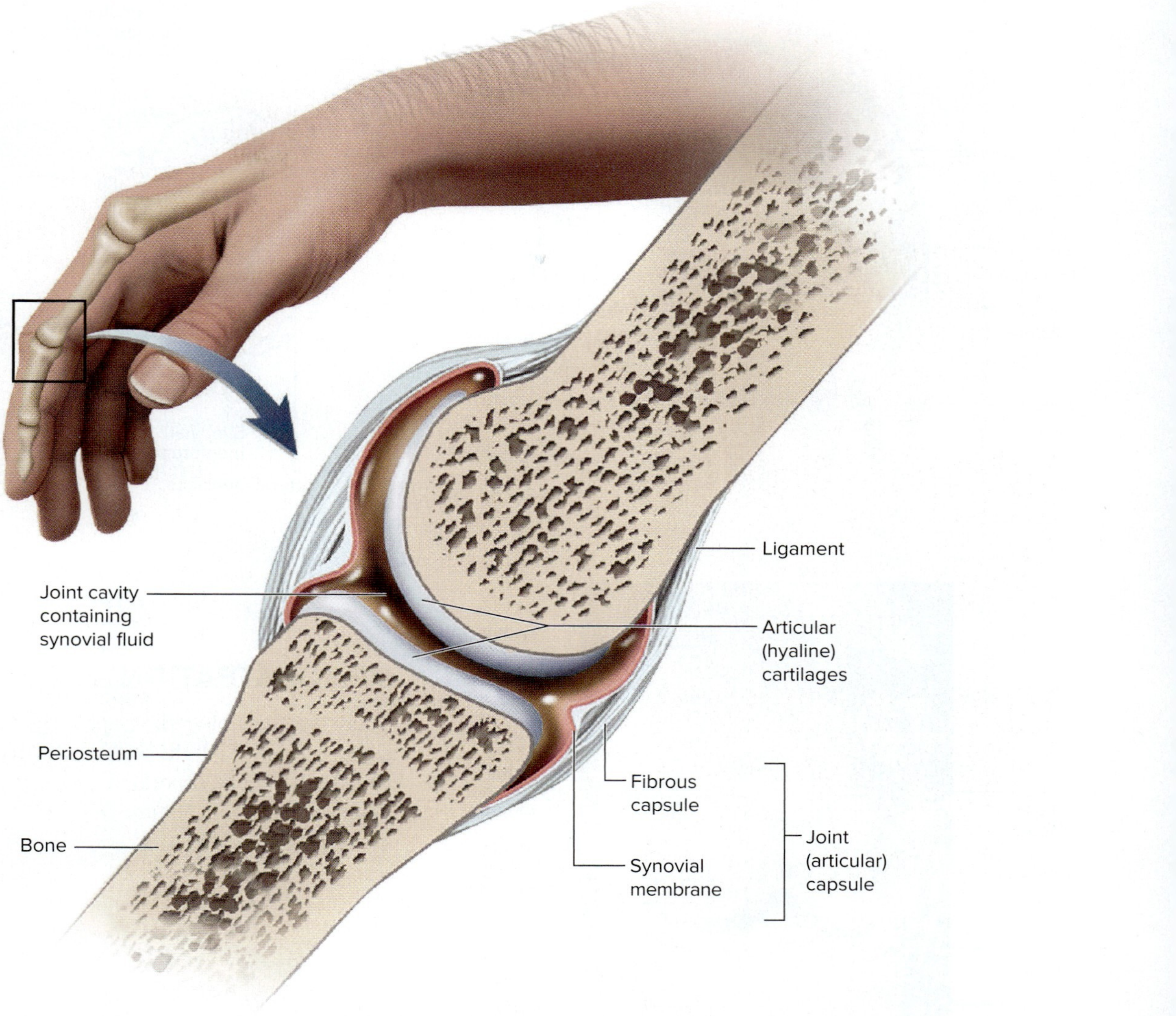

FIGURE 19.3 Basic structure of a synovial joint. APR

PROCEDURE C: Joint Movements

1. Examine figures 19.5 and 19.6.
2. Most joints and body parts are extended and adducted when the body is positioned in anatomical position. Skeletal muscle action involves the more movable end (*insertion*) being pulled toward the more stationary end (*origin*). In the limbs, the origin is usually proximal to the insertion; in the trunk, the origin is usually medial to the insertion. Use these concepts as reference points as you move joints. Move various parts of your body to demonstrate the synovial joint movements described in tables 19.2 and 19.3.
3. Complete Part E of the laboratory assessment.

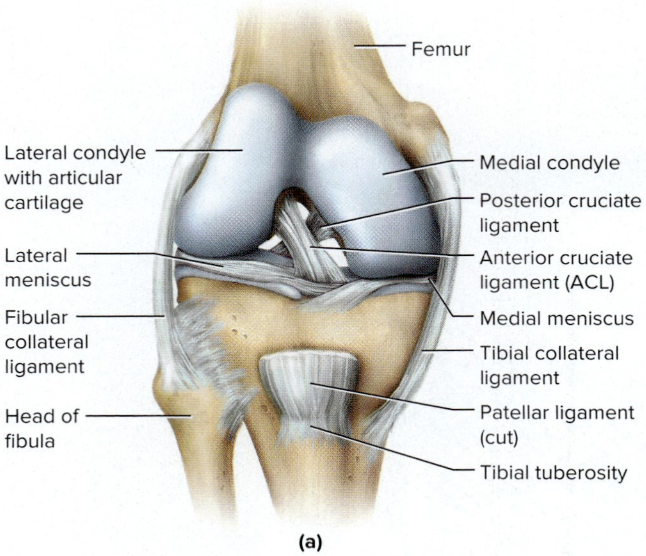

(a)

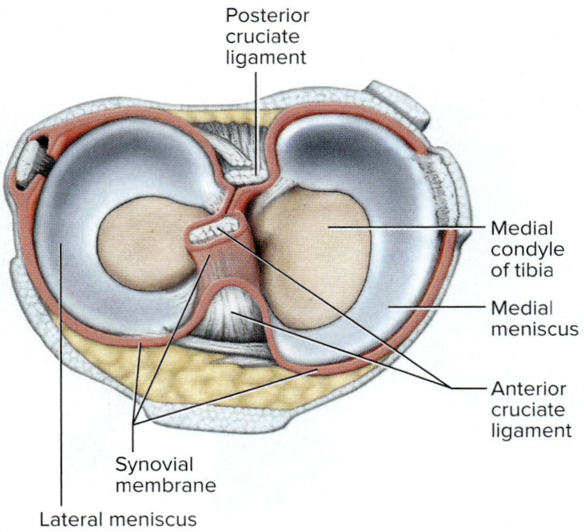

(c)

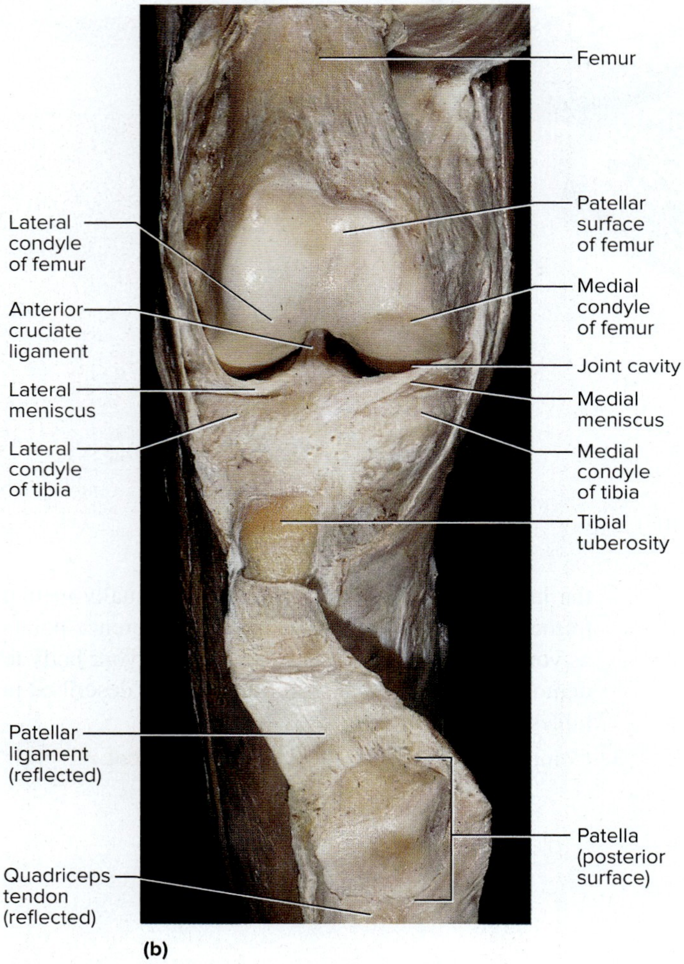

(b)

DEMONSTRATION ACTIVITY

Study the available radiographs of joints by holding the films in front of a light source. Identify the type of joint and the bones incorporated in the joint. Also identify other major visible features.

FIGURE 19.4 Right knee joint: (a) anterior view; (b) anterior view of a cadaver; (c) superior view of tibial end.
(b) Rebecca Gray/McGraw Hill APR

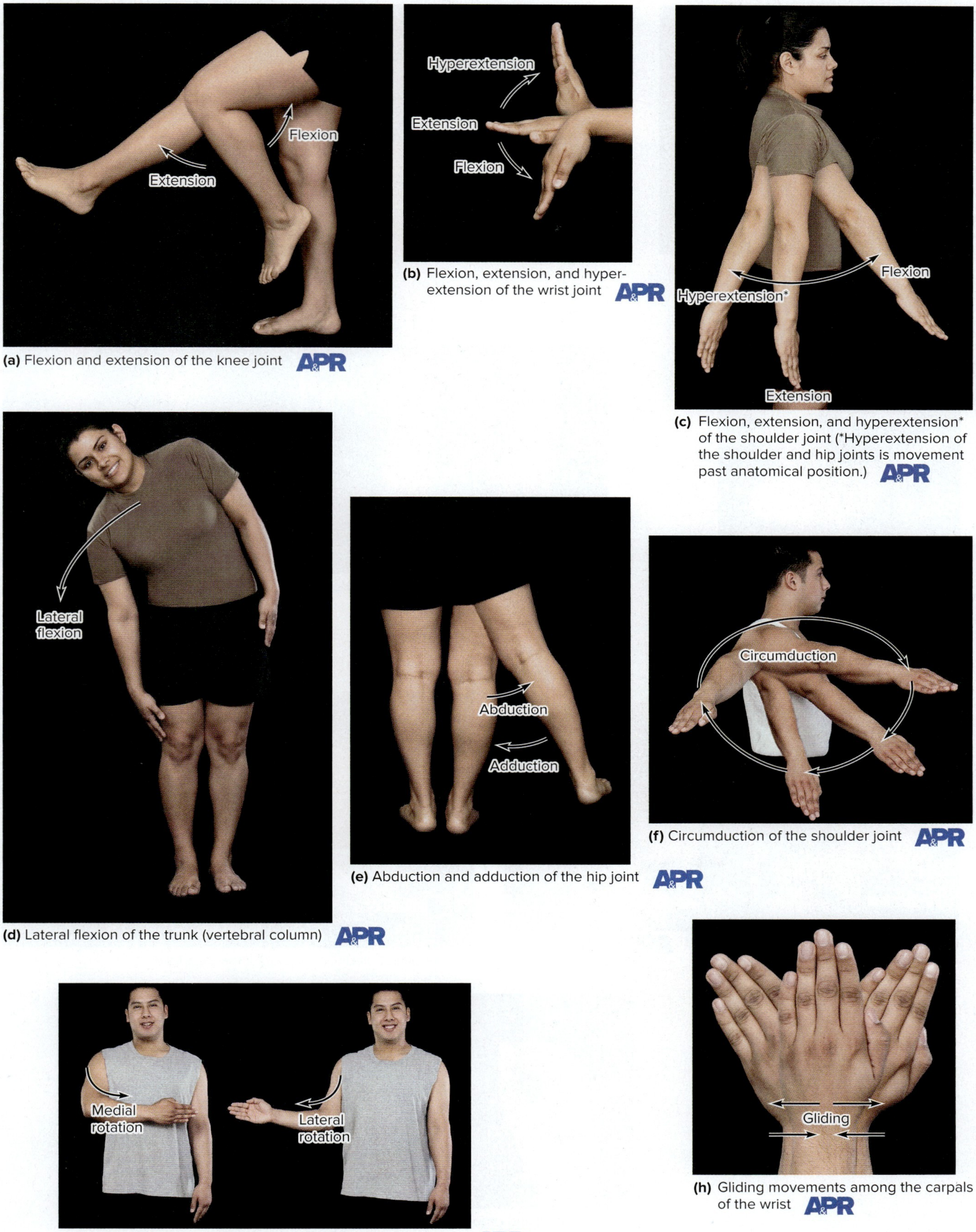

(a) Flexion and extension of the knee joint APR

(b) Flexion, extension, and hyperextension of the wrist joint APR

(c) Flexion, extension, and hyperextension* of the shoulder joint (*Hyperextension of the shoulder and hip joints is movement past anatomical position.) APR

(d) Lateral flexion of the trunk (vertebral column) APR

(e) Abduction and adduction of the hip joint APR

(f) Circumduction of the shoulder joint APR

(g) Medial and lateral rotation of the arm at the shoulder joint APR

(h) Gliding movements among the carpals of the wrist APR

FIGURE 19.5 Examples of angular, rotational, and gliding movements of synovial joints (a)–(h). ©J & J Photography

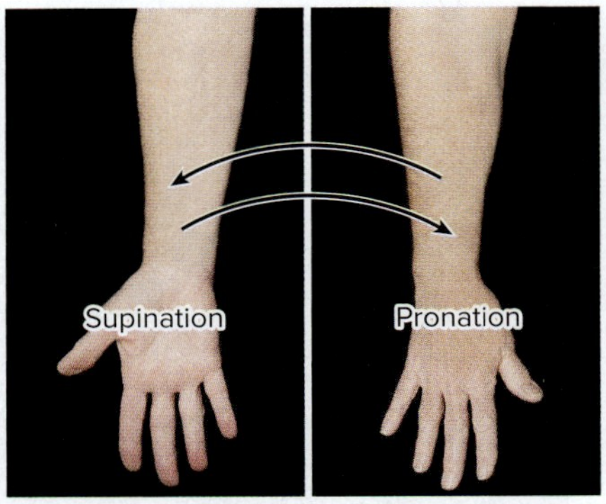

(a) Supination and pronation of the forearm and hand involving movement at the radioulnar joint

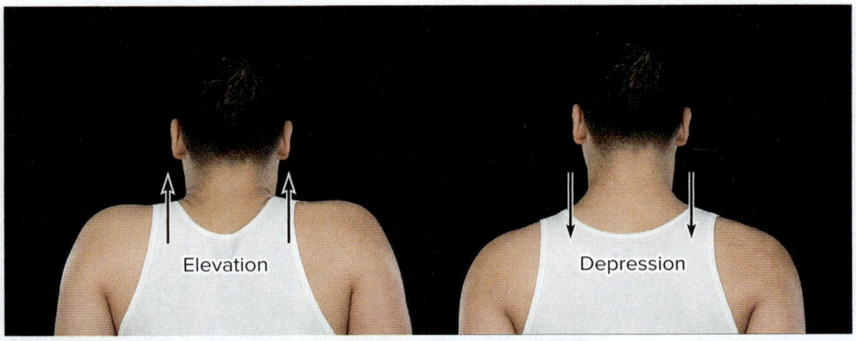

(b) Elevation and depression of the shoulder (scapula)

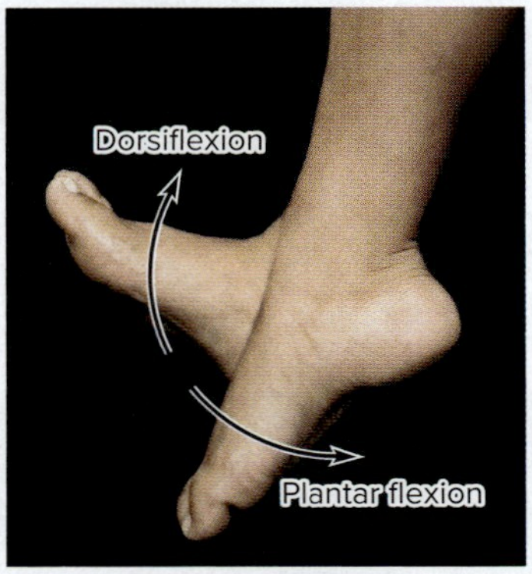

(c) Dorsiflexion and plantar flexion of the foot at the ankle joint

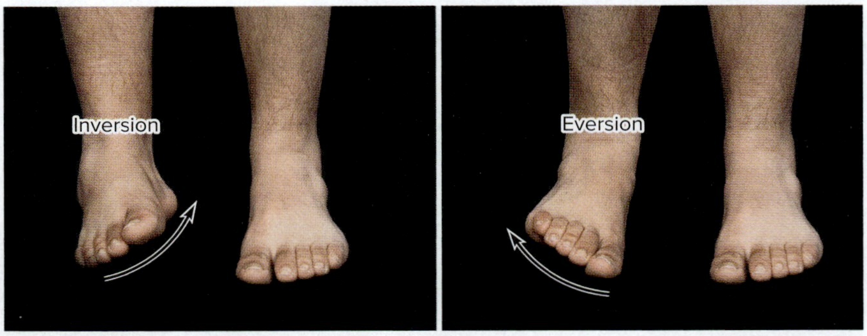

(d) Inversion and eversion of the right foot at the ankle joint (The left foot is unchanged for comparison.)

(e) Protraction and retraction of the head

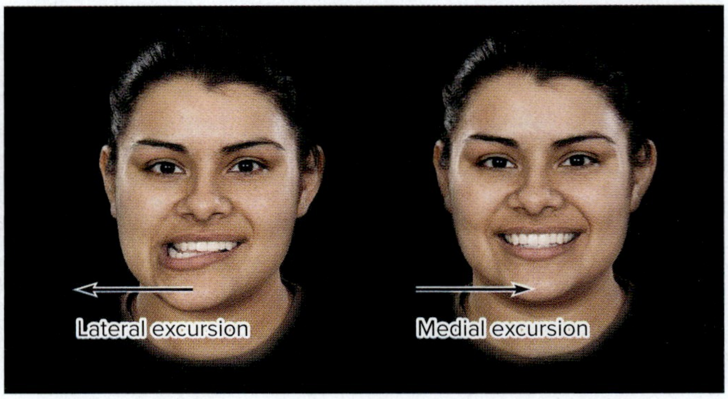

(f) Lateral and medial excursion of the mandible

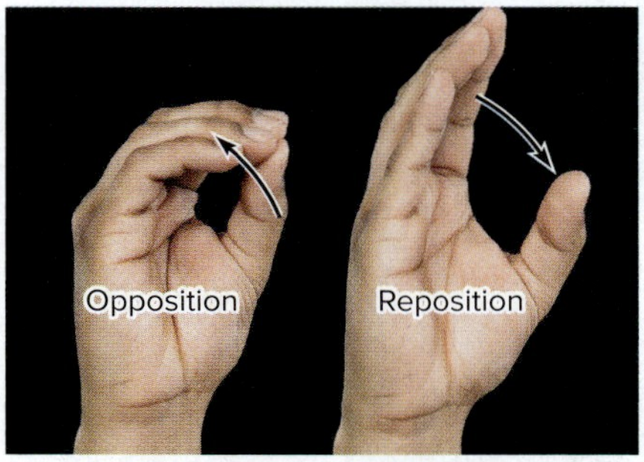

(g) Opposition and reposition of the thumb at the carpometacarpal (saddle) joint

FIGURE 19.6 Special movements of synovial joints (a)–(g). ©J & J Photography

Table 19.2 Angular, Rotational, and Gliding Movements

Movement	Description
Flexion	Decrease of an angle (usually in the sagittal plane)
Lateral flexion	Bending the trunk (vertebral column) to the side
Extension	Increase of an angle (usually in the sagittal plane)
Hyperextension	Extension beyond anatomical position
Abduction	Movement of a body part away from the midline (usually in the frontal plane)
Adduction	Movement of a body part toward the midline (usually in the frontal plane)
Circumduction	Circular movement (combines flexion, abduction, extension, and adduction)
Rotation	Movement of part around its long axis
Medial (internal)	Inward rotation
Lateral (external)	Outward rotation
Gliding	Back-and-forth and side-to-side sliding movements of plane joints

Table 19.3 Special Movements (Pertain to Specific Joints)

Movement	Description
Supination	Rotation of forearm and palm of hand anteriorly or upward
Pronation	Rotation of forearm and palm of hand posteriorly or downward
Elevation	Movement of body part upward
Depression	Movement of body part downward
Dorsiflexion	Movement of ankle joint so dorsum (superior) of foot becomes closer to anterior surface of leg (as standing on heels)
Plantar flexion	Movement of ankle joint so the plantar surface of foot becomes closer to the posterior surface of leg (as standing on toes)
Inversion (called supination in some health professions)	Medial movement of sole of foot at ankle joint
Eversion (called pronation in some health professions)	Lateral movement of sole of foot at ankle joint
Protraction	Anterior movement in the transverse plane
Retraction	Posterior movement in the transverse plane
Excursion	Special movements of mandible when grinding food
Lateral	Movement of mandible laterally
Medial	Movement of mandible medially
Opposition	Movement of the thumb to touch another finger of the same hand
Reposition	Return of thumb to anatomical position

CRITICAL THINKING ACTIVITY

Have your laboratory partner do some of the preceding movements and see if you can correctly identify the movements made.

CRITICAL THINKING ACTIVITY

Describe a body position that can exist when all major body parts are flexed.________________________

____

____

____

NOTES

LABORATORY
ASSESSMENT 19

Name ____________________

Date ____________________

Section ____________________

The (A) corresponds to the indicated Learning Outcome(s) (O) found at the beginning of the Laboratory Exercise

Joint Structure and Movements

PART A: Assessments

Match the terms in column A with the descriptions in column B. Place the letter of your choice in the space provided. A1 A2

Column A		Column B
a. Gomphosis	______	1. Immovable joint between flat bones of the skull united by a thin layer of dense connective tissue
b. Suture	______	2. Fibrocartilage fills the slightly movable joint
c. Symphysis	______	3. Temporary joint in which bones are united by bands of hyaline cartilage
d. Synchondrosis	______	4. Slightly movable joint in which bones are united by interosseous membrane
e. Syndesmosis	______	5. Joint formed by union of tooth root in bony socket

PART B: Assessments

Identify the types of structural and functional joints numbered in figure 19.7. A3

	Structural Classification	Functional Classification
1.	____________________	____________________
2.	____________________	____________________
3.	____________________	____________________
4.	____________________	____________________
5.	____________________	____________________
6.	____________________	____________________
7.	____________________	____________________
8.	____________________	____________________
9.	____________________	____________________

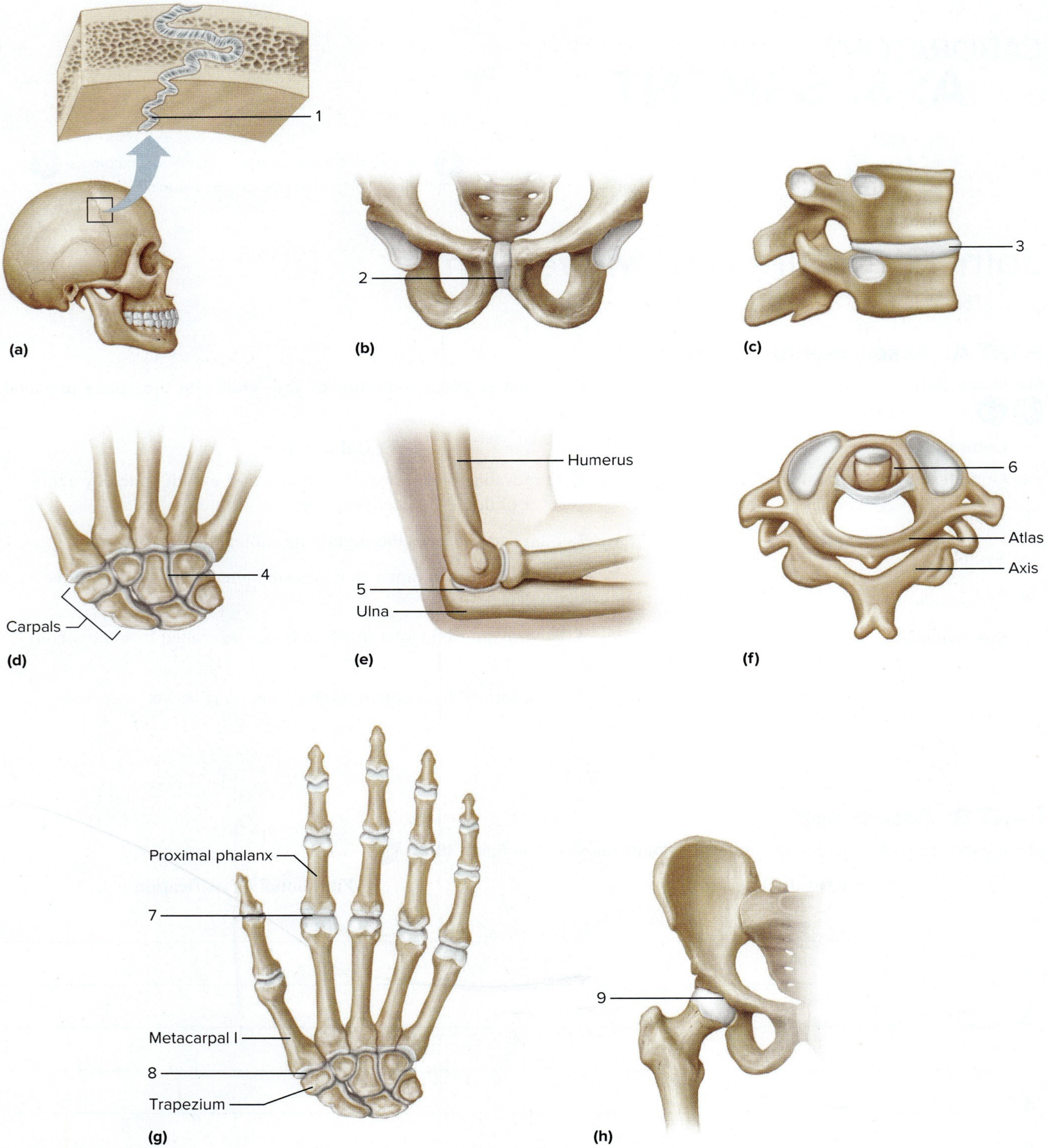

FIGURE 19.7 Identify the types of structural and functional joints numbered in these illustrations (a)–(h).

PART C: Assessments

Match the types of synovial joints in column A with the examples in column B. Place the letter of your choice in the space provided. (Some answers may be used more than once.) A3

Column A	Column B
a. Ball-and-socket	_____ 1. Hip joint
b. Condylar (ellipsoid)	_____ 2. Metacarpal–phalanx
c. Hinge	_____ 3. Proximal radius–ulna
d. Pivot	_____ 4. Humerus–ulna of the elbow joint
e. Plane (gliding)	_____ 5. Phalanx–phalanx
f. Saddle	_____ 6. Shoulder joint
	_____ 7. Knee joint between femur and tibia (main movement)
	_____ 8. Carpal–metacarpal of the thumb
	_____ 9. Carpal–carpal
	_____ 10. Tarsal–tarsal

PART D: Assessments

1. Identify the features of the knee joint indicated in figure 19.8.

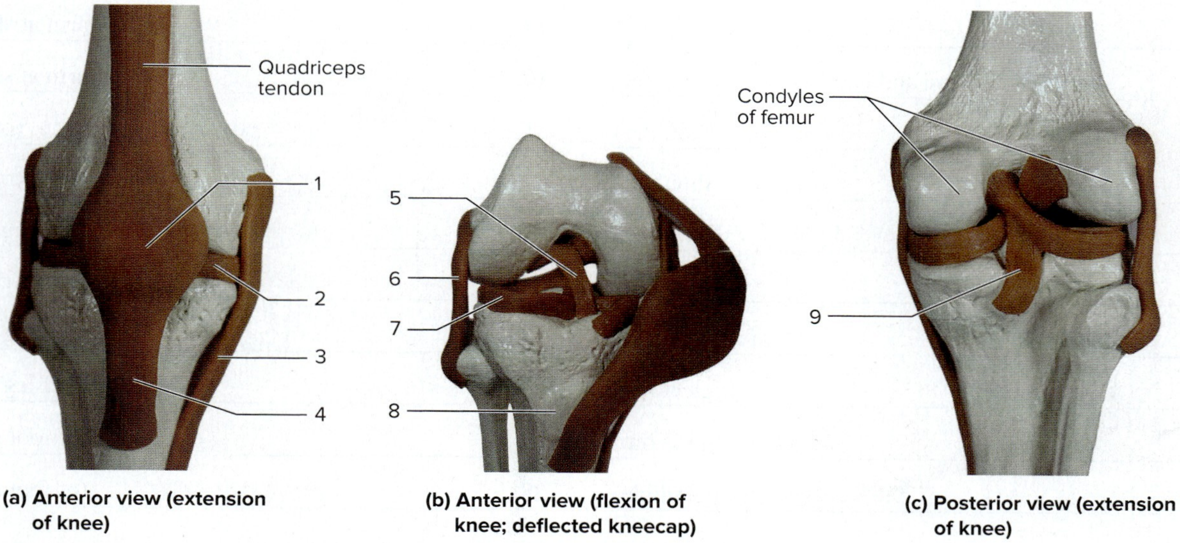

Terms:

_____Anterior cruciate ligament
_____Fibular collateral ligament
_____Medial meniscus
_____Lateral meniscus
_____Patella
_____Patellar ligament
_____Posterior cruciate ligament
_____Tibial collateral ligament
_____Tibial tuberosity

FIGURE 19.8 Label the features of the knee joint in (a) anterior view with knee joint in extension, (b) anterior view with knee joint in flexion and kneecap deflected, and (c) posterior view with knee joint in extension by placing the correct numbers in the spaces provided. A4 © 2017 Denoyer-Geppert Science Company

2. Complete the missing components of the following table: A4 A5

Name of Joint	Type of Joint	Bones Included	Types of Movement Possible
Shoulder joint		Humerus, scapula	
Elbow joint	Hinge, plane, pivot		Flexion and extension between humerus and ulna; twisting between radius and humerus; rotation between head of radius and ulna
Hip joint			Movements in all planes and rotation
Knee joint	Hinge (modified), condylar, plane		

PART E: Assessments

Identify the types of joint movements numbered in figure 19.9. A5

1. ______________________ (of head)
2. ______________________ (of shoulder)
3. ______________________ (of shoulder)
4. ______________________ (of forearm and hand at radioulnar joint)
5. ______________________ (of forearm and hand at radioulnar joint)
6. ______________________ (of arm at shoulder)
7. ______________________ (of arm at shoulder)
8. ______________________ (of wrist)
9. ______________________ (of wrist)
10. ______________________ (of thigh at hip)
11. ______________________ (of thigh at hip)
12. ______________________ (of lower limb at hip)
13. ______________________ (of chin/mandible)
14. ______________________ (of chin/mandible)
15. ______________________ (of vertebral column)
16. ______________________ (of vertebral column)
17. ______________________ (of head and neck)
18. ______________________ (of head and neck)
19. ______________________ (of arm at shoulder)
20. ______________________ (of arm at shoulder)
21. ______________________ (of elbow)
22. ______________________ (of elbow)
23. ______________________ (of thigh at hip)
24. ______________________ (of thigh at hip)
25. ______________________ (of leg at knee)
26. ______________________ (of leg at knee)
27. ______________________ (of foot at ankle)
28. ______________________ (of foot at ankle)

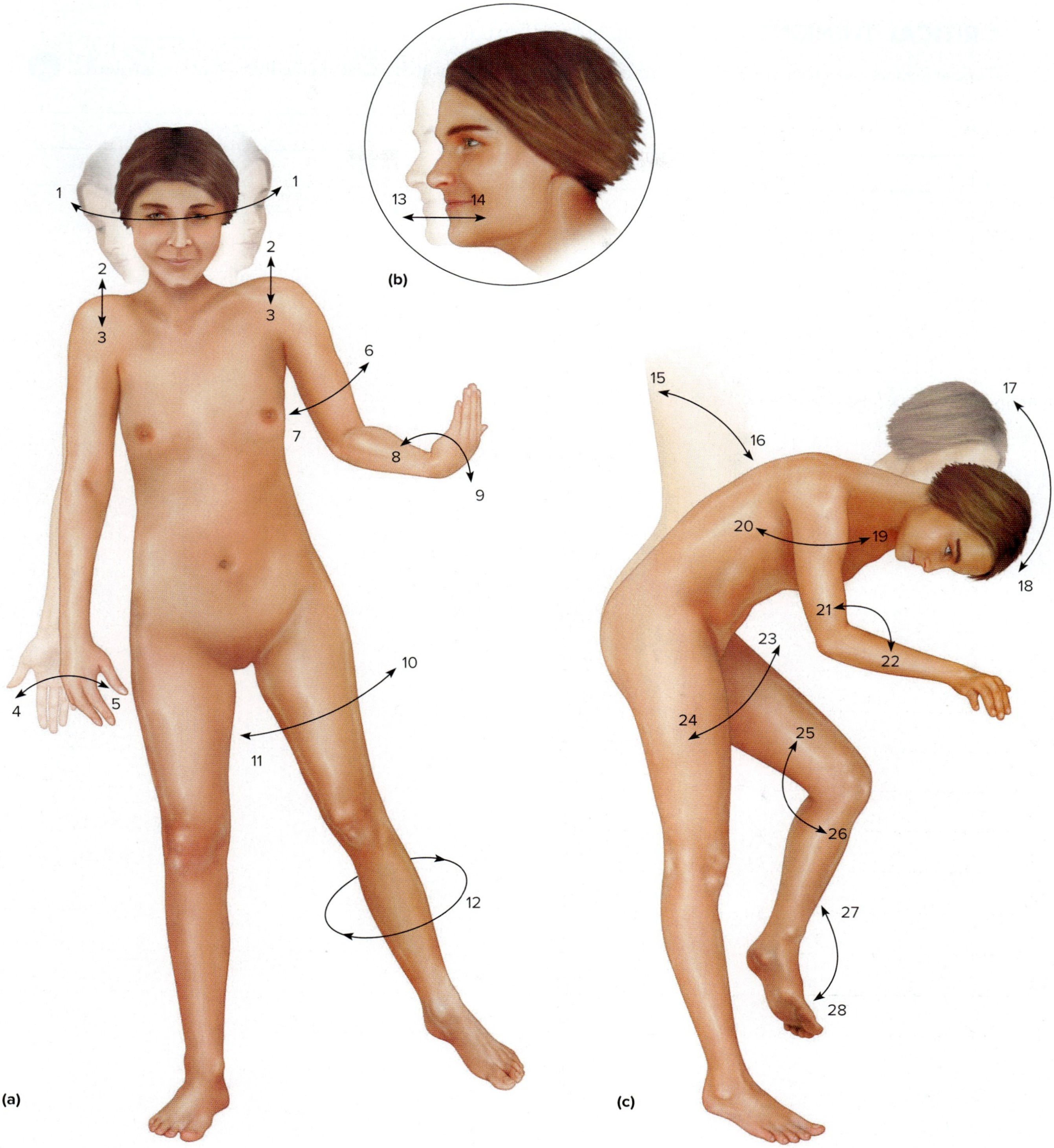

FIGURE 19.9 Identify each of the types of movements numbered and illustrated: (a) anterior view; (b) lateral view of head; and (c) lateral view. Adapted from original concept of illustration drawn by Ross Martin.

CRITICAL THINKING ASSESSMENT—PART E

Choose a common activity in everyday life and describe the various movements at the joints that are involved. A5

____

____

____

LABORATORY
EXERCISE 20

Skeletal Muscle Structure and Function

Purpose of the Exercise

To study the structure and general function of skeletal muscles as cells and as organs.

MATERIALS NEEDED

Compound light microscope

Prepared microscope slide of skeletal muscle tissue (longitudinal section and cross section)

Human muscular model

Model of skeletal muscle fiber

For Demonstration Activity:

Fresh round beefsteak

SAFETY

- Review all the safety guidelines in Appendix 1 of your laboratory manual.
- Wear disposable gloves and protective eyewear when handling the fresh beefsteak.
- Wash your hands before leaving the laboratory.

Chapter Opener: ©101cats/Getty Images

Learning Outcomes APR

After completing this exercise, you should be able to:

O1 Locate the structures of a skeletal muscle fiber (cell).

O2 Describe how connective tissue is associated with muscle tissue within a skeletal muscle.

O3 Distinguish between the origin and insertion of a muscle.

O4 Describe and demonstrate the general actions of agonists, antagonists, synergists, and fixators.

The O corresponds to the assessments A indicated in the Laboratory Assessment for this Exercise.

Pre-Lab

Carefully read the introductory material and examine the entire lab. Be familiar with skeletal muscle tissue and muscle structure and function from lecture or the textbook. Answer the pre-lab questions.

Pre-Lab Questions Select the correct answer for each of the following questions:

1. The outermost layer of connective tissue of a muscle is the
a. fascicle. **b.** endomysium.
c. epimysium. **d.** perimysium.

2. The thick myofibril filament of a sarcomere is composed of the protein
a. myosin. **b.** actin.
c. titin. **d.** sarcolemma.

3. The muscle primarily responsible for an action is the
a. synergist. **b.** agonist.
c. antagonist. **d.** origin.

4. The neuron and the muscle fibers it innervates is called the
a. neural impulse. **b.** muscle fiber.
c. motor neuron. **d.** motor unit.

5. The functional contractile unit of a muscle fiber (cell) is the
a. sarcoplasm. **b.** sarcolemma.
c. sarcomere. **d.** sarcoplasmic reticulum.

6. Bundles of muscle fibers are called
a. fascicles. **b.** cells.
c. myofibrils. **d.** myofilaments.

7. The action of a particular muscle is always the same.
a. True **b.** False

8. A synergistic muscle contraction assists the agonist.
a. True **b.** False

9. The plasma membrane of a muscle fiber (cell) is called the sarcolemma.

a. True b. False

A skeletal muscle represents an organ of the muscular system and is composed of several types of tissues. These tissues include skeletal muscle tissue, nervous tissue, and various connective tissues. It is called skeletal muscle because many of these muscles attach directly or indirectly to bones. Some skeletal muscles attach to connective tissues of other muscles or to the dermis of the skin.

Each skeletal muscle is encased within and permeated by connective tissue sheaths. The connective tissues surround and extend into the structure of a muscle and separate it into compartments. The entire muscle is encased by the *epimysium*. The *perimysium* covers bundles of cells (*fascicles*), within the muscle. The deepest connective tissue surrounds each individual muscle fiber (cell) as a thin *endomysium*. The connective tissues provide support and reinforcement during muscular contractions. The connective tissue often projects beyond the end of a muscle, providing an attachment to other muscles or to bones as a flattened, broad sheath called an *aponeurosis*, or a rope-like *tendon*. Some collagen fibers of the connective tissue are continuous with the tendon and embed through the periosteum, making for a strong structural continuity.

The *origin* of a skeletal muscle is the site of bony attachment that is the more stationary end, while the *insertion* is the attachment site that is more movable. This terminology is imperfect, however, because the origin and insertion may change depending on the *action* of a muscle, which is the movement that the muscle produces or prevents. The *shape* of a muscle influences the direction in which the muscle fibers pull and the strength of the muscle, due to the number of muscle fibers. For example, triangular muscles (e.g., pectoralis major) are generally stronger than parallel muscles (e.g., rectus abdominis).

Muscles are named according to their location, size, shape, action, attachments, number of origins, or the direction of the fibers. Examples of how muscles are named include: gluteus maximus (location and size); adductor longus (action and size); sternocleidomastoid (attachments); serratus anterior (shape and location); biceps brachii (two origins and location); and orbicularis oculi (direction of fibers and location).

Skeletal muscles are composed of many muscle fibers (cells). These muscle fibers are innervated by somatic motor neurons of the nervous system. Anatomically, each motor neuron is arranged to extend to a specific number of muscle fibers. (Note that each muscle fiber is innervated by only one motor neuron.) The motor neuron and the collection of muscle fibers it innervates is called a *motor unit*. When the motor neuron relays a neural impulse (action potential) to the muscle, all the muscle fibers that it innervates (i.e., the motor unit) will be stimulated to contract. If only a few motor units are stimulated, the muscle as a whole exerts little force. If a larger number of motor units are stimulated, the muscle will exert a greater force. Thus, to lift a heavy object like a suitcase requires the activation of more motor units in the biceps brachii than the lifting of a lighter object like a book.

PROCEDURE: Skeletal Muscle Structure and Function

1. Study a human muscular model and a skeletal muscle fiber model along with figure 20.1.
2. Examine the microscopic structure of a longitudinal section of skeletal muscle by observing a prepared microscope slide of this tissue. Use figures 20.2*a* and 20.3*a* of skeletal muscle tissue to locate the following features:

   skeletal muscle fiber (cell)

   nuclei

   striations (alternating light and dark)

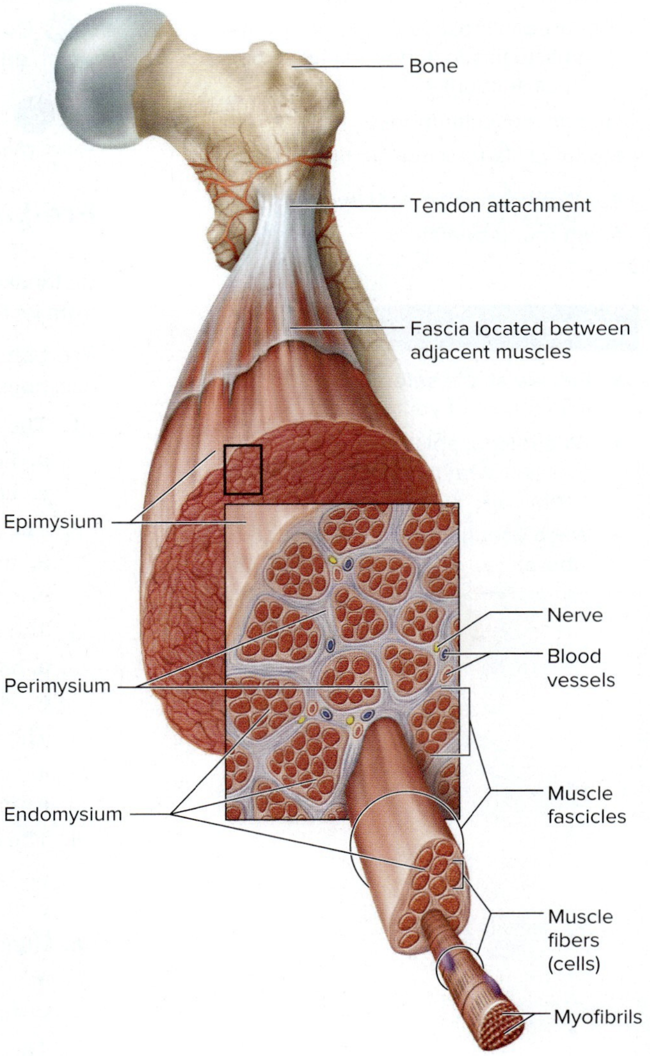

FIGURE 20.1 Skeletal muscle structure from the gross anatomy to the microscopic arrangement. Note the distribution pattern of the epimysium, perimysium, and endomysium. APR

3. Skeletal muscle cells are stimulated by impulses (action potentials) over cellular processes (axons) of motor neurons (fig. 20.2*a*). The *neuromuscular junction* is an area where the axon of a motor neuron terminates and synapses with a skeletal muscle fiber. The motor impulse results in a release of a chemical messenger, called a *neurotransmitter,* that allows the motor neuron to communicate with the muscle fiber (fig. 20.2*b*).
4. Study figures 20.1 and 20.3. Note the arrangement of muscle fibers in relation to the connective tissues of the

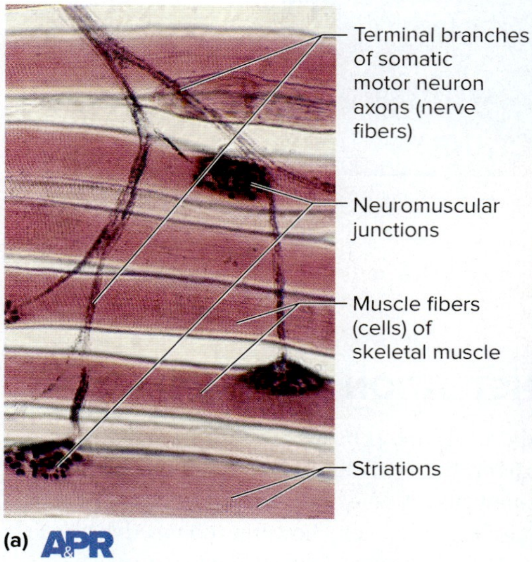

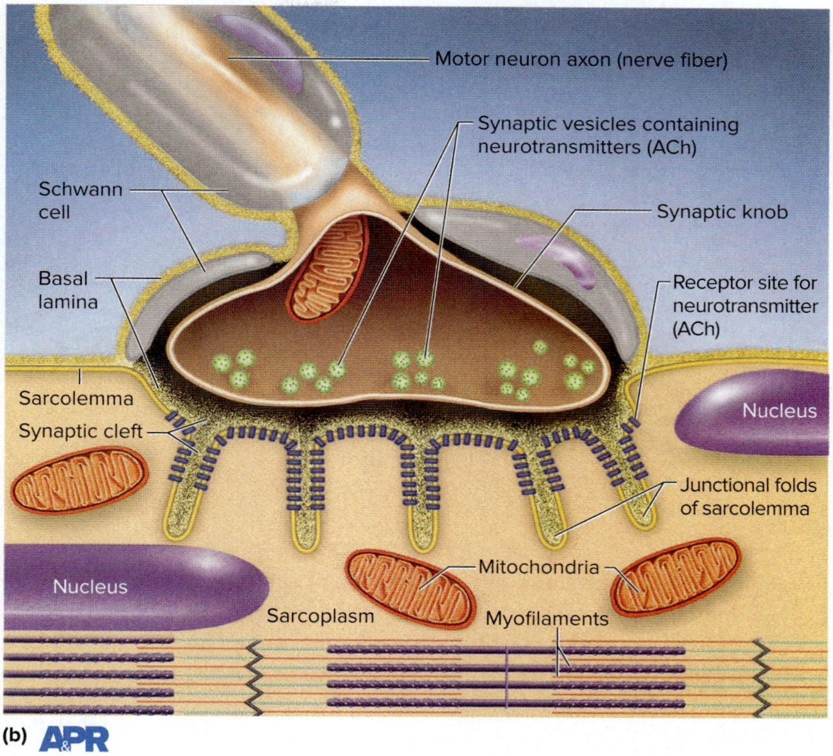

FIGURE 20.2 Skeletal muscle fibers with neuromuscular junctions: (a) micrograph of a longitudinal section (400×); (b) an illustration of one neuromuscular junction. (a) Dr. Thomas Caceci

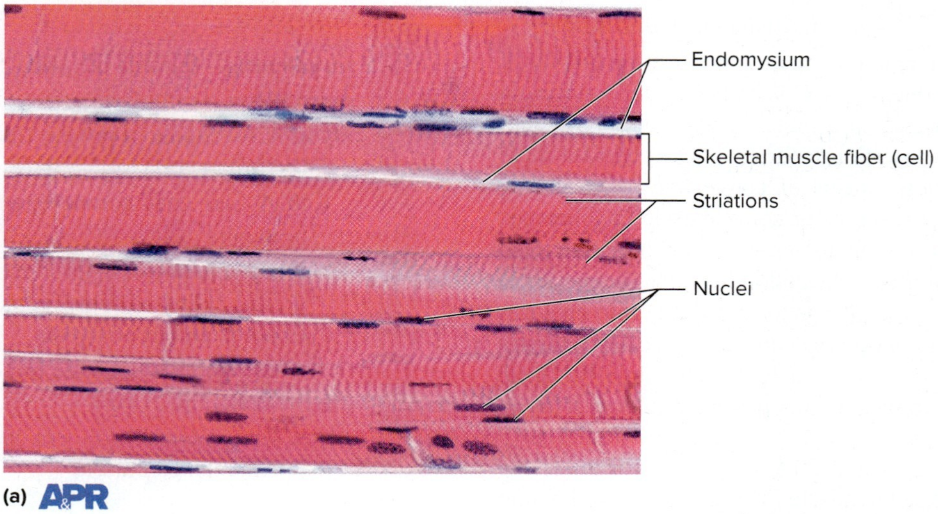

FIGURE 20.3 Micrograph of (a) longitudinal section of skeletal muscle fibers (400x), and (b) cross section of a fascicle and associated connective tissues (800×). Al Telser/McGraw Hill

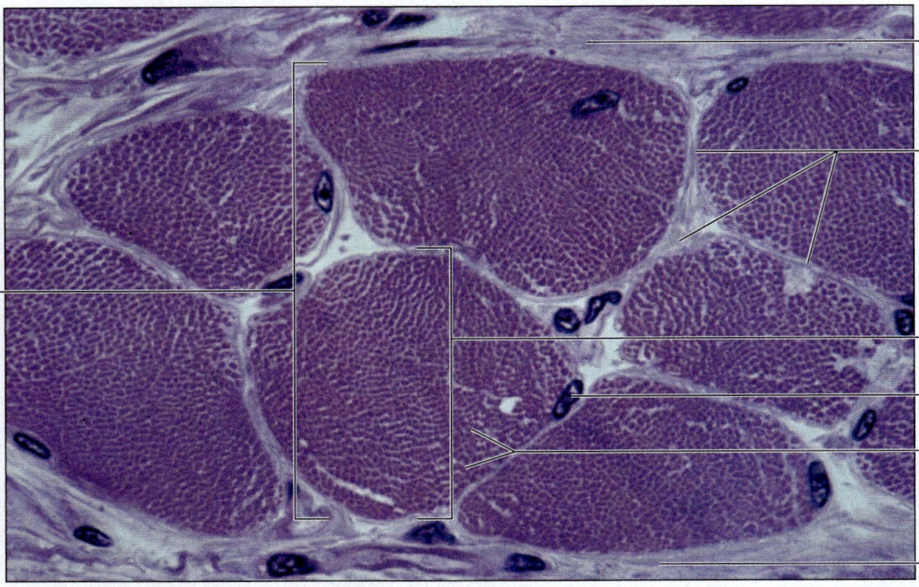

(b)

FIGURE 20.3 *Continued.* Ed Reschke/Getty Images

epimysium, perimysium, and endomysium of an individual muscle. Examine a prepared slide of the microscopic structure of a cross section of skeletal muscle tissue along with figure 20.3*b* and locate the following features:

connective tissue sheaths

- epimysium—encases entire muscle
- perimysium—surrounds fascicles
- endomysium—covers individual fibers (cells)

fascicle—bundle of fibers within a muscle

skeletal muscle fiber

- nuclei—near plasma membrane (sarcolemma)
- myofibrils—protein filaments within a fiber

5. Examine the human torso model and locate examples of tendons and aponeuroses. One end of the tendon (origin tendon) is attached to a more fixed location, whereas the other end (insertion tendon) is attached to a more movable location. Sheets of connective tissue, called aponeuroses, also serve for some muscle attachments. An example of a large aponeurosis visible on the torso or a muscle chart is in the abdominal area. It appears as a broad white sheet of connective tissue that encloses or connects the abdominal muscles. Locate examples of cordlike tendons in your body. The large calcaneal tendon is easy to palpate in a posterior ankle just above the heel.
6. Using figures 20.4 and 20.5 as a guide, examine the model of the skeletal muscle fiber. Locate the following:

sarcolemma—plasma membrane of a muscle fiber

sarcoplasm—cytoplasm of a muscle fiber

myofibril—bundle of protein filaments (myofilaments)

- thick filament—composed of contractile protein myosin
- thin filament—composed mostly of contractile protein actin and some regulatory proteins troponin and tropomyosin
  - troponin—bound to tropomyosin; receives calcium
  - tropomyosin—blocks active sites on actin in relaxed muscle

DEMONSTRATION ACTIVITY

Examine the fresh round beefsteak. It represents a cross section through the beef thigh muscles. Note the white lines of connective tissue that separate the individual skeletal muscles. Also note how the connective tissue extends into the structure of a muscle and separates it into small compartments of muscle tissue. Locate the epimysium and the perimysium of an individual muscle.

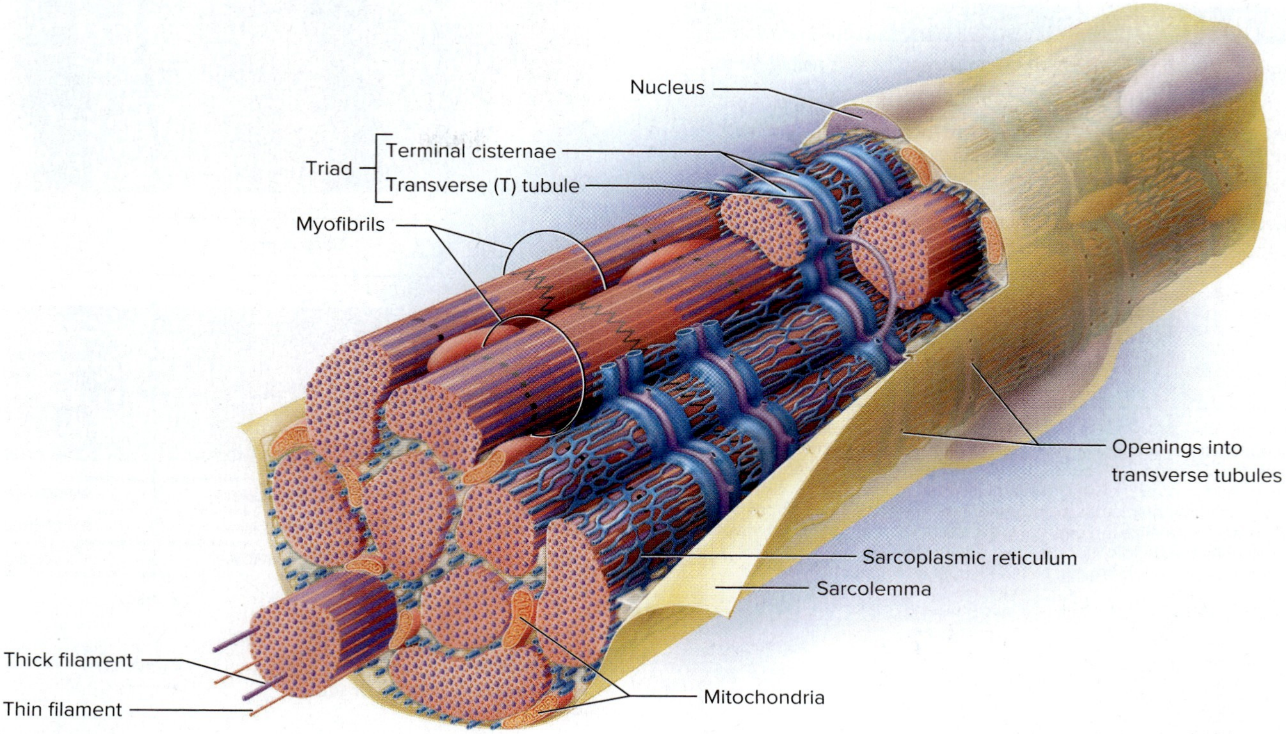

FIGURE 20.4 Structures of a segment of a muscle fiber (cell).

- elastic filament—composed of the springy protein titin

sarcoplasmic reticulum (SR)—specialized smooth ER of muscle fiber that stores calcium

- terminal cisternae—extended ends of SR next to transverse tubules

transverse (T) tubules—inward extensions of sarcolemma through the muscle fiber

sarcomere—functional contractile unit within muscle fiber (fig. 20.5)

- A (anisotropic) band—dark band of thick filaments with overlapping thin filaments
- I (isotropic) band—light band of thin filaments and bisecting Z disc
- H zone (band)—light band in middle of A band
- M line—dark line in middle of H zone
- Z disc (line)—thin and elastic filaments anchored at ends of sarcomere

7. Complete Parts A and B of Laboratory Assessment 20.

8. Study figure 20.6 and table 20.1.

9. Locate the biceps brachii, brachialis, and triceps brachii and their origins and insertions in the human torso model and in your body.

10. Make various movements with your upper limb at the shoulder and elbow. For each movement, determine which muscles are functioning as *agonists* and as *antagonists.* When an agonist contracts (shortens) and a joint moves, its antagonist relaxes (lengthens). Antagonistic muscle pairs pull from opposite sides of the same body region. As an example, consider flexion of the forearm at the elbow versus extension of the forearm at the same elbow joint. The agonist and antagonist change, depending upon which movement is occurring. *Synergistic muscles* often supplement the contraction force of an agonist, or by acting as *fixators,* they might also stabilize nearby joints. The role of a muscle as an agonist, antagonist, or synergist depends upon the movement under consideration, as their roles change.

11. Complete Part C of the laboratory assessment.

Table 20.1 Various Roles of Muscles

Functional Category	Description
Agonist (prime mover)*	Muscle primarily responsible for the action (movement)
Antagonist	Muscle responsible for action in the opposite direction of an agonist or for resistance to an agonist
Synergist	Muscle that assists an agonist, often by supplementing the contraction force
Fixator	Special type of synergist muscle; muscle contraction will stabilize a joint so another contracting muscle exerts a force on something else

*The functional category agonist can be interchangeable with prime mover. Most actions (movements) involve several synergistic muscles participating in the force. The agonist responsible for the greatest amount of the force is often called the prime mover. An example would be the rotator cuff muscles (supraspinatus, infraspinatus, teres minor, and subscapularis) whose tendons act to stabilize the head of the humerus at the glenoid cavity of the scapula while the deltoid muscle lifts (abducts) the arm. These S.I.T.S. muscles are special synergists, called fixators, to the agonist (deltoid).

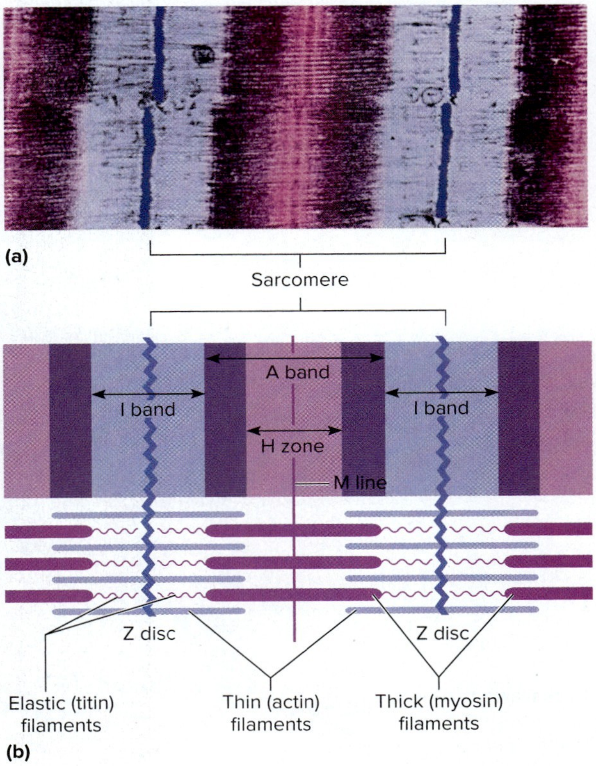

FIGURE 20.5 Sarcomere contractile unit of a relaxed muscle fiber: (a) micrograph (16,000×); (b) illustration of the repeating pattern of striations (bands). (a) H.E. Huxley APR

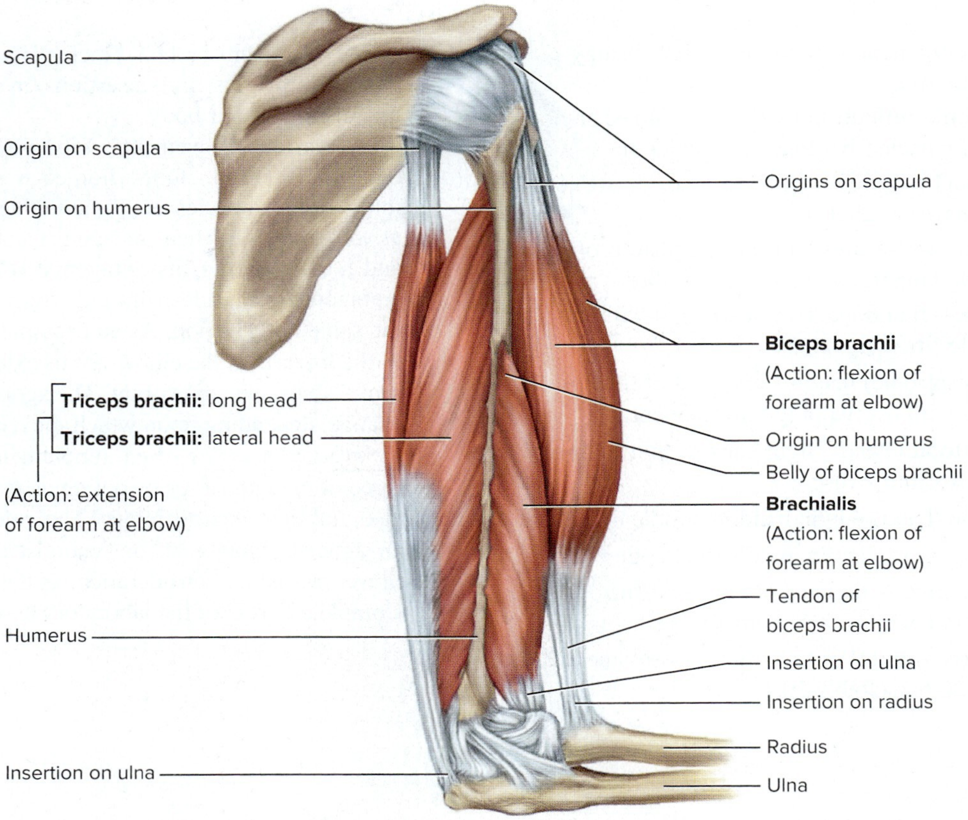

FIGURE 20.6 Antagonistic muscle pairs (in bold) are located on opposite sides of the same body region as shown in an arm (posterior view). The muscle acting as the agonist depends upon the movement that occurs.

LABORATORY
ASSESSMENT 20

Name ____________________

Date ____________________

Section ____________________

The (A) corresponds to the indicated Learning Outcome(s) (O) found at the beginning of the Laboratory Exercise.

Skeletal Muscle Structure and Function

PART A: Assessments

Match the terms in column A with the definitions in column B. Place the letter of your choice in the space provided. (A1) (A2)

Column A	Column B
a. Endomysium	____ 1. Protein that comprises part of the thin filament with actin
b. Epimysium	____ 2. Cytoplasm of a muscle fiber
c. Fascia	____ 3. Connective tissue located between adjacent muscles
d. Fascicle	____ 4. Layer of connective tissue that separates a muscle into small bundles called fascicles
e. Myosin	____ 5. Plasma membrane of a muscle fiber
f. Perimysium	____ 6. Layer of connective tissue that surrounds a skeletal muscle
g. Sarcolemma	____ 7. Unit of alternating light and dark striations between Z discs (lines)
h. Sarcomere	____ 8. Layer of connective tissue that surrounds an individual muscle fiber
i. Sarcoplasm	____ 9. Cellular organelle in muscle fiber corresponding to the endoplasmic reticulum
j. Sarcoplasmic reticulum	____ 10. Cordlike part that attaches a muscle to a bone
k. Tendon	____ 11. Protein found within thick filament
l. Troponin	____ 12. Small bundle of muscle fibers within a muscle

PART B: Assessments

Provide the labels for the electron micrograph in figure 20.7.

___ A band (dark)
___ H zone
___ I band (light)
___ M line
___ Sarcomere
___ Z disc

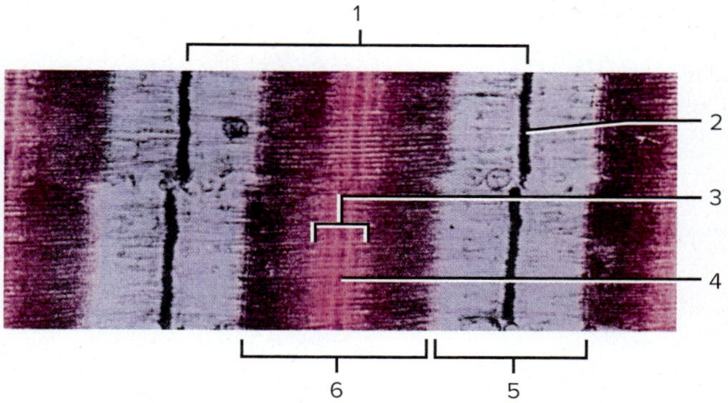

FIGURE 20.7 Label this transmission electron micrograph (16,000×) of a relaxed sarcomere by placing the correct numbers in the spaces provided. H.E. Huxley (A1)

CRITICAL THINKING ASSESSMENT—PART B

When stimulated to contract, the sarcomeres shorten ⟶ myofibrils shorten ⟶ muscle fiber shortens ⟶ the muscle shortens. Complete the table by indicating the changes (if any) in two adjacent sarcomeres when they move from a relaxed state to a contracted state. A1

Contractile Unit	Changes that Occur in Relaxed ⟶ Contracted Sarcomeres
A band	
I band	
H zone	
Z discs	

PART C: Assessments

Complete the following statements:

1. The ________________________ of a muscle is attached to a more fixed location. A3
2. The ________________________ of a muscle is attached to a more movable location. A3
3. A muscle responsible for most of an action is called a(n) ________________________. A4
4. Assisting muscles of an agonist are called ________________________. A4
5. Antagonists are muscles that resist the actions of ________________________ and cause movement in the opposite direction. A4
6. When the forearm is extended at the elbow joint, the ________________________ muscle acts as the agonist. A4
7. When the brachialis acts as the agonist in forearm flexion, the ________________________ muscle assists as a synergist. A4
8. A ________________________ is a synergistic muscle that will stabilize a joint when another contracting muscle exerts a force on something else. A4

LABORATORY
EXERCISE 21

Electromyography: BIOPAC® Exercise

Purpose of the Exercise

To measure and calculate fist clench strength from electromyograms obtained using the BIOPAC system and to listen to the EMG "sounds"; to measure and calculate the force generated during motor unit recruitment and to observe muscle fatigue.

MATERIALS NEEDED

Computer system [Mac OS X 10.7—10.15 (Catalina), or PC running Windows 10, 8, or 7]

Biopac Student Lab software version 4.0 or above

BIOPAC Acquisition Unit MP36, MP35, or MP46/45

BIOPAC electrode lead set (SS2L)

BIOPAC disposable vinyl electrodes (EL503), six electrodes per subject

BIOPAC electrode gel (GEL1) and abrasive pad (ELPAD), or alcohol wipes, or another type of skin cleanser

BIOPAC Headphones (OUT1/OUT1A for MP36/35 or 40 HP for MP46/45) optional (*Note:* OUT1A works only with the MP36)

BIOPAC Hand Dynamometer (SS25LA, SS25LB, or SS25L), or Hand Clench Force Pump Bulb (SS56L) for Part 2. *Notes:* If you are using the SS56L, the Clench Force Transducer Preference must be set before the calibration step. The SS25LB is compatible only with BSL 4.1 or higher.

SAFETY

- The BIOPAC electrode lead set is safe and easy to use and should be used only as described in the procedures section of the laboratory exercise.
- The electrode lead clips are color-coded. Make sure they are connected to the properly placed electrodes as demonstrated in figure 21.6.
- The vinyl electrodes are disposable and meant to be used only once. Each subject should use a new set.

Learning Outcomes

After completing this exercise, you should be able to:

O1 Record the maximum clench strength for dominant and nondominant hands.

O2 Examine, record, and correlate motor unit recruitment with increased power of skeletal muscle contraction.

O3 Listen to EMG "sounds" and correlate sound intensity with motor unit recruitment.

O4 Examine, record, and calculate motor unit recruitment with increased power of skeletal muscle contraction.

O5 Determine the maximum clench force between dominant and nondominant hands (and male and female, if possible).

O6 Record the force produced by clenched muscles during fatigue.

O7 Calculate the time it takes clenched muscles to fatigue.

The O corresponds to the assessments A indicated in the Laboratory Assessment for this Exercise.

Pre-Lab

Carefully read the introductory material and examine the entire lab. Be familiar with motor unit recruitment and muscle fatigue from lecture or the textbook.

This is the first of a series of laboratory exercises that will utilize the BIOPAC computer-based data acquisition and analysis system. This is a complete system consisting of hardware and software, which provides data similar to that obtained by a physiograph. Once the software package has been loaded onto your computer, each lesson will use a different set of hardware. Connections and directions for using the instrumentation will be explained in each lesson.

The hardware includes the MP36/35 or MP46/45 Data Acquisition Unit, connection cables, transformers, transducers, electrodes and electrode cables, and other accessories. The electrodes and other accessories often contain sensors that pick up a signal and send it to the MP36/35 or MP46/45 Data Acquisition Unit. The software picks up electrical signals coming into the MP36/35 or MP46/45 Data Acquisition Unit, which are then displayed on the screen as a waveform. The software then guides the user through the lesson and manages data saving and review.

The BIOPAC computer-based data acquisition and analysis system works in a manner similar to other recording equipment in sensing electrical signals in the body. See figure 21.1 for the basic setup of the Biopac Student Lab System. The major advantage of a computerized system such as the BIOPAC system is that it performs dynamic calibration. With other equipment such as a physiograph, the instructor or student has to manually adjust the dials and settings to obtain a good signal for the recording. With the Biopac Student Lab, the adjustments are made automatically for each subject during the calibration procedure in each lesson.

Before you start any lesson, the software should already be installed on the computer, and you should read through the lesson completely. The Biopac Student Laboratory Manual that comes with the system has a full description of the lessons and some troubleshooting information. The lessons as described in this manual are shortened versions of these. *Note:* In the most recent upgrades of the Biopac Student Lab System, versions 4.0 and 4.1, brief instructions for lab procedures are displayed in the Journal area below the data window. As you perform the lab, you may follow along with the brief instructions in the Journal, but you will have to refer to this lab manual for more detailed instructions. The instructions for data analysis are not provided in the Journal area, so you will have to refer to the lab manual for them. Figure 21.2 and table 21.1 have been included here as a quick reference guide to the BIOPAC Display Tools for analysis as you proceed through the BIOPAC lessons.

In Laboratory Exercise 21, you will investigate the electrical properties of skeletal muscle using *electromyography.* This is a recording of skin-surface voltage that is produced by underlying skeletal muscle contraction. The voltage produced by contraction is weak, but it can be picked up by surface sensors to produce the recording, referred to as an *electromyogram (EMG).*

Skeletal muscle is one of three types of muscle tissue in the body. Skeletal muscle is mainly attached to the various bones of the skeleton; it helps provide a framework and support for the body, as well as the ability to move in response to willing an action to occur. (Cardiac muscle is found only in the heart, where it originates and spreads the heartbeats that keep the blood circulating around the body. Smooth muscle is located mostly in the tubular organs of our internal systems; it accomplishes many types of internal movement, such as the churning action of the stomach and the transport of urine from the kidney to the bladder.)

Each skeletal muscle in our body is composed of thousands of skeletal muscle fibers (cells). Skeletal muscle fibers are unique among the three types of muscle tissue in that they contract only in response to stimulation by a motor neuron. The response of a single skeletal muscle fiber to stimulation by its motor neuron is called a *twitch.* A twitch consists of a rapid period of tension development and contraction, followed by a decrease in tension and relaxation of the muscle.

Each motor neuron associated with skeletal muscle fibers innervates (stimulates, controls) a certain number of fibers,

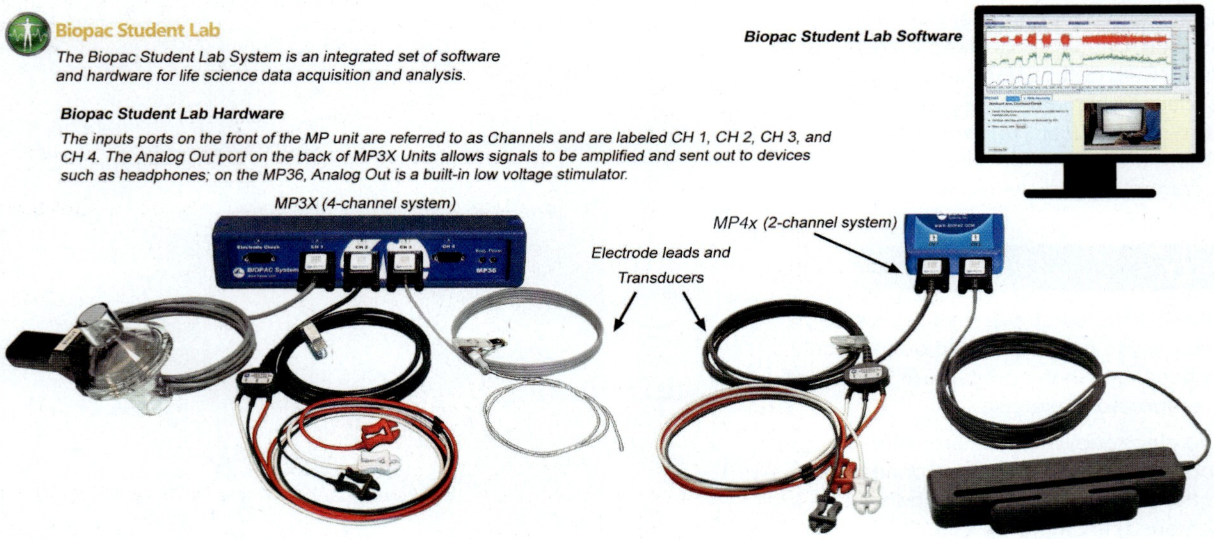

FIGURE 21.1 The basic setup of the Biopac Student Lab System includes a computer, an MP36/35 Data Acquisition Unit, and various transducers.

Courtesy of and © *BIOPAC* Systems, Inc. www.biopac.com

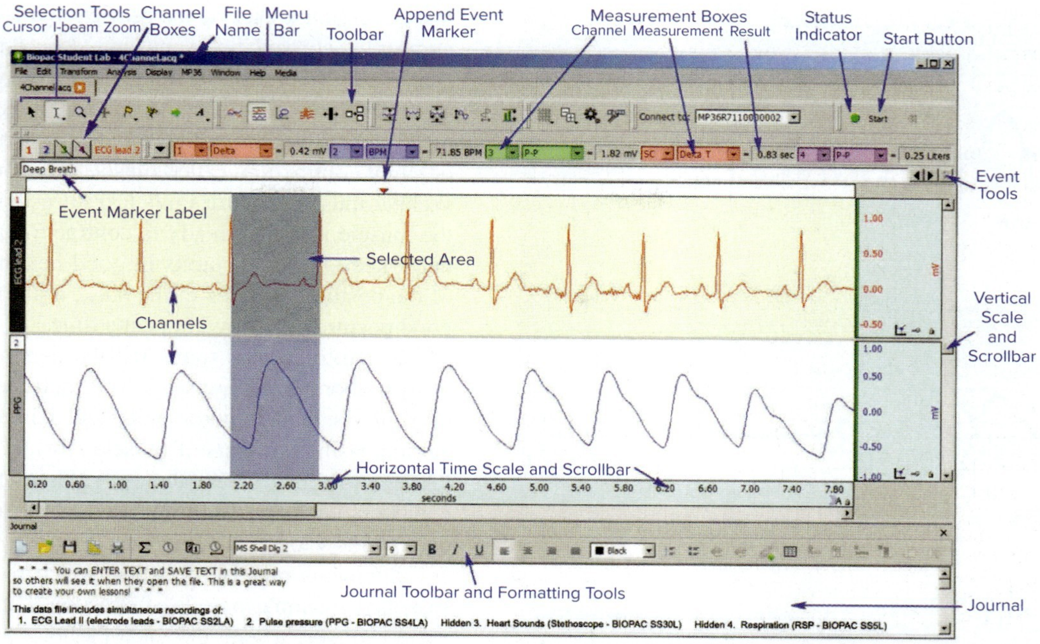

FIGURE 21.2 Basic setup and icons of the BIOPAC display window.

Courtesy of and © *BIOPAC* Systems, Inc. www.biopac.com

Table 21.1 Display Tools for Analysis

Viewing Tool	Overview
Editing and Selection Tools	The **Editing and Selection tools** icons are located in the upper left corner of the display window, just above the channel boxes. The **Arrow Tool** is a general-purpose cursor, used for selecting waveforms and scrolling through data.
I-Beam Tool	The **I-beam** icon is located in the upper left corner of the display window. The **I-beam Tool** is used to select a point or an area for measurement. To use it for a point measurement, click on the **I beam,** and click on the point that is to be used for a measurement. This will bring up a flashing line, and the measurement will be taken. To activate it for an area of measurement, click on it and move the mouse such that the cursor is positioned at the beginning of the region that you want to select. The cursor *must* be positioned within the data window. Hold the mouse button down and move (drag) the cursor to the second position. Release the mouse button to complete the selection. The selected area should remain darkened. Measurements will apply to the selected area.
Zoom Tool	The **Zoom** icon is in the upper left corner of the display window. Once selected, the **Zoom Tool** allows you to define an area of the waveform by click-hold-drag-releasing the mouse over the desired section. This draws a rectangle around the selected area. The horizontal (time) and vertical (amplitude) scales will expand to display just the selected section.
Zoom Back	Click on the **Display** menu to access this option. After you have used the **Zoom Tool, Zoom Back** will revert the display to the previous scale settings.
Horizontal (time) Scroll Bar	The **Horizontal Scroll Bar** is located below the horizontal time scale. It only works when you are viewing just a portion of the waveform data. To move the display's time position, click on the scroll box arrows or hold the mouse button down on the scroll box and drag the box left (earlier) or right (later).
Vertical (amplitude) Scroll Bar	The **Vertical Scroll Bar** is located to the right of the vertical amplitude (mV) scale. To move the selected channel's vertical position, click on the scroll box or hold the mouse button down on the scroll box and drag the box up or down.
Autoscale waveforms	Click on the **Display** menu to access this option. Optimizes the vertical (amplitude) scale so that all the data will be shown on the screen.
Autoscale horizontal	Click on the **Display** menu to access this option. Compresses or expands the horizontal (time) scale so the entire recording will fit on the screen.
Grids	Grids provide a quick visual guide to determine amplitude and duration. To turn the grids display on or off, click on the **Show Grids** or **Hide Grids** button below the Menu Bar.
Adjust Baseline button	The **Adjust Baseline** button only appears in Laboratory Exercise 46, ECG. Allows you to position the waveform up or down in small increments so that the baseline can be set to exactly zero. This is not needed to get accurate amplitude measurements, but it may be desired before making a printout or when using grids. When the **Adjust Baseline** button is pressed, **Up** and **Down** buttons will appear. Simply click on these to move the waveform up or down.

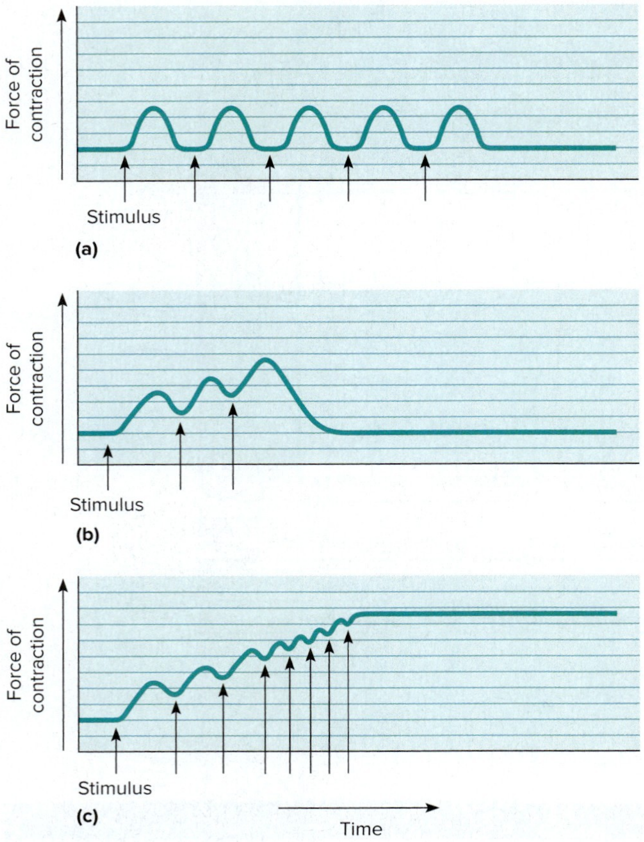

FIGURE 21.3 Myograms of (a) isolated muscle twitches, (b) temporal summation, and (c) a tetanic contraction.

ranging from fewer than 10 to thousands. A single motor neuron plus all of the muscle fibers it controls is called a *motor unit.* In a motor unit, whenever the motor neuron is activated, all of its muscle fibers are commanded to contract simultaneously. Therefore, in real life, entire motor units, rather than single muscle fibers, are contracting to accomplish body movements.

When a single muscle fiber is stimulated by its motor neuron to contract, then relaxes completely, and then contracts again, the subsequent contractions all produce the same amount of muscle tension (fig. 21.3*a*). However, if a muscle fiber is stimulated again, before it has time to relax completely, then the second twitch will generate more tension than the first. This is called *temporal (wave) summation,* and it is the body's way of increasing the force generated by muscles (fig. 21.3*b*). When a muscle fiber is stimulated to contract by several stimuli in rapid succession, and there is no opportunity for muscle relaxation between stimuli, the tension increases until it reaches the maximum possible force for that muscle fiber, and then remains at that level until the stimulation stops; this type of sustained contraction is called *tetanus,* or a *tetanic contraction* (fig. 21.3*c*).

Although there are many motor units in a skeletal muscle, not all of them are stimulated during every muscle contraction. For example, if a person decides to pick up a 1 lb. weight, just a small number of motor units will be activated to contract. If, however, that person picks up a 10 lb. weight instead, many more motor units will be stimulated, in order to generate enough muscle tension to pick up the heavier object. An increase in the number of activated motor units in a skeletal muscle, in order to increase the tension in the muscle, is called *motor unit recruitment.* See figure 21.4 for an illustration of motor unit recruitment.

When skeletal muscles are at rest, they maintain a state of slight contraction, called muscle tone, or *tonus.* In resting skeletal muscles, motor units take turns contracting, so that the muscle is always ready to contract on short notice. This helps keep the motor units in good health, and in the case of the postural muscles in the back, it allows us to maintain good posture for many hours at a time.

Recruitment and tonus will be demonstrated in Part 1 of this laboratory exercise; as you increase the clench force of your fist, more motor units will be recruited to generate increasing amounts of muscle tension. You will be able to observe tonus in the short periods of time between fist clenches. The periods of tonus will show up in the BIOPAC data window as very low mV readings.

Strenuous exercise or work, which requires repeated or sustained maximal muscle contraction, sometimes results in *muscle fatigue,* a state in which the force of muscle contraction decreases, even when the muscle is still being stimulated. Several factors appear to contribute to muscle fatigue. One of the main causes seems to be a buildup of lactic acid from the anaerobic respiration of glucose, the resulting decrease in blood pH in the vicinity, and decreased responsiveness of the muscle at low pH. In other cases, muscle fatigue can be attributed to conduction failure, in which an accumulation of K^+ ions in the transverse tubules of skeletal muscle fibers eventually leads to an inability to produce subsequent action potentials. Part 2 of this lab exercise will demonstrate muscle fatigue. You will use a *hand dynamometer* or hand clench force pump bulb to generate a recording called a *dynagram.* See figure 21.5 for the appearance and proper grip positions of two different types of dynamometers and the pump bulb. Muscle fatigue will be induced by a sustained, maximum fist clench, and will appear on the dynagram as a steady decline in muscle tension.

PART 1: Electromyography: Standard and Integrated EMG

PROCEDURE A: Setup

1. With your computer turned **ON** and the BIOPAC MP36/35 unit turned **OFF,** plug the electrode lead set (SS2L) into Channel 1 and plug the headphones into the back of the unit if desired. If you are using the MP46/45, the USB cable must be connected, and the ready light must be ON. Plug the BIOPAC SS25LA/LB or SS25 Hand Dynamometer or pump bulb into Channel 2 only if you are going to do Part 2 of this lab exercise. The dynamometer or pump bulb is not used in Part 1.
2. Turn on the MP36/35 Data Acquisition Unit.
3. Attach the electrodes to both forearms of the subject, as shown in figure 21.6. To help to ensure a good contact, make sure the area to be in contact with the electrodes is clean by wiping with the abrasive pad (ELPAD), skin cleanser, or an alcohol wipe. A small amount of BIOPAC

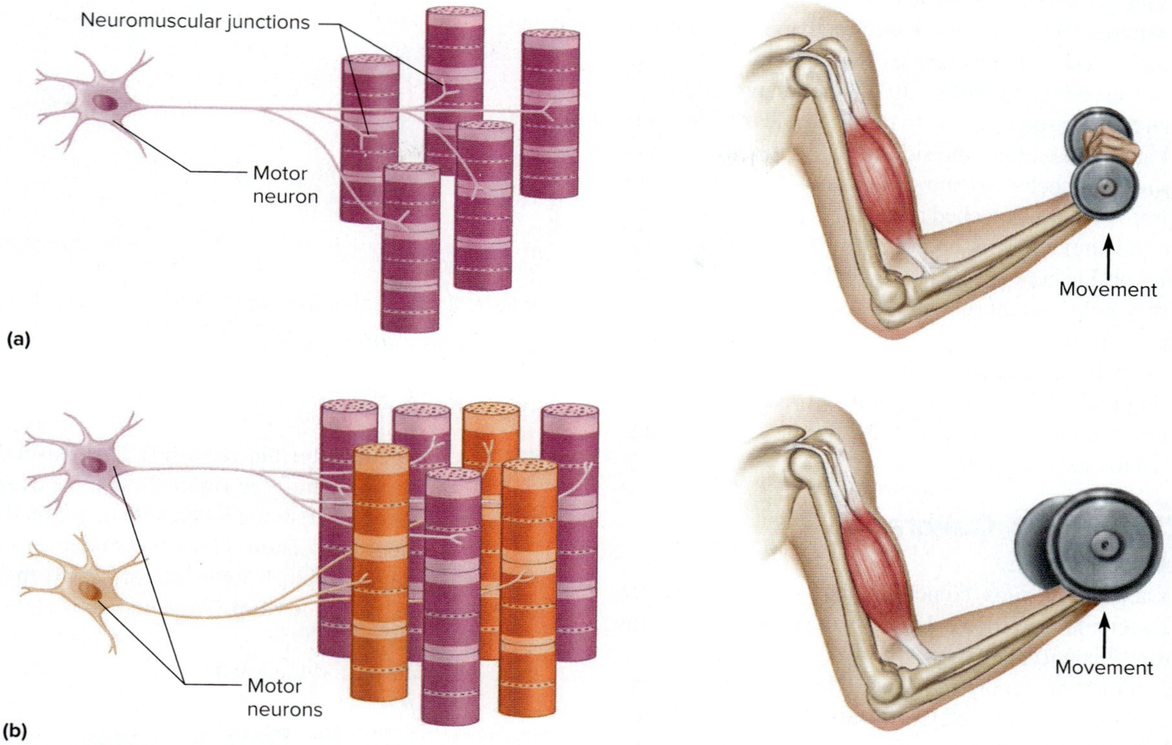

FIGURE 21.4 Motor unit recruitment. (a) A small number of activated motor units only generates enough force to pick up a lightweight object. (b) Recruitment of more motor units has increased the force to the extent that a heavier object can be lifted.

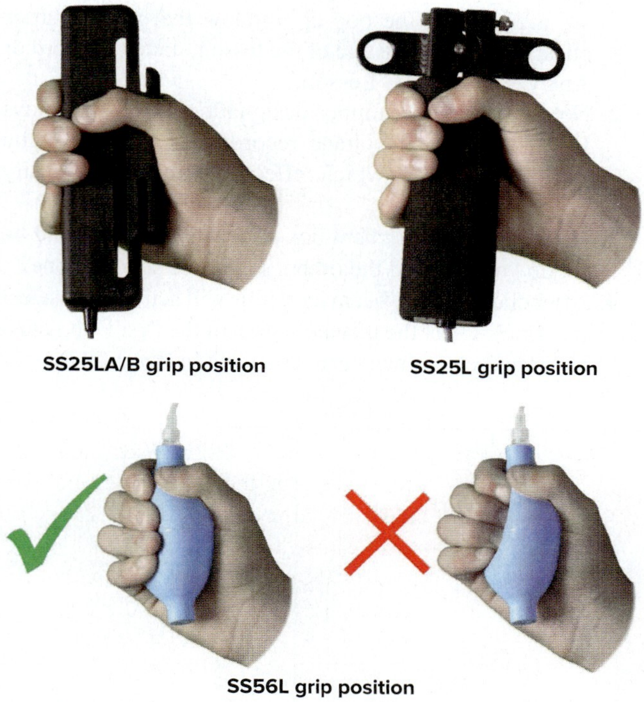

FIGURE 21.5 These are the various types of hand dynamometers and/or hand clench force pump bulb that may be available in your lab. Notice the proper grip position for each type.

Courtesy of and © *BIOPAC* Systems, Inc. www.biopac.com

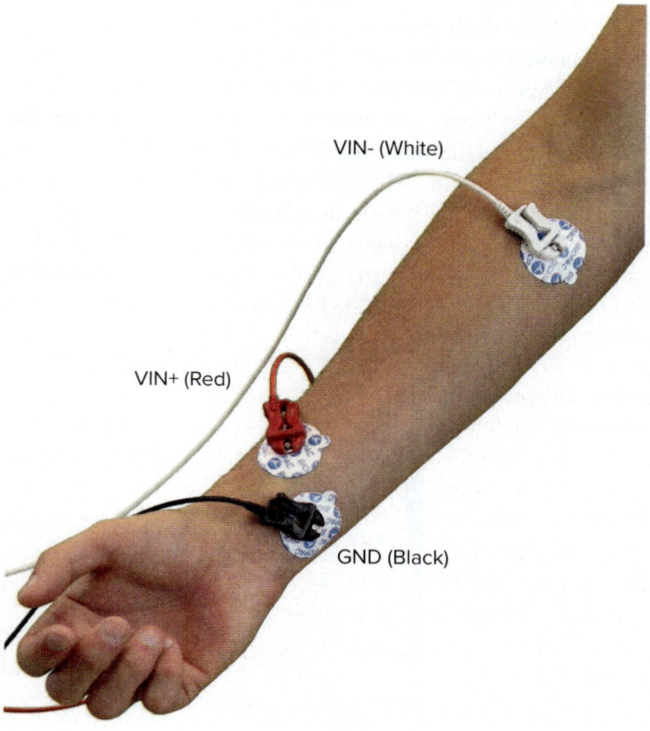

FIGURE 21.6 Electrode lead attachment.

Courtesy of and © *BIOPAC* Systems, Inc. www.biopac.com

electrode gel (GEL1) may also be used to make better contact between the sensor in the electrode and the skin. For optimal adhesion, the electrodes should be placed on the skin at least 5 minutes before the calibration procedure.

4. Attach the electrode lead set to the electrodes on the dominant arm (most often, the side of the writing hand), via the pinch connectors as shown in figure 21.6. Make sure the proper color is attached to the appropriate electrode.
5. Have the subject sit in a chair, with the dominant arm resting on his/her thigh, electrodes facing up. The subject may hold a small rubber ball to clench during Part 1 of the exercise.
6. Start the Biopac Student Lab Program for Lesson 1, by clicking on L01-Electromyography (EMG) I.
7. Designate a unique filename to be used to save the subject's data and click OK.

PROCEDURE B: Calibration

1. Click on **Calibrate.**
2. Wait about 2 seconds, clench the fist as hard as possible for 2 seconds, and then release. Wait for the calibration to stop (this will take about 8 seconds).
3. The calibration recording should look similar to figure 21.7. If it does, click **Continue.** If it does not, click **Redo Calibration.** If the baseline is not at zero, the subject did not wait 2 seconds before clenching; in this case, you should redo the calibration. Also, if the recording shows a flat line, check all connections and redo the calibration.

PROCEDURE C: Recording

1. Fill out the subject profile in Laboratory Assessment 21. Click on **Record** and have the subject conduct a series of four clench-release-wait cycles, holding for 2 seconds during the clench and wait phases. During the first cycle the clench should be gentle; then increase the grip with each successive cycle so that the fourth clench is of maximum force.
2. Click **Suspend.** The recording should be similar to figure 21.8. If it is not, click **Redo** and repeat step 1.
3. Switch the pinch connectors of the electrode lead set to the nondominant arm. Click Continue, then click **Record,** and repeat from step 1 of Procedure C: Recording.

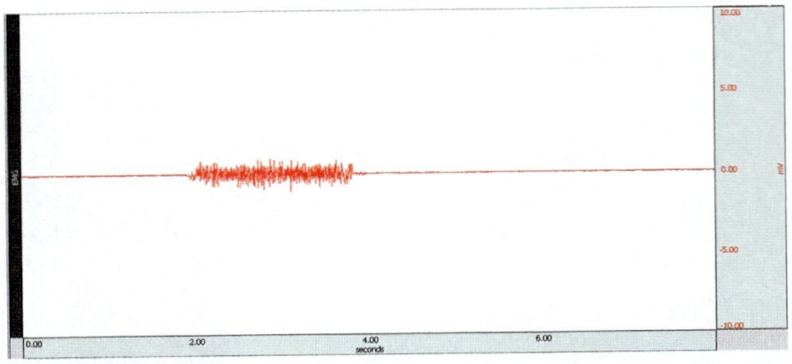

FIGURE 21.7 Calibration recording, part 1 (single clench).

Courtesy of and © *BIOPAC* Systems, Inc. www.biopac.com

4. Click **Suspend** after the recording has finished. If the recording looks similar to figure 21.8, click **Stop.**
5. If you want to listen to the EMG signal, put on the headphones and click **Listen.** Have the subject experiment by changing the clench force as you watch the screen. When finished, click **Stop.** These data will not be saved for analysis.
6. Click **Done** to end the lesson.

PROCEDURE D: Data Analysis

1. You can either click **Analyze current date file** now, to proceed directly to the data analysis, or save your file and do the analysis later, using the **Review Saved Data** mode and selecting the correct file. Both options are available in the pop-up window that appears after clicking **Done** at the end of the lesson, under "Record or Analyze a BIOPAC Lesson."
2. Note the channel number designations: **CH 1** displays **EMG** (the actual voltage recording) and **CH 40** the **Integrated EMG** (which reflects the absolute intensity of the voltage).
3. Notice the measurement box settings: **CH 40** should be set on "mean," and the others should be set on "none."
4. After clicking the I-beam icon (this will activate the select function), select the plateau region of the first EMG cluster from the dominant arm, as shown in figure 21.9.

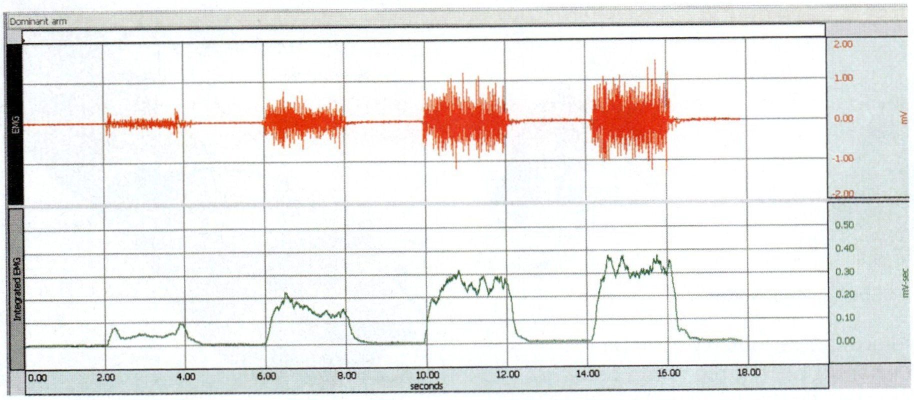

FIGURE 21.8 Sample recording of four clench-release-wait cycles of increasing strength.

Courtesy of and © *BIOPAC* Systems, Inc. www.biopac.com

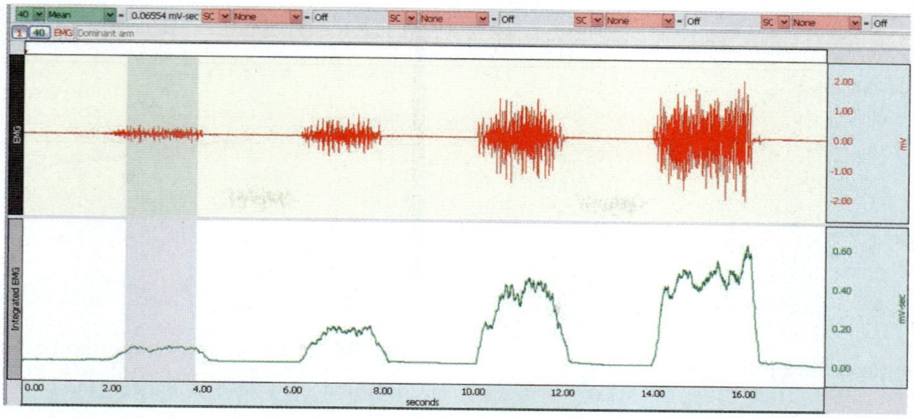

FIGURE 21.9 Proper selection of EMG data for one cluster.

Courtesy of and © *BIOPAC* Systems, Inc. www.biopac.com

5. The values in the measurement boxes should be recorded in the table in Part IA of Laboratory Assessment 21.
6. Repeat for the other three clusters and record the data in Part IA of the laboratory assessment. Then record the data for all four clusters from the nondominant arm.
7. Complete Part I of the laboratory assessment.
8. If you are not going to complete Part 2, then exit the program by clicking **File,** and **Quit.**

PART 2: Electromyography: Motor Unit Recruitment and Fatigue

PROCEDURE A: Setup

If continuing with the same subject used in Part 1, move to setup step 5 below.

1. With your computer turned **ON** and the BIOPAC MP36/35 unit turned **OFF,** plug the electrode lead set (SS2L) into Channel 1 and BIOPAC Hand Dynamometer (SS25L or SS25LA/LB), or BIOPAC Hand Clench Force Pump Bulb (SS56L) into Channel 2. If you are using the MP46/45, the USB cable must be connected, and the ready light must be ON. Plug the headphones into the back of the unit if desired.
2. Turn on the MP36/35 Data Acquisition Unit.
3. If you did not complete Part 1 of the lab, attach the electrodes to both forearms of the subject, as shown in figure 21.6. To help to ensure a good contact, make sure the area to be in contact with the electrodes is clean by wiping with the abrasive pad (ELPAD) or alcohol wipe. A small amount of BIOPAC electrode gel (GEL1) may also be used to make better contact between the sensor in the electrode and the skin. For optimal adhesion, the electrodes should be placed on the skin at least 5 minutes before the calibration procedure.
4. Attach the electrode lead set to the electrodes on the dominant arm (most often, the side of the writing hand), via the pinch connectors, as shown in figure 21.6. Make sure the proper color is attached to the appropriate electrode.
5. Have the subject sit in a chair, with the dominant arm resting on his/her thigh, electrodes facing up.
6. Start the Biopac Student Lab Program for Electromyography II, lesson L02-EMG-2.
7. Designate a unique filename to be used to save the subject's data, then click **OK.**
8. Remember that if you are using the pump bulb, you will have to click on **File,** and **Lesson Preferences,** and set the preference for the SS56L before the calibration step.

PROCEDURE B: Calibration

1. Click **Calibrate.** Set the hand dynamometer or pump bulb on the table and click **OK.** This establishes a zero-force calibration. This portion of the calibration will run for only 1–2 seconds.
2. Grasp the hand dynamometer or pump bulb with the dominant hand when prompted. See figure 21.5 for proper grip positions. Note the position of your hand and try to replicate this during the recording phase.
3. Follow the instructions on the two successive pop-up windows and click **OK** when ready for each.
4. Wait 2 seconds, then clench the hand dynamometer or pump bulb as hard as possible, then release.
5. Wait for the calibration to stop. This will take about 8 seconds.
6. If the calibration recording resembles figure 21.10, click **Continue.** If it does not, click **Redo Calibration.**
7. Before proceeding to the recording procedure, the software will determine the force increments that you will use for the recruitment recordings. The increments will be shown on the right side of the data window. Using the maximum force from the calibration of a hand dynamometer, the increments will be set as follows: for 0–25 kg, 5 kg increments will be used; for 25–50 kg, 10 kg increments will be used; for 50–75 kg, 15 kg increments will be used, and for >75 kg, 20 kg increments will be used. If you are using a pump bulb, these increments will also be set by

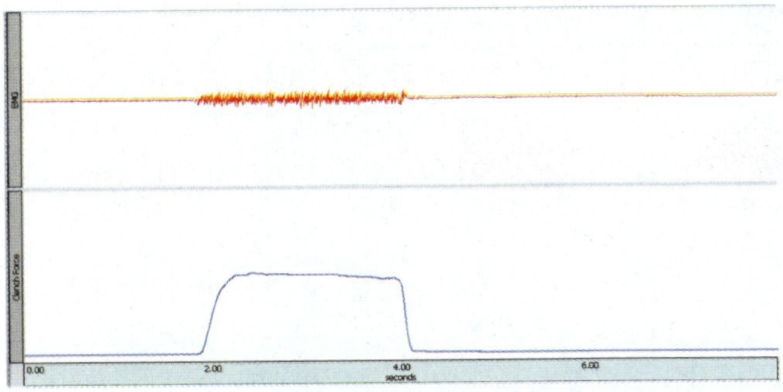

FIGURE 21.10 Sample calibration recording, Part 2 (single clench).

Courtesy of and © *BIOPAC* Systems, Inc. www.biopac.com

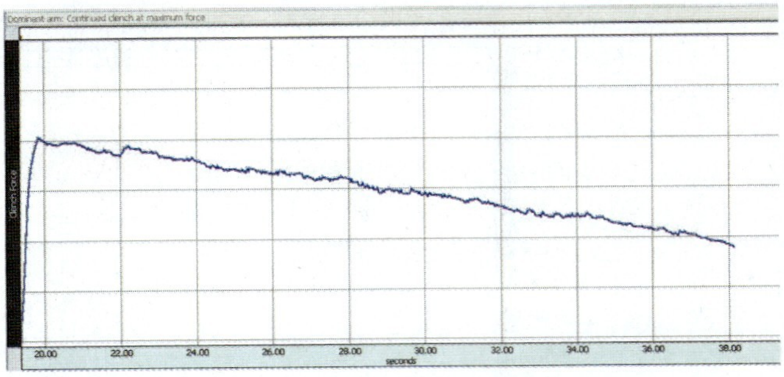

FIGURE 21.12 Sample recording of muscle fatigue.

Courtesy of and © *BIOPAC* Systems, Inc. www.biopac.com

the software: for 0–5,000 kgf/m^2, increments of 1,000; for 5,000–7,500, increments of 1,500; for 7,500–10,000, increments of 2,000; for 10,000–12,500, increments of 2,500; and for >12,500, increments of 3,000 will be used.

PROCEDURE C: Recording

1. Click on **Record** and have the subject:
   - Clench-release-wait and repeat with increasing clench force by the increments determined in step 7 above until you reach your maximum clench force. (Increase your clench force by 1 grid line each time, as shown in fig. 21.11.) This will demonstrate motor unit recruitment.
   - Each clench should be held for 2 seconds, and there should be a 2-second interval between clenches.
2. Click **Suspend.** The recording should resemble figure 21.11 for motor unit recruitment. If it does not, click **Redo** and repeat step 1.
3. Click **Continue,** and then **Record.**
4. To demonstrate muscle fatigue, clench the hand dynamometer or pump bulb with your maximum force. Note the force and try to maintain it.
5. When the maximum clench force has decreased by 50%, click **Suspend.** The recording should resemble figure 21.12 for muscle fatigue. If it does not, click **Redo** and repeat steps 3–5.
6. If correct, click **Continue.**
7. Switch the pinch connectors of the electrode lead set to the nondominant forearm. Grasp the hand dynamometer or pump bulb, and click **Record.** Proceed as with the dominant arm for recording recruitment and fatigue data.
8. When finished, click **Stop,** and then **Done.**

PROCEDURE D: Data Analysis

1. You can either click **Analyze current date file** now, to proceed directly to the data analysis, or save your file and do the analysis later using the **Review Saved Data** mode.
2. Note the channel number designations: **CH 1** displays EMG (hidden); **CH 40** displays the **Integrated EMG** (mV); and **CH 41** displays clench force.
3. Note the channel/measurement boxes as follows:

   CH 41: Mean; CH 40: Mean; CH 41: Value; and CH 40: DeltaT

4. Use the I-beam cursor to select an area on the plateau phase of the first clench of the dominant arm, as shown in figure 21.13. Record the measurements in Laboratory Assessment 21 in the table for Motor Unit Recruitment (Part II A1).
5. Repeat step 4 on the plateau for each successive clench.
6. Scroll to the second recording segment (for fatigue).

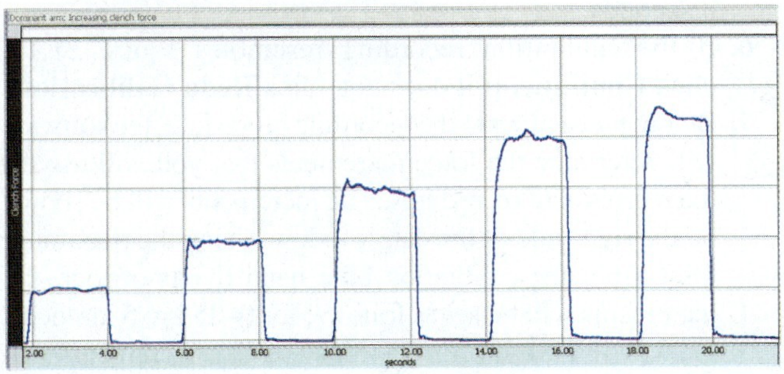

FIGURE 21.11 Sample recording of motor unit recruitment.

Courtesy of and © *BIOPAC* Systems, Inc. www.biopac.com

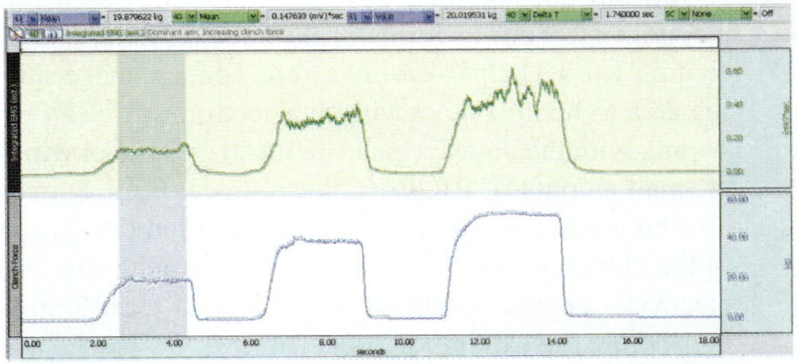

FIGURE 21.13 Proper selection of the plateau area of the first clench.

Courtesy of and © *BIOPAC* Systems, Inc. www.biopac.com

7. Use the I-beam cursor to select a point of maximal clench force at the beginning of the muscle fatigue segment. Record the measurements in the laboratory assessment in the table for fatigue (Part II A2).
8. Calculate 50% of the maximum clench force from step 7. Record the calculation in the laboratory assessment in the table for fatigue (Part II A2).
9. Select the area from the point of 50% clench force to the maximum clench force by dragging the I-beam cursor. See figure 21.14 for proper selection of the area. Note the time to fatigue (CH 40/Δ T) and record in the laboratory assessment in the table for fatigue (Part II A2).
10. Repeat from step 2 for the nondominant forearm.
11. Complete Part II of the laboratory assessment.
12. Exit the program by clicking **File** and **Quit.**

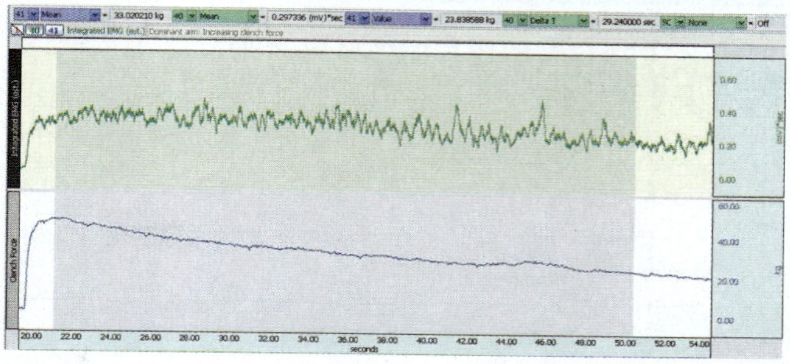

FIGURE 21.14 Proper selection of the area between 50% clench force and maximum clench force.

Courtesy of and © *BIOPAC* Systems, Inc. www.biopac.com

NOTES

LABORATORY
ASSESSMENT
21

Name ______________________________

Date ______________________________

Section ______________________________

The (A) corresponds to the indicated Learning Outcome(s) (O) found at the beginning of the Laboratory Exercise.

Electromyography: BIOPAC® Exercise

Subject Profile

Name ______________________ Height ____________

Age ______________________ Weight ____________

Gender: Male/Female Dominant Forearm: Right/Left

I. Standard and Integrated EMG

PART A: Data and Calculations Assessments

1. EMG Measurements (A1) (A2)

Cluster #	DOMINANT ARM	NONDOMINANT ARM
	40 Mean	40 Mean
1		
2		
3		
4		

Note: "Clusters" are the EMG bursts associated with each clench.

2. Use the mean measurement from the preceding table to compute the percentage increase in EMG activity recorded between the weakest clench and the strongest clench of the dominant arm.

Calculation:

Answer: __________ %

PART B: Assessments

Complete the following:

1. If you tested both forearms on the same subject, are the mean measurements for the right and left maximum grip EMG cluster the same? Which one suggests the greater grip strength? Explain. (A1)

____

____

____

2. What factors in addition to gender contribute to observed differences in clench strength? A2

3. Explain the source of signals detected by the EMG electrodes. A3

4. What is meant by the term *motor unit recruitment?*

5. Define electromyography.

II. Motor Unit Recruitment and Fatigue

PART A: Data and Calculations Assessments

1. Increasing Clench Force (Recruitment) Data: A4 A5

Peak #	Assigned Force Increment SS25L/LA/LB = kg SS56L = kgf/m^2	DOMINANT FOREARM		NONDOMINANT FOREARM	
		Force at Peak	Integrated EMG (mV)	Force at Peak	Integrated EMG (mV)
		41 Mean	40 Mean	41 Mean	40 Mean
1					
2					
3					
4					
5					
6					
7					
8					

2. Maximum Clench Force (Muscle Fatigue) Data: A6 A7

DOMINANT FOREARM			NONDOMINANT FOREARM		
Maximum Clench Force	50% of Max Clench Force	Time* to Fatigue	Maximum Clench Force	50% of Max Clench Force	Time* to Fatigue
41 Value	*Calculate*	40 Delta T	**41 Value**	*Calculate*	40 Delta T

**Note:* You do not need to indicate the Delta T (time to fatigue) polarity. The polarity of the Delta T measurement reflects the direction the I-beam cursor was dragged to select the data. Data selected left to right will have a positive ("+") polarity, while data selected right to left will have a negative ("–") polarity.

PART B: Assessments

Complete the following:

1. Compare the strength of the dominant forearm to that of the nondominant forearm. A5

____

____

____

2. Is there a difference in the absolute values of force generated by males and females in your class? What might explain the difference? A5

____

____

____

3. Is there a difference in fatigue times between the two forearms? Why might this happen?

____

____

____

4. Define muscle fatigue, and its possible causes.

____

____

____

5. Define dynamometry.

____

____

____

NOTES

LABORATORY
EXERCISE 22

Muscles of the Head and Neck

Purpose of the Exercise

To review the locations, actions, origins, insertions, and innervations of the muscles of the head and neck.

MATERIALS NEEDED

Human muscular model
Human skull
Human skeleton, articulated

For Learning Extension Activity:
Long rubber bands

Learning Outcomes APR

After completing this exercise, you should be able to:

O1 Locate and identify the muscles of facial expression, the muscles of mastication, the muscles that move the head and neck, and the muscles that move the hyoid bone and larynx.

O2 Describe and demonstrate the action of each of these muscles.

O3 Locate the origin and insertion and identify the innervation of each of these muscles in a human skeleton and the musculature of the human torso model.

The O corresponds to the assessments A indicated in the Laboratory Assessment for this Exercise.

Pre-Lab

Carefully read the introductory material and examine the entire lab. Be familiar with the locations, actions, origins, insertions, and innervations of the muscles of the head and neck from lecture or the textbook. Answer the pre-lab questions.

Pre-Lab Questions Select the correct answer for each of the following questions:

1. Muscles of mastication are all inserted on the
   a. mandible. **b.** maxilla.
   c. tongue. **d.** teeth.
2. Muscles of mastication include the following *except* the
   a. lateral pterygoid. **b.** masseter.
   c. temporalis. **d.** platysma.
3. Facial expressions are so variable partly because many facial muscles are inserted
   a. on the mandible. **b.** in the skin.
   c. on the maxilla. **d.** on facial bones.
4. Which of the following muscles does *not* move the hyoid bone?
   a. sternocleidomastoid **b.** mylohyoid
   c. sternohyoid **d.** thyrohyoid
5. Which of the following is *not* a facial expression muscle?
   a. zygomaticus major **b.** buccinator
   c. orbicularis oris **d.** scalenes
6. Flexion of the head and neck is an action of the _____ muscle.
   a. semispinalis capitis **b.** splenius capitis
   c. sternocleidomastoid **d.** trapezius
7. Which of the following muscles is the most visible from an anterior view of a person in anatomical position?
   a. splenius capitis **b.** nasalis
   c. temporalis **d.** medial pterygoid

Chapter Opener: ©101cats/Getty Images

8. The muscles of facial expression are innervated by the _____ nerve.
   a. mandibular **b.** phrenic
   c. accessory **d.** facial
9. Which of the following is *not* a muscle that moves the head and neck?
   a. sternocleidomastoid **b.** masseter
   c. semispinalis capitis **d.** splenius capitis

The skeletal muscles of the head include the muscles of facial expression. Human facial expressions are more variable than those of other mammals because many of the muscles have insertions in the dermis of the skin, rather than on bones, allowing great versatility of movements and emotions.

The muscles of mastication (chewing) are all inserted on the mandible. Muscles for chewing allow the movements of depression and elevation of the mandible to open and close the jaw. However, jaw movements also involve protraction, retraction, lateral excursion, and medial excursion needed to bite off pieces of food and grind the food.

The muscles that move the head are located in the neck. The actions of flexion, extension, and hyperextension of the head and the cervical vertebral column occur when right and left paired muscles work simultaneously. Additional actions, such as lateral flexion and rotation, result from muscles on one side only contracting or alternating muscles contracting.

A group of neck muscles that have attachments on the hyoid bone and larynx assist in swallowing and speech. Some of the neck muscles that move the head and neck have their attachments in the thorax.

PROCEDURE: Muscles of the Head and Neck

The muscle lists and tables in this laboratory exercise reflect muscle groupings using a combination of related locations and functions. Study and master one group at a time as a separate task. Frequently refer to the illustrations, tables, skeletons, and models. Consider working with a partner or a small group to study and review.

1. Study figure 22.1 and table 22.1.
2. Locate the following muscles in the human muscular model and in your body whenever possible:

   muscles of facial expression
   - epicranius (occipitofrontalis)
     - frontal belly (frontalis)
     - occipital belly (occipitalis)
   - nasalis
   - orbicularis oculi
   - orbicularis oris
   - risorius
   - depressor anguli oris
   - depressor labii inferioris
   - zygomaticus major
   - zygomaticus minor
   - levator labii superioris
   - levator anguli oris
   - mentalis
   - buccinator
   - platysma
3. Demonstrate the actions of these muscles in your body.

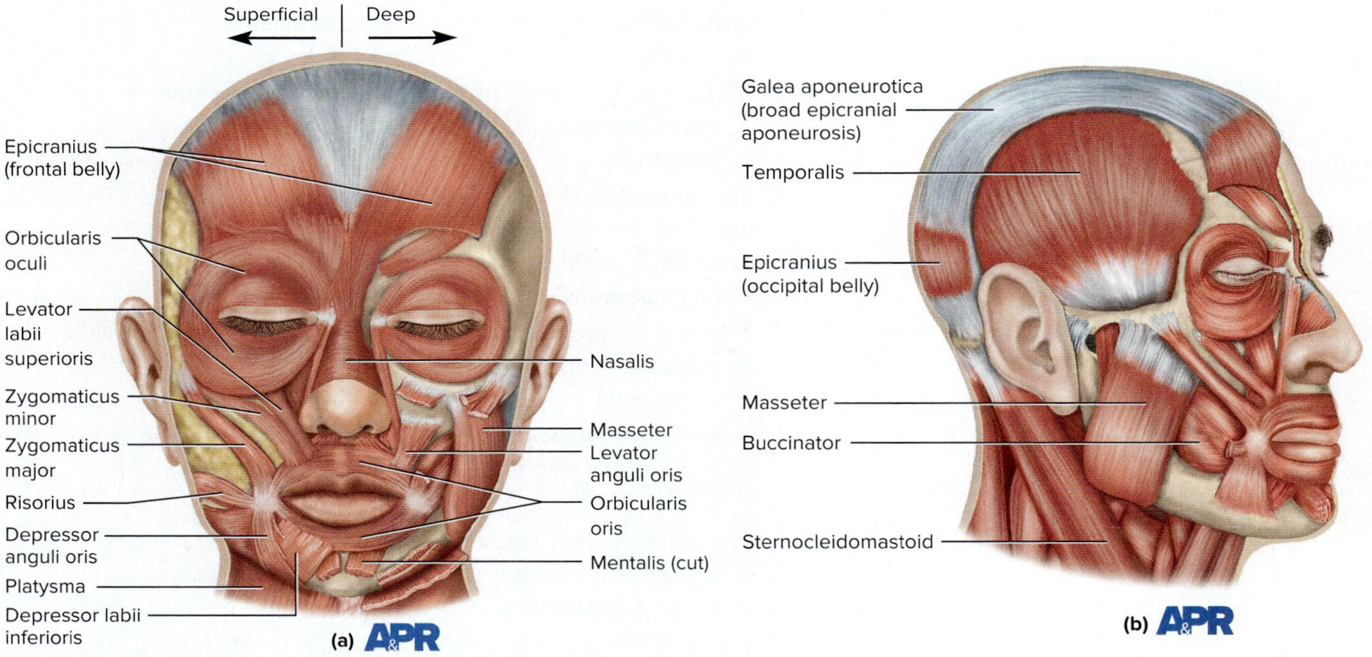

FIGURE 22.1 Muscles of facial expression and mastication: (a) anterior view; (b) lateral view.

Table 22.1 Muscles of Facial Expression

Muscle	Origin	Insertion	Action	Innervation
Epicranius	Occipital bone; galea aponeurotica	Skin of eyebrows; galea aponeurotica	Raises eyebrows; retracts scalp	Facial nerve
Nasalis	Maxilla lateral to nose	Bridge of nose	Widens nostrils	Facial nerve
Orbicularis oculi	Maxillary and frontal bones	Skin around eye	Closes eyes as in blinking	Facial nerve
Orbicularis oris	Muscles near the mouth	Skin of lips	Closes lips; protrudes lips as for kissing	Facial nerve
Risorius	Fascia near ear	Corner of mouth	Draws corner of mouth laterally as when expression of horror	Facial nerve
Depressor anguli oris	Mandible body inferior margin	Corner of mouth	Draws corner of mouth laterally and downward to open mouth and expression of sadness	Facial nerve
Depressor labii inferioris	Mandible near chin	Skin of lower lip	Draws lower lip downward and laterally in chewing and expression of doubt	Facial nerve
Zygomaticus major	Zygomatic bone	Corner of mouth	Raises corner of mouth as when smiling and laughing	Facial nerve
Zygomaticus minor	Zygomatic bone	Corner of mouth	Raises corner of mouth as when smiling and laughing	Facial nerve
Levator labii superioris	Zygomatic bone and maxilla inferior to orbit of eye	Upper lip muscles	Opens lips; raises and everts upper lip; expression of sadness	Facial nerve
Levator anguli oris	Maxilla below infraorbital foramen	Corner of mouth	Raises corner of mouth as when smiling and laughing	Facial nerve
Mentalis	Mandible near inferior incisors	Skin of chin	Raises and protrudes lower lip in drinking, pouting, and expression of doubt; raises and wrinkles skin of chin	Facial nerve
Buccinator	Lateral surfaces of maxilla and mandible	Orbicularis oris	Compresses cheeks inward as when blowing air	Facial nerve
Platysma	Fascia in upper chest	Lower border of mandible and skin at corner of mouth	Draws angle of mouth downward as when pouting or expression of horror	Facial nerve

4. Locate the origins and insertions of these muscles in the human skull and skeleton, and identify their innervations.
5. Study figures 22.1 and 22.2 and table 22.2.
6. Locate the following muscles in the human torso model and in your body whenever possible:

   muscles of mastication
   - masseter
   - temporalis
   - lateral pterygoid
   - medial pterygoid
7. Demonstrate the actions of these muscles in your body.
8. Locate the origins and insertions of these muscles in the human skull and skeleton, and identify their innervations.
9. Study figures 22.3 to 22.5 and table 22.3.
10. Locate the following muscles in the human torso model and in your body whenever possible:

    muscles that move head and neck
    - sternocleidomastoid
    - trapezius (superior part)
    - scalenes (anterior, middle, and posterior)
    - splenius capitis
    - semispinalis capitis

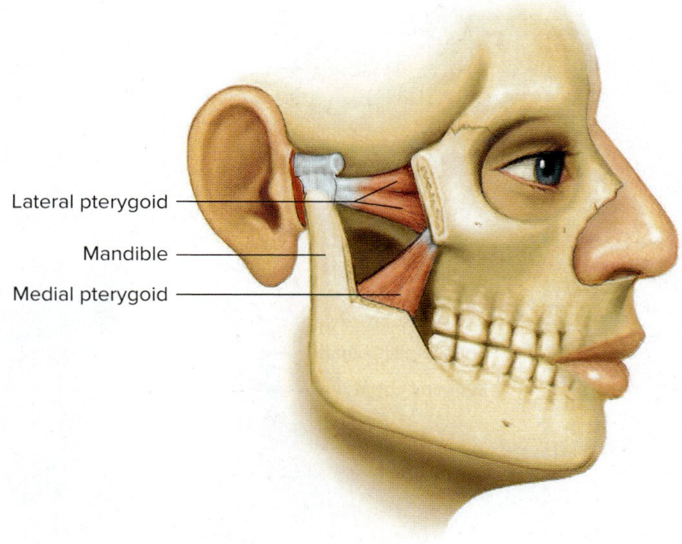

FIGURE 22.2 Deep muscles of mastication. APR

Table 22.2 Muscles of Mastication

Muscle	Origin	Insertion	Action	Innervation
Masseter	Lower border of zygomatic arch	Lateral surface of mandible	Elevates mandible; closes jaw	Trigeminal nerve
Temporalis	Temporal bone	Coronoid process and anterior ramus of mandible	Elevates and retracts mandible; closes jaw	Trigeminal nerve
Lateral pterygoid	Sphenoid bone	Anterior surface of mandibular condyle	Depresses and protracts mandible and moves it from side to side as when grinding food	Trigeminal nerve
Medial pterygoid	Sphenoid, palatine, and maxillary bones	Medial surface of mandible	Elevates mandible and moves it from side to side as when grinding food	Trigeminal nerve

11. Demonstrate the actions of these muscles in your body.
12. Locate the origins and insertions of these muscles in the human skull and skeleton, and identify their innervations.
13. Complete Part A of Laboratory Assessment 22.
14. Study figure 22.6 and table 22.4.
15. Locate the following muscles in the human torso model:

muscles that move hyoid bone and larynx

- suprahyoid muscles
    - digastric (2 parts)
    - stylohyoid
    - mylohyoid
- infrahyoid muscles
    - sternohyoid
    - omohyoid (2 parts)
    - sternothyroid
    - thyrohyoid

16. Demonstrate the actions of these muscles in your body.
17. Locate the origins and insertions of these muscles in the human skull and skeleton, and identify their innervations.
18. Complete Parts B, C, and D of the laboratory assessment.

LEARNING EXTENSION ACTIVITY

A long rubber band can be used to simulate muscle locations, origins, insertions, and actions on the human muscular model, the skeleton, or a laboratory partner. Hold one end of the rubber band firmly on the origin location of a muscle; then slightly stretch the rubber band and hold the other end on the insertion site. Allow the insertion end to slowly move toward the origin end to simulate the contraction and action of the muscle. For example, hold one end of the rubber band firmly on the zygomatic arch (the origin) and the other end on the mandible (the insertion); then move the mandible closer to the zygomatic arch to observe the action of the masseter muscle (elevate mandible).

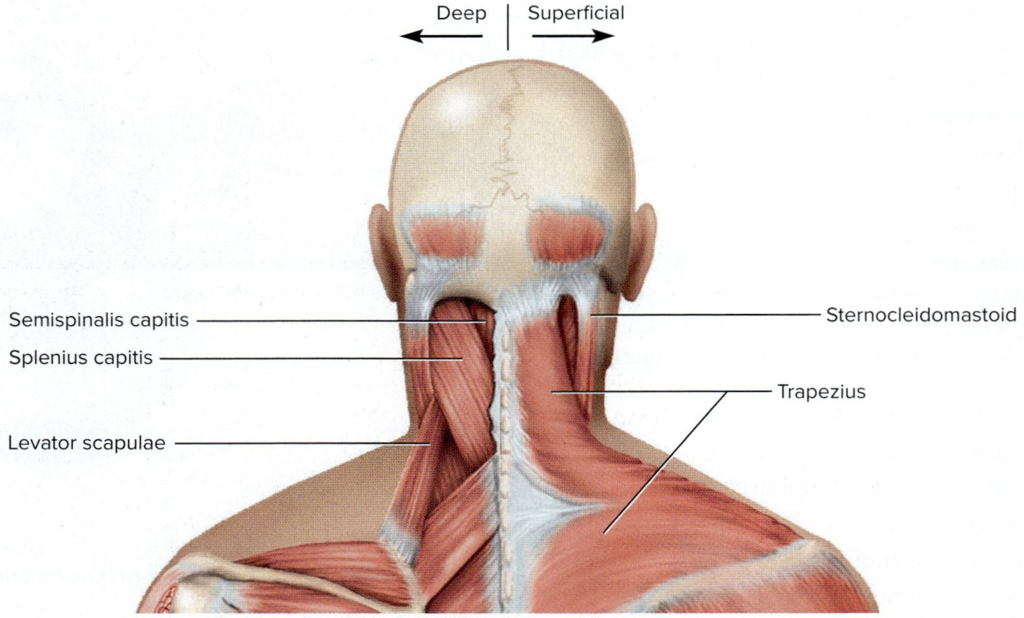

FIGURE 22.3 Posterior muscles that move the head and neck. APR

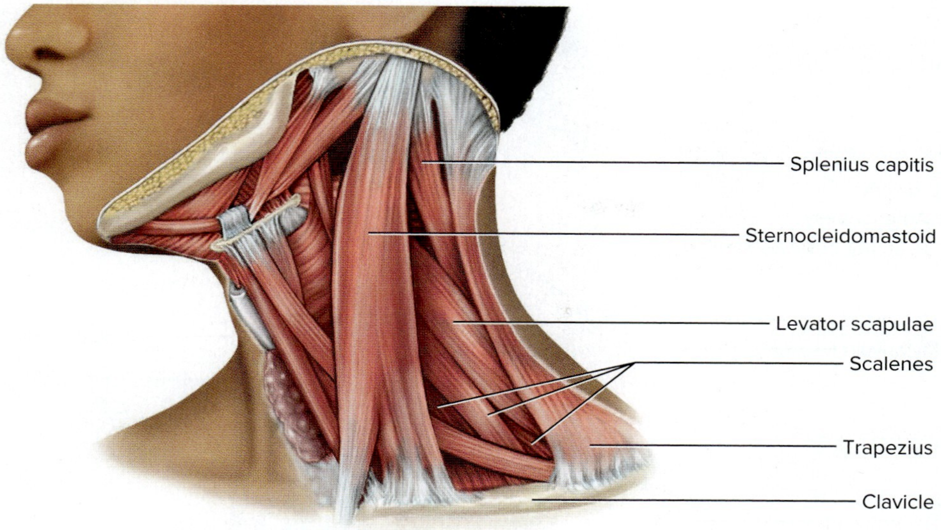

FIGURE 22.4 Muscles that move the head and neck, lateral view. APR

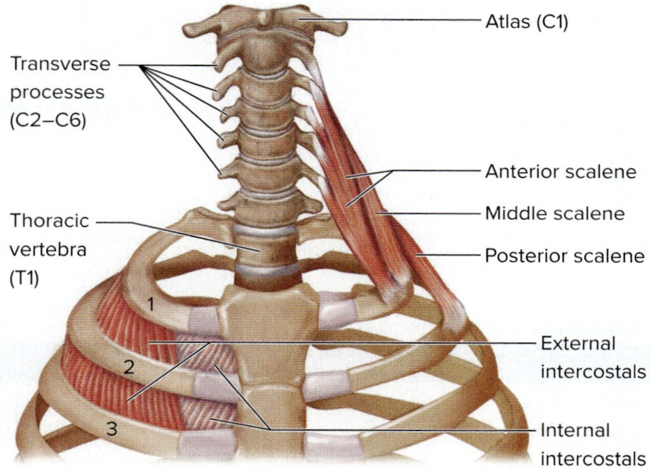

FIGURE 22.5 Three scalene muscles located in the neck and upper thorax, anterior view. The scalenes assist in neck movements and respiration. APR

Table 22.3 Muscles that Move Head and Neck

Muscle	Origin	Insertion	Action	Innervation
Sternocleidomastoid	Manubrium of sternum and medial clavicle	Mastoid process of temporal bone	Flexion of head and neck; rotation of head to left or right	Accessory nerve; spinal nerves C2–C4
Trapezius (superior part)	Occipital bone and spinous processes of C7 and several thoracic vertebrae	Clavicle and the spine and acromion of scapula	Extends head (as a synergist) (primary actions on scapula are covered in another lab)	Accessory nerve
Scalenes (anterior, middle, and posterior)	Transverse processes of C2–C6 cervical vertebrae	Ribs 1–2	Elevates ribs 1–2; flexion and rotation of neck when rib 1 is fixed	Anterior rami of spinal nerves C3–C8
Splenius capitis	Spinous processes of C7–T6	Mastoid process and occipital bone	Extends head; rotates head	Posterior rami of middle cervical nerves
Semispinalis capitis	Processes of inferior cervical and superior thoracic vertebrae	Occipital bone	Extends head; rotates head	Posterior rami of cervical and thoracic nerves

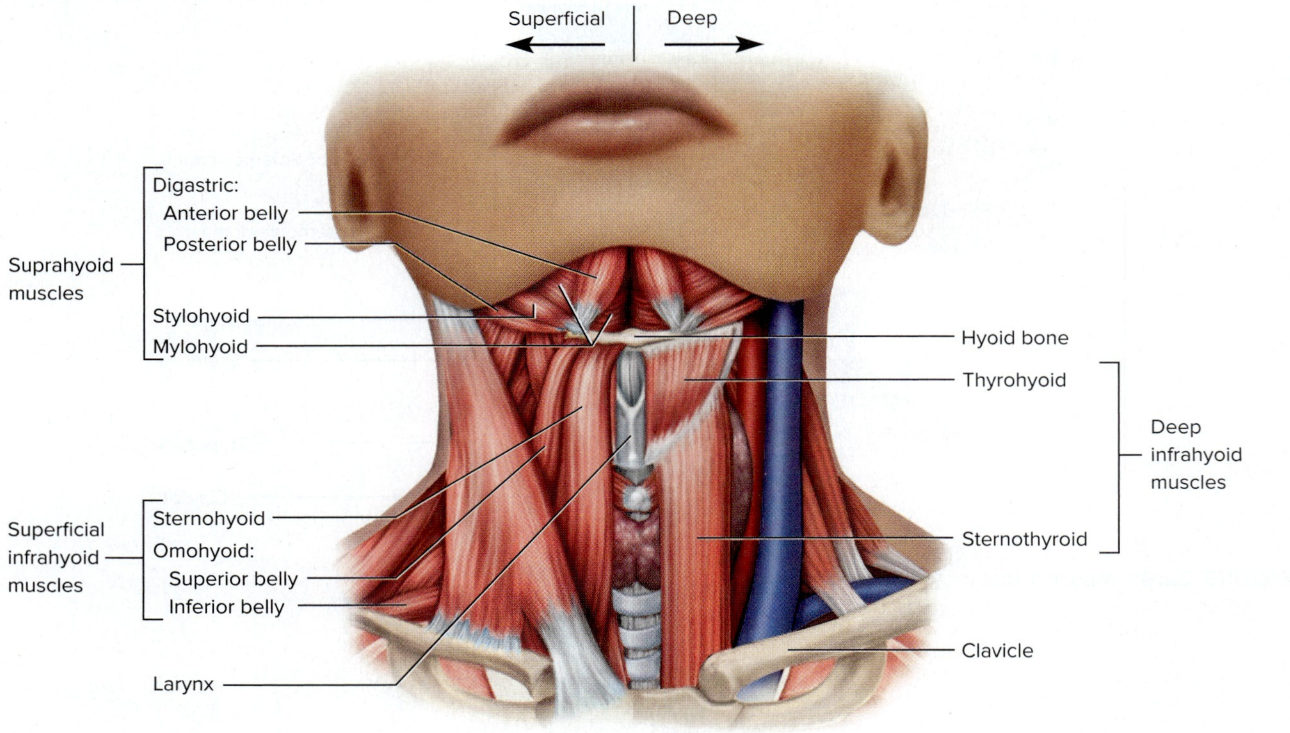

FIGURE 22.6 Muscles that move the hyoid bone and larynx assist in swallowing and speech and are grouped into suprahyoid and infrahyoid muscles. APR

Table 22.4 **Muscles that Move Hyoid Bone and Larynx**

Muscle	Origin	Insertion	Action	Innervation
Digastric (2 parts)	Inferior mandible (anterior belly) and mastoid process (posterior belly)	Hyoid bone	Opens mouth; depresses mandible; elevates hyoid bone	Facial nerve (posterior); trigeminal nerve (anterior)
Stylohyoid	Styloid process of temporal bone	Hyoid bone	Retracts and elevates hyoid bone	Facial nerve
Mylohyoid	Mandible	Hyoid bone	Elevates hyoid bone during swallowing	Trigeminal nerve
Sternohyoid	Manubrium and medial clavicle	Hyoid bone	Depresses hyoid bone	Ansa cervicalis
Omohyoid (2 parts)	Superior border of scapula	Hyoid bone	Depresses hyoid bone	Ansa cervicalis
Sternothyroid	Manubrium	Thyroid cartilage of larynx	Depresses larynx	Ansa cervicalis
Thyrohyoid	Thyroid cartilage of larynx	Hyoid bone	Depresses hyoid bone; elevates larynx	Spinal nerve C1 via hypoglossal nerve

Name ______________________________

Date ______________________________

Section ______________________________

The **A** corresponds to the indicated Learning Outcome(s) **O** found at the beginning of the Laboratory Exercise.

LABORATORY ASSESSMENT 22

Muscles of the Head and Neck

PART A: Assessments

Identify the muscles indicated in the head and neck in figures 22.7 and 22.8.

Terms:

- Depressor labii inferioris
- Frontalis
- Levator anguli oris
- Masseter
- Nasalis
- Orbicularis oculi
- Orbicularis oris
- Platysma
- Risorius
- Zygomaticus major
- Zygomaticus minor

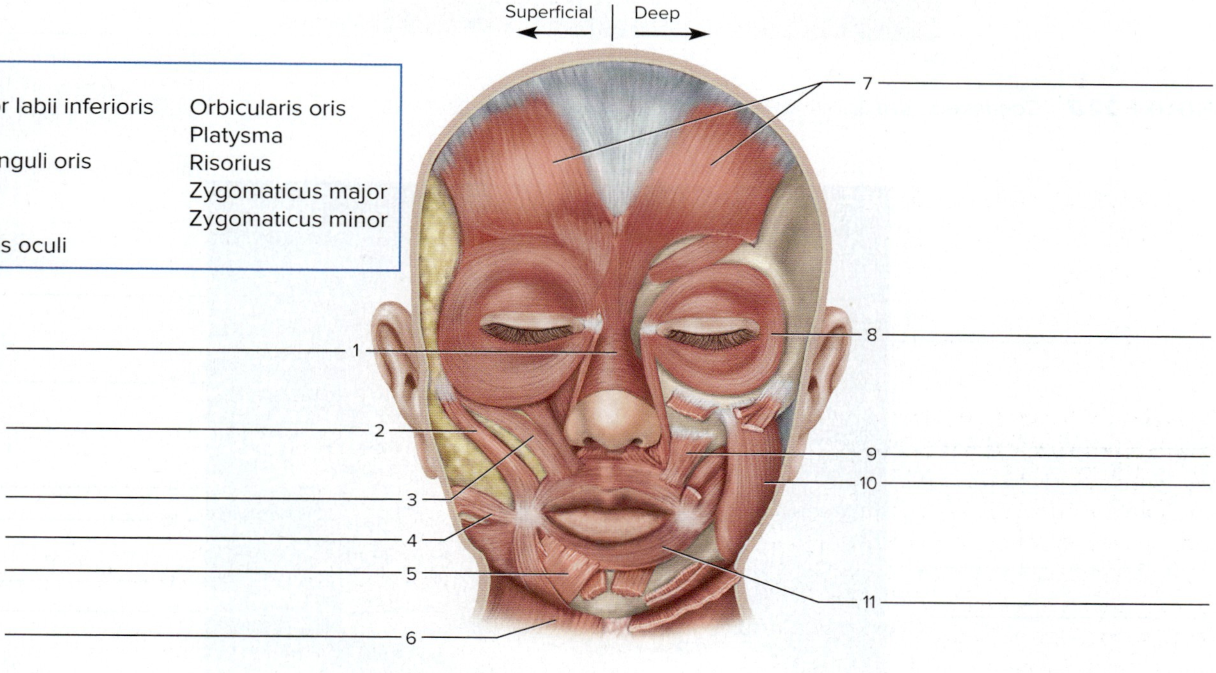

(a)

FIGURE 22.7 Label the anterior muscles of the head and neck, (a) an illustration and (b) a model, using the terms provided for each. APR A1

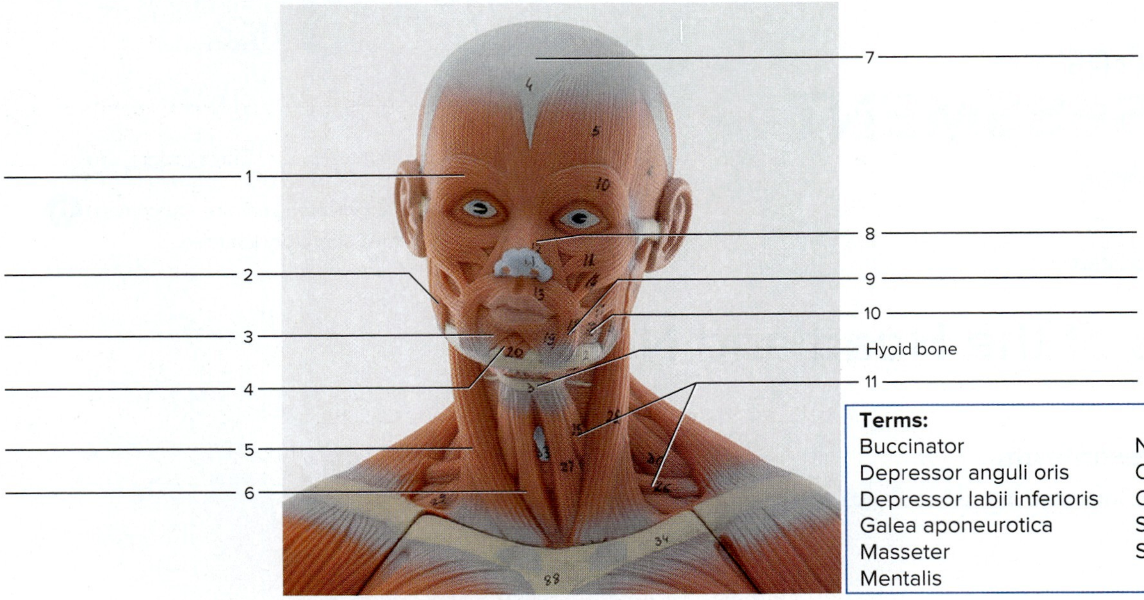

Terms:
Buccinator
Depressor anguli oris
Depressor labii inferioris
Galea aponeurotica
Masseter
Mentalis
Nasalis
Omohyoid
Orbicularis oculi
Sternocleidomastoid
Sternohyoid

(b)

FIGURE 22.7 *Continued.* ©3B Scientific GmbH, Germany, 2017 www.3bscientific.com

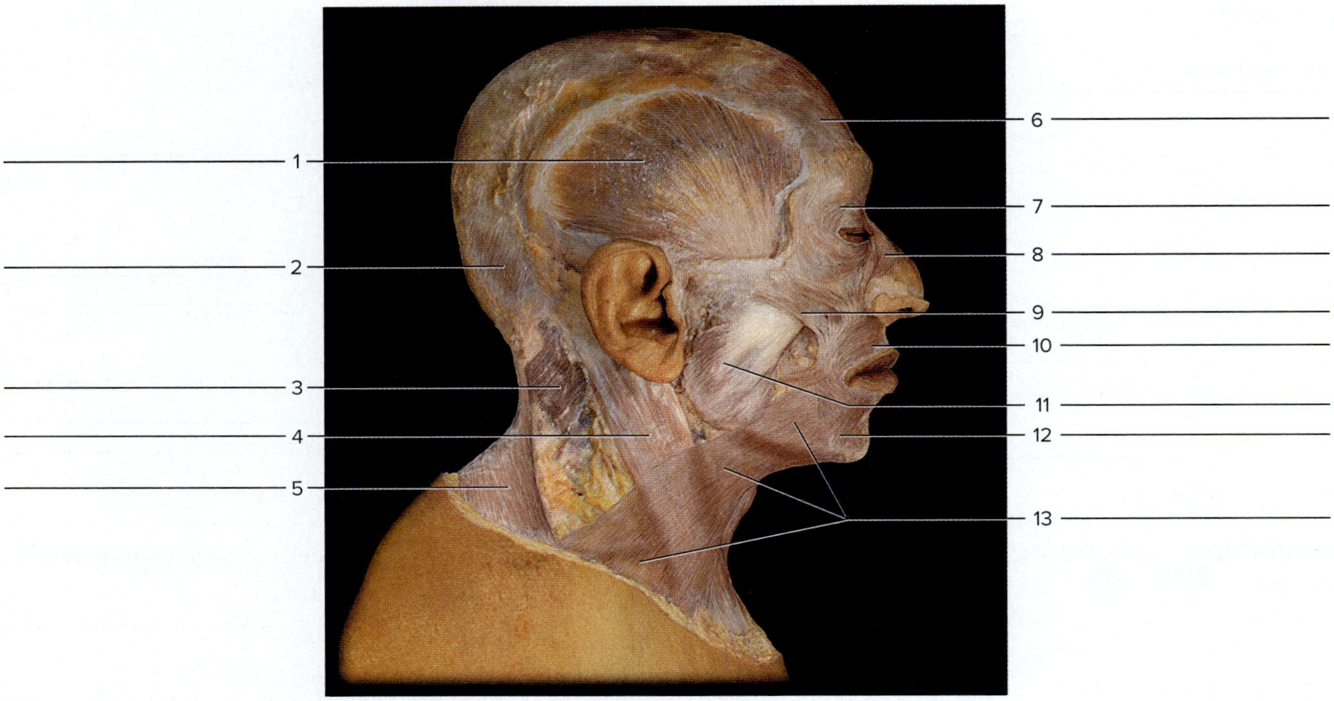

Terms:
Depressor anguli oris
Frontal belly (of epicranius)
Masseter
Nasalis
Occipital belly (of epicranius)
Orbicularis oculi
Orbicularis oris
Platysma
Splenius capitis
Sternocleidomastoid
Temporalis
Trapezius (superior part)
Zygomaticus major

FIGURE 22.8 Identify the muscles of the head and neck in this lateral view of a cadaver, using the terms provided. APR/McGraw Hill APR A1

PART B: Assessments

Complete the following statements:

1. When the ______________________ contracts, the corner of the mouth is drawn upward and laterally when laughing. A2
2. The ______________________ acts to compress the wall of the cheeks when air is blown out of the mouth. A2
3. The ______________________ causes the lips to close and pucker during kissing, whistling, and speaking. A2
4. The temporalis acts to ____. A2
5. The ______________________ pterygoid can close the jaw and pull it sideways. A2
6. The ______________________ pterygoid can protrude the jaw, pull the jaw sideways, and open the mouth. A2
7. The ____________________ wrinkles the skin of the chin, and raises and protrudes the lower lip when drinking, pouting, or expressing doubt. A2
8. The ______________________ can close the eye, as in blinking. A2
9. The ______________________ can pull the head toward the chest. A2
10. The muscle used to pout and to express horror is the ______________________. A2
11. The muscle used to widen the nostrils is the ______________________. A2
12. The muscle used to elevate the hyoid bone during swallowing is the ______________________. A2
13. The ______________________ raises the eyebrows and moves the scalp. A2
14. The splenius capitis and ______________________ muscles act to extend and rotate the head. A2

PART C: Assessments

Name the muscle indicated by the following combinations of origin, insertion, and innervation. A3

	Origin	Insertion	Innervation	Muscle
1.	Manubrium of sternum	Thyroid cartilage of larynx	Ansa cervicalis	______________________
2.	Zygomatic arch	Lateral surface of mandible	Trigeminal nerve	______________________
3.	Sphenoid bone	Anterior surface below mandibular condyle	Trigeminal nerve	______________________
4.	Manubrium of sternum and medial clavicle	Mastoid process of temporal bone	Accessory nerve; spinal nerves C2–C4	______________________
5.	Lateral surfaces of mandible and maxilla	Orbicularis oris	Facial nerve	______________________
6.	Fascia in upper chest	Lower border of mandible and skin around corner of mouth	Facial nerve	______________________
7.	Temporal bone	Coronoid process and anterior ramus of mandible	Trigeminal nerve	______________________
8.	Spinous processes of cervical and thoracic vertebrae (C7–T6)	Mastoid process of temporal bone and occipital bone	Posterior rami of middle cervical nerves	______________________
9.	Styloid process of temporal bone	Hyoid bone	Facial nerve	______________________

10. Transverse processes of cervical vertebrae (C2–C6)	Ribs 1–2	Anterior rami of spinal nerves (C3–C8)	____________
11. Frontal and maxillary bones	Skin around eye	Facial nerve	____________
12. Medial clavicle and manubrium	Hyoid bone	Ansa cervicalis	____________
13. Zygomatic bone and maxilla inferior to orbit of eye	Upper lip muscles	Facial nerve	____________

PART D: Assessments

CRITICAL THINKING ASSESSMENT—PART D

Identify the muscles of various facial expressions in the photographs of figure 22.9.

1. ______________________________
2. ______________________________
3. ______________________________
4. ______________________________
5. ______________________________

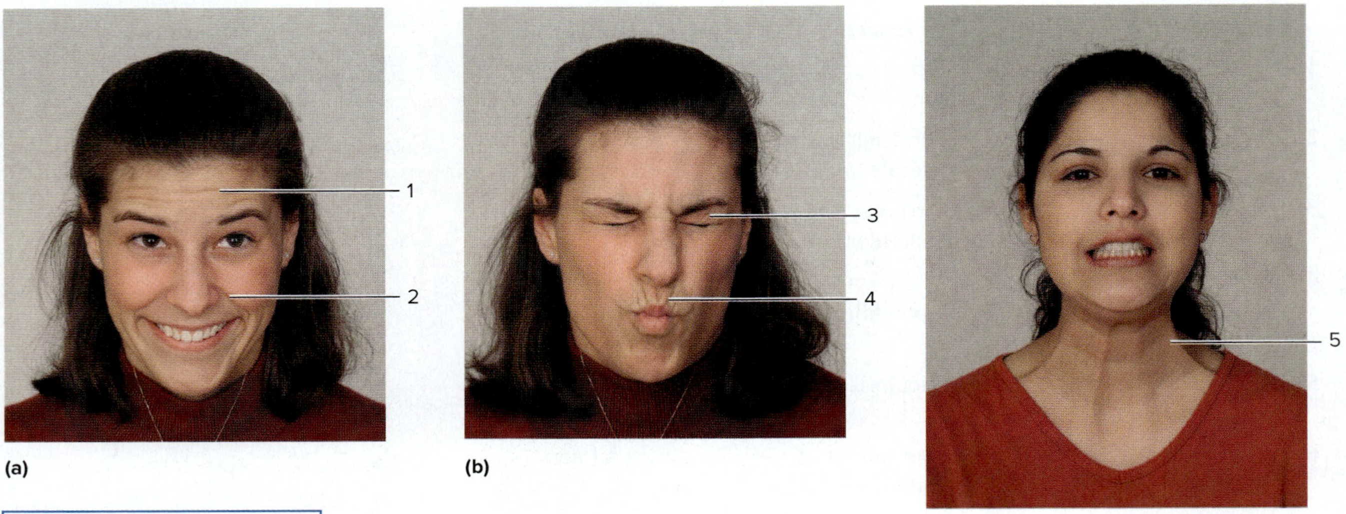

(a) (b) (c)

Terms:
Epicranius (frontal belly)
Orbicularis oculi
Orbicularis oris
Platysma
Zygomaticus major

FIGURE 22.9 Identify the muscles of facial expression being contracted in each of these photographs (a)–(c), using the terms provided. ©J & J Photography A1 A2

LABORATORY
EXERCISE 23

Muscles of the Chest, Shoulder, and Upper Limb

Purpose of the Exercise

To review the locations, actions, origins, insertions, and innervations of the muscles in the chest, shoulder, and upper limb.

MATERIALS NEEDED

Human torso model
Human skeleton, articulated
Muscular models of the upper limb

For Learning Extension Activity:
Long rubber bands

Learning Outcomes APR

After completing this exercise, you should be able to:

O1 Locate and identify the muscles of the chest, shoulder, and upper limb.

O2 Describe and demonstrate the action of each of these muscles.

O3 Locate the origin and insertion and identify the innervation of each of these muscles in a human skeleton and on the muscular models.

The O corresponds to the assessments A indicated in the Laboratory Assessment for this Exercise.

Pre-Lab

Carefully read the introductory material and examine the entire lab. Be familiar with the locations, actions, origins, insertions, and innervations of the muscles of the chest, shoulder, and upper limb from lecture or the textbook. Answer the pre-lab questions.

Pre-Lab Questions Select the correct answer for each of the following questions:

1. Chest and shoulder muscles that move the arm at the shoulder joint have insertions on the
a. scapula. **b.** clavicle and ribs.
c. humerus. **d.** radius or ulna.

2. Muscles located in the arm that move the forearm at the elbow joint have insertions on the
a. scapula and clavicle. **b.** humerus.
c. radius or ulna. **d.** carpals, metacarpals, and phalanges.

3. The major muscles of respiration include all of the following *except* the _____.
a. diaphragm **b.** external intercostals
c. internal intercostals **d.** rectus abdominis

4. Rotator cuff muscles all have origins on the
a. scapula. **b.** clavicle.
c. humerus. **d.** radius.

5. Which of the following is *not* a rotator cuff muscle?
a. supraspinatus **b.** infraspinatus
c. teres major **d.** teres minor

6. The belly of the muscle is usually _____ the region pulled.
a. located on **b.** opposite
c. distal to **d.** proximal to

Chapter Opener: ©101cats/Getty Images

7. The belly of the flexor digitorum profundus muscle is located in the
   a. posterior superficial region of the forearm.
   b. anterior deep region of the forearm.
   c. anterior superficial region of the forearm.
   d. posterior superficial region of the arm.
8. There are more muscles that move the forearm than move the hand.
   a. True b. False
9. The action of the brachialis muscle is flexion of the elbow joint.
   a. True b. False
10. Flexor and extensor muscles of the forearm act to move the hand.
   a. True b. False

Many of the chest and shoulder muscles insert on the scapula and provide various movements of the pectoral girdle. Some chest and shoulder muscles crossing the shoulder joint have insertions on the humerus and are responsible for moving the arm in various ways at the shoulder joint. Several muscles of the chest are involved in respiration during inspiration and expiration. Muscles located in the arm having insertions on the radius and ulna are responsible for forearm movements at the elbow joint (flexion and extension), as well as rotation. Muscles located in the forearm with insertions on carpals, metacarpals, or phalanges are responsible for various movements of the hand.

The muscles of the upper limb vary from large sizes that provide powerful actions to numerous smaller muscles in the forearm and hand that allow finer motor skills. The human hand, with its opposable thumb, enables us to grasp objects and provides great dexterity for tasks such as writing, drawing, and playing musical instruments.

As you demonstrate the actions of muscles in an upper limb, use the standard anatomical regions to describe them. The *arm* refers to shoulder to elbow; the *forearm* refers to elbow to wrist; and the *hand* includes the wrist, palm, and fingers. Also remember that the belly of a muscle responsible for a joint movement is usually proximal to the region being pulled. If a muscle crosses two joints, the actions allowed are even more diverse.

PROCEDURE A: Muscles of the Chest and Shoulder

The muscle lists and tables in this laboratory exercise reflect muscle groupings using a combination of related locations and functions. Study and master one group at a time as a separate task. Frequently refer to the illustrations, tables, skeletons, and models. Consider working with a partner or a small group to study and review.

1. Study the major muscles of respiration in figures 23.1 and 23.2 and table 23.1. The major muscles of inspiration include the diaphragm and the external intercostals; the internal intercostals are the primary muscles for forced expiration. Other muscles, such as the sternocleidomastoid, scalenes, pectoralis minor, rectus

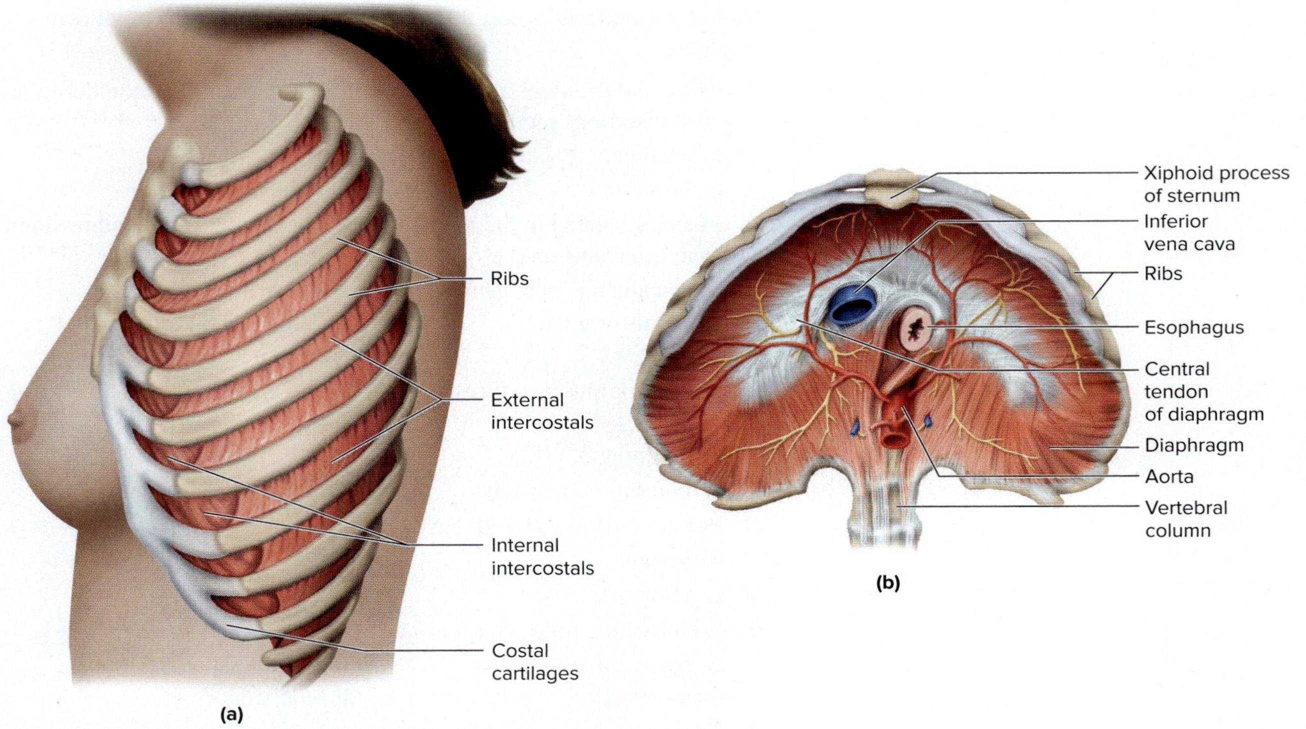

FIGURE 23.1 Major muscles of respiration: (a) lateral view of intercostals; (b) inferior view of diaphragm. (See fig. 51.1 for anterior views of the major muscles of respiration as well as additional muscles that assist in respiration.) APR

abdominis, and external oblique, serve as accessory muscles in respiration. Their actions in respiration are covered in more detail in Laboratory Exercise 51 as part of the respiratory system.

2. Locate the following muscles in the human torso model:

   diaphragm

   external intercostals

   internal intercostals

3. Demonstrate the actions of these muscles in your body.
4. Locate the origins and insertions of these muscles in the human skeleton, and identify their innervations.
5. Study figures 23.2 and 23.3 and table 23.2.
6. Locate the following muscles in the human torso model. Also locate in your body as many of the muscles as you can.

muscles that move the pectoral girdle

- anterior muscles
  - pectoralis minor
  - serratus anterior
- posterior muscles
  - trapezius
  - rhomboid major
  - rhomboid minor
  - levator scapulae

7. Demonstrate the actions of these muscles in your body.
8. Locate the origins and insertions of these muscles in the human skeleton, and identify their innervations.
9. Study figures 23.2 to 23.4, and table 23.3.

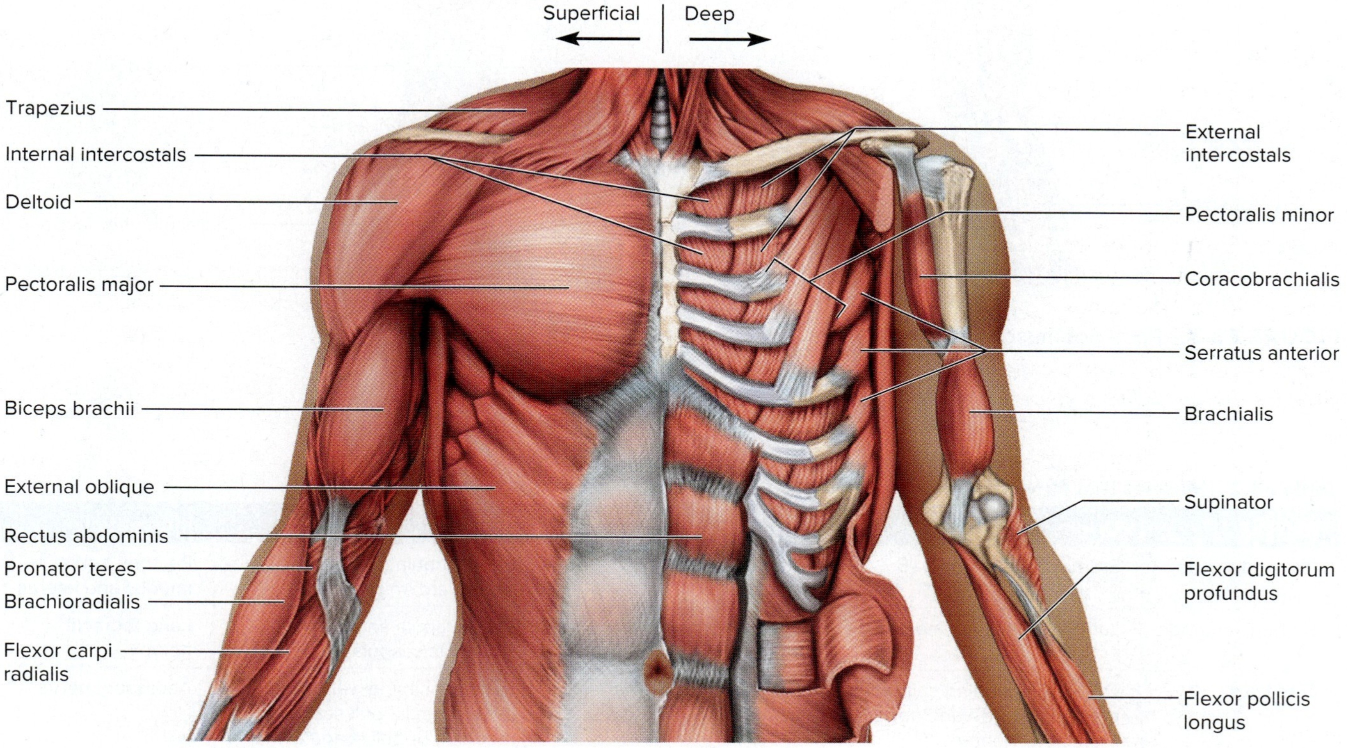

FIGURE 23.2 Anterior muscles of the chest, shoulder, and arm. Some regional muscles are also labeled. APR

Table 23.1 Major Muscles of Respiration

Muscle	Origin	Insertion	Action	Innervation
Diaphragm	Costal cartilages and ribs 7–12, xiphoid process, and lumbar vertebrae	Central tendon of diaphragm	Flattens during contraction, which expands thoracic cavity and compresses abdominal viscera during inspiration	Phrenic nerves
External intercostals	Inferior border of superior rib	Superior border of inferior rib	Elevates ribs, which expands thoracic cavity during inspiration	Intercostal nerves
Internal intercostals	Superior border of inferior rib	Inferior border of superior rib	Depresses ribs, which compresses thoracic cavity during forced expiration	Intercostal nerves

Note: The external intercostals extend from the tubercles of the ribs posteriorly to the costal cartilages anteriorly, and the internal intercostals extend from the sternum anteriorly to the curve (angle) of the ribs.

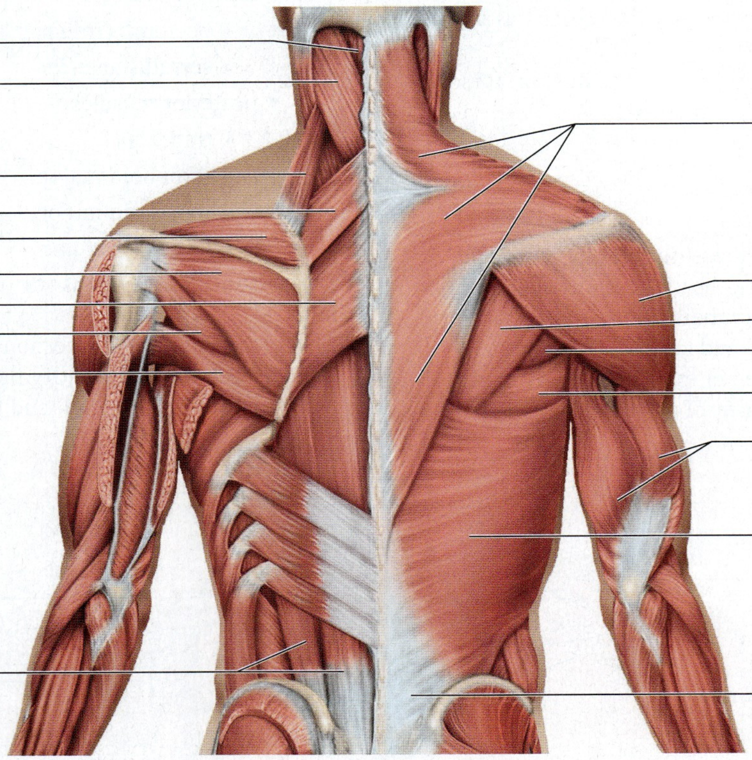

FIGURE 23.3 Posterior muscles of the shoulder, back, and arm. APR

Table 23.2 Muscles that Move the Pectoral Girdle

Muscle	Origin	Insertion	Action	Innervation
Pectoralis minor	Sternal ends of ribs 3–5	Coracoid process of scapula	Pulls scapula forward and downward; raises ribs	Pectoral nerves (medial and lateral)
Serratus anterior	Lateral surfaces of most ribs	Medial border of scapula	Pulls scapula anteriorly (protracts scapula)	Long thoracic nerve
Trapezius	Occipital bone and spinous processes of C7 and upper thoracic vertebrae	Clavicle and the spine and acromion of scapula	Rotates scapula; various fibers raise scapula, pull scapula medially, or pull scapula and shoulder downward	Accessory nerve
Rhomboid major	Spinous processes of thoracic vertebrae (T2–T5)	Medial border of scapula	Retracts and elevates scapula	Posterior scapular nerve
Rhomboid minor	Spinous processes of cervical vertebra C7 and thoracic vertebra T1	Medial border of scapula	Retracts and elevates scapula	Posterior scapular nerve
Levator scapulae	Transverse processes of cervical vertebrae (C1–C4)	Medial border of scapula	Elevates scapula	Spinal nerves C3–C4 and C5 by posterior scapular nerve

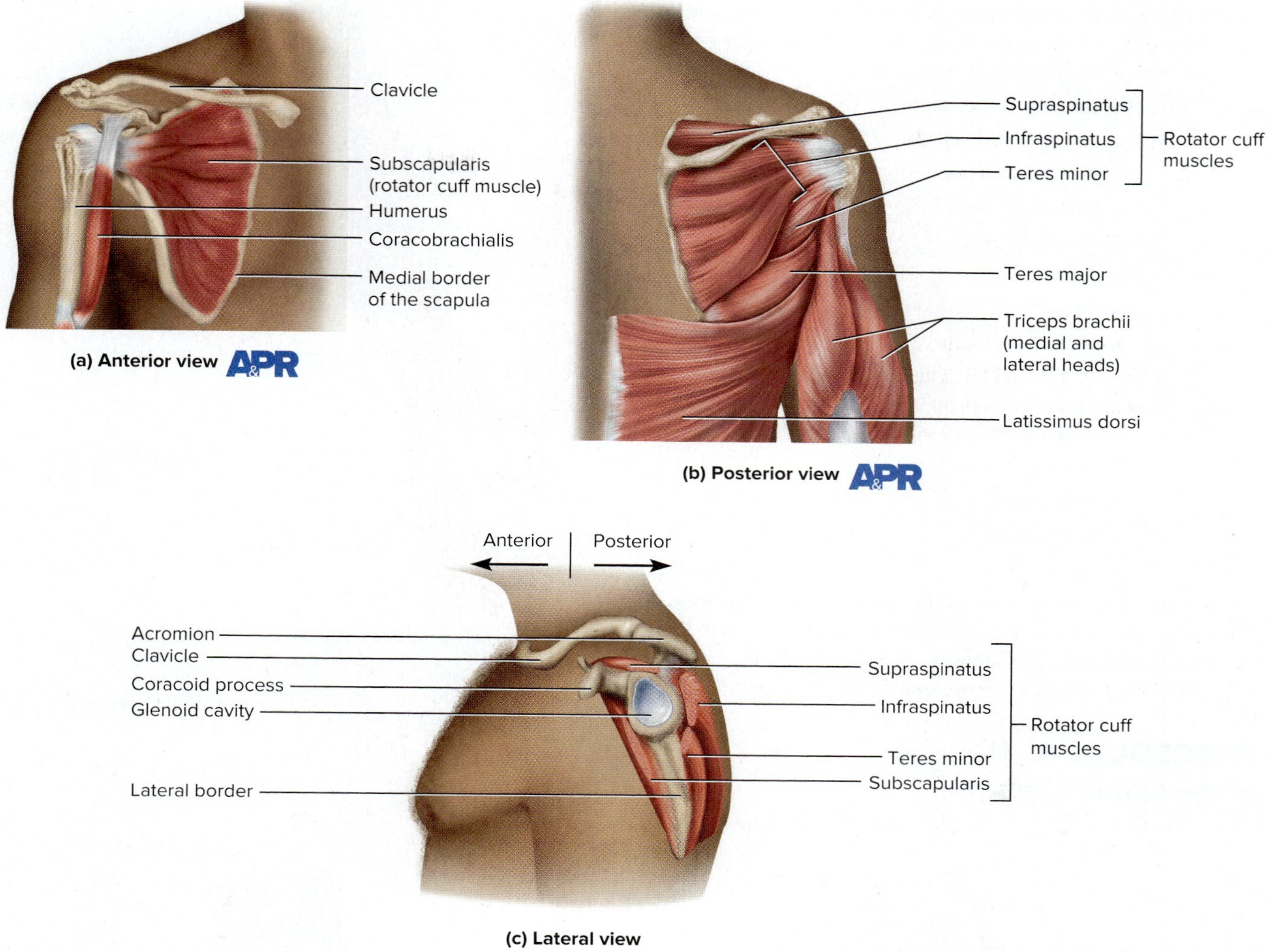

FIGURE 23.4 Rotator cuff muscles: (a) anterior view; (b) posterior view; (c) lateral view. Some regional bones, features, and muscles are also labeled.

Table 23.3 Muscles that Move the Arm

Muscle	Origin	Insertion	Action	Innervation
Pectoralis major	Clavicle, sternum, and costal cartilages of ribs 1–7	Intertubercular sulcus of humerus	Flexes, adducts, and rotates arm medially	Pectoral nerves (medial and lateral)
Latissimus dorsi	Spinous processes of lumbar and lower thoracic vertebrae, iliac crest, and lower ribs	Intertubercular sulcus of humerus	Extends, adducts, and rotates the arm medially	Thoracodorsal nerve
Deltoid	Acromion and spine of the scapula and the clavicle	Deltoid tuberosity of humerus	Abducts, extends, and flexes arm	Axillary nerve
Teres major	Lateral border of scapula	Intertubercular sulcus of humerus	Extends, adducts, and rotates arm medially	Lower subscapular nerve
Coracobrachialis	Coracoid process of scapula	Shaft of humerus	Flexes and adducts the arm	Musculocutaneous nerve
Supraspinatus	Supraspinous fossa of scapula	Greater tubercle of humerus	Abducts the arm	Suprascapular nerve
Infraspinatus	Infraspinous fossa of scapula	Greater tubercle of humerus	Rotates arm laterally	Suprascapular nerve
Teres minor	Lateral border of scapula	Greater tubercle of humerus	Rotates arm laterally	Axillary nerve
Subscapularis	Subscapular fossa of scapula	Lesser tubercle of humerus	Rotates arm medially	Upper and lower subscapular nerves

10. Locate the following muscles in the human torso model and models of the upper limb. Also locate in your body as many of the muscles as you can.

muscles that move the arm

- origins on axial skeleton
  - pectoralis major
  - latissimus dorsi
- origins on scapula
  - deltoid
  - teres major
  - coracobrachialis
- rotator cuff (SITS) muscles (origins also on scapula)—SITS is an acronym that corresponds to the first letter of each muscle in the group
  - supraspinatus
  - infraspinatus
  - teres minor
  - subscapularis

11. Demonstrate the actions of these muscles in your body.
12. Locate the origins and insertions of these muscles in the human skeleton, and identify their innervations.
13. Complete Part A of Laboratory Assessment 23.

PROCEDURE B: Muscles of the Upper Limb

1. Study figures 23.2, 23.3, 23.5, and 23.6 and table 23.4.

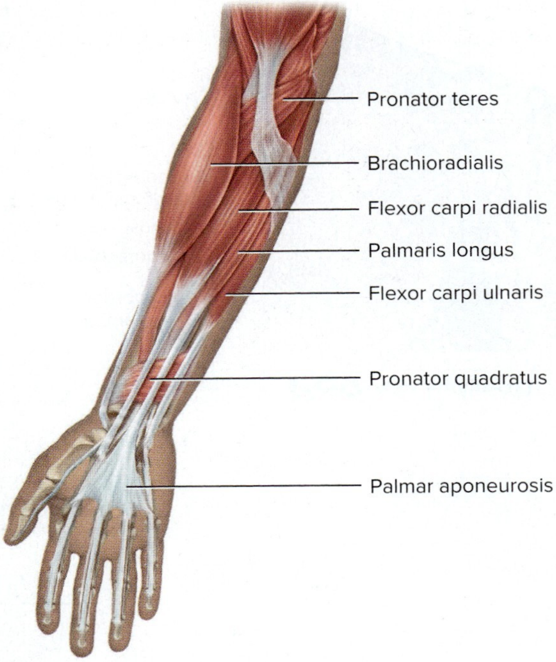

(a) Anterior muscles: superficial

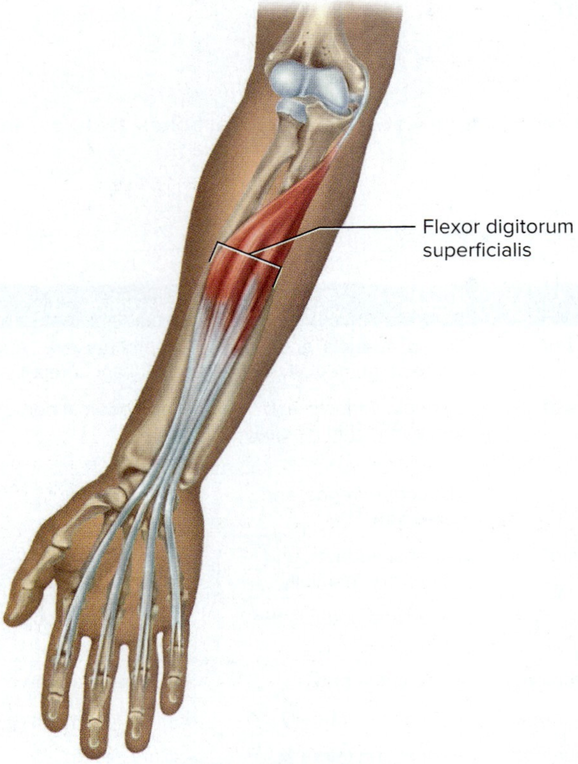

(b) Anterior muscles: intermediate

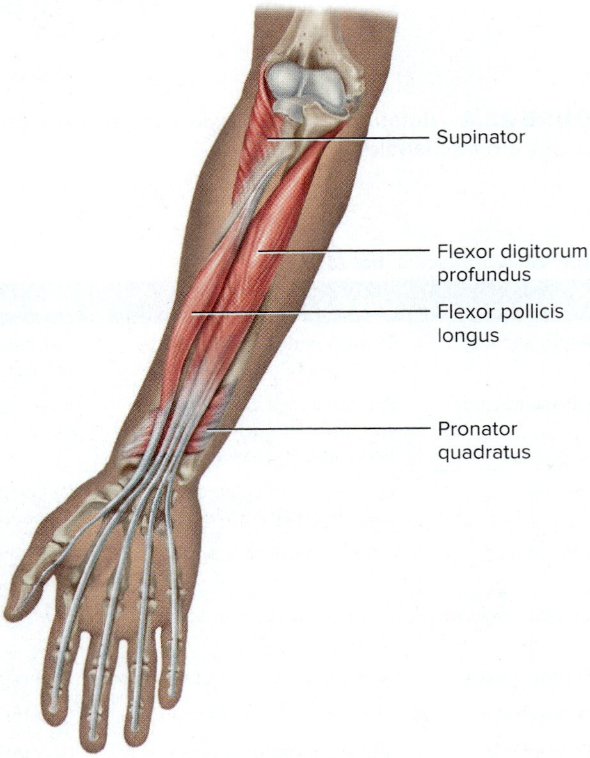

(c) Anterior muscles: deep

FIGURE 23.5 The (a) superficial, (b) intermediate, and (c) deep anterior forearm muscles. APR

2. Locate the following muscles in the human torso model and models of the upper limb. Also locate in your body as many of the muscles as you can.

muscles that move the forearm

- muscle bellies in arm
  - biceps brachii
  - brachialis
  - triceps brachii
- muscle bellies in forearm
  - brachioradialis
  - anconeus
  - supinator
  - pronator teres
  - pronator quadratus

3. Demonstrate the actions of these muscles in your body.
4. Locate the origins and insertions of these muscles in the human skeleton, and identify their innervations.

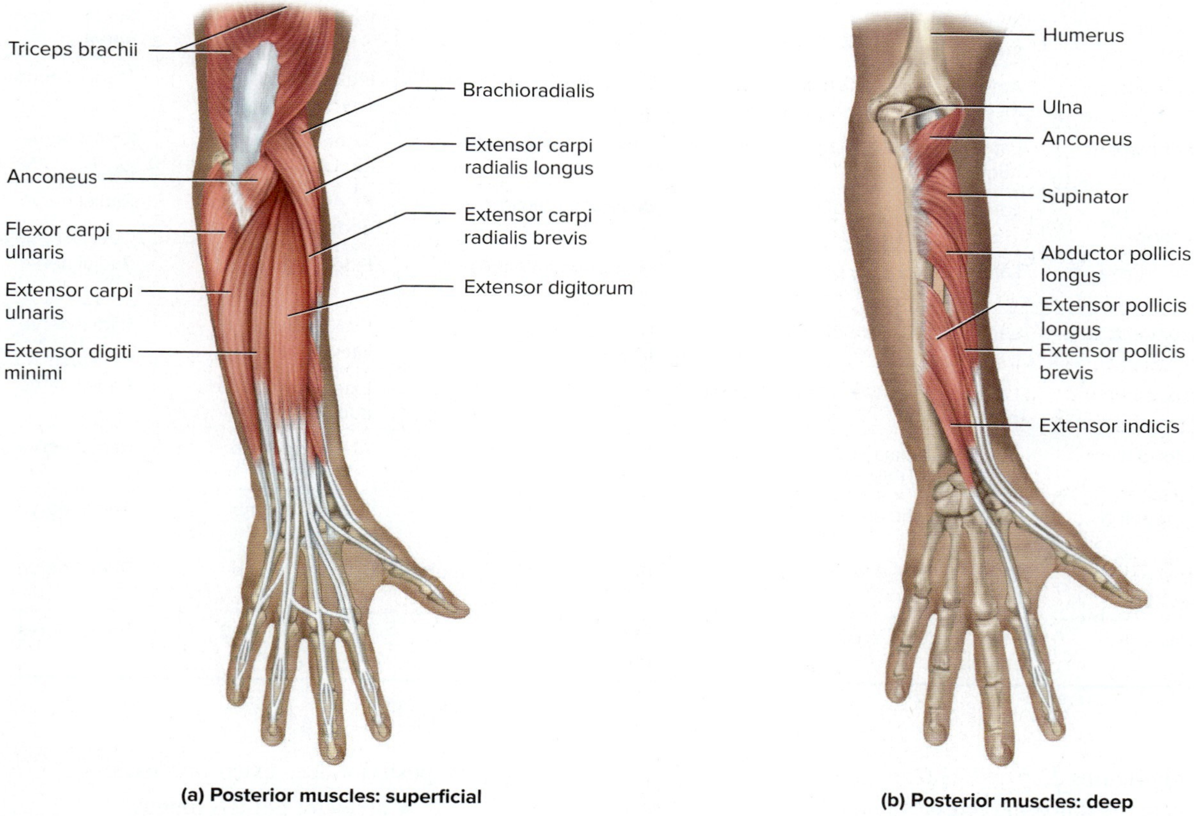

FIGURE 23.6 The (a) superficial and (b) deep posterior forearm muscles. APR

Table 23.4 Muscles that Move the Forearm

Muscle	Origin	Insertion	Action	Innervation
Biceps brachii	Coracoid process (short head) and tubercle above glenoid cavity of scapula (long head)	Radial tuberosity	Flexes elbow and supinates forearm and hand	Musculocutaneous nerve
Brachialis	Distal anterior shaft of humerus	Coronoid process of ulna	Flexes elbow	Musculocutaneous nerve; radial nerve
Triceps brachii	Tubercle below glenoid cavity and lateral and medial surfaces of humerus	Olecranon process of ulna	Extends elbow	Radial nerve
Brachioradialis	Distal lateral end of humerus	Lateral surface of radius near styloid process	Flexes elbow	Radial nerve
Anconeus	Lateral epicondyle of humerus	Olecranon process of ulna	Extends elbow	Radial nerve
Supinator	Lateral epicondyle of humerus and ulna distal to radial notch	Lateral surface of proximal radius	Supinates forearm and hand	Radial nerve
Pronator teres	Medial epicondyle of humerus and coronoid process of ulna	Lateral surface of radius	Pronates forearm and hand	Median nerve
Pronator quadratus	Anterior distal end of ulna	Anterior distal end of radius	Pronates forearm and hand	Median nerve

Table 23.5 Muscles that Move the Hand

Muscle	Origin	Insertion	Action	Innervation
Palmaris longus (sometimes absent)	Medial epicondyle of humerus	Palmar aponeurosis	Flexes wrist	Median nerve
Flexor carpi radialis	Medial epicondyle of humerus	Base of metacarpals II–III	Flexes wrist and abducts hand	Median nerve
Flexor carpi ulnaris	Medial epicondyle of humerus and olecranon process of ulna	Carpals (pisiform and hamate) and metacarpal V	Flexes wrist and adducts hand	Ulnar nerve
Flexor digitorum superficialis	Medial epicondyle of humerus, coronoid process of ulna, and radius	Middle phalanges of fingers 2–5	Flexes fingers and wrist	Median nerve
Flexor digitorum profundus	Anterior ulna and interosseous membrane	Distal phalanges of fingers 2–5	Flexes fingers 2–5	Median nerve; ulnar nerve
Flexor pollicis longus	Anterior radius and interosseous membrane	Distal phalanx of thumb	Flexes thumb	Median nerve
Extensor carpi radialis longus	Lateral supracondylar ridge of humerus	Base of metacarpal II	Extends wrist and abducts hand	Radial nerve
Extensor carpi radialis brevis	Lateral epicondyle of humerus	Base of metacarpal III	Extends wrist and abducts hand	Radial nerve
Extensor digitorum	Lateral epicondyle of humerus	Posterior phalanges of fingers 2–5	Extends fingers	Radial nerve
Extensor digiti minimi	Lateral epicondyle of humerus	Proximal phalanx of finger 5	Extends finger 5 (little finger)	Radial nerve
Extensor carpi ulnaris	Lateral epicondyle of humerus and posterior ulna	Base of metacarpal V	Extends wrist and adducts hand	Radial nerve
Abductor pollicis longus	Posterior radius and ulna and interosseous membrane	Metacarpal I and trapezium	Abducts thumb	Radial nerve
Extensor pollicis longus	Posterior ulna and interosseous membrane	Distal phalanx of thumb	Extends thumb	Radial nerve
Extensor pollicis brevis	Posterior radius and interosseous membrane	Proximal phalanx of thumb	Extends thumb	Radial nerve
Extensor indicis	Posterior ulna and interosseous membrane	Middle and distal phalanges of finger 2	Extends finger 2 (index finger)	Radial nerve

5. Study figures 23.5 and 23.6 and table 23.5.
6. Locate the following muscles in the human torso model and models of the upper limb. Also locate in your body as many of the muscles as you can.

   muscles that move the hand

   - anterior superficial flexor muscles
     - palmaris longus (absent sometimes)
     - flexor carpi radialis
     - flexor carpi ulnaris
     - flexor digitorum superficialis (intermediate depth)
   - anterior deep flexor muscles
     - flexor digitorum profundus
     - flexor pollicis longus
   - posterior superficial extensor muscles
     - extensor carpi radialis longus
     - extensor carpi radialis brevis
     - extensor digitorum
     - extensor digiti minimi
     - extensor carpi ulnaris
   - posterior deep extensor muscles
     - abductor pollicis longus
     - extensor pollicis longus
     - extensor pollicis brevis
     - extensor indicis
7. Demonstrate the actions of these muscles in your body.
8. Locate the origins and insertions of these muscles in the human skeleton, and identify their innervations.
9. Complete Parts B, C, D, and E of the laboratory assessment.

LEARNING EXTENSION ACTIVITY

A long rubber band can be used to simulate muscle locations, origins, insertions, and actions on muscular models, the skeleton, or a laboratory partner. Hold one end of the rubber band firmly on the origin location of a muscle; then slightly stretch the rubber band and hold the other end on the insertion site. Allow the insertion end to slowly move toward the origin end to simulate the contraction and action of the muscle.

LABORATORY
ASSESSMENT 23

Name ______________________

Date ______________________

Section ______________________

The (A) corresponds to the indicated Learning Outcome(s) (O) found at the beginning of the Laboratory Exercise.

Muscles of the Chest, Shoulder, and Upper Limb

PART A: Assessments

1. Identify the chest and respiratory muscles shown in figures 23.7 and 23.8.

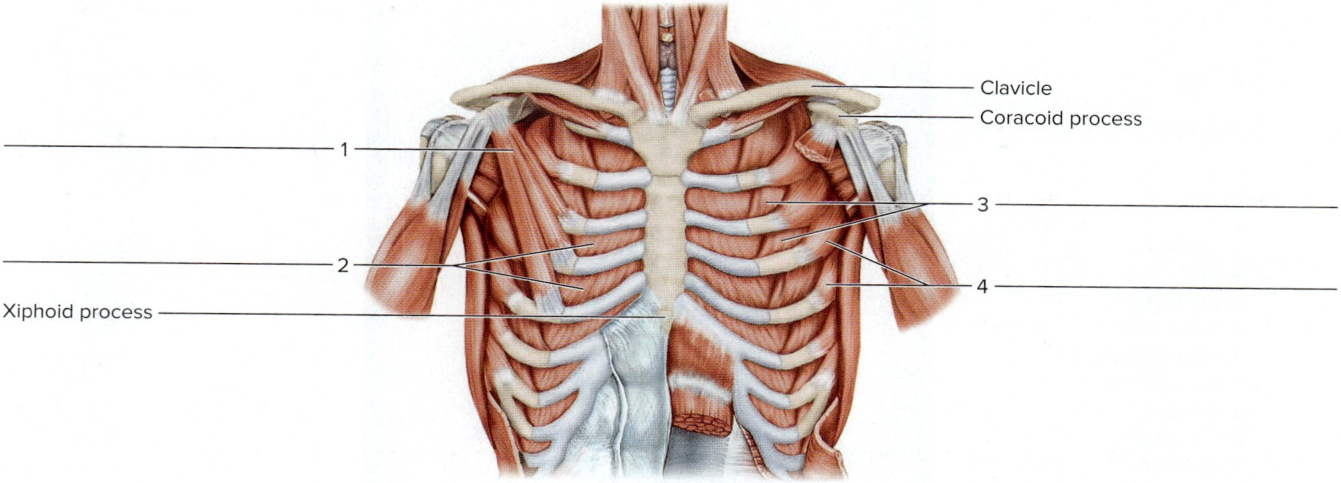

FIGURE 23.7 Label the deep anterior muscles of the chest. APR A1

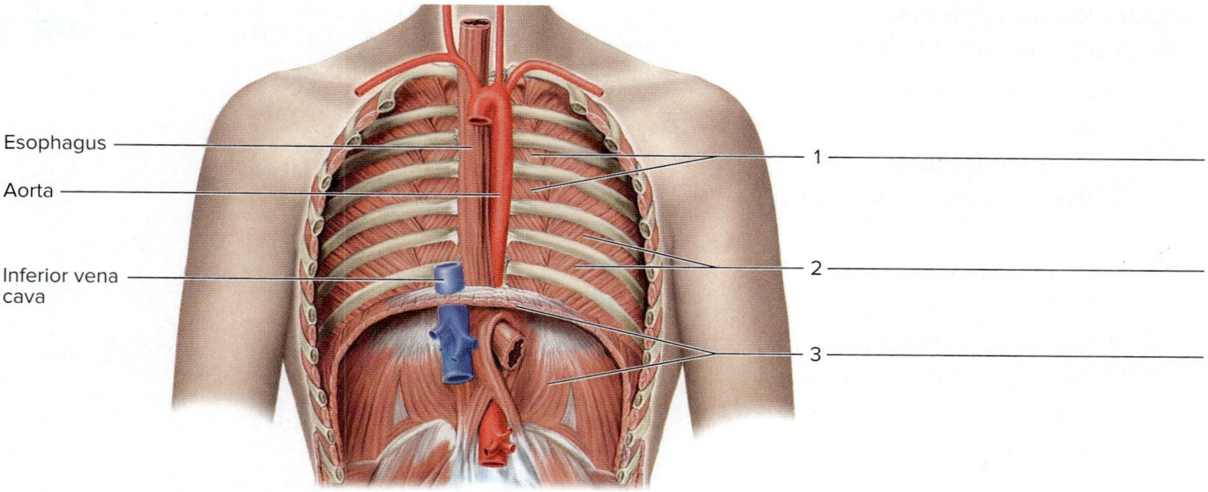

FIGURE 23.8 Label the muscles of respiration of the posterior body wall (anterior view). The heart and lungs are removed. APR A1

CRITICAL THINKING ASSESSMENT—PART A

The prime mover (agonist) of inspiration is the ______________ muscle. During normal, quiet breathing, expiration is a passive process as this muscle relaxes into its original dome-shape. Provide conditions when expiration becomes an active process, or forced. Explain. A2 ____

____.

2. Identify the muscles of the shoulder, back, and arm shown in figure 23.9.

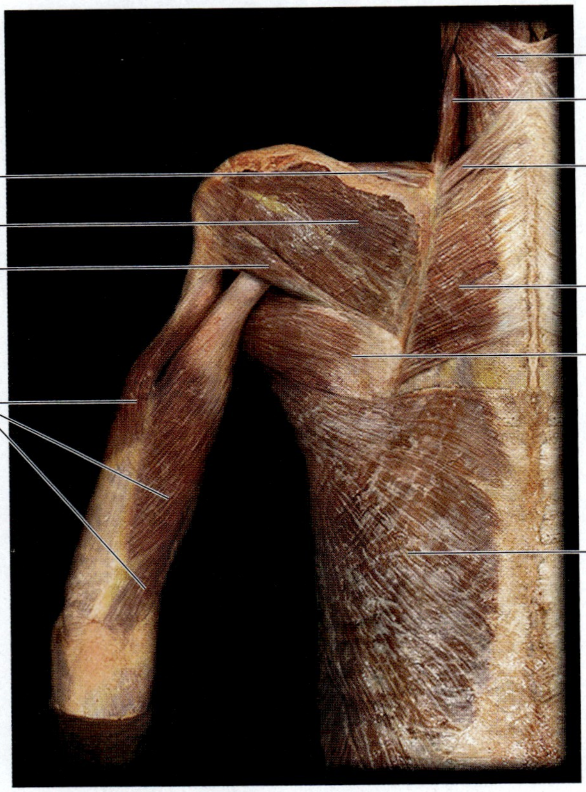

Terms:
Infraspinatus
Latissimus dorsi
Levator scapulae
Rhomboid major
Rhomboid minor
Supraspinatus
Teres major
Teres minor

(a)

FIGURE 23.9 Identify the posterior muscles of the left shoulder, back, and arm of (a) a cadaver and (b) a model. The terms are provided for (a) only. *Note:* The deltoid and trapezius muscles have been removed in (a). APR/McGraw Hill APR A1

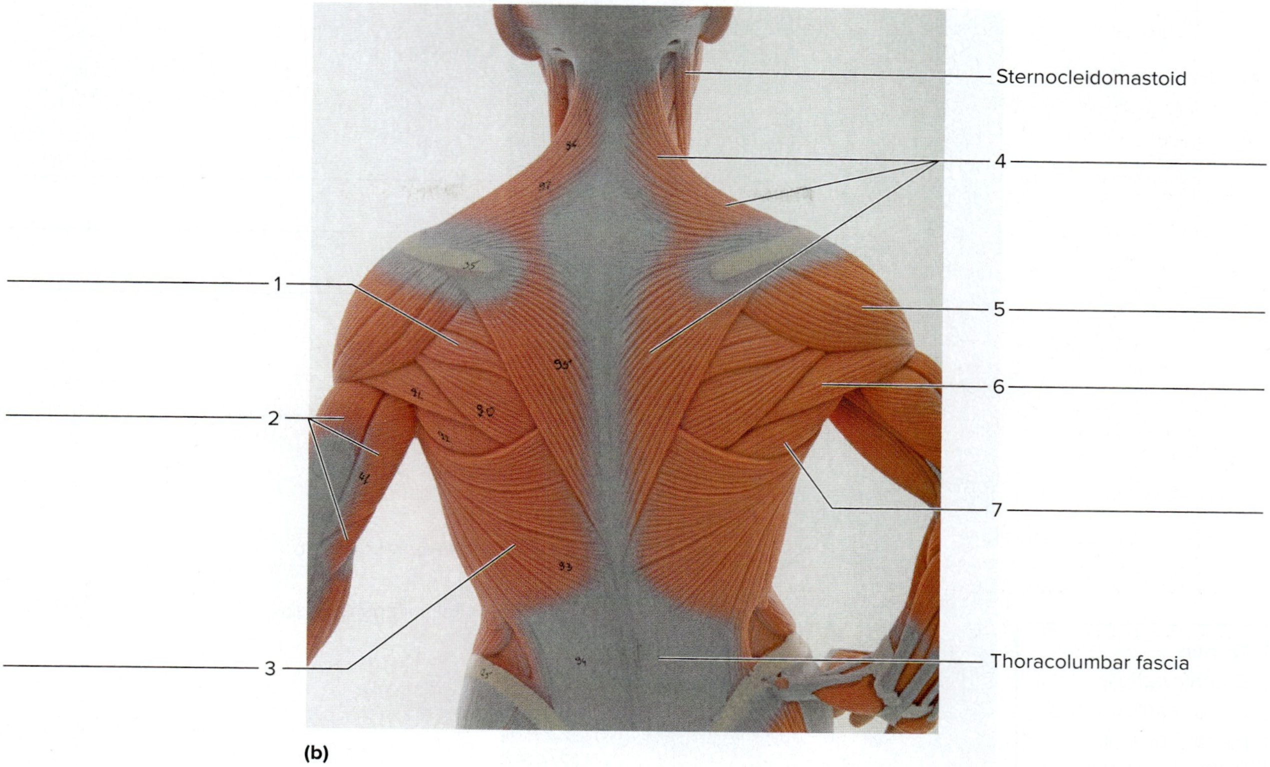

FIGURE 23.9 *Continued.* B59 [1000212] 1/3 Life-Size Muscle Figure, 2-part © 3B Scientific GmbH, Germany, 2017 www.3bscientific.com

PART B: Assessments

Identify the muscles indicated in the chest, shoulder, arm, and forearm in figures 23.10 and 23.11.

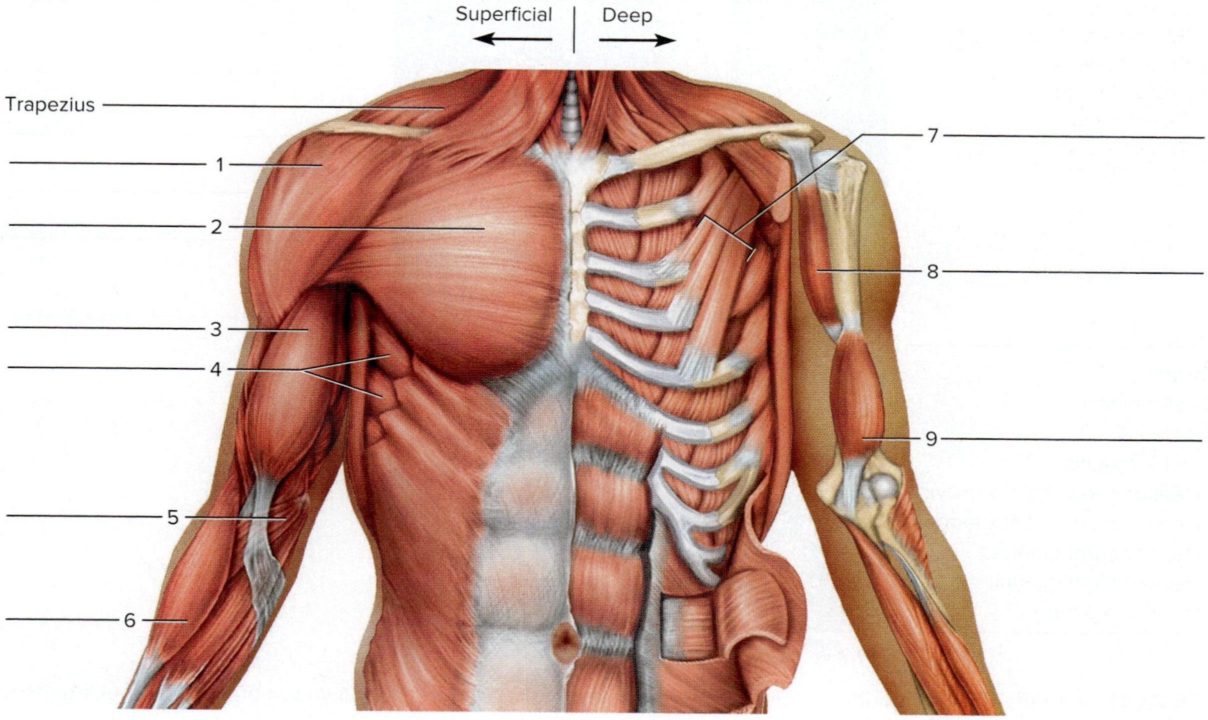

FIGURE 23.10 Label the anterior muscles of the chest, shoulder, and upper limb. APR A1

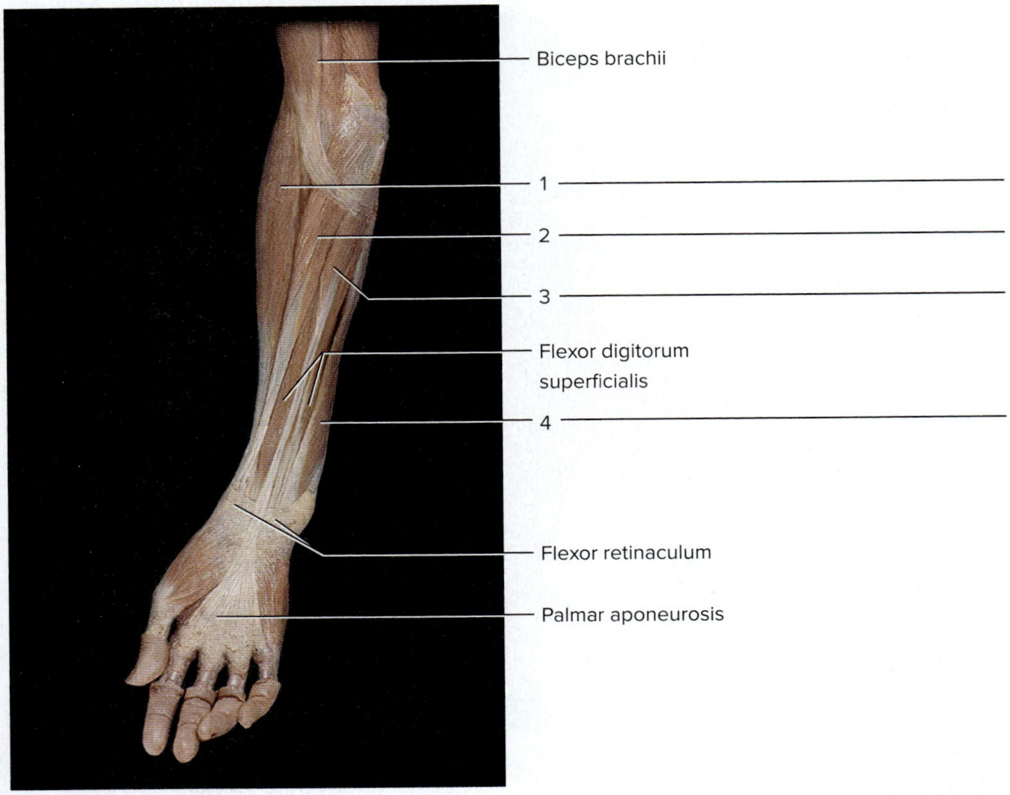

Terms:
Brachioradialis
Flexor carpi radialis
Flexor carpi ulnaris
Palmaris longus

(a) APR

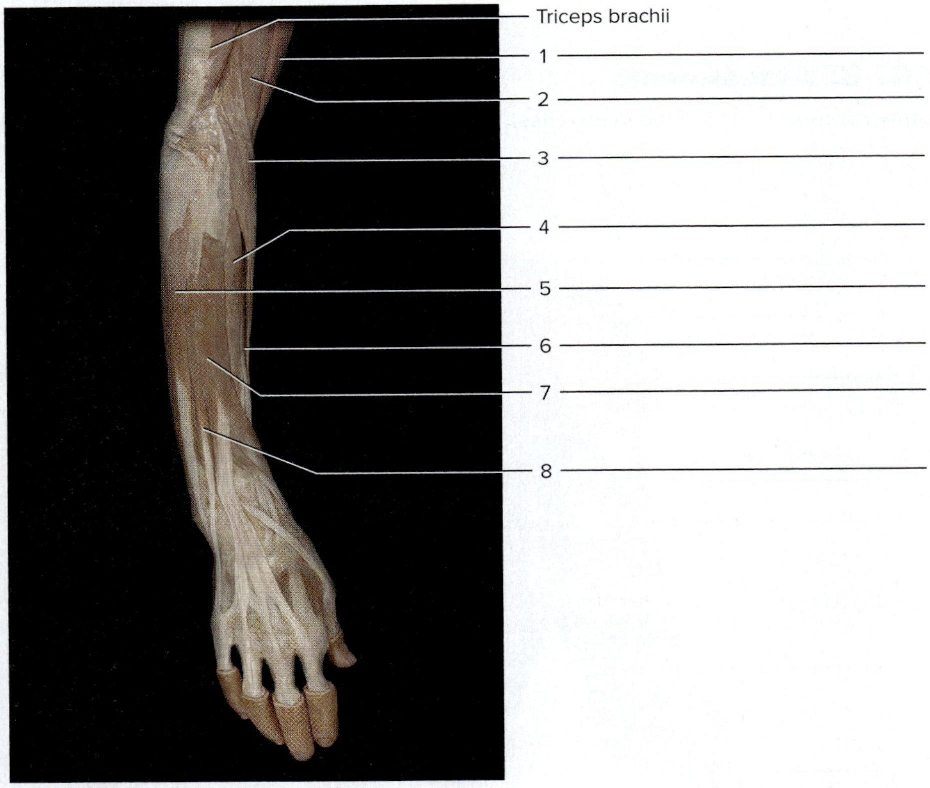

Terms:
Biceps brachii
Brachialis
Brachioradialis
Extensor carpi radialis brevis
Extensor carpi radialis longus
Extensor carpi ulnaris
Extensor digiti minimi
Extensor digitorum

(b) APR

FIGURE 23.11 Identify the (a) anterior muscles and (b) posterior muscles of the right forearm of a cadaver, using the terms provided. APR/McGraw Hill A1

PART C: Assessments

Match the muscles in column A with the actions in column B. Place the letter of your choice in the space provided. A2

Column A		Column B
a. Brachialis	______	1. Abducts, extends, and flexes arm
b. Coracobrachialis	______	2. Pulls arm anteriorly (flexion) and across chest (adduction) and rotates arm medially
c. Deltoid	______	3. Flexes wrist and adducts hand
d. Extensor carpi ulnaris	______	4. Raises and adducts (retracts) scapula
e. Flexor carpi ulnaris	______	5. Raises ribs in forceful inhalation or pulls scapula forward and downward
f. Infraspinatus	______	6. Used to thrust shoulder anteriorly (protraction), as when pushing something
g. Pectoralis major	______	7. Flexes the forearm at the elbow
h. Pectoralis minor	______	8. Flexes and adducts arm at the shoulder along with pectoralis major
i. Rhomboid major	______	9. Extends the forearm at the elbow
j. Serratus anterior	______	10. Extends, adducts, and rotates arm medially
k. Teres major	______	11. Extends and adducts hand at the wrist
l. Triceps brachii	______	12. Rotates arm laterally

PART D: Assessments

Name the muscle indicated by the following combinations of origin, insertion, and innervation. A3

Origin	Insertion	Innervation	Muscle
1. Spinous processes of thoracic vertebrae	Medial border of scapula	Posterior scapular nerve	______________
2. Lateral surfaces of most ribs	Medial border of scapula	Long thoracic nerve	______________
3. Sternal ends of ribs 3–5	Coracoid process of scapula	Medial and lateral pectoral nerves	______________
4. Coracoid process of scapula	Shaft of humerus	Musculocutaneous nerve	______________
5. Lateral border of scapula	Intertubercular sulcus of humerus	Lower subscapular nerve	______________
6. Subscapular fossa of scapula	Lesser tubercle of humerus	Upper and lower subscapular nerves	______________
7. Lateral border of scapula	Greater tubercle of humerus	Axillary nerve	______________
8. Distal anterior shaft of humerus	Coronoid process of ulna	Musculocutaneous nerve; radial nerve	______________
9. Medial epicondyle of humerus	Palmar aponeurosis	Median nerve	______________
10. Anterior radius and interosseous membrane	Distal phalanx of thumb	Median nerve	______________
11. Posterior ulna and interosseous membrane	Middle and distal phalanges of finger 2	Radial nerve	______________
12. Clavicle and spine and acromion of scapula	Deltoid tuberosity of humerus	Axillary nerve	______________
13. Ribs 7–12 and xiphoid process	Central tendon	Phrenic nerves	______________
14. Superior border of inferior rib	Inferior border of superior rib	Intercostal nerves	______________

PART E: Assessments

CRITICAL THINKING ASSESSMENTS—PART E

1. Identify the muscles indicated in figure 23.12.

____ Biceps brachii
____ Deltoid
____ External oblique
____ Pectoralis major
____ Rectus abdominis
____ Serratus anterior
____ Sternocleidomastoid
____ Trapezius

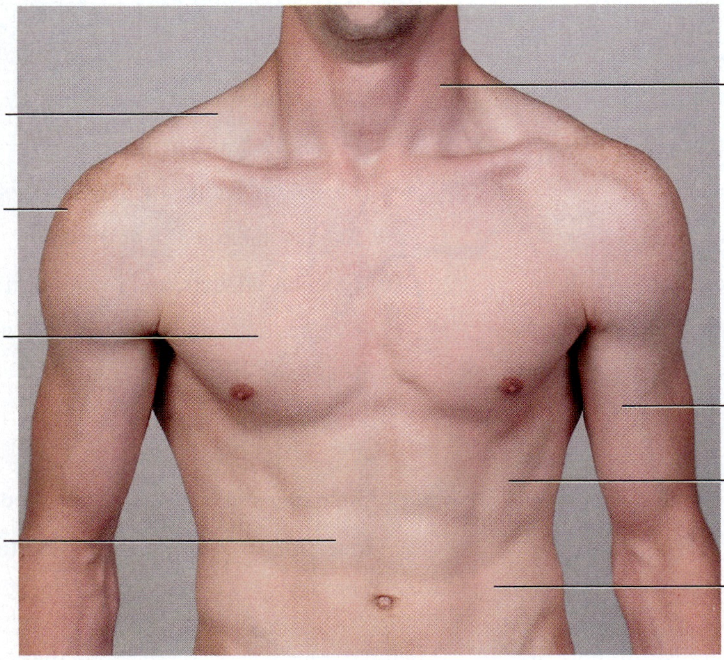

(a)

____ Biceps brachii
____ Deltoid
____ Infraspinatus
____ Latissimus dorsi
____ Trapezius
____ Triceps brachii

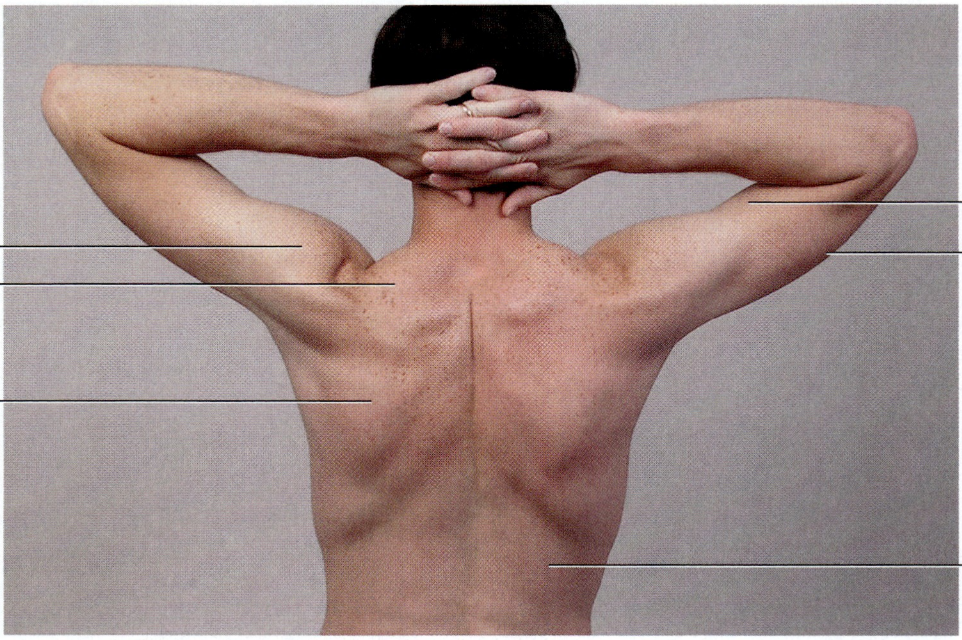

(b)

FIGURE 23.12 Label these muscles that appear as body surface features in these photographs [(a)–(c)] by placing the correct numbers in the spaces provided. ©J & J Photography A1

____ Biceps brachii
____ Brachioradialis
____ Deltoid
____ Pectoralis major
____ Serratus anterior
____ Trapezius
____ Triceps brachii

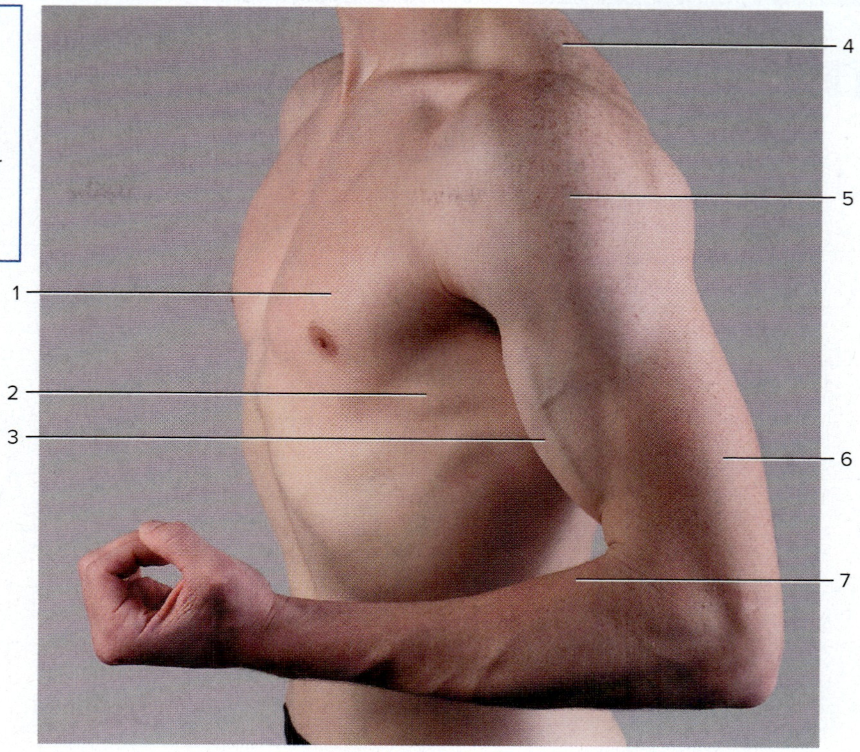

(c)

FIGURE 23.12 *Continued.* ©J & J Photography

2. Tendons of the rotator cuff muscles hold the head of the humerus to the glenoid cavity of the scapula. What agonist (prime mover) of arm abduction are these muscles assisting? ______________________
Why is their synergistic action important, particularly in sports where there is repetitive throwing or rotation of the arm (e.g., pitching, cricket, swimming)? A2

____

____

NOTES

LABORATORY
EXERCISE 24

Muscles of the Vertebral Column, Abdominal Wall, and Pelvic Floor

Purpose of the Exercise

To review the locations, actions, origins, insertions, and innervations of the muscles of the vertebral column, abdominal wall, and pelvic floor.

MATERIALS NEEDED

Human torso model with musculature
Human skeleton, articulated
Muscular models of male and female pelves

Chapter Opener: ©101cats/Getty Images

Learning Outcomes APR

After completing this exercise, you should be able to:

O1 Locate and identify the muscles of the vertebral column, abdominal wall, and pelvic floor.

O2 Describe the action of each of these muscles.

O3 Locate the origin and insertion and identify the innervation of each of these muscles in a human skeleton or on muscular models.

The O corresponds to the assessments A indicated in the Laboratory Assessment for this Exercise.

Pre-Lab

Carefully read the introductory material and examine the entire lab. Be familiar with the locations, actions, origins, insertions, and innervations of the muscles of the vertebral column, abdominal wall, and the pelvic floor from lecture or the textbook. Answer the pre-lab questions.

Pre-Lab Questions Select the correct answer for each of the following questions:

1. Which of the following muscles is *not* part of the erector spinae group?
   a. iliocostalis **b.** longissimus
   c. spinalis **d.** quadratus lumborum
2. The abdominal wall muscles possess the following functions *except*
   a. extend the spine.
   b. compress the abdominal visceral organs.
   c. assist in forceful exhalation.
   d. help maintain posture.
3. The abdominal muscle nearest the midline is the
   a. external oblique. **b.** internal oblique.
   c. rectus abdominis. **d.** transversus abdominis.
4. Which of the following muscles would be most important in relation to a vaginal delivery?
   a. bulbospongiosus **b.** quadratus lumborum
   c. spinalis **d.** erector spinae
5. The four muscles of the abdominal wall
   a. contain muscle fibers in the same direction.
   b. contain muscle fibers in four different directions.
   c. are responsible for trunk extension.
   d. fully overlap each other.

6. The anterior urogenital triangle and posterior anal triangle make up the _____.
   a. perineum   b. abdominal wall
   c. diaphragm   d. semispinalis group
7. Which of the following muscles functions to extend and rotate the head and neck?
   a. quadratus lumborum   b. serratus posterior
   c. semispinalis capitis   d. multifidus
8. The iliocostalis represents the lateral muscle group of the erector spinae.
   a. True   b. False
9. The pelvic floor muscles are all arranged in a single superficial layer.
   a. True   b. False

The deep muscles of the back extend the vertebral column. Because the muscles have many origins, insertions, and subgroups, the muscles overlap each other. The deep back muscles can extend the spine when contracting as a group but also help to maintain posture and normal spine curvatures.

The anterior and lateral walls of the abdomen contain broad, flattened muscles arranged in layers. These muscles connect the rib cage and vertebral column to the pelvic girdle. The abdominal wall muscles compress the abdominal visceral organs, help maintain posture, assist in forceful exhalation, and contribute to trunk flexion and waist rotation.

The muscles in the region of the perineum form the pelvic floor (outlet) and can be divided into two triangles: a urogenital triangle and an anal triangle. The anterior urogenital triangle contains superficial and deeper muscle layers associated with the external genitalia. The posterior anal triangle contains the pelvic diaphragm and the anal sphincter. The pelvic floor is penetrated by the urethra, vagina, and anus in a female; thus, pelvic floor muscles are important in obstetrics.

PROCEDURE A: Muscles of the Vertebral Column and Abdominal Wall

The muscles that move the vertebral column are located deep in the back from the pelvic girdle to the skull. The erector spinae group includes three major columns of muscles. Each group is composed of several muscles (table 24.1). For the purpose of this exercise, the entire erector spinae group is considered to have the common action of extending the vertebral column when they contract bilaterally; if they contract unilaterally, lateral flexion of the trunk occurs. A familiar lifting injury, known as a muscle strain (pulled muscle), can occur especially in the lumbar region of the erector spinae group. The three muscles of the semispinalis group have actions similar to the erector spinae group. The quadratus lumborum is another major muscle of the inferior vertebral column and functions in extension and lateral flexion of the vertebral column.

The splenius capitis and the semispinalis capitis muscles, located in the posterior neck extending from the vertebral column to the skull, are large muscles associated with the vertebral column and act in extensions and rotations of the head and neck. Smaller muscles of the vertebral column, including the serratus posterior and multifidus, assist in either breathing, extension, or rotation functions.

The abdominal wall is composed of four muscles situated as broad, flat muscle layers. Their muscle fibers are oriented in different directions much like plywood. This provides additional strength to support and compress the abdominal viscera. The muscle layers attach to structures such as the linea alba, inguinal ligament, or aponeuroses, since the abdominal region lacks skeletal attachments for these muscles.

1. Study figure 24.1 and table 24.1.
2. Locate the following muscles in the human torso model:

   erector spinae group
   - iliocostalis group (lateral group)
   - longissimus group (intermediate group)
   - spinalis group (medial group)

   semispinalis group

   quadratus lumborum
3. Demonstrate the actions of these muscles in your body.
4. Locate the origins and insertions of these muscles in the human skeleton, and identify their innervations.
5. Study figures 24.2 and 24.3, and table 24.2.
6. Locate the following muscles in the human torso model:

   external oblique

   internal oblique

   transversus abdominis (transverse abdominal)

   rectus abdominis

Table 24.1 Muscles of the Vertebral Column

Muscle	Origin	Insertion	Action	Innervation
Erector spinae (three columns of muscles: iliocostalis group, longissimus group, and spinalis group)	Various sites including cervical; thoracic; and lumbar vertebrae, ribs, and sacrum	Various sites including occipital bone, mastoid process, cervical and thoracic vertebrae	Maintains posture; extends vertebral column; lateral flexion when unilateral contractions occur	Posterior rami of cervical to lumbar spinal nerves
Semispinalis group (semispinalis capitis, cervicis, and thoracis)	Transverse processes of cervical and thoracic vertebrae	Spinous processes of cervical and thoracic vertebrae and occipital bone	Extends head and vertebral column; lateral flexion of neck and vertebral column when unilateral contractions occur	Posterior rami of cervical and thoracic spinal nerves
Quadratus lumborum	Iliac crest	Transverse processes of L1–L4 and rib 12	Extends vertebral column; lateral flexion when unilateral contractions occur	Anterior rami of spinal nerves T12–L4

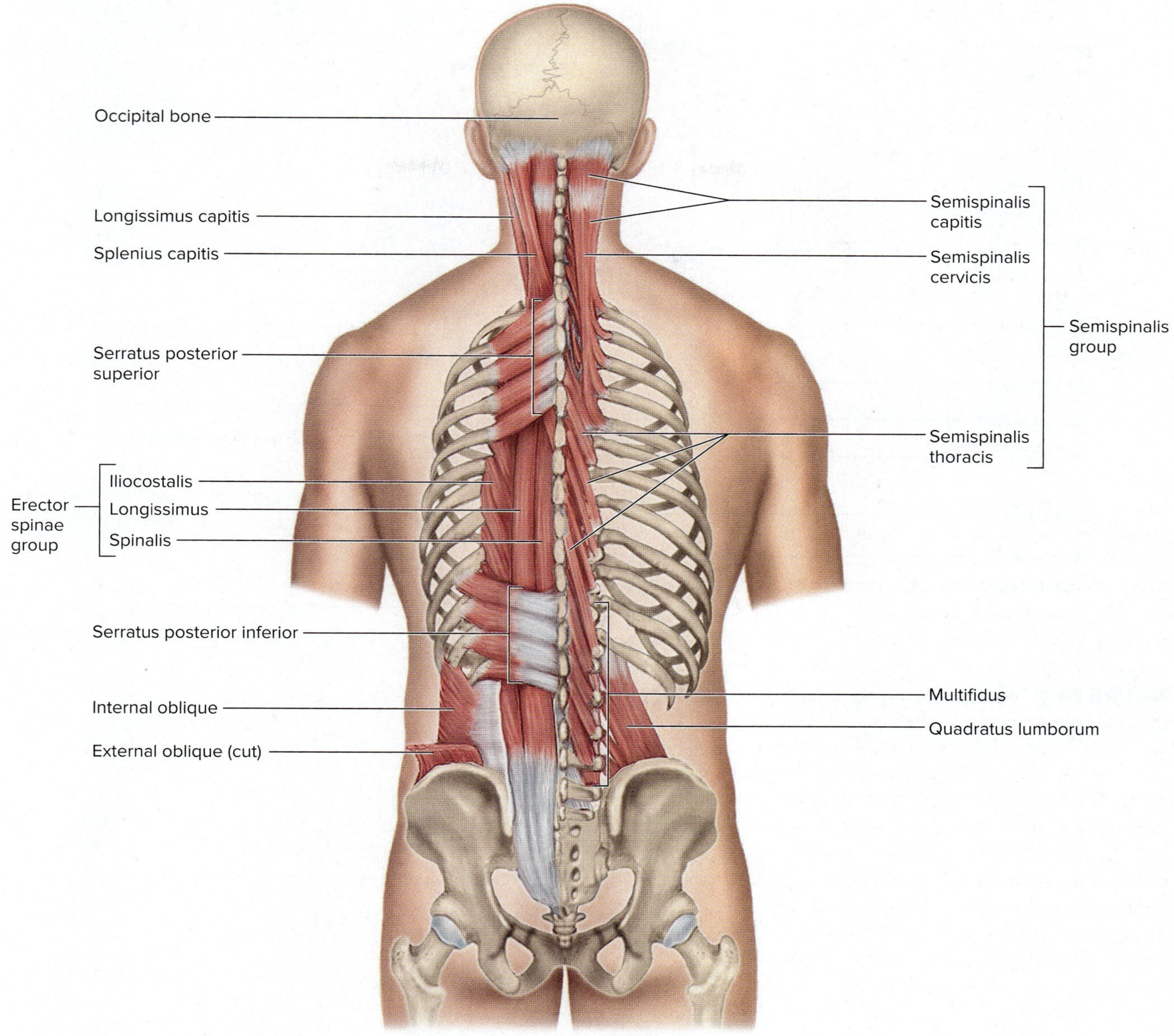

FIGURE 24.1 Muscles associated with the vertebral column. Muscles shown on the right side are deeper than those on the left side. (*Note:* Superficial muscles are not included on this figure.) APR

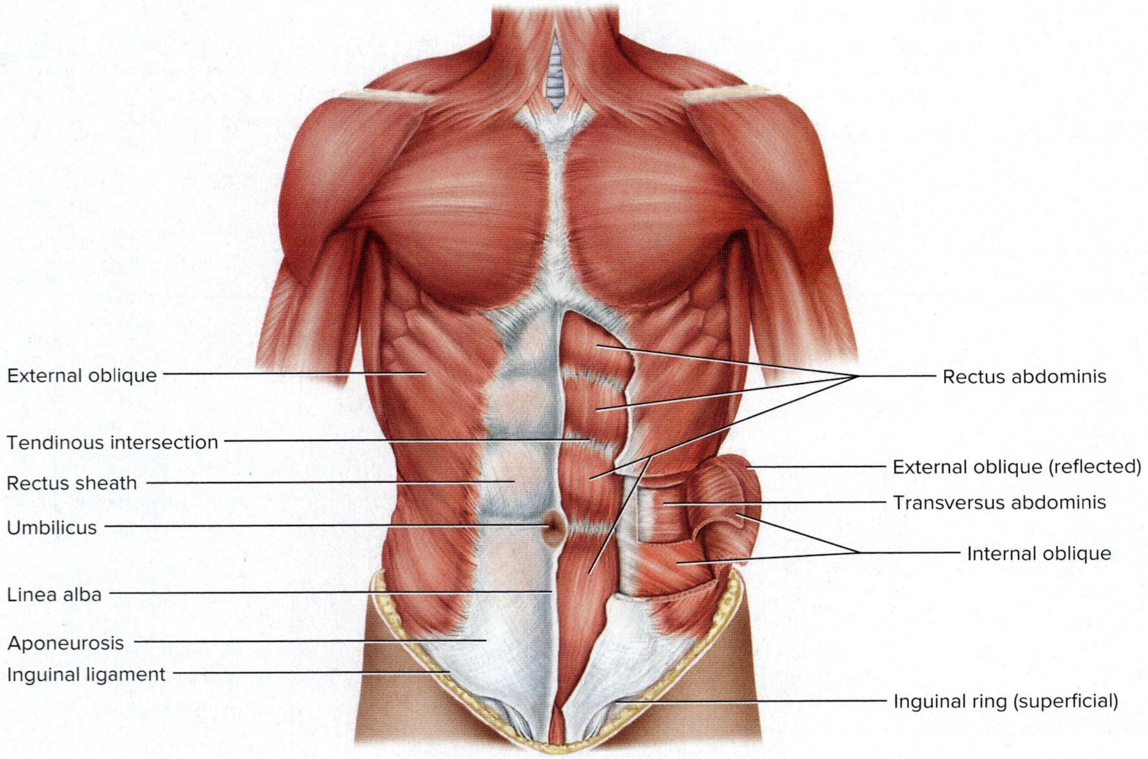

FIGURE 24.2 Muscles of the abdominal wall. APR

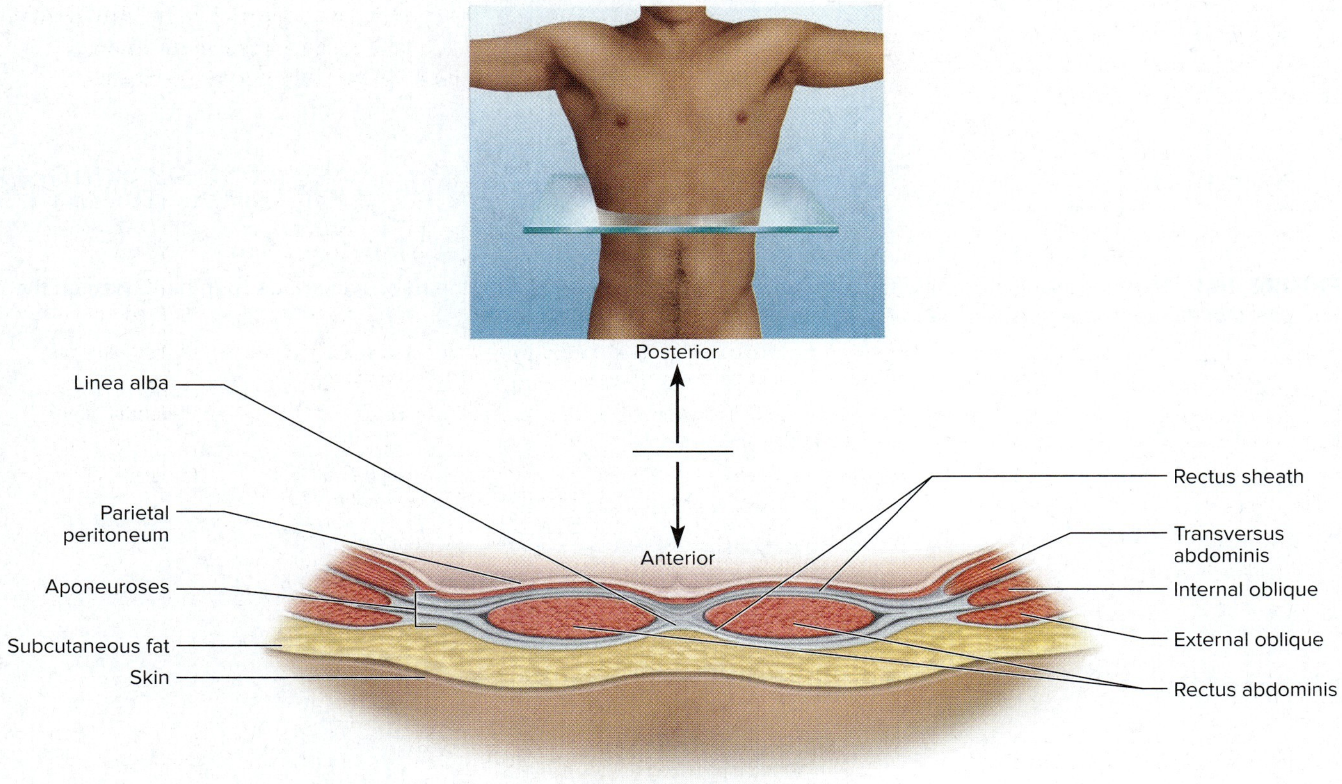

FIGURE 24.3 Transverse view of the midsection of the abdominal wall.

Table 24.2 Muscles of the Abdominal Wall

Muscle	Origin	Insertion	Action	Innervation
External oblique	Anterior surfaces of ribs 5–12	Anterior iliac crest and linea alba by aponeurosis	Tenses abdominal wall; compresses abdominal contents; aids in trunk rotation	Anterior rami of spinal nerves T7–T12
Internal oblique	Iliac crest and inguinal ligament	Cartilages of lower ribs, linea alba, and crest of pubis	Same as external oblique	Anterior rami of spinal nerves T7–L1
Transversus abdominis	Costal cartilages of lower ribs, iliac crest, and inguinal ligament	Linea alba and pubic crest	Compresses abdominal contents	Anterior rami of spinal nerves T7–L1
Rectus abdominis	Pubic crest and pubic symphysis	Xiphoid process of sternum and costal cartilages of ribs 5–7	Flexes vertebral column; compresses abdominal contents	Anterior rami of spinal nerves T6–T12

7. Demonstrate the actions of these muscles in your body.
8. Locate the origins and insertions of these muscles in the human skeleton.
9. Complete Parts A and B of Laboratory Assessment 24.

PROCEDURE B: Muscles of the Pelvic Floor

The diamond-shaped region between the lower limbs is the *perineum*. It is located between the ischial tuberosities on the lateral sides, the anterior pubic symphysis, and the posterior coccyx. The perineum possesses an anterior *urogenital triangle* and a posterior *anal triangle*. The urogenital triangle includes the external genitalia, the urogenital openings, and associated muscles. The anal triangle includes the anus as the digestive opening and associated muscles. The pelvic floor is composed of three layers of muscles. The term *diaphragm* is used in conjunction with some of these muscle layers, forming a type of partition or wall, and referred to as a urogenital diaphragm and a pelvic diaphragm.

1. Study figure 24.4 and table 24.3.
2. Locate the following muscles in the models of the male and female pelves:

   pelvic floor (pelvic diaphragm)
   - levator ani (group of muscles)
   - coccygeus

   deep perineal muscles (urogenital diaphragm)
   - deep transverse perineal muscle
   - external urethral sphincter

   superficial perineal muscles of perineum
   - superficial transverse perineal muscle
   - bulbospongiosus
   - ischiocavernosus
   - external anal sphincter
3. Locate the origins and insertions of these muscles in the human skeleton, and identify their innervations.
4. Complete Part C of the laboratory assessment.

Table 24.3 Muscles of the Pelvic Floor

Muscle	Origin	Insertion	Action	Innervation
Levator ani	Pubis and ischial spine	Coccyx and midline from anus through perineum (median raphe)	Supports pelvic viscera and provides sphincterlike action in anal canal and vagina	Pudendal nerve; spinal nerves S2–S4
Coccygeus	Ischial spine	Sacrum and coccyx	Supports pelvic viscera	Pudendal nerve
External anal sphincter	Coccyx and central tendon	Attached to connective tissues that encircle anal canal opening	Constricts opening of anus until voluntarily relax to defecate	Pudendal nerve
Deep transverse perineal muscle	Ischial and pubic ramus	Central tendon of perineum (perineal body)	Supports pelvic viscera	Pudendal nerve
External urethral sphincter	Ischial and pubic rami	Midline of urogenital diaphragm (median raphe)	Constricts opening of urethra; also encircles vagina in females	Pudendal nerve
Superficial transverse perineal muscle	Ischial tuberosity	Central tendon of perineum (perineal body)	Supports pelvic viscera and helps stabilize the central tendon of perineum	Pudendal nerve
Bulbospongiosus	Central tendon of perineum (perineal body)	Males: Ensheaths root of penis into corpus cavernosa Females: Ensheaths vagina and root of clitoris into corpus cavernosa	Males: Assists emptying of urethra and assists in erection of penis Females: Constricts vagina and assists in erection of clitoris	Pudendal nerve
Ischiocavernosus	Ischial tuberosity and ramus	Pubic symphysis near base and crus of corpus cavernosa of penis or clitoris	Males: Erects penis Females: Erects clitoris	Pudendal nerve

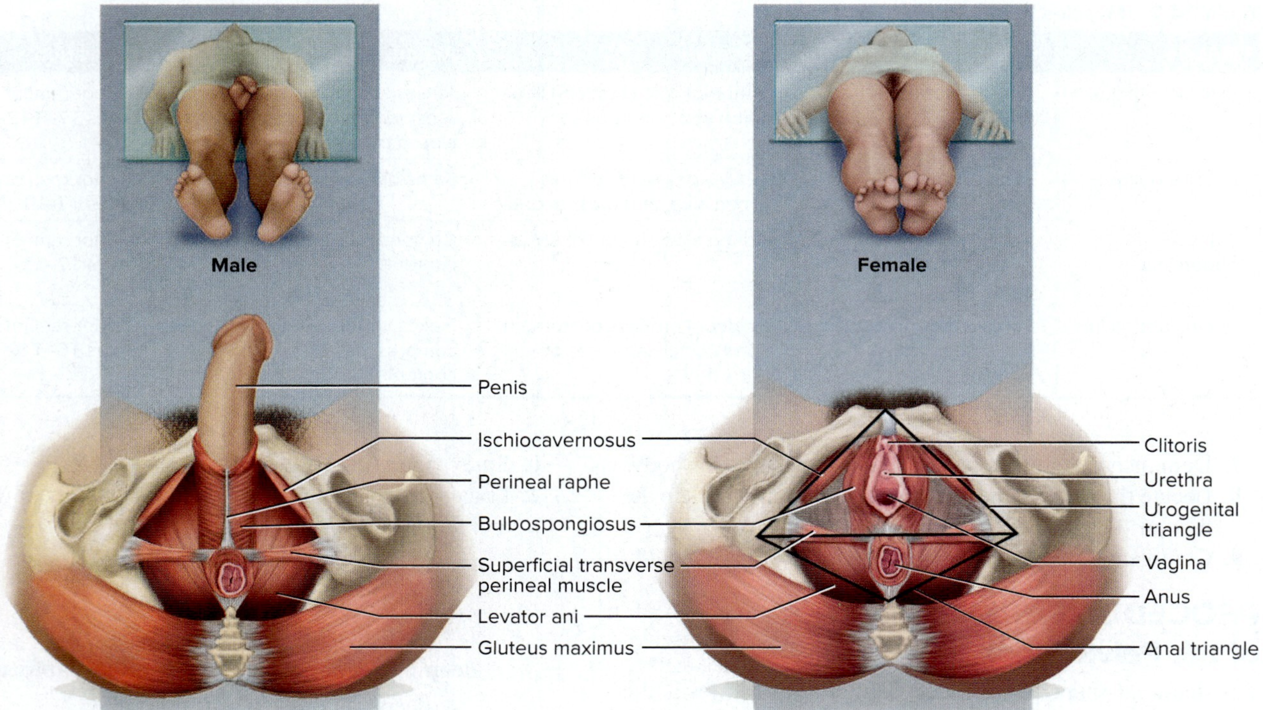

(a) Superficial region of perineum, inferior view

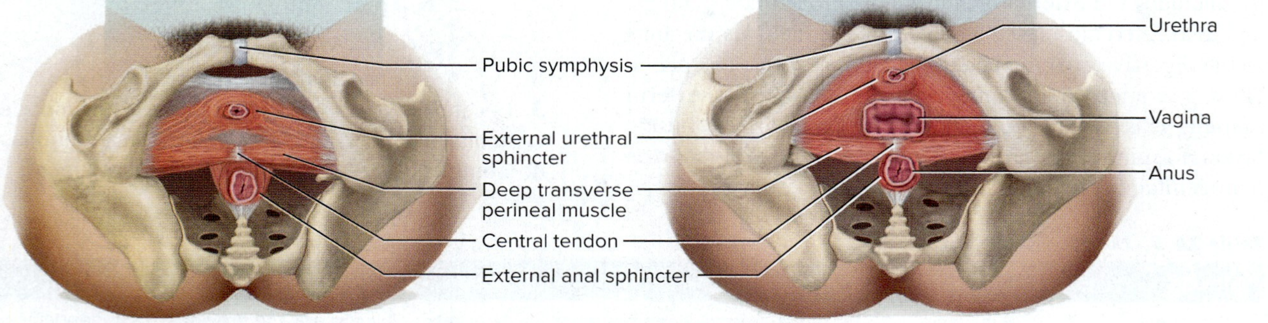

(b) Urogenital diaphragm of perineum, inferior view

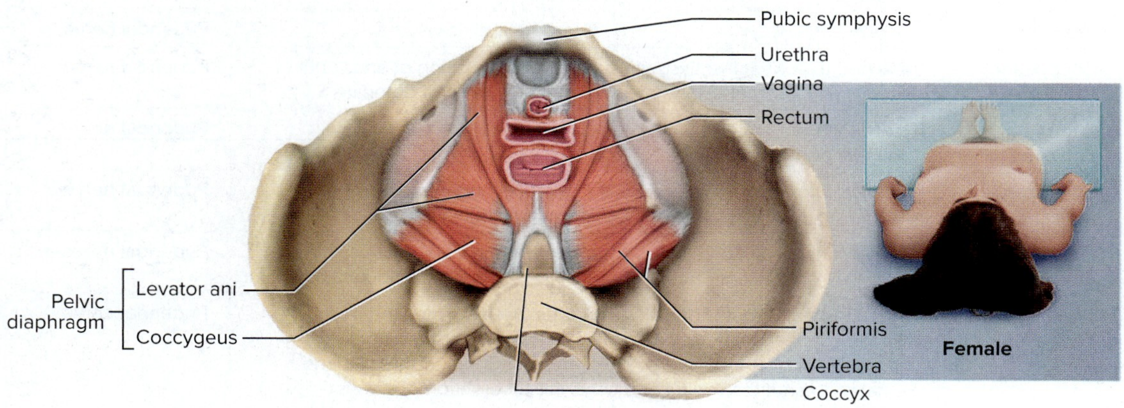

(c) Female pelvic diaphragm, superior view APR

FIGURE 24.4 Muscles of the pelvic floor: (a) superficial perineal muscles of male and female; (b) urogenital diaphragm (next deeper perineal muscles) of male and female; (c) pelvic diaphragm (deepest layer) of female.

LABORATORY
ASSESSMENT 24

Name ___________________________

Date ___________________________

Section ___________________________

The **A** corresponds to the indicated Learning Outcome(s) **O** found at the beginning of the Laboratory Exercise.

Muscles of the Vertebral Column, Abdominal Wall, and Pelvic Floor

PART A: Assessments

Identify the muscles indicated in figures 24.5 and 24.6.

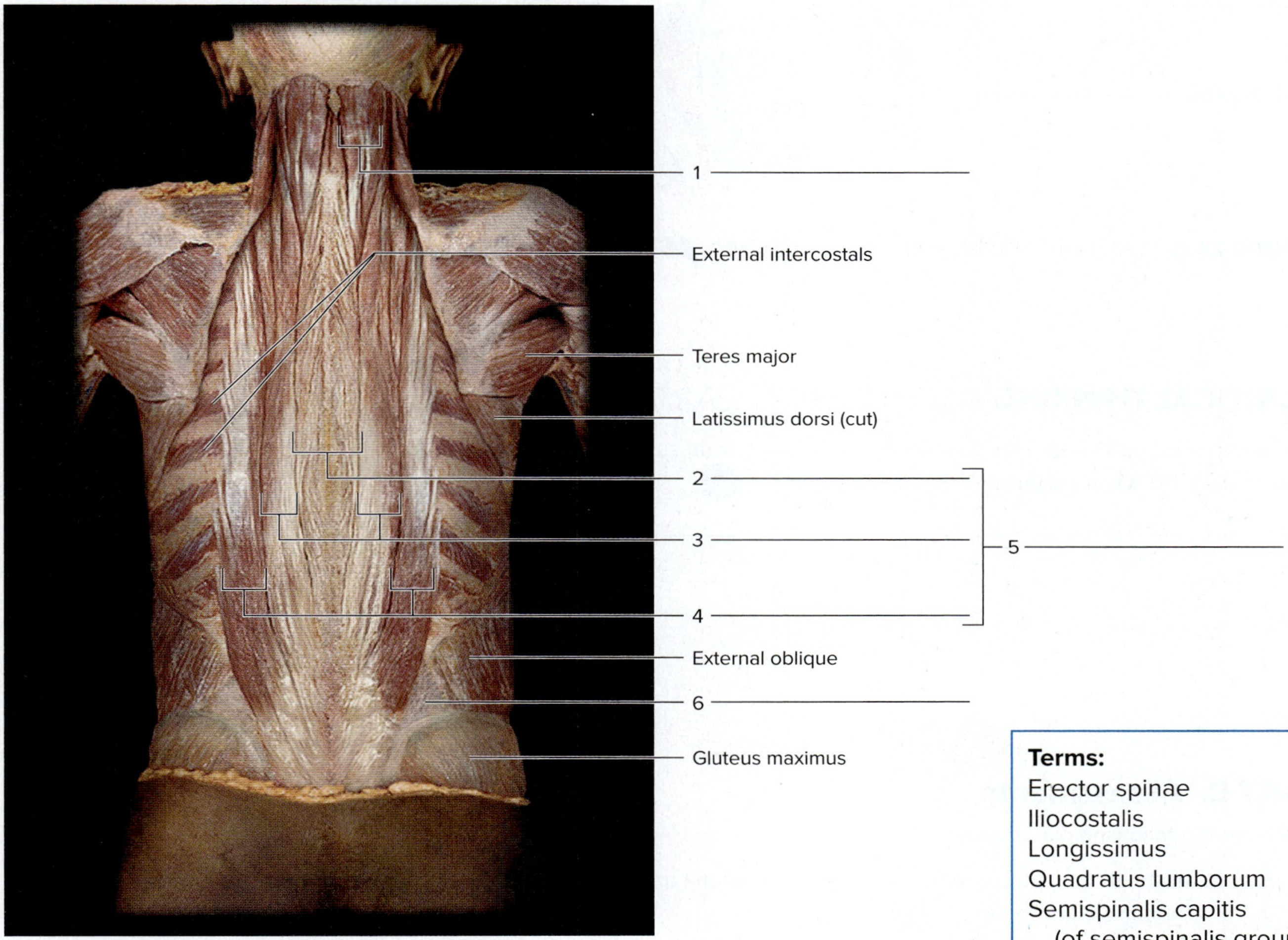

Terms:
Erector spinae
Iliocostalis
Longissimus
Quadratus lumborum
Semispinalis capitis (of semispinalis group)
Spinalis

FIGURE 24.5 Label the deep back muscles of a cadaver, using the terms provided. APR/McGraw Hill APR A1

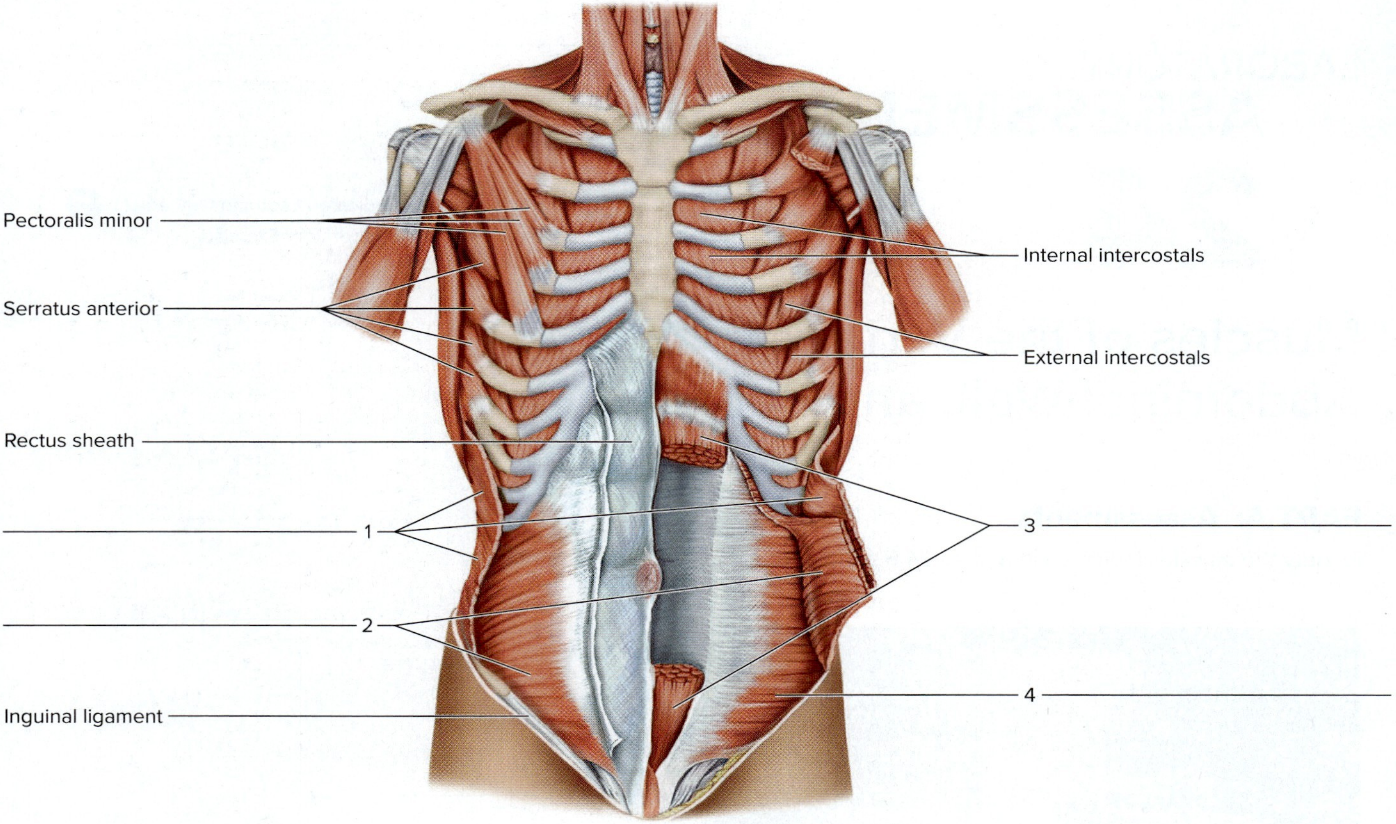

FIGURE 24.6 Label the muscles of the abdominal wall. APR A1

CRITICAL THINKING ASSESSMENT—PART A

Name the abdominal muscles a surgeon would incise from superficial (1) to deep (3) when performing an appendectomy. A1

(1) ______________________________

(2) ______________________________

(3) ______________________________

PART B: Assessments

Complete the following statements:

1. A band of tough connective tissue in the midline of the anterior abdominal wall called the ______________________ serves as a muscle attachment. A1
2. The ______________________ muscle spans from the costal cartilages of ribs 5–7 and the xiphoid process of the sternum to the pubic bones. A3

3. The ________________________ forms the third layer (deepest layer) of the abdominal wall muscles. A1

4. The action of the external oblique muscle is to ________________________. A2

5. The action of the rectus abdominis is to ________________________. A2

6. The iliocostalis, longissimus, and spinalis muscles together form the ________________________ group. A1

7. The ________________________ has the origin on the iliac crest and the insertion on rib 12 and L1–L4 vertebrae. A3

8. The origin of the ________________________ is on the iliac crest and inguinal ligament. A3

9. The insertion of the ________________________ is on the xiphoid process of the sternum and costal cartilages. A3

10. Name four muscles that compress the abdominal contents. A2

________________________ ________________________

________________________ ________________________

11. The actions of the ________________________ group of vertebral column muscles are similar to the erector spinae group and the quadratus lumborum. A2

12. Name the three muscles included in the semispinalis group. A1

________________________ ________________________

____

PART C: Assessments

1. Identify the muscles in figure 24.7.

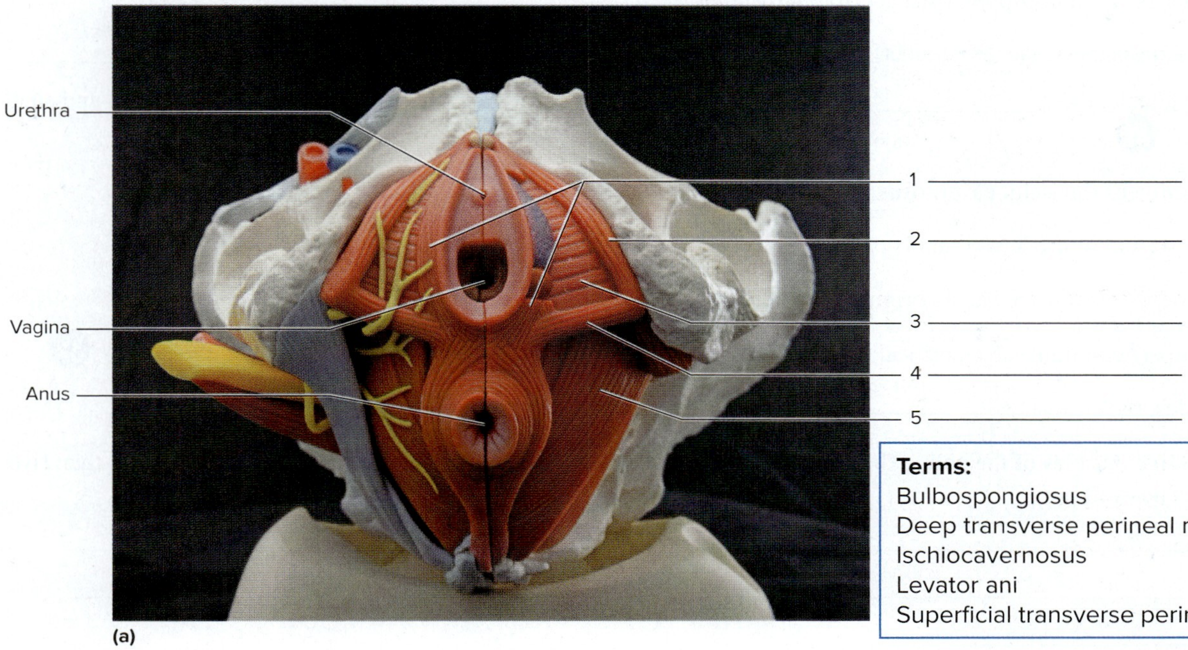

Terms:
Bulbospongiosus
Deep transverse perineal muscle
Ischiocavernosus
Levator ani
Superficial transverse perineal muscle

FIGURE 24.7 Label the muscles of the female pelvic floor in (a) an inferior view of a model of the superficial and urogenital diaphragm muscles of the perineurium, and a superior view of (b) a model and (c) a cadaver of the pelvic diaphragm, using the terms provided. ©3B Scientific GmbH, Germany, 2017 www.3bscientific.com

Terms:
Coccygeus
Levator ani

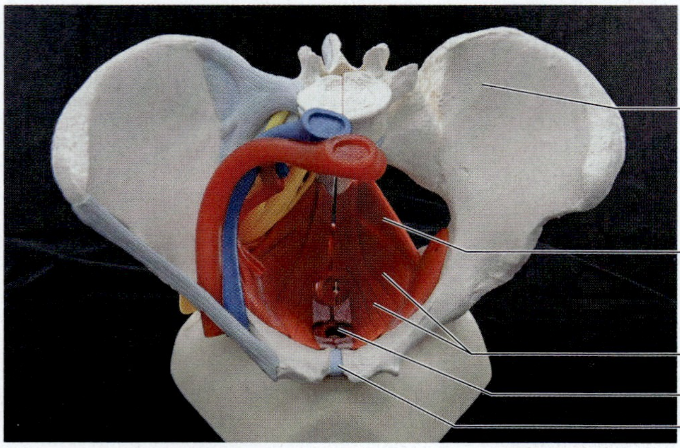

(b)

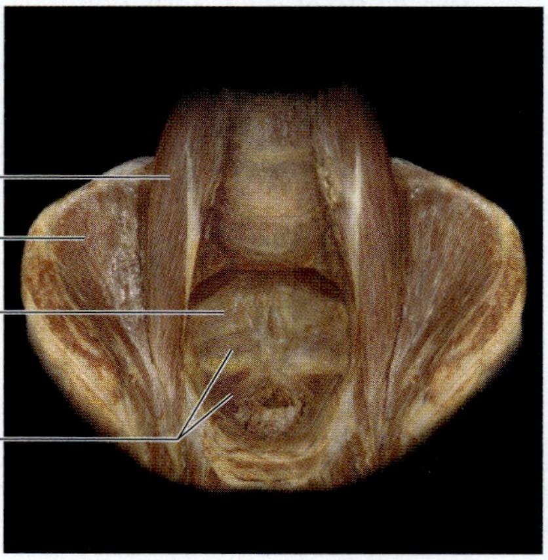

(c)

FIGURE 24.7 *Continued.* (b) ©3B Scientific GmbH, Germany, 2017, www.3bscientific.com; (c) APR/McGraw Hill

2. Complete the following statements:

a. The levator ani and coccygeus together form the ______________________________ diaphragm. A1

b. The posterior portion of the perineum is called the ______________________________ triangle. A1

c. The ______________________________ assists in closing the anal opening. A2

d. The muscles of the pelvic floor are innervated by the ______________________________ nerve. A3

e. The ______________________________ ensheaths the root of the penis. A1

f. In females, the bulbospongiosus acts to constrict the ______________________________. A2

g. In the female, the pelvic floor (outlet) is penetrated by the urethra, vagina, and ______________________________. A1

h. In the female, the ______________________________ muscles are separated by the vagina, urethra, and anal canal. A1

i. The action of the ischiocavernosus is to ______________________________. A2

j. The insertion of the levator ani is on the ______________________________ bone. A3

k. The coccygeus muscle has an origin on the ______________________________. A3

l. The superficial transverse perineal muscle acts to support the ______________________________. A2

m. In males, the ______________________________ acts to help empty the urethra. A2

n. Name five muscles of the pelvic floor that males and females both have in common. A1 (There are more than five correct answers.)

______________________________ ______________________________

______________________________ ______________________________

____

LABORATORY
EXERCISE 25

Muscles of the Hip and Lower Limb

Purpose of the Exercise

To review the locations, actions, origins, insertions, and innervations of the muscles that move the thigh, leg, and foot.

MATERIALS NEEDED

Human muscular model

Human skeleton, articulated

Muscular models of the lower limb

For Learning Extension Activity:

Long rubber bands

Learning Outcomes APR

After completing this exercise, you should be able to:

O1 Locate and identify the muscles that move the thigh, leg, and foot.

O2 Describe and demonstrate the actions of each of these muscles.

O3 Locate the origin and insertion and identify the innervation of each of these muscles in a human skeleton and on muscular models.

The O corresponds to the assessments A indicated in the Laboratory Assessment for this Exercise.

Pre-Lab

Carefully read the introductory material and examine the entire lab. Be familiar with the locations, actions, origins, insertions, and innervations of the muscles of the hip and lower limb from lecture or the textbook. Answer the pre-lab questions.

Pre-Lab Questions Select the correct answer for each of the following questions:

1. Muscles that move the thigh at the hip joint have origins on the
- **a.** pelvis.
- **b.** femur.
- **c.** tibia.
- **d.** patella.

2. Anterior thigh muscles serve as prime movers of
- **a.** flexion of the leg at the knee joint.
- **b.** abduction of the thigh at the hip joint.
- **c.** adduction of the thigh at the hip joint.
- **d.** extension of the leg at the knee joint.

3. The muscles of the lower limb _____ than those of the upper limb.
- **a.** are more numerous
- **b.** are larger
- **c.** conduct actions with more precision
- **d.** are less powerful

4. Which of the following muscles is *not* part of the quadriceps femoris group?
- **a.** rectus femoris
- **b.** vastus lateralis
- **c.** semitendinosus
- **d.** vastus medialis

5. Which of the following muscles is *not* part of the hamstring group?
- **a.** rectus femoris
- **b.** biceps femoris
- **c.** semitendinosus
- **d.** semimembranosus

6. Muscles located in the lateral leg include actions of
- **a.** extension of the knee joint.
- **b.** eversion of the foot.
- **c.** flexion of the toes.
- **d.** extension of the toes.

7. Posterior muscles of the leg act in _____ of the foot.
- **a.** plantar flexion
- **b.** dorsiflexion
- **c.** eversion
- **d.** extension

8. Contractions of the sartorius muscle involve movements across two different joints.
   a. True **b.** False
9. The insertions of the four quadriceps femoris muscles are on the femur.
   a. True **b.** False

The muscles that move the thigh at the hip joint have their origins on some part of the pelvic girdle and insertions usually on the femur. Those muscles attached on the anterior pelvis primarily act to flex the thigh at the hip joint; those attached on the posterior pelvis primarily act to extend the thigh at the hip joint; those attached on the lateral side act as abductors and rotators of the thigh at the hip joint; and those attached on the medial side act as adductors of the thigh at the hip joint.

The muscles that move the leg at the knee joint have their origins on the femur or the hip bone. The anterior thigh muscles have their insertions on the proximal tibia and serve as prime movers of extension of the leg at the knee joint. The posterior thigh muscles have their insertions on the tibia or fibula and act as prime movers of flexion of the leg at the knee joint.

The muscles that move the foot have their origins on the tibia, fibula, or femur. The anterior leg muscles have actions of dorsiflexion and extension of the toes; the posterior leg muscles have actions of plantar flexion and flexion of the toes; and the lateral leg muscles have actions of eversion of the foot.

In contrast to the upper limb muscles, those in the lower limb tend to be much larger and are involved in powerful contractions for walking or running as well as isometric contractions in standing. Although the lower limb has more stability than the upper limb, it has a more limited range of motion. As compared to the forearm, the leg has fewer muscles because the foot is not as involved in actions that require precision movements as the hand.

As you demonstrate the actions of muscles in the lower limb, refer to standard anatomical regions of the limb. The *thigh* refers to hip to knee; the *leg* refers to knee to ankle; and the *foot* refers to ankle, metatarsal area, and toes. Several muscles in the hip and lower limb cross more than one joint and involve movements of both joints. Examples of muscles crossing more than one joint include the sartorius and gastrocnemius.

PROCEDURE: Muscles of the Hip and Lower Limb

1. Study figures 25.1 to 25.4, and table 25.1.
2. Locate the following muscles in the human torso model and in the lower limb models. Also locate as many of them as possible in your body.

   muscles that move the thigh/hip joint
   - anterior hip muscles
     - iliopsoas group
       - psoas major
       - iliacus
   - posterior and lateral hip muscles
     - gluteus maximus
     - gluteus medius
     - gluteus minimus
     - tensor fasciae latae
     - piriformis
     - quadratus femoris
   - medial adductor muscles
     - pectineus
     - adductor longus
     - adductor magnus
     - adductor brevis
     - gracilis
3. Demonstrate the actions of these muscles in your body.
4. Locate the origins and insertions of these muscles in the human skeleton.
5. Study figures 25.1, 25.3, and 25.5 and table 25.2.
6. Locate the following muscles in the human torso model and in the lower limb models. Also locate as many of them as possible in your body.

   muscles that move the leg/knee joint
   - anterior thigh muscles
     - sartorius
     - quadriceps femoris group
       - rectus femoris
       - vastus lateralis
       - vastus medialis
       - vastus intermedius
   - posterior thigh muscles
     - hamstring group
       - biceps femoris
       - semitendinosus
       - semimembranosus
7. Demonstrate the actions of these muscles in your body.
8. Locate the origins and insertions of these muscles in the human skeleton.
9. Complete Part A of Laboratory Assessment 25.
10. Study figures 25.6 to 25.8, and table 25.3.
11. Locate the following muscles in the human torso model and in the lower limb models. Also locate as many of them as possible in your body.

    muscles that move the foot
    - anterior leg muscles
      - tibialis anterior
      - extensor digitorum longus

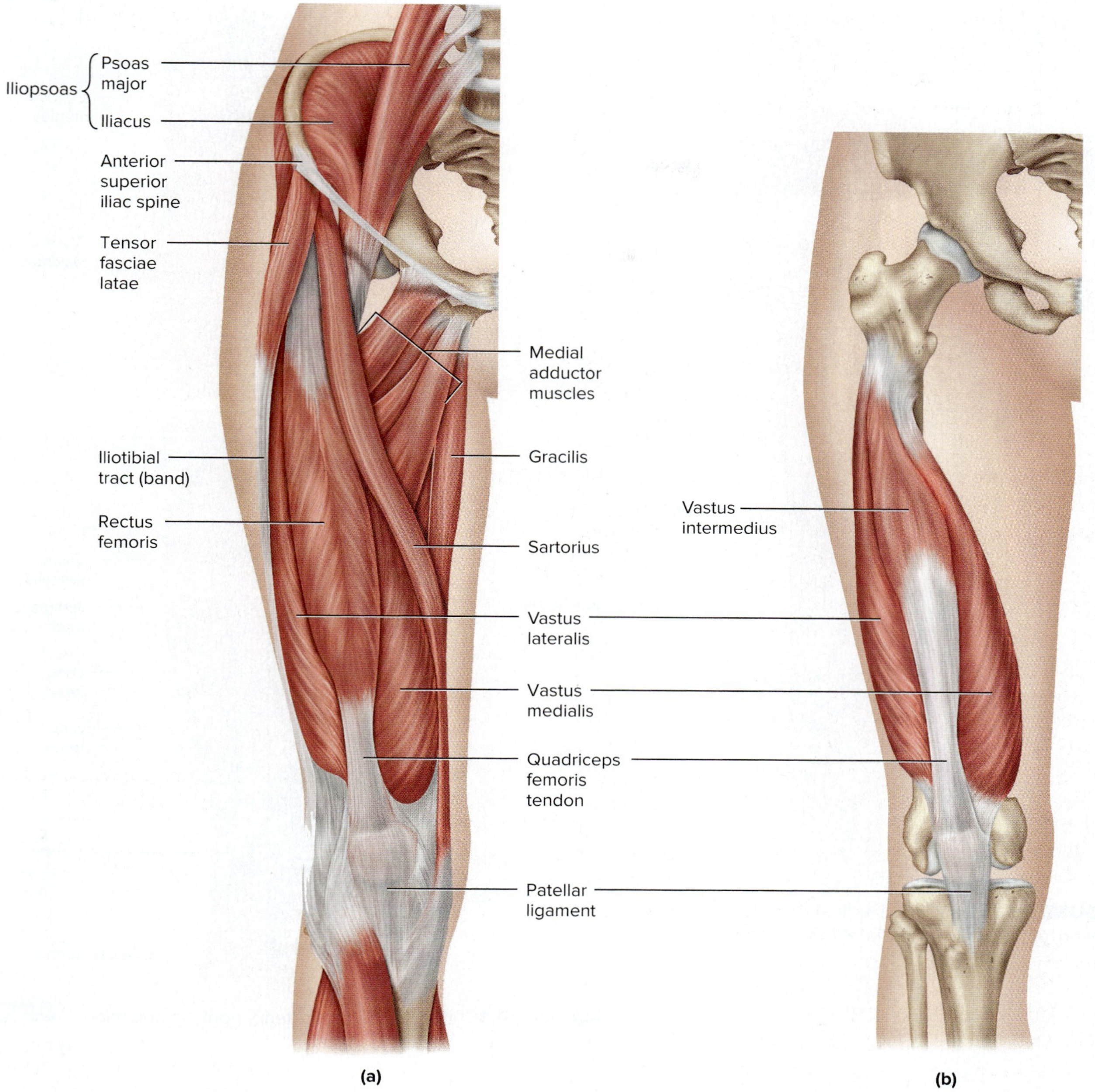

FIGURE 25.1 (a) Muscles of the anterior right hip and thigh; (b) selected quadriceps femoris muscles of the anterior right thigh. The rectus femoris is removed to expose the deeper muscles and tendons of the quadriceps femoris group. APR

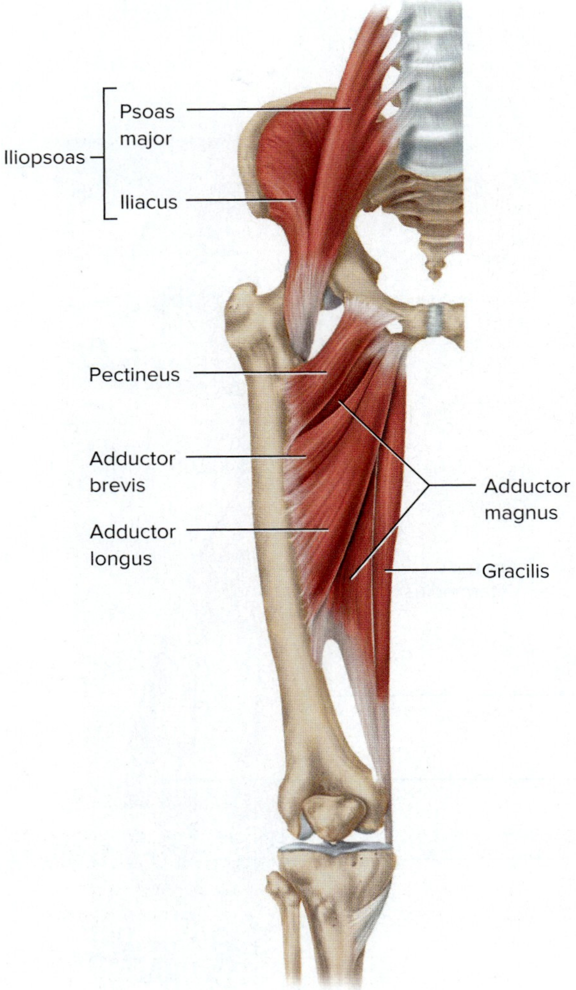

FIGURE 25.2 Selected individual muscles of the anterior right hip and medial thigh. APR

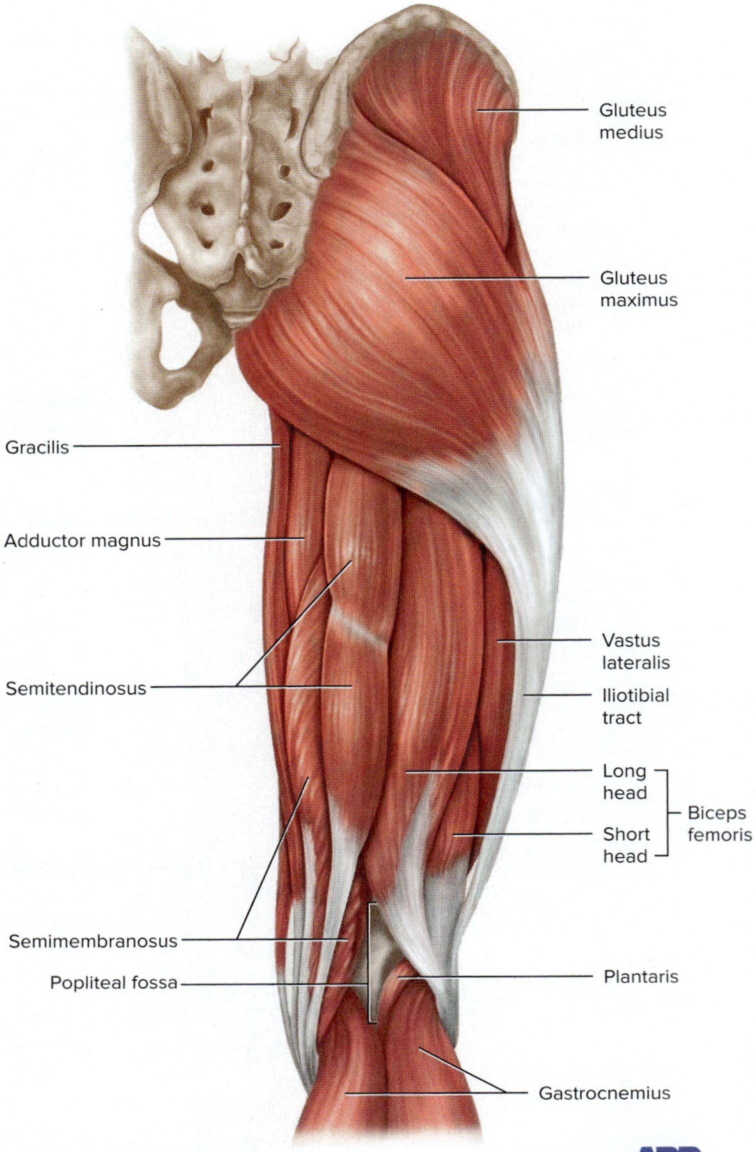

FIGURE 25.3 Muscles of the posterior right hip and thigh. APR

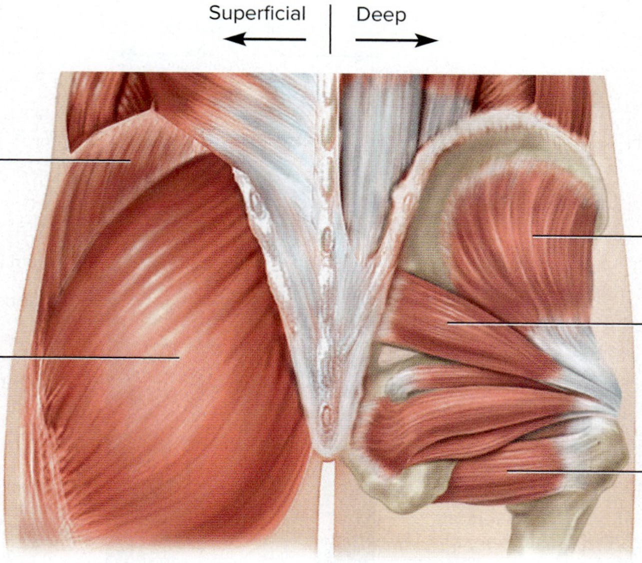

FIGURE 25.4 Superficial and deep posterior hip muscles. APR

Table 25.1 Muscles that Move the Thigh

Muscle	Origin	Insertion	Action	Innervation
Psoas major	Intervertebral discs, bodies, and transverse processes of T12–L5	Lesser trochanter of femur	Flexes thigh at hip; flexes trunk at hip (when thigh is fixed)	Anterior rami of lumbar spinal nerves
Iliacus	Iliac fossa of ilium	Lesser trochanter of femur	Same as psoas major	Femoral nerve
Gluteus maximus	Sacrum, coccyx, and posterior iliac crest	Gluteal tuberosity of femur and iliotibial tract	Extends thigh at hip	Inferior gluteal nerve
Gluteus medius	Lateral surface of ilium	Greater trochanter of femur	Abducts and medially rotates thigh	Superior gluteal nerve
Gluteus minimus	Lateral surface of ilium	Greater trochanter of femur	Same as gluteus medius	Superior gluteal nerve
Tensor fasciae latae	Iliac crest and anterior superior iliac spine	Iliotibial tract	Abducts and medially rotates thigh	Superior gluteal nerve
Piriformis	Anterior surface of sacrum	Greater trochanter of femur	Abducts and laterally rotates thigh	Spinal nerves L5–S2
Quadratus femoris	Ischial tuberosity	Trochanter crest of femur	Laterally rotates thigh	Nerve to quadratus femoris
Pectineus	Pubis	Femur distal to lesser trochanter	Adducts and flexes thigh	Femoral nerve
Adductor longus	Pubis near pubic symphysis	Linea aspera of femur	Adducts and flexes thigh	Obturator nerve
Adductor magnus	Pubis and ischial tuberosity	Linea aspera of femur	Adducts thigh; posterior portion extends thigh and anterior portion flexes thigh	Obturator nerve; tibial nerve
Adductor brevis	Pubis	Linea aspera of femur	Adducts thigh	Obturator nerve
Gracilis	Pubis	Medial surface of tibia	Adducts thigh and flexes leg at knee	Obturator nerve

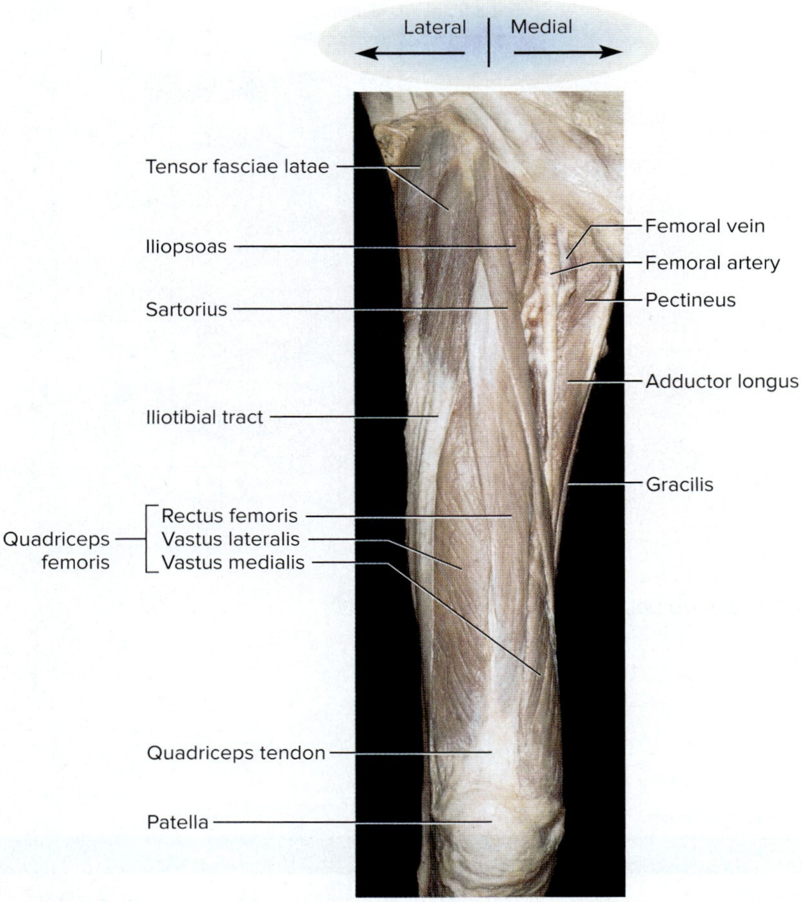

FIGURE 25.5 Anterior thigh muscles include the quadriceps femoris muscle, which contains four heads: rectus femoris, vastus lateralis, vastus medialis, and vastus intermedius (not shown in this figure). Rebecca Gray/McGraw Hill APR

Table 25.2 Muscles that Move the Leg

Muscle	Origin	Insertion	Action	Innervation
Sartorius	Anterior superior iliac spine	Proximal medial surface of tibia	Flexes, abducts, and laterally rotates thigh at hip; flexes leg at knee	Femoral nerve
Quadriceps Femoris Group				
Rectus femoris	Anterior inferior iliac spine and superior margin of acetabulum	Patella by common quadriceps tendon, which continues as patellar ligament to tibial tuberosity	Extends leg at knee; flexes thigh at hip	Femoral nerve
Vastus lateralis	Greater trochanter and linea aspera of femur		Extends leg at knee	Femoral nerve
Vastus medialis	Linea aspera of femur		Extends leg at knee	Femoral nerve
Vastus intermedius	Anterior and lateral surfaces of femur		Extends leg at knee	Femoral nerve
Hamstring Group				
Biceps femoris	Ischial tuberosity (long head) and linea aspera of femur (short head)	Head of fibula and lateral condyle of tibia	Flexes leg at knee; rotates leg laterally; extends thigh	Tibial nerve; common fibular nerve
Semitendinosus	Ischial tuberosity	Proximal medial surface of tibia	Flexes leg at knee; rotates leg medially; extends thigh	Tibial nerve
Semimembranosus	Ischial tuberosity	Posterior medial condyle of tibia	Flexes leg at knee; rotates leg medially; extends thigh	Tibial nerve

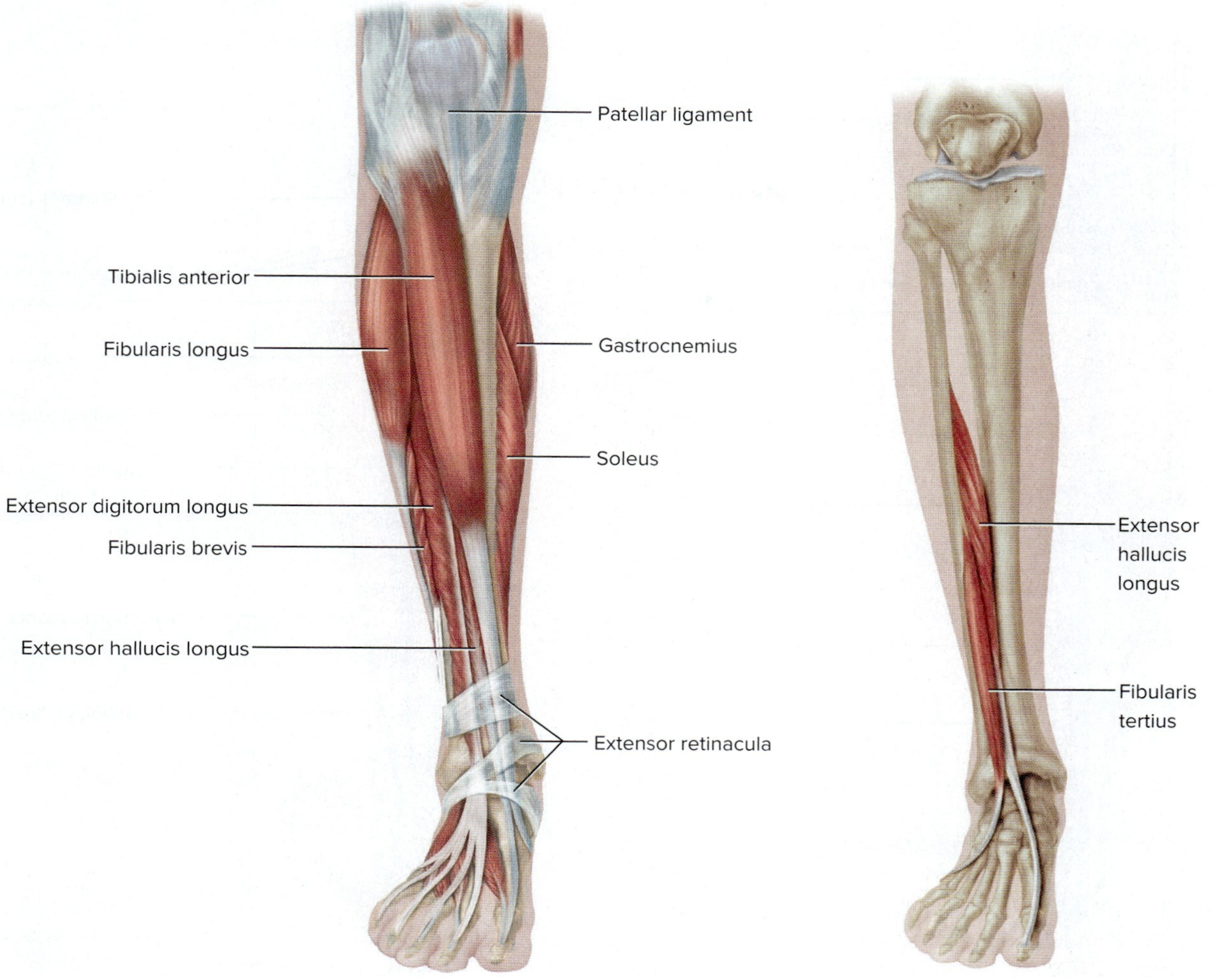

(a) Superficial view

(b) Selected individual muscles

FIGURE 25.6 Muscles of the anterior right leg (a) and (b). APR

- extensor hallucis longus
- fibularis (peroneus) tertius

- posterior leg muscles
  - superficial group
    - gastrocnemius
    - soleus
    - plantaris
  - deep group
    - popliteus
    - tibialis posterior
    - flexor digitorum longus
    - flexor hallucis longus
- lateral leg muscles
  - fibularis (peroneus) longus
  - fibularis (peroneus) brevis

12. Demonstrate the actions of these muscles in your body.
13. Locate the origins and insertions of these muscles in the human skeleton.
14. Complete Parts B, C, and D of the laboratory assessment.

LEARNING EXTENSION ACTIVITY

A long rubber band can be used to simulate muscle locations, origins, insertions, and actions on muscular models, the skeleton, or a laboratory partner. Hold one end of the rubber band firmly on the origin location of a muscle; then slightly stretch the rubber band and hold the other end on the insertion site. Allow the insertion end to slowly move toward the origin end to simulate the contraction and action of the muscle.

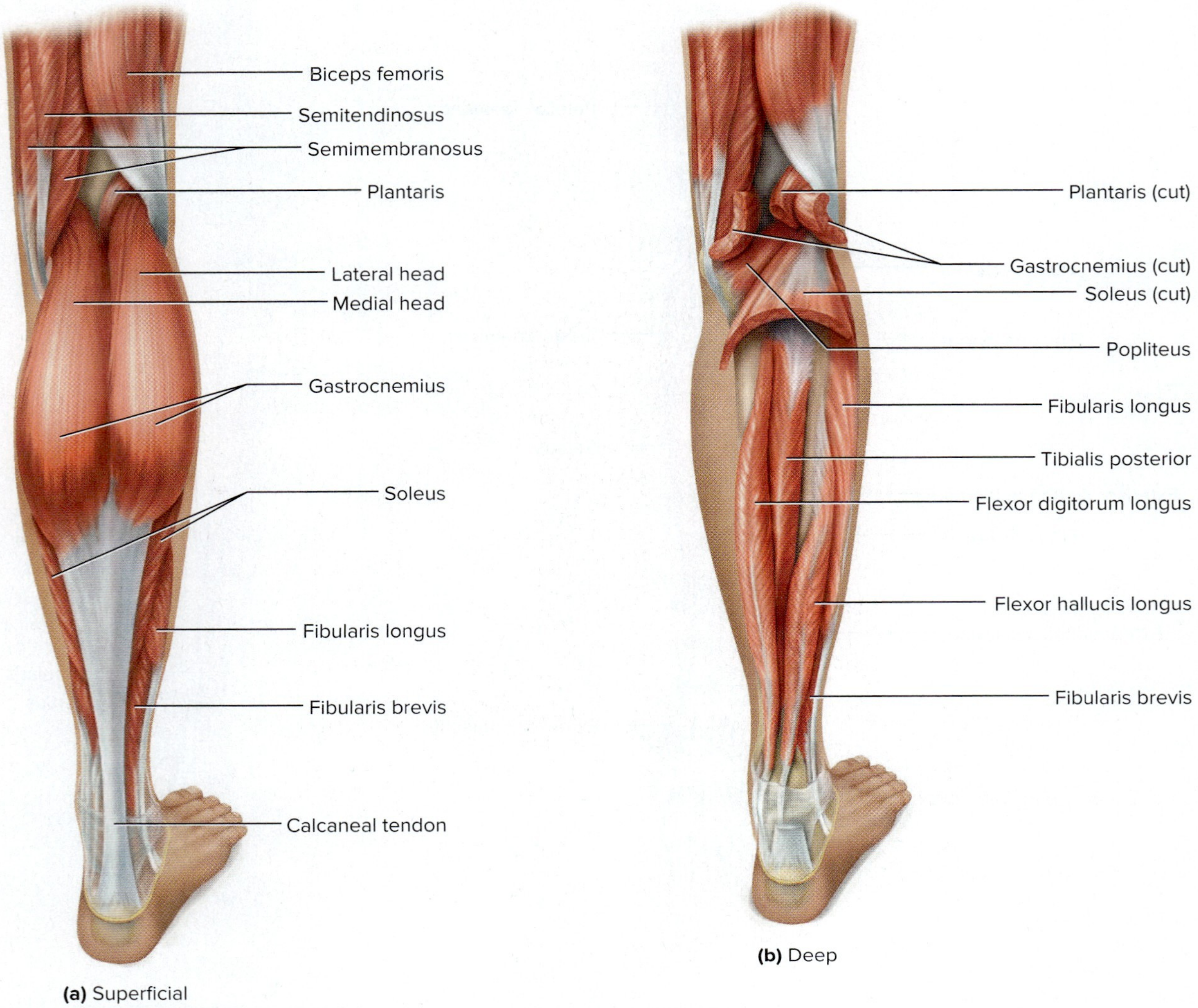

FIGURE 25.7 Muscles of the posterior right leg (a) and (b). APR

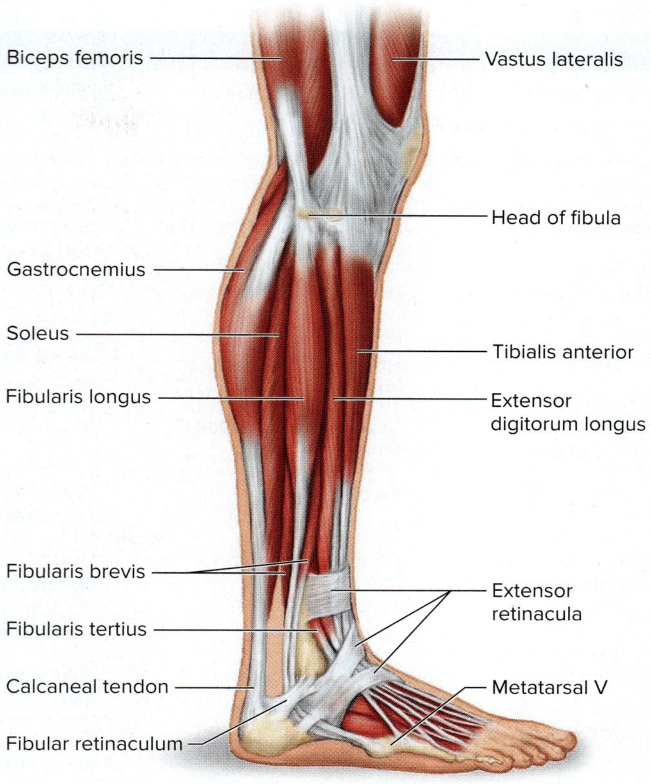

FIGURE 25.8 Muscles of the lateral right leg.

Table 25.3 Muscles that Move the Foot

Muscle	Origin	Insertion	Action	Innervation
Tibialis anterior	Lateral condyle and proximal tibia	Medial cuneiform and metatarsal I	Dorsiflexion and inversion of foot	Deep fibular (peroneal) nerve
Extensor digitorum longus	Lateral condyle of tibia and anterior surface of fibula	Dorsal surfaces of second and third phalanges of toes 2–5	Extends toes 2–5 and dorsiflexion	Deep fibular (peroneal) nerve
Extensor hallucis longus	Anterior surface of fibula	Distal phalanx of the great toe	Extends great toe and dorsiflexion	Deep fibular (peroneal) nerve
Fibularis tertius	Anterior distal surface of fibula	Dorsal surface of metatarsal V	Dorsiflexion and eversion of foot	Deep fibular (peroneal) nerve
Gastrocnemius	Lateral and medial condyles of femur	Calcaneus via calcaneal tendon	Plantar flexion of foot and flexes knee	Tibial nerve
Soleus	Head and shaft of fibula and posterior surface of tibia	Calcaneus via calcaneal tendon	Plantar flexion of foot	Tibial nerve
Plantaris	Superior to lateral condyle of femur	Calcaneus	Plantar flexion of foot and flexes knee	Tibial nerve
Popliteus	Lateral condyle of femur; lateral meniscus and joint capusule	Posterior upper tibia	Unlocks knee to allow flexion; may prevent femur dislocation in forward direction during crouching	Tibial nerve
Tibialis posterior	Posterior tibia and fibula	Tarsals (several) and metatarsals II–IV	Plantar flexion and inversion of foot	Tibial nerve
Flexor digitorum longus	Posterior surface of tibia	Distal phalanges of toes 2–5	Flexes toes 2–5 and plantar flexion and inversion of foot	Tibial nerve
Flexor hallucis longus	Posterior distal fibula	Distal phalanx of great toe	Flexes great toe and plantar flexes foot	Tibial nerve
Fibularis longus	Head and shaft of fibula and lateral condyle of tibia	Medial cuneiform and metatarsal I	Plantar flexion and eversion of foot; also supports arch	Superficial fibular (peroneal) nerve
Fibularis brevis	Distal fibula	Metatarsal V	Plantar flexion and eversion of foot	Superficial fibular (peroneal) nerve

Name ______________________

Date ______________________

Section ______________________

The (A) corresponds to the indicated Learning Outcome(s) (O) found at the beginning of the Laboratory Exercise.

LABORATORY ASSESSMENT 25

Muscles of the Hip and Lower Limb

PART A: Assessments

Identify the muscles indicated in the hip and thigh in figures 25.9 and 25.10.

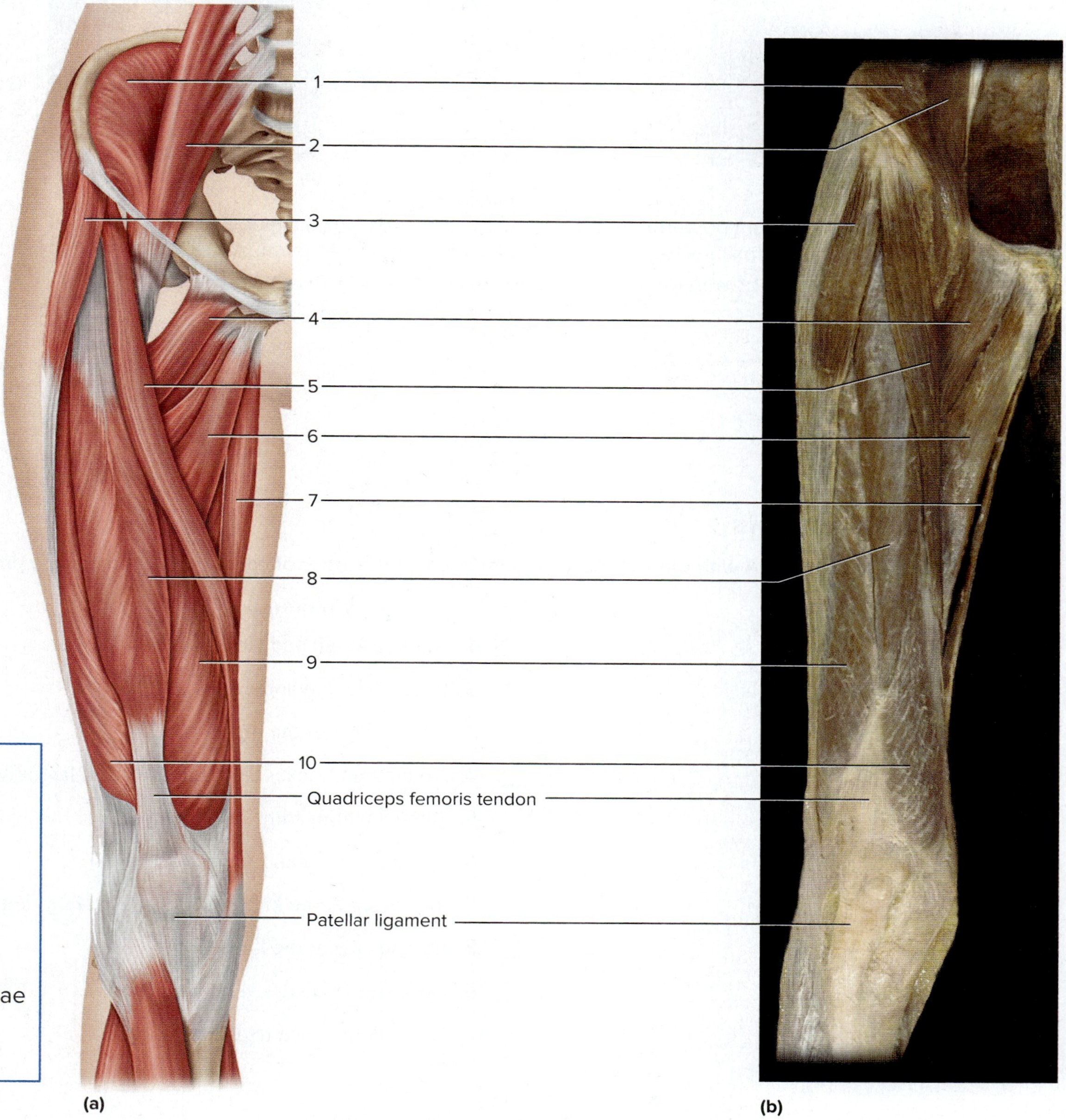

Terms:
Adductor longus
Gracilis
Iliacus
Pectineus
Psoas major
Rectus femoris
Sartorius
Tensor fasciae latae
Vastus lateralis
Vastus medialis

FIGURE 25.9 Label the right anterior muscles of the hip and thigh shown in (a) an illustration and (b) a cadaver, using the terms provided. (b) APR/McGraw Hill APR A1

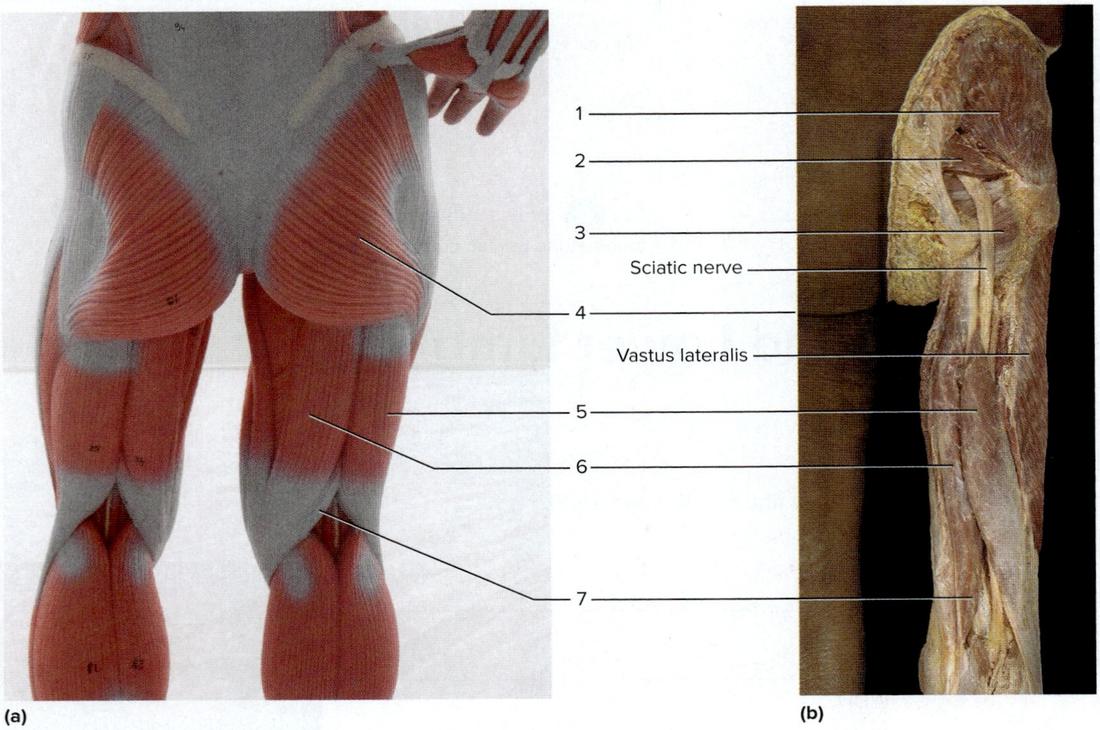

Terms:
Biceps femoris (long head)
Gluteus maximus
Gluteus medius
Pirformis
Quadratus femoris
Semimembranosus
Semitendinosus

FIGURE 25.10 Label the right posterior muscles of the hip and thigh in (a) a model and (b) a cadaver, using the terms provided. *Note:* The gluteus maximus has been removed in (b) to show deeper muscles. (a) B59 [1000212] 1/3 Life-Size Muscle Figure, 2-part © 3B Scientific GmbH, Germany, 2017 www.3bscientific.com; (b) APR/McGraw Hill APR A1

PART B: Assessments

Match the muscles in column A with the actions in column B. Place the letter of your choice in the space provided. A2

Column A	Column B
a. Biceps femoris	______ **1.** Adducts thigh and flexes knee
b. Fibularis (peroneus) longus	______ **2.** Plantar flexion and eversion of foot
c. Gluteus medius	______ **3.** Flexes thigh at the hip
d. Gracilis	______ **4.** Abducts and flexes thigh and rotates it laterally; flexes leg at knee
e. Popliteus	______ **5.** Abducts thigh and rotates it medially
f. Psoas major and iliacus	______ **6.** Plantar flexion and inversion of foot
g. Quadriceps femoris group	______ **7.** Flexes leg at the knee and laterally rotates leg
h. Sartorius	______ **8.** Extends leg at the knee
i. Tibialis anterior	______ **9.** Dorsiflexion and inversion of foot
j. Tibialis posterior	______ **10.** Unlocks the knee to allow flexion

PART C: Assessments

Name the muscle indicated by the following combinations of origin, insertion, and innervation. A3

	Origin	Insertion	Innervation	Muscle
1.	Lateral surface of ilium	Greater trochanter of femur	Superior gluteal nerve	______________
2.	Anterior superior iliac spine	Medial surface of proximal tibia	Femoral nerve	______________
3.	Lateral and medial condyles of femur	Calcaneus	Tibial nerve	______________
4.	Iliac crest and anterior superior iliac spine	Iliotibial tract	Superior gluteal nerve	______________
5.	Greater trochanter and linea aspera of femur	Patella to tibial tuberosity	Femoral nerve	______________
6.	Ischial tuberosity	Medial surface of proximal tibia	Tibial nerve	______________
7.	Linea aspera of femur	Patella to tibial tuberosity	Femoral nerve	______________
8.	Posterior surface of tibia	Distal phalanges of four lateral toes	Tibial nerve	______________
9.	Lateral condyle and proximal tibia	Medial cuneiform and first metatarsal	Deep fibular (peroneal) nerve	______________
10.	Ischial tuberosity and pubis	Linea aspera of femur	Obturator nerve; tibial nerve	______________
11.	Anterior sacrum	Greater trochanter of femur	Spinal nerves L5–S2	______________

PART D: Assessments

Identify the muscles in the leg in figure 25.11.

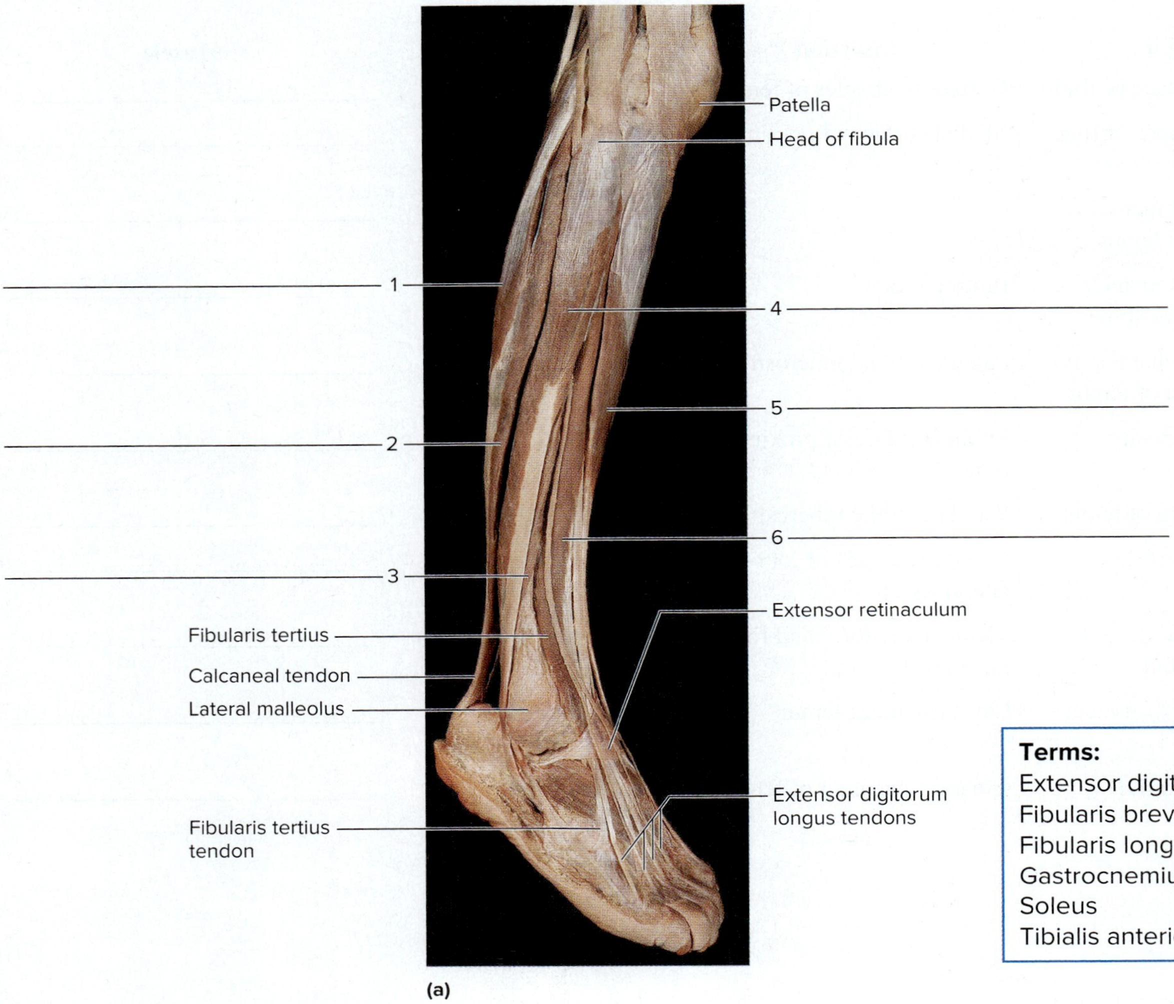

Terms:
Extensor digitorum longus
Fibularis brevis
Fibularis longus
Gastrocnemius
Soleus
Tibialis anterior

FIGURE 25.11 Label lateral views of (a) the right leg muscles of a cadaver and (b) the left leg muscles of a model, using the terms provided. Christine Eckel/McGraw Hill A1

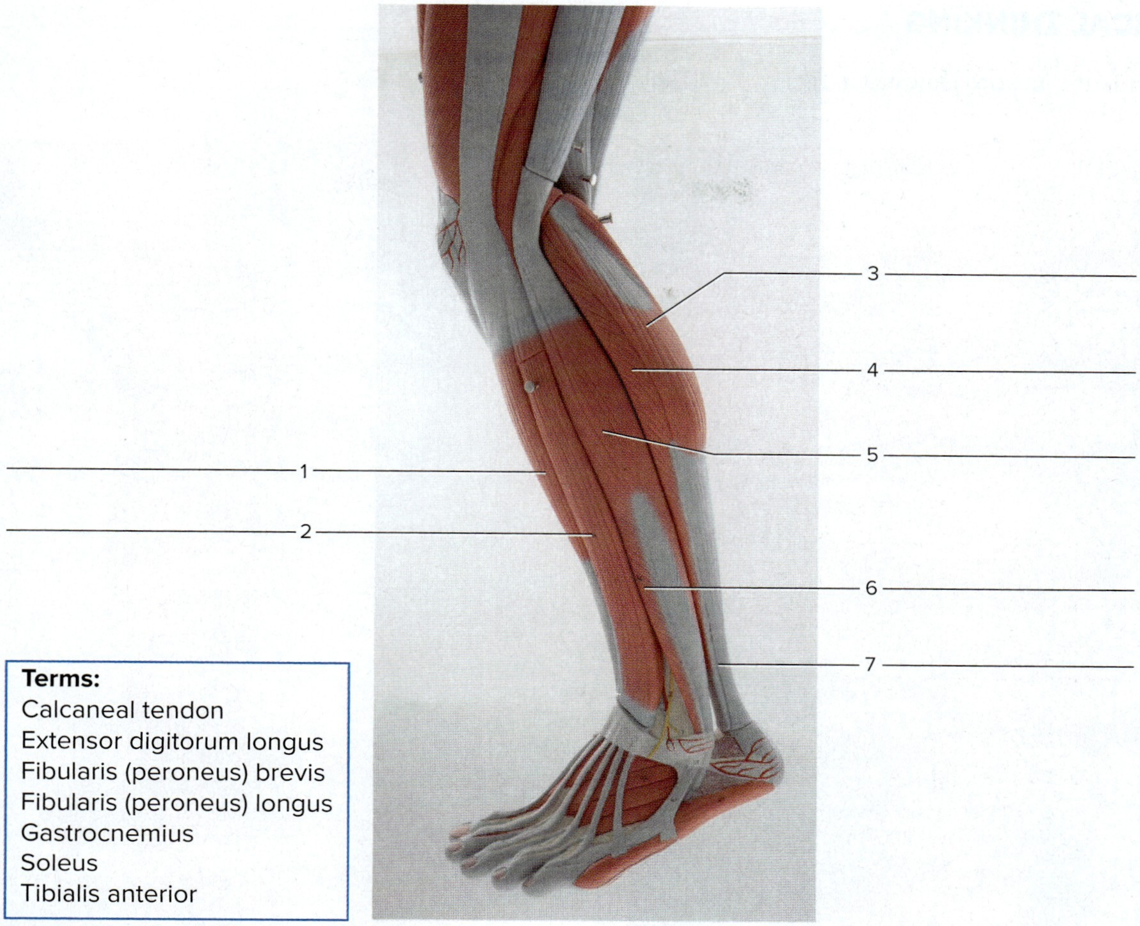

Terms:
Calcaneal tendon
Extensor digitorum longus
Fibularis (peroneus) brevis
Fibularis (peroneus) longus
Gastrocnemius
Soleus
Tibialis anterior

(b)

FIGURE 25.11 *Continued.* 3B Scientific GmbH A1

CRITICAL THINKING ASSESSMENTS—PART D

1. Identify the muscles indicated in figure 25.12.

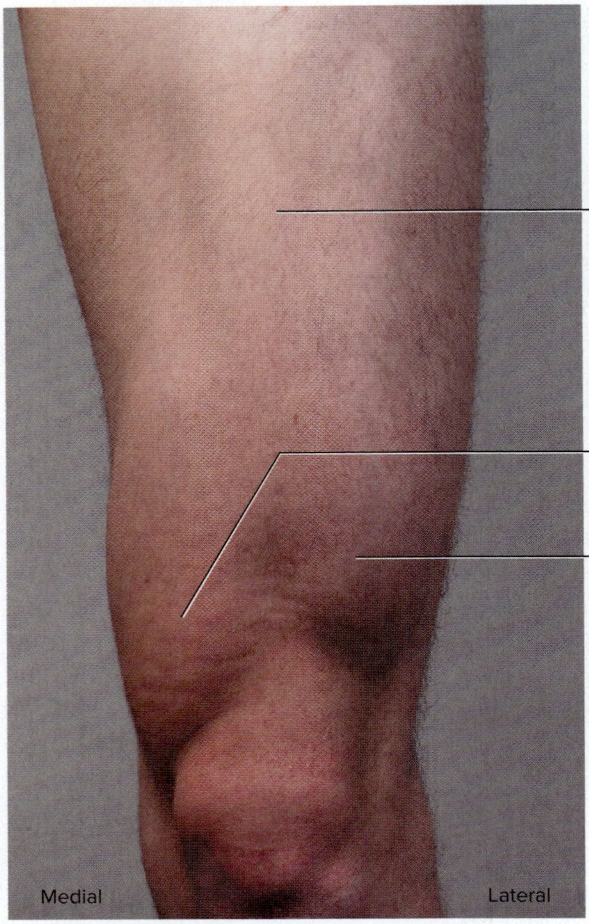

(a) Left thigh, anterior view

____ Rectus femoris
____ Vastus lateralis
____ Vastus medialis

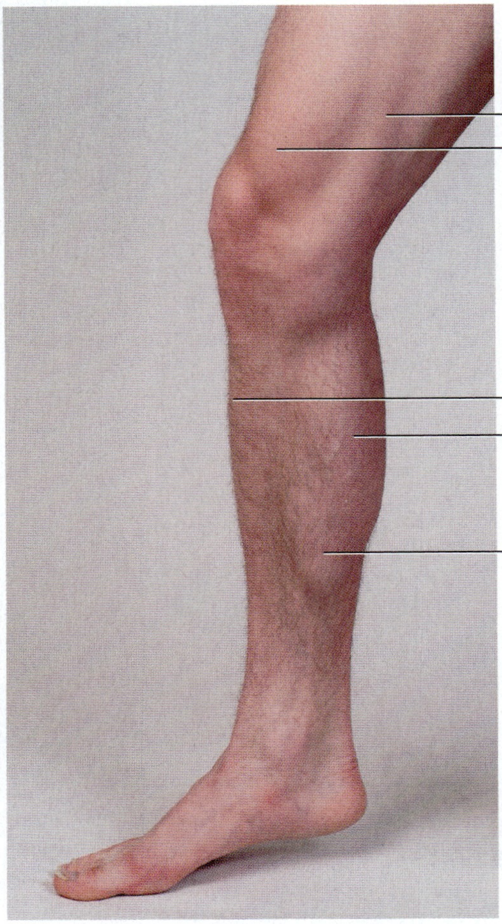

(b) Right lower limb, medial view

____ Gastrocnemius
____ Sartorius
____ Soleus
____ Tibialis anterior
____ Vastus medialis

FIGURE 25.12 Label these muscles that appear as lower limb surface features in these photographs (a) and (b), by placing the correct numbers in the spaces provided. ©J & J Photography A1

2. Wesley tore his calcaneal tendon playing a pick-up game of basketball. What two muscles of the leg would be directly affected by this injury, and what would be the immediate consequence in Wesley's mobility? Explain. ____________

____

____. A2 A3

LABORATORY
EXERCISE 26

Surface Anatomy

Purpose of the Exercise

To examine the surface features of the human body and the terms used to describe them.

MATERIALS NEEDED

Small round stickers
Colored pencils (black and red)

Learning Outcomes

After completing this exercise, you should be able to:

O1 Locate and identify major surface features of the human body.

O2 Arrange surface features by body region.

O3 Distinguish between surface features that are bony landmarks and soft tissue.

O4 Associate surface features with deeper structures and possible abnormalities.

The O corresponds to the assessments A indicated in the Laboratory Assessment for this Exercise.

Pre-Lab

Carefully read the introductory material and examine the entire lab. Be familiar with body regions, the skeletal system, and the muscular system from lecture or the textbook. Answer the pre-lab questions.

Pre-Lab Questions Select the correct answer for each of the following questions:

1. The technique of examining surface features with hands and fingers is called
 a. exploration. **b.** touching.
 c. palpation. **d.** feeling.

2. Soft tissues include the following *except*
 a. bone. **b.** muscle.
 c. ligament. **d.** skin.

3. Which of the following is a bony feature of the upper torso?
 a. pectoralis major **b.** anterior axillary fold
 c. jugular notch **d.** deltoid

4. Which of the following is a soft tissue feature of the lower limb?
 a. tibial tuberosity **b.** head of the fibula
 c. medial malleolus **d.** vastus lateralis

5. Palpation enables us to determine all of the following characteristics of a structure *except*
 a. location. **b.** function.
 c. size. **d.** texture.

6. The laryngeal prominence can be palpated in the anterior neck.
 a. True **b.** False

7. The linea alba is a feature of the posterior torso (trunk).
 a. True **b.** False

8. It is important that healthcare professionals have knowledge of surface anatomy and practice in palpation.
 a. True **b.** False

Chapter Opener: ©101cats/Getty Images

External landmarks, called surface anatomy, located on the human body provide an opportunity to examine surface features and to help locate other internal structures. This exercise will focus on bony landmarks and superficial soft tissue structures, primarily related to the skeletal and muscular systems. The technique used for the examination of surface features is called *palpation* and is accomplished by touching with the hands or fingers. This enables us to determine the location, size, and texture of the underlying anatomical structure. Some of the surface features can be visible without the need of palpation, especially on body builders and very thin individuals. If the individual involved has considerable subcutaneous adipose tissue, additional pressure might be necessary to palpate some of the structures.

These surface features will help us to locate additional surface and deeper structures during the study of the organ systems. Surface features will help us to determine the pulse locations of superficial arteries, to assess injection sites, to describe pain and injury sites, to insert examination tubes, and to listen to heart, lung, and bowel sounds with a stethoscope. For those who choose careers in emergency medical services, nursing, medicine, physical education, physical therapy, chiropractic, and massage therapy, surface anatomy is especially valuable.

Many clues to a person's overall health can be revealed from the body surface. The general nature of the surface texture and coloration can be observed and examined quickly. Palpation of various surface features can provide additional clues to possible ailments. With the use of different pressures during palpations, clues to the depth of organs and possible abnormalities become more evident. An external examination is an important aspect of a physical exam. The usefulness of this laboratory exercise will become even more apparent as you continue the study of anatomy and physiology and related careers.

PROCEDURE: Surface Anatomy

1. Review the earlier laboratory manual exercises and figures on body regions, skeletal system, and muscular system.
2. Examine the surface features labeled in figures 26.1 through 26.8. Features in **boldface** are *bony features*. Other labeled features (not boldface) are *soft tissue features*. A bony feature is a part of a bone and the skeletal system. Soft tissue features are composed of tissue other than bone, such as a muscle, tendon, ligament, cartilage, or skin. Palpate the labeled surface features to distinguish the texture and degree of firmness of the bony features and the soft tissue features. Most of the surface features included in this laboratory exercise can be palpated on your own body.

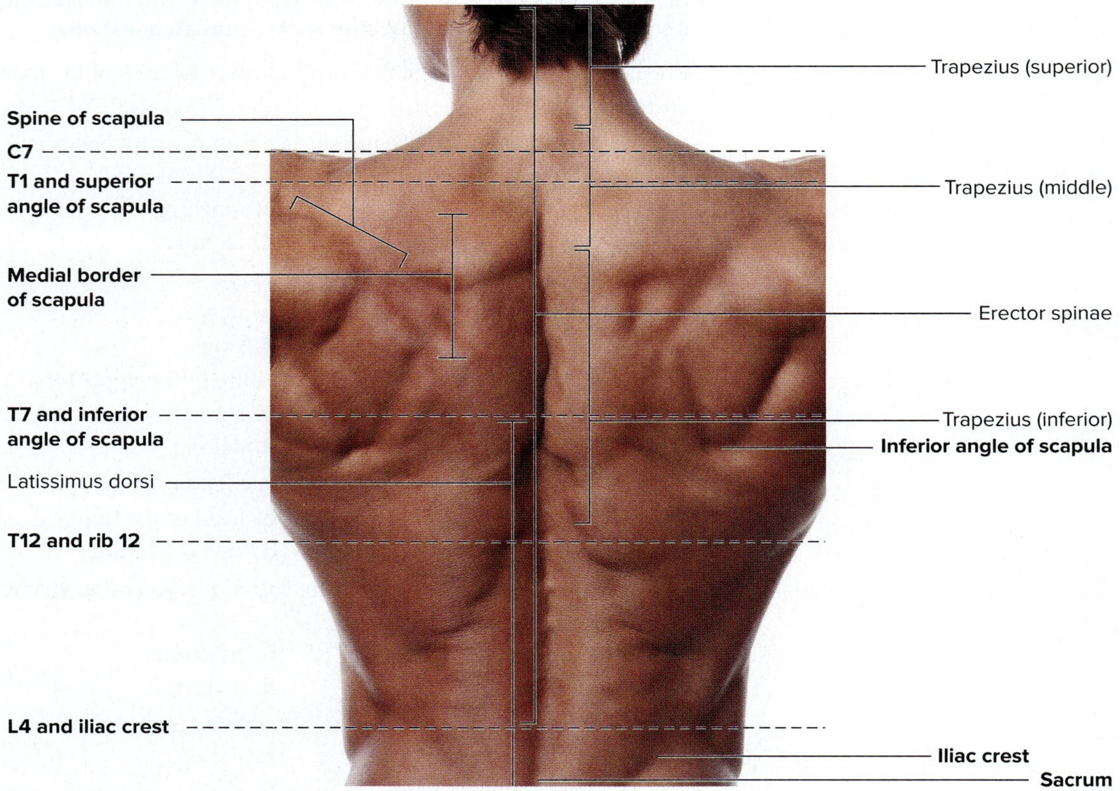

FIGURE 26.1 Surface features of the posterior shoulder and torso. (**Boldface** indicates bony features; not boldface indicates soft tissue features.) BLACKDAY/Shutterstock APR

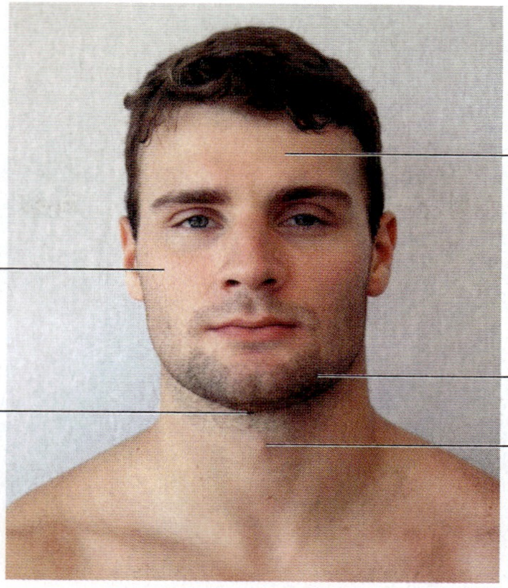

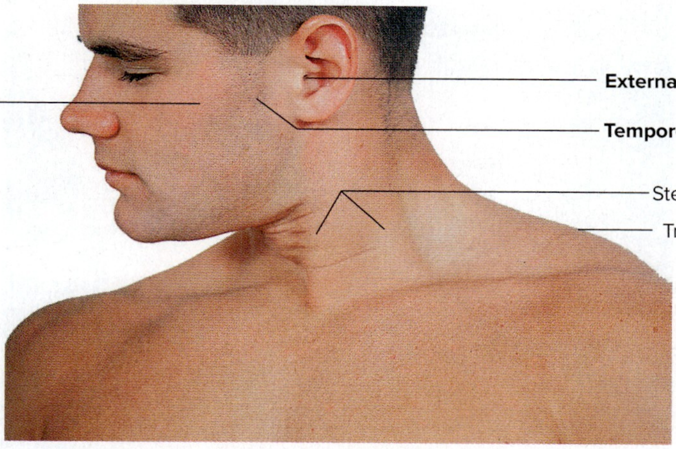

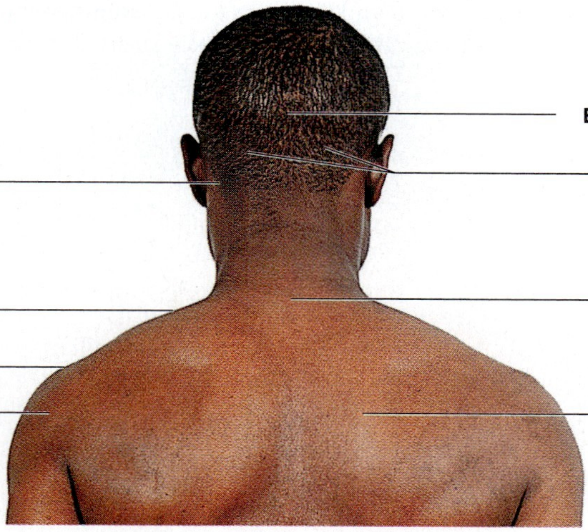

FIGURE 26.2 Surface features of (a) the anterior head and neck, (b) the lateral head and neck, and (c) the posterior head and neck. (**Boldface** indicates bony features; not boldface indicates soft tissue features.) (a) Tara McDermott/McGraw Hill; (b, c) Eric Wise

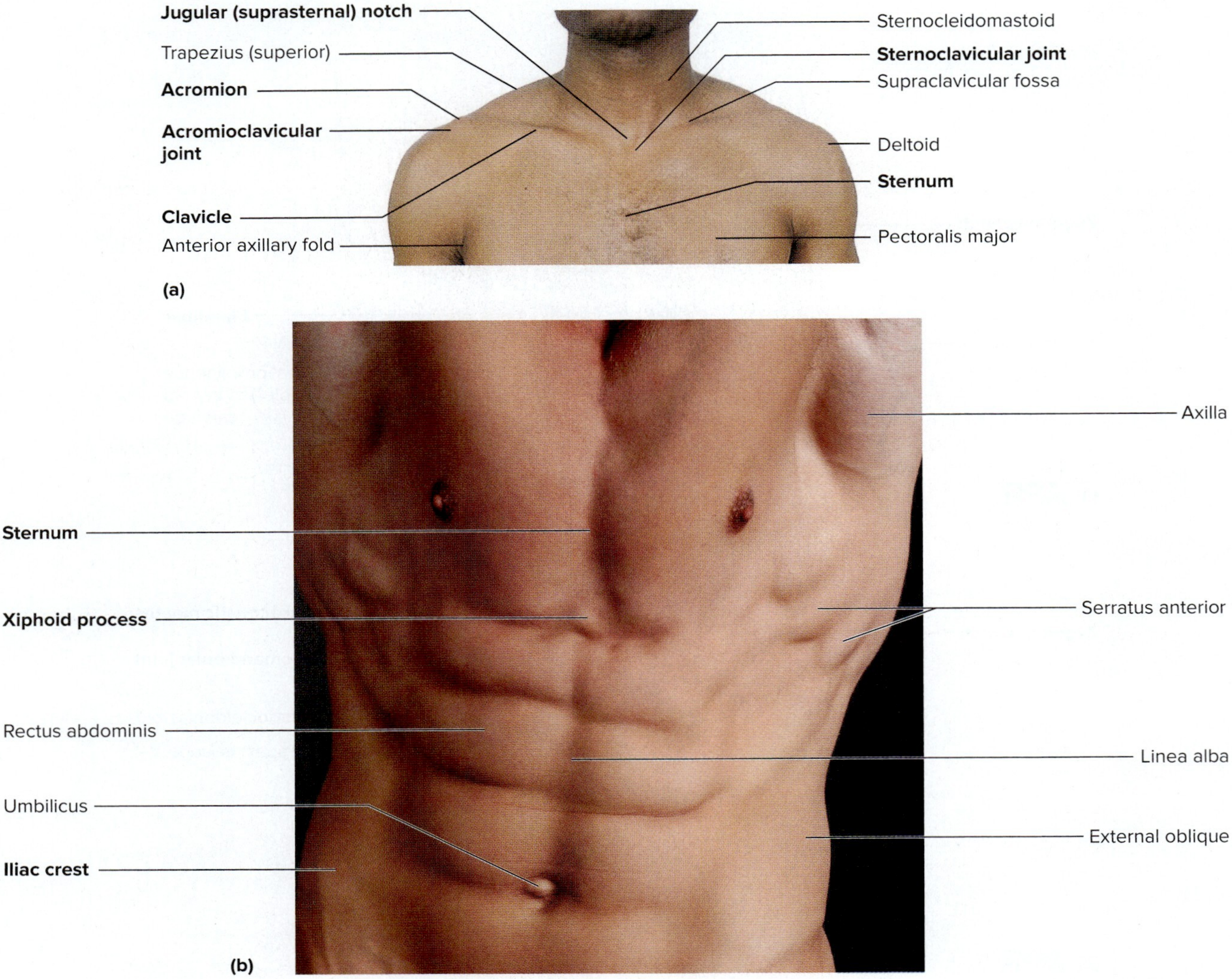

FIGURE 26.3 Anterior surface features of (a) upper torso and (b) lower torso. (**Boldface** indicates bony features; not boldface indicates soft tissue features.) (a) Eric Wise/McGraw Hill; (b) Stocked House Studio/Shutterstock APR

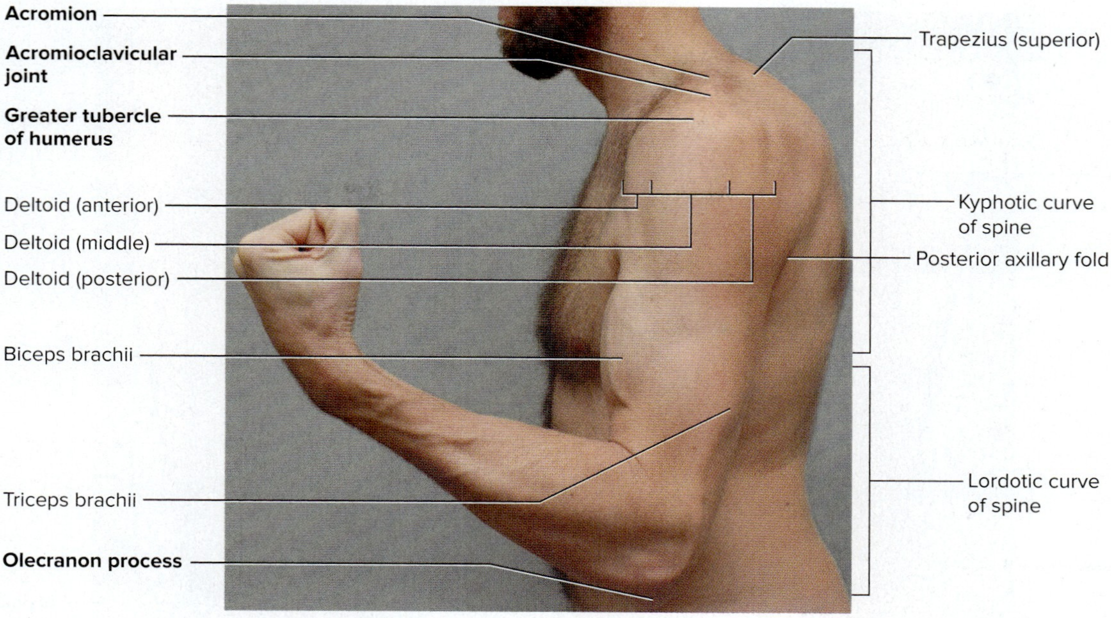

FIGURE 26.4 Surface features of the lateral shoulder and upper limb. (**Boldface** indicates bony features; not boldface indicates soft tissue features.) Krystal Rose Faust APR

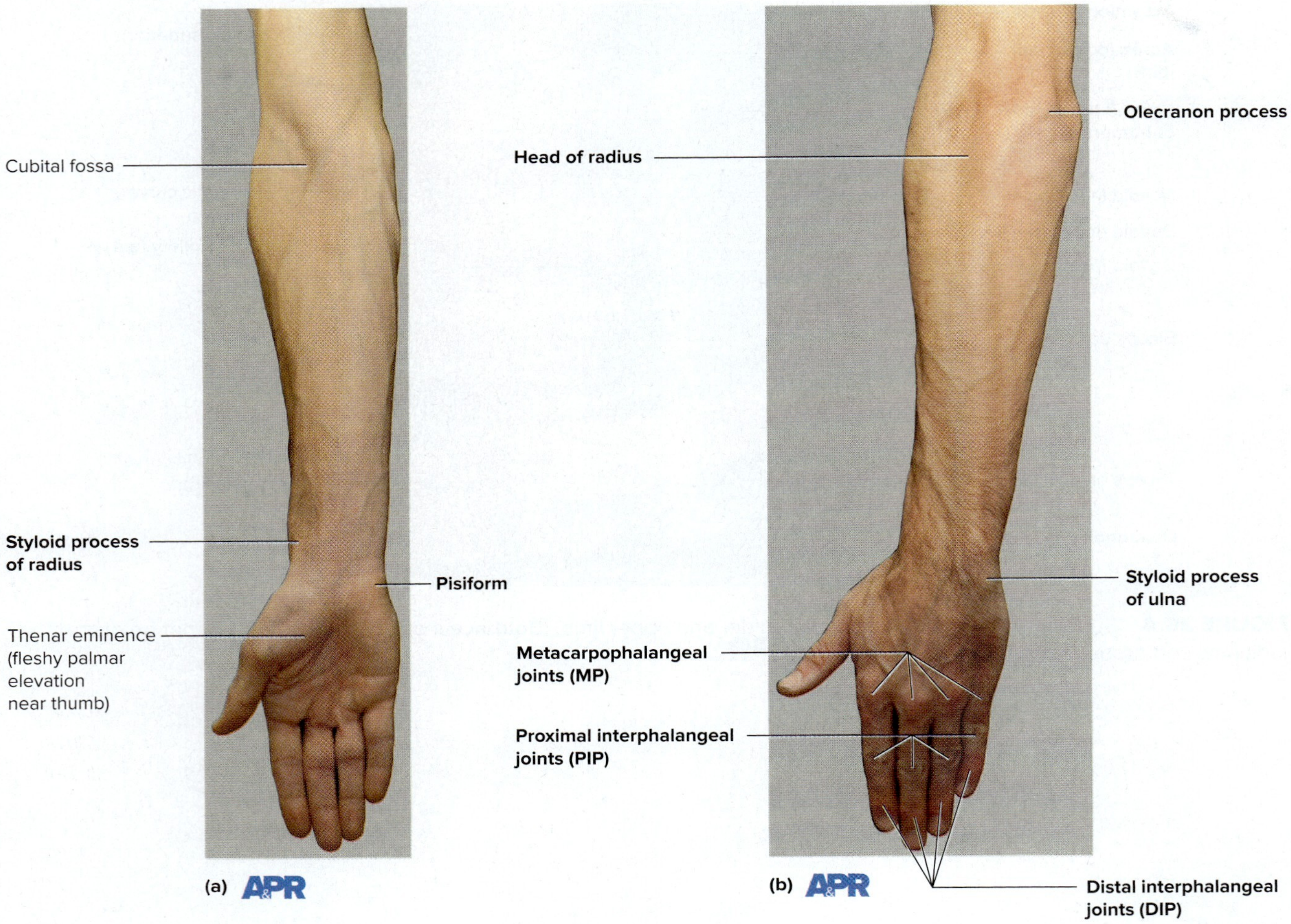

FIGURE 26.5 Surface features of the upper limb: (a) anterior view; (b) posterior view. (**Boldface** indicates bony features; not boldface indicates soft tissue features.) Krystal Rose Faust

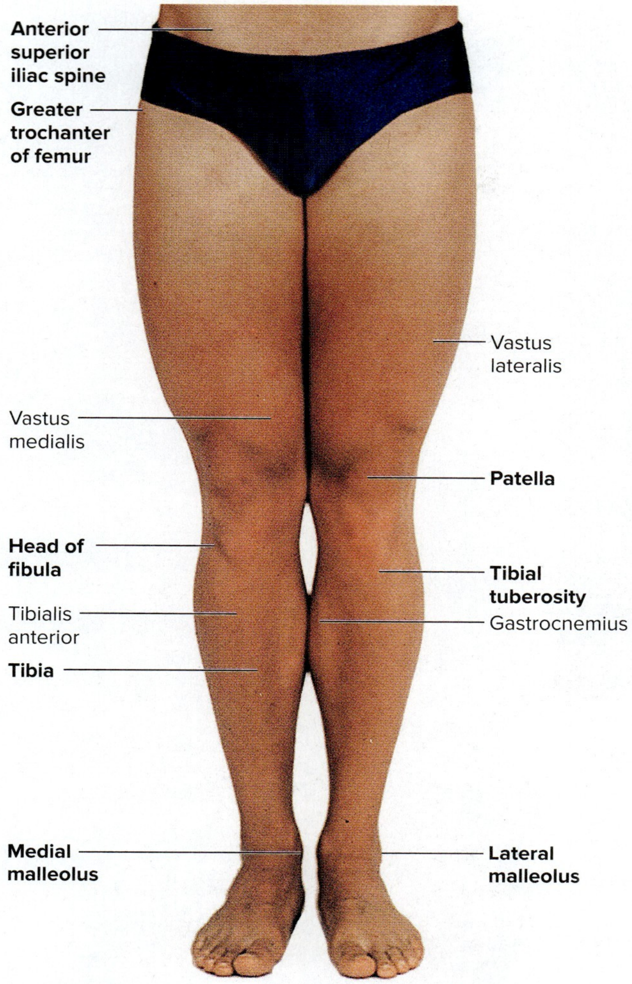

FIGURE 26.6 Surface features of the anterior view of the body. (**Boldface** indicates bony features; not boldface indicates soft tissue features.) Eric Wise APR

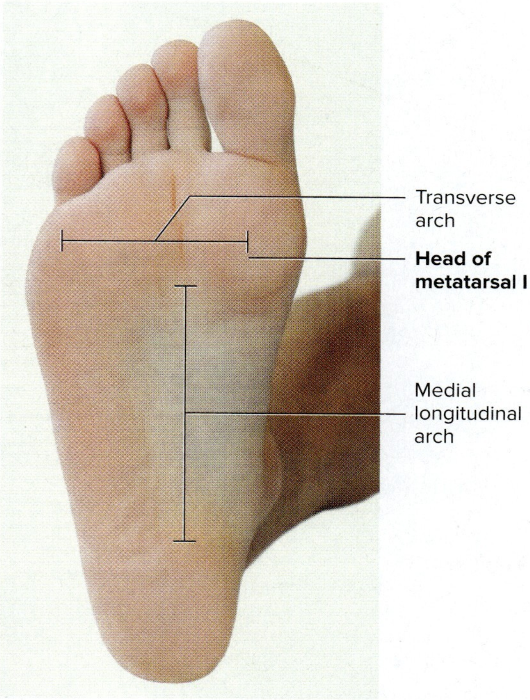

FIGURE 26.7 Surface features of the plantar surface of the foot. (**Boldface** indicates bony features; not boldface indicates soft tissue features.) Tara McDermott/McGraw Hill APR

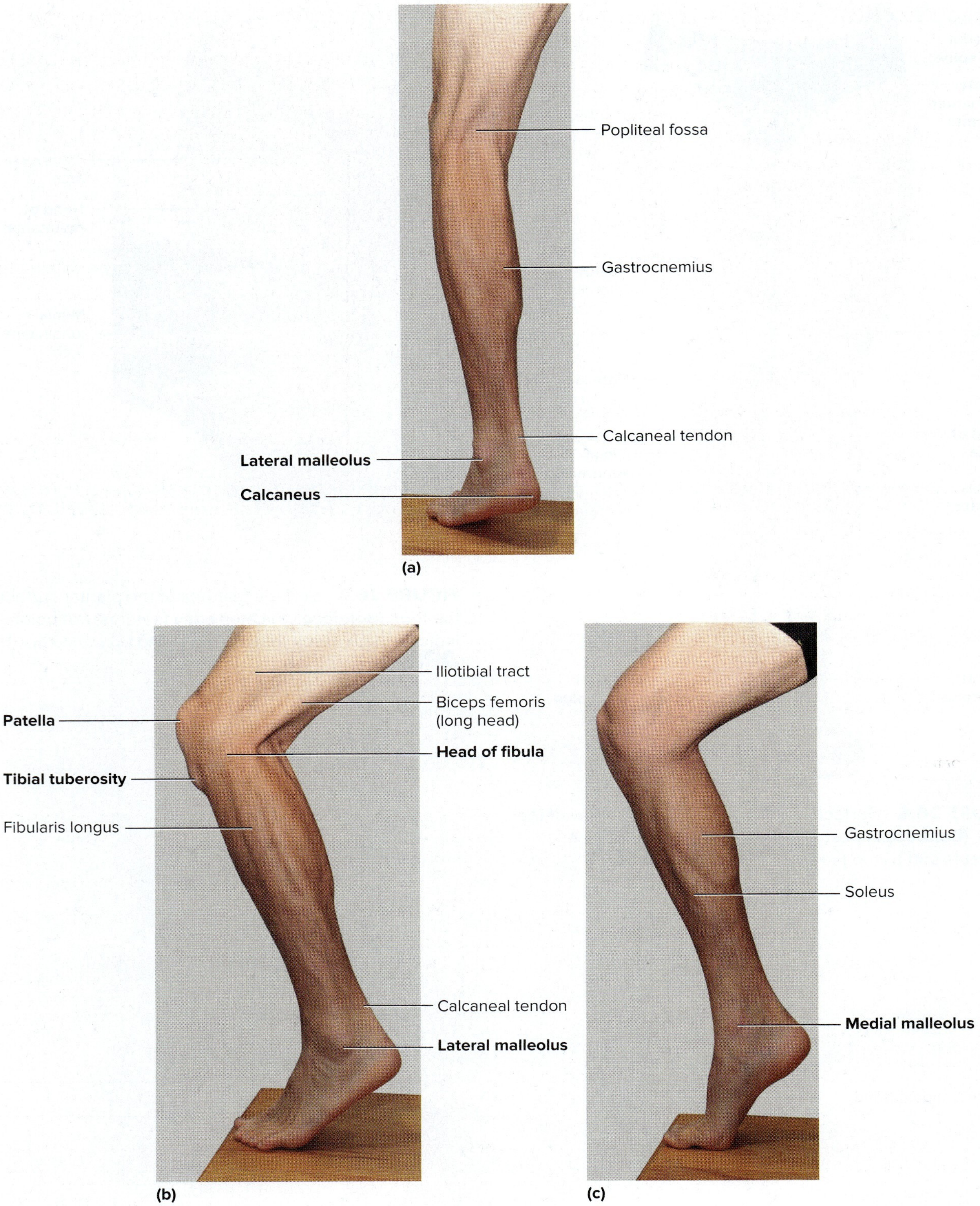

FIGURE 26.8 Surface features of the (a) posterior lower limb, (b) lateral lower limb, and (c) medial lower limb. (**Boldface** indicates bony features; not boldface indicates soft tissue features.) Krystal Rose Faust APR

3. Complete Parts A and B of Laboratory Assessment 26.
4. Use figures 26.1 through 26.8 to determine which items listed in table 26.1 are bony features and which items are soft tissue features. Examine the list of bony and soft tissue surface features in the first column. If the feature listed is a bony surface feature, place an "X" in the second column; if the feature listed is a soft tissue feature, place an "X" in the third column.
5. Complete Part C of the laboratory assessment.

LEARNING EXTENSION ACTIVITY

Select fifteen surface features presented in the laboratory exercise figures. Using small, round stickers, write the name of a surface feature on each sticker. Working with a lab partner, locate the selected surface features on yourself and on your lab partner. Use the stickers to accurately "label" your partner.

- Were you able to find all the features on both of you? ____________________
- Was it easier to find the surface features on yourself or your lab partner? ____________________

Table 26.1 **Representative Bony and Soft Tissue Surface Features**

(*Note:* After indicating your "X" in the appropriate column, use the extra space in columns 2 and 3 to add personal descriptions of the nature of the bony feature or the soft tissue feature palpation.)

Bony and Soft Tissues	Bony Features	Soft Tissue Features
Biceps brachii		
Biceps femoris		
Calcaneal tendon		
Deltoid		
Fibularis longus		
Gastrocnemius		
Greater trochanter		
Head of fibula		
Iliac crest		
Iliotibial tract		
Lateral malleolus		
Mandible		
Mastoid process		
Medial border of scapula		
Rectus abdominis		
Sacrum		
Serratus anterior		
Sternocleidomastoid		
Sternum		
Thenar eminence		
Tibialis anterior		
Tibial tuberosity		
Zygomatic arch		

NOTES

LABORATORY
ASSESSMENT 26

Name ______________________________

Date ______________________________

Section ______________________________

The A corresponds to the indicated Learning Outcome(s) O found at the beginning of the Laboratory Exercise.

Surface Anatomy

PART A: Assessments

Match the body regions in column A with the surface features in column B. Place the letter of your choice in the space provided. (Answers may be used more than once.) A1 A2

Column A	Column B
a. Head	______ **1.** Umbilicus
b. Trunk (torso, including shoulder and hip)	______ **2.** Medial malleolus
c. Upper limb	______ **3.** Iliac crest
d. Lower limb	______ **4.** Transverse arch
	______ **5.** Spine of scapula
	______ **6.** External occipital protuberance
	______ **7.** Cubital fossa
	______ **8.** Sternum
	______ **9.** Olecranon process
	______ **10.** Zygomatic arch
	______ **11.** Mastoid process
	______ **12.** Thenar eminence
	______ **13.** Popliteal fossa
	______ **14.** Metacarpophalangeal joints

PART B: Assessments

Label figure 26.9 with the surface features provided.

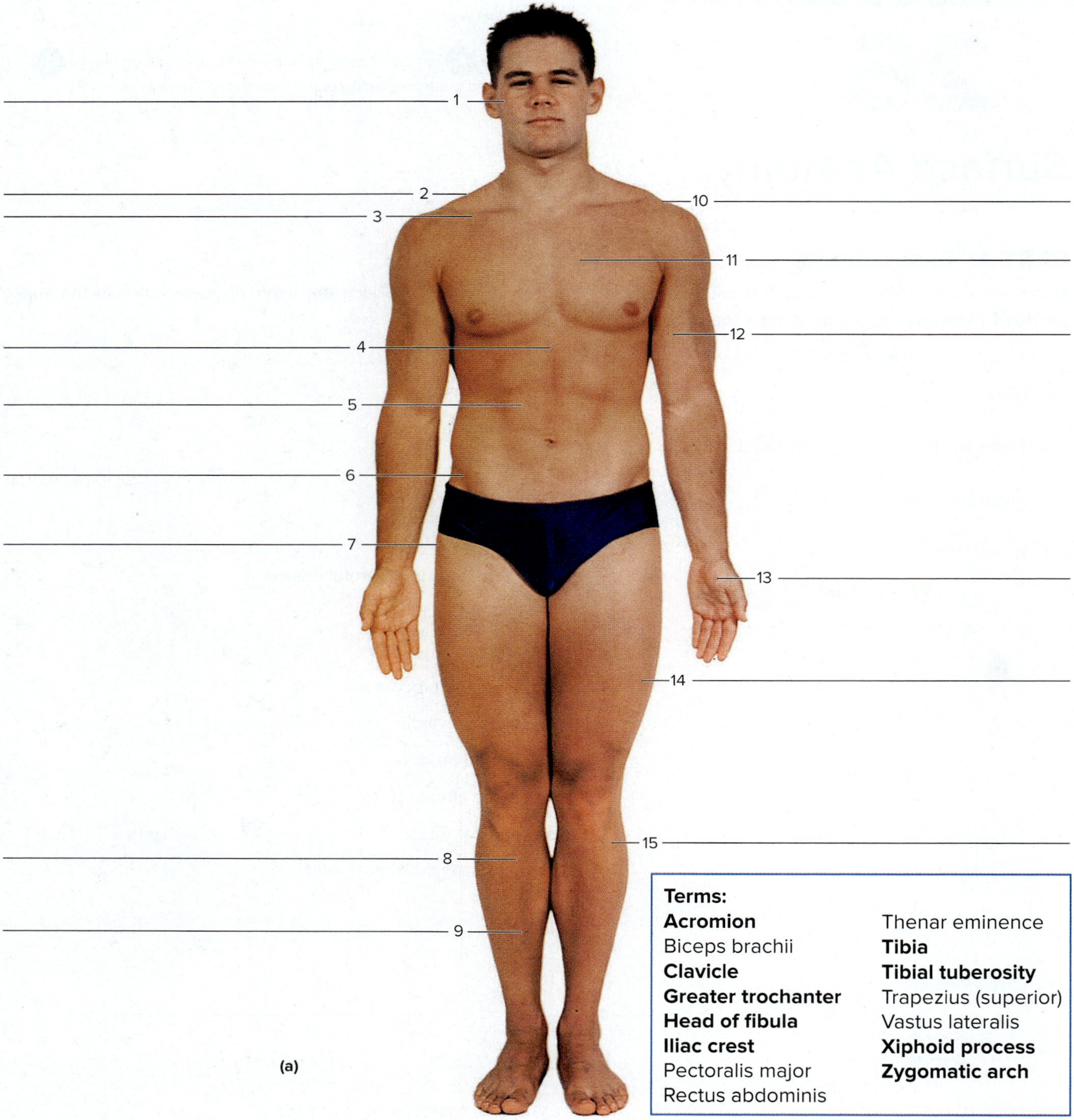

Terms:

Acromion	Thenar eminence
Biceps brachii	**Tibia**
Clavicle	**Tibial tuberosity**
Greater trochanter	Trapezius (superior)
Head of fibula	Vastus lateralis
Iliac crest	**Xiphoid process**
Pectoralis major	**Zygomatic arch**
Rectus abdominis	

FIGURE 26.9 Using the terms provided, label the surface features of the (a) anterior view of the body and (b) posterior view of the body. (**Boldface** indicates bony features; not boldface indicates soft tissue features.) Eric Wise A1

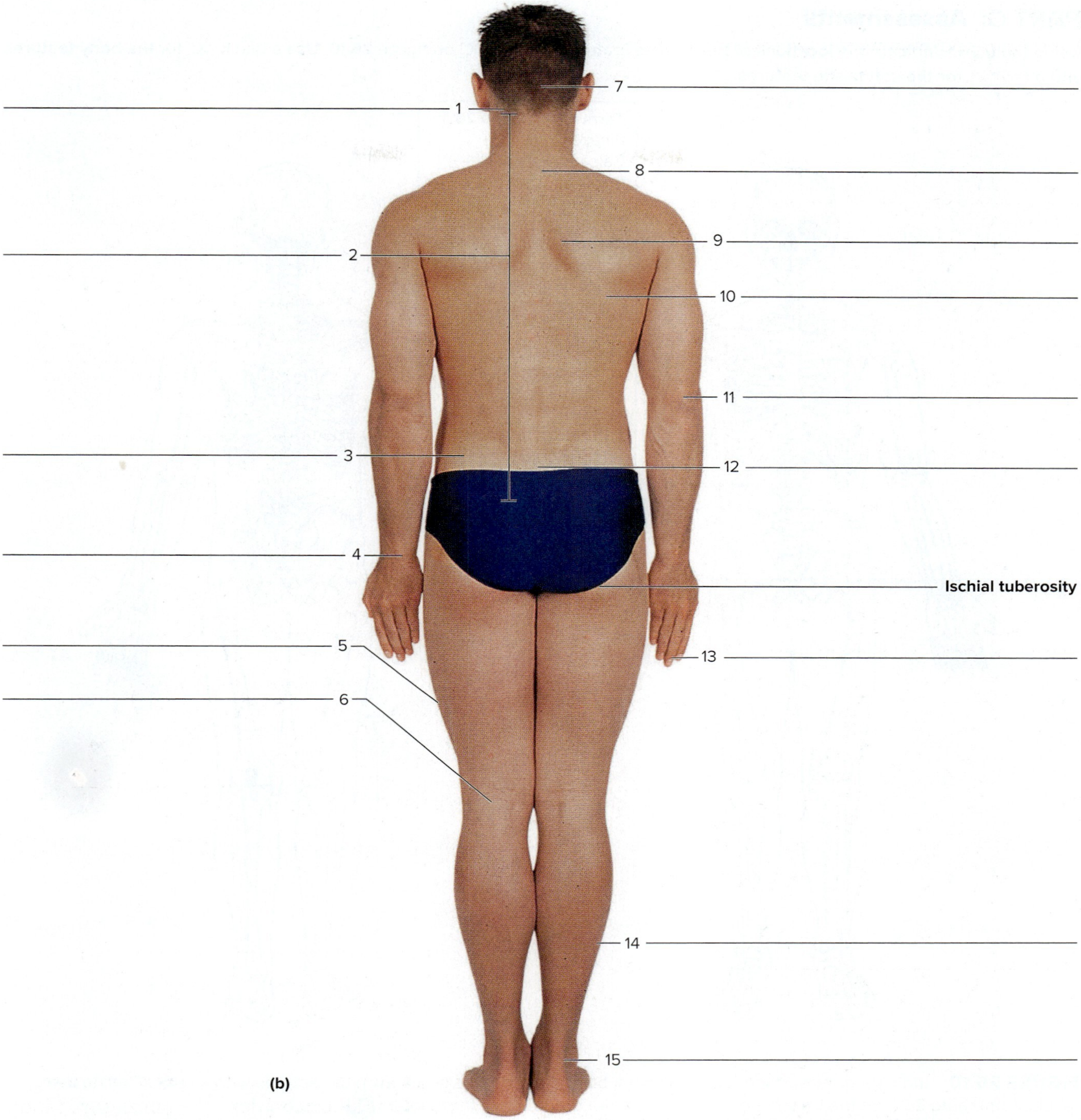

Terms:

C7	**Iliac crest**	**Olecranon process**
Calcaneal tendon	Iliotibial tract	Popliteal fossa
Distal interphalangeal joint (DIP)	**Inferior angle of scapula**	**Sacrum**
Erector spinae	**Mastoid process**	Soleus
External occipital protuberance	**Medial border of scapula**	**Styloid process of ulna**

FIGURE 26.9 *Continued.* Eric Wise

PART C: Assessments

Using table 26.1, indicate the locations of the surface features with an "X" on figure 26.10. Use a black "X" for the bony features and a red "X" for the soft tissue features.

Terms from table 26.1:
Biceps brachii
Calcaneal tendon
Deltoid
Gastrocnemius
Greater trochanter
Head of fibula
Iliac crest
Iliotibial tract
Lateral malleolus
Mandible
Mastoid process
Medial border of scapula
Rectus abdominis
Sacrum
Serratus anterior
Sternocleidomastoid
Sternum
Thenar eminence
Tibialis anterior
Zygomatic arch

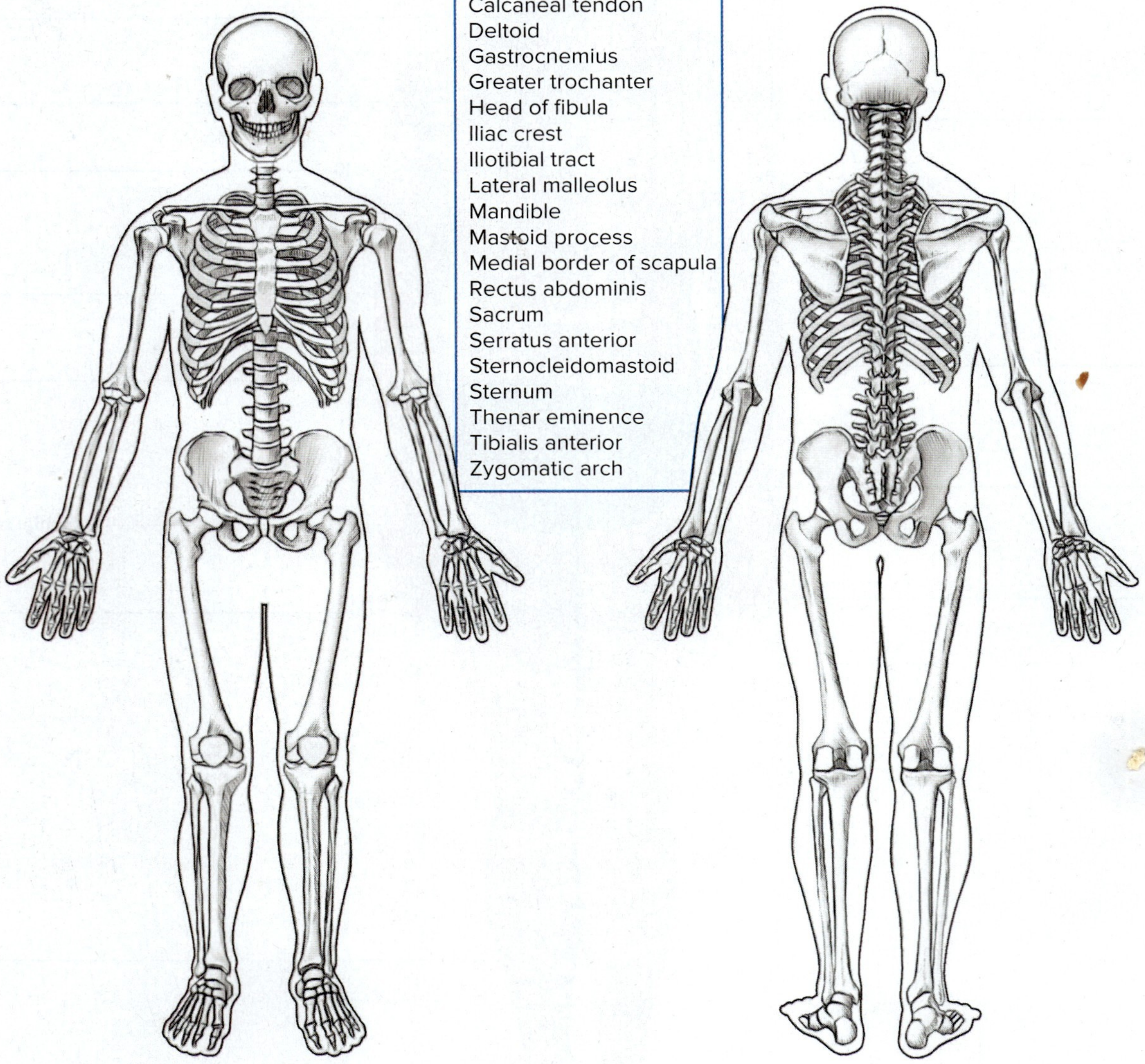

(a) (b)

FIGURE 26.10 Indicate bony surface features with a black "X" and soft tissue surface features with a red "X" using the results from table 26.1, on the (a) anterior and (b) posterior diagrams of the body. A3

CRITICAL THINKING ASSESSMENT—PART C

What is being palpated when taking a radial pulse? A4 ______________________

When a patient is suspected of having strep throat, what common area would a physician palpate when doing an examination? A4 ______________________

Palpation of the breasts may detect lumps or changes in texture or size that may be a sign of what illness? A4

____

Identify some organs of the abdominal cavity that a physician may palpate when doing an examination. A4

____

LABORATORY

EXERCISE 27

Nervous Tissue and Nerves

Purpose of the Exercise

To review the characteristics of nervous tissue and to observe neurons, neuroglia, and various features of a nerve.

MATERIALS NEEDED

Compound light microscope
Prepared microscope slides of the following:
- Spinal cord (smear)
- Posterior root ganglion (section)
- Neuroglia (astrocytes)
- Peripheral nerve (cross section and longitudinal section)

Neuron model

For Learning Extension Activity:
Prepared microscope slide of Purkinje cells from cerebellum

Chapter Opener: ©101cats/Getty Images

Learning Outcomes APR

After completing this exercise, you should be able to:

O1 Describe the location and characteristics of nervous tissue.

O2 Distinguish structural and functional characteristics between neurons and neuroglia.

O3 Identify and sketch the major structures of a neuron and a nerve.

The O corresponds to the assessments A indicated in the Laboratory Assessment for this Exercise.

Pre-Lab

Carefully read the introductory material and examine the entire lab. Be familiar with the structures and functions of nervous tissue and nerves from lecture or the textbook. Answer the pre-lab questions.

Pre-Lab Questions Select the correct answer for each of the following questions:

1. The cell body of a neuron contains the
   a. nucleus. **b.** dendrites.
   c. axon. **d.** neuroglia.
2. A multipolar neuron contains
   a. one dendrite and many axons.
   b. many dendrites and one axon.
   c. one dendrite and one axon.
   d. a single process with the dendrite and axon.
3. Impulse conduction between neurons within the CNS is the function of
   a. interneurons. **b.** motor neurons.
   c. sensory neurons. **d.** efferent neurons.
4. Neuroglia that produce myelin insulation in the CNS are
   a. microglia. **b.** astrocytes.
   c. ependymal cells. **d.** oligodendrocytes.
5. The PNS contains
   a. 12 pairs of cranial nerves only.
   b. 31 pairs of spinal nerves only.
   c. 12 pairs of cranial nerves and 31 pairs of spinal nerves.
   d. 43 pairs of spinal nerves.
6. Schwann cells
   a. are only in the brain.
   b. are only in the spinal cord.
   c. are throughout the CNS.
   d. have a myelin sheath and neurilemma.

7. A _____ neuron is the most common structural neuron in the brain and spinal cord.
   a. multipolar   b. tripolar
   c. bipolar   d. unipolar
8. Sensory neurons conduct impulses from the spinal cord to a muscle or a gland.
   a. True   b. False
9. Astrocytes have contacts between blood vessels and neurons in the CNS.
   a. True   b. False

Nervous tissue, which occurs in the brain, spinal cord, and peripheral nerves, contains *neurons* (nerve cells) and *neuroglia* (neuroglial cells; glial cells). Neurons are irritable (excitable) and easily respond to stimuli, resulting in the conduction of an *action potential* (*impulse; nerve impulse; nerve signal*). A neuron contains a cell body with a nucleus and most of the cytoplasm, and elongated cell processes (typically, several dendrites and a single axon) along which impulse conductions occur. Neurons can be classified according to structural variations of their cell processes: *multipolar* with many dendrites and one axon (nerve fiber), *bipolar* with one dendrite and one axon, and *unipolar* with a single process where the dendrite leads directly into the axon. Functional classifications are used according to the direction of the impulse conduction: *sensory (afferent) neurons* conduct impulses from receptors toward the central nervous system (CNS), *interneurons* (*association neurons*) conduct impulses within the CNS, and *motor* (*efferent*) *neurons* conduct impulses away from the CNS to effectors (muscles or glands).

Neuroglia are located in close association with neurons and generally perform supportive and protective functions. Four types of neuroglia are found in the CNS: *astrocytes, microglia, oligodendrocytes,* and *ependymal cells.* Astrocytes are numerous, and their branches support neurons and blood vessels. Microglia can phagocytize microorganisms and nerve tissue debris. Oligodendrocytes produce the myelin insulation in the CNS. Ependymal cells line the brain ventricles and secrete cerebrospinal fluid (CSF), and the cilia on their apical surfaces aid the circulation of the CSF.

Two types of neuroglia are located in the peripheral nervous system (PNS): *Schwann cells* and *satellite cells.* Schwann cells surround nerve fibers numerous times. The wrappings nearest the nerve fiber represent the myelin sheath and serve to insulate and increase the impulse speed, while the last wrapping, called the neurilemma, contains most of the cytoplasm and the nucleus and functions in nerve fiber regeneration in PNS neurons. Satellite cells surround and support the cell body regions, called ganglia, of peripheral neurons.

The PNS contains 12 pairs of *cranial nerves* that arise from the brain and 31 pairs of *spinal nerves* that arise from the spinal cord, all containing parallel axons of neurons and neuroglia representing the nervous tissue components. Most nerves are mixed in that they contain both sensory and motor neurons, but some contain only sensory or motor components. The nerves represent an organ structure as they also contain small blood vessels and fibrous connective tissue. The fibrous connective tissue around each nerve fiber and the Schwann cell is the *endoneurium;* a bundle of nerve fibers, representing a *fascicle,* is surrounded by a *perineurium;* and the entire nerve is surrounded by an *epineurium.* The fibrous connective components of a nerve provide protection and stretching during body movements.

PROCEDURE A: Nervous Tissue APR

In this procedure you will be asked to identify the locations and characteristics of structures in nervous tissue. Two tables of nervous tissue include the structural and functional

Table 27.1 Structural and Functional Types of Neurons

CLASSIFIED BY STRUCTURE		
Type	**Structural Characteristics**	**Location**
Multipolar neuron	Cell body with one axon and multiple dendrites	Most common type of neuron in the brain and spinal cord
Bipolar neuron	Cell body with one axon and one dendrite	In receptor parts of the eyes, nose, and ears; rare type
Unipolar neuron	Cell body with a single process that divides into two branches and functions as an axon; only the receptor ends of the peripheral (distal) process function as dendrites	Most sensory neurons
CLASSIFIED BY FUNCTION		
Type	**Functional Characteristics**	**Structural Characteristics**
Sensory (afferent) neuron	Conducts impulses from receptors in peripheral body parts into the brain or spinal cord	Most unipolar; some bipolar
Interneuron	Transmits impulses between neurons in the brain and spinal cord	Multipolar
Motor (efferent) neuron	Conducts impulses from the brain or spinal cord out to muscles or glands (effectors)	Multipolar

characteristics of neurons, and neuroglia within the CNS and the PNS. Diagrams of neurons and neuroglia are provided as sources of comparison for your microscopic observations using the compound microscope. When you make sketches in the laboratory assessment, include all the observed structures with labels.

1. Examine tables 27.1 and 27.2 and figures 27.1 to 27.3.

Table 27.2 Types of Neuroglia in the CNS and PNS

Type	Structural Characteristics	Functions
CNS (BRAIN AND SPINAL CORD)		
Astrocytes	Star-shaped cells contacting neurons and blood vessels; most abundant type	Structural support; formation of scar tissue to replace damaged neurons; transport of substances between blood vessels and neurons; communicate with one another and with neurons; remove excess ions and neurotransmitters
Microglia	Small cells with slender cellular processes	Structural support; phagocytosis of microorganisms and damaged tissue
Oligodendrocytes	Shaped like astrocytes, but with fewer cellular processes; processes wrap around axons	Form myelin sheaths in the brain and spinal cord
Ependyma	Cuboidal cells lining cavities of the brain and spinal cord	Secrete and assist in circulation of cerebrospinal fluid (CSF)
PNS (PERIPHERAL NERVES)		
Schwann cells	Cells that wrap tightly around the axons of peripheral neurons	Form myelin sheaths in PNS and speed neurotransmission
Satellite cells	Flattened cells that surround cell bodies of neurons in ganglia	Support and protect ganglia

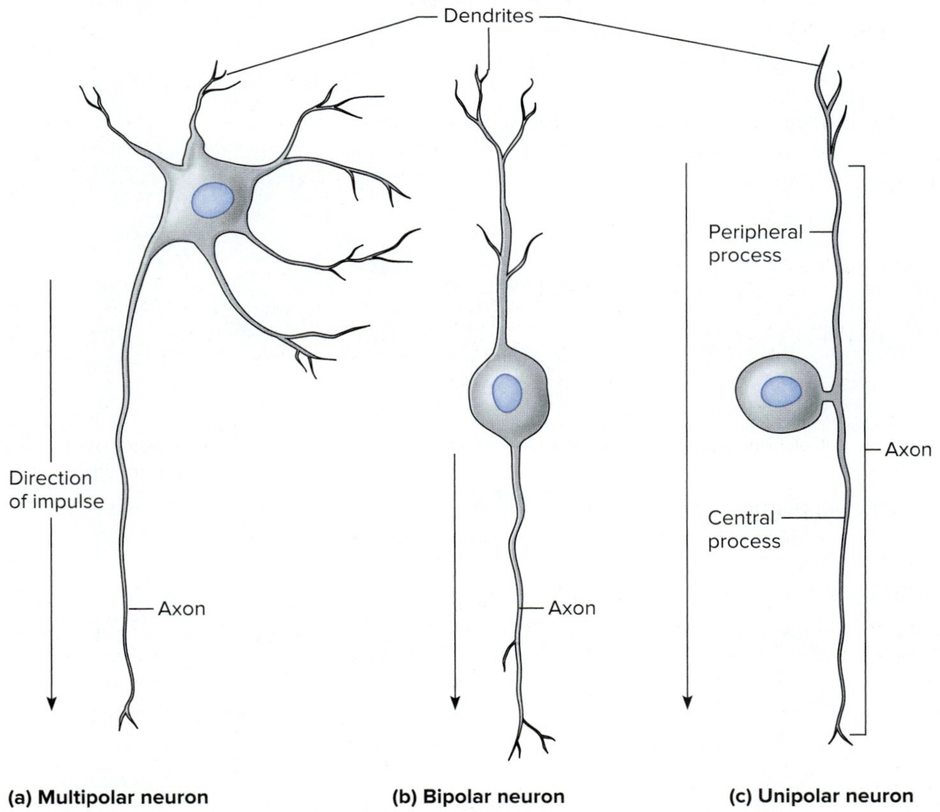

FIGURE 27.1 Structural types of neurons: (a) multipolar neuron; (b) bipolar neuron; (c) unipolar neuron.

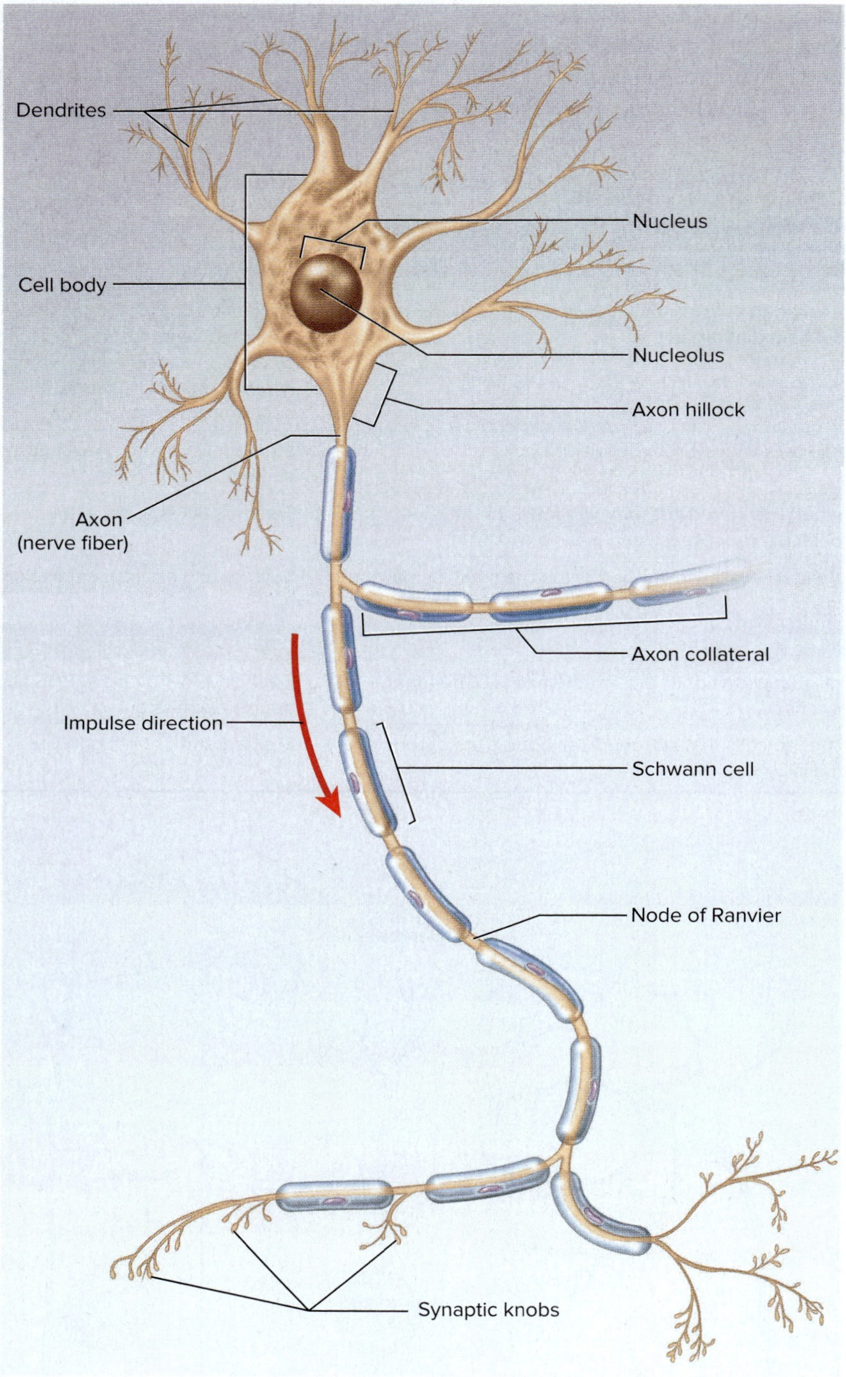

FIGURE 27.2 Diagram of a multipolar motor neuron. APR

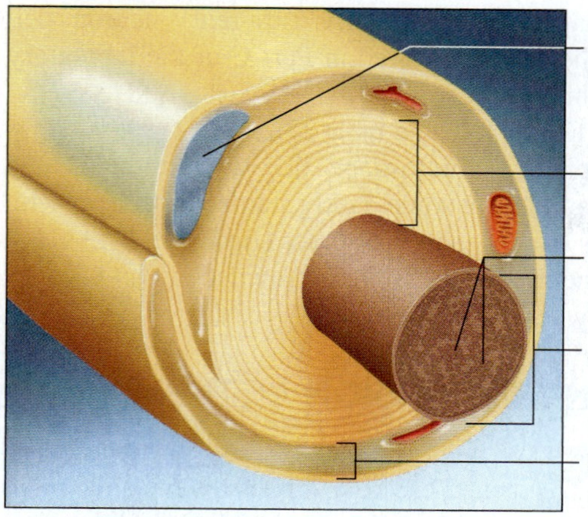

FIGURE 27.3 Diagram of a cross section of a myelinated axon (nerve fiber) of a spinal nerve. APR

2. Complete Parts A and B of Laboratory Assessment 27.
3. Obtain a prepared microscope slide of a spinal cord smear. Using low-power magnification, search the slide and locate the relatively large, deeply stained cell bodies of multipolar motor neurons.
4. Observe a single multipolar motor neuron, using high-power magnification (fig. 27.4), and note the following features:

cell body (soma)
- nucleus
- nucleolus
- Nissl bodies (chromatophilic substance)—a type of rough ER in neurons
- neurofibrils—threadlike structures extending into axon
- axon hillock—origin region of axon

dendrites—conduct impulses toward cell body

axon (nerve fiber)—conducts impulse away from cell body

Compare the slide to the neuron model and to figure 27.4. Also note small, darkly stained nuclei of neuroglia around the motor neuron.

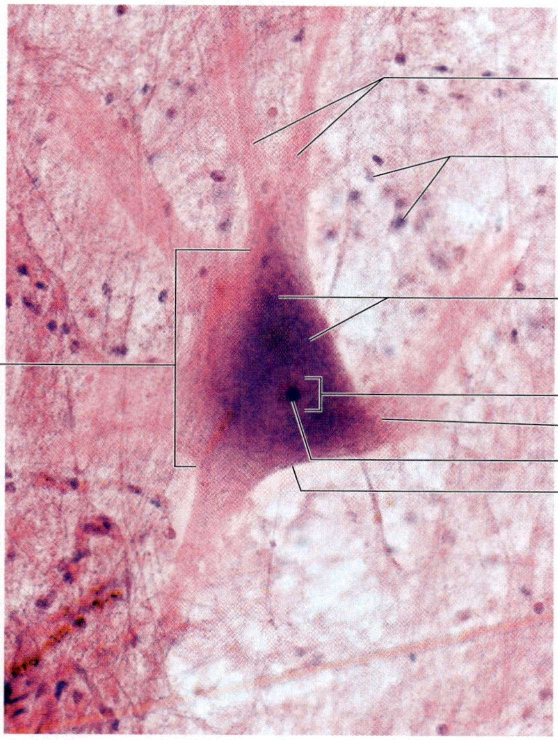

FIGURE 27.4 Micrograph of a multipolar neuron and neuroglia from a spinal cord smear (350×). Alvin Telser/McGraw Hill APR

5. Sketch and label a motor (efferent) neuron in the space provided in Part C of the laboratory assessment.
6. Obtain a prepared microscope slide of a posterior root ganglion. Search the slide and locate a cluster of sensory neuron cell bodies. You also may note bundles of nerve fibers passing among groups of neuron cell bodies (fig. 27.5).

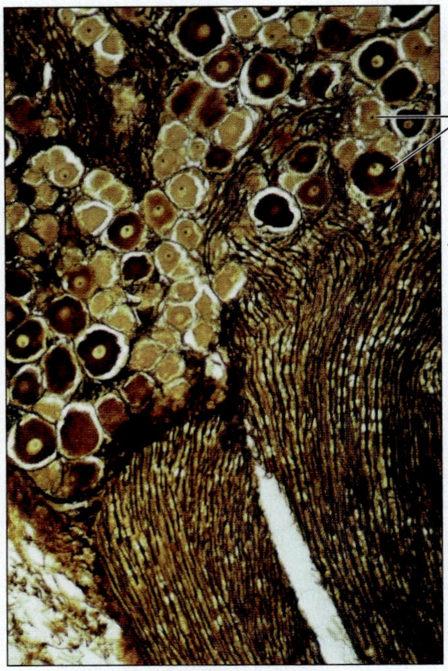

FIGURE 27.5 Micrograph of a posterior root ganglion containing cell bodies of sensory (afferent) neurons (100×).
Ed Reschke/Photolibrary/Getty Images

7. Sketch and label a sensory (afferent) neuron cell body in the space provided in Part C of the laboratory assessment.
8. Examine figure 27.6 and table 27.2.
9. Obtain a prepared microscope slide of neuroglia. Search the slide and locate some darkly stained astrocytes with numerous long, slender processes (fig. 27.7).
10. Sketch a neuroglia in the space provided in Part C of the laboratory assessment.

LEARNING EXTENSION ACTIVITY

Obtain a prepared microscope slide of Purkinje cells (fig. 27.8). To locate these neurons, search the slide for large, flask-shaped cell bodies. Each cell body has one or two large, thick dendrites that give rise to extensive branching networks of dendrites. These large cells are located in a particular region of the brain (cerebellar cortex). Their axons synapse with neurons that connect to the brainstem.

PROCEDURE B: Nerves

In this procedure you will compare an illustration of a peripheral spinal nerve with actual microscopic views of nerves. A nerve is an organ of the nervous system and contains various tissues in combination with the nervous tissue. The nerve contains the nerve fibers (axons) of the neurons and the Schwann cells representing some neuroglia in the PNS.

1. Study figure 27.9 of a cross section of a spinal nerve. Examine the fibrous connective tissue pattern including the epineurium around the entire nerve, the perineurium around fascicles, and the endoneurium around an

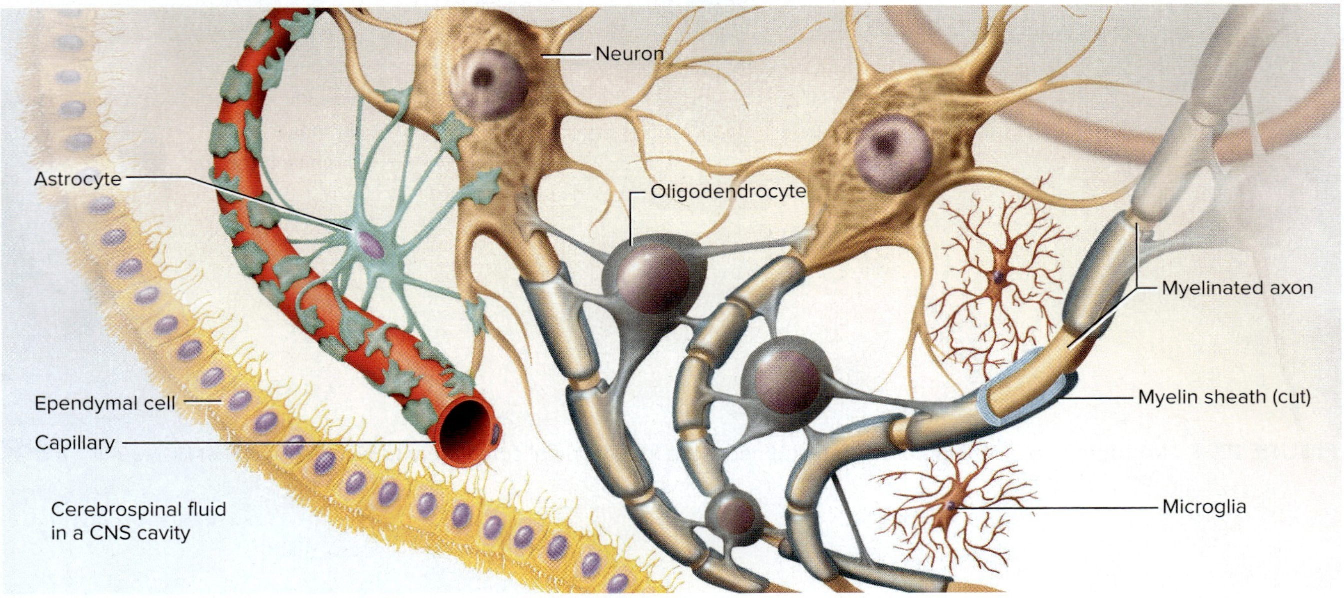

FIGURE 27.6 Types of neuroglia in the CNS.

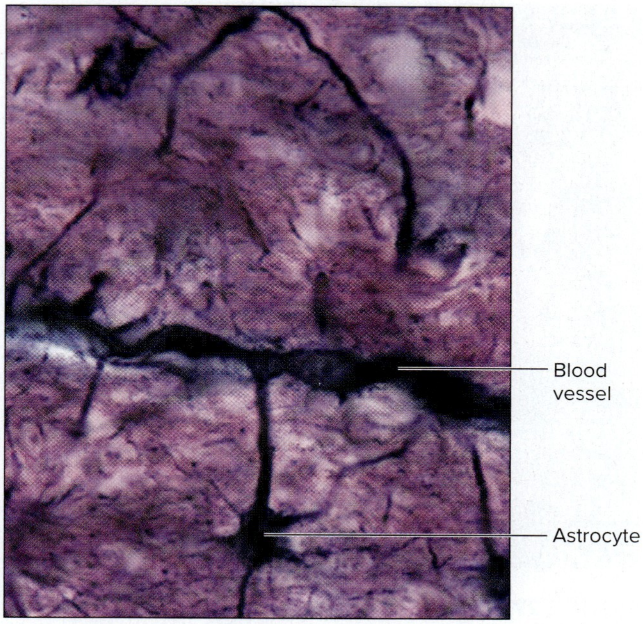

FIGURE 27.7 Micrograph of astrocytes (1,000×). Ed Reschke/Photolibrary/Getty Images

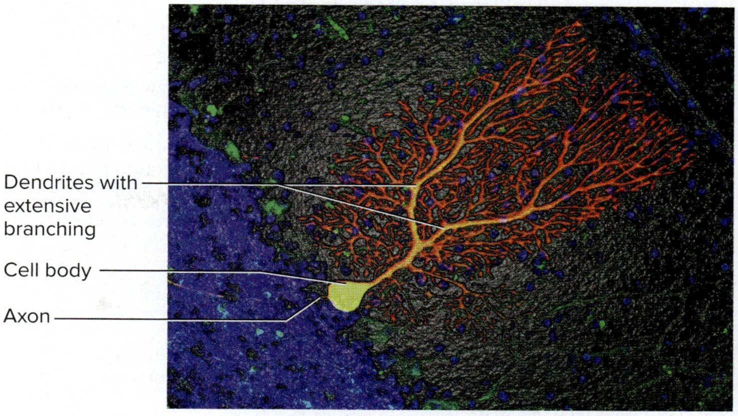

FIGURE 27.8 Large multipolar Purkinje cell from cerebellum of the brain (400×). David Becker/Science Source

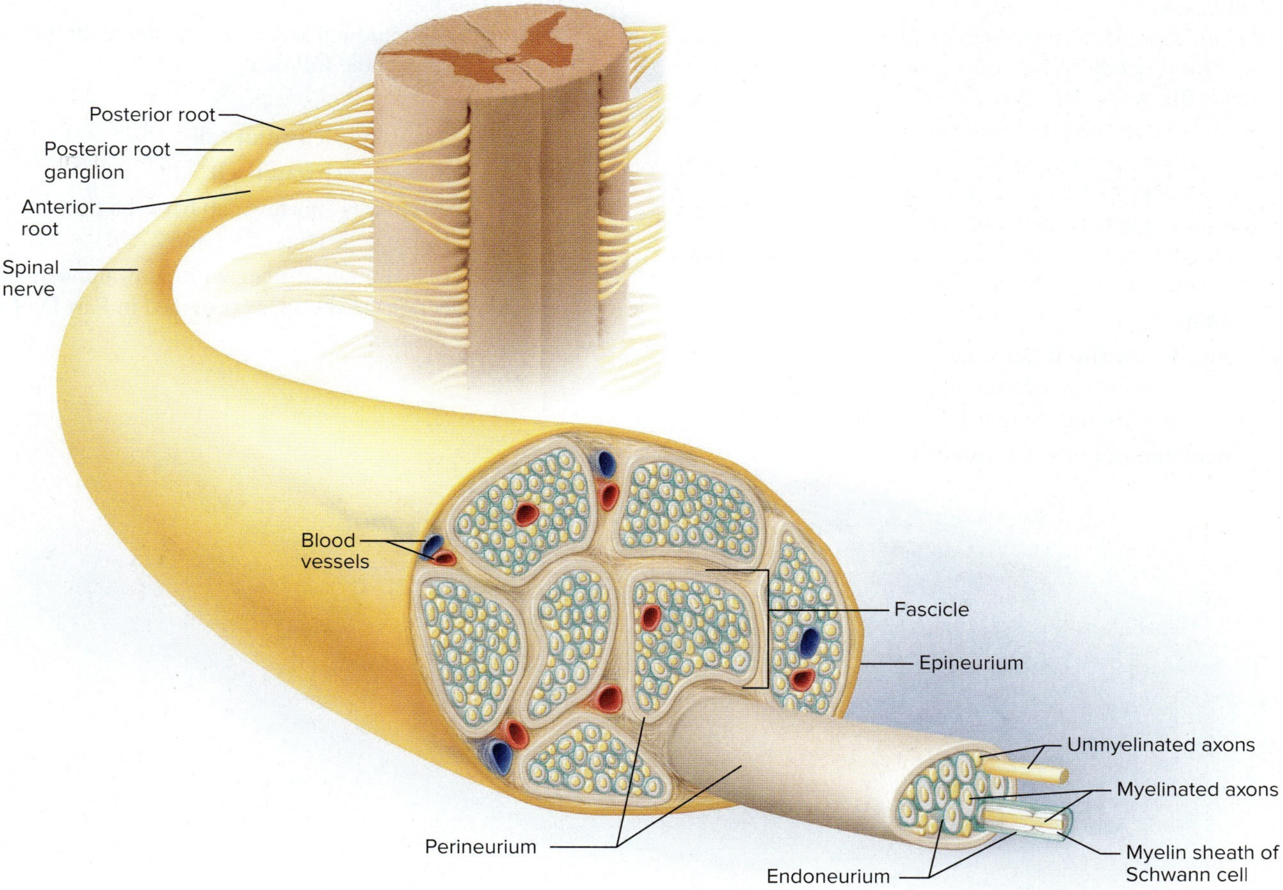

FIGURE 27.9 Diagram of a cross section of a peripheral spinal nerve. APR

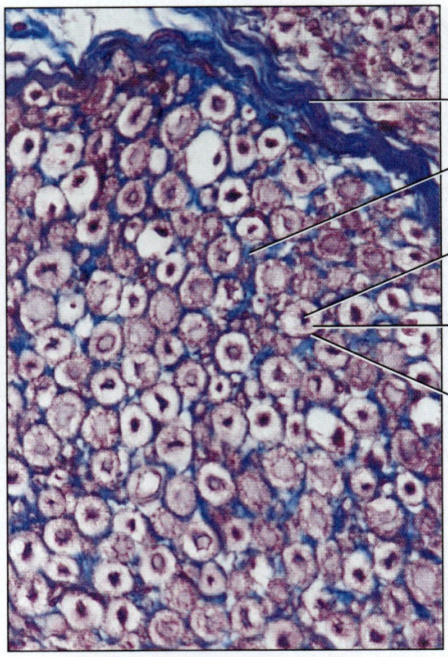

FIGURE 27.10 Cross section of a bundle of axons within a fascicle of a nerve (400×). Ed Reschke

FIGURE 27.11 Longitudinal section of a nerve (2,000×). Ed Reschke/Photolibrary/Getty Images APR

individual axon and Schwann cell. Reexamine figure 27.3 to note additional cross-sectional detail of the Schwann cell layers.

2. Obtain a prepared microscope slide of a nerve. Locate the cross section of the nerve and note the many round nerve fibers inside. Also note the dense layer of connective tissue (perineurium) that encircles a fascicle of nerve fibers and holds them together in a bundle. The individual nerve fibers are surrounded by a layer of more delicate connective tissue (endoneurium) (fig. 27.10).
3. Using high-power magnification, observe a single nerve fiber and note the following features:

   axon

   myelin sheath of Schwann cell around the axon (most of the myelin may have been dissolved and lost during the slide preparation)

   neurilemma of Schwann cell
4. Sketch and label a nerve fiber with Schwann cell (cross section) in the space provided in Part D of the laboratory assessment.
5. Locate the longitudinal section of the nerve on the slide (fig. 27.11). Note the following:

   axon

   myelin sheath of Schwann cell

   neurilemma of Schwann cell

   node of Ranvier—narrow gap between the Schwann cells (figs. 27.2 and 27.11)
6. Sketch and label a nerve fiber with Schwann cell (longitudinal section) in the space provided in Part D of the laboratory assessment.

LABORATORY
ASSESSMENT 27

Name ______________________________

Date ______________________________

Section ______________________________

The A corresponds to the indicated Learning Outcome(s) O found at the beginning of the Laboratory Exercise.

Nervous Tissue and Nerves

PART A: Assessments

Match the terms in column A with the descriptions in column B. Place the letter of your choice in the space provided. A1 A2

Column A	Column B
a. Astrocyte	_____ 1. Sheath of Schwann cell containing cytoplasm and nucleus that encloses myelin
b. Axon	_____ 2. Corresponds to rough endoplasmic reticulum in other cells
c. Bipolar neuron	_____ 3. Network of threadlike structures within cell body and extending into axon
d. Collateral	_____ 4. Substance of Schwann cell composed of lipoprotein that insulates axons and increases impulse speed
e. Dendrite	_____ 5. Neuron process with many branches that conducts an action potential (impulse) toward the cell body
f. Myelin	_____ 6. Branch of an axon
g. Neurilemma	_____ 7. Star-shaped neuroglia between neurons and blood vessels
h. Neurofibrils	_____ 8. Nerve fiber arising from a slight elevation of the cell body that conducts an action potential (impulse) away from the cell body
i. Nissl bodies (chromatophilic substance)	_____ 9. Possesses a single process from the cell body
j. Unipolar neuron	_____ 10. Found in the eyes, nose, and ears

PART B: Assessments

Match the terms in column A with the descriptions in column B. Place the letter of your choice in the space provided. A1 A2

Column A	Column B
a. Effector	_____ 1. Transmits impulse from sensory to motor neuron within central nervous system
b. Ependymal cell	_____ 2. Transmits impulse out of the brain or spinal cord to effectors (muscles and glands)
c. Ganglion	_____ 3. Transmits impulse into brain or spinal cord from receptors
d. Interneuron (association neuron)	_____ 4. Myelin-forming neuroglia in brain and spinal cord
e. Microglia	_____ 5. Phagocytic neuroglia
f. Motor (efferent) neuron	_____ 6. Structure capable of responding to motor impulse
g. Oligodendrocyte	_____ 7. Specialized mass of neuron cell bodies outside the brain or spinal cord
h. Sensory (afferent) neuron	_____ 8. Cells that line cavities of the brain and secrete cerebrospinal fluid

PART C: Assessments

Each circle below represents the field of view as seen through the microscope. In each circle, sketch the indicated cells. Label any of the cellular structures observed, and indicate the magnification of each sketch. Alongside your sketch, add notes that would assist in your understanding of the structures and functions. A2 A3

Motor neuron (______×)

Sensory neuron cell body (______×)

Neuroglia (______×)

PART D: Assessments

Each circle below represents the field of view as seen through the microscope. In each circle, sketch the indicated view of a nerve fiber (axon). Label any structures observed, and indicate the magnification of each sketch. Alongside your sketch, add notes that would assist in your understanding of the structures and functions. A2 A3

Nerve fiber cross section with Schwann cell (______×)

Nerve fiber longitudinal section with Schwann cell (______×)

CRITICAL THINKING ASSESSMENT—PART D

Nutrient deficiencies throughout infancy and early childhood can significantly alter the composition and formation of the myelin sheath during this crucial time of rapid neurological development. What macronutrients are important in the proper development and maintenance of the myelin sheath and what neuroglia will be utilizing these nutrients? What might be the effect of early life malnourishment? A2

NOTES

LABORATORY
EXERCISE 28

Meninges, Spinal Cord, and Spinal Nerves

Purpose of the Exercise

To review the characteristics of the meninges, spinal cord, spinal nerves, and nerve plexuses and to observe the major anatomical features of these structures.

MATERIALS NEEDED

Spinal cord model with meninges
Vertebral column model with spinal nerves
Compound light microscope
Prepared microscope slide of a spinal cord cross section with spinal nerve roots
Limb models with plexuses and nerves
For Demonstration Activity:
Preserved spinal cord with meninges intact

Chapter Opener: ©101cats/Getty Images

Learning Outcomes APR

After completing this exercise, you should be able to:

O1 Describe the arrangement of the spinal cord meninges and the structure of each.

O2 Identify the major anatomical features and functions of the spinal cord.

O3 Locate the distribution and functions of the spinal nerves and plexuses.

The O corresponds to the assessments A indicated in the Laboratory Assessment for this Exercise.

Pre-Lab

Carefully read the introductory material and examine the entire lab. Be familiar with the structures and functions of the spinal cord, spinal nerves, and meninges from lecture or the textbook. Answer the pre-lab questions.

Pre-Lab Questions Select the correct answer for each of the following questions:

1. The _____ is the most superficial membrane of the meninges.
 a. subarachnoid space **b.** pia mater
 c. arachnoid mater **d.** dura mater

2. The inferior end of the adult spinal cord terminates
 a. just inferior to L1. **b.** inferior to L5.
 c. in the sacrum. **d.** in the coccyx.

3. The posterior root of spinal nerves contains
 a. interneurons. **b.** sensory neurons.
 c. motor neurons. **d.** sensory and motor neurons.

4. Which of the following is *not* part of the gray matter of the spinal cord?
 a. gray commissure **b.** posterior horn
 c. lateral funiculus **d.** lateral horn

5. The central canal of the spinal cord is located within the
 a. white matter. **b.** epidural space.
 c. gray commissure. **d.** subarachnoid space.

6. The cerebrospinal fluid (CSF) flows within the
 a. epidural space. **b.** subarachnoid space.
 c. subdural space. **d.** dural sheath.

7. The brachial plexus is formed from components of spinal nerves C5–T1.
 a. True **b.** False

8. There are eight pairs of cervical spinal nerves.
 a. True **b.** False

9. The major ascending (sensory) and descending (motor) tracts compose the gray matter of the spinal cord.
 a. True **b.** False

Three membranes, or *meninges*, surround the entire central nervous system (CNS). They consist of three protective layers of fibrous connective tissue membranes located between the bones of the skull and vertebral column and the soft tissues of the CNS. The most superficial layer, the *dura mater,* is a tough membrane with an epidural space containing blood vessels, loose connective tissue, and adipose tissue between the membrane and the vertebrae. A thin, weblike *arachnoid mater* adheres to the inside of the dura mater. The *subarachnoid space,* located beneath the arachnoid mater, contains *cerebrospinal fluid* (*CSF*) and serves as a protective cushion for the spinal cord and brain. The delicate innermost membrane, the *pia mater,* adheres to the surface of the spinal cord and brain. The *denticulate ligaments,* extending from the pia mater to the dura mater, anchor the spinal cord. An inferior extension of the pia mater, the *filum terminale,* anchors the spinal cord to the coccyx.

The spinal cord is a column of nerve fibers and supporting glial cells that extends from the foramen magnum through the vertebral canal to about the level of lumbar vertebra L1. Together with the brain, it makes up the CNS. In the cervical and lumbar regions of the spinal cord, enlargements give rise to spinal nerves to the upper and lower limbs, respectively. The terminal end of the spinal cord, the *conus medullaris,* is located just inferior to L1 in the adult.

There are thirty-one pairs of spinal nerves attached to the spinal cord. At close proximity to the cord, the spinal nerve has two branches: a *posterior (dorsal) root* containing *sensory neurons* and *posterior root ganglion,* and an *anterior (ventral) root* containing *motor neurons.* (Recall that a ganglion is a cluster of cell bodies (somas) of sensory neurons outside the CNS.) The spinal nerves emerge through the nearby intervertebral foramina; however, most lumbar and sacral nerves extend inferiorly through the vertebral canal as the *cauda equina* (resembling a horse's tail) to emerge in their respective regions of the vertebral column.

The spinal cord has a central region, the *gray matter,* with paired *posterior* and *anterior horns* connected by a *gray commissure* containing the *central canal.* Paired *lateral horns* are seen from thoracic vertebra T2 through L1. The gray matter is a processing center for spinal reflexes and synaptic integration. The posterior horns receive sensory nerve fibers, often synapsing in the horn with interneurons. The anterior horns contain cell bodies of motor neurons whose axons will lead to and innervate skeletal muscle. The lateral horns contain cell bodies of neurons of the sympathetic nervous system. The *white matter* of the spinal cord, represented by paired *posterior, lateral,* and *anterior funiculi* (*columns*), contains *ascending* (*sensory*) and *descending* (*motor*) *tracts.* (Recall that a tract is a collection of nerve fibers (axons) in the CNS.) At various levels of the spinal cord, many of the tracts cross over (decussate) to the opposite side of the spinal cord or the brainstem.

PROCEDURE A: Meninges APR

Three membranes, or meninges, surround the entire CNS. The superficial, tough dura mater of the spinal cord meninges has an epidural space between the membrane and the vertebrae. A subarachnoid space exists between the middle arachnoid mater and the pia mater that contains CSF. The delicate pia mater closely adheres to the surface of the brain and spinal cord. Collectively, the meninges enclose and provide physical protective layers around the delicate brain and spinal cord. The CSF provides an additional protective cushion from sudden jolts to the head or the back. The cranial meninges, which are more directly associated with the brain and skull, are covered in Laboratory Exercise 30.

1. Study figures 28.1 to 28.3.

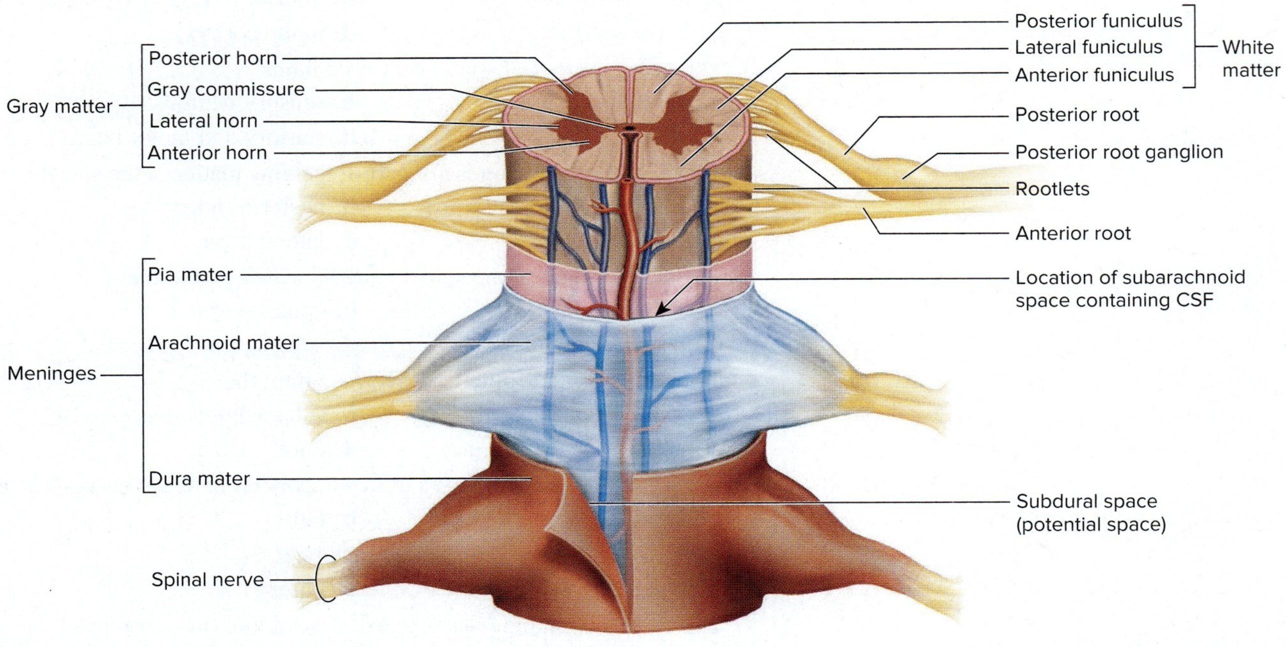

FIGURE 28.1 The meninges associated near the spinal cord and spinal nerves, anterior view.

FIGURE 28.2 Posterior view of a segment of the cervical spinal cord. The three meninges surround the spinal cord. Rootlets of posterior and anterior spinal nerves enter the spinal cord. Posterior portions of the cervical vertebrae have been removed to expose all the labeled structures. Dr. Robert A. Chase, M.D. APR

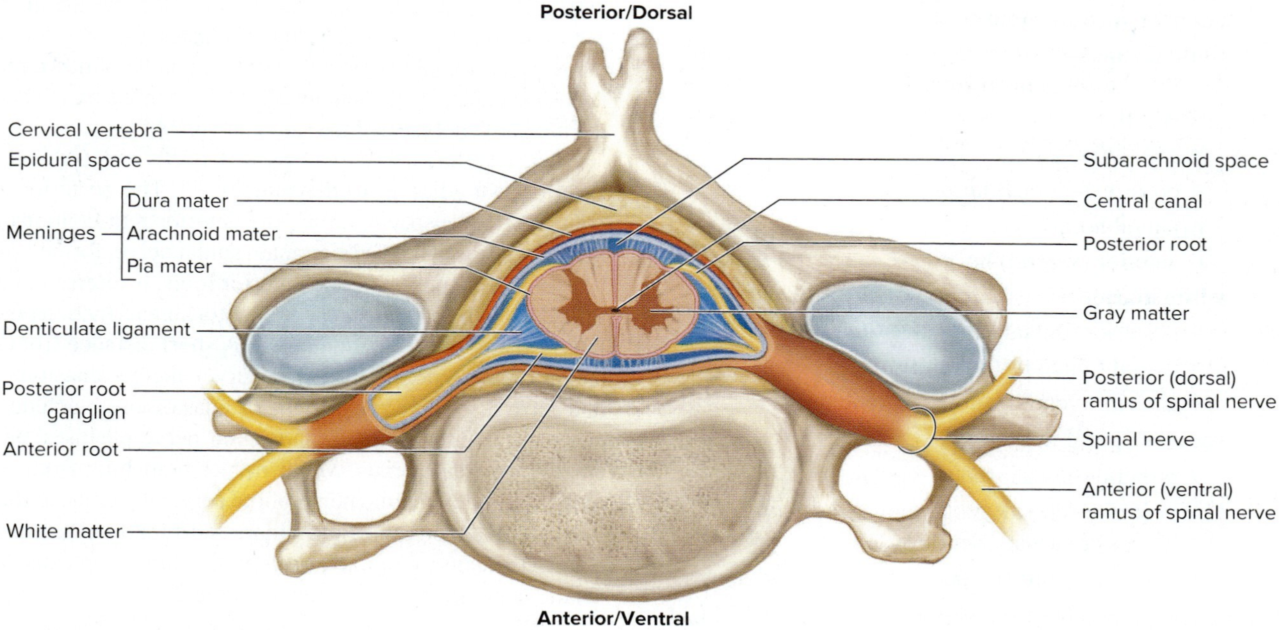

FIGURE 28.3 Features of the spinal cord cross section and surrounding structures. APR

2. Observe the spinal cord model with meninges and locate the following features from superficial to deep:
   - **epidural space**—between vertebra and dura mater; contains connective tissues and blood vessels; site where epidural anesthesia is administered.
   - **dura mater (dural sheath)**—most superficial of the three meninges
   - **subdural space**—a narrow potential space with a thin film of fluid
   - **arachnoid mater (membrane)**—thin, weblike middle layer of meninges
   - **subarachnoid space**—contains protective CSF
   - **denticulate ligament**—extensions of pia mater to dura mater to suspend and anchor the spinal cord
   - **pia mater**—adheres closely to spinal cord
3. Complete Part A of Laboratory Assessment 28.

PROCEDURE B: Structure of the Spinal Cord

1. Study figures 28.1, 28.3, and 28.4, along with the cross-section model of the spinal cord.

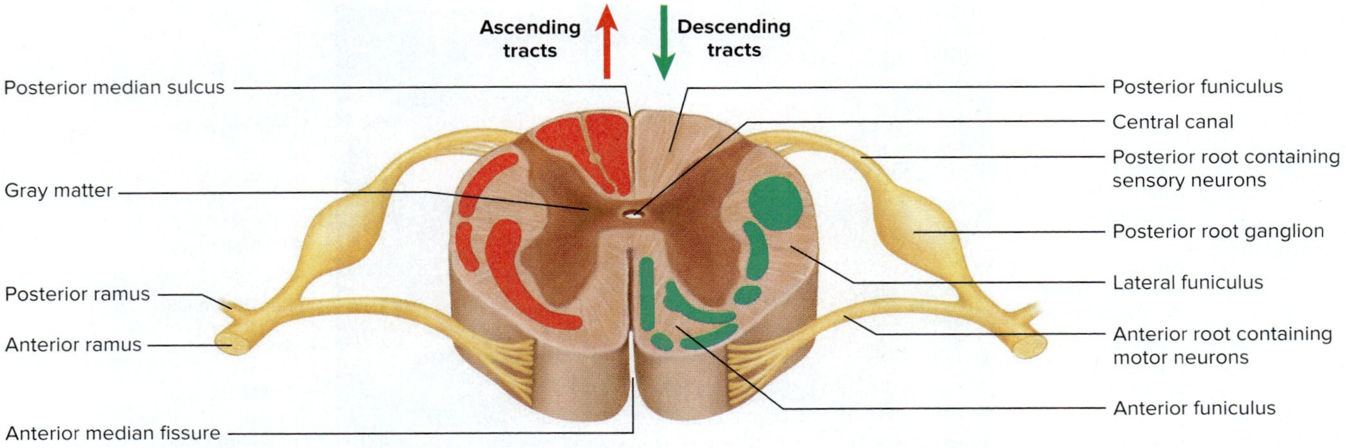

FIGURE 28.4 Cross section of the spinal cord showing ascending (sensory) tracts (pathways) carrying sensory signals up the spinal cord on one side and the descending (motor) tracts (pathways) sending motor signals down the spinal cord on the other side. Ascending and descending tracts are present on both sides (bilateral) of the spinal cord in the white matter.

2. Obtain a prepared microscope slide of a spinal cord cross section. Use the scan and low power of the microscope to locate the following features:

   posterior median sulcus

   anterior median fissure

   central canal—within gray commissure; collapsed in most areas in an adult

   gray matter

   - gray commissure
   - posterior (dorsal) horn
   - lateral horn
   - anterior (ventral) horn

   white matter

   - posterior (dorsal) funiculus (column)
   - lateral funiculus (column)
   - anterior (ventral) funiculus (column)

   roots of spinal nerve

   - posterior (dorsal) roots—contain sensory neurons
   - posterior (dorsal) root ganglia—contain cell bodies of sensory neurons
   - anterior (ventral) roots—contain motor neurons

3. Observe the model of the vertebral column with spinal nerves along with figure 28.5. Note the length of the spinal cord from the foramen magnum to the conus medullaris. Note also the thirty-one pairs of spinal nerves, the cauda equina, and the filum terminale.
4. Complete Part B of the laboratory assessment.

PROCEDURE C: Spinal Nerves and Nerve Plexuses

There are thirty-one pairs of spinal nerves that emerge between bones of the vertebral column. Cervical spinal nerve pair C1 emerges between the foramen magnum of the occipital bone and the atlas. The other cervical nerves emerge inferior to the seven cervical vertebrae, resulting in eight pairs of cervical nerves. The rest of the spinal nerves emerge inferiorly to the corresponding vertebrae, resulting in twelve pairs of thoracic nerves, five pairs of lumbar nerves, five pairs of sacral nerves, and one coccygeal pair. Recall that the single sacrum of the adult is a result of fusions of five sacral vertebrae. Because the inferior end of the spinal cord of the adult terminates just inferior to L1, many of the spinal nerves give rise to a bundle of nerve roots, the cauda equina, extending inferiorly through the remainder of the lumbar vertebrae and the sacrum (fig. 28.5).

Each spinal nerve possesses a posterior root and an anterior root adjacent to the spinal cord. The posterior root contains the sensory neurons; the anterior root contains the motor neurons. The resulting main spinal nerve, formed from a fusion of the posterior and anterior roots, is referred to as a mixed nerve because sensory and motor signals (action potentials) exist within the same nerve. A short distance from the spinal cord, some of the spinal nerves merge (anastomose) and form a weblike plexus. Major anastomoses include the cervical, brachial, lumbar, and sacral nerve plexuses, and a small coccygeal plexus. Major nerves branching from each plexus include: phrenic nerve of the cervical plexus; axillary, musculocutaneous, median, ulnar, and radial nerves of the brachial plexus; femoral nerve of the lumbar plexus; and sciatic nerve of the sacral plexus (table 28.1).

1. Examine figures 28.2 and 28.5.
2. Observe a vertebral column model with spinal nerves extending from the intervertebral foramina and limb models with plexuses and nerves. Compare the emerging locations of the spinal nerves with the names of the vertebrae.
3. The *posterior* and *anterior roots* have multiple *rootlets* attached to the spinal cord (figs. 28.1–28.3). All of the thirty-one pairs of spinal nerves are mixed, containing sensory and motor nerve fibers. The actual spinal nerve is very short as it divides after emerging from its foramen, where it branches into a *posterior (dorsal) ramus,* an *anterior (ventral) ramus,* and a *meningeal branch.* The tiny meningeal branch reenters the vertebral canal and

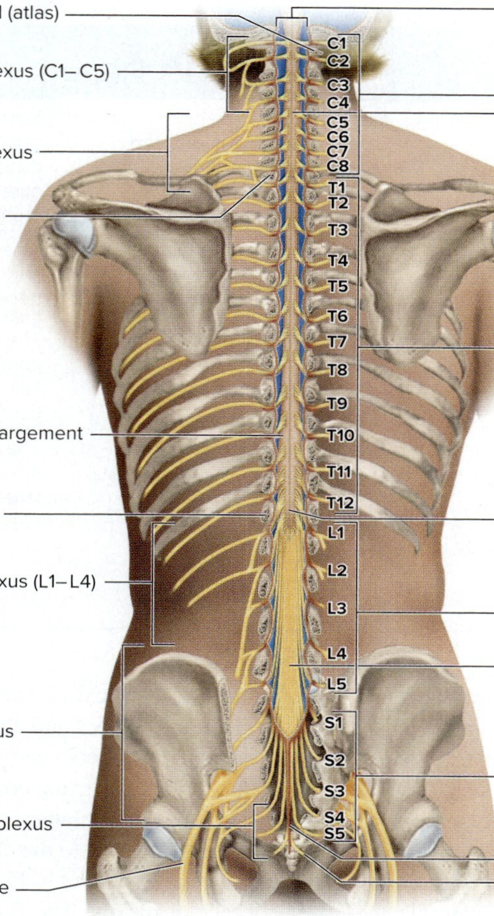

(a)

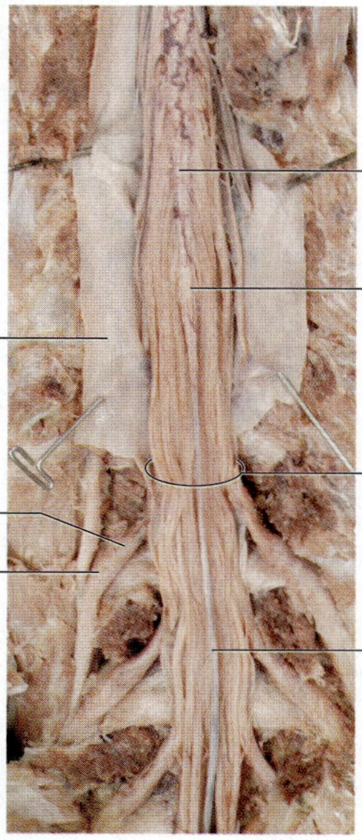

(b)

FIGURE 28.5 (a) Posterior view of the origins and categories of the thirty-one pairs of spinal nerves on the right. The spinal cord begins at the foramen magnum and ends at the conus medullaris. The plexuses formed by various spinal nerves are illustrated on the left. (b) Cadaver dissection photo of the conus medullaris and cauda equina. (b) Christine Eckel/McGraw Hill

Table 28.1 Nerve Plexuses

Plexus	Spinal Nerves Contributing Fibers	Most Significant Nerves of Plexus	Area Served	Comments
Cervical (figs. 28.7 and 28.8)	C1–C5	Phrenic (C3–C5), Ansa cervicalis	Diaphragm, omohyoid, sternohyoid, and sternothyroid muscles, other neck and shoulder muscles, and portions of the head	Some do not include C5 as part of cervical plexus; only a few fibers of C5 contribute to the phrenic nerve; the phrenic nerve (C3–C5) is essential in breathing
Brachial (figs. 28.9 and 28.10)	C5–T1	Musculocutaneous, axillary, radial, median, ulnar	Shoulder, arm, forearm, and hand	Small contribution of fibers from C4 and T2; a very complex network of nerves
Lumbar (fig. 28.11)	T12 and L1–L4	Femoral, obturator, genitofemoral, ilioinguinal, iliohypogastric	Inferior abdominopelvic and anterior thigh, male cremaster muscle	The lumbar and sacral plexuses have connecting fibers and are sometimes called the lumbosacral plexus; minor contributions from T12
Sacral (fig. 28.12)	L4–L5 and S1–S4	Sciatic (common fibular and tibial), superior gluteal, inferior gluteal (figs. 28.11 and 28.12)	Buttock (gluteals), posterior thigh, leg, and foot	The sciatic nerve has fibers from five spinal nerves (L4–L5 and S1–S3); the large sciatic nerve is located deep to the gluteus maximus and posterior thigh and is vulnerable to injuries and sciatica pain
Coccygeal (fig. 28.12)	S4–S5 and Co1	Pudendal	Pelvic floor (perineum)	Tiny plexus; fibers from sacral plexus contribute to pudendal nerve

innervates the meninges and vertebrae, while the relatively large anterior ramus has communicating rami that connect to the autonomic fibers of the *sympathetic trunk ganglion* (*sympathetic chain ganglion*) (fig. 28.6).

4. The posterior rami innervate muscles of the spine and skin in the back regions. In many areas along the spinal cord, the anterior rami form *nerve plexuses* with complex networks of branches and fibers, comparable to an electrical junction box, primarily serving the limbs (table 28.1 and fig. 28.5). In some areas, some fibers intermingle with an adjacent plexus. The nerve plexuses include: cervical plexus (figs. 28.7 and 28.8); brachial plexus (figs. 28.9 and 28.10); lumbar plexus (fig. 28.11); and sacral and coccygeal plexuses (fig. 28.12).
5. The brachial plexus, considered the most complex, can be palpated just lateral to the sternocleidomastoid muscle and superior to the clavicle (fig. 28.10). The anterior rami of C5 through T1 represent the five *roots* of the brachial plexus. The five roots unite into three *trunks;* each trunk divides into *anterior* and *posterior divisions.* The anterior and posterior divisions give rise to three bundles called *lateral, medial,* and *posterior cords.*

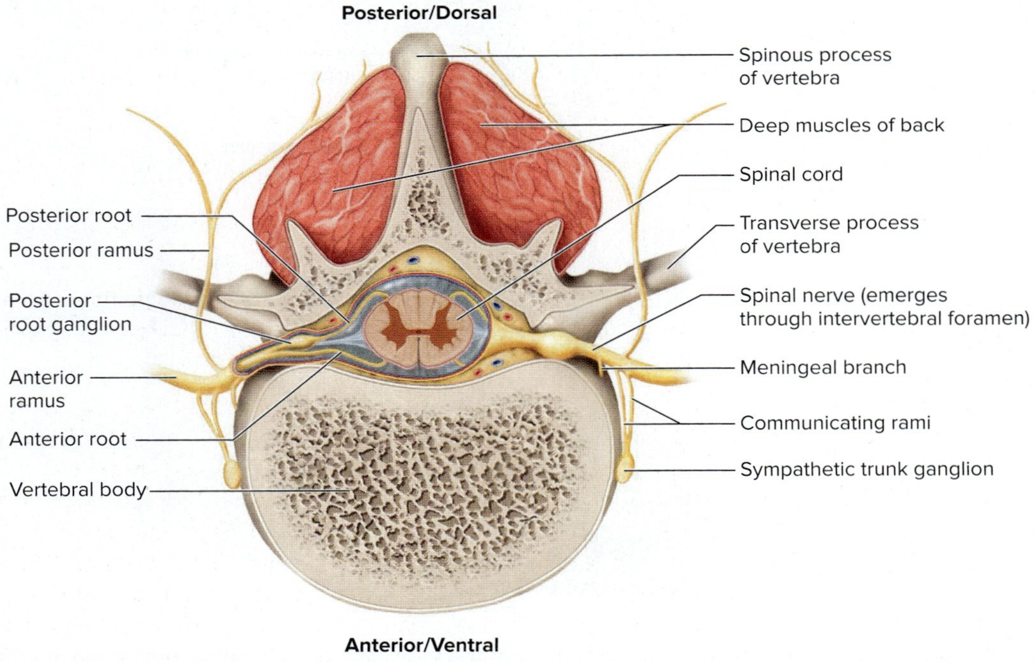

FIGURE 28.6 Transverse view of the spinal cord and a pair of spinal nerves with its branches associated with a vertebra.

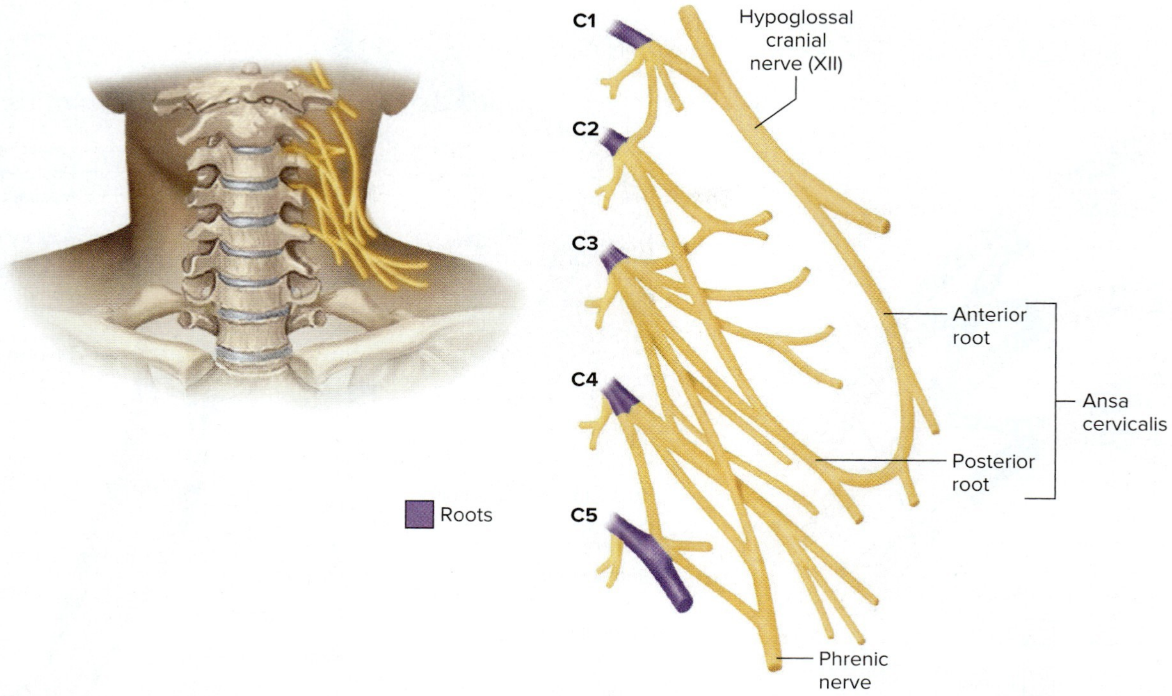

FIGURE 28.7 The cervical plexus (C1–C5). See table 28.1 and figure 28.8, the cervical plexus of the cadaver.

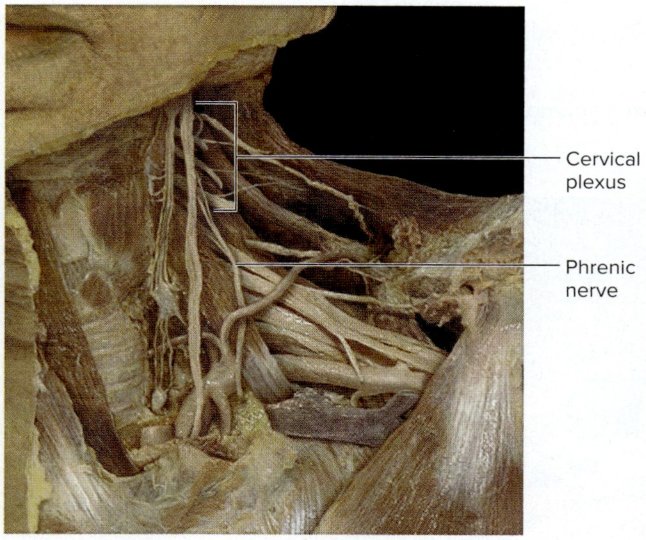

FIGURE 28.8 The cervical plexus of a cadaver, anterior view. The phrenic nerve, containing some fibers from C3, C4, and C5 innervates the diaphragm, the skeletal muscle essential in breathing. See table 28.1. APR/McGraw Hill APR

Arising from the three cords in the axillary region, the five most significant nerves of the brachial plexus emerge (figs. 28.9 and 28.10). The brachial plexus has a characteristic **W** or **M** shape when viewed in a careful dissection of a cadaver (fig. 28.10).

6. The posterior root near the spinal cord possesses sensory neurons that conduct nerve impulses from peripheral regions of the body. This sensory input correlates to specific areas of the skin. A specific area of the skin that a particular cutaneous branch of a spinal nerve innervates is called a *dermatome.* Figure 28.13 represents a *dermatome map* of anterior and posterior cutaneous regions of the body innervated by specific spinal nerves. Testing various dermatomes by light pinpricks or pokes can be used to diagnose possible spinal cord injuries and spinal nerve damage. The specific spinal nerve damaged may not be precisely identified because there is some overlap of neurons with close neighboring dermatomes.
7. Complete Parts C and D of the laboratory assessment.

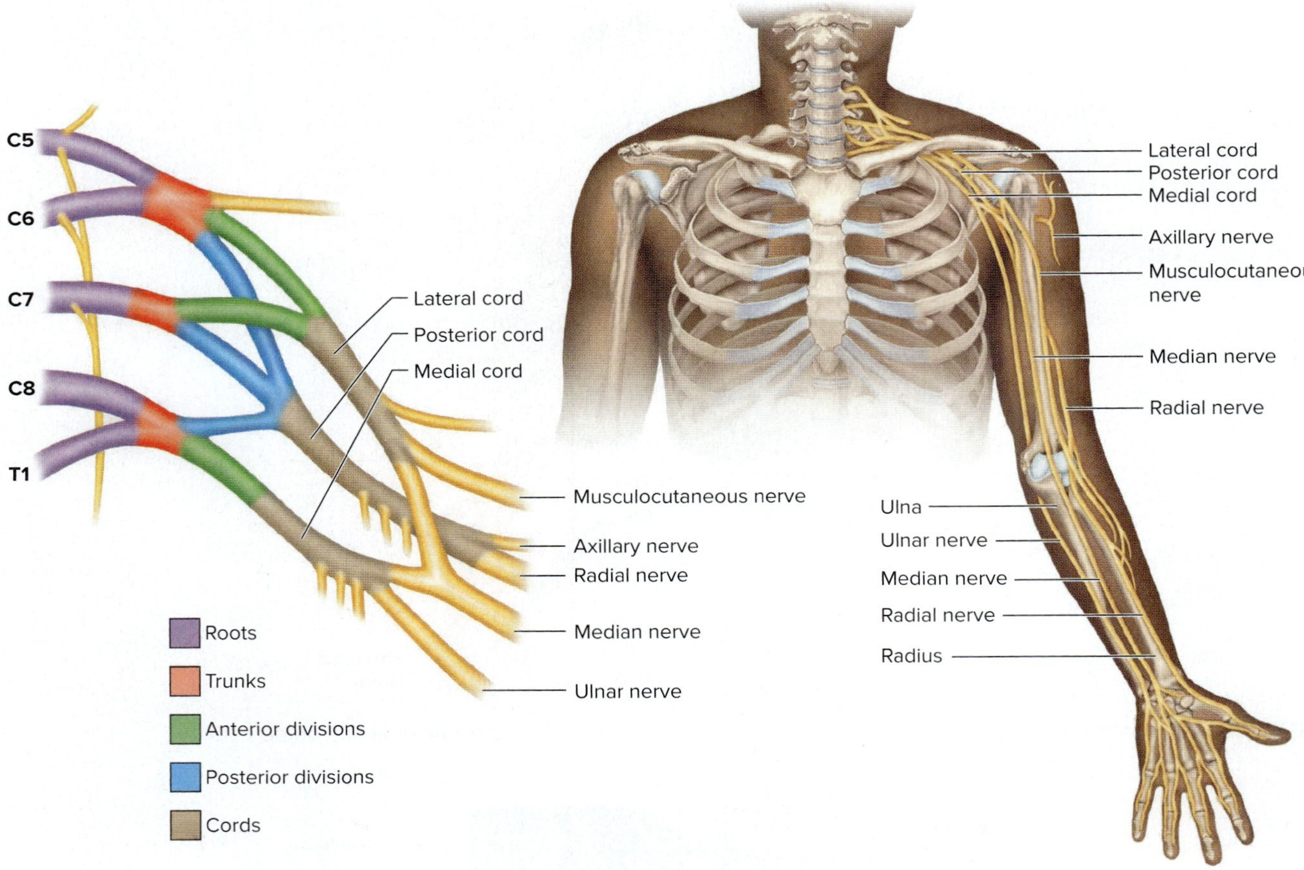

FIGURE 28.9 The brachial plexus anterior view (C5–T1). See table 28.1 and figure 28.10, the brachial plexus of the cadaver. APR

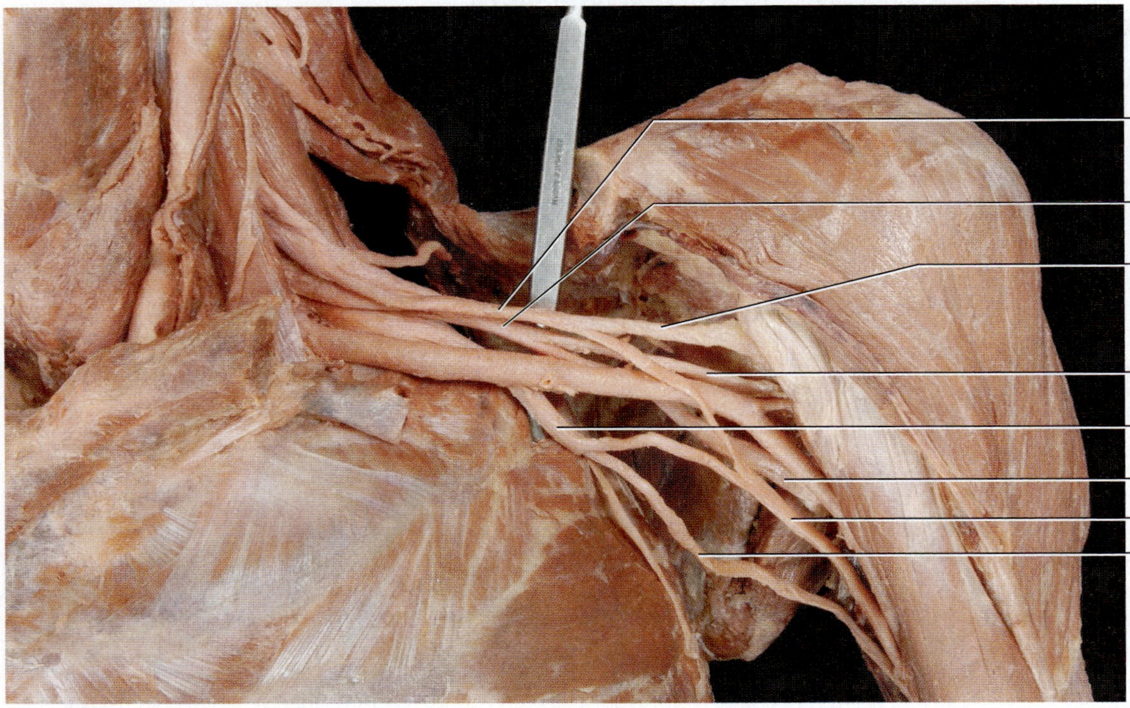

FIGURE 28.10 The brachial plexus of the left shoulder of the cadaver. The five most significant nerves of the plexus are exposed in this anterior view. See table 28.1. Christine Eckel/McGraw Hill APR

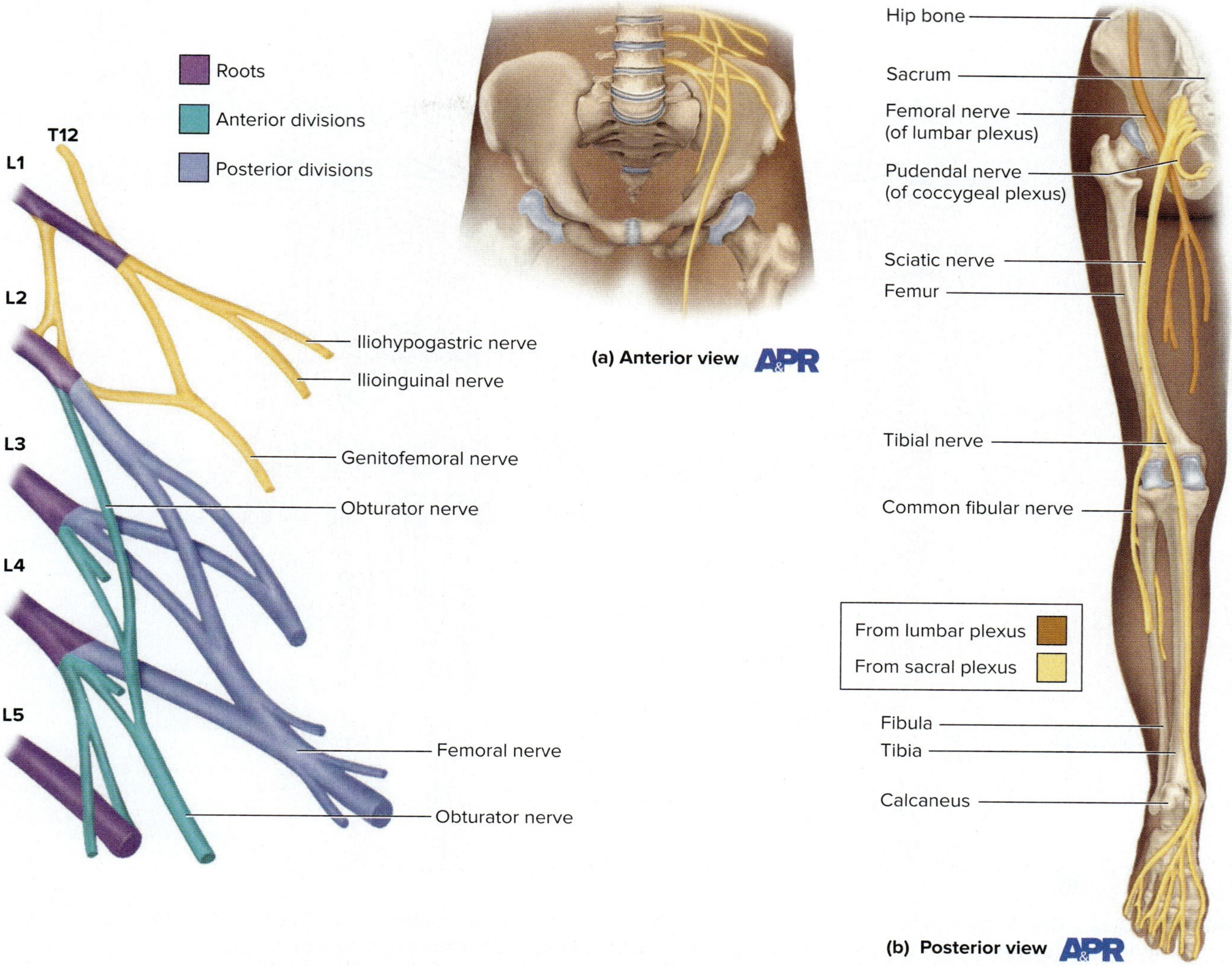

FIGURE 28.11 An anterior view of the lumbar plexus (T12 and L1–L4) in (a), and a posterior view of the sacral plexus (L4–L5 and S1–S4) and the femoral nerve of the lumbar plexus shown in (b). See table 28.1.

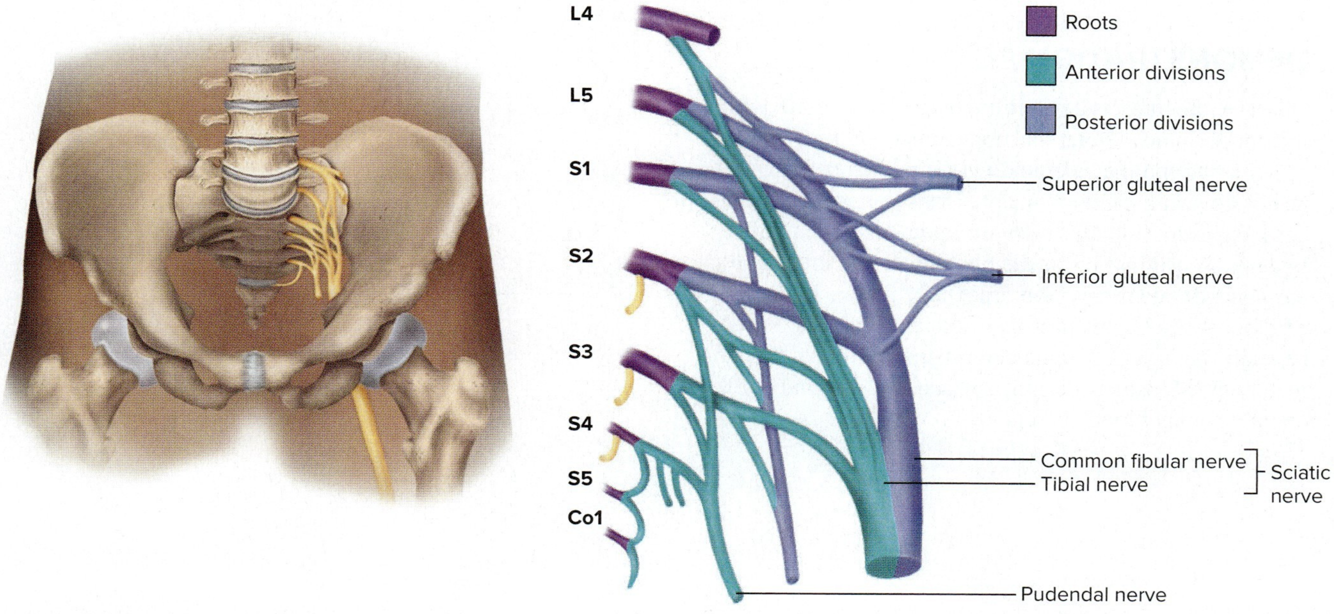

FIGURE 28.12 The anterior view of the sacral plexus (L4–L5 and S1–S4) and the coccygeal plexus (S4–S5 and Co1). See table 28.1 and the posterior view of figure 28.11*b*.

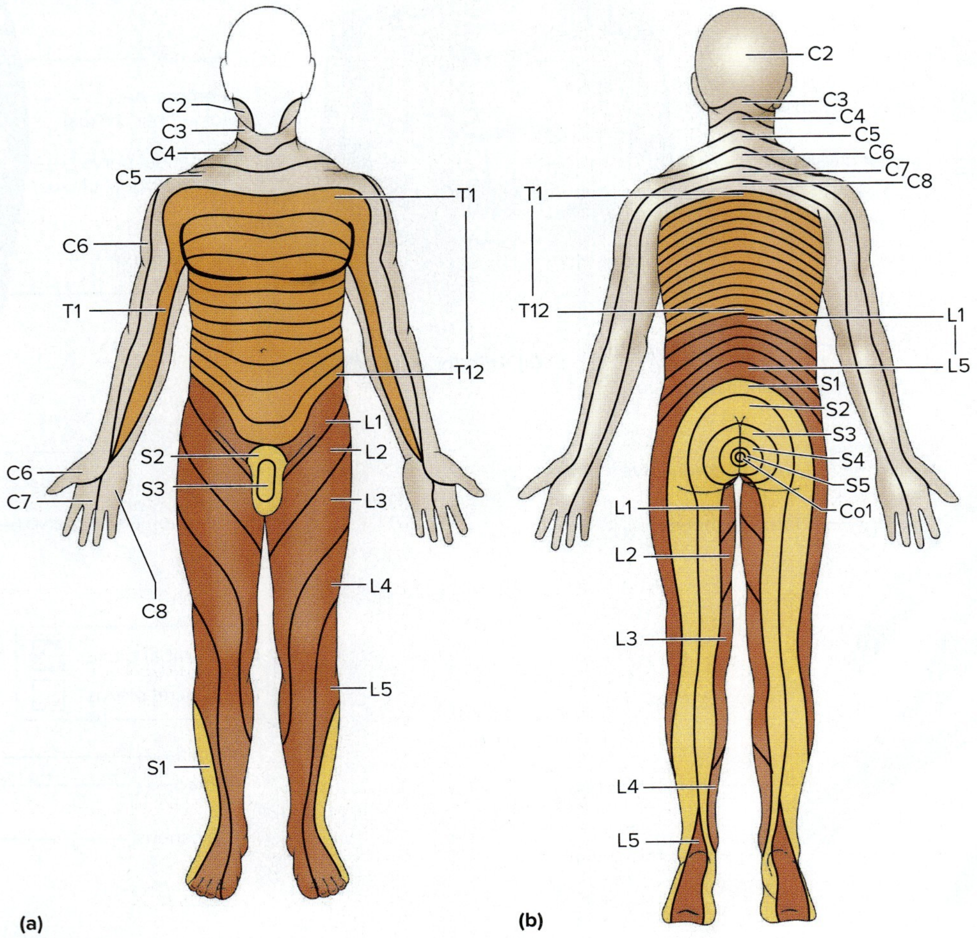

FIGURE 28.13 Dermatome map of the body surface: (a) anterior; (b) posterior. Spinal nerve C1 does not innervate the skin, but all of the other spinal nerves possess sensory branches to a zone indicated on the dermatome map.

DEMONSTRATION ACTIVITY

Observe the preserved section of a spinal cord with meninges intact. Note the appearance of the central gray matter and the surrounding white matter. Note the heavy covering of dura mater, firmly attached to the cord on each side by a set of ligaments (denticulate ligaments) originating in the pia mater. The intermediate layer of meninges, the arachnoid mater, is devoid of blood vessels, but in a live human being, the space beneath this layer contains cerebrospinal fluid. The pia mater, closely attached to the surface of the spinal cord, contains many blood vessels.

LABORATORY
ASSESSMENT 28

Name ______________________________

Date ______________________________

Section ______________________________

The A corresponds to the indicated Learning Outcome(s) O found at the beginning of the Laboratory Exercise.

Meninges, Spinal Cord, and Spinal Nerves

PART A: Assessments

Match the terms in column A with the descriptions in column B. Place the letter of your choice in the space provided. A1

Column A		Column B
a. Arachnoid mater	______	**1.** Connections from pia mater to dura mater that anchor the spinal cord
b. Denticulate ligaments	______	**2.** Inferior continuation of pia mater to the coccyx
c. Dura mater	______	**3.** Outermost layer of meninges
d. Epidural space	______	**4.** Follows irregular contours of spinal cord surface
e. Filum terminale	______	**5.** Contains a protective cushion of cerebrospinal fluid
f. Pia mater	______	**6.** Thin, weblike middle membrane
g. Subarachnoid space	______	**7.** Separates dura mater from bone of vertebra
h. Subdural space	______	**8.** Potential narrow space with a thin film of fluid

PART B: Assessments

Identify the features indicated in the spinal cord cross section of figures 28.14 and 28.15.

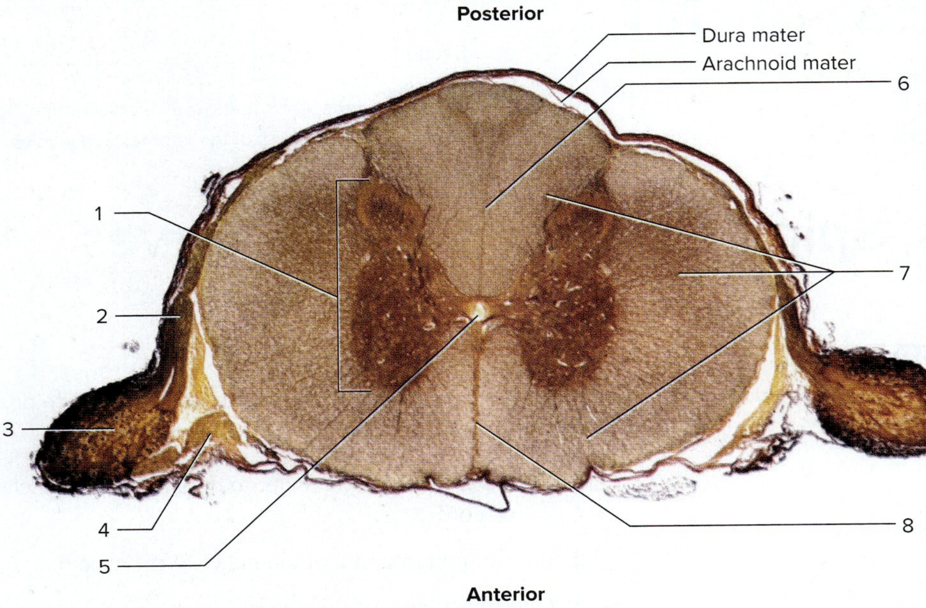

_____ Anterior median fissure
_____ Anterior root of spinal nerve
_____ Central canal
_____ Gray matter
_____ Posterior median sulcus
_____ Posterior root ganglion
_____ Posterior root of spinal nerve
_____ White matter

FIGURE 28.14 Micrograph of a spinal cord cross (transverse) section with spinal nerve roots (7.5×). Label the features by placing the correct numbers in the spaces provided. Ed Reschke/Photolibrary/Getty Images A2

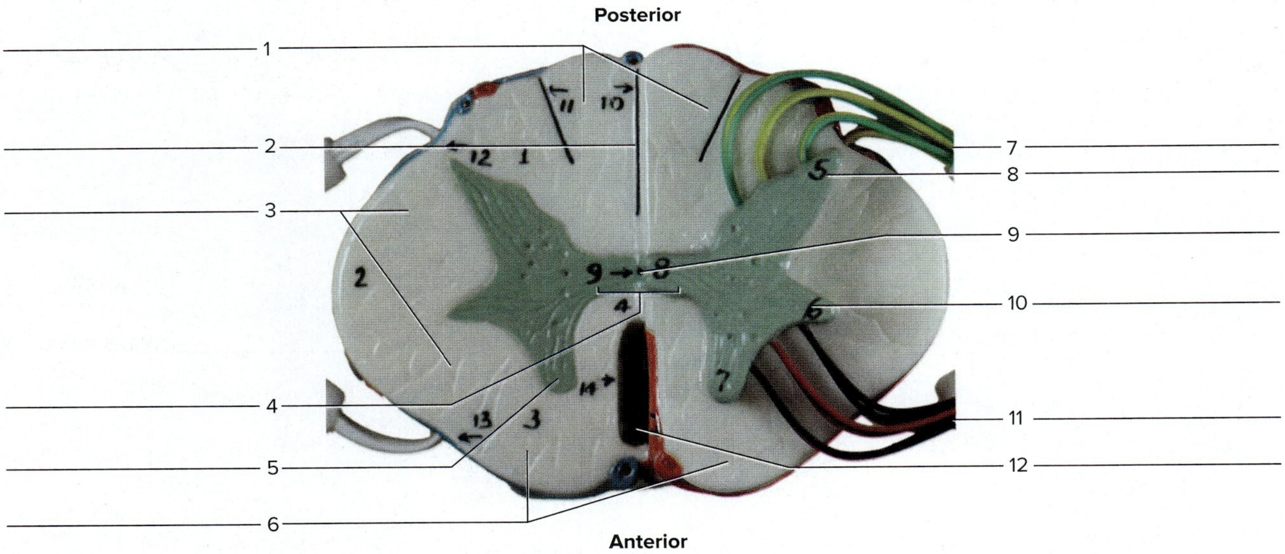

FIGURE 28.15 Label the features indicated on the image of a cross-section model of a spinal cord with spinal nerve roots. ©2017 Denoyer-Geppert Science Company A2

PART C: Assessments

1. Identify the nerves and plexi indicated in the forearm and leg of figure 28.16.

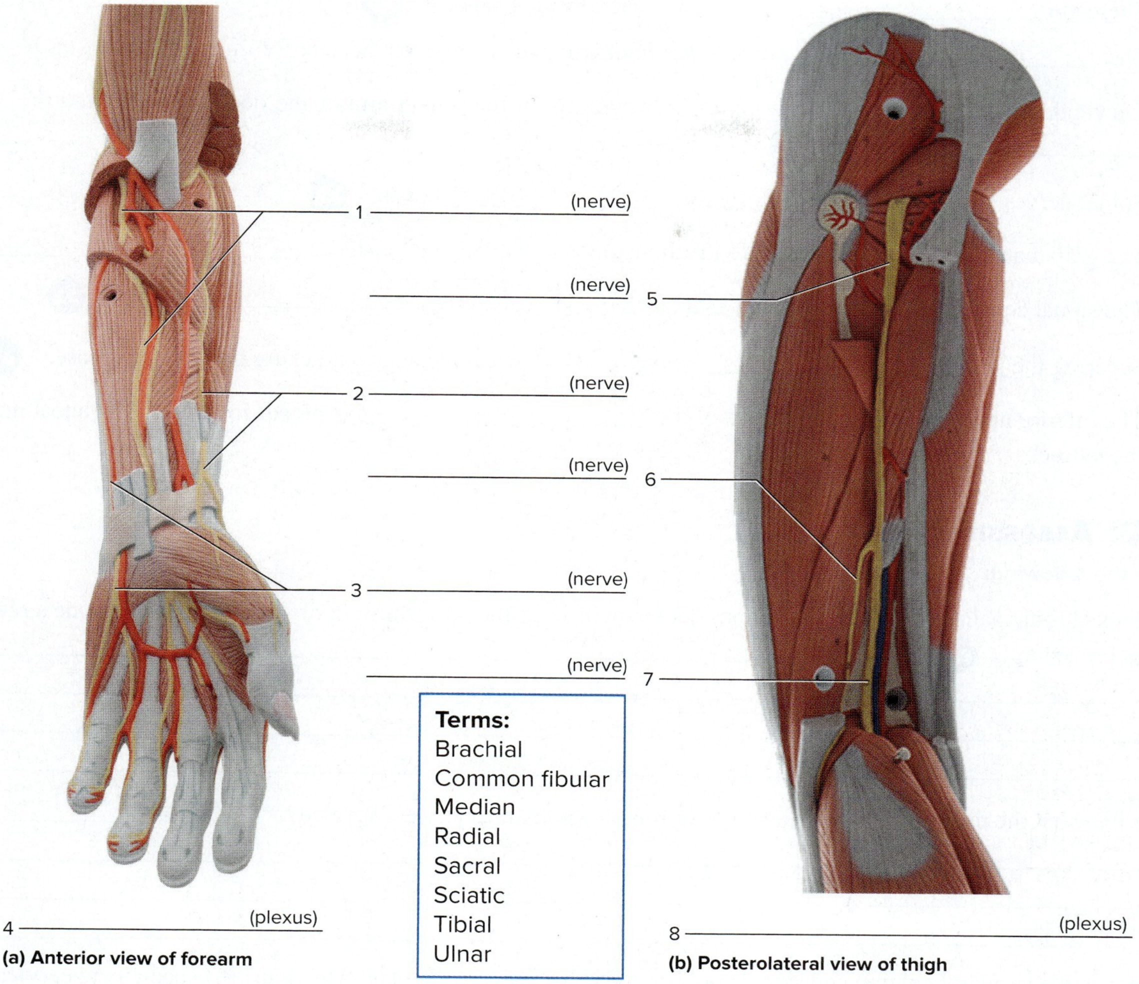

FIGURE 28.16 Label the nerves and their plexus of the (a) anterior forearm and (b) posterolateral thigh, using the terms provided. [*Note:* The superficial flexor muscles have been removed in (a) and the gluteus maximus and part of the biceps femoris muscles have been removed in (b).] @3B Scientific GmbH, Germany, 2017, www.3bscientific.com A3

2. Complete the following statements:

a. The spinal cord gives rise to thirty-one pairs of ______________________. A2 A3

b. The bulge in the spinal cord that gives off nerves to the upper limbs is called the ________________ enlargement. A2

c. The bulge in the spinal cord that gives off nerves to the lower limbs is called the ________________ enlargement. A2

d. The ______________________ is a groove that extends the length of the spinal cord posteriorly. A2

e. In a spinal cord cross section, the posterior ______________________ of the gray matter resemble the upper wings of a butterfly. A2

f. The axons of motor neurons are found in the ______________________ roots of spinal nerves. A3

g. The ______________________ connects the gray matter on the left and right sides of the spinal cord. A2

h. The ______________________ in the gray commissure of the spinal cord contains cerebrospinal fluid and is continuous with the cavities of the brain. A2

i. The white matter of the spinal cord is divided into anterior, lateral, and posterior ______________________ that contain ascending and descending tracts. A2

j. There are ________________ pairs of cervical spinal nerves. A3

k. There are ________________ pairs of sacral spinal nerves. A3

l. The ________________ nerve of the lumbar plexus innervates the anterior thigh muscles.

m. Cervical spinal nerve pair C1 originates between the foramen magnum of the occipital bone and the ________________. A3

n. Spinal nerves L4 through S4 form a ________________ plexus. A3

o. The gray matter of the spinal cord is divided into anterior, lateral, and posterior ________________. A2

p. The spinal cord ends just inferior to L1 in a tapered point called the ________________. A2

q. Severing the ________________ nerves of the cervical plexus would cause breathing to cease. A3

r. The inferior and superior gluteal nerves of the ________________ plexus innervate the gluteal muscles of the buttocks. A3

PART D: Assessments

Complete the following:

1. To block chronic pain in a patient, sometimes doctors will sever the posterior root of a spinal nerve. Provide a rationale for such a procedure. A2 A3 ________________

2. Describe what the result would be if the anterior root of a spinal nerve were severed? A2 A3 ________________

3. Injuries to the vertebral column and spinal cord are always of concern. Explain why injuries, especially to cervical regions, are of the greatest concern. Include specific cervical areas as part of your answer. A2 A3 ________________

CRITICAL THINKING ASSESSMENTS—PART D

1. Why are intramuscular injections given in the gluteus medius muscle rather than in the larger gluteus maximus muscle in the buttock region? Explain. A3

2. Incorrectly using a crutch may cause impaired sensory or motor function to the arm, forearm, and hand as nerves are compressed, creating a condition called crutch paralysis. A3

   What nerve plexus is involved? ________________

   How might a person who uses a crutch avoid this injury? ________________

LABORATORY
EXERCISE 29

Reflex Arc and Somatic Reflexes

Purpose of the Exercise

To review the components of reflex arcs and characteristic reflexive behavior and to demonstrate several common somatic reflexes that occur in the human body.

MATERIALS NEEDED

Rubber percussion hammer

Reflex arc model

Chapter Opener: ©101cats/Getty Images

Learning Outcomes APR

After completing this exercise, you should be able to:

O1 Demonstrate and record results of stretch reflexes that occur under normal laboratory conditions and from mental distractions.

O2 Describe the components of a reflex arc.

O3 Analyze the components and patterns of stretch reflexes.

The O corresponds to the assessments A indicated in the Laboratory Assessment for this Exercise.

Pre-Lab

Carefully read the introductory material and examine the entire lab. Be familiar with reflexes from lecture or the textbook. Answer the pre-lab questions.

Pre-Lab Questions Select the correct answer for each of the following questions:

1. The impulse over a motor neuron will lead to
- **a.** an interneuron.
- **b.** the spinal cord.
- **c.** a receptor.
- **d.** an effector.

2. An effector of somatic reflexes would be
- **a.** smooth muscle.
- **b.** cardiac muscle.
- **c.** skeletal muscle.
- **d.** glands.

3. Stretch reflex receptors are called
- **a.** effectors.
- **b.** muscle spindles.
- **c.** interneurons.
- **d.** motor neurons.

4. Stretch reflexes include all of the following *except* the _____ reflex.
- **a.** withdrawal
- **b.** patellar
- **c.** calcaneal
- **d.** biceps

5. A withdrawal reflex could occur from
- **a.** striking the patellar ligament.
- **b.** striking the calcaneal tendon.
- **c.** striking the triceps tendon.
- **d.** touching a hot object.

6. The calcaneal reflex response is
- **a.** pain interpretation.
- **b.** flexion of the leg at the knee joint.
- **c.** plantar flexion of the foot.
- **d.** a separation of toes.

7. The quadriceps femoris is the effector muscle of the patellar reflex.
- **a.** True
- **b.** False

8. The posterior roots of spinal nerves contain the axons of the motor neurons.
- **a.** True
- **b.** False

9. The normal patellar reflex response involves extension of the leg at the knee joint.
- **a.** True
- **b.** False

A reflex arc represents the simplest type of nerve pathway found in the nervous system. This pathway begins with a *receptor* at the dendrite end of a *sensory* (*afferent*) *neuron.* The sensory neuron leads into the central nervous system (CNS) (*integration center*) and may communicate with one or more *interneurons.* Some of these interneurons, in turn, communicate with *motor* (*efferent*) *neurons,* whose axons (nerve fibers) lead outward to *effectors.* Thus, when a sensory receptor is stimulated by a change occurring inside or outside the body, impulses may pass through a *reflex arc,* and, as a result, effectors may respond. Such an automatic, subconscious response is called a *reflex.*

Skeletal muscles are involved in *somatic reflexes.* Somatic reflexes include *stretch reflexes* and *withdrawal reflexes.* Some reflexes involve a single synapse (monosynaptic) between a sensory and motor neuron; more complex arcs or circuits involve two or more synapses (polysynaptic) and some interneurons. A reflex arc includes five basic components:

1. **Receptor**—location of the stimulus at some site in the PNS
2. **Sensory neuron**—conducts impulse from receptor to the CNS (brain or spinal cord)
3. **Integration center**—the brain or spinal cord where synapses occur with some interneurons
4. **Motor neuron**—conducts impulse from the integration center to the effector
5. **Effector**—a muscle or a gland that responds with a contraction or a secretion

Skeletal muscles function as the effectors for the somatic stretch reflexes included in this laboratory exercise. Withdrawal reflex examples are described and illustrated in this laboratory exercise, but they will not be demonstrated for safety reasons.

Smooth muscle, cardiac muscle, and glands are the effectors of *autonomic reflexes.* Some examples of autonomic reflexes are included in Laboratory Exercise 36.

A stretch reflex involves a single synapse (monosynaptic) between a sensory and a motor neuron within the gray matter of the spinal cord. A branch of a sensory fiber is also involved in an inhibitory pathway with an interneuron and motor neuron to an antagonist (fig. 29.1). Examples of stretch reflexes include the patellar, calcaneal, biceps, triceps, and plantar reflexes. Other more complex withdrawal reflexes involve interneurons in combination with sensory and motor neurons; thus, they are polysynaptic. Examples of withdrawal reflexes include responses to touching hot objects or stepping on sharp objects (fig. 29.2).

Reflexes demonstrated in this laboratory exercise are stretch reflexes. When a muscle is stretched by a tap over its

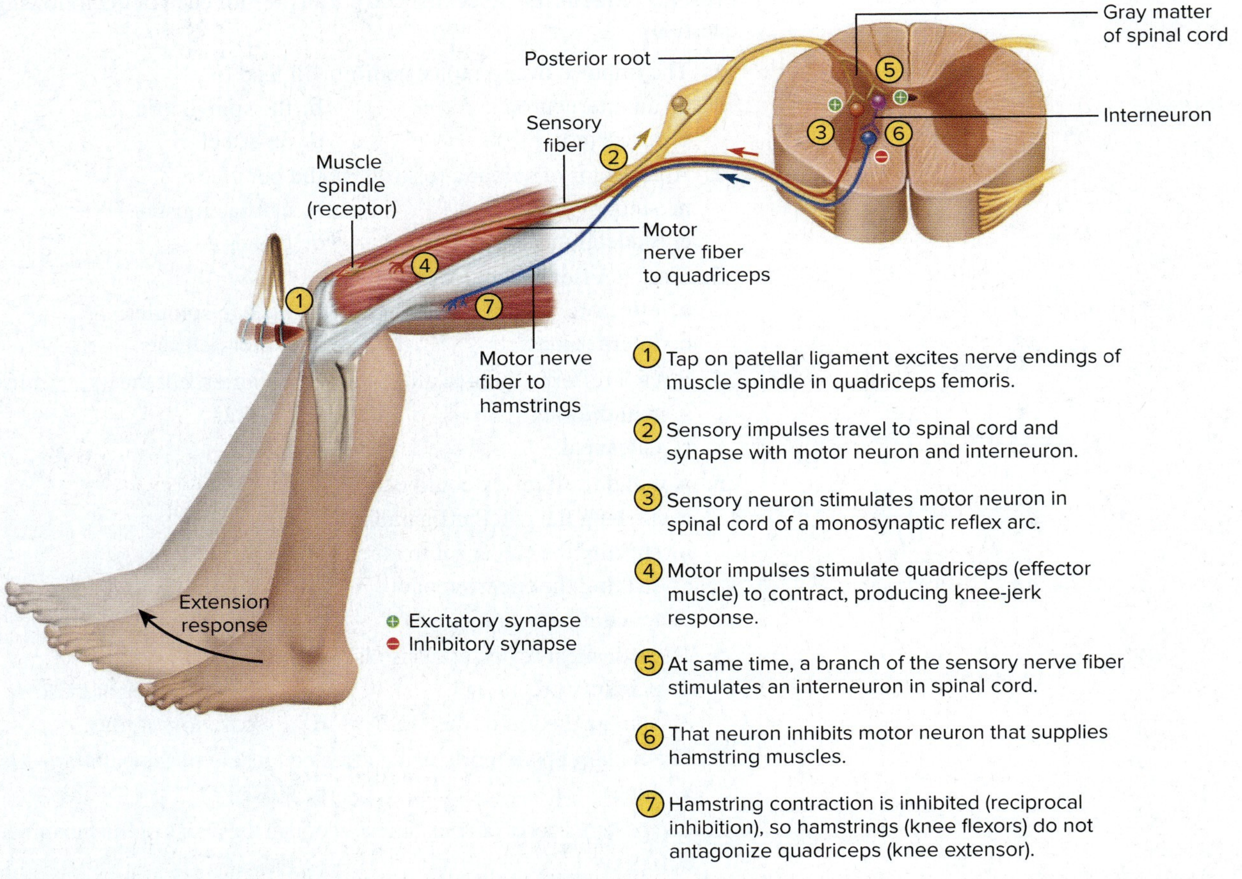

FIGURE 29.1 Diagram and pathway of a stretch (monosynaptic) reflex arc, and the reciprocal inhibition of its antagonistic muscle. The patellar reflex represents a specific example of a stretch reflex. APR

tendon, stretch receptors (proprioceptors) called *muscle spindles* are stretched within the muscle, which initiates an impulse over a reflex arc. A sensory neuron conducts an impulse from the muscle spindle into the gray matter of the spinal cord, where it synapses with a motor neuron, which conducts the impulse to the effector muscle. The stretched muscle responds by contracting to resist or reverse further stretching. These stretch reflexes are important to maintaining proper posture, balance, and movements. For example, if a person is trying to stand up straight and starts to sway, resulting in a potential loss of balance, the stretched muscle will trigger a reflex arc, causing the stretched muscle to contract to correct the stance. Observations of many of these reflexes in clinical tests on patients may indicate damage to a level of the spinal cord or peripheral nerves of the particular reflex arc. Indications of certain diseases can be diagnosed if a reflex is absent, hypoactive (diminished) or hyperactive (exaggerated).

PROCEDURE: Reflex Arc and Somatic Reflexes

1. Study figure 29.1 as an example of a stretch reflex. Compare the stretch reflex with the withdrawal reflex shown in figure 29.2. Examine a model of a reflex arc.
2. Work with a laboratory partner to demonstrate each of the reflexes listed. (See fig. 29.3*a–e* also.) *It is important that muscles involved in the reflexes be totally relaxed to observe proper responses.* If a person is trying too hard

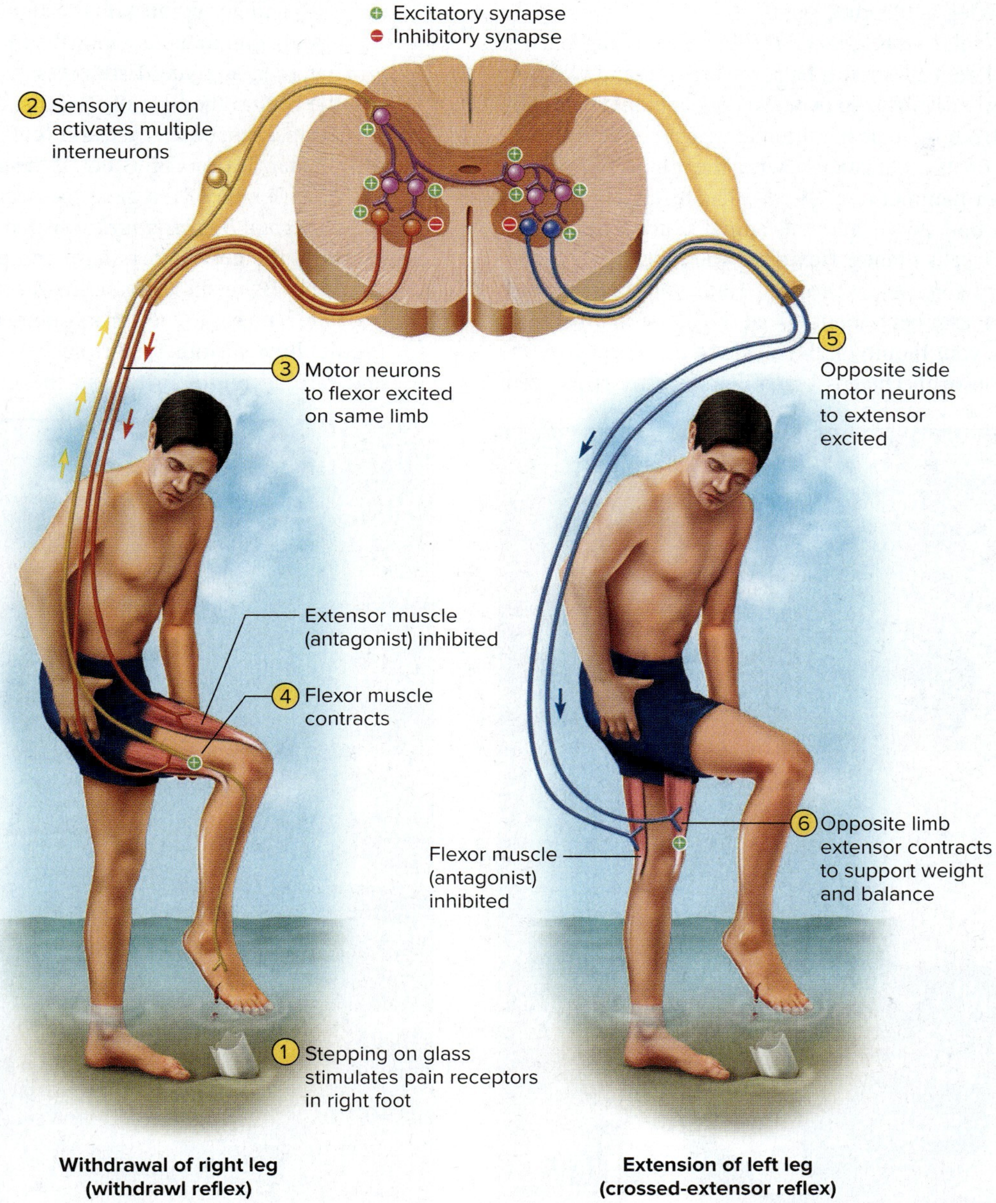

FIGURE 29.2 Polysynaptic withdrawal reflex along with crossed-extensor reflex on opposite limb. Note that when the agonist muscle contracts, the antagonist muscle is inhibited. For safety reasons, withdrawal reflexes are not tested in this laboratory exercise.

to experience the reflex or is trying to suppress the reflex, assign a multitasking activity while the stimulus with the rubber percussion hammer occurs. For example, assign a physical task with upper limbs along with a complex mental activity during the patellar reflex demonstration. After each demonstration, record your observations in the table provided in Part A of Laboratory Assessment 29.

a. *Patellar (knee-jerk) reflex.* Have your laboratory partner sit on a table (or sturdy chair) with legs relaxed and hanging freely over the edge without touching the floor. Gently strike your partner's patellar ligament (just below the patella) with the blunt side of a rubber percussion hammer (fig. 29.3*a*). The reflex arc involves the femoral nerve and the spinal cord. The normal response is a moderate extension of the leg at the knee joint.

b. *Calcaneal (ankle-jerk) reflex.* Have your partner kneel on a chair with back toward you and with feet slightly dorsiflexed over the edge and relaxed. Gently strike the calcaneal tendon (just above its insertion on the calcaneus) with the blunt side of the rubber hammer (fig. 29.3*b*). The reflex arc involves the tibial nerve and the spinal cord. The normal response is plantar flexion of the foot.

c. *Biceps (biceps-jerk) reflex.* Have your partner place a bare arm bent about 90° at the elbow on the table. Press your thumb on the inside of the elbow over the tendon of the biceps brachii, and gently strike your thumb with the rubber hammer (fig. 29.3*c*). The reflex arc involves the musculocutaneous nerve and the spinal cord. Watch the biceps brachii for a response. The response might be a slight twitch of the muscle or flexion of the forearm at the elbow joint.

d. *Triceps (triceps-jerk) reflex.* Have your partner lie supine with an upper limb bent about 90° across the abdomen. Gently strike the tendon of the triceps brachii near its insertion just proximal to the olecranon process at the tip of the elbow (fig. 29.3*d*). The reflex arc involves the radial nerve and the spinal cord. Watch the triceps brachii for a response. The response might be a slight twitch of the muscle or extension of the forearm at the elbow joint.

e. *Plantar reflex.* Have your partner remove a shoe and sock and lie supine with the lateral surface of the foot resting on the table. Draw the metal tip of the rubber hammer, applying firm pressure, over the sole from the heel to the base of the large toe (fig. 29.3*e*). The normal response is flexion (curling) of the toes and plantar flexion of the foot. If the toes spread apart and dorsiflexion of the great toe occurs, the reflex is the abnormal *Babinski reflex* response (normal in infants until the nerve fibers have complete myelinization). If the Babinski reflex occurs later in life, it may indicate damage to the corticospinal tract of the CNS.

3. Predict how various factors such as mental distractions and fatigue might influence the degree of the reflex

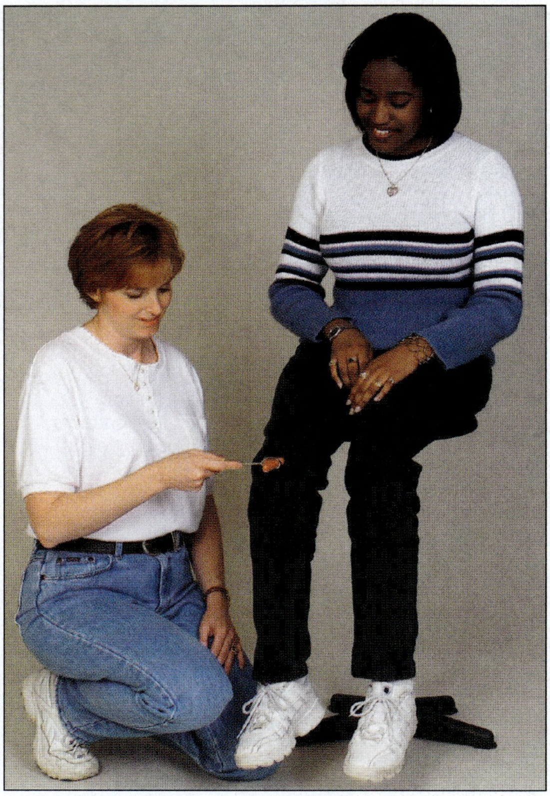
(a) Patellar reflex

(b) Calcaneal reflex

FIGURE 29.3 Demonstrate each of the following reflexes: (a) patellar reflex; (b) calcaneal reflex; (c) biceps reflex; (d) triceps reflex; and (e) plantar reflex. ©J & J Photography reaction. Test the mental distraction factor by repeating each of the reflex demonstrations while someone is purposely distracting the subject. To test the fatigue factor, have your partner exercise and then repeat all of the reflex demonstrations again. Record your observations in the table in Part A of the laboratory assessment.

4. Complete Part B of the laboratory assessment.

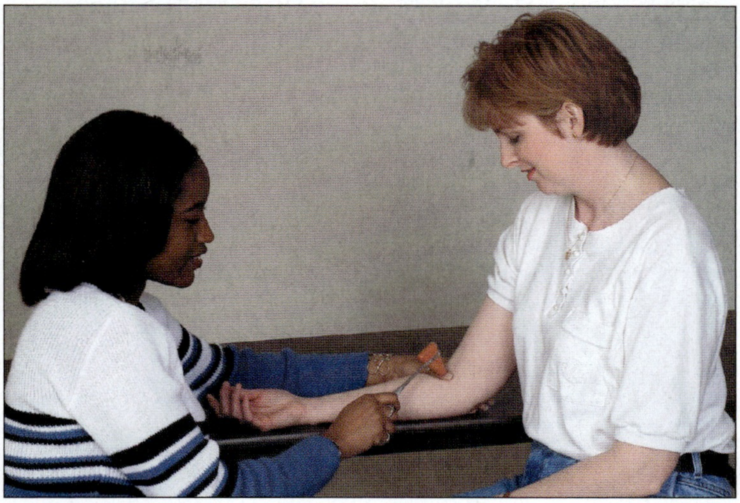

(c) Biceps reflex

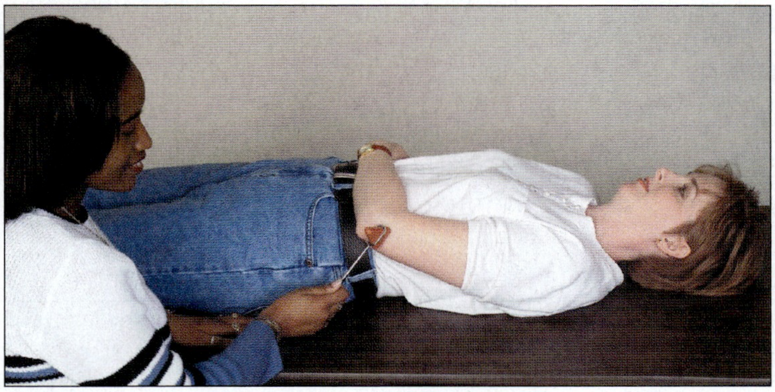

(d) Triceps reflex

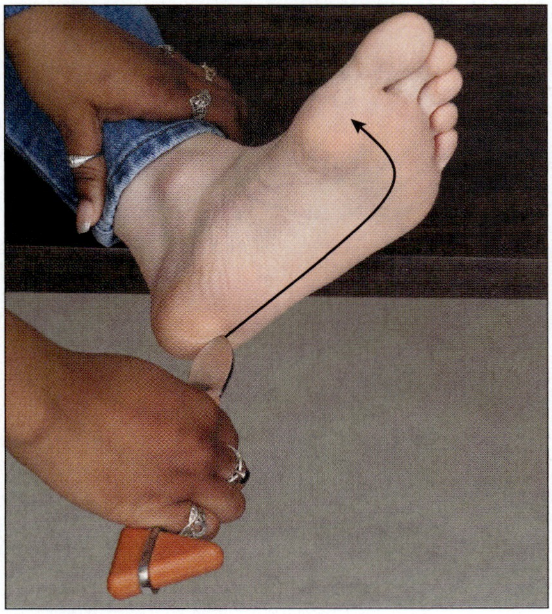

(e) Plantar reflex

FIGURE 29.3 *Continued.* ©J & J Photography

NOTES

LABORATORY
ASSESSMENT 29

Name ______________________________

Date ______________________________

Section ______________________________

The **A** corresponds to the indicated Learning Outcome(s) **O** found at the beginning of the Laboratory Exercise.

Reflex Arc and Somatic Reflexes

PART A: Assessments

1. Complete the following table: **A1**

Reflex Tested	Response Observed	Effector Muscle Involved	Response During Mental Distraction or Fatigue (No Change, More, or Less)
Patellar			
Calcaneal			
Biceps			
Triceps			
Plantar			

2. What conclusions can you make when comparing results of somatic reflexes under typical laboratory conditions to the results under conditions of mental distraction or fatigue? How did these results compare to your original prediction? **A1**

____

____

____

PART B: Assessments

Complete the following statements:

1. A withdrawal reflex employs ________________________ in conjunction with sensory and motor neurons. A2
2. Interneurons in a withdrawal reflex are located in the ________________________. A2
3. A reflex arc begins with the stimulation of a ________________________ at the dendrite end of a sensory neuron. A2
4. Effectors of somatic reflex arcs are ________________________ muscles. A2
5. A monosynaptic reflex arc involves a synapse between a ____________________ neuron and a ____________________ neuron. A2
6. The effector muscle of the patellar reflex is the ________________________. A2
7. The sensory stretch receptors (muscle spindles) of the patellar reflex are located in the ________________________ muscle. A2
8. The posterior root of a spinal nerve contains the ________________________ neurons. A2
9. The normal plantar reflex results in ________________________ of toes. A2
10. Stroking the sole of the foot in infants results in dorsiflexion of the great toe and toes that spread apart, called the ________________________ reflex. A2
11. In a stretch reflex arc, when the agonist muscle contracts, reciprocal inhibition occurs in its ________________________ muscle. A2
12. Synapses, whether excitatory or inhibitory, occur in the ________________________ matter of the spinal cord. A2
13. Effectors of autonomic reflexes include glands, smooth muscle, and ________________________. A2
14. List the major events that occur in the patellar reflex, from the striking of the patellar ligament to the resulting response. A2 A3

____

____

____

CRITICAL THINKING ASSESSMENT—PART B

What characteristics do the reflexes you demonstrated have in common? A3

____

____

LABORATORY EXERCISE 30

Brain and Cranial Nerves

Purpose of the Exercise

To review the structural and functional characteristics of the human brain and cranial nerves.

MATERIALS NEEDED

Dissectible model of the human brain
Human brain ventricles model
Human brain model with cranial nerves
Preserved human brain
Anatomical charts of the human brain

For testing cranial nerves:
Aromatic substances
Snellen eye chart
Pen flashlight
Sugar
Salt
Unsweetened lemon juice
Tuning fork

Learning Outcomes APR

After completing this exercise, you should be able to:

- 01 Identify the locations and functions of the cranial meninges.
- 02 Locate and describe the structural and functional regions of the brain.
- 03 Identify the twelve pairs of cranial nerves.
- 04 Differentiate the functions of each cranial nerve.
- 05 Distinguish the cranial nerve possibly damaged from an observed impaired function.

The O corresponds to the assessments A indicated in the Laboratory Assessment for this Exercise.

Pre-Lab

Carefully read the introductory material and examine the entire lab. Be familiar with the brain and cranial nerves from lecture or the textbook. Answer the pre-lab questions.

Pre-Lab Questions Select the correct answer for each of the following questions:

1. Each hemisphere of the cerebrum regulates
  a. motor functions on the opposite side of the body.
  b. motor functions on the same side of the body.
  c. only functions within the brain.
  d. only functions within the spinal cord.

2. There are _____ pairs of cranial nerves.
  a. 2 **b.** 12
  c. 31 **d.** 43

3. Which of the following is *not* part of the brainstem?
  a. midbrain **b.** pons
  c. medulla oblongata **d.** thalamus

4. The _____ is the deep lobe of the cerebrum.
  a. temporal **b.** insula
  c. parietal **d.** occipital

5. The _____ separates the precentral and postcentral gyrus.
  a. lateral sulcus **b.** parieto-occipital sulcus
  c. central sulcus **d.** longitudinal fissure

6. The falx cerebri extends into the _____ that separates the two cerebral hemispheres.
  a. central sulcus **b.** lateral sulcus
  c. transverse fissure **d.** longitudinal fissure

7. The cranial meninges include the dura mater, arachnoid mater, and pia mater.
  a. True **b.** False

Chapter Opener: ©101cats/Getty Images

8. Ventricles of the brain contain cerebrospinal fluid.
   a. True b. False
9. Functions of the cerebellum include reasoning, memory, and regulation of body temperature.
   a. True b. False

The brain, the largest and most complex part of the nervous system, contains centers associated with sensory perceptions and integration. It issues motor commands to skeletal muscles and carries on higher mental activities. The brain coordinates muscular movements, and it contains centers and neural pathways necessary to regulate the functions of internal organs. The cranial meninges surround the brain and are continuous with the spinal cord meninges.

The cerebral cortex, comprised of gray matter with billions of interneurons, represents areas for conscious awareness and decision-making processes. Sensory areas receive information from various receptors, association areas interpret sensory input, and motor areas involve planning and controlling muscle movements. All of these functional regions are influenced and integrated together in making complex decisions. Each hemisphere primarily interprets sensory and regulates motor functions on the opposite (contralateral) side of the body.

Twelve pairs of cranial nerves arise from the base (inferior surface) of the brain and are designated by number and name. The cranial nerves are part of the peripheral nervous system (PNS); most arise from the brainstem region of the brain. Although most of these nerves conduct both sensory and motor impulses, some contain only sensory fibers associated with special sense organs. Others are primarily composed of motor fibers and are involved with the activities of muscles and glands.

PROCEDURE A: Cranial Meninges and Brain APR

The cranial meninges, as well as the spinal cord meninges, include the *dura mater, arachnoid mater,* and *pia mater.* The thick, opaque, outermost layer, the dura mater, adheres to the inside of the skull. The weblike, semitransparent, middle layer, the arachnoid mater, spans the fissures of the brain. A *subarachnoid space* containing protective *cerebrospinal fluid* (*CSF*) is located between the arachnoid mater and the innermost pia mater and provides chemical stability, buoyancy, and protection. The thin, delicate pia mater adheres tightly to the surfaces of the ridges (*gyri*) and grooves (*sulci*) on the surface of the brain (fig. 30.1).

The superficial dura mater is a single meningeal layer around the spinal cord, but it becomes a double layer in the cranial cavity due to the fusion of the meningeal layer with the periosteal layer of the skull (fig. 30.1). The two layers of the dura mater separate in some cranial areas, forming *dural sinuses.* The blood-filled dural sinuses are modified pathways for collecting venous blood of the brain. The large dural sinus, the *superior sagittal sinus,* is shown in figures 30.1 and 30.2.

In certain areas, the meningeal layer of the dura mater forms *dural septa,* membranous partitions separating various parts of the brain. These partitions function to support and stabilize the brain within the cranial cavity. The largest partition, the *falx cerebri,* extends into the longitudinal fissure separating the cerebral hemispheres (figs. 30.1 and 30.2). Other dural septa include one that separates the cerebellum from the cerebrum (*tentorium cerebelli*) and a partial septum separating the inferior cerebellar hemispheres (*falx cerebelli*) (fig. 30.2).

The cranial subarachnoid space contains CSF as the protective cushion of the brain. The periosteal layer of the

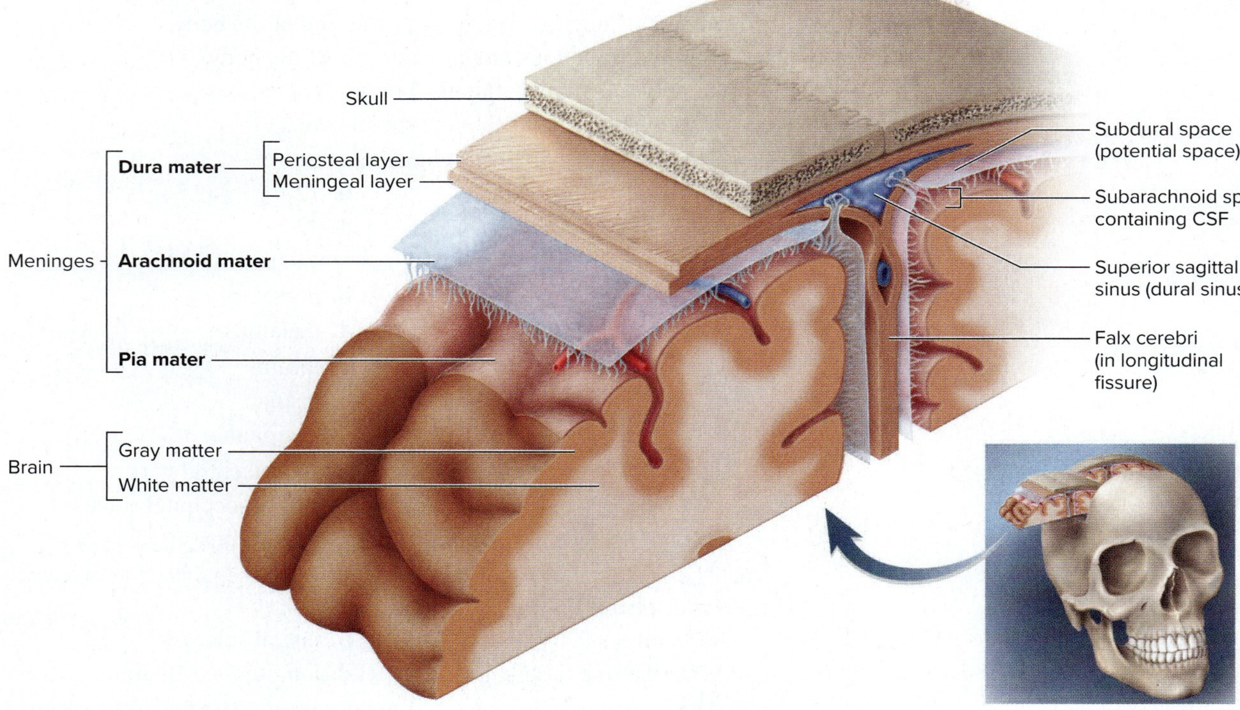

FIGURE 30.1 Cranial meninges associated with the brain. APR

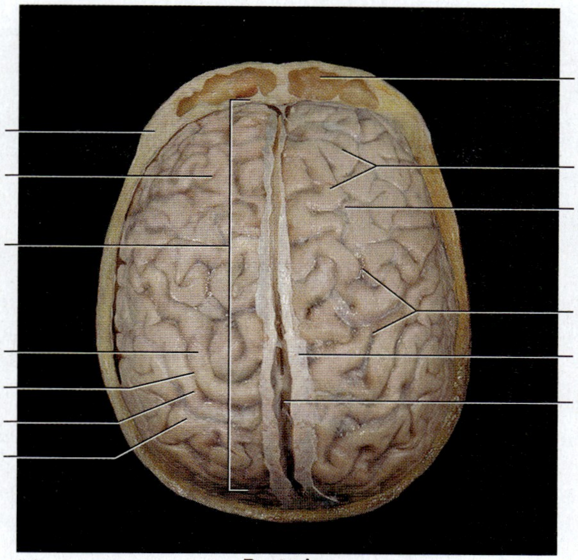

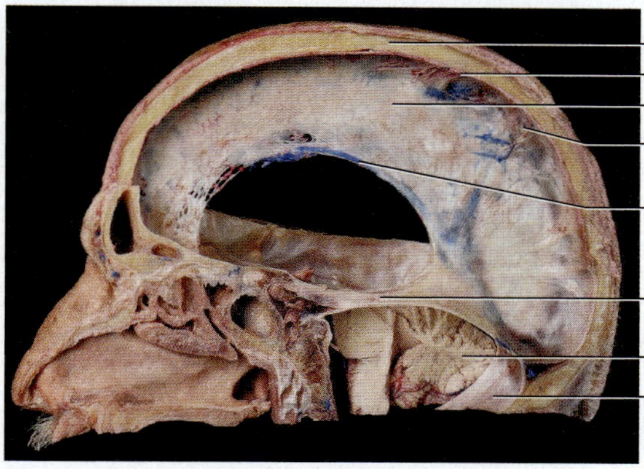

FIGURE 30.2 (a) Superior view of the surface of the brain within the skull of a cadaver. A portion of the dura mater, the outer meningeal layer, forms the falx cerebri within the longitudinal fissure. (b) Midsagittal view of the partitions of the dural septa. (a) APR/McGraw Hill; (b) Christine Eckel/McGraw Hill APR

dura mater adheres tightly to the inside of the skull, and therefore there is no cranial epidural space as seen in the spinal cord. An epidural space and the subdural space are potential spaces in the cranial cavity. Head injuries can result in accumulation of blood in the epidural or subdural potential spaces, referred to as an epidural hematoma or a subdural hematoma. Such head injuries can result in compression on brain tissue and possible death.

The study of the brain will include several regional categories: the ventricles, external surface features, cerebral hemispheres, diencephalon, brainstem, and cerebellum. During each regional study, examine available anatomical charts, dissectible models, and a human brain.

1. Examine figures 30.3 and 30.4 illustrating the ventricles of the brain. The four ventricles contain a clear CSF secreted by blood capillaries named choroid plexuses. The CSF flows from the lateral ventricles into the third ventricle and the fourth ventricle and then into the central canal of the spinal cord and through pores into the subarachnoid space. Locate each of the following features using a human brain ventricles model and dissectible brain models:

   ventricles

   - lateral ventricles—the largest ventricles; one located in each cerebral hemisphere, separated by thin membrane called septum pellucidum
   - third ventricle—located within the diencephalon inferior to the corpus callosum
   - fourth ventricle—located between the pons and the cerebellum

2. Examine figures 30.2 and 30.5 for a study of the external surface features of the brain. Locate the following features using dissectible models and the human brain:

   gyri—elevated surface ridges

   - precentral gyrus
   - postcentral gyrus

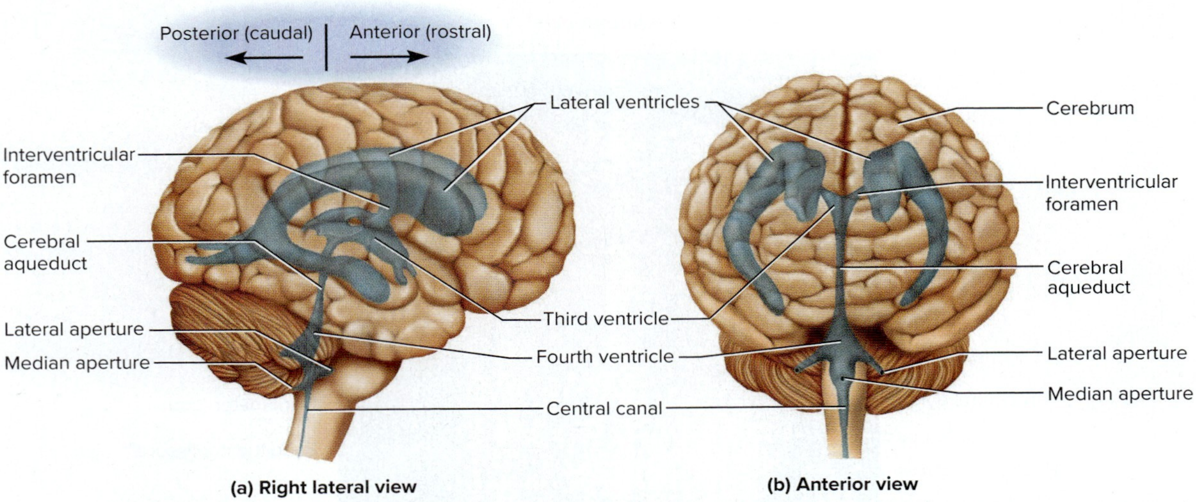

FIGURE 30.3 Four ventricles of the brain and their connections: (a) right lateral view; (b) anterior view. APR

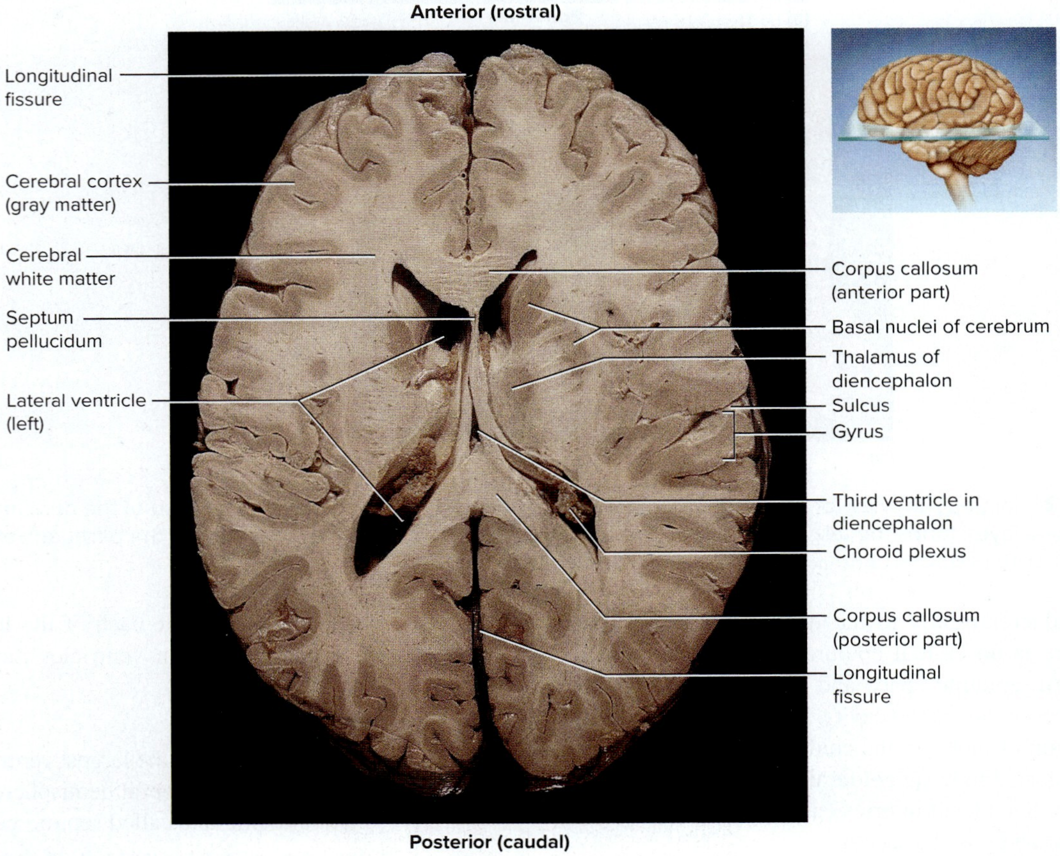

FIGURE 30.4 Transverse section of the human brain showing some ventricles, surface features, and internal structures of the cerebrum (superior view). This section exposes the anterior and posterior horns of both lateral ventricles. Rebecca Gray/Don Kincaid/McGraw Hill

sulci—shallow grooves

- central sulcus—divides the frontal from the parietal lobe
- lateral sulcus (fissure)—divides the temporal from the parietal lobe
- parieto-occipital sulcus—divides the occipital from the parietal lobe

fissures—deep grooves

- longitudinal fissure—separates the cerebral hemispheres; contains falx cerebri, a dural septum
- transverse fissure—separates the cerebrum from the cerebellum; contains tentorium cerebelli, a dural septum

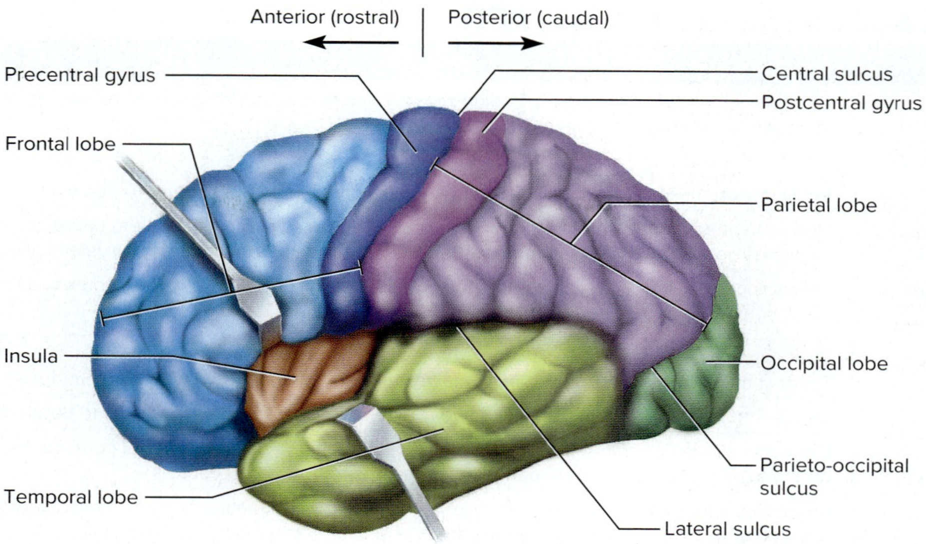

FIGURE 30.5 Lobes of the cerebrum. Retractors are used to expose the deep insula. APR

lobes of cerebrum—names associated with bones of the cranium

- frontal lobe
- parietal lobe
- temporal lobe
- occipital lobe
- insula (insular lobe)—deep within cerebrum; not visible on surface

3. Examine figures 30.2, 30.4 to 30.6, and table 30.1 during the study of the cerebral hemispheres. The cerebrum is the largest part of the brain and has two hemispheres connected by a white matter tract called the corpus callosum. The lobes of the cerebrum were included in the surface features because they are also visible on the surface, except the deep insula. The functions associated with these lobes will be included

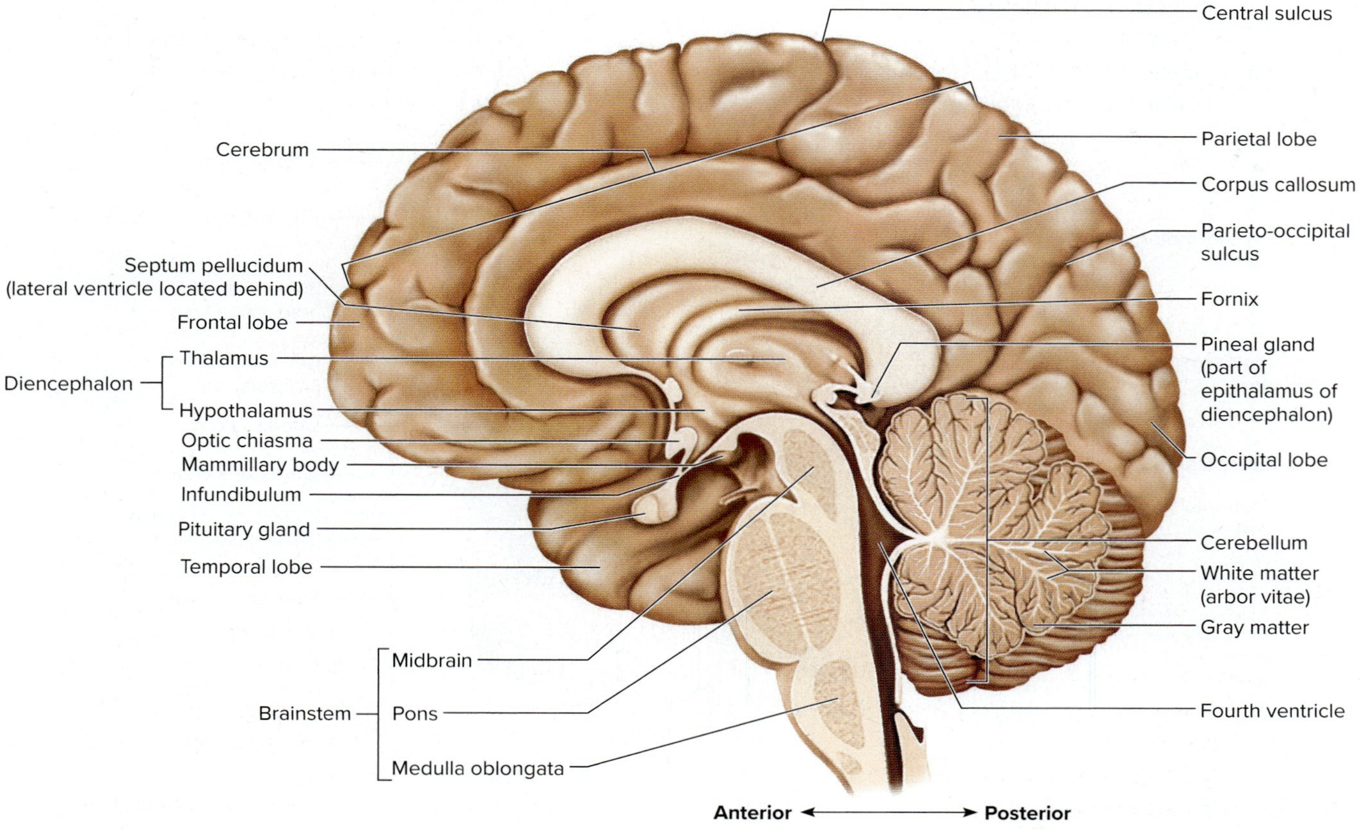

FIGURE 30.6 Diagram of a sagittal (median) section of the brain. APR

Table 30.1 Major Regions and Functions of the Brain APR

Region	Functions
1. Cerebrum	Controls higher brain functions, including sensory perception, storing memory, reasoning, and determining intelligence; initiates voluntary muscle movements
2. Diencephalon	
a. Thalamus	Relay station for sensory impulses ascending from other parts of the nervous system to the cerebral cortex
b. Hypothalamus	Helps maintain homeostasis by regulating visceral activities and by linking the nervous and endocrine systems; regulates body temperature, sleep cycles, emotions, and autonomic nervous system control
c. Epithalamus	Contains pineal gland that secretes the hormone melatonin; contains habenula connecting limbic system and midbrain
3. Brainstem	
a. Midbrain	Contains visual reflex centers that move the eyes and head and auditory reflex centers; contains fiber tracts that connect the pons with the cerebrum; contains aqueduct between third and fourth ventricles where CSF flows
b. Pons	Relays impulses between higher and lower brain regions; helps regulate rate and depth of breathing
c. Medulla oblongata	Conducts ascending and descending impulses between the brain and spinal cord; contains cardiac, vasomotor, and respiratory control centers and various nonvital reflex control centers
4. Cerebellum	Processes information from other parts of the CNS by nerve tracts; integrates sensory information concerning the position of body parts; and coordinates muscle activities and maintains posture

in step 4. Locate the following features using dissectible models and a human brain:

cerebral cortex—thin surface layer of gray matter containing very little myelin; area of conscious awareness and processing information

cerebral white matter—largest portion of the cerebrum containing myelinated nerve fibers; transmits impulses between cerebral areas and lower brain centers; an example is the fornix, a thin tract that connects the hippocampus with other limbic system structures in the diencephalon

basal nuclei—masses of gray matter deep within the white matter; sometimes called basal ganglia as a clinical term; relay motor impulses from the cerebral cortex to the brainstem and spinal cord; the main structures include the caudate nucleus, putamen, and globus pallidus

4. Examine figures 30.5 and 30.7 to compare the structural lobes of the cerebrum with the functional regions of the lobes. Functional areas are not visible as distinct parts on an actual human brain. Examine the labeled areas in figure 30.7 that represent the following functional regions of the cerebrum:

sensory areas

- primary somatosensory cortex—receives information from skin receptors and proprioceptors

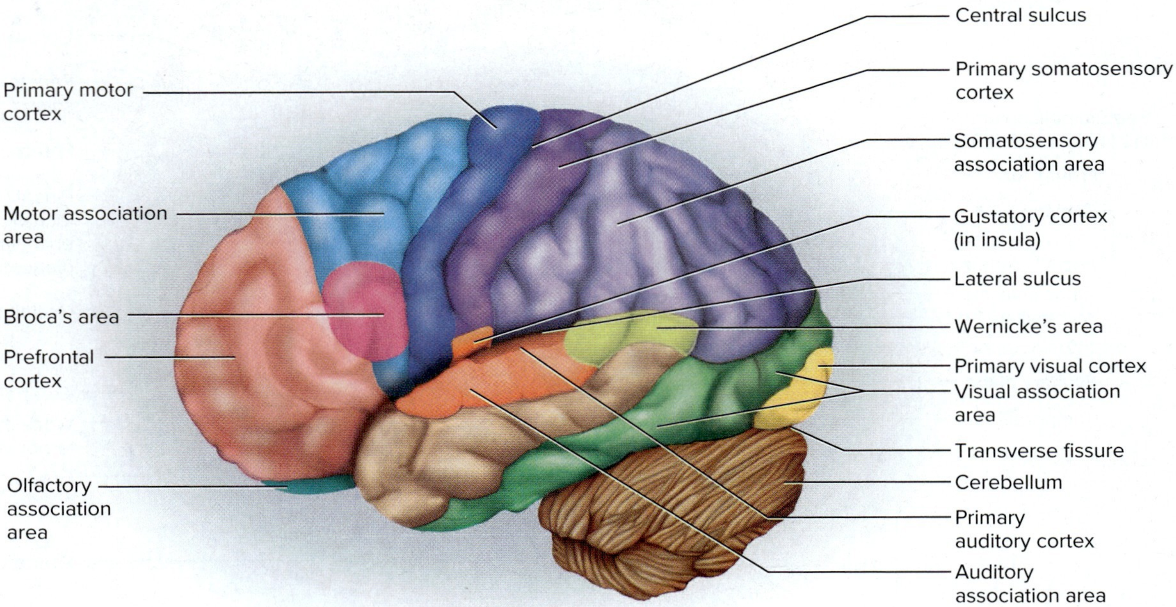

FIGURE 30.7 Some structural and functional areas of the left cerebral hemisphere. (*Note:* These areas are not visible as distinct parts of the brain.)

- somatosensory association cortex—integrates sensory information from primary cortex
- Wernicke's area—recognizes and comprehends spoken and written language; usually within left hemisphere only
- visual areas—consists of a primary and association (interpretation) area for vision
- auditory areas—consist of a primary and association area for hearing
- olfactory association area—interpretation of odors
- gustatory cortex—perceptions of taste

motor areas

- primary motor cortex (somatic motor area)—controls skeletal muscles
- motor association (premotor) area—planning body movements
- Broca's area (motor speech area)—planning speech movements; most often found in left inferolateral portion of frontal lobe

5. Examine figures 30.6 and 30.8 and table 30.1 during the study of the diencephalon, brainstem, and cerebellum. Locate the following features using anatomical charts, dissectible models, and a human brain:

diencephalon

- thalamus—largest portion
- hypothalamus—inferior portion
- optic chiasma (chiasm)—optic nerves meet
- mammillary bodies—pair of small humps; part of limbic system
- pineal gland—formed from epithalamus; secretes hormone melatonin

brainstem

- midbrain—superior region of brainstem
  - cerebral peduncles—connect pons to cerebrum
  - corpora quadrigemina—four bulges [two superior colliculi (visual reflexes) and two inferior colliculi (auditory reflexes)]
  - cerebral aqueduct—CSF flows from third to fourth ventricle
- pons—bulge on underside of brainstem
- medulla oblongata—inferior region of brainstem

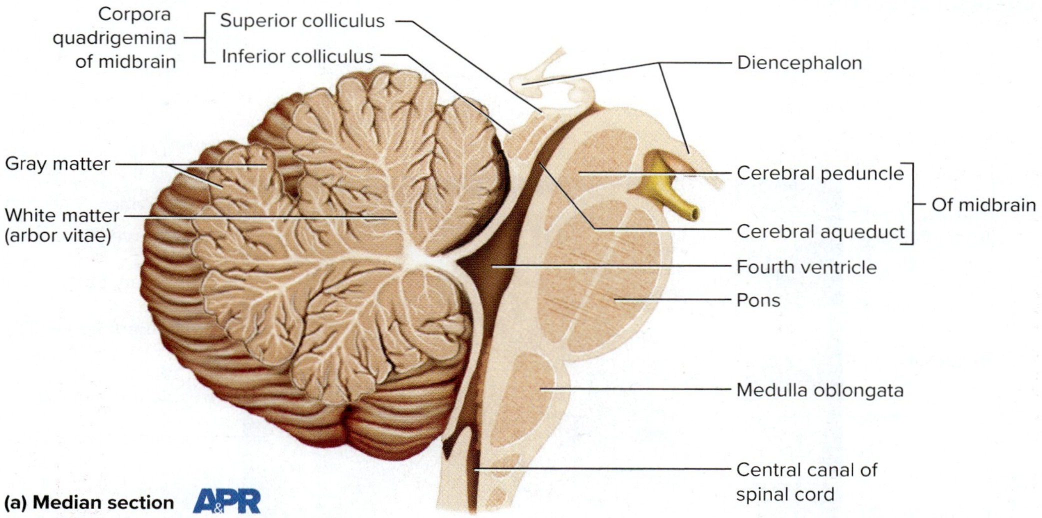

(a) Median section APR

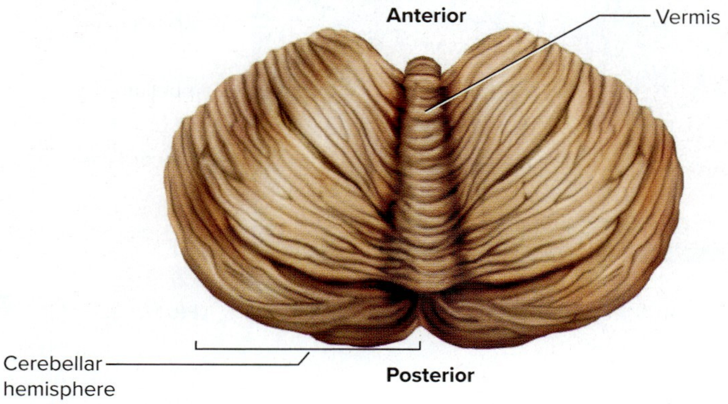

(b) Superior view

FIGURE 30.8 Cerebellum and brainstem: (a) median section; (b) superior view of cerebellum.

cerebellum—cauliflower like appearance

- right and left hemispheres
- vermis—connects the two hemispheres
- cerebellar cortex—gray matter portion
- arbor vitae—deeper branching pattern of white matter

6. Complete Parts A, B, and C of Laboratory Assessment 30.

PROCEDURE B: Cranial Nerves

1. The cranial nerves are part of the PNS. Examine figures 30.9 to 30.12 and table 30.2.
2. Observe the brain model with cranial nerves and preserved specimen of the human brain, and locate as many of the following cranial nerves as possible as you differentiate their associated functions:

olfactory nerves (I)
optic nerves (II)
oculomotor nerves (III)
trochlear nerves (IV)
trigeminal nerves (V)
abducens nerves (VI)
facial nerves (VII)
vestibulocochlear nerves (VIII)
glossopharyngeal nerves (IX)
vagus nerves (X)
accessory nerves (XI)
hypoglossal nerves (XII)

The following mnemonic device will help you learn the twelve pairs of cranial nerves in the proper order:

Old **Op**ie **oc**casionally **tr**ies **trig**onometry,
and **f**eels **ve**ry **glo**omy, **vag**u**e**, **a**nd **hypo**active.[1]

---

3. Examine table 30.3, which contains symptoms of a particular cranial nerve damaged, and the basic clinical laboratory tests to detect the damaged nerve.
4. With a laboratory partner, conduct each of the twelve clinical tests of the cranial nerves indicated in the last column of table 30.3. Summarize your results of the basic clinical tests in Part F of the laboratory assessment.
5. Complete Parts D, E, and F of the laboratory assessment.

[1]From *HAPS-Educator,* Winter 2002. An official publication of the Human Anatomy & Physiology Society (HAPS).

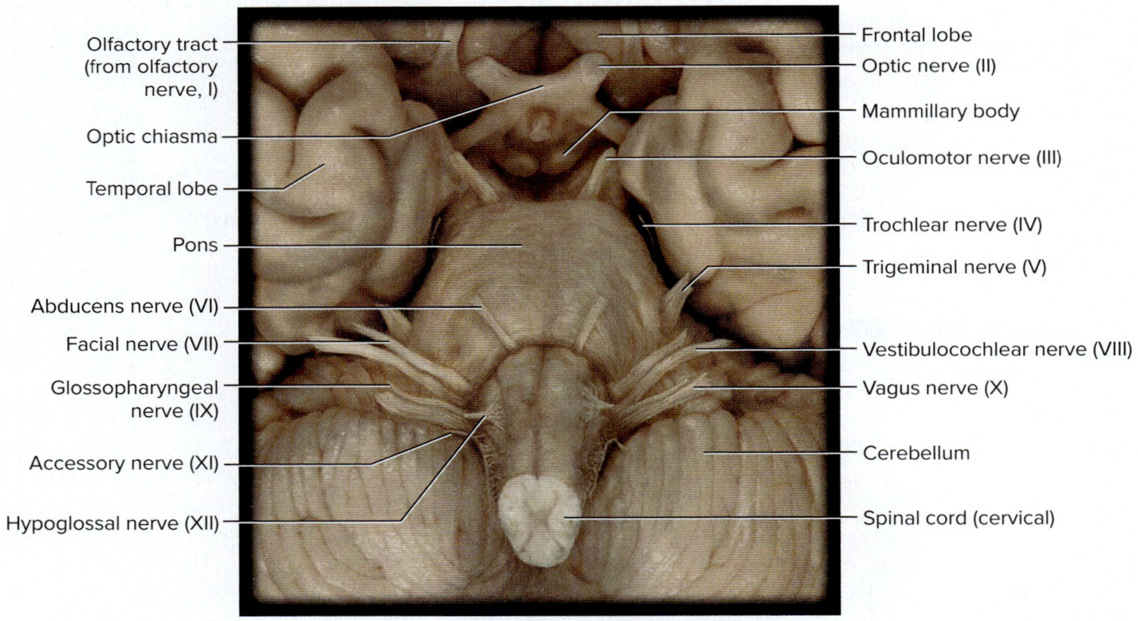

FIGURE 30.9 Photograph of the cranial nerves attached to the base of the human brain. APR/McGraw Hill APR

Table 30.2 The Cranial Nerves and Functions

Number and Name		Type	Function
I	Olfactory	Sensory	Sensory impulses associated with smell
II	Optic	Sensory	Sensory impulses associated with vision
III	Oculomotor	Primarily motor*	Motor impulses to superior, inferior, and medial rectus and inferior oblique muscles that move the eyes, adjust the amount of light entering the eyes (pupil constriction), focus the lenses, and raise the eyelids
IV	Trochlear	Primarily motor*	Motor impulses to superior oblique muscles that move the eyes
V	Trigeminal	Mixed	
	Ophthalmic division		Sensory impulses from the surface of the eyes, tear glands, scalp, forehead, and upper eyelids
	Maxillary division		Sensory impulses from the upper teeth, upper gum, upper lip, lining of the palate, and skin of the face
	Mandibular division		Sensory impulses from the scalp, skin of the jaw, lower teeth, lower gum, and lower lip Motor impulses to muscles of mastication and to muscles in the floor of the mouth
VI	Abducens	Primarily motor*	Motor impulses to lateral rectus muscles that move the eyes laterally
VII	Facial	Mixed	Sensory impulses associated with taste receptors of the anterior tongue Motor impulses to muscles of facial expression, tear glands, and salivary glands
VIII	Vestibulocochlear	Sensory	
	Vestibular branch		Sensory impulses associated with equilibrium
	Cochlear branch		Sensory impulses associated with hearing
IX	Glossopharyngeal	Mixed	Sensory impulses from the pharynx, tonsils, posterior tongue, and carotid arteries Motor impulses to salivary glands and to muscles of the pharynx used in swallowing
X	Vagus	Mixed	Somatic motor impulses to muscles associated with speech and swallowing; autonomic motor impulses to the viscera of the thorax and abdomen Sensory impulses from the pharynx, larynx, esophagus, and viscera of the thorax and abdomen
XI	Accessory	Primarily motor*	
	Cranial branch		Motor impulses to muscles of the soft palate, pharynx, and larynx
	Spinal branch		Motor impulses to muscles of the neck and shoulder
XII	Hypoglossal	Primarily motor*	Motor impulses to muscles that move the tongue

*These nerves contain a small number of sensory impulses from proprioceptors.

Table 30.3 Clinical Tests to Detect Possible Cranial Nerve Damage

Cranial Nerve	Symptoms of Nerve Damage	Basic Clinical Tests to Detect Damaged Cranial Nerve
Olfactory I (fig. 30.10)	Impaired smell or loss of smell	Sniff and determine if aromatic substances such as clove oil, perfume, and coffee are recognized
Optic II	Impaired vision or loss of vision	Read the time on the wall clock; identify the letters on a Snellen eye chart
Oculomotor III	Drooping eyelid; impaired eye movements; dilated pupil	Follow a finger without moving head; use a pen flashlight to gently shine light into the eye to check if pupils constrict
Trochlear IV	Impaired eye movements	Follow a finger and check if eye can be rotated
Trigeminal V (fig. 30.11)	Loss of sensations of touch, hot, cold, and pain in facial areas; impaired jaw movements	Touch the face in areas corresponding to the branches of the nerve to see if touch is detected; open and close mouth and move jaw in various directions
Abducens VI	Loss of lateral eye movements	Follow a finger to test for lateral eye movements
Facial VII (fig. 30.12)	Distorted taste sensations on anterior region of tongue; dry eye and mouth; impaired facial muscle expressions	Place grains of sugar (or salt) on tip of tongue for detection of sweet (or salty) sensation; test ability to smile and whistle
Vestibulocochlear VIII	Loss of balance; loss of hearing	Try to walk a straight line to check balance (vestibular branch); use a tuning fork to test hearing ability (cochlear branch)
Glossopharyngeal IX	Dry mouth; impaired swallowing; impaired taste sensation on posterior tongue	Check for ability to swallow and cough; check posterior tongue with unsweetened lemon juice for detection of sour sensation
Vagus X	Impaired swallowing; difficulty in speaking and experiencing hoarseness	Check for swallowing impairments (cranial nerves IX and X innervate pharynx muscles); check for monotone and hoarseness in voice
Accessory XI	Impaired shoulder and rotating head movements	Examine ability to "shrug" shoulders and turn head against some resistance
Hypoglossal XII	Deviations or inability to stick out tongue; speech and swallowing difficulties	Check for ability to protrude and retract tongue without any deviations in position

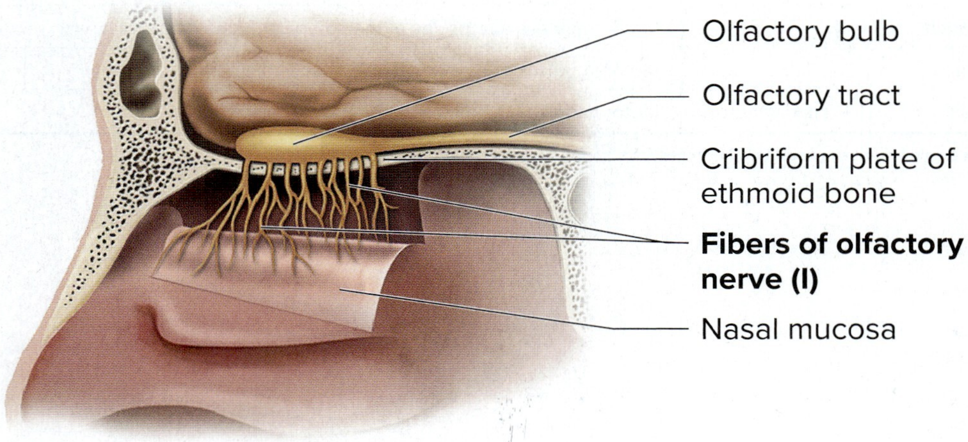

FIGURE 30.10 Olfactory cranial nerve. APR

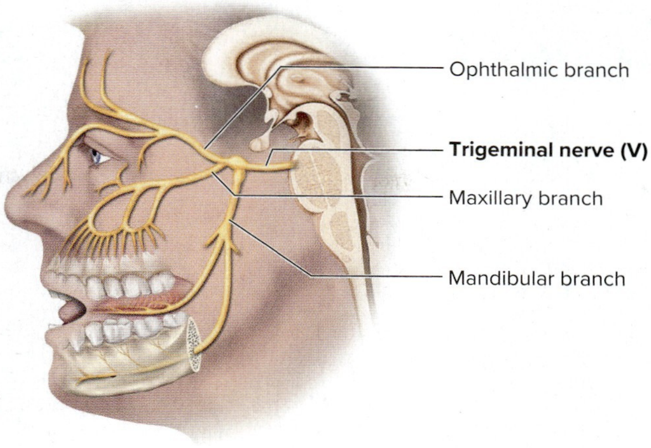

FIGURE 30.11 Trigeminal cranial nerve and its three major branches. APR

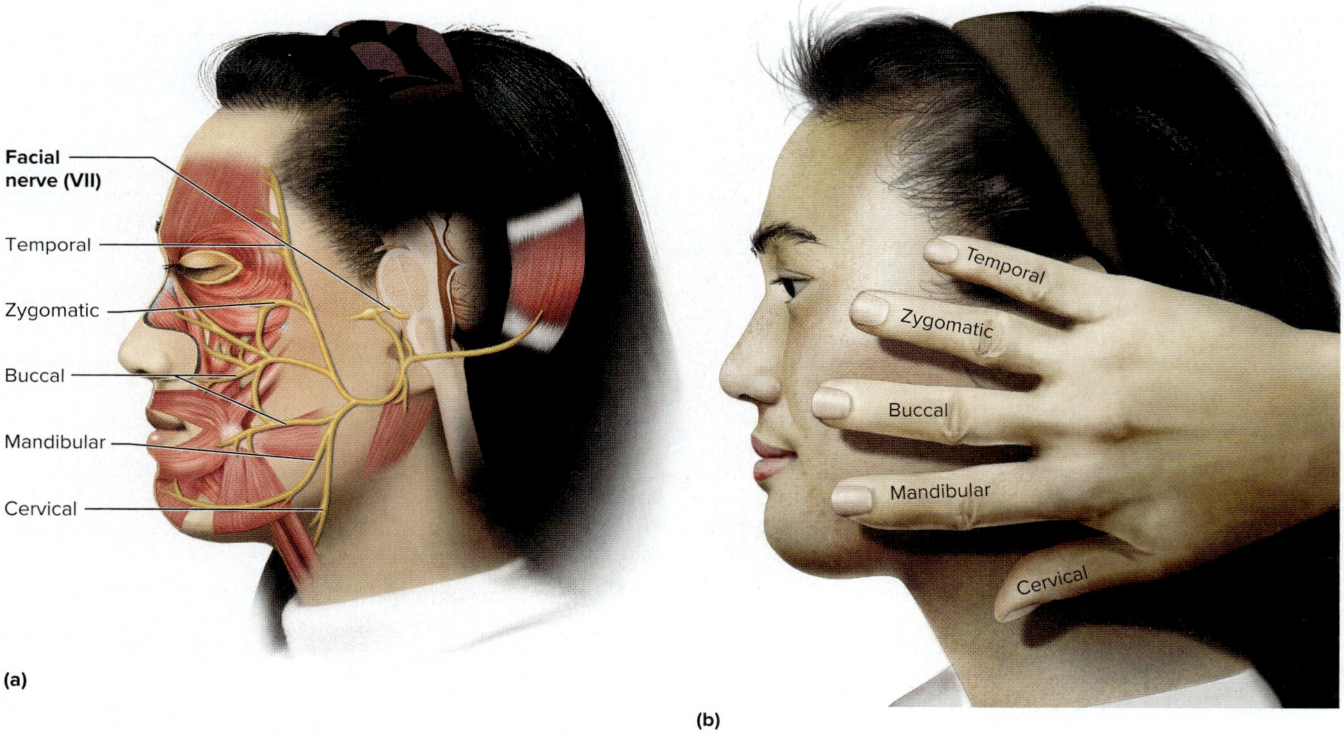

FIGURE 30.12 Facial cranial nerve: (a) illustration of facial nerve and its five major branches; (b) illustration to help remember the facial areas served by the five branches. APR

NOTES

LABORATORY
ASSESSMENT 30

Name ______________________________

Date ______________________________

Section ______________________________

The (A) corresponds to the indicated Learning Outcome(s) (O) found at the beginning of the Laboratory Exercise.

Brain and Cranial Nerves

PART A: Assessments

Match the terms in column A with the descriptions in column B. Place the letter of your choice in the space provided. A1 A2 A3

Column A		Column B
a. Central sulcus	______	**1.** Structure formed by the crossing-over of nerve fibers from optic nerves
b. Cerebral cortex	______	**2.** Part of diencephalon that forms lower walls and floor of third ventricle
c. Corpus callosum	______	**3.** Cone-shaped gland in the upper posterior portion of diencephalon
d. Gyrus	______	**4.** Connects cerebral hemispheres
e. Hypothalamus	______	**5.** Ridge on surface of cerebrum with pia mater closely attached
f. Insula	______	**6.** Separates frontal and parietal lobes
g. Medulla oblongata	______	**7.** Part of brainstem between diencephalon and pons
h. Midbrain	______	**8.** Rounded bulge on underside of brainstem
i. Optic chiasma	______	**9.** Part of brainstem continuous with the spinal cord
j. Pineal gland	______	**10.** Internal brain chamber filled with CSF
k. Pons	______	**11.** Cerebral lobe located deep within lateral sulcus
l. Ventricle	______	**12.** Thin layer of gray matter on surface of cerebrum

PART B: Assessments

Complete the following statements:

1. The ____________________ and the ____________________ together compose the dura mater in the cranial cavity. A1
2. The subarachnoid space contains a protective ____________________. A1
3. The ____________________, a dural septum, is located within the longitudinal fissure between the cerebral hemispheres. A1
4. The superior sagittal sinus collects and contains ____________________. A1
5. The cerebral cortex contains ____________________ matter. A2
6. Grooves on the surface of the brain are sulci; ridges on the surface are ____________________. A2
7. The auditory areas of the brain are part of the ____________________ lobe. A2
8. The vision areas of the brain are part of the ____________________ lobe. A2
9. The left cerebral hemisphere primarily controls the ____________________ side of the body. A2

10. The brainstem includes the pons, the midbrain, and the ________________________. A2

11. The delicate ________________________ membrane is located on the surface of the brain. A1

12. The ________________________ fissure separates the two cerebral hemispheres. A2

13. The primary motor cortex is located within the ________________________ gyrus. A2

14. Arbor vitae and vermis are components of the ________________________. A2

15. The ________________________ ventricle is located between the pons and the cerebellum. A2

16. The ________________________ connects the two hemispheres of the cerebellum. A2

PART C: Assessments

Identify the features indicated in the median section of the human brain in figures 30.13 and 30.14.

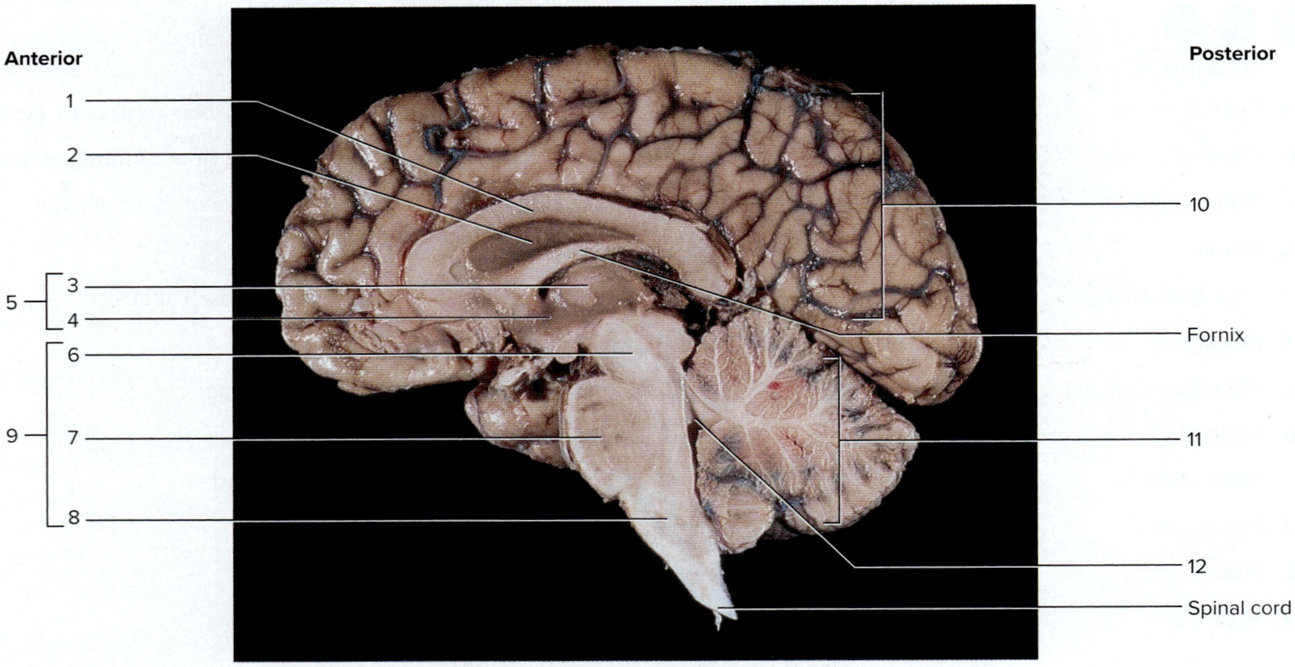

____ Brainstem	____ Diencephalon	____ Midbrain
____ Cerebellum	____ Fourth ventricle	____ Pons
____ Cerebrum	____ Hypothalamus	____ Septum pellucidum
____ Corpus callosum	____ Medulla oblongata	____ Thalamus

FIGURE 30.13 Label the features on this median section of the right half of the human brain by placing the correct numbers in the spaces provided. Rebecca Gray/Dennis Strete/McGraw Hill APR A2

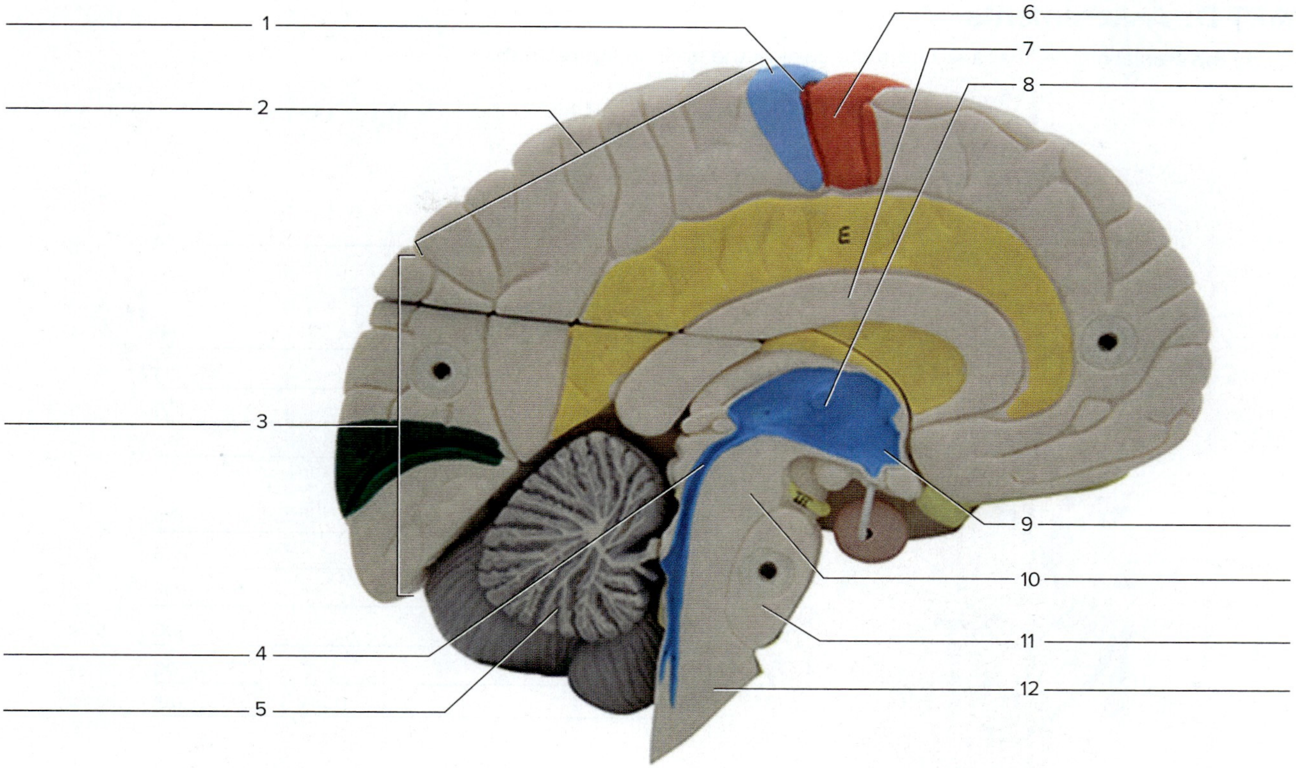

Terms:

Central sulcus	Hypothalamus	Pons
Cerebellum	Medulla oblongata	Precentral gyrus
Cerebral aqueduct	Midbrain	Occipital lobe
Corpus callosum	Parietal lobe	Thalamus

FIGURE 30.14 Label the features on this median section of a model of the left half of the human brain, using the terms provided. ©3B Scientific GmbH, Germany, 2017 www.3bscientific.com

PART D: Assessments

Identify the cranial nerves that arise from the base of the brain in figure 30.15.

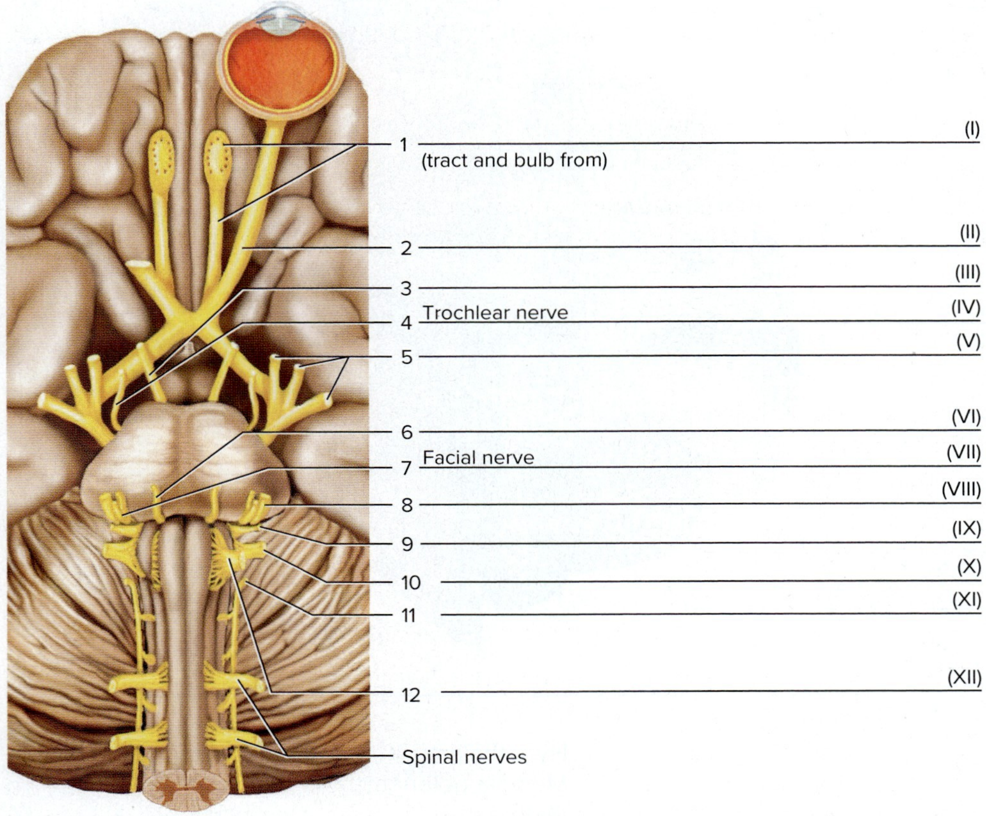

FIGURE 30.15 Complete the labeling of the twelve pairs of cranial nerves as viewed from the base of the brain. The Roman numerals indicated are also often used to reference a cranial nerve. APR A3

PART E: Assessments

Match the cranial nerves in column A with the associated functions in column B. Place the letter of your choice in the space provided. A4

Column A	Column B
a. Abducens	______ **1.** Regulates thoracic and abdominal viscera
b. Accessory	______ **2.** Equilibrium and hearing
c. Facial	______ **3.** Stimulates superior oblique muscle of eye
d. Glossopharyngeal	______ **4.** Sensory impulses from teeth and face
e. Hypoglossal	______ **5.** Adjusts light entering eyes and eyelid opening
f. Oculomotor	______ **6.** Smell
g. Olfactory	______ **7.** Controls neck and shoulder movements
h. Optic	______ **8.** Controls tongue movements
i. Trigeminal	______ **9.** Vision
j. Trochlear	______ **10.** Stimulates lateral rectus muscle of eye
k. Vagus	______ **11.** Sensory from anterior tongue and controls salivation and secretion of tears
l. Vestibulocochlear	______ **12.** Sensory from posterior tongue and controls salivation and swallowing

PART F: Assessments

1. Summarize the results of the basic clinical tests performed to detect any possible damage to a particular cranial nerve. A4 A5

   Name the cranial nerves that indicated normal functional results based upon the clinical tests performed. ______________
   ____
   ____
   ____

   Name any of the cranial nerves that had indications of impaired functional results based upon the clinical tests performed, and describe the specific impaired function observed. ______________
   ____
   ____
   ____

2. Place the number and name of the cranial nerve best associated with each of the following normal abilities or impaired functions. All twelve cranial nerves are represented for the answers. A4 A5

   ______________ **a.** Experiences motion sickness and seems intoxicated
   ______________ **b.** Unable to rotate eyeball inferolaterally
   ______________ **c.** Able to depress shoulder joint, but unable to elevate shoulder
   ______________ **d.** Identifies tissues using the compound microscope
   ______________ **e.** Experiences difficulty with tongue movements when talking and swallowing
   ______________ **f.** Abducts the eyeballs
   ______________ **g.** Detects the odor of burning fall leaves
   ______________ **h.** Detects taste sensation of a medicine placed on the back of the tongue
   ______________ **i.** Moves jaw from side to side and chews food
   ______________ **j.** One pupil of an eye has a different shape and size
   ______________ **k.** Hoarseness experienced in voice and difficulty speaking
   ______________ **l.** Experiences facial muscle paralysis and sagging corner of mouth (Bell palsy)

NOTES

LABORATORY EXERCISE 31A

Reaction Time: BIOPAC® Exercise

Purpose of the Exercise

To measure reaction time in response to random-interval and fixed-interval stimuli using the Biopac Student Lab system, to observe differences in reaction time between the dominant and nondominant hands, and to conduct a brief statistical analysis of the reaction times of a group of subjects.

MATERIALS NEEDED

Computer system [Mac OS X 10.7–10.15 (Catalina), or PC running Windows 10, 8, or 7]

Biopac Student Lab software, version 4.0 or above

BIOPAC Data Acquisition Unit MP36, MP35

BIOPAC Hand Switch (SS10L)

BIOPAC Headphones (OUT1 or OUT1A)
Note: OUT1A is compatible only with the MP36

SAFETY

- The BIOPAC system is safe and easy to use. There are no special safety considerations in this lab exercise, but subjects should wash their hands before and after holding the hand switch, to reduce the chances of spreading infection.

Learning Outcomes

After completing this exercise, you should be able to:

O1 Explain the effects of learning on reaction time in response to random-interval and fixed-interval stimuli.

O2 Compare the reaction times between the dominant and nondominant hands.

O3 Explain individual differences in reaction time.

O4 Analyze the reaction times of a group of subjects through a series of statistical calculations of the group mean, variance, and standard deviation.

The O corresponds to the assessments A indicated in the Laboratory Assessment for this Exercise.

Pre-Lab

Carefully read the introductory material and examine the entire lab. Be familiar with the structure and function of synapses, neuromuscular junctions, and neural pathways from lecture or the textbook. Also, if this is the first BIOPAC lab that you are performing, please read the general description of the Biopac Student Lab system in Exercise 21.

Chapter Opener: ©101cats/Getty Images

In order to maintain homeostasis, we have to be able to respond to changes in our internal and external environments. Some of these responses are involuntary (automatic, not under conscious control), while others are voluntary (willed to occur, under conscious control). For example, if a person walks out of a warm house on a cold morning, the cold temperature will be the stimulus for both involuntary and voluntary responses. Involuntary (reflex) responses include the redistribution of blood flow, with more blood being sent toward the core of the body, inactivation of the sweat glands, and perhaps shivering. Voluntary reactions to the same stimulus may include zipping up your jacket, putting on warmer clothes, or simply going back into the house.

The neural pathways for voluntary and involuntary responses are similar, but not identical. Figure 31A.1 illustrates the basic pathway for a neural response to a stimulus.

Both voluntary and involuntary responses require stimulation of specific receptors, the transmission of nerve impulses over a series of neurons (one or more in the central nervous system), and the activation of effectors that carry out the response. However, involuntary reflex responses occur more quickly than voluntary reactions; they involve fewer neurons and do not require input from the cerebral cortex. Therefore, the reaction time for involuntary reflex responses is shorter than that of voluntary reactions.

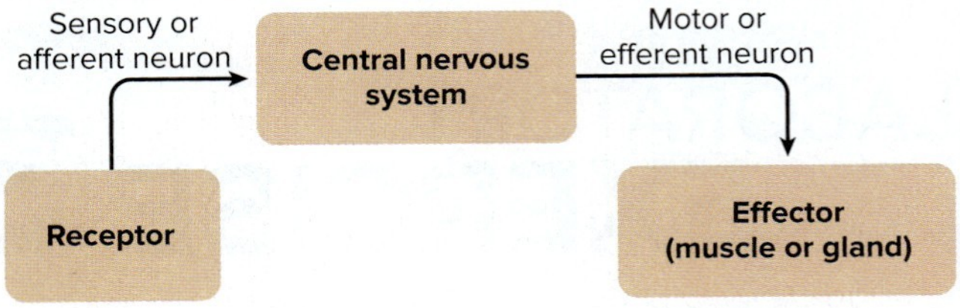

FIGURE 31A.1 A general neural pathway for a voluntary or involuntary response to a stimulus.

The neural pathway leading to an involuntary reflex response is called a *reflex arc*. Figure 31A.2 shows the components of a spinal reflex arc:

1. *Sensory receptor,* which detects a specific type of change or stimulus
2. *Sensory (afferent) neuron,* which transmits nerve impulses from a receptor into the central nervous system (CNS), in this case, the spinal cord
3. *Interneuron (association neuron),* which integrates and coordinates the information received from sensory neurons (present in some types of reflex arcs, but not others; the processing is done in the CNS, whether or not an interneuron is present)
4. *Motor (efferent) neuron,* which transmits a motor command from an interneuron (or sensory neuron, in some cases) to an effector
5. *Effector,* a muscle or gland that carries out the reflex response, according to the instructions from a motor neuron

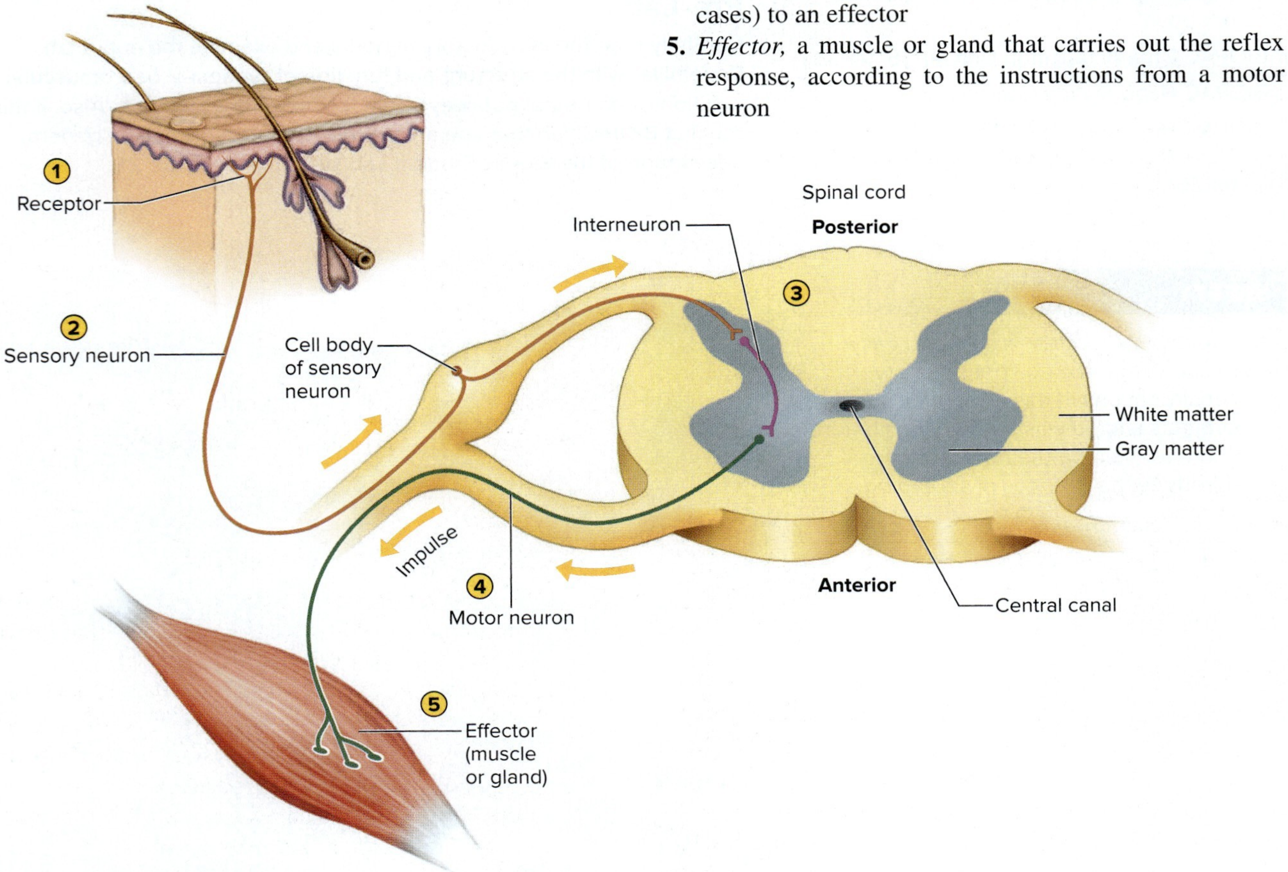

FIGURE 31A.2 The components of a spinal reflex arc (involuntary).

The response to laying a hand on the surface of a hot stove is conducted through a reflex arc such as the one shown in figure 31A.2. Since it causes a person to withdraw a limb from a painful stimulus, it is called a *withdrawal reflex*. When a hand is placed on the hot stove, several types of receptors are stimulated: warm receptors, pain receptors, touch receptors, and pressure receptors, but the pain receptors elicit the reflex response. Each receptor is connected to a sensory neuron, which transmits sensory information to an interneuron in the spinal cord. The interneuron coordinates the information from the various pain receptors and formulates a motor command, which it sends to a motor neuron. The motor neuron then sends the command to the effectors of the reflex, which in this case are flexor muscles of the arm. As a result of muscle contraction, the hand is lifted away from the painful stimulus, reducing or eliminating tissue damage. Although the brain is made aware of the reflex response, it is not a necessary component of the reflex arc.

In a voluntary reaction, however, a person chooses to respond in a certain way to a particular stimulus. This laboratory exercise is an example of a voluntary reaction to a stimulus. In order to measure reaction time, subjects are asked to hold a hand switch, and press a button in response to hearing a click. Since this type of reaction requires a conscious choice and involvement of the brain, its neural pathway is more complex than the reflex arc of an involuntary response. The neural pathway for the voluntary reaction to an auditory stimulus, which is the basis for this lab exercise, is shown in the following flowchart: Auditory stimulus (click) → sensory (hearing) receptors in inner ear → sensory neurons in vestibulocochlear nerve → interneurons in brainstem → interneurons in thalamus → interneurons in auditory cortex of cerebrum → descending pathways to spinal cord → motor neurons to proper muscles to move the thumb → pressing of hand switch.

In this lab exercise, you will test your reaction time by listening for clicks through headphones, and pressing a button on a hand switch as soon as possible after hearing each click. Reaction times will be tested in your dominant and nondominant hands, in response to random-interval and fixed-interval stimuli. You will also gather data from other students in your class and observe individual differences, and then determine data variability through calculation of variance and standard deviation. The BIOPAC program will record the reaction times and print them in the Journal area below the data window. You will then be guided through the statistical calculations.

PROCEDURE A: Setup

1. Before starting the lab exercise, you should designate a **Subject** to perform the experiment, a **Director,** who will inform the subject of each step in the experiment, and a **Recorder,** who will run the computer.
2. With your computer turned **ON** and the BIOPAC MP35/36 unit turned **OFF,** plug the hand switch (SS10L) into Channel 1 and the headphones (OUT1 or OUT1A) into the back of the Data Acquisition Unit. (*Note:* The OUT1A headphones can only be used with the MP36.)
3. Turn on the MP36/35.
4. Start the Biopac Student Lab program, and click on **L11-Reaction Time I;** then click **OK.**
5. Designate a filename for the subject's data, and click **OK.**

PROCEDURE B: Calibration

1. Have the subject get ready for the calibration recording by sitting in a chair in a relaxed position, with eyes closed. The subject should put the headphones on and hold the hand switch upright in the dominant hand, with the thumb ready to press the button on top of the hand switch.
2. Click **Calibrate.** The calibration recording will last for 8 seconds and will stop automatically. About 4–5 seconds into the calibration, the subject will hear a click through the headphones. The subject should firmly press the button on the hand switch as soon as possible after hearing the click, and then quickly release it.
3. After the calibration stops, check the recording. If it looks similar to figure 31A.3, proceed to the data recording section, by clicking **Continue.** If not, check all connections, and then click **Redo Calibration.**

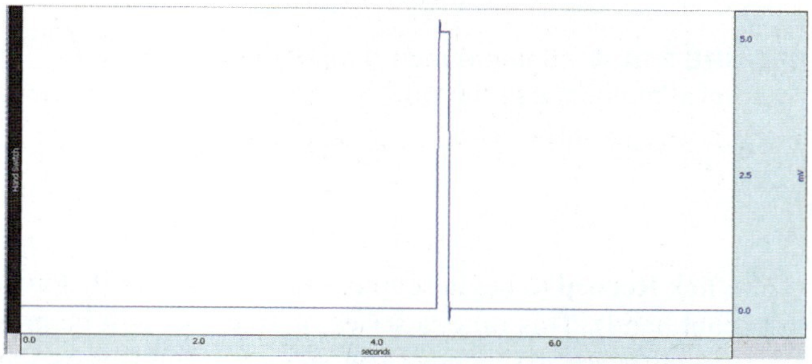

FIGURE 31A.3 Sample calibration recording.

Courtesy of and © *BIOPAC* Systems, Inc. www.biopac.com

PROCEDURE C: Recording

1. The subject should remain seated and relaxed, with eyes closed, holding the hand switch upright in the dominant hand.
2. All members of the group should read and become familiar with the procedure, before starting to record data. The recording portion of lab exercise 31A consists of four segments, in the following order:
   A. Stimuli (clicks) presented at random intervals (1–10 seconds); dominant hand
   B. Stimuli presented at fixed intervals (every 4 seconds); dominant hand

C. Stimuli presented at random intervals (1–10 seconds); nondominant hand

D. Stimuli presented at fixed intervals (every 4 seconds); nondominant hand

3. Click **Record** to begin segment A (random-interval, dominant hand). A series of 10 stimuli (clicks) will be presented to the subject at intervals ranging from 1 to 10 seconds. The subject should press the button on the hand switch as soon as possible after hearing each click, and then release the button quickly.
4. The recording will stop automatically after 10 clicks. The data window should look similar to figure 31A.4, with a reaction pulse displayed after each event marker (a triangle at the top of the data window). If similar, click **Continue** to proceed to step 5. If not, click **Redo.** In some cases, the subject may not have pressed the button hard enough for the system to record a pulse. If no response has been recorded for one or more clicks, you should redo the segment.

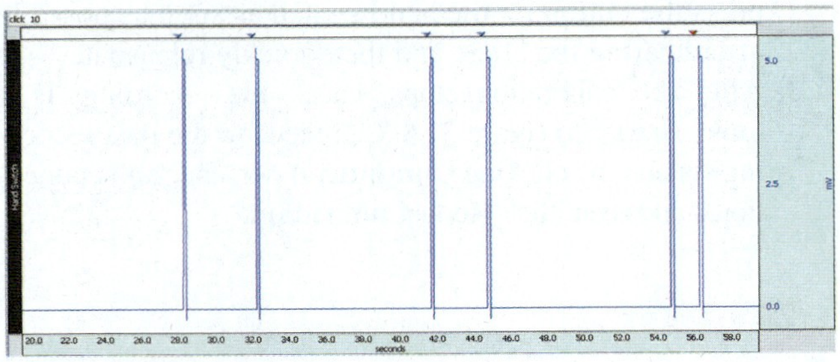

FIGURE 31A.4 Sample recording of data for random-interval stimulus presentation.

Courtesy of and © *BIOPAC* Systems, Inc. www.biopac.com

5. Click **Record** to begin segment B (fixed-interval, dominant hand). This time, a series of 10 clicks will be presented every 4 seconds. The subject should press the button on the hand switch as in step 3.
6. The recording will stop automatically after 10 clicks. The data window should display a reaction pulse for each of the 10 stimuli, and it should look similar to figure 31A.4. If it does, click **Continue** to proceed to step 7. If not, click **Redo.**
7. The subject should now transfer the hand switch to the nondominant hand.
8. Click **Record** to begin segment C (random-interval, nondominant hand). Press the button on the hand switch as above. Click **Redo** if necessary; otherwise, click **Continue** to proceed to step 9.
9. Click **Record** to begin segment D (fixed-interval, nondominant hand). Again, press the button on the hand switch as above. Click **Redo** if necessary; otherwise click **Done** to end the recording portion of the lab exercise.
10. A pop-up window will now appear with several options. Click either **Analyze current data file,** if you are going to conduct your data analysis now, or **Quit,** if you are going to analyze the file at a later time. Then click **OK.** The data from the lab exercise will be saved by the BIOPAC program.
11. When you are sure that you have obtained complete data for the lab exercise, you can unplug the hand switch and headphones.

PROCEDURE D: Data Analysis

1. You can either **Analyze the current data file** now, or conduct the analysis later, by using the **Review Saved Data** mode. When you begin the data analysis, the data window should look similar to figure 31A.5.

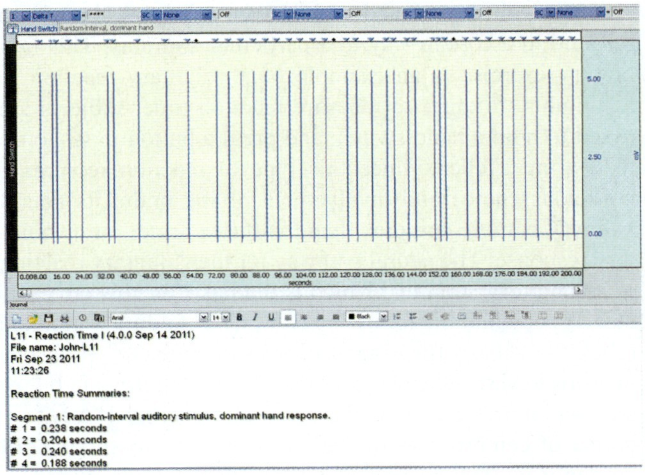

FIGURE 31A.5 Sample reaction time data and journal.

Courtesy of and © *BIOPAC* Systems, Inc. www.biopac.com

2. Note the channel number designation: CH 1—hand switch.
3. Across the top of the recording, you will see three boxes for each channel: channel number, measurement type, and result. Note the setting for the measurement box: CH 1 should be set on **Delta T,** which measures the change in time between the beginning and end of the selected area.
4. Set up your data window so that you are only viewing the first event marker and the first reaction pulse of Segment A. You can spread out the data, or "zoom in" on the first pulse, by clicking anywhere in the time line at the bottom of the data window. This will bring up a text box, in which you can choose the "start" and "end" times to display in the window. After choosing these times (perhaps from 5 to 8 seconds), click **OK.**

5. Your data window should now look similar to figure 31A.6. This view will be used to manually calculate the reaction time between the first event marker and the first reaction pulse. Using the I-beam cursor, select the area between the first event marker and the beginning of the first reaction pulse. See figure 31A.7 for proper selection of the area. The event marker represents the stimulus (the click), and the beginning of the reaction pulse represents the point at which the button was pressed. Record this reaction time (**Delta T**) measurement in question 1 in Part A of Laboratory Assessment 31A.

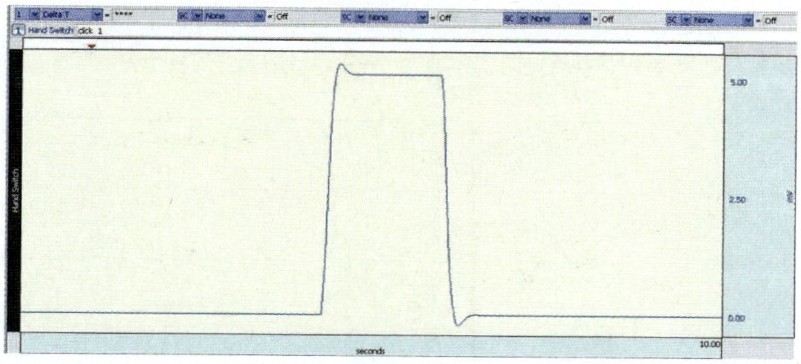

FIGURE 31A.6 Data window displaying the first event marker and reaction pulse.

Courtesy of and © *BIOPAC* Systems, Inc. www.biopac.com

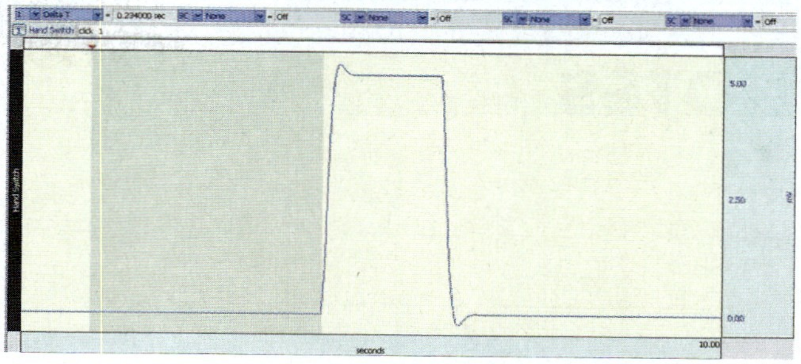

FIGURE 31A.7 Selection of the area for manual calculation of reaction time.

Courtesy of and © *BIOPAC* Systems, Inc. www.biopac.com

6. Compare the manually calculated reaction time (**Delta T**) to the first reaction time shown in the journal. The two numbers should be close. This manual calculation is conducted in order for you to confirm that the journal results are correct.
7. Copy the data from the journal into Table 31A.1 in the laboratory assessment. *Note:* The reaction times provided in the journal are shown in seconds. When you enter the data into the table, enter them in milliseconds (msec). For example, 0.215 seconds = 215 msec.
8. Then proceed to fill out the rest of the tables, following the directions provided in the laboratory assessment.

NOTES

LABORATORY
ASSESSMENT 31A

Name ______________________

Date ______________________

Section ______________________

The (A) corresponds to the indicated Learning Outcome(s) (O) found at the beginning of the Laboratory Exercise.

Reaction Time: BIOPAC® Exercise

Subject Profile

Name ______________ Height ______________

Age ______________ Weight ______________

Gender: Male/Female

PART A: Data and Calculations Assessments

1. Manual Calculation of Reaction Time for the First Stimulus

   Measure the reaction time for the first stimulus of the lab exercise. **Delta T =** __________ msec

2. Reaction Time Results for One Subject (copy out of journal)

Table 31A.1 Reaction Times for One Subject A1 A2

Stimulus Number	REACTION TIMES (MSEC)			
	Dominant Hand		Nondominant Hand	
	Random-Interval	Fixed-Interval	Random-Interval	Fixed-Interval
1				
2				
3				
4				
5				
6				
7				
8				
9				
10				
Mean				

3. Comparison of Reaction Time to Stimulus Presentation Number in Five Subjects

Copy the appropriate data for your subject from Table 31A.1 into line 1 of Table 31A.2. This table will be filled out with data from segments A (random-interval, dominant hand) and B (fixed-interval, dominant hand) of the lab exercise. Next, obtain corresponding data from four other subjects to complete the table. Then calculate the means for each column of data.

Table 31A.2 Reaction Times by Stimulus Presentation Number A1 A2

Name of Subject	RANDOM-INTERVAL REACTION TIMES (MSEC) (DOMINANT HAND)			FIXED-INTERVAL REACTION TIMES (MSEC) (DOMINANT HAND)		
	Stimulus #1	Stimulus #5	Stimulus #10	Stimulus #1	Stimulus #5	Stimulus #10
1.						
2.						
3.						
4.						
5.						
Mean						

4. Group Summary of Mean Reaction Times in Five Subjects

Fill in line 1 of Table 31A.3 with your subject's mean reaction times from the last row of Table 31A.1 (*Note:* The data in Table 31A.1 are not in the same order as the data in Table 31A.3; be careful to copy the data into the correct column.) Next, obtain data from the same four other subjects to complete the table. Then calculate the means for each column of data.

Table 31A.3 Mean Reaction Times for a Group of Five Subjects A1 A2

Name of Subject	RANDOM-INTERVAL REACTION TIMES (MSEC)		FIXED-INTERVAL REACTION TIMES (MSEC)	
	Dominant Hand	Nondominant Hand	Dominant Hand	Nondominant Hand
1.				
2.				
3.				
4.				
5.				
Mean				

5. Variance and Standard Deviation Calculations

The formulas for variance and standard deviation follow. You will be able to complete these statistical calculations by following the directions in Tables 31A.4 and 31A.5.

$$Variance = \frac{1}{n-1} \sum_{j=1}^{n} (X_j - \bar{X})^2$$

Definition of variables:

n = number of subjects

X_j = mean reaction time for each subject

$\bar{X}$ = group mean (same for all subjects)

$\sum_{j=1}^{n}$ = sum of all subjects' data

$$Standard\ Deviation = \sqrt{\text{Variance}}$$

Calculate the variance and standard deviation for the group of five subjects, using the data from Table 31A.3. To fill out Table 31A.4, copy the data from Table 31A.3, column 3 (random-interval, nondominant hand), into Table 31A.4, column 2 (Mean Reaction Time for Subject, X_j). Also copy the group mean into column 3. Then complete the calculations as indicated. When you calculate the deviation from the mean in column 4, enter absolute values; negative signs should be eliminated. These statistical calculations will indicate the variability of reaction times obtained from the five subjects in the random-interval, nondominant hand segment of the lab exercise.

Table 31A.4 Calculation of Variance and Standard Deviation for Random-Interval Reaction Times for the Nondominant Hand in a Group of Five Subjects A4

	ENTER	ENTER	CALCULATE	CALCULATE
Name of Subject	Mean Reaction Time for Subject X_j	Group Mean $\bar{X}$	Deviation from Mean $X_j - \bar{X}$	Deviation from Mean² $(X_j - \bar{X})^2$
1.				
2.				
3.				
4.				
5.				
Sum of data for all five subjects: (Add the five numbers in column 5) $\sum_{j=1}^{n} (X_j - \bar{X})^2$				
Variance = Sum × 0.25 or $\frac{1}{n-1}$				
Standard Deviation = Square Root of Variance				

To fill in Table 31A.5, copy the data from Table 31A.3, column 5 (fixed-interval, nondominant hand), into Table 31A.5, column 2 (Mean Reaction Time for Subject, X_j). Then complete the calculations as indicated. These statistical calculations will indicate the reaction time variability obtained from the five subjects in the fixed-interval, nondominant hand segment of the lab exercise.

Table 31A.5 Calculation of Variance and Standard Deviation for Fixed-Interval Reaction Times for the Nondominant Hand in a Group of Five Subjects A4

	ENTER	ENTER	CALCULATE	CALCULATE
Name of Subject	Mean Reaction Time for Subject X_j	Group Mean $\bar{X}$	Deviation from Mean $X_j - \bar{X}$	Deviation from Mean² $(X_j - \bar{X})^2$
1.				
2.				
3.				
4.				
5.				
Sum of data for all five subjects: (Add the five numbers in column 5) $\sum_{j=1}^{n} (X_j - \bar{X})^2$				
Variance = Sum × 0.25 or $\frac{1}{n-1}$				
Standard Deviation = Square Root of Variance				

PART B: Assessments

Complete the following:

1. **a.** What are the components of a typical reflex arc? List them in order. ______________________

   b. How does a neural pathway for a voluntary reaction differ from a spinal reflex arc? A1 ______________________

2. What is the difference between a reflex response and a voluntary reaction to a stimulus? Which would you expect to have a shorter response time? Why? A1 ______________________

3. From the results in Table 31A.1, what was the range of your subject's reaction time? A2 ______________________

4. From the results in Table 31A.1, which condition showed the shorter mean reaction time in your dominant hand, fixed-interval or random-interval? Did you expect these results? Why or why not? A1 ______________________________

____

____

5. From the results in Table 31A.1, which hand showed the shorter mean reaction time, the dominant or nondominant hand? Did you expect these results? Why or why not? A2 ______________________________

____

____

6. From the results in Table 31A.2, did the reaction times get longer or shorter as the stimuli progressed from #1 to #10? Did you expect these results? Why or why not? A1 ______________________________

____

____

____

7. **a.** From the group means in Table 31A.3, which condition showed the shorter reaction time, random-interval or fixed-interval? A1 A3 ______________________________

b. Which hand had the shorter reaction time, dominant or nondominant? A2 A3 ______________________________

c. Did the group results agree with the results from your subject? A3 ______________________________

8. From the calculations of standard deviation in Tables 31A.4 and 31A.5, which condition showed the greater variability in reaction times, random-interval or fixed-interval? Did you expect these results? Why or why not? A3 A4 __________

____

____

NOTES

LABORATORY
EXERCISE 31B

Electroencephalography I: BIOPAC® Exercise

Purpose of the Exercise

To record an EEG from an awake, resting subject with eyes open and closed, and to identify and examine alpha, beta, delta, and theta components of an EEG.

MATERIALS NEEDED

- Computer system [MAC OS X 10.7–10.15 (Catalina), or PC running Windows 10, 8, or 7]
- Biopac Student Lab software version 4.0 or above
- BIOPAC Data Acquisition Unit MP36, MP35, or MP46/45
- BIOPAC electrode lead set (SS2L)
- BIOPAC disposable vinyl electrodes (EL503), three electrodes per subject
- BIOPAC electrode gel (GEL1) and abrasive pad (ELPAD) or alcohol wipes
- Suggested: Lycra swim cap or supportive wrap (such as 3M Coban™ self-adhering support wrap) to press electrodes against the head for improved contact

SAFETY

- The BIOPAC electrode lead set is safe and easy to use and should be used only as described in the procedures section of the laboratory exercise.
- The electrode lead clips are color-coded. Make sure they are connected to the properly placed electrode as demonstrated in figure 31B.2.
- The vinyl electrodes are disposable and meant to be used only once. Each subject should use a new set.

Learning Outcomes

After completing this exercise, you should be able to:

O1 Identify alpha, beta, delta, and theta components of an EEG.

O2 Distinguish variations in these components due to differences in the level of stimulation (whether the eyes are closed or open).

The O corresponds to the assessments A indicated in the Laboratory Assessment for this Exercise.

Pre-Lab

Carefully read introductory material and examine the entire lab. Be familiar with structures and functions of the brain, and the physiology of action potentials from lecture or the textbook. Also, if this is the first BIOPAC lab that you are performing, please read the general description of the Biopac Student Lab system in Exercise 21.

Nerve impulses generated in the neurons of the nervous system are responsible for controlling many vital functions in the body, such as thought processes, responses to stimuli, higher mental functions (such as problem solving), and the control of muscle contraction. Nerve impulses are electrical events that transmit information from neurons to other neurons, muscle cells, or glandular cells. This information is then used to accomplish various functional tasks throughout the body.

Neurons respond to certain types of stimuli by generating *action potentials,* which are accomplished by transporting sodium (Na^+) and potassium (K^+) ions across their plasma membranes in a particular sequence. This neural response leads to changes in the charge both inside the neurons and in the extracellular fluid surrounding them. The movement of ions in the extracellular fluid of the cerebral cortex can be detected by placing electrodes on the surface of the scalp and measuring the voltage between them. A recording of electrical activity in the cerebral cortex, via electrodes placed on the scalp, is called an *electroencephalogram (EEG).* An EEG is a safe, noninvasive, and painless method of recording the sum of electrical activity occurring simultaneously in many thousands of neurons in the cerebral cortex. It is important to realize that an EEG is not a record of action potentials occurring in an individual neuron.

An EEG can be used to assess states of consciousness (awake, concentrating, asleep) and various brain disorders, and to establish brain death. The major brain waves found in EEGs are of four types, called *alpha, beta, delta,* and *theta* waves. Each type of brain wave is characterized by a specific range of frequencies and amplitudes. Figure 31B.1 shows the typical appearance of the four types of brain waves.

The *frequency* of a brain wave is the number of complete wave cycles that occurs each second; it is measured in Hertz (Hz, or cycles per second). Frequency is an indication of the state of alertness; therefore, the higher the wave frequency, the more alert and attentive the person is.

The *amplitude* of a brain wave is the height, or magnitude, of the wave; it is measured in microvolts (μV, or one millionth of a volt). Amplitude indicates the degree of synchrony, or simultaneous firing of many neurons underneath the recording electrodes. Therefore, the higher the amplitude, the more neurons are firing simultaneously (in synchrony). It is important to understand that synchrony is not a measure of alertness; in fact, the highest amplitudes typically are found in delta waves, which occur mainly during the deeper stages of sleep. The lowest amplitudes are usually found in beta waves, which are dominant during high levels of attentiveness. This can be explained by the fact that asynchronous firing of neurons leads to a "canceling out" of positive and negative voltages occurring in different neurons at the same time.

All four types of brain waves can usually be observed in the EEG of a normal adult, although different wave forms become predominant in the various states of consciousness, including the different stages of sleep. There is also quite a bit of individual variation in the characteristics of brain waves. This variation is due to a variety of factors, including gender, age, personality, the type of activity a person is engaged in, the area of the brain being used for the EEG, and various disease states of the brain. Table 31B.1 shows the average frequencies and amplitudes of the four types of brain waves, along with a short description of the state(s) of consciousness in which they are predominant.

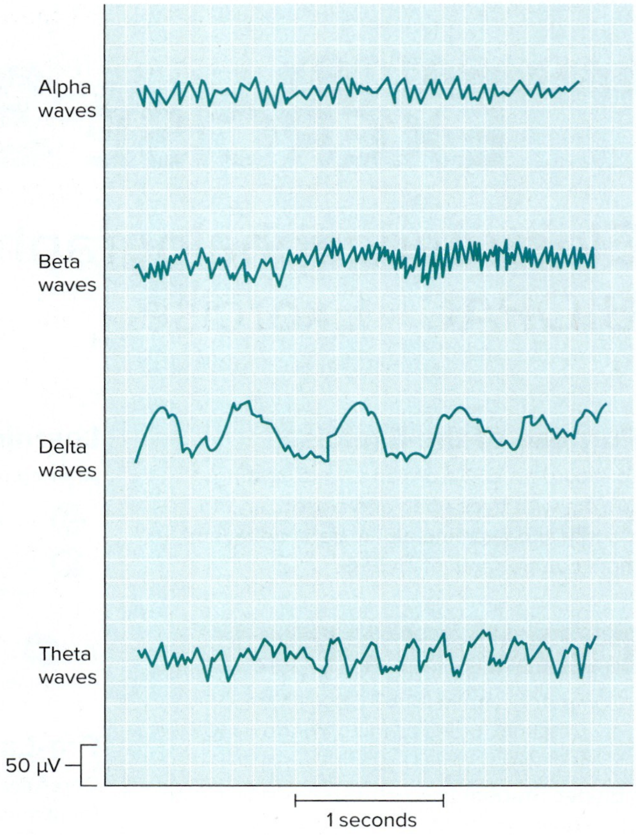

FIGURE 31B.1 The four types of brain waves of the EEG.

In this laboratory exercise, you will be using the bipolar recording method. In this method, two electrodes are placed over the same region of the cerebral cortex, perhaps 2–3" apart, and the difference in the electrical potential (voltage) between the electrodes is measured. A third electrode is placed over the mastoid process as a "ground," or reference point for the body's baseline voltage.

The BIOPAC program will separate your EEG into the four types of brain waves. You will be able to observe all four types and measure the frequencies of the waves, as well as the standard deviation (variability) of wave amplitudes. You will also demonstrate *alpha block,* a replacement of alpha waves with lower-amplitude beta waves that occurs when closed eyes are suddenly opened.

Table 31B.1 Frequencies, Amplitudes, and State(s) of Consciousness Associated with the Four Types of Brain Waves

Brain Wave	Average Frequency (Hz)	Average Amplitude (μV)	State(s) of Consciousness
Alpha	8–13	~50 (can vary from 20 to 200)	Awake, resting, with eyes closed; decrease when concentrating
Beta	14–30 (sometimes higher)	5–10 (can be as high as 30)	Awake, alert, with eyes open, paying attention or concentrating
Delta	1–4	~100 (can vary from 20 to 200)	Deep stages of sleep; some forms of severe brain disease
Theta	4–7	5–10 (can be as high as 60–70)	More commonly observed in children, but often seen in adults, especially while concentrating or under stress, or during sleep; some forms of brain disease

PROCEDURE A: Setup

1. With your computer turned **ON** and the BIOPAC MP36/35 unit turned **OFF,** plug the electrode lead set (SS2L) into Channel 1. If you are using the MP46/45, the USB cable must be connected, and the ready light must be on.
2. Turn on the MP36/35 Data Acquisition Unit.
3. Selection of the subject is important; choose someone with easy access to his/her scalp so that you can be sure to make good contact between the skin and the electrode. With the subject sitting in a relaxed position, place the electrodes on the same side of the head, as in figure 31B.2. The ground electrode (black) is attached over the mastoid process. To help ensure good contact, make sure the area to be in contact with the electrodes is clean by wiping with the abrasive pad (ELPAD) or alcohol wipe. A small amount of BIOPAC electrode gel (GEL1) may also be used to make better contact if necessary. For optimal adhesion, hold the electrodes against the skin for about a minute after placement. Wrap the subject's head with a swim cap or supportive wrap to secure the electrodes in place.
4. Start the Biopac Student Lab Program for *L03-Electroencephalography (EEG) I,* then click **OK.**
5. Designate a filename for the subject's data and click **OK.**

PROCEDURE B: Calibration

1. The subject should be in a sitting position with eyes closed. Click on **Calibrate.** The calibration will stop automatically after 8 seconds.
2. The calibration recording should look similar to figure 31B.3. If it does, click on **Continue** to proceed to the EEG recording. If not, click **Redo Calibration.**

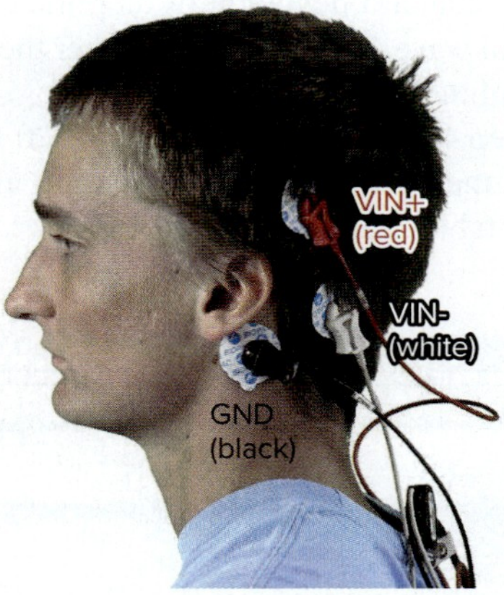

FIGURE 31B.2 Electrode placement for EEG experiment.

Courtesy of and © *BIOPAC* Systems, Inc. www.biopac.com

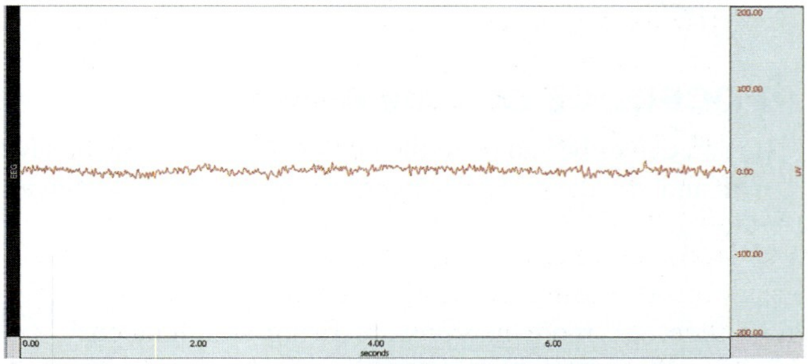

FIGURE 31B.3 Sample calibration recording.

Courtesy of and © *BIOPAC* Systems, Inc. www.biopac.com

PROCEDURE C: Recording

1. Before beginning the recording phase, you should designate a **Director** (who will instruct the subject to open or close eyes) and a **Recorder** (who will insert markers by pressing the F4 key when eyes are opened, and the F5 key when eyes are closed.)
2. The data for this exercise will be recorded over a 1-minute period. During the first 20 seconds, the subject will have his/her eyes closed. During the next 20 seconds, the subject should have eyes open, and for the last 20 seconds, eyes closed again. Make sure all group members understand the procedure before proceeding to step 3.
3. Click on **Record.** The **Director** should then instruct the subject to remain relaxed and change eye position as follows:
   a. 0–20 seconds, eyes closed
   b. 21–40 seconds, eyes open
   c. 41–60 seconds, eyes closed
   The **Recorder** should insert markers at these times: 21 seconds when eyes are opened (F4 key), and 41 seconds when eyes are closed again (F5 key).
4. Click on **Suspend.** If the recording is similar to figure 31B.4, click **Done,** and go to step 5. If there are large fluctuations or baseline drift in the recording, or if the EEG is flat, click on **Redo.** This will erase the current data. Check electrode placement, and make sure that the subject does not make any excessive movements during the new recording. After completing the new recording, click **Suspend,** and then **Done.**
5. A pop-up window will appear with several options. Click either **Analyze current data file,** if you are going to conduct your data analysis now, or **Quit,** if you are going to analyze the file at a later time. Then click **OK.** The data will be saved by the BIOPAC program.

PROCEDURE D: Data Analysis

1. You can either analyze the current data file now or save it and do the analysis later, using the **Review Saved Data** mode. When you enter the analysis mode, the program will separate the EEG into alpha, beta, delta, and theta waves, and display them in the data window. The window should look similar to figure 31B.5. If it is similar, the subject may now remove the electrodes.
2. Note the channel number designations: CH 1—EEG (hidden), CH 40—alpha, CH 41—beta, CH 42—delta, CH 43—theta.
3. Across the top of the recordings you will see three boxes for each channel: channel number, measurement type, and the result. Note the settings for the measurement boxes: CH 40–CH 43 should all be set on **Stddev** and SC on **Freq.** The program will automatically calculate the standard deviation or frequency for the selected area.
4. Using the I-beam cursor, select the area from time 0 to 20 seconds (to the first marker). See figure 31B.6 for an illustration of how to select the area for the first 20 seconds. Record the standard deviation for each channel in table 31B.2 of Part A of Laboratory Assessment 31B. The standard deviation of amplitude will not provide actual wave amplitudes, but rather the variability in those amplitudes.
5. Repeat step 4 for the time period from 20 to 40 seconds (between the first and second markers) and record the results in table 31B.2.

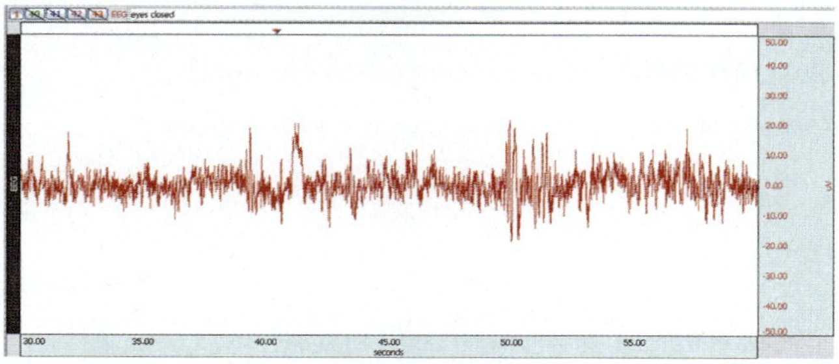

FIGURE 31B.4 Sample EEG recording.

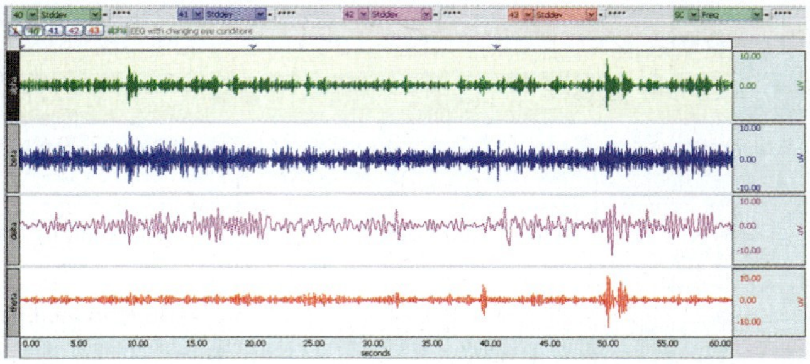

FIGURE 31B.5 EEG recording showing the four different types of brain waves.

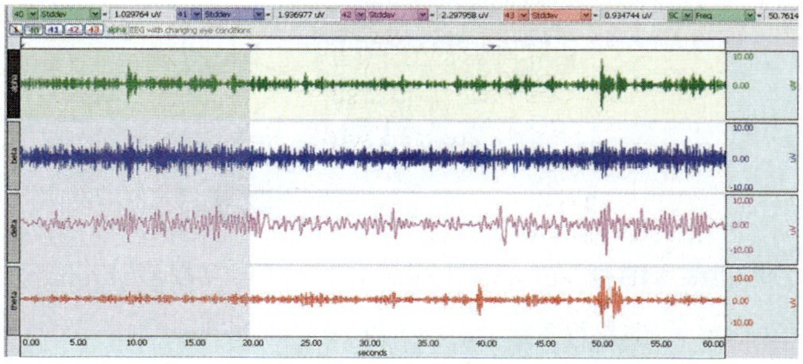

FIGURE 31B.6 Proper selection of the area from 0 to 20 seconds for the standard deviation measurements.

6. Repeat step 4 for the time period from 40 to 60 seconds (from the second marker to the end) and record the results in table 31B.2.
7. Now set up the data window, so you can just see a 4- to 5-second interval in the first 20 seconds of the recording. This can be done by clicking anywhere in the horizontal scale (the numbers just above the word "seconds" at the bottom of the data window). A box will pop up in the middle of the screen, called **Set Screen Horizontal Axis.** Change the start time to 2 and the end time to 6. Then click **OK.** This will spread out the data so you can see individual waves and measure their frequency.
8. Now begin to measure wave frequencies. In order for the program to measure **FREQ** of each type of wave form, you have to select each type and measure it individually. First select alpha waves for **FREQ** measurement by clicking on the word "alpha" along the left side of the data window. Now **SC (Selected Channel)** will display only alpha wave frequencies.
9. Use the I-beam cursor to select an area that represents one cycle of the alpha wave. One cycle runs from one peak to the next peak (refer to fig. 31B.7). Record the result in table 31B.3 of the laboratory assessment.
10. Repeat for two other nearby alpha cycles and record the results. Calculate the mean and record it in table 31B.3.
11. Repeat steps 9 and 10 for each of the other waveforms (beta, delta, and theta) and record the results in table 31B.3. (Remember to click on the word "beta," delta," or "theta" before beginning to measure **FREQ** for each of these wave types.)

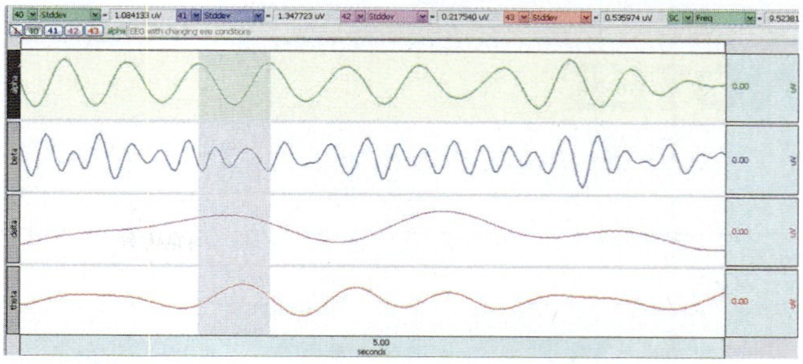

FIGURE 31B.7 Proper selection of one alpha wave for the frequency measurement.

Courtesy of and © *BIOPAC* Systems, Inc. www.biopac.com

12. If you would like to view the entire recording after completing **FREQ** measurements, open the **Display** menu. Click on **Autoscale Horizontal** and then **Autoscale Waveforms.** Now the entire 1-minute recording will be visible in the data window.
13. When finished, you can save the data on the hard drive, a disk, or a network drive, and then exit the system by clicking **File, Quit.**
14. Complete Part B of the laboratory assessment.

NOTES

Name ______________________________

Date ______________________________

Section ______________________________

The (A) corresponds to the indicated Learning Outcome(s) (O) found at the beginning of the Laboratory Exercise.

LABORATORY ASSESSMENT 31B

Electroencephalography I: BIOPAC® Exercise

Subject Profile

Name ____________________ Height ____________________

Age ____________________ Weight ____________________

Gender: Male/Female

PART A: Data and Calculations Assessments

1. EEG Variability in Amplitude Measurements (A1) (A2)

Table 31B.2 Standard Deviation [stddev]

Rhythm	CH Measurement	Eyes Closed	Eyes Open	Eyes Re-closed
Alpha	40 Stddev			
Beta	41 Stddev			
Delta	42 Stddev			
Theta	43 Stddev			

2. EEG Frequency Measurements (A1) (A2)

Table 31B.3 Frequency [Hz] Measurements

Rhythm	CH Measurement	Cycle 1	Cycle 2	Cycle 3	Mean
Alpha	SC Freq				
Beta	SC Freq				
Delta	SC Freq				
Theta	SC Freq				

PART B: Assessments

Complete the following:

1. Examine the alpha and beta waveforms for changes between the "eyes closed" state and the "eyes open" state. A2

   a. Did desynchronization of the alpha rhythm occur when the eyes were open? How do you know?

   ____
   ____
   ____
   ____

   b. Did the beta rhythm change in the "eyes open" state? If so, how did it change? A2

   ____
   ____
   ____
   ____

2. Examine the delta and theta rhythms. Was there a change in delta and theta activity when the eyes were open? Explain your observation. A1

   ____
   ____
   ____

3. Define the following terms: A1

   a. Alpha rhythm

   ____
   ____

   b. Beta rhythm

   ____
   ____

   c. Delta rhythm

   ____
   ____

   d. Theta rhythm

   ____
   ____

LABORATORY
EXERCISE 32

Dissection of the Sheep Brain

Purpose of the Exercise

To observe the major features of the sheep brain and to compare these features with those of the human brain.

MATERIALS NEEDED

Dissectible model of human brain
Preserved sheep brain
Dissecting tray
Dissection instruments
Long knife
Disposable gloves

For Demonstration Activity:
Frontal sections of sheep brains

SAFETY

- Review all the safety guidelines in Appendix 1 of your laboratory manual.
- Wear disposable gloves and protective eyewear when handling the sheep brains.
- Save or dispose of the brains as instructed.
- Wash your hands before leaving the laboratory.

Chapter Opener: ©101cats/Getty Images

Learning Outcomes APR

After completing this exercise, you should be able to:

O1 Examine the major structures of the sheep brain.
O2 Summarize differences and similarities between the sheep brain and the human brain.
O3 Locate the most developed cranial nerves of the sheep brain.

The O corresponds to the assessments A indicated in the Laboratory Assessment for this Exercise.

Pre-Lab

Carefully read the introductory material and examine the entire lab. Be familiar with basic structures and functions of the meninges, brain, and cranial nerves from the previous laboratory exercises. Answer the pre-lab questions.

Pre-Lab Questions Select the correct answer for each of the following questions:

1. The spinal cord of the sheep has a _____ orientation.
a. vertical **b.** transverse
c. horizontal **d.** sagittal

2. The gyri and sulci of the sheep brain possess the closely attached
a. pia mater. **b.** arachnoid mater.
c. subarachnoid space. **d.** dura mater.

3. By bending the cerebellum and medulla oblongata downward, the _____ should be visible.
a. vagus nerve **b.** optic nerve
c. pituitary gland **d.** pineal gland

4. The sheep brain has _____ pairs of attached cranial nerves.
a. 2 **b.** 10
c. 12 **d.** 31

5. The superior and inferior colliculi compose the
a. pineal gland. **b.** corpora quadrigemina.
c. cerebrum. **d.** cerebellum.

6. The thin covering seen on the surface of the sheep brain is the
a. pia mater. **b.** arachnoid mater.
c. dura mater. **d.** dural septa.

7. Sulci and gyri are surface features of the cerebrum of the sheep brain.
a. True **b.** False

8. The cerebellum of the sheep brain has deep gray matter and superficial white matter.
a. True **b.** False

Mammalian brains have many features in common. Human brains may not be available, so sheep brains often are dissected as an aid to understanding mammalian brain structure. However, the adaptations of the sheep differ from the adaptations of the human, so comparisons of their structural features may not be precise. The sheep is a quadruped; therefore, the spinal cord is horizontal, unlike the vertical orientation in a bipedal human. Preserved sheep brains have a different appearance and are firmer than those that are removed directly from the cranial cavity; this is caused by the preservatives used.

PROCEDURE: Dissection of the Sheep Brain

As you examine a sheep brain, contemplate any differences and similarities between the sheep brain and the human brain. Several questions in the laboratory assessment address comparisons of sheep and human brains.

1. Obtain a preserved sheep brain and rinse it thoroughly in water to remove as much of the preserving fluid as possible.
2. Examine the surface of the brain for the presence of meninges. (The outermost layers of these membranes may have been mostly lost during removal of the brain from the cranial cavity.) Locate the following:

   dura mater—the thick, opaque outer layer that adheres to the inside of the skull

   arachnoid mater—the weblike, semitransparent middle layer spanning the fissures of the brain

   pia mater—the thin, delicate, innermost layer that adheres to the gyri and sulci of the surface of the brain
3. Remove any remaining dura mater by cutting and pulling it gently from the surface of the brain.
4. Position the brain with its ventral surface down in the dissecting tray. Study figure 32.1, and locate the following structures on the specimen:

   longitudinal fissure

   cerebral hemispheres

   gyri

   sulci

   frontal lobe

   parietal lobe

   temporal lobe

   occipital lobe

   cerebellum

   medulla oblongata

   spinal cord
5. Gently separate the cerebral hemispheres along the longitudinal fissure and expose the transverse band of white fibers (tracts) within the fissure that connects the hemispheres. This band is the *corpus callosum.*
6. Bend the cerebellum and medulla oblongata slightly downward and away from the cerebrum (fig. 32.2). This will expose the *pineal gland* in the upper midline and the *corpora quadrigemina*, which consists of four rounded structures, called *colliculi*, associated with the midbrain.

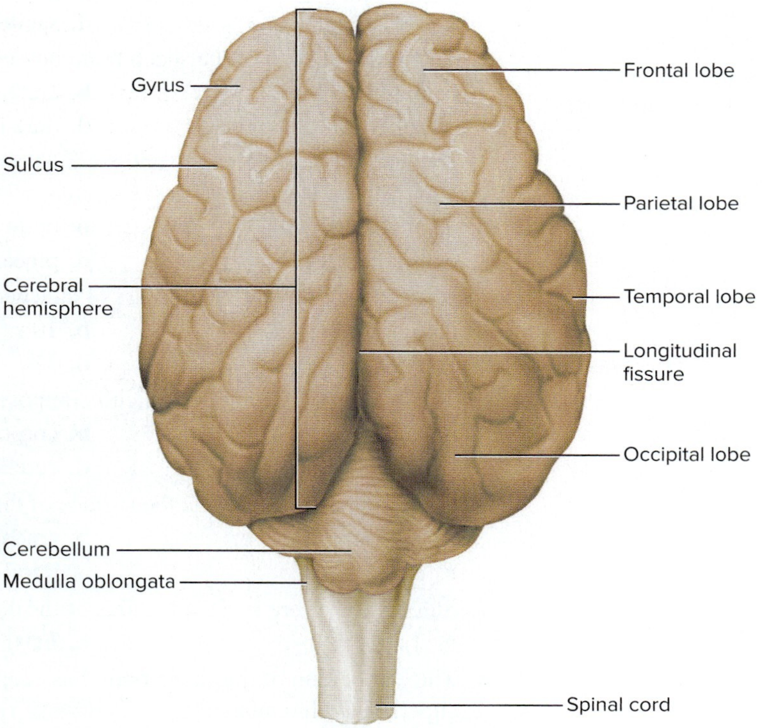

FIGURE 32.1 Dorsal surface of the sheep brain.

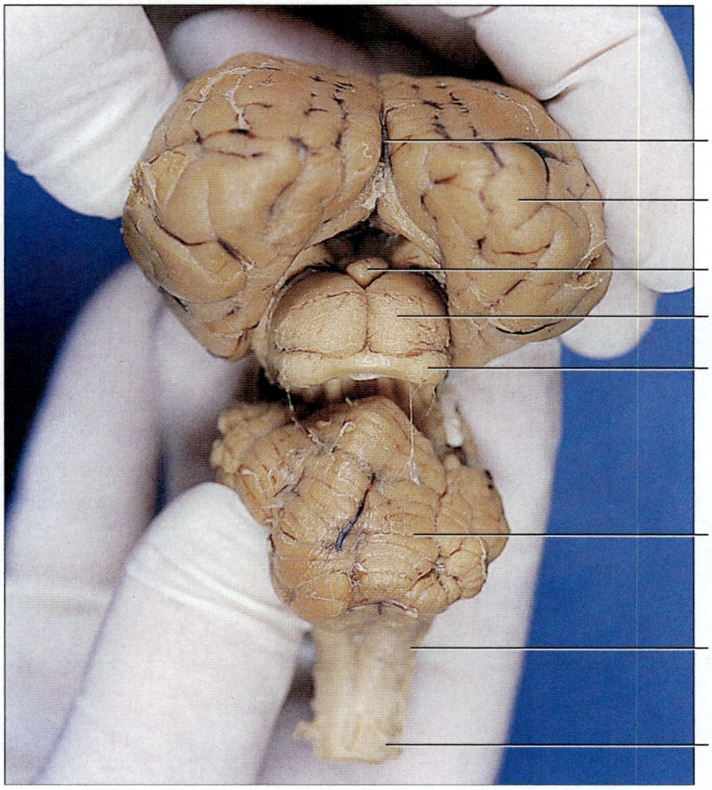

FIGURE 32.2 Gently bend the cerebellum and medulla oblongata away from the cerebrum to expose the pineal gland of the diencephlon and the corpora quadrigemina of the midbrain. ©J & J Photography

7. Position the brain with its ventral surface upward. Study figures 32.3 and 32.4, and locate the following structures on the specimen:

    longitudinal fissure
    olfactory bulbs
    optic nerves
    optic chiasma
    optic tract
    mammillary body
    infundibulum (pituitary stalk)
    midbrain
    pons

8. Although some of the cranial nerves may be missing or are quite small and difficult to find, locate as many of the following as possible, using figures 32.3 and 32.4 as references:

    oculomotor nerves
    trochlear nerves
    trigeminal nerves
    abducens nerves
    facial nerves
    vestibulocochlear nerves
    glossopharyngeal nerves
    vagus nerves
    accessory nerves
    hypoglossal nerves

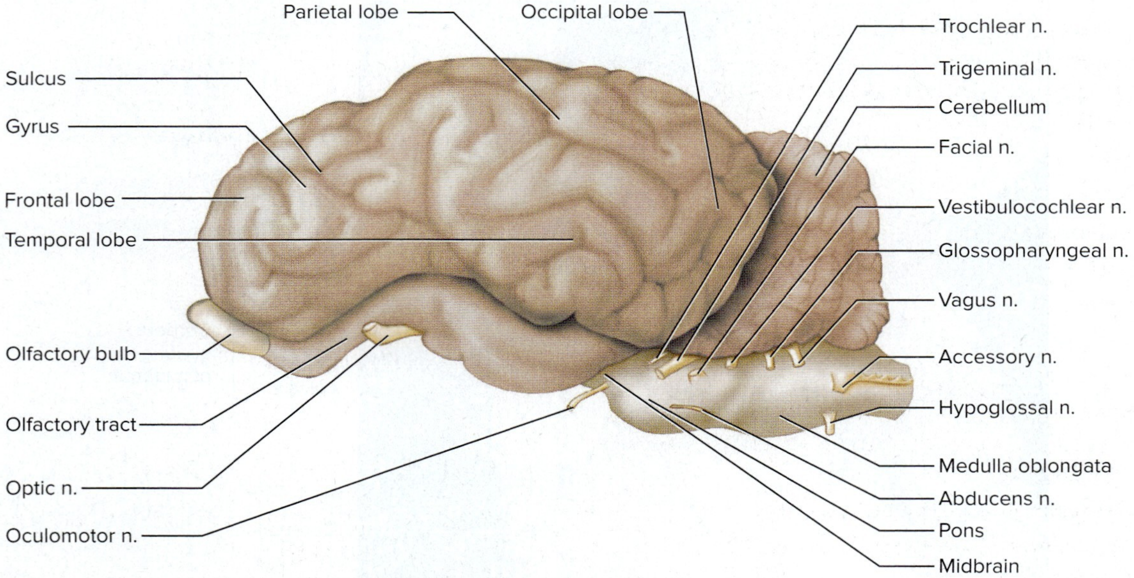

FIGURE 32.3 Lateral surface of the sheep brain. (The "n." stands for nerve.)

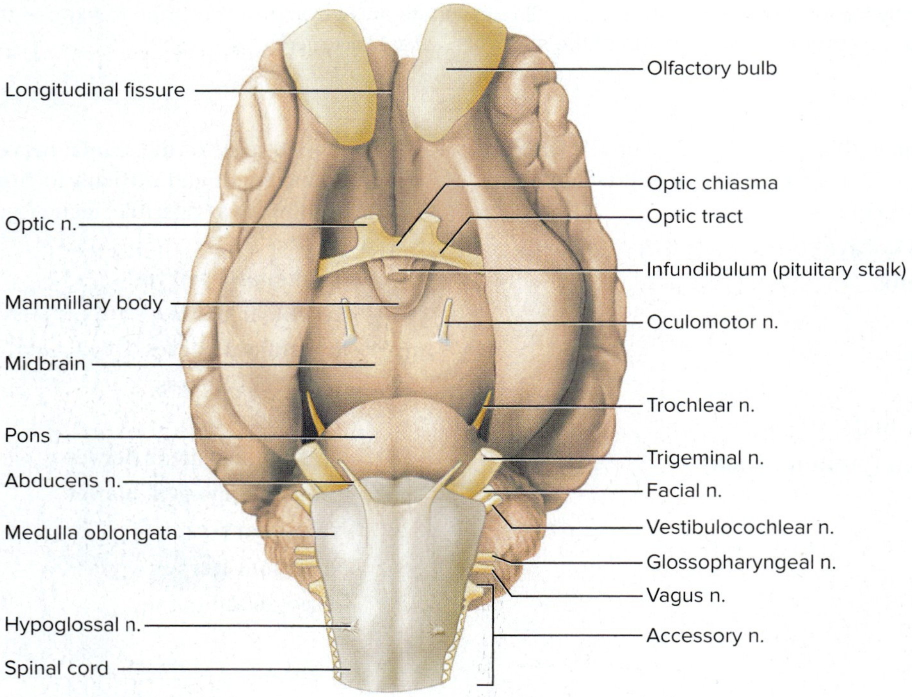

FIGURE 32.4 Ventral surface of the sheep brain. (The "n." stands for nerve.)

9. Using a long, sharp knife or scalpel, cut the sheep brain along the midline to produce a median section. Study figures 32.2 and 32.5, and locate the following structures on the specimen:

cerebrum

olfactory bulb

corpus callosum

fornix

cerebellum

- white matter
- gray matter

lateral ventricle—one in each cerebral hemisphere, separated by thin membrane called septum pellucidum

third ventricle—within diencephalon

fourth ventricle—between brainstem and cerebellum

choroid plexus—capillary networks; form cerebrospinal fluid (CSF)

diencephalon

- optic chiasma
- infundibulum—this structure may be missing
- pituitary gland—this structure may be missing
- mammillary body—single in sheep
- thalamus
- hypothalamus
- pineal gland of epithalmus

midbrain

- corpora quadrigemina
    - superior colliculus
    - inferior colliculus
- cerebral peduncle
- cerebral aqueduct

pons

medulla oblongata

DEMONSTRATION ACTIVITY

Observe a frontal section from a sheep brain (fig. 32.6). Note the longitudinal fissure, gray matter, white matter, corpus callosum, lateral ventricles, third ventricle, and thalamus. Compare this to the image of the frontal section of the human brain (fig. 32.7). Note differences and similarities between the two.

10. Dispose of the sheep brain as directed by the laboratory instructor.
11. Complete Parts A, B, and C of Laboratory Assessment 32.

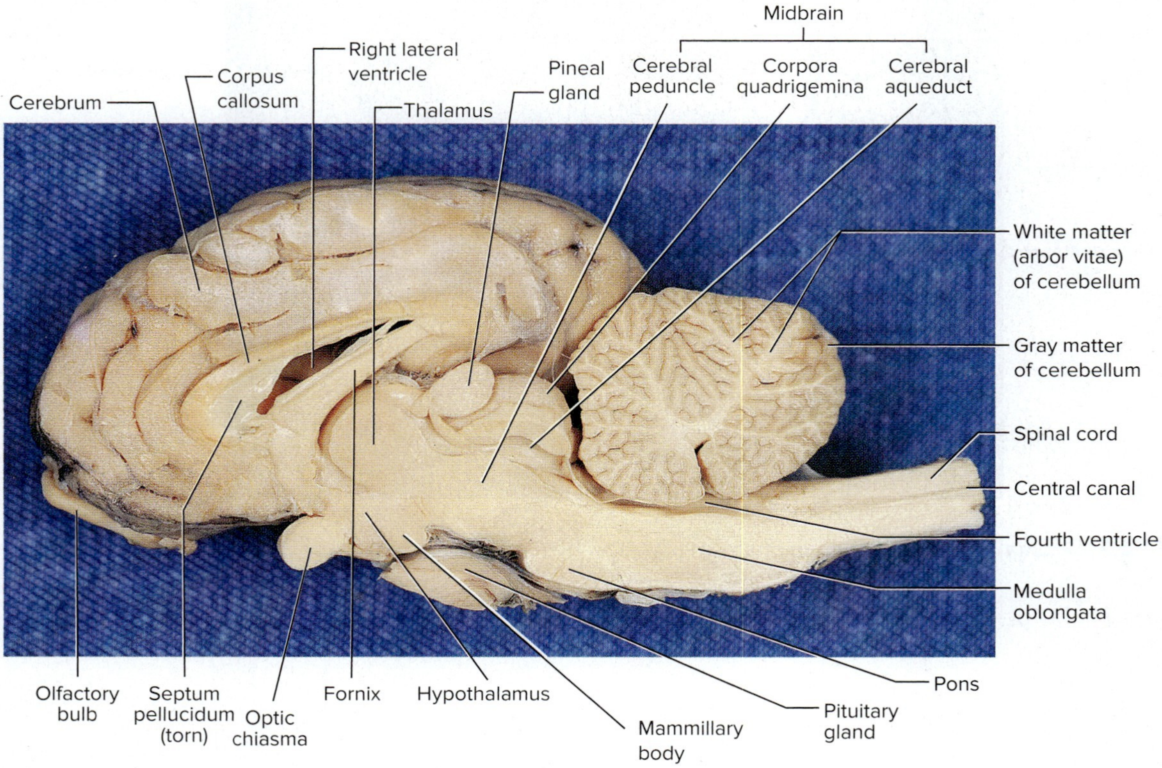

FIGURE 32.5 Median section of the right half of the sheep brain dissection. ©J & J Photography

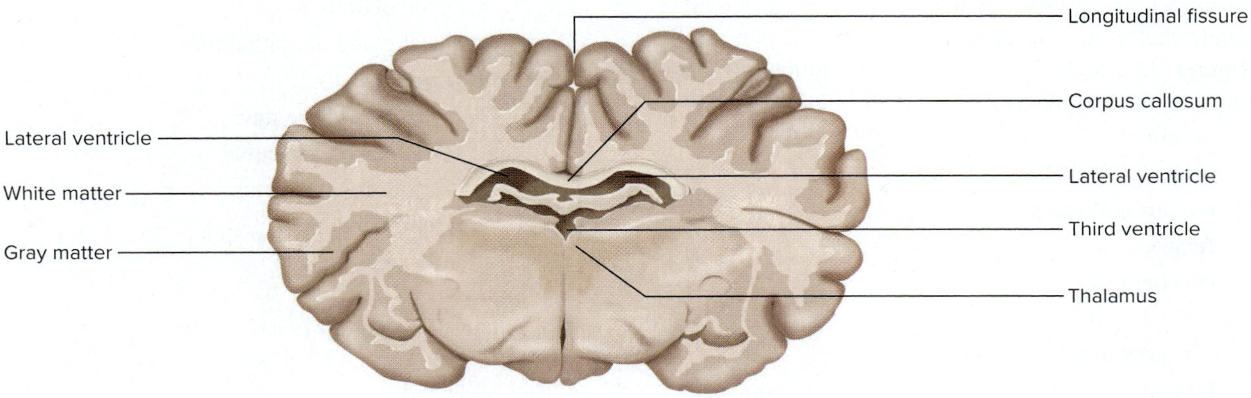

FIGURE 32.6 Frontal section of a sheep brain.

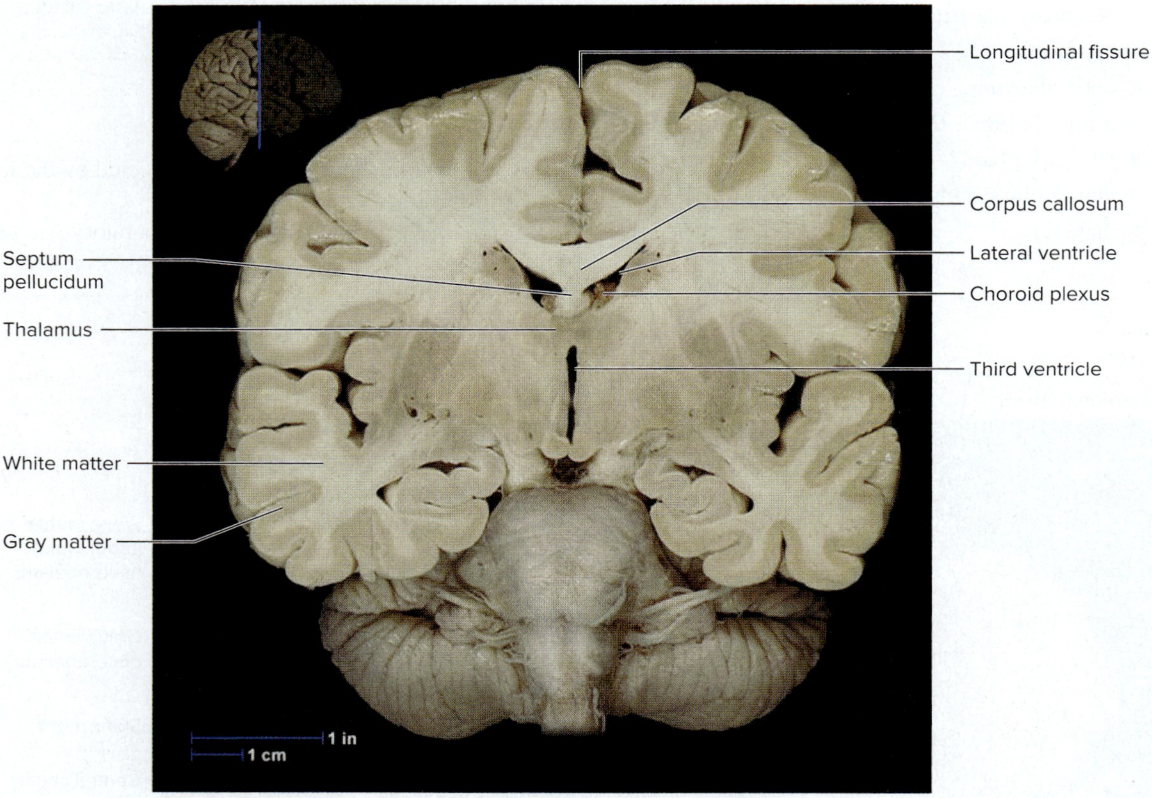

FIGURE 32.7 Frontal section of a human brain. APR/McGraw Hill APR

LABORATORY
ASSESSMENT 32

Name ____________________

Date ____________________

Section ____________________

The (A) corresponds to the indicated Learning Outcome(s) (O) found at the beginning of the Laboratory Exercise.

Dissection of the Sheep Brain

PART A: Assessments

Answer the following questions as you compare the sheep brain and human brain (model or cadaver):

1. Describe the location of any meninges observed to be associated with the sheep brain. A1 ____________________

2. How do the relative sizes of the sheep and human cerebral hemispheres differ? A2 ____________________

3. How do the gyri and sulci of the sheep cerebrum compare with the human cerebrum in numbers? A2 ____________________

4. What is the significance of the differences you noted in your answers for questions 2 and 3? A2 ____________________

5. What difference did you note in the structures of the sheep cerebellum and the human cerebellum? A2 ____________________

6. How do the sizes of the olfactory bulbs of the sheep brain compare with those of the human brain? A2 ____________________

7. Based on their relative sizes, which of the cranial nerves seems to be most highly developed in the sheep brain? A3 ____________________

CRITICAL THINKING ASSESSMENT—PART A

Since the nineteenth-century, it was hypothesized that humans have an inferior sense of smell compared to other mammals that rely on smell. What do you think this hypothesis was based on?

____

Recent evidence, however, suggests that humans have a similar number of olfactory neurons in their olfactory bulbs and the orbitofrontal cortex of the brain is more complex for interpreting odors compared to other mammals, such as sheep. There is also evidence that odors influence human behavior and are more powerful than previously thought. Evaluate the validity of the nineteenth-century hypothesis and provide supporting evidence. A2

____

PART B: Assessments

CRITICAL THINKING ASSESSMENT—PART B

Prepare a list of at least six features to illustrate ways in which the brains of sheep and humans are similar. A2

1. ______________________________
2. ______________________________
3. ______________________________
4. ______________________________
5. ______________________________
6. ______________________________

Interpret the significance of these similarities. A2 ______________________________

____

PART C: Assessments

Identify the features of the sheep brain in figures 32.8 to 32.11.

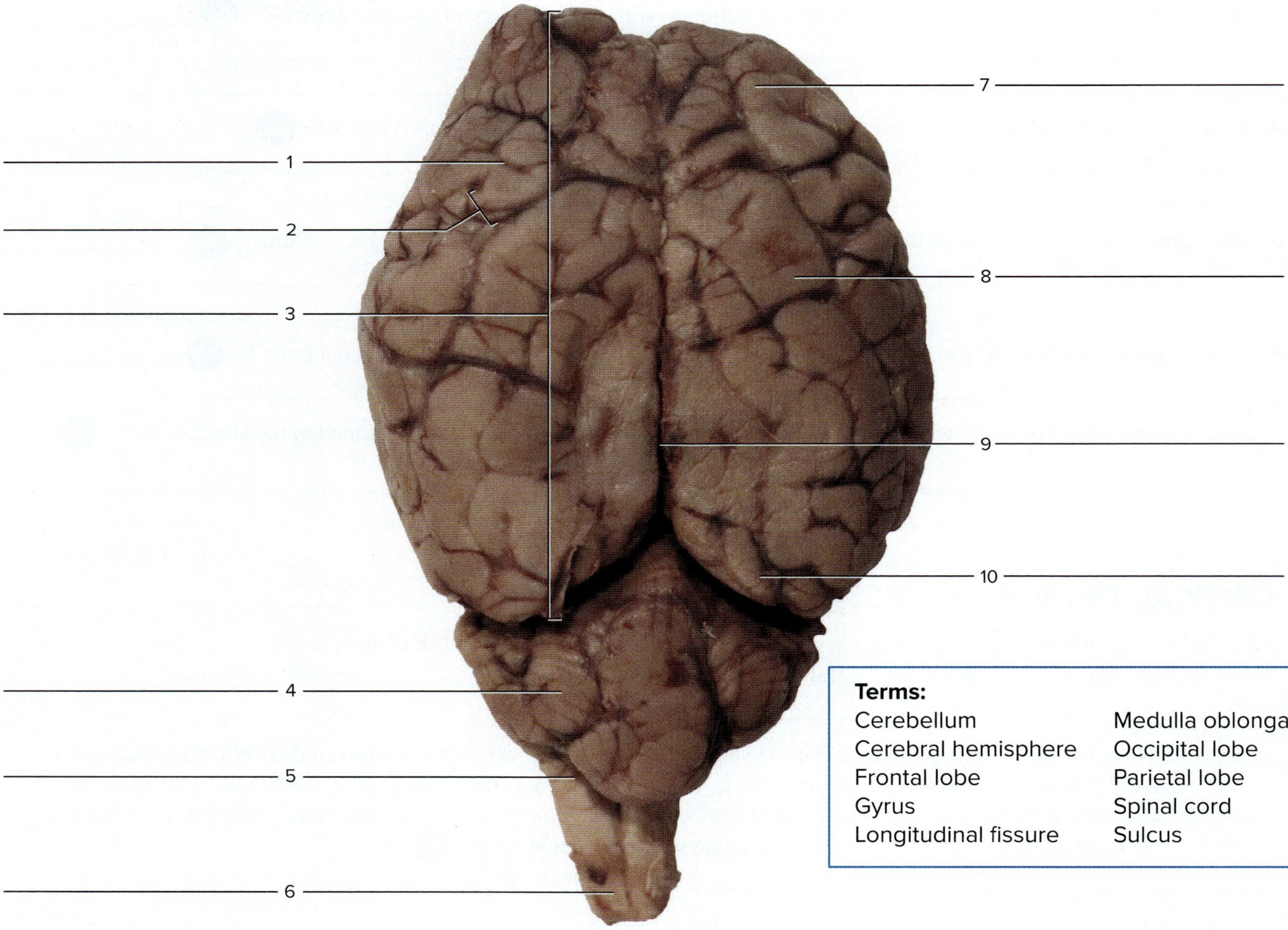

Terms:

Cerebellum	Medulla oblongata
Cerebral hemisphere	Occipital lobe
Frontal lobe	Parietal lobe
Gyrus	Spinal cord
Longitudinal fissure	Sulcus

FIGURE 32.8 Label the features of this dorsal surface of the sheep brain, using the terms provided. Cynthia Prentice-Craver/McGraw Hill A1

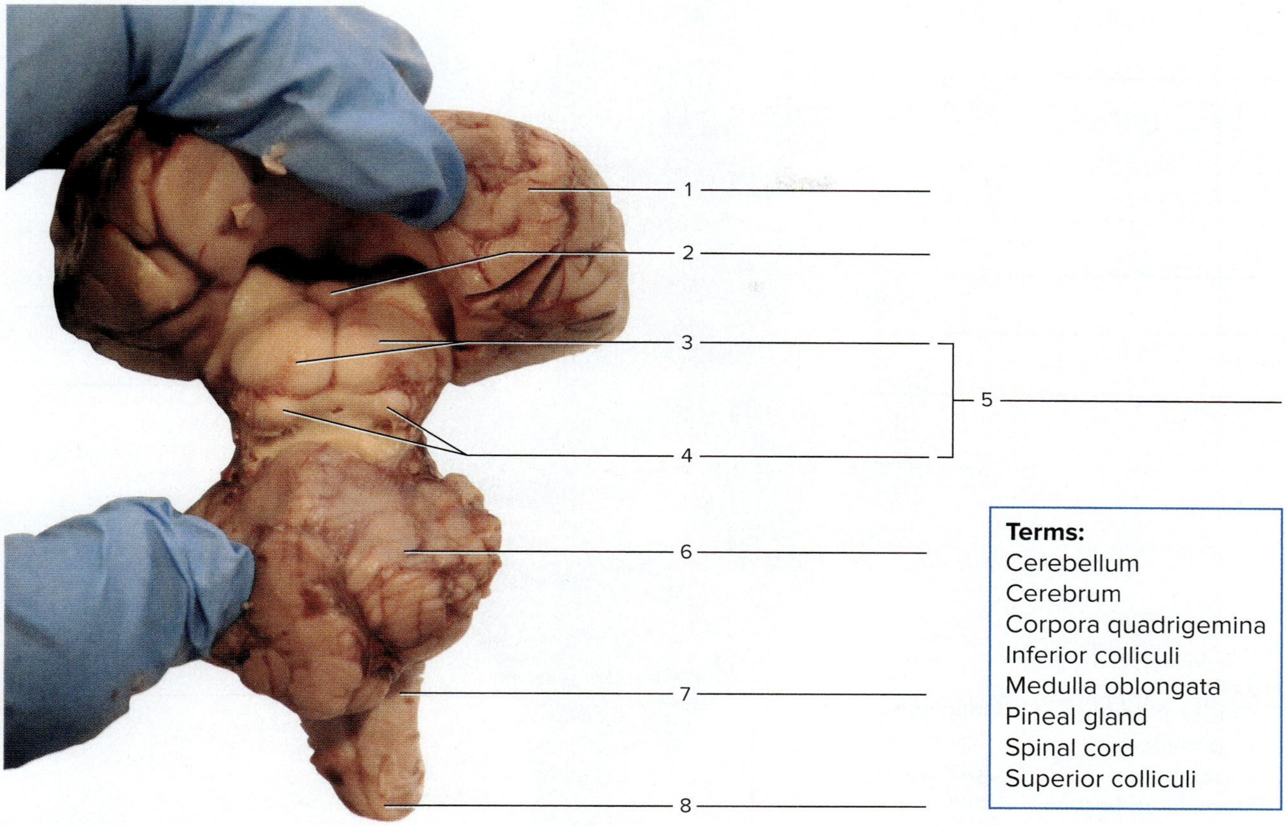

FIGURE 32.9 Label the features of this dorsal view that is showing some of the internal anatomy of the sheep brain, using the terms provided. Cynthia Prentice-Craver/McGraw Hill A1

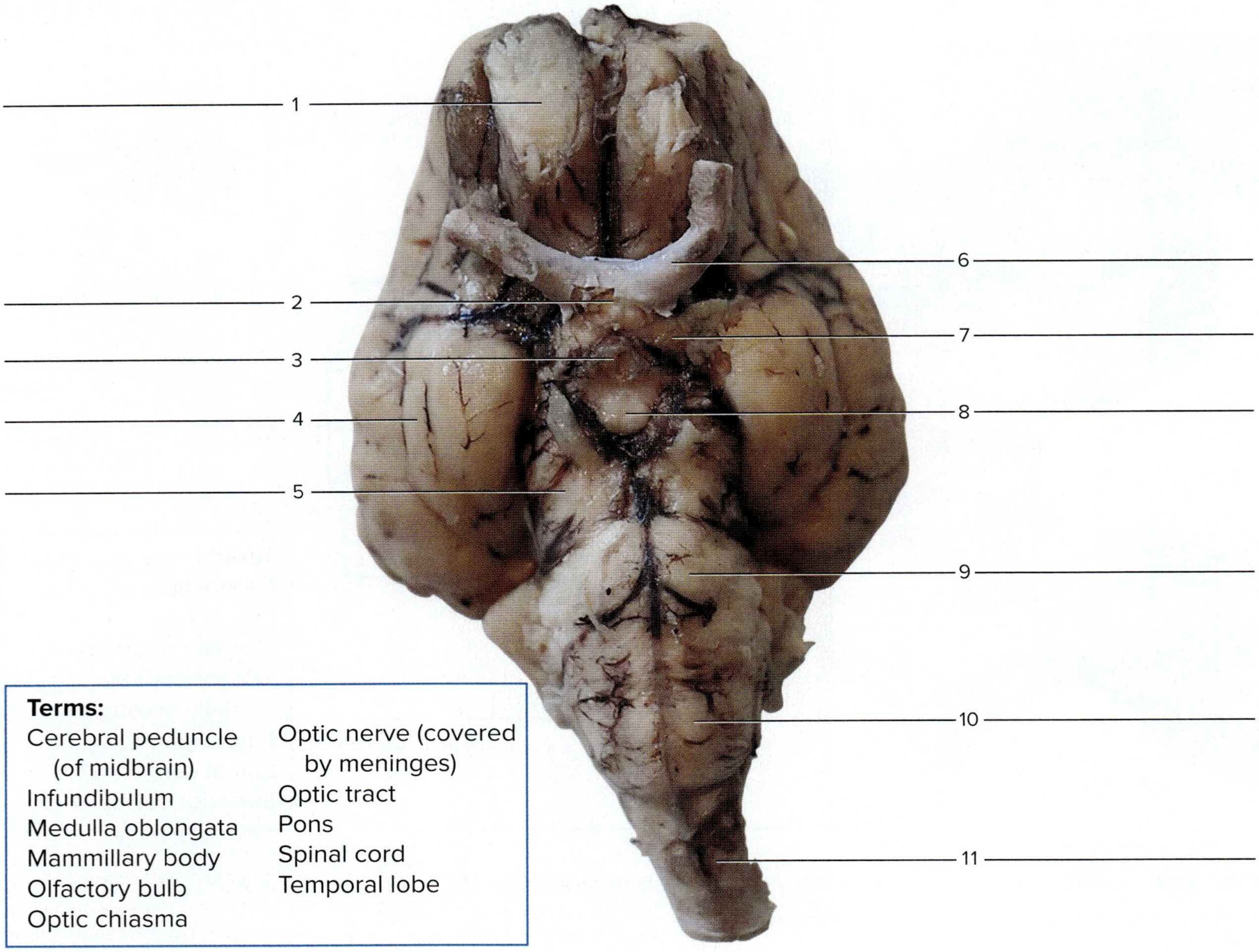

Terms:

Cerebral peduncle (of midbrain)	Optic nerve (covered by meninges)
Infundibulum	Optic tract
Medulla oblongata	Pons
Mammillary body	Spinal cord
Olfactory bulb	Temporal lobe
Optic chiasma	

FIGURE 32.10 Label the features of this ventral view that is showing some of the external anatomy of the sheep brain, using the terms provided. (*Note:* The pituitary gland has been removed.) Cynthia Prentice-Craver/McGraw Hill **A1**

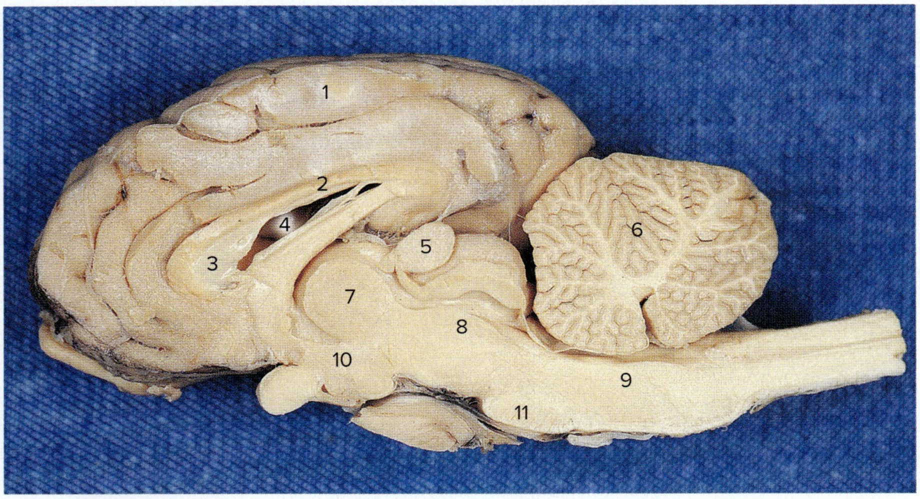

____ Cerebellum	____ Lateral ventricle	____ Pons
____ Cerebrum	____ Medulla oblongata	____ Septum pellucidum
____ Corpus callosum	____ Midbrain	____ Thalamus
____ Hypothalamus	____ Pineal gland	

FIGURE 32.11 Label the features of this median section of the sheep brain by placing the correct numbers in the spaces provided. ©J & J Photography **A1**

LABORATORY EXERCISE 33

General Senses

Purpose of the Exercise

To review the characteristics of sensory receptors and general senses and to demonstrate some of the general senses associated with the skin.

MATERIALS NEEDED

Marking pens in variety of colors (washable)

Millimeter ruler

Forceps with fine points or another two-point discriminator device

Blunt metal probes

Three beakers (250 mL)

Warm tap water or 45°C (113°F) water bath

Cold water (ice water)

Thermometer

For Demonstration Activity:

Prepared microscope slides of tactile (Meissner's) and lamellated (Pacinian) corpuscles

Compound light microscope

Learning Outcomes APR

After completing this exercise, you should be able to:

O1 Associate types of sensory receptors with general senses throughout the body.

O2 Determine and record a person's tactile localization ability relative to the distribution of mechanoreceptors in various regions of the skin.

O3 Measure the two-point threshold of various regions of the skin.

O4 Determine and record the distribution of thermoreceptors in various regions of the skin.

The O corresponds to the assessments A indicated in the Laboratory Assessment for this Exercise.

Pre-Lab

Carefully read the introductory material and examine the entire lab. Be familiar with basic structures and functions of the receptors associated with general senses from the lecture or the textbook. Answer the pre-lab questions.

Pre-Lab Questions Select the correct answer for each of the following questions:

1. General senses include all of the following *except*
 a. touch. **b.** vision.
 c. temperature. **d.** pain.

2. Thermoreceptors are associated with
 a. deep pressure. **b.** light touch.
 c. tissue trauma. **d.** temperature changes.

3. When receptors are continuously stimulated, the sensations may fade away; this phenomenon is known as
 a. tolerance. **b.** perception.
 c. adaptation. **d.** fatigue.

4. Encapsulated nerve endings include
 a. tactile corpuscles. **b.** pain receptors.
 c. cold receptors. **d.** warm receptors.

5. A lamellated corpuscle is stimulated by
 a. light touch. **b.** deep pressure.
 c. warm temperatures. **d.** cold temperatures.

6. Determining the area of the body that has been touched is called
 a. touch localization. **b.** two-point threshold.
 c. proprioception. **d.** discrimination.

7. Free nerve endings function as pain, warm, and cold receptors.
 a. True **b.** False

8. Lamellated corpuscles are located in the epidermis of the skin.
 a. True **b.** False

Sensory receptors are sensitive to changes that occur within the body and its surroundings. Each type of receptor is particularly sensitive to a distinct kind of environmental change and is much less sensitive to other forms of stimulation. When receptors are stimulated, they initiate nerve impulses that travel into the central nervous system. The raw form in which these receptors send information to the brain is called *sensation*. The way our brains interpret this information is called *perception*.

The sensory receptors found widely distributed throughout skin, muscles, joints, and visceral organs are associated with **general senses.** These senses include touch, pressure, temperature, pain, and the senses of muscle movement and body position. Receptors (muscle spindles) associated with muscle movements were included as part of Laboratory Exercise 29.

The general senses associated with the body surface can be classified according to the stimulus type. *Mechanoreceptors* include tactile (Meissner's) corpuscles, which are stimulated by light touch, stretch, or vibration, and lamellated (Pacinian) corpuscles, which are stimulated by deep pressure, stretch, or vibration. *Thermoreceptors* include those associated with temperature changes. Warm receptors are most sensitive to temperatures between 25°C (77°F) and 45°C (113°F). Cold receptors are most sensitive to temperatures between 10°C (50°F) and 20°C (68°F). *Nociceptors* are pain receptors that respond to tissue trauma, which may include a cut or pinch, extreme heat or cold, or excessive pressure. Tests for pain receptors are not included as part of this laboratory exercise.

Receptors possess structural differences at the dendrite ends of sensory neurons. Those with unencapsulated free nerve endings include pain, cold, and warm receptors; those with encapsulated nerve endings include tactile and lamellated corpuscles.

A sensation may fade away when receptors are continuously stimulated. This is known as *sensory adaptation,* like adjusting to a room temperature or an odor of our environment. However, sensory adaptation to pain is not as prevalent because impulses may continue into the CNS for longer periods of time. The importance of perceiving pain not only alerts us to a change in our environment and potential injury, but it also allows us to detect body abnormalities so that proper adjustments can be taken.

Sensory receptors that are more specialized and confined to the head are associated with **special senses.** Laboratory Exercises 34 through 38 describe the special senses.

PROCEDURE A: Receptors and General Senses

1. Reexamine the introduction to this laboratory exercise and study table 33.1.
2. Complete Part A of Laboratory Assessment 33.

DEMONSTRATION ACTIVITY

Observe the tactile (Meissner's) corpuscle with the microscope set up by the laboratory instructor. This type of receptor is abundant in the superficial dermis in outer regions of the body, such as in the fingertips, soles, lips, and external genital organs. It is responsible for the sensation of light touch. (See fig. 33.1.)

Observe the lamellated (Pacinian) corpuscle in the second demonstration microscope. This corpuscle is composed of many layers of connective tissue cells and has a nerve fiber in its central core. Lamellated corpuscles are numerous in the hands, feet, joints, and external genital organs. They are responsible for the sense of deep pressure (fig. 33.2). How are tactile and lamellated corpuscles similar?

____

____

How are they different?

____

____

Table 33.1 Receptors Associated with General Senses of Skin

Receptor Category	Respond to	Receptor Examples
Mechanoreceptors	Touch, stretch, vibration, pressure	Encapsulated tactile (Meissner's) corpuscle; encapsulated lamellated (Pacinian) corpuscle
Thermoreceptors	Temperature changes	Free nerve endings for warm temperatures; free nerve endings for cold temperatures
Nociceptors	Intense mechanical, chemical, or temperature stimuli; tissue trauma	Free nerve endings for pain

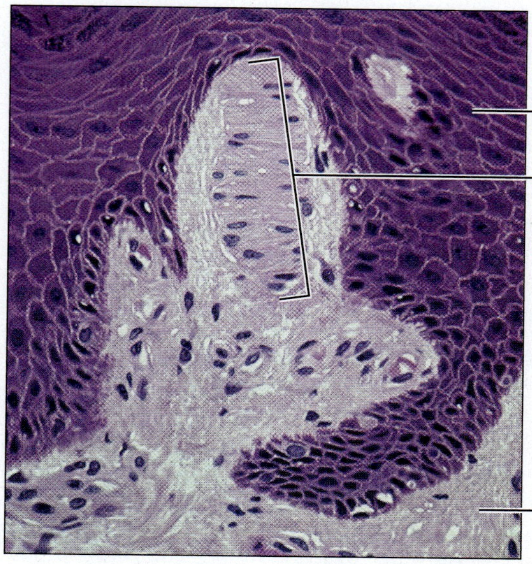

FIGURE 33.1 Tactile (Meissner's) corpuscles, such as this one, are responsible for the sensation of light touch (250×). Ed Reschke/Photolibrary/Getty Images

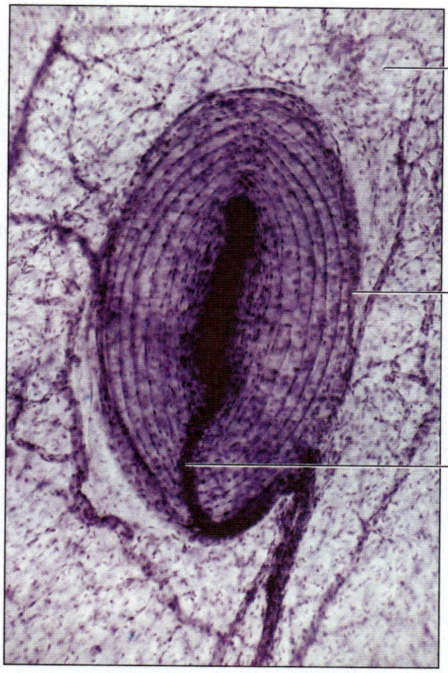

FIGURE 33.2 Lamellated (Pacinian) corpuscles, such as this one, are responsible for the sensation of deep pressure (100×). Ed Reschke/Photolibrary/Getty Images

PROCEDURE B: Touch (Tactile) Localization

1. Certain areas of the body have numerous touch receptors, making touch (tactile) localization more precise, while other areas have few touch receptors, resulting in less precision. The brain interprets the touch receptor field in the primary somatosensory cortex. Investigate the ability of your partner to determine where the skin has been touched. To do this, follow these steps:
   - **a.** Ask your partner to relax and sit down in a comfortable chair with wrists resting on thighs. Give your partner a pointed marking pen. Your partner should keep eyes closed through the remainder of this procedure.
   - **b.** Using a different colored pointed marking pen, touch your partner's fingertip. Then, ask your partner to use his/her pointed marker to touch the exact location where you touched.
   - **c.** Use a millimeter ruler to measure in millimeters (mm) the distance between your touch and that of your partner; thus, you are measuring the error of localization. Record the first test measurement in the table in Part B of the laboratory assessments.
   - **d.** Do a second test of the fingertip as you did before, and then average the two measurements. Record these measurements in Part B of the laboratory assessment.
2. Repeat steps a to d to test your partner's palm, back of the hand, back of the neck, forearm, and leg.
3. Answer the questions in Part B of the laboratory assessment.

PROCEDURE C: Two-Point Threshold

1. Test your partner's ability to recognize the difference between one and two points of skin being stimulated simultaneously. To do this, follow these steps:
   - **a.** Have your partner place a hand with the palm up on the table and close his or her eyes.
   - **b.** Hold the tips of a forceps tightly together and gently touch the skin on your partner's fingertip (fig. 33.3).
   - **c.** Ask your partner to report if it feels like one or two points are touching the finger.
   - **d.** Allow the tips of the forceps to spread so they are 1 mm apart, press both points against the skin simultaneously, and ask your partner to report as before.
   - **e.** Repeat this procedure, allowing the tips of the forceps to spread more each time until your partner can feel both tips being pressed against the skin. The minimum distance between the tips of the forceps when both can be felt is called the *two-point threshold.* As soon as you are able to distinguish two points, two separate receptors are being stimulated instead of only one receptor (fig. 33.3). Areas of skin with a small threshold are more sensitive than areas with large two-point thresholds.
   - **f.** Record the two-point threshold for the skin of a fingertip in Part C of the laboratory assessment.

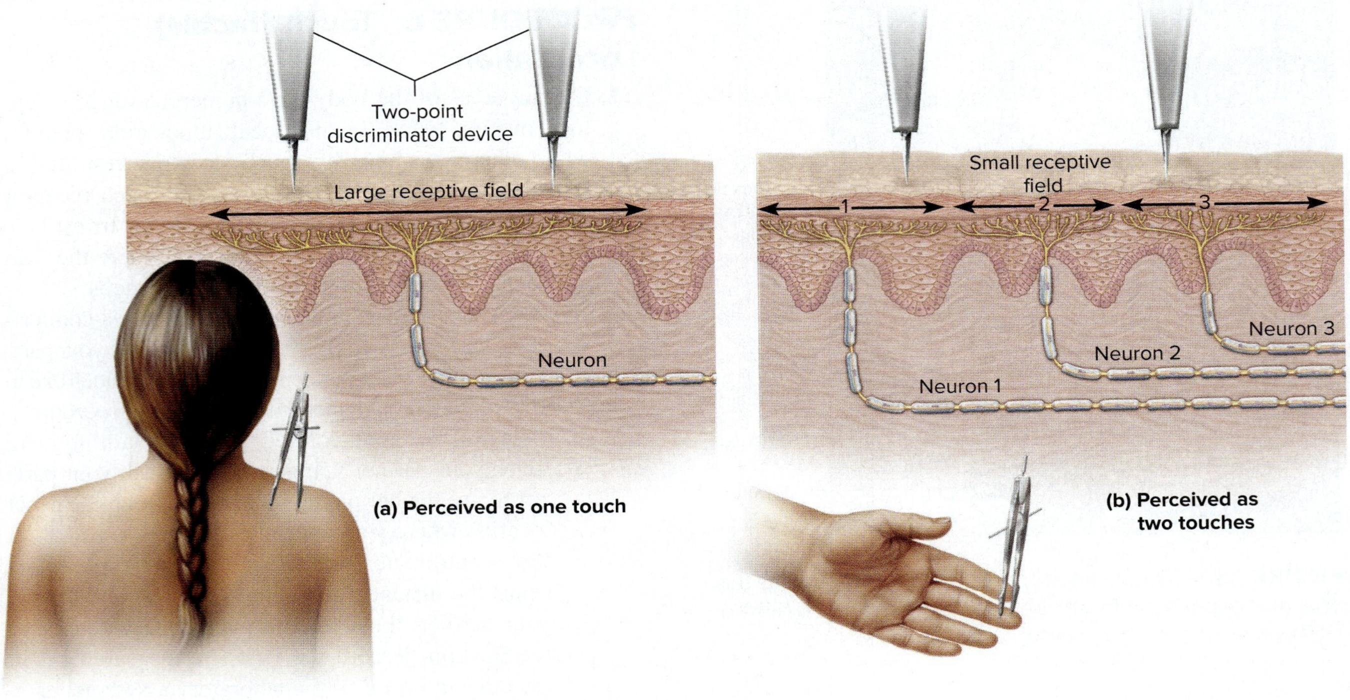

FIGURE 33.3 Two-point threshold determination test. Example (a) stimulates a single sensory neuron and (b) stimulates two different sensory neurons.

2. Repeat this procedure to determine the two-point threshold of the palm, the back of the hand, the back of the neck, the forearm, and the leg. Record the results in Part C of the laboratory assessment.
3. Answer the questions in Part C of the laboratory assessment.

PROCEDURE D: Sense of Temperature

1. Investigate the distribution of *warm thermoreceptors* in your partner's skin. To do this, follow these steps:
   a. Mark a square with 2.5 cm sides on your partner's palm.
   b. Prepare a grid by dividing the square into smaller squares, 0.5 cm on a side.
   c. Have your partner rest the marked palm on the table and close his or her eyes.
   d. Heat a blunt metal probe by placing it in a beaker of warm (hot) water (about 40–45°C/ 104–113°F) for a minute or so. (*Be sure the probe does not get so hot that it burns the skin.*) Use a thermometer to monitor the appropriate warm water from the tap or the water bath.
   e. Wipe the probe dry and touch it to the skin on some part of the grid.
   f. Ask your partner to report if the probe feels warm. Then record the results in Part D of the laboratory assessment.
   g. Keep the probe warm, and repeat the procedure until you have randomly tested several different locations on the grid.
2. Investigate the distribution of *cold receptors* by repeating the procedure. Use a blunt metal probe that has been cooled by placing it in ice water for a minute or so. Record the results in Part D of the laboratory assessment.
3. Answer the questions in Part D of the laboratory assessment.

LEARNING EXTENSION ACTIVITY

Prepare three beakers of water of different temperatures. One beaker should contain warm water (about 40°C/104°F), one should be room temperature (about 22°C/72°F), and one should contain cold water (about 10°C/50°F). Place the index finger of one hand in the warm water and, at the same time, place the index finger of the other hand in the cold water for about 2 minutes. Then, simultaneously move both index fingers into the water at room temperature. What temperature do you sense with each finger? How do you explain the resulting perceptions?

____

____

____

____

____

____

LABORATORY
ASSESSMENT 33

Name ______________________________

Date ______________________________

Section ______________________________

The A corresponds to the indicated Learning Outcome(s) O found at the beginning of the Laboratory Exercise.

General Senses

PART A: Receptors and General Senses Assessments

Complete the following statements:

1. Whenever tissues are damaged, ______________________ receptors are likely to be stimulated. A1
2. Receptors that are sensitive to temperature changes are called ______________________. A1
3. A sensation may seem to fade away when receptors are continuously stimulated as a result of ______________________ adaptation. A1
4. Tactile (Meissner's) corpuscles are responsible for the sense of light ______________________. A1
5. Lamellated (Pacinian) corpuscles are responsible for the sense of deep ______________________. A1
6. ______________________ receptors are most sensitive to temperatures between 25°C (77°F) and 45°C (113°F). A1
7. ______________________ receptors are most sensitive to temperatures between 10°C (50°F) and 20°C (68°F). A1
8. Widely distributed sensory receptors throughout the body are associated with ______________________ senses in contrast to special senses. A1

PART B: Touch (Tactile) Localization Assessments

1. Record the localization error for each area of the body you tested. A2

AREA OF BODY TESTED	ERROR OF LOCALIZATION (MM)		
	First Test	Second Test	Average
Fingertip			
Palm			
Back of hand			
Back of neck			
Forearm			
Leg			

2. Answer the following questions:

   a. What body area tested had the smallest error of localization? ______________________

   What body area tested has the largest error of localization? ______________________

   What is your interpretation of these results? ______________________

   ____ A2

b. Compare your measurement results of the first test to those of the second test of each body area. Did your ability to localize the touch improve or deteriorate? ____________________

____

Predict the outcome if you were to do a third test on each body area and explain your reasoning. ____________________

____

____________________. A2

PART C: Two-Point Threshold Assessments

1. Record the two-point threshold in millimeters for skin in each of the following regions: A3

Area of Body Tested	Two-Point Threshold (mm)
Fingertip	
Palm	
Back of hand	
Back of neck	
Forearm	
Leg	

2. Answer the following questions:

a. What region of the skin tested has the greatest ability to discriminate two points? ____________________

What region of the skin tested has the least sensitivity to this test? ____________________

What is the significance of these observations? ____________________

____

____________________ A3

b. Predict how a two-point threshold measurement of your lips would compare to the areas of the body you tested and recorded in #1. ____________________

____________________ A3

PART D: Sense of Temperature Assessments

1. Record a + to indicate where warm was felt and a *0* to indicate where it was not felt. A4

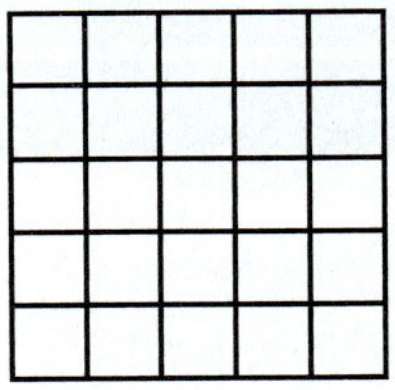

Skin of palm

2. Record a + to indicate where cold was felt and a *0* to indicate where it was not felt. A4

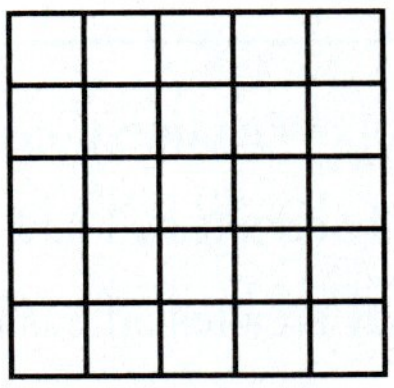

Skin of palm

3. Answer the following questions:

a. How do temperature receptors appear to be distributed in the skin of the palm? A4 ______________________

____

b. Compare the distribution and concentration of warm and cold receptors in the skin of the palm. A4 __________

____

____

NOTES

LABORATORY
EXERCISE 34

Smell and Taste

Purpose of the Exercise

To review the structures of the organs of smell and taste and to investigate the abilities of smell and taste receptors to discriminate various chemical substances.

MATERIALS NEEDED

For Procedure A: Sense of Smell (Olfaction)

Compound light microscope

Prepared microscope slide of olfactory epithelium

Set of substances in stoppered bottles: cinnamon, sage, vanilla, garlic powder, oil of clove, oil of wintergreen, and perfume

For Procedure B: Sense of Taste (Gustation)

Compound light microscope

Prepared microscope slide of the tongue surface with taste buds

Paper cups (small)

Cotton swabs (sterile; disposable)

5% sucrose solution (sweet)

5% NaCl solution (salt)

1% acetic acid or unsweetened lemon juice (sour)

0.5% quinine sulfate solution or 0.1% Epsom salt solution (bitter)

1% monosodium glutamate (MSG) solution (umami)

For Learning Extension Activity:

Pieces of apple, potato, carrot, and onion or packages of mixed flavors of LifeSavers

SAFETY

- Review all the safety guidelines in Appendix 1 of your laboratory manual.
- Be aware of possible food allergies when selecting test solutions and foods to taste.
- Prepare fresh solutions for use in Procedure B.
- Wash your hands before starting the taste experiment.
- Wear disposable gloves and protective eyewear when performing taste tests on your laboratory partner.
- Use a clean cotton swab for each test. Do not dip a used swab into a test solution.
- Dispose of used cotton swabs and paper towels as directed.
- Wash your hands before leaving the laboratory.

Chapter Opener: ©101cats/Getty Images

Learning Outcomes APR

After completing this exercise, you should be able to:

O1 Identify and describe the anatomical structures and functions associated with the sense of smell.

O2 Indicate the cranial nerves and the regions of the brain involved with the perception of smell and taste.

O3 Record recognized odors and the time needed for olfactory sensory adaptation to occur.

O4 Identify and describe the anatomical structures and functions associated with the sense of taste.

O5 Locate the distribution of taste receptors on the surface of the tongue and mouth cavity.

The O corresponds to the assessments A indicated in the Laboratory Assessment for this Exercise.

Pre-Lab

Carefully read the introductory material and examine the entire lab. Be familiar with the basic structures and functions of the receptors associated with smell and taste from the lecture or the textbook. Answer the pre-lab questions.

Pre-Lab Questions Select the correct answer for each of the following questions:

1. Receptor cells for taste are located

a. only on the tongue.
b. on the tongue, oral cavity, and pharynx.
c. in the oral cavity except the tongue.
d. on the tongue and nasal passages.

2. The primary olfactory interpretation center is located in the

a. oral cavity.
b. inferior nose.
c. temporal lobe of the cerebrum.
d. brainstem.

3. Which of the following is *not* considered a recognized taste?
   a. mint b. salty
   c. sweet d. umami
4. Taste interpretation occurs in the _____ of the cerebrum.
   a. frontal b. temporal
   c. occipital d. insula
5. Sour sensations are produced from
   a. acids. b. sugars.
   c. ionized inorganic salts. d. alkaloids.
6. Sense of taste is called _____.
   a. gustation b. olfaction
   c. sensation d. adaptation
7. The facial, glossopharyngeal, and vagus cranial nerves conduct impulses from taste receptors to the brain.
   a. True b. False
8. Olfactory nerves are located in the foramina of the inferior nasal concha.
   a. True b. False

The senses of smell (olfaction) and taste (gustation) are dependent upon modified sensory neurons, called *chemoreceptors,* that are stimulated by various chemicals dissolved in liquids. The receptors of smell are found in the olfactory organs, which are located in the superior parts of the nasal cavity and in a portion of the nasal septum. The receptors of taste occur in the taste buds, which are sensory organs primarily found on the surface of the tongue. Chemicals are considered odorless and tasteless if receptor sites for them are absent.

Olfactory receptor cells are actually neurons surrounded by pseudostratified columnar epithelial cells, called supporting cells. Their apical ends are covered with cilia, also called olfactory hairs, embedded in the mucus of the superior nasal cavity. These neurons form the *olfactory nerve* (cranial nerve I). In order to detect an odor, the molecules must first dissolve in the mucus before they bind to the receptor sites of the cilia (olfactory hairs). When receptor cells temporarily bind to an odorant molecule, it results in an action potential over the olfactory neurons passing through the foramina of the cribriform plate (of the ethmoid bone) and nerve fibers in the olfactory bulb. While several areas of the brain are involved as interpreting centers for smell, the *primary olfactory cortex* is located in the *temporal lobe* of the cerebrum. Although humans have the ability to distinguish thousands of different odors, coded by genes, various odors can result from combinations of the receptor cells stimulated. We become sensory adapted to an odor very quickly. *Sensory adaptation* occurs when there is a decreased response or interpretation of a certain stimulus. Exposure to a different substance can be quickly noticed.

Taste receptor cells are located in *taste buds* on the tongue, but receptor cells are also distributed in other areas of the oral cavity and pharynx. A taste bud contains *taste* (*gustatory*) *cells* (receptor cells), *basal cells,* and *supporting cells.* Taste cells are frequently damaged by friction and temperature changes of foods in the oral cavity, and they have a life span of only 7 to 10 days. Fortunately, the basal cells are stem cells that replace the taste receptor cells that have died. Taste cells possess terminal microvilli, called *taste hairs,* that project through a taste pore on the epithelium of the tongue. The taste hairs function as sensory receptor surfaces for substances we are able to taste. Taste sensations are grouped into five recognized categories: sweet, sour, salt, bitter, and umami. The *sweet* sensation is produced from sugars, the *sour* sensation from acids, the *salt* sensation from ionized inorganic salts, the *bitter* sensation from alkaloids and spoiled foods, and the *umami* sensation from aspartic and glutamic acids or a derivative such as monosodium glutamate. Three cranial nerves conduct impulses from the taste buds. The *facial nerve* (*VII*) conducts sensory impulses from the anterior two-thirds of the tongue; the *glossopharyngeal nerve* (*IX*) from the posterior tongue; and the *vagus nerve* (*X*) from the pharyngeal region. When a molecule binds to a receptor cell, a neurotransmitter is secreted and an action potential occurs over a sensory neuron. The sensory information continues through the medulla oblongata and the thalamus to the *insula* of the cerebrum. The insula serves as the *primary gustatory cortex* for the perception of taste. Sensory adaptation to taste sensations occurs rather quickly, which helps explain the reason the first few bites of a particular food have the most vivid flavor.

The senses of smell and taste function closely together, because substances that are tasted often are smelled at the same moment, and they play important roles in the selection of foods. The taste and aroma of foods are also influenced by appearance, texture, temperature, and the person's mood.

PROCEDURE A: Sense of Smell (Olfaction)

1. Obtain a prepared microscope slide of an olfactory epithelium and observe it under high-power magnification. The olfactory receptor cells are spindle-shaped, bipolar neurons with spherical nuclei. They also have six to eight cilia (olfactory hairs) at their apical ends. The supporting cells are pseudostratified columnar epithelial cells. However, in this region the tissue lacks goblet cells (fig. 34.1).
2. Reexamine the introduction to this laboratory exercise and study figure 34.2.
3. Complete Part A of Laboratory Assessment 34.
4. Test your laboratory partner's ability to recognize the odors of the bottled substances available in the laboratory. To do this, follow these steps:
   a. Have your partner keep their eyes closed.
   b. Remove the stopper from one of the bottles, and hold it about 4 cm under your partner's nostrils for about 2 seconds.
   c. Ask your partner to identify the odor, and then replace the stopper.
   d. Record your partner's response in Part B of the laboratory assessment.
   e. Repeat steps b to d for each of the bottled substances.
5. Repeat the preceding procedure, using the same set of bottled substances, but present them to your partner in

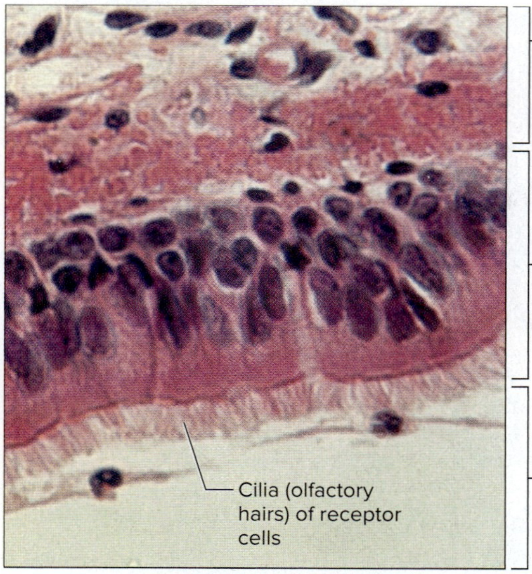

FIGURE 34.1 Micrograph of olfactory epithelium (250×). Olfactory receptors have cilia (olfactory hairs) at their apical ends. Biophoto Associates/Science Source APR

a different sequence. Record the results in Part B of the laboratory assessment.

6. Wait 10 minutes and then determine the time it takes for your partner to experience olfactory sensory adaptation. To do this, follow these steps:
   a. Ask your partner to breathe in through the nostrils and exhale through the mouth.
   b. Remove the stopper from one of the bottles, and hold it about 4 cm under your partner's nostrils.
   c. Keep track of the time that passes until your partner is no longer able to detect the odor of the substance.
   d. Record the result in Part B of the laboratory assessment.
   e. Wait 5 minutes and repeat this procedure, using a different bottled substance.
   f. Test a third substance in the same manner.
   g. Record the results as before.
7. Complete Part B of the laboratory assessment.

PROCEDURE B: Sense of Taste (Gustation)

1. Obtain a prepared microscope slide of the tongue surface with taste buds and observe it under low-power and high-power magnifications. Observe the oval-shaped taste bud and the surrounding epithelial cells. The taste pore, an opening into the taste bud, may be filled with taste hairs (microvilli). Within the taste bud there are supporting cells and thinner taste-receptor cells, which often have lightly stained nuclei (figs. 34.3 and 34.4).
2. Reexamine the introduction to this laboratory exercise and study figure 34.5.
3. Complete Part C of the laboratory assessment.
4. Identify the distribution of the receptors for the primary taste sensations on your partner's tongue. To do this, follow these steps:
   a. Ask your partner to rinse his or her mouth with water and then partially dry the surface of the tongue with a paper towel.
   b. Moisten a clean cotton swab with 5% sucrose solution, and touch several regions of your partner's tongue with the swab.
   c. Each time you touch the tongue, ask your partner to report what sensation is experienced.
   d. Test the tip, sides, and back of the tongue in this manner.

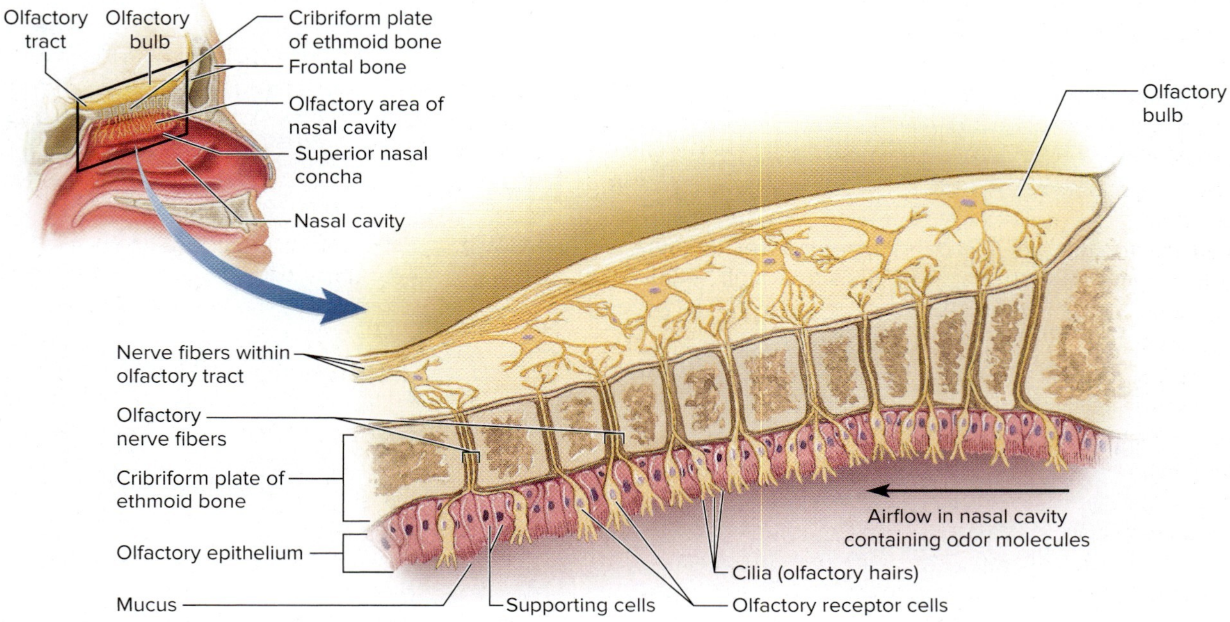

FIGURE 34.2 Structures associated with smell. APR

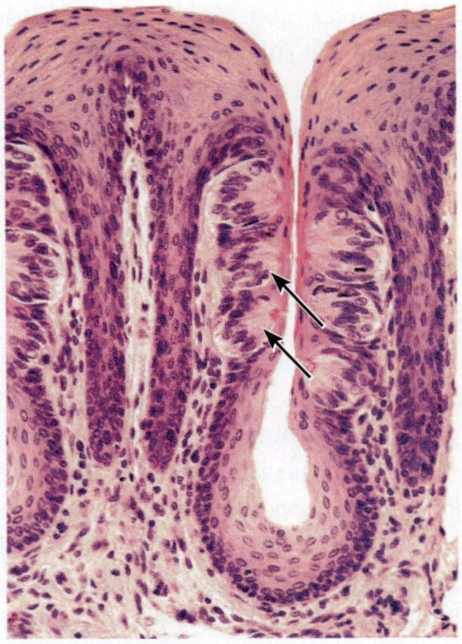

FIGURE 34.3 Micrograph of tongue surface with taste buds (225x). The arrows mark two of several taste buds viewed in this segment of the tongue. Jose Luis Calvo/ Shutterstock APR

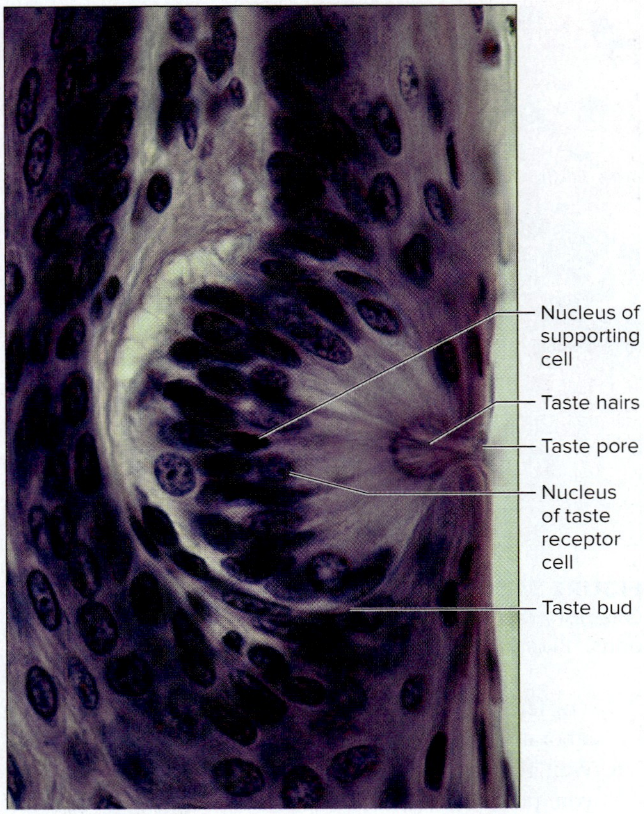

FIGURE 34.4 Micrograph of a taste bud (400x). Taste receptors are found in taste buds. Ed Reschke/Photolibrary/ Getty Images

   e. Test some other representative areas inside the oral cavity such as the cheek, gums, and roof of the mouth (hard and soft palate) for a sweet sensation.
   f. Record your partner's responses in Part D of the laboratory assessment.
   g. Have your partner rinse his or her mouth and dry the tongue again, and repeat the preceding procedure, using each of the other four test solutions—NaCl, acetic acid, quinine or Epsom salt, and MSG solution. Be sure to use a fresh swab for each test substance and dispose of used swabs and paper towels as directed.

5. Complete Part D of the laboratory assessment.

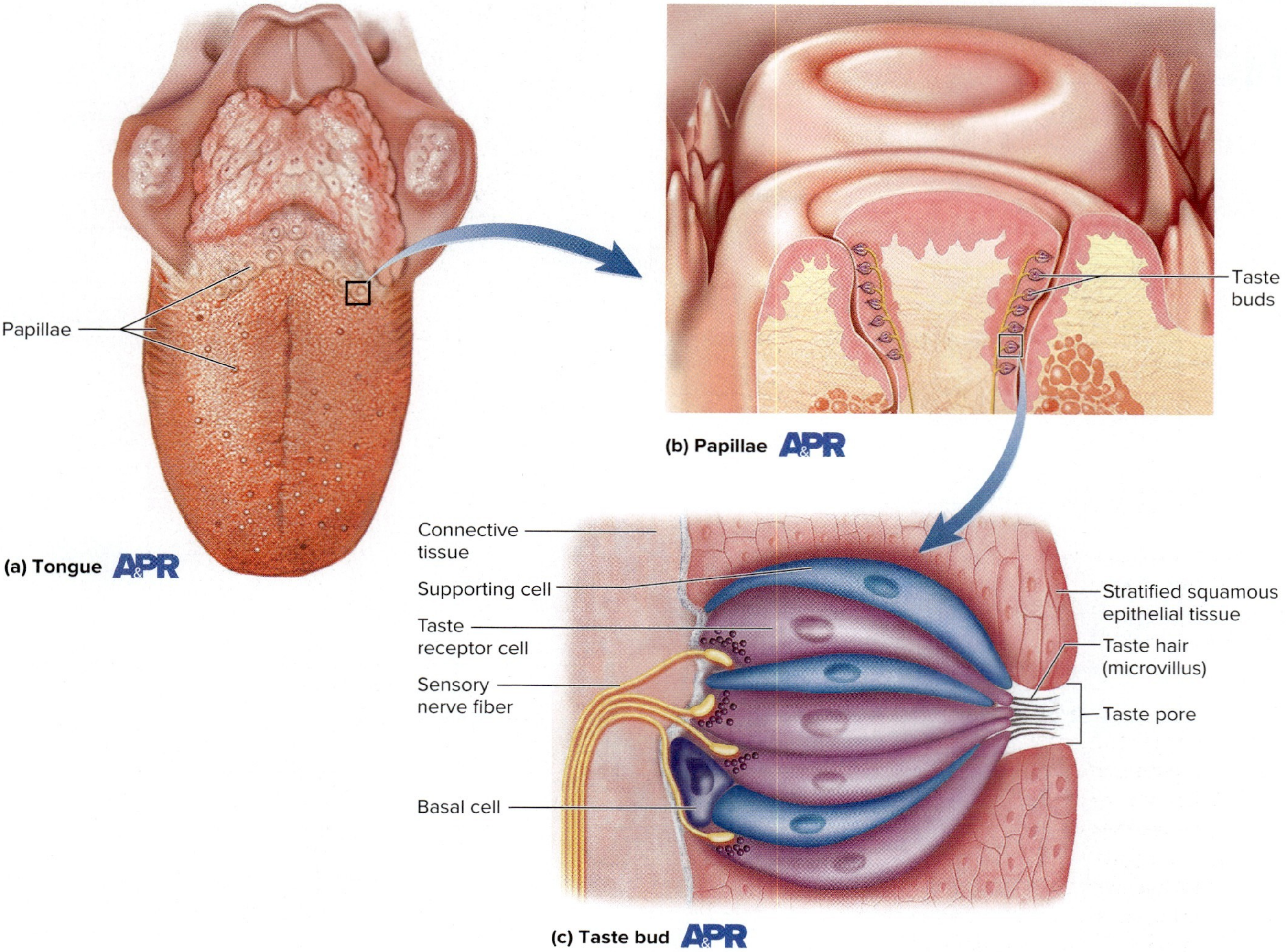

FIGURE 34.5 Structures associated with taste: (a) tongue, (b) papillae, and (c) taste bud.

LEARNING EXTENSION ACTIVITY

Test your laboratory partner's ability to recognize the tastes of apple, potato, carrot, and onion. (A package of mixed flavors of LifeSavers is a good alternative.) To do this, follow these steps:

1. Have your partner close their eyes and hold the nostrils shut.
2. Place a small piece of one of the test substances on your partner's tongue.
3. Ask your partner to identify the substance without chewing or swallowing it.
4. Repeat the procedure for each of the other substances.

How do you explain the results of this experiment?

____

____

____

NOTES

Name ______________________________

Date ______________________________

Section ______________________________

The (A) corresponds to the indicated Learning Outcome(s) (O) found at the beginning of the Laboratory Exercise.

LABORATORY ASSESSMENT 34

Smell and Taste

PART A: Assessments

Complete the following statements:

1. The distal ends of the olfactory neurons are covered with hairlike ________________. A1
2. Before gaseous substances can stimulate the olfactory receptors, they must be dissolved in ________________ that surrounds the cilia. A1
3. The axons of olfactory receptors pass through small openings in the ________________ of the ethmoid bone. A2
4. The primary olfactory cortex for interpreting smell is located in the ________________ of the cerebrum. A2
5. Olfactory sensations usually fade rapidly as a result of ________________. A1
6. A chemical would be considered ________________ if a person lacks a particular receptor site on the cilia of the olfactory neurons. A1

PART B: Sense of Smell Assessments

1. Record the results (as +, if recognized; as *0*, if unrecognized) from the tests of odor recognition in the following table: A3

Substance Tested	Odor Reported	
	First Trial	Second Trial

2. Record the results of the olfactory sensory adaptation time in the following table: A3

Substance Tested	Adaptation Time in Seconds

3. Complete the following:

   a. How do you describe your partner's ability to recognize the odors of the substances you tested? A1

   ____

   ____

   b. Compare your experimental results with those of others in the class. Did you find any evidence to indicate that individuals may vary in their ability to recognize odors? Explain your answer. A1

   ____

   ____

   ____

CRITICAL THINKING ASSESSMENT—PART B

Does the time it takes for sensory adaptation to occur seem to vary with the substances tested? Explain your answer. A3

____

____

____

PART C: Assessments

Complete the following statements:

1. Taste is interpreted in the ________________________ of the cerebrum. A2
2. The opening to a taste bud is called a ________________________. A4
3. The ________________________ of a taste cell are its sensitive part. A4
4. The facial, ________________________, and vagus cranial nerves conduct impulses related to the sense of taste. A2
5. Before the taste of a substance can be detected, the substance must be dissolved in ________________________. A4
6. Basal cells are ________________________ cells that multiply to replace dead taste cells. A4
7. Substances that stimulate taste cells bind with ________________________ sites on the surfaces of taste hairs. A4
8. Sour receptors are mainly stimulated by ________________________. A4
9. Salt receptors are mainly stimulated by ionized inorganic ________________________. A4
10. Alkaloids usually have a ________________________ taste. A4

PART D: Sense of Taste Assessments

1. *Taste receptor distribution.* Record a + to indicate where a taste sensation seemed to originate and a *0* if no sensation occurred when the spot was stimulated. A5

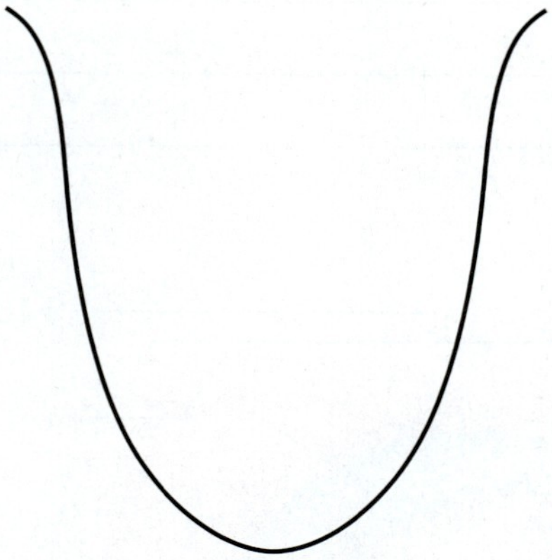

Sucrose (sweet sensation)

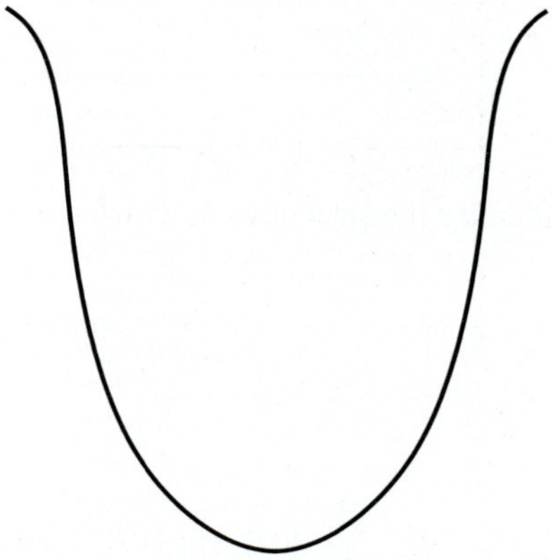

NaCl (salty sensation)

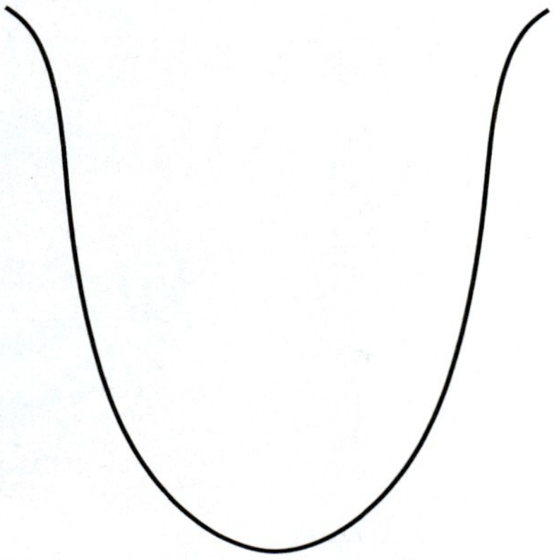

MSG (umami sensation)

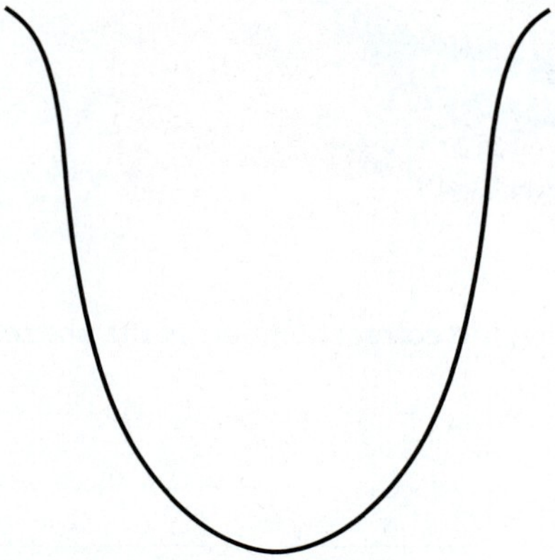

Acetic acid (sour sensation)

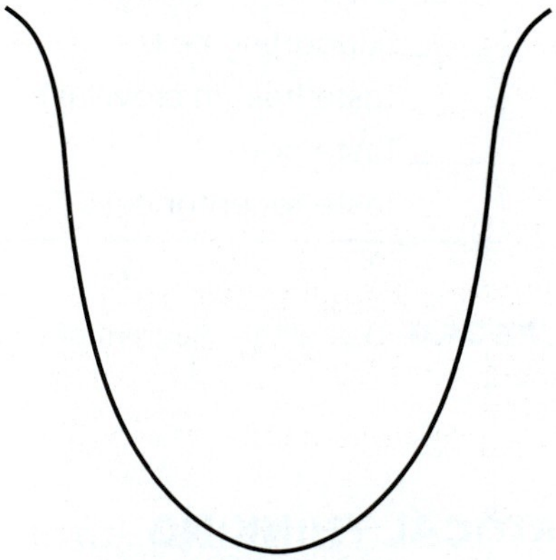

Quinine (bitter sensation)

2. Complete the following:

   a. Describe how each type of taste receptor is distributed on the surface of your partner's tongue. A5

   b. Describe other locations inside the mouth where any sensations of sweet, salty, sour, bitter, or umami were located. A5

c. How does your taste receptor distribution on the tongue compare to those of other students in the class? A5

____

____

____

____

3. Identify the structures associated with a taste bud in figure 34.6. A4

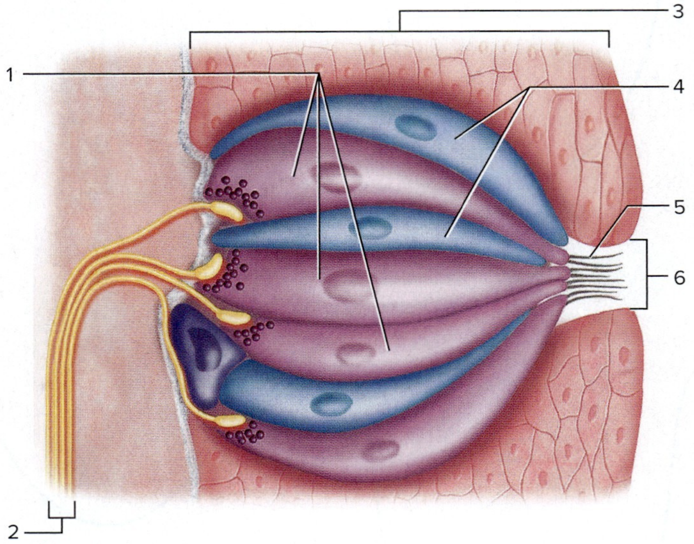

____ Epithelial tissue of tongue
____ Sensory nerve fibers
____ Supporting cells
____ Taste hair (microvillus)
____ Taste pore
____ Taste receptor cells

FIGURE 34.6 Label this diagram of structures associated with a taste bud by placing the correct numbers in the spaces provided.

CRITICAL THINKING ASSESSMENT—PART D

Why would an individual who has an upper respiratory infection with nasal congestion have difficulty sensing smell and taste? A1 A4 ____

____

____

LABORATORY
EXERCISE 35

Eye Structure

Purpose of the Exercise

To review the structure and function of the eye and to dissect a mammalian eye.

MATERIALS NEEDED

- Dissectible eye model
- Compound light microscope
- Prepared microscope slide of a mammalian eye (sagittal section)
- Sheep or cow eye (fresh or preserved)
- Dissecting tray
- Dissecting instruments—forceps, sharp scissors, and dissecting needle
- Disposable gloves

For Learning Extension Activity:
Ophthalmoscope

SAFETY

- Review all the safety guidelines in Appendix 1 of your laboratory manual.
- Do not allow light to reach the macula lutea during the eye exam for longer than 1-second intervals.
- Wear disposable gloves and protective eyewear when working on the eye dissection.
- Dispose of tissue remnants and gloves as instructed.
- Wash the dissecting tray and instruments as instructed.
- Wash your laboratory table.
- Wash your hands before leaving the laboratory.

Chapter Opener: ©101cats/Getty Images

Learning Outcomes APR

After completing this exercise, you should be able to:

O1 Identify the major structures of an eye.

O2 Describe the functions of the structures of an eye.

O3 Trace the structures through which light passes as it travels from the cornea to the retina.

O4 Dissect a mammalian eye and locate its major features.

The O corresponds to the assessments A indicated in the Laboratory Assessment for this Exercise.

Pre-Lab

Carefully read the introductory material and examine the entire lab. Be familiar with the basic structures and functions of parts of the eye from the lecture or the textbook. Answer the pre-lab questions.

Pre-Lab Questions Select the correct answer for each of the following questions:

1. The cornea and the sclera compose the _____ layer.
   a. outer **b.** middle
   c. inner **d.** gel
2. We are able to see color because the eye contains
   a. melanin. **b.** an optic disc.
   c. rods. **d.** cones.
3. The perception of vision occurs in the
   a. optic nerve. **b.** occipital lobe of the cerebrum.
   c. retina. **d.** frontal lobe of the cerebrum.
4. Which of the following is *not* part of the middle eye layer?
   a. choroid **b.** conjunctiva
   c. ciliary body **d.** iris
5. The area of our eye where visual acuity is best is the
   a. macula lutea. **b.** optic disc.
   c. fovea centralis. **d.** pupil.
6. Which of the following extrinsic skeletal muscles rotates the eyeball superiorly and medially?
   a. superior rectus **b.** superior oblique
   c. inferior rectus **d.** inferior oblique
7. The conjunctiva covers the superficial surface of the cornea.
   a. True **b.** False
8. Tears from the lacrimal gland eventually flow through a nasolacrimal duct into the nasal cavity.
   a. True **b.** False

The special sense of vision involves the complex structure of the eye, accessory structures associated with the orbit, the optic nerve, and the brain. The eyebrow, eyelids, conjunctiva, and lacrimal apparatus protect the eyes. The vascular *conjunctiva,* a mucous membrane, covers all of the anterior surface of the eyeball except the cornea, and it lines the eyelids. Its mucus prevents the eye from drying out, and the lacrimal fluid from the *lacrimal* (*tear*) *gland* cleanses the eye and prevents infections due to the antibacterial agent lysozyme in its secretions. Two voluntary muscles involve control of the eyelids, while six extrinsic muscles allow voluntary eye movements in multiple directions.

Three layers (tunics) compose the main structures of the eye. The *outer* (*fibrous*) *layer* includes the protective white *sclera* and a transparent anterior portion, the *cornea,* which allows light to enter the eye cavities. The *middle* (*vascular*) *layer* includes the heavily pigmented *choroid* that supports the inner layer by supplying nourishment and removing wastes, the *ciliary body* that controls the shape of the *lens,* and the *iris* with its smooth muscle fibers that control the amount of light passing through its central opening, the *pupil.* The inner layer includes pigmented and neural components of the *retina* and the optic nerve, which contains sensory neurons that conduct nerve impulses from the rods and cones. A large posterior cavity is filled with a transparent gel, the *vitreous humor.* A smaller anterior cavity has anterior and posterior chambers separated at the iris, which is filled with a clear *aqueous humor.*

The photoreceptor cells, the *rods* and *cones* of the retina, can absorb light to produce visual images. The more numerous rods can be stimulated in dim light and are most abundant in the peripheral retina. The cones are stimulated only in brighter light, but they allow us to perceive sharper images. They allow us to see images in color because there are three types of cones, with each type most sensitive to wavelengths peaking in the blue, green, or red portion of the visible light spectrum. Located at the posterior center of the retina is a region containing mostly cones, called the *macula lutea* ("yellow spot"). A tiny pit in the center of the macula lutea, the *fovea centralis,* contains only cones and represents our best image location. Slightly to the medial side of the macula lutea is the *optic disc,* where the nerve fibers of the retina leave the eye to become parts of the optic nerve. It is also referred to as the "blind spot" because rods and cones are absent at this location. Sensory impulses from the optic nerves pass through the thalamus and eventually arrive at the primary visual cortex in the occipital lobe of the cerebrum, where the perception of visual images occurs.

PROCEDURE A: Structure and Function of the Eye APR

1. Reexamine the introduction to this laboratory exercise.
2. Use figures 35.1 to 35.3, and table 35.1 as references as you progress through this procedure.
3. Examine the dissectible model of the eye and locate the following features:

   eyelid (palpebra)—anterior protection of eyes

   conjunctiva—a mucous membrane

   tarsal glands—secrete oily lubricant

   orbicularis oculi—closes eyelids

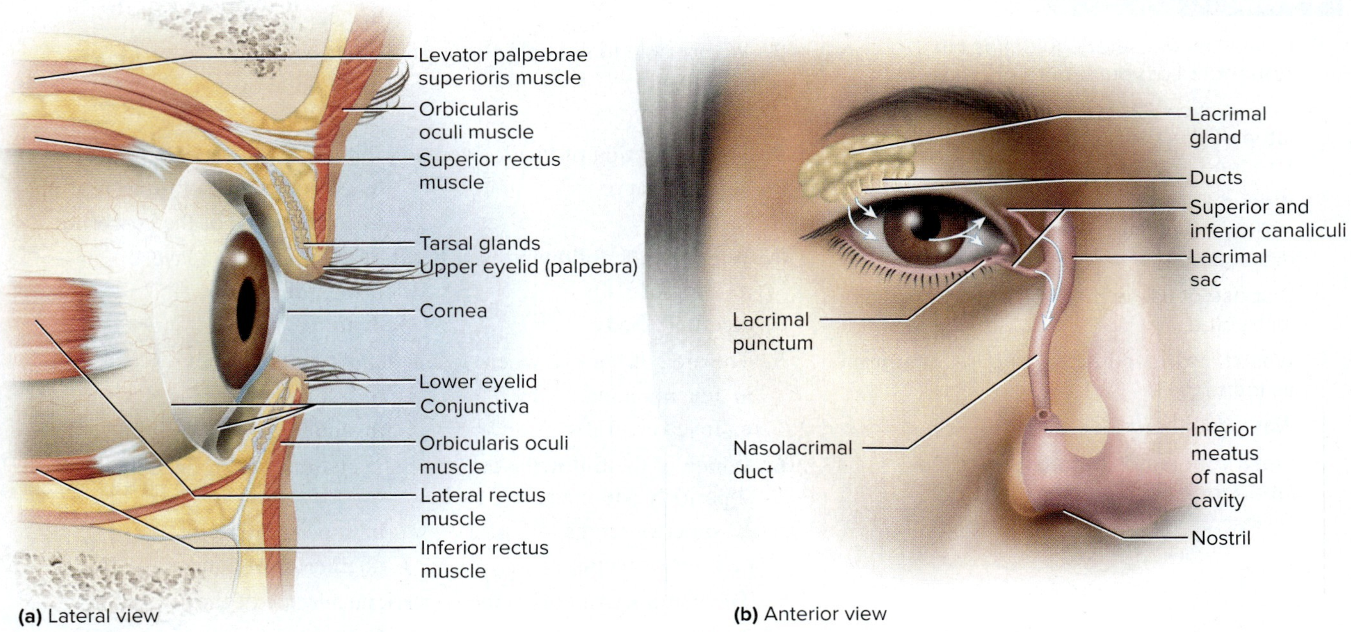

FIGURE 35.1 Accessory structures of the eye: (a) sagittal section of orbit; (b) lacrimal apparatus.

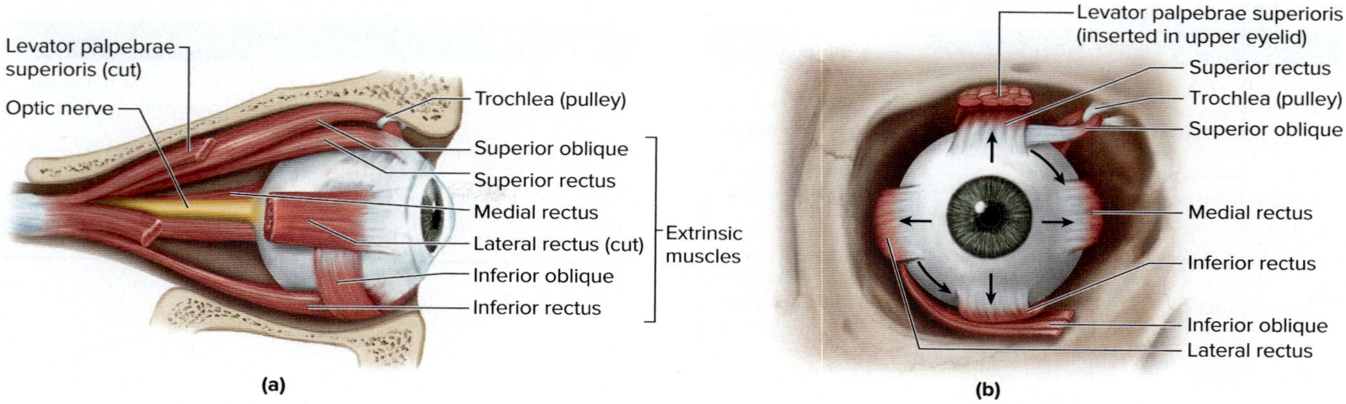

FIGURE 35.2 Extrinsic muscles of the right eyeball: (a) lateral view; (b) anterior view. The arrows indicate the direction of the pull by each of the six eyeball muscles.

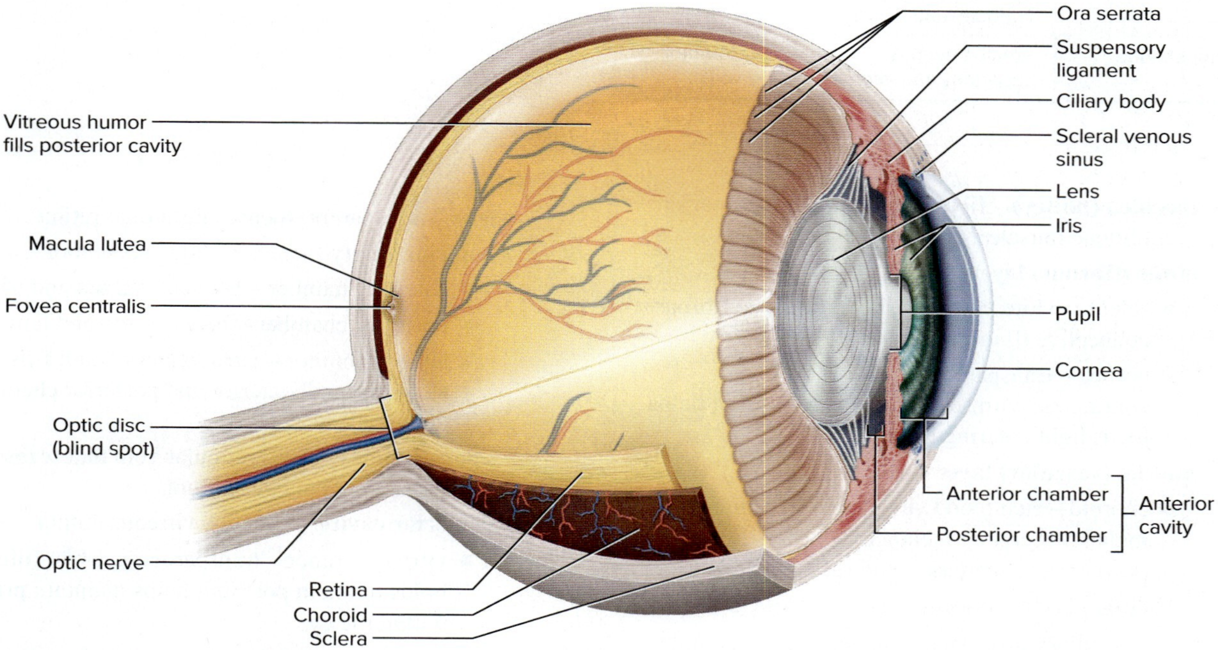

FIGURE 35.3 Structures of the eye (sagittal section). APR

levator palpebrae superioris—elevates upper eyelid

lacrimal apparatus (fig. 35.1*b*)

- lacrimal gland—secretes tears
- lacrimal puncta—two small pores into canaliculi
- canaliculi—passageways to lacrimal sac
- lacrimal sac—collects tears from cannaliculi
- nasolacrimal duct—drainage to nasal cavity

extrinsic muscles (fig. 35.2 and table 35.1)

- superior rectus
- inferior rectus
- medial rectus
- lateral rectus
- superior oblique
- inferior oblique

Table 35.1 Muscles Associated with the Eyelids and Eyes

Skeletal Muscles	Action	Innervation
Muscles of the eyelids		
Orbicularis oculi	Closes eye	Facial nerve (VII)
Levator palpebrae superioris	Opens eye	Oculomotor nerve (III)
Extrinsic muscles of the eyeballs		
Superior rectus	Rotates eyeball superiorly and medially	Oculomotor nerve (III)
Inferior rectus	Rotates eyeball inferiorly and medially	Oculomotor nerve (III)
Medial rectus	Rotates eyeball medially	Oculomotor nerve (III)
Lateral rectus	Rotates eyeball laterally	Abducens nerve (VI)
Superior oblique	Rotates eyeball inferiorly and laterally	Trochlear nerve (IV)
Inferior oblique	Rotates eyeball superiorly and laterally	Oculomotor nerve (III)

Smooth Muscles	Action	Innervation
Ciliary muscles	Relax suspensory ligaments	Oculomotor nerve (III) parasympathetic fibers
Iris, circular muscles	Constrict pupil	Oculomotor nerve (III) parasympathetic fibers
Iris, radial muscles	Dilate pupil	Sympathetic fibers

trochlea (pulley)—fibrocartilage ring for superior oblique muscle

outer (fibrous) layer (tunic)

- sclera—white protective layer of dense fibrous connective tissue
- cornea—transparent anterior portion continuous with sclera; bulges anteriorly; helps focus light entering the eye

middle (vascular) layer (tunic)

- choroid—rich blood vessel area; contains melanin-secreting melanocytes; absorbs light to prevent light scattering and reflecting
- ciliary body—thickest part of middle layer
  - ciliary processes—secrete aqueous humor
  - ciliary body (muscle)—change lens shape
  - suspensory ligaments—hold lens in position; attached to ciliary body
- iris—smooth muscles change pupil diameter
- pupil—opening in center of iris allowing light passage

inner layer (tunic)

- retina—sensory layer with rods and cones
  - macula lutea—high cone density area
  - fovea centralis—area of best visual acuity
  - optic disc—location where optic nerve leaves eye
  - orra serrata—serrated anterior margin of sensory retina
- optic nerve—contains axons of sensory neurons

lens—transparent; focuses light onto retina

anterior cavity—contains aqueous humor

- anterior chamber—between cornea and iris
- posterior chamber—between iris and lens
- aqueous humor—clear, watery fluid; fills and flows through anterior and posterior chambers and pupil

scleral venous sinus—circular vein that reabsorbs circulating aqueous humor

posterior cavity—contains vitreous humor

- vitreous humor—transparent, gel-like filler; holds retina in position; helps maintain pressure within the eye

4. Obtain a microscope slide of a mammalian eye section, and locate as many of the features in step 3 as possible.
5. Using low-power magnification, observe the posterior portion of the eye wall, and locate the sclera, choroid, and retina (fig. 35.4).
6. Using high-power magnification, examine the retina portion of the eye wall in more detail and note its layered structure. Locate the following:

   nerve fibers leading to the optic nerve—innermost layer of the retina

   layer of ganglion cells

   layer of bipolar neurons

   nuclei of rods and cones

   receptor ends of rods and cones

   pigmented epithelium—outermost layer of the retina
7. Complete Part A of Laboratory Assessment 35.

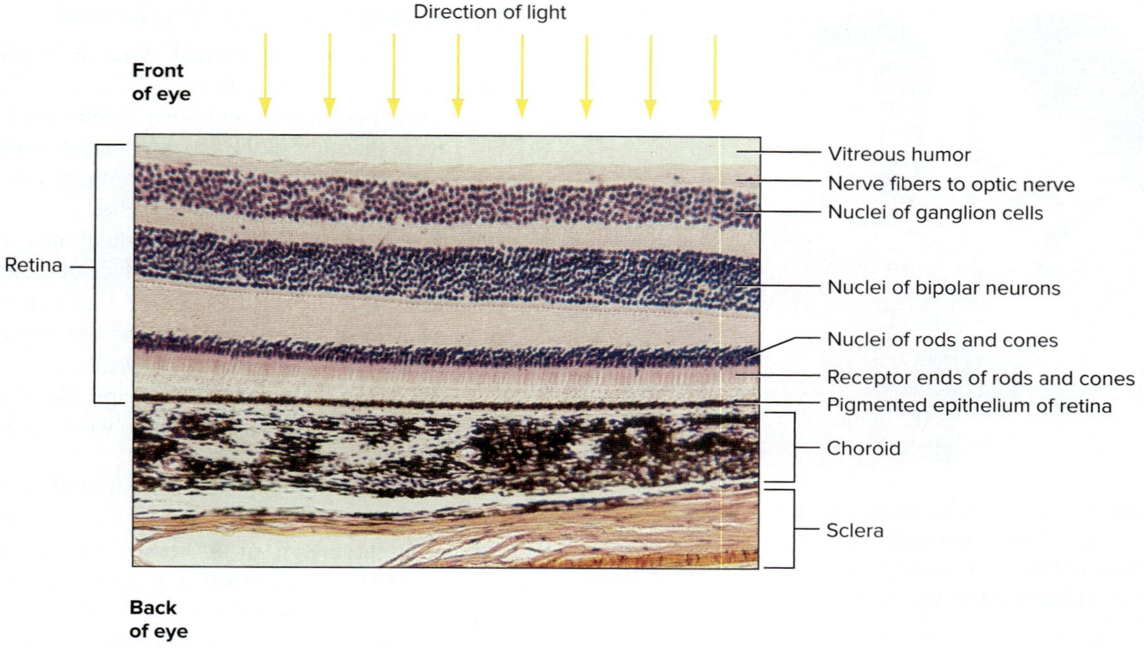

FIGURE 35.4 The cells of the retina are arranged in distinct layers (75×). ©Per Kjeldsen APR

LEARNING EXTENSION ACTIVITY

Use an ophthalmoscope to examine the interior of your laboratory partner's eye. This instrument consists of a set of lenses held in a rotating disc, a light source, and some mirrors that reflect the light into the test subject's eye (fig. 35.5).

The examination should be conducted in a dimly lit room. Have your partner seated and staring straight ahead at eye level. Move the rotating disc of the ophthalmoscope so that the *O* appears in the lens selection window. Hold the instrument in your right hand with the end of your index finger on the rotating disc. Direct the light at a slight angle from a distance of about 15 cm into the subject's right eye (fig. 35.6). The light beam should pass along the inner edge of the pupil. Look through the instrument, and you should see a reddish, circular area—the interior of the eye. Rotate the disc of lenses to higher values until sharp focus is achieved.

Move the ophthalmoscope to within about 5 cm of the eye being examined, *being very careful that the instrument does not touch the eye,* and again rotate the lenses to sharpen the focus (fig. 35.6). Locate the optic disc and the blood vessels that pass through it. Also locate the macula lutea by having your partner stare directly into the light of the instrument (fig. 35.7). *Do not allow light to reach the macula lutea region for longer than 1-second intervals.*

Examine the subject's iris by viewing it from the side and by using a lens with a +15 or +20 value.

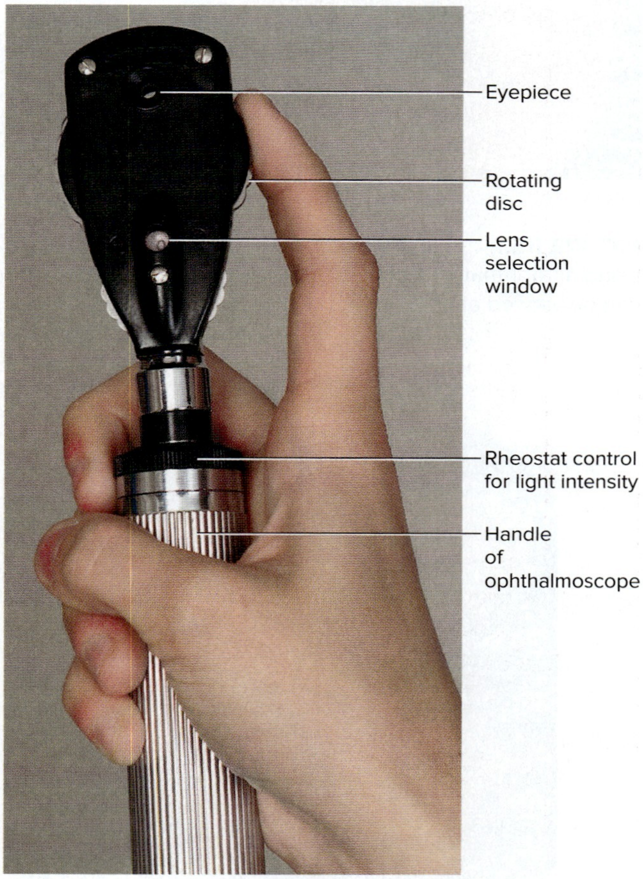

FIGURE 35.5 The ophthalmoscope is used to examine the interior of the eye. ©J & J Photography

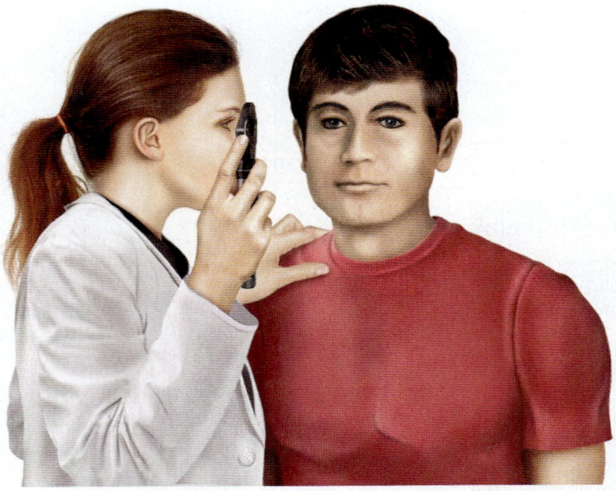

FIGURE 35.6 Examine the retina with the use of an ophthalmoscope. This is the proper position in which to examine the right eye of the subject. Follow the directions in the Learning Extension Activity to examine the left eye.

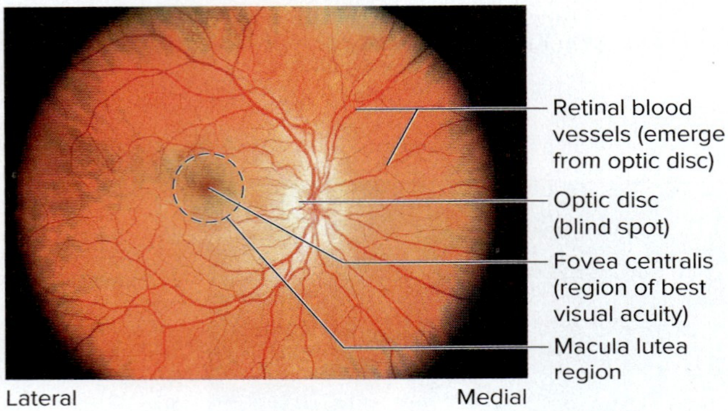

FIGURE 35.7 Right retina as viewed through the pupil using an ophthalmoscope. Only small portions of the retina can be viewed at a time. ©Lisa Klancher

PROCEDURE B: Eye Dissection

1. Obtain a mammalian eyeball, place it in a dissecting tray, and dissect it as follows:
   a. Trim away the fat and other connective tissues but leave the stubs of the *extrinsic muscles* and the *optic nerve* (fig. 35.8). This nerve projects outward from the posterior region of the eyeball.
   b. The *conjunctiva* lines the inside of the eyelid and is reflected over the anterior surface of the eye, except for the cornea. Lift some of this thin membrane away from the eye with forceps and examine it.
   c. Locate and observe the *cornea, sclera,* and *iris.* Also note the *pupil* and its shape. The cornea from a fresh eye is transparent; when preserved, it becomes opaque.
   d. Use sharp scissors to make a frontal section of the eye. To do this, cut through the wall about 1 cm from the margin of the cornea and continue all the way around the eyeball. Try not to damage the internal structures of the eye (fig. 35.9).
   e. Gently separate the eyeball into anterior and posterior portions. Usually, the jellylike vitreous humor will remain in the posterior portion, and the lens may adhere to it. Place the parts in the dissecting tray with their contents facing upward (fig. 35.10).
   f. Examine the anterior portion of the eye, and locate the *ciliary body,* which appears as a dark, circular structure. Also note the *iris* and the *lens* if it remained in the anterior portion. The lens is normally attached to the ciliary body by many suspensory ligaments, which appear as delicate, transparent threads that detach when a lens is removed.
   g. Use a dissecting needle to gently remove the lens, and examine it. If the lens is still transparent, hold it up and look through it at something in the distance and note that the lens inverts the image. The lens of

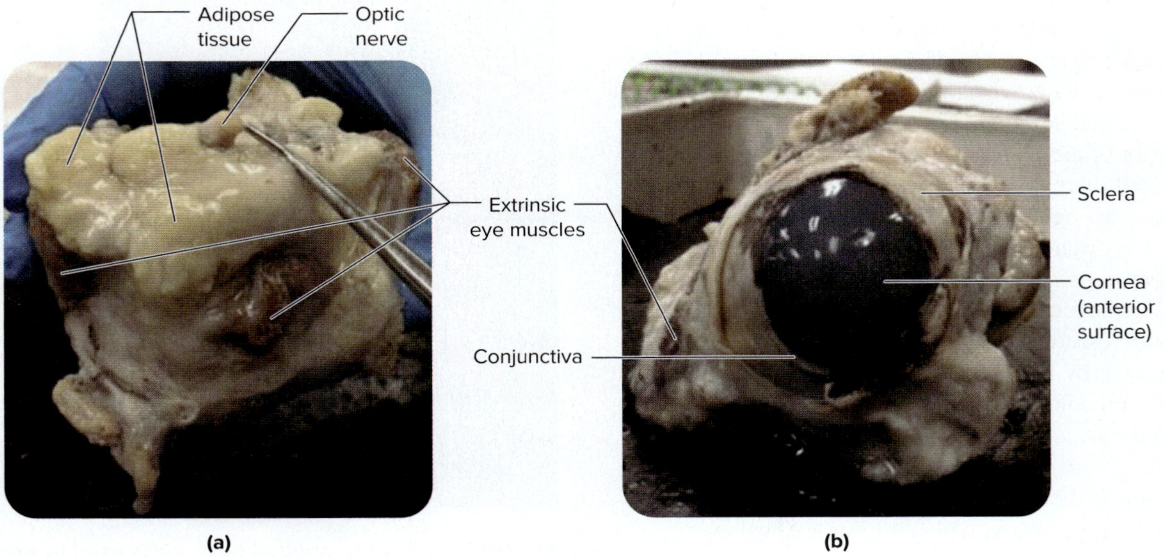

FIGURE 35.8 External structures of a preserved cow eye; (a) posterior view; (b) anterior view. Cynthia Prentice-Craver/McGraw Hill

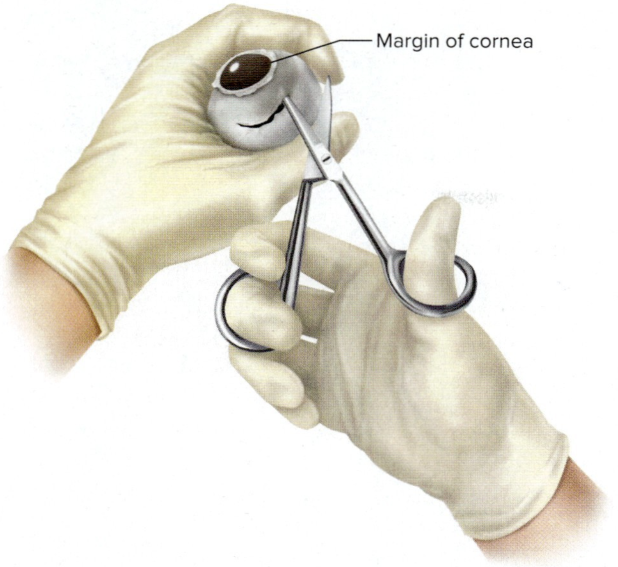

FIGURE 35.9 Prepare a frontal section of the eye.

a preserved eye is usually too opaque for this experience. If the lens of the human eye becomes opaque, the defect is called a cataract (fig. 35.11).

h. Examine the posterior portion of the eye. Note the *vitreous humor.* This jellylike mass helps to hold the lens in place anteriorly and helps to hold the *retina* against the choroid.

i. Carefully remove the vitreous humor and examine the retina. This layer will appear as a thin, nearly colorless to cream-colored membrane that detaches easily from the choroid. Compare the structures identified to figure 35.10.

j. Locate the *optic disc*—the point where the retina is attached to the posterior wall of the eyeball and where the optic nerve originates. There are no receptor cells in the optic disc, so this region is also called the "blind spot."

k. Note the iridescent area of the choroid beneath the retina. This colored surface in ungulates (mammals having hoofs) is called the *tapetum fibrosum* (an anatomical modification of the *tapetum lucidum* found in nonhoofed mammals). It reflects light back through the retina, an action thought to aid the night vision of some animals. The tapetum fibrosum is lacking in the human eye.

2. Discard the tissues of the eye as directed by the laboratory instructor.

3. Complete Parts B and C of the laboratory assessment.

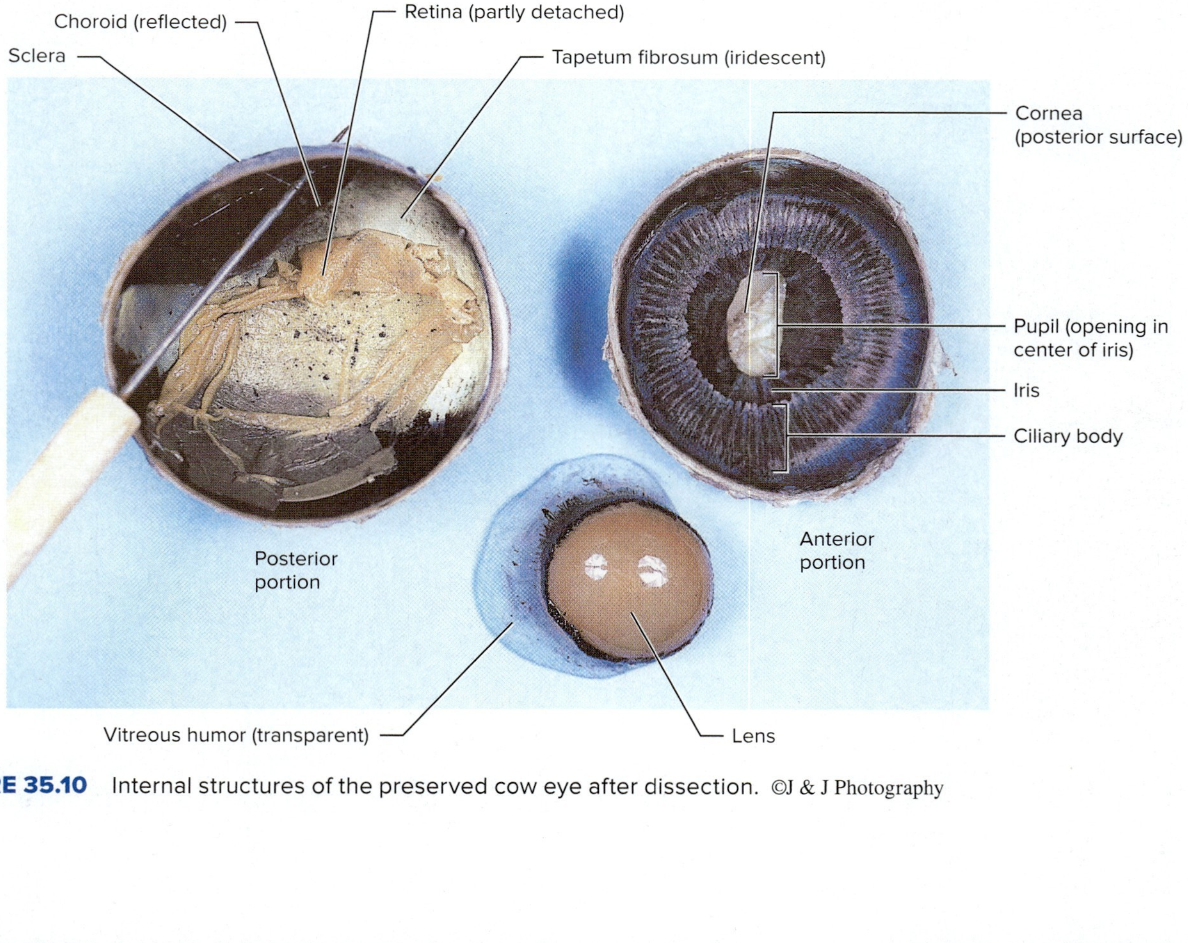

FIGURE 35.10 Internal structures of the preserved cow eye after dissection. ©J & J Photography

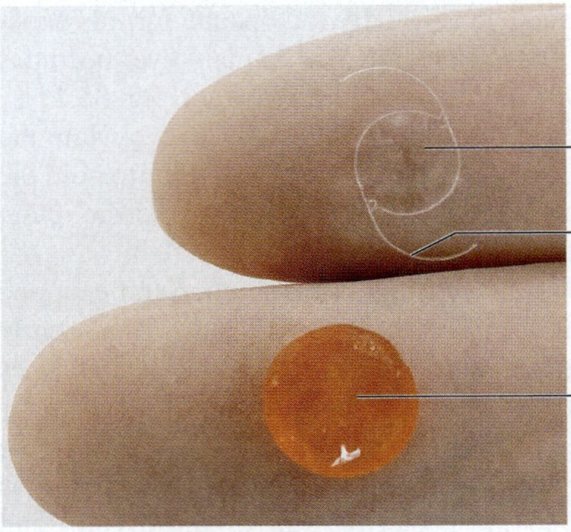

FIGURE 35.11 Human lens showing a cataract and one type of lens implant (intraocular lens replacement) on tips of fingers to show their relative size. A normal lens is transparent. ©J & J Photography

Name ______________________

Date ______________________

Section ______________________

The **A** corresponds to the indicated Learning Outcome(s) **O** found at the beginning of the Laboratory Exercise.

LABORATORY ASSESSMENT 35

Eye Structure

PART A: Assessments

Identify the features of the eye indicated in figures 35.12 to 35.15.

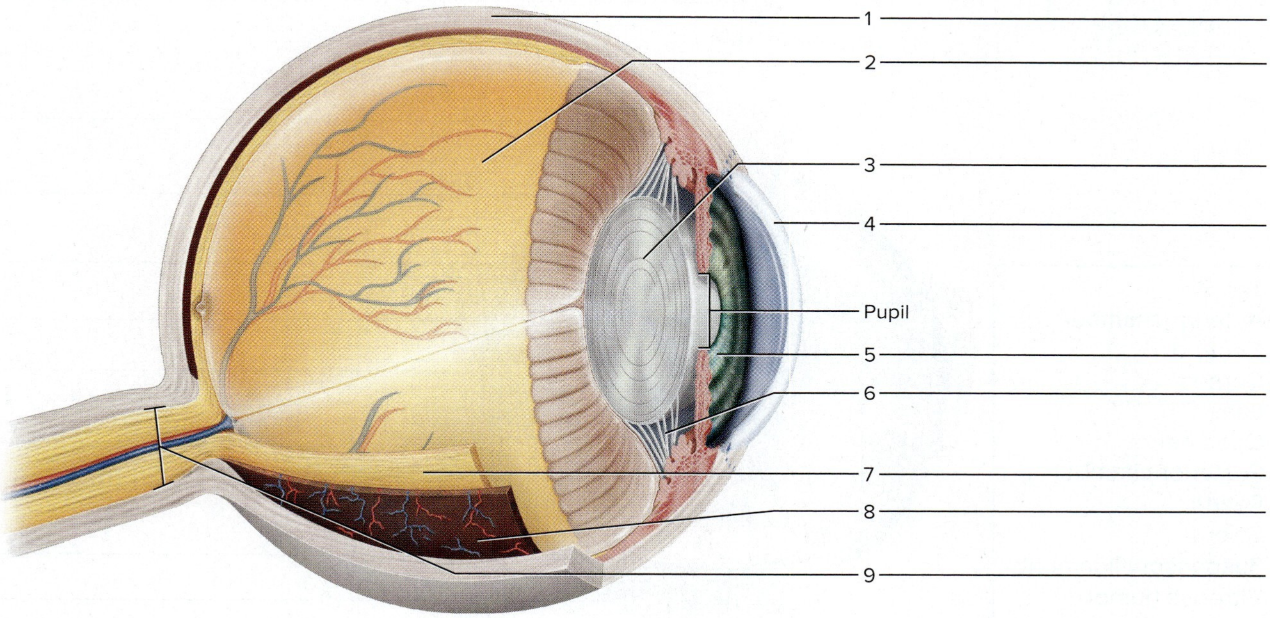

FIGURE 35.12 Label the structures in the sagittal section of the eye. **A1**

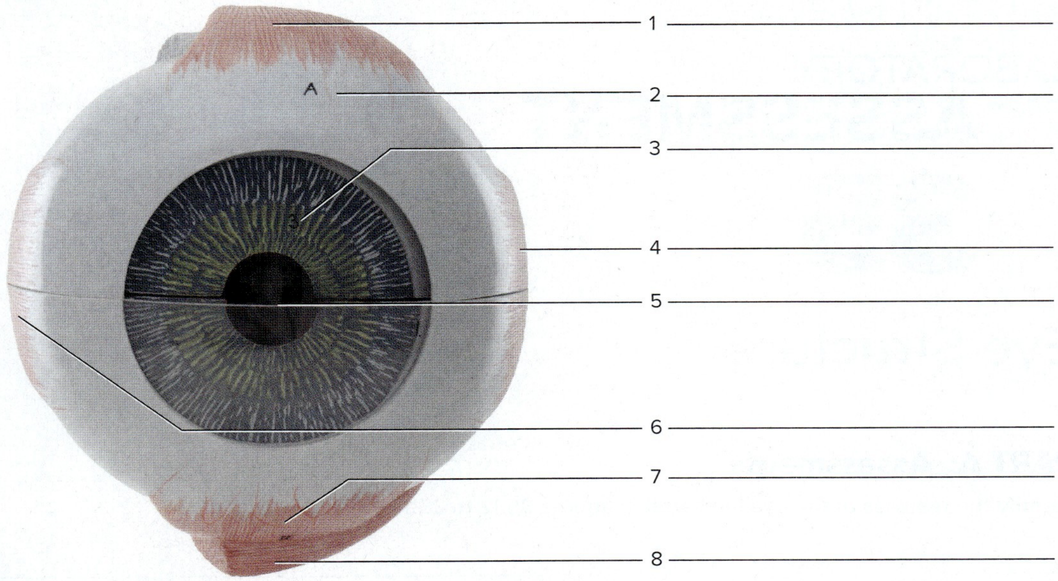

Terms:
Inferior oblique
Inferior rectus
Iris
Lateral rectus
Medial rectus
Pupil
Sclera
Superior rectus

(a) Anterior view

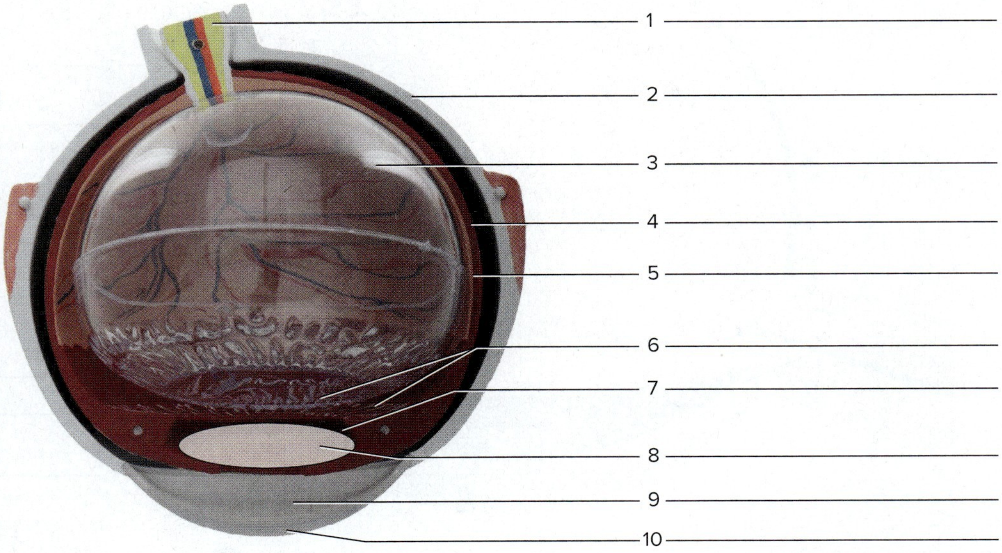

Terms:
Anterior chamber
Choroid
Cornea
Lens
Optic nerve
Posterior chamber
Retina
Sclera
Suspensory ligaments
Vitreous humor
(in posterior cavity)

(b) Superior view

FIGURE 35.13 Label the features of the (a) anterior view image of the left eye model and (b) superior view image of the eye model, using the terms provided for each. ©3B Scientific GmbH, Germany, 2017 www.3bscientific.com A1

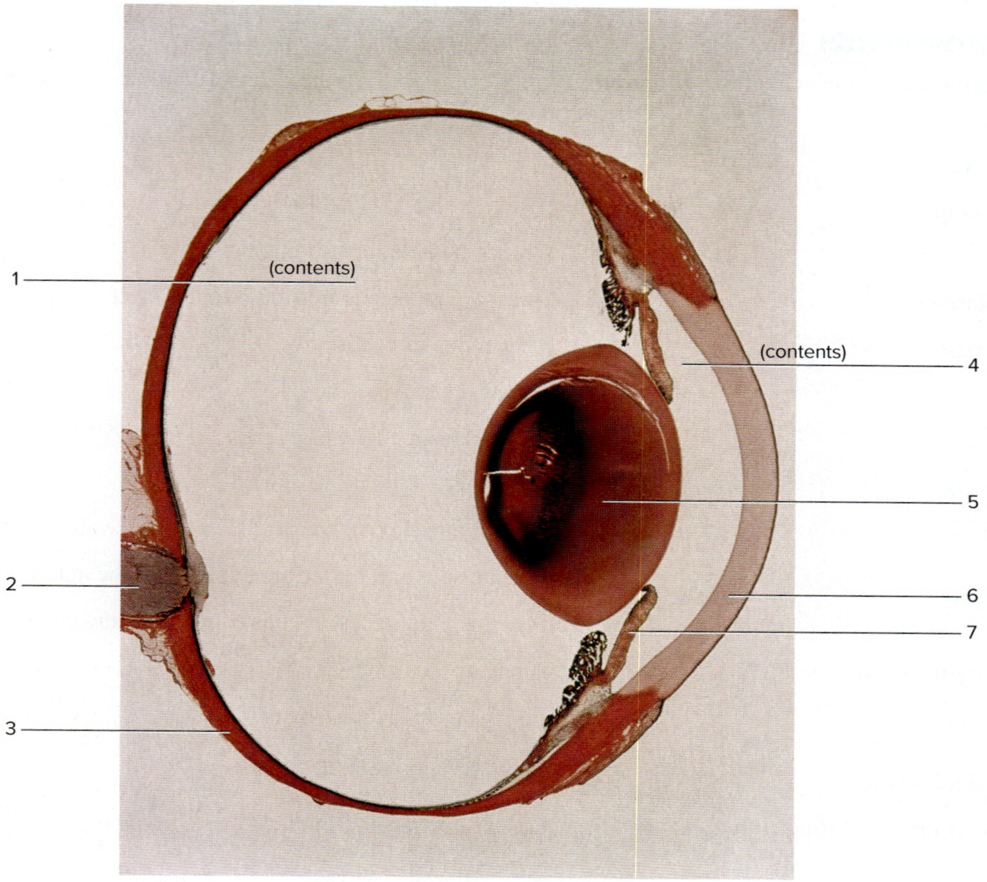

_____ Aqueous humor	_____ Optic nerve
_____ Cornea	_____ Sclera/outer layer
_____ Iris	_____ Vitreous humor
_____ Lens	

FIGURE 35.14 Sagittal section of the eye (5×). Identify the numbered features by placing the correct numbers in the spaces provided. Manfred Kage/Science Source A1 APR

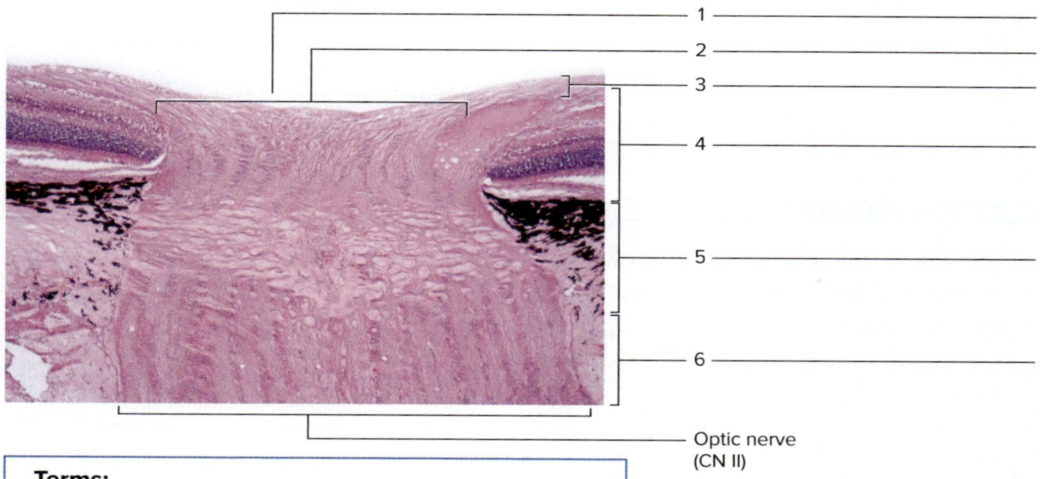

Terms:

Choroid	Optic disc	Sclera
Nerve fibers (to optic nerve)	Retina	Vitreous humor

FIGURE 35.15 Identify the tunics and features of the eye in the micrograph, using the terms provided. Victor P. Eroschenko

PART B: Assessments

1. Match the terms in column A with the descriptions in column B. Place the letter of your choice in the space provided. A1 A2

Column A		Column B
a. Aqueous humor	______	1. Posterior five-sixths of middle (vascular) layer
b. Choroid	______	2. White part of outer (fibrous) layer
c. Ciliary muscles	______	3. Transparent anterior portion of outer layer
d. Conjunctiva	______	4. Inner lining of eyelid
e. Cornea	______	5. Secretes tears
f. Iris	______	6. Fills posterior cavity of eye
g. Lacrimal gland	______	7. Area where optic nerve exits the eye
h. Optic disc	______	8. Smooth muscle that controls the pupil size and light entering the eye
i. Retina	______	9. Fills anterior and posterior chambers of the anterior cavity of the eye
j. Sclera	______	10. Contains photoreceptor cells called rods and cones
k. Suspensory ligament	______	11. Connects lens to ciliary body
l. Vitreous humor	______	12. Cause lens to change shape

2. Complete the following:

a. List the structures and fluids through which light passes as it travels from the cornea to the retina. A3 ______________________________

____

____

____

b. List three ways in which rods and cones differ in structure or function. A2 ______________________________

____

____

____

____

CRITICAL THINKING ASSESSMENT—PART B

Sally, a 64-year old female, complained to an ophthalmologist about experiencing blurriness of both eyes, particularly in the central area of vision. The physician diagnosed Sally with dry macular degeneration. Explain this diagnosis and the symptoms she may be experiencing. A1 A2 ______________________________

____

____

____.

PART C: Assessments

1. Identify the features of the cow eye indicated in figures 35.16 and 35.17. A4

Terms:
Choroid
Ciliary body
Iris
Lens
Pupil
Retina
Sclera
Vitreous humor

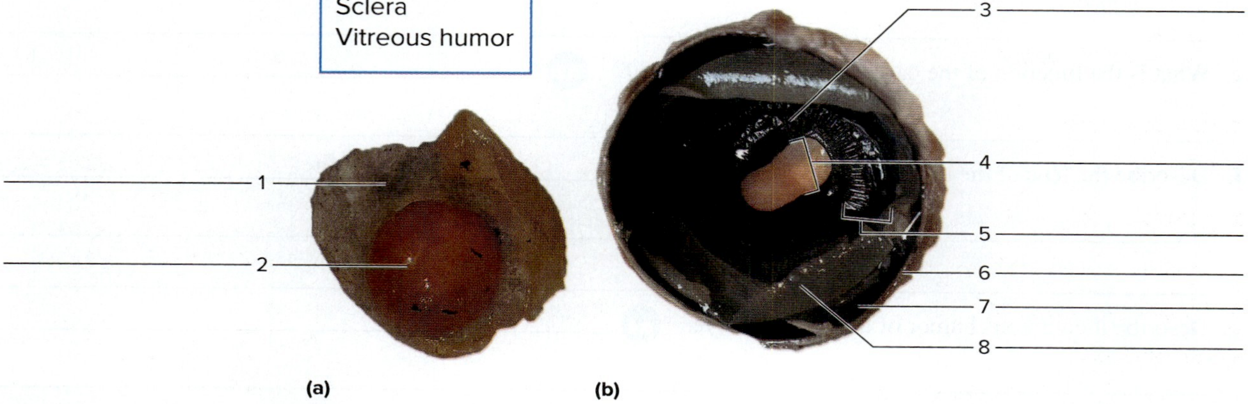

FIGURE 35.16 Frontal cut of dissected cow eye. Label the numbered internal structures of (a) and (b), using the terms provided. Cynthia Prentice-Craver/McGraw Hill

Terms:
Cornea
Retina
Sclera
Tapetum fibrosum (of choroid)

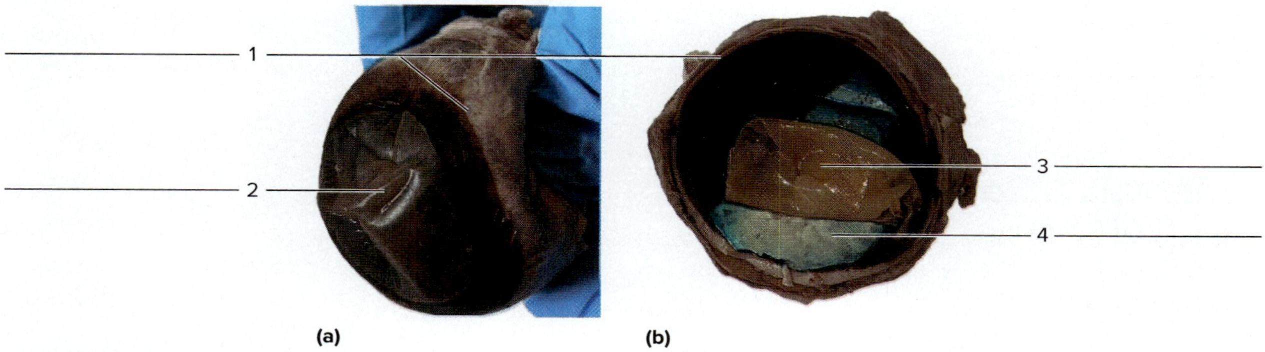

FIGURE 35.17 Label the numbered external (a) and internal frontal section (b) tunics and features of the cow eye, using the terms provided. Cynthia Prentice-Craver/McGraw Hill

2. Complete the following questions about the cow eye dissection.

   a. Which layer/tunic of the eye was the most difficult to cut? A4 ______________________________

   b. What kind of tissue do you think is responsible for this quality of toughness? A4 ______________________________

   c. How do you compare the shape of the pupil in the dissected eye with your own pupil? A4 ______________________________

   ____

   ____

   d. Where was the aqueous humor in the dissected eye? A4 ______________________________

   ____

   e. What is the function of the dark pigment in the choroid? A2 ______________________________

   ____

   f. Describe the lens of the dissected eye. A4 ______________________________

   ____

   ____

   g. Describe the vitreous humor of the dissected eye. A4 ______________________________

   ____

   ____

CRITICAL THINKING ASSESSMENT—PART C

A strong blow to the head might cause the retina to detach. From observations made during the eye dissection, explain why this can happen. A1 A4 ______________________________

____

______________________________.

LABORATORY
EXERCISE 36

Visual Tests and Demonstrations

Purpose of the Exercise

To conduct and interpret the tests for visual acuity, astigmatism, accommodation, and color vision, and to demonstrate the blind spot and reflexes of the eye.

MATERIALS NEEDED

Snellen eye chart
3" × 5" card (plain)
3" × 5" card with any word typed in center
Astigmatism chart
Meterstick
Metric ruler
Pen flashlight
Ichikawa's or Ishihara's color plates for color blindness test

Chapter Opener: ©101cats/Getty Images

Learning Outcomes

After completing this exercise, you should be able to:

- **O1** Conduct and interpret the tests used to evaluate visual acuity, astigmatism, the ability to accommodate for close vision, and color vision.
- **O2** Describe four conditions that can lead to defective vision.
- **O3** Demonstrate and describe the blind spot, photopupillary reflex, accommodation pupillary reflex, and convergence reflex.
- **O4** Summarize the results of the performed visual tests and demonstrations.

The **O** corresponds to the assessments **A** indicated in the Laboratory Assessment for this Exercise.

Pre-Lab

Carefully read the introductory material and examine the entire lab. Be familiar with common eye tests and defects from lecture or the textbook. Answer the pre-lab questions.

Pre-Lab Questions Select the correct answer for each of the following questions:

1. As a person ages, the elasticity of the lens
  a. decreases. **b.** increases.
  c. remains unchanged. **d.** improves.

2. Visual acuity is measured using a
  a. metric ruler. **b.** pen flashlight.
  c. Snellen eye chart. **d.** astigmatism chart.

3. Color blindness is a characteristic in _____ of males.
  a. 0% **b.** 0.4%
  c. 7% **d.** 10%

4. The blind spot is located at the _____ of the eye.
  a. pupil **b.** vitreous humor
  c. aqueous humor **d.** optic disc

5. Astigmatism results from a defect in the
  a. control of the pupil size. **b.** curvature of the cornea or lens.
  c. rods of the retina. **d.** cones of the retina.

6. Only males can inherit the color blindness condition.
  a. True **b.** False

7. Nearsightedness is also called myopia.
  a. True **b.** False

8. A convex lens can be used to correct hyperopia.
  a. True **b.** False

Normal vision (emmetropia) results when light rays from objects in the external environment are refracted by the cornea and lens of the eye and focused onto the photoreceptors of the retina. If a person has *nearsightedness (myopia),* the image focuses in front of the retina because the eyeball is too long; nearby objects are clear, but distant objects are blurred. If a person has *farsightedness (hyperopia),* the image focuses behind the retina because the eyeball is too short; distant objects are clear, but nearby objects are blurred. A concave lens that causes light to diverge somewhat before entering the eye is required to correct vision for nearsightedness; a convex lens that causes light to converge somewhat before entering the eye is required to correct vision for farsightedness (fig. 36.1).

If a defect occurs from unequal curvatures of the cornea or lens, a condition called *astigmatism* results. Focusing in different planes cannot occur simultaneously. Corrective cylindrical lenses allow proper refraction of light onto the retina to compensate for the unequal curvatures.

As part of a natural aging process, the lens's elasticity decreases, which degrades our ability to focus on nearby objects. The near point of accommodation test is used to determine the ability to accommodate. Often a person will use "reading glasses" to compensate for this condition.

Hereditary defects in color vision result from a lack of certain cones needed to absorb certain wavelengths of light. Special color test plates are used to diagnose color blindness.

If you wear glasses, perform the tests with and without the corrective lenses. If you wear contact lenses, it is not necessary to run the tests under both conditions; however, indicate in the laboratory assessment that all tests were performed with contact lenses in place.

PROCEDURE A: Visual Tests

Perform the following visual tests using your laboratory partner as a test subject. If your partner usually wears glasses, test each eye with and without the glasses.

1. *Visual acuity test.* Visual acuity (sharpness of vision) can be measured by using a Snellen eye chart (fig. 36.2). This chart consists of several sets of letters in different sizes printed on a white card. The letters near the top of the chart are relatively large, and those in each lower set become smaller. At one end of each set of letters is an acuity value in the form of a fraction. One of the sets near the bottom of the chart, for example, is marked 20/20. The normal eye can clearly see these letters from the standard distance of 20 feet and thus is said to have 20/20 vision. The letter at the top of the chart is marked 20/200. The normal eye can read letters of this size from a distance of 200 feet. Thus, an eye able to read only the top letter of the chart from a distance of 20 feet is said

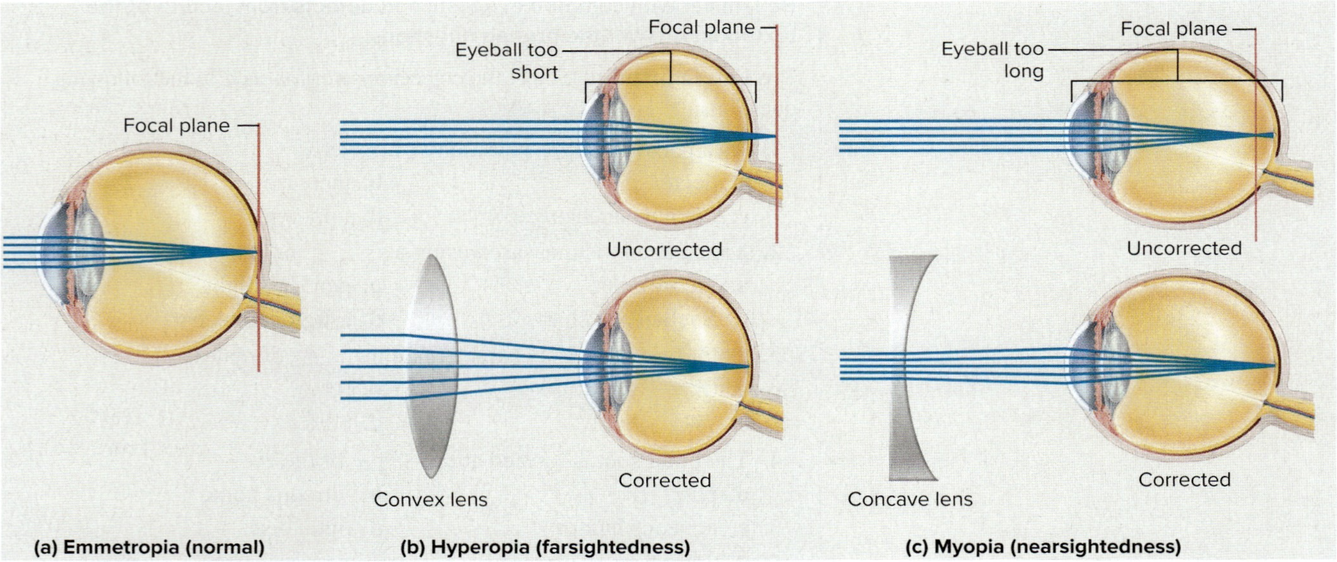

FIGURE 36.1 Normal eye compared to two common refractive defects. (a) The focal point is on the retina in the normal eye (emmetropia); (b) the focal point is behind the retina in hyperopia; and (c) the focal point is before the retina in myopia. A convex lens can correct hyperopia; a concave lens can correct myopia.

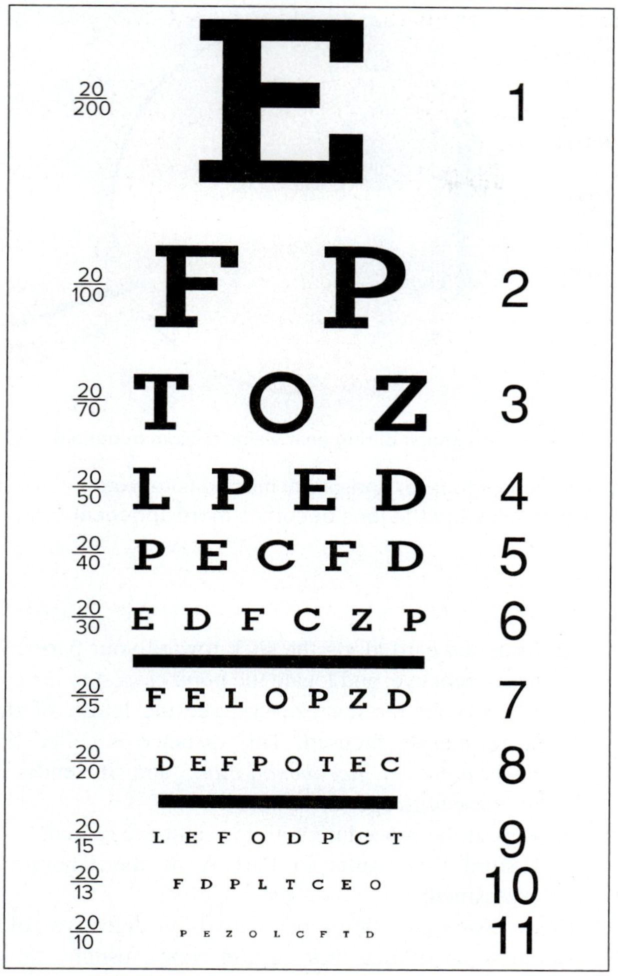

FIGURE 36.2 The Snellen eye chart looks similar to this but is somewhat larger.

to have 20/200 vision. This person has less than normal vision. A line of letters near the bottom of the chart is marked 20/15. The normal eye can read letters of this size from a distance of 15 feet, but a person might be able to read it from 20 feet. This person has better than normal vision.

To conduct the visual acuity test, follow these steps:

a. Hang the Snellen eye chart on a well-illuminated wall at eye level.

b. Have your partner stand 20 feet in front of the chart, gently cover your partner's left eye with a 3" × 5" card, and ask your partner to read the smallest set of letters possible using the right eye.

c. Record the visual acuity value for that set of letters in Part A of Laboratory Assessment 36.

d. Repeat the procedure, using the left eye for the test.

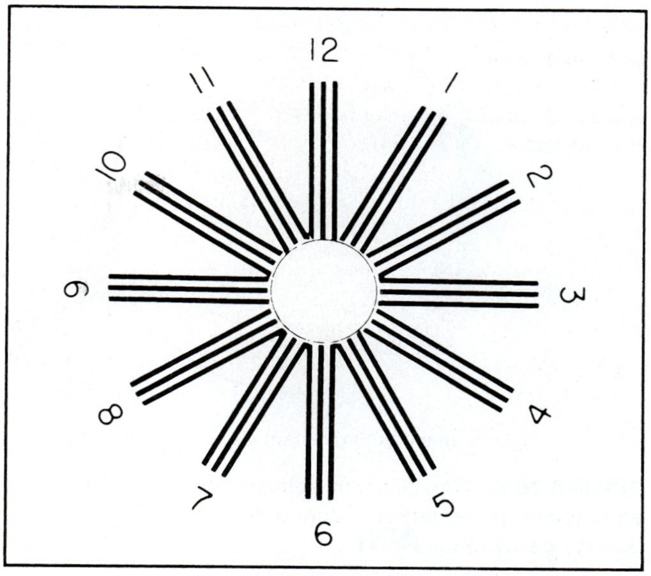

FIGURE 36.3 Astigmatism is evaluated using a chart such as this one.

2. *Astigmatism test.* Astigmatism is a condition that results from a defect in the curvature of the cornea or lens. As a consequence, some portions of the image projected on the retina are sharply focused, but other portions are blurred. Astigmatism can be evaluated by using an astigmatism chart (fig. 36.3). This chart consists of sets of black lines radiating from a central spot like the spokes of a wheel. To a normal eye, these lines appear sharply focused and equally dark; however, if the eye has an astigmatism some sets of lines appear sharply focused and dark, whereas others are blurred and less dark.

To conduct the astigmatism test, follow these steps:

a. Hang the astigmatism chart on a well-illuminated wall at eye level.

b. Have your partner stand 20 feet in front of the chart, and gently cover your partner's left eye with a 3" × 5" card. Ask your partner to focus on the spot in the center of the radiating lines using the right eye and report which lines, if any, appear more sharply focused and darker.

c. Repeat the procedure, using the left eye for the test.

d. Record the results in Part A of the laboratory assessment.

3. *Accommodation test.* The changing of the shape of the lens that occurs when the normal eye is focused for near (close) vision is called *accommodation.* It involves a reflex in which muscles of the ciliary body are stimulated to contract, releasing tension on the suspensory ligaments that are fastened to the lens capsule. This allows the

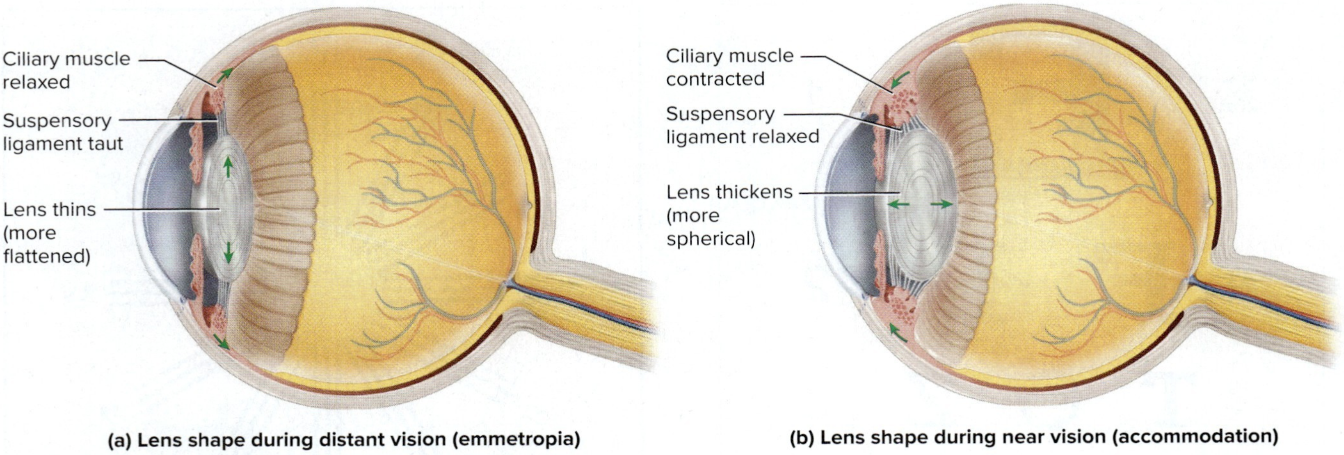

FIGURE 36.4 The ciliary muscle and the lens are involved during accommodation. (a) The ciliary muscle is relaxed and the lens more flattened during distant vision; (b) the ciliary muscle contracts and the lens becomes more spherical during near vision. APR

capsule to rebound elastically, causing the surface of the lens to become more spherical (convex) (fig. 36.4).

The ability to accommodate is likely to decrease with age because the tissues involved tend to lose their elasticity. This common aging condition, called *presbyopia,* can be corrected with reading glasses or bifocal lenses.

To evaluate the ability to accommodate, follow these steps:

a. Hold the end of a meterstick against your partner's chin so that the stick extends outward at a right angle to the plane of the face (fig. 36.5).
b. Have your partner close the left eye.
c. Hold a 3" × 5" card with a word typed in the center at the distal end of the meterstick.
d. Slide the card along the stick toward your partner's open right eye, and locate the *point closest to the eye* where your partner can still see the letters of the word sharply focused. This distance is called the *near point of accommodation,* and it tends to increase with age (table 36.1).
e. Repeat the procedure with the right eye closed.
f. Record the results in Part A of the laboratory assessment.

4. *Color vision test.* Some people exhibit defective color vision because they lack certain cones, usually those sensitive to the reds or greens. This trait is an X-linked (sex-linked) inheritance, so the condition is more prevalent in males (7%) than in females (0.4%). People who

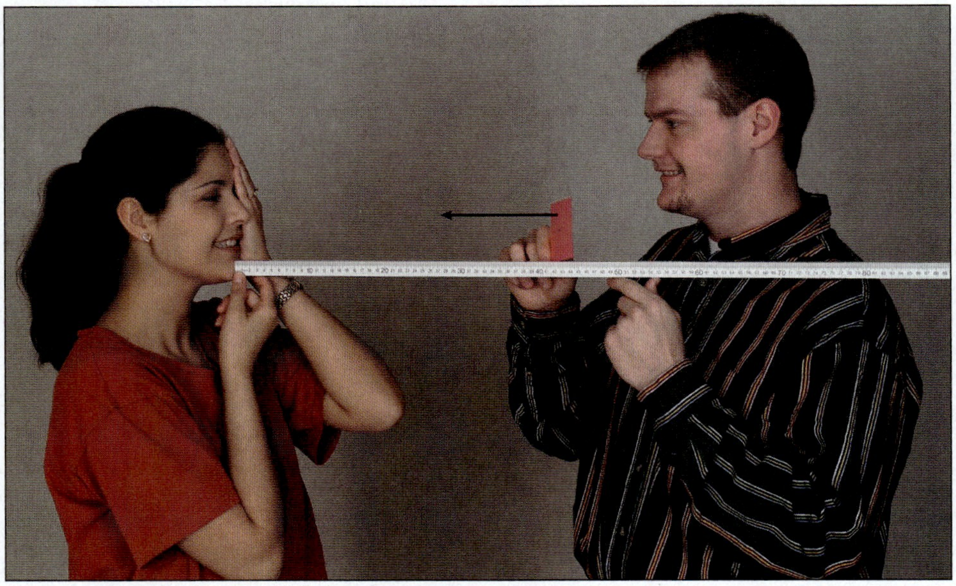

FIGURE 36.5 To determine the near point of accommodation, slide the 3" × 5" card along the meterstick toward your partner's open eye until it reaches the closest location where your partner can still see the word sharply focused. ©J & J Photography

Table 36.1 Near Point of Accommodation

Age (years)	Average Near Point (cm)
10	7
20	10
30	13
40	20
50	45
60	90

lack or possess decreased sensitivity to the red-sensitive cones possess protanopia color blindness; those who lack or possess decreased sensitivity to green-sensitive cones possess deuteranopia color blindness. The color blindness condition is often more of a deficiency or a weakness than one of blindness. Laboratory Exercise 62 describes the genetics for color blindness.

To conduct the color vision test, follow these steps:

a. Examine the color test plates in Ichikawa's or Ishihara's book to test for any red-green color vision deficiency. Also examine figure 36.6.

b. Hold the test plates approximately 30 inches from the subject in bright light. All responses should occur within 3 seconds.

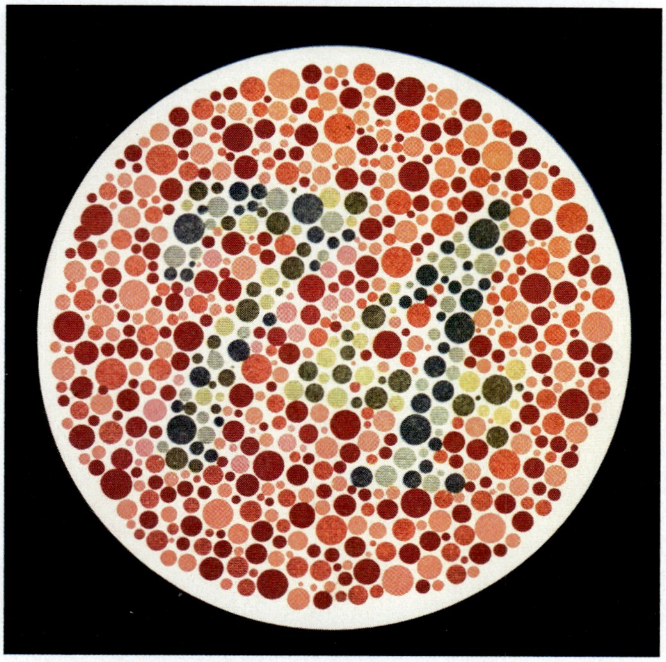

FIGURE 36.6 Sample test for red-green color blindness. A person with normal vision sees a number 74. A person with red-green color blindness does not see a 74 as the red and green colors are too similar. For accurate testing, the original plates should be used. Steve Allen/Brand X Pictures/Getty Images

c. Compare your responses with the correct answers in Ichikawa's or Ishihara's book. Determine the percentage of males and females in your class who exhibit any color-deficient vision. If an individual exhibits color-deficient vision, determine if the condition is protanopia or deuteranopia.

d. Record the class results in Part A of the laboratory assessment.

5. Complete Part A of the laboratory assessment.

PROCEDURE B: Visual Demonstrations

Perform the following demonstrations with the help of your laboratory partner.

1. *Blind-spot demonstration.* There are no photoreceptors in the optic disc, located where the nerve fibers of the retina leave the eye and enter the optic nerve. Consequently, this region of the retina is commonly called the *blind spot.*

To demonstrate the blind spot, follow these steps:

a. Close your left eye, hold figure 36.7 about 34 cm away from your face, and stare at the "+" sign in the figure with your right eye.

b. Move the figure closer to your face as you continue to stare at the "+" until the dot on the figure suddenly disappears. This happens when the image of the dot is focused on the optic disc. Measure the right eye distance using a metric ruler or a meterstick.

c. Repeat the procedures with your right eye closed. This time stare at the dot with your left eye, and the "+" will disappear when the image falls on the optic disc. Measure the distance.

d. Record the results in Part B of the laboratory assessment.

CRITICAL THINKING ACTIVITY

Explain why small objects are not lost from our vision under normal visual circumstances.

____

____

____

____

FIGURE 36.7 Blind-spot demonstration.

2. *Photopupillary reflex.* The smooth muscles of the iris control the size of the pupil. For example, when the intensity of light entering the eye increases, a photopupillary reflex is triggered, and the circular muscles of the iris are stimulated to contract. As a result, the size of the pupil decreases, and less light enters the eye. This reflex involves the autonomic nervous system (ANS) and protects the photoreceptor cells of the retina from damage. In this autonomic reflex, the retina is the *receptor;* the *optic nerve* conducts the sensory impulses to the *control center* in the brain; the *oculomotor nerve* conducts motor impulses to the circular, smooth muscle fibers of the iris (the *effector*). The response, constriction of the pupil, is a result of impulses from the *parasympathetic division* of the ANS. (In contrast, during dim light situations, impulses over the *sympathetic division* of the ANS result in stimulation of radial, smooth muscle fibers and dilation of the pupil.)

   To demonstrate this photopupillary reflex, follow these steps:

   a. Ask your partner to sit with his or her hands thoroughly covering his or her eyes for 2 minutes.
   b. Position a pen flashlight close to one eye with the light shining on the hand that covers the eye.
   c. Ask your partner to remove the hand quickly.
   d. Observe the pupil and note any change in its size.
   e. Have your partner remove the other hand, but keep that uncovered eye shielded from extra light.
   f. Observe both pupils and note any difference in their sizes.

3. *Accommodation pupillary reflex.* The pupil constricts as a normal accommodation reflex response to focusing on close objects. This reduces peripheral light rays and allows the refraction of light to occur closer to the center portion of the lens, resulting in a clearer image.

   To demonstrate the accommodation pupillary reflex, follow these steps:

   a. Have your partner stare for several seconds at a dimly illuminated object in the room that is more than 20 feet away.
   b. Observe the size of the pupil of one eye. Then hold a pencil about 25 cm in front of your partner's face, and have your partner stare at it.
   c. Note any change in the size of the pupil.

4. *Convergence reflex.* The eyes converge as a normal convergence response to focusing on close objects. The convergence reflex orients the eyeballs toward the object so the image falls on the fovea centralis of both eyes.

   To demonstrate the convergence reflex, follow these steps:

   a. Repeat the procedure outlined for the accommodation pupillary reflex.
   b. Note any change in the position of the eyeballs as your partner changes focus from the distant object to the pencil.

5. Complete Part B of the laboratory assessment.

LABORATORY
ASSESSMENT 36

Name ______________________

Date ______________________

Section ______________________

The (A) corresponds to the indicated Learning Outcome(s) (O) found at the beginning of the Laboratory Exercise.

Visual Tests and Demonstrations

PART A: Assessments

1. Visual acuity test results: A1

Eye Tested	Acuity Values
Right eye	
Right eye with glasses (if applicable)	
Left eye	
Left eye with glasses (if applicable)	

2. Astigmatism test results: A1

Eye Tested	Darker Lines
Right eye	
Right eye with glasses (if applicable)	
Left eye	
Left eye with glasses (if applicable)	

3. Accommodation test results: A1

Eye Tested	Near Point (cm)
Right eye	
Right eye with glasses (if applicable)	
Left eye	
Left eye with glasses (if applicable)	

4. Color vision test results: A1

Condition	Males			Females		
	Class Number	Class Percentage	Expected Percentage	Class Number	Class Percentage	Expected Percentage
Normal color vision			93			99.6
Deficient red-green color vision			7			0.4
Protanopia (lack red-sensitive cones)			Less-frequent type			Less-frequent type
Deuteranopia (lack green-sensitive cones)			More-frequent type			More-frequent type

5. Complete the following:

a. What is meant by 20/70 vision? A2 ______

b. What is meant by 20/10 vision? A2 ______

c. What visual problem is created by astigmatism? A2 ______

d. Why does the near point of accommodation often increase with age? A2 ______

e. Describe the eye defect that causes color-deficient vision. A2 ______

PART B: Assessments

1. Blind-spot results: A3

a. Right eye distance ______

b. Left eye distance ______

2. Complete the following:

a. Explain why an eye has a blind spot. A3 ______

b. Describe the photopupillary reflex. A3 ______

c. What difference did you note in the size of the pupils when one eye was exposed to bright light and the other eye was shielded from the light? **A3** ___________

____

____

d. Describe the accommodation pupillary reflex. **A3** ___________

____

____

e. Describe the convergence reflex. **A3** ___________

____

____

3. Summarize the vision of the person tested based upon the visual tests and demonstrations conducted in this laboratory exercise. Include information on suspected structural defects, genetic disorders, and aging conditions. **A4** ___________

____

____

____

____

____

NOTES

LABORATORY EXERCISE 37

Ear and Hearing

Purpose of the Exercise

To review the structural and functional characteristics of the ear and to conduct some ordinary hearing tests.

MATERIALS NEEDED

Dissectible ear model
Watch that ticks
Tuning fork (128 or 256 cps)
Rubber hammer
Cotton
Meterstick

For Demonstration Activities:
Compound light microscope
Prepared microscope slide of cochlea (section)
Audiometer

Learning Outcomes APR

After completing this exercise, you should be able to:

O1 Identify the major structures of the ear.

O2 Trace the pathway of sound vibrations from the tympanic membrane to the hearing receptors.

O3 Describe the functions of the structures of the ear.

O4 Conduct four ordinary hearing tests and summarize the results.

The O corresponds to the assessments A indicated in the Laboratory Assessment for this Exercise.

Pre-Lab

Carefully read the introductory material and examine the entire lab. Be familiar with the basic structures and functions of the ear associated with hearing from lecture or the textbook. Answer the pre-lab questions.

Pre-Lab Questions Select the correct answer for each of the following questions:

1. Hearing is interpreted in the _____ lobe of the cerebrum.
   a. frontal **b.** occipital
   c. parietal **d.** temporal

2. Sound loudness is measured in
   a. pitch. **b.** frequencies.
   c. decibels. **d.** hertz.

3. The middle ear bones articulate from tympanic membrane to oval window in which of the following sequences?
   a. malleus, incus, stapes **b.** incus, stapes, malleus
   c. stapes, malleus, incus **d.** malleus, stapes, incus

4. The _____ test is done to assess possible conduction deafness by comparing bone and air conduction.
   a. Weber **b.** Rinne
   c. auditory acuity **d.** sound localization

5. Which of the following structures is part of the inner ear?
   a. tympanic membrane **b.** cochlea
   c. stapes **d.** pharyngotympanic tube

6. The pharyngotympanic tube connects the outer ear to the inner ear.
   a. True **b.** False

7. The cochlear nerve serves as the hearing branch of the vestibulocochlear nerve.
   a. True **b.** False

8. Endolymph is located within the cochlear duct of the cochlea.
   a. True **b.** False

The ear is composed of outer (external), middle, and inner (internal) parts. The large external *auricle* gathers sound waves (vibrations) of air and directs them into the *external acoustic meatus* to the *tympanic membrane* (eardrum). The middle ear *auditory ossicles* (small bones) conduct and amplify the waves from the tympanic membrane to the *oval window.* To allow the same pressure in the outer and middle ear, the pharyngotympanic (auditory) tube connects the middle ear to the pharynx. Equalizing pressure is important for proper vibration of the eardrum. The sound waves are further conducted through the fluids (perilymph and endolymph) of the *cochlea* in the inner ear. Perilymph is located within the scala vestibuli and the scala tympani; endolymph occupies the cochlear duct, which contains the *spiral organ.* Finally, some hair cells of the spiral organ are stimulated. As certain hair cells (receptor cells) are stimulated, these receptors initiate nerve impulses that are conducted over the *cochlear branch* of the *vestibulocochlear nerve* into the brain. When nerve impulses arrive at the *primary auditory cortex* of the temporal lobe of the cerebrum, the impulses are interpreted and the sensations of hearing are created (fig. 37.1).

The human hearing range is approximately 20 to 20,000 waves per second, known as *hertz* (*Hz*). We perceive these different frequencies as pitch. Different frequencies stimulate particular hair cells in specific regions of the spiral organ, allowing us to interpret different pitches. An additional quality of sound is loudness. The intensity of the sound is measured in units called *decibels* (*dB*). Permanent damage can happen to the hair cells of the spiral organ if frequent and prolonged exposures occur over 90 decibels.

Although the ear is considered here as a hearing organ, it also has important functions for equilibrium (balance). The inner ear semicircular ducts and vestibule have equilibrium functions. The ear and equilibrium will be studied in Laboratory Exercise 38.

PROCEDURE A: Structure and Function of the Ear APR

1. Use figures 37.2 to 37.4 as references as you progress through this procedure.
2. Examine the dissectible model of the ear and locate the following features:

outer (external) ear

- auricle (pinna)—funnel for sound waves
- external acoustic meatus (auditory canal)—2.5 cm S-shaped tube; contains ceruminous glands; passageway for sound waves
- tympanic membrane (eardrum)—semitransparent; vibrates upon sound wave arrival

middle ear

- tympanic cavity—air-filled chamber in temporal bone
- auditory ossicles (fig. 37.5)—transmit vibrations to oval window
  - malleus (hammer)—first ossicle to vibrate
  - incus (anvil)—middle ossicle to vibrate
  - stapes (stirrup)—final ossicle to vibrate at oval window
- oval window—opening at lateral wall of inner ear
- tensor tympani—skeletal muscle attached to the malleus
- stapedius—skeletal muscle attached to the stapes
- round window—membrane-covered opening of inner ear wall
- pharyngotympanic tube (auditory tube; eustachian tube)—connects middle ear to pharynx; removes secretions and debris from middle ear; helps equalizes tympanic cavity pressure with pressure in external auditory canal

inner (internal) ear

- bony labyrinth—bony chambers filled with perilymph; includes the cochlea, semicircular canals, and vestibule
- membranous labyrinth—membranous chambers filled with endolymph; follows pattern of bony labyrinth
- cochlea—converts sound waves to nerve impulses at spiral organ
- semicircular ducts—equilibrium organ
- vestibule—equilibrium organ with two chambers
  - utricle
  - saccule

vestibulocochlear nerve VIII—cranial nerve

- vestibular nerve—balance branch
- cochlear nerve—hearing branch

3. Complete Parts A, B, and C of Laboratory Assessment 37.

DEMONSTRATION ACTIVITY

Observe the section of the cochlea in the demonstration microscope. Locate one of the turns of the cochlea, and using figures 37.4 and 37.6 as a guide, identify the *scala vestibuli, cochlear duct* (*scala media*), *scala tympani, vestibular membrane, basilar membrane,* and the *spiral organ* (*organ of Corti*).

PROCEDURE B: Hearing Tests

Perform the following tests in a quiet room, using your laboratory partner as the test subject.

1. *Auditory acuity test.* To conduct this test, follow these steps:
   a. Have the test subject sit with eyes closed.
   b. Pack one of the subject's ears with cotton.
   c. Hold a ticking watch close to the open ear, and slowly move it straight out and away from the ear.

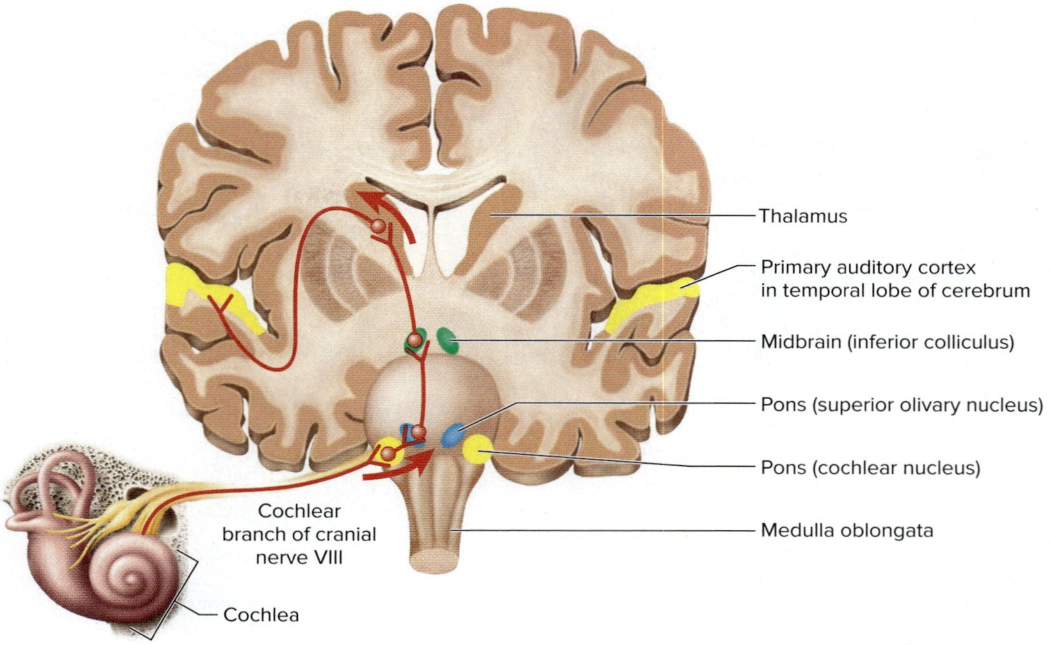

FIGURE 37.1 Auditory pathways. The sensory receptors are hair cells of the spiral organ in the cochlea. The sensory information continues through the medulla oblongata, pons, midbrain, thalamus, and to the primary auditory cortex in the temporal lobe, where sound is perceived.

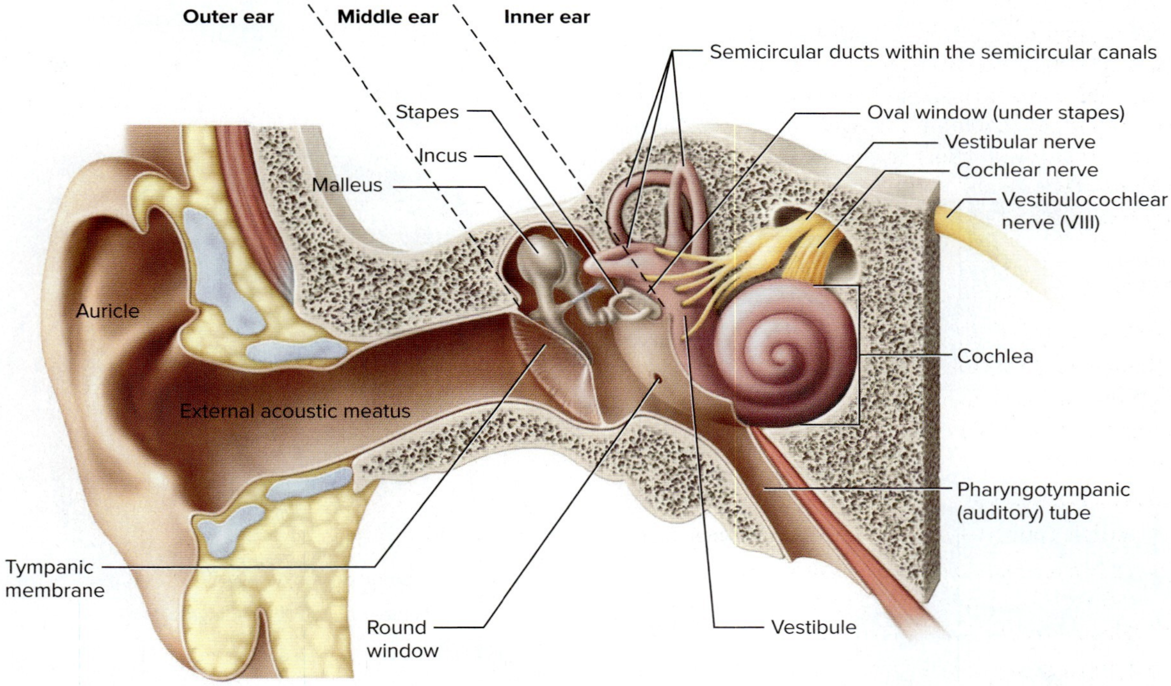

FIGURE 37.2 Major structures of the ear. The three divisions of the ear are roughly indicated by the dashed lines. APR

d. Have the subject indicate when the sound of the ticking can no longer be heard.

e. Use a meterstick to measure the distance in centimeters from the ear to the position of the watch.

f. Repeat this procedure to test the acuity of the other ear.

g. Record the test results in Part D of the laboratory assessment.

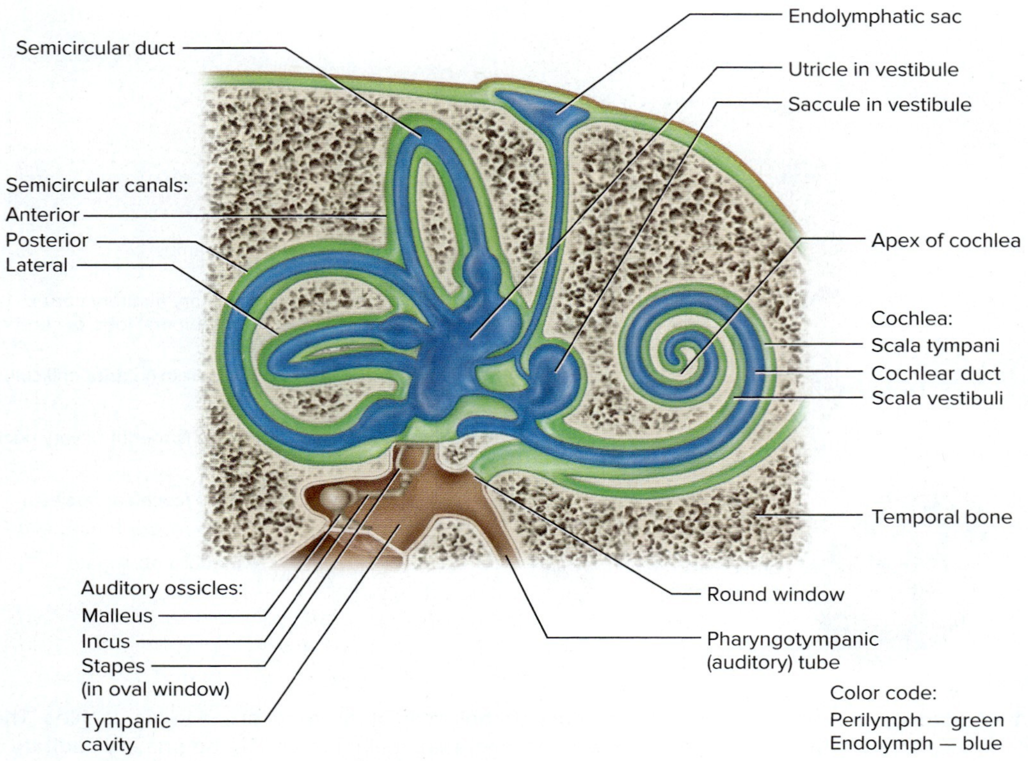

FIGURE 37.3 Structures of the middle and inner ear.

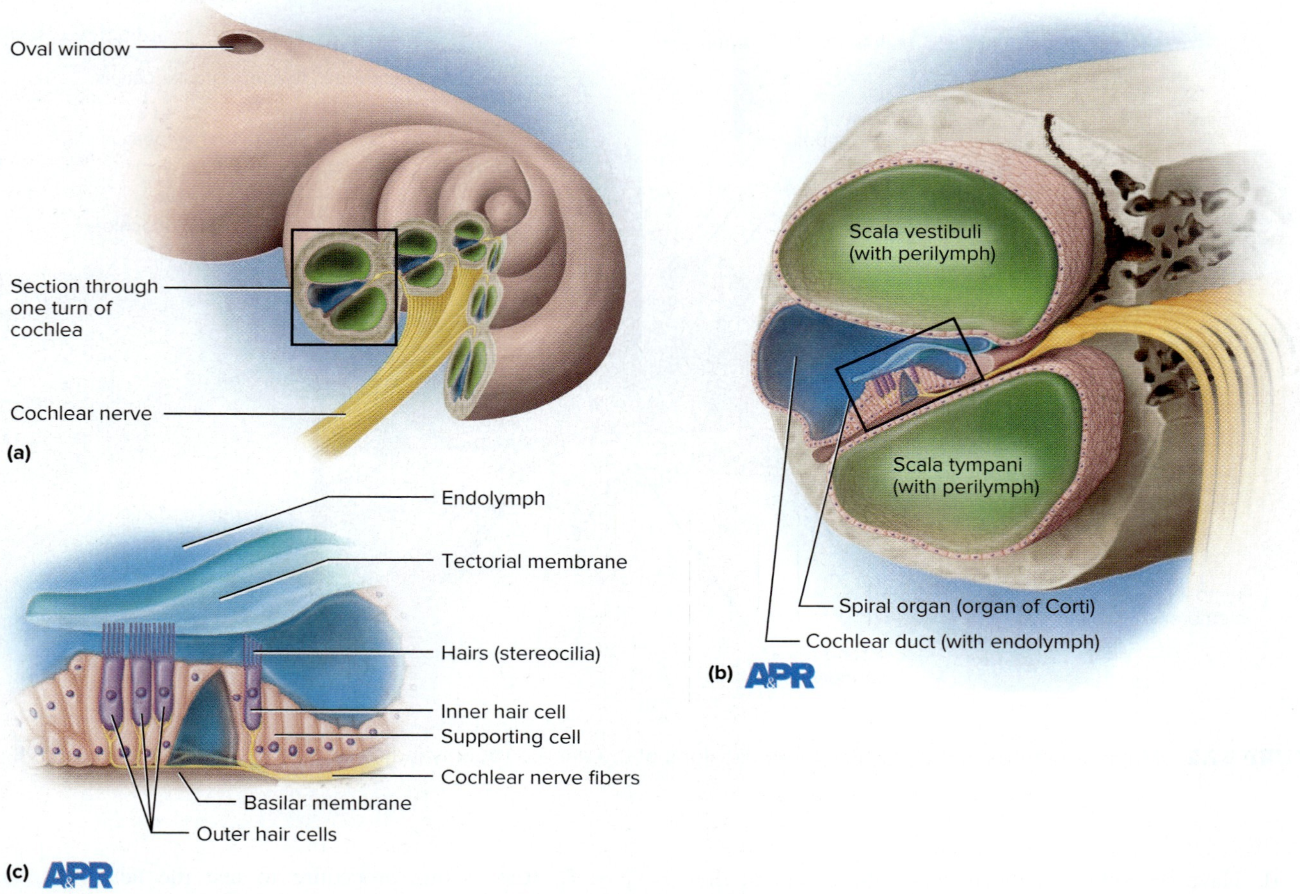

FIGURE 37.4 Anatomy of the cochlea: (a) section through cochlea; (b) view of one turn of cochlea; and (c) view of spiral organ (organ of Corti).

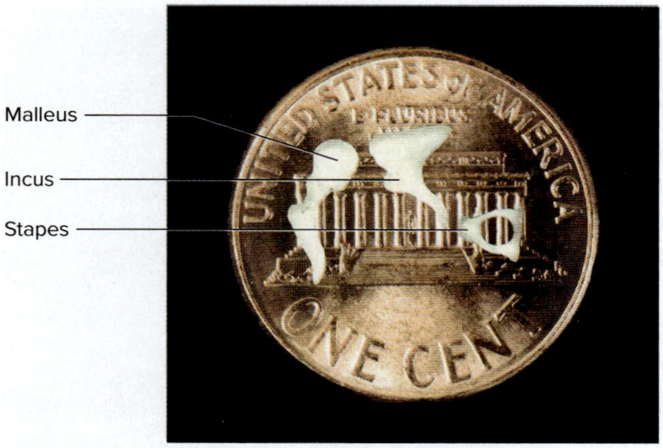

FIGURE 37.5 Middle ear bones (auditory ossicles) superimposed on a penny to show their relative size. The arrangement of the malleus, incus, and stapes in the enlarged photograph resembles the articulations in the middle ear. ©J & J Photography

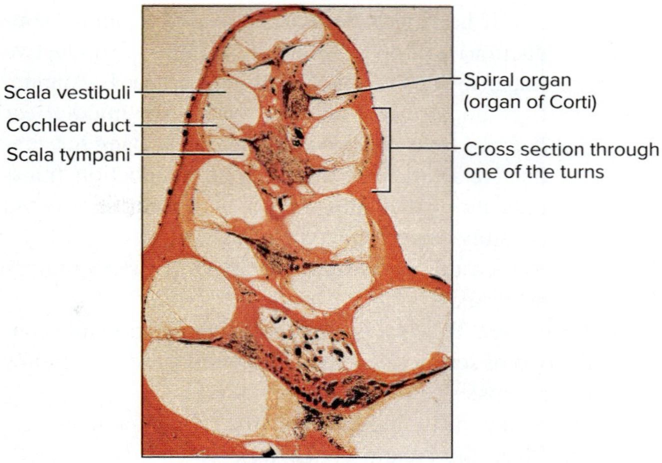

FIGURE 37.6 A section through the cochlea (22×). Biophoto Associates/Science Source APR

2. *Sound localization test.* To conduct this test, follow these steps:
   a. Have the subject sit with eyes closed.
   b. Hold the ticking watch somewhere within the audible range of the subject's ears, and ask the subject to point to the watch.
   c. Move the watch to another position and repeat the request. In this manner, determine how accurately the subject can locate the watch when it is in each of the following positions: in front of the head, behind the head, above the head, on the right side of the head, and on the left side of the head.
   d. Record the test results in Part D of the laboratory assessment.

3. *Rinne test.* This test is done to assess possible conduction deafness by comparing bone and air conduction. To conduct this test, follow these steps:
   a. Obtain a tuning fork and strike it with a rubber hammer, or on the heel of your hand, causing it to vibrate.
   b. Place the end of the fork's handle against the subject's mastoid process behind one ear. Have the prongs of the fork pointed downward and away from the ear, and be sure nothing is touching them (fig. 37.7*a*). The sound sensation is that of bone conduction. If no sound is experienced, nerve deafness exists.
   c. Ask the subject to indicate when the sound is no longer heard.
   d. Then quickly remove the fork from the mastoid process and position it in the air close to the opening of the nearby external acoustic meatus (fig. 37.7*b*).

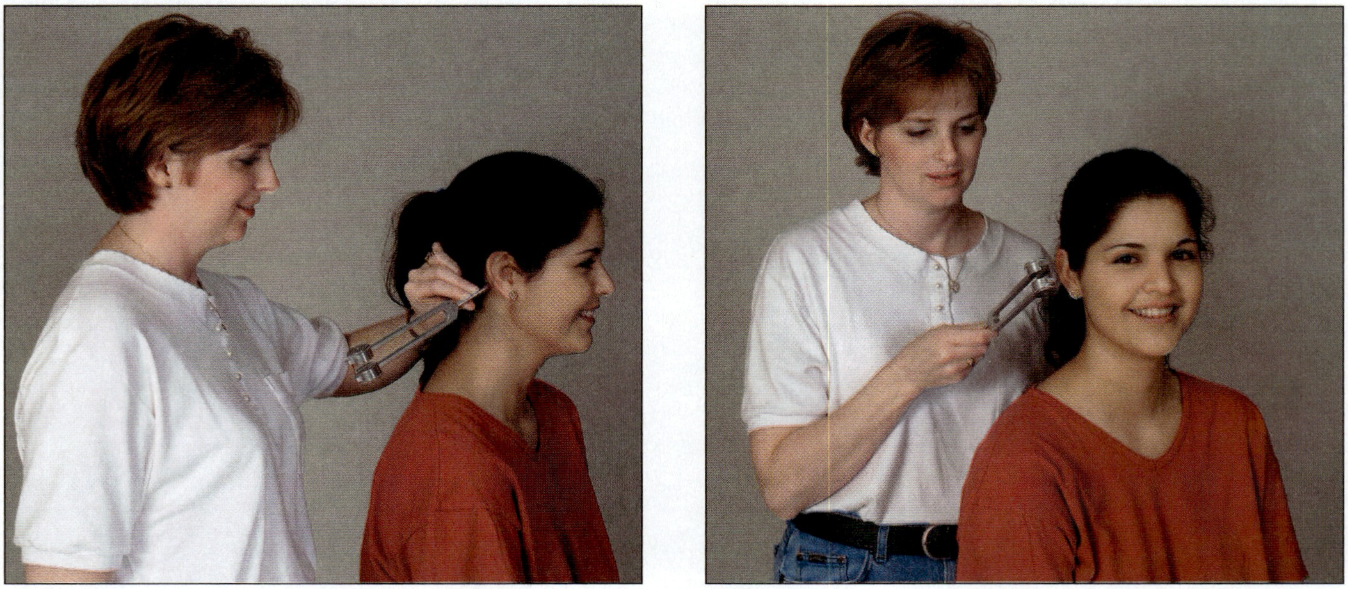

(a) (b)

FIGURE 37.7 Rinne test: (a) first placement of vibrating tuning fork until sound is no longer heard; (b) second placement of tuning fork to assess air conduction. ©J & J Photography

If hearing is normal, the sound (from air conduction) will be heard again; if there is conductive impairment, the sound will not be heard. Conductive impairment involves outer or middle ear defects. Hearing aids can improve hearing for conductive deafness because bone conduction transmits the sound into the inner ear. Surgery could possibly correct this type of defect.

   e. Record the test results in Part D of the laboratory assessment.

4. *Weber test.* This test is used to distinguish possible conduction or sensory deafness. To conduct this test, follow these steps:
   a. Strike the tuning fork with the rubber hammer.
   b. Place the handle of the fork against the subject's forehead in the midline (fig. 37.8).
   c. Ask the subject to indicate if the sound is louder in one ear than in the other or if it is equally loud in both ears.

      If hearing is normal, the sound will be equally loud in both ears. If there is conductive impairment, the sound will appear louder in the affected ear. If some degree of sensory (nerve) deafness exists, the sound will be louder in the normal ear. The impairment involves the spiral organ or the cochlear nerve. Hearing aids will not improve sensory deafness.
   d. Have the subject experience the effects of conductive impairment by packing one ear with cotton and repeating the Weber test. Usually, the sound appears louder in the plugged (impaired) ear because extraneous sounds from the room are blocked out.
   e. Record the test results in Part D of the laboratory assessment.

5. Complete Part D of the laboratory assessment.

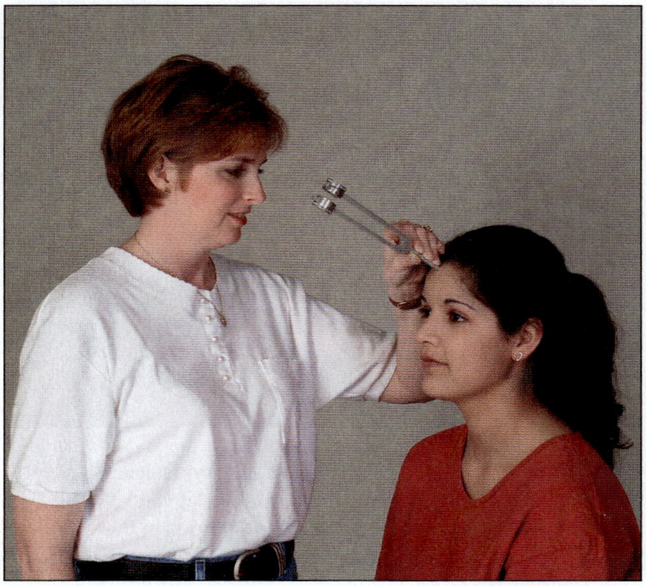

FIGURE 37.8 Weber test. ©J & J Photography

CRITICAL THINKING ACTIVITY

Ear structures from the outer ear into the inner ear are progressively smaller. Using results obtained from the hearing tests, explain this advantage.

____

____

____

DEMONSTRATION ACTIVITY

Ask the laboratory instructor to demonstrate the use of the audiometer. This instrument produces sound vibrations of known frequencies transmitted to one or both ears of a test subject through earphones. The audiometer can be used to determine the threshold of hearing for different sound frequencies, and, in the case of hearing impairment, it can be used to determine the percentage of hearing loss for each frequency.

LABORATORY
ASSESSMENT 37

Name ______________________________

Date ______________________________

Section ______________________________

The (A) corresponds to the indicated Learning Outcome(s) (O) found at the beginning of the Laboratory Exercise.

Ear and Hearing

PART A: Assessments

1. Identify the features of the ear indicated in figures 37.9 to 37.11.

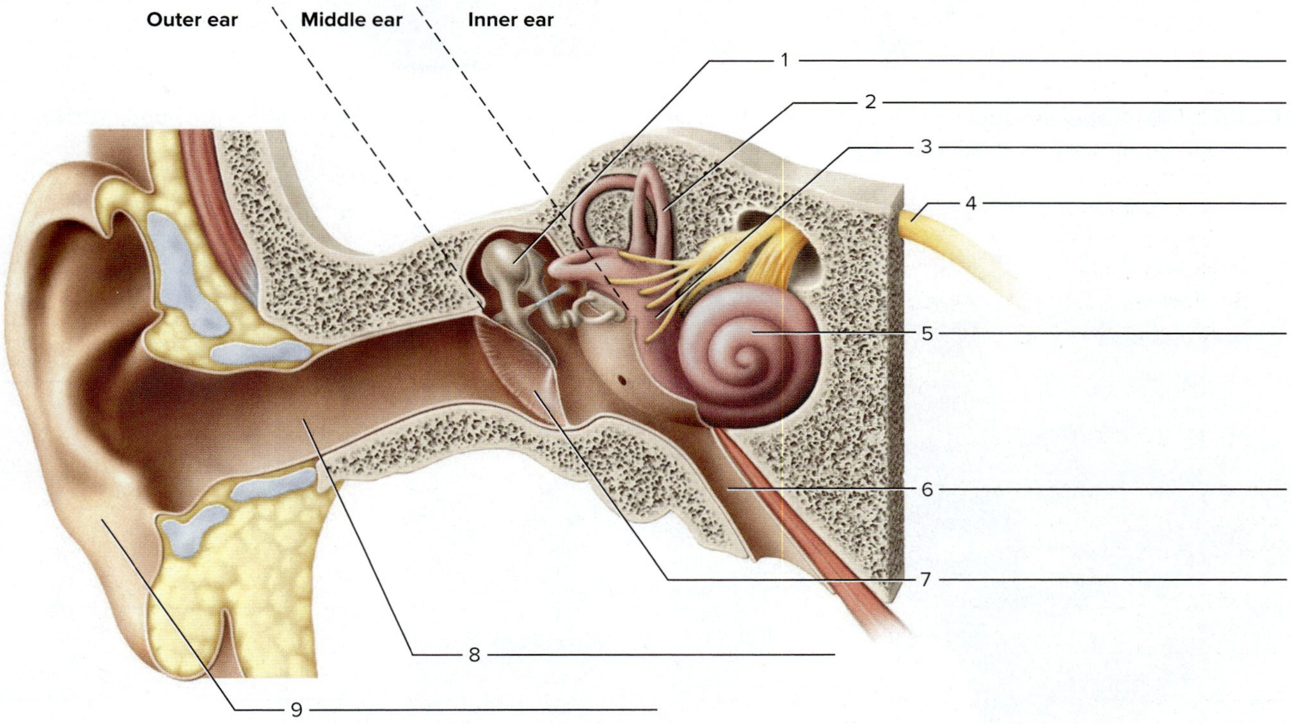

FIGURE 37.9 Label the structures associated with the right ear. (A1)

Terms:
Auditory tube
Auricle
Cochlea
External acoustic meatus
Incus
Malleus
Semicircular ducts
Stapes
Tympanic membrane
Vestibule

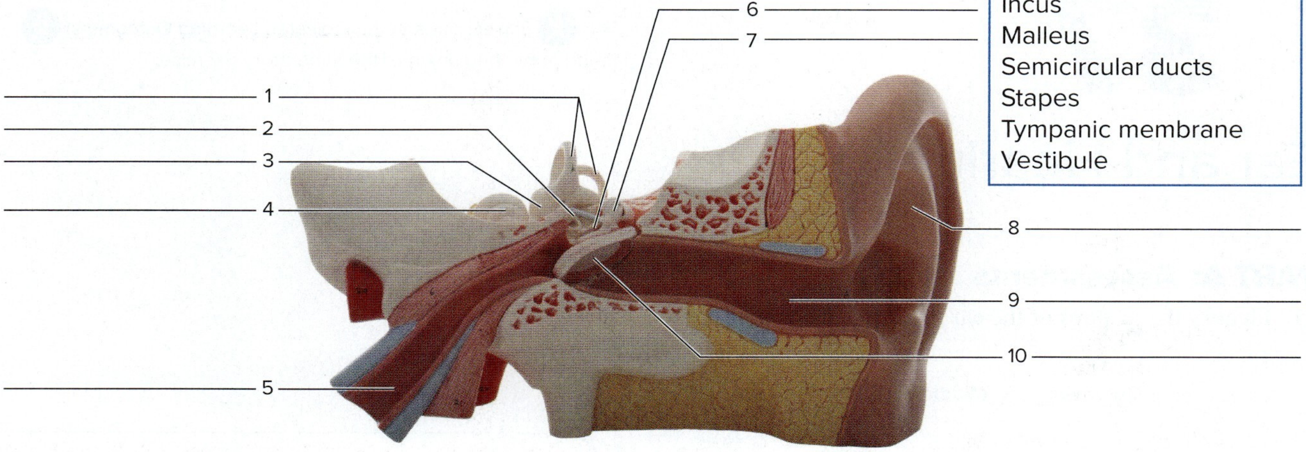

FIGURE 37.10 Label the structures of the anterior view image of the left ear model, using the terms provided. ©3B Scientific GmbH, Germany, 2017 www.3bscientific.com A1

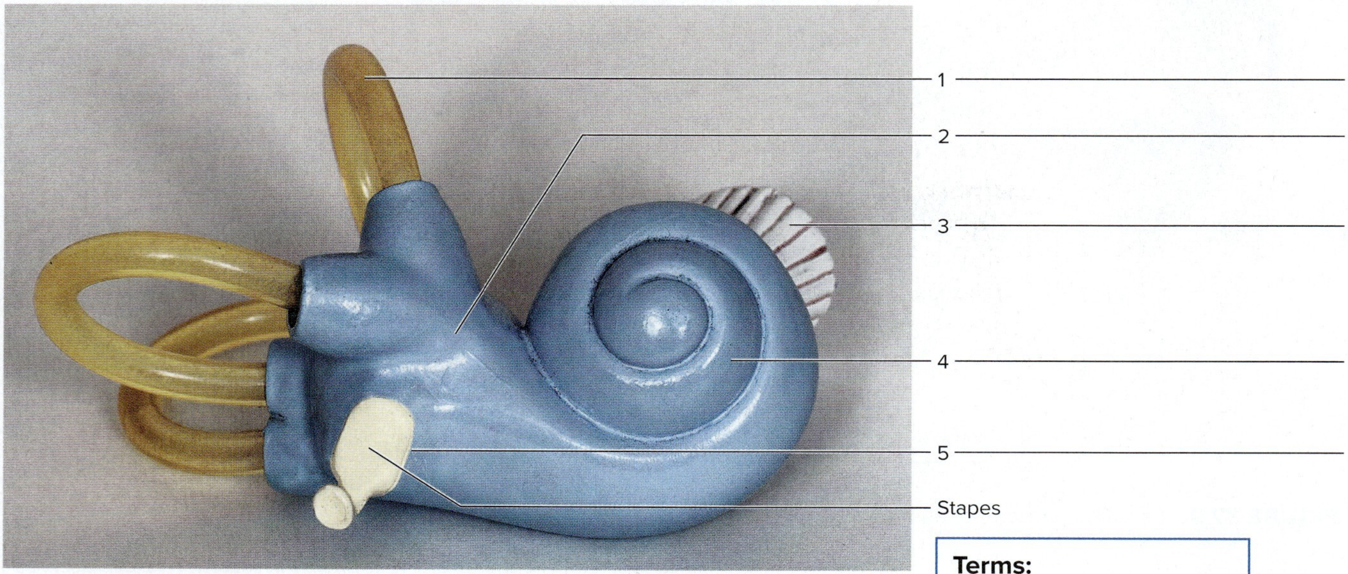

Terms:
Cochlea
Oval window
Semicircular duct
Vestibule
Vestibulocochlear nerve

FIGURE 37.11 Identify the features indicated on this removable part of an ear model, using the terms provided. ©J & J Photography A1

2. Label the structures indicated in the micrograph of the spiral organ in figure 37.12.

Terms:
Basilar membrane
Cochlear duct
Hair cells
Scala tympani
Tectorial membrane

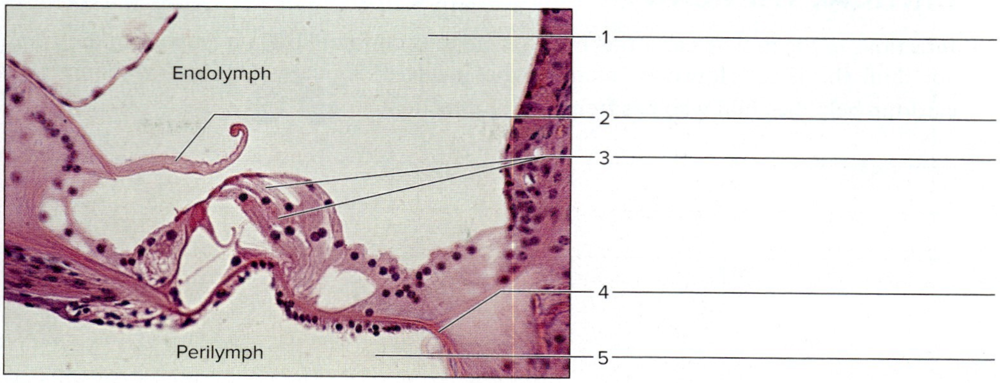

FIGURE 37.12 Label the structures associated with this spiral organ region of a cochlea, using the terms provided (400×).
Ed Reschke/Getty Images APR A1

PART B: Assessments

Number the following structures (1–9) to indicate their respective positions in relation to the pathway of the sound vibrations. Assign number 1 to the most superficial portion of the outer ear. A2

_____ Auricle (air vibrations within)

_____ External acoustic meatus (air vibrations within)

_____ Basilar membrane of spiral organ within cochlea

_____ Hair cells of spiral organ

_____ Incus

_____ Malleus

_____ Oval window

_____ Stapes

_____ Tympanic membrane

PART C: Assessments

Match the terms in column A with the descriptions in column B. Place the letter of your choice in the space provided. A1 A3

Column A		Column B
a. Bony labyrinth	_____	**1.** Auditory ossicle attached to tympanic membrane
b. Cochlear duct	_____	**2.** Air-filled space containing auditory ossicles within middle ear
c. External acoustic meatus	_____	**3.** Contacts hairs of hearing receptors
d. Malleus	_____	**4.** Leads from oval window to apex of cochlea and contains perilymph
e. Membranous labyrinth	_____	**5.** S-shaped tube leading to tympanic membrane
f. Pharyngotympanic (auditory) tube	_____	**6.** Tube within cochlea containing spiral organ and endolymph
g. Scala tympani	_____	**7.** Cone-shaped, semitransparent membrane attached to malleus
h. Scala vestibuli	_____	**8.** Auditory ossicle attached to oval window
i. Stapes	_____	**9.** Chambers containing endolymph within bony labyrinth
j. Tectorial membrane	_____	**10.** Bony chambers of inner ear in temporal bone
k. Tympanic cavity	_____	**11.** Connects middle ear and pharynx
l. Tympanic membrane (eardrum)	_____	**12.** Extends from apex of cochlea to round window and contains perilymph

CRITICAL THINKING ASSESSMENT—PART C

Infections of the middle ear (otitis media) tend to occur more often in young children compared to adults. Why do you think this is so? How does treatment that involves surgically putting a ventilation (tympanostomy) tube in the eardrum help the child who has frequent ear infections? A1 A3

____

____

____

PART D: Assessments

1. Results of auditory acuity test: A4

Ear Tested	Audible Distance (cm)
Right	
Left	

2. Results of sound localization test: A4

Actual Location	Reported Location
Front of the head	
Behind the head	
Above the head	
Right side of the head	
Left side of the head	

3. Results of experiments using tuning forks: A4

Test	Left Ear (Normal or Impaired)	Right Ear (Normal or Impaired)
Rinne		
Weber		

4. Summarize the results of the hearing tests you conducted on your laboratory partner. A4

____

____

____

____

____

____

____

LABORATORY
EXERCISE 38

Ear and Equilibrium

Purpose of the Exercise

To review the structure and function of the organs of equilibrium and to conduct some tests of equilibrium.

MATERIALS NEEDED

Swivel chair
Bright light

For Demonstration Activity:
Compound light microscope
Prepared microscope slide of semicircular duct (cross section through ampulla)

SAFETY

- Do not pick subjects who have frequent motion sickness.
- Have four people surround the subject in the swivel chair in case the person falls from vertigo or loss of balance.
- Stop your experiment if the subject becomes nauseated.

Learning Outcomes

After completing this exercise, you should be able to:

O1 Locate the organs of static and dynamic equilibrium and describe their functions.

O2 Explain the role of vision in the maintenance of equilibrium.

O3 Conduct and record the results of the Romberg and Bárány tests of equilibrium.

The **O** corresponds to the assessments **A** indicated in the Laboratory Assessment for this Exercise.

Pre-Lab

Carefully read the introductory material and examine the entire lab. Be familiar with the basic structures and functions of the ear associated with equilibrium from lecture or the textbook. Answer the pre-lab questions.

Pre-Lab Questions Select the correct answer for each of the following questions:

1. The sense organs associated with equilibrium are within the
a. outer ear. **b.** middle ear.
c. inner ear. **d.** tympanic membrane.

2. The _____ nerve conducts impulses associated with equilibrium.
a. trochlear **b.** trigeminal
c. facial **d.** vestibulocochlear

3. Hair cell receptors associated with dynamic equilibrium are located within the
a. utricle. **b.** crista ampullaris.
c. saccule. **d.** cochlea.

4. Otoliths are composed of
a. calcium carbonate. **b.** carbon.
c. bone tissue. **d.** stereocilia.

5. Which of the following is *not* associated with rotational movements?
a. crista ampullaris **b.** otoliths
c. ampulla **d.** semicircular duct

6. Eye twitching movements characteristic during rotational movements are called nystagmus.
a. True **b.** False

7. Otoliths are located within the semicircular ducts of the inner ear.
a. True **b.** False

8. Sense of equilibrium depends on conscious awareness of body position and movements.
a. True **b.** False

The sense of equilibrium involves two sets of sensory organs. One set helps to maintain the stability of the head and body when they are motionless or during linear acceleration and produces a sense of static (gravitational) equilibrium. The other set is concerned with balancing the head and body when angular acceleration produces a sense of dynamic (rotational) equilibrium.

The sense organs associated with the sense of *static equilibrium* are located within the *vestibules* of the inner ears. Two chambers, the *utricle* and *saccule,* contain receptors called *maculae.* Each macula is composed of hair cells embedded within a gelatinous otolithic membrane. Embedded within the otolithic membrane are numerous tiny calcium carbonate "ear stones" called *otoliths.* As we change our head position, gravitational forces allow the shift of the otoliths to bend the stereocilia of the *hair cells* of the macula, resulting in stimulation of sensory vestibular neurons. Stimulation of the maculae also occurs during linear acceleration. Horizontal acceleration (as occurs when riding in a car) involves the utricle; vertical acceleration (as occurs when riding in an elevator) involves the saccule. As a result of static equilibrium, recognition of movements such as falling are detected and postural adjustments are accomplished.

The sense organs associated with the sense of *dynamic equilibrium* are located within the *ampullae* of the three *semicircular ducts* of the inner ear. Each membranous semicircular duct is located within a *semicircular canal* of the temporal bone. A small elevation within each ampulla possesses the *crista ampullaris.* Each crista ampullaris contains *hair cells* with stereocilia embedded within a gelatinous cap called the *cupula.* During angular acceleration, the cupula is bent, which stimulates hair cells and sensory neurons of the vestibular nerve. Because the three semicircular ducts are in different planes, rotational movements in any direction result in stimulation of the associated hair cells. Impulses from vestibular neurons of the semicircular ducts result in reflex movements of the eye. During rotational movements, characteristic twitching movements of the eyes called *nystagmus* occur, often accompanied by dizziness (vertigo).

Impulses from inner ear receptors travel over the vestibular neurons of the vestibulocochlear nerve and include processing destinations in the brainstem, cerebellum, thalamus, and cerebrum (fig. 38.1). The sense of equilibrium works subconsciously to initiate appropriate corrections to body position and movements. Additional senses work in conjunction with the inner ear and equilibrium. Stretch receptors in muscles and tendons, touch, and vision complement the inner ear for equilibrium and proper body adjustments.

PROCEDURE A: Organs of Equilibrium

1. Re-examine the introduction to this laboratory exercise.
2. Study figures 38.2 and 38.3, which present the structures and functions of static equilibrium.
3. Study figures 38.4 and 38.5, which present the structures and functions of dynamic equilibrium.
4. Complete Part A of Laboratory Assessment 38.

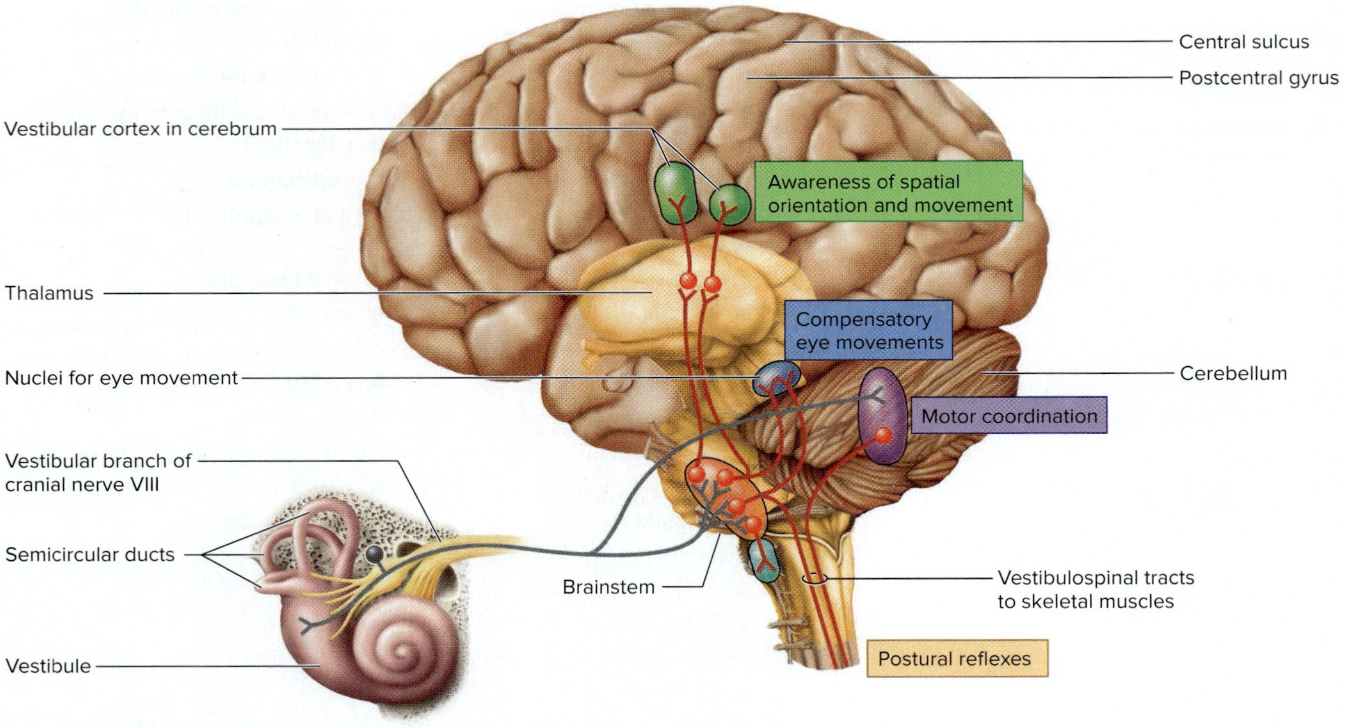

FIGURE 38.1 Vestibular pathways. The sensory receptors are hair cells within the vestibule and semicircular ducts. Sensory information over vestibular neurons continues through the brainstem, cerebellum, thalamus, and cerebrum of the brain. Conscious awareness of the body position and movement is processed in the cerebrum.

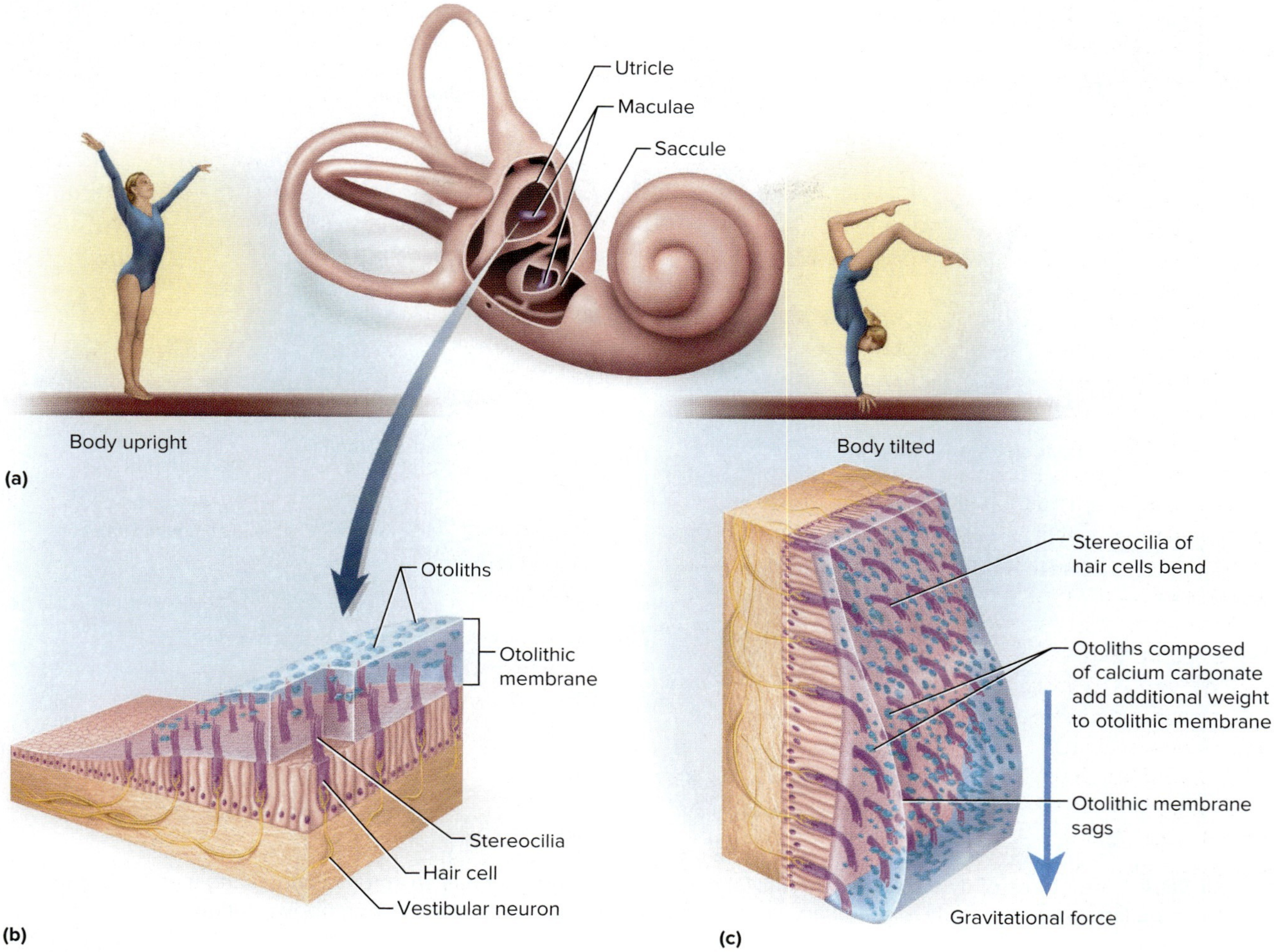

FIGURE 38.2 Structures of static equilibrium: (a) utricle and saccule each contain a macula; (b) macula and otolithic membrane orientation when body is upright; and (c) macula and otolithic membrane changes when the body is tilted.

DEMONSTRATION ACTIVITY

Observe the cross section of the semicircular duct through the ampulla in the demonstration microscope. Note the crista ampullaris (fig. 38.6) projecting into the lumen of the membranous labyrinth, which in a living person is filled with endolymph. The space between the membranous and bony labyrinths is filled with perilymph.

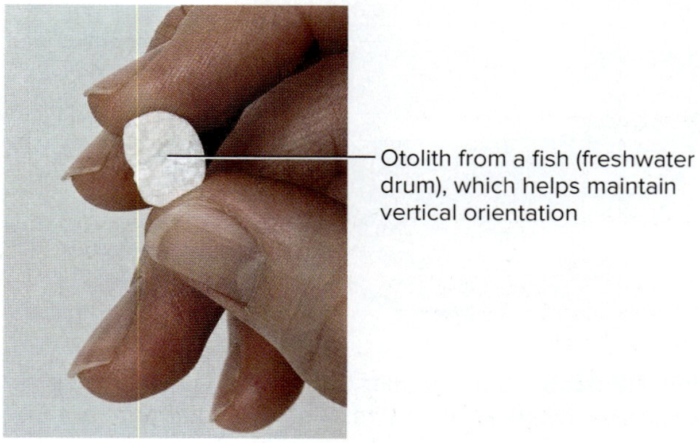

FIGURE 38.3 Particularly large otoliths are found in a freshwater drum (*Aplodinotus grunniens*). They have been used for lucky charms and jewelry. Human otoliths are microscopic size. ©J & J Photography

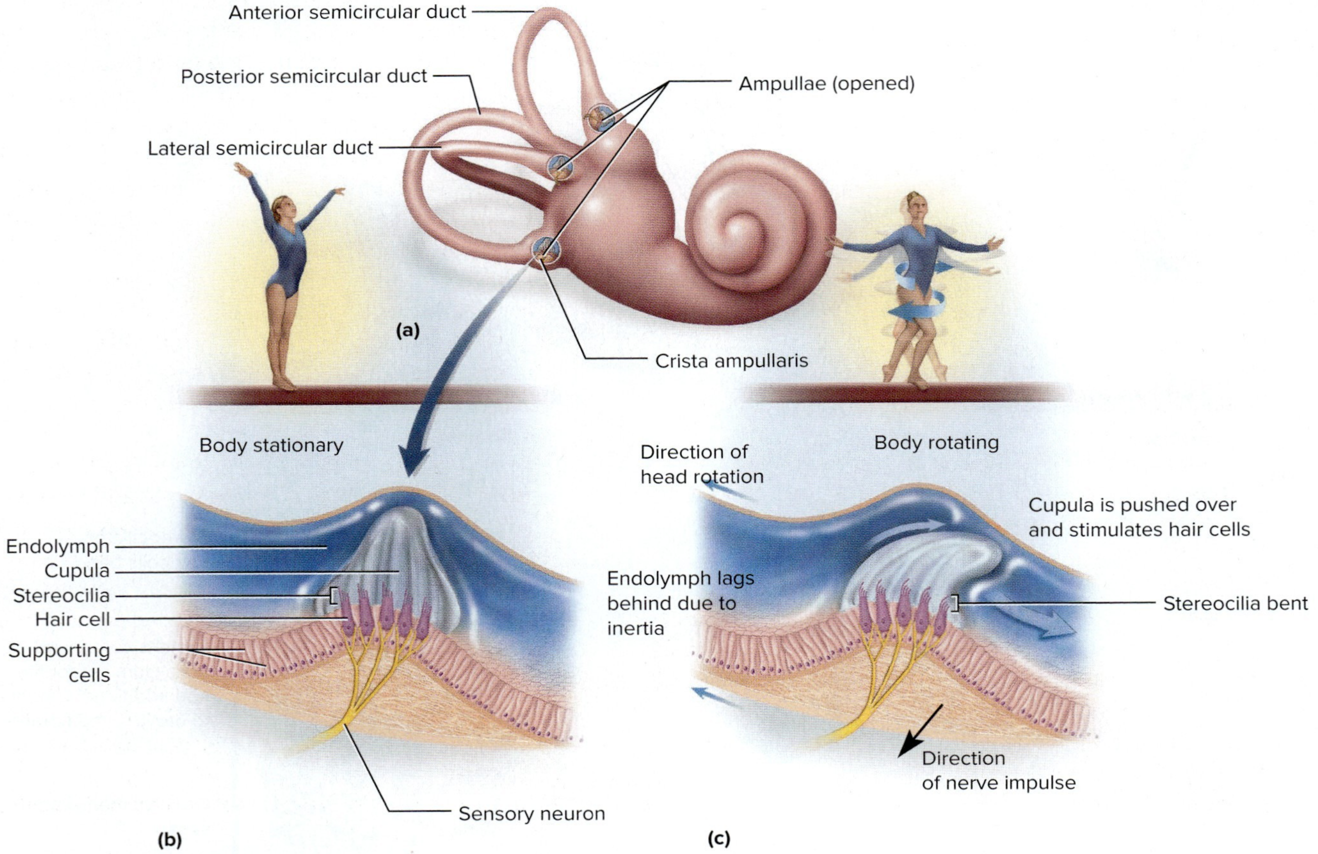

FIGURE 38.4 Structures of dynamic equilibrium: (a) each semicircular duct has an ampulla containing a crista ampullaris; (b) crista ampullaris when the body is stationary; and (c) changes in crista ampullaris during body rotation.

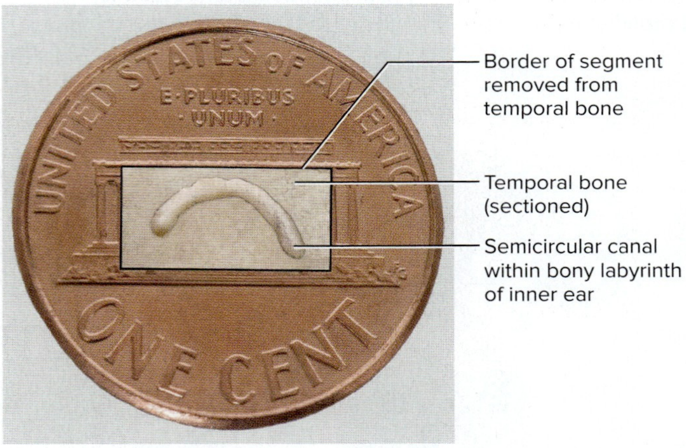

FIGURE 38.5 Semicircular canal superimposed on a penny to show its relative size. A semicircular duct of the same shape would occupy the semicircular canal.
©J & J Photography

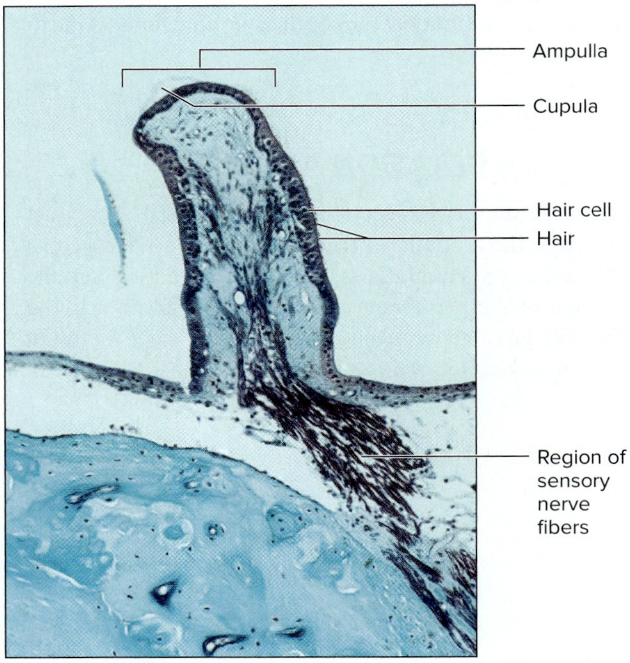

FIGURE 38.6 A micrograph of a crista ampullaris (400×). The crista ampullaris is located within the ampulla of each semicircular duct. Biophoto Associates/Science Source

PROCEDURE B: Tests of Equilibrium

Use figures 38.2 and 38.4 as references as you progress through the various tests of equilibrium. Perform the following tests, using a person as a test subject who is not easily disturbed by dizziness or rotational movement. Also have some other students standing close by to help prevent the test subject from falling during the tests. *The tests should be stopped immediately if the test subject begins to feel uncomfortable or nauseated.*

1. *Vision and equilibrium test.* To demonstrate the importance of vision in the maintenance of equilibrium, follow these steps:
   a. Have the test subject stand erect on one foot for 1 minute with his or her eyes open.
   b. Observe the subject's degree of unsteadiness.
   c. Repeat the procedure with the subject's eyes closed. *Be prepared to prevent the subject from falling.*
   d. Answer the questions related to the vision and equilibrium test in Part B of the laboratory assessment.
2. *Romberg test.* The purpose of this test is to evaluate how the organs of static equilibrium in the vestibule enable one to maintain balance (fig. 38.2). To conduct this test, follow these steps:
   a. Position the test subject close to a chalkboard with the back toward the board.
   b. Place a bright light in front of the subject so that a shadow of the body is cast on the board.
   c. Have the subject stand erect with feet close together and eyes staring straight ahead for 3 minutes.
   d. During the test, make marks on the chalkboard along the edge of the shadow of the subject's shoulders to indicate the range of side-to-side swaying.
   e. Measure the maximum sway in centimeters and record the results in Part B of the laboratory assessment.
   f. Repeat the procedure with the subject's eyes closed.
   g. Position the subject so one side is toward the chalkboard.
   h. Repeat the procedure with the eyes open.
   i. Repeat the procedure with the eyes closed.

   The Romberg test is used to evaluate a person's ability to integrate sensory information from proprioceptors and receptors within the organs of static equilibrium and to relay appropriate motor impulses to postural muscles. A person who shows little unsteadiness when standing with feet together and eyes open, but who becomes unsteady when the eyes are closed, has a positive Romberg test.
3. *Bárány test.* The purpose of this test is to evaluate the effects of rotational acceleration on the semicircular ducts and dynamic equilibrium (fig. 38.4). To conduct this test, follow these steps:
   a. Have the test subject sit on a swivel chair with his or her eyes open and focused on a distant object, the head tilted forward about 30°, and the hands gripped firmly to the seat. Position four people around the chair for safety. *Be prepared to prevent the subject and the chair from tipping over.*
   b. Rotate the chair ten rotations within 20 seconds.
   c. Abruptly stop the movement of the chair. The subject will still have the sensation of continuous movement and might experience some dizziness (vertigo).
   d. Have the subject look forward and immediately note the nature of the eye movements and their direction. (Such reflex eye twitching movements are called nystagmus.) Also note the time it takes for the nystagmus to cease. Nystagmus will continue until the cupula is returned to an original position.
   e. Record your observations in Part B of the laboratory assessment.
   f. Allow the subject several minutes of rest, then repeat the procedure with the subject's head tilted nearly 90° onto one shoulder.
   g. After another rest period, repeat the procedure with the subject's head bent forward so that the chin is resting on the chest.

   In this test, when the head is tilted about 30°, the lateral semicircular ducts receive maximal stimulation, and the nystagmus is normally from side to side. When the head is tilted at 90°, the superior ducts are stimulated, and the nystagmus is up and down. When the head is bent forward with the chin on the chest, the posterior ducts are stimulated, and the nystagmus is rotary.
4. Complete Part B of the laboratory assessment.

NOTES

LABORATORY
ASSESSMENT 38

Name ______________________________

Date ______________________________

Section ______________________________

The (A) corresponds to the indicated Learning Outcome(s) (O) found at the beginning of the Laboratory Exercise.

Ear and Equilibrium

PART A: Assessments

Complete the following statements:

1. The organs of static equilibrium are located within two expanded chambers within the vestibule called the ______________ and the saccule. A1
2. All of the balance organs are found within the ______________________ bone of the skull. A1
3. Otoliths are small grains composed of ______________________. A1
4. Linear acceleration changes of the head, such as when a person tilts his/her head downward to pet a dog, is detected by organs of ______________________ equilibrium. A1
5. Sensory impulses travel from the organs of equilibrium to the brain on vestibular neurons of the ______________________ nerve. A1
6. The sensory organ of a semicircular duct lies within a small elevation called the ______________________. A1
7. The sensory organ within the ampulla of a semicircular duct is called a(an) ______________________. A1
8. The ______________________ of this sensory organ consists of a dome-shaped gelatinous cap. A1
9. The vestibular cortex in the ______________________ of the brain processes awareness of body position and movement. A1
10. In an angular acceleration, when the head first rotates, the fluid (endolymph) inside the membranous portion of the semicircular duct tends to lag behind due to ______________________. A1

PART B: Tests of Equilibrium Assessments

1. Vision and equilibrium test results:

   a. When the eyes are open, what sensory organs provide information needed to maintain equilibrium? A2

   ____

   ____

   b. When the eyes are closed, what sensory organs provide such information? A2

   ____

   ____

2. Romberg test results:

a. Record the test results in the following table: A3

Conditions	Maximal Movement (cm)
Back toward board, eyes open	
Back toward board, eyes closed	
Side toward board, eyes open	
Side toward board, eyes closed	

b. Did the test subject's unsteadiness increase when the eyes were closed? ____________ What is the significance of this observation? A2 ____

____

____

c. Why would you expect a person with impairment of the organs of equilibrium to become more unsteady when the eyes are closed? A2 ____

____

____

3. Bárány test results:

a. Record the test results in the following table: A3

Position of Head	Description of Eye Movements	Time for Movement to Cease
Tilted 30° forward		
Tilted 90° onto shoulder		
Tilted forward, chin on chest		

b. Summarize the results of this test. A3

____

____

____

CRITICAL THINKING ASSESSMENT—PART B

What additional sensory information would you expect persons with impairment of organs of equilibrium to use to supplement their relative lack of some sensory information? A2

____

____

____

LABORATORY
EXERCISE 39

Endocrine Structure and Function

Purpose of the Exercise

To examine and identify anatomically and microscopically the major endocrine glands and tissues, and to specify the hormones secreted and their general functions.

MATERIALS NEEDED

Human torso model

Compound light microscope

Prepared microscope slides of the following:

- Pituitary gland
- Thyroid gland
- Parathyroid gland
- Adrenal gland
- Pancreas
- Ovary
- Testis

For Learning Extension Activity:

Water bath equipped with temperature control mechanism set at 37°C (98.6°F)

Laboratory thermometer

Learning Outcomes APR

After completing this exercise, you should be able to:

O1 Name and locate the major endocrine glands.

O2 Sketch and label tissue sections from the pituitary gland, thyroid gland, parathyroid glands, adrenal glands, pancreas, ovary, and testis.

O3 Associate the principal hormones secreted by each of the major endocrine glands.

O4 Describe the general physiological function of the principal hormones.

The O corresponds to the assessments A indicated in the Laboratory Assessment for this Exercise.

Pre-Lab

Carefully read the introductory material and examine the entire lab. Be familiar with basic structures and functions of the pituitary, thyroid, parathyroid, adrenal, and pancreas glands, and ovary and testis from lecture or the textbook. Answer the pre-lab questions.

Pre-Lab Questions Select the correct answer for each of the following questions:

1. Endocrine glands secrete _____ into the blood.
a. neurotransmitters **b.** true hormones
c. paracrine substances **d.** autocrine substances

2. True hormones influence
a. blood cells. **b.** the secreting cells.
c. neighboring cells. **d.** target cells around the body.

3. Hormones bind to _____ on target cells to bring about an effect.
a. phospholipids **b.** synapses
c. receptors **d.** vesicles

4. Which of the following endocrine glands is also an exocrine (ductile secretion) gland?
a. pancreas **b.** thyroid
c. thymus **d.** adrenal

5. Which two endocrine glands have the closest proximity to each other?
a. pancreas; thymus **b.** thymus; thyroid
c. thyroid; parathyroid **d.** pituitary; thyroid

6. Which endocrine gland stores hormones synthesized by the hypothalamus?
a. posterior pituitary **b.** anterior pituitary
c. adrenal cortex **d.** adrenal medulla

7. The endocrine system most often establishes homeostasis through
a. negative feedback. **b.** paracrine secretions.
c. positive feedback. **d.** autocrine secretions.

Chapter Opener: ©101cats/Getty Images

8. The hypothalamus, pineal gland, and pituitary gland are located within the cranial cavity.
   a. True b. False
9. Insulin is secreted by beta cells of the pancreatic islets.
   a. True b. False
10. The adrenal medulla secretes the hormones aldosterone and cortisol.
   a. True b. False

The endocrine system consists of *ductless glands* that act together with parts of the nervous system to help control body activities. The endocrine glands secrete regulatory molecules, called *hormones,* into the internal environment of interstitial fluid. Hormones are then transported in the *blood* and influence *target cells* around the body. A target cell is any cell that expresses a receptor that will bind the hormone. In most cases, hormones bind to specific *receptors* on the cell membranes of target cells, resulting in the activation of processes leading to functional changes within the target cells. In this way, hormones influence the rate of metabolic reactions, the transport of substances through cell membranes, the regulation of water and electrolyte balances, and many other functions.

Some of the glands included in this laboratory exercise have ducts and are known as exocrine glands. These glands are mixed endocrine and exocrine. The pancreas, ovaries, and testes have ducts and function within systems other than the endocrine system. Additional secreted substances that are regulatory, although not considered true hormones, include paracrine and autocrine secretions. *Paracrine secretions* influence only neighboring cells; *autocrine secretions* influence only the secreting cell itself.

The nervous system and the endocrine system are both regulatory and involve chemical secretions and receptor sites. The neurons of the nervous system secrete neurotransmitters. Receptor sites for neurotransmitters are on the postsynaptic cell, resulting in rapid responses of a brief duration unless additional stimulation occurs. In contrast, the endocrine cells secrete hormones transported by the blood to receptor sites on target cells. The responses from the hormones last longer and might even continue for hours to days even without additional hormonal secretions. By controlling cellular processes, endocrine glands play important roles in the maintenance of homeostasis.

The relationship between the hypothalamus, anterior pituitary, and thyroid gland demonstrates the role of the endocrine system in establishing homeostasis through negative feedback. When you walk outside on a cold winter day, your body temperature will drop, triggering specific physiological processes to respond to restore normal body temperature. Temperature receptors in the skin stimulate neural signals to the hypothalamus, which results in release of thyrotropin-releasing hormone (TRH). TRH in the blood travels to the anterior pituitary, stimulating release of thyroid-stimulating hormone (TSH) (also called thyrotropin). TSH in the blood travels to the thyroid gland, stimulating release of thyroid hormones (T_3 and T_4). Thyroid hormone increases the metabolic rate of nearly all cells, especially neurons, by stimulating them to increase their number of membrane Na^+/K^+ pumps. With more Na^+/K^+ pumps, more energy is converted from the chemical energy of ATP to the mechanical energy of pumping Na^+ out of the cells and K^+ into the cells against their concentration gradients; thus, more heat is produced (second law of thermodynamics) to maintain normal body temperature. As soon as homeostasis is reached, the feedback on the hypothalamus inhibits TRH release, the anterior pituitary will not be stimulated to release TSH, and therefore the thyroid gland will not release thyroid hormone.

PROCEDURE: Endocrine Gland Histology

1. Study the location and function(s) of the major endocrine glands indicated in figure 39.1 and tables 39.1 and 39.2.
2. Examine the human torso model and locate the following:

   hypothalamus
   pituitary stalk (infundibulum)
   pituitary gland (hypophysis)
   - anterior lobe (adenohypophysis)
   - posterior lobe (neurohypophysis)

   thyroid gland
   parathyroid glands

LEARNING EXTENSION ACTIVITY

The secretions of endocrine glands are usually controlled by negative feedback systems. As a result, the concentrations of hormones in body fluids remain relatively stable, although they will fluctuate slightly within a normal range.

Similarly, the mechanism used to maintain the temperature of a laboratory water bath involves negative feedback. In this case, a temperature-sensitive thermostat in the water allows a water heater to operate whenever the water temperature drops below the thermostat's set point. Then, when the water temperature reaches the set point, the thermostat causes the water heater to turn off (a negative effect), and the water bath begins to cool again.

Use a laboratory thermometer to monitor the temperature of the water bath in the laboratory. Measure the temperature at regular intervals until you have recorded ten readings. What was the lowest temperature you recorded? ____________ The highest temperature? ____________ What was the average temperature of the water bath? ____________
How is the water bath temperature control mechanism similar to a hormonal control mechanism in the body?

____

____

____

____

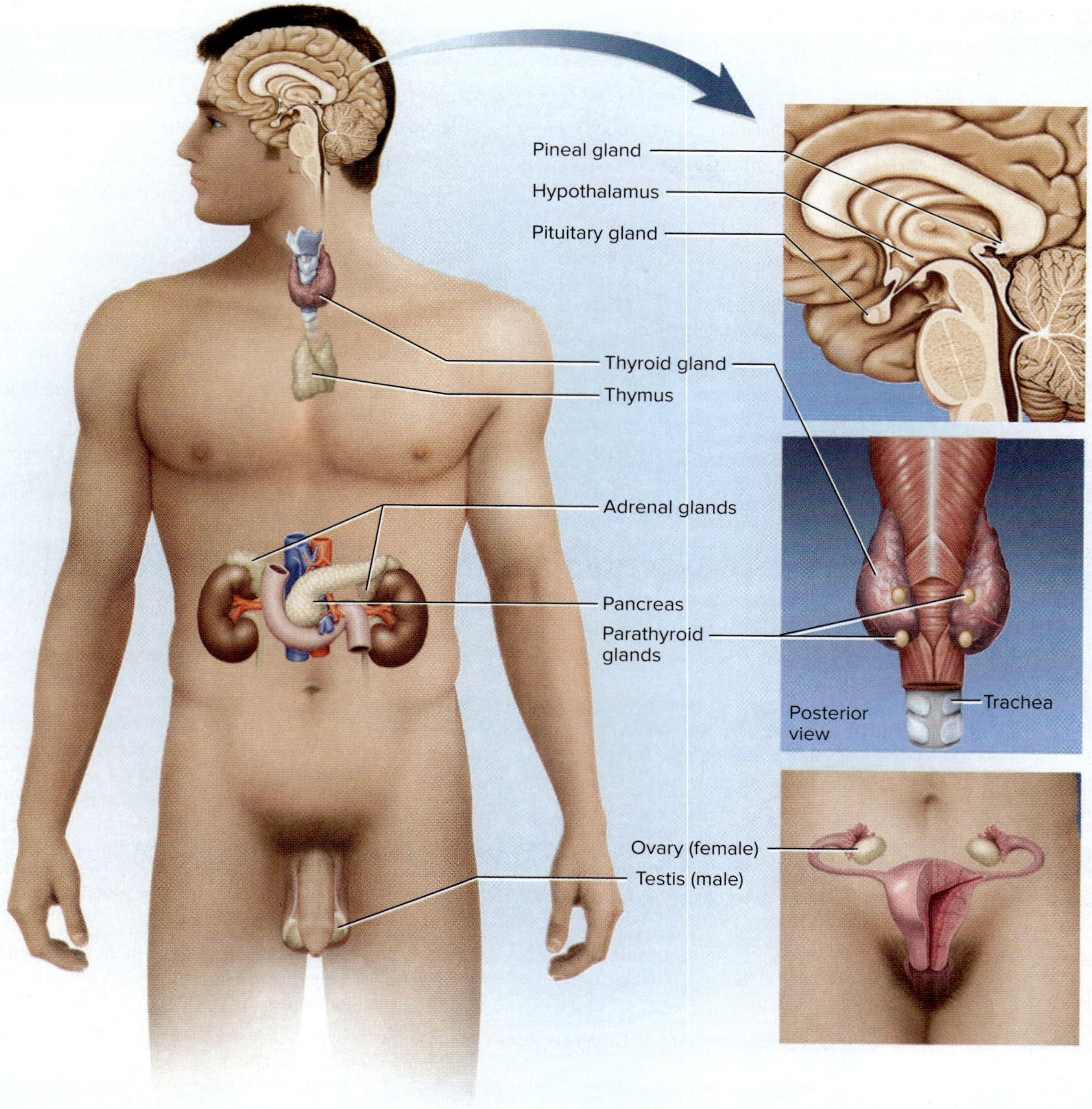

FIGURE 39.1 Major endocrine glands. APR

Table 39.1 Releasing and Inhibiting Hormones of the Hypothalamus

Hormone	General Function(s)
Thyrotropin-releasing hormone (TRH)	Stimulates anterior pituitary to secrete thyroid-stimulating hormone (TSH) and prolactin (PRL)
Corticotropin-releasing hormone (CRH)	Stimulates anterior pituitary to secrete adrenocorticotropic hormone (ACTH)
Gonadotropin-releasing hormone (GnRH)	Stimulates anterior pituitary to secrete follicle-stimulating hormone (FSH) and luteinizing hormone (LH)
Growth hormone-releasing hormone (GHRH)	Stimulates anterior pituitary to secrete growth hormone (GH)
Prolactin-inhibiting hormone (PIH)	Inhibits anterior pituitary secretion of prolactin (PRL)
Somatostatin (Growth hormone-inhibiting hormone (GHIH))	Inhibits anterior pituitary secretion of growth hormone (GH) and thyroid-stimulating hormone (TSH)

adrenal glands (suprarenal glands)

- adrenal medulla
- adrenal cortex

pancreas

pineal gland—secretes melatonin

thymus

ovaries

testes

Table 39.2 Endocrine Hormones and Their Functions

Endocrine Gland or Tissue	Hormone(s)	General Function(s)
	PITUITARY GLAND	
Anterior lobe	Thyroid-stimulating hormone (TSH)	Thyroid gland growth and secretion of thyroid hormone
	Adrenocorticotropic hormone (ACTH)	Adrenal cortex growth and secretion of glucocorticoids
	Prolactin (PRL)	Synthesis of milk in females and increases sensitivity of LH in males
	Growth hormone (GH)	Mitosis, differentiation, and growth of tissues (especially muscle, cartilage, bone, and fat)
	Follicle-stimulating hormone (FSH)	Follicle growth and secretion of estrogen in ovaries and sperm production in testes
	Luteinizing hormone (LH)	Ovulation and maintaining corpus luteum in ovaries and secretion of testosterone in testes
Posterior lobe	Antidiuretic hormone (ADH)	Water retention by kidneys to help prevent dehydration
	Oxytocin (OT)	Feelings of sexual satisfaction and emotional bonding; labor contractions; milk ejection in lactating mothers
	THYROID GLAND	
Follicle cells	Thyroid hormone (T_3 and T_4)	Increases metabolic rate; accelerates breakdown of food for fuel; alertness
C (parafollicular) cells	Calcitonin	Responds to high blood calcium by stimulating osteoblast activity to deposit calcium and form bone
	PARATHYROID GLAND	
Chief cells	Parathyroid hormone (PTH)	Responds to low blood calcium by indirectly stimulating increased osteoclast activity to release calcium from bone; acts on the kidney cells to reabsorb calcium so less is elimated in urine; stimulates final step in calcitriol synthesis by the kidneys
	ADRENAL CORTEX	
Zona glomerulosa	Mineralocorticoids–aldosterone	Sodium retention in body fluids and excretion of potassium in urine
Zona fasciculata	Glucocorticoids–cortisol	Fat and protein breakdown; synthesis of glucose from noncarbohydrates (gluconeogenesis); help body adapt to stress; anti-inflammatory effect
Zona reticularis	Sex steroids–androgens and estrogens	Growth of secondary sexual characteristic (pubic and axillary hair, apocrine scent glands, libido)
	ADRENAL MEDULLA	
Chromaffin cells	Epinephrine and norepinephrine (NE)	Increase alertness; mobilize fuels for energy; raise heart rate, respiratory rate, metabolic rate
	PANCREAS	
Alpha cells	Glucagon	Raises glucose in the blood; breakdown glycogen in liver; amino acid absorption for gluconeogenesis; fat breakdown
Beta cells	Insulin	Lowers glucose in the blood; absorption of glucose, fatty acids, amino acids by cells for use, storage, or synthesis
Delta cells	Somatostatin	Inhibits stomach acid secretion
	OVARY	
Granulosa cells of follicle	Estrogen (estradiol)	Development and growth of reproductive organs; regulation of menstrual cycle and pregnancy; mammary glands prepared for lactation
Corpus luteum	Progesterone (and estrogen)	Regulation of menstrual cycle and pregnancy; mammary glands prepared for lactation
	TESTIS	
Interstitial cells	Testosterone	Development and growth of reproductive organs; growth of skeleton and muscle; sperm production; libido

3. Examine the microscopic tissue sections of the following glands, and identify the features described:

Pituitary gland. APR The pituitary gland is located in the sella turcica of the sphenoid bone. To examine the pituitary tissue, follow these steps:

a. Observe the tissues using scanning and then low-power magnification (fig. 39.2).
b. Locate the *pituitary stalk* (infundibulum), the *anterior lobe* (the largest part of the gland), and the *posterior lobe.* The anterior lobe synthesizes hormones and appears glandular; the posterior lobe stores and releases hormones synthesized by the hypothalamus and has the appearance of nervous tissue.
c. Observe an area of the anterior lobe using high-power magnification (fig. 39.3). Locate a cluster of relatively large cells and identify some *acidophil cells,* which contain pink-stained granules, and some *basophil cells,* which contain blue-stained granules. These acidophil and basophil cells are the hormone-secreting cells. The hormones produced and secreted include TSH, adrenocorticotropic hormone (ACTH), prolactin (PRL), growth hormone (GH), follicle-stimulating hormone (FSH), and luteinizing hormone (LH). The *chromophobes* are depleted of hormones so they stain weakly and appear pale; some are thought to be stem cells. The general functions of these hormones are listed in table 39.2.
d. Observe an area of the posterior lobe using high-power magnification (fig. 39.4). Note the numerous unmyelinated nerve fibers present in this lobe. Also locate some *pituicytes,* a type of glial cell, scattered among the nerve fibers. The posterior lobe stores and releases antidiuretic hormone (ADH) and oxytocin (OT), which are synthesized in the hypothalamus. The general functions of these hormones are listed in table 39.2.
e. Prepare labeled sketches of representative portions of the anterior and posterior lobes of the pituitary gland in Part A of Laboratory Assessment 39.

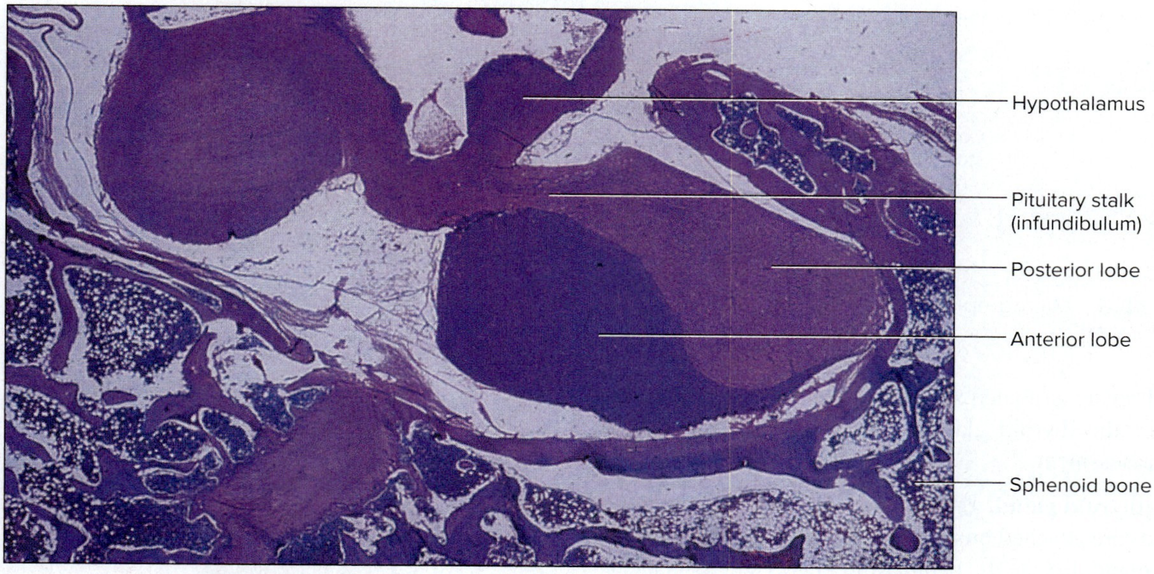

FIGURE 39.2 Micrograph of the pituitary gland (6×). Biophoto Associates/Science Source APR

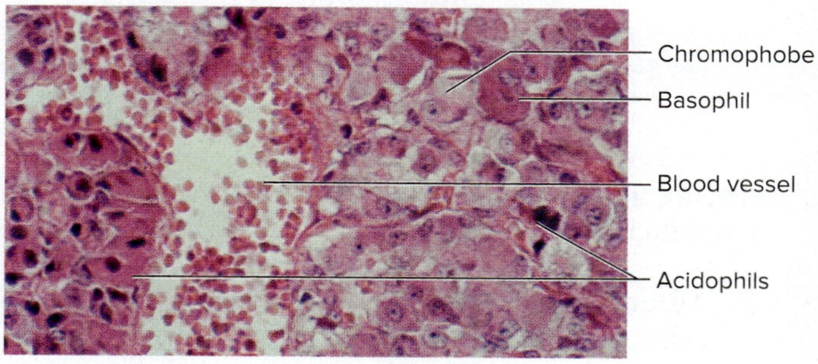

FIGURE 39.3 Micrograph of the anterior lobe of the pituitary gland (400×). Cynthia Prentice-Craver/McGraw Hill APR

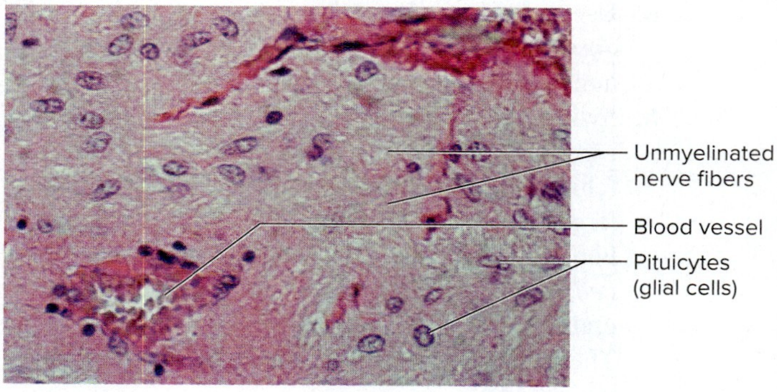

FIGURE 39.4 Micrograph of the posterior lobe of the pituitary gland (400×). Cynthia Prentice-Craver/McGraw Hill APR

Thyroid gland. APR Locate the thyroid gland and associated structures (fig. 39.5). To examine the thyroid tissue, follow these steps:

a. Use scanning and then low-power magnification to observe the tissue (fig. 39.6). Note the numerous *follicles,* each of which consists of a layer of cells surrounding a colloid-filled cavity.

b. Observe the tissue using high-power magnification. The follicular cells forming the wall of a follicle are simple cuboidal epithelial cells. These cells secrete thyroid hormone (T_3 and T_4). The parafollicular cells outside the follicle secrete calcitonin. The general functions of these hormones are listed in table 39.2.

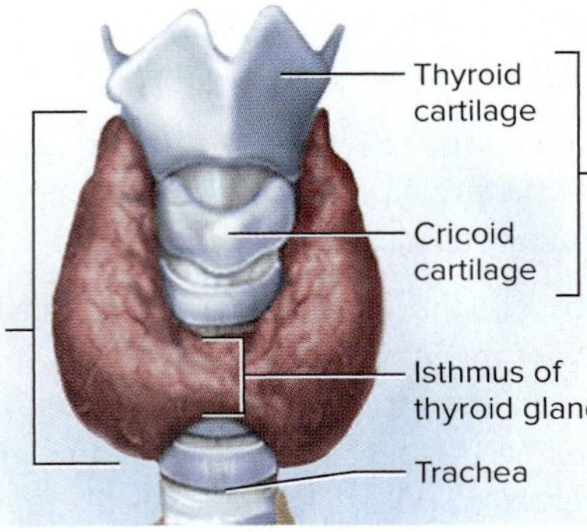

FIGURE 39.5 Anterior view of thyroid gland and associated structures. APR

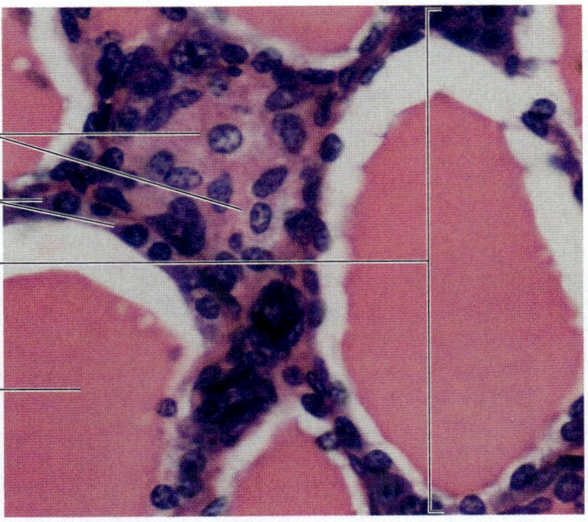

FIGURE 39.6 Micrograph of thyroid gland (400×). Cynthia Prentice-Craver/McGraw Hill APR

c. Prepare a labeled sketch of a representative portion of the thyroid gland in Part A of the laboratory assessment.

Parathyroid gland. APR Typically, four parathyroid glands are attached on the posterior surface of the thyroid gland. Locate the parathyroid glands and associated structures (fig. 39.7). To examine the parathyroid tissue, follow these steps:

a. Use scanning and then low-power magnification to observe the tissue (fig. 39.8). The gland consists of numerous tightly packed secretory cells.

b. Switch to high-power magnification and locate two types of cells—a smaller, more abundant form (chief cells) arranged in cordlike patterns and a larger form (oxyphil cells) that have distinct cell boundaries and are present in clusters. The *chief cells* secrete parathyroid hormone (PTH). The general function of this hormone is listed in table 39.2. The function of the *oxyphil cells* is not well understood.

c. Prepare a labeled sketch of a representative portion of the parathyroid gland in Part A of the laboratory assessment.

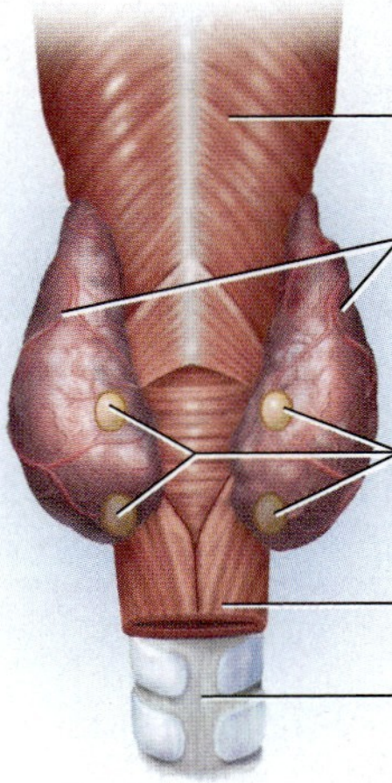

FIGURE 39.7 Posterior view of parathyroid glands and associated structures.

Adrenal gland. APR Locate an adrenal gland and associated structures (fig. 39.9). To examine the adrenal tissue, follow these steps:

a. Use scanning and then low-power magnification to observe the tissue (fig. 39.10). Note the thin capsule

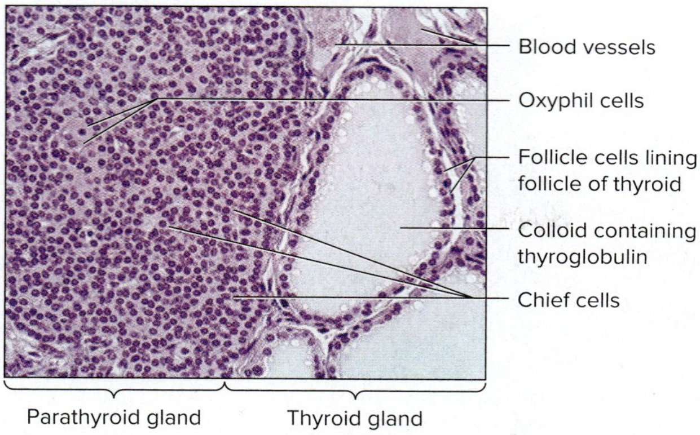

FIGURE 39.8 Micrograph of the parathyroid gland adjacent to the thyroid gland (400×). Victor P. Eroschenko APR

that covers the gland. Just beneath the capsule, there is a relatively thick *adrenal cortex*. The central portion of the gland is the *adrenal medulla*. The cells of the cortex are in three poorly defined layers. Those of the outer layer (*zona glomerulosa*) are arranged irregularly in small rounded clusters and secrete a group of hormones called mineralocorticoids; those of the thicker middle layer (*zona fasciculata*) are in long cords and primarily secrete a group of hormones called glucocorticoids; and those of the inner layer (*zona reticularis*) are arranged in an interconnected network of cords and primarily secrete a group of hormones called sex steroids. Together, the adrenal cortex hormones are considered *corticosteroids* of which cholesterol is the precursor. The most noted corticosteroids secreted from the cortex include aldosterone from the zona glomerulosa, cortisol from the zona fasciculata, and androgens and estrogens from the zona reticularis. The cells of the medulla (chromaffin cells) are relatively large, irregularly shaped, and often occur in clusters. The medulla cells secrete epinephrine and norepinephrine. The general functions of these hormones are listed in table 39.2.

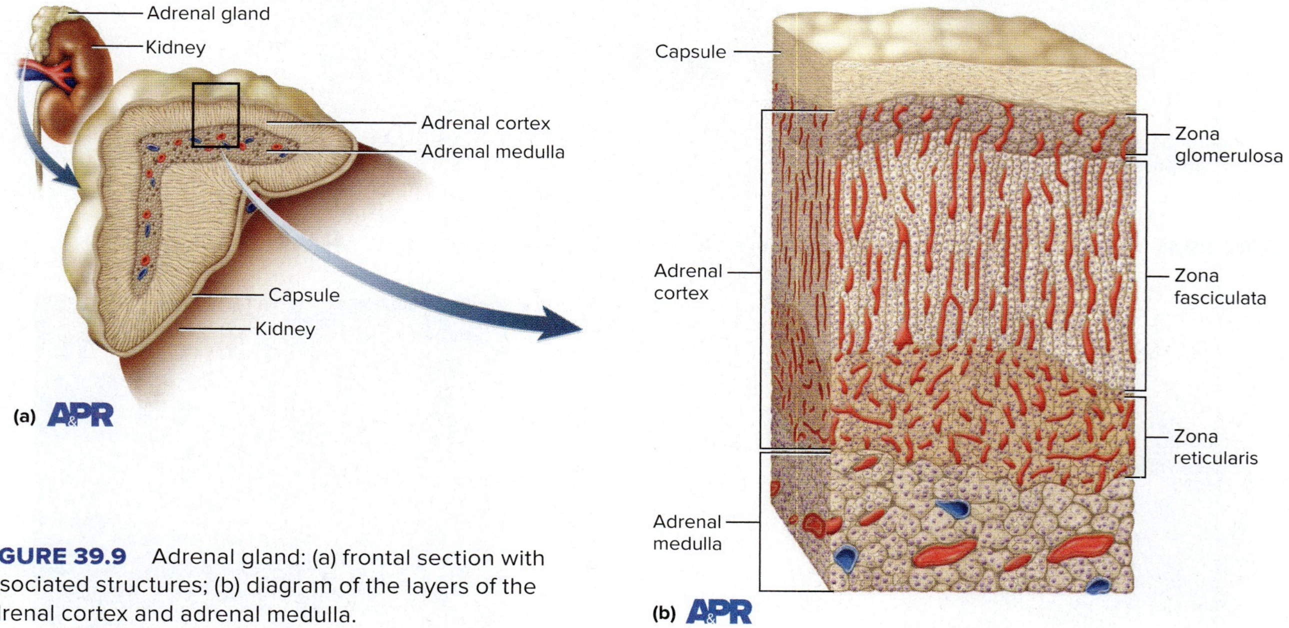

FIGURE 39.9 Adrenal gland: (a) frontal section with associated structures; (b) diagram of the layers of the adrenal cortex and adrenal medulla.

b. Using high-power magnification, observe each of the layers of the cortex and the cells of the medulla.

c. Prepare labeled sketches of representative portions of the adrenal cortex and medulla in Part A of the laboratory assessment.

Pancreas. APR Locate the pancreas and associated structures (fig. 39.11). To examine the tissues of the pancreas, follow these steps:

a. Use scanning and then low-power magnification to observe the tissue (fig. 39.12). The gland largely consists of deeply stained exocrine cells (acinar cells) arranged in clusters around secretory ducts. These exocrine cells secrete pancreatic juice rich in

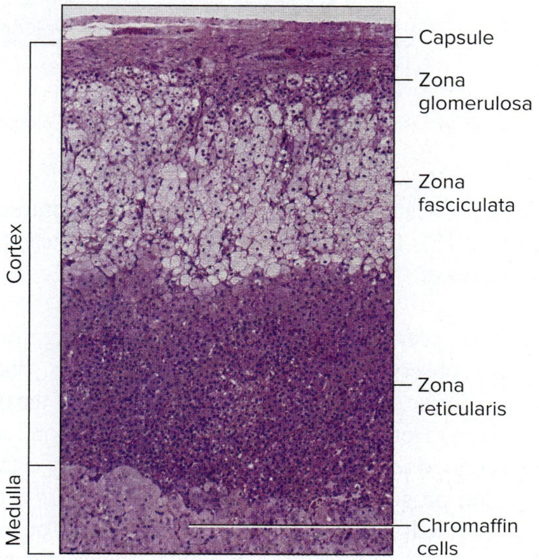

FIGURE 39.10 Micrograph of the adrenal cortex and the adrenal medulla (75×). Ed Reschke/Getty Images APR

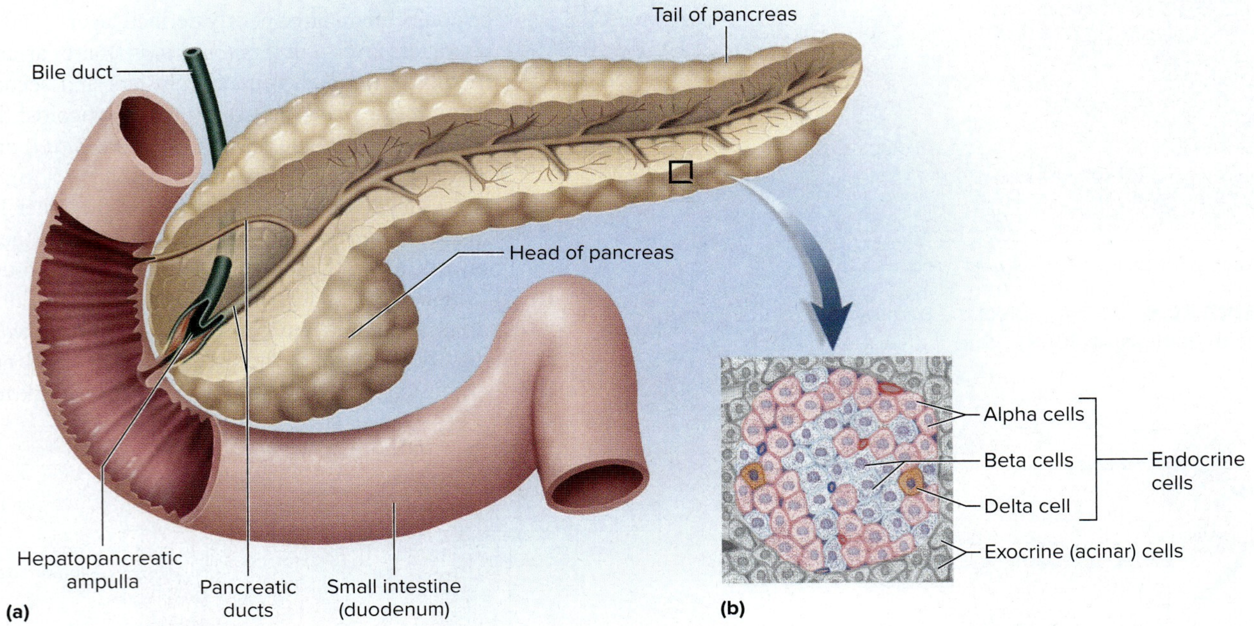

FIGURE 39.11 (a) Pancreas and associated structures. (b) Diagram of a pancreatic islet. APR

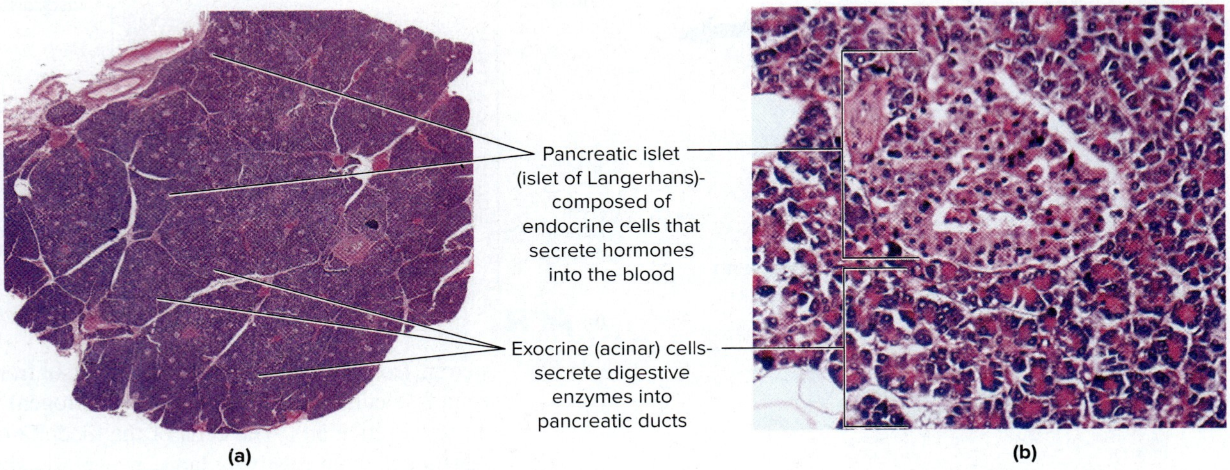

FIGURE 39.12 Micrographs of the pancreas at (a) 40× and (b) 400×. (a) Dr. Thomas Caceci, Virginia-Maryland Regional College of Veterinary Medicine; (b) Cynthia Prentice-Craver/McGraw Hill APR

digestive enzymes. There are circular masses of differently stained cells scattered throughout the gland. These clumps of cells constitute the *pancreatic islets (islets of Langerhans),* and they represent the endocrine portion of the pancreas. The *beta cells* secrete insulin, the *alpha cells* secrete glucagon, and the *delta cells* secrete somatostatin (growth hormone-inhibiting hormone). The general functions of these hormones are listed in table 39.2.

b. Examine an islet, using high-power magnification. Unless special stains are used, it is not possible to distinguish alpha, beta, and delta cells.

c. Prepare a labeled sketch of a representative portion of the pancreas in Part A of the laboratory assessment.

Ovary. Locate the ovary and associated structures (fig. 39.13). To examine the tissues of the ovary, follow these steps:

a. Use scanning and then low-power magnification to observe the tissue. The ovary has both an exocrine secretion, which is the ejection of the oocyte (egg) from the mature follicle at ovulation, and an endocrine secretion of estrogen (primarily estradiol) and progesterone. In the outer cortex of the ovary, you will see follicles that each contain an oocyte (egg) at various developmental stages. Primordial follicles, the most abundant of the follicles, consist of a single layer of follicular cells surrounding an

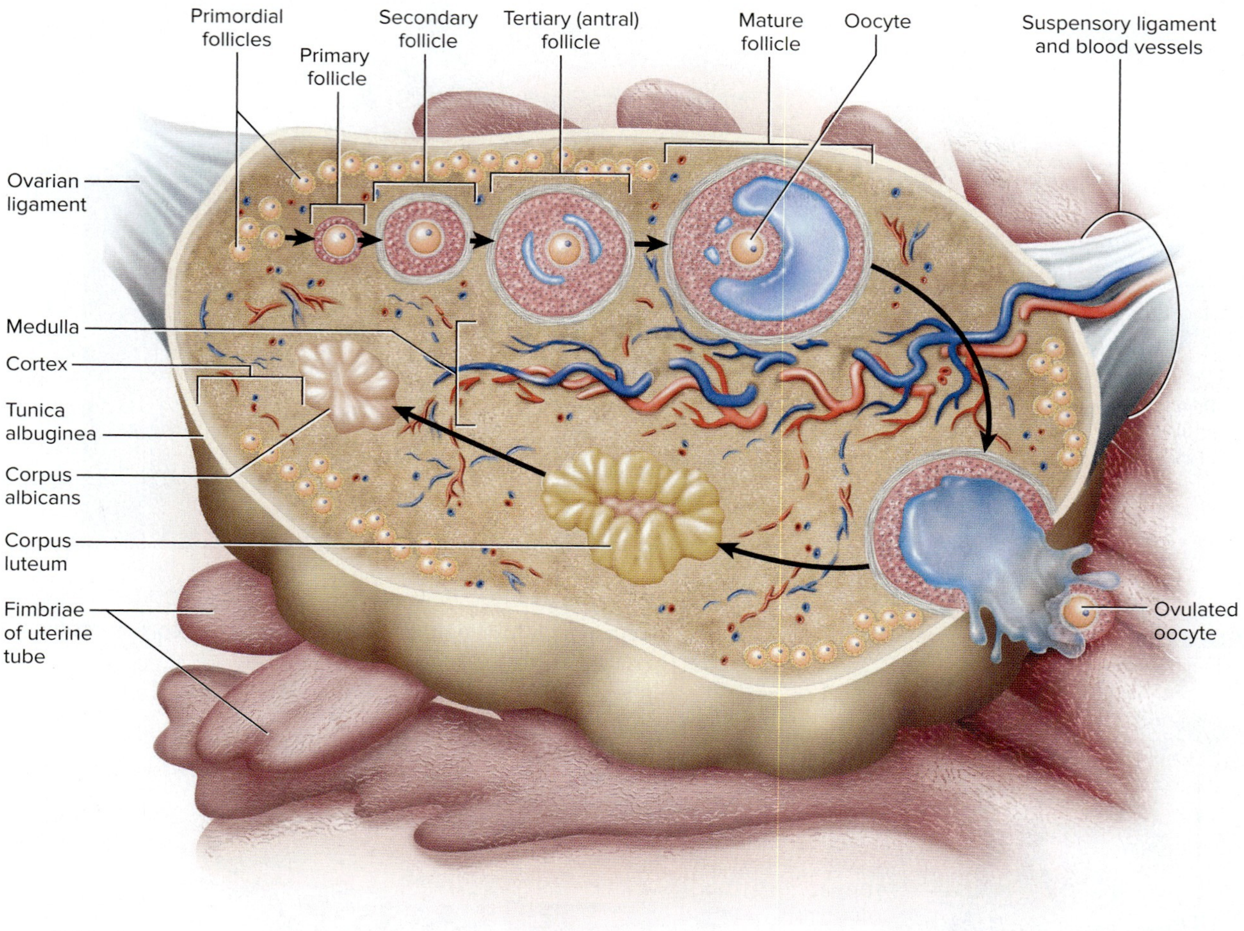

FIGURE 39.13 Ovary structure, with arrows showing the sequence of follicle development (folliculogenesis), and the corpus luteum that formed after ovulation. (*Note:* The sequential arrows do not imply that follicles migrate around the ovary as depicted in this single illustration; instead, they occur in a particular location of the ovary during a corresponding ovarian cycle.) APR

oocyte; these are formed before birth, and many persist throughout adulthood. A cohort of primordial follicles may continue growth into primary follicles, and eventually, stimulated by FSH from the anterior pituitary, one *mature follicle* will form. Surrounding the mature follicle is an outer theca folliculi whose cells make steroid hormones (androgens) that will diffuse to the inner *granulosa cells* to be converted to estrogen (estradiol) (fig. 39.14*a*). After ovulation, the ruptured follicle implodes and theca and granulosa cells fill the antrum, forming the *corpus luteum* (fig. 39.14*b*). Cells of the corpus luteum are initially stimulated by LH from the anterior pituitary to grow and secrete progesterone and estradiol. The general functions of these hormones are listed in table 39.2.

b. Examine a mature follicle under high-power magnification. Locate the corpus luteum (if it is present on your prepared slide) to compare the cells (now called lutein cells) of the corpus luteum to the granulosa cells found lining the mature follicle.

c. Prepare a labeled sketch of a mature ovarian follicle and a corpus luteum in Part A of the laboratory assessment.

Testis. Locate the testis and associated structures (fig. 39.15).

a. Use scanning and then low-power magnification to observe the tissue. Each testis contains numerous *seminiferous tubules* that nurture sperm (spermatozoon) production, an exocrine secretion. Germ cells (spermatagonia) in the outer portion of the tubule differentiate into sperm, each with half the number of chromosomes as the original germ cell, by the time they reach the lumen. LH from the anterior pituitary stimulates *interstitial cells* (Leydig cells), located between the seminiferous tubules, to synthesize and secrete testosterone, an endocrine secretion. The general functions of this hormone are listed in table 39.2. With the help of FSH from the anterior pituitary and testosterone from the interstitial cells, sperm production (spermatogenesis) is stimulated.

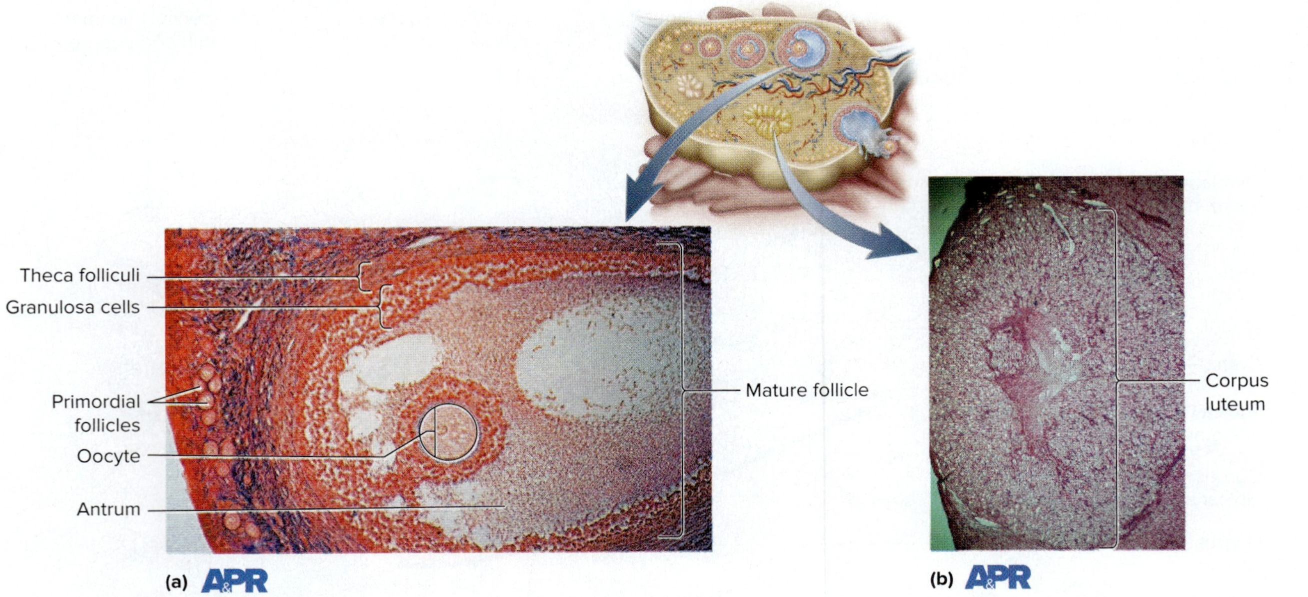

FIGURE 39.14 Ovary: (a) micrograph of primordial follicles and a mature follicle (100x); (b) micrograph of a corpus luteum (100x). Cynthia Prentice-Craver/McGraw Hill

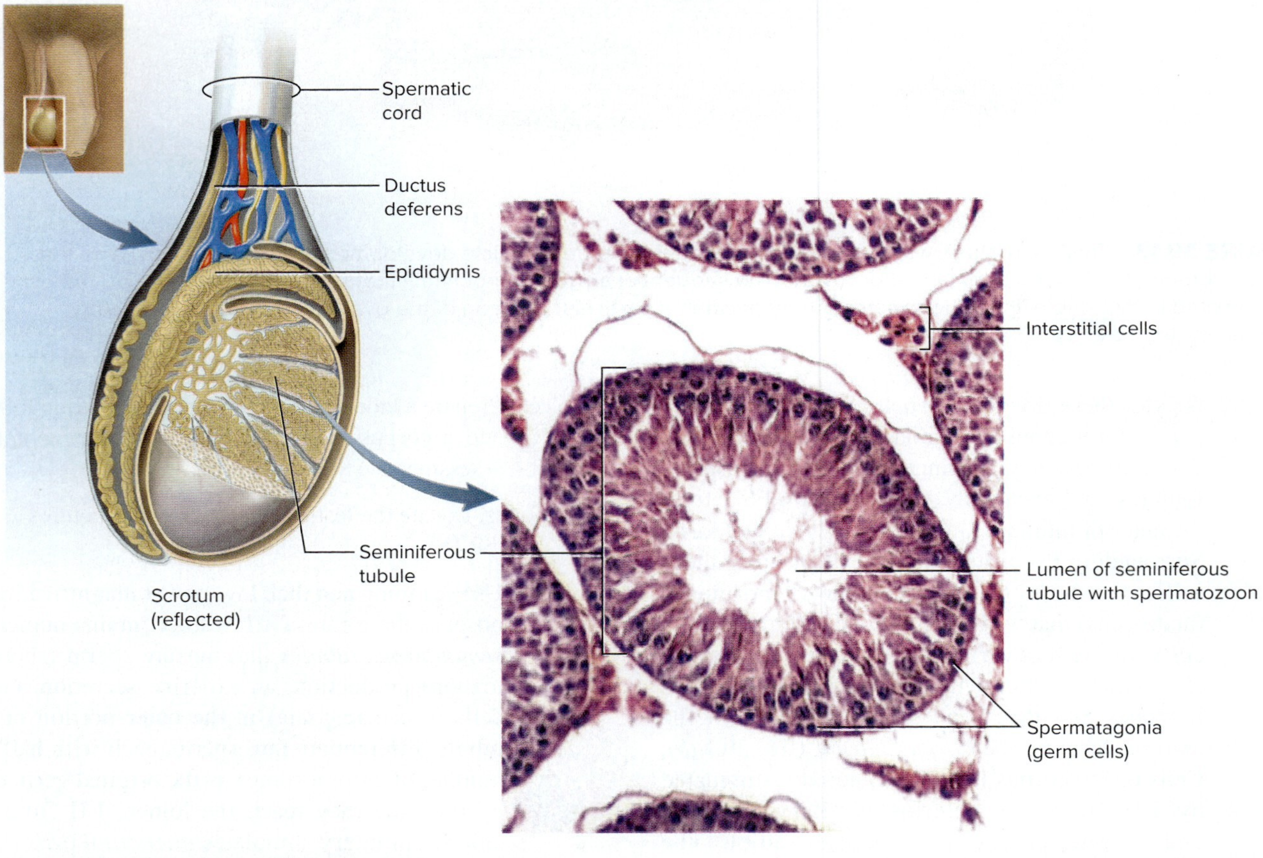

FIGURE 39.15 Structures within the testis, and micrograph of the seminiferous tubules and interstitial cells of the testes (100x). Cynthia Prentice-Craver/McGraw Hill APR

b. Examine the testis under high-power magnification. Locate the seminiferous tubules and recognize the cells that are developing into sperm (spermatogenic cells), as well as the spermatogonia (germ cells). Identify the interstitial cells.

c. Prepare a labeled sketch of a testis that includes seminiferous tubules and interstitial cells in Part A of the laboratory assessment.

4. Complete Parts B, C, and D of the laboratory assessments.

LABORATORY
ASSESSMENT 39

Name ____________________

Date ____________________

Section ____________________

The A corresponds to the indicated Learning Outcome(s) O found at the beginning of the Laboratory Exercise.

Endocrine Structure and Function

PART A: Assessments

Each circle below represents the microscopic field of view. In each circle, sketch and label representative portions of the indicated endocrine gland. A2

Pituitary gland and hypothalamus (_____×)

Pituitary gland (anterior lobe) (_____×)

Pituitary gland (posterior lobe) (_____×)

Thyroid gland (_____×)

Parathyroid gland (_____×)

Adrenal gland (cortex and medulla) (_____×)

Pancreas (______×)

Ovary (______×)

Testis (______×)

PART B: Assessments

1. Label the endocrine organs shown in the images of models in figure 39.16. A1

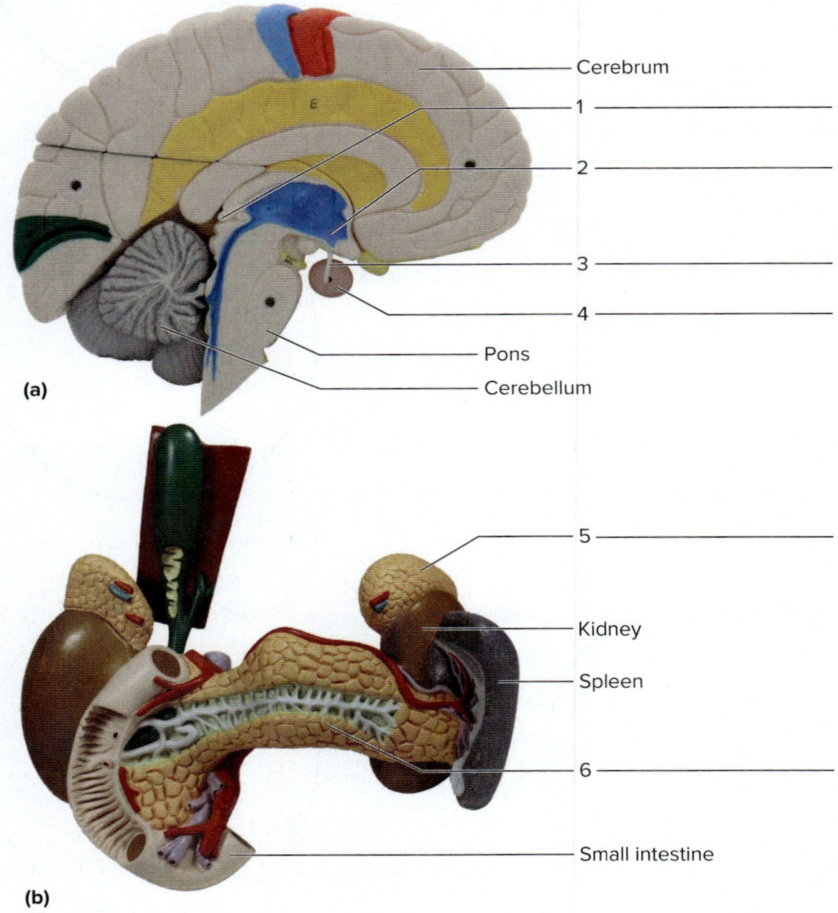

FIGURE 39.16 Major endocrine organs shown on model images of (a) a median section of the brain and (b) an anterior view of abdominal organs. (a) ©3B Scientific GmbH, Germany, 2017 www.3bscientific.com; (b) Christine Eckel/Denoyer-Geppert

2. Match the endocrine gland in column A with a characteristic of the gland in column B. Place the letter of your choice in the space provided. A1

Column A		Column B
a. Adrenal cortex	_______	1. Contains alpha, beta, and delta cells
b. Adrenal medulla	_______	2. Attached to pituitary gland by a stalk
c. Hypothalamus	_______	3. Contains seminiferous tubules and interstitial cells
d. Ovary	_______	4. Attached to the posterior surface of thyroid gland
e. Pancreatic islets	_______	5. Contains three layers that secrete corticosteroids
f. Parathyroid gland	_______	6. Located in mediastinum
g. Pituitary gland	_______	7. Contains colloid-filled follicles and parafollicular (C) cells
h. Testis	_______	8. Located in center portion of gland superior to kidneys
i. Thymus	_______	9. Contains developing follicles and corpus luteum
j. Thyroid gland	_______	10. Rests in sella turcica of sphenoid bone

PART C: Assessments

Complete the following:

1. Name six hormones secreted by the anterior lobe of the pituitary gland. A3 ________________________________

____

2. Name two hormones secreted by the posterior lobe of the pituitary gland. A3 ________________________

3. Name two hormones secreted by the corpus luteum within the ovary. A3 ________________________

4. Name two thyroid hormones secreted from the follicular cells. A3 ________________________

5. What hormone is secreted by alpha cells of the pancreatic islets? A3 ________________________

6. What hormone is secreted by beta cells of the pancreatic islets? A3 ________________________

7. Name the hormone secreted from the parafollicular cells of the thyroid gland. A3 ________________________

8. Name two hormones secreted by the adrenal medulla. A3 ________________________

9. Name the most noted glucocorticoid secreted by the zona fasciculata cells of the adrenal cortex. A3 ____________

10. Name the hormone secreted by interstitial cells of the testes. A3 ________________________

11. Where are the target cells for the releasing and inhibiting hormones of the hypothalamus located? A3 ____________

PART D: Assessments

In each of the following statements, circle the underlined choice that makes the statement about hormone function correct: A4

1. Calcitonin is released by the parafollicular (C) cells of the thyroid gland when blood calcium levels fall above/below normal levels in the blood.
2. Insulin is a hypoglycemic/hyperglycemic hormone because it acts to lower blood glucose levels.
3. The "stress hormone" released by cells of the zona fasciculata is aldosterone/cortisol.
4. Parathyroid hormone indirectly activates osteoclasts in bone to raise/lower blood calcium levels.
5. The hormone responsible for libido and sperm production is luteinizing hormone/testosterone.
6. When blood sodium (Na^+) is too low or blood potassium (K^+) is too high, the hormone aldosterone/cortisol is released by the zona glomerulosa.

7. The hormone responsible for milk ejection in lactating mothers is oxytocin/prolactin.

8. Thyroid hormone secreted by the thyroid gland and epinephrine and norepinephrine secreted by chromaffin cells of the adrenal medulla act to increase fatigue/alertness.

CRITICAL THINKING ASSESSMENTS—PART D

1. Name two hormones of the endocrine system that function oppositely to maintain homeostasis. ______________________________ Explain your reasoning. A3 A4 ______________________________

2. For each group of hormones, circle the one hormone that does *not* belong to that group and then, explain your reasoning. A3 A4

Group 1:

Hormones:	Growth hormone	Oxytocin	Thyroid-stimulating hormone	Adrenocorticotropic hormone
Explain your reasoning:				

Group 2:

Hormones:	Epinephrine	Cortisol	Estradiol	Aldosterone
Explain your reasoning:				

Group 3:

Hormones:	Follicle-stimulating hormone	Luteinizing hormone	Testosterone	Insulin
Explain your reasoning:				

LABORATORY

EXERCISE 40

Diabetic Physiology

Purpose of the Exercise

To understand the disease of diabetes mellitus, to observe behavioral changes that occur during insulin shock and recovery from insulin shock, and to examine microscopically the pancreatic islets.

MATERIALS NEEDED

500 mL beakers
Live small fish (1"–1.5" goldfish, guppy, rosy red feeder, or other)
Small fish net
Insulin (regular U-100) (HumulinR in 10 mL vials has 100 units/mL; store in refrigerator—do not freeze)
Syringes for U-100 insulin
10% glucose solution
Clock with second hand or timer
Compound light microscope
Prepared microscope slides of the following:
- Normal human pancreas (stained for alpha and beta cells)
- Human pancreas of a diabetic

SAFETY

- Review all the safety guidelines in Appendix 1 of your laboratory manual.
- Wear disposable gloves and protective eyewear when handling the fish and syringes.
- Dispose of the used syringe and needle in the puncture-resistant container.
- Wash your hands before leaving the laboratory.

Chapter Opener: ©101cats/Getty Images

Learning Outcomes APR

After completing this exercise, you should be able to:

O1 Identify the cells of the pancreatic islets and the hormones they produce and secrete.

O2 Compare the causes, symptoms, and treatments for type 1 and type 2 diabetes mellitus.

O3 Examine and record the behavioral changes caused by insulin shock.

O4 Examine and record the recovery from insulin shock when sugar is provided to cells.

O5 Distinguish tissue sections of normal pancreatic islets from those indicating diabetes mellitus, and sketch both types.

The O corresponds to the assessments A indicated in the Laboratory Assessment for this Exercise.

Pre-Lab

Carefully read the introductory material and examine the entire lab. Be familiar with diabetes mellitus from lecture or the textbook. Answer the pre-lab questions.

Pre-Lab Questions Select the correct answer for each of the following questions:

1. The _____ cells of pancreatic islets produce insulin.
a. alpha **b.** beta
c. delta **d.** PP

2. Type 1 diabetes mellitus usually
a. occurs in those over 40. **b.** afflicts 90% of diabetics.
c. has a gradual onset. **d.** has a rapid onset.

3. Most of the insulin-secreting cells of a pancreas are
a. scattered evenly throughout the pancreatic islet.
b. near the periphery of the pancreatic islet.
c. near the central portion of the pancreatic islet.
d. among the exocrine cells of the pancreas.

4. Beta cells of a pancreatic islet occupy about _____ of the islet.
a. 1% **b.** 5%
c. 15% **d.** 80%

5. Which of the following has the *least* influence in the onset of type 2 diabetes mellitus?
a. sleep patterns **b.** obesity
c. heredity **d.** lack of exercise

6. Insulin levels are highest in the blood
a. during sleep. **b.** directly after a meal.
c. between meals. **d.** when glucose levels are low.

7. Type 1 diabetes mellitus occurs in more people than type 2 diabetes mellitus.
   a. True **b.** False
8. Obesity and lack of exercise increase the risk for the onset of type 2 diabetes mellitus.
   a. True **b.** False

The primary "fuel" for cells is sugar (glucose). Insulin produced by the beta cells of the pancreatic islets has a primary regulation role in the transport of blood sugar across the plasma membranes of most cells of the body. Some individuals do not produce enough insulin; others might not have adequate transport of glucose into the body cells when these target cells lose insulin receptors. This rather common disease is called *diabetes mellitus*. Diabetes mellitus is a functional disease that can have its onset either during childhood or later in life. (A different functional disease, *diabetes insipidus,* results from a deficiency of antidiuretic hormone with symptoms of copious urine production, but not hyperglycemia.)

The protein hormone insulin has several functions: it stimulates the liver and skeletal muscles to synthesize glycogen from glucose; it inhibits the conversion of noncarbohydrates into glucose; it promotes the facilitated diffusion of glucose across plasma membranes of target cells possessing insulin receptors (adipose tissue, cardiac muscle, and skeletal muscle); it decreases blood sugar (glucose); it increases protein synthesis; and it promotes fat storage in adipose cells. Normally, as a person's blood sugar increases during nutrient absorption after a meal, the rising glucose levels directly stimulate beta cells of the pancreas to secrete insulin. Insulin prevents sudden surges in blood glucose (hyperglycemia) by promoting glycogen synthesis in the liver and increasing glucose entry into muscle and adipose cells.

Between meals and during sleep, blood glucose levels decrease and insulin decreases. When insulin concentration decreases, more glucose is available to enter cells that lack insulin receptors. Brain cells, liver cells, kidney cells, and red blood cells either lack insulin receptors or express limited insulin receptors, and they absorb and utilize glucose without the need for insulin. These cells depend upon blood glucose concentrations to enable cellular respiration and ATP production. When insulin is given as a medication for diabetes mellitus, blood glucose levels may drop drastically (hypoglycemia) if the individual does not eat and/or has been vigorously exercising. This can have significant negative effects on the nervous system, resulting in *insulin shock.*

Type 1 diabetes mellitus usually has a rapid onset relatively early in life, but it can occur in adults. Because this type of diabetes mellitus occurs when insulin is no longer produced, it is sometimes referred to as *insulin-dependent diabetes mellitus* (*IDDM*). Affecting about 10% to 15% of diabetics, this autoimmune disorder occurs when antibodies from the immune system destroy beta cells, resulting in a decrease in insulin production. The treatment for type 1 diabetes mellitus is to administer insulin, usually by injection.

Type 2 diabetes mellitus has a gradual onset, usually diagnosed in people over age 40. It is sometimes called *non insulin-dependent diabetes mellitus (NIDDM).* This disease results as body cells become less sensitive to insulin or lose insulin receptors and thus cannot respond to insulin even if insulin levels remain normal, a condition called *insulin resistance.* Risk factors for the onset of type 2 diabetes mellitus, which afflicts about 85% to 90% of diabetics, include heredity, obesity, and lack of exercise. Obesity is a major indicator for predicting type 2 diabetes; a weight loss of a mere 10 pounds will increase the body's sensitivity to insulin as a means of prevention. The treatments for type 2 diabetes mellitus include a diet that avoids foods that stimulate insulin production, weight control, exercise, and medications. (The causes and treatments for type 2 diabetes closely resemble those for coronary artery disease.)

PROCEDURE A: Insulin Shock

If too much insulin is present relative to the amount of glucose available to the cells, insulin shock can occur. In this procedure, insulin shock will be created in a fish by introducing insulin into the water with a fish. Insulin will be absorbed into the blood of the gills of the fish. When insulin in the blood reaches levels above normal, rapid hypoglycemia occurs from the increased insulin absorption; consequently, the brain cells obtain less of the much-needed glucose, and epinephrine is secreted. The cumulative effect of complications during insulin shock is a rapid heart rate, anxiety, sweating, mental disorientation, impaired vision, dizziness, convulsions, and possible unconsciousness. Although these complications take place in humans, some of them are visually noticeable in a fish that is in insulin shock. The most observable components of insulin shock in the fish are related to behavioral changes, including rapid and irregular movements of the entire fish and increased gill cover and mouth movements.

1. Reexamine the introduction to this laboratory exercise.
2. Complete the table in Part A of Laboratory Assessment 40.
3. A fish can be used to observe normal behavior, induced insulin shock, and recovery from insulin shock. The pancreatic islets are located in the pyloric ceca or other scattered regions in fish. The behavioral changes of fish when they experience insulin shock and recovery mimic those of humans. Observe a fish in 200 mL of aquarium water in a 500 mL beaker. Make your observations of swimming, gill cover (operculum), and mouth movements for 5 minutes. Record your observations in Part B of the laboratory assessment.
4. Add 200 units of room-temperature insulin slowly, using the syringe, into the water with the fish to induce insulin shock. Record the time or start the timer when the insulin is added. Watch for changes of the swimming, gill cover, and mouth movements as insulin diffuses into the blood at the gills. Record your observations of any

behavioral changes and the time of the onset in Part B of the laboratory assessment.

5. After definite behavioral changes are observed, use the small fish net to move the fish into another 500 mL beaker containing 200 mL of a 10% room-temperature glucose solution. Record the time when the fish is transferred into this container. Make observations of any recovery of normal behaviors, including the amount of time involved. Record your observations in Part B of the laboratory assessment.
6. When the fish appears to be fully recovered, return it to its normal container. Dispose of the syringe according to directions from your laboratory instructor.
7. Complete Part B of the laboratory assessment.

PROCEDURE B: Pancreatic Islets

1. Examine a normal stained pancreas tissue under low-power and high-power magnification (fig. 40.1). Locate the clumps of cells that constitute the *pancreatic islets (islets of Langerhans)* that represent the endocrine portions of the organ. A pancreatic islet contains four distinct cells that secrete different hormones: *alpha cells,* which secrete glucagon, *beta cells,* which secrete insulin, *delta cells,* which secrete somatostatin, and *PP cells,* which secrete pancreatic polypeptide. Some of the specially stained pancreatic islets allow for distinguishing the different types of cells (figs. 40.2 and 40.3). The normal pancreatic islet has about 80% beta cells concentrated in the central region of the islet, about 15% alpha cells near the periphery of an islet, about 5% delta cells, and a small number of F cells. There are complex interactions among the four hormones affecting blood sugar, but in general, glucagon will increase blood sugar, whereas insulin will lower blood sugar; thus, these two hormones are considered antagonistic in their function. Somatostatin acts as a paracrine secretion to inhibit alpha and beta cell secretions, and pancreatic polypeptide inhibits secretions from delta cells and digestive enzyme secretions from the pancreatic acini.
2. Prepare a labeled sketch of a normal pancreatic islet in Part C of the laboratory assessment. Include labels for alpha cells and beta cells.

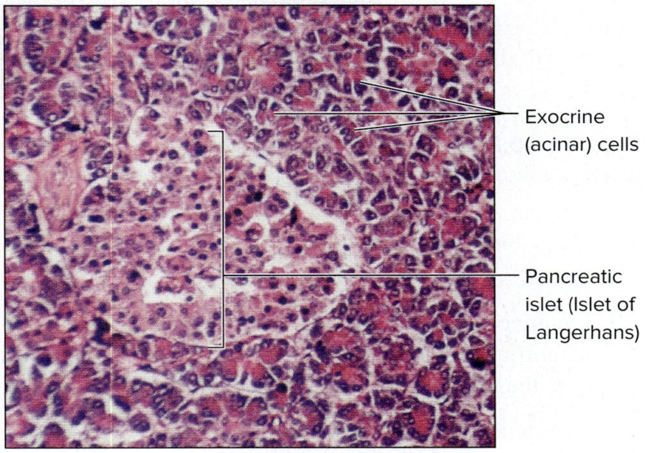

FIGURE 40.1 Micrograph of a stained normal pancreas (400×). Cynthia Prentice-Craver/McGraw Hill APR

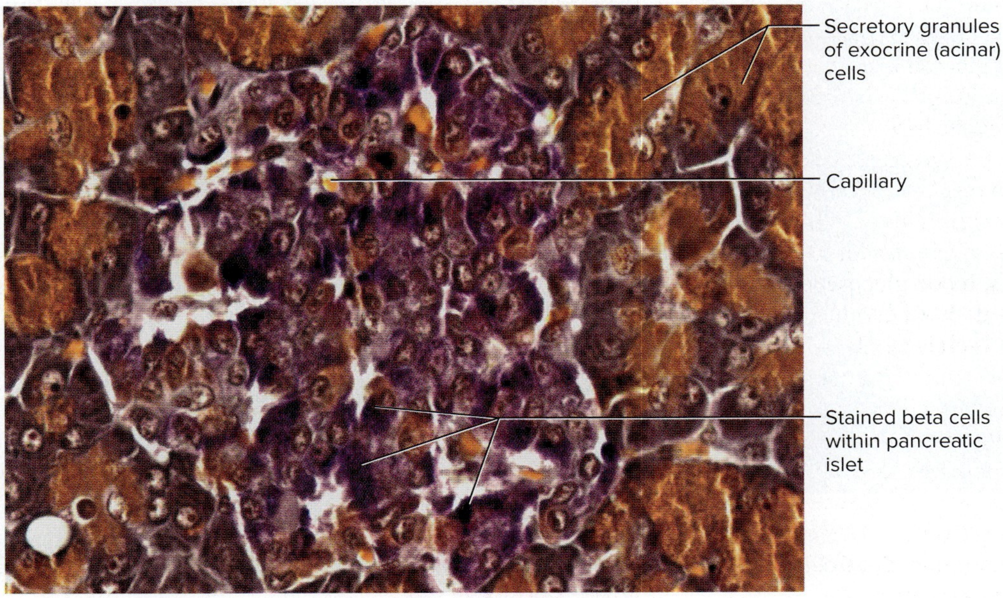

FIGURE 40.2 Micrograph of a specially stained pancreatic islet surrounded by exocrine (acinar) cells (500×). The insulin-containing beta cells are stained dark purple. Al Telser/McGraw Hill APR

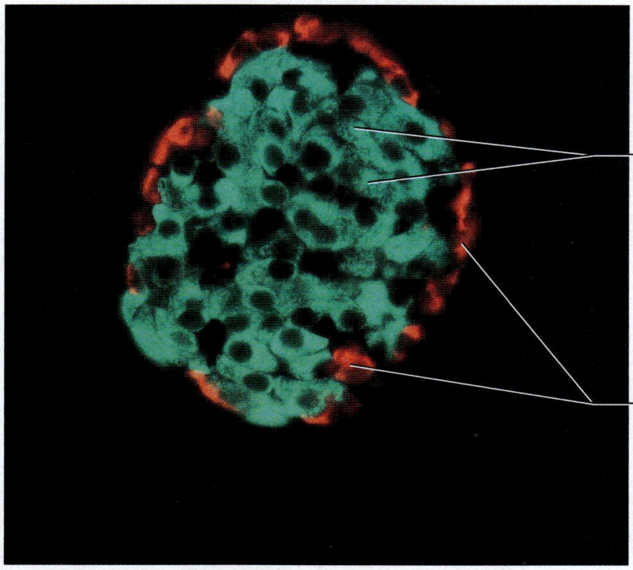

FIGURE 40.3 Micrograph of a pancreatic islet using immunofluorescence stains to visualize alpha and beta cells. Only the areas of interest are visible in the fluorescence microscope. Al Telser/McGraw Hill APR

3. Examine the pancreas of a diabetic under high-power magnification (fig. 40.4). Locate a pancreatic islet and note the number of beta cells as compared to a normal number of beta cells in the normal pancreatic islet.
4. Prepare a labeled sketch of a pancreatic islet that shows some structural changes due to diabetes mellitus. Place the sketch in Part C of the laboratory assessment.
5. Complete Part C of the laboratory assessment.

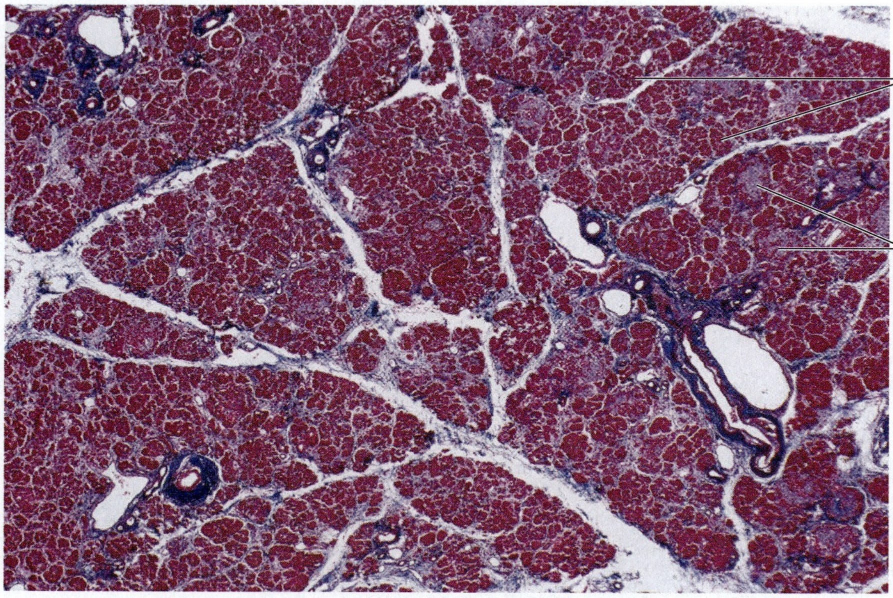

FIGURE 40.4 Micrograph of pancreas showing indications of changes from diabetes mellitus (100×). The changes from diabetes mellitus include areas of fibrosis and amyloid replacement of islet cells. Michael Abbey/Science Source APR

LABORATORY
ASSESSMENT 40

Name ______________________________

Date ______________________________

Section ______________________________

The (A) corresponds to the indicated Learning Outcome(s) (O) found at the beginning of the Laboratory Exercise.

Diabetic Physiology

PART A: Assessments

Complete the missing parts of the table on diabetes mellitus: A2

Characteristic	Type 1 Diabetes	Type 2 Diabetes
Onset age	Early age or adult	
Onset of symptoms		Slow
Percentage of diabetics		85–90%
Natural insulin levels	Below normal	
Beta cells of pancreatic islets		Not destroyed
Pancreatic islet cell antibodies	Present	
Risk factors of having the disease	Heredity	
Typical treatments	Insulin administration	
Untreated blood sugar levels		Hyperglycemia

PART B: Assessments

1. Record your observations of normal behavior of the fish.

a. Swimming movements: ______________________________

____

b. Gill cover movements: ______________________________

____

c. Mouth movements: ______________________________

____

2. Record your observations of the fish after insulin was added to the water. A3

a. Recorded time that insulin was added: ______________________________

b. Swimming movements: ______________________________

____

c. Gill cover movements: ______________________________

____

d. Mouth movements: ______________________________

____

e. Elapsed time until the insulin shock symptoms occurred: ______________________________

3. Record your observations after the fish was transferred into a glucose solution. **A4**

a. Recorded time that fish was transferred into the glucose solution: ____________________

b. Describe any changes in the behavior of the fish that indicate a recovery from insulin shock. ____________________

____

____

____

c. Elapsed time until indications of a recovery from insulin shock occurred: ____________________

PART C: Assessments

1. Match the pancreatic islet hormone in column A to its proper description in column B. Place the letter of your choice in the space provided. (Answers will be used more than once.) **A1**

Column A	Column B
a. Insulin	______ **1.** Inhibits pancreatic digestive enzyme secretion
b. Glucagon	______ **2.** Released in response to low blood sugar levels
c. Somatostatin	______ **3.** Released in response to high blood sugar levels
d. Pancreatic polypeptide	______ **4.** Inhibits or modulates alpha and beta cell secretions
	______ **5.** Secreted by beta cells
	______ **6.** Secreted by delta cells
	______ **7.** Secreted by alpha cells
	______ **8.** Secreted by PP cells
	______ **9.** Hyperglycemic hormone
	______ **10.** Hypoglycemic hormone

2. The circles below represent the microscopic field of view. In each circle, sketch and label representative portions of the indicated pancreatic islets: **A5**

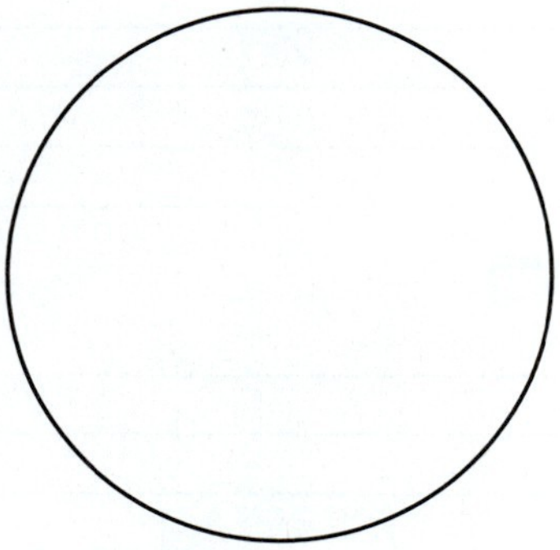

Pancreatic islet (______×)
(normal)

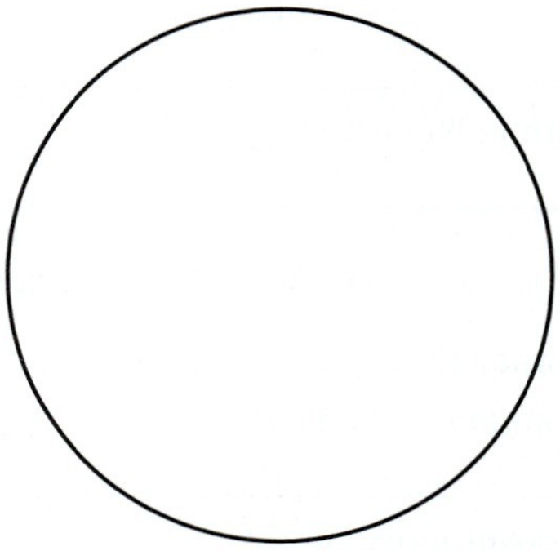

Pancreatic islet (______×)
(showing changes from diabetes mellitus)

3. Describe the differences that you observed between a normal pancreatic islet and a pancreatic islet of a person with diabetes mellitus. A5 ____

____

____

____

CRITICAL THINKING ASSESSMENTS—PART C

1. Justify the importance for a type 1 diabetic to regulate insulin administration, meals, and exercise. A2

____

____

____

____

____

2. A patient came to the hospital complaining of excessive urination and intense thirst and hunger. Clinical tests of the blood and urine were conducted to confirm diagnosis of diabetes mellitus. Explain what these clinical tests would show and why. A2

____

____

____

____

____

NOTES

LABORATORY EXERCISE 41

Blood Cells

Purpose of the Exercise

To review the characteristics of blood cells, to examine them microscopically, and to perform a differential white blood cell count.

MATERIALS NEEDED

Compound light microscope
Prepared microscope slides of human blood (Wright's stain)
Colored pencils

For Demonstration Activity:

Mammal blood other than human or contaminant-free human blood is suggested as a substitute for collected blood
Microscope slides (precleaned)
Sterile disposable blood lancets
Alcohol swabs (wipes)
Slide staining rack and tray
Wright's stain
Distilled water
Disposable gloves

For Learning Extension Activity:

Prepared slides of pathological blood, such as eosinophilia, leukocytosis, leukopenia, and lymphocytosis

SAFETY

- Review all the safety guidelines in Appendix 1 of your laboratory manual.
- It is important that students learn and practice correct procedures for handling body fluids. Consider using either mammal blood other than human or contaminant-free blood that has been tested and is available from various laboratory supply houses. Some of the procedures might be accomplished as demonstrations only. If student blood is used, it is important that students handle only their own blood.
- Use an appropriate disinfectant to wash the laboratory tables before and after the procedures.
- Wear disposable gloves and protective eyewear when handling blood samples.
- Clean the end of a finger with an alcohol swab before the puncture is performed.
- Use the sterile blood lancet only once.
- Dispose of used lancets and blood-contaminated items in an appropriate container (never use the wastebasket).
- Wash your hands before leaving the laboratory.

Learning Outcomes APR

After completing this exercise, you should be able to:

O1 Identify and sketch red blood cells, five types of white blood cells, and platelets.

O2 Describe the structure and function of red blood cells, white blood cells, and platelets.

O3 Perform and interpret the results of a differential white blood cell count.

The O corresponds to the assessments A indicated in the Laboratory Assessment for this Exercise.

Pre-Lab

Carefully read the introductory material and examine the entire lab. Be familiar with RBCs, WBCs, and platelets from lecture or the textbook. Answer the pre-lab questions.

Pre-Lab Questions Select the correct answer for each of the following questions:

1. Which of the following have significant functions mainly during bleeding?
a. erythrocytes **b.** leukocytes
c. platelets **d.** plasma

2. Which of the following is among the agranulocytes?
a. monocyte **b.** neutrophil
c. eosinophil **d.** basophil

3. Which white blood cell has the greatest nuclear variations?
a. monocyte **b.** neutrophil
c. eosinophil **d.** basophil

4. A(n) _____ lacks a nucleus.
a. erythrocyte **b.** lymphocyte
c. monocyte **d.** basophil

Chapter Opener: ©101cats/Getty Images

5. Which cell has a large nucleus that fills most of the cell?
   a. erythrocyte
   b. platelet
   c. eosinophil
   d. lymphocyte
6. Which leukocyte is the most abundant in a normal differential count?
   a. basophil
   b. monocyte
   c. neutrophil
   d. lymphocyte
7. Eosinophil numbers typically increase during allergic reactions.
   a. True
   b. False
8. Erythrocytes are also called granulocytes because granules are visible in their cytoplasm when using Wright's stain.
   a. True
   b. False

WARNING

Because of the possibility of blood infections being transmitted from one student to another if blood slides are prepared in the classroom, it is suggested that commercially prepared blood slides be used in this exercise. The instructor, however, may wish to demonstrate the procedure for preparing such a slide. Observe all safety procedures for this lab.

DEMONSTRATION ACTIVITY

To prepare a stained blood slide, follow these steps:

1. Obtain two precleaned microscope slides. Avoid touching their flat surfaces.
2. Thoroughly wash hands with soap and water and dry them with paper towels. Don disposable gloves except on the hand of the person with the finger to be lanced.
3. Cleanse the end of the middle finger with an alcohol swab and let the finger dry in the air.
4. Remove a sterile disposable blood lancet from its package without touching the sharp end.
5. Puncture the skin on the side near the tip of the middle finger with the lancet and properly discard the lancet.
6. Wipe away the first drop of blood with the alcohol swab. Place a drop of blood about 2 cm from the end of a clean microscope slide. Cover the lanced finger location with a bandage.
7. Use a second slide to spread the blood across the first slide, as illustrated in figure 41.1. Discard the slide used for spreading the blood in the appropriate container.
8. Place the blood slide on a slide staining rack and let it dry in the air.
9. Put enough Wright's stain on the slide to cover the smear but not overflow the slide. Count the number of drops of stain that are used.
10. After 2–3 minutes, add an equal volume of distilled water to the stain and let the slide stand for 4 minutes. From time to time, gently blow on the liquid to mix the water and stain.
11. Flood the slide with distilled water until the blood smear appears light blue.
12. Tilt the slide to pour off the water, and let the slide dry in the air.
13. Examine the blood smear with low-power magnification, and locate an area where the blood cells are well distributed. Observe these cells, using high-power magnification and then an oil immersion objective if one is available.

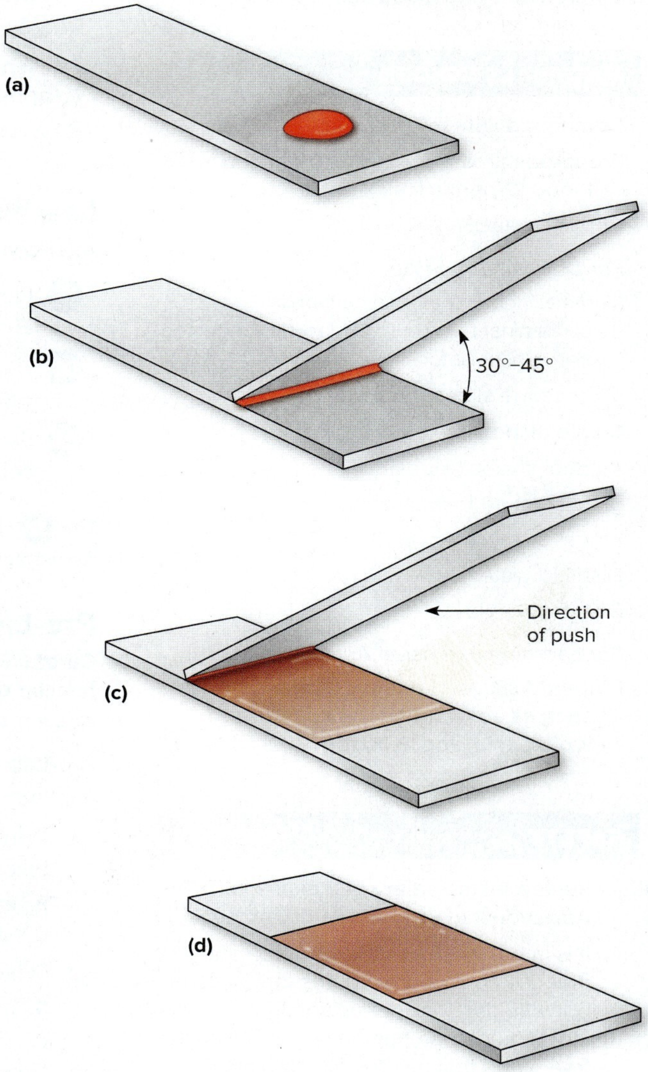

FIGURE 41.1 To prepare a blood smear: (a) place a drop of blood about 2 cm from the end of a clean slide; (b) hold a second slide at about a 30°–45° angle to the first one, allowing the blood to spread along its edge; (c) push the second slide over the surface of the first so that it pulls the blood with it; (d) observe the completed blood smear. The ideal smear should be 1.5 inches in length, be evenly distributed, and contain a smooth, feathered edge.

Blood is a type of connective tissue whose cells are suspended in a liquid extracellular matrix called *plasma*. Plasma is composed of water, proteins, nutrients, electrolytes, hormones, wastes, and gases. Within the plasma are *formed elements* that include *erythrocytes* (*red blood cells; RBCs*), *leukocytes* (*white blood cells*; *WBCs*), and cellular fragments called *platelets* (*thrombocytes*). Platelets are formed from huge cells called megakaryocytes. Erythrocytes, leukocytes, and megakaryocytes are produced in red bone marrow. The formed elements compose about 45% of the total blood volume; the plasma composes approximately 55% of the blood volume.

Erythrocytes contain hemoglobin and transport gases (oxygen and carbon dioxide) between the body cells and the lungs, leukocytes defend the body against infections, and platelets play an important role in stoppage of bleeding (hemostasis).

Laboratories in modern hospitals and clinics use updated hematology analyzers for evaluations of the blood characteristics (fig. 41.2). The more traditional procedures that you perform in Laboratory Exercises 41, 42, and 43 will help you better understand the methodology and purpose of measuring these blood characteristics. On occasion, a doctor might question a blood test result from the hematology blood analyzer and request a retest of the blood. Additional verification of a test result might be performed using traditional procedures. A self-diagnosis should never be made as a result of a test conducted in the biology laboratory. Always obtain proper medical exams and treatments from medical personnel.

PROCEDURE A: Types of Blood Cells

1. Refer to figures 41.3 to 41.5 as an aid in identifying the various types of blood cells. Study the functions of the blood cells listed in table 41.1.

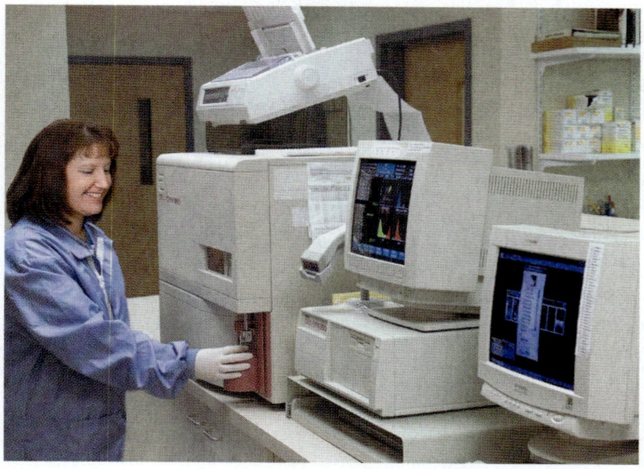

FIGURE 41.2 Modern hematology analyzer being used in the laboratory of a clinic. ©J & J Photography

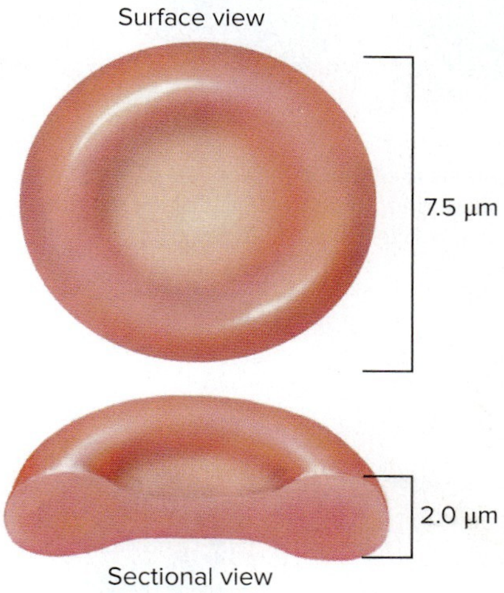

FIGURE 41.3 Red blood cells (erythrocytes) have a biconcave shape and function in the transport of oxygen and carbon dioxide.

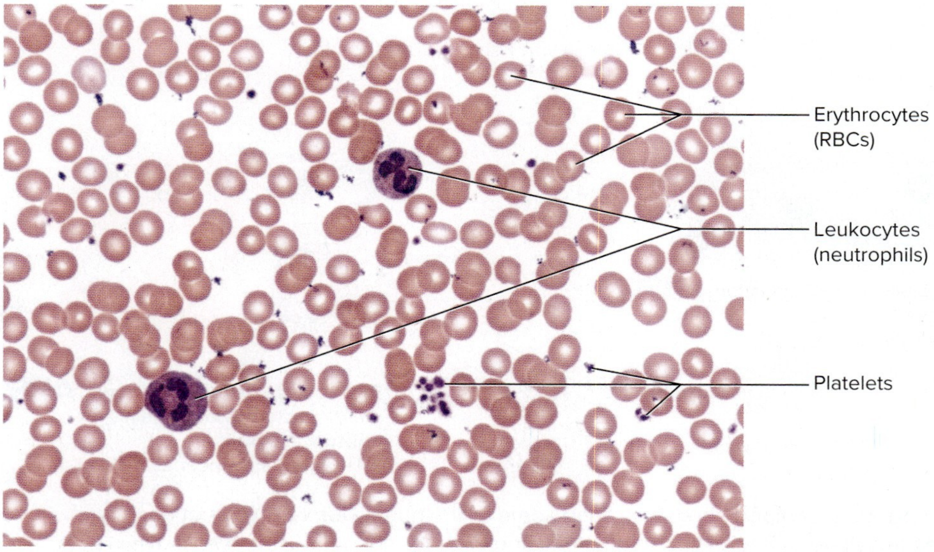

FIGURE 41.4 Micrograph of a blood smear using Wright's stain (500×). Al Telser/McGraw Hill APR

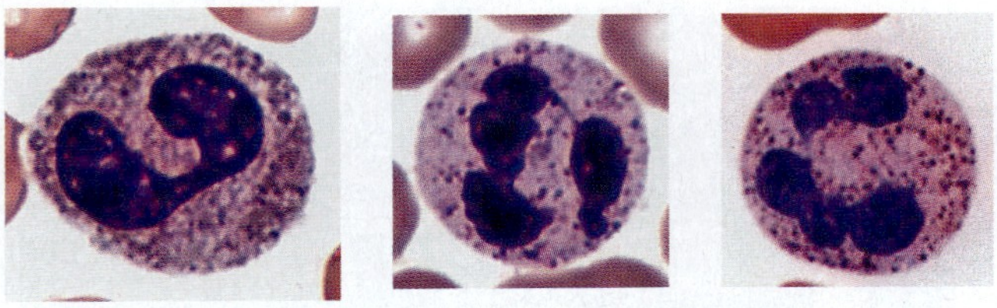

Neutrophils (3 of many variations)
- Fine light-purple granules
- Nucleus single to five lobes (highly variable)
- Immature neutrophils, called bands, have a single C-shaped nucleus
- Mature neutrophils, called segs, have a multilobed nucleus
- Often called polymorphonuclear leukocytes when older

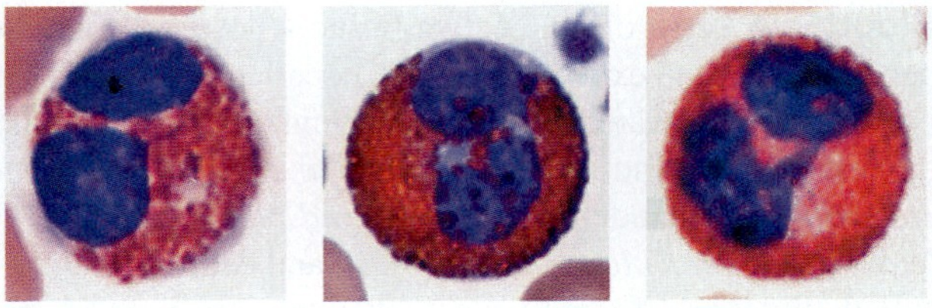

Eosinophils (3 of many variations)
- Coarse reddish granules
- Nucleus usually bilobed

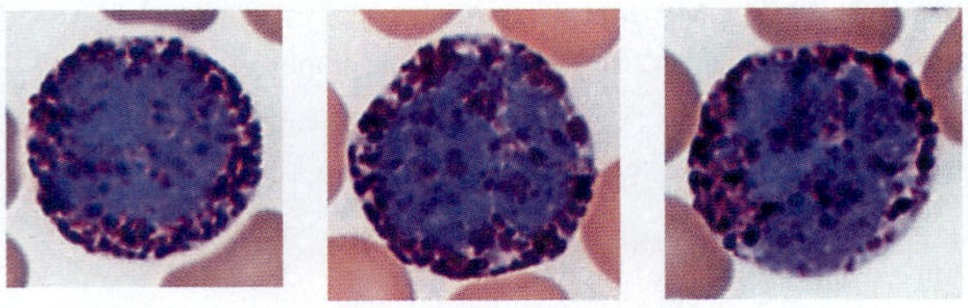

Basophils (3 of many variations)
- Coarse deep blue to almost black granules
- Nucleus often almost hidden by granules

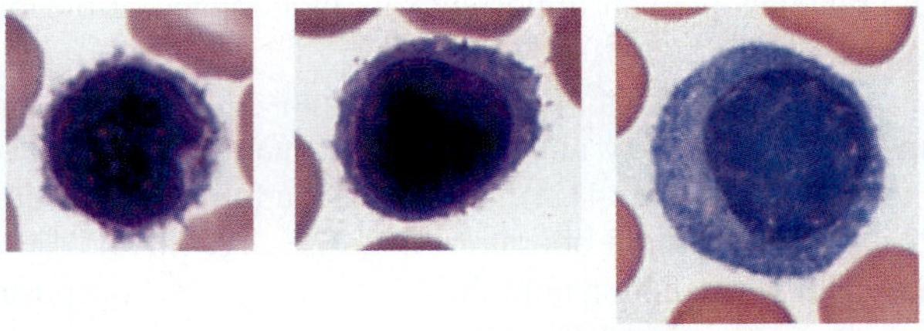

Lymphocytes (3 of many variations)
- Slightly larger than RBCs
- Thin rim of nearly clear cytoplasm
- Nearly round nucleus appears to fill most of cell in smaller lymphocytes
- Larger lymphocytes hard to distinguish from monocytes

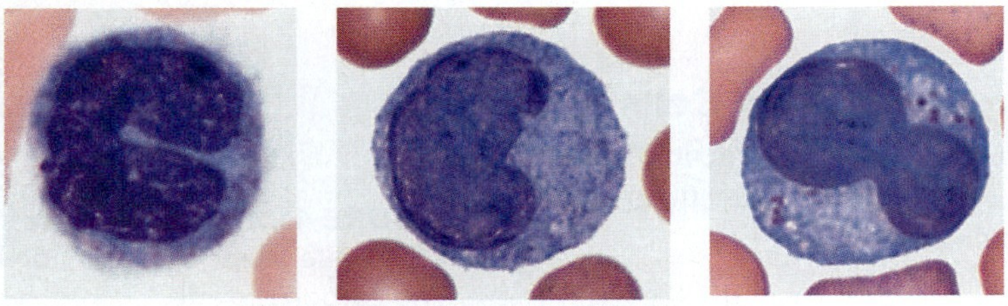

Monocytes (3 of many variations)
- Largest WBC; 2–3x larger than RBCs
- Cytoplasm nearly clear
- Nucleus round, kidney-shaped, oval, or lobed

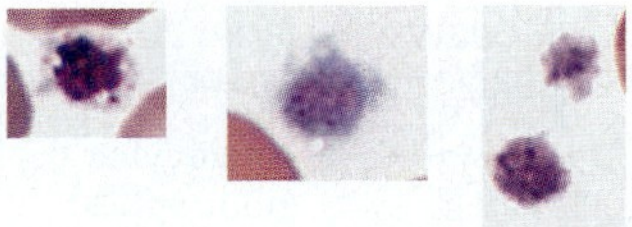

Platelets (several variations)
- Cell fragments
- Single to small clusters

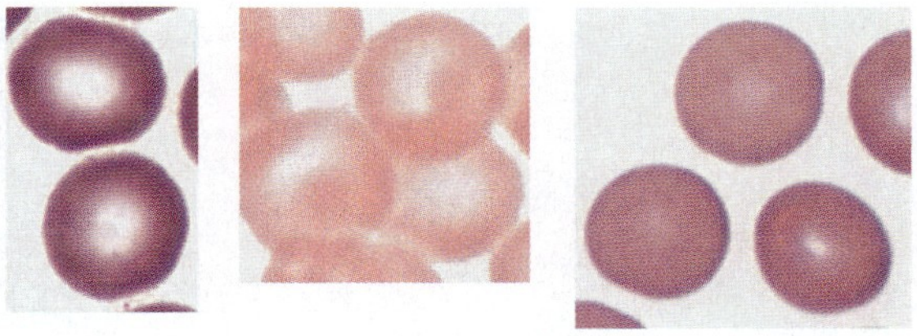

Erythrocytes (several variations)
- Lack nucleus (mature cell)
- Biconcave discs
- Thin centers appear almost hollow

FIGURE 41.5 Micrographs of blood cells illustrating some of the numerous variations of each type. Appearance characteristics for each cell pertain to a thin blood film using Wright's stain (1,000×). ©Alvin Telser, Ph.D.

Use the prepared slide of blood and locate each of the following:

red blood cell (erythrocyte)

white blood cell (leukocyte)

- granulocytes
  - neutrophil
  - eosinophil
  - basophil
- agranulocytes
  - lymphocyte
  - monocyte

platelet (thrombocyte)

2. In Part A of Laboratory Assessment 41, prepare sketches of single blood cells to illustrate each type. Pay particular attention to the relative size, nuclear shape, and color of granules in the cytoplasm (if present). The sketches should be accomplished using either the high-power objective or the oil immersion objective of the compound light microscope.
3. Complete Part B of the laboratory assessment.

Table 41.1 Cellular Components of Blood

Component	Function
Red blood cell (erythrocyte)	Contains hemoglobin (Hb) that transports oxygen and carbon dioxide
White blood cell (leukocyte)	Destroys pathogenic microorganisms and parasites, removes worn cells, and provides immunity
Granulocytes—have granular cytoplasm	
1. Neutrophil	Phagocytizes bacteria
2. Eosinophil	Destroys parasites and helps control inflammation and allergic reactions
3. Basophil	Releases heparin (an anticoagulant) and histamine (a blood vessel dilator)
Agranulocytes—lack granular cytoplasm	
1. Monocyte	Phagocytizes dead or dying cells and microorganisms
2. Lymphocyte	Provides immunity; produces antibodies; destroys foreign cells and infected cells with viruses
Platelet (thrombocyte)	Helps control blood loss from injured blood vessels; secretes chemicals important in blood clotting, tissue repair, inflammatory response, and aggregation of platelets

LEARNING EXTENSION ACTIVITY

Obtain a prepared slide of pathological blood that has been stained with Wright's stain. Perform a differential white blood cell count, using this slide, and compare the results with the values for normal blood listed in table 41.2. What differences do you note?

____

____

____

____

PROCEDURE B: Differential White Blood Cell Count

A differential white blood cell count is performed to determine the percentage of each of the various types of white blood cells present in a blood sample. The test is useful because the relative proportions of white blood cells may change in particular diseases, as indicated in table 41.2. Neutrophils, for example, usually increase during bacterial infections, whereas eosinophils may increase during certain parasitic infections and allergic reactions.

1. To make a differential white blood cell count, follow these steps:
   a. Using high-power magnification or an oil immersion objective, focus on the cells at one end of a prepared blood slide where the cells are well distributed.
   b. Slowly move the blood slide back and forth, following a path that avoids passing over the same cells twice (fig. 41.6).
   c. Each time you encounter a white blood cell, identify its type and record it in Part C of the laboratory assessment.
   d. Continue searching for and identifying white blood cells until you have recorded 100 cells in the data table. *Percent* means "parts of 100" for each type of white blood cell, so the total number observed is equal to its percentage in the blood sample.
2. Complete Part C of the laboratory assessment.

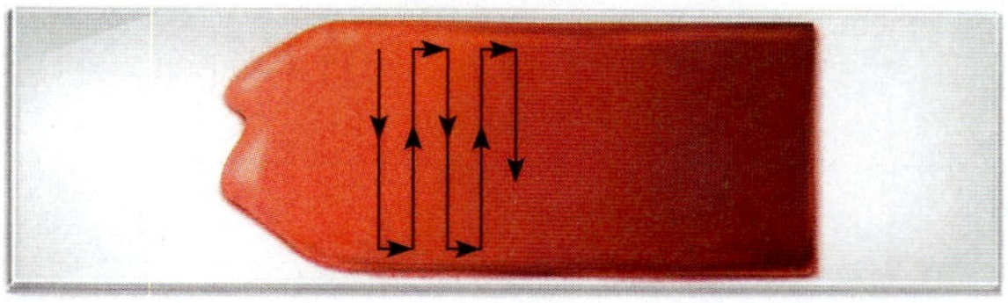

FIGURE 41.6 Move the blood slide back and forth to avoid passing the same cells twice.

Table 41.2 Differential White Blood Cell Count

Cell Type	Normal Value (percent)	Elevated Levels May Indicate
Neutrophil	50–70	Bacterial infections, stress
Lymphocyte	25–33	Mononucleosis, whooping cough, viral infections
Monocyte	3–9	Malaria, tuberculosis, fungal infections
Eosinophil	1–3	Allergic reactions, autoimmune diseases, parasitic worms
Basophil	<1	Cancers, chicken pox, hypothyroidism

NOTES

LABORATORY ASSESSMENT 41

Name ____________________

Date ____________________

Section ____________________

The **A** corresponds to the indicated Learning Outcome(s) **O** found at the beginning of the Laboratory Exercise.

Blood Cells

PART A: Assessments

Sketch a single blood cell of each type in the following circles that represent the microscope field of view. Use colored pencils to represent the stained colors of the cells. Label any features that can be identified. **A1**

Red blood cell (____ x)

Neutrophil (____ x)

Lymphocyte (____ x)

Monocyte (____ x)

Eosinophil (____ x)

Basophil (____ x)

Platelet (____ x)

PART B: Assessments

Complete the following statements:

1. Red blood cells (RBCs) are also called ____________________. A2
2. The shape of a red blood cell can be described as a ____________________ disc. A2
3. The functions of red blood cells are ____. A2
4. ____________________ is the oxygen-carrying substance in a red blood cell. A2
5. A mature red blood cell cannot reproduce because it lacks a ____________________ that was extruded during late development. A2
6. White blood cells are also called ____________________. A2
7. White blood cells with granular cytoplasm are called ____________________. A2
8. White blood cells lacking granular cytoplasm are called ____________________. A2
9. Polymorphonuclear leukocyte is another name for a ____________________ with a segmented nucleus. A2
10. Normally, the most numerous white blood cells are ____________________. A2
11. White blood cells with coarse reddish cytoplasmic granules are called ____________________. A2
12. ____________________ are normally the least abundant of the white blood cells. A2
13. ____________________ are the largest in size of the white blood cells. A2
14. ____________________ are small agranulocytes that have relatively large, round nuclei with thin rims of cytoplasm. A2
15. Small cell fragments that function to prevent blood loss from an injury site are called ____________________. A2

PART C: Assessments

1. *Differential White Blood Cell Count Data Table*. As you identify white blood cells, record them in the table by using a tally system, such as 卌 ||. Place tally marks in the "Number Observed" column and total each of the five WBCs when the differential count is completed. Obtain a total of all five WBCs counted to determine the percentage of each WBC type. **A3**

Type of WBC	Number Observed	Total	Percent
Neutrophil			
Lymphocyte			
Monocyte			
Eosinophil			
Basophil			
		Total of column	

2. How do the results of your differential white blood cell count compare with the normal values listed in table 41.2? **A3**

3. Identify the blood cells indicated in figure 41.7.

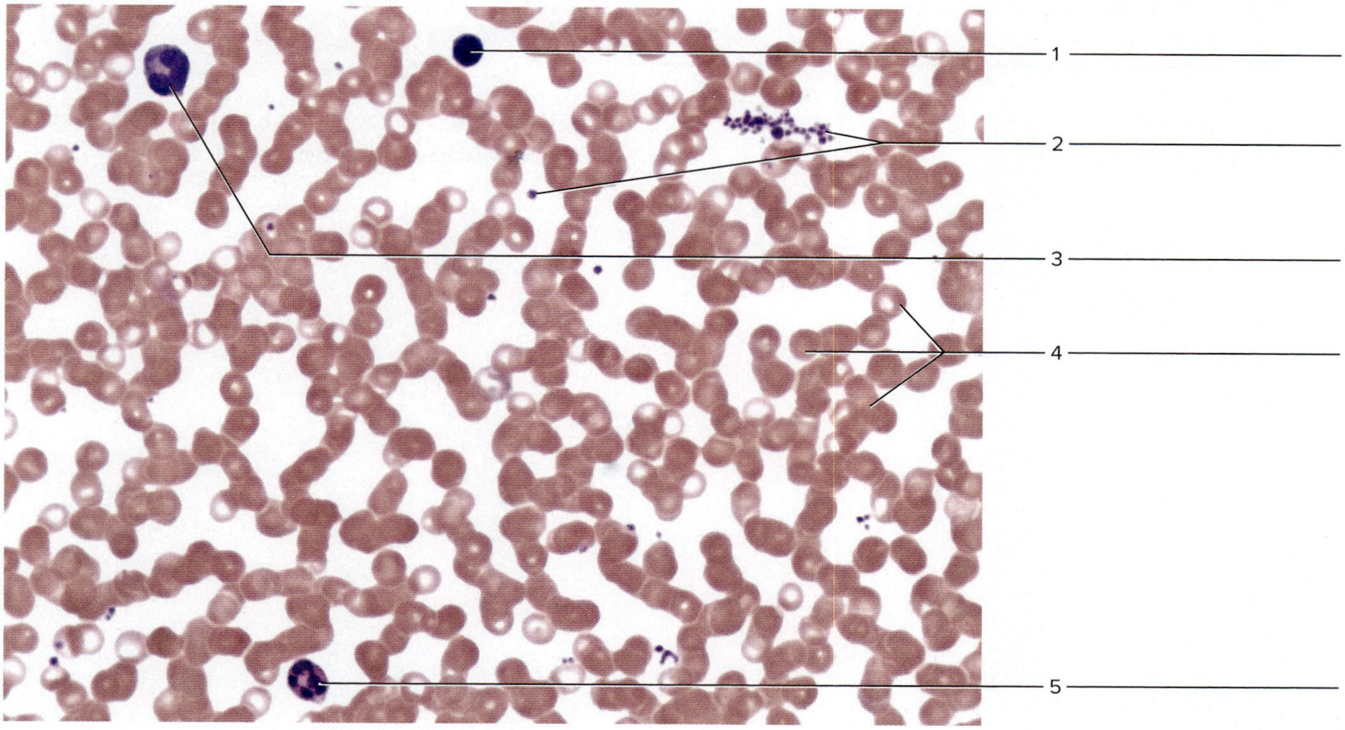

FIGURE 41.7 Label the specific blood cells on this micrograph of a stained blood smear (400×). Al Telser/McGraw Hill **A1**

CRITICAL THINKING ASSESSMENTS—PART C

1. Explain the difference between a differential white blood cell count and a total white blood cell count. A3

____

____

____

2. Which leukocyte type would likely be elevated in a patient who has strep throat? ____________

   Explain your reasoning. ____________.

   Which leukocyte type would likely be elevated in a patient who has influenza? ____________

   Explain your reasoning. ____________.

   Which leukocyte type would likely be elevated in a patient who has tapeworm? ____________

   Explain your reasoning. ____________.

3. Examine the blood smear shown in the micrograph in figure 41.8. How do you distinguish a leukocyte from an erythrocyte? A2

   ____

   ____

   What leukocyte(s) is/are not shown in the blood smear? ________

   ____

   Explain how you came to this conclusion. ____________

   ____

   ____

   ____.

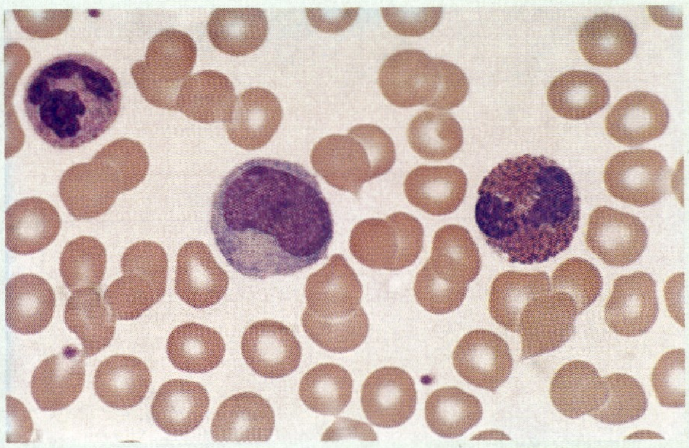

FIGURE 41.8 Micrograph of a normal blood smear.
Biophoto Associates/Science Source A1

LABORATORY EXERCISE 42

Blood Testing

Purpose of the Exercise

To perform and interpret the blood tests used to determine hematocrit, hemoglobin content, coagulation, and cholesterol level.

MATERIALS NEEDED

Mammal blood other than human or contaminant-free human blood is suggested as a substitute for collected blood

Sterile disposable blood lancets

Alcohol swabs (wipes)

Disposable gloves

For Procedure A:

Heparinized microhematocrit capillary tube

Sealing clay (or Critocaps)

Microhematocrit centrifuge

Microhematocrit reader

For Procedure B:

Tallquist test kit

Hemoglobinometer

Lens paper

Hemolysis applicator

For Procedure C:

Small triangular file

Capillary tube (nonheparinized)

Timer

For Procedure D:

Simulated blood kit for cholesterol determination*

* Kit contains all materials needed and is available from several biological supply companies.

SAFETY

- Review all the safety guidelines in Appendix 1 of your laboratory manual.
- It is important that students learn and practice correct procedures for handling body fluids. Consider using either mammal blood other than human or contaminant-free blood that has been tested and is available from various laboratory supply houses. Some of the procedures might be accomplished as demonstrations only. If student blood is used, it is important that students handle only their own blood.
- Use an appropriate disinfectant to wash the laboratory tables before and after the procedures.
- Wear disposable gloves and protective eyewear when handling blood samples.
- Clean the end of a finger with alcohol swabs before the puncture is performed.
- Use the sterile blood lancet only once.
- Dispose of used lancets and blood-contaminated items in an appropriate container (never use the wastebasket).
- Wash your hands before leaving the laboratory.

Chapter Opener: ©101cats/Getty Images

Learning Outcomes

After completing this exercise, you should be able to:

O1 Determine the hematocrit, hemoglobin, coagulation, and cholesterol level in a blood sample.

O2 Evaluate the results and significance of the blood tests compared to normal values.

O3 Identify the blood tests performed in this exercise that could indicate anemia.

O4 Identify some variables that affect the percent saturation of hemoglobin.

The O corresponds to the assessments A indicated in the Laboratory Assessment for this Exercise.

Pre-Lab

Carefully read the introductory material and examine the entire lab. Be familiar with hematocrit, hemoglobin, coagulation, and cholesterol from lecture or the textbook. Answer the pre-lab questions.

Pre-Lab Questions Select the correct answer for each of the following questions:

1. The _____ is the percentage of red blood cells in whole blood.
   a. hemoglobin **b.** hematocrit **c.** buffy coat **d.** coagulation
2. In a normal sample of whole blood, the _____ has/have the lowest total volume percentage.
   a. plasma **b.** red blood cells
   c. buffy coat **d.** hematocrit
3. The specific part of the hemoglobin molecule where oxygen (O_2) binds is the
   a. globin portion. **b.** amino acids of the beta subunits.
   c. iron (Fe) of the heme. **d.** amino acids of the alpha subunits.

4. Oxyhemoglobin is formed when oxygen binds to
   a. the hematocrit.
   b. the red blood cell membrane.
   c. globin of hemoglobin.
   d. iron of hemoglobin.
5. When there is a "shift left" on the oxygen dissociation curve,
   a. more O_2 is bound to hemoglobin.
   b. the O_2 bound to hemoglobin is unaffected.
   c. less O_2 is bound to hemoglobin.
   d. hemoglobin has no O_2 bound to it.
6. In order to determine the hematocrit, blood is placed in a
   a. coagulation tube. b. hemoglobinometer.
   c. centrifuge. d. movable slide.
7. Coagulation culminates in the formation of insoluble threads of
   a. fibrin. b. fibrinogen.
   c. thrombin. d. platelets.
8. Coagulation should normally occur within
   a. 10 seconds. b. 1 minute.
   c. 8 minutes. d. 1 hour.
9. A normal blood total cholesterol level is less than 200 mg/dL.
   a. True b. False
10. The high-density lipoproteins (HDLs) are sometimes considered the "bad" type of cholesterol.
   a. True b. False

As an aid in identifying various disease conditions, tests are often performed on blood to determine how its composition compares with normal values. These tests commonly include measuring the hematocrit [red blood cell (RBC) percentage], hemoglobin content, coagulation, and cholesterol.

When a blood sample is collected in a heparinized (to prevent coagulation) capillary tube and left standing, the heavier cellular components settle to the bottom of the tube. Spinning the tube in a centrifuge accelerates this process. The red layer at the bottom portion of the tube represents the compacted RBCs. The percentage of compacted RBCs in the blood sample is called the *hematocrit,* normally about 45% of the entire volume. A thin, whitish layer, known as the *buffy coat,* forms on the top of the compacted RBC layer. This contains the white blood cells and platelets, and it normally represents less than 1% of the total volume. The remaining straw-colored upper portion of the contents contains the plasma, the remaining approximately 55% of the volume.

Molecules of the protein *hemoglobin (Hb)* compose about one-third of the volume within the RBC. When you take a breath of air into your lungs, the inspired oxygen diffuses across the respiratory membrane of the alveoli and capillary beds of your lungs into your blood, where it enters a RBC and binds to a hemoglobin molecule. The hemoglobin molecule is composed of four protein subunits (two alpha and two beta), each containing a heme molecule with an atom of iron (Fe). Each iron binds to one oxygen molecule (O_2). As a result, each hemoglobin molecule can bind to as many as four oxygen molecules. When oxygen binds to hemoglobin, it forms a bright red *oxyhemoglobin* and is transported to systemic capillaries, where it is released from the hemoglobin to diffuse into your body tissues. When oxygen detaches (dissociates) from the hemoglobin in the capillaries of the tissues, it becomes a darker red and is called *deoxyhemoglobin,* which can appear bluish when viewed through the skin's blood vessel walls.

The amount of oxygen bound to hemoglobin molecules is expressed as the *percent (%) saturation of hemoglobin* (i.e., the percentage of the iron-binding sites of hemoglobin bound with oxygen). The percent saturation of hemoglobin is dependent on a number of variables, the most important being the partial pressure of oxygen (similar to concentration of oxygen in the air). The relationship of percent saturation of hemoglobin to partial pressure of oxygen can be graphed to produce an *oxygen dissociation curve.* (The term *dissociate* means "to separate"; thus, oxygen separates from hemoglobin.) Other variables that influence percent saturation of hemoglobin include temperature, carbon dioxide levels, and pH. The change of any of these variables will trigger a "shift" in the oxygen dissociation curve, as shown in figure 42.1 with the effect of temperature and hemoglobin saturation. A "shift right" indicates that for any given partial pressure of oxygen, there is a decrease in the percent saturation of hemoglobin. A "shift left" indicates that for any given partial pressure of oxygen, there is an increase in the percent saturation of hemoglobin.

A *complete blood count (CBC)* measures white blood cell count, RBC count (including average RBC size), hematocrit, and hemoglobin concentration (including Hb amount per RBC). A CBC is used to detect or monitor various health conditions. An increased hematocrit may be a result of dehydration, living at higher altitudes, smoking, emphysema,

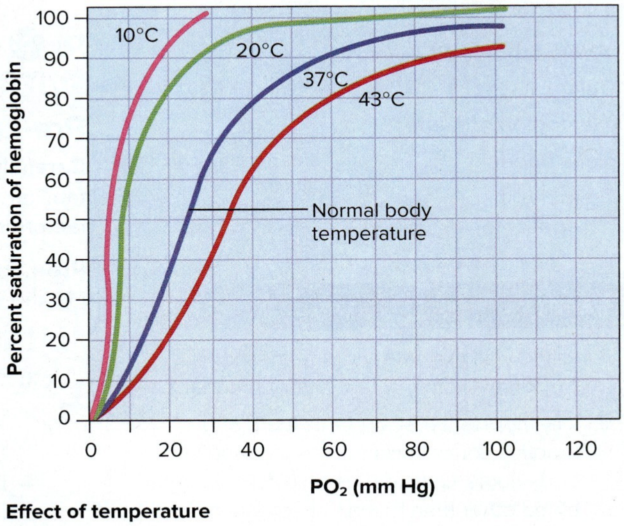

FIGURE 42.1 Effects of temperature on oxyhemoglobin dissociation. At higher temperatures, hemoglobin unloads more oxygen for a given PO_2; seen as a "shift right" as hemoglobin saturation decreases. At lower temperatures, oxygen remains loaded on the hemoglobin; seen as a "shift left" as hemoglobin saturation increases.

or the abuse of erythropoietin (EPO) by athletes for blood doping purposes. Decreased values of hematocrits or hemoglobin levels could be attributed to anemia, hemorrhage, dietary deficiencies, kidney disease where EPO production is impaired, infections, heredity, bone marrow cancer, or exposure to ionizing radiation. Anemia is a condition that has many causes, but the consequences (low blood oxygen levels, low blood osmolarity, and low blood viscosity) put more stress on the cardiovascular system and may result in cardiac failure.

Stoppage of bleeding, called *hemostasis,* involves three defense mechanisms: blood vessel spasm, platelet plug formation, and *coagulation*. The coagulation mechanism involves a series of chain reactions that result in a blood clot. The final stage of blood clot formation occurs when thrombin converts a soluble plasma protein called fibrinogen into insoluble threads of *fibrin*. The fibrin mesh traps cellular components of the blood, stopping the blood loss. Once blood clotting is initiated, it usually takes 3–8 minutes for fibrin to form. Clotting deficiencies could be attributed to low platelet count, leukemia, hemophilia, liver disease, malnutrition, exposure to ionizing radiation, or anticoagulant drugs.

Blood cholesterol is an important component needed for the structure of cellular membranes and the formation of steroid hormones and bile components. However, blood cholesterol higher than recommended levels may increase the chances of cardiovascular diseases and risks of heart attacks and strokes. Hereditary factors are the primary determinant of total blood cholesterol levels. High-cholesterol diets and lack of exercise are among factors that contribute to high cholesterol levels. Low-cholesterol diets, an increase in exercise, and various medications are some ways used to lower the levels of cholesterol.

The clinical significance or conclusive diagnosis of a disease is often difficult to achieve. Most often, several blood and urine characteristics along with other symptoms are assessed collectively in order to make a determination of the abnormality and its potential cause. Any diagnosis should always be obtained by proper medical exams and treatments from a medical professional.

WARNING

Because of the possibility of blood infections being transmitted from one student to another during blood-testing procedures, it is suggested that the following procedures be performed as demonstrations by the instructor. Observe all safety procedures listed for this lab.

PROCEDURE A: Hematocrit

To determine the hematocrit (percentage of RBCs) of a whole blood sample, the cells must be separated from the liquid plasma. This separation can be quickly accomplished by placing a tube of blood in a centrifuge. The force created by the spinning motion of the centrifuge causes the cells to be packed into the lower end of the tube. The quantities of cells and plasma can then be measured, and the hematocrit can be calculated. The normal range for males is 42–52%; the normal range for females is 37–47%.

1. Review all of the safety procedures in Appendix 1 of your laboratory manual and for Laboratory Exercise 42 related to using human blood. To determine the hematocrit in a blood sample, follow these steps:
   - **a.** Thoroughly wash hands with soap and water and dry them with paper towels. Don disposable gloves except on the hand of the person with the finger to be lanced. (If purchased blood is used for this procedure, skip to step 1e.)
   - **b.** Cleanse the end of the middle finger with an alcohol swab and let the finger dry in the air.
   - **c.** Remove a sterile disposable blood lancet from its package without touching the sharp end.
   - **d.** Puncture the skin on the side near the tip of the middle finger with the lancet and properly discard the lancet. Wipe away the first drop of blood with the alcohol swab.
   - **e.** Touch a drop of blood with the colored end of a heparinized capillary tube. Hold the tube tilted slightly upward so that the blood will easily move into it by capillary action (fig. 42.2*a*). To prevent an air bubble, keep the tip in the blood until filled.
   - **f.** Allow the blood to fill about two-thirds of the length of the tube. Cover the lanced finger location with a bandage.
   - **g.** Hold a finger over the tip of the dry end so that blood will not drain out while you seal the blood end. Plug the blood end of the tube by pushing it with a rotating motion into sealing clay or by adding a plastic Critocap (fig. 42.2*b*).
   - **h.** Place the sealed tube into one of the numbered grooves of a microhematocrit centrifuge. The tube's sealed end should point outward from the center and should touch the rubber lining on the rim of the centrifuge (fig. 42.2*c*).
   - **i.** The centrifuge should be balanced by placing specimen tubes on opposite sides of the moving head, the inside cover should be tightened, and the outside cover should be securely fastened.
   - **j.** Run the centrifuge for 3 to 5 minutes.
   - **k.** After the centrifuge has stopped, remove the specimen tube. The RBCs have been packed into the bottom of the tube. The clear liquid on top of the cells is plasma.
   - **l.** Use a microhematocrit reader to determine the percentage of RBCs in the tube. If a microhematocrit reader is not available, measure the total length of the blood column in millimeters (red cells plus plasma) and the length of the RBC column alone in millimeters. Divide the RBC length by the total blood column length and multiply the answer by 100 to calculate the percentage of RBCs.
2. Record the test result in Part A of Laboratory Assessment 42.

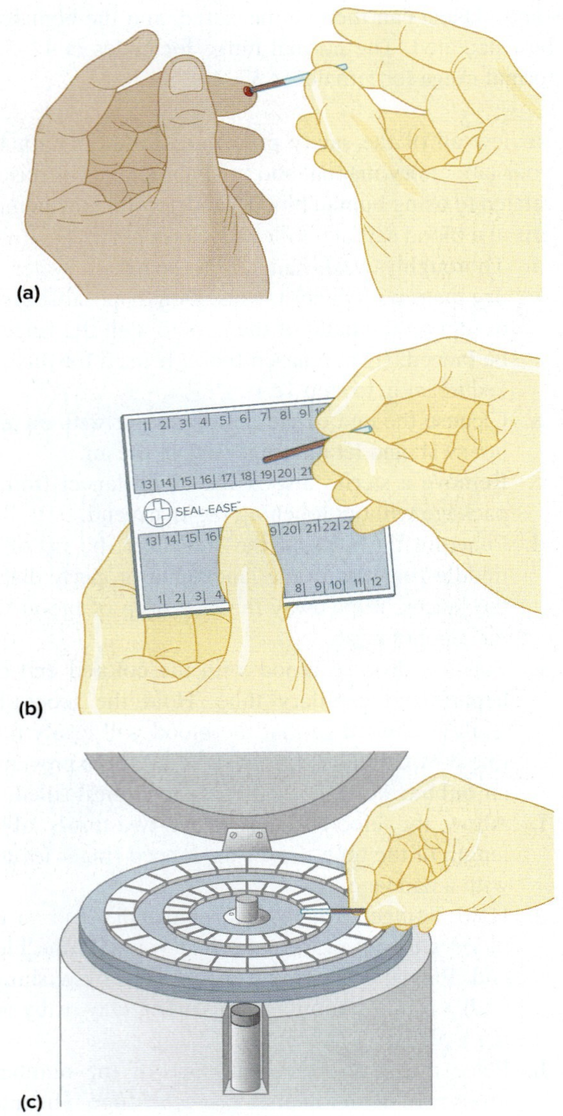

FIGURE 42.2 Steps of the hematocrit (red blood cell percentage) procedure: (a) load a heparinized capillary tube with blood; (b) plug the blood end of the tube with sealing clay; (c) place the tube in a microhematocrit centrifuge.

PROCEDURE B: Hemoglobin Content

Hemoglobin (Hb) is responsible for binding oxygen in the RBC. It is possible that anemia may be a result of low Hb even if the RBC count is normal (normocytic, hypochromic anemia). The hemoglobin content of a blood sample can be measured in several ways, including the *Tallquist method* and the *hemoglobinometer method,* both included in this laboratory exercise.

The Tallquist method uses a special test paper and is a simple, less expensive test that can be done more quickly than the hemoglobinometer method. The hemoglobinometer instrument is designed to compare the color of light passing through a hemolyzed blood sample with a standard color. The results of these tests are expressed in grams of hemoglobin per 100 mL (g/dL). The normal range for males is 13–18 g/dL; the normal range for females is 12–16 g/dL.

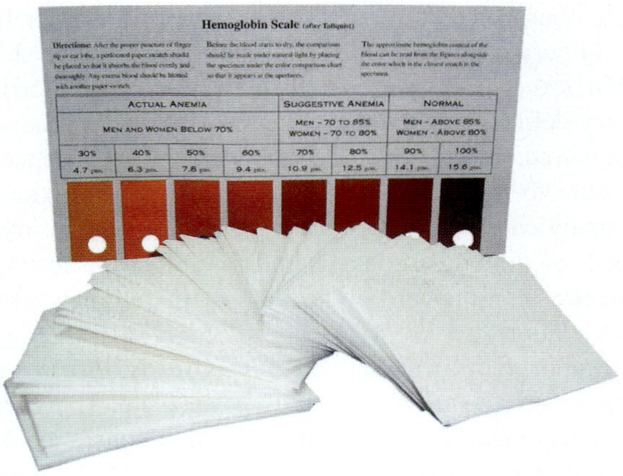

FIGURE 42.3 A simple method of measuring the amount of hemoglobin blood is to compare a piece of Tallquist paper that has been saturated with a sample of blood with a Tallquist color scale. ©Kemtec Science

Tallquist Method

1. To measure the hemoglobin content of a blood sample, follow these steps:
   a. Gather one piece of paper from the Tallquist booklet/kit, a Tallquist hemoglobin scale, and a blood sample.
   b. Obtain a drop of blood from a finger by following the directions in Procedure A.
   c. Place a large drop of blood on the center of the Tallquist paper so that the drop of blood is larger than the holes in the Tallquist hemoglobin scale.
   d. Let the blood dry only long enough to lose its glossy appearance. (It should not dry to a brown color.)
   e. Match the color, using natural light, of the blood on the Tallquist paper with the most accurate match of the color scale provided with the kit (fig. 42.3).
2. Record the test result in Part A of the laboratory assessment.

Hemoglobinometer Method

1. To measure the hemoglobin content of a blood sample, follow these steps:
   a. Obtain a hemoglobinometer and remove the blood chamber from the slot in its side.
   b. Separate the pieces of glass from the metal clip and clean them with alcohol swabs and lens paper. One of the pieces of glass has two broad, U-shaped areas surrounded by depressions. The other piece is flat on both sides.
   c. Obtain a large drop of blood from a finger, by following the directions in Procedure A.
   d. Place the drop of blood on one of the U-shaped areas of the blood chamber glass (fig. 42.4*a*).

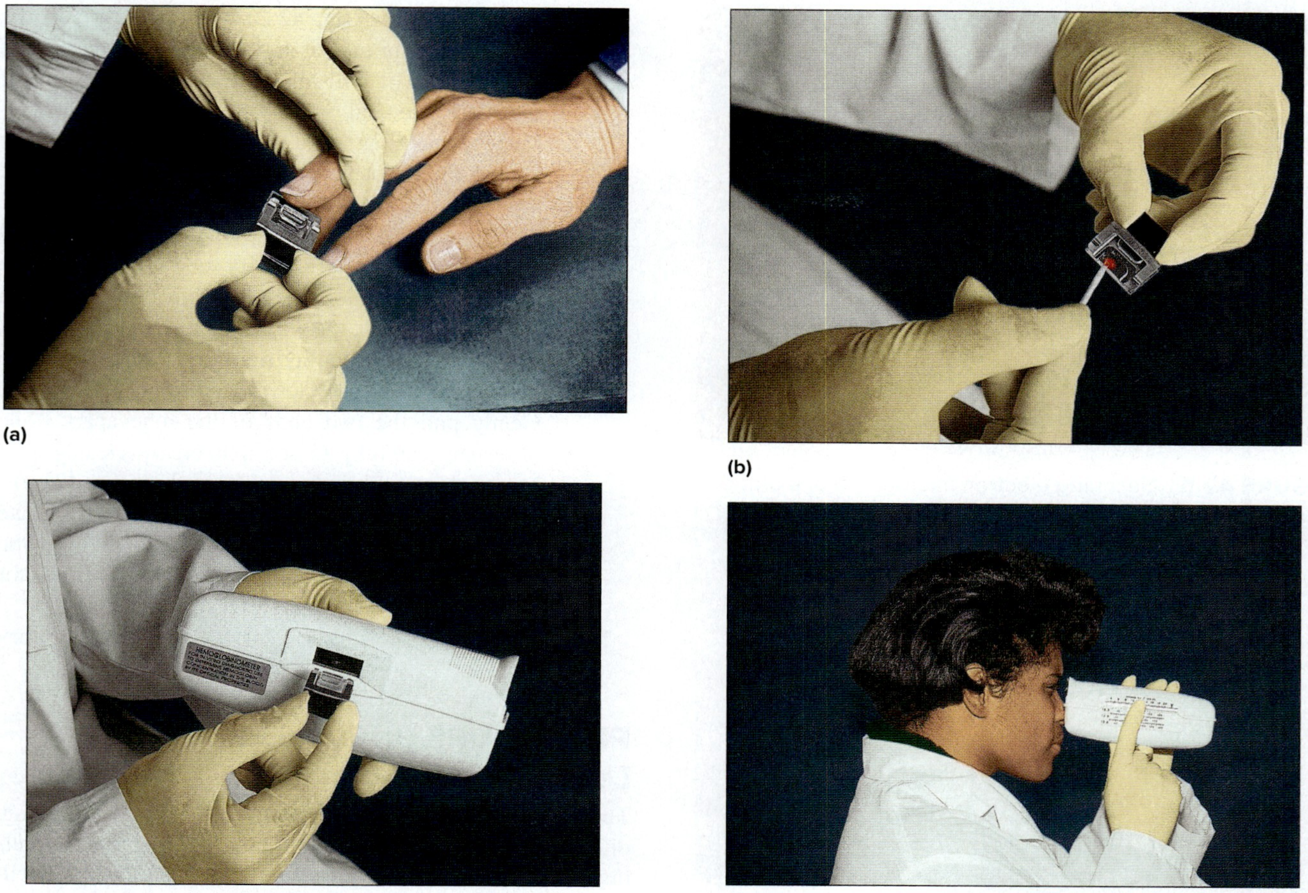

(a) (b) (c) (d)

FIGURE 42.4 Steps of the hemoglobin content procedure: (a) load the blood chamber with blood; (b) stir the blood with a hemolysis applicator; (c) place the blood chamber in the slot of the hemoglobinometer; (d) match the colors in the green area by moving the slide on the side of the instrument. James Schaffer/McGraw Hill

- **e.** Stir the blood with the tip of a hemolysis applicator until the blood appears clear rather than cloudy. This usually takes about 45 seconds (fig. 42.4*b*).
- **f.** Place the flat piece of glass on top of the blood plate and slide both into the metal clip of the blood chamber.
- **g.** Push the blood chamber into the slot on the side of the hemoglobinometer, making sure that it is in all the way (fig. 42.4*c*).
- **h.** Hold the hemoglobinometer in the left hand with the thumb on the light switch on the underside (fig. 42.4*d*).
- **i.** Look into the eyepiece and note the green area split in half.
- **j.** Slowly move the slide on the side of the instrument back and forth with the right hand until the two halves of the green area look the same.
- **k.** Note the value in the upper scale (grams of hemoglobin per 100 mL of blood), indicated by the mark in the center of the movable slide.

2. Record the test result in Part A of the laboratory assessment.

PROCEDURE C: Coagulation

Coagulation time, often called clotting time, is the time from the onset of bleeding until the clot is formed. Injured tissues and platelets release chemicals that trigger the clotting cascade, which ultimately results in thrombin converting soluble fibrinogen into insoluble fibrin. Fibrin is a mesh that traps RBCs and forms the clot (fig. 42.5). Clotting time normally ranges from 3 to 8 minutes. This process is prolonged if the person has clotting deficiencies or is being treated with anticoagulants such as heparin (some is produced by basophils and mast cells), warfarin (Coumadin), apixaban (Eliquis), or aspirin. In this laboratory exercise we will try to determine the time to the nearest minute.

1. To determine coagulation time, follow these steps:

- **a.** Prepare the finger to be lanced by following the directions in Procedure A. Lance the end of a finger to obtain a drop of blood. Wipe away the first drop of blood with the alcohol swab.
- **b.** Touch a drop of blood with one end of a nonheparinized capillary tube. Hold the tube tilted slightly upward so that the blood will easily move into it by

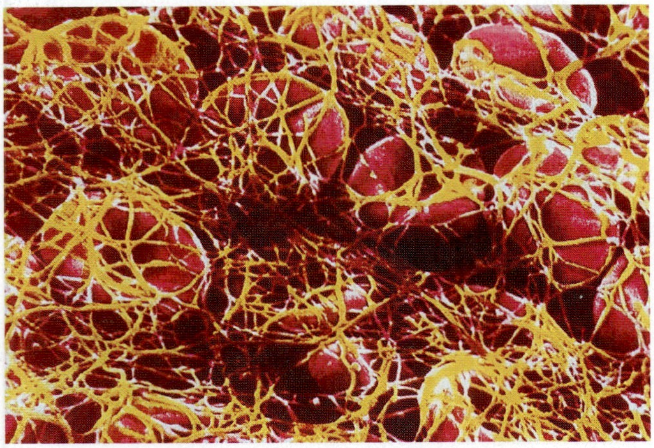

FIGURE 42.5 Scanning electron micrograph of a blood clot (2,800×). Note that the red blood cells are trapped within a mesh of fibrin threads. CNRI/Science Source

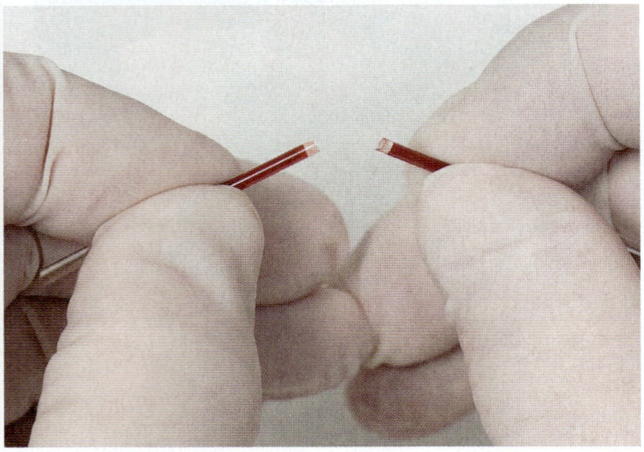

(a)

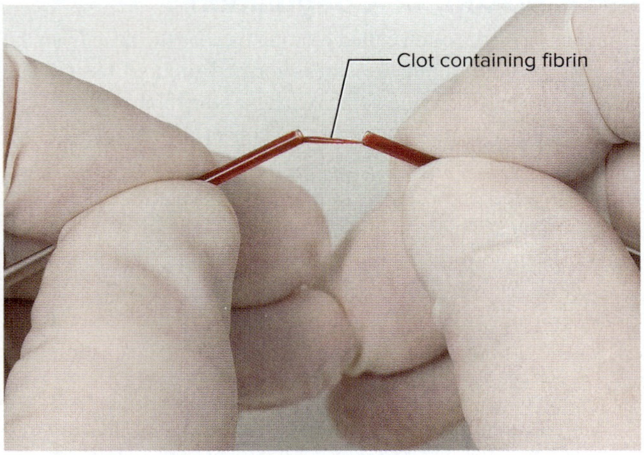

(b)

FIGURE 42.6 Steps of the coagulation procedure: (a) clean break of capillary tube before any fibrin formed; (b) fibrin spans between the broken capillary tube segments when coagulation occurs. ©J & J Photography

capillary action (fig. 42.2*a*). Keep the tip in the blood until the tube is nearly filled. If the tube is nearly filled, it will allow enough tube length for breaking it several times.

c. Place the capillary tube on a paper towel. Cover the lanced finger location with a bandage. Record the time: ______________

d. At 1-minute intervals, use the small triangular file and make a scratch on the capillary tube starting near one end of the tube. Hold the tube with fingers on each side of the scratch, the weakened location of the tube, and break the tube away from you, being careful to keep the two pieces close together after the break. Gently pull the two ends of the tube apart while observing carefully to see if it breaks cleanly. If it breaks cleanly, fibrin has not formed yet (fig. 42.6*a*).

e. Continue breaking the capillary tube each minute until fibrin is noted spanning the two parts of the capillary tube (fig. 42.6*b*). Note the time for coagulation.

2. Record the test results in Part A of the laboratory assessment.

PROCEDURE D: Blood Cholesterol Level Determination

The recommended total cholesterol level is less than 200 mg/dL [dL = deciliter; 1 dL = 100 milliliters (mL)], because levels higher than this have been correlated with increased probability of cardiovascular disease. Total blood cholesterol of less than 100 (hypocholesterolemia) is associated with such possible conditions as malnutrition, hyperthyroidism, and depression. Since cholesterol is not water soluble, it is transported in the plasma incorporated with a protein as a low-density lipoprotein (LDL) and a high-density lipoprotein (HDL). The **H**DLs are transported to the liver, where they are degraded and removed from the body, and are therefore sometimes referred to as the "**H**ealthy/good" form. The LDLs are transported to our tissue cells for cellular needs, but if the levels are excessive, there is an increased occurrence of atherosclerosis. Atherosclerosis is the deposition of fatty material within blood vessel walls, resulting in impeded blood flow and increased work by the heart. **L**DLs are sometimes referred to as the "**L**ousy/bad" cholesterol. The recommended levels of HDLs are >40 mg/dL; recommended levels of LDLs are <130 mg/dL.

1. Each kit contains cholesterol samples of simulated blood, test strips, color charts, and directions for the particular kit being used. Some available kits contain samples that reflect the use of cholesterol-lowering medications, while others might contain samples that represent diets of various cholesterol levels. Dip a test strip into a sample; then compare the strip to a color chart to determine the cholesterol level of the sample.
2. Record the test results in Part A of the laboratory assessment.
3. Complete Part B of the laboratory assessment that pertains to all procedures performed in this laboratory exercise.

LABORATORY
ASSESSMENT 42

Name ____________________

Date ____________________

Section ____________________

The A corresponds to the indicated Learning Outcome(s) O found at the beginning of the Laboratory Exercise.

Blood Testing

PART A: Assessments

Blood test data: A1

Blood Test	Test Results	Normal Values
Hematocrit (mL per 100 mL blood)		Males: 42–52% Females: 37–47%
Hemoglobin content (g per 100 mL blood; g/dL)		Males: 13–18 g/dL Females: 12–16 g/dL
Coagulation		3–8 minutes
Cholesterol	Before reduction program: After reduction program:	Below 200 mg/dL

CRITICAL THINKING ASSESSMENT—PART A

Predict how moving from Portland, Oregon to Denver, Colorado would affect a person's hematocrit. Explain. A2

____

____.

PART B: Assessments

Complete the following:

1. How does the hematocrit from the blood test compare with the normal value? If the test results were abnormal, what condition would it possibly indicate and why? A2 ____________________

____

____

2. How does the hemoglobin content from the blood test compare with the normal value? If the test results were abnormal, what condition would it possibly indicate and why? A2 ____________________

____

____

3. How does the coagulation time from the blood test compare with the normal value? If the test results were abnormal, what condition would it possibly indicate and why? A2 ____________________

____

____

4. Assess the most beneficial method of lowering cholesterol from samples provided in the simulated blood kit. Explain your reasoning. A2 ____

____

____

CRITICAL THINKING ASSESSMENTS—PART B

1. A patient who is taking warfarin (Coumadin) must go off this medicine a few days prior to having surgery. Referencing your laboratory activity on coagulation, explain why this protocol is important. A2

____

____

____.

2. Which blood tests performed in this lab could be used to determine possible anemia? Explain your reasoning. A3

____

____

____.

3. Predict what direction the oxygen dissociation curve will shift during physical activity. Provide your reasons why. A4

____

____

____.

LABORATORY
EXERCISE 43

Blood Typing

Purpose of the Exercise

To determine and explain the significance of the ABO blood type of a blood sample and an Rh blood-typing test.

MATERIALS NEEDED

For Procedure A:

ABO blood-typing kit

Simulated blood-typing kits are suggested as a substitute for collected blood

For Procedure B:

Microscope slide

Alcohol swabs (wipes)

Sterile blood lancet

Disposable gloves

Toothpicks

Anti-D serum

Slide warming table (Rh blood-typing box)

SAFETY

- Review all the safety guidelines in Appendix 1 of your laboratory manual.
- It is important that students learn and practice correct procedures for handling body fluids. Consider using simulated blood-typing kits or contaminant-free blood that has been tested and is available from various laboratory supply houses. Some of the procedures might be accomplished as demonstrations only. If student blood is used, it is important that students handle only their own blood.
- Use an appropriate disinfectant to wash the laboratory tables before and after the procedures.
- Wear disposable gloves and protective eyewear when handling blood samples.
- Clean the end of a finger with an alcohol swab before the puncture is performed.
- Use the sterile blood lancet only once.
- Dispose of used lancets and blood-contaminated items in an appropriate container (never use the wastebasket).
- Wash your hands before leaving the laboratory.

Learning Outcomes

After completing this exercise, you should be able to:

O1 Explain the fundamentals of the ABO blood group and possible adverse reactions from transfusions.

O2 Interpret the ABO type of a blood sample.

O3 Analyze the basis and importance of the Rh blood factor.

O4 Interpret the Rh type of a blood sample.

The O corresponds to the assessments O indicated in the Laboratory Assessment for this Exercise.

Pre-Lab

Carefully read the introductory material and examine the entire lab. Be familiar with the ABO blood group and the Rh blood group from lecture or the textbook. Answer the pre-lab questions.

Pre-Lab Questions Select the correct answer for each of the following questions:

1. The antigens related to the ABO blood group are located
   a. on the red blood cell membrane.
   b. within the red blood cell nucleus.
   c. within the red blood cell cytosol.
   d. on the red blood cell ribosome.
2. When B antibodies (anti-B) react with B antigens, _____ occurs.
   a. coagulation **b.** agglutination
   c. transfusion **d.** proliferation
3. The D antigen related to the Rh factor is present in about _____ of Americans.
   a. 4% **b.** 38%
   c. 47% **d.** 85%
4. Blood type _____ is sometimes considered the universal donor within the ABO blood group.
   a. A **b.** B
   c. AB **d.** O
5. Blood type _____ is sometimes considered the universal recipient within the ABO blood group.
   a. A **b.** B
   c. AB **d.** O

Chapter Opener: ©101cats/Getty Images

6. In a cross-matching test, if there is no agglutination,
   a. a transfusion reaction will occur.
   b. a transfusion can safely be done.
   c. a transfusion cannot be safely done.
   d. a transfusion will cause RBC hemolysis.
7. Hemolytic disease of the newborn could be of concern when an
   a. Rh-positive fetus and an Rh-positive mother condition occur.
   b. Rh-positive fetus and an Rh-negative mother condition occur.
   c. Rh-negative fetus and an Rh-positive mother condition occur.
   d. Rh-negative fetus and an Rh-negative mother condition occur.
8. Blood type B is the most common blood type found in the United States population.
   a. True b. False
9. An individual with blood type O lacks both RBC antigens A and B.
   a. True b. False

Blood typing involves identifying protein substances called *antigens* present on the outer surface of red blood cell membranes. There are many different antigens associated with human red blood cells, but only a few of them are of clinical importance. These include the antigens of the *ABO group* and those of the *Rh group*.

To determine which antigens are present, a blood sample is mixed with blood-typing sera that contain known types of *antibodies*. If a particular antibody contacts a corresponding antigen, a reaction occurs, and the red blood cells clump together (*agglutination*). Thus, if red blood cells are mixed with serum containing anti-A (antibodies A) that react with antigen A and the cells agglutinate, antigen A must be present on the outer surface of those cells.

Although there are many antigens on the red blood cell membranes, the ABO group and the Rh group antigens are of the greatest concern during *transfusions*. Most other antigens cause little, if any, transfusion reaction. As a final check on blood compatibility before transfusions, an important cross-matching test occurs. In this test, small samples of the blood from the donor and the recipient are mixed to be sure agglutination does not create a transfusion reaction, which would result in red blood cell destruction (hemolysis) and other life-threatening conditions. If surgery is elective, autologous transfusions are promoted, allowing some of the person's own predonated blood from storage to be used during the surgery.

An antigen commonly found on the human RBC membrane was first identified in rhesus monkeys, and it became known as the *Rh factor*. Although many different types of antigens are related to the Rh factor, because it is the most reactive, only the *D antigen* is checked using an anti-D reagent. About 85% of Americans possess the D antigen and are therefore considered to be *Rh-positive* (*Rh*$^+$). Those lacking the D antigen are considered *Rh-negative* (*Rh*$^-$). If an Rh-negative woman is pregnant carrying an Rh-positive fetus, the mother might obtain some RBCs from the fetus during the birth process or during a miscarriage. As a result, the Rh-negative mother would begin producing anti-D antibodies, creating complications for future pregnancies. This condition is known as *hemolytic disease* of the fetus and newborn (erythroblastosis fetalis). This condition can be prevented with the proper administration of *RhoGAM,* which prevents the mother from producing anti-D antibodies. Antibodies are not formed until a Rh$^-$ person is exposed to Rh$^+$ blood, but as a precaution a pregnant Rh$^-$ woman is typically given RhoGAM in every pregnancy.

WARNING

Because of the possibility of blood infections being transmitted from one student to another if blood testing is performed in the classroom, it is suggested that commercially prepared blood-typing kits containing virus-free human blood be used for ABO blood typing and for the Rh blood typing. The instructor may wish to demonstrate Rh blood typing. Observe all of the safety procedures listed for this lab.

PROCEDURE A: ABO Blood Typing

1. Reexamine the introductory material and study figures 43.1 to 43.3 and table 43.1.
2. Figure 43.1 illustrates the antigens on the RBC membranes that are compatible with various antibodies in the plasma of the ABO blood group. If an antigen of the RBC is introduced into a person with an incompatible plasma antibody, an adverse reaction occurs between the antigen and the antibody, illustrated as an example in figure 43.2. This is most likely to occur if blood from an incompatible donor was used for the transfusion. Table 43.1 indicates the preferred donors, permissible donors (in limited amounts), and incompatible donors for transfusions. If blood is used from an incompatible donor for a transfusion, clumps of RBCs form rather quickly that can block circulation through smaller blood vessels, resulting in serious consequences (fig. 43.3).

   Table 43.1 labels blood type AB a *universal recipient* and blood type O a *universal donor.* This terminology can help you to understand the concept of possible clumping (agglutination) reactions during transfusions. Blood type AB is sometimes referred to as a universal recipient (acceptor) because AB blood lacks anti-A and anti-B antibodies and will not clump with other blood types when used for transfusions. Similarly, blood type O is theoretically referred to as a universal donor as it lacks both A and B antigens (agglutinogens). However, it is not reliable to refer to a blood type as a universal recipient or a universal donor. Under certain conditions, an adverse reaction could

Table 43.1 Antigens and Antibodies of the ABO Blood Group and Preferred, Permissible, and Incompatible Donors

Blood Type	RBC Antigens (Agglutinogens)	Plasma Antibodies (Agglutinins)	Preferred Donor Type	Permissible Donor Type in Limited Amounts	Incompatible Donor
A	A	Anti-B	A	O	B, AB
B	B	Anti-A	B	O	A, AB
AB (universal recipient)	A and B	Neither anti-A nor anti-B	AB	A, B, O	None
O (universal donor)	Neither A nor B	Both anti-A and anti-B	O	No alternative types	A, B, AB

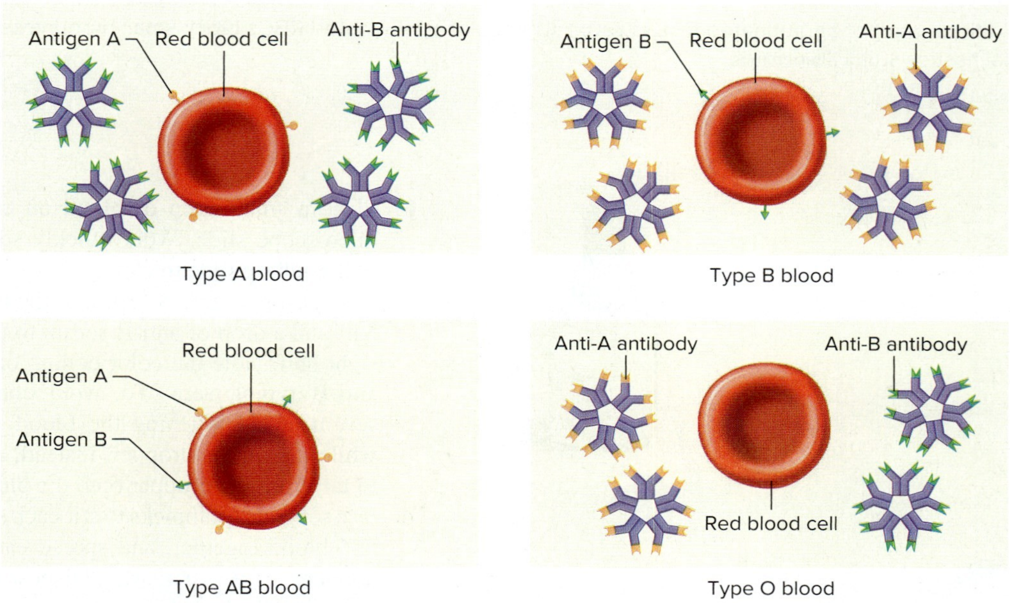

FIGURE 43.1 Illustration of four possible combinations of the RBC antigens and plasma antibodies within the ABO blood group.

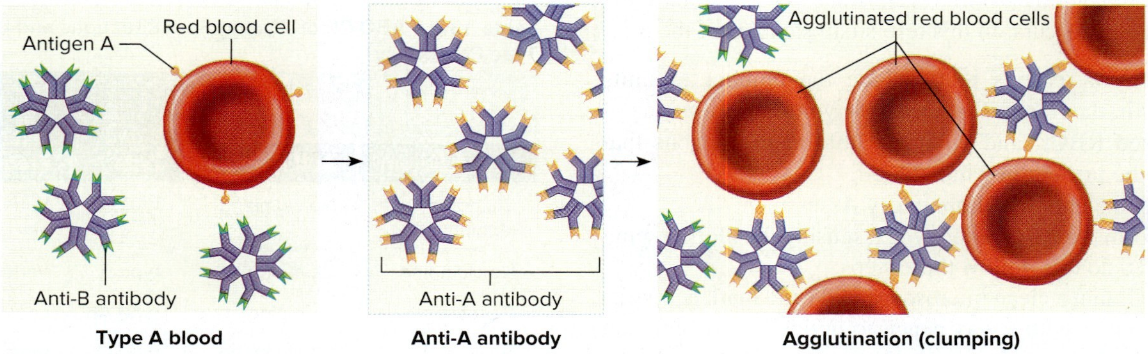

FIGURE 43.2 Illustration of an agglutination (clumping) reaction when antigen A reacts with anti-A antibody in an agglutination test.

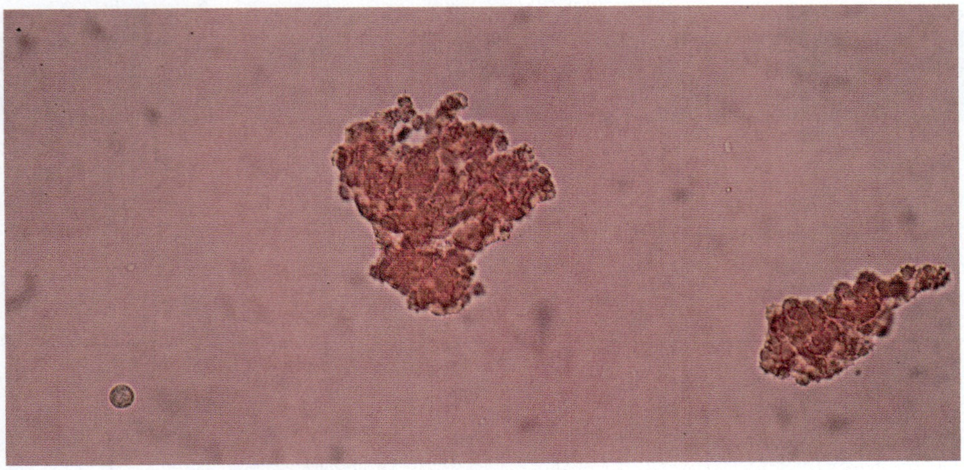

FIGURE 43.3 Micrograph of agglutinated RBCs (220x). Clumps of agglutinated RBCs block small blood vessels and circulation to vital tissue. Schira/Shutterstock

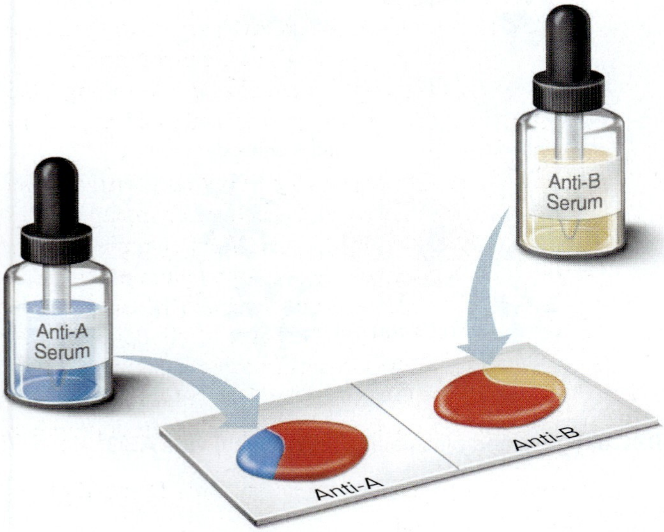

FIGURE 43.4 Slide prepared for ABO blood typing.

b. Place a small drop of blood on each half of the microscope slide. Work quickly so that the blood will not have time to clot.

c. Add a drop of anti-A serum to the blood on the left half and a drop of anti-B serum to the blood on the right half. Note the color coding of the anti-A and anti-B typing sera. To avoid contaminating the serum, avoid touching the blood with the serum while it is in the dropper; instead, allow the serum to fall from the dropper onto the blood.

d. Use separate toothpicks to stir each sample of serum and blood together, and spread each over an area about as large as a quarter. Dispose of toothpicks in an appropriate container.

e. Examine the samples for clumping of red blood cells (agglutination) after 2 minutes.

f. See table 43.2 and figure 43.5 for aid in interpreting the test results. The frequency of occurrence of

Table 43.2 ABO Blood-Typing Sera Reactions and U.S. Blood Type Frequency

POSSIBLE REACTIONS		Blood Type	Overall U.S. Frequency
Anti-A Serum	Anti-B Serum		
Clumping (agglutination)	No clumping	Type A	Varies
No clumping	Clumping	Type B	Varies
Clumping	Clumping	Type AB	Least common
No clumping	No clumping	Type O	Most common

Note: The inheritance of the ABO blood groups is described in Laboratory Exercise 62.

occur for reasons that include the number of units used in the transfusions, plasma volumes, infusions of packed RBCs, and other possible agglutinogens that may be present in whole blood.

3. Compete Part A of Laboratory Assessment 43.
4. Perform the ABO blood type test using the blood-typing kit. To do this, follow these steps:
   a. Obtain a clean microscope slide and mark across its center with a wax pencil to divide it into right and left halves. Also write "Anti-A" near the edge of the left half and "Anti-B" near the edge of the right half (fig. 43.4).

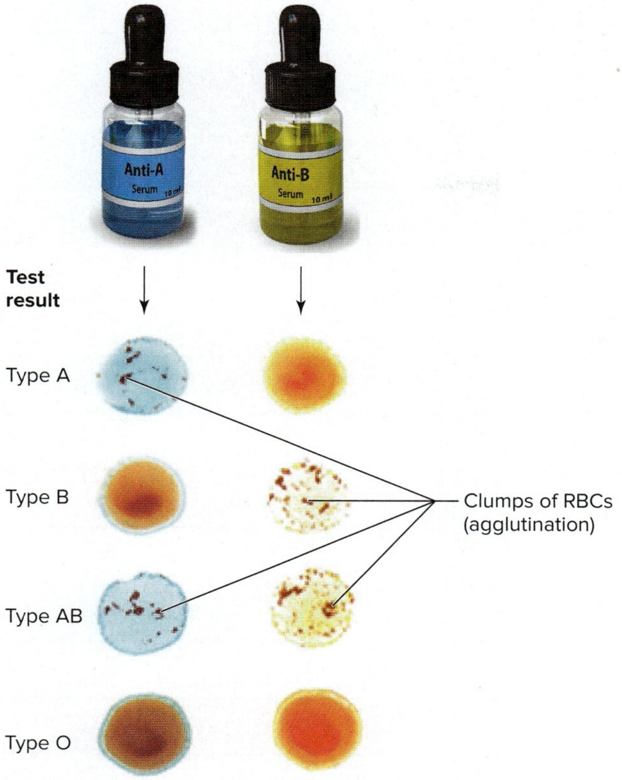

FIGURE 43.5 Four possible results of the ABO test.
McGraw Hill

ABO blood types varies substantially between ethnic groups. Although there is a very diverse population in the United States, blood type O is the most common and blood type AB the least common.

g. Discard contaminated materials as directed by the laboratory instructor.

5. Record your results and complete Part B of the laboratory assessment.

CRITICAL THINKING ACTIVITY

Judging from your observations of the blood-typing results, suggest which components in the anti-A and anti-B sera caused clumping (agglutination).

____

____

____

PROCEDURE B: Rh Blood Typing

1. Reexamine the introductory material.
2. Complete Part C of the laboratory assessment.
3. To determine the Rh blood type of a blood sample, follow these steps:
   a. Lance the tip of a finger. (See the demonstration procedures in Laboratory Exercise 41 for directions.) Place a small drop of blood in the center of a clean microscope slide. Cover the lanced finger location with a bandage.
   b. Add a drop of anti-D (anti-Rh) serum to the blood and mix them together with a clean toothpick.
   c. Place the slide on the plate of a slide warming table (Rh blood-typing box) prewarmed to 45°C (113°F) (fig. 43.6).
   d. Slowly rock the table (box) back and forth to keep the mixture moving and watch for clumping (agglutination) of the blood cells. When clumping occurs in anti-D serum, the clumps usually are smaller than those that appear in anti-A or anti-B sera, so they may be less obvious. However, if clumping occurs, the blood is called Rh positive; if no clumping occurs *within 2 minutes,* the blood is called Rh negative.
   e. Discard all contaminated materials in appropriate containers.
4. Record your results and complete Parts D and E of the laboratory assessment.

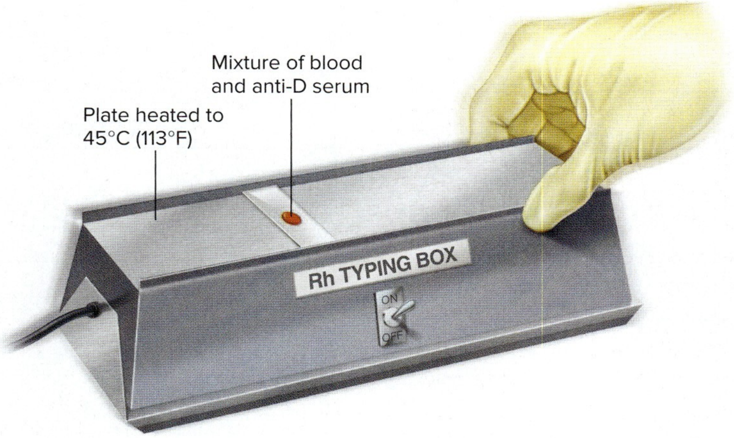

FIGURE 43.6 Slide warming table (box) used for Rh blood typing.

NOTES

LABORATORY
ASSESSMENT 43

Name ______________________________

Date ______________________________

Section ______________________________

The (A) corresponds to the indicated Learning Outcome(s) (O) found at the beginning of the Laboratory Exercise.

Blood Typing

PART A: Assessments

Complete the following statements:

1. The antigens of the ABO blood group are located on the red blood cell ________________. A1
2. The blood of every person contains one of (how many possible?) ________________ combinations of antigens. A1
3. Type A blood contains antigen ________________. A1
4. Type B blood contains antigen ________________. A1
5. Type A blood contains ________________ antibody in the plasma. A1
6. Type B blood contains ________________ antibody in the plasma. A1
7. Persons with ABO blood type ________________ are sometimes called universal recipients. A1
8. Persons with ABO blood type ________________ are sometimes called universal donors. A1
9. Assume a person with blood type B receives a transfusion of blood type AB. Would the transfusion most likely have successful results? ________________ Explain your answer. A1

____

____

PART B: Assessments

1. Record your results in the table. (*Note:* If you conducted more than one ABO blood-typing test, complete the table accordingly.) [Record a (+) or (−) to indicate an agglutination reaction with each serum.]

Name of Individual	Anti-A Serum	Anti-B Serum	ABO Blood Type
1.			
2.			
3.			
4.			

2. Complete the following:
[*Note:* If more than one ABO blood-typing test was conducted, choose one individual from your results in the table above (or as directed by your instructor) to answer these questions.]

   a. What was the ABO type of the blood tested? A2 ________________
   b. What ABO antigens are present on the red blood cells of this blood type? A2 ________________
   c. What ABO antibodies are present in the plasma of this blood type? A2 ________________
   d. If a person with this blood type needed a blood transfusion, what ABO type(s) of blood could be received safely? A1

____

e. If a person with this blood type donated blood, what ABO blood type(s) could receive the blood safely? A1

____

PART C: Assessments

Complete the following statements:

1. The Rh blood group was named after the ______________. A3
2. Of the antigens in the Rh group, the most important is ______________. A3
3. If red blood cells lack Rh antigens, the blood is called ______________. A3
4. If an Rh-negative person who is sensitive to Rh-positive blood receives a transfusion of Rh-positive blood, the donor's cells are likely to ______________. A3
5. An Rh-negative woman who might be carrying a(an) ______________ fetus is given an injection of RhoGAM to prevent hemolytic disease of the fetus and newborn (erythroblastosis fetalis). A3

PART D: Assessments

1. Record your results in the table. (*Note:* If you conducted more than one Rh blood-typing test, complete the table accordingly.) [Record a (+) or (−) to indicate an agglutination reaction with anti-Rh (anti-D) serum.]

Name of Individual	Anti-Rh Serum	Rh Blood Type (Rh-positive or Rh-negative)
1.		
2.		
3.		
4.		

2. Complete the following:
[*Note:* If more than one Rh blood-typing test was conducted, choose one individual from your results in the table above (or as directed by your instructor) to answer these questions.]

a. What was the Rh type of the blood tested? A4 ______________

b. What Rh antigen is present on the red blood cells of this type of blood? A4 ______________

c. If a person with this blood type needed a blood transfusion, what blood type could be received safely? A3

____

d. If a person with this blood type donated blood, a person with what blood type could receive the blood safely? A3

____

PART E: Assessments

Examine the test results shown in figure 43.7 from an individual being tested for both ABO type and Rh type. Observe whether clumping (agglutination) occurred to answer the questions that follow.

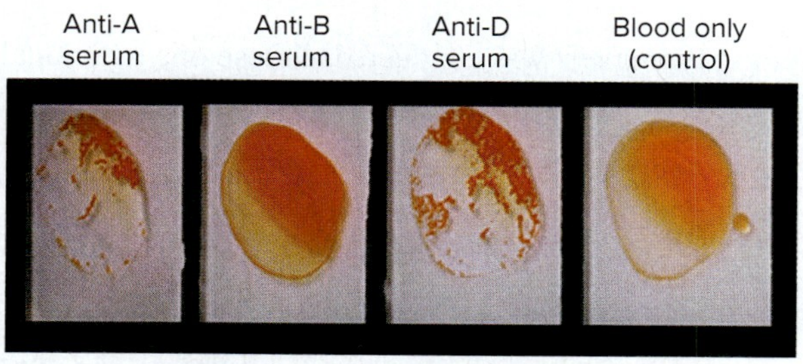

FIGURE 43.7 Blood test results after adding anti-A serum, anti-B serum, and anti-D serum to separate drops of blood. An unaltered drop of blood is shown as a control; it does not indicate any clumping (agglutination) reaction. Image Source/Getty Images

1. This individual has ABO type ______________ because ____. A2

2. This individual has Rh type ______________ because ____. A4

CRITICAL THINKING ASSESSMENT—PART E

A patient with unknown blood type is in need of a transfusion. When a blood typing test is done, the results show the patient's blood agglutinates when mixed with anti-B and with anti-D. Answer the following questions: A2 A4

What is this patient's blood type? ____

What blood type(s) can this patient safely receive in a transfusion? ______________________________

At a later date when the patient is well, what blood type(s) could this patient safely donate? ______________

____.

NOTES

LABORATORY
EXERCISE 44

Heart Structure

Purpose of the Exercise

To review the structural characteristics of the human heart and to examine the major features of a mammalian heart.

MATERIALS NEEDED

Dissectible human heart model
Dissectible human torso model
Preserved sheep or other mammalian heart
Dissecting tray
Dissecting instruments
Disposable gloves
Long knife
Millimeter ruler

For Learning Extension Assessment:
Colored pencils (red and blue)

SAFETY

- Review all the safety guidelines in Appendix 1 of your laboratory manual.
- Wear disposable gloves and protective eyewear when working on the heart dissection.
- Save or dispose of the dissected heart as instructed.
- Wash the dissecting tray and instruments as instructed.
- Wash your laboratory table.
- Wash your hands before leaving the laboratory.

Learning Outcomes APR

After completing this exercise, you should be able to:

O1 Identify and label the major structural features of the human heart and closely associated blood vessels.

O2 Trace blood flow through the heart chambers and the systemic and pulmonary circuits.

O3 Describe locations and functions of heart structures.

O4 Compare the features of the human heart with those of another mammal.

The O corresponds to the assessments A indicated in the Laboratory Assessment for this Exercise.

Pre-Lab

Carefully read the introductory material and examine the entire lab. Be familiar with the structure of the heart from lecture or the textbook. Answer the pre-lab questions.

Pre-Lab Questions Select the correct answer for each of the following questions:

1. The _____ is the inferior end of the heart that is bluntly pointed.
a. base **b.** apex
c. auricle **d.** sulcus

2. Oxygen-rich blood is located in the
a. left-side chambers. **b.** right-side chambers.
c. superior chambers. **d.** inferior chambers.

3. The two superior heart chambers are the
a. left atrium and left ventricle.
b. right atrium and right ventricle.
c. atria.
d. ventricles.

4. Which of the following is an atrioventricular (AV) valve?
a. aortic **b.** pulmonary
c. semilunar **d.** mitral

5. The _____ lines the heart chambers.
a. endocardium **b.** fibrous pericardium
c. serous pericardium **d.** epicardium

6. Which heart valve is located between the left atrium and left ventricle?
a. aortic **b.** mitral
c. tricuspid **d.** pulmonary

7. The _____ valves regulate blood flow from the ventricles into the large arteries.
a. mitral **b.** tricuspid
c. atrioventricular **d.** semilunar

8. Chordae tendineae connect the cusps of the AV valves to papillary muscles of the ventricles.
   a. True **b.** False
9. The systemic circuit delivers blood to the lungs and back to the heart.
   a. True **b.** False
10. The right and left coronary arteries containing oxygen-rich blood originate from the base of the aorta.
   a. True **b.** False

The heart is a muscular pump located within the mediastinum and resting upon the diaphragm. It is enclosed by the lungs, thoracic vertebrae, and sternum, and attached at its superior end (the *base*) are several large blood vessels. Its inferior end extends downward to the left and terminates as a bluntly pointed *apex.*

The heart and the proximal ends of the attached blood vessels are enclosed by a double-layered *pericardium.* The innermost layer of this membrane (visceral layer of serous pericardium) consists of a thin covering closely applied to the surface of the heart, whereas the outer layer (parietal layer of serous pericardium with fibrous pericardium) forms a tough, protective sac surrounding the heart. Between the parietal and visceral layers of the pericardium is a narrow space, the *pericardial cavity,* that contains a small volume of *serous (pericardial) fluid.*

Functioning as a muscular pump of blood, the heart consists of four chambers: two superior *atria* and two inferior *ventricles.* The *interatrial septum* and the *interventricular septum* prevent oxygen-rich (oxygenated) blood in the left-side chambers from mixing with oxygen-poor (deoxygenated) blood in the right chambers. *Atrioventricular (AV) valves (mitral and tricuspid)* are located between each atrium and ventricle, while *semilunar valves (aortic and pulmonary)* are at the exits of the ventricles into the two large arteries. The heart valves only allow blood flow in one direction.

PROCEDURE A: The Human Heart APR

As you study the human heart, use a combination of laboratory manual figures and heart models available in the laboratory to locate structures. This laboratory exercise covers basic information that will be very important as you work on the rest of the cardiovascular system exercises.

1. Study figures 44.1 to 44.6.

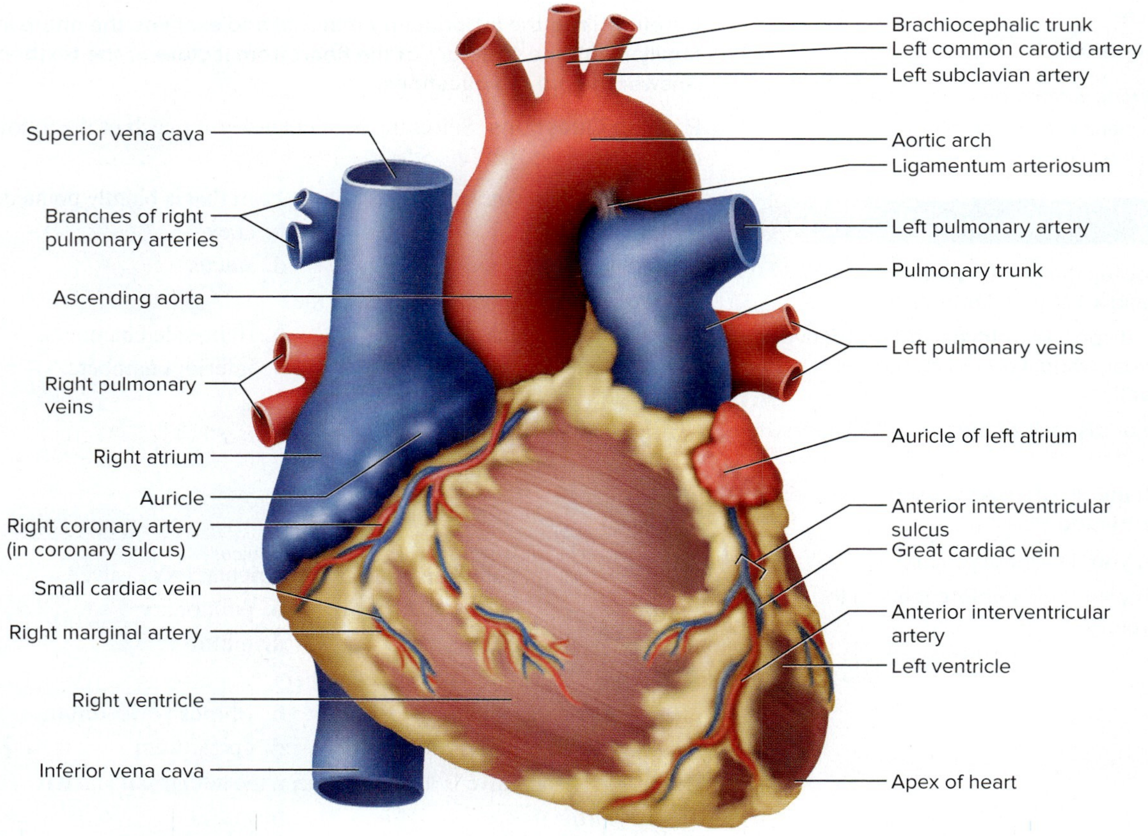

FIGURE 44.1 Anterior view of the human heart. APR

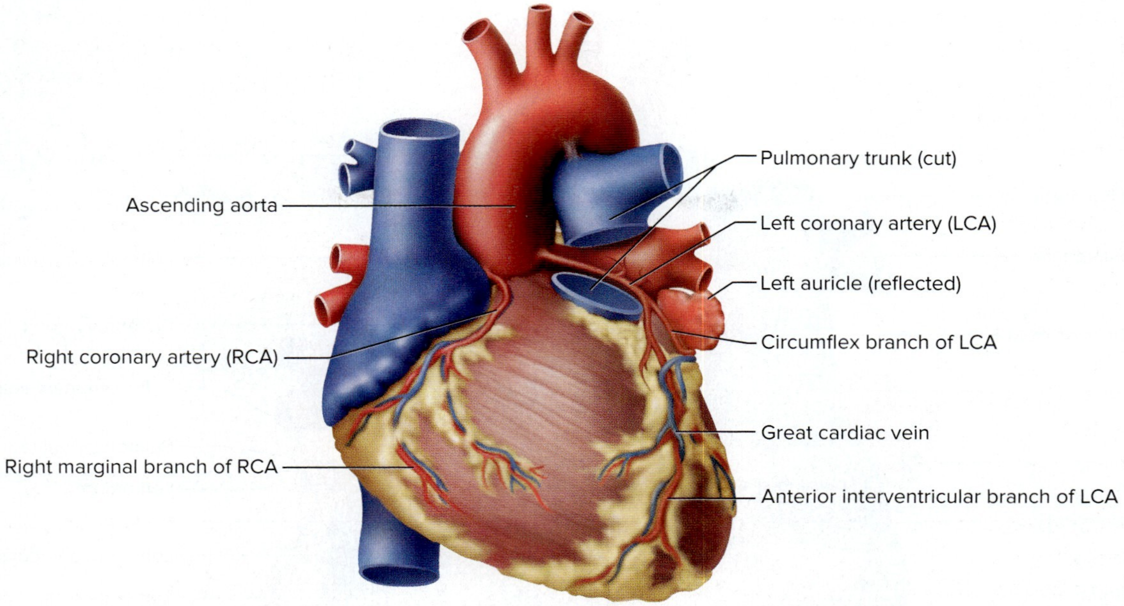

FIGURE 44.2 Anterior view of the two coronary arteries and their major branches. The coronary arteries originate from the base of the aorta and carry oxygen-rich blood to the cardiac muscle tissue. APR

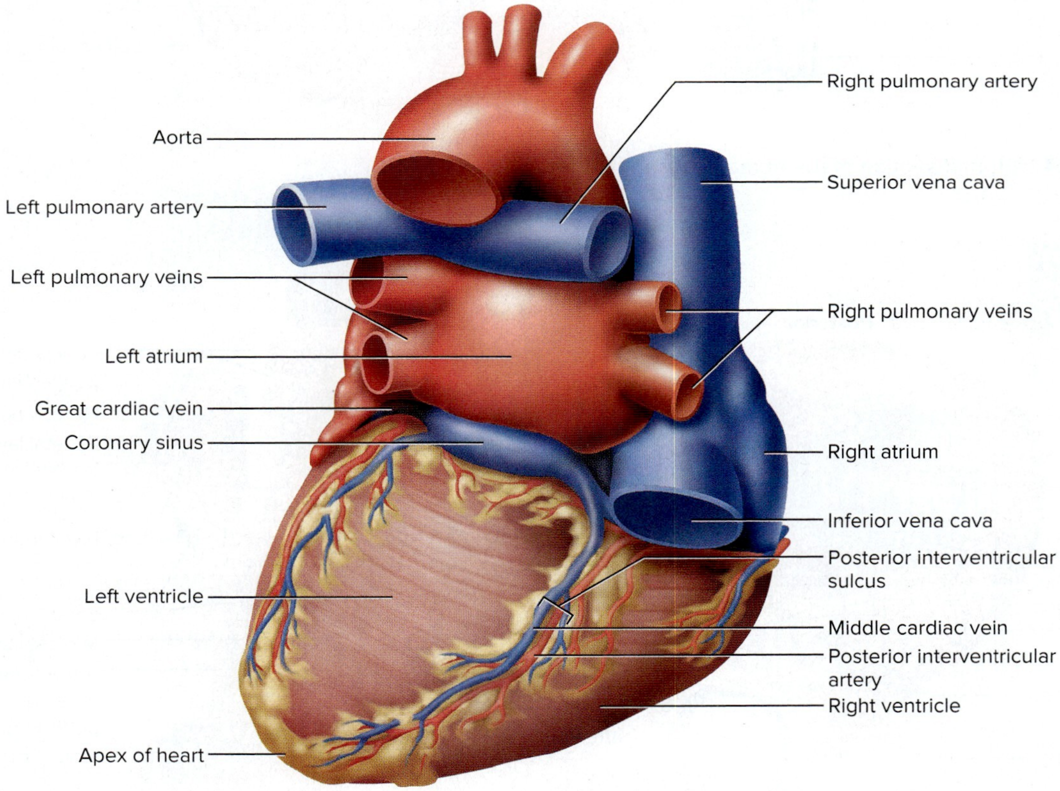

FIGURE 44.3 Posterior view of the human heart. APR

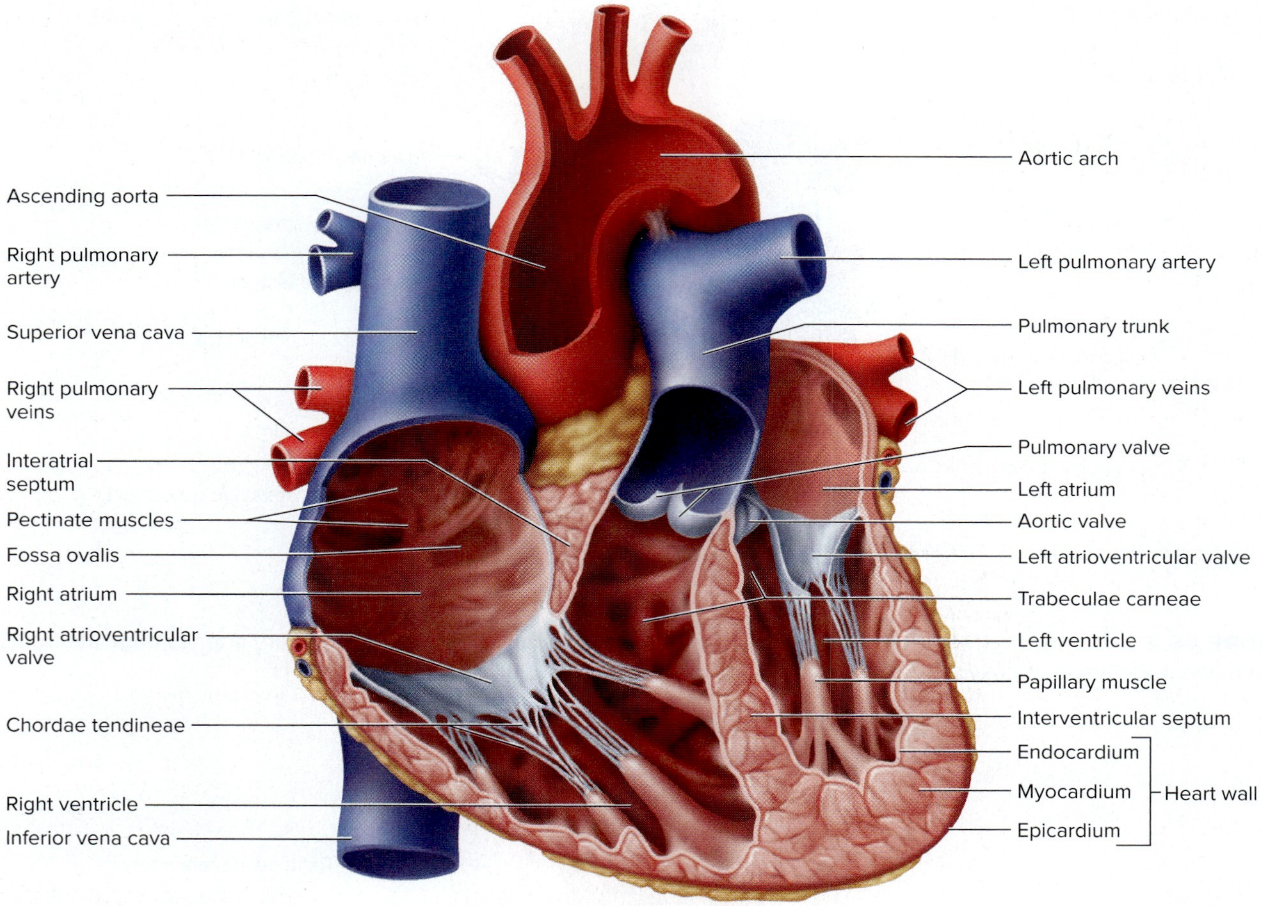

FIGURE 44.4 Frontal section of the human heart. APR

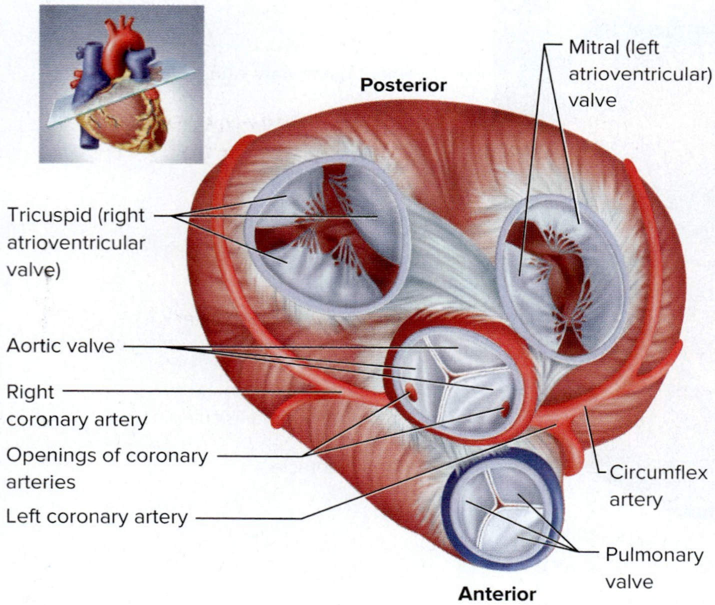

FIGURE 44.5 Superior view of the four heart valves (atria removed).

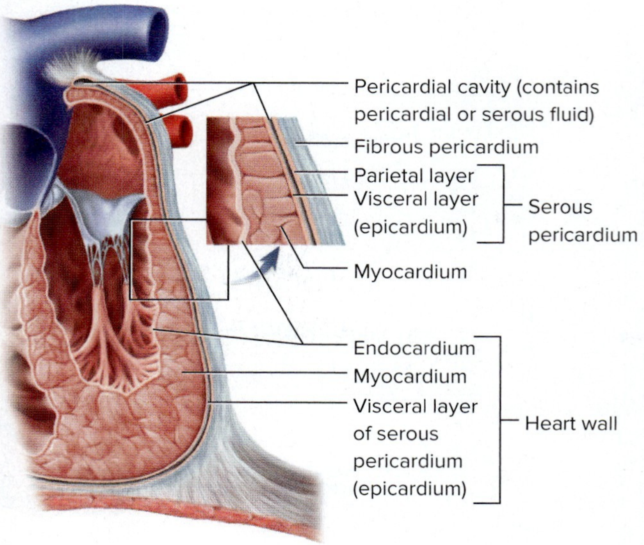

FIGURE 44.6 Frontal section of the heart wall and pericardium.

2. Examine the human heart model and locate the following features:

heart

- base of heart—superior region where blood vessels emerge
- apex of heart—inferior, rounded end; tilts slightly to the left

3. Examine figure 44.6 of the pericardium and the heart wall. Observe the arrangement of the structures and note the descriptions of the following features:

pericardium (pericardial sac)

- fibrous pericardium—outer layer
- serous pericardium
  - parietal layer—inner fused lining of fibrous pericardium
  - visceral layer (epicardium)

pericardial cavity—between parietal and visceral layers of serous pericardial membranes; contains serous (pericardial) fluid

myocardium—composed of cardiac muscle tissue (fig. 44.7)

endocardium—lines heart chambers

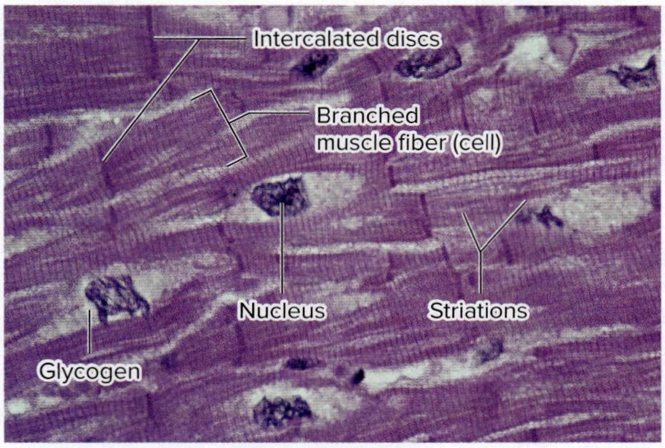

Cardiac muscle tissue (400×)

FIGURE 44.7 Cardiac muscle tissue forms the myocardium of the heart chamber walls. It also composes the muscular ridges (trabeculae carneae of ventricles and pectinate muscles of portions of atria) and papillary muscles of ventricles. Ed Reschke/Stone/Getty Images APR

4. Use the human heart model and examine the structures, locations, and descriptions of the following:

atria—two superior chambers

- right atrium—receives blood from systemic veins and from venous drainage of the heart tissue
- left atrium—receives blood from pulmonary veins
- interatrial septum—muscular wall separating two atria
- auricles—earlike extensions of atria allowing greater expansions for blood volumes
- pectinate muscles—ridges of myocardium found within right atrium and both auricles

ventricles—two inferior chambers

- right ventricle—pumps blood into pulmonary circuit
- left ventricle—pumps blood into systemic circuit; has thicker myocardium than right ventricle
- interventricular septum—thick muscular wall separating two ventricles
- trabeculae carneae—ridges of myocardium found within ventricles

atrioventricular orifices—openings between atria and ventricles and guarded by AV valves

atrioventricular valves (AV valves)—located between atria and ventricles

- right atrioventricular valve (tricuspid valve)
- left atrioventricular valve (mitral valve)

semilunar valves—regulate flow into two large arteries that exit from ventricles

- pulmonary valve—has three pocketlike cusps at entrance of pulmonary trunk
- aortic valve—has three pocketlike cusps at entrance of aorta

chordae tendineae (tendinous cords)—connect AV valve cusps to papillary muscles

papillary muscles—projections from inferior floor of ventricles; anchor AV valve cusps during ventricular contraction

coronary sulcus (atrioventricular sulcus or groove)—encircles heart between atria and ventricles

interventricular sulci—grooves with blood vessels along septum between ventricles

- anterior sulcus
- posterior sulcus

5. Inspect figure 44.8 as a reference to trace the flow of blood through the heart chambers and the systemic and pulmonary circuits. The arrows within the figure indicate the direction of the blood flow within the heart chambers and the two circuits. Use a dissectible human torso model and the heart model and locate the major blood vessels near the heart. Follow the arrows on figure 44.8 to trace the pathway of blood. Observe the figures and the models to take notice of the use of the red and blue colors that pertain to the relative oxygen levels of the blood within those blood vessels and heart chambers. The red color indicates oxygen-rich (oxygenated) blood; the blue color indicates oxygen-poor (deoxygenated) blood. Compare the two circuits in relation to the oxygen content in the arteries and veins. *Arteries carry blood away from the heart, while veins return blood to the heart. Notice that the arteries of the systemic circuit contain oxygen-rich blood; the arteries of the pulmonary circuit contain oxygen-poor blood.*

The layers of the heart wall do not obtain nutrients or oxygen from the blood flowing within its chambers; therefore, the heart has its own blood supply, called the *coronary circulation*, to deliver blood to (and from) the heart tissue. Locate the following blood vessels associated with the heart and those serving the myocardium:

superior vena cava—returns blood to right atrium from superior part of systemic circuit

inferior vena cava—returns blood to right atrium from inferior part of systemic circuit

pulmonary trunk—main artery of pulmonary circuit; divides into left and right pulmonary arteries

pulmonary arteries—deliver oxygen-poor blood to lungs

pulmonary veins—return oxygen-rich blood from lungs into left atrium

aorta—main artery of systemic circuit

left coronary artery—originates from base of aorta; has two main branches (figs. 44.2 and 44.5)

- circumflex artery
- anterior interventricular (descending) artery

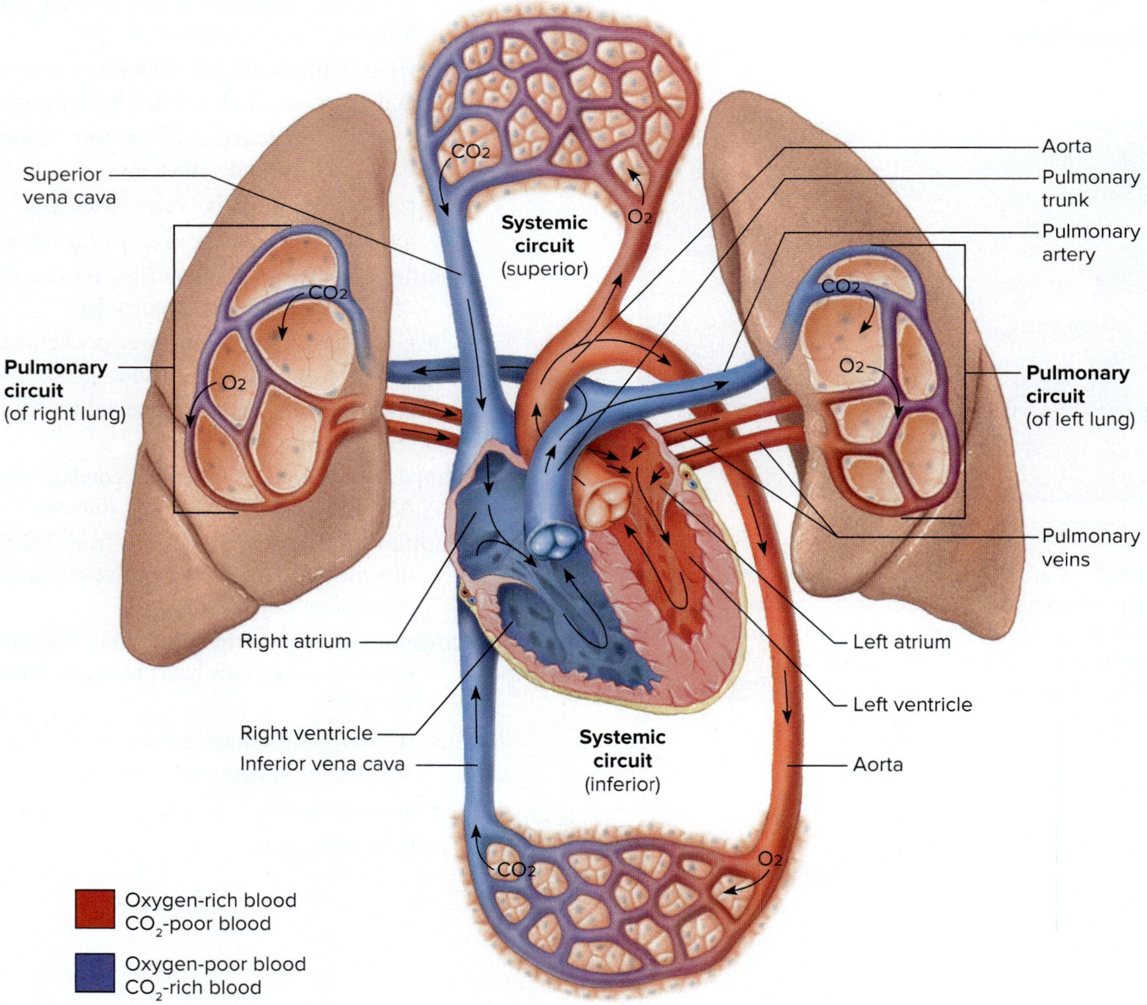

FIGURE 44.8 The heart pumps blood into two major circuits. The left heart side pumps blood into the systemic circuit, serving all body tissues except the lungs. The right heart side pumps blood into the pulmonary circuit, serving both lungs. Red indicates oxygen-rich blood; blue indicates oxygen-poor blood. The arrows indicate the direction of blood flow. APR

right coronary artery—originates from base of aorta; has two main branches (figs. 44.2, 44.3, and 44.5)

- posterior interventricular artery
- right marginal artery

cardiac veins—drain the myocardium into the coronary sinus; include several main veins (figs. 44.1–44.3)

- great cardiac vein
- middle cardiac vein
- small cardiac vein

coronary sinus—for return of blood from cardiac veins into right atrium (fig. 44.3)

6. Complete Parts A and B of Laboratory Assessment 44.

PROCEDURE B: Dissection of a Sheep Heart

The directions and the figures for this procedure are for a sheep heart. The structures of a sheep heart are similar to those of a human heart and will therefore be used for futher exploration. Although sheep are quadrupeds, human terminology will be used for the sheep heart. The directions and the figures would be comparable if another mammal heart were used for the dissection. Other mammal hearts that might be used include pig, cow, and deer. A cadaver observation would enhance this laboratory experience if it could be arranged.

1. Obtain a preserved sheep heart. Rinse it in water thoroughly to remove as much of the preservative as possible. Also run water into the large blood vessels to force any blood clots out of the heart chambers.
2. Place the heart in a dissecting tray with its anterior surface up (fig. 44.9), and proceed as follows:
   a. Although the relatively thick *pericardial sac* probably is missing, look for traces of this membrane around the origins of the large blood vessels. Review figure 44.6 to help you locate some pericardial sac structures.
   b. Locate the *visceral pericardium (epicardium),* which appears as a thin, transparent layer on the surface of the heart. Use a scalpel to remove a portion of this layer and expose the *myocardium* beneath. Also note the abundance of adipose tissue (fat) along the paths of various blood vessels, helping to anchor and protect them. This adipose tissue is found within the loose connective tissue that underlies the visceral pericardium in these areas.
   c. Identify the following:
      right atrium
      right ventricle
      left atrium
      left ventricle
      coronary (atrioventricular) sulcus
      anterior interventricular sulcus
   d. Carefully remove the fat from the anterior interventricular sulcus and expose the blood vessels that pass along this groove. They include a branch of the *left coronary artery* (*anterior interventricular artery*) and a *cardiac vein.*

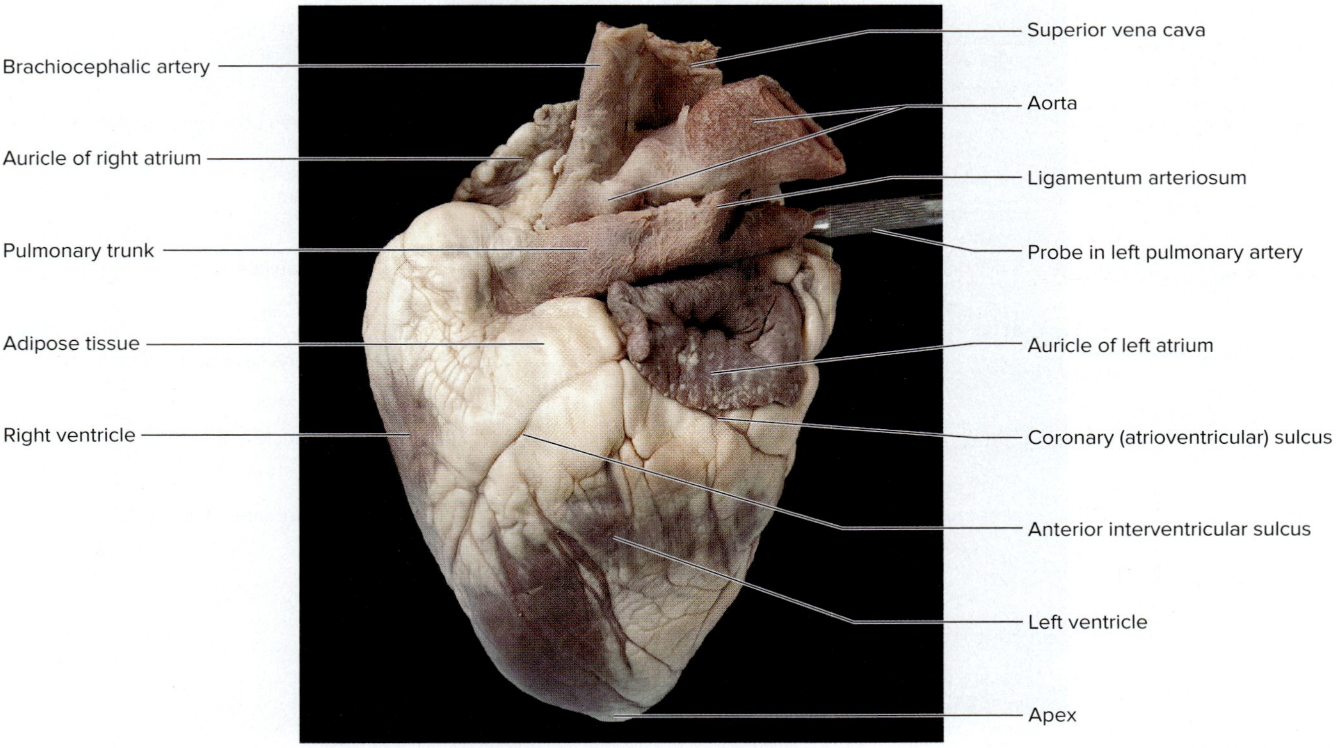

FIGURE 44.9 Anterior surface of sheep heart. The probe passes through the left pulmonary artery and the pulmonary trunk. ©J & J Photography APR

3. Examine the posterior surface of the heart (fig. 44.10), and proceed as follows:
   a. Identify the *coronary sulcus* and the *posterior interventricular sulcus.* The posterior interventricular sulcus contains a *posterior interventricular artery* and a *cardiac vein.*
   b. Locate the stumps of two relatively thin-walled veins that enter the right atrium. Demonstrate this connection by passing a slender probe through them. The upper vessel is the *superior vena cava,* and the lower one is the *inferior vena cava* (fig. 44.10).
4. In order to view the internal heart chambers and heart valves, a frontal (coronal) section of the heart will be prepared using a long knife or a scalpel. Place the base of the heart down in a dissecting tray with the apex pointing upward and the anterior part of the heart facing toward your own body. Make a cut from the apex toward the base while attempting the cut near the center of the heart to separate the heart into equal anterior and posterior portions (fig. 44.11). This cut should expose the four chambers, interventricular septum, and heart valves.
5. Place the anterior portion in the dissecting tray with the anterior surface down, which will allow internal viewing of structures of the anterior portion (section) of the heart (fig. 44.11*a*). Identify as many of the following structures as possible:

   right atrium

   right ventricle

   left atrium

   left ventricle

   interventricular septum

   atrioventricular valves (AV valves)—mitral and tricuspid
   - cusps of valves
   - chordae tendineae
   - papillary muscles

   semilunar valves—aortic and pulmonary
6. Next place the posterior portion in the dissecting tray with the posterior surface down, which will allow internal viewing of structures of the posterior portion (section) of the heart (fig. 44.11*b*). Many of the structures listed in step 5 of Procedure B should also be located. Depending on the major blood vessels that remained on the heart for the frontal section, portions of the stumps of sectioned major blood vessels such as the aorta, pulmonary trunk, brachiocephalic artery, and superior vena cava can be identified (fig. 44.11). Observe the thickness of the wall of the aorta compared to the wall of the pulmonary trunk.
7. Note the muscular ridges and pitted pattern of the *trabeculae carneae* inside the ventricle chambers (fig. 44.11*a*). Examine the muscular ridges of the *pectinate muscles* inside the right atrium (fig. 44.11*b*).
8. Obtain another sheep heart that has not been dissected for making a transverse cut. Place the heart in the dissecting tray with the posterior of the heart down and the anterior of the heart facing upward. Make a transverse cut starting

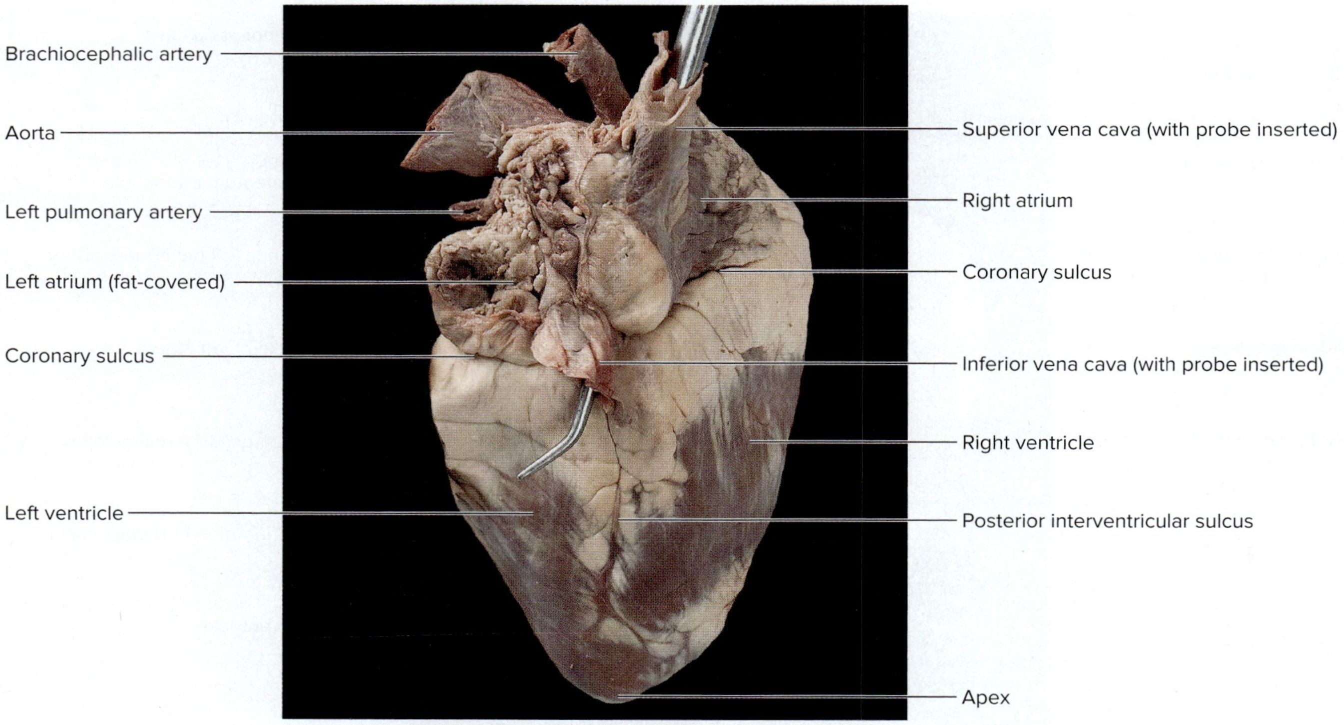

FIGURE 44.10 Posterior surface of sheep heart. The probe passes through the right atrium and the vena cavae.

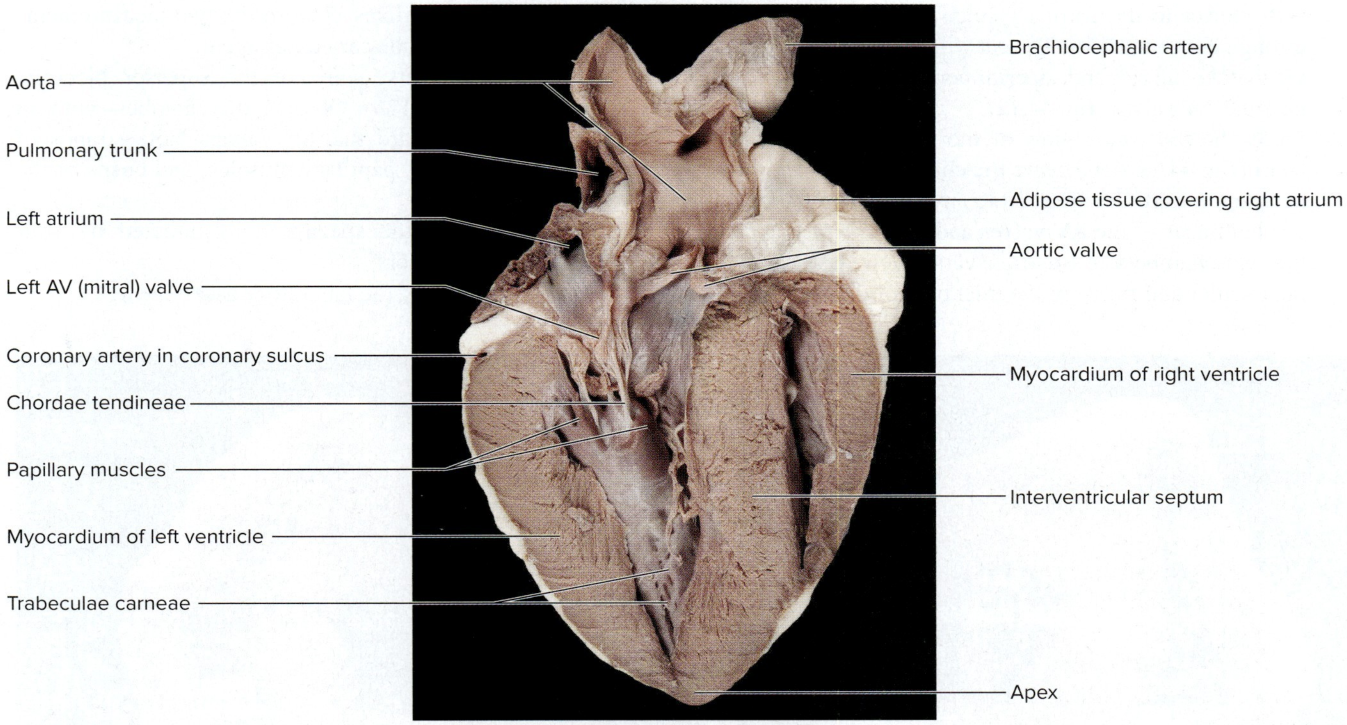

(a) Internal view of anterior portion

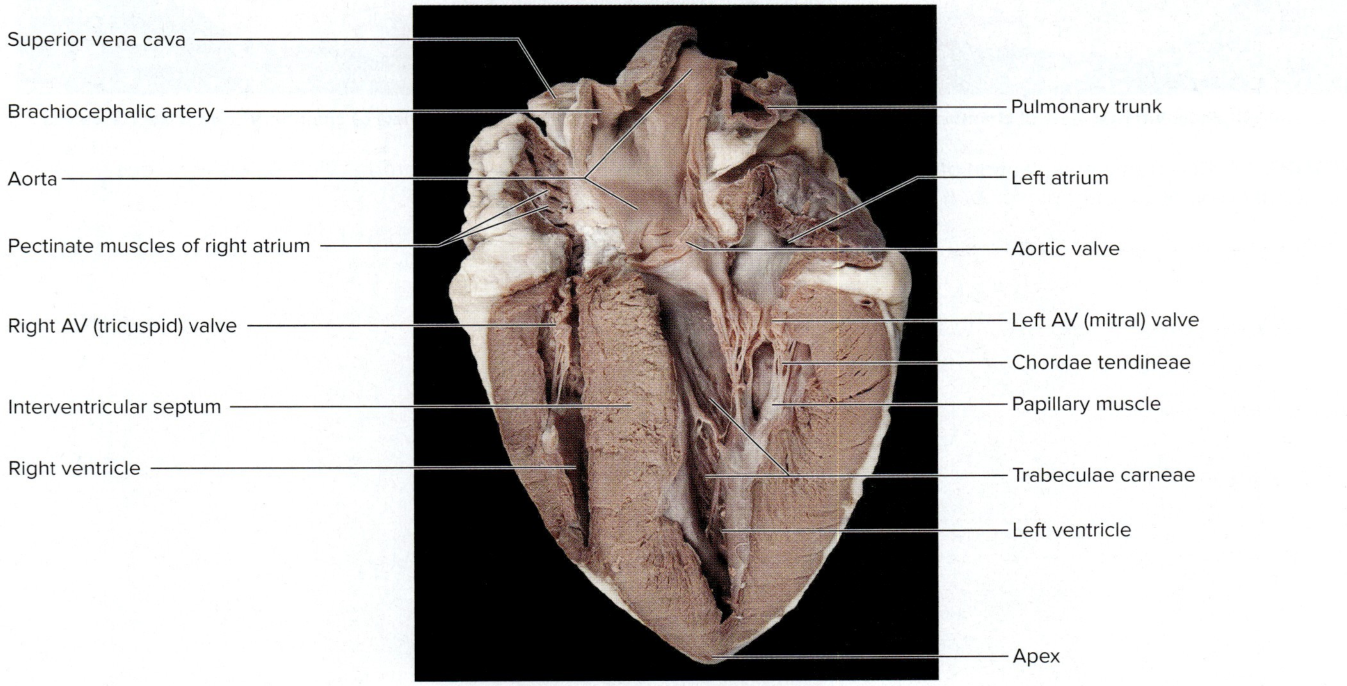

(b) Internal view of posterior portion

FIGURE 44.11 Frontal section of sheep heart with internal views of the chambers, valves, and major blood vessels: (a) anterior portion; (b) posterior portion. ©J & J Photography just inferior to the coronary sulci and continue the cut through both ventricles. The cut will produce a transverse view of the inferior and superior portions of the ventricles near the AV valves (fig. 44.12).

9. Study the transverse view of the inferior piece of the heart (fig. 44.12*a*). Observe the chambers with an internal view toward the apex. Note any strands of the chordae tendineae of the AV valves and the general shape of the two chambers in this transverse view. Use a millimeter ruler and measure the thickness of the wall of the left and right ventricles. Record the two measurements in Part C of the laboratory assessment.
10. Study the transverse view of the superior piece of the heart (fig. 44.12*b*). Observe the chambers with an internal view toward the AV valves. Note strands of chordae tendineae, papillary muscles, and cusps of the AV valves.
11. Discard or save the specimens as directed by your laboratory instructor.
12. Complete Part C of the laboratory assessment.

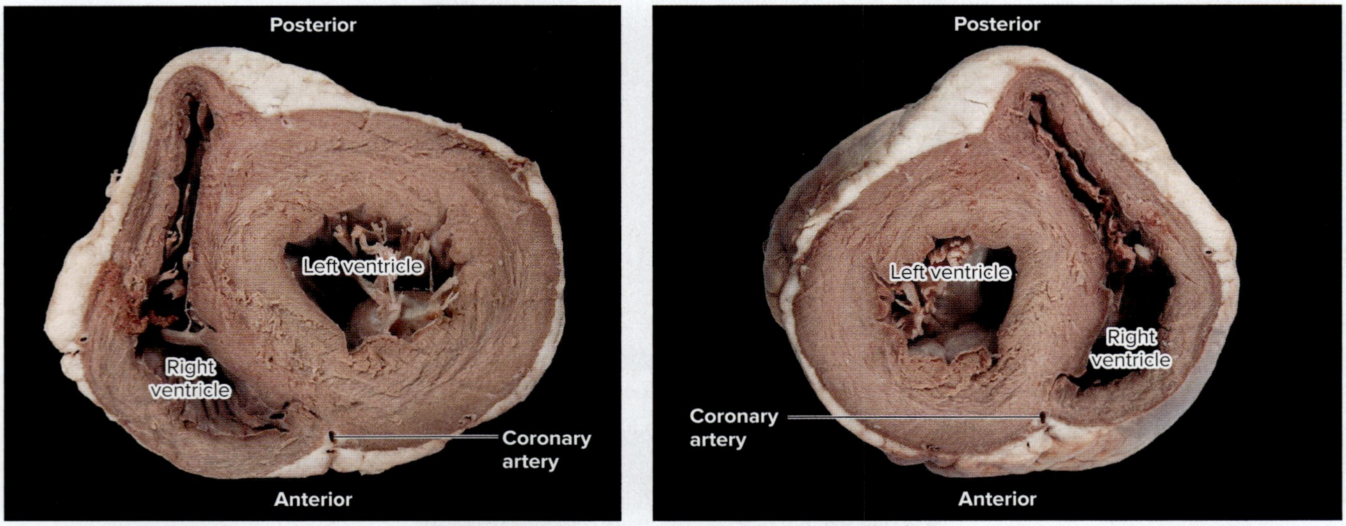

(a) Inferior piece: internal view of chambers toward the apex **(b) Superior piece: internal view of chambers toward the AV valves**

FIGURE 44.12 Transverse section of ventricles of sheep heart near the AV valves: (a) internal view of inferior piece; (b) internal view of superior piece. ©J & J Photography

LABORATORY
ASSESSMENT 44

Name ____________________

Date ____________________

Section ____________________

The (A) corresponds to the indicated Learning Outcome(s) (O) found at the beginning of the Laboratory Exercise.

Heart Structure

PART A: Assessments

Examine the labeled features and identify the numbered features in figures 44.13 to 44.16 of the heart.

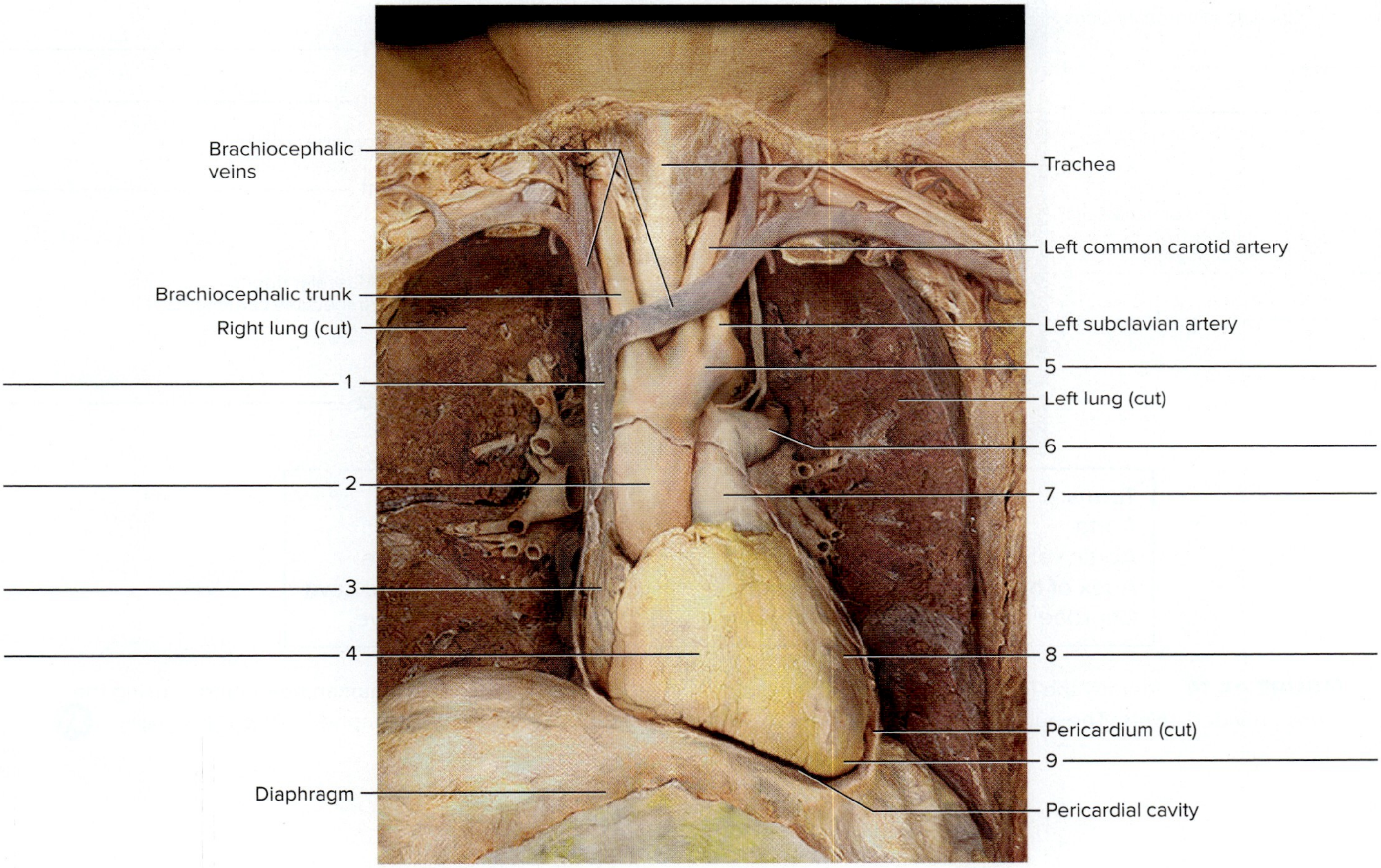

Terms:

Aortic arch	Pulmonary trunk
Apex of heart	Right atrium
Ascending aorta	Right ventricle
Left pulmonary artery	Superior vena cava
Left ventricle	

FIGURE 44.13 Identify the features on this anterior view of the heart region of a cadaver, using the terms provided. APR/McGraw Hill

APR (A1)

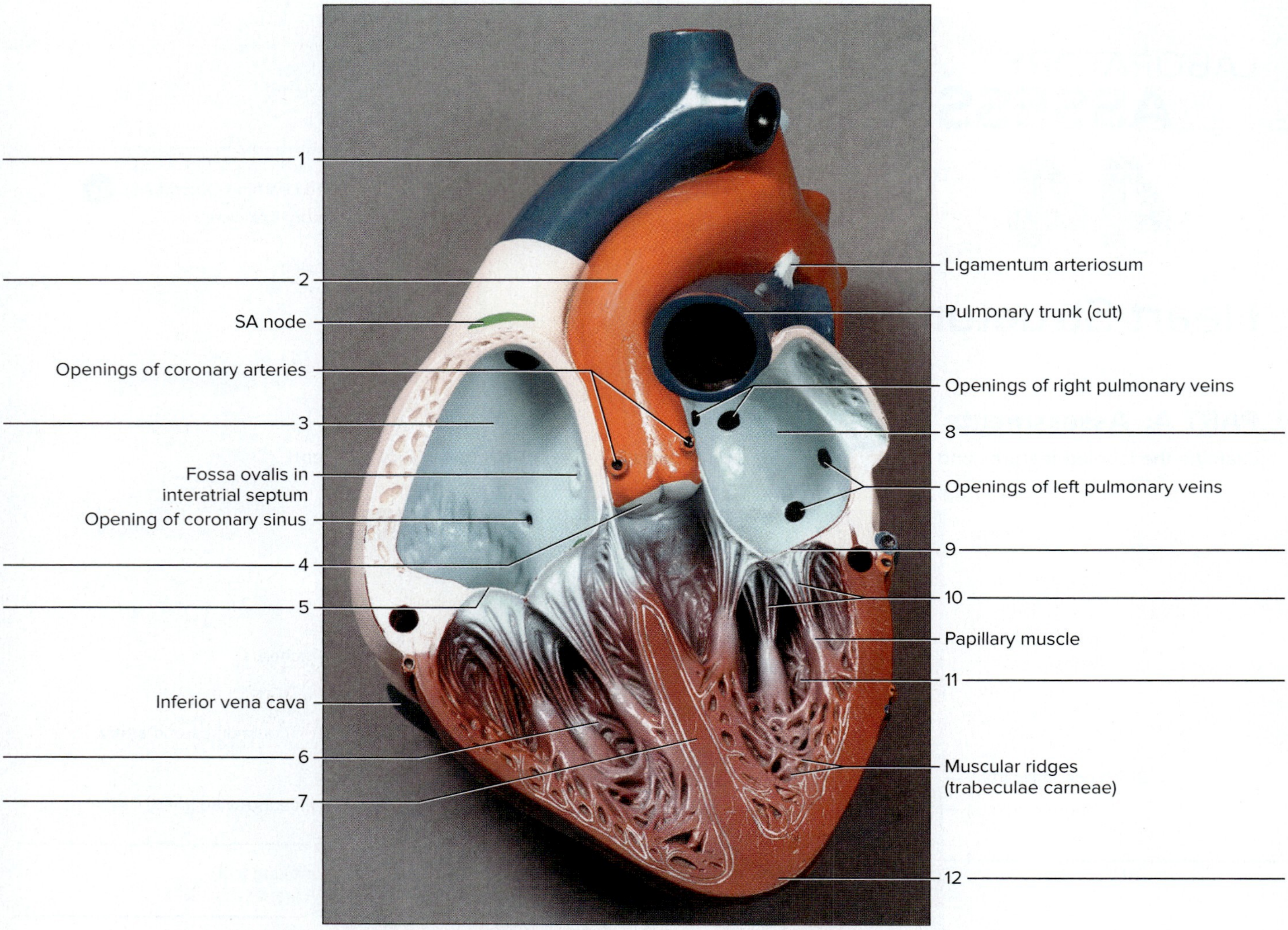

Terms:

Aorta	Interventricular septum	Right atrium
Aortic valve	Left atrium	Right ventricle
Apex of heart	Left ventricle	Superior vena cava
Chordae tendineae	Mitral valve	Tricuspid valve

FIGURE 44.14 Identify the features indicated on this anterior view of a frontal section of a human heart model, using the terms provided. (*Note:* The pulmonary valve is not shown on the portion of the model photographed.) ©J & J Photography A1

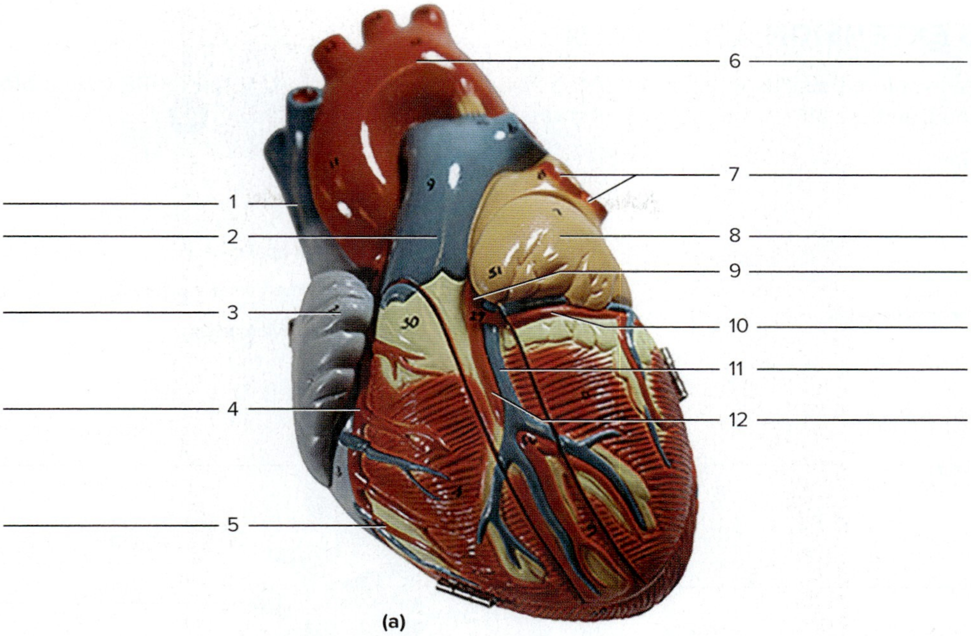

(a)

Terms:

Anterior interventricular artery	Left atrium	Right auricle
Aortic arch	Left coronary artery	Right coronary artery
Circumflex artery	Left pulmonary veins	Right marginal artery
Great cardiac vein	Pulmonary trunk	Superior vena cava

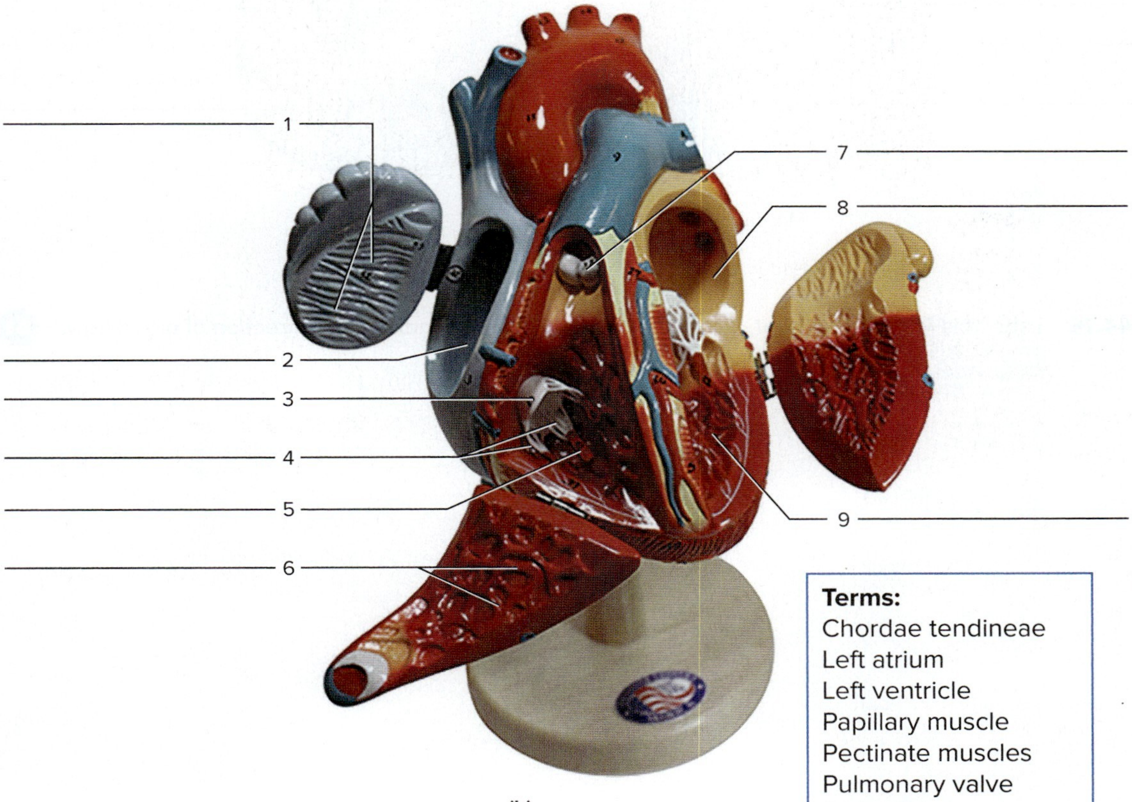

(b)

Terms:
Chordae tendineae
Left atrium
Left ventricle
Papillary muscle
Pectinate muscles
Pulmonary valve
Right atrium
Trabeculae carneae
Tricuspid valve

FIGURE 44.15 Identify the features indicated on this image of a heart model in (a) anterior view of external features and (b) anterior view of internal features, using the terms provided for each. ©2017 Denoyer-Geppert Science Company A1

LEARNING EXTENSION ASSESSMENT

Use red- and blue-colored pencils to color the blood vessels in figure 44.16. Use red to illustrate a blood vessel with oxygen-rich blood, and use blue to illustrate a blood vessel with oxygen-poor blood. A2

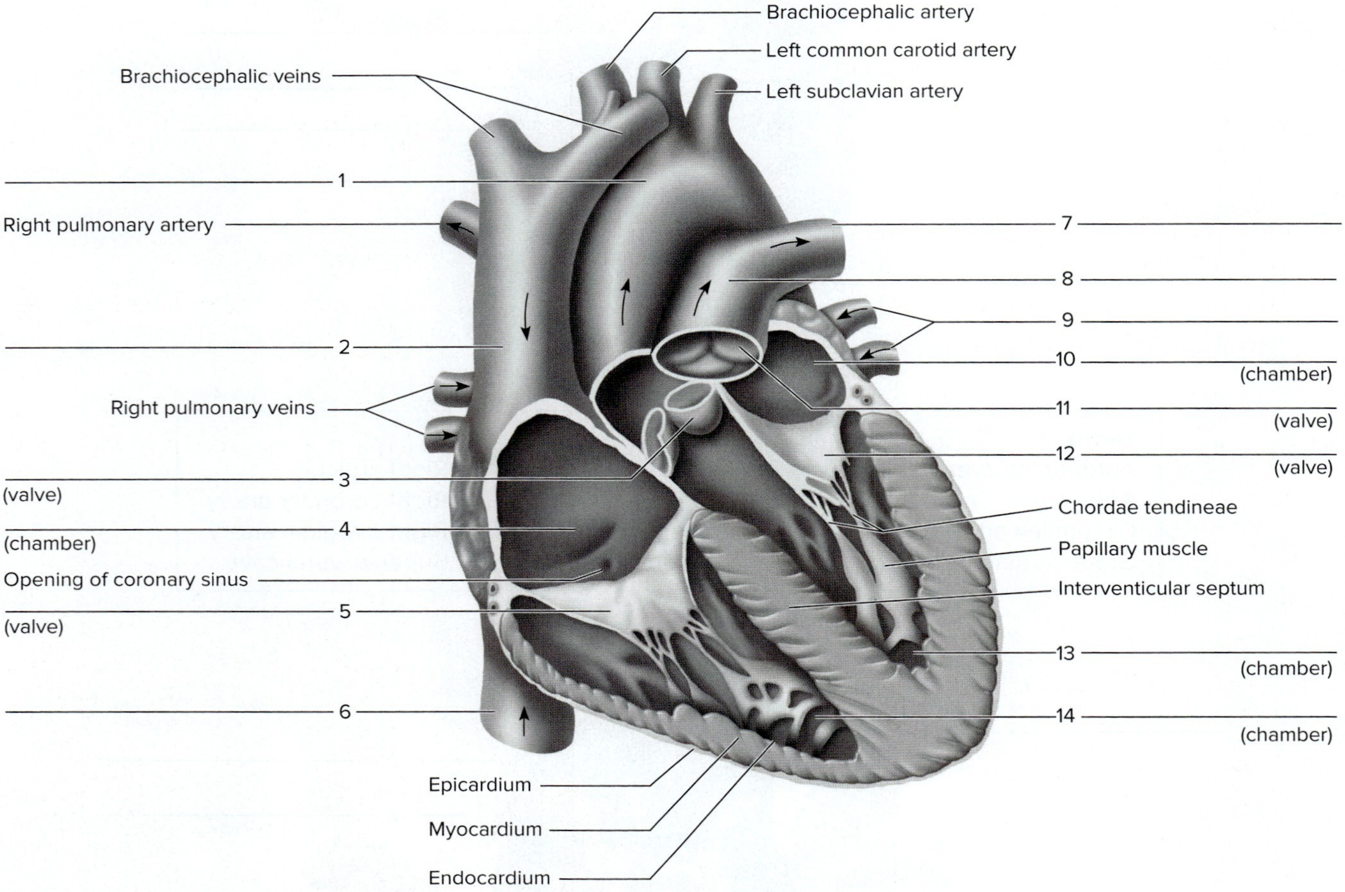

FIGURE 44.16 Label this frontal section of the human heart. The arrows indicate the direction of blood flow. A1

PART B: Assessments

Match the terms in column A with the descriptions in column B. Place the letter of your choice in the space provided. A2 A3

Column A	Column B
a. Aorta	_____ **1.** Structure from which chordae tendineae originate
b. Cardiac vein	_____ **2.** Prevents blood movement from right ventricle to right atrium
c. Coronary artery	_____ **3.** Membranes around heart
d. Coronary sinus	_____ **4.** Prevents blood movement from left ventricle to left atrium
e. Endocardium	_____ **5.** Gives rise to left and right pulmonary arteries
f. Mitral valve	_____ **6.** Drains blood from myocardium into right atrium
g. Myocardium	_____ **7.** Inner lining of heart chamber
h. Papillary muscle	_____ **8.** Layer largely composed of cardiac muscle tissue
i. Pericardial cavity	_____ **9.** Space containing serous fluid to reduce friction during heartbeats
j. Pericardial sac	_____ **10.** Drains blood from myocardial capillaries
k. Pulmonary trunk	_____ **11.** Supplies blood to heart muscle
l. Tricuspid valve	_____ **12.** Distributes blood to body organs (systemic circuit) except lungs

CRITICAL THINKING ASSESSMENTS—PART B

1. Disease of the aortic valve flaps (cusps) (fig. 44.17) causes *stenosis,* the narrowing of the opening through which blood flows. A2 A3

Which ventricle will be most affected, and why? ____

____.

How might the structure of this ventricle change as a result?

____

____.

What symptoms do you expect a person with aortic stenosis to have? ____

____.

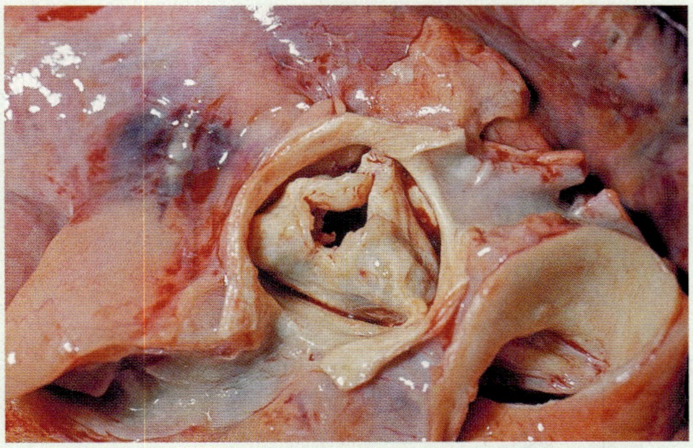

FIGURE 44.17 Diseased aortic valve. Dr. Edwin P. Ewing, Jr./CDC

2. Blood flows down a pressure gradient from high-to-low pressure. Changes in pressure control the opening and closing of the heart valves. Circle the underlined choices that make each statement describing the location and function of the heart valves correct: A2 A3

a. When pressure in the right/left ventricle is greater than the pressure in the right/left atrium, the tricuspid valve opens/closes.

b. When pressure in the right/left ventricle is greater than the pressure in the aorta/pulmonary trunk, the aortic valve opens/closes.

c. When pressure in the right/left atrium is greater than the pressure in the right/left ventricle, the mitral valve opens/closes.

d. When pressure in the aorta/pulmonary trunk is greater than the pressure in the right/left ventricle, the pulmonary valve opens/closes.

PART C: Assessments

Complete the following:

1. Compare the structure of the left atrioventricular valve with that of the aortic valve. A4 ______________________________

2. Describe the functions of the chordae tendineae and the papillary muscles. A4 ______________________________

3. What is the functional significance of the difference in thickness between the wall of the aorta and the wall of the pulmonary trunk? A4 ______________________________

4. List the correct pathway through which blood must flow in relation to regions of the heart. Assume blood is currently in the vena cava and will eventually enter the aorta. Include the major blood vessels (arteries and veins) attached directly to a heart chamber, the four heart chambers, and the four heart valves in your list. A2 A4 ______________________________

5. Describe the general overall shape of the left and right chambers of the ventricles as observed in a transverse section of the heart. A4 ______________________________

6. What was the measured thickness of the left ventricle wall? ______________________________
   What was the measured thickness of the right ventricle wall? A4 ______________________________

CRITICAL THINKING ASSESSMENT—PART C

Explain the functional significance of the different thicknesses of the ventricular walls. A4 ______________________________.

LABORATORY
EXERCISE 45

Cardiac Cycle

Purpose of the Exercise

To review the events of a cardiac cycle, to become acquainted with normal heart sounds, and to record an electrocardiogram.

MATERIALS NEEDED

For Procedure A—Heart Sounds:

Stethoscope

Alcohol swabs

For Procedure B—Electrocardiogram:

Electrocardiograph (or other instrument for recording an ECG)

Cot or table

Alcohol swabs

Electrode paste (gel)

Electrodes and cables

SAFETY

- Review all the safety guidelines in Appendix 1 of your laboratory manual.
- Wear proper personal protection equipment when working with others.

Learning Outcomes APR

After completing this exercise, you should be able to:

O1 Interpret the major events of a cardiac cycle.

O2 Associate the sounds produced during a cardiac cycle with the valves closing.

O3 Correlate the structures and components of a normal ECG pattern with the phases of a cardiac cycle.

O4 Record and interpret an electrocardiogram.

O5 Integrate the electrical changes and the heart sounds with the cardiac cycle.

The O corresponds to the assessments A indicated in the Laboratory Assessment for this Exercise.

Pre-Lab

Carefully read the introductory material and examine the entire lab. Be familiar with the cardiac cycle, heart sounds, cardiac conduction system, and electrocardiograms from lecture or the textbook. Answer the pre-lab questions.

Pre-Lab Questions Select the correct answer for each of the following questions:

1. The _____ of the conduction system is known as the pacemaker.
a. SA node **b.** AV node
c. AV bundle **d.** left bundle branch

2. The _____ of the conduction system is/are located throughout the ventricular walls.
a. AV node **b.** left bundle branch
c. right bundle branch **d.** Purkinje fibers

3. The first heart sound (S1; lubb) occurs when the
a. AV valves open. **b.** AV valves close.
c. semilunar valves open. **d.** semilunar valves close.

4. One cardiac cycle would consist of
a. left chamber contractions followed by right chamber contractions.
b. right chamber contractions followed by left chamber contractions.
c. atrial chamber contractions followed by ventricular chamber contractions.
d. ventricular chamber contractions followed by atrial chamber contractions.

5. The sinoatrial (SA) node of the heart is located in the
a. right atrium. **b.** left atrium.
c. left ventricle. **d.** right ventricle.

6. The depolarization of ventricular fibers is indicated by the _____ of an ECG.
   a. P wave b. QRS complex
   c. T wave d. P-Q interval
7. During ventricular _____, the chambers fill with blood.
   a. diastole b. systole
   c. contraction d. ejection
8. The dupp sound occurs when the semilunar valves are closing during ventricular diastole.
   a. True b. False
9. The P wave of an ECG occurs during the repolarization of the atria.
   a. True b. False

A set of atrial contractions while the ventricular walls relax, followed by ventricular contractions while the atrial walls relax, constitutes a *cardiac cycle*. Each cycle generates blood pressure changes within the heart chambers, resulting in the ejection of blood. Movement only occurs in one direction due to the one-way valves found between the superior and inferior chambers (atrioventricular valves) and between the ventricles and the large arteries of the circulatory circuits (semilunar valves). As the valves close, blood hits the tissues and the vibrations create the sounds associated with the heartbeat. If backflow of blood occurs when a heart valve is closed, it creates a turbulence noise known as a murmur.

The regulation and coordination of the cardiac cycle involves the *cardiac (intrinsic) conduction system*. The pathway of electrical signals originates from the *sinoatrial (SA) node* located in the right atrium near the entrance of the superior vena cava. Because the stimulation of the heartbeat and the heart rate originate from the SA node, it is often called the *pacemaker*. As the signals pass through the atrial walls toward the *atrioventricular (AV) node*, contractions of the atria followed by relaxations take place. When the AV node has been signaled, the rapid continuation of the electrical signals occurs through the *AV bundle* and the *right and left bundle branches* within the interventricular septum and terminates via the *subendocardial conducting network (Purkinje fibers)* throughout the ventricular walls. Once the myocardium of the ventricles has been stimulated, ventricular contractions followed by relaxations occur, completing one cardiac cycle. During relaxation, the ventricles fill with blood for the next cycle.

The sympathetic and parasympathetic subdivisions of the autonomic nervous system influence the activity of the pacemaker under various conditions. An increased rate results from sympathetic responses; a decreased rate results from parasympathetic responses.

A number of electrical changes also occur in the myocardium as it contracts and relaxes. These changes can be detected by using electrodes placed on the skin surface and an instrument called an *electrocardiograph.* The recording produced by the instrument is an *electrocardiogram,* or *ECG (EKG). Depolarization* and *repolarization* electrical events of the cardiac cycle can be observed and interpreted from the ECG graphic recording. Clinicians can compare a subject's ECG recording to normal ECG patterns to get a noninvasive view of cardiac function. The heart rate can be determined by counting the number of QRS complexes on the ECG during 1 minute.

PROCEDURE A: Heart Sounds APR

1. Listen to your heart sounds. To do this, follow these steps:
   a. Obtain a stethoscope and clean its earpieces and the diaphragm using alcohol swabs.
   b. Fit the earpieces into your ear canals so that the angles are positioned in the forward direction.
   c. Firmly place the diaphragm (bell) of the stethoscope on the chest over the fifth intercostal space near the apex of the heart (fig. 45.1) and listen to the sounds. This is a good location to hear the first sound (*lubb*) of a cardiac cycle when the AV valves are closing, which occurs during ventricular *systole* (contraction).
   d. Move the diaphragm to the second intercostal space, just to the left of the sternum, and listen to the sounds from this region. You should be able to hear the second sound (*dupp*) of the cardiac cycle clearly when the semilunar valves are closing, which occurs during ventricular *diastole* (relaxation).
   e. It is possible to hear sounds associated with the aortic and pulmonary valves by listening from the second intercostal space on either side of the sternum. The aortic valve sound comes from the right and the pulmonary valve sound comes from the left. The sound associated with the mitral valve can be heard from the fifth intercostal space at the nipple line on the left. The sound of the tricuspid valve can be heard at the fifth intercostal space just to the left of the sternum (fig. 45.1). The four points for listening to (auscultating) the four heart valve sounds when they close are not directly superficial to the valve locations. This is because the sounds associated with the closing valves travel in a slanting direction to the surface of the chest wall where the stethoscope placements occur.
2. Inhale slowly and deeply and exhale slowly while you listen to the heart sounds from each of the locations as before. Note any changes that have occurred in the sounds.
3. Exercise moderately outside the laboratory for a few minutes so that other students listening to heart sounds will not be disturbed. After the exercise period, listen to the heart sounds and note any changes that have

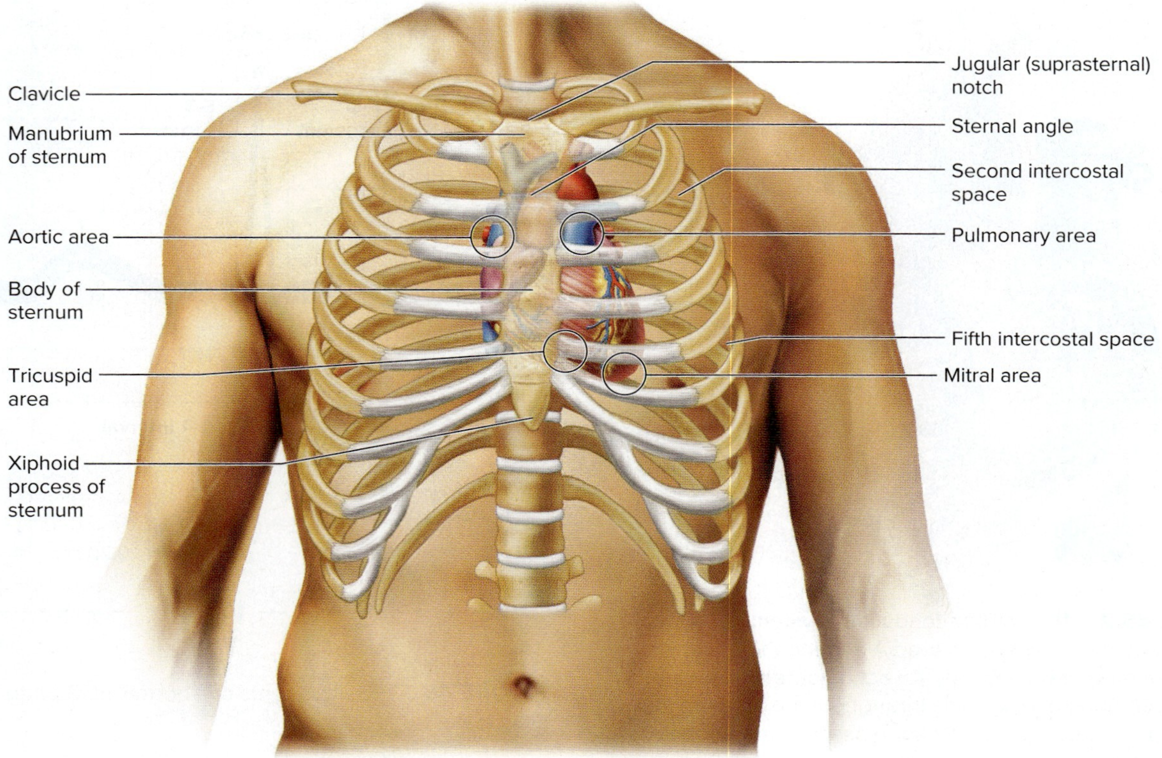

FIGURE 45.1 The first sound (lubb) of a cardiac cycle can be heard by placing the diaphragm of a stethoscope over the fifth intercostal space near the apex of the heart. The second sound (dupp) can be heard over the second intercostal space, just left of the sternum. The thoracic regions circled indicate where the sounds of each heart valve are most easily heard.

occurred in them. *This exercise should be avoided by anyone with health risks.*

4. Complete Parts A and B of Laboratory Assessment 45.

PROCEDURE B: Electrocardiogram

A single heartbeat is a sequence of precisely timed contractions and relaxations of the four heart chambers initiated from the cardiac conduction system (fig. 45.2). The basic rhythm is controlled from the SA node (pacemaker), but it can be modified by the autonomic nervous system. During atrial contraction (systole), the ventricles are in a state of relaxation (diastole) and are filling with blood forced into them from the atria. Atrial contraction is soon followed by ventricular contraction (systole), during which time atrial relaxation (diastole) occurs. As a result of the synchronized ventricular contractions, blood is forced into the two large arteries that carry blood away from the heart. The aorta receives blood from the left ventricle into the systemic circuit; the pulmonary trunk receives blood from the right ventricle into the pulmonary circuit. This entire sequence of events of a single heartbeat is the cardiac cycle.

Due to the action of the cardiac conduction system, the heart contracts in a coordinated manner. The events of the cardiac conduction system and the cardiac cycle form the basis for the electrocardiogram. The appearance of a normal ECG pattern for one heartbeat is shown in figure 45.3. The ECG components, normal time durations, and corresponding significance of the waves and intervals are placed in table 45.1. An illustration of the relationship of the ECG components with the depolarization and repolarization of the atria followed by the ventricles is presented in figure 45.4.

1. Study figures 45.2 to 45.4, and table 45.1.
2. Complete Parts C and D of the laboratory assessment.
3. The laboratory instructor will demonstrate the proper adjustment and use of the instrument available to record an electrocardiogram.

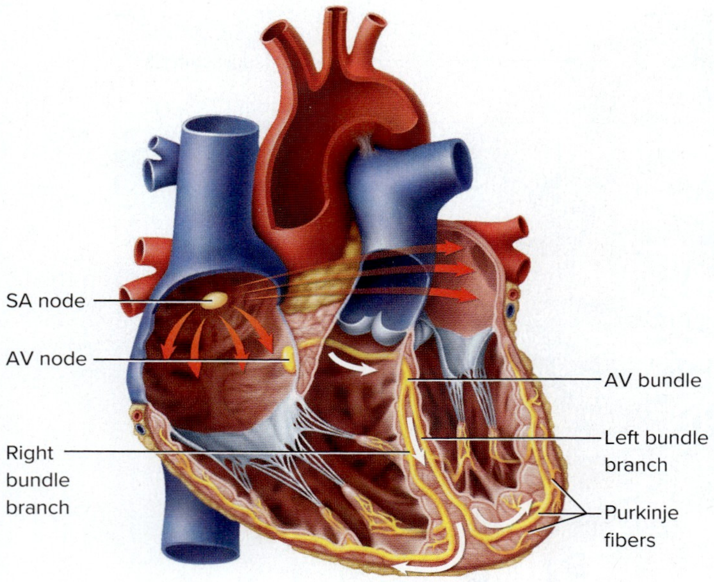

FIGURE 45.2 The cardiac conduction system pathway, indicated by the arrows, from the SA node to the Purkinje fibers (frontal section of heart). An action potential originates from the SA node and spreads throughout the atria to the AV node. Conduction from the AV node passes through the AV bundle and spreads along bundle branches within the interventricular septum to the network of Purkinje fibers within the ventricle walls. APR

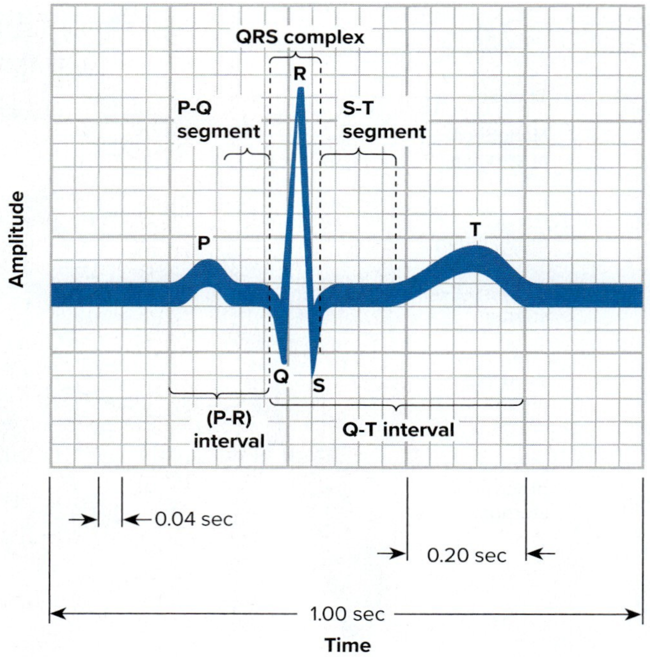

FIGURE 45.3 Components of a normal ECG pattern with a time scale.

Table 45.1 ECG Components, Durations, and Significance

ECG Component	Duration in Seconds	Corresponding Significance
P wave	0.06–0.11	Depolarization of atrial fibers
P-Q segment	0.08	Time for atria to contract
P-R interval	0.12–0.20	Time for cardiac impulse from SA node through AV node
QRS complex	<0.12	Depolarization of ventricular fibers
S-T segment	0.12	Time for ventricles to contract and eject blood; corresponds to myocardial plateau of action potential
Q-T interval	0.36–0.44	Time from ventricular depolarization to end of ventricular repolarization; length varies according to heart rate
T wave	0.16	Repolarization of ventricular fibers (ends pattern)

Note: Atrial repolarization is not observed in the ECG because it occurs at the same time that ventricular fibers depolarize. Therefore, it is obscured by the QRS complex.

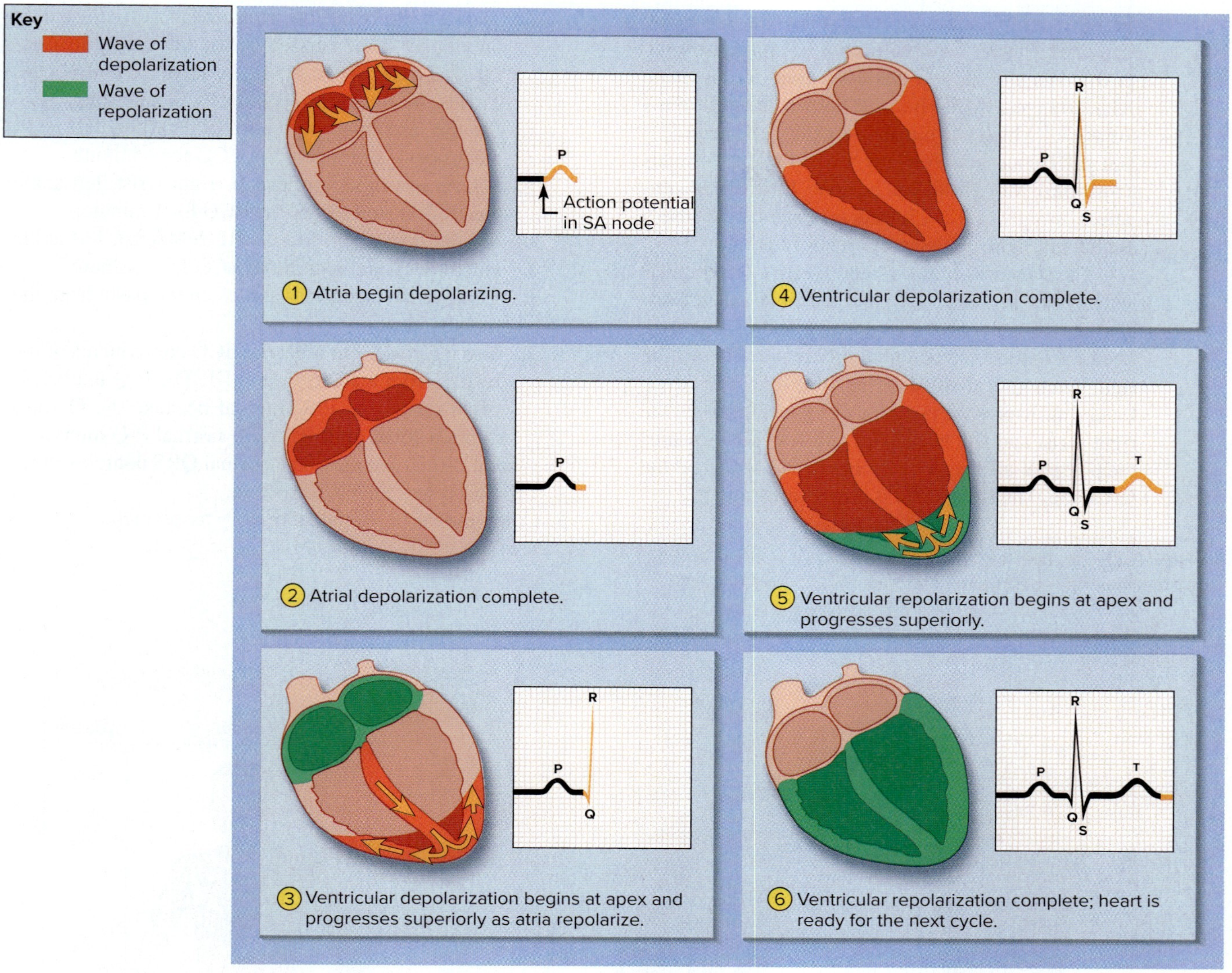

FIGURE 45.4 Depolarization waves (red) followed by repolarization waves (green) of the atria and ventricles during a cardiac cycle, along with the representative segment of the ECG pattern. The yellow arrows indicate the direction the wave is traveling.

4. Record your laboratory partner's ECG. To do this, follow these steps:
   a. Have your partner lie on a cot or table close to the electrocardiograph, remaining as relaxed and still as possible.
   b. Scrub the electrode placement locations with alcohol swabs (fig. 45.5). Apply a small quantity of electrode paste (if necessary) to the skin on the insides of the wrists and ankles. (Any jewelry on the wrists or ankles should be removed.)
   c. Spread some electrode paste (if necessary) over the inner surfaces of four plate electrodes and attach one to each of the prepared skin areas (fig. 45.5). Make sure there is good contact between the skin and the metal of the electrodes. The electrode on the right ankle is the grounding system.
   d. Attach the electrodes to the corresponding cables of the ECG recording instrument. When an ECG recording is made, only two electrodes are used at a time, and the instrument selector switch allows

various combinations of electrodes (leads) to be activated. Three standard limb leads placed on the two wrists and the left ankle are used for an ECG. This arrangement has become known as *Einthoven's triangle,*[1] which permits the recording of the potential difference between any two of the electrodes (fig. 45.6). More leads can be done. In a clinical situation, usually a 12–15 lead arrangement is used, which gives the clinician more information.

The standard leads I, II, and III are called bipolar leads because they are the potential difference between two electrodes (a positive and a negative). Lead I measures the potential difference between the right wrist (negative) and the left wrist (positive). Lead II measures the potential difference between the right wrist and the left ankle, and lead III measures the potential difference between the left wrist and the left ankle. The right ankle is always the ground.

[1]Willem Einthoven (1860–1927), a Dutch physiologist, received the Nobel prize for Physiology or Medicine for his work with electrocardiograms.

e. Turn on the recording instrument and adjust it as previously demonstrated by the laboratory instructor. The paper speed should be set at 2.5 cm/second. This is the standard speed for ECG recordings.

f. Set the instrument to lead I (right wrist, left wrist electrodes), and record the ECG for 1 minute.

g. Set the instrument to lead II (right wrist, left ankle electrodes), and record the ECG for 1 minute.

h. Set the instrument to lead III (left wrist, left ankle electrodes), and record the ECG for 1 minute.

i. Remove the electrodes and clean the paste from the metal and skin.

j. Use figure 45.3 to label the ECG components of the results from leads I, II, and III. The P-Q interval is often called the P-R interval because the Q wave is often small or absent. The normal P-Q interval is 0.12–0.20 seconds. The normal QRS complex duration is less than 0.10 second.

5. Complete Part E of the laboratory assessment.

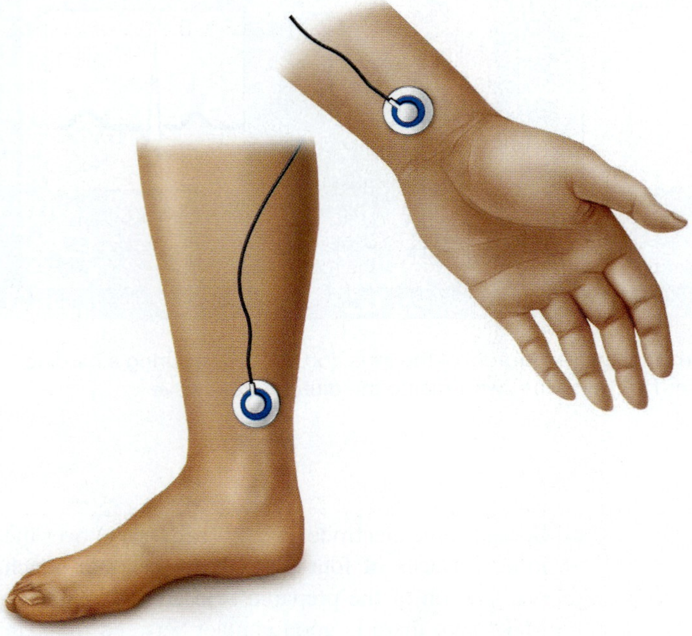

FIGURE 45.5 To record an ECG, attach electrodes to the wrists and ankles.

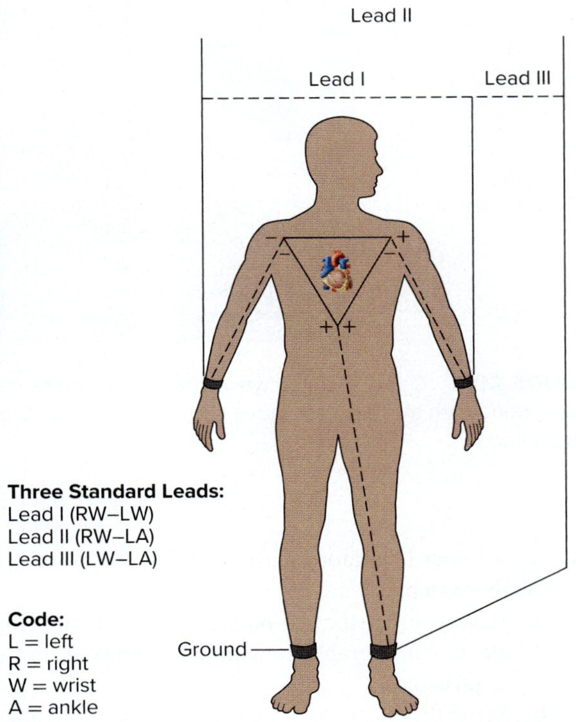

FIGURE 45.6 Standard limb leads for electrocardiograms.

LABORATORY
ASSESSMENT 45

Name ______________________________

Date ______________________________

Section ______________________________

The (A) corresponds to the indicated Learning Outcome(s) (O) found at the beginning of the Laboratory Exercise.

Cardiac Cycle

PART A: Assessments

Complete the following statements:

1. The period during which a heart chamber is contracting is called ______________________. A1
2. The period during which a heart chamber is relaxing is called ______________________. A1
3. During ventricular contraction, the AV valves (tricuspid and mitral valves) are ______________________. A1
4. During ventricular relaxation, the AV valves are ______________________. A1
5. The pulmonary and aortic valves open when the pressure in the ______________________ exceeds the pressure in the pulmonary trunk and aorta. A1
6. Heart sounds are due to ______________________ as blood hits the heart tissues when the valves close. A2
7. The first sound of a cardiac cycle occurs when the ______________________ are closing. A2
8. The second sound of a cardiac cycle occurs when the ______________________ are closing. A2
9. The sound created when blood leaks back through an incompletely closed valve is called a(an) ______________________. A2

PART B: Assessments

Complete the following:

1. What changes did you note in the heart sounds when you inhaled deeply? A2

____

____

2. What changes did you note in the heart sounds following the exercise period? A2

____

____

PART C: Assessments

Complete the following statements:

1. Normally, the ______________________ node serves as the pacemaker of the heart. A3
2. Stimulation of the SA node by the ______________________ nervous system can modify the heart rate. A3
3. The ______________________ node is located in the inferior portion of the interatrial septum. A3

4. The network of fibers that carry cardiac impulses from the interventricular septum into the myocardium are called ____________________ fibers. A3

5. A(n) ____________________ is a recording of electrical changes occurring in the myocardium during a cardiac cycle. A3

6. The P wave corresponds to depolarization of the muscle fibers of the ____________________. A3

7. The QRS complex corresponds to depolarization of the muscle fibers of the ____________________. A3

8. The T wave corresponds to repolarization of the muscle fibers of the ____________________. A3

9. The time when the ventricles contract and eject blood is the ____________________. A3

10. Why is atrial repolarization not observed in the ECG? A3

____

____

CRITICAL THINKING ASSESSMENT—PART C

If a person has a heart murmur caused by the improper opening and closing of a heart valve, the ECG *or* heart sounds (choose one) will be unusual as the blood is forced through a more narrow opening (stenotic valve), or the blood backflows because the valve does not close properly (incompetent valve). A2

Explain your reasoning for the answer choice. ____

____

____.

PART D: Assessments

Identify the heart chambers and conduction system structures in the frontal section of the heart in figure 45.7.

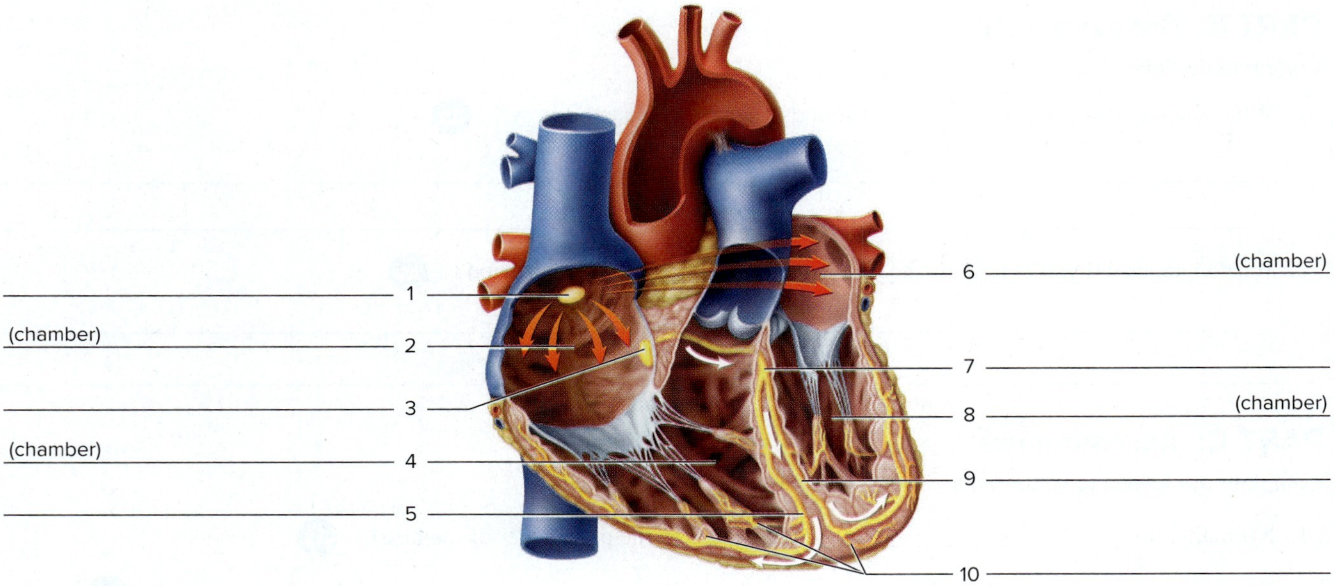

FIGURE 45.7 Label the indicated heart chambers and conduction system structures (anterior view). A3

PART E: Assessments

1. Attach a short segment of the ECG recording from each of the three leads you used, and label the waves of each. A4

Lead I

Lead II

Lead III

2. What differences do you find in the ECG patterns of these leads? A4 ________________________________

____

____

____

____

____

3. How much time passed from the beginning of the P wave to the beginning of the QRS complex (P-Q interval, or P-R interval) in the ECG from lead I? A4 ________________________________

4. What is the significance of this P-R interval? A4 ________________________________

____

____

____

____

5. How can you determine heart rate from an electrocardiogram? A4 ________________________________

____

____

____

6. What was your heart rate as determined from the ECG? A4 ________________________________

CRITICAL THINKING ASSESSMENTS—PART E

1. If a person's heart rate was 72 beats per minute, determine the number of QRS complexes that would have appeared on an ECG during the first 30 seconds. A4 ________________________________

________________________________.

2. Answer the following questions regarding heart sounds and an ECG:

   a. As blood in the ventricles surges back against the closed AV valves, the ______________ heart sound ("lubb") can be heard. Why does this heart sound occur near the middle of the QRS complex in an ECG? Explain. A5

   ____

   ____

   ________________________________.

   b. As blood in the great arteries ricochets back against the closed semilunar valves, the ______________ heart sound ("dupp") can be heard. Why does this heart sound occur toward the end of the T wave in an ECG? Explain. A5

   ____

   ____

   ________________________________.

LABORATORY
EXERCISE 46

Electrocardiography: BIOPAC® Exercise

Purpose of the Exercise

To record an ECG and calculate heart rate and rhythm changes that occur with changes in body position and breathing.

MATERIALS NEEDED

Computer system [Mac OS X 10.7–10.15 (Catalina), or PC running Windows 10, 8, or 7]

Biopac Student Lab software version 4.0 or above

BIOPAC Acquisition Unit MP36, MP35, or MP46/45

BIOPAC electrode lead set (SS2L)

BIOPAC disposable vinyl electrodes (EL503), three electrodes per subject

BIOPAC electrode gel (GEL1) and abrasive pad (ELPAD) or alcohol wipes

Exercise mat, cot, or lab table with pillow

Chair or stool

SAFETY

- The BIOPAC electrode lead set is safe and easy to use and should be used only as described in the procedures section of the laboratory exercise.
- The electrode lead clips are color-coded. Make sure they are connected to the properly placed electrodes as demonstrated in figure 46.3.
- The vinyl electrodes are disposable and are meant to be used only once. Each subject should use a new set.

Learning Outcomes

After completing this exercise, you should be able to:

O1 Record the electrical components of an electrocardiogram.

O2 Correlate electrical events as displayed on the ECG with the mechanical events that occur during the cardiac cycle.

O3 Interpret rate and rhythm changes in the ECG associated with changes in body position, activity level, and breathing.

The O corresponds to the assessments A indicated in the Laboratory Assessment for this Exercise.

Pre-Lab

Carefully read the introductory material and examine the entire lab content. Be familiar with the cardiac cycle, cardiac conduction system, and electrocardiogram from lecture or the textbook. Also, if this is the first BIOPAC lab that you are performing, please read the general description of the Biopac Student Lab system in Exercise 21.

The beating of the heart is one of the basic requirements for maintaining life, pumping blood that contains vital oxygen and nutrients to all the active cells of the body. A single heartbeat is a coordinated sequence of events, involving precisely timed contractions and relaxations of the four chambers of the heart. During a heartbeat, the atria, the superior chambers of the heart, contract simultaneously. This action squeezes the blood from the atria into the ventricles, the inferior chambers of the heart. During atrial contraction (*systole*), the ventricles are in a state of relaxation (*diastole*) and are filling with blood from the contracting atria. Atrial contraction is soon followed by ventricular contraction, the simultaneous contraction of the two large pumping chambers of the heart. After a short period in which all four heart chambers are relaxed, these events repeat. This entire sequence of events occurs during every heartbeat and is called the *cardiac cycle*.

The heart is a large, cone-shaped pump, consisting mainly of cardiac muscle, one of the three types of muscle tissue in the body. Cardiac muscle differs from skeletal muscle in that it is controlled by specialized cardiac muscle cells within the heart, rather than by somatic motor neurons. The action of cardiac muscle can be modified by autonomic nerve input, but its basic rhythm is set by specialized cells of the heart, called *pacemaker cells,* found in the small clumps (nodes) of the cardiac conduction system.

The *cardiac conduction system* is a specialized group of cardiac muscle cells that generates and spreads cardiac impulses throughout the heart. Following are the major components of the cardiac conduction system:

1. **Sinoatrial (SA) node:** This small clump of specialized cardiac muscle cells in the upper wall of the right atrium is called the *pacemaker* of the heart. The SA node sets the rhythm of the heart, since it has the fastest innate firing rate of any conducting cells of the heart. It spreads its impulses across the atria to initiate atrial contraction, and to the next component of the cardiac conduction system, the AV node.
2. **Atrioventricular (AV) node:** This tiny clump of conducting cardiac muscle cells in the inferior portion of the right atrium receives impulses from the SA node and transports them to the next structure of the conduction system, the AV bundle. The AV node serves to delay the signal from the SA node for about 0.1 seconds before sending it on to the AV bundle. This gives the atria time to complete their contraction and squeeze the remaining blood into the ventricles before the ventricles begin to contract.
3. **Atrioventricular bundle (AV bundle; bundle of His):** This group of conducting cells at the superior end of the interventricular septum transports impulses from the AV node to the next structures of the conduction system, the left and right bundle branches. This structure distributes impulses toward the left and right ventricles.
4. **Left and right bundle branches:** The bundle branches transport cardiac impulses down the interventricular septum toward the apex of the heart. From these branches, the Purkinje fibers (subendocardial conducting network) diverge.
5. **Purkinje fibers:** Purkinje fibers are large cardiac muscle fibers that split off from the bundle branches and transport cardiac impulses throughout the entire myocardium of the ventricles. These fibers are considered the final component of the cardiac conduction system. From here, the impulses are spread from one contractile cardiac muscle cell to the next via gap junctions in their cell membranes.

Figure 46.1 shows the arrangement of the cardiac conduction system and the locations of its components.

Due to the action of the cardiac conduction system, the heart contracts in a coordinated manner. The atria contract just before the ventricles, providing the ventricles with time to fill completely, before starting their own contraction. This increases the efficiency of the heart as a pump and ensures that the heart contracts as a functional unit almost every time.

The events of the cardiac conduction system and the cardiac cycle form the basis for this lab exercise. An *electrocardiogram* (ECG or EKG) is a recording of the electrical activity derived from the heart during the cardiac cycle. Similar to the EMG and EEG that you studied in previous lab exercises, the ECG is derived from the sum of the ionic movements that occur during the action potentials of many cardiac muscle cells at the same time. It is not a recording of action potentials in a single cardiac muscle cell. The movement of ions in the extracellular fluids generates current flow, which can be detected by electrodes placed on the skin. ECGs are used not only to observe normal heart action, but

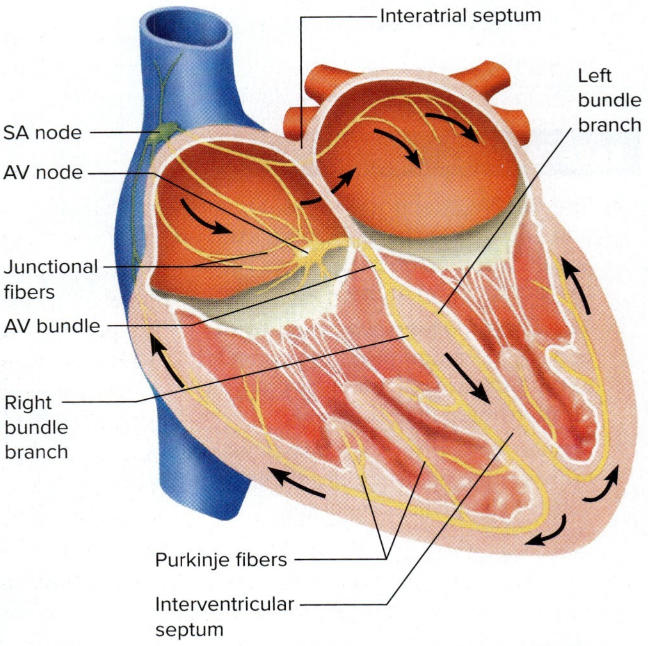

FIGURE 46.1 The components of the cardiac conduction system.

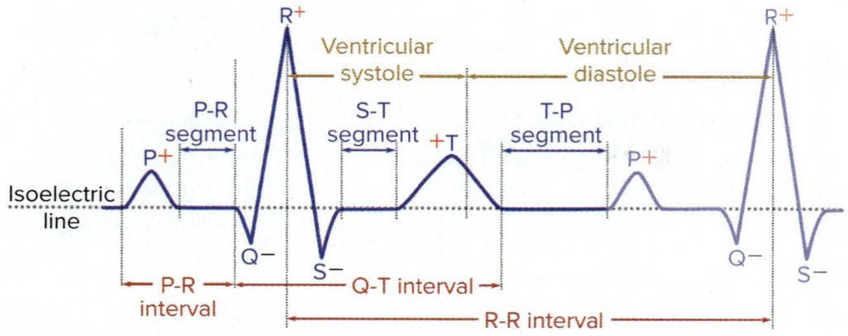

FIGURE 46.2 Components of the ECG.

Courtesy of and ©BIOPAC Systems, Inc. www.biopac.com

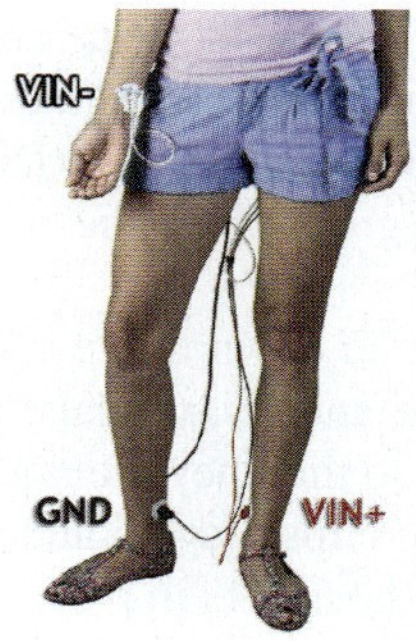

FIGURE 46.3 Electrode attachments for lead II.

Courtesy of and ©BIOPAC Systems, Inc. www.biopac.com

also to help in the diagnosis of many disorders of the cardiac conduction system and some types of heart disease.

The ECG is recorded in units of mV (millivolts). The typical appearance of an ECG for a single heartbeat is shown in figure 46.2. It consists of three deflections from the isoelectric line (the baseline, or zero-voltage line, which occurs during the short periods in which no voltage is being recorded):

1. **P wave:** The P wave is a positive (+) deflection from the isoelectric line that results from atrial depolarization. This depolarization is caused by the firing of the SA node, and it leads to the immediate contraction (systole) of the atria.
2. **QRS complex:** The QRS complex is a large deflection that results from ventricular depolarization. This depolarization of the ventricles is immediately followed by ventricular contraction (systole). The electrical events of atrial repolarization occur simultaneously with those of ventricular depolarization. Therefore, atrial repolarization is "hidden" inside the QRS complex and cannot be detected separately on an ECG.
3. **T wave:** The T wave is a positive deflection that results from ventricular repolarization. Repolarization of the ventricles is immediately followed by ventricular relaxation (diastole).

In addition to the wave components of the ECG, various segments and intervals can also be measured, all with their own significance in the cardiac cycle. A *segment* is a period of time between the end of a wave and the beginning of the next wave; it lies entirely along the isoelectric line and is associated with a plateau in the action potential of myocardial cells. For example, the P-R segment runs from the end of the P wave to the beginning of the QRS complex; it represents the plateau when myocardial cells of the atria are contracting. An *interval* is a period of time that contains at least one wave and one segment along the isoelectric line. For example, the P-R interval runs from the beginning of the P wave to the beginning of the QRS complex; it represents the time between the beginning of atrial depolarization and the beginning of ventricular depolarization.

ECGs can be conducted by placing electrodes in a variety of positions on the limbs and chest and measuring the voltage between them. There are several possible combinations of electrodes that can measure the ECG, and the results will vary with electrode placement. The particular arrangement of positive and negative electrodes, along with a ground electrode, that is used to conduct the ECG is called a *lead*. In this lab exercise, you are going to use lead II, in which the positive electrode is on the left ankle, the negative electrode on the right wrist, and the ground electrode on the right ankle. Proper electrode placement for lead II is shown in figure 46.3. Typical lead II ECG values are shown in table 46.1.

PROCEDURE A: Setup

1. With your computer turned **ON** and the BIOPAC MP36/35 unit turned **OFF,** plug the electrode lead set (SS2L) into Channel 1. If you are using the MP46/45, the USB cable must be connected, and the ready light must be on.
2. Turn on the MP36/35 Data Acquisition Unit.
3. Attach the electrodes to the subject as shown in figure 46.3. To help to ensure good contact, make sure the area to be in contact with the electrodes is clean by wiping with the abrasive pad (ELPAD) or alcohol wipe. A small amount of BIOPAC electrode gel (GEL1) may also be used to provide better contact between the sensor in the electrode and the skin. For optimal adhesion, the electrodes should be placed on the skin at least 5 minutes before the calibration procedure.
4. Attach the electrode lead set with the pinch connectors as shown in figure 46.3. Make sure the proper color

Table 46.1 Normal Lead II ECG Values

Phase	Duration (seconds)	Amplitude (millivolts)
P wave	0.06–0.11	<0.25
P-R interval	0.12–0.20	
P-R segment	0.08	
QRS complex (R)	<0.12	0.8–1.2
S-T segment	0.12	
Q-T interval	0.36–0.44	
T wave	0.16	<0.5

electrode cable is attached to the appropriate electrode: *white lead* on the anterior medial surface of the right wrist, *red lead* on the medial surface of the left ankle, and *black lead* (ground) on the medial surface of the right ankle.

5. Have the subject lie down in the supine position on the mat, cot, or table and relax. Make sure to position the electrode cables so that they are not pulling on the electrodes. The cable clips may be attached to the subject's clothing to hold the cables in place.
6. Start the Biopac student lab program for L05-Electrocardiography (ECG) 1, and click **OK.**
7. Designate a unique filename to be used to save the subject's data and click **OK.**

PROCEDURE B: Calibration

1. Make sure the subject is relaxed throughout the calibration procedure. *For all procedures, the subject must be still—no laughing or talking.* Click on **Calibrate.** The calibration will stop automatically after 8 seconds.
2. The calibration recording should look similar to figure 46.4. If it does, click **Continue.** If it does not, check connections and electrode adhesion, and click **Redo Calibration.** If the baseline is not flat (which is called baseline drift), or there are any large spikes or jitters, click **Redo Calibration.**

PROCEDURE C: Recording

This lab exercise consists of recording the ECG under four different conditions (segments): lying down in a supine position, seated (immediately after sitting up), performing deep breathing, and after exercise. In addition to the subject, you will have to designate a **Director** to coordinate the group members and a **Recorder** to insert event markers and time the exercise.

Segment 1—Supine (Lying Down)

1. With the subject lying down and relaxed, click **Record.** A marker labeled "Supine" will come up automatically. Record for 20 seconds.
2. Click **Suspend.** The recording should be similar to figure 46.5. If it is not, click **Redo** and repeat step 1.
3. If correct, click **Continue** to proceed to Segment 2.

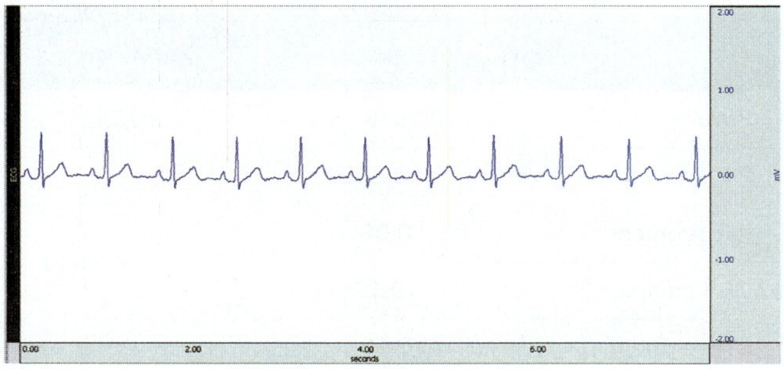

FIGURE 46.4 Sample calibration recording.

Courtesy of and ©BIOPAC Systems, Inc. www.biopac.com

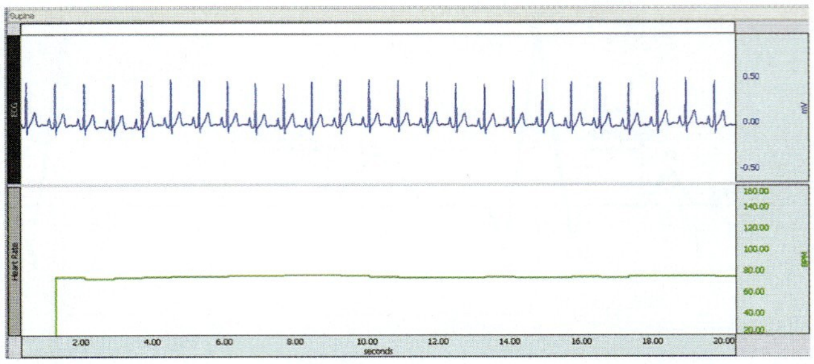

FIGURE 46.5 Segment 1 ECG—supine (lying down).

Courtesy of and ©BIOPAC Systems, Inc. www.biopac.com

Segment 2—Seated (After Sitting Up)

4. Have the subject quickly get up and sit in a chair, with forearms resting comfortably on the chair armrests or thighs. As soon as the subject is seated, relaxed, and fairly still, click **Record.** A marker labeled "Seated" will come up automatically.
5. Record for 20 seconds (seconds 21–40).
6. Click **Suspend.** The recording should be similar to figure 46.6. If it is not, click **Redo** and repeat steps 4 and 5.
7. If correct, click **Continue** to proceed to Segment 3.

Segment 3—Deep Breathing

8. Have the subject remain seated and prepare to take five deep breaths during the Segment 3 recording. Allow the subject to breathe at his/her own resting rate.
9. Click **Record.** A marker labeled "Deep Breathing" will come up automatically. Have the subject inhale and exhale deeply for five cycles (approximately 30 seconds; seconds 41–70). At the beginning of each inhalation, the Recorder should press the F4 key. At the beginning of each exhalation, the Recorder should press the F5 key.
10. After five breaths, click **Suspend.** The recording should be similar to figure 46.7. If it is not, click **Redo** and repeat step 9.
11. If correct, click **Continue** to proceed to Segment 4.

Segment 4—After Exercise

12. Have the subject perform moderate exercise to elevate his/her heart rate, such as jumping jacks or running in place.

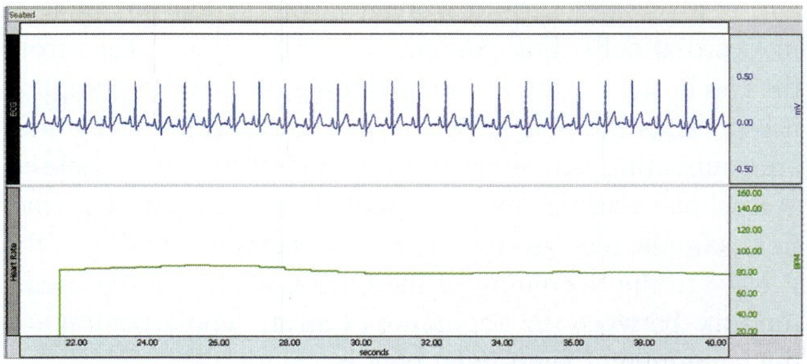

FIGURE 46.6 Segment 2 ECG—seated (after sitting up).

Courtesy of and ©BIOPAC Systems, Inc. www.biopac.com

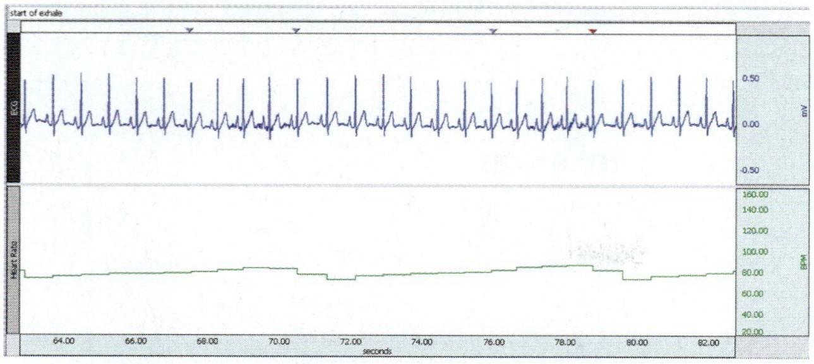

FIGURE 46.7 Segment 3 ECG—deep breathing.

Courtesy of and ©BIOPAC Systems, Inc. www.biopac.com

The electrode cable pinch connectors may be removed during exercising, but do not remove the electrodes. Immediately after the subject stops exercising, quickly reattach the pinch connectors; then move on to step 13.

13. Click **Record.** A marker labeled "After Exercise" will come up automatically.
14. Record for 30 seconds (seconds 71–100).
15. Click **Suspend.** The recording should be similar to figure 46.8. If it is not, click **Redo** and repeat steps 12–14.
16. If correct, click **Done.**
17. When a pop-up message asks you if you are sure you are done recording, click **Yes.** Remove the electrodes.

PROCEDURE D: Data Analysis

1. You can either analyze the current file now or save it and do the analysis later, using the **Review Saved Data** mode and selecting the correct file. Note the channel number designations: **CH 1, ECG,** and **CH 40, Heart Rate.** The data window should look similar to figure 46.9.

Heart Rate

2. Note the settings for the measurement boxes across the top of the data window as follows:
   - CH 40—Value
   - CH 1—Delta T (Δ T)
   - CH 1—P-P
   - CH 1—BPM

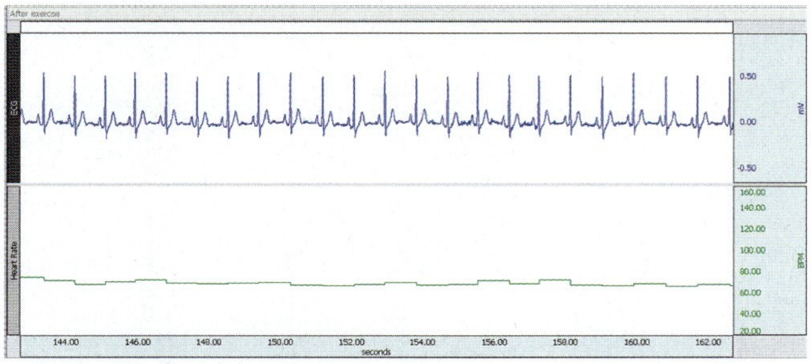

FIGURE 46.8 Segment 4 ECG—after exercise.

Courtesy of and ©BIOPAC Systems, Inc. www.biopac.com

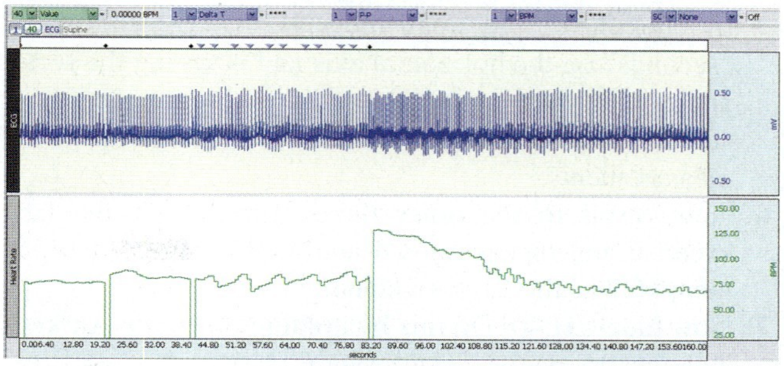

FIGURE 46.9 Data window for the entire ECG exercise.

Courtesy of and ©BIOPAC Systems, Inc. www.biopac.com

These will provide you with the following measurements:

Value—BPM or beats per minute (heart rate) in the preceding R-R interval

Δ T—the difference in time between the beginning and the end of the selected area

P-P (Peak to Peak)—measures the difference in amplitude between the maximum and minimum values of the selected area

BPM—the beats per minute; time difference between the beginning and the end of the selected area divided into 60 seconds/minute

3. Set up your display window for optimal viewing of three to five successive beats from Segment 1 (0–20 seconds). Do this by clicking anywhere in the horizontal axis (the numbers just above the word "seconds" at the bottom of the data window). A box will pop up in the middle of the screen, called **Set Screen Horizontal Axis.** Set the start time to 2 seconds and the end time to 8 seconds. Then click **OK.** This will spread out the data so you can see ECG recordings of individual heartbeats.
4. To measure heart rate, click on the I-beam and select any point within the R-R interval, as shown in figure 46.10. The result will be displayed by CH 40 value. Record the heart rate in table 46.2 of Part A of Laboratory Assessment 46.
5. Measure heart rate for two more heartbeats within the 0–20 second range. You can scroll through the segment

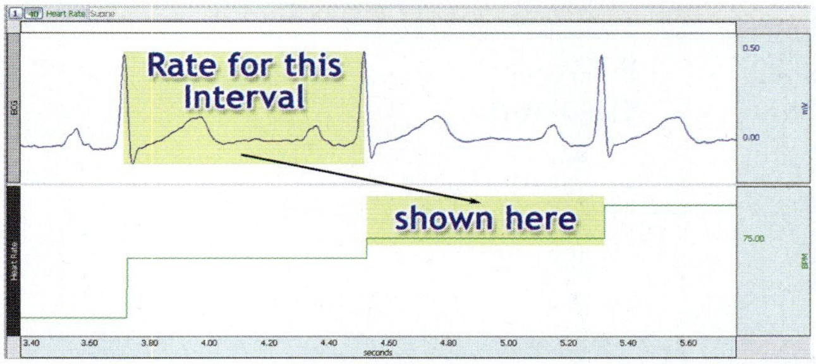

FIGURE 46.10 Proper selection of a data point for measuring heart rate.

Courtesy of and ©BIOPAC Systems, Inc. www.biopac.com

by clicking on the small rectangle beneath the word "seconds" on the horizontal axis and dragging the rectangle with your cursor. Record these heart rate measurements in table 46.2. Then take the mean of the three measurements.

6. Now scroll to the other three segments of the lab exercise and repeat steps 4 and 5. Record the data in table 46.2 of the lab assessment.
7. Now hide CH 40. Do this by holding down the Alt key and clicking on the CH 40 square.

Ventricular Systole and Diastole

8. Zoom in on a single cardiac cycle from Segment 1. Using the I-beam tool, measure the duration of ventricular systole and diastole by selecting the appropriate areas, as shown in figure 46.2. To measure ventricular systole, select the area between the peak of an R wave and the point one-third of the way down the descending portion of the T wave. To measure ventricular diastole, select the area between the end point of the systole measurement and the peak of the next R wave. The result will be displayed by CH 1, Delta T. Record your results in table 46.3.
9. Now scroll to Segment 4, "After Exercise," Repeat step 8 for Segment 4, and record your results in table 46.3.

Waves, Intervals, and Segments of the ECG

10. Scroll back to Segment 1 (Supine) and observe the ECG of a single cardiac cycle (heartbeat).
11. Using the I-beam tool, measure the durations of the various waves, intervals, and segments of the ECG, as required to fill out table 46.4. See figure 46.11 for an example of how to measure the duration of a P wave. See figure 46.12 for measuring P-R interval. The results for duration measurements will be displayed by CH 1, Delta T. (Refer to fig. 46.2 for proper selection of the areas to measure the intervals and segments.)
12. Measure the durations for all waves, intervals, and segments of two more heartbeats in Segment 1 and record the results in table 46.4. Then calculate the means as required in the table.
13. Measure the amplitudes of the P wave, QRS complex, and T wave for three heartbeats in Segment 1. To measure the amplitude of a P wave, for example, use the I-beam tool to select the area between the beginning of the P wave deflection and the highest point of the P wave. The result will be displayed by CH 1, P-P. (The P-P measurement calculates the difference between the maximum and minimum values in the selected area. Therefore, your selection does not have to be perfect; you can simply include the maximum and minimum points in your selection of the P wave, and the P-P measurement will calculate the amplitude correctly.) Record your results in table 46.4, and calculate the means.
14. Scroll to the beginning of Segment 4 (After Exercise).
15. Measure the durations of all waves, intervals, and segments for a single heartbeat, close to the beginning of Segment 4. Record your results in table 46.5.
16. Measure the amplitudes of the P wave, QRS complex, and T wave for a single heartbeat in Segment 4. Record your results in table 46.5.
17. Exit the program.
18. Complete Part B of the laboratory assessment.

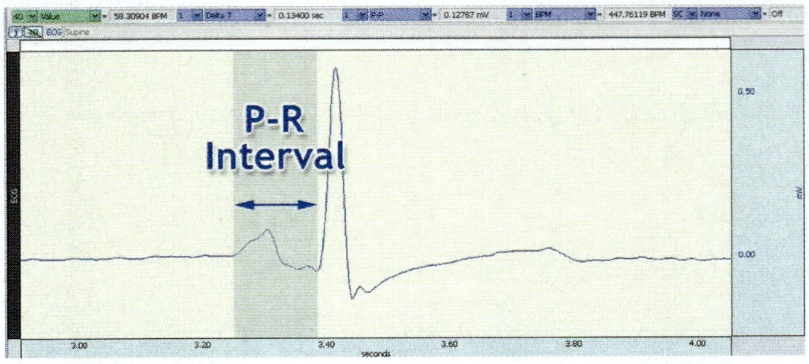

FIGURE 46.12 Proper selection of area for measuring the duration of the P-R interval.

Courtesy of and ©BIOPAC Systems, Inc. www.biopac.com

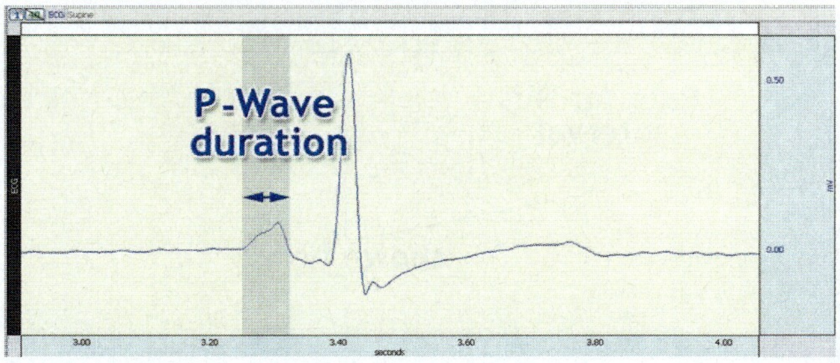

FIGURE 46.11 Proper selection of area for measuring the duration of a P wave.

Courtesy of and ©BIOPAC Systems, Inc. www.biopac.com

LABORATORY
ASSESSMENT 46

Name ______________________

Date ______________________

Section ______________________

The (A) corresponds to the indicated Learning Outcome(s) (O) found at the beginning of the Laboratory Exercise.

Electrocardiography: BIOPAC® Exercise

PART A: Data and Calculations Assessments

Subject Profile

Name ______________________ Height ______________

Age ______________________

Gender: Male / Female

Complete the following tables with the lesson data indicated and calculate the mean as appropriate. (A1)

1. Heart Rate Measurement

Table 46.2 Measurements of Heart Rate

Segment Condition	Cardiac Cycle 40 Value			Mean (Calculate)
	1	2	3	
1: Supine				
2: Seated				
3: Start of inhale				
3: Start of exhale				
4: After exercise				

2. Ventricular Systole and Diastole Measurements

Table 46.3 Measurements of Ventricular Systole and Ventricular Diastole Duration

Segment	Duration (sec) 1 Delta T	
	Ventricular Systole	Ventricular Diastole
1: Supine		
4: After exercise		

3a. Waves, Intervals, and Segments of the ECG (Segment 1, Supine) A2 A3

Table 46.4 Measurements from Three Cardiac Cycles of the Supine Recording

Component of the ECG	Normal Values (Based on resting heart rate of 75 BPM)		Duration (sec) 1 Delta T				Amplitude (mV) 1 P-P			
			Cycle 1	Cycle 2	Cycle 3	Mean (Calc.)	Cycle 1	Cycle 2	Cycle 3	Mean (Calc.)
Waves	Dur. (sec)	Amp. (mV)								
P	0.07–0.18	< 0.20								
QRS complex	0.06–0.12	0.10–1.5								
T	0.10–0.25	< 0.5								
Intervals	Duration (sec)									
P-R	0.12–0.20									
Q-T	0.32–0.36									
R-R	~ 0.80									
Segments	Duration (sec)									
P-R	0.02–0.10									
S-T	< 0.20									
T-P	0–0.40									

3b. Waves, Intervals, and Segments of the ECG (Segment 4, After Exercise)

Table 46.5 Measurements from One Cardiac Cycle of the After Exercise Recording

Component of the ECG	Normal Values (Based on resting heart rate of 75 BPM)		Duration (sec) 1 Delta T	Amplitude (mV) 1 P-P
Waves	Dur. (sec)	Amp. (mV)		
P	0.07–0.18	< 0.20		
QRS complex	0.06–0.12	0.10–1.5		
T	0.10–0.25	< 0.5		
Intervals	Duration (sec)			
P-R	0.12–0.20			
Q-T	0.32–0.36			
R-R	~ 0.80			
Segments	Duration (sec)			
P-R	0.02–0.10			
S-T	< 0.20			
T-P	0–0.40			

PART B: Assessments

Complete the following:

1. Was there a change in heart rate when the subject went from the supine position (Segment 1) to sitting (Segment 2)?

 ___

 Describe the physiological mechanisms causing this change. A3

 ___

2. Was there a difference between the heart rates when the subject was sitting at rest (Segment 2) and just after exercise (beginning of Segment 4)? Explain any differences. A3

 ___

3. What changes occurred in the duration of ventricular systole between the supine position (Segment 1) and postexercise (beginning of Segment 4)? Explain any differences. A3

 ___

4. What changes occurred in the duration of ventricular diastole between the supine position (Segment 1) and postexercise (beginning of Segment 4)? Explain any differences. A3

 ___

5. What changes occurred in the cardiac cycle during the respiratory cycle? In other words, did you observe a difference in heart rate between inhaling and exhaling (Segment 3)? Explain. A3

 ___

NOTES

LABORATORY
EXERCISE 47

Blood Vessel Structure, Arteries, and Veins

Purpose of the Exercise

To review the structure and functions of blood vessels and the circulatory pathways, and to locate major arteries and veins.

MATERIALS NEEDED

- Compound light microscope
- Prepared microscope slides:
  - Artery cross section
  - Vein cross section
- Live frog or goldfish
- Frog Ringer's solution
- Frog board or heavy cardboard (with a 1-inch hole cut in one corner)
- Paper towel
- Rubber bands
- Masking tape
- Dissecting pins
- Thread
- Dissectible human heart model
- Human torso model
- Anatomical charts of the cardiovascular system

For Learning Extension Activity:

- Ice
- Thermometer
- Hot plate

SAFETY

- Review all the safety guidelines in Appendix 1 of your laboratory manual.
- Wear disposable gloves and protective eyewear when handling the live frogs.
- Return the frogs to the location indicated after the experiment.
- Wash your hands before leaving the laboratory.

Chapter Opener: ©101cats/Getty Images

Learning Outcomes APR

After completing this exercise, you should be able to:

O1 Distinguish and sketch the three major layers in the wall of an artery and a vein.

O2 Examine and identify the types of blood vessels in the web of a frog's foot.

O3 Distinguish and trace the pulmonary and systemic circuits.

O4 Locate the major arteries in the circuits on a diagram, chart, or model.

O5 Locate the major veins in the circuits on a diagram, chart, or model.

The O corresponds to the assessments A indicated in the Laboratory Assessment for this Exercise.

Pre-Lab

Carefully read the introductory material and examine the entire lab. Be familiar with the structure of blood vessels and the major arteries and veins of the systemic and pulmonary circuits from lecture or the textbook. Answer the pre-lab questions.

Pre-Lab Questions Select the correct answer for each of the following questions:

1. The middle layer of an artery and vein contains mostly
   a. connective tissue. **b.** simple squamous epithelium.
   c. smooth muscle. **d.** nervous tissue.

2. Arteries carry blood
   a. away from the heart.
   b. toward the heart.
   c. both away from and toward the heart.
   d. only in the systemic circuit.

3. Blood vessels containing only an endothelium and a basement membrane are
   a. veins. **b.** arterioles.
   c. venules. **d.** capillaries.

4. The _____ has the thickest wall.
   a. left atrium **b.** right atrium
   c. left ventricle **d.** right ventricle

5. Which of the following arteries is *not* a direct branch of the aorta?
   a. coronary artery **b.** brachiocephalic trunk
   c. left subclavian artery **d.** brachial artery

6. Which of the following veins is *not* located in the neck?
   a. vertebral vein **b.** brachiocephalic vein
   c. external jugular vein **d.** internal jugular vein

7. Which of the following arteries is part of the cerebral arterial circle (circle of Willis)?
   a. anterior communicating artery
   b. external carotid artery
   c. facial artery
   d. popliteal artery
8. Blood pressure in the systemic veins is higher than in the systemic arteries.
   a. True   b. False
9. The left ventricle pumps blood into the aorta of the systemic circuit.
   a. True   b. False
10. The pulmonary trunk is an artery.
   a. True   b. False

The blood vessels form a closed system of tubes that carry blood to and from the heart, lungs, and body cells. These tubes include *arteries* and *arterioles* that transport blood away from the heart; *capillaries* in which exchanges of substances occur between the blood and surrounding tissues; and *venules* and *veins* that return blood to the heart.

Arteries and veins are composed of three layers or tunics. The inner layer, *tunica interna*, is composed of an endothelium of simple squamous epithelial cells and the basement membrane. This establishes a smooth lining next to the lumen of the blood vessel. The middle layer, *tunica media*, is composed of smooth muscle. The contraction and relaxation of the smooth muscle modify the size of the lumen, which influences the amount of blood flow through a particular area of the body. The outer layer, *tunica externa*, is composed of connective tissues rich in collagen, providing some support and protection. In general, the arteries have a thicker middle layer and elastic laminae, which provide additional support and elasticity to withstand the higher blood pressures and reduce extreme fluctuations in blood pressure during systole and diastole of the ventricles. In contrast, veins have much lower blood pressure, contain a thinner middle layer, and have occasional valves to prevent backflow. Veins appear collapsed when empty, and they depend on mechanisms (such as skeletal muscles) to create pressure gradients that help blood return to the heart. Relatively small arteries are called arterioles; small veins are termed venules. Capillaries are often found connecting arterioles to venules, and they retain only the endothelium and basement membrane.

The blood vessels of the cardiovascular system can be divided into two major pathways—the *pulmonary circuit* and the *systemic circuit*. Within each circuit, arteries transport blood away from the heart. After exchanges of gases, nutrients, and wastes have occurred between the blood and the surrounding tissues, veins return the blood to the heart.

Variations exist in anatomical structures among humans, especially in arteries and veins. The illustrations in this laboratory manual represent typical (meaning the most common variation) anatomy.

PROCEDURE A: Blood Vessel Structure

1. Reexamine the introductory material. Compare the structure of an artery and a vein, illustrated in figure 47.1. Note that arteries and veins possess the same three

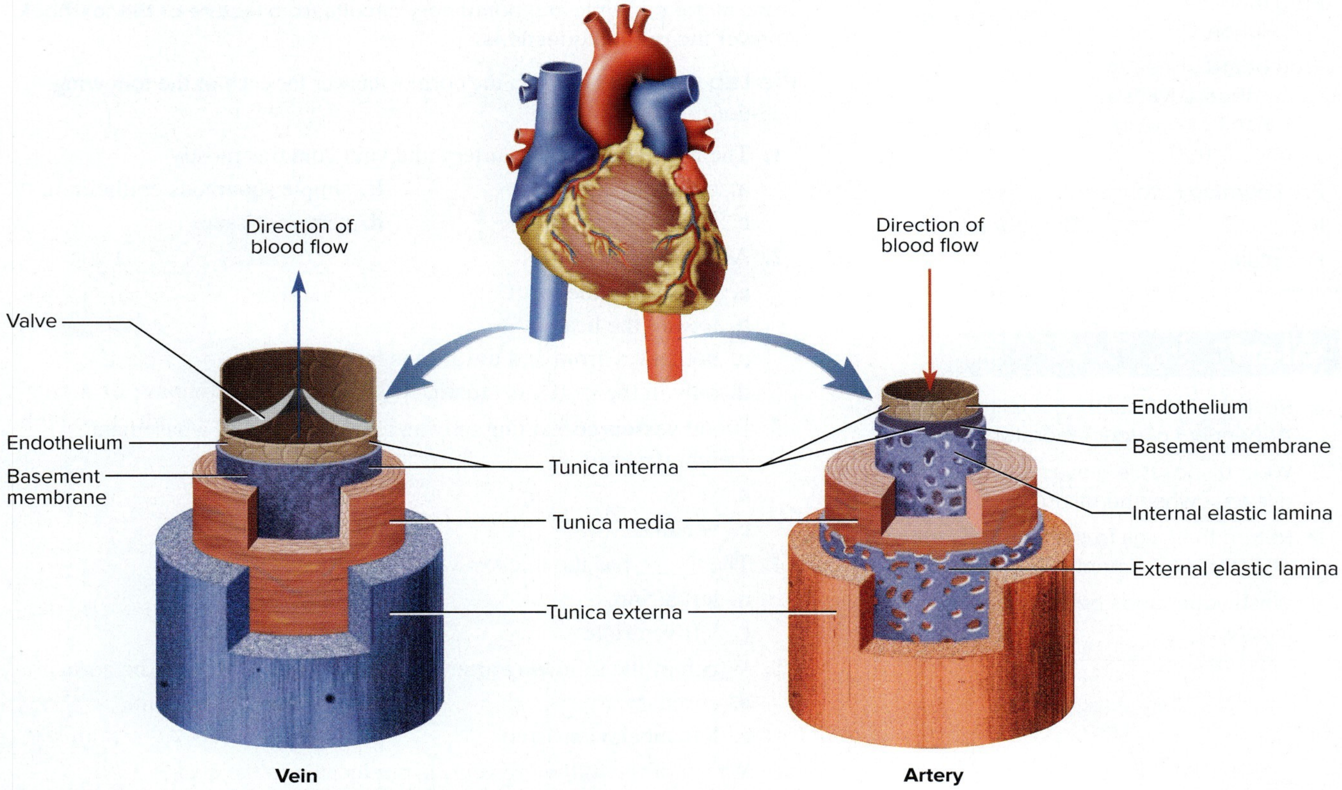

FIGURE 47.1 Structure of the wall of an artery and a vein.

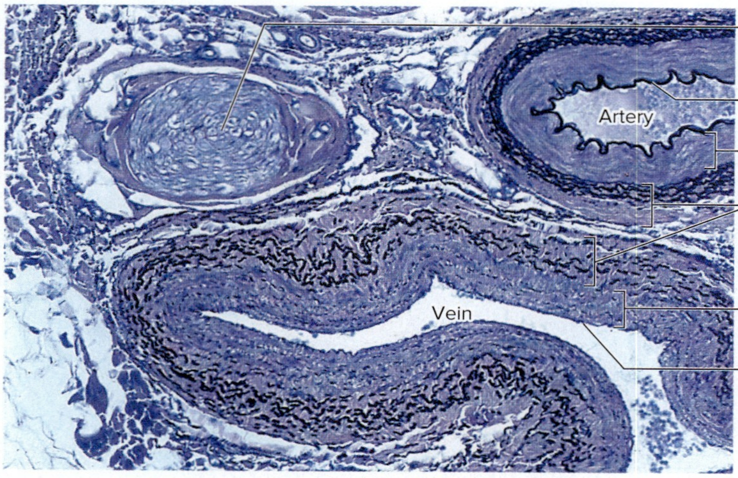

(a)

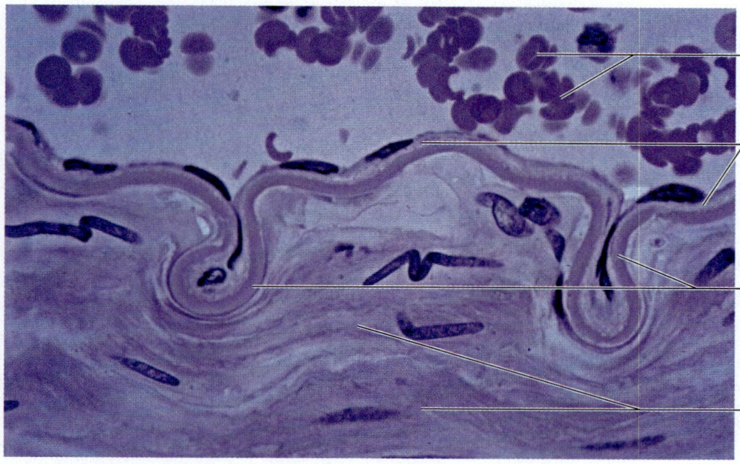

(b)

FIGURE 47.2 Micrograph of (a) a neurovascular bundle, showing a small peripheral nerve and labeled tunics of a small artery and small vein (100×), and (b) the tunica interna and part of the tunica media of a small artery (400×). (a) Dennis Strete/McGraw Hill; (b) Ed Reschke/Getty Images APR

layers or tunics, but an artery has thicker middle tunic and also contains two elastic laminae. The vein has a thinner middle tunic, lacks the two elastic laminae, and has periodic valves.

2. Obtain a microscope slide of an artery or arteriole cross section, and examine it using scanning, low-power, and then high-power magnification (figs. 47.2 and 47.3). Identify the three distinct layers (tunics) of the arterial wall. The inner layer (tunica interna) is composed of an endothelium (simple squamous epithelium) and appears as a wavy line due to an abundance of elastic fibers that have recoiled just beneath it. The middle layer (tunica media) consists of numerous concentrically arranged smooth muscle cells with elastic fibers scattered among them. The outer layer (tunica externa) contains connective tissue rich in collagen fibers.
3. Prepare a labeled sketch of the arterial wall in Part A of Laboratory Assessment 47.
4. Obtain a slide of a vein or venule cross section and examine it as you did the artery or arteriole cross section (figs. 47.2 and 47.3). Note the thinner wall and larger lumen relative to an artery of comparable locations. Identify the three layers of the wall and prepare a labeled sketch in Part A of the laboratory assessment.
5. Complete Part A of the laboratory assessment.
6. Observe the blood vessels in the webbing of a frog's foot. (As a substitute for a frog, a live goldfish could be used to observe circulation in the tail.) To do this, follow these steps:
   a. Obtain a live frog. Wrap its body in a moist paper towel, leaving one foot extending outward. Secure the towel with rubber bands, but be careful not to wrap the animal so tightly that it could be injured. Try to keep the nostrils exposed.
   b. Place the frog on a frog board or on a piece of heavy cardboard with the foot near the hole in one corner.
   c. Fasten the wrapped body to the board with masking tape.
   d. Carefully spread the web of the foot over the hole and secure it to the board with dissecting pins and

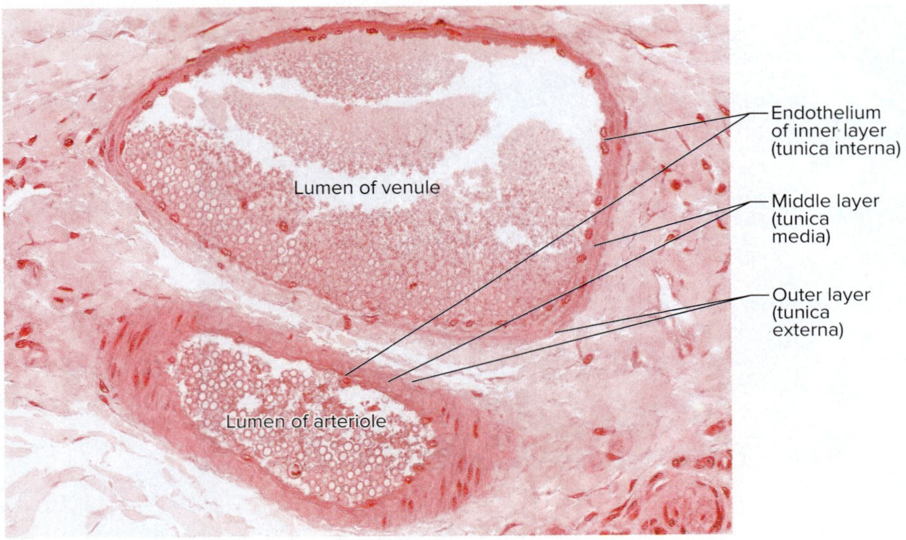

FIGURE 47.3 Cross section of an arteriole and a venule (200×). Al Telser/McGraw Hill APR

thread (fig. 47.4). Keep the web moist with frog Ringer's solution.

e. Secure the board on the stage of a microscope with heavy rubber bands and position it so that the web is beneath the objective lens.

f. Focus on the web, using low-power magnification, and locate some blood vessels. Note the movement of the blood cells and the direction of the blood flow. You might notice that red blood cells of frogs are nucleated. Identify an arteriole, a capillary, and a venule.

g. Examine each of these vessels with high-power magnification.

h. When finished, return the frog to the location indicated by your instructor. The microscope lenses and stage will likely need cleaning after the experiment.

7. Complete Part B of the laboratory assessment.

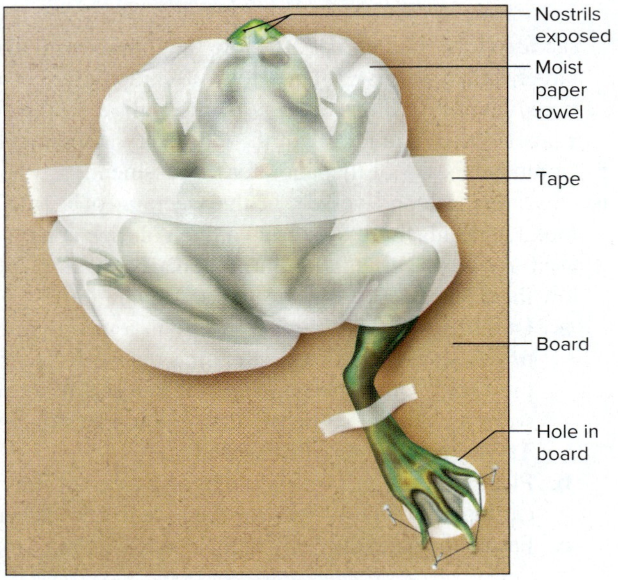

FIGURE 47.4 Spread the web of the foot over the hole and secure it to the board with pins and thread.

LEARNING EXTENSION ACTIVITY

Investigate the effect of temperature change on the blood vessels of the frog's foot by flooding the web with a small quantity of ice water. Observe the blood vessels with low-power magnification and note any changes in their diameters or the rate of blood flow. Remove the ice water and replace it with water heated to about 35°C (95°F). Repeat your observations. What do you conclude from this experiment?

____

____

____

PROCEDURE B: Paths of Circulation

1. Study figures 47.5 and 47.6, which illustrate the pathway of circulation in the pulmonary and systemic circuits. Examine the major blood vessels of the two circuits that are associated near and connected to the heart chambers.
2. Locate the following blood vessels on the available anatomical charts, the dissectible human heart model, and the human torso model:

pulmonary circuit (figs. 47.5 and 47.6)

- pulmonary trunk 1
- pulmonary arteries 2
- pulmonary veins 4

systemic circuit (fig. 47.5)

- aorta 1
- superior vena cava 1
- inferior vena cava 1

PROCEDURE C: The Arterial System

The *pulmonary trunk,* connected to the right ventricle, is the main artery of the pulmonary circuit. The *aorta,* connected to

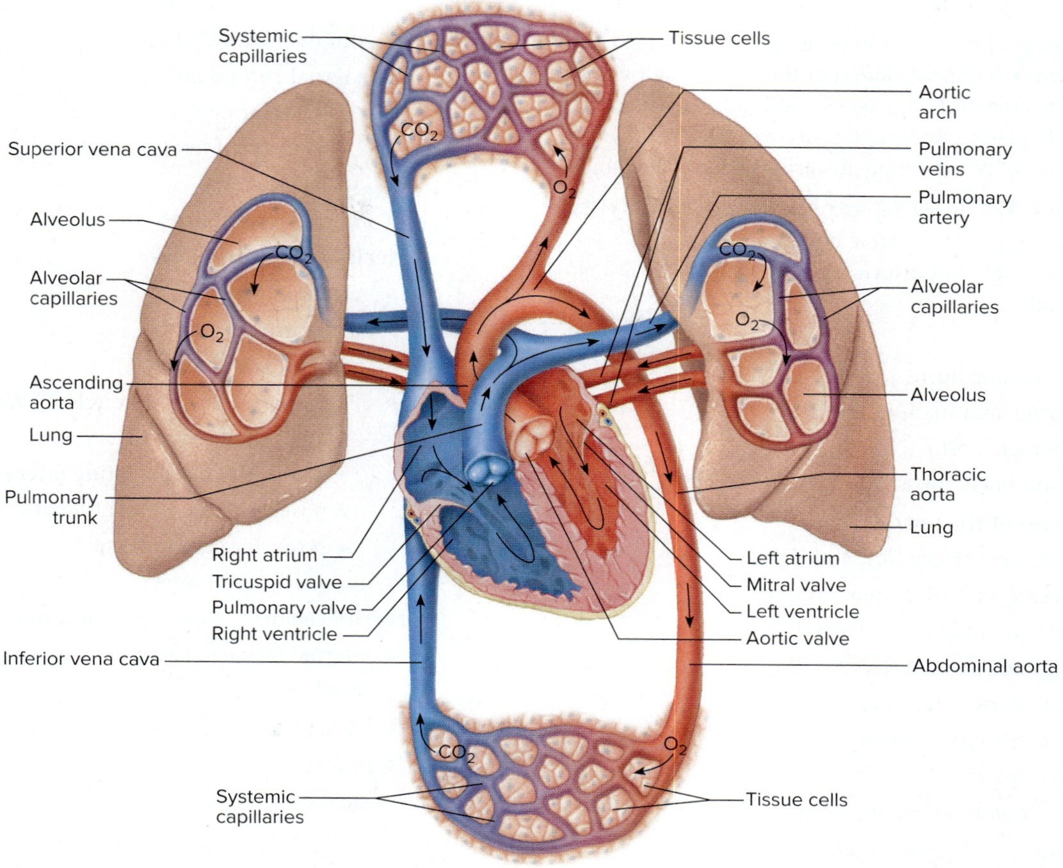

FIGURE 47.5 The major blood vessels associated with the pulmonary and systemic circuits. The red colors indicate locations of oxygen-rich (oxygenated) and CO_2-poor blood. The blue colors indicate locations of oxygen-poor (deoxygenated) and CO_2-rich blood. APR

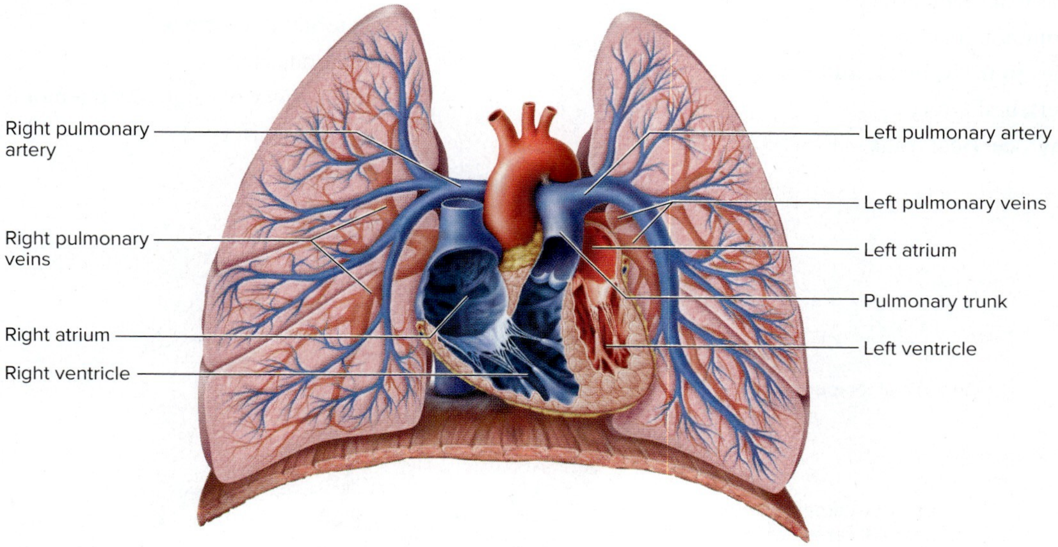

FIGURE 47.6 Blood vessels of the pulmonary circuit. The pulmonary arteries carry oxygen-poor blood from the right ventricle to the lung capillaries; the pulmonary veins carry oxygen-rich blood from the lung capillaries to the left atrium. Exchange of carbon dioxide and oxygen occurs at the pulmonary alveoli.

the left ventricle, is the main artery of the systemic circuit. The aorta has a thicker wall than the pulmonary trunk. This allows the aorta to withstand the greater pressure created from the contraction of the very thick myocardium of the left ventricle compared to the right ventricle. In this procedure, trace the pathway of blood vessels as outlined in the various sections. The first branches of the aorta near the aortic valve are the two coronary arteries, which deliver oxygen-rich blood to the myocardium. The brachiocephalic trunk, the left common carotid artery, and the left subclavian artery branch from the aortic arch. One pathway of particular importance to note is the *cerebral arterial circle* (*circle of Willis*) of the arteries at the base of the brain. This *arterial anastomosis* of blood vessels encircles the pituitary gland and the optic chiasma, which allows for alternative routes of blood supply to various parts of the brain.

1. Use figures 47.7 to 47.12 as references to identify the major arteries of the systemic circuit.
2. Locate the following arteries on the charts and human torso model:

aorta

- ascending aorta
- aortic arch (arch of aorta)
- thoracic aorta
- abdominal aorta

branches of the aorta

- coronary artery (left and right)
- brachiocephalic trunk (artery)
- left common carotid artery
- left subclavian artery
- celiac trunk (artery)
    - left gastric artery
    - splenic artery
    - common hepatic artery
- superior mesenteric artery
- renal artery
- gonadal artery
- inferior mesenteric artery
- median sacral artery
- common iliac artery

arteries to neck, head, and brain

- vertebral artery
- thyrocervical trunk
- common carotid artery
- external carotid artery
    - superior thyroid artery
    - superficial temporal artery
    - facial artery
- internal carotid artery

arteries of base of brain

- vertebral artery
- basilar artery
- internal carotid artery
- cerebral arterial circle (circle of Willis)
    - anterior cerebral artery
    - anterior communicating artery
    - posterior communicating artery
    - posterior cerebral artery
- middle cerebral artery

arteries to chest, shoulder, and upper limb

- internal thoracic (mammary) artery
- subclavian artery
- axillary artery
- brachial artery
- deep brachial artery
- ulnar artery
- radial artery

arteries to pelvis and lower limb

- common iliac artery
- internal iliac artery
- external iliac artery
- femoral artery
- deep artery of thigh (deep femoral artery)
- popliteal artery

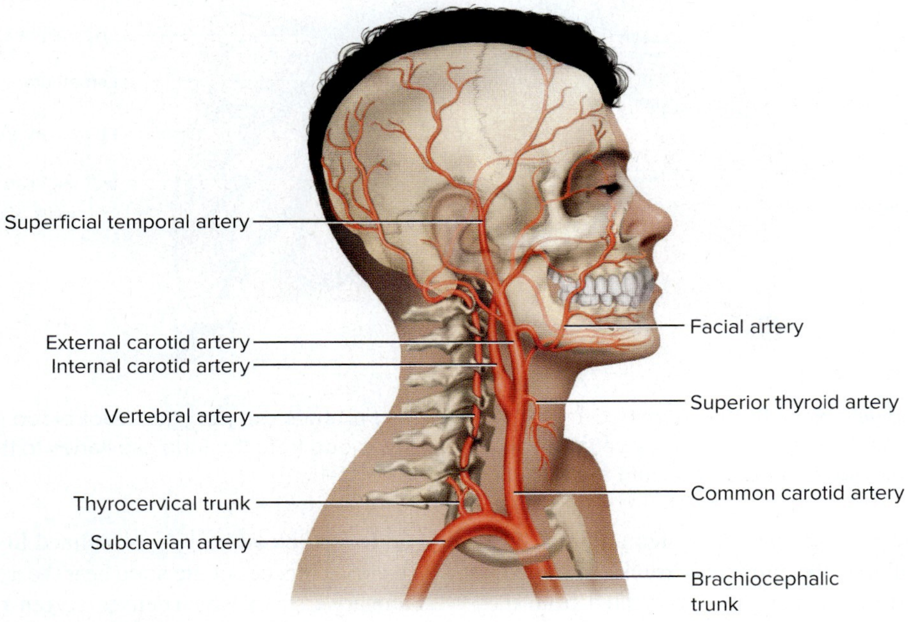

FIGURE 47.7 Arteries supplying the right side of the neck and head. (The clavicle has been removed.) APR

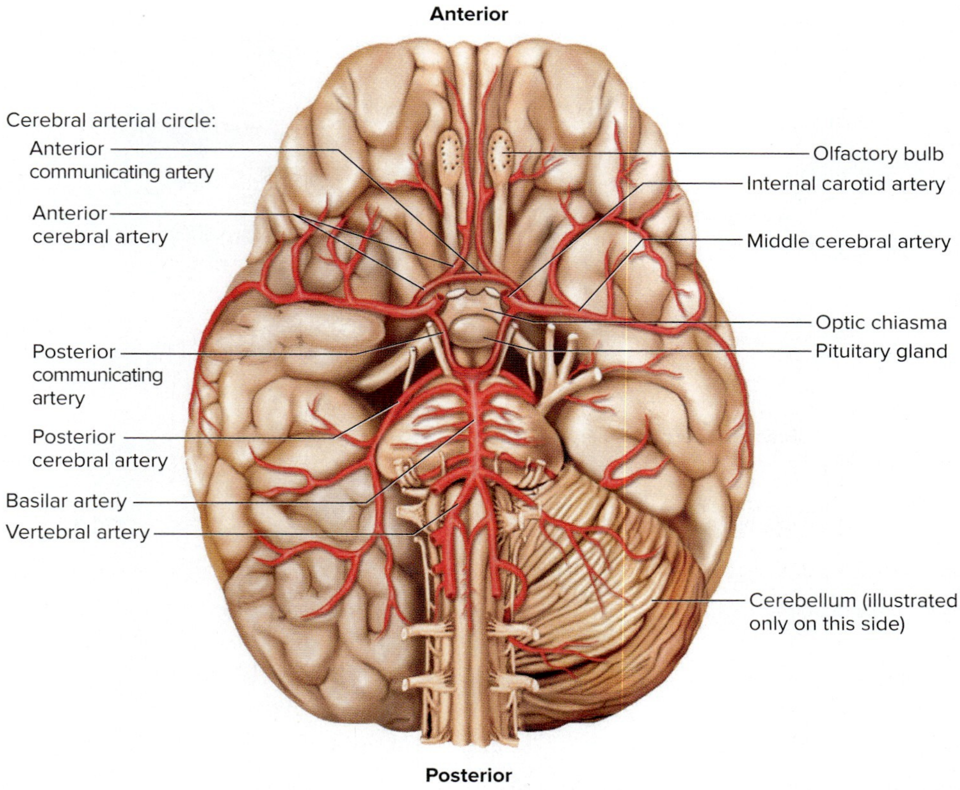

FIGURE 47.8 Inferior view of the brain showing the major arteries of the base of the brain, including the cerebral arterial circle. APR

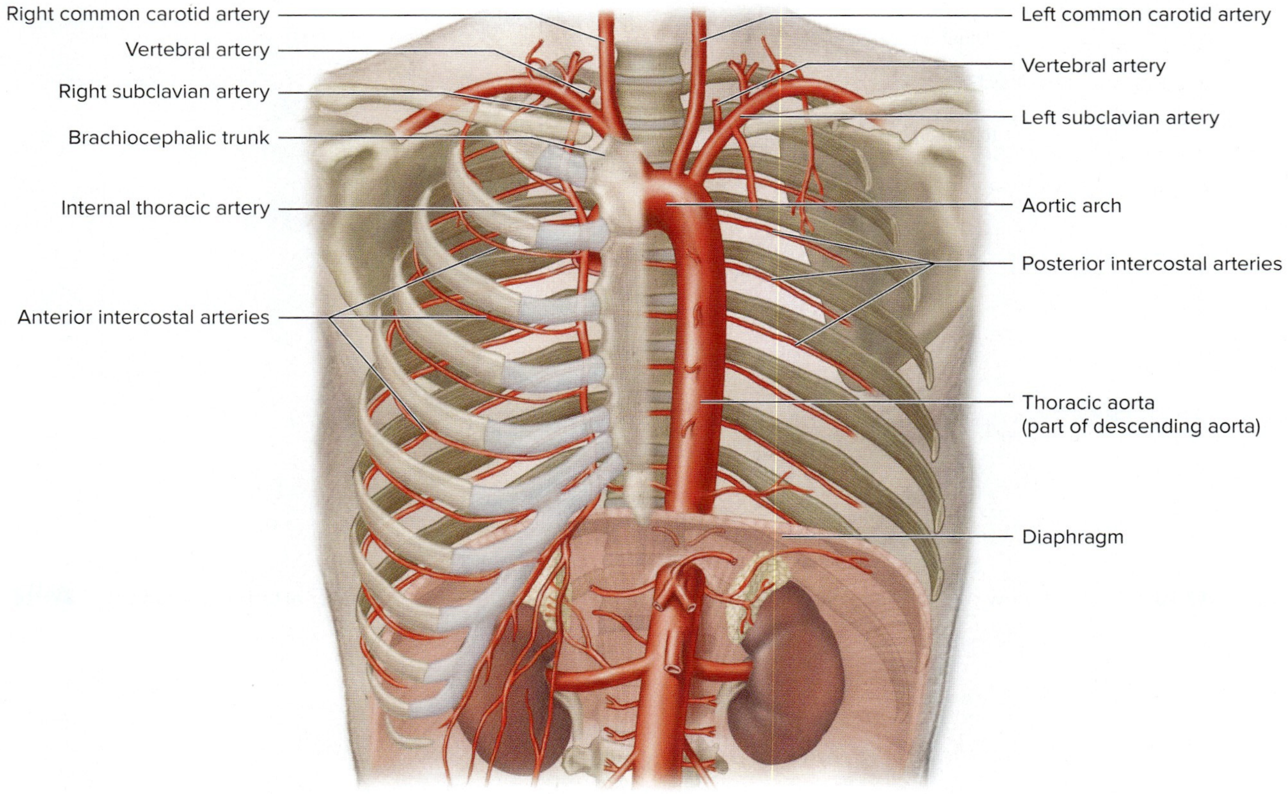

FIGURE 47.9 Anterior view of the major arteries of the chest (thoracic) wall and neck. APR

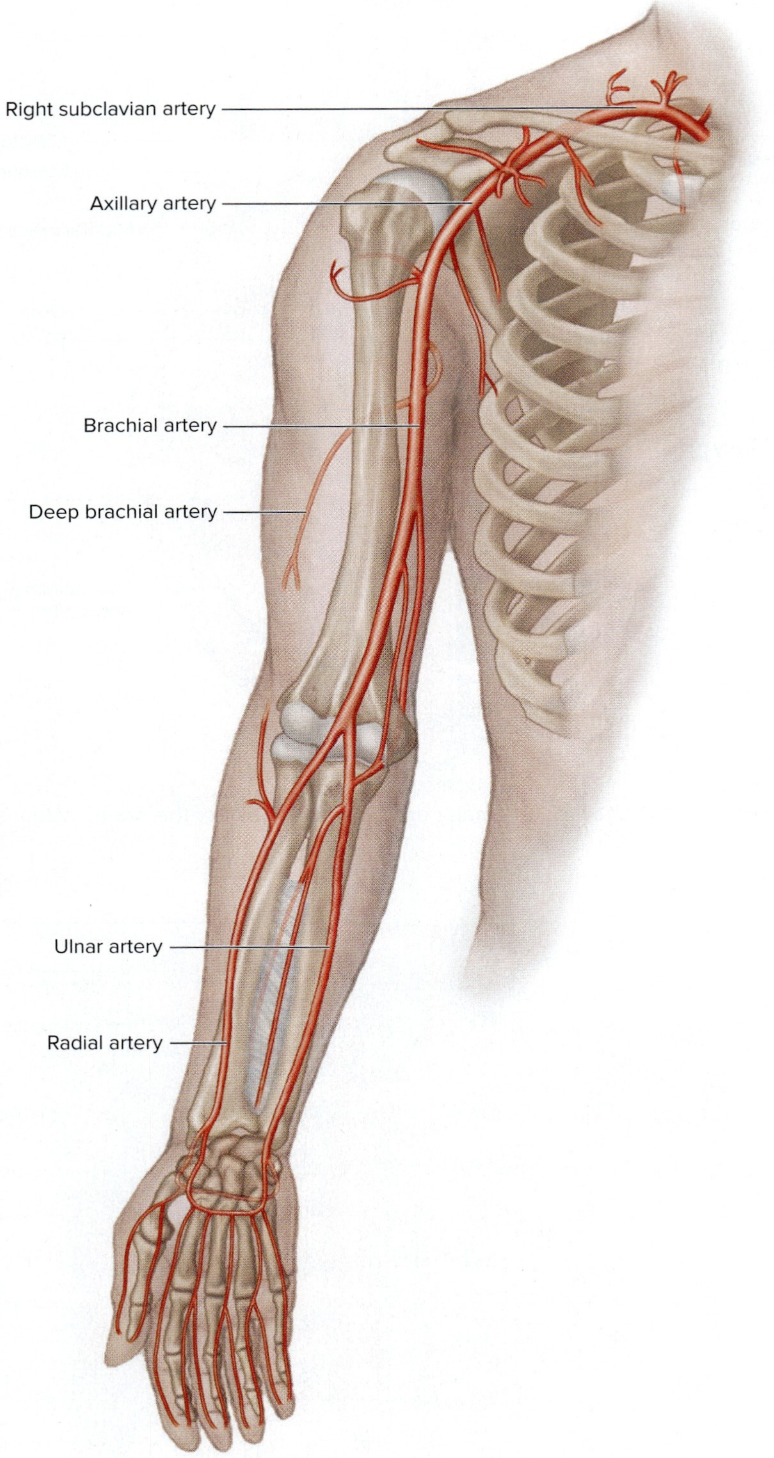

FIGURE 47.10 Anterior view of major arteries of the right shoulder and upper limb (posterior arteries are faded). APR

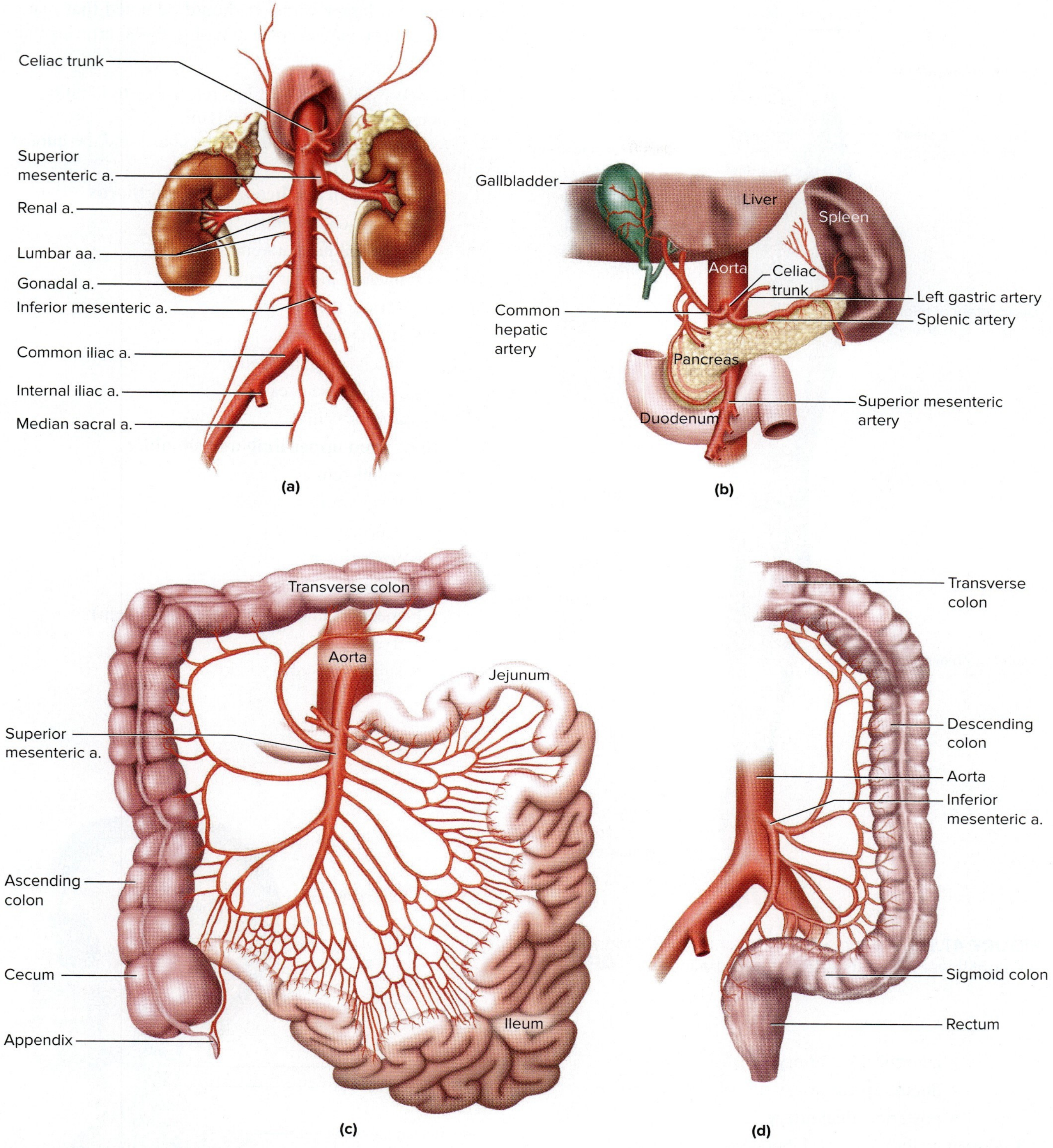

FIGURE 47.11 Anterior views of (a) major abdominal, mesenteric, renal, gonadal, and pelvic arteries; (b) major branches of the celiac trunk (the stomach has been removed); (c) superior mesenteric artery and its distribution; and (d) inferior mesenteric artery and its distribution. (a. = artery; aa. = arteries) APR

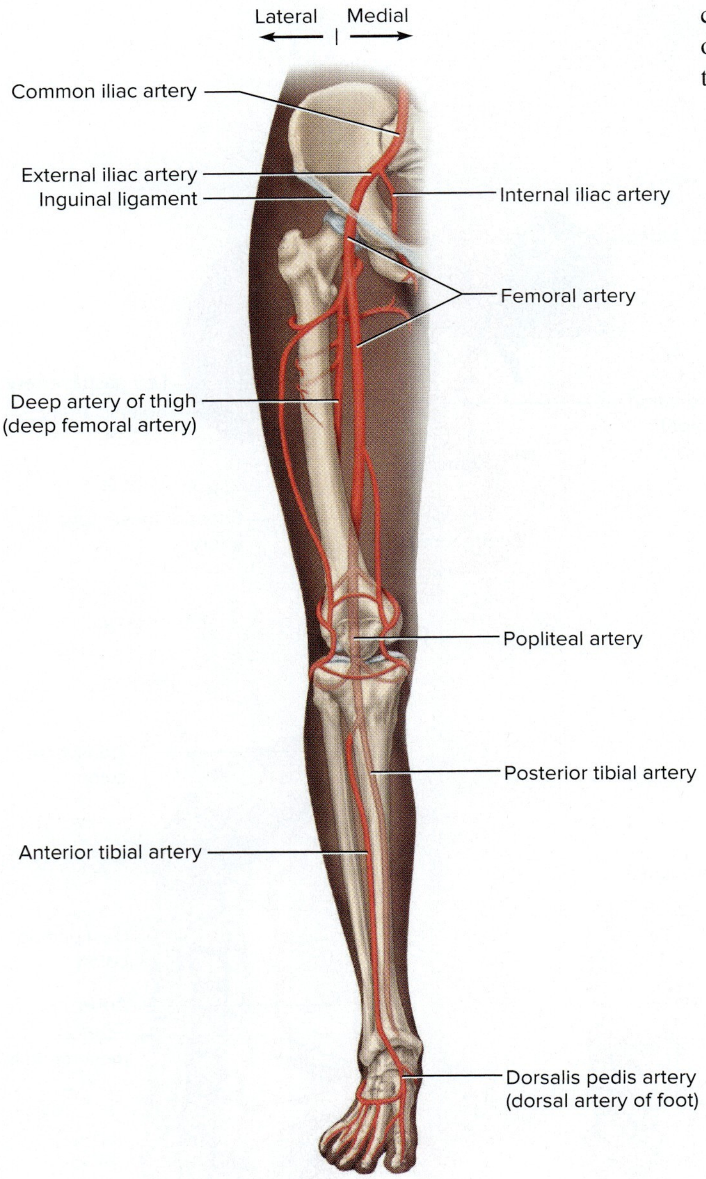

FIGURE 47.12 Anterior view of major arteries supplying the right lower limb (posterior arteries are faded). APR

- anterior tibial artery
- dorsalis pedis artery (dorsal artery of foot)
- posterior tibial artery

3. Complete Part C of the laboratory assessment.

PROCEDURE D: The Venous System

The two venae cavae represent the very large veins returning blood from the superior and inferior portions of the systemic circuit into the right atrium. In this procedure, trace the veins as outlined in the various sections, keeping in mind that the flow of blood from smaller veins converges into larger veins. It should be noted that many of the veins have names corresponding to the arteries that they parallel.

1. Use figures 47.13 to 47.17 as references to identify the major veins of the systemic circuit.
2. Locate the following veins on the charts and the human torso model:

veins from brain, head, neck, and thorax

- dural venous sinus
- external jugular vein
- internal jugular vein
- vertebral vein
- subclavian vein
- brachiocephalic vein
- superior vena cava
- azygos vein

veins from upper limb and shoulder

- radial vein
- ulnar vein
- brachial vein
- basilic vein
- cephalic vein
- median cubital vein (antecubital vein)
- axillary vein
- subclavian vein

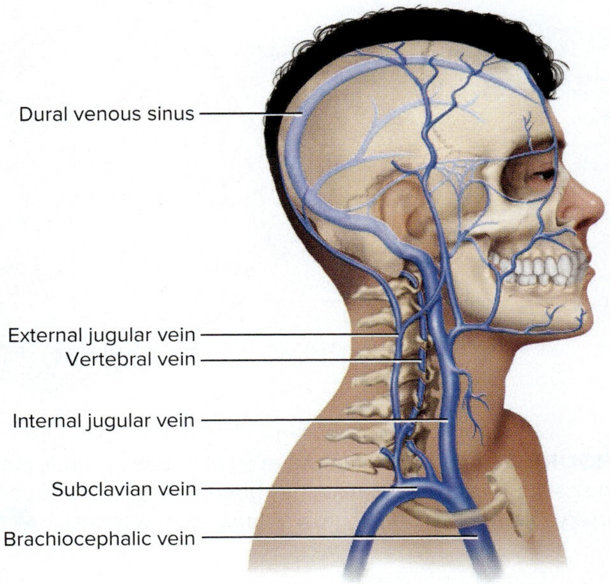

FIGURE 47.13 Major veins associated with the right side of the head and neck. (The clavicle has been removed.) APR

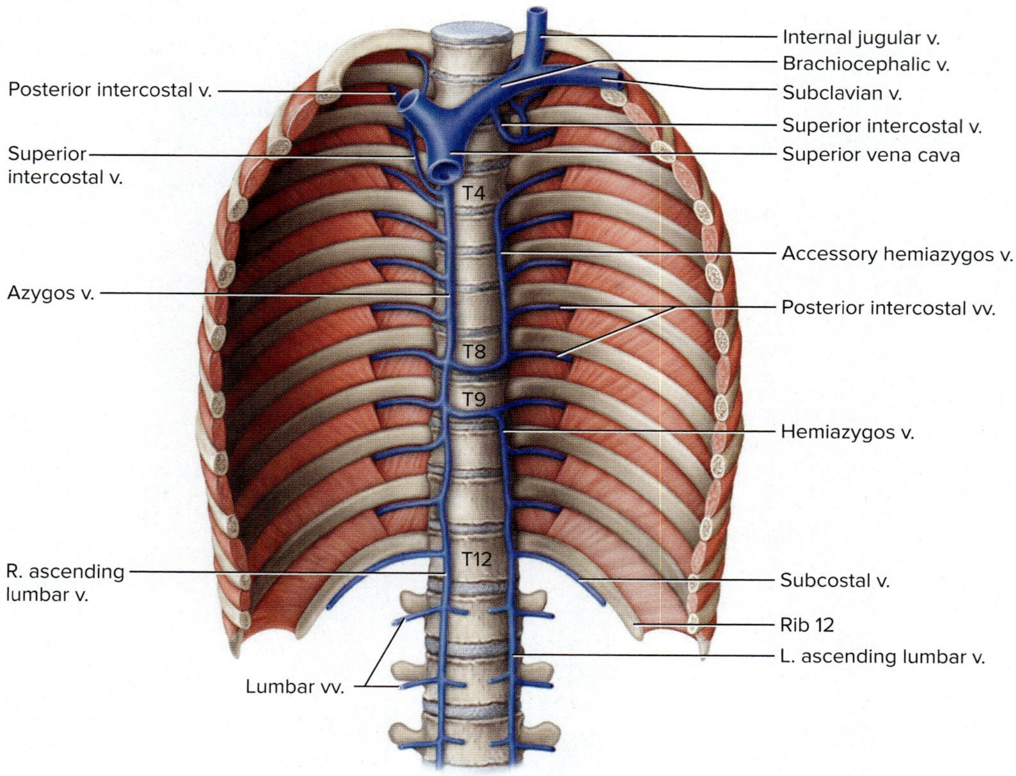

FIGURE 47.14 Anterior view of the azygos vein in the thoracic wall and the veins it drains. (R. = Right; L. = Left; v. = vein; vv. = veins; m. = muscle) APR

veins of the abdominal viscera

- hepatic portal vein
- gastric vein
- superior mesenteric vein
- splenic vein
- inferior mesenteric vein
- hepatic vein
- renal vein
- gonadal vein

veins from lower limb and pelvis

- anterior tibial vein
- posterior tibial vein
- popliteal vein
- femoral vein
- great (long) saphenous vein
- small (short) saphenous vein
- external iliac vein
- internal iliac vein
- common iliac vein
- inferior vena cava

3. Complete Parts D and E of the laboratory assessment.

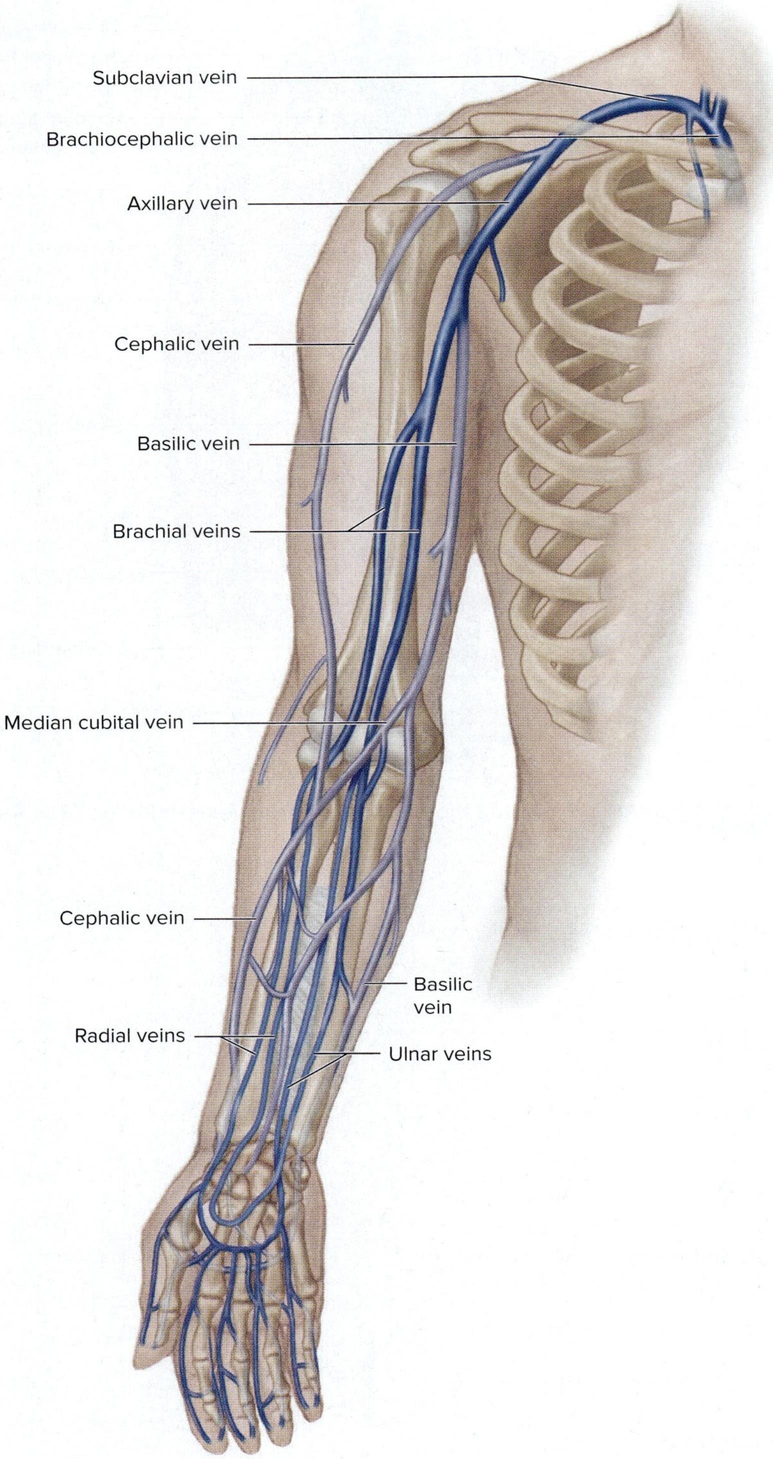

FIGURE 47.15 Anterior view of the veins of the right upper limb. The superficial veins are light purple and the deep veins are dark blue. (*Note:* The ulnar and radial veins travel in pairs around their corresponding arteries.) APR

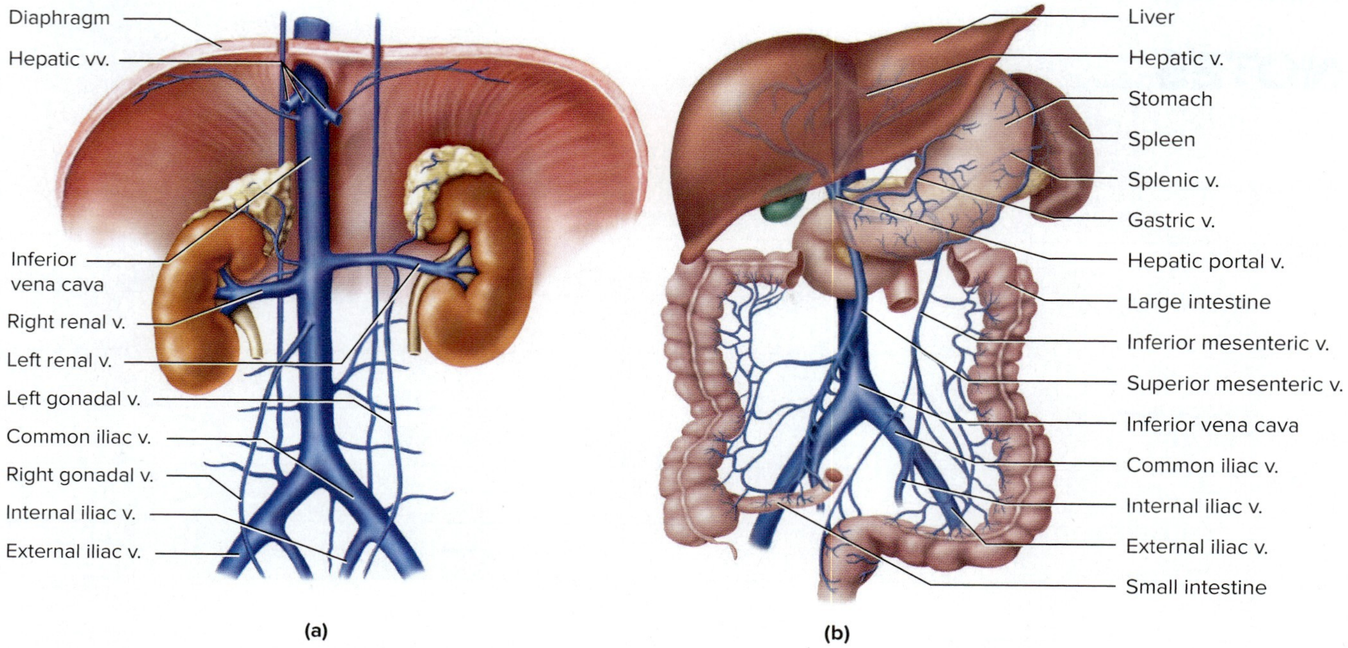

FIGURE 47.16 Anterior view of (a) major tributaries of the inferior vena cava, and (b) major abdominal, hepatic portal, and pelvic veins. (Most of small intestine and part of transverse colon of large intestine have been removed; v. = vein and vv. = veins.) APR

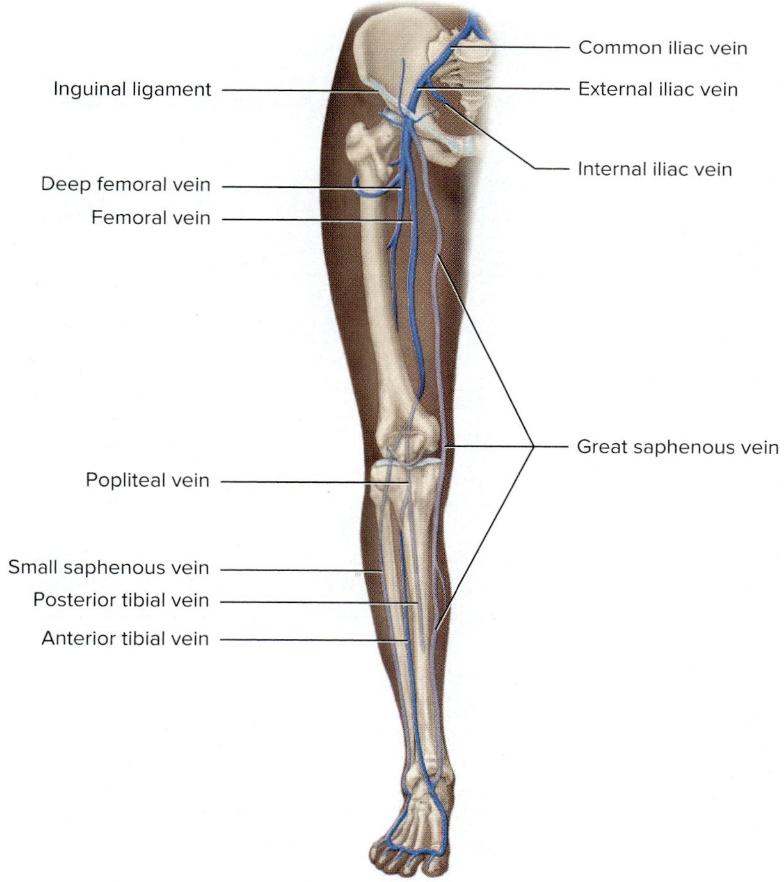

FIGURE 47.17 Anterior view of major veins of the right pelvis and lower limb. The superficial veins are light purple color and the deep veins are dark blue (posterior veins are faded). APR

NOTES

LABORATORY
ASSESSMENT 47

Name ______________________________

Date ______________________________

Section ______________________________

The (A) corresponds to the indicated Learning Outcome(s) (O) found at the beginning of the Laboratory Exercise.

Blood Vessel Structure, Arteries, and Veins

PART A: Assessments

1. Sketch and label a section of an arterial wall next to a section of a venous wall in each circle that represents the field of view as seen through the microscope. (A1)

Artery wall (_____×)   Vein wall (_____×)

2. Describe the differences you noted in the structures of the arterial and venous walls. Mention each of the three layers of the wall. (A1) ______________________________

____

____

CRITICAL THINKING ASSESSMENT—PART A

Explain the functional significance of the differences you noted in the structures of the arterial and venous walls. (A1)

____

____

PART B: Assessments

Complete the following:

1. How did you distinguish between arterioles and venules when you observed the vessels in the web of the frog's foot? (A2)

____

____

2. How did you recognize capillaries in the web? A2 ____

____

3. What differences did you note in the rate of blood flow through the arterioles, capillaries, and venules? A2

____

____

PART C: Assessments

Provide the name of the missing artery in each of the following sequences: A3

1. Brachiocephalic trunk, ____________________, right axillary artery
2. Ascending aorta, ____________________, descending thoracic aorta
3. Abdominal aorta, ____________________, ascending colon (right side of large intestine)
4. Brachiocephalic trunk, ____________________, right external carotid artery
5. Axillary artery, ____________________, radial artery
6. Common iliac artery, ____________________, femoral artery
7. Pulmonary trunk, ____________________, lungs
8. Vertebral artery, ____________________, circle of Willis

PART D: Assessments

Provide the name of the missing vein or veins in each of the following sequences: A3

1. Right subclavian vein, ____________________, superior vena cava
2. Anterior tibial vein, ____________________, femoral vein
3. Internal iliac vein, ____________________, inferior vena cava
4. Vertebral vein, ____________________, brachiocephalic vein
5. Popliteal vein, ____________________, external iliac vein
6. Lungs, ____________________, left atrium
7. Kidney, ____________________, inferior vena cava
8. Superior mesenteric vein, ____________________, liver

CRITICAL THINKING ASSESSMENTS—PART D

1. Why are varicose veins more common in the lower limb? Which veins are most likely to have varicosities? Explain. A3 ____

____

____.

2. The hepatic portal system drains blood from the gastrointestinal tract and the spleen, gallbladder, and pancreas. What constituents would you expect to be in the blood that enters the liver and what might the liver do with what it receives? Explain. A3 A5 ____

____

____

____.

PART E: Assessments

Label the major arteries and veins indicated in figures 47.18 to 47.20.

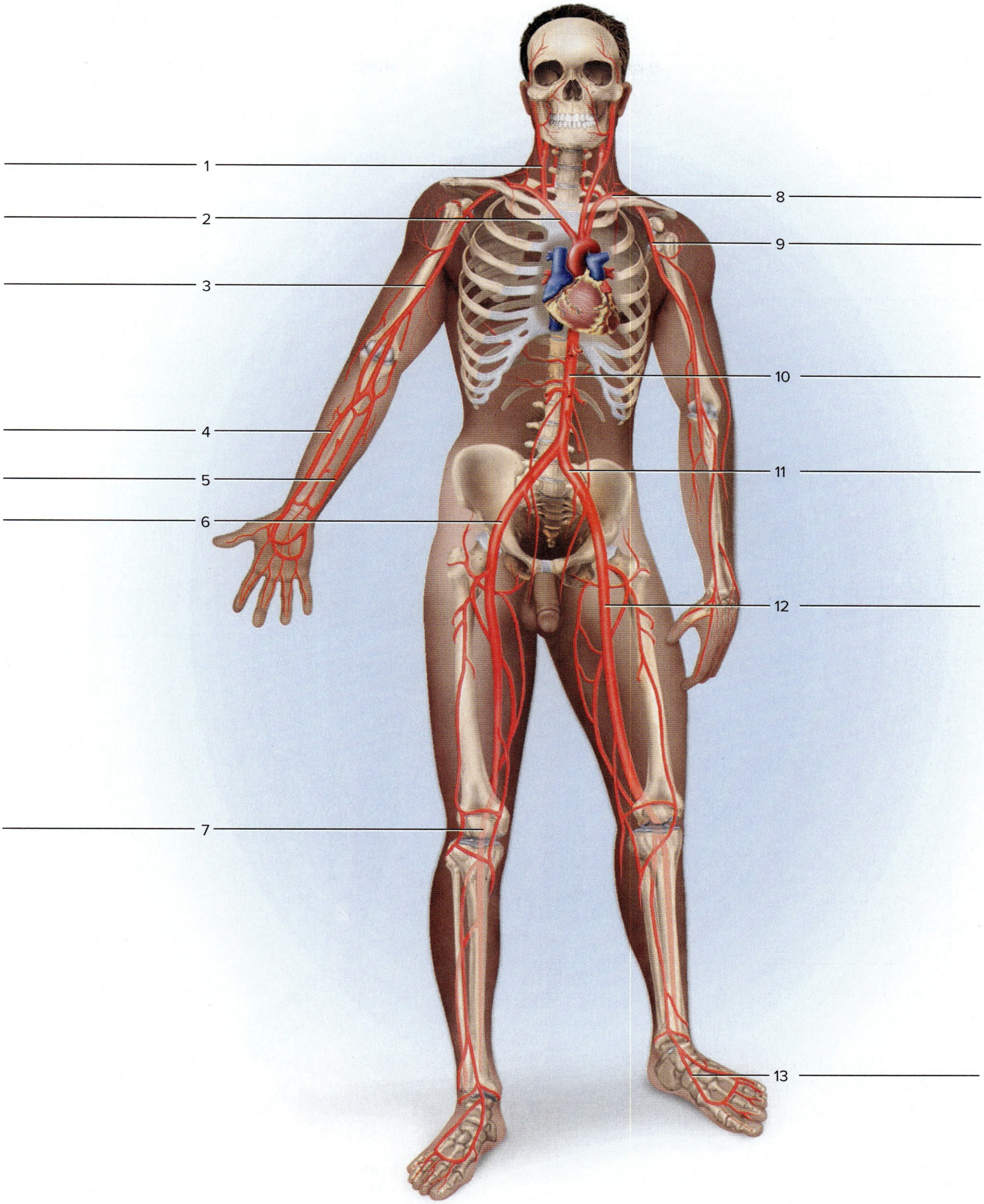

FIGURE 47.18 Label the major systemic arteries (posterior arteries are faded). A4

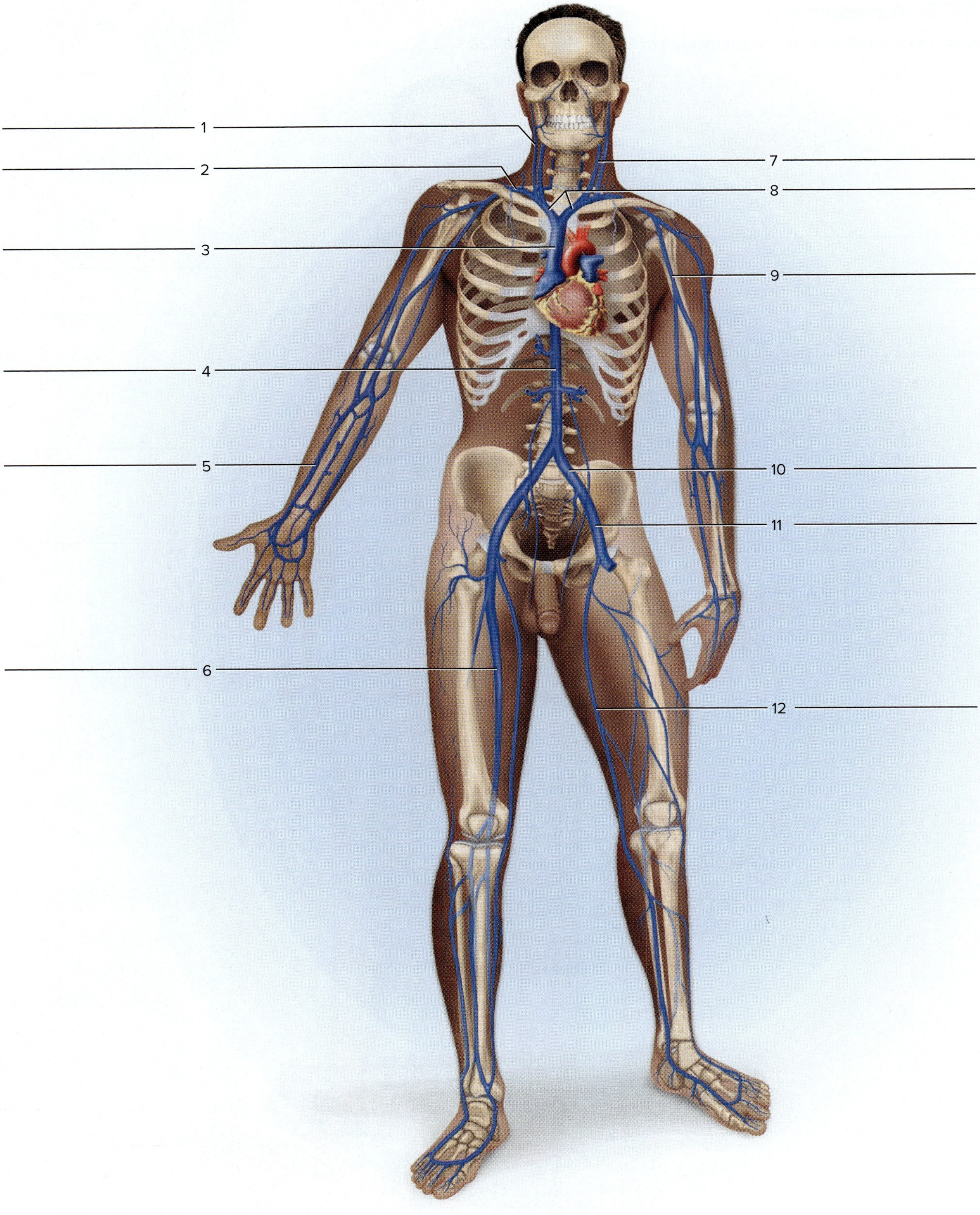

FIGURE 47.19 Label the major systemic veins (posterior veins are faded). A5

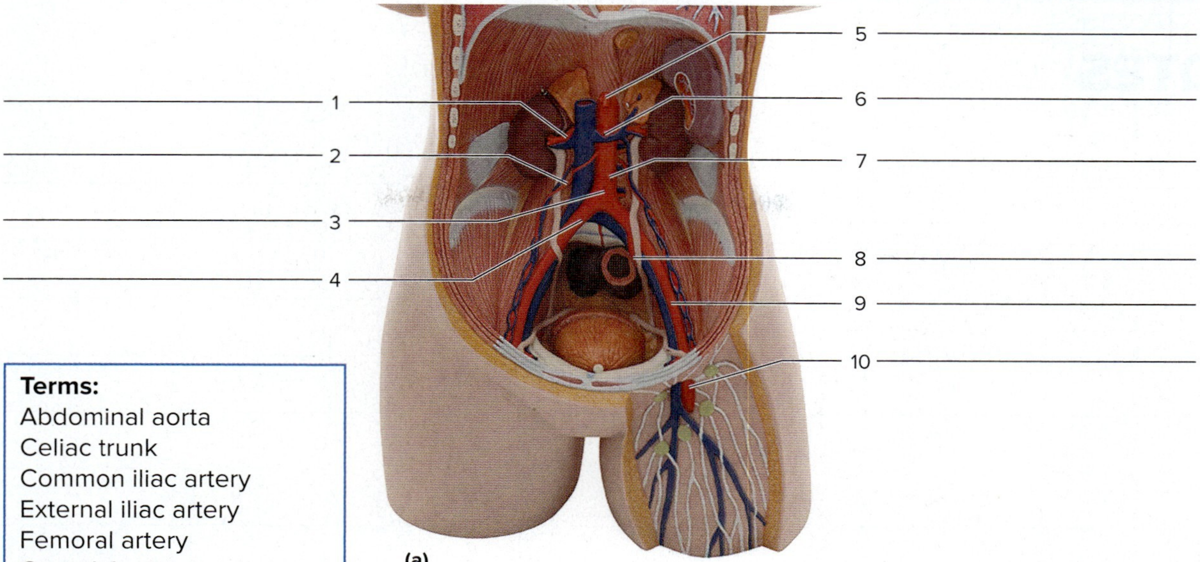

(a)

Terms:
Abdominal aorta
Celiac trunk
Common iliac artery
External iliac artery
Femoral artery
Gonadal artery
Inferior mesenteric artery
Internal iliac artery
Superior mesenteric artery
Renal artery

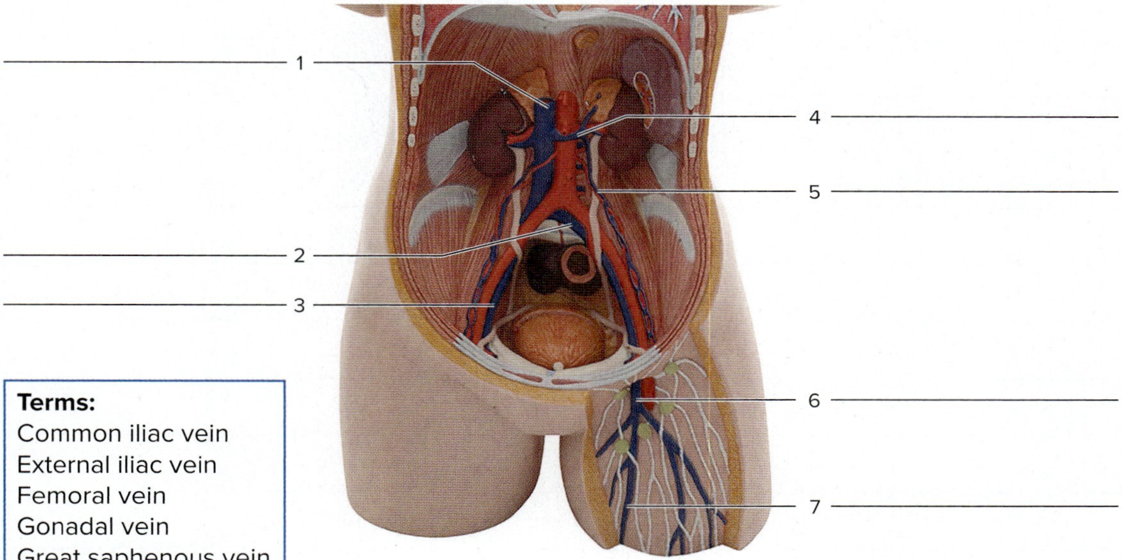

(b)

Terms:
Common iliac vein
External iliac vein
Femoral vein
Gonadal vein
Great saphenous vein
Inferior vena cava
Renal vein

FIGURE 47.20 Anterior view of torso model image showing major (a) arteries and (b) veins. ©3B Scientific GmbH, Germany, 2017 www.3bscientific.com A4 A5

NOTES

LABORATORY
EXERCISE 48

Pulse Rate and Blood Pressure

Purpose of the Exercise

To examine the pulse, to determine the pulse rate, to measure blood pressure, and to investigate the effects of body position and exercise on pulse rate and blood pressure.

MATERIALS NEEDED

Clock with second hand
Sphygmomanometer
Stethoscope
Alcohol swabs (wipes)

For Demonstration Activity:

Pulse pickup transducer or plethysmograph
Physiological recording apparatus

Chapter Opener: ©101cats/Getty Images

Learning Outcomes

After completing this exercise, you should be able to:

O1 Determine pulse rate and correlate pulse characteristics.

O2 Test the effects of various factors on pulse rate.

O3 Measure and record blood pressure using a sphygmomanometer and determine pulse pressure.

O4 Test the effects of various factors on blood pressure.

O5 Distinguish components of pulse and blood pressure measurements.

The O corresponds to the assessments A indicated in the Laboratory Assessment for this Exercise.

Pre-Lab

Carefully read the introductory material and examine the entire lab. Be familiar with characteristics of pulse and blood pressure from lecture or the textbook. Answer the pre-lab questions.

Pre-Lab Questions Select the correct answer for each of the following questions:

1. The _____ is responsible for the pulse wave in the systemic circuit.
a. right atrium **b.** right ventricle
c. left atrium **d.** left ventricle

2. Which of the following would *not* be palpated on the surface of the body?
a. facial artery **b.** dorsalis pedis artery
c. common iliac artery **d.** common carotid artery

3. The _____ is the most common artery palpated to determine heart rate.
a. temporal artery **b.** radial artery
c. popliteal artery **d.** posterior tibial artery

4. The _____ is the standard artery used to determine blood pressure.
a. brachial artery **b.** common carotid artery
c. femoral artery **d.** posterior tibial artery

5. Which of the following resting blood pressures would be considered hypertension?
a. 100/60 mm Hg **b.** 110/76 mm Hg
c. 120/80 mm Hg **d.** 140/90 mm Hg

6. The pulse of the _____ can be palpated in the upper medial thigh.
a. popliteal artery **b.** posterior tibial artery
c. common iliac artery **d.** femoral artery

7. The thumb should be used to sense the pulse when palpating an artery.
a. True **b.** False

8. Pulse pressure is the difference between the systolic and diastolic pressures.
a. True **b.** False

9. The blood pressure in a capillary is higher than in a small artery.

a. True b. False

10. The stroke volume is the amount of blood pumped by a ventricle in 1 minute.

a. True b. False

The sudden surge of blood that enters the arteries each time the ventricles of the heart contract (systole) causes the elastic walls of these vessels to expand. Then, as the ventricles relax (diastole), the arterial walls recoil. This alternate expanding and recoiling of an arterial wall can be palpated (felt) as a *pulse* in vessels that are near the surface of the body. Because most of the major arteries are deep in the body, there are a limited number of possible locations to palpate the pulse. If any artery is superficial and firm tissue is just beneath the artery, a pulse can be palpated at that location. The number of pulse expansions per minute correlates with the heart rate or cardiac cycles. The left ventricle is the systemic pumping chamber, and it is therefore responsible for the pulse wave and for the blood pressure in the arteries selected for these assessments.

The force exerted by the blood pressing against the inner walls of arteries creates *blood pressure.* This systemic arterial pressure reaches its maximum, called the *systolic pressure,* during contraction of the left ventricle. Because the left ventricle relaxes when it fills with blood again, the blood pressure then drops to its lowest level, called the *diastolic pressure.* Blood pressure is measured in millimeters of mercury (mm Hg) and is expressed as a systolic pressure over diastolic pressure. A normal adult resting blood pressure is considered to be less than 120/80 mm Hg. If pressures during resting conditions are too high, the condition is called hypertension; if pressures are too low, the condition is called hypotension.

The pulse and the systolic and diastolic blood pressures have some close relationships. The difference between the systolic pressure and the diastolic pressure is called the *pulse pressure,* which is related to the natural elasticity and recoil of the arteries. Obtain the pulse pressure by subtracting the diastolic pressure from the systolic pressure. For example, a person with a blood pressure of 120/80 mm Hg would have a normal pulse pressure of 40 mm Hg. The pulse pressure indicates the force exerted upon the arteries from ventricular contraction and indicates the general condition of the cardiovascular system. An expanded pulse pressure could be an indication of atherosclerosis and hypertension. There are many factors that influence blood pressure, so it is important to realize that a single measurement does not provide enough information to make conclusions about health.

After blood has passed through the capillaries, the blood pressure established by the heart is no longer sufficient to return blood to the heart via the veins. The skeletal muscle pump of the limbs and the respiratory pump of the torso assist in venous return. When the skeletal muscles of the limbs contract, they squeeze in on the blood in the veins of the limbs, forcing the blood into the torso (valves in the veins prevent the backflow of blood). In the torso there are two cavities, the thoracic cavity and abdominopelvic cavity, divided by the diaphragm. When the diaphragm contracts (along with the external intercostal muscles), the thoracic cavity volume increases and the pressure decreases. Simultaneously, the volume of the abdominopelvic cavity decreases and the pressure increases. This difference in pressure between the abdominopelvic cavity and the thoracic cavity assists the movement of blood into the veins of the thoracic cavity and back into the heart. As a result, more blood enters the heart, and the heart rate increases to pump the additional blood. Less blood enters the heart during expiration, since the thoracic cavity volume decreases and its pressure increases.

PROCEDURE A: Pulse Rate

1. Identify some of the pulse locations indicated in figure 48.1.

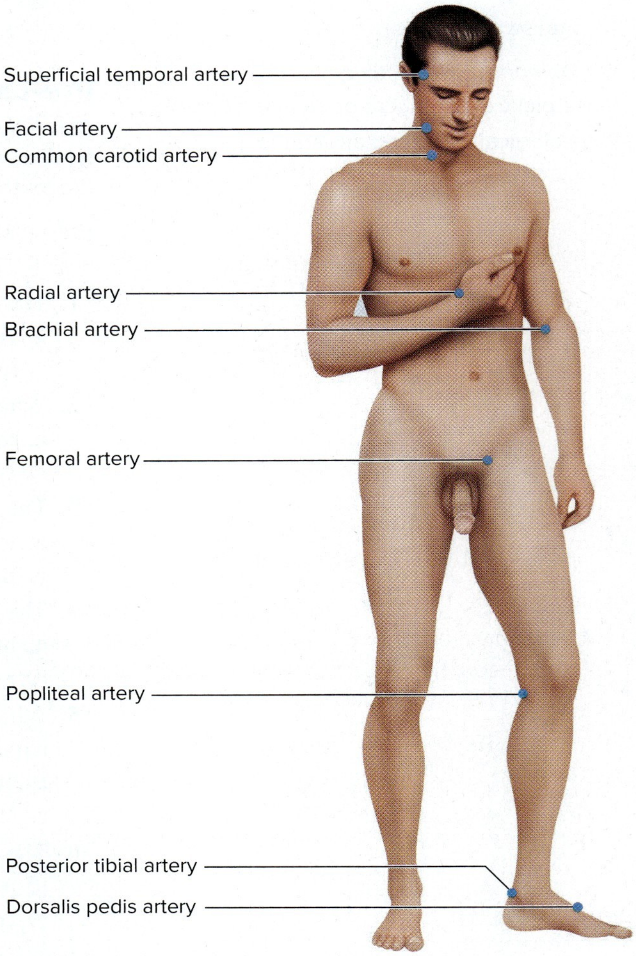

FIGURE 48.1 Locations where an arterial pulse can be palpated.

2. Examine your laboratory partner's radial pulse. To do this, follow these steps:
   a. Have your partner sit quietly, remaining as relaxed as possible.
   b. Locate the pulse by placing your index and middle fingers over the radial artery on the anterior surface near the lateral side of the wrist (fig. 48.2). Do not use your thumb for sensing the pulse because you may feel a pulse coming from an artery in the thumb. The radial artery is most often used because it is easy and convenient to locate.
   c. To determine the pulse rate, count the number of pulses that occur in 1 minute. This can be accomplished by counting pulses in 30 seconds and multiplying that number by 2. (A pulse count for a full minute, although taking a little longer, would give a better opportunity to detect pulse characteristics and irregularities.)
   d. Note the characteristics of the pulse. That is, can it be described as regular or irregular, strong or weak, hard or soft? The pulse should be regular and the amplitude (magnitude) will decrease as the distance from the left ventricle increases. The strength of the pulse reveals some indications of blood pressure. Under high blood pressure, the pulse feels very hard and strong; under low blood pressure, the pulse feels weak and can be easily compressed.
   e. In Part A of Laboratory Assessment 48, record the pulse rate and pulse characteristics determined while sitting.

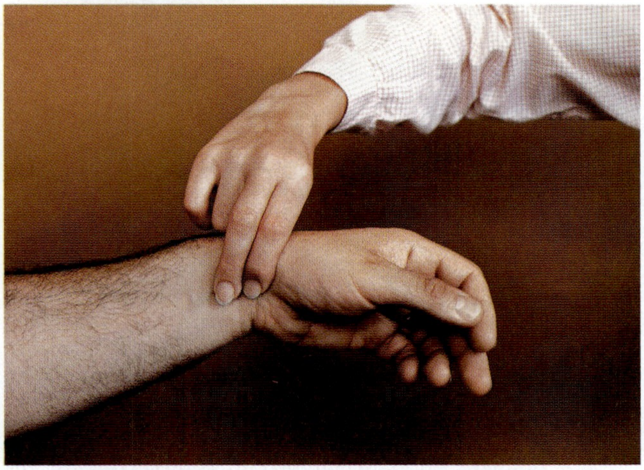

FIGURE 48.2 Taking a pulse rate from the radial artery.
Image Source/Getty Images

LEARNING EXTENSION ACTIVITY

Determine the pulse rate and pulse characteristics in two additional locations on your laboratory partner. Locate and record the pulse rate from the common carotid artery and the dorsalis pedis artery (fig. 48.1). A1

Common carotid artery pulse rate ______________

Dorsalis pedis artery pulse rate ______________

Compare the amplitude of the pulse characteristic between these two pulse locations, and interpret any of the variations noted. ______________

____

____

DEMONSTRATION ACTIVITY

If the equipment is available, the laboratory instructor will demonstrate how a photoelectric pulse pickup transducer or plethysmograph can be used together with a physiological recording apparatus to record the pulse. Such a recording allows an investigator to analyze certain characteristics of the pulse more precisely than is possible using a finger to examine the pulse. For example, the pulse rate can be determined very accurately from a recording, and the heights of the pulse waves provide information about the blood pressure.

3. Repeat the procedure and determine the pulse rate and pulse characteristics in each of the following conditions:
   a. immediately after lying down;
   b. 3–5 minutes after lying down;
   c. immediately after standing;
   d. 3–5 minutes after standing quietly;
   e. immediately after 3 minutes of moderate exercise (*omit if the person has health problems*);
   f. 3–5 minutes after exercise has ended.
   g. Record the pulse rates and pulse characteristics in Part A of the laboratory assessment.
4. Complete Part A of the laboratory assessment.

PROCEDURE B: Blood Pressure

Reexamine the introductory material related to blood pressure. Several basic structural and functional factors influence arterial pressure, including *cardiac output, blood volume,* and *peripheral resistance.* Cardiac output is the volume of blood discharged from a ventricle in a minute. The blood ejected from a single ventricle contraction is about 70 mL and is called the *stroke volume.* A 75 beats/minute resting heart rate and a stroke volume of 70 mL would calculate as a cardiac output of 5,250 mL/minute, which is about the average adult blood volume. If either heart rate or stroke volume increases, cardiac output would increase as would blood pressure. Blood volume influences blood pressure. Because blood pressure is directly proportional to blood volume, any volume change can alter blood pressure. Peripheral resistance refers to the amount of friction between the blood and the walls of the blood vessels. Although blood viscosity and blood vessel length influence peripheral resistance, changes in blood vessel diameter as a result of contraction (vasoconstriction) or relaxation (vasodilation) of smooth muscle of the middle tunic (tunica media) are the most significant. Certain conditions, such as obesity, excess salt intake, inactive lifestyle, stress, smoking, certain medications, and others can aggravate blood pressures. Be aware of these factors and conditions influencing blood pressure as you take a blood pressure measurement using a sphygmomanometer and a stethoscope.

1. Although blood pressure is present in all blood vessels, the standard location to record blood pressure is the brachial artery. Examine figure 48.3, which indicates various pressure relationships throughout the systemic circuit.
2. Measure your laboratory partner's arterial blood pressure. To do this, follow these steps:
   a. Obtain a sphygmomanometer and a stethoscope.
   b. Clean the earpieces and the diaphragm of the stethoscope with alcohol swabs.
   c. Have your partner sit quietly with upper limb resting on a table at heart level. Have the person remain as relaxed as possible.
   d. Locate the brachial artery at the antecubital space. Wrap the cuff of the sphygmomanometer around the arm so that its lower border is about 2.5 cm above the bend of the elbow. Center the bladder of the cuff in line with the *brachial pulse* (fig. 48.4).
   e. Palpate the *radial pulse*. Close the valve on the neck of the rubber bulb connected to the cuff and pump air from the bulb into the cuff. Inflate the cuff while watching the sphygmomanometer and note the pressure when the pulse disappears. (This is a rough estimate of the systolic pressure.) Immediately deflate the cuff. *Do not leave the cuff inflated for more than 1 minute.*
   f. Position the stethoscope over the brachial artery. Reinflate the cuff to a level 30 mm Hg higher than the point where the pulse disappeared during palpation (fig. 48.5). When the cuff is inflated above the systolic pressure, no blood flows distal to the cuff, and no sound is heard.
   g. Slowly open the valve of the bulb until the pressure in the cuff drops at a rate of about 2 or 3 mm Hg per second.
   h. Listen for sounds (*Korotkoff sounds*) from the brachial artery. When the systolic pressure becomes greater than the cuff's pressure, the brachial artery is forced open. When the first loud tapping sound is heard, record the reading as the systolic pressure. This indicates the pressure exerted against the arterial wall during systole.
   i. Continue to listen to the sounds as the pressure drops and note the level when the last sound is heard. Record this reading as the diastolic pressure, which measures the constant arterial resistance. When sounds are no longer heard, the cuff's pressure no longer compresses the brachial artery enough to restrict blood flow to the upper limb.
   j. Release all of the pressure from the cuff.

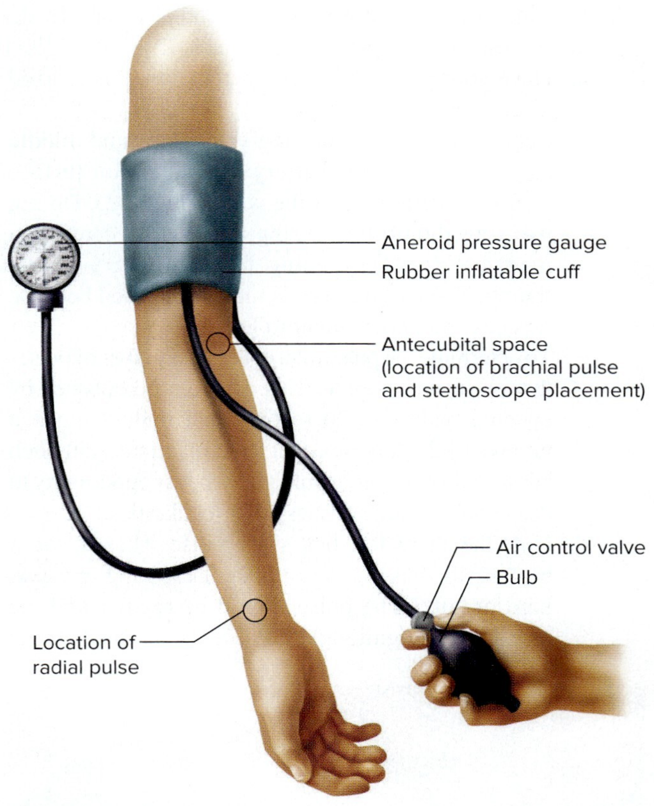

FIGURE 48.4 Blood pressure is commonly measured by using a sphygmomanometer (blood pressure cuff). The use of a column of mercury is the most accurate measurement, but due to environmental concerns, it has been replaced by alternative gauges and digital readouts.

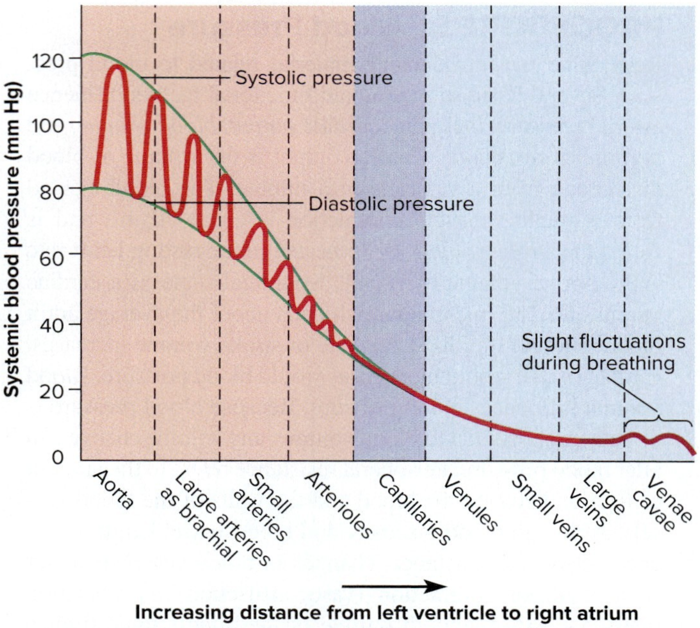

FIGURE 48.3 Pressure changes in blood vessels of the systemic circuit.

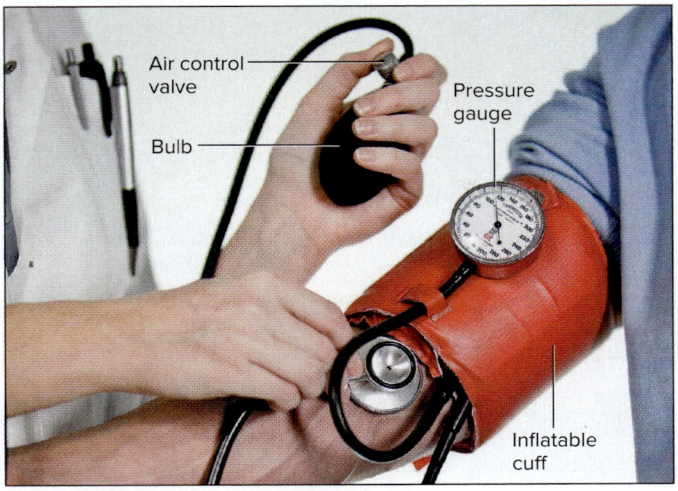

FIGURE 48.5 Sphygmomanometer and stethoscope placement on arm while taking blood pressure. Johner Images/SuperStock

   k. Repeat the procedure until you have two blood pressure measurements from each arm, allowing 2–3 minutes of rest between readings.
   l. Average your readings and enter them in the table in Part B of the laboratory assessment.

3. Measure your partner's blood pressure in each of the following conditions:
   a. 3–5 minutes after lying down;
   b. 3–5 minutes after standing quietly;
   c. immediately after 3 minutes of moderate exercise *(omit if the person has health problems);*
   d. 3–5 minutes after exercise has ended.
   e. Record the blood pressures in Part B of the laboratory assessment.

4. Complete Parts B and C of the laboratory assessment.

NOTES

LABORATORY
ASSESSMENT 48

Name ______________________________

Date ______________________________

Section ______________________________

The A corresponds to the indicated Learning Outcome(s) O found at the beginning of the Laboratory Exercise.

Pulse Rate and Blood Pressure

PART A: Assessments

1. Enter your observations of pulse rates and pulse characteristics in the table. A1 A2

Test Subject	Pulse Rate (Beats/min)	Pulse Characteristics
Sitting		
Lying down		
3–5 minutes later		
Standing		
3–5 minutes later		
After exercise		
3–5 minutes later		

2. Summarize the effects of body position and exercise on the pulse rates and pulse characteristics. A2 ______________________________

____

____

____

PART B: Assessments

1. Enter the initial measurements of blood pressure from a relaxed sitting position in the table. A3

Reading	Blood Pressure in Right Arm	Blood Pressure in Left Arm
First		
Second		
Average		

2. What was the average pulse pressure calculated from the blood pressure recorded from the arms? A3 ______

3. Enter your test results in the table. A4

Test Subject	Blood Pressure
3–5 minutes after lying down	
3–5 minutes after standing	
After 3 minutes of moderate exercise	
3–5 minutes later	

4. Summarize the effects of body position and exercise on blood pressure. A4 ______

5. Summarize any correlations between pulse rate and blood pressure from any of the experimental conditions. A5 ______

PART C: Assessments

Complete the following statements:

1. The maximum pressure achieved during ventricular contraction is called ________________ pressure. A5

2. The lowest pressure that remains in the arterial system during ventricular relaxation is called ________________ pressure. A5

3. The pulse rate is equal to the ________________ rate. A5

4. A pulse that feels full and is not easily compressed is produced by an elevated ________________. A5

5. The instrument commonly used to measure systemic arterial blood pressure is called a ________________. A5

6. Blood pressure is expressed in units of ________________. A5

7. The upper number of the fraction used to record blood pressure indicates the ________________ pressure. A5

8. The ________________ artery in the arm is the standard systemic artery in which blood pressure is measured. A5

9. A person with a blood pressure of 134/82 mm Hg would have a pulse pressure of ________________ mm Hg. A5

CRITICAL THINKING ASSESSMENT—PART C

When a pulse is palpated and counted, which pressure (systolic or diastolic) is characteristic at that moment? Explain your answer. A1 A5 ____

____

____

____.

NOTES

LABORATORY EXERCISE 49

Lymphatic System

Purpose of the Exercise

To review the anatomical structures and features of the lymphatic system and to observe the microscopic structure and function of a lymph node, the thymus, the spleen, and a tonsil.

MATERIALS NEEDED

Human torso model
Anatomical chart of the lymphatic system
Compound light microscope
Prepared microscope slides:
- Lymph node section
- Human thymus section
- Human spleen section
- Human tonsil section

Chapter Opener: ©101cats/Getty Images

Learning Outcomes APR

After completing this exercise, you should be able to:

O1 Trace the major lymphatic pathways and identify the components of the lymphatic system.

O2 Locate and sketch the major microscopic structures of a lymph node, the thymus, the spleen, and a tonsil.

O3 Describe the structure and function of a lymphatic vessel, a lymph node, the thymus, the spleen, and a tonsil.

The O corresponds to the assessments A indicated in the Laboratory Assessment for this Exercise.

Pre-Lab

Carefully read the introductory material and examine the entire lab. Be familiar with the basic structures and functions of lymphatic pathways, lymph nodes, thymus, spleen, and tonsils from lecture or the textbook. Answer the pre-lab questions.

Pre-Lab Questions Select the correct answer for each of the following questions:

1. Lymph finally flows into lymphatic _____ just before connecting to a blood vessel.
a. capillaries **b.** collecting ducts
c. vessels **d.** trunks

2. The right lymphatic duct drains only the _____ quadrant of the entire body.
a. right upper **b.** left upper
c. right lower **d.** left lower

3. Which of the following areas would contain the fewest lymph nodes?
a. cervical **b.** axillary
c. inguinal **d.** plantar

4. The _____ is the largest lymphatic organ.
a. thymus **b.** pharyngeal tonsil
c. spleen **d.** popliteal lymph node

5. The thoracic duct drains into the
a. right internal jugular vein. **b.** right subclavian vein.
c. left internal jugular vein. **d.** left subclavian vein.

6. Which of the following is *not* a lymphatic organ?
a. tonsil **b.** thymus
c. thyroid **d.** spleen

7. Lymph flows faster during periods of exercise.
a. True **b.** False

8. The thymus continues to enlarge as a person ages.
   a. True b. False
9. Lymphatic capillaries and blood capillaries are open at both ends.
   a. True b. False

The lymphatic system is closely associated with the cardiovascular system and includes a network of vessels that assist in the circulation of body fluids. Fluid in the lymphatic vessels is *lymph.* As blood flows through the arterial ends of capillary beds, fluids and some dissolved biochemicals continually filter from the blood plasma, with most of the plasma proteins left behind in the blood. Most of the fluids are reabsorbed back into the blood at the venous ends of the capillary beds; however, the amount not reabsorbed becomes part of the interstitial fluids among the tissue cells. Lymphatic vessels provide necessary pathways through which this excess tissue fluid can be carried away from the interstitial spaces and returned to the bloodstream (fig. 49.1). Without the lymphatic system, this fluid would accumulate in tissue spaces, producing *edema.*

The *lymphatic capillaries* are microscopic, blind-ended vessels, in which excess interstitial fluid first enters the lymphatic system (fig. 49.2). These capillaries drain into *lymphatic collecting vessels,* then into larger *lymphatic trunks,* and finally into one of two *collecting ducts* just before the connections with the subclavian veins (fig. 49.1). As a result of the orientation of valves, only one-way flow of lymph occurs (fig. 49.3). This happens during skeletal muscle contraction and breathing, causing compression upon the vessels that forces fluid through the lymphatic vessels. The larger *thoracic duct* provides drainage for the entire body except for the right upper quadrant. The thoracic duct connects to the left subclavian vein, while a *right lymphatic duct* drains only the right upper quadrant into the right subclavian vein (fig. 49.4).

Along the pathway of the lymphatic vessels, *lymph nodes* are positioned; they are especially concentrated in

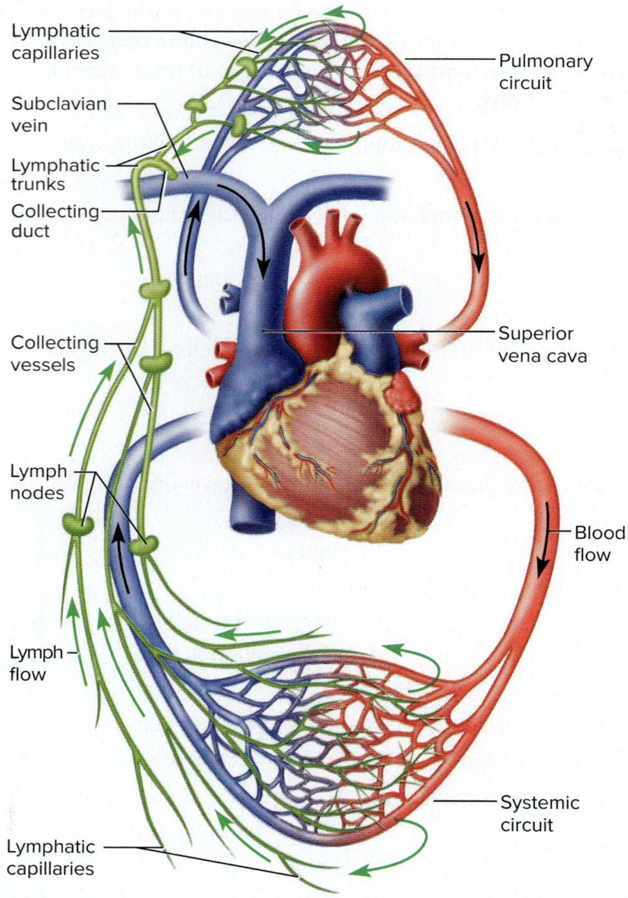

FIGURE 49.1 Schematic relationship of fluid movements between the cardiovascular and lymphatic systems. APR

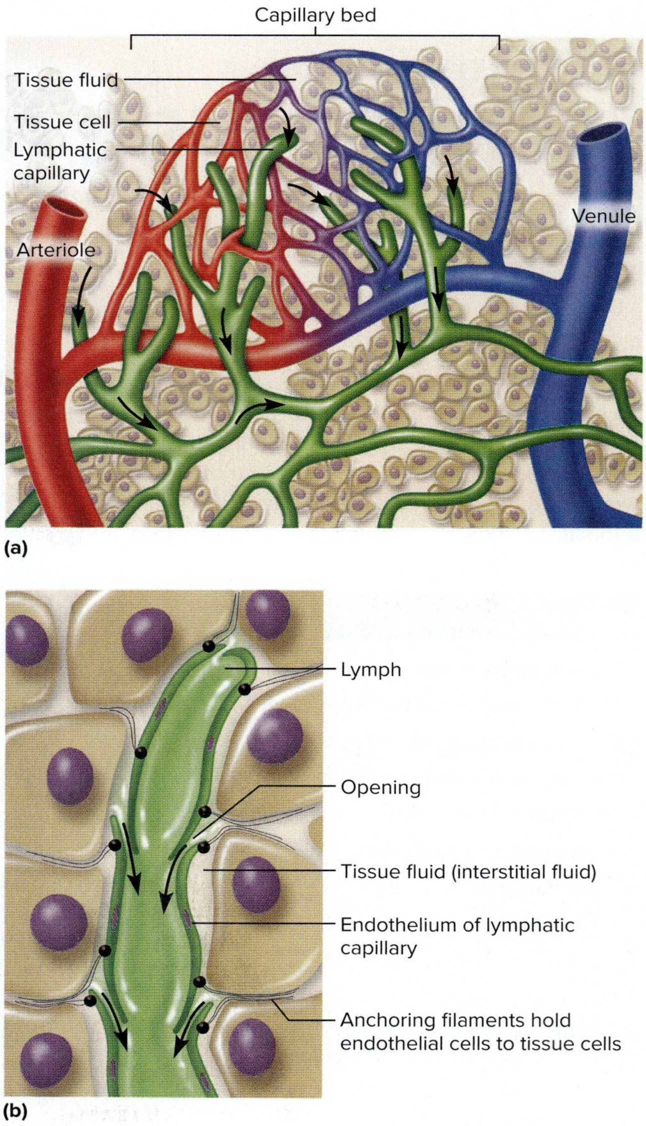

FIGURE 49.2 (a) Blind-ended lymphatic capillaries among tissue cells and blood vessel capillaries; (b) uptake of tissue fluid between overlapping endothelial cells of lymphatic capillary.

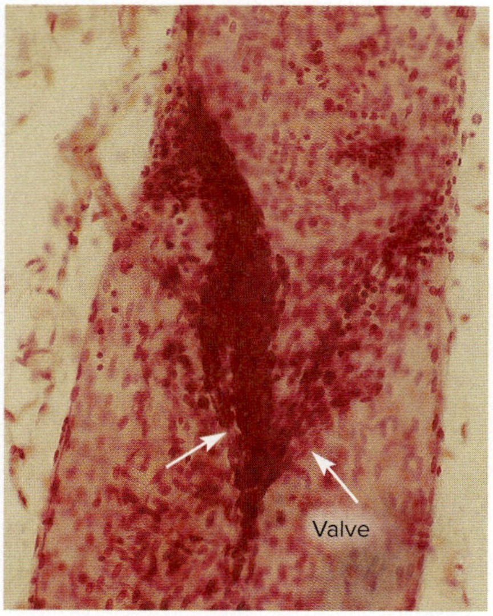

FIGURE 49.3 Micrograph of a valve within a lymphatic vessel (60x). The arrows indicate the flaplike valve that helps prevent backflow of lymph. Dennis Strete/McGraw Hill

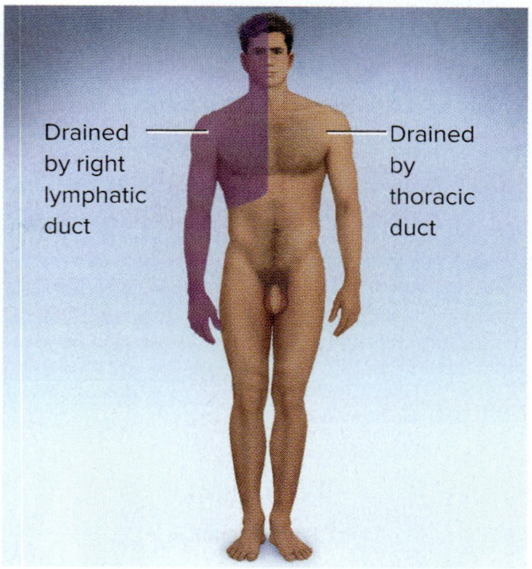

FIGURE 49.4 Regions of the body drained by the right lymphatic duct are shaded purple. The unshaded regions of the body are drained by the larger thoracic duct into the left subclavian vein.

areas of major joints like axillary and inguinal regions. The lymph nodes contain concentrations of lymphocytes and macrophages that cleanse the lymph and activate an immune response. Swollen and sore lymph nodes represent areas of infection that occur as our bodies fight off the microorganisms.

Other lymphatic organs are part of the lymphatic system. The red bone marrow is a major production site of white blood cells. The *thymus,* located within the mediastinum, is a location of T lymphocyte maturation. The thymus is proportionately very large in a fetus and in a child, but it begins to atrophy around puberty, and this continues as we age. The *spleen,* the largest lymphatic organ, is located in the left upper quadrant of the abdominopelvic cavity. The spleen is highly vascular and contains large numbers of red blood cells, lymphocytes, and macrophages. Palatine, pharyngeal, and lingual *tonsils* are located in regions of the entrance into the pharynx (throat). The pharyngeal tonsils are also called the adenoids. Numerous lymphatic nodules within the tonsils contain concentrations of lymphocytes. Tissues lining certain passages open to outside the body, such as the gastrointestinal and respiratory tracts, contain scattered lymphocytes called mucosa-associated lymphatic tissue (MALT) in the submucosal membrane that help protect the mucosa. The organs of the lymphatic system help to defend the tissues against infections by filtering particles from the lymph and by supporting the activities of lymphocytes that furnish immunity against specific disease-causing agents or pathogens.

PROCEDURE A: Lymphatic Pathways

1. Study figures 49.5 and 49.6.
2. Observe the human torso model and the anatomical chart of the lymphatic system, and locate the following features:

   lymphatic vessels—collect lymph from microscopic lymphatic capillaries

   lymph nodes—bean-shaped organs that cleanse lymph and activate immune response

   lymphatic trunks—collect lymph from lymphatic vessels

   - lumbar trunk
   - intestinal trunk
   - bronchomediastinal trunk
   - subclavian trunk
   - jugular trunk

   cisterna chyli—inferior expanded sac of thoracic duct where fatty intestinal lymph collects

   collecting ducts—largest lymphatic vessels; collect lymph from lymphatic trunks

   - thoracic duct—connects to left subclavian vein near left internal jugular vein
   - right lymphatic duct—connects to right subclavian vein near right internal jugular vein

   internal jugular veins

   subclavian veins—receive lymph from collecting ducts
3. Complete Part A of Laboratory Assessment 49.

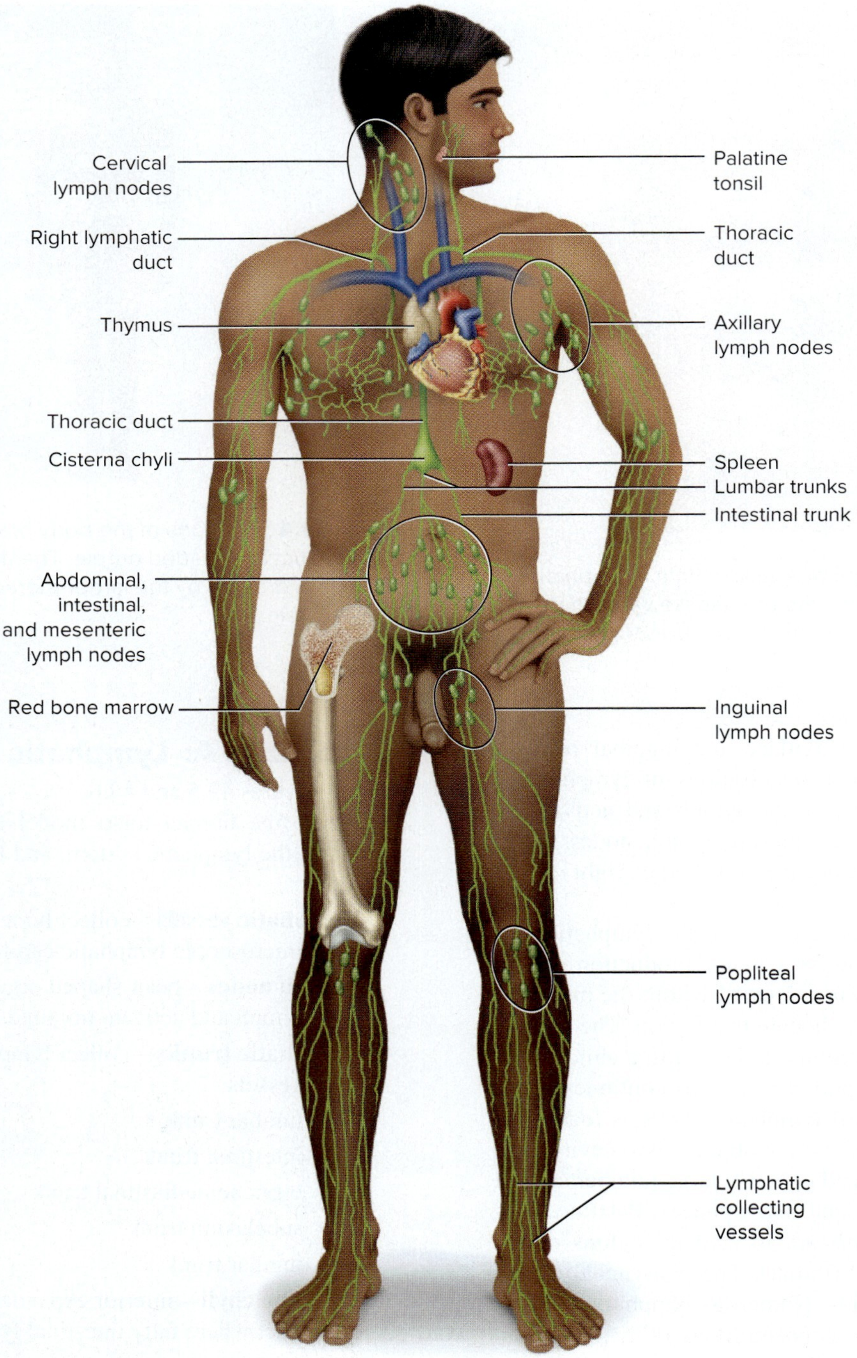

FIGURE 49.5 Lymphatic vessels, lymph nodes, and lymphatic organs. APR

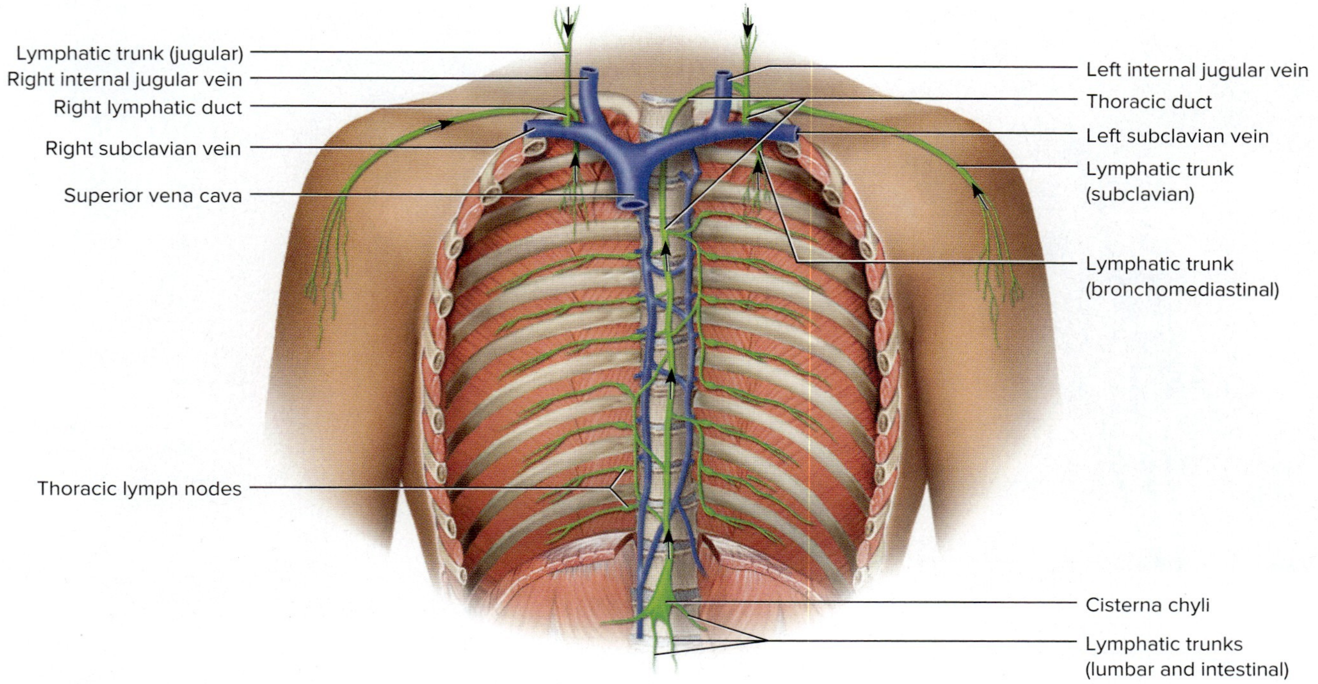

FIGURE 49.6 Lymphatic pathways into the subclavian veins. The arrows indicate the direction of lymph drainage.

PROCEDURE B: Lymph Nodes

1. Study the lymph node locations in figures 49.5 and 49.6.
2. Observe the anatomical chart of the lymphatic system and the human torso model, and locate the clusters of lymph nodes in the following regions:
   cervical region
   axillary region
   popliteal region
   inguinal region
   pelvic cavity
   abdominal cavity
   thoracic cavity
3. Palpate the lymph nodes in your cervical region. They are located along the lower border of the mandible and between the ramus of the mandible and the sternocleidomastoid muscle. They feel like small, firm lumps.
4. Study figures 49.7 and 49.8.
5. Obtain a prepared microscope slide of a lymph node and observe it, using scanning and then low-power magnification (fig. 49.8*c*). Identify the *capsule* surrounding the node that is mainly composed of collagen fibers, and the *lymphatic nodules* that appear as dense masses near the surface of the node. Each nodule contains lymphocytes, macrophages, plasma cells, reticular cells, and reticular fibers as well as a germinal center with proliferating B lymphocytes that differentiate to plasma cells when combating a pathogen. There are *lymphatic sinuses* that appear as narrow spaces where lymph circulates between the nodules and the capsule.

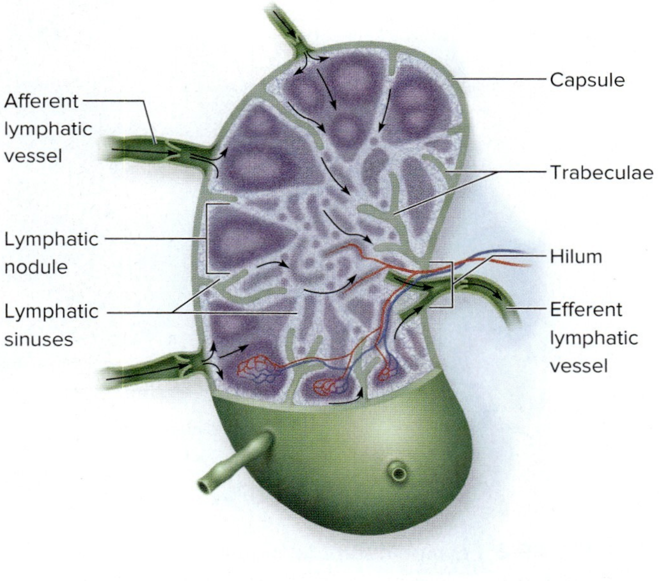

FIGURE 49.7 Diagram of a lymph node showing internal structures, attached lymphatic vessels, and arrows indicating lymph flow. APR

6. Using high-power magnification, examine a nodule within the lymph node. The nodule contains densely packed *lymphocytes*.
7. Prepare a labeled sketch of a representative section of a lymph node in Part B of the laboratory assessment.
8. Complete Part C of the laboratory assessment.

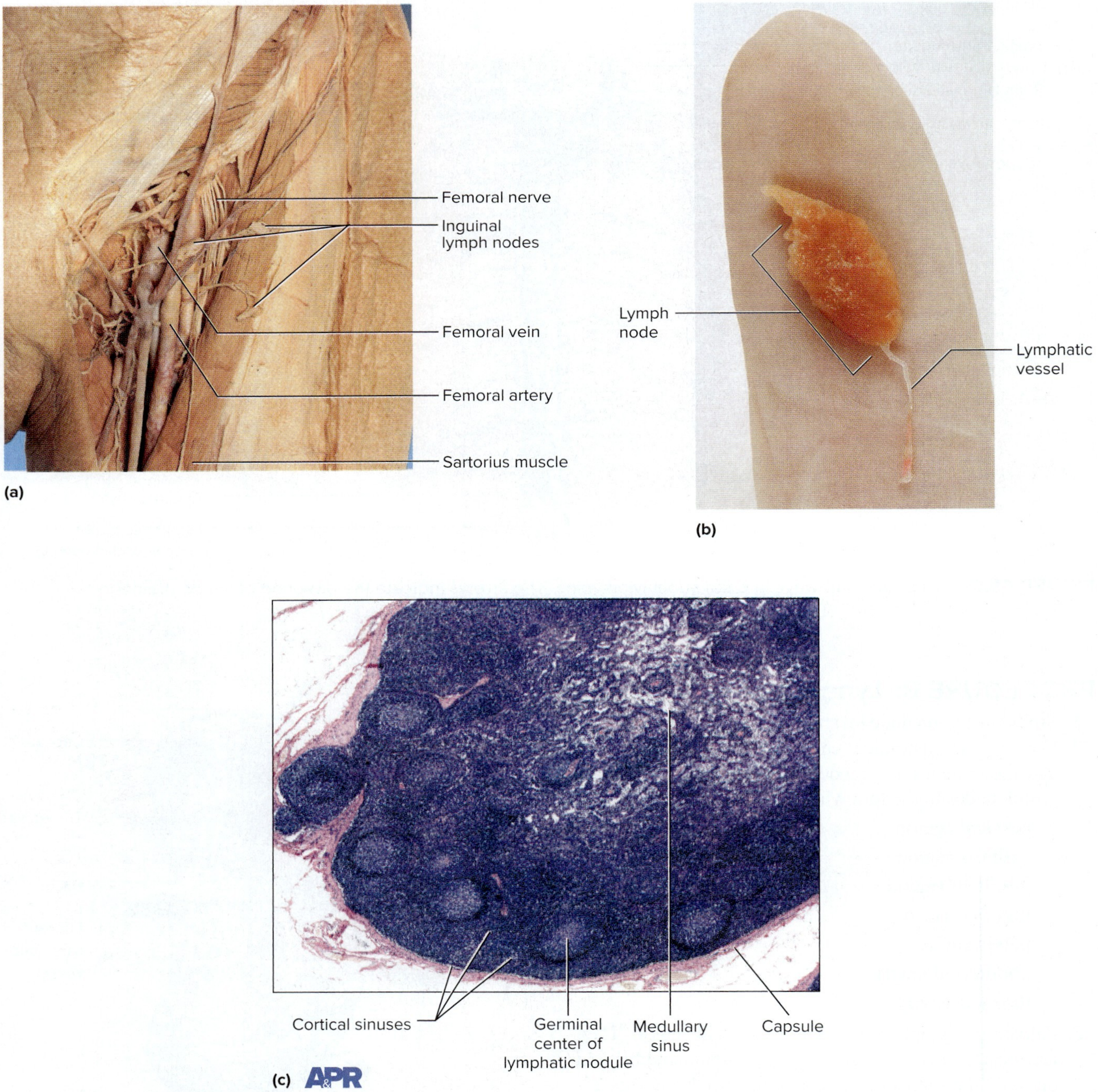

FIGURE 49.8 Lymph nodes: (a) inguinal lymph nodes in a cadaver; (b) photograph of a human lymph node on tip of finger; (c) micrograph of a lymph node (20×). (a) Rebecca Gray/McGraw Hill; (b) ©J & J Photography; (c) Alvin Telser/McGraw Hill

PROCEDURE C: Thymus, Spleen, and Tonsil

1. Locate the thymus, spleen, and tonsils in the anatomical chart of the lymphatic system and on the human torso model.
2. Obtain a prepared microscope slide of human thymus and observe it, using scanning and then low-power magnification (fig. 49.9). Note how the thymus is subdivided into *lobules* by *septa* of connective tissue that contain blood vessels. Identify the capsule of loose connective tissue that surrounds the thymus, the outer *cortex* of a lobule composed of densely packed cells and deeply stained, and the inner *medulla* of a lobule composed of loosely packed mature T lymphocytes and epithelial cells and lightly stained.

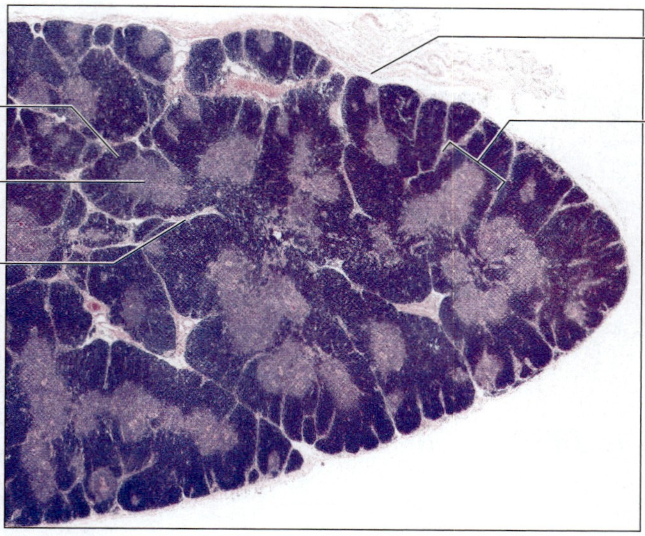

FIGURE 49.9 Micrograph of a section of the thymus (10×). Jose Luis Calvo/Shutterstock APR

3. Examine the cortex tissue of a lobule using high-power magnification. The cells of the cortex are composed of densely packed developing *T lymphocytes* among some epithelial cells and *macrophages*. Some of these cortical cells may be undergoing mitosis, so their chromosomes may be visible.
4. Prepare a labeled sketch of a representative section of the thymus in Part B of the laboratory assessment.
5. Obtain a prepared slide of the human spleen and observe it, using scanning and then low-power magnification (fig. 49.10). Identify the *capsule* of dense connective tissue that surrounds the spleen. The tissues of the spleen include circular *nodules* of *white pulp* enclosed in a matrix of *red pulp*.
6. Using high-power magnification, examine a nodule of white pulp and red pulp. The cells of the white pulp are mainly lymphocytes. Also, there may be an arteriole centrally located in the nodule. The cells of the red pulp are mostly red blood cells with many lymphocytes and macrophages. The macrophages engulf and destroy aging red blood cells, cellular debris, and bacteria.
7. Prepare a labeled sketch of a representative section of the spleen in Part B of the laboratory assessment.
8. Obtain a prepared microscope slide of a tonsil and observe it, using scanning and then low-power magnification (fig. 49.11*b*). Identify the surface epithelium, deep invaginated pits called *tonsillar crypts,* and numerous lymphatic nodules. Although the tonsillar crypts often harbor bacteria and food debris, infections are usually prevented within these lymphoid organs.
9. Prepare a labeled sketch of a representative section of a tonsil in Part B of the laboratory assessment.
10. Complete Part D of the laboratory assessment.

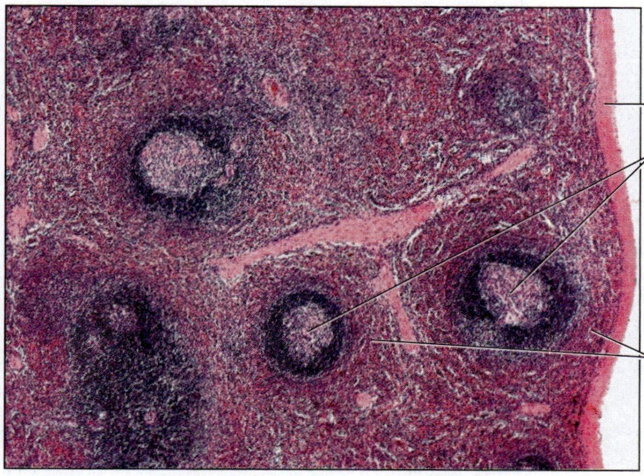

FIGURE 49.10 Micrograph of a section of the spleen (40×). Al Telser/McGraw Hill APR

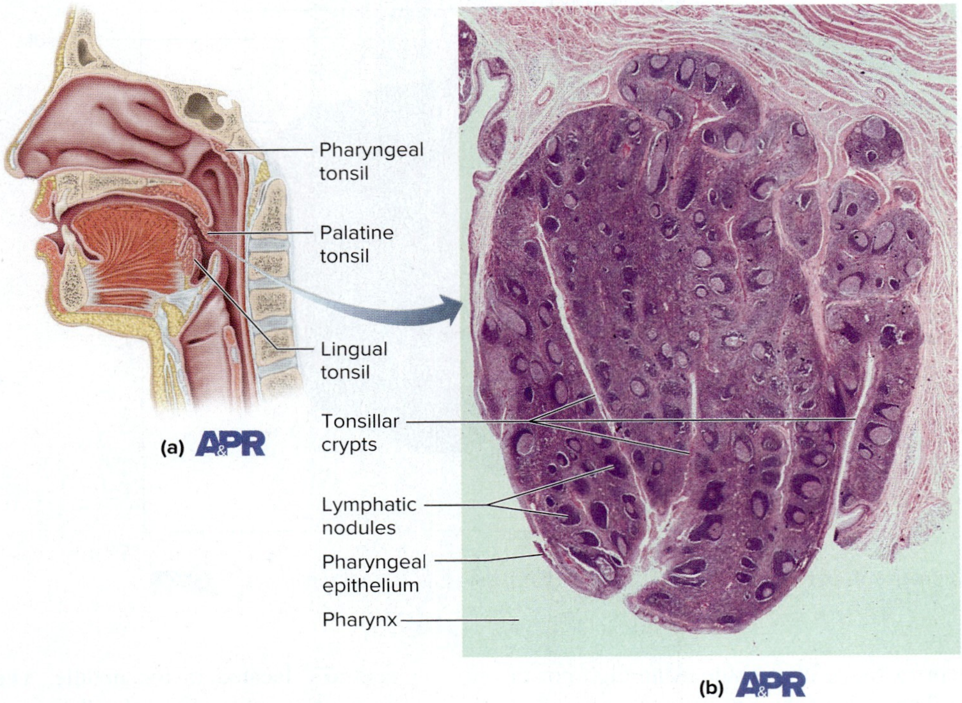

FIGURE 49.11 (a) Location of tonsils in sagittal section illustration of head; (b) micrograph of a section of the palatine tonsil (5x). (b) Biophoto Associates/Science Source

LABORATORY
ASSESSMENT 49

Name ____________________

Date ____________________

Section ____________________

The (A) corresponds to the indicated Learning Outcome(s) (O) found at the beginning of the Laboratory Exercise.

Lymphatic System

PART A: Assessments

Complete the following statements:

1. Lymphatic pathways begin as lymphatic ____________________ that merge to form lymphatic vessels. A1
2. Tissue fluid enters lymphatic capillary openings between overlapping ____________________ cells. A1
3. Lymph drainage from collecting ducts enters the ____________________ veins. A1
4. Once tissue (interstitial) fluid is inside a lymphatic capillary, the fluid is called ____________________. A1
5. Lymphatic vessels contain ____________________ that help prevent the backflow of lymph. A1
6. Lymphatic vessels usually lead to lymph ____________________ that filter the fluid being transported. A1
7. The cisterna chyli collects ____________________ lymph from the small intestine after a meal. A1
8. The lymphatic trunk that receives lymph from lymphatic vessels of the head and neck is called the ____________________ trunk. A1
9. The ____________________ duct is the larger and longer of the two lymphatic collecting ducts. A1
10. The ____________________ duct receives lymph from the intestinal and lumbar trunks. A1

PART B: Assessments

Each circle below represents the field of view as seen through the microscope. In each circle, sketch and label the microscopic structures of the indicated organs: A2

Lymph node (______×)

Thymus (______×)

Spleen (______×)

Tonsil (______×)

PART C: Assessments

Complete the following statements:

1. Lymph nodes contain large numbers of white blood cells called ______________________ and macrophages that fight invading microorganisms. A3
2. The indented region where blood vessels and nerves join a lymph node is called the ______________________. A3
3. Lymphatic ______________________ that contain germinal centers are the structural units of a lymph node. A3
4. The spaces within a lymph node are called lymphatic ______________________ through which lymph circulates. A3
5. Lymph enters a node through a(an) ______________________ lymphatic vessel. A3
6. The lymph nodes in the neck region that are often palpated when a person has a sore throat are the ______________________ lymph nodes. A3
7. The lymph nodes associated with the lymphatic vessels that drain the lower limbs are located in the ______________________ region. A1
8. The lymph nodes associated with the lymphatic vessels that drain the upper limbs are located in the ______________________ region. A3

CRITICAL THINKING ASSESSMENTS—PART C

1. Predict what would happen if a lymphatic vessel were blocked or lymph nodes were removed. Explain. A1 A3

____

____.

2. An active bacterial or viral infection may result in "swollen glands," which are inflamed ________________. Why are they swollen? A3 ____

____.

PART D: Assessments

Complete the following statements:

1. The thymus is located in the ________________________, anterior to the aortic arch. A1
2. The ________________________ is very large in infancy and early childhood and tends to decrease in size (atrophy) in advanced age. A3
3. Although these originate from precursor cells in the red bone marrow, ________________________ continue their development in the thymus. A3
4. The ________________________ is the largest of the lymphatic organs. A1
5. A ________________________ of connective tissue surrounds the spleen. A3
6. The tiny islands (nodules) of tissue within the spleen that contain many lymphocytes constitute the ________________________ pulp. A3
7. The ________________________ pulp of the spleen contains large numbers of red blood cells, lymphocytes, and macrophages. A3
8. ________________________ within the spleen engulf and destroy foreign particles and cellular debris. A3
9. The lymphoid organs in the pharynx (throat) that are possible infection sites are collectively called ________________________. A3

NOTES

LABORATORY
EXERCISE 50

Respiratory Organs

Purpose of the Exercise

To review the structure and function of the respiratory organs and to examine the tissues of some of these organs.

MATERIALS NEEDED

Human skull (sagittal section)
Human torso model
Larynx model
Thoracic organs model
Compound light microscope
Prepared microscope slides of the following:
  Trachea (cross section)
  Lung, human (normal)

For Demonstration Activities:
Animal lung with trachea (fresh or preserved)
Bicycle pump
Prepared microscope slides of the following:
  Lung tissue (smoker)
  Lung tissue (emphysema)

SAFETY

- Review all the safety guidelines in Appendix 1 of your laboratory manual.
- Wear disposable gloves and protective eyewear when working on the fresh or preserved animal lung demonstration.
- Wash your hands before leaving the laboratory.

Chapter Opener: ©101cats/Getty Images

Learning Outcomes APR

After completing this exercise, you should be able to:

01 Locate the major organs and structural features of the respiratory system.
02 Describe the functions of these organs.
03 Sketch and label features of tissue sections of the trachea and lung.

The O corresponds to the assessments A indicated in the Laboratory Assessment for this Exercise.

Pre-Lab

Carefully read the introductory material and examine the entire lab. Be familiar with the basic structures and functions of the respiratory organs from lecture or the textbook. Answer the pre-lab questions.

Pre-Lab Questions Select the correct answer for each of the following questions:

1. The right lung has ____ lobes; the left lung has ____ lobes.
   a. 3; 3 **b.** 2; 2
   c. 3; 2 **d.** 2; 3
2. Paranasal sinuses are within the following bones *except* the
   a. maxillary bones. **b.** mandible.
   c. frontal bone. **d.** sphenoid bone.
3. The alveoli are composed of
   a. simple squamous epithelium.
   b. hyaline cartilage.
   c. smooth muscle.
   d. epithelium with cilia.
4. The _____ adheres to the surface of the lung.
   a. pericardium **b.** pleural cavity
   c. parietal pleura **d.** visceral pleura
5. The _____ is the most inferior cartilage of the larynx.
   a. epiglottis **b.** cricoid cartilage
   c. thyroid cartilage **d.** corniculate cartilage
6. Which of the following structures increases the surface area and air turbulence the most during breathing?
   a. nasal meatuses **b.** nasal septum
   c. nasal conchae **d.** nares (nostrils)
7. Which of the following airway tubes have the smallest lumens?
   a. alveolar ducts **b.** segmental bronchi
   c. lobar bronchi **d.** main bronchi
8. The epiglottic cartilage is composed of hyaline cartilage.
   a. True **b.** False

The respiratory system involves the movement of oxygen from the external environment into our lungs upon inspiration and eventually to our cells, and the movement of carbon dioxide produced by our cells until it exits our body upon expiration. Breathing (pulmonary ventilation) involves the nose, nasal cavity, paranasal sinuses, pharynx, larynx, trachea, and bronchial tree, all serving as passageways for the movement of gases into and out of the lungs. The exchange of oxygen and carbon dioxide in the lungs is called *external respiration;* the exchange of oxygen and carbon dioxide in the tissues is called *internal respiration.* The blood transports gases to and from the alveolar sacs and the body cells. The cells utilize the oxygen and produce carbon dioxide in *cellular respiration.*

Larger bronchial tubes possess supportive cartilaginous rings and plates so that they do not collapse during breathing. Smooth muscle composes part of the walls of these bronchial tubes. If the smooth muscle of these tubes relaxes, the air passages dilate, which allows a greater volume of air movement. If the smooth muscle of the respiratory tubes contracts, a reduced volume of oxygen enters the lungs. The epithelial lining of the respiratory tubes includes occasional *goblet cells,* which secrete protective mucus that catches dust and potential pathogens. Numerous epithelial cells possess *cilia* that extend into the mucus. The action of the cilia creates a current of mucus that moves any entrapped debris away from the lungs toward the pharynx, which helps prevent respiratory infections. The mucus with any entrapped particles is normally swallowed.

The right lung has three lobes; the left lung has two lobes. A right main bronchus branches into three lobar bronchi, each extending into one of the three right lobes. A left main bronchus branches into two lobar bronchi, each extending into a left lobe. The *hilum* of the lung represents an area where a main bronchus and major blood vessels are located. A *visceral pleura* adheres to the surface of each lung, including the fissures between the lobes. A *parietal pleura* covers the internal surface of the thoracic wall and the superior surface of the diaphragm. A narrow potential space, called the *pleural cavity,* is located between the two pleurae and contains pleural fluid secreted by these membranes. This pleural fluid is a serous fluid that serves as lubrication during breathing movements, and due to the surface tension, it resists separation of the pleurae.

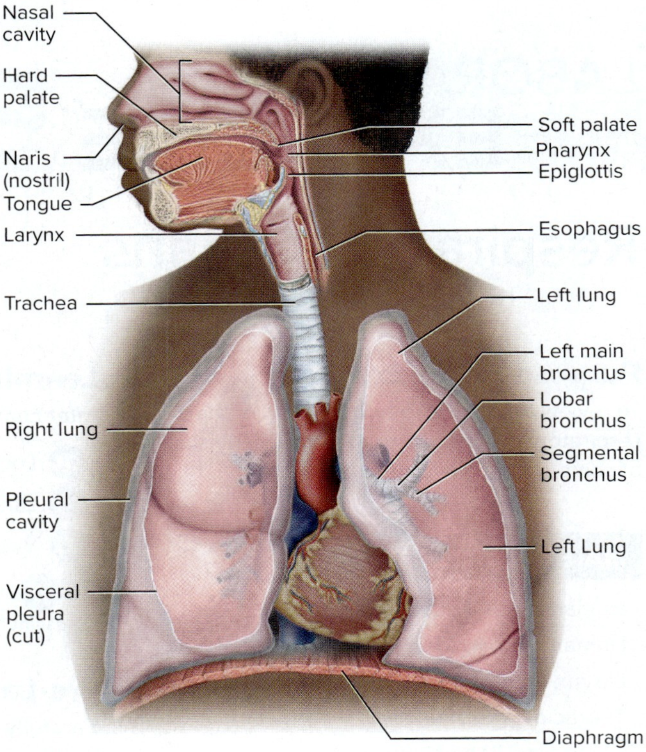

FIGURE 50.1 The major features of the respiratory system. APR

PROCEDURE A: Respiratory Organs

1. Study figures 50.1 and 50.2.
2. Examine the sagittal section of the human skull and the human torso model. Locate the following features:

   nasal cavity—hollow space behind nose
   - naris (nostril)—external opening into nasal cavity
   - nasal septum—divides nasal cavity into right and left portions
   - nasal conchae—increase surface area and air turbulence during breathing
   - nasal meatuses—air passageways along nasal conchae

   paranasal sinuses—air-filled spaces that open into nasal cavity; lined with mucous membrane
   - maxillary sinus
   - frontal sinus
   - ethmoidal sinus
   - sphenoidal sinus
3. Examine figures 50.1 to 50.5.
4. Observe the larynx model, the thoracic organs model, and the human torso model. Palpate the larynx as you swallow. Locate the following features:

   pharynx—often called throat
   - nasopharynx—air passageway only
   - oropharynx—air and food passageway
   - laryngopharynx—air and food passageway

   larynx—airway enlargement superior to trachea; voicebox
   - vocal folds—consist of two sets of folds
     - vestibular folds (false vocal cords)—upper folds but do not produce sounds
     - vocal folds (true vocal cords)—vibrate when exhaling to produce sounds
   - thyrohyoid ligament—hangs inferiorly from hyoid bone

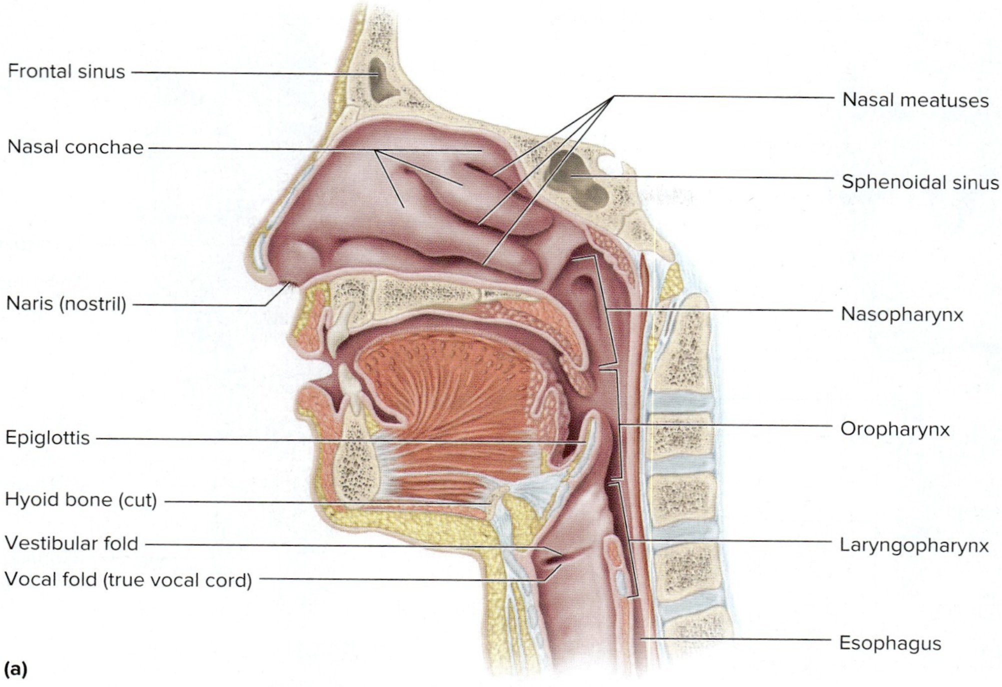

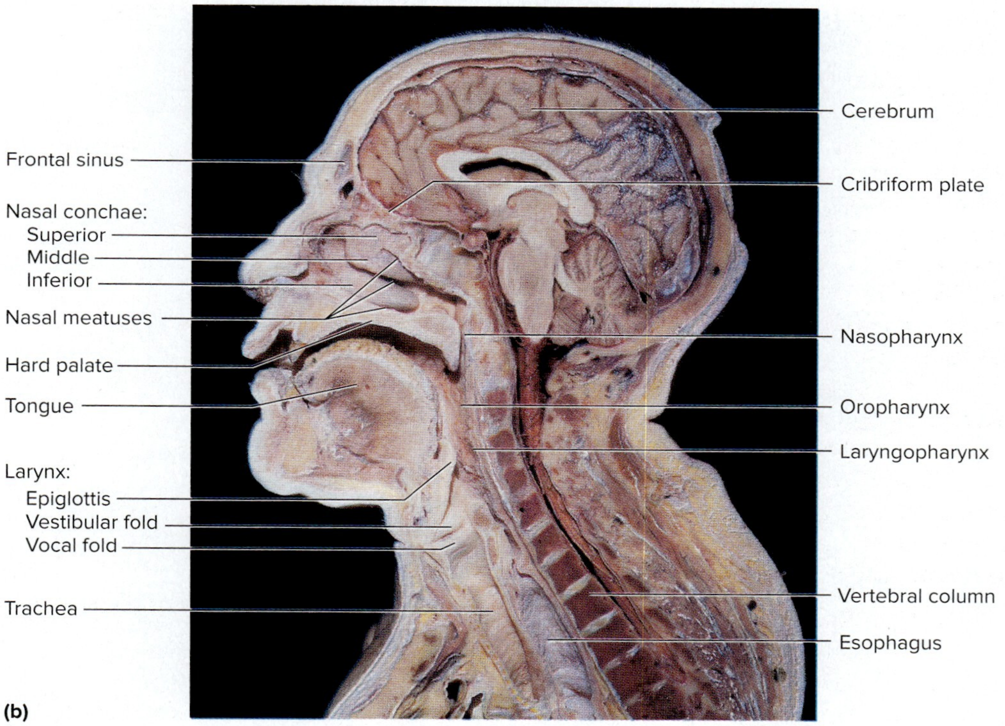

FIGURE 50.2 A sagittal section of the features of the upper respiratory system in (a) an illustration and (b) a cadaver.
(b) Rebecca Gray/McGraw Hill APR

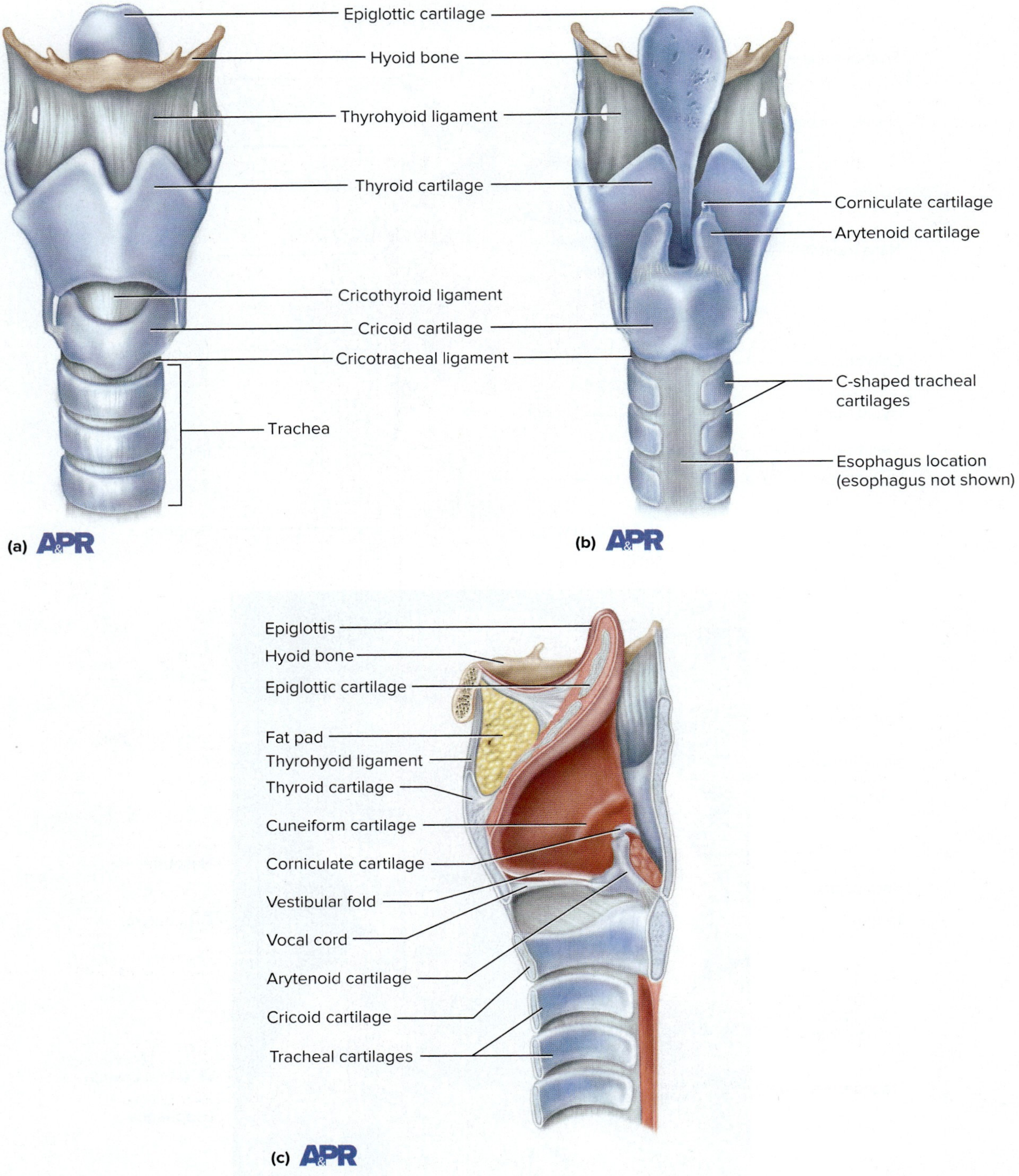

FIGURE 50.3 Major features of the larynx: (a) anterior view; (b) posterior view; (c) sagittal view.

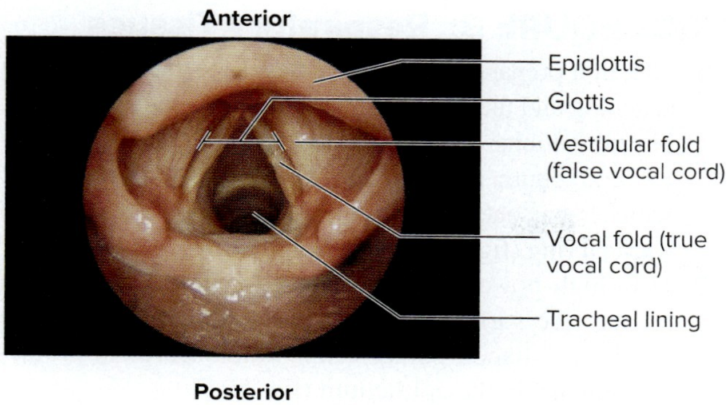

FIGURE 50.4 Superior view of larynx with the glottis open as seen using a laryngoscope. CNRI/Science Source

DEMONSTRATION ACTIVITY

Observe the animal lung and the attached trachea. Identify the larynx, major laryngeal cartilages, trachea, and incomplete cartilaginous rings of the trachea. Open the larynx and locate the vocal folds. Examine the visceral pleura on the surface of a lung. A bicycle pump can be used to demonstrate lung inflation. Section the lung and locate some bronchioles and alveoli. Squeeze a portion of a lung between your fingers. How do you describe the texture of the lung?

___

___

___

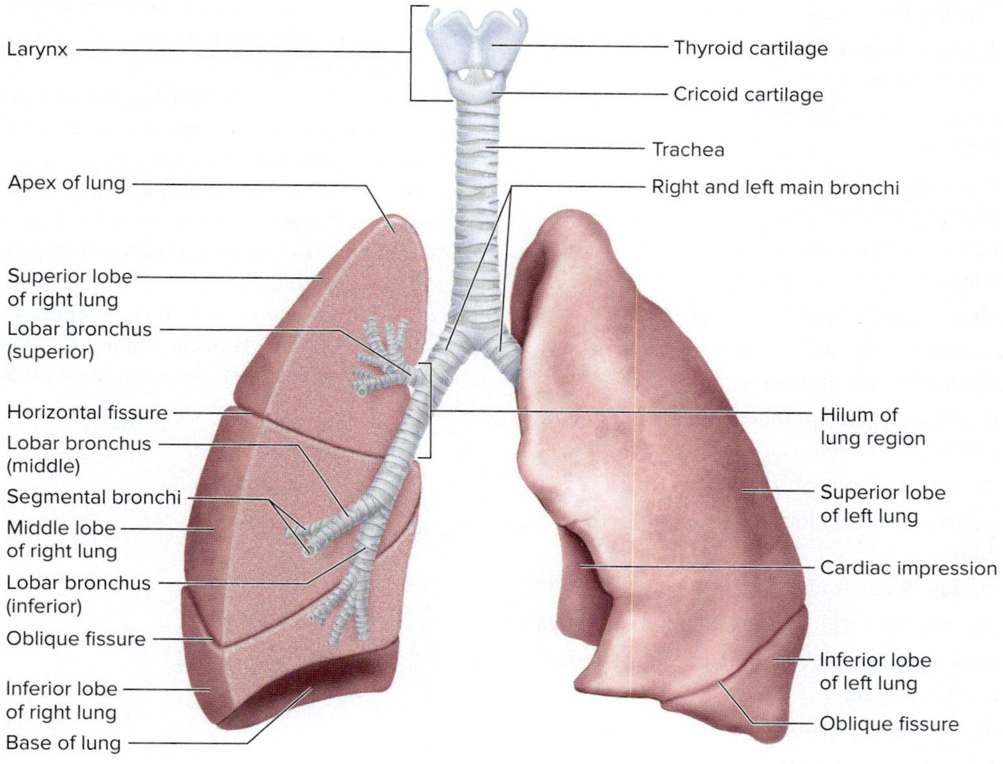

FIGURE 50.5 Features of the lower respiratory system (anterior view). APR

- thyroid cartilage ("Adam's apple")—largest hyaline cartilage of larynx; covered by part of thyroid gland; more pronounced in males
- cricoid cartilage—complete ring of hyaline cartilage of inferior larynx
- cricothyroid ligament—connects thyroid and cricoid cartilages
- epiglottis—covers part of larynx opening during swallowing
- epiglottic cartilage—elastic cartilage supporting epiglottis
- cuneiform cartilages—support soft tissues inferior to the epiglottis
- arytenoid cartilages—vocal cords attached to these small cartilages
- corniculate cartilages—small cartilages superior to arytenoid cartilages
- glottis—vocal folds and opening between them

trachea—windpipe anterior to esophagus

bronchial tree—branched airways extending throughout lungs

- right and left main (primary) bronchi—arise from trachea
- lobar (secondary) bronchi—branches of main bronchi into lobes of lungs
- segmental (tertiary) bronchi—branches of lobar bronchi into segments of lungs

bronchioles—smaller branches of airways

lung—location of external respiration between alveoli and blood capillaries

- hilum of lung—medial indentation where main bronchus, blood vessels, and nerves enter lung
- lobes—large regions of lungs
  - superior (upper) lobe—located in both lungs
  - middle lobe—only in right lung
  - inferior (lower) lobe—located in both lungs
- fissures—horizontal (only in right lung) and oblique (both lungs)
- lobules—smallest visible subdivisions of a lung

visceral pleura—serous membrane on surface of lung

parietal pleura—serous membrane on inside of thoracic wall and superior surface of diaphragm

pleural cavity—small space between pleurae containing lubricating serous fluid

5. Complete Part A of Laboratory Assessment 50.

PROCEDURE B: Respiratory Tissues

1. Obtain a prepared microscope slide of a trachea, and use scanning and then low-power magnification to examine it. Notice the inner lining of ciliated pseudostratified columnar epithelium and the deep layer of hyaline cartilage, which represents a portion of an incomplete (C-shaped) tracheal ring (figs. 50.6 and 50.7*a*).
2. Use high-power magnification to observe the cilia on the free surface of the epithelial lining. Locate the wineglass-shaped goblet cells, which secrete a protective mucus, in the epithelium (fig. 50.7*b*).
3. Prepare a labeled sketch of a representative portion of the tracheal wall in Part B of the laboratory assessment.
4. Study the internal structures of a lung (fig. 50.8).
5. Obtain a prepared microscope slide of human lung tissue. Examine it, using scanning and then low-power magnification, and note the many open spaces of the air sacs (alveoli). Look for a bronchiole—a tube with a relatively thick wall and a wavy inner lining. Locate the smooth muscle tissue in the wall of this tube (fig. 50.9). You also may see a section of cartilage as part of the bronchiole wall.
6. Use high-power magnification to examine the alveoli (fig. 50.10). Their walls are composed of simple squamous epithelium formed by type I alveolar cells. The thin walls of the alveoli and the associated capillaries allow efficient gas exchange of oxygen and carbon dioxide. You also may see sections of blood vessels containing blood cells.
7. Prepare a labeled sketch of a representative portion of the lung in Part B of the laboratory assessment.
8. Complete Part C of the laboratory assessment.

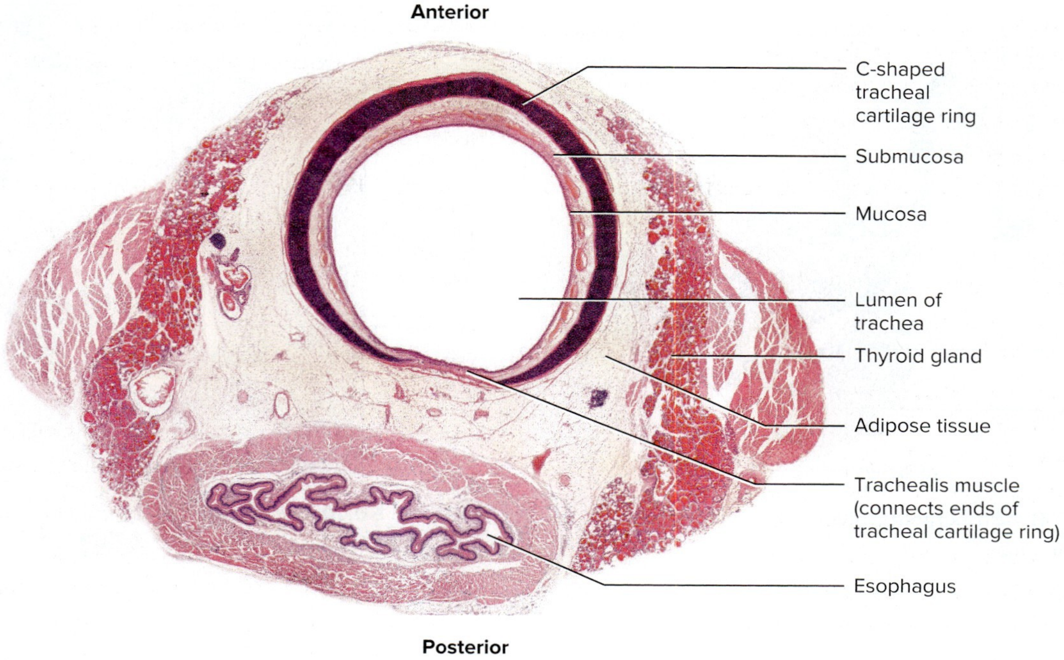

FIGURE 50.6 Micrograph of a section of the trachea and surrounding organs and tissues (10×). Jose Luis Calvo/Shutterstock

DEMONSTRATION ACTIVITY

Examine the prepared microscope slides of the lung tissue of a smoker and a person with emphysema, using low-power magnification. How does the smoker's lung tissue compare with that of the normal lung tissue that you examined previously?

____

____

How does the emphysema patient's lung tissue compare with the normal lung tissue?

____

____

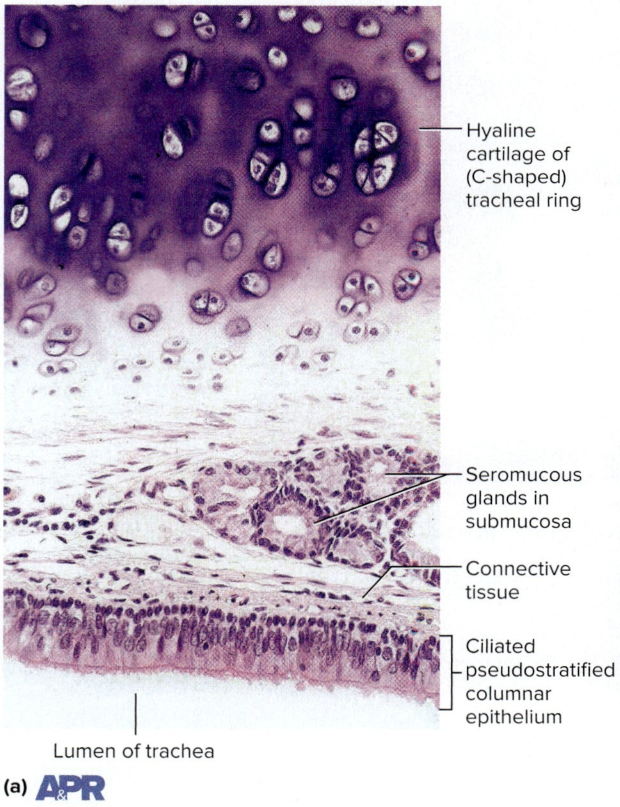

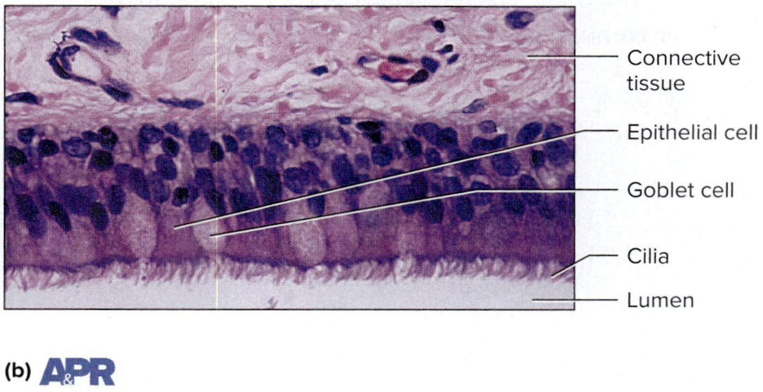

FIGURE 50.7 Micrograph of (a) a section of the tracheal wall (60×) and (b) ciliated pseudostratified columnar epithelium in the respiratory tract (275×). (a) Victor P. Eroschenko; (b) Biophoto Associates/Science Source

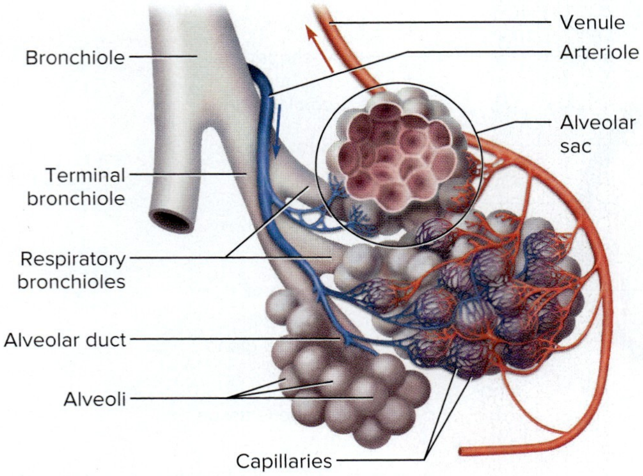

FIGURE 50.8 Bronchioles, alveoli, and blood vessel network in a lung.

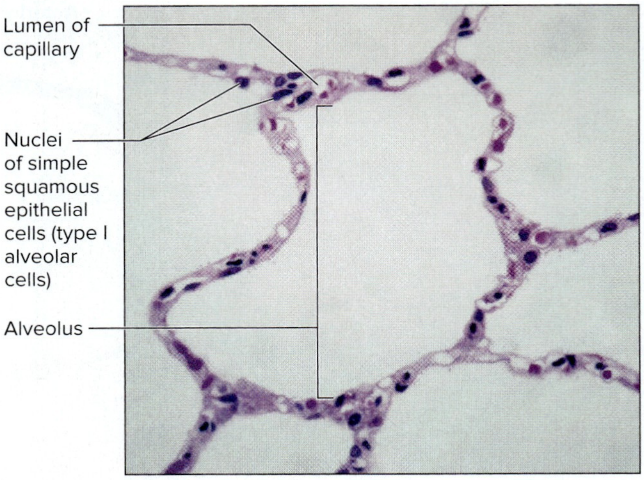

FIGURE 50.10 Micrograph of human lung tissue (250×). Ed Reschke/Getty Images APR

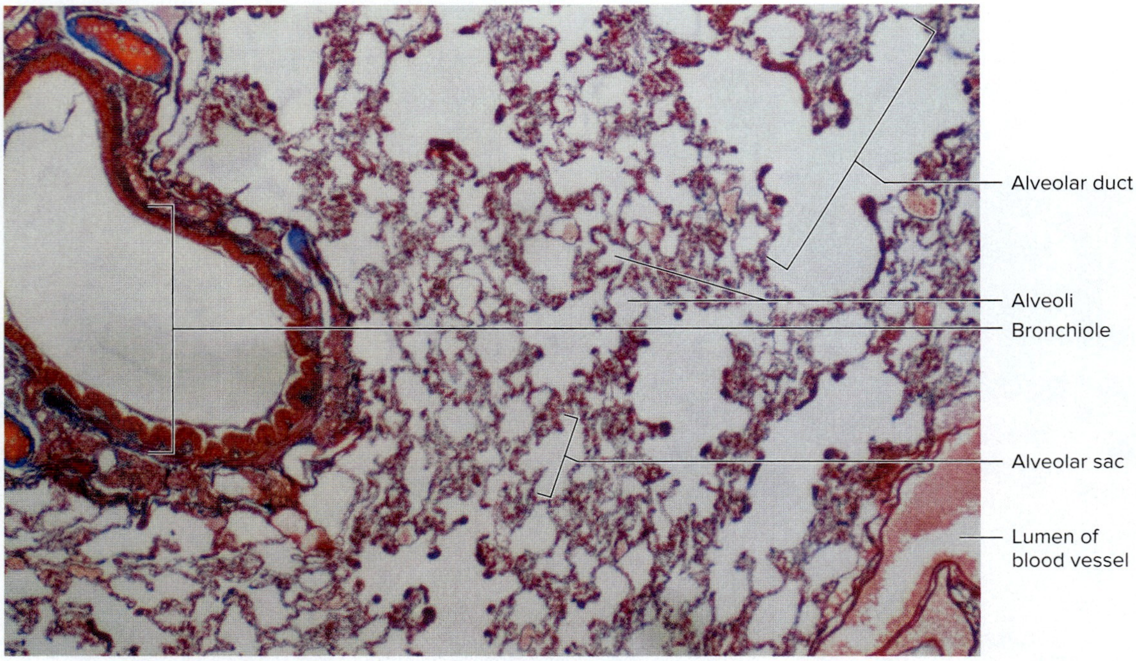

FIGURE 50.9 Micrograph of human lung tissue (40×). Cynthia Prentice-Craver/McGraw Hill APR

LABORATORY ASSESSMENT

50

Name ______________________

Date ______________________

Section ______________________

The (A) corresponds to the indicated Learning Outcome(s) (O) found at the beginning of the Laboratory Exercise.

Respiratory Organs

PART A: Assessments

1. Match the terms in column A with the descriptions in column B. Place the letter of your choice in the space provided. A1 A2

Column A		Column B
a. Alveolus	______	**1.** Potential space between visceral and parietal pleurae
b. Cricoid cartilage	______	**2.** Most inferior portion of larynx
c. Epiglottis	______	**3.** Air-filled space in skull bone that opens into nasal cavity
d. Glottis	______	**4.** Microscopic air sac for gas exchange
e. Lung	______	**5.** Consists of large lobes
f. Nasal concha	______	**6.** Vocal folds, including the opening between them
g. Pharynx	______	**7.** Fold of mucous membrane containing elastic fibers responsible for sounds
h. Pleural cavity	______	**8.** Increases surface area of nasal mucous membrane
i. Sinus (paranasal sinus)	______	**9.** Passageway for air and food
j. Vocal fold (true vocal cord)	______	**10.** Partially covers opening of larynx during swallowing

2. Label the structures indicated in figures 50.11 to 50.13.

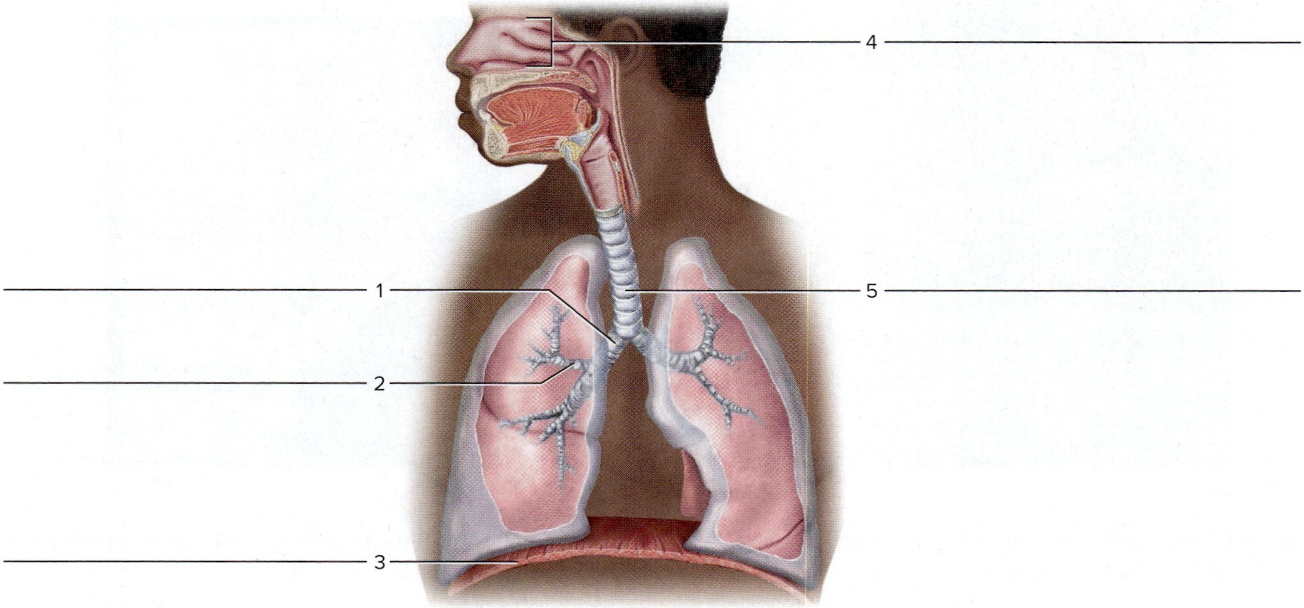

(a)

FIGURE 50.11 Label the features of the respiratory system on (a) an illustration and (b) a model image (anterior view), using the terms provided [for (b) only]. A1

Terms:
Diaphragm
Horizontal fissure
Larynx
Lobar (secondary) bronchus
Main (primary) bronchus
Oblique fissure
Right inferior lobe
Right middle lobe
Right superior lobe
Segmental (tertiary) bronchus
Trachea

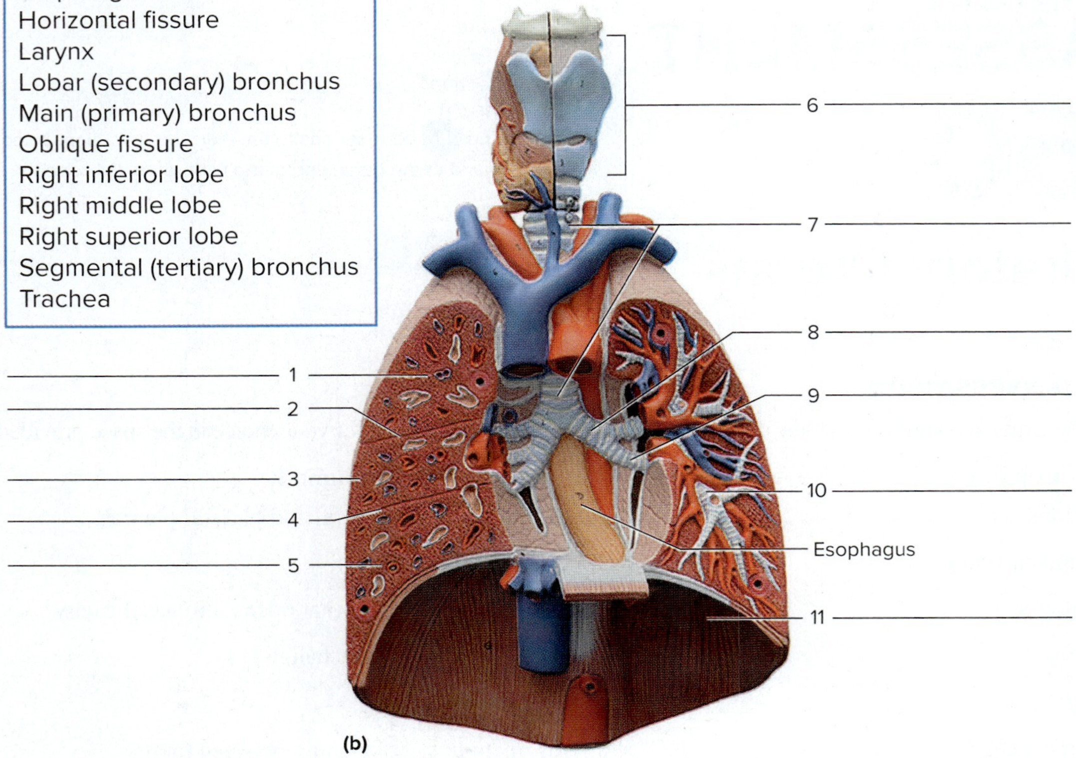

FIGURE 50.11 *Continued.* ©3B Scientific GmbH, Germany, 2017 www.3bscientific.com A1

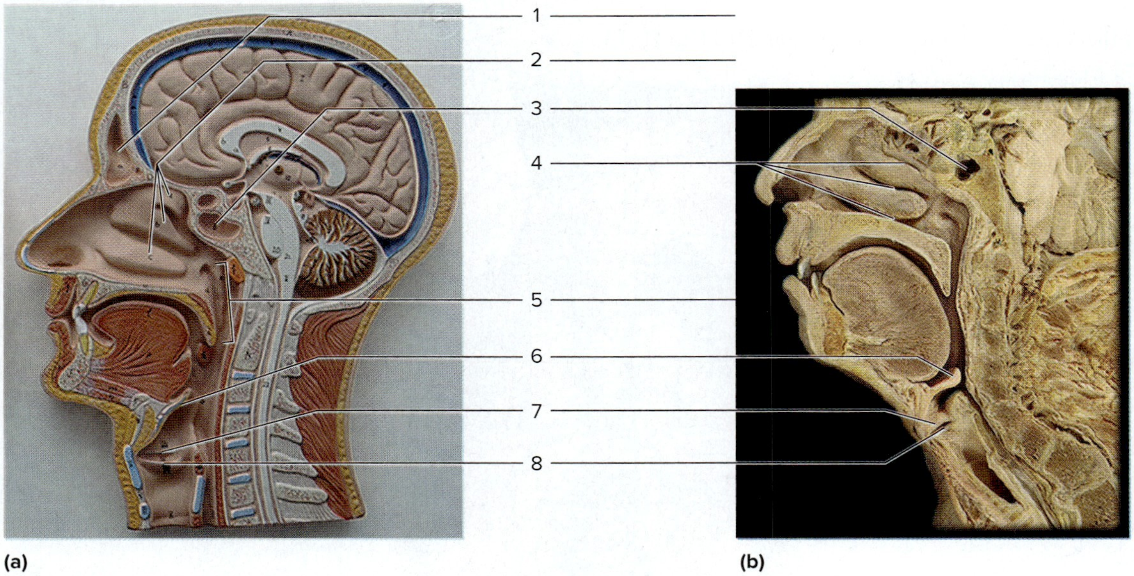

FIGURE 50.12 Label the features of the upper respiratory system on the (a) model and (b) cadaver images (sagittal section).
(a) ©3B Scientific GmbH, Germany, 2017 www.3bscientific.com; (b) APR/McGraw Hill A1

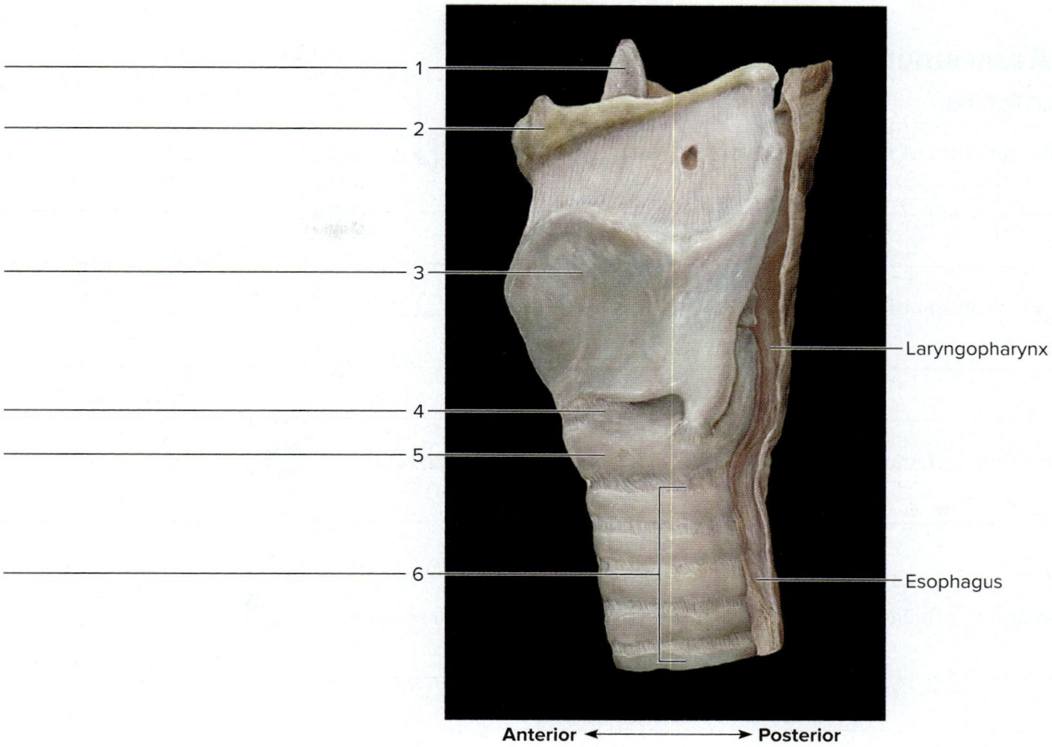

FIGURE 50.13 Label the features of the larynx region of a cadaver (lateral view). APR/McGraw Hill A1

1. ________________________
2. ________________________
3. ________________________
4. ________________________
5. ________________________
6. ________________________

PART B: Assessments

Each circle below represents the microscopic field of view. In each circle, sketch and label a portion of the tracheal wall and a portion of lung tissue. A3

Tracheal wall (______×)

Lung tissue (_____×)

CRITICAL THINKING ASSESSMENT—PART B

Examine the micrograph shown in figure 50.14. What are your observations? A2 A3 ____

____

____.

Based on your observations, is this a healthy lung tissue? Explain. A2 A3

____

____

____.

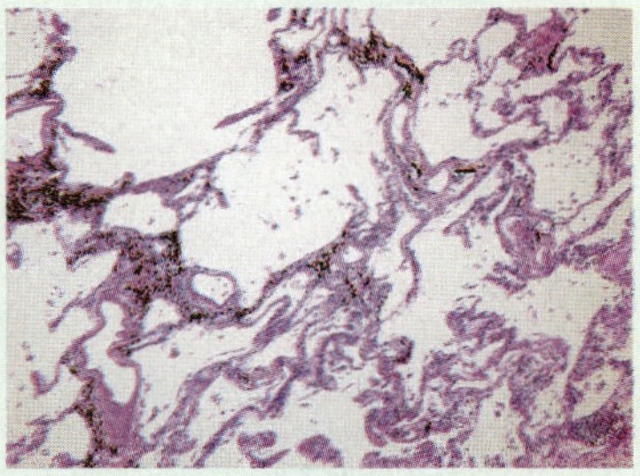

FIGURE 50.14 Micrograph of lung tissue. Dennis Strete

PART C: Assessments

Complete the following:

1. What is the function of the mucus secreted by the goblet cells? A2 ______________________________

2. Describe the function of the cilia in the respiratory tubes. A2 ______________________________

3. How is breathing affected if the smooth muscle of the bronchial tree relaxes? A2 ______________________________

4. How is breathing affected if the smooth muscle of the bronchial tree contracts? A2 ______________________________

CRITICAL THINKING ASSESSMENTS—PART C

1. What is the functional advantage of the alveolar walls being so thin? A2

______________________________.

What affect would pulmonary edema have on this function? A2

______________________________.

2. Describe the airway of a patient who is having an asthma attack. What would be the desired effect of treating this patient with medications such as beta-2 agonists and corticosteroids? A2

______________________________.

LABORATORY EXERCISE 51

Breathing and Respiratory Volumes

Purpose of the Exercise

To review the mechanisms of breathing and to measure or calculate and analyze certain respiratory air volumes and respiratory capacities.

MATERIALS NEEDED

Lung function model
Spirometer, handheld (dry portable)
Alcohol swabs (wipes)
Disposable mouthpieces
Nose clips
Meterstick

For Learning Extension Activity:
Clock or watch with seconds timer

SAFETY

- Review all the safety guidelines in Appendix 1 of your laboratory manual.
- Wipe the spirometer with an alcohol swab (wipe) before use.
- Place a new disposable mouthpiece on the stem of the spirometer before each use.
- Dispose of the alcohol swabs and mouthpieces according to directions from your laboratory instructor.

CAUTION

The risk of contracting SARS-CoV-2 (the COVID 19 virus) or other pathogen that may affect the respiratory system is a concern for this lab. The equipment used to measure respiratory volumes must be sterilized after **each student** uses the equipment. Your instructor will determine how the exercises will be accomplished, whether by demonstration, by conducting a virtual lab, or by individual students as indicated in the lab exercises if proper safety protocols are in place to perform these exercises in the safest possible way. Under no circumstances should nonsterile equipment be used when completing these exercises. The following Centers for Disease Control and Prevention (CDC) website contains recommendations for the sterilization of equipment: https://www.cdc.gov/infectioncontrol/pdf/guidelines/disinfection-guidelines-H.pdf

Learning Outcomes APR

After completing this exercise, you should be able to:

O1 Differentiate between the mechanisms and muscles responsible for inspiration and expiration.

O2 Measure respiratory volumes using a spirometer and analyze results.

O3 Calculate respiratory capacities using the data obtained from respiratory volumes and analyze results.

O4 Match the respiratory volumes and respiratory capacities with their definitions.

The O corresponds to the assessments A indicated in the Laboratory Assessment for this Exercise.

Pre-Lab

Carefully read the introductory material and examine the entire lab. Be familiar with inspiration, expiration, and respiratory volumes from lecture or the textbook. Answer the pre-lab questions.

Pre-Lab Questions Select the correct answer for each of the following questions:

1. The size of the thoracic cavity is increased by contractions of all of the following muscles *except* the
   a. diaphragm. **b.** external intercostals.
   c. pectoralis minor. **d.** external oblique.
2. A _____ is an instrument to measure air volumes during breathing.
   a. flow meter **b.** spirometer
   c. lung function model **d.** capacity meter
3. The _____ is the maximum volume of air that can be exhaled after taking the deepest breath possible.
   a. tidal volume **b.** expiratory reserve volume
   c. vital capacity **d.** total lung capacity
4. Tidal volume is estimated to be about
   a. 500 mL. **b.** 1,200 mL.
   c. 3,600 mL. **d.** 4,800 mL.
5. A normal resting breathing rate is about _____ breaths per minute.
   a. 5–10 **b.** 12–15
   c. 16–20 **d.** 21–30

6. The contraction of the diaphragm increases the size of the thoracic cavity.
   a. True   b. False
7. Vital capacity is the total of tidal volume, expiratory reserve volume, and residual volume.
   a. True   b. False
8. Vital capacities gradually decrease as a person continues to age.
   a. True   b. False

Breathing, or pulmonary ventilation, involves the movement of air from outside the body through the bronchial tree and into the alveoli and the reversal of this air movement to allow gas (oxygen and carbon dioxide) exchange between air and blood. These movements are caused by changes in the size of the thoracic as a result of skeletal muscle contractions. The size of the thoracic cavity during *inspiration* is increased by contractions of the diaphragm, external intercostals, internal intercostals (intercartilaginous part), pectoralis minor, sternocleidomastoid, and scalenes. *Expiration* is aided by contractions of the internal intercostals (interosseous part), rectus abdominis, and external oblique, and from the elastic recoil (elasticity) of stretched tissues (fig. 51.1 and table 51.1).

Within the respiratory system passageway is the conducting division, where there is only airflow and no gas exchange. This division includes the nasal cavity, pharynx, larynx, trachea, bronchi, and major bronchioles. The respiratory division, which is the area where gas exchange occurs, includes the respiratory bronchioles, alveolar ducts, alveolar sacs, and the most important component, the alveoli.

The volumes of air that move into and out of the lungs during various phases of breathing are called *respiratory (pulmonary) volumes* and *capacities*. A *volume* is a single measurement, whereas *capacity* represents two or more combined volumes. Respiratory volumes can be measured by using an instrument called a *spirometer*. Respiratory capacities can be determined by using various combinations of

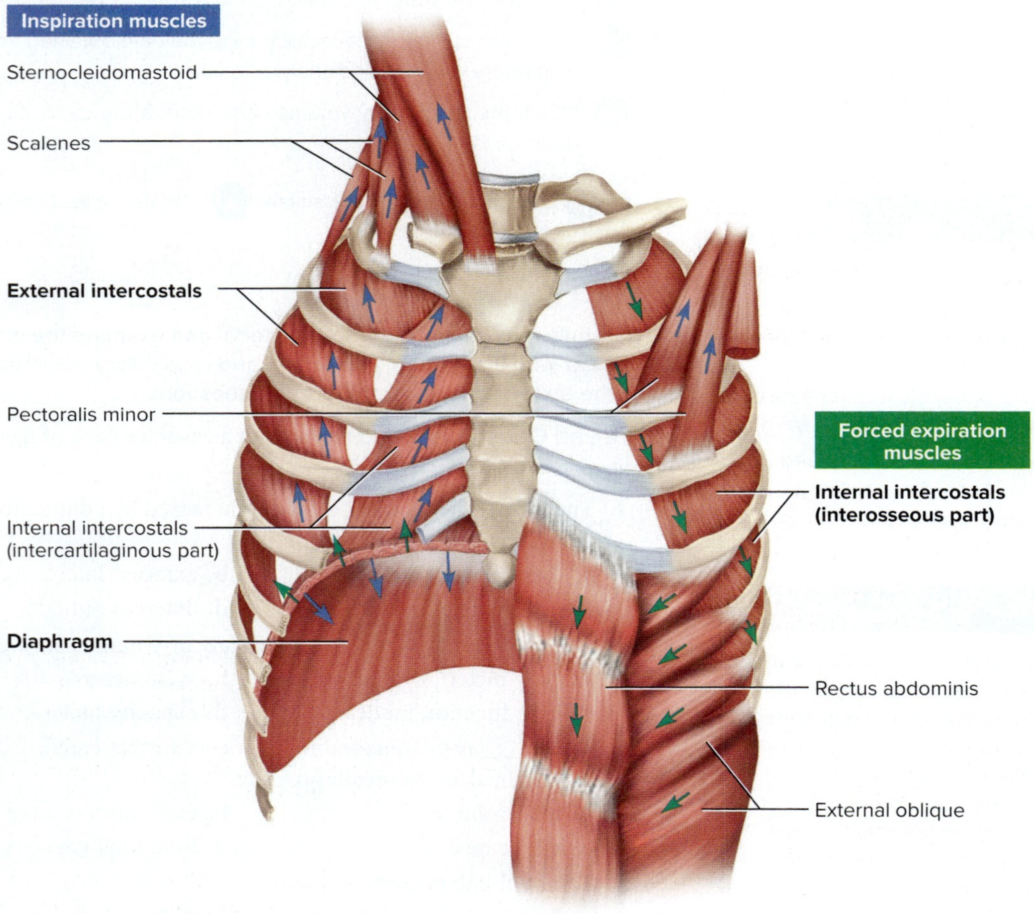

FIGURE 51.1 Respiratory muscles involved in inspiration and forced expiration (anterior view). Boldface indicates the primary muscles increasing or decreasing the size of the thoracic cavity. Blue arrows indicate the direction of contraction during inspiration that increases the capacity of the thoracic cavity; the green arrows indicate the direction of contraction during forced expiration that decreases the capacity of the thoracic cavity. (The green arrows associated with the diaphragm indicate its ascending direction primarily from the elastic recoil of the stretched tissues during quiet breathing.) APR

Table 51.1 Muscles of Respiration

Muscles of Inspiration	Origin	Insertion	Inspiration Action
Diaphragm	Costal cartilages and ribs 7–12, xiphoid process, and lumbar vertebrae	Central tendon of diaphragm	Flattens (descends) during contraction; this expands thoracic cavity and compresses abdominal viscera during inspiration
External intercostals	Inferior border of superior rib	Superior border of inferior rib	Elevates ribs; this expands thoracic cavity during inspiration
Sternocleidomastoid	Manubrium of sternum and medial clavicle	Mastoid process of temporal bone	Elevates sternum
Scalenes	Transverse processes of cervical vertebrae C2–C6	Ribs 1 and 2	Elevates ribs 1 and 2
Pectoralis minor	Sternal ends of ribs 3–5	Coracoid process of scapula	Elevates ribs 3–5
Internal intercostals (intercartilaginous part)	Superior border of inferior costal cartilages	Inferior border of superior costal cartilages	Assists in elevating ribs
Muscles of (Forced) Expiration	**Origin**	**Insertion**	**Expiration Action**
Internal intercostals (interosseous part)	Superior border of inferior rib	Inferior border of superior rib	Depresses ribs; this compresses thoracic cavity during forced expiration
Rectus abdominis	Pubic crest and pubic symphysis	Xiphoid process of sternum and costal cartilages of ribs 5–7	Compresses inferior ribs and abdominal contents, helping to raise diaphragm
External oblique	Anterior surfaces of ribs 5–12	Anterior iliac crest and linea alba by aponeurosis	Compresses abdominal contents, helping to raise diaphragm

respiratory volumes. The values obtained vary with a person's age, sex, height, weight, stress, and physical fitness.

The volume of air taken in or out in a normal breath is called the *tidal volume (TV)*. The average TV is 500 mL (imagine the equivalent of a 1/2 L bottle). This volume of air is moved in through the nose and mouth and through the conducting and respiratory divisions to the alveoli. A certain amount of that air never reaches the alveoli; instead, it remains within the respiratory system passageway. This air is said to be in the *anatomic dead space*. The average volume of anatomic dead space is 150 mL. If the TV is 500 mL and the volume of the anatomic dead space is 150 mL, what is the volume of air reaching the alveoli? Understanding the difference between TV and the volume of air that reaches the lungs is important because *only* the air that reaches the alveoli provides oxygen for gas exchange with the blood.

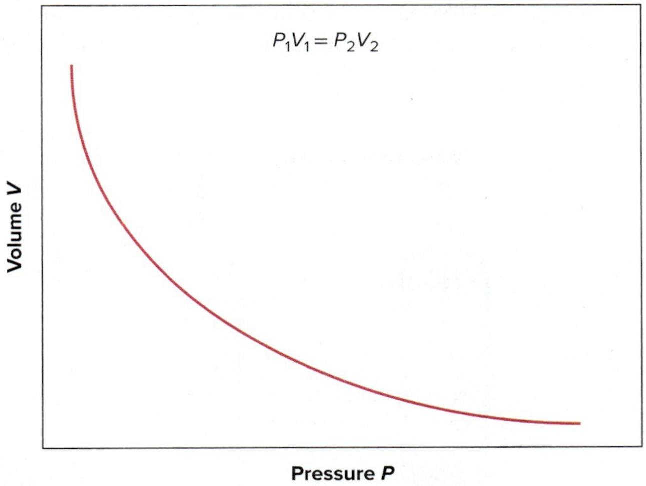

FIGURE 51.2 Formula for Boyle's Law. P_1 and V_1 represent the original pressure and volume, respectively, and P_2 and V_2 represent the second pressure and volume. The pressure of a gas is inversely proportional to its volume, assuming a constant temperature. The graph shows this relationship.

PROCEDURE A: Breathing Mechanisms

The size of the thoracic cavity increases during inspiration upon contractions of inspiration muscle. Contractions of the diaphragm results in increased depth of the thoracic cavity, and contractions of the external intercostals widens the thoracic cavity. A combination of synergistic contractions of the sternocleidomastoid, scalenes, pectoralis minor, and the intercartilaginous part of the internal intercostals elevates the sternum and rib cage. During forced expiration, the interosseous part of the internal intercostals depresses the rib cage, narrowing the thoracic cavity. The contractions of the rectus abdominis and the external oblique compress the abdominal viscera, forcing the diaphragm to ascend and thus reducing the size of the thoracic cavity (fig. 51.1).

Air enters the lungs during inspiration and exits the lungs during expiration based upon characteristics of *Boyle's law*. According to Boyle's law, the pressure of a gas is inversely proportional to its volume, assuming a constant temperature (fig. 51.2). In other words, when the volume increases in a container such as the thoracic cavity during inspiration, pressure decreases and air flows into the lungs. In contrast, when volume decreases in a container such as the thoracic cavity during expiration, pressure increases and air flows out of the lungs. Air moves into and out of the lungs in much the same way as can be demonstrated using a syringe (fig. 51.3). When you pull back on the plunger of the syringe, the volume increases in the barrel of the syringe, which decreases the pressure, resulting in air rushing into the barrel of the syringe. In contrast, pushing the plunger into the barrel of the syringe causes the pressure to increase in the barrel of the syringe, and air flows out of the container.

1. Observe the mechanical lung function model (fig. 51.4). It consists of a heavy plastic bell jar with rubber sheeting clamped over its wide open end. Its narrow upper opening is plugged with a rubber stopper through which a Y tube is passed. Small rubber balloons are fastened to the arms of the Y. What happens to the balloons when the rubber sheeting is pulled downward?

   ____

   ____

   What happens when the sheeting is pushed upward?

   ____

   ____

2. Compare the lung function model with figures 51.1 and 51.3. Note the muscles of inspiration that increase the size of the thoracic cavity, and the muscles of expiration that decrease the size of the thoracic cavity. Relate these volume changes to pressure changes and airflow into and out of the lungs.
3. Complete Part A of Laboratory Assessment 51.

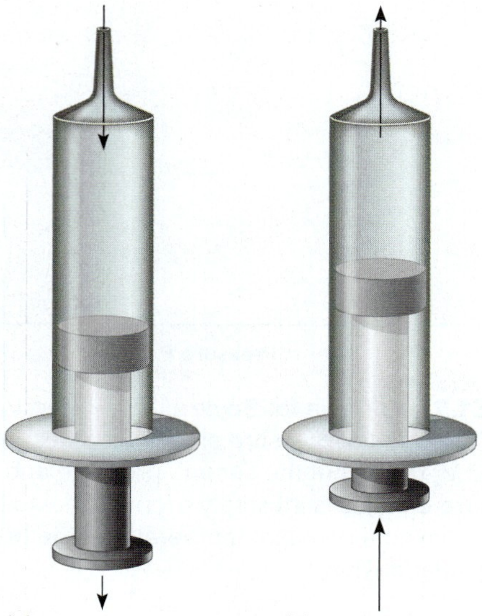

FIGURE 51.3 Moving the plunger of a syringe as in (a) results in air movements into the barrel of the syringe; and (b) results in air movements out of the barrel of the syringe. Air movements in and out of the lung function model and the lungs during breathing occur for similar reasons. APR

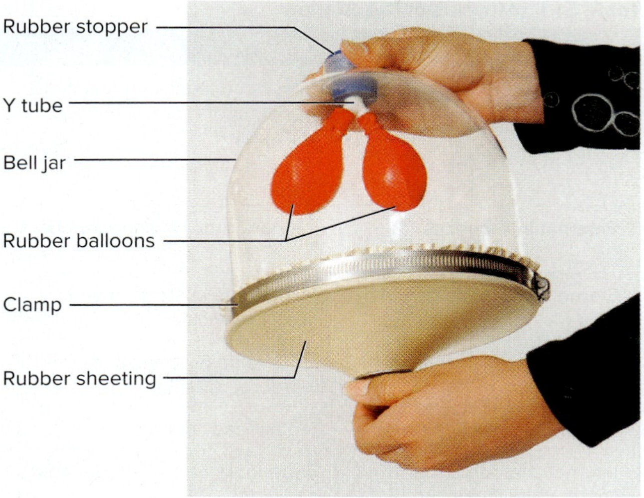

FIGURE 51.4 A lung function model. ©J & J Photography

PROCEDURE B: Respiratory Volumes and Capacities

1. Get a handheld spirometer (fig. 51.5) that has been sterilized. Point the needle to zero by rotating the adjustable dial. Before using the instrument, wipe it with an alcohol swab and place a new disposable mouthpiece over its stem. The instrument should be held with the dial upward, and *air should be blown only into the disposable mouthpiece* (fig. 51.6). If air tends to exit the nostrils during exhalation, you can prevent it from doing so by using a nose clip or by pinching your nose when exhaling into the spirometer. Movement of the needle indicates the air volume that leaves the lungs. The exhalation should be slowly and forcefully performed. Too rapid an exhalation can result in erroneous data or damage to the spirometer.
2. *Tidal volume (TV)* (about 500 mL) is the volume of air that enters (or leaves) the lungs during a *respiratory cycle* (one inspiration plus the following expiration). *Resting tidal volume* is the volume of air that enters (or leaves)

WARNING

If you begin to feel dizzy or light-headed while performing Procedure B, stop the exercise and breathe normally.

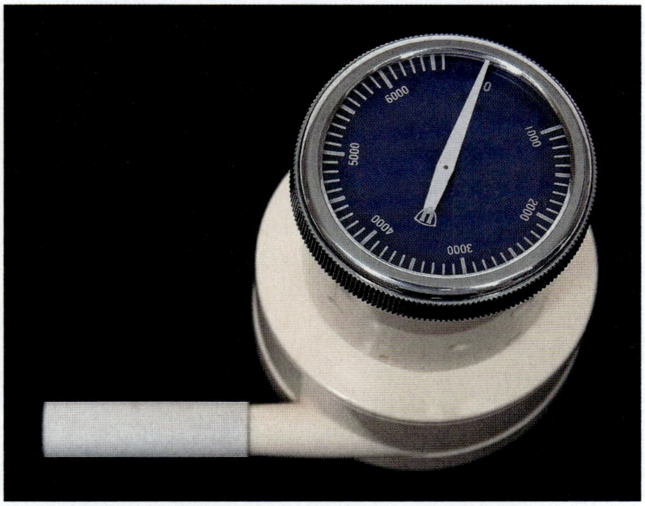

FIGURE 51.5 A handheld spirometer can be used to measure respiratory volumes. ©J & J Photography

the lungs during normal, quiet breathing (fig. 51.7). To measure this volume, follow these steps:

a. Sit quietly for a few moments.

b. Position the spirometer dial so that the needle points to zero.

c. Place the mouthpiece between your lips and exhale three ordinary expirations into it after inhaling through the nose each time. *Do not force air out of your lungs; exhale normally.*

d. Divide the total value indicated by the needle by 3 and record this amount as your resting TV on the table in Part B of the laboratory assessment.

3. *Expiratory reserve volume (ERV)* (about 1,200 mL) is the volume of air in addition to the TV that leaves the lungs during forced expiration (fig. 51.7). Chronic obstructive pulmonary disease (COPD) reduces pulmonary ventilation; COPD occurs in emphysema and chronic bronchitis, which are often caused by smoking and inhaling other pollutants. With COPD, the ERV is greatly reduced during physical exertion, and the person becomes easily exhausted from expending more energy just to breathe. To measure this volume, follow these steps:

a. Breathe normally for a few moments. Set the needle to zero.

b. At the end of an ordinary expiration, place the mouthpiece between your lips and exhale all of the air you can force from your lungs through the spirometer. Use a nose clip or pinch your nose to prevent any air from exiting your nostrils.

FIGURE 51.6 Demonstration of use of a handheld spirometer. Air should be blown slowly and forcefully only into a disposable mouthpiece. *Use a nose clip or pinch your nose when measuring expiratory reserve volume and vital capacity volume if air exits the nostrils.* ©J & J Photography

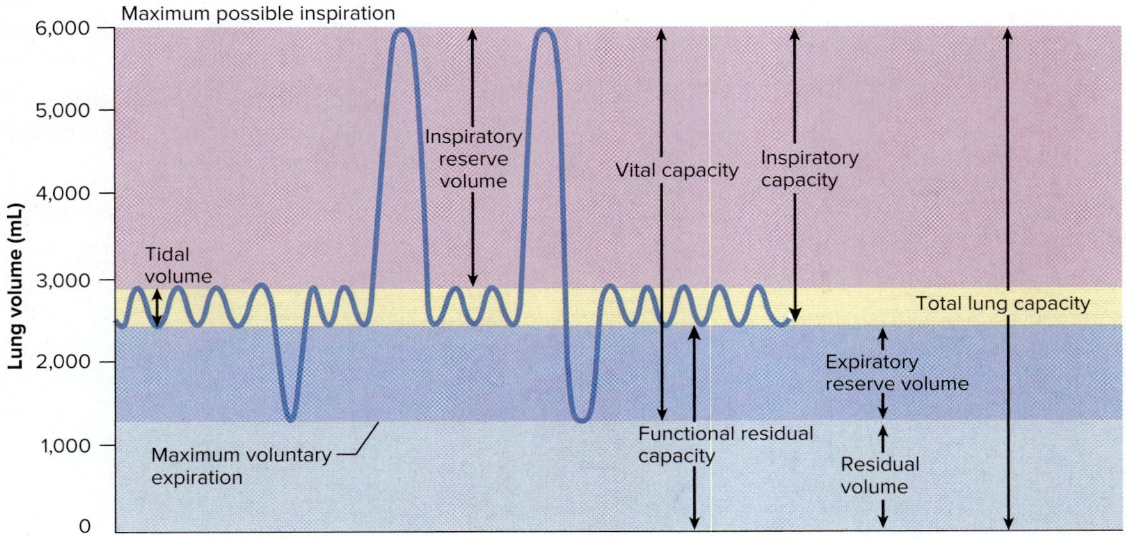

FIGURE 51.7 Graphic representation (spirogram) of respiratory volumes and capacities.

c. Record the results as your ERV in Part B of the laboratory assessment.

4. *Vital capacity (VC)* (about 4,700 mL) is the maximum volume of air that can be exhaled after taking the deepest breath possible (fig. 51.7). To measure this volume, follow these steps:

   a. Breathe normally for a few moments. Set the needle at zero.

   b. Breathe in and out deeply a couple of times; then take the deepest breath possible.

   c. Place the mouthpiece between your lips and exhale all the air out of your lungs, slowly and forcefully. Use a nose clip or pinch your nose to prevent any air from exiting your nostrils.

   d. Record the value as your VC in Part B of the laboratory assessment. Compare your result with that expected for a person of your sex, age, and height listed in tables 51.2 and 51.3. Use the meterstick to determine your height in centimeters if necessary or multiply your height in inches times 2.54 to calculate your height in centimeters. Considerable individual variations from the expected will be noted due to parameters other than sex, age, and height, which can include fitness level, health, medications, and others.

5. *Inspiratory reserve volume (IRV)* (about 3,000 mL) is the volume of air in addition to the TV that enters the lungs during forced inspiration (fig. 51.7). Calculate your IRV by subtracting your TV and your ERV from your VC:

$$IRV = VC - (TV + ERV)$$

6. *Inspiratory capacity (IC)* (about 3,500 mL) is the maximum volume of air a person can inhale following exhalation of the TV (fig. 51.7). Calculate your IC by adding your TV and your IRV:

$$IC = TV + IRV$$

7. *Functional residual capacity (FRC)* (about 2,500 mL) is the volume of air that remains in the lungs following exhalation of the TV (fig. 51.7). Calculate your FRC by adding your ERV and your residual volume (RV), which you can assume is 1,300 mL:

$$FRC = ERV + 1{,}300$$

8. *Residual volume (RV)* (about 1,300 mL) is the volume of air that always remains in the lungs after the most forceful expiration (fig. 51.7). Although it is part of *total lung capacity* (about 6,000 mL), it cannot be measured with a spirometer. The residual air allows gas exchange and the alveoli to remain open during the respiratory cycle.

9. Complete Parts B and C of the laboratory assessment.

CRITICAL THINKING ACTIVITY

It can be noted from the data in tables 51.2 and 51.3 that vital capacities gradually decrease with age. Propose an explanation for this normal correlation.

____

____

____

____

LEARNING EXTENSION ACTIVITY

Determine your *minute respiratory volume.* To do this, follow these steps:

1. Sit quietly for a while, and then to establish your breathing rate, count the number of times you breathe in 1 minute. This might be inaccurate because conscious awareness of breathing rate can alter the results. You might ask a laboratory partner to record your breathing rate sometime when you are not expecting it to be recorded. A normal resting breathing rate is about 12–15 breaths per minute.
2. Calculate your minute respiratory volume by multiplying your breathing rate by your tidal volume:

   __________ × __________ = ______________

   (breathing rate) (tidal volume) (minute respiratory volume)

3. This value indicates the total volume of air that moves into your respiratory passages during each minute of ordinary breathing.

Table 51.2 Predicted Vital Capacities (in Milliliters) for Females

	HEIGHT IN CENTIMETERS																								
Age	146	148	150	152	154	156	158	160	162	164	166	168	170	172	174	176	178	180	182	184	186	188	190	192	194
16	2950	2990	3030	3070	3110	3150	3190	3230	3270	3310	3350	3390	3430	3470	3510	3550	3590	3630	3670	3715	3755	3800	3840	3880	3920
18	2920	2960	3000	3040	3080	3120	3160	3200	3240	3280	3320	3360	3400	3440	3480	3520	3560	3600	3640	3680	3720	3760	3800	3840	3880
20	2890	2930	2970	3010	3050	3090	3130	3170	3210	3250	3290	3330	3370	3410	3450	3490	3525	3565	3605	3645	3695	3720	3760	3800	3840
22	2860	2900	2940	2980	3020	3060	3095	3135	3175	3215	3255	3290	3330	3370	3410	3450	3490	3530	3570	3610	3650	3685	3725	3765	3800
24	2830	2870	2910	2950	2985	3025	3065	3100	3140	3180	3220	3260	3300	3335	3375	3415	3455	3490	3530	3570	3610	3650	3685	3725	3765
26	2800	2840	2880	2920	2960	3000	3035	3070	3110	3150	3190	3230	3265	3300	3340	3380	3420	3455	3495	3530	3570	3610	3650	3685	3725
28	2775	2810	2850	2890	2930	2965	3000	3040	3070	3115	3155	3190	3230	3270	3305	3345	3380	3420	3460	3495	3535	3570	3610	3650	3685
30	2745	2780	2820	2860	2895	2935	2970	3010	3045	3085	3120	3160	3195	3235	3270	3310	3345	3385	3420	3460	3495	3535	3570	3610	3645
32	2715	2750	2790	2825	2865	2900	2940	2975	3015	3050	3090	3125	3160	3200	3235	3275	3310	3350	3385	3425	3460	3495	3535	3570	3610
34	2685	2725	2760	2795	2835	2870	2910	2945	2980	3020	3055	3090	3130	3165	3200	3240	3275	3310	3350	3385	3425	3460	3495	3535	3570
36	2655	2695	2730	2765	2805	2840	2875	2910	2950	2985	3020	3060	3095	3130	3165	3205	3240	3275	3310	3350	3385	3420	3460	3495	3530
38	2630	2665	2700	2735	2770	2810	2845	2880	2915	2950	2990	3025	3060	3095	3130	3170	3205	3240	3275	3310	3350	3385	3420	3455	3490
40	2600	2635	2670	2705	2740	2775	2810	2850	2885	2920	2955	2990	3025	3060	3095	3135	3170	3205	3240	3275	3310	3345	3380	3420	3455
42	2570	2605	2640	2675	2710	2745	2780	2815	2850	2885	2920	2955	2990	3025	3060	3100	3135	3170	3205	3240	3275	3310	3345	3380	3415
44	2540	2575	2610	2645	2680	2715	2750	2785	2820	2855	2890	2925	2960	2995	3030	3060	3095	3130	3165	3200	3235	3270	3305	3340	3375
46	2510	2545	2580	2615	2650	2685	2715	2750	2785	2820	2855	2890	2925	2960	2995	3030	3060	3095	3130	3165	3200	3235	3270	3305	3340
48	2480	2515	2550	2585	2620	2650	2685	2715	2750	2785	2820	2855	2890	2925	2960	2995	3030	3060	3095	3130	3160	3195	3230	3265	3300
50	2455	2485	2520	2555	2590	2625	2655	2690	2720	2755	2785	2820	2855	2890	2925	2955	2990	3025	3060	3090	3125	3155	3190	3225	3260
52	2425	2455	2490	2525	2555	2590	2625	2655	2690	2720	2755	2790	2820	2855	2890	2925	2955	2990	3020	3055	3090	3125	3155	3190	3220
54	2395	2425	2460	2495	2530	2560	2590	2625	2655	2690	2720	2755	2790	2820	2855	2885	2920	2950	2985	3020	3050	3085	3115	3150	3180
56	2365	2400	2430	2460	2495	2525	2560	2590	2625	2655	2690	2720	2755	2790	2820	2855	2885	2920	2950	2980	3015	3045	3080	3110	3145
58	2335	2370	2400	2430	2460	2495	2525	2560	2590	2625	2655	2690	2720	2750	2785	2815	2850	2880	2920	2945	2975	3010	3040	3075	3105
60	2305	2340	2370	2400	2430	2460	2495	2525	2560	2590	2625	2655	2685	2720	2750	2780	2810	2845	2875	2915	2940	2970	3000	3035	3065
62	2280	2310	2340	2370	2405	2435	2465	2495	2525	2560	2590	2620	2655	2685	2715	2745	2775	2810	2840	2870	2900	2935	2965	2995	3025
64	2250	2280	2310	2340	2370	2400	2430	2465	2495	2525	2555	2585	2620	2650	2680	2710	2740	2770	2805	2835	2865	2895	2920	2955	2990
66	2220	2250	2280	2310	2340	2370	2400	2430	2460	2495	2525	2555	2585	2615	2645	2675	2705	2735	2765	2800	2825	2860	2890	2920	2950
68	2190	2220	2250	2280	2310	2340	2370	2400	2430	2460	2490	2520	2550	2580	2610	2640	2670	2700	2730	2760	2795	2820	2850	2880	2910
70	2160	2190	2220	2250	2280	2310	2340	2370	2400	2425	2455	2485	2515	2545	2575	2605	2635	2665	2695	2725	2755	2780	2810	2840	2870
72	2130	2160	2190	2220	2250	2280	2310	2335	2365	2395	2425	2455	2480	2510	2540	2570	2600	2630	2660	2685	2715	2745	2775	2805	2830
74	2100	2130	2160	2190	2220	2245	2275	2305	2335	2360	2390	2420	2450	2475	2505	2535	2565	2590	2620	2650	2680	2710	2740	2765	2795

Source: E. DeF. Baldwin and E. W. Richards, Jr., "Pulmonary Insufficiency 1, Physiologic Classification, Clinical Methods of Analysis, Standard Values in Normal Subjects," *Medicine*, vol. 27, p. 243.

Table 51.3 Predicted Vital Capacities (in Milliliters) for Males

	HEIGHT IN CENTIMETERS																							
Age	146	148	150	152	154	156	158	160	162	164	166	168	170	172	174	176	178	180	182	184	186	188	190	192	194
16	3765	3820	3870	3920	3975	4025	4075	4130	4180	4230	4285	4335	4385	4440	4490	4540	4590	4645	4695	4745	4800	4850	4900	4955	5005
18	3740	3790	3840	3890	3940	3995	4045	4095	4145	4200	4250	4300	4350	4405	4455	4505	4555	4610	4660	4710	4760	4815	4865	4915	4965
20	3710	3760	3810	3860	3910	3960	4015	4065	4115	4165	4215	4265	4320	4370	4420	4470	4520	4570	4625	4675	4725	4775	4825	4875	4930
22	3680	3730	3780	3830	3880	3930	3980	4030	4080	4135	4185	4235	4285	4335	4385	4435	4485	4535	4585	4635	4685	4735	4790	4840	4890
24	3635	3685	3735	3785	3835	3885	3935	3985	4035	4085	4135	4185	4235	4285	4330	4380	4430	4480	4530	4580	4630	4680	4730	4780	4830
26	3605	3655	3705	3755	3805	3855	3905	3955	4000	4050	4100	4150	4200	4250	4300	4350	4395	4445	4495	4545	4595	4645	4695	4740	4790
28	3575	3625	3675	3725	3775	3820	3870	3920	3970	4020	4070	4115	4165	4215	4265	4310	4360	4410	4460	4510	4555	4605	4655	4705	4755
30	3550	3595	3645	3695	3740	3790	3840	3890	3935	3985	4035	4080	4130	4180	4230	4275	4325	4375	4425	4470	4520	4570	4615	4665	4715
32	3520	3565	3615	3665	3710	3760	3810	3855	3905	3950	4000	4050	4095	4145	4195	4240	4290	4340	4385	4435	4485	4530	4580	4625	4675
34	3475	3525	3570	3620	3665	3715	3760	3810	3855	3905	3950	4000	4045	4095	4140	4190	4225	4285	4330	4380	4425	4475	4520	4570	4615
36	3445	3495	3540	3585	3635	3680	3730	3775	3825	3870	3920	3965	4010	4060	4105	4155	4200	4250	4295	4340	4390	4435	4485	4530	4580
38	3415	3465	3510	3555	3605	3650	3695	3745	3790	3840	3885	3930	3980	4025	4070	4120	4165	4210	4260	4305	4350	4400	4445	4495	4540
40	3385	3435	3480	3525	3575	3620	3665	3710	3760	3805	3850	3900	3945	3990	4035	4085	4130	4175	4220	4270	4315	4360	4410	4455	4500
42	3360	3405	3450	3495	3540	3590	3635	3680	3725	3770	3820	3865	3910	3955	4000	4050	4095	4140	4185	4230	4280	4325	4370	4415	4460
44	3315	3360	3405	3450	3495	3540	3585	3630	3675	3725	3770	3815	3860	3905	3950	3995	4040	4085	4130	4175	4220	4270	4315	4360	4405
46	3285	3330	3375	3420	3465	3510	3555	3600	3645	3690	3735	3780	3825	3870	3915	3960	4005	4050	4095	4140	4185	4230	4275	4320	4365
48	3255	3300	3345	3390	3435	3480	3525	3570	3615	3655	3700	3745	3790	3835	3880	3925	3970	4015	4060	4105	4150	4190	4235	4280	4325
50	3210	3255	3300	3345	3390	3430	3475	3520	3565	3610	3650	3695	3740	3785	3830	3870	3915	3960	4005	4050	4090	4135	4180	4225	4270
52	3185	3225	3270	3315	3355	3400	3445	3490	3530	3575	3620	3660	3705	3750	3795	3835	3880	3925	3970	4010	4055	4100	4140	4185	4230
54	3155	3195	3240	3285	3325	3370	3415	3455	3500	3540	3585	3630	3670	3715	3760	3800	3845	3890	3930	3975	4020	4060	4105	4145	4190
56	3125	3165	3210	3255	3295	3340	3380	3425	3465	3510	3550	3595	3640	3680	3725	3765	3810	3850	3895	3940	3980	4025	4065	4110	4150
58	3080	3125	3165	3210	3250	3290	3335	3375	3420	3460	3500	3545	3585	3630	3670	3715	3755	3800	3840	3880	3925	3965	4010	4050	4095
60	3050	3095	3135	3175	3220	3260	3300	3345	3385	3430	3470	3500	3555	3595	3635	3680	3720	3760	3805	3845	3885	3930	3970	4015	4055
62	3020	3060	3110	3150	3190	3230	3270	3310	3350	3390	3440	3480	3520	3560	3600	3640	3680	3730	3770	3810	3850	3890	3930	3970	4020
64	2990	3030	3080	3120	3160	3200	3240	3280	3320	3360	3400	3440	3490	3530	3570	3610	3650	3690	3730	3770	3810	3850	3900	3940	3980
66	2950	2990	3030	3070	3110	3150	3190	3230	3270	3310	3350	3390	3430	3470	3510	3550	3600	3640	3680	3720	3760	3800	3840	3880	3920
68	2920	2960	3000	3040	3080	3120	3160	3200	3240	3280	3320	3360	3400	3440	3480	3520	3560	3600	3640	3680	3720	3760	3800	3840	3880
70	2890	2930	2970	3010	3050	3090	3130	3170	3210	3250	3290	3330	3370	3410	3450	3480	3520	3560	3600	3640	3680	3720	3760	3800	3840
72	2860	2900	2940	2980	3020	3060	3100	3140	3180	3210	3250	3290	3330	3370	3410	3450	3490	3530	3570	3610	3650	3680	3720	3760	3800
74	2820	2860	2900	2930	2970	3010	3050	3090	3130	3170	3200	3240	3280	3320	3360	3400	3440	3470	3510	3550	3590	3630	3670	3710	3740

Source: E. DeF. Baldwin and E. W. Richards, Jr., "Pulmonary Insufficiency 1, Physiologic Classification, Clinical Methods of Analysis, Standard Values in Normal Subjects," *Medicine*, vol. 27, p. 243.

Name ______________________________

Date ______________________________

Section ______________________________

The **A** corresponds to the indicated Learning Outcome(s) **O** found at the beginning of the Laboratory Exercise.

LABORATORY
ASSESSMENT 51

Breathing and Respiratory Volumes

PART A: Assessments

Complete the following statements:

1. Breathing is also called ______________________________. **A1**

2. When using the lung function model, what part of the respiratory system is represented by: **A1**

   a. The rubber sheeting? ______________________________ b. The bell jar? ______________________________

   c. The Y tube? ______________________________ d. The balloons? ______________________________

3. When the diaphragm contracts, the size of the thoracic cavity ______________________________. **A1**

4. The ribs are raised primarily by contraction of the ______________________________ muscles, which increases the size of the thoracic cavity. **A1**

5. The force responsible for normal resting expiration comes from ______________________________ of stretched tissues. **A1**

6. The primary muscles that help to force out more than the normal volume of air by pulling the ribs downward and inward are the ______________________________. **A1**

7. The diaphragm can be forced to ascend higher than normal by the contraction of ______________________________ wall muscles. **A1**

8. We inhale when the diaphragm ______________________________. **A1**

9. We exhale when the diaphragm ______________________________. **A1**

CRITICAL THINKING ASSESSMENT—PART A

Predict the effect that injury to a phrenic nerve would have on pulmonary ventilation. **A1**

____

____.

PART B: Assessments

1. Test results for respiratory air volumes and capacities: A2 A3

Respiratory Volume or Capacity	Expected Value* (Approximate)	Test Result	Percent of Expected Value (Test Result/Expected Value × 100)
Tidal volume (resting) (TV)	500 mL		
Expiratory reserve volume (ERV)	1,200 mL		
Vital capacity (VC)	4,700 mL ________ (enter yours from table 51.2 or 51.3)		
Inspiratory reserve volume (IRV)	3,000 mL		
Inspiratory capacity (IC)	3,500 mL		
Functional residual capacity (FRC)	2,500 mL		

*The values listed are most characteristic for a healthy, tall, young adult male. In general, adult females have smaller bodies and therefore smaller lung volumes and capacities. If your expected value for vital capacity is considerably different than 4,700 mL, your other values will vary accordingly.

2. Complete the following: A2 A3

   a. How do your test results compare with the expected values? ______________________________

   ____

   ____

   ____

   b. How does your vital capacity compare with the average value for a person of your sex, age, and height?

   ____

   ____

   c. What measurement in addition to vital capacity is needed before you can calculate your total lung capacity?

   ____

3. If your experimental results are considerably different than the predicted vital capacities, propose reasons for the differences. As you write this paragraph, consider factors such as smoking, physical fitness, respiratory disorders, stress, and medications. (Your instructor might have you make some class correlations from class data.)

____

____

____

____

PART C: Assessments

Match the air volumes in column A with the definitions in column B. Place the letter of your choice in the space provided. A4

Column A

a. Expiratory reserve volume

b. Functional residual capacity

c. Inspiratory capacity

d. Inspiratory reserve volume

e. Residual volume

f. Tidal volume

g. Total lung capacity

h. Vital capacity

Column B

_____ **1.** Volume of air in addition to tidal volume that leaves the lungs during forced expiration

_____ **2.** Vital capacity plus residual volume

_____ **3.** Volume of air that remains in lungs after the most forceful expiration

_____ **4.** Volume of air that enters or leaves lungs during a respiratory cycle

_____ **5.** Volume of air in addition to tidal volume that enters lungs during forced inspiration

_____ **6.** Maximum volume of air a person can exhale after taking the deepest possible breath

_____ **7.** Maximum volume of air a person can inhale following exhalation of the tidal volume

_____ **8.** Volume of air remaining in the lungs following exhalation of the tidal volume

CRITICAL THINKING ASSESSMENT—PART C

To determine possible obstruction of the airway, a forced expiratory volume (FEV) may be measured using a spirometer. Under normal circumstances, a healthy adult can expel 75–85% of the vital capacity in 1.0 second (FEV_1). Predict how this would change in a person who has asthma, and explain your reasoning. A2 A3 A4

____

____

____.

NOTES

LABORATORY
EXERCISE 52

Spirometry: BIOPAC® Exercise

Purpose of the Exercise

To measure and calculate pulmonary volumes and capacities using the BIOPAC airflow transducer system.

MATERIALS NEEDED

Computer system [Mac OS X 10.7–10.15 (Catalina), or PC running Windows 10, 8, or 7]

Biopac Student Lab software, version 4.0 or above

BIOPAC Data Acquisition Unit (MP36, MP35, or MP46/45)

BIOPAC Airflow Transducer (SS11LA, with removable/autoclavable head, or SS11LB, with version 4.1.1 or higher)

BIOPAC Disposable Mouthpiece (AFT2), optional Autoclavable Mouthpiece (AFT8), or Mouthpiece/filter Unit (AFT36), only with SS11LB

BIOPAC Disposable Nose Clip (AFT3), or SS11LB, with version 4.1.1 or higher

BIOPAC Bacteriological Filter (AFT1), if necessary

BIOPAC Calibration Syringe, 0.6 L (AFT6 or AFT6A + AFT11A) or 2 L (AFT26)

SAFETY

- If your lab autoclaves the removable head of the airflow transducer (SS11LA), and you use the autoclavable mouthpieces (AFT8), you will need to use a bacteriological filter (AFT1) only for the calibration step.
- If your lab does not autoclave the removable head of the airflow transducer (SS11LA), and you use the disposable mouthpieces (AFT2), you will need a bacteriological filter (AFT1) for the calibration step, and a clean and sterilized one for each subject.
- If your lab uses the mouthpiece/filter unit (AFT36), which works only with the SS11LB airflow transducer and version 4.1.1 or higher, you will not need to add a bacteriological filter (AFT1).
- Each subject will need a disposable mouthpiece (AFT2), a clean, autoclaved mouthpiece (AFT8), or a mouthpiece/filter unit (AFT36).
- Each subject will need a disposable nose clip (AFT3).

Learning Outcomes

After completing this exercise, you should be able to:

O1 Examine experimentally and record and/or calculate selected respiratory volumes and capacities.

O2 Compare the measured volumes and capacities with average values.

O3 Compare the volumes and capacities of subjects differing in gender, age, height, and weight.

The O corresponds to the assessments A indicated in the Laboratory Assessment for this Exercise.

Pre-Lab

Carefully read the introductory material and examine the entire lab content. Be familiar with inspiration, expiration, and respiratory volumes from lecture or the textbook. Also, if this is your first BIOPAC lab, please read the general description of the Biopac Student Lab system in Exercise 21.

Breathing (pulmonary ventilation), one of the most vital processes of life, consists of two phases: inspiration and expiration. *Inspiration* consists of the inhalation of air into the respiratory passages and lungs, which provides oxygen (O_2) to the cells of the body. Oxygen is required for the aerobic cellular respiration of food, to break down nutrients into their building blocks, and to generate ATP for energy. Expiration involves the exhalation of air from the lungs into the atmosphere, which rids the body of carbon dioxide (CO_2). Carbon dioxide is produced as a waste product of cellular respiration.

One way in which the breathing process can be studied is through the use of a spirometer. There are various types of spirometers, from small handheld dry spirometers, to wet bell varieties, to computer-linked spirometers. But all spirometers work on the same principle—they all measure certain respiratory volumes, based on different degrees of effort in the breathing process. The respiratory volumes used in spirometry are as follows:

1. *Tidal volume (TV):* The volume of air inspired or expired in one normal, quiet breath (about 500 mL).
2. *Inspiratory reserve volume (IRV):* The volume of air that can be inspired forcefully above a tidal inspiration (about 3,000 mL).
3. *Expiratory reserve volume (ERV):* The volume of air that can be expired forcefully beyond a tidal expiration (about 1,200 mL).
4. *Residual volume (RV):* The volume of air remaining in the lungs after forceful expiration (about 1,300 mL).

Tidal respiration does not require any extra effort, but volumes such as IRV and ERV are defined by the extra effort used to measure them. For example, if you are preparing to blow up a balloon, and you inhale the maximum possible amount of air, the amount you inhale above your tidal inspiration will be your IRV. When you actually blow up the balloon, and you exhale the maximum possible amount of air, the amount you exhale beyond your tidal expiration will be your ERV. The RV cannot be measured with a spirometer, since it always remains in the lungs; for the purpose of this lab exercise, the default setting for RV by the BIOPAC software is 1.0 L.

In addition to respiratory volumes, several respiratory capacities can be defined; these can be measured in the lab and/or calculated. Respiratory capacities can be calculated by adding two or more volumes together, as follows:

1. *Inspiratory capacity (IC):* The maximal volume of air that can be inspired after a tidal expiration, IC = TV + IRV (about 3,500 mL).
2. *Expiratory capacity (EC):* The maximal volume of air that can be expired after a tidal inspiration, EC = TV + ERV (about 1,700 mL).
3. *Functional residual capacity (FRC):* The volume of air remaining in the lungs after a tidal expiration, FRC = ERV + RV (about 2,500 mL).
4. *Vital capacity (VC):* The volume of air that can be forcefully expired after a maximal inspiration, VC = TV + IRV + ERV (about 4,700 mL).
5. *Total lung capacity (TLC):* The total volume of air that the lungs and respiratory passages can hold, TLC = TV + IRV + ERV + RV or TLC = VC + RV (about 6,000 mL).

It is important to realize that there is quite a bit of individual variation in respiratory volumes and capacities. This variation is due to differences in gender, height, weight, age, and exercise history. Normal averages for a young, healthy, adult male are given above, but any reading within 20% of the average is still considered to be in the normal range. For example, even though there are several factors determining the vital capacity (VC) of an individual, VC can be estimated fairly accurately based on gender, height (H) in cm, and age (A) in years, with the following equations:

$$\text{Male:} \quad VC = 0.052\ H - 0.022\ A - 3.60$$

$$\text{Female:} \quad VC = 0.041\ H - 0.018\ A - 2.69$$

This BIOPAC exercise uses a computer-linked spirometry system to measure respiratory volumes. Each subject will breathe into an airflow transducer; the BIOPAC software will then convert airflow into volume. Several volumes and capacities will be measured and/or calculated in this lab exercise. VC will be both measured and estimated from one of the equations given above; the two values will then be compared. Again, remember that RV cannot be measured through spirometry. Figure 52.1 shows a record of volume changes over time for various respiratory volumes and capacities; this is called a *spirogram.*

Spirometry is used in medicine to help in the diagnosis of specific respiratory disorders, determine the extent of a respiratory illness or disease, and assess the respiratory status of a person recovering from an illness, such as pneumonia or an asthma attack. Certain illnesses are associated with characteristic changes in specific respiratory volumes and/or capacities. For example, in people with obstructive lung disease, such as chronic bronchitis, the FRC and RV may be elevated due to an inability to expel the normal amount of air during expiration and excess inflation of the lungs. In cases of restrictive lung disease, such as pulmonary fibrosis, decreased lung compliance makes it difficult to expand the lungs, which results in a decrease in IC and VC.

PROCEDURE A: Setup

1. With your computer **ON** and the BIOPAC MP36/35 unit turned **OFF,** plug the airflow transducer (SS11LA or SS11LB) into Channel 1. If you are using the MP46/45, the USB cable must be connected, and the ready light must be on.
2. Turn on the MP36/35 Data Acquisition Unit.
3. Start the Biopac student lab program for **L12—Pulmonary Function I,** and click **OK.**
4. Designate a filename to be used to save the subject's data, and click **OK.**
5. After typing the filename, a window will pop up, asking for subject details (gender, age, and height). Fill in the information, and click **OK.**

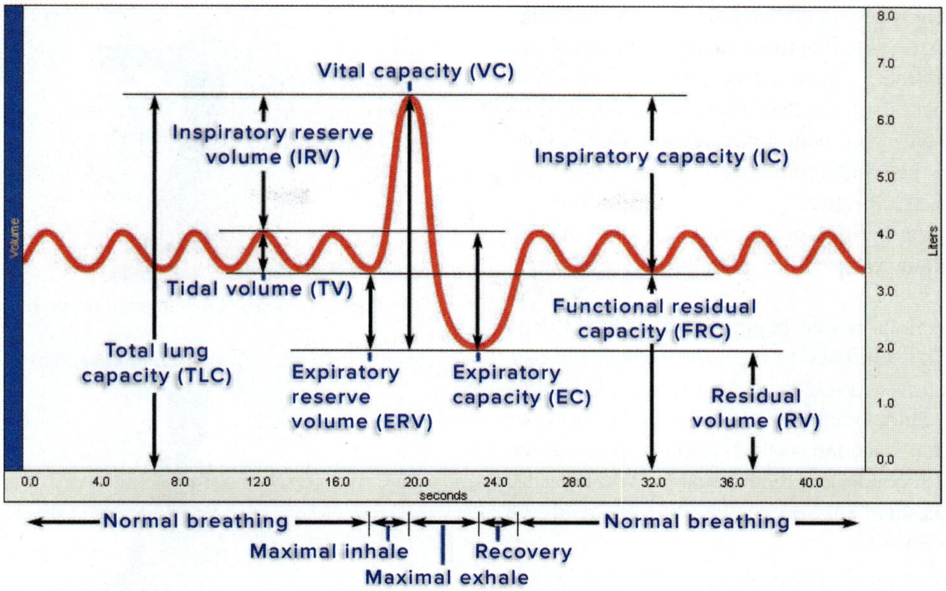

FIGURE 52.1 Example of a spirogram, showing respiratory volumes and capacities.

PROCEDURE B: Calibration

1. The calibration will be conducted in two stages. For the first stage, simply hold the airflow transducer still and upright, and click **Calibrate.** This calibration will run for 4–8 seconds and then stop automatically. The recording should look similar to figure 52.2. If it does, click **Continue.** If it does not, click **Redo Calibration.**
2. To prepare for the second calibration stage, connect a bacteriological filter to the calibration syringe, and then connect the calibration syringe/filter assembly to the airflow transducer, as shown in figure 52.3. If you are using the AFT36 mouthpiece/filter unit with the SS11LB

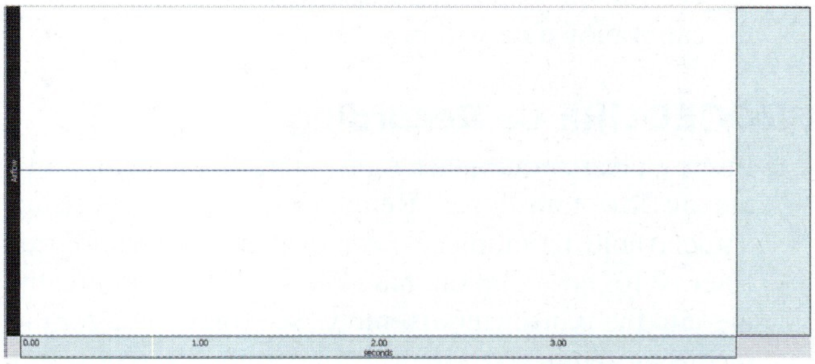

FIGURE 52.2 Sample recording for first calibration stage.

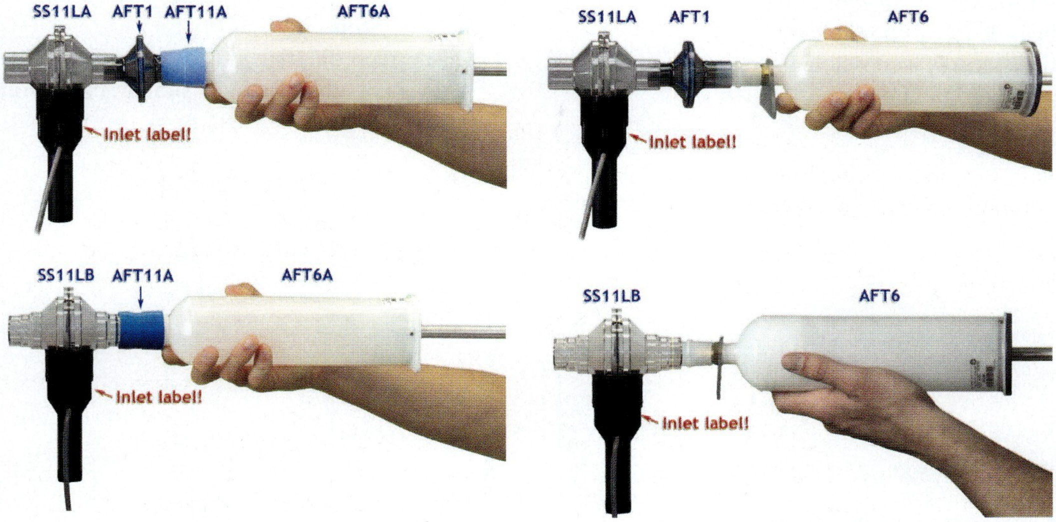

FIGURE 52.3 Setup for second calibration stage.

airflow transducer, you will not have to add a bacteriological filter (AFT1). Be sure that the assembly is inserted into the side labeled "Inlet." Once you have put the calibration assembly together, always support the calibration syringe by holding it up with your nondominant hand. Do not hold the assembly by the handle of the airflow transducer; this could damage the calibration syringe.

3. Pull the calibration syringe plunger all the way out, and hold the calibration syringe/filter assembly still and parallel to the floor.
4. Read these directions before beginning the second calibration stage. You will have to push the syringe plunger in and out five times (a series of ten strokes). This is to mimic normal, quiet breathing, and should be done at this rate: push for 1 second, wait 2 seconds, pull out for 1 second, wait 2 seconds, and then repeat the whole cycle. When you understand this procedure, and are ready to begin, click **Calibrate.**
5. After completing five push-and-pull cycles, click **End Calibration.** If your calibration data resembles figure 52.4, remove the calibration syringe, and click **Continue** to proceed to the **Recording** section. If it does not, click **Redo Calibration.** (If you have to redo the calibration, you will have to redo both stages; all of the calibration data will be erased.)

PROCEDURE C: Recording

1. Insert a clean mouthpiece (and filter, if applicable) into the airflow transducer. Remember, if you are using autoclavable mouthpieces, remove the bacteriological filter. With nose clip on, have the subject begin breathing into the transducer assembly. See figure 52.5 for the recording setup.
2. Have the subject breathe normally for about 20 seconds, making sure to close his/her mouth completely around the mouthpiece. Then click **Record** and have the subject:
   a. Take five normal breaths, each consisting of an inspiration followed by an expiration.
   b. Inhale as deeply as possible once.
   c. Exhale as deeply as possible once.
   d. Take five normal breaths.

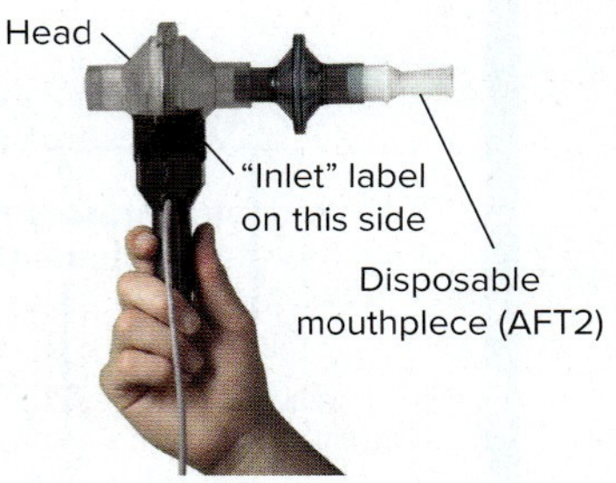

SS11LA with unsterilized head

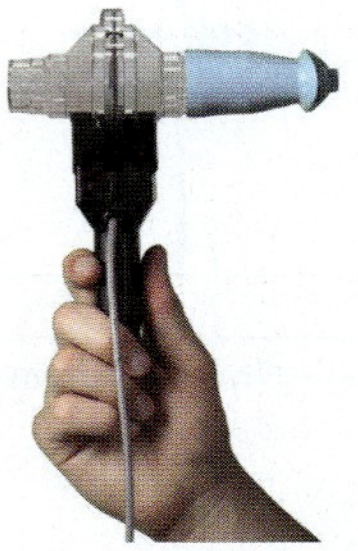

SS11LA/LB with sterilized head

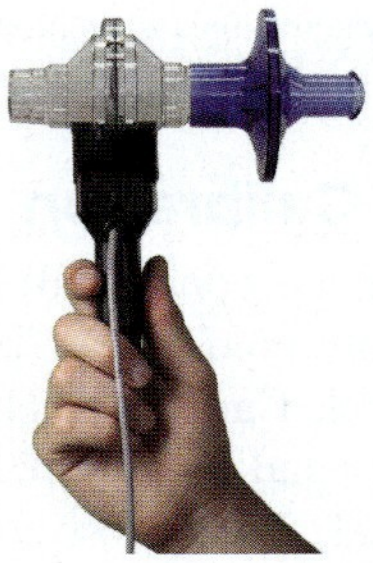

SS11LB with reusable filter/mouthpiece combination

FIGURE 52.5 Recording setups, showing the various airflow transducers, mouthpieces, and filters that your lab may have.

Courtesy of and ©BIOPAC Systems, Inc. www.biopac.com

3. Click **Stop.**
4. The recording should look similar to figure 52.6. If it does not, click **Redo.** If similar, click **Done.**
5. Click **Yes.**

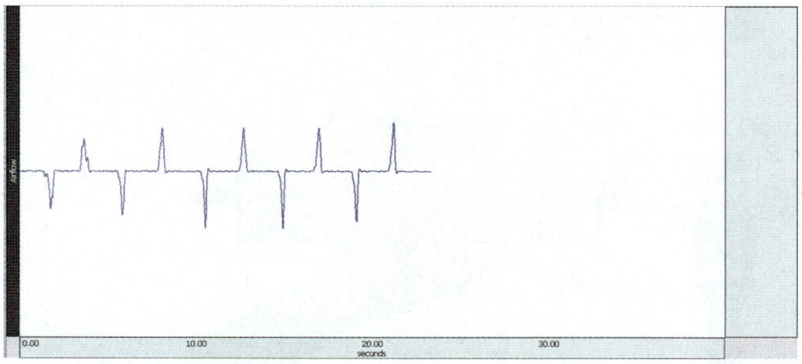

FIGURE 52.4 Sample recording for second calibration stage.

Courtesy of and ©BIOPAC Systems, Inc. www.biopac.com

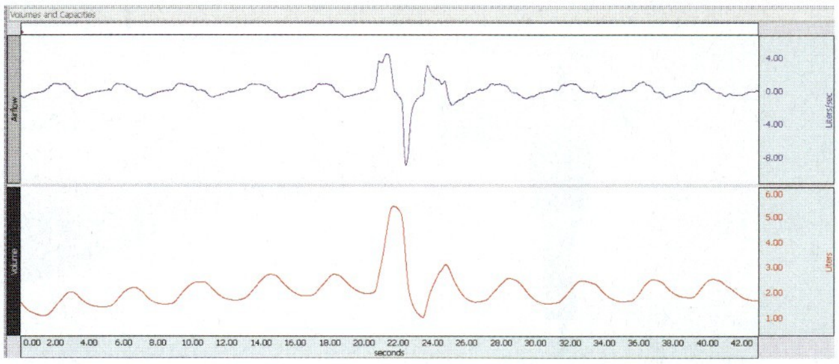

FIGURE 52.6 Sample recording for spirometry exercise.

Courtesy of and ©BIOPAC Systems, Inc. www.biopac.com

PROCEDURE D: Data Analysis

1. You can either analyze the current data file at this time or save it and do the analysis later, using the **Review Saved Data** mode. Note the channel number designations: CH 1: Airflow (hidden) and CH 2: Volume.
2. Note the settings for the measurement boxes, across the top of the data window: CH 2: P–P, CH 2: Max, CH 2: Min, and CH 2: Delta. Following are explanations of these settings:

P–P:	The difference between the maximum and minimum points in the selected area
Max:	Shows the maximum point within the selected area
Min:	Shows the minimum point within the selected area
Delta:	Calculates the amplitude difference between the last and first points in the selected area

3. Measure the subject's *vital capacity* (VC). Click on the I-beam tool. Using the I-beam cursor, select the area from the highest point of the deep inspiration to the lowest point of the deep expiration. Selection of the P–P area does not have to be perfect; it just has to include the highest and lowest points. See figure 52.7 for an example of proper selection of the area for VC measurement. The result will be displayed by CH 2: P–P. Record your result in Part A of Laboratory Assessment 52.
4. Measure the *tidal volume* (TV), during both normal inspiration and expiration. Use the I-beam tool to make these measurements. To measure the TV during inspiration, select the area between the lowest point before the third normal breath and the highest point of the third breath (valley to peak; see fig. 52.8). For TV during expiration, select the area from the highest point of the third breath to the lowest point at the end of the third breath (peak to valley; see fig. 52.9). Then repeat the inspiration and expiration measurements for the fourth breath. The results will be displayed by CH 2: P–P. Record your results in table 52.1 of the laboratory assessment.
5. Measure the *inspiratory reserve volume* (IRV). Using the I-beam cursor, select the area from the peak of the fifth normal breath to the peak of the deep inspiration. See figure 52.10. The result will be displayed by CH 2: Delta. Select the area carefully; the Delta measurement must be exact in order to be accurate. Record your result in table 52.1.
6. Measure the *expiratory reserve volume* (ERV). Using the I-beam cursor, select the area from the lowest point of the deep expiration to the lowest point at the end of the normal breath following it (valley to valley). The result will

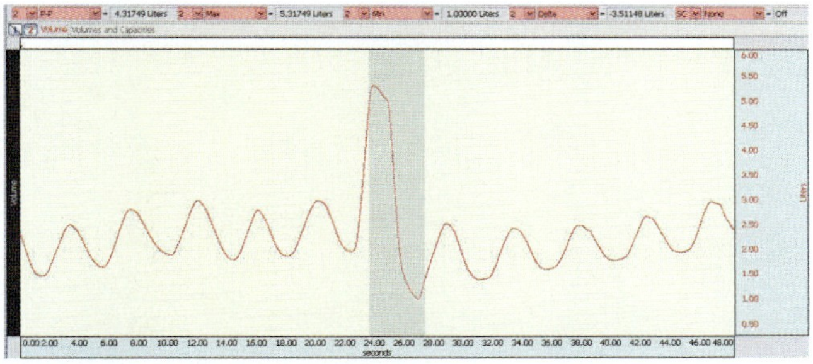

FIGURE 52.7 Proper selection of area for measurement of vital capacity (VC).

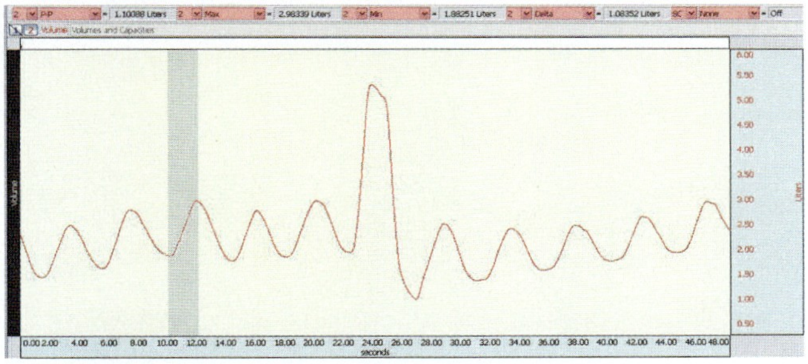

FIGURE 52.8 Proper selection of area for measurement of tidal volume (TV) during inspiration.

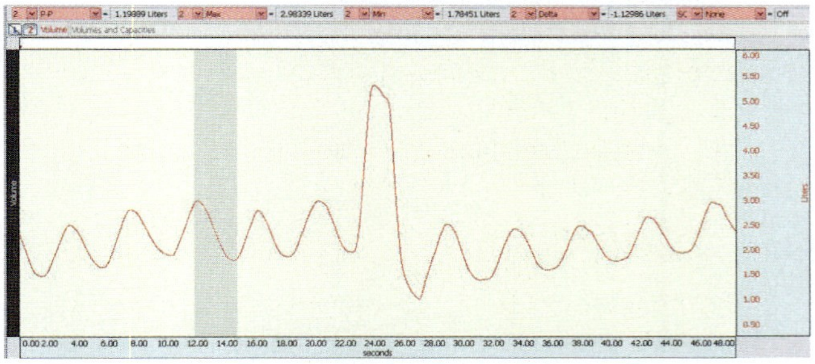

FIGURE 52.9 Proper selection of area for measurement of tidal volume (TV) during expiration.

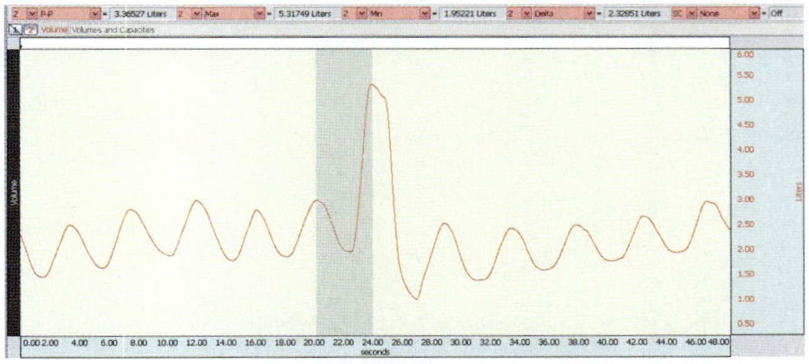

FIGURE 52.10 Proper selection of area for measurement of inspiratory reserve volume (IRV).

be displayed by CH 2: Delta. As in step 5, select the area carefully. Record your result in table 52.1.

7. Measure the *residual volume* (RV). Using the I-beam cursor, select a small area that includes the lowest point of the deep expiration. The result will be displayed by CH 2: Min. Remember that spirometers cannot actually measure RV. Since the BIOPAC software uses a default setting of 1.0 L for RV, your RV result may be preset for 1.0 L. If this is the case, just record 1.0 L as your measurement result for RV in table 52.1. *Note:* Because the RV cannot be measured, it is not possible to measure functional residual capacity (FRC) either. It will be calculated in table 52.1 by adding RV + ERV.
8. Now measure the *inspiratory capacity* (IC). Using the I-beam cursor, select the area from the lowest point at the end of the fifth normal breath to the highest point of the deep inspiration (valley to peak). The result will be displayed by CH 2: Delta. Record your result in table 52.1.
9. Now measure the *expiratory capacity* (EC). Using the I-beam cursor, select the area from the lowest point of the deep expiration to the highest point of the normal breath following it (valley to peak). The result will be displayed by CH 2: Delta. Record your result in table 52.1.
10. Finally, measure the *total lung capacity* (TLC). Using the I-beam cursor, select a small area that includes the highest point of the deep inspiration. The result will be displayed by CH 2: Max. Record your result in table 52.1.
11. Now turn to Part A of the laboratory assessment. Complete the calculations for estimated VC and follow the directions to compare measured VC with estimated VC.
12. Complete the calculations for table 52.1 using the measured results you just obtained. It will be interesting to see how close the measured and calculated results are.
13. Complete table 52.2, to see how the subject's measured TV, IRV, and ERV compare with the normal averages.
14. When finished, you can save your data on the hard drive or a network drive and then **Exit** the system or **Quit** the file.
15. Answer the questions in Part B of the laboratory assessment.

LABORATORY
ASSESSMENT 52

Name ______________________________

Date ______________________________

Section ______________________________

The A corresponds to the indicated Learning Outcome(s) O found at the beginning of the Laboratory Exercise.

Spirometry: BIOPAC® Exercise

PART A: Assessments

1. Vital capacity A1

a. Record the subject's measured vital capacity (VC) as instructed in step 3 of the Data Analysis:

Subject's measured VC = __________ L.

b. Estimate the subject's vital capacity (VC) by using the appropriate equation below:

Male: VC = 0.052 H – 0.022A – 3.60

Female: VC = 0.041 H – 0.018 A – 2.69

(H = height in cm, and A = age in years)

Subject's estimated VC = __________ L.

c. Perform the following calculation to determine how close the subject's measured VC came to the estimated VC:

Measured VC / Estimated VC = __________ x 100 = __________ %

2. Measurements and calculations of respiratory volumes and capacities: A1

Table 52.1 Measurements and Calculations of Subject's Respiratory Volumes and Capacities

Title	Measurement Result			Calculation
Tidal volume (TV)	a = 2	P–P	Cycle 3 inhale:	(a + b + c + d) / 4 =
	b = 2	P–P	Cycle 3 exhale:	
	c = 2	P–P	Cycle 4 inhale:	
	d = 2	P–P	Cycle 4 exhale:	
Inspiratory reserve volume (IRV)	2	Delta		
Expiratory reserve volume (ERV)	2	Delta		
Residual volume (RV)	2	Min		Default = 1.0 L *(Preference setting)*
Inspiratory capacity (IC)	2	Delta		TV + IRV =
Expiratory capacity (EC)	2	Delta		TV + ERV =
Functional residual capacity (FRC)				ERV + RV =
Total lung capacity (TLC)	2	Max		IRV + TV + ERV + RV =

3. Comparison of subject's measured volumes with average volumes: A2

Table 52.2 Comparison of Subject's Measured Respiratory Volumes with Average Volumes

Volume Title		Average Volume	Measured Volume
Tidal volume	TV	Resting subject, normal breathing: TV is approximately 500 mL During exercise: TV can be more than 3 L	Greater than Equal to Less than
Inspiratory reserve volume	IRV	Resting IRV for young adults is males = approximately 3,300 mL females = approximately 1,900 mL	Greater than Equal to Less than
Expiratory reserve volume	ERV	Resting ERV for young adults is males = approximately 1,000 mL females = approximately 700 mL	Greater than Equal to Less than

PART B: Assessments

Complete the following:

1. How do the experimental results compare to the expected or normal values presented in the introduction? A2

____

____

____

2. Why is the predicted vital capacity dependent upon the height of the subject? A3

____

____

____

3. Why might vital capacity and expiratory reserve decrease with age? A3

____

____

____

LABORATORY EXERCISE 53

Control of Breathing

Purpose of the Exercise

To review the muscles that control breathing and to describe the mechanisms that regulate the rate and depth of breathing.

MATERIALS NEEDED

Clock or watch with seconds timer
Paper bags, small

For Demonstration Activity:
Flasks
Glass tubing
Rubber stoppers, two-hole
Calcium hydroxide solution (limewater)

For Learning Extension Activity:
Pneumograph
Physiological recording apparatus

Learning Outcomes APR

After completing this exercise, you should be able to:

O1 Locate the respiratory areas in the brainstem.

O2 Describe the mechanisms that control and influence breathing.

O3 Select the respiratory muscles involved in inspiration and forced expiration.

O4 Test and record the effect of various factors on the rate and depth of breathing.

The O corresponds to the assessments A indicated in the Laboratory Assessment for this Exercise.

Pre-Lab

Carefully read the introductory material and examine the entire lab. Be familiar with control of breathing from lecture or the textbook. Answer the pre-lab questions.

Pre-Lab Questions Select the correct answer for each of the following questions:

1. Respiratory areas of the brain include all of the following *except* the
a. brainstem. **b.** pons.
c. medulla oblongata. **d.** pineal gland.

2. A quiet, more relaxed breathing is called _____.
a. dyspnea **b.** hyperpnea
c. eupnea **d.** apnea

3. Which of the following muscles is *not* involved in forced expiration?
a. internal intercostal muscles **b.** rectus abdominis muscles
c. sternocleidomastoid muscles **d.** external oblique muscles

4. Peripheral chemoreceptors sensitive to low blood oxygen levels are located in the
a. heart. **b.** aortic arch and carotid arteries.
c. aorta only. **d.** carotid arteries only.

5. Normal blood pH is
a. 6.4. **b.** 8.0.
c. 6.8–8.0. **d.** 7.35–7.45.

6. Breathing rate increases when blood _____ levels increase.
a. carbon dioxide and hydrogen ion
b. oxygen
c. carbon dioxide and oxygen
d. oxygen and hydrogen ion

7. Inhaled dust and pollutants stimulate _____ to elicit a protective response of bronchoconstriction and coughing.
a. chemoreceptors **b.** stretch receptors
c. proprioceptors **d.** irritant receptors

8. An increase in the duration of inspirations is the normal response from the inflation reflex.
   a. True **b.** False
9. The ventral respiratory group of the brainstem respiratory centers is involved with stimulation of the contractions of the diaphragm.
   a. True **b.** False

Breathing is controlled from regions of the *brainstem respiratory centers,* which control both inspiration and expiration. These areas initiate and regulate nerve impulses that travel to various skeletal muscles of breathing, causing rhythmic contractions that can be adjusted to the rate and depth of breathing needed to meet various cellular needs.

PROCEDURE A: Control of Breathing

The brainstem respiratory centers include two bilateral groups of neurons, the *ventral respiratory group* (*VRG*) and the *dorsal respiratory group* (*DRG*), in the medulla oblongata, and another bilateral group of neurons, the *pontine respiratory group* (*PRG*), in the pons. The VRG is the primary generator of impulses for the respiratory rhythm during quiet breathing. A quiet, more relaxed breathing is called *eupnea*. Impulses from the VRG pass through spinal cord integrating centers and then to the diaphragm via the phrenic nerves and to the intercostal muscles via the intercostal nerves to expand the thoracic cavity volume. An inspiration from contractions of the diaphragm and external intercostal muscles takes about 2 seconds, while expiration takes about 3 seconds. This produces a quiet (resting) breathing rate of about 12–15 breaths per minute. During inspiration, the expiratory neurons of the VRG are inhibited. When inspiratory neurons stop firing, expiratory neurons of the VRG conduct impulses to the muscles for expiration, and the inspiratory muscles relax. There is a delicate balance between inspiratory and expiratory muscles being stimulated and inhibited from the VRG during each respiratory breathing cycle (fig. 53.1).

The DRG receives input from various sources, including the pons, central chemoreceptors, peripheral chemoreceptors, stretch receptors, and irritant receptors. As a result of input into the DRG, messages are relayed to the VRG to adjust the rate and depth of the respiratory rhythm during various changing conditions (fig. 53.1). In essence, the DRG modulates the activity of the VRG.

Neurons in the pons compose the PRG (formerly the pneumotaxic center). Input from higher brain centers to the PRG includes the cerebral cortex, the limbic system, and the hypothalamus. As a result of various messages into the PRG, messages are conducted to the DRG and the VRG, and modifications can occur to the respiratory rhythm. Such modifications to a respiratory rhythm occur during varying situations such as exercise, sleep, and emotional responses (fig. 53.1).

Various factors can influence the respiratory areas and thus affect the rate and depth of breathing. These factors include stretch of the lung tissues, emotional state, and the presence of certain chemicals in the blood, such as carbon dioxide, hydrogen ions, and oxygen. The breathing rate increases primarily as the blood concentration of carbon dioxide or hydrogen ions increases, and much less often as the concentration of oxygen decreases.

The medulla oblongata possesses *central chemoreceptors* that are sensitive to changes in the levels of carbon dioxide and hydrogen ions in the blood and cerebrospinal fluid (CSF). If the levels of carbon dioxide or hydrogen ions increase, the chemoreceptors relay this information to the brainstem respiratory centers. As a result, the rate and depth of breathing increase and more carbon dioxide is exhaled. When the levels decline to a more normal range, breathing rate decreases to normal. Exercise, breath holding, and hyperventilation are examples that considerably alter the blood levels of carbon dioxide and hydrogen ions (fig. 53.1).

Low blood oxygen levels are detected by *peripheral chemoreceptors* located in *carotid bodies* in the carotid arteries and *aortic bodies* in the aortic arch. This information is relayed to the respiratory centers in the brainstem, and the breathing rate increases. In order to trigger this response, the blood oxygen levels must be very low. Therefore, the increased blood levels of carbon dioxide and hydrogen ions have a much greater influence upon the peripheral chemoreceptors for the rate and depth of breathing (fig. 53.2).

The depth of breathing is also influenced by the *inflation reflex*. If the lungs are greatly expanded during forceful breathing, *stretch receptors* within the bronchial tree and the visceral pleura are stimulated. Sensory impulses are conducted to the DRG via the vagus nerves and, as a result, there is a decrease in the duration of inspirations.

Irritant receptors, located in the respiratory bronchial tree, are stimulated by inhaled dust and various pollutants. When receptors are stimulated, sensory impulses are conducted to the DRG via the vagus nerves. Coughing and respiratory bronchial tree constrictions occur as protective responses to the irritants.

The muscles involved with inspiration and expiration are all voluntary skeletal muscles. Therefore, a person has some cerebral conscious influence over the rate and depth of breathing as in talking, singing, and breath holding. However, respiratory centers of the brainstem, chemoreceptors, and certain reflexes represent the primary mechanism of subconscious, automatic control of breathing. The complex coordination of breathing involves considerable integration between respiratory areas of the brain and receptors; however, some details of the control of breathing are still obscure.

1. Reexamine figures 53.1 and 53.2. Notice that breathing involves input and responses of the nervous, respiratory, muscular, and skeletal systems. Various environmental factors can also alter the rate and depth of respiratory cycles.
2. Study figure 53.3 and table 53.1. Several skeletal muscles contract during breathing. The principal muscles involved during inspiration are the diaphragm and external intercostals. The sternocleidomastoid, scalenes, and pectoralis minor are synergistic during more forceful inhalation, resulting in greater volumes of air inhaled. During quiet respiration, there is minimal

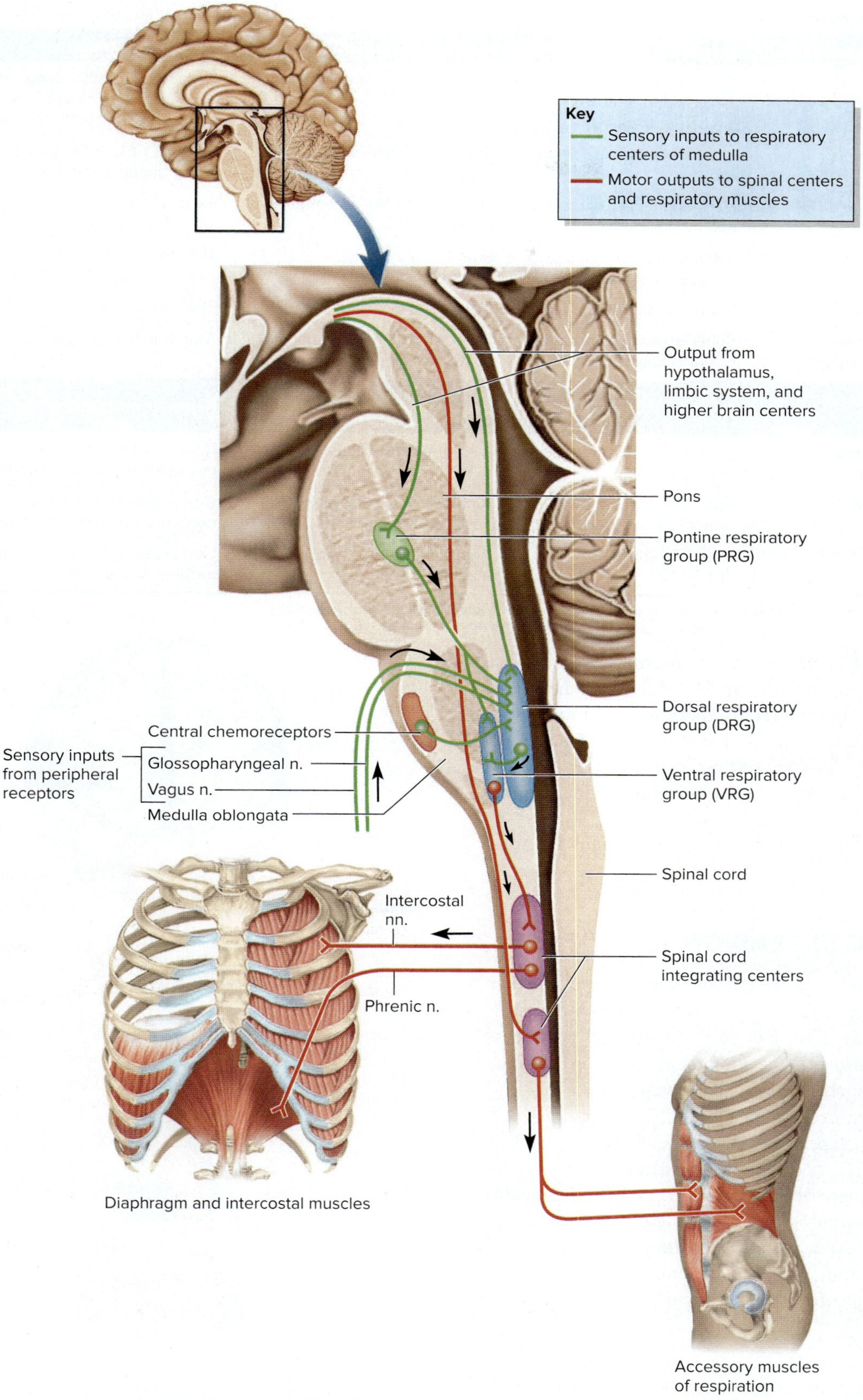

FIGURE 53.1 Components of respiratory control. The interactions of the brainstem respiratory centers (VRG, DRG, and PRG) control the respiratory rhythm of inspiration and expiration. Additional inputs from central chemoreceptors, higher brain centers, and peripheral receptors can result in alterations to the rate and depth of breathing. (n. = nerve; nn. = nerves)

Table 53.1 Muscles of Respiration

Muscles of Inspiration	Origin	Insertion	Inspiration Action
Diaphragm	Costal cartilages and ribs 7–12, xiphoid process, and lumbar vertebrae	Central tendon of diaphragm	Flattens (descends) during contraction; this expands thoracic cavity and compresses abdominal viscera during inspiration
External intercostals	Inferior border of superior rib	Superior border of inferior rib	Elevates ribs; this expands thoracic cavity during inspiration
Sternocleidomastoid	Manubrium of sternum and medial clavicle	Mastoid process of temporal bone	Elevates sternum
Scalenes	Transverse processes of cervical vertebrae C2–C6	Ribs 1 and 2	Elevates ribs 1 and 2
Pectoralis minor	Sternal ends of ribs 3–5	Coracoid process of scapula	Elevates ribs 3–5
Internal intercostals (intercartilaginous part)	Superior border of inferior costal cartilages	Inferior border of superior costal cartilages	Assists in elevating ribs
Muscles of (Forced) Expiration	**Origin**	**Insertion**	**Expiration Action**
Internal intercostals (interosseous part)	Superior border of inferior rib	Inferior border of superior rib	Depresses ribs; this compresses thoracic cavity during forced expiration
Rectus abdominis	Pubic crest and pubic symphysis	Xiphoid process of sternum and costal cartilages of ribs 5–7	Compresses inferior ribs and abdominal contents, helping to raise diaphragm
External oblique	Anterior surfaces of ribs 5–12	Anterior iliac crest and linea alba by aponeurosis	Compresses abdominal contents, helping to raise diaphragm

Note: Boldface muscles indicate the principle muscles increasing and decreasing the thoracic cavity.

involvement of the expiratory muscles. During forceful expiration, the principal muscles are the internal intercostals, but the rectus abdominis and the external oblique muscles can provide extra force. Contractions of accessory muscles during inspiration or expiration result in extra forces and volumes of air into or out of the lungs.

3. Complete Part A of Laboratory Assessment 53.

LEARNING EXTENSION ACTIVITY

A *pneumograph* is a device that can be used together with some type of recording apparatus to record breathing movements. The laboratory instructor will demonstrate the use of this equipment to record various movements, such as those that accompany coughing, laughing, yawning, and speaking.

Devise an experiment to test the effect of some factor, such as hyperventilation, rebreathing air, or exercise, on the length of time a person can hold the breath. *After the laboratory instructor has approved your plan,* carry out the experiment, using the pneumograph and recording equipment. What conclusion can you draw from the results of your experiment?

____

____

____

____

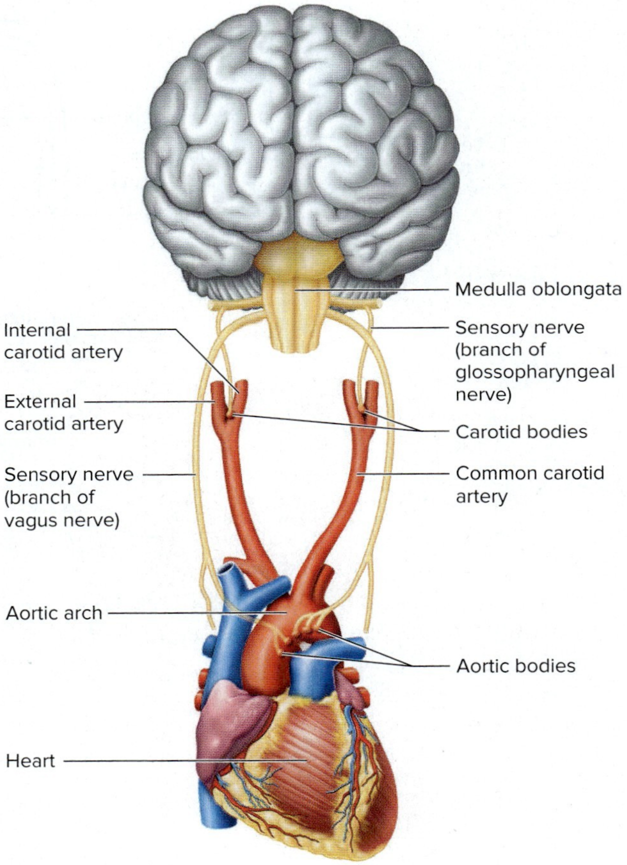

FIGURE 53.2 The peripheral chemoreceptors involved in control of breathing. Aortic bodies in the aortic arch and carotid bodies in the carotid arteries are sensitive to increased levels of blood carbon dioxide and hydrogen ions as well as decreased blood oxygen levels.

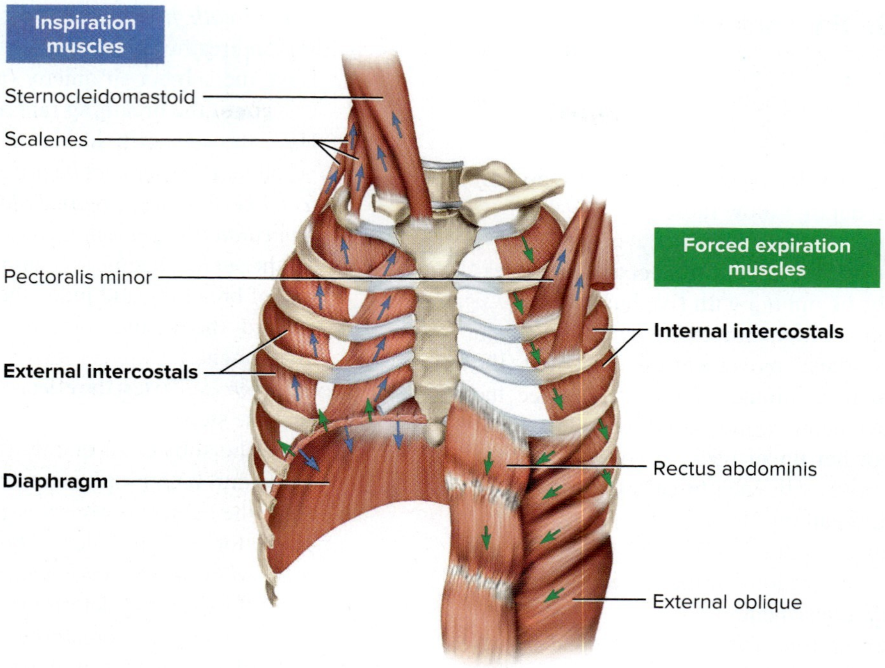

FIGURE 53.3 Respiratory muscles involved in inspiration and forced expiration. Boldface indicates the principle muscles increasing and decreasing the size of the thoracic cavity. The blue arrows indicate the direction of muscle contraction during inspiration; the green arrows indicate the direction of muscle contraction during forced expiration. Boldface indicates the primary muscles involved in breathing. APR

DEMONSTRATION ACTIVITY

When a solution of calcium hydroxide is exposed to carbon dioxide, a chemical reaction occurs, and a white precipitate of calcium carbonate is formed as indicated by the following reaction:

$$Ca(OH)_2 + CO_2 \rightarrow CaCO_3 + H_2O$$

Thus, a clear water solution of calcium hydroxide (limewater) can be used to detect the presence of carbon dioxide because the solution becomes cloudy if this gas is bubbled through it.

The laboratory instructor will demonstrate this test for carbon dioxide by drawing some air through limewater in an apparatus such as that shown in figure 53.4. Then the instructor will blow an equal volume of expired air through a similar apparatus. (*Note:* The apparatus should be cleaned and sterilized and a new sterile mouthpiece should be used each time the apparatus is demonstrated.) Watch for the appearance of a precipitate that causes the limewater to become cloudy. Was there any carbon dioxide in the atmospheric air drawn through the limewater?

____

If so, how did the amount of carbon dioxide in the atmospheric air compare with the amount in the expired air?

____

____

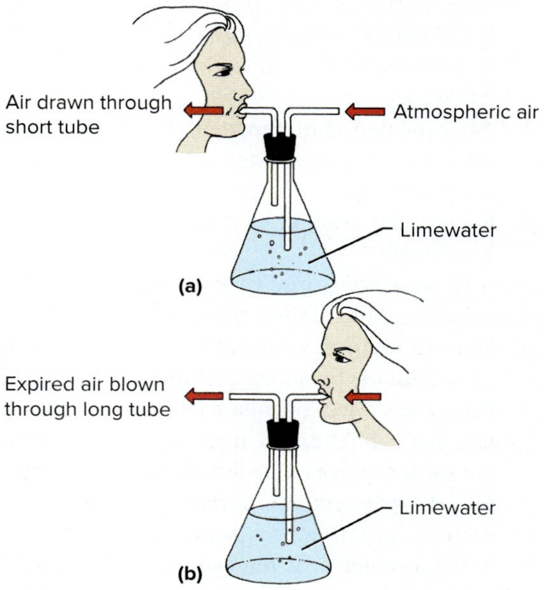

FIGURE 53.4 Apparatus used to demonstrate the presence of carbon dioxide in air: (a) atmospheric air is drawn through limewater; (b) expired air is blown through limewater.

PROCEDURE B: Factors Affecting Breathing

Perform each of the following tests, using your laboratory partner as a test subject.

1. *Normal breathing.* To determine the subject's normal breathing rate and depth, follow these steps:
   a. Have the subject sit quietly for a few minutes.
   b. After the rest period, ask the subject to count backward mentally, beginning with five hundred.
   c. While the subject is distracted by counting, watch the subject's chest movements, and count the breaths taken in a minute. If you cannot see the subject's movements while breathing, place your hand on his or her upper back so you can feel the upper body move with each breath. Use this value as the normal breathing rate (breaths per minute).
   d. Note the relative depth of the breathing movements.
   e. Record your observations in the table in Part B of the laboratory assessment.
2. *Effect of hyperventilation.* To test the effect of hyperventilation on breathing, follow these steps:
   a. Seat the subject and *guard to prevent the possibility of the subject falling over.*
   b. Have the subject breathe rapidly and deeply for a maximum of 20 seconds. *If the subject begins to feel dizzy, the hyperventilation should be halted immediately to prevent the subject from fainting from complications of alkalosis. The increased blood pH causes vasoconstriction of cerebral arterioles, which decreases circulation and oxygen to the brain.*
   c. After the period of hyperventilation, determine the subject's breathing rate and judge the breathing depth as before.
   d. Record the results in Part B of the laboratory assessment.
3. *Effect of rebreathing air.* To test the effect of rebreathing air on breathing, follow these steps:
   a. Have the subject sit quietly (approximately 5 minutes) until the breathing rate returns to normal.
   b. Have the subject breathe deeply into a small paper bag that is held tightly over the nose and mouth. *If the subject begins to feel light-headed or like fainting, the rebreathing air should be halted immediately to prevent further acidosis and fainting.*
   c. After 2 minutes of rebreathing air, determine the subject's breathing rate and judge the depth of breathing.
   d. Record the results in Part B of the laboratory assessment.
4. *Effect of breath holding.* To test the effect of breath holding on breathing, follow these steps:
   a. Have the subject sit quietly (approximately 5 minutes) until the breathing rate returns to normal.
   b. Have the subject hold his or her breath as long as possible. *If the subject begins to feel light-headed or feel like fainting, breath holding should be halted immediately to prevent further acidosis and fainting.*
   c. As the subject begins to breathe again, determine the rate of breathing and judge the depth of breathing.
   d. Record the results in Part B of the laboratory assessment.
5. *Effect of exercise.* To test the effect of exercise on breathing, follow these steps:
   a. Have the subject sit quietly (approximately 5 minutes) until breathing rate returns to normal.
   b. Have the subject exercise by moderately running in place for 3–5 minutes. *This exercise should be avoided by anyone with health risks.*
   c. After the exercise, determine the breathing rate and judge the depth of breathing.
   d. Record the results in Part B of the laboratory assessment.
6. The importance of the respiratory rate and blood pH relationship is illustrated in figure 53.5. A condition of respiratory acidosis or alkalosis, leading to possible death, can occur from changes in the depth or rate of breathing.
7. Complete Part B of the laboratory assessment.

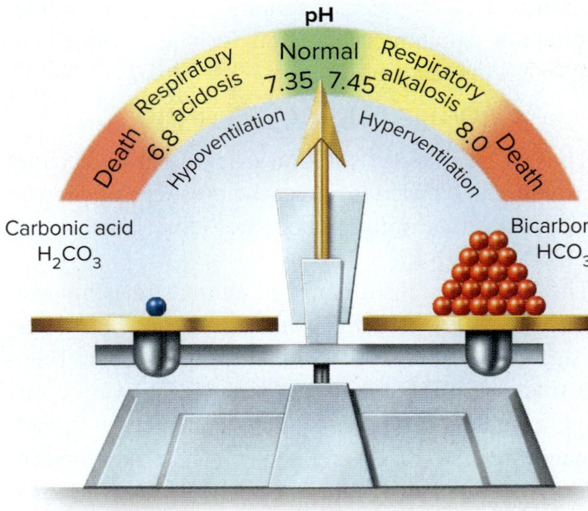

FIGURE 53.5 The relationship of respiratory rate and blood pH. The normal breathing rate of 12–15 breaths per minute relates to a blood pH within the normal range.

LABORATORY

ASSESSMENT

53

Name ______________________________

Date ______________________________

Section ______________________________

The A corresponds to the indicated Learning Outcome(s) O found at the beginning of the Laboratory Exercise.

Control of Breathing

PART A: Assessments

Complete the following statements:

1. The respiratory centers are widely scattered throughout the ______________ and medulla oblongata of the brainstem. A1
2. The ______________________ respiratory group relays messages to the VRG and the DRG to modify the respiratory rhythm. A1
3. The ______________________ respiratory group within the medulla oblongata generates the basic rhythm of quiet breathing. A1
4. The ______________________ respiratory group within the medulla oblongata receives input from chemoreceptors, stretch receptors, and irritant receptors. A1
5. Central chemoreceptors are sensitive to changes in the blood concentrations of hydrogen ions and ______________. A2
6. As the blood concentration of carbon dioxide increases, the breathing rate ______________________. A2
7. As a result of increased breathing, the blood concentration of carbon dioxide is ______________________. A2
8. Peripheral chemoreceptors include aortic bodies and the ______________________. A2
9. The principal muscles of forced expiration are the ______________________. A3
10. The principal muscles of inspiration are the ______________________ and the external intercostal muscles. A3

PART B: Assessments

1. Record the results of your breathing tests in the table. A4

Factor Tested	Breathing Rate (Breaths/Minute)	Breathing Depth (+, + +, + + +)
Normal		
Hyperventilation (shortly after)		
Rebreathing air (shortly after)		
Breath holding (shortly after)		
Exercise (shortly after)		

2. Explain the reason for the changes in breathing that occurred in each of the following cases: A2

a. Hyperventilation

____

____

____

____

b. Rebreathing air

____

____

____

____

c. Breath holding

____

____

____

____

d. Exercise

____

____

____

____

3. Complete the following: A2

a. Why is it important to distract a person when you are determining the normal rate of breathing?

____

____

b. How can the depth of breathing be measured accurately?

____

____

CRITICAL THINKING ASSESSMENTS—PART B

1. Use the following carbonic acid–bicarbonate buffer system equation to identify the direction (*to the right* or *to the left*) the equation would move as a corrective homeostatic response by the respiratory system to acidosis:
$CO_2 + H_2O \longleftrightarrow H_2CO_3 \longleftrightarrow HCO_3^- + H^+$ ______________________ Explain. **A2**

____

____.

2. Why is it dangerous for a swimmer to hyperventilate in order to hold the breath for a longer period of time? **A2**

____

____

____

____.

NOTES

LABORATORY
EXERCISE 54

Digestive Organs

Purpose of the Exercise

To review the structure and function of the digestive organs and to examine the tissues of these organs.

MATERIALS NEEDED

- Human torso model
- Head model, sagittal section
- Skull with teeth
- Teeth, sectioned
- Tooth model, sectioned
- Paper cup
- Compound light microscope
- Prepared microscope slides of the following:
    - Sublingual gland (salivary gland)
    - Esophagus
    - Stomach (fundus)
    - Pancreas (exocrine portion)
    - Liver
    - Small intestine (jejunum)
    - Large intestine

Chapter Opener: ©101cats/Getty Images

Learning Outcomes APR

After completing this exercise, you should be able to:

- O1 Sketch and label digestive organ structures in tissue (histological) sections.
- O2 Locate the major organs and structural features of the digestive system.
- O3 Match digestive organs with their descriptions.
- O4 Describe the function(s) of each digestive organ.

The O corresponds to the assessments A indicated in the Laboratory Assessment for this Exercise.

Pre-Lab

Carefully read the introductory material and examine the entire lab. Be familiar with the basic structures and functions of the digestive organs from lecture or the textbook. Answer the pre-lab questions.

Pre-Lab Questions Select the correct answer for each of the following questions:

1. Which of the following digestive organs possesses villi?
   a. esophagus **b.** stomach
   c. small intestine **d.** large intestine
2. Which is a major function of the oral cavity?
   a. absorb water **b.** mechanical digestion
   c. secrete hydrochloric acid **d.** emulsify fat
3. Bile is produced in the
   a. small intestine. **b.** pancreas.
   c. gallbladder. **d.** liver.
4. The _____ is the exposed part of a tooth.
   a. crown **b.** neck
   c. root **d.** periodontal ligament
5. The _____ sphincter is the muscular valve at the exit of the stomach.
   a. lower esophageal **b.** pyloric
   c. hepatopancreatic **d.** ileocecal
6. The stomach can stretch enough to hold the contents of a large meal because it has
   a. the pyloric antrum. **b.** a fundus region.
   c. gastric folds. **d.** a greater curvature.
7. The _____ lobe is the largest lobe of the liver.
   a. left **b.** quadrate
   c. caudate **d.** right
8. Which of the following digestive organs does *not* secrete enzymes that carry out chemical digestion?
   a. stomach **b.** large intestine
   c. pancreas **d.** salivary glands

9. The appendix, cecum, and ascending colon are on the left side of the body.
   a. True **b.** False
10. The anal sphincter muscles include an external voluntary sphincter and an internal involuntary sphincter.
   a. True **b.** False

The digestive system includes the organs associated with the *alimentary canal* and several accessory structures. The alimentary canal, a muscular tube, passes through the body from the opening of the mouth to the anus. It includes the mouth, pharynx, esophagus, stomach, small intestine, and large intestine. The canal is adapted to move substances throughout its length. It is specialized in various regions to store, digest, and absorb food materials and to eliminate the residues. The accessory organs, which include the salivary glands, liver, gallbladder, and pancreas, secrete products into the alimentary canal that aid digestive functions.

The *oral cavity* is a primary area for mechanical digestion where food is physically broken down, involving the mastication (chewing) of food using the teeth and jaw muscles. Three pairs of salivary glands secrete mucus and salivary amylase into the oral cavity, where chemical digestion of starch begins. Chemical digestion uses enzymes to hydrolyze larger molecules into smaller absorbable molecules. The *pharynx* and *esophagus* secrete mucus and serve as passageways for the food and liquids to reach the stomach. The contents are forced along by peristaltic waves.

Within the *stomach,* swallowed contents mix with gastric juice, which contains hydrochloric acid and the enzyme pepsin. The hydrochloric acid creates a pH of about 2, serving to destroy most ingested bacteria, and it activates pepsin to begin the chemical digestion of proteins. The gastric folds (rugae) allow the stomach to hold large quantities of food. Only minimal absorption occurs in the stomach. The partially digested content, called chyme, enters the small intestine periodically through the muscular pyloric sphincter.

Within the *small intestine,* chyme is mixed with bile and pancreatic juice. Bile is produced in the liver and then stored in the gallbladder. Bile emulsifies fat, breaking it into smaller particles to allow the enzyme lipase to act upon it more easily. Pancreatic juice contains bicarbonate ions necessary to neutralize acidic chyme, and a variety of enzymes for the chemical digestion of carbohydrates, fats, and proteins. The small intestine is nearly 6 meters (21 feet) long and possesses microvilli, villi, and circular folds that help to increase its surface area. Chemical digestion is completed and most nutrient absorption occurs before the contents reach the large intestine.

Contents not absorbed enter the *large intestine* through the ileocecal valve. Within the large intestine, mucus is secreted and water and electrolytes are absorbed. Bacteria that inhabit the large intestine break down some remaining residues, producing some vitamins that are absorbed. Feces composed of water, undigested substances, mucus, and bacteria are formed and stored until elimination.

PROCEDURE A: Oral Cavity and Salivary Glands

1. Study figure 54.1 and the list of structures and descriptions. Examine the head model (sagittal section) and a skull. Locate the following structures:

   oral cavity (mouth)—location of mechanical digestion, some chemical digestion

   vestibule—narrow space between teeth and lips and cheeks

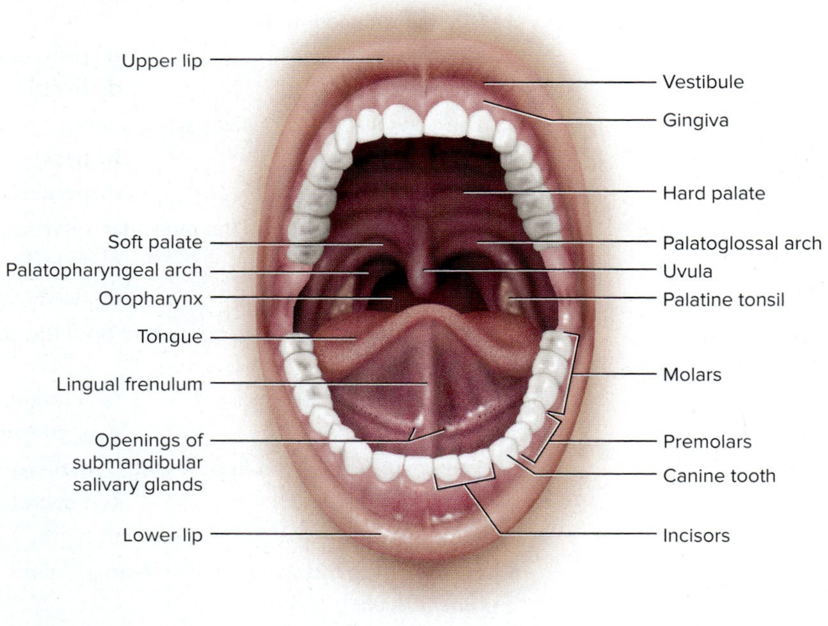

FIGURE 54.1 The features of the oral cavity. APR

tongue—muscular organ for food manipulation
- lingual frenulum (frenulum of tongue)—membranous fold connecting tongue and floor of oral cavity
- papillae—contain taste buds

palate—roof of oral cavity
- hard palate—anterior part formed by portions of maxillary and palatine bones
- soft palate—posterior part composed of skeletal muscle and glandular tissue; includes uvula; without bone
- uvula—cone-shaped projection; along with soft palate, helps prevent ingested contents from entering nasal regions during swallowing
- palatoglossal arches—located anteriorly
- palatopharyngeal arches—located posteriorly

palatine tonsils—lymphatic tissues on lateral walls between the arches

gingivae (gums)—soft tissue that surrounds neck of tooth and alveolar processes

teeth
- incisors
- canines (cuspids)
- premolars (bicuspids)
- molars

2. Study figure 54.2 and the list of structures and descriptions. Examine a sectioned tooth and a tooth model. Locate the following features:

crown—tooth portion projection beyond gingivae
- enamel—hardest substance of body on surface of crown
- dentin—living connective tissue forming most of tooth

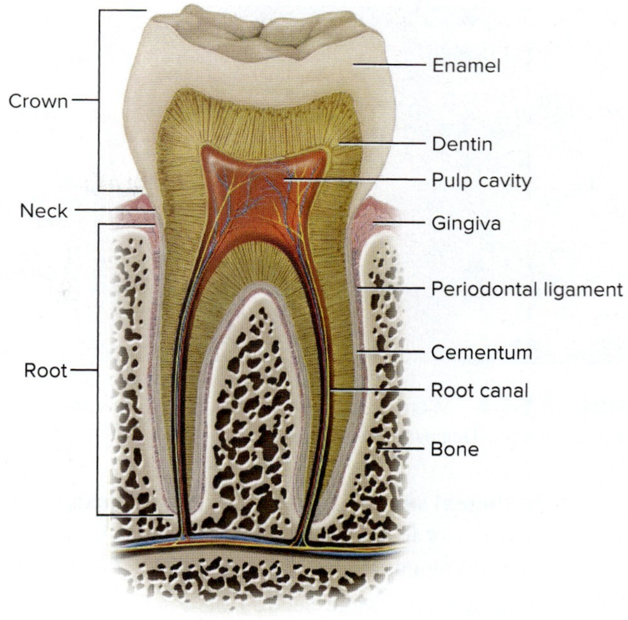

FIGURE 54.2 Longitudinal section of a molar.

neck—junction between crown and root

root—deep portion of tooth beneath surface
- pulp cavity—central portion of tooth
- cementum (cement)—thin surface area of bonelike material
- root canal—contains blood vessels and nerves

3. Study figures 54.1 and 54.3. Observe the head of the human torso model and locate the following:

parotid salivary glands—largest glands anterior to ear
- parotid duct (Stensen's duct)

submandibular salivary glands—along mandible
- submandibular duct (Wharton's duct)

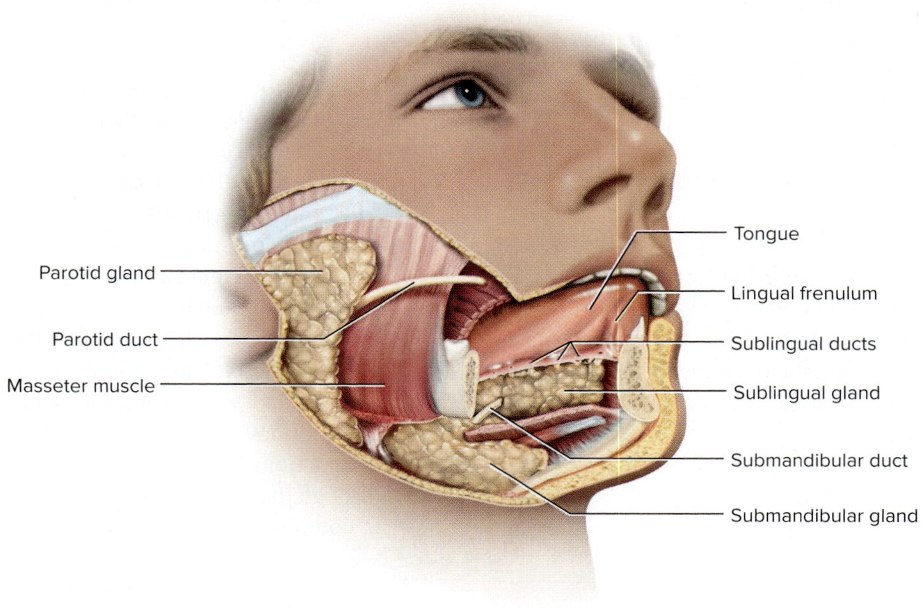

FIGURE 54.3 The features associated with the salivary glands. APR

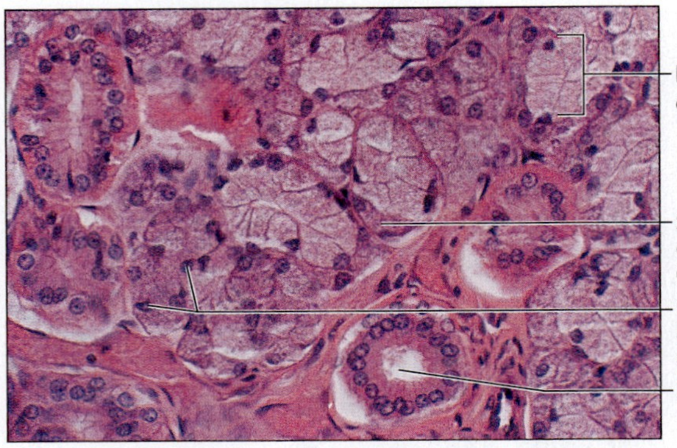

FIGURE 54.4 Micrograph of the sublingual salivary gland (300×). Dennis Strete/McGraw Hill

sublingual salivary glands—smallest glands inferior to tongue

- sublingual ducts (10–12 ducts)

4. Examine a microscopic section of a sublingual gland using scanning, low-power, and then high-power magnification. Note that the mucous cells produce mucus and the serous cells produce enzymes. Serous cells sometimes form caps (demilunes) around mucous cells. Also note any larger secretory duct surrounded by cuboidal epithelial cells (fig. 54.4).
5. Prepare a labeled sketch of a representative section of a salivary gland in Part A of Laboratory Assessment 54.

PROCEDURE B: Pharynx and Esophagus

1. Reexamine figure 50.2 in Laboratory Exercise 50.
2. Observe the human torso model and locate the following features:

pharynx—connects nasal and oral cavities with larynx

- nasopharynx—superior to soft palate; without digestive functions
- oropharynx—posterior to oral cavity
- laryngopharynx—extends from inferior oropharynx to esophagus

epiglottis—deflects food and fluids into esophagus when swallowing

esophagus—straight collapsible tube from pharynx to stomach

lower esophageal sphincter (cardiac sphincter; gastroesophageal sphincter)—thickened smooth muscle at junction with stomach

3. Have your laboratory partner take a swallow from a cup of water. Carefully watch the movements in the anterior region of the neck. What steps in the swallowing process did you observe?

____

____

____

____

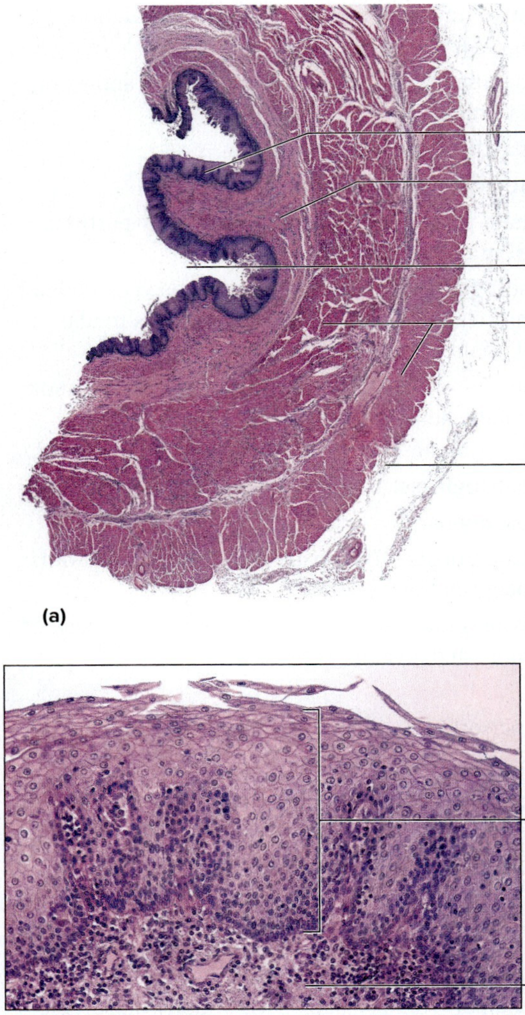

FIGURE 54.5 Micrograph of a cross section of the (a) layers that make up the wall of the esophagus (10×), and (b) the mucosa (400×). Jose Luis Calvo/Shutterstock APR

4. Examine a microscopic section of esophagus wall using scanning and then low-power magnification (fig. 54.5). The inner mucosa layer is composed of nonkeratinized stratified squamous epithelium. Locate some mucous glands in the submucosa layer. They appear as clusters of lightly stained cells. There are two layers of muscle tissue in the muscularis externa, an inner circular layer and an outer longitudinal layer. In this layer, the upper third of the esophagus contains skeletal muscle, the middle third contains a mix of skeletal and smooth muscle, and the lower third contains smooth muscle. Peristaltic waves occur from contractions of the muscular layers that move contents through the alimentary canal. The outer adventitia layer is composed of a fibrous connective tissue that melds with surrounding structures.
5. Prepare a labeled sketch of the esophagus wall in Part A of the laboratory assessment.

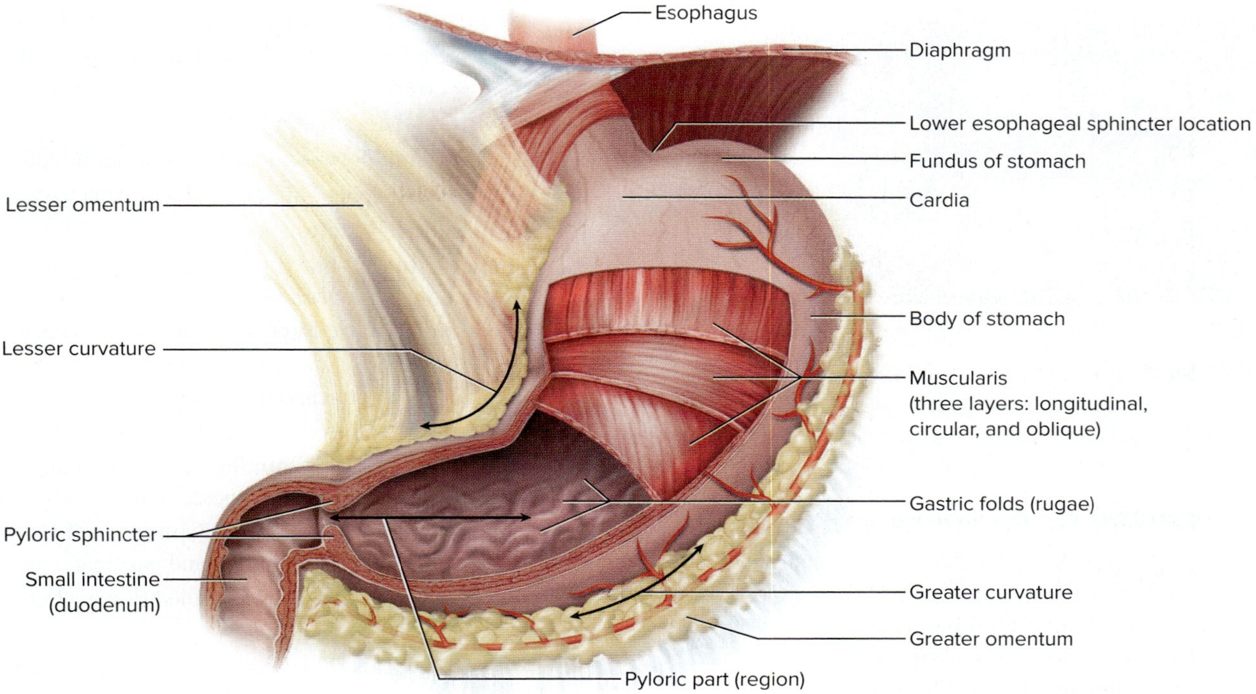

FIGURE 54.6 The major features of the stomach and associated structures (frontal section of internal structures). APR

PROCEDURE C: Stomach

1. Study figure 54.6 and the list of structures and descriptions. Observe the human torso model and locate the following features of the stomach:

   gastric folds (rugae)—allow stomach to expand to hold food and liquid contents

   cardia (cardiac region)—region near lower esophageal sphincter

   fundus of stomach (fundic region)—dome-shaped region superior to opening from esophagus

   body of stomach (body region)—large main region

   pyloric part (pyloric region)—region near passage into small intestine

   - pyloric antrum—funnel-like portion
   - pyloric canal—narrow region
   - pylorus—terminal passage into duodenum

   pyloric sphincter—thick, smooth muscle valve that regulates chyme passage into duodenum

   lesser curvature—on superior surface; mesentery called the lesser omentum attaches here

   greater curvature—on lateral and inferior surface; mesentary called the greater omentum attaches here and drapes down over the small intestine

2. Examine a microscopic section of stomach wall using scanning and then low-power magnification. Note how the inner mucosa layer of simple columnar epithelium dips inward to form gastric pits. The gastric glands are tubular structures that open into the gastric pits. Near the deep ends of these glands, locate the intensely stained (bluish) chief cells and the lightly stained (pinkish) parietal cells (fig. 54.7). The parietal cells secrete hydrochloric acid; the chief cells produce pepsinogen, the inactive form of pepsin.

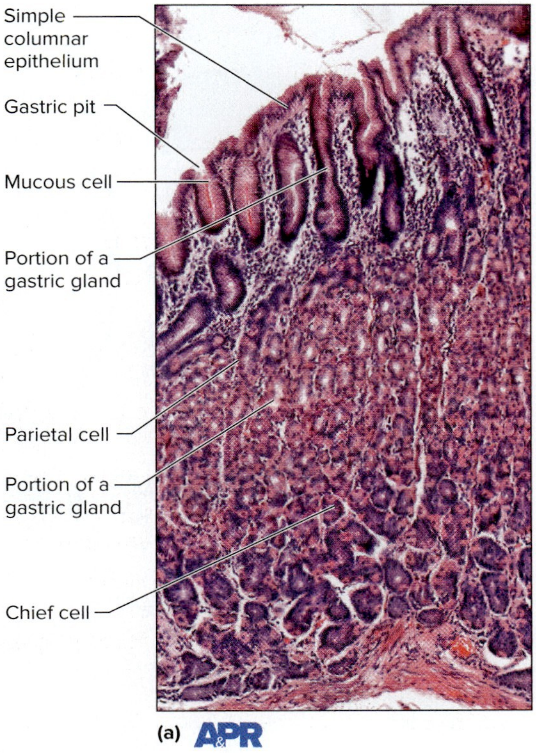

FIGURE 54.7 Micrograph of the mucosa of the (a) stomach wall (60×), and (b) chief cells and parietal cells (400×).
Ed Reschke

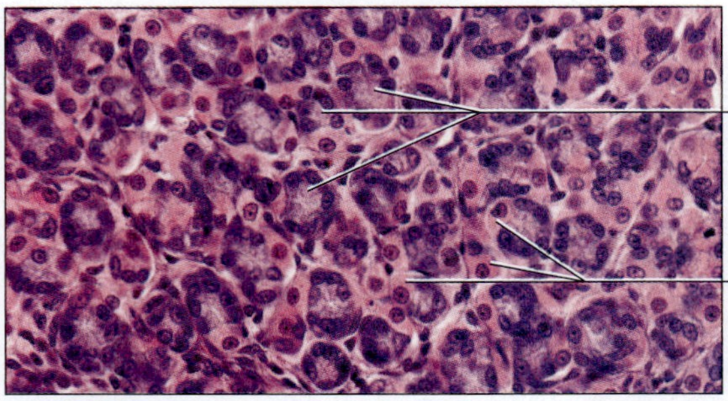

FIGURE 54.7 *Continued.* Jose Luis Calvo/Shutterstock

3. Prepare a labeled sketch of a representative section of the stomach wall in Part A of the laboratory assessment.

PROCEDURE D: Pancreas and Liver

1. Study figures 54.8 and 54.9 and the list of structures and descriptions of the pancreas and the liver. Observe the human torso model and locate the following structures:

   pancreas—a retroperitoneal organ; secretes an alkaline mixture of digestive enzymes and bicarbonate ions (fig. 54.8)
   - tail of pancreas—part farthest from duodenum
   - body of pancreas
   - head of pancreas—part closest to duodenum
   - pancreatic duct—main duct
   - accessory pancreatic duct

   liver—has four lobes; produces bile (fig. 54.9)
   - right lobe—largest lobe; bare area portion attached to diaphragm
   - left lobe—smaller than right lobe
   - quadrate lobe—minor lobe near gallbladder
   - caudate lobe—minor lobe near vena cava

   gallbladder—stores and concentrates bile

   hepatic ducts—bile ducts from inferior liver to common hepatic duct

   common hepatic duct—bile duct from liver to cystic duct

   cystic duct—attached to gallbladder

   bile duct—duct from cystic duct to duodenum

   hepatopancreatic ampulla—enlarged area where bile duct and main pancreatic duct merge

   hepatopancreatic sphincter (sphincter of Oddi)—regulates passage of bile and pancreatic juice

   major duodenal papilla—folded tissue where ampulla ends

2. Examine the pancreas slide using scanning and then low-power magnification. Then, examine the tissue using high-power magnification. Observe the exocrine (acinar) cells that secrete pancreatic juice into pancreatic ducts (fig. 54.10). The pancreatic duct secretes

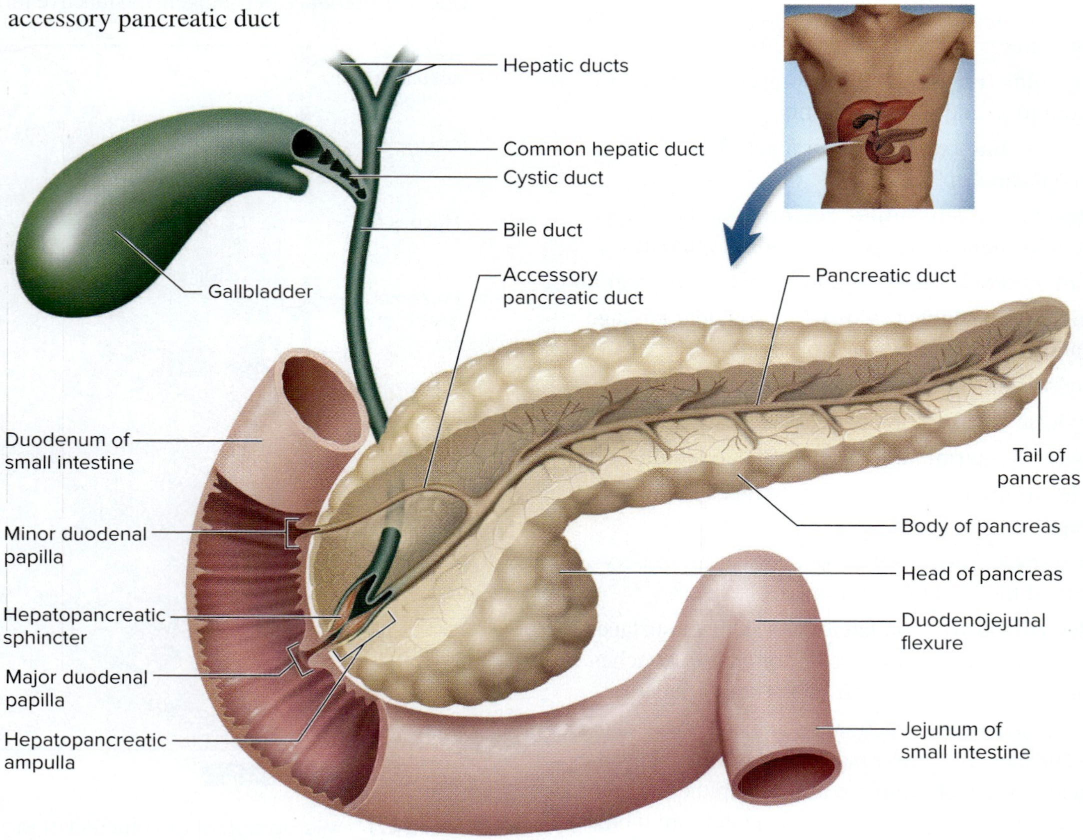

FIGURE 54.8 The major features of the pancreas and the gallbladder associated with the liver and the duodenum of the small intestine.

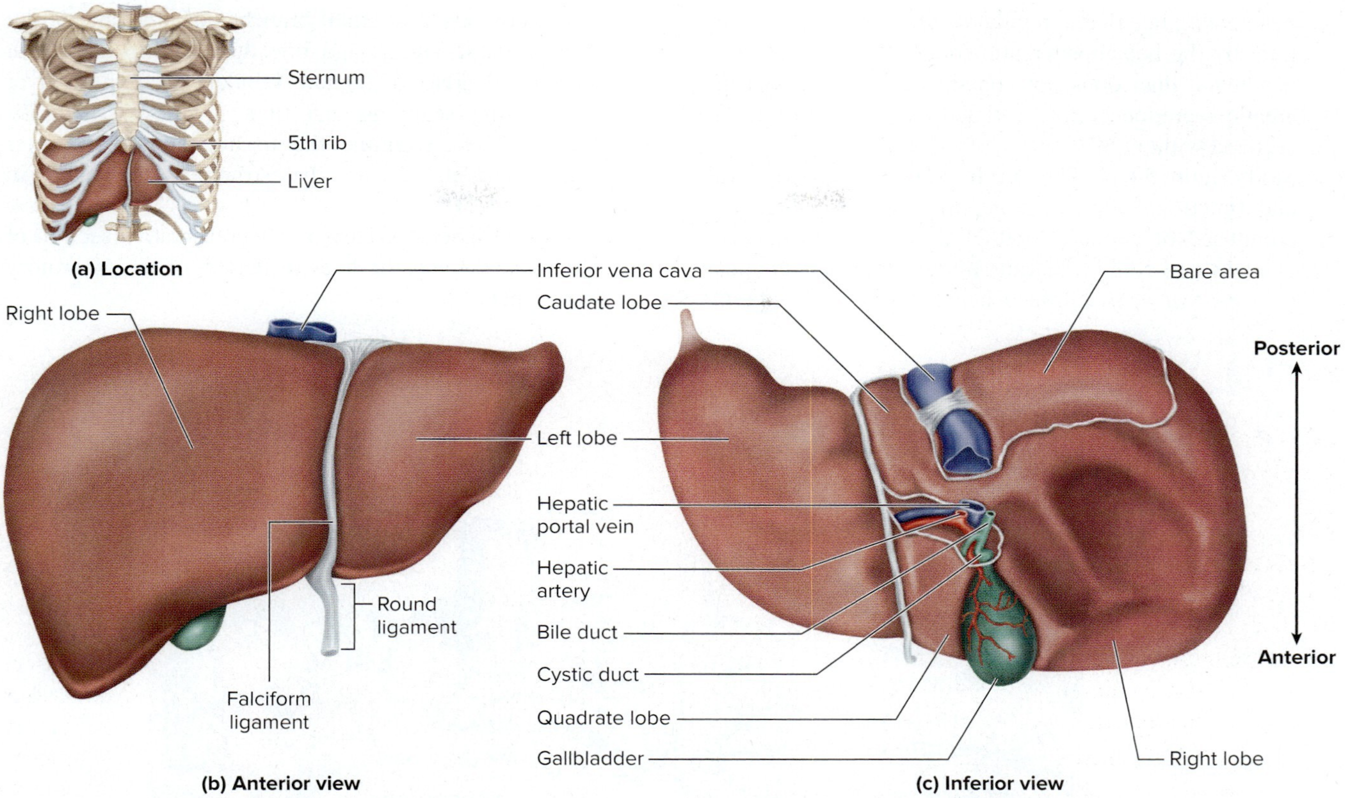

FIGURE 54.9 Features of the liver: (a) liver location; (b) anterior view; (c) inferior view. APR

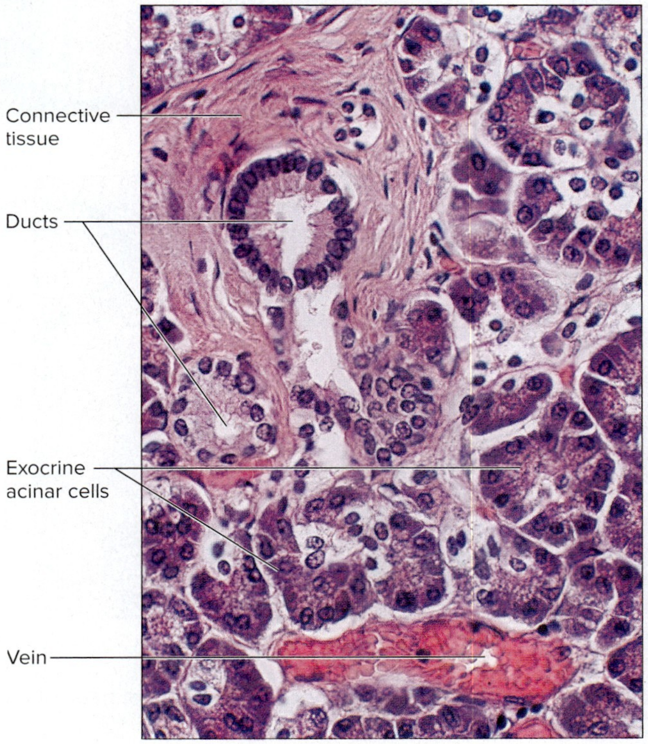

FIGURE 54.10 Micrograph of exocrine portion of pancreas showing acinar cells and small pancreatic ducts (200×). The acinar cells secrete an alkaline pancreatic juice into small ducts that converge into the larger pancreatic duct. Dennis Strete/McGraw Hill APR

pancreatic juice that is regulated, along with bile secretion, by the hepatopancreatic sphincter. The accessory pancreatic duct does not contain a sphincter, so pancreatic secretion is not regulated and is allowed to be secreted without bile.

3. Study figure 54.11. The four liver lobes have structural and functional units, the *hepatic lobules,* which are composed of cuboidal epithelial cells called *hepatocytes.* Blood-filled radiating *hepatic sinusoids* connect to a *central vein* within a hepatic lobule. A *hepatic triad,* composed of small branches of a hepatic portal vein, hepatic artery, and bile ductule, are located at peripheral areas of hepatic lobules. Examine a liver slide using scanning and then low-power magnification. Then, examine the tissue using high-power magnification. Locate the structures indicated in figures 54.11*b* and *c*.
4. Prepare a labeled sketch of a representative section of the pancreas and the liver in Part A of the laboratory assessment.

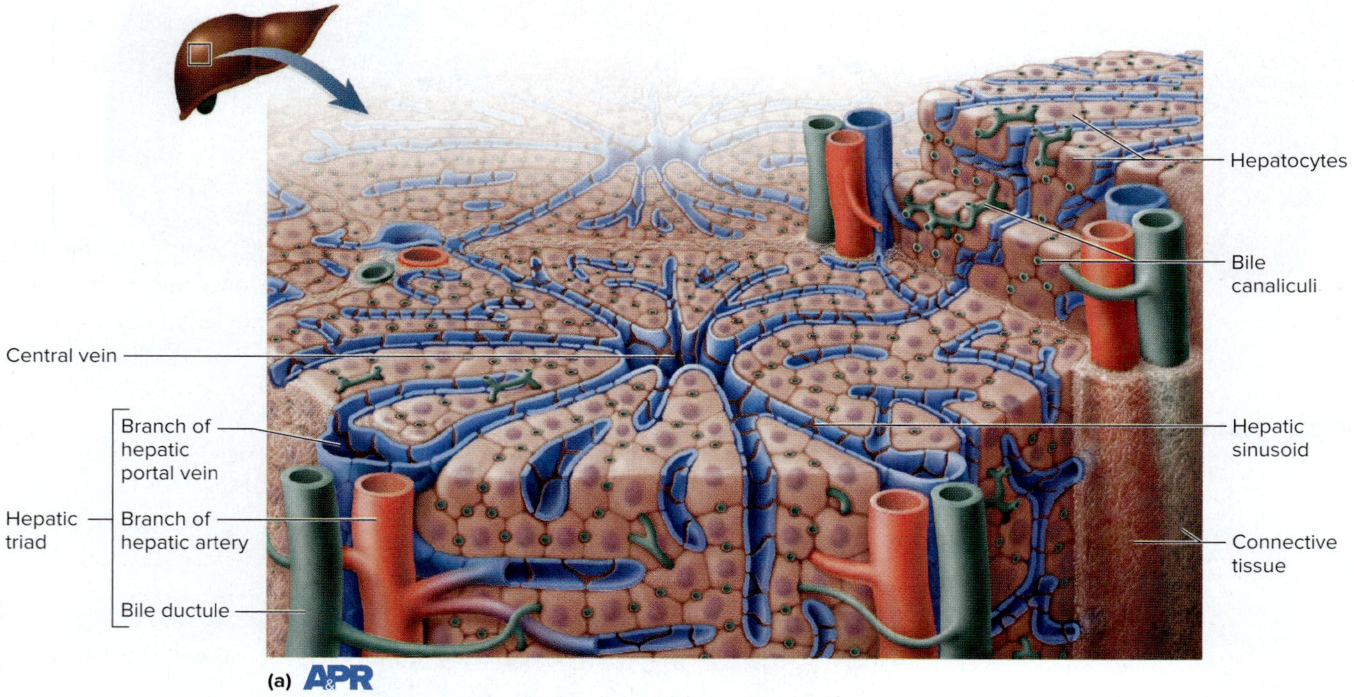

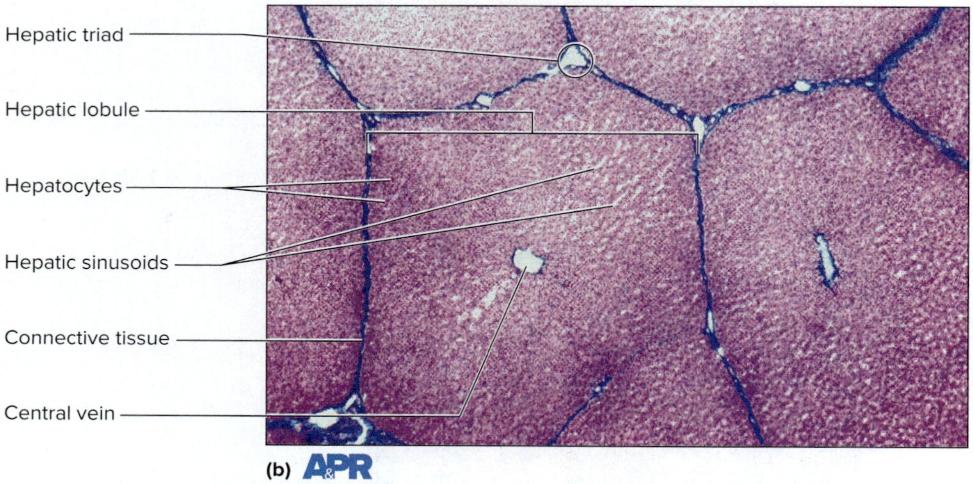

FIGURE 54.11 Microscopic structure of the liver: (a) illustration; (b) micrograph of hepatic lobules (40×); (c) micrograph of hepatic triad (400×). (b) Victor P. Eroschenko

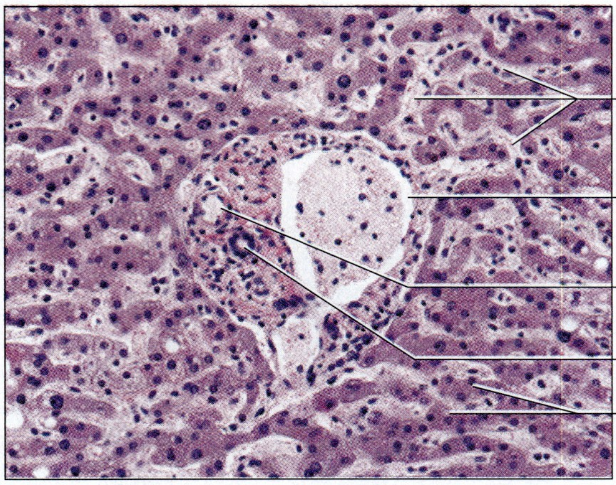

(c)

FIGURE 54.11 *Continued.* Jose Luis Calvo/Shutterstock

PROCEDURE E: Small and Large Intestines

1. Study figures 54.12 to 54.14 and the list of structures and descriptions. Observe the human torso model and locate each of the following features:

small intestine—tube with small diameter and many loops and coils where chemical digestion and absorption of nutrients occur (fig. 54.12)

- duodenum—first 25 cm (10 inches); shortest portion; located retroperitoneal
- jejunum—next nearly 2.5 m (8 feet); most digestion and absorption occurs here
- ileum—last 3.5 m (12 feet); joins with large intestine

mesentery—fold of peritoneal membrane that suspends and supports abdominal viscera

ileocecal valve (sphincter)—regulates contents entering large intestine

large intestine—tube with large diameter nearly 1.5 m (5 feet) long (fig. 54.13)

- large intestinal wall—has three unique features
  - taeniae coli—three longitudinal bands of smooth muscle
  - haustra—pouches created by muscle tone of taeniae coli
  - epiploic (omental) appendages—fatty pouches on outer surfaces
- cecum—blind pouch inferior to ileocecal valve
- appendix (vermiform appendix)—2–7 cm wormlike blind pouch attached to cecum (fig. 54.14); contains abundant immune cells
- ascending colon—course on right side from cecum to transverse colon

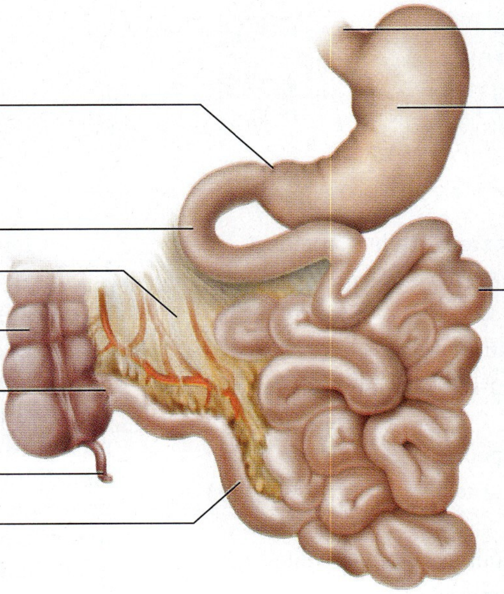

FIGURE 54.12 The features of the small intestine and associated structures (anterior view). (*Note:* The small intestine is pulled aside to expose the ileocecal junction.) APR

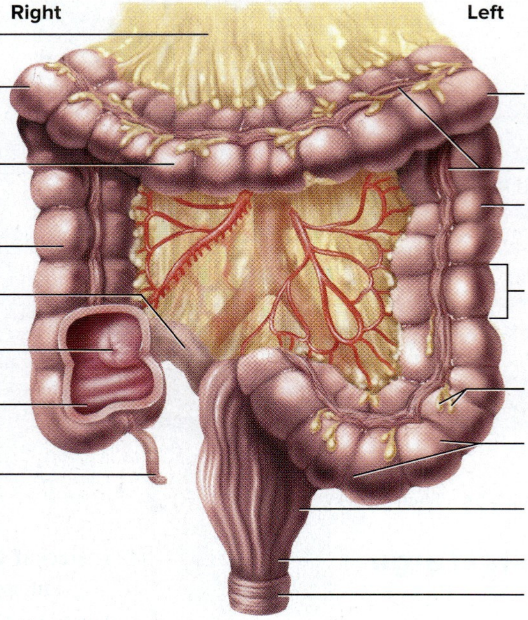

FIGURE 54.13 The features of the large intestine (anterior view). APR

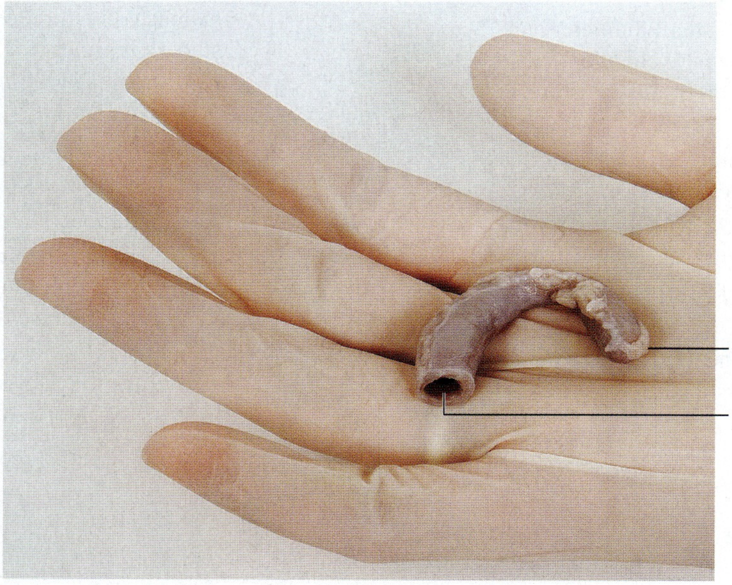

FIGURE 54.14 Normal appendix. ©J & J Photography

- right colic (hepatic) flexure—right-angle turn
- transverse colon—course from ascending to descending colon
- left colic (splenic) flexure—right-angle turn
- descending colon—course from transverse to sigmoid colon
- sigmoid colon—S-shaped course to rectum
- rectum—course anterior to sacrum; retains feces until defecation
- anal canal—last few centimeters of large intestine

anal sphincter muscles—encircle anus

- external anal sphincter—the voluntary sphincter of skeletal muscle
- internal anal sphincter—the involuntary sphincter of smooth muscle

anus—external opening

2. Using scanning and then low-power magnification, examine a microscopic section of small intestine wall. Identify the mucosa, submucosa, muscularis, and serosa. Note the villi that extend into the lumen of the tube. Study a single villus using high-power magnification. Note the core of connective tissue and the covering of simple columnar epithelium that contains some lightly stained goblet cells and microvilli (brush border) projecting from the apical ends of the cells (fig. 54.15). The microvilli and villi projections greatly increase the surface area for absorption of digestive products. Note

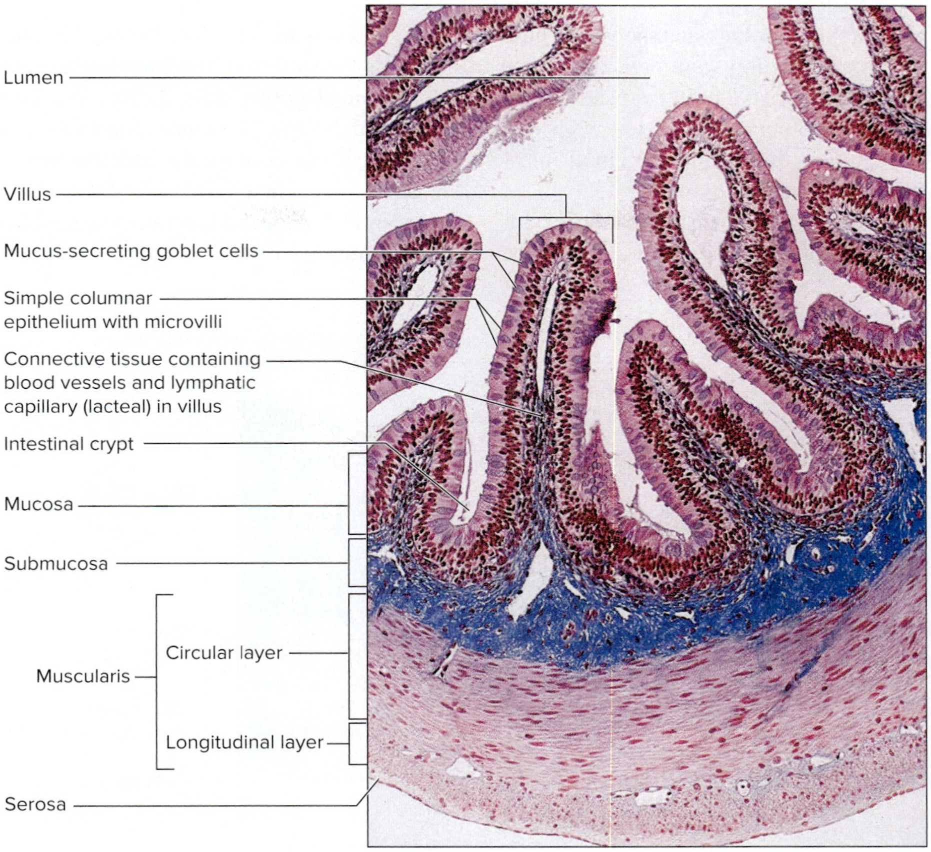

(a)

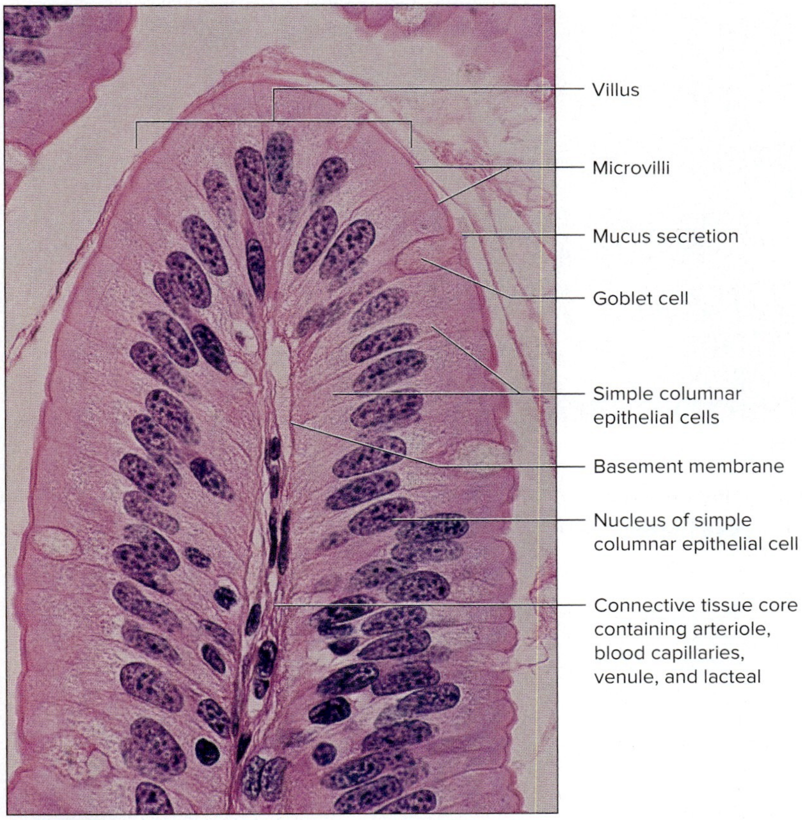

(b)

FIGURE 54.15 Micrograph of (a) the small intestine wall (40×), and (b) a single villus (200×). (a) Dennis Strete; (b) Ed Reschke/Getty Images the intestinal crypts on the floor between the villi. These glands contain abundant dividing stem cells and some Paneth cells that secrete defensive proteins used to protect the mucosa from microbial infection.

3. Prepare a labeled sketch of the wall of the small intestine in Part A of the laboratory assessment.
4. Using scanning and then low-power magnification, examine a microscopic section of large intestine wall. Note the lack of villi. Also note the tubular mucous glands that open on the surface of the inner lining and the numerous lightly stained goblet cells. Locate the four layers of the wall (fig. 54.16). The mucus functions as a lubricant and holds the particles of fecal matter together.
5. Prepare a labeled sketch of the wall of the large intestine in Part A of the laboratory assessment.
6. Complete Parts B, C, and D of the laboratory assessment.

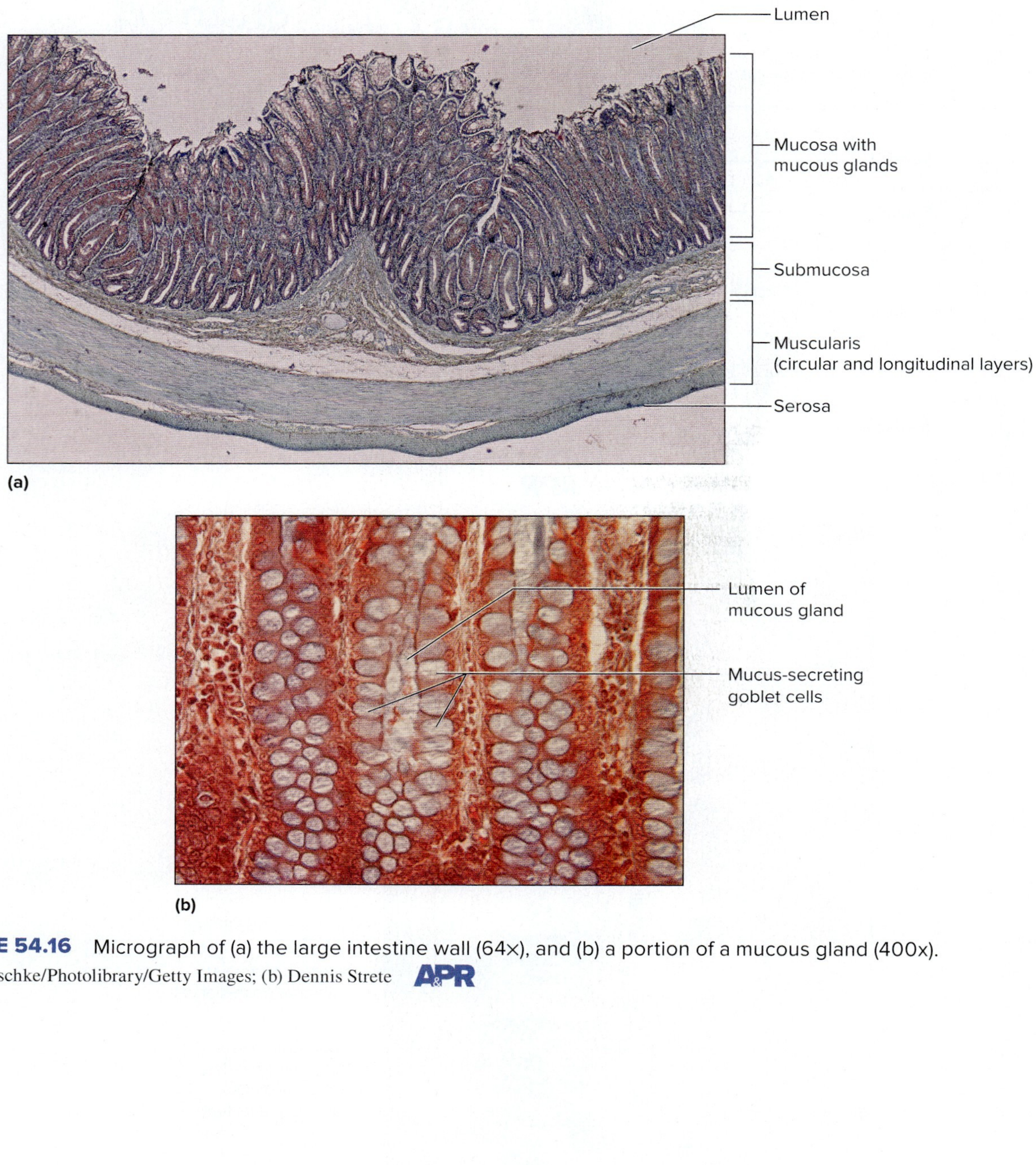

FIGURE 54.16 Micrograph of (a) the large intestine wall (64×), and (b) a portion of a mucous gland (400×).
(a) Ed Reschke/Photolibrary/Getty Images; (b) Dennis Strete APR

LABORATORY
ASSESSMENT 54

Name ______________________________

Date ______________________________

Section ______________________________

The (A) corresponds to the indicated Learning Outcome(s) (O) found at the beginning of the Laboratory Exercise.

Digestive Organs

PART A: Assessments

Each circle below represents the microscopic field of view. In each circle, sketch a representative area of the organ indicated. Label any of the structures observed and indicate the magnification used for each sketch. Add any personal notes in any of the extra space on the sides of your sketch that would assist your future ability to recognize the organ, its features, and to interpret its function. (A1)

Salivary gland (______×)

Esophagus (______×)

Stomach wall (______×)

Pancreas (______×)

Small intestine wall (______×)

Large intestine wall (______×)

Liver (______×)

CRITICAL THINKING ASSESSMENT—PART A

Describe how the structure of the small intestine is better adapted for absorption than that of the large intestine. A1

____

____

____

PART B: Assessments

Identify the numbered features in figures 54.17 to 54.22.

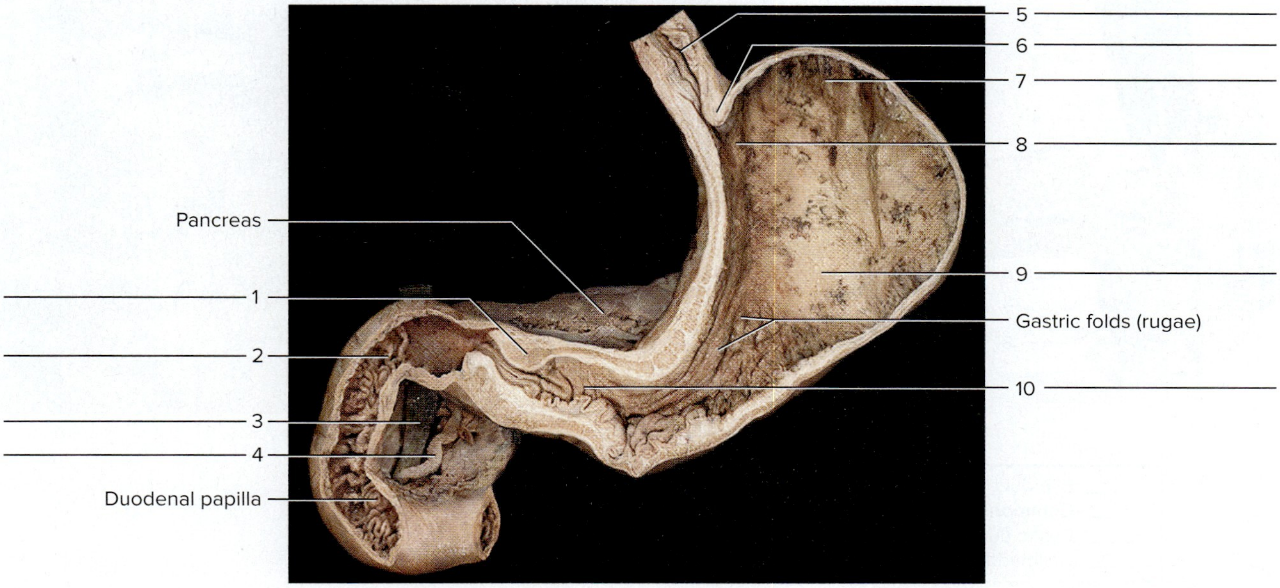

FIGURE 54.17 Label the features of the stomach and nearby regions in this frontal section of a cadaver (anterior view).
APR/McGraw Hill APR A2

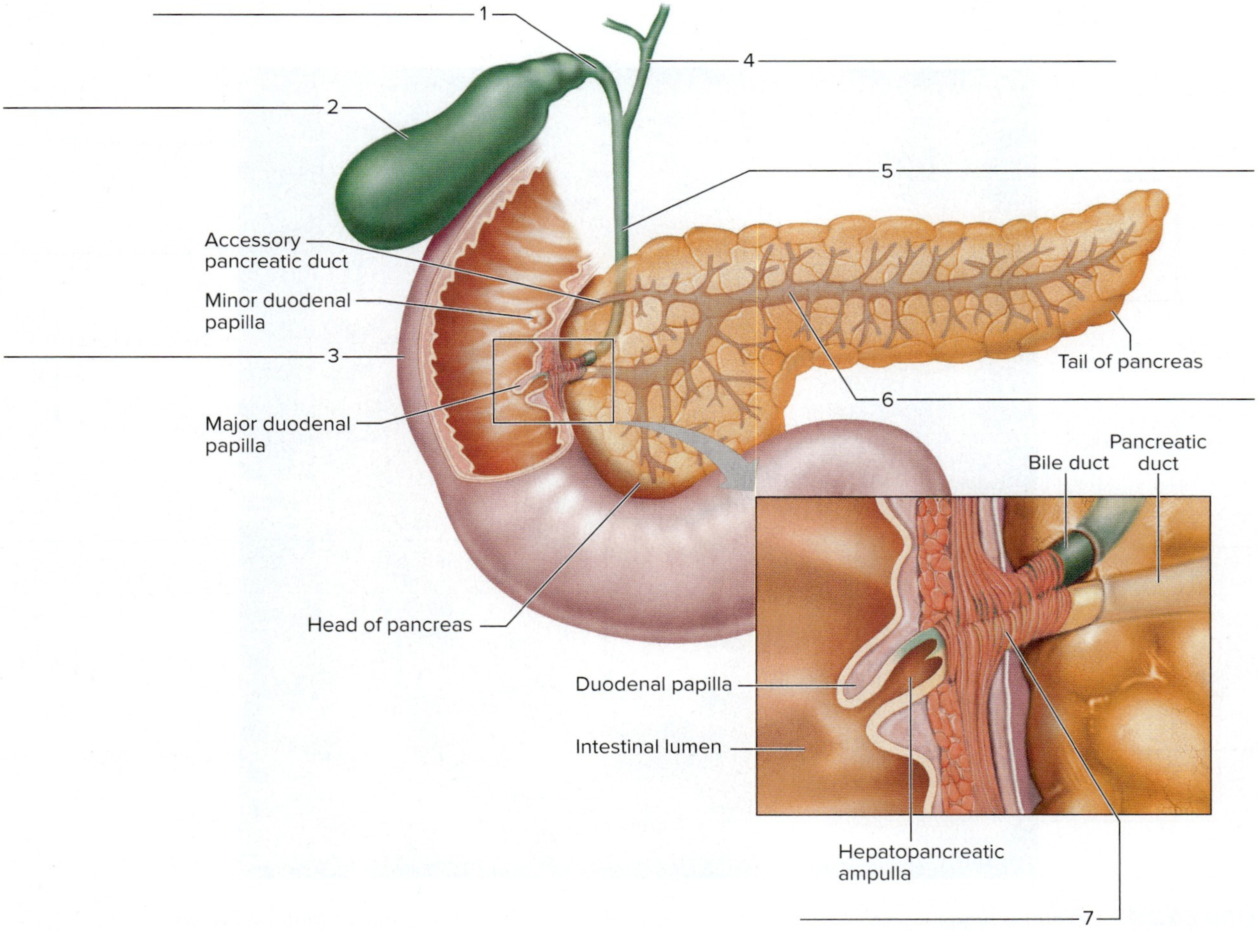

FIGURE 54.18 Label the features associated with the liver and pancreas (liver is removed). APR A2

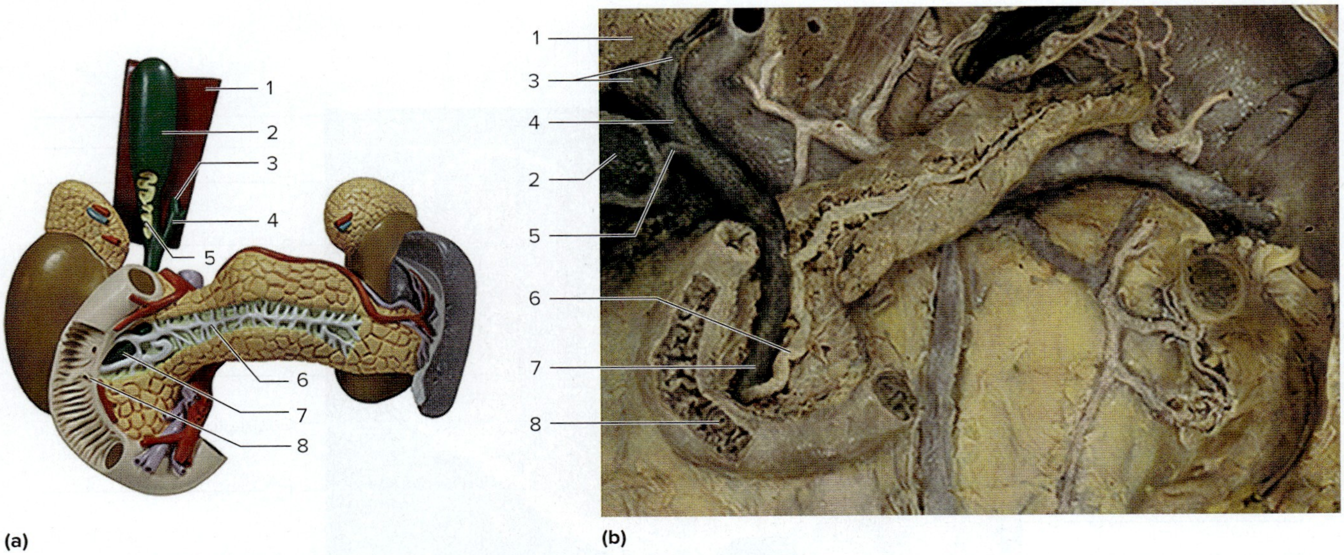

_____ Bile duct	_____ Hepatic ducts
_____ Common hepatic duct	_____ Liver
_____ Cystic duct	_____ Major duodenal papilla
_____ Gallbladder	_____ Pancreatic duct

FIGURE 54.19 Identify the features indicated on the images of abdominal digestive organs and corresponding ducts in an anterior view: (a) a model; (b) a cadaver by placing the correct numbers in the spaces provided. (a) Christine Eckel/Denoyer-Geppert; (b) APR/McGraw Hill A2

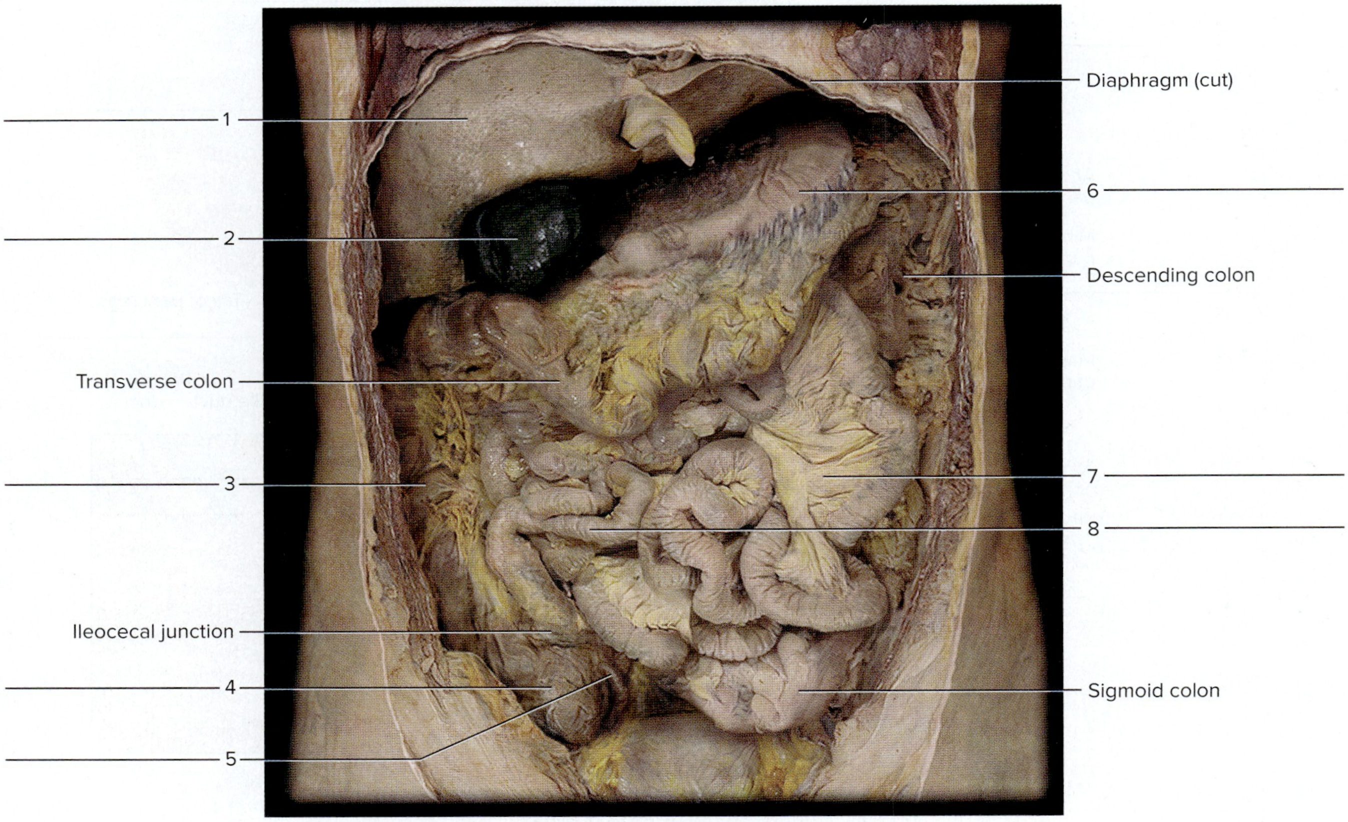

FIGURE 54.20 Label the digestive structures of this abdominopelvic cavity of a cadaver (anterior view).
APR/McGraw Hill APR A2

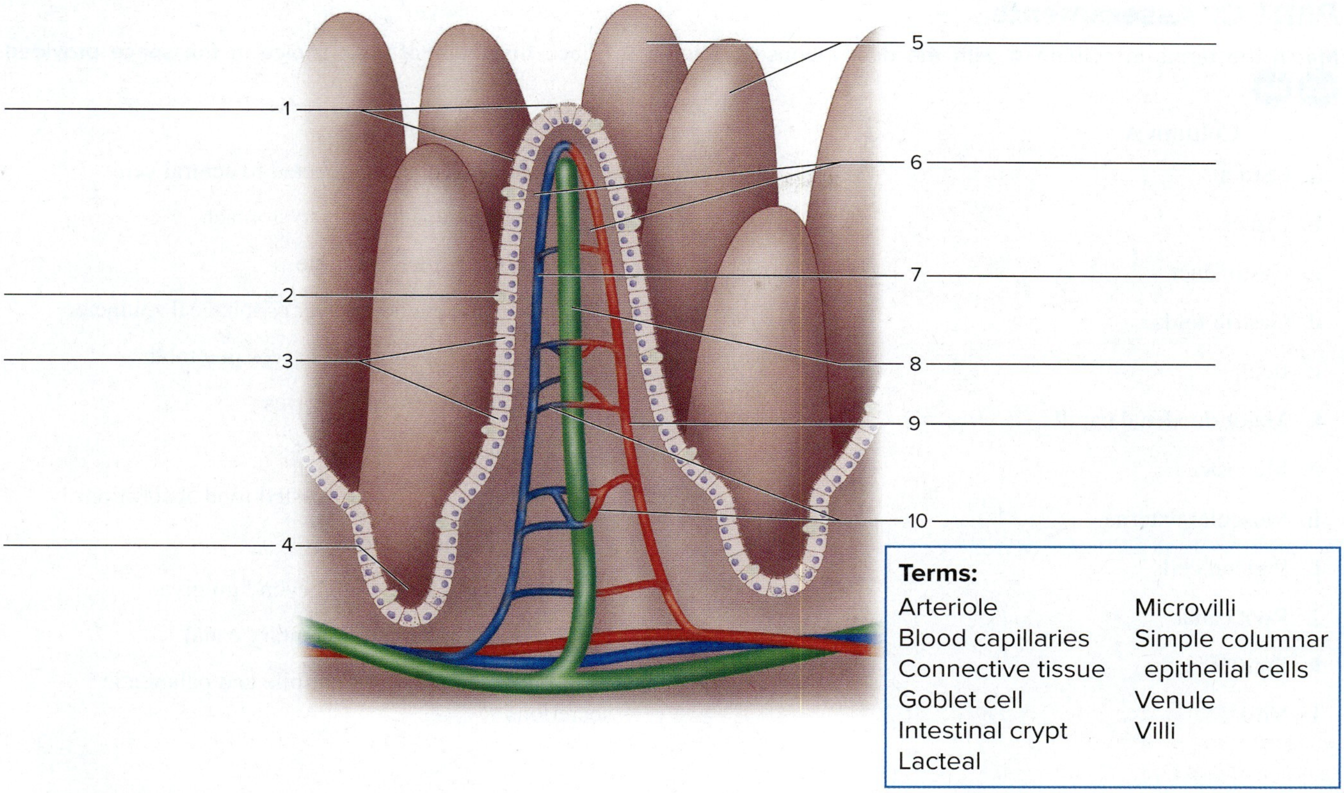

FIGURE 54.21 Identify the features indicated on the simplified illustration of a section of the small intestine mucosa, using the terms provided. A2

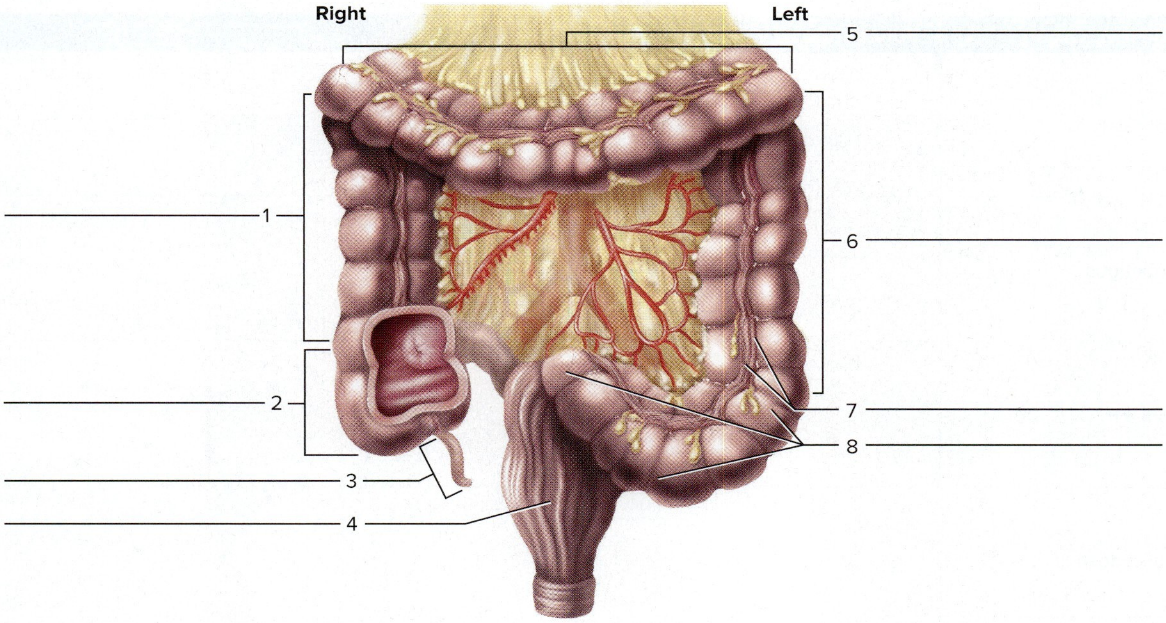

FIGURE 54.22 Label the features of the large intestine. A2

PART C: Assessments

Match the terms in column A with the descriptions in column B. Place the letter of your choice in the space provided. A3 A4

Column A

a. Cardia
b. Crown
c. Cystic duct
d. Gastric folds
e. Ileum
f. Major duodenal papilla
g. Mucosa
h. Muscular externa
i. Parietal cells
j. Root canal
k. Sinusoids
l. Villi

Column B

______ **1.** Blood-filled channels connected to central vein
______ **2.** Secrete hydrochloric acid into stomach
______ **3.** Last section of small intestine
______ **4.** Region of stomach near lower esophageal sphincter
______ **5.** Contains blood vessels and nerves in a tooth
______ **6.** Responsible for peristaltic waves
______ **7.** Allow stomach to expand
______ **8.** Increase surface area for digested food absorption
______ **9.** Attached to gallbladder
______ **10.** Portion of tooth projecting beyond gingivae
______ **11.** Layer nearest lumen of alimentary canal
______ **12.** Common opening region for bile and pancreatic secretions

PART D: Assessments

Complete the following:

1. Summarize the function(s) of each digestive organ: A4

Digestive Organ	Function(s)
Oral cavity	
Esophagus	
Stomach	
Small intestine	
Large intestine	

2. Name the valve (sphincter) located between the stomach and the duodenum. A2 ______________________

What is the function of this valve? A4 ____

____

3. Name the valve located between the small and the large intestines. A2 ______________________

What is the function of this valve? A4 ____

____

CRITICAL THINKING ASSESSMENT—PART D

The liver receives blood from the ______________________ artery and the ______________________ vein. Explain why the liver receives blood from these two separate sources. A4

____

____

____.

NOTES

LABORATORY EXERCISE 55

Action of a Digestive Enzyme

Purpose of the Exercise

To investigate the action of amylase and the effect of heat on its enzymatic activity.

MATERIALS NEEDED

- 0.5% amylase solution*
- Beakers (50 and 500 mL)
- Distilled water
- Funnel
- Pipettes (1 and 10 mL)
- Pipette rubber bulbs
- 0.5% starch solution
- Graduated cylinder (10 mL)
- Test tubes
- Test-tube clamps
- Wax marker
- Iodine-potassium-iodide solution
- Medicine dropper
- Ice
- Water bath, 37°C (98.6°F)
- Porcelain test plate
- Benedict's solution
- Hot plates
- Test-tube rack
- Thermometer

For Alternative Procedure Activity:

- Small, disposable cups
- Distilled water

*The amylase should be free of sugar for best results; a low-maltose solution of amylase yields good results. See Appendix 1.

SAFETY

- Review all the safety guidelines in Appendix 1 of your laboratory manual.
- Use only a mechanical pipetting device (never your mouth). Use pipettes with rubber bulbs or dropping pipettes.
- Wear protective eyewear when working with acids and when heating test tubes.
- Use test-tube clamps when handling hot test tubes.
- If an open flame is used for heating the test solutions, keep clothes and hair away from the flame.
- If student saliva is used as a source of amylase, it is important that students wear disposable gloves and handle only their own materials.
- Use an appropriate disinfectant to wash the laboratory tables before and after the procedures.
- Dispose of chemicals according to appropriate directions.
- Wash your hands before leaving the laboratory.

Chapter Opener: ©101cats/Getty Images

Learning Outcomes APR

After completing this exercise, you should be able to:

- **O1** Test a solution for the presence of starch or the presence of sugar.
- **O2** Explain the action of amylase.
- **O3** Test the effects of varying temperatures on the activity of amylase.

The **O** corresponds to the assessments **A** indicated in the Laboratory Assessment for this Exercise.

Pre-Lab

Carefully read the introductory material and examine the entire lab. Be familiar with enzyme structure and activity from lecture or the textbook. Answer the pre-lab questions.

Pre-Lab Questions Select the correct answer for each of the following questions:

1. The enzyme amylase accelerates the reaction of changing
- **a.** polysaccharides into monosaccharides.
- **b.** disaccharides into monosaccharides.
- **c.** starch into disaccharides.
- **d.** disaccharides into glucose.

2. Amylase originates from
- **a.** salivary glands and stomach.
- **b.** salivary glands and pancreas.
- **c.** esophagus and stomach.
- **d.** pancreas and intestinal mucosa.

3. The optimum pH for amylase activity is
- **a.** 0–14.
- **b.** 2.
- **c.** 6.8–7.0.
- **d.** 7–8.

4. After an enzyme reaction is completed, the enzyme
- **a.** is still available.
- **b.** was structurally changed.
- **c.** was denatured.
- **d.** is part of the enzyme-substrate complex.

5. As temperatures decrease, enzyme activity
- **a.** increases.
- **b.** remains unchanged.
- **c.** ceases.
- **d.** decreases.

6. Starch digestion begins in the _____.
   a. stomach b. small intestine
   c. mouth d. esophagus
7. The change of the unique three-dimensional shape of a protein, such as an enzyme, due to extreme temperature or pH conditions is called denaturation.
   a. True b. False
8. Salivary amylase does not function in chemical digestion in the stomach.
   a. True b. False

This laboratory exercise is designed to examine the relationship between enzymes and chemical digestion. *Enzymes* serve as *biological catalysts.* Acting as a catalyst, an enzyme will modify and rapidly increase the rate of a particular chemical reaction without being consumed in the reaction, which enables the enzyme to function repeatedly. A particular enzyme will catalyze a specific reaction. Various metabolic pathways and chemical digestive processes are accelerated by numerous enzymatic reactions. The **lock-and-key model** of enzyme activity involves an *enzyme* and *substrate* (substance acted upon by the enzyme) temporarily combining as an *enzyme-substrate complex* and ending with a reaction *product* and the original *enzyme* (fig. 55.1).

Most enzymes are globular proteins with a three-dimensional structure determined by various chemical bonds including hydrogen bonds. For the enzyme to function as the catalyst, the unique conformational shape is critical at the enzyme's active site for the enzyme-substrate complex to form properly. If the environmental temperature, including the body temperature, rises to a certain level, the three-dimensional shape of the enzyme is modified as hydrogen bonds start breaking and the enzyme becomes *denatured.* Although the primary structure of the amino acid sequence remains unaltered, the secondary and tertiary structures have been altered. A denatured protein could regain its shape unless the three-dimensional shape is destroyed when conditions become too extreme, and the enzyme becomes irreversibly denatured.

The digestive enzyme in salivary secretions that acts on starch (amylose) is called *salivary amylase.* Pancreatic amylase is secreted among several other pancreatic enzymes. A bacterial extraction of amylase is available for laboratory experiments. This enzyme catalyzes the reaction of hydrolyzing starch molecules into sugar (disaccharide) molecules, which is the first step in the digestion of complex carbohydrates.

As in the case of most other enzymes, amylase is a protein catalyst. Its activity is affected by exposure to certain environmental factors, including various temperatures, pH, radiation, and electricity. As temperatures increase, faster chemical reactions occur as the collisions of molecules happen at a greater frequency. Eventually, temperatures increase to a point that the enzyme becomes denatured, and the rate of the enzyme activity rapidly declines. As temperatures decrease, enzyme activity also decreases due to fewer collisions of the molecules; however, colder temperatures do not denature the enzyme. Normal body temperature provides an environment for enzyme activity near the optimum for enzymatic reactions.

The pH of the environment where enzymes are secreted also has a major influence on the enzyme reactions. The optimum pH for amylase activity is between 6.8 and 7.0, typical of salivary secretions. When salivary amylase arrives in the stomach, hydrochloric acid deactivates the enzyme, diminishing any further chemical digestion of remaining starch in the stomach. However, pancreatic amylase is secreted into the small intestine, where an optimum pH for amylase is provided once again. Other enzymes in the digestive system have different optimum pH ranges of activity compared to amylase. For example, pepsin from stomach secretions has an optimum activity around pH 2, whereas trypsin from pancreatic secretions operates best around pH 7–8.

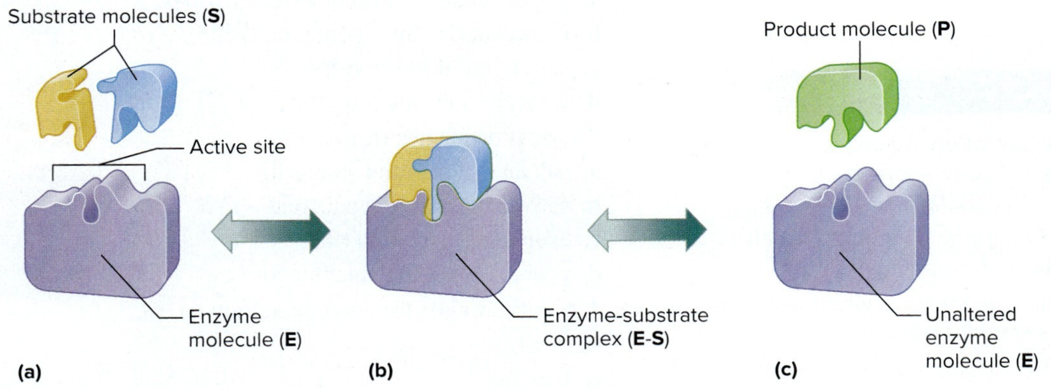

FIGURE 55.1 Lock-and-key model of an enzyme-catalyzed reaction **(E + S → E–S → P + E)**. (Many enzyme-catalyzed reactions, as depicted here, are reversible.) In the forward reaction (dark-shaded arrows), (a) the shapes of the substrate molecules (the "keys") fit the shape of the enzyme's active site (the "lock"). (b) When the substrate molecules temporarily combine with the enzyme, a chemical reaction occurs. (c) The result is a product molecule and an unaltered enzyme. The active site of the enzyme changes shape somewhat as the substrate binds, such that formation of the enzyme-substrate complex is more like a hand fitting into a glove, which has some flexibility that is not seen in a key fitting into a lock. APR

PROCEDURE A: Amylase Activity

1. Study the amylase action on starch digestion (fig. 55.2).
2. Examine the locations where starch is digested (fig. 55.3).
3. Mark three clean test tubes as *tubes 1, 2,* and *3,* and prepare the tubes as follows (fig. 55.4):

   Tube 1: **Add 6 mL of amylase solution.**

   Tube 2: **Add 6 mL of starch solution.**

   Tube 3: **Add 5 mL of starch solution and 1 mL of amylase solution.**

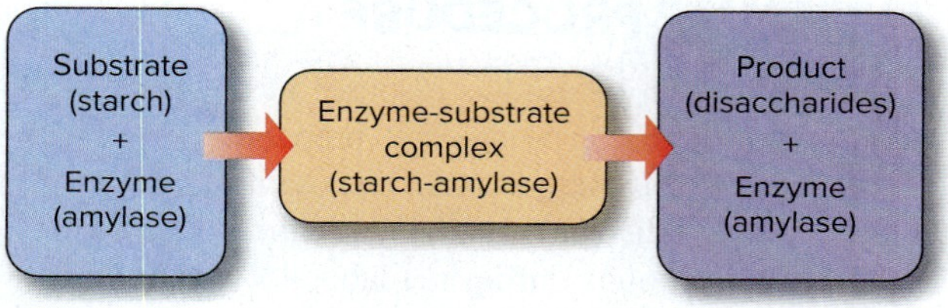

FIGURE 55.2 Action of amylase on starch digestion.

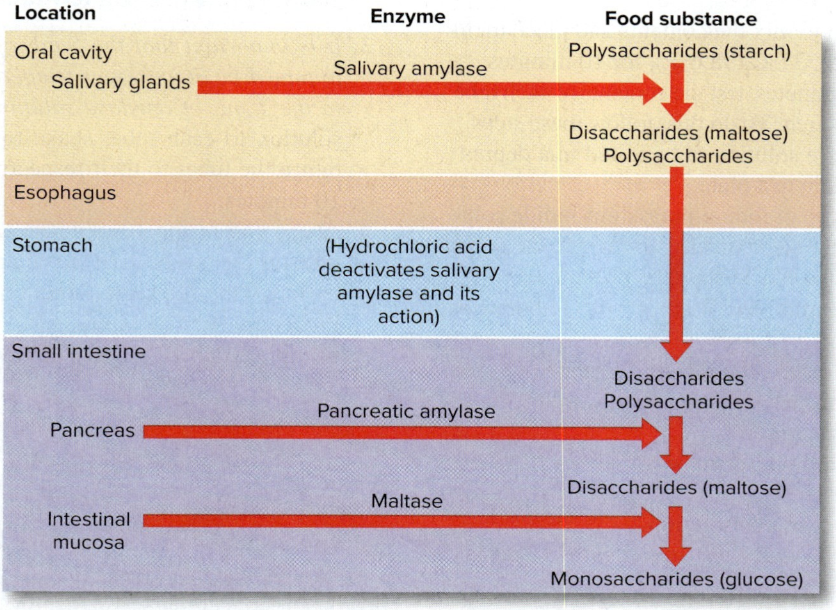

FIGURE 55.3 Flowchart of starch digestion.

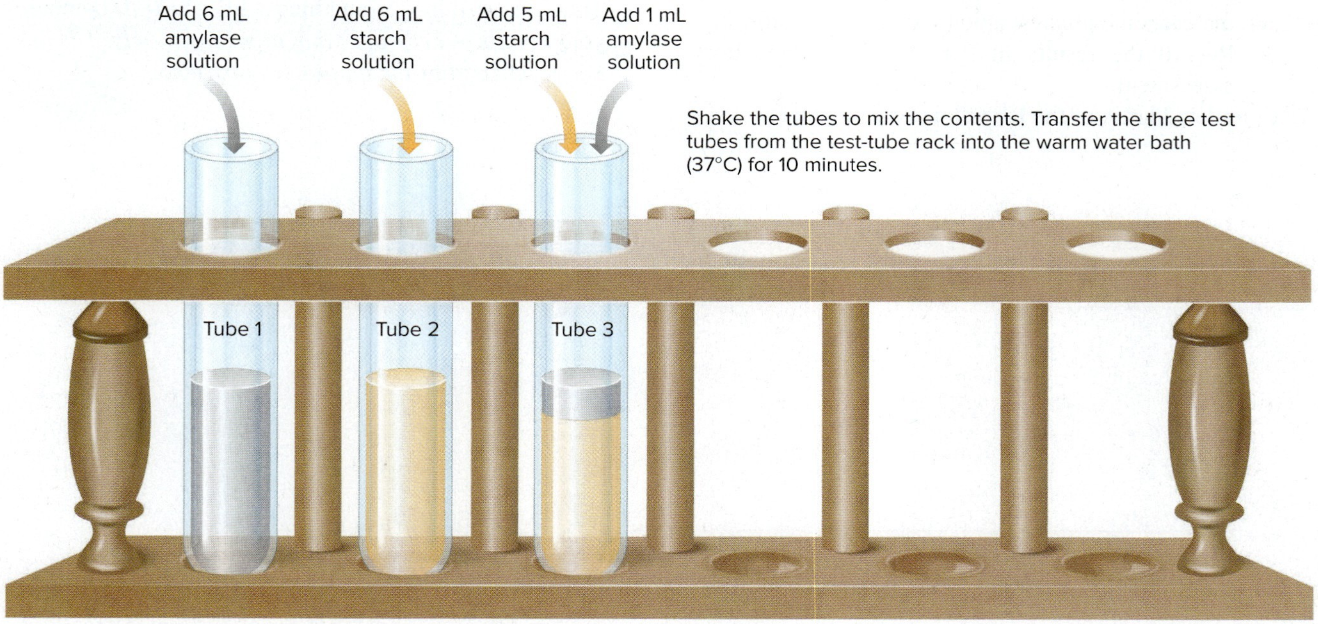

FIGURE 55.4 Test tubes prepared for testing amylase activity.

ALTERNATE PROCEDURE ACTIVITY

Human saliva can be used as a source of amylase solutions instead of bacterial amylase preparations. Collect about 5 mL of saliva into a small, disposable cup. Add an equal amount of distilled water and mix them together for the amylase solutions during the laboratory procedures. Be sure to follow all of the safety guidelines.

4. Shake the tubes well to mix the contents and place them in a warm water bath, 37°C (98.6°F), for 10 minutes.
5. At the end of the 10 minutes, test the contents of each tube for the presence of starch. To do this, follow these steps:
   a. Place 1 mL of the solution to be tested in a depression of a porcelain test plate.
   b. Next add one drop of iodine-potassium-iodide solution and note the color of the mixture. If the solution becomes blue-black, starch is present.
   c. Record the results in Part A of Laboratory Assessment 55.
6. Test the contents of each tube for the presence of sugar (disaccharides, in this instance). To do this, follow these steps:
   a. Place 1 mL of the solution to be tested in a clean test tube.
   b. Add 1 mL of Benedict's solution.
   c. Place the test tube with a test-tube clamp in a beaker of boiling water for 2 minutes.
   d. Note the color of the liquid. If the solution becomes green, yellow, orange, or red, sugar is present. Blue indicates a negative test, whereas green indicates a positive test with the least amount of sugar, and red indicates the greatest amount of sugar present.
   e. Record the results in Part A of the laboratory assessment.
7. Complete Part A of the laboratory assessment.

PROCEDURE B: Effect of Heat

1. Mark three clean test tubes as *tubes 4, 5,* and *6.*
2. Add 1 mL of amylase solution to each of the tubes and expose each solution to a different test temperature for 3 minutes as follows:

   Tube 4: **Place in beaker of ice water (about 0°C [32°F]).**

   Tube 5: **Place in warm water bath (about 37°C [98.6°F]).**

   Tube 6: **Place in beaker of boiling water (about 100°C [212°F]). Use a test-tube clamp.**
3. *It is important that the 5 mL of starch solution added to tube 4 be at ice-water temperature before it is added to the 1 mL of amylase solution.* Add 5 mL of starch solution to each tube, shake to mix the contents, and return the tubes to their respective test temperatures for 10 minutes.
4. At the end of the 10 minutes, test the contents of each tube for the presence of starch and the presence of sugar by following the same directions as in steps 5 and 6 in Procedure A.
5. Complete Part B of the laboratory assessment.

LEARNING EXTENSION ACTIVITY

Devise an experiment to test the effect of some other environmental factor on amylase activity. For example, you might test the effect of a strong acid by adding a few drops of concentrated hydrochloric acid to a mixture of starch and amylase solutions. Be sure to include a control in your experimental plan. That is, include a tube containing everything except the factor you are testing. Then you will have something with which to compare your results. *Carry out your experiment only if it has been approved by the laboratory instructor.*

LABORATORY
ASSESSMENT 55

Name ______________________

Date ______________________

Section ______________________

The **A** corresponds to the indicated Learning Outcome(s) **O** found at the beginning of the Laboratory Exercise.

Action of a Digestive Enzyme

PART A: Assessments

1. Test results: **A1**

Tube	Starch	Sugar
1 Amylase solution		
2 Starch solution		
3 Starch-amylase solution		

2. Complete the following:

a. Explain the reason for including tube 1 in this experiment. **A2** ______________________

b. What is the importance of tube 2? **A2** ______________________

c. Explain why these test tubes were placed in a 37°C water bath. **A2** ______________________

d. What do you conclude from the results of this experiment? **A2** ______________________

PART B: Assessments

1. Test results: A3

Tube	Starch	Sugar
4 0°C (32°F)		
5 37°C (98.6°F)		
6 100°C (212°F)		

2. Complete the following:

   a. What do you conclude from the results of this experiment? A3

   ____

   ____

   ____

   ____

   ____

   b. If digestion failed to occur in one of the tubes in this experiment, how can you tell whether the amylase was destroyed by the factor being tested or the amylase activity was simply inhibited by the test treatment? A3

   ____

   ____

   ____

   ____

   ____

CRITICAL THINKING ASSESSMENT—PART B

What test result would occur if the amylase you used contained sugar? ______________________ Would your results be as valid? Explain your answer. A2

____

____

____

____

LABORATORY
EXERCISE 56

Metabolism

Purpose of the Exercise

To develop an understanding of metabolism, to observe and experience a technique of measuring metabolism, and to relate metabolic rate to body functions and actions.

MATERIALS NEEDED

Respiratory chambers (assembly required)

- Chamber—medium-sized mason (canning) jar with mouth approximately 6 cm in diameter, to fit #12 two-hole rubber stopper; 1-gallon pickle jar with mouth approximately 10 cm in diameter, to fit #15 two-hole rubber stopper (Recommend: 4–5 medium-size chambers and 1–2 large chamber)
- Rubber stoppers—#12 two-hole for medium chamber; #15 two-hole stopper for large chamber
- Glass tubing—6 mm diameter glass tubing, one straight piece approximately 6 cm in length; bent piece approximately 30 cm in length (bent to shape and fire polished in-house)
- Rubber tubing—4–5 mm inside dimension latex approximately 5–6 cm in length per chamber
- Hoffman metal clamp attached to rubber tubing
- 10 cc syringe attached to hypodermic needle pushed through stopper; tip blunted
- CO_2 absorbent ($Ca(OH)_2$ = soda lime)—place inside empty tea bag and then seal the bag
- Rubber tubing (thick)—12 mm outside dimension, 4 mm inside dimension, approximately 27 cm heavy rubber tubing to fit snug inside jar, using short piece of glass tubing to join ends to form a ring inside chamber to keep galvanized mesh screening in place over CO_2 absorbent
- Galvanized mesh screening—6 mm × 6 mm cut to fit inside bottom of chamber, keeping animal from reaching CO_2 absorbent

Colored water

Triple-beam balance with metal basket *or* electronic balance 500 g with readability 0.1 g (and container for animal)

Calculator

Laboratory animals (rats, mice, gerbils, hamsters)—need weight range of approximately 12 animals

Leather gloves for handling rodents

Dropper bottle of isopropyl alcohol

Colored pencils

Markers

SAFETY

- Review all the safety guidelines in Appendix 1 of your laboratory manual.
- Wear leather gloves when handling rodents.
- Take precautions when handling rodents.

Learning Outcomes

After completing this exercise, you should be able to:

O1 Define metabolism and how it can be measured.

O2 Explain the relationship of oxygen consumption in determining metabolic rate.

O3 Calculate and graph metabolic rate relative to body weight and interpret its significance to body temperature regulation.

O4 Integrate the concepts of body size (surface area to volume ratio), oxygen consumption, and metabolism, and explain their application to the human body.

O5 Identify factors that influence the rate of metabolism.

The O corresponds to the assessments A in the Laboratory Assessment for this Exercise.

Pre-Lab

Carefully read the introductory material and examine the entire lab. Be familiar with metabolism and metabolic rate from lecture or the textbook. Answer the pre-lab questions.

Pre-Lab Questions Select the correct answer for each of the following questions:

1. The type of metabolic reactions that release energy are _____.
   a. catabolism **b.** anabolism
   c. synthesis **d.** active transport
2. The reactants of glucose catabolism are _____.
   a. oxygen and carbon dioxide **b.** water and glucose
   c. glucose and oxygen **d.** carbon dioxide and water

Chapter Opener: ©101cats/Getty Images

3. Most heat loss occurs through _____.
   a. the skin's surface   b. breathing
   c. defecation   d. urination
4. Which of the following is *not* a factor that can influence the rate of metabolism?
   a. hair color   b. physical activity
   c. thyroid hormone   d. age
5. A measurement of the minimal amount of energy needed to maintain vital body functions is _____.
   a. catabolism   b. anabolism
   c. basal metabolic rate   d. calorimetry
6. Direct calorimetry measures _____.
   a. oxygen consumption   b. heat expended
   c. carbon dioxide produced   d. food consumption
7. Basal metabolic rate (BMR) is best measured when a person has just finished exercising and is eating a meal.
   a. True   b. False
8. In general, as an organism decreases in size, the ratio of surface area to volume increases.
   a. True   b. False
9. Carbohydrates, proteins, and fats can be used to synthesize ATP.
   a. True   b. False

***Metabolism* refers to all** the chemical reactions within cells that use or release energy. Reactions known as *anabolism* synthesize complex molecules from simple molecules and use energy in the process, whereas reactions known as *catabolism* break down complex molecules into simple molecules and release energy in the process. Measuring metabolism provides information on how an organism obtains cellular energy and how quickly and efficiently it is used.

Metabolism is difficult to measure because there are so many chemical reactions that occur in the body. The second law of thermodynamics states that for every chemical reaction, some energy is always lost as heat. Therefore, one way to monitor anabolism and catabolism processes is to determine the amount of heat energy the body releases (fig. 56.1), which makes up the vast majority of expended energy.

Cells of the body utilize fuel from the foods that are eaten and oxygen (O_2) from the air that is inhaled to synthesize adenosine triphosphate (ATP). Carbon dioxide (CO_2), water (H_2O), and energy in the form of ATP and heat are produced in this reaction (fig. 56.2). The greater the number of cells or the more metabolically active the cells, the more heat is produced. Warm-blooded animals rely on heat production, as well as mechanisms to allow heat to escape, to keep their body temperature within normal limits (homeostasis). Most heat is lost through the surface of the skin. Thus, a larger surface area of the skin in proportion to body volume allows more heat to escape toward the outside environment. For a given volume of cells, the metabolic rate of the cells is established and maintained to offset the heat that is lost by the skin surface (within limits). The more heat that is lost at the skin surface, the more metabolically active the cells must be to produce the heat required to keep the body warm and maintain homeostasis.

In addition to heat and surface area to volume ratio, there are many other factors that can influence the rate of metabolism, such as age, sex, diet, absorptive or postabsorptive conditions, health, endocrine function, genetics, and varying levels of physical activity (table 56.1). For this reason, variables need to be controlled if meaningful comparisons of metabolic rates are to be made. One way to approach this goal is to measure the *basal metabolic rate (BMR)*. BMR is a measurement of the minimal amount of energy needed to maintain vital body functions. This is best measured in humans when the person is in a relaxed and awake state, reclined in a dark room at a comfortable temperature, having had nothing to eat for 12–14 hours, after a restful night's sleep, and having had no exercise for at least 1 hour.

To take a valid measure of metabolism when the organism is not doing work, a *direct calorimeter* can be used to measure the amount of heat given off by an organism, usually expressed in units of heat energy such as calories or kilocalories per body weight per hour (kcal/body weight/hour). However, this device requires a closed water-filled chamber that is tedious and difficult to operate. An easier and more commonly used method of measurement

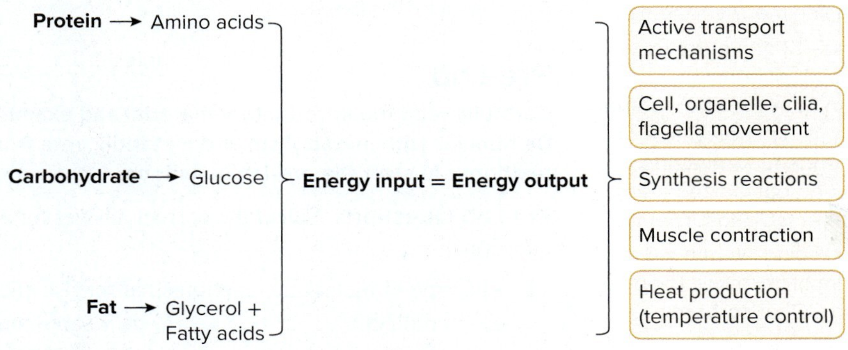

FIGURE 56.1 Foods we eat contain proteins, carbohydrates, and fats that are broken down by hydrolysis in the digestive system, releasing energy in the process, and then absorbed into the body's cells. Energy is utilized by the cells to do work and help maintain homeostasis.

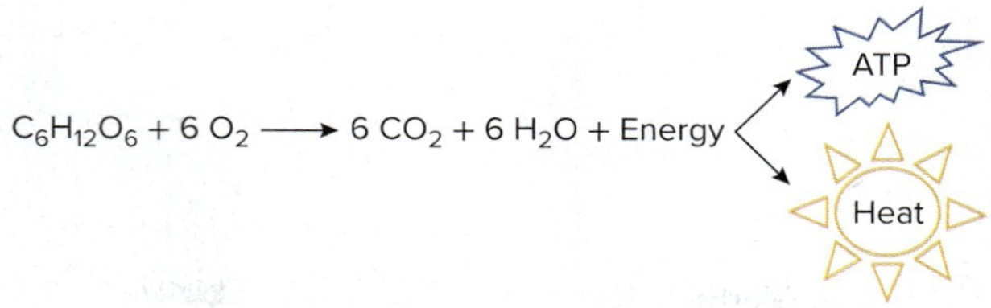

FIGURE 56.2 Summary equation of glucose catabolism through aerobic respiration. Glucose is obtained by glycogen hydrolysis or by ingested foods and oxygen is inhaled, forming carbon dioxide, water, and energy in the form of ATP and heat. APR

Table 56.1 Energy Consumed [in Kilocalories (kcal) per Minute] in Different Types of Activities Given a Person's Weight in Pounds

Activity	Weight in Pounds			
	105–115	127–137	160–170	182–192
Bicycling				
10 mph	5.41	6.16	7.33	7.91
Stationary, 10 mph	5.50	6.25	7.41	8.16
Calisthenics	3.91	4.50	7.33	7.91
Dancing				
Aerobic	5.83	6.58	7.83	8.58
Square	5.50	6.25	7.41	8.00
Gardening, Weeding, and Digging	5.08	5.75	6.83	7.50
Jogging				
5.5 mph	8.58	9.75	11.50	12.66
6.5 mph	8.90	10.20	12.00	13.20
8.0 mph	10.40	11.90	14.10	15.50
9.0 mph	12.00	13.80	16.20	17.80
Rowing, Machine				
Easily	3.91	4.50	5.25	5.83
Vigorously	8.58	9.75	11.50	12.66
Skiing				
Downhill	7.75	8.83	10.41	11.50
Cross-country, 5 mph	9.16	10.41	12.25	13.33
Cross-country, 9 mph	13.08	14.83	17.58	19.33
Swimming, Crawl				
20 yards per minute	3.91	4.50	5.25	5.83
40 yards per minute	7.83	8.91	10.50	11.58
55 yards per minute	11.00	12.50	14.75	16.25
Walking				
2 mph	2.40	2.80	3.30	3.60
3 mph	3.90	4.50	6.30	6.80
4 mph	4.50	5.20	6.10	6.80

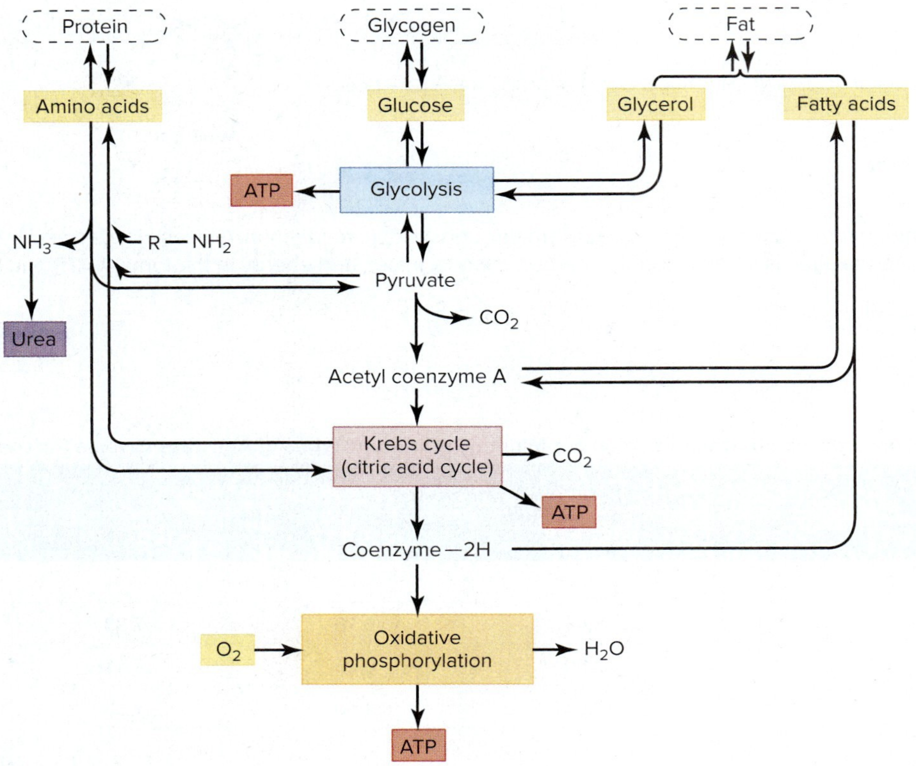

FIGURE 56.3 Pathways of metabolism of protein, carbohydrate (glycogen), and fat (triglyceride) and their relationships. APR

is *indirect calorimetry,* using an apparatus called a *respirometer,* where the rate of oxygen consumption is measured. The products of digestion from complex organic molecules of carbohydrates, fats, and proteins found in food are utilized in aerobic respiration to synthesize adenosine triphosphate (ATP). Given that this process is aerobic, oxygen is consumed during the process and carbon dioxide is released as a by-product (fig. 56.3). Approximately 4.82 kcal of energy is released from organic nutrients per liter of oxygen (*or* 0.004825 kcal per milliliters of oxygen).

PROCEDURE: Measuring and Calculating Metabolic Rate

In this laboratory exercise, the respirometer that will be used to measure oxygen consumption of rodents of various sizes will be a respiration chamber as shown in figure 56.4. Within this chamber is a carbon dioxide compound that is used to absorb the carbon dioxide released by the test animal. As the test animal consumes oxygen and releases carbon dioxide, the carbon dioxide is absorbed quickly, and consequently the amount of gas in the respiratory chamber will decrease. The amount of oxygen consumed will be monitored and measured using the attached manometer.

To compare the size of the rodents, the weight of the animal can be determined using a triple-beam balance with a metal basket (fig. 56.5), or a similar reliable device such as an electronic balance, measuring in mass units of grams (g). (*Note:* Weight and mass are different concepts; however, in this lab, we refer to body mass as the "weight" of the animal.) Although it would be more accurate to assess the size by determining the ratio of surface area to volume of the animal, this is not easily measured, so it will not be done in this laboratory exercise. In general, though, as an organism increases in size, the ratio of surface area to volume decreases. Another way of saying this is that smaller organisms have a higher ratio of surface area to volume than larger organisms.

1. Work in groups of 3–4 individuals. Set up the respiratory chamber as illustrated in figure 56.4 with the valve (C) open.
2. Make sure the triple-beam balance is at zero and balanced, with the metal basket and lid in place.
3. Wearing leather gloves, carefully take the animal from its cage and place it inside the metal basket of the triple-beam balance to determine its weight in grams. Record the value in Part A of Laboratory Assessment 56.
4. Carefully place the test animal in the respiratory chamber.
5. Firmly place the stopper (A) on the respiratory chamber so that there is a tight seal, making sure the valve (C) is open.
6. Because the animal's body will release heat and warm the air and walls of the respiratory chamber, the volume of air will be affected. In order to wait until this change is accomplished, allow at least 3 minutes for

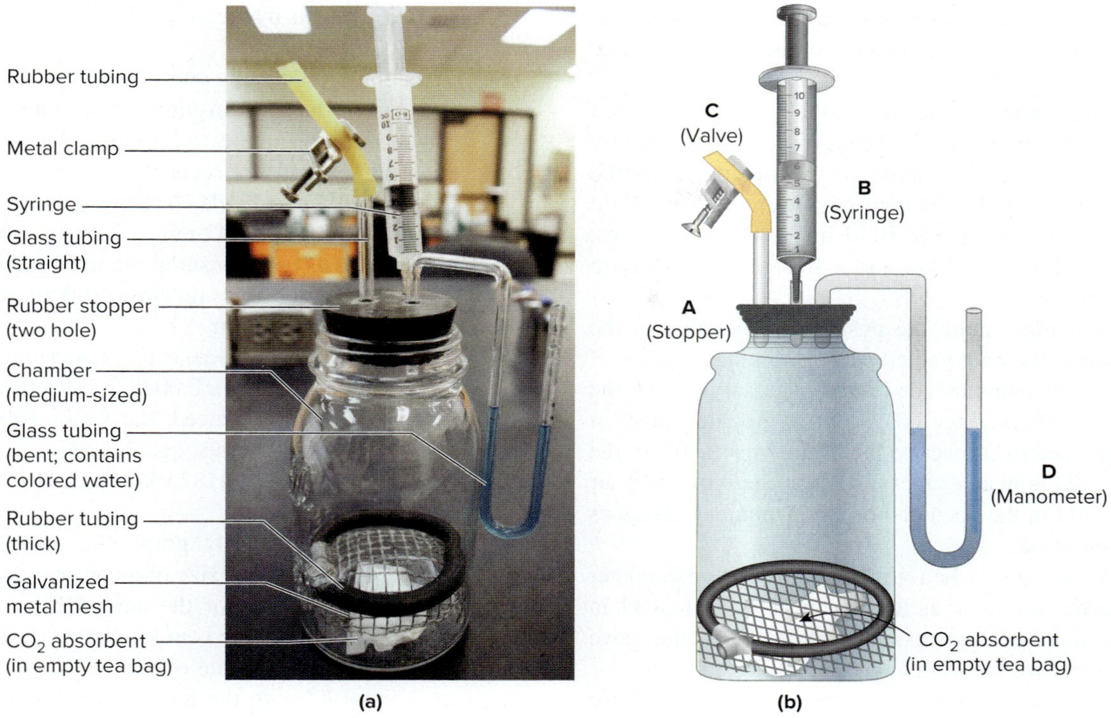

FIGURE 56.4 (a) Respiratory chamber photo and (b) respiratory chamber illustration, each with labeled parts. Cynthia Prentice-Craver/McGraw Hill

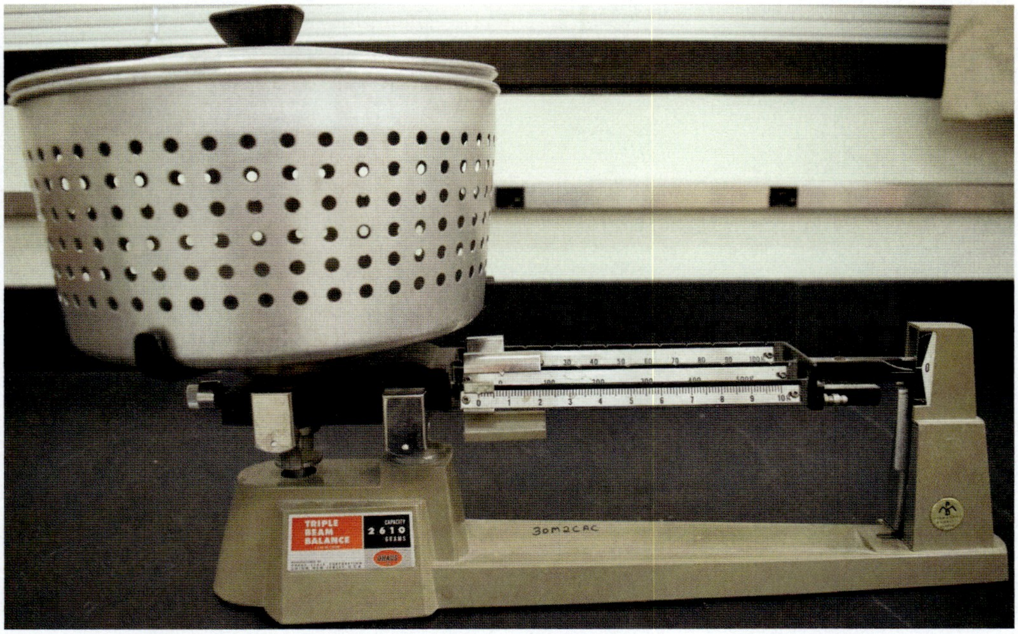

FIGURE 56.5 Triple-beam balance with metal basket to hold animal. Cynthia Prentice-Craver/McGraw Hill

temperature equilibration. As you are waiting, pull the plunger back on the syringe (B) to just under the 10 cc mark.

7. Close the clamp so that the valve (C) is closed and readjust the plunger on the syringe (B) to exactly the 10 cc mark. The fluid in the open end of the manometer (D) should now be lower than the fluid at the closed end. If the fluid in the open end rises, check for leaks or allow more time for temperature equilibration.
8. Using a marker, mark the position of the fluid in the open end of the manometer and start timing.
9. At 1-minute intervals (or less if necessary), push the plunger of the syringe (B) to bring the fluid level in the manometer (D) back to its original mark. When the syringe (B) is nearly empty, push in the remaining air so the fluid in the open end of the manometer (D) goes above the mark.
10. Record the time when the fluid in the manometer (D) reaches the mark as Run #1 for Test Animal #1 in Part A of Laboratory Assessment 56. Open the valve (C) and remove the stopper (A) to provide fresh air.
11. After a few minutes, repeat steps 5–10 as Run #2 for Test Animal #1, and record the results in Part A of Laboratory Assessment 56. Return Test Animal #1 to its proper cage.
12. Using a different test animal that has a significantly different body size, repeat steps 2–11 and record your results as Test Animal #2 in Part A of Laboratory Assessment 56. Return Test Animal #2 to its proper cage.
13. Use the following example to help calculate the metabolic rate of each test animal and run. Show your calculation results by recording them in Part A of Laboratory Assessment 56.

    Calculation of metabolic rate:

    a. Calculate oxygen consumption per minute. Convert seconds into minutes by dividing by 60 (example: 4 minutes and 25 seconds = 4 min + 25/60 = 4.42 min). Then divide 10 mL by the time in minutes (example: 10 mL/4.42 min = 2.26 mL/min).

    b. Calculate oxygen consumption rate by using the following formula: (O_2 mL/min) (60 min/hr)/animal weight = O_2 mL/g/hr

    c. Calculate metabolic rate expressed in units of energy production. The factor 0.004825 converts mL O_2 into kcal of energy produced; the factor 1000 converts grams (g) to kilograms (kg).

    O_2 mL/g/hr × 0.004825 kcal/mL × 1000 g/kg = kcal/kg/hr.
14. In order to more easily graph the data to show a correlation between the size of an animal and its metabolic rate, the weight of the animal is in grams (g), while the metabolic rate is in kilograms (kg). Share the weight and metabolic rate of your Test Animal #1 and Test Animal #2 with the rest of the class. Record the class data in the table provided in Part B of Laboratory Assessment 56.
15. Use the graph page provided in Part B of Laboratory Assessment 56 to graph your data using a colored pencil, and then graph the rest of the class data using a regular pencil. After plotting the data, draw an "average curve" rather than connecting each data plot.
16. Complete Part B, Part C, and Part D of Laboratory Assessment 56.

Name ______________________________

Date ______________________________

Section ______________________________

The **A** corresponds to the indicated Learning Outcome(s) **O** found at the beginning of the Laboratory Exercise.

LABORATORY ASSESSMENT 56

Metabolism

PART A: Assessments

1. Enter your calculations and test results as you complete each test run for animal #1 and animal #2: **A1** **A2** **A3**

	Run #1	Run #2
Animal #1 Weight: ______ g	Time: ________ min ________ sec O_2 consumed: 10 mL/________ min O_2 consumption: ________ mL/min O_2 consumption rate: ________ mL/g/hr Metabolic rate: ________ kcal/kg/hr	Time: ________ min ________ sec O_2 consumed: 10 mL/________ min O_2 consumption: ________ mL/min O_2 consumption rate: ________ mL/g/hr Metabolic rate: ________ kcal/kg/hr
Animal #2 Weight: ______ g	Time: ________ min ________ sec O_2 consumed: 10 mL/________ min O_2 consumption: ________ mL/min O_2 consumption rate: ________ mL/g/hr Metabolic rate: ________ kcal/kg/hr	Time: ________ min ________ sec O_2 consumed: 10 mL/________ min O_2 consumption: ________ mL/min O_2 consumption rate: ________ mL/g/hr Metabolic rate: ________ kcal/kg/hr

2. What variables were controlled in this experiment? **A1** ______________________________

____

3. What do you suggest could be done to make this experiment and its test results more reliable? **A1** ______________

____

4. Based on your calculations and test results of these two animals, what can you infer about the relationship between oxygen consumption and metabolic rate? **A2**

____

____

PART B: Assessments

1. Gather the test results from your classmates to complete the following class data:

Class Data:

Weight (g)	Metabolic Rate (kcal/kg/hr)

Weight (g)	Metabolic Rate (kcal/kg/hr)

2. Use the class data from #1 above to complete the graph on the next page where the x-axis is the animal weight (g) and the y-axis is the metabolic rate (kcal/kg/hr). A3

3. Analyze the "average curve" result that you sketched in your graph. What can you conclude from these results? ____ A3

PART C: Assessments

Complete the following statements:
(*Note:* Where appropriate, circle the underlined word to make the statement correct.)

1. The polymerization of glycogen from glucose molecules is *catabolism/anabolism*, and energy is *used/released* in the process. A1

2. The sum total of all chemical reactions that occur in the body is ________________. A1

3. In glycolysis, the break down of glucose to form pyruvate is *catabolism/anabolism*, and energy is *used/released* in the process. A1

4. In this experiment, metabolic rate was measured indirectly by measuring the amount of ________________ consumed. A1

5. The cells of the body utilize oxygen in the process of ________________, which produces carbon dioxide, water, and energy in the form of ATP and heat. A2

6. Smaller organisms have a ________________ metabolism than larger organisms. A3

7. Smaller organisms have a ________________ ratio of surface area to volume than larger organisms. A3

8. The greater the ratio of surface area to volume of an organism, the ________________ heat is expended (lost). A3

9. Observe the equation in figure 56.2. To maintain normal body temperature, if an animal is losing heat, the animal must consume more ________________ and ________________ to produce more heat to make up for what has been lost. A4

Graph of Class Data Comparing Animal Size and Metabolic Rate

- Graph your own data using colored pencil
- Graph class data using regular pencil
- Draw an "average curve"

Body weight (g)

Metabolic rate (kcal/kg/hr)

CRITICAL THINKING ASSESSMENT—PART C

If Bill, who weighs 160 pounds, consumed 2,025 mL of O_2 during a 6-minute period, what would be his metabolic rate (in units of kcal/kg/hr)? (*Hint:* 1 kg = 2.2 lbs.) A4

Assume that, on an average diet, people generate 4.825 kcal of energy for every liter of oxygen they consume. How many calories (kcal) does Bill need per day? (*Hint:* No need to include Bill's weight in this calculation.) A4

PART D: Assessments

For each of the following factors, indicate whether it would *increase* or *decrease* a person's metabolic rate: A5

____________________ **1.** Experiencing anxiety and stress.

____________________ **2.** Being physically active.

____________________ **3.** A larger percent body fat compared to muscle mass.

____________________ **4.** Hypothyroid condition.

____________________ **5.** Older person as compared to younger person (i.e., aging).

____________________ **6.** Increase in the ratio of surface area to volume.

____________________ **7.** Experiencing a fever.

____________________ **8.** Being in an absorptive state, having just eaten a meal.

LABORATORY

EXERCISE 57

Urinary Organs

Purpose of the Exercise

To review the structure and general function of urinary organs, including the gross anatomy of a kidney as well as the microscopic nephron.

MATERIALS NEEDED

Human torso model
Kidney model
Preserved pig (or sheep) kidney
Dissecting tray
Dissecting instruments
Disposable gloves
Long knife
Compound light microscope
Prepared microscope slide of the following:
- Kidney section
- Ureter, cross section
- Urinary bladder
- Urethra, cross section

SAFETY

- Review all the safety guidelines in Appendix 1 of your laboratory manual.
- Wear disposable gloves and protective eyewear when working on the kidney dissection.
- Dispose of the kidney and gloves as directed by your laboratory instructor.
- Wash the dissecting tray and instruments as instructed.
- Wash your laboratory table.
- Wash your hands before leaving the laboratory.

Chapter Opener: ©101cats/Getty Images

Learning Outcomes APR

After completing this exercise, you should be able to:

O1 Identify the structural features of a kidney and associated urinary organs and their general functions.
O2 Identify and sketch the structures of a nephron.
O3 Trace the path of filtrate through a renal nephron.
O4 Trace the path of blood through the renal blood vessels.
O5 Identify and sketch the structures of the wall of a ureter, urinary bladder, and urethra in tissue (histological) sections.
O6 Trace the path of urine flow through the urinary system.

The O corresponds to the assessments A indicated in the Laboratory Assessment for this Exercise.

Pre-Lab

Carefully read the introductory material and examine the entire lab. Be familiar with the structures and functions of the urinary organs from lecture or the textbook. Answer the pre-lab questions.

Pre-Lab Questions Select the correct answer for each of the following questions:

1. When comparing the position of the two kidneys,
a. they are at the same level.
b. the right kidney is slightly superior to the left kidney.
c. the right kidney is slightly inferior to the left kidney.
d. the right kidney is anterior to the left kidney.

2. Which of the following is *not* a function of the kidneys?
a. secrete erythropoietin **b.** regulate pH of blood
c. produce bile **d.** remove metabolic wastes from blood

3. Cortical nephrons represent about _____ of the nephrons.
a. 50% **b.** 100%
c. 15% **d.** 85%

4. Which of the following does *not* represent one of the processes in urine formation?
a. secretion of renin **b.** glomerular filtration
c. tubular reabsorption **d.** tubular secretion

5. The _____ arteries and veins are located in the corticomedullary junction.
a. arcuate **b.** renal
c. cortical radiate **d.** peritubular

6. The _____ is the tube located from the kidney to the urinary bladder.
   a. urethra **b.** ureter
   c. renal pelvis **d.** renal column
7. The trigone is a triangular, funnel-like region of the
   a. renal cortex. **b.** renal medulla.
   c. urethra. **d.** urinary bladder.
8. The external urethral sphincter is composed of involuntary smooth muscle.
   a. True **b.** False
9. Contractions of the detrusor muscle provide the force during micturition (urination).
   a. True **b.** False

The kidneys are the primary organs of the urinary system. They are located in the abdominal cavity, against the posterior wall and behind the parietal peritoneum (*retroperitoneal*). Masses of adipose tissue associated with the kidneys hold them in place at a vertebral level between T12 and L3. The right kidney is slightly inferior (lower) than the left kidney due to the large mass of the liver near its superior border. There are a variety of functions of the kidneys. They remove metabolic wastes from the blood; help regulate blood volume, blood pressure, and pH of blood; control water and electrolyte concentrations; convert vitamin D to its active form; and secrete renin and erythropoietin.

The movements of muscular walls of the *ureters* force urine by means of *peristaltic waves* into the *urinary bladder,* which temporarily stores urine. The *urethra* conveys urine to the outside of the body.

Each *kidney* contains about 1 million *nephrons,* which serve as the basic structural and functional units of the kidney. A *glomerular capsule, proximal convoluted tubule, nephron loop,* and *distal convoluted tubule* compose the microscopic, multicellular structure of a relatively long nephron tubule. Several nephrons merge and drain into a common *collecting duct.* Approximately 85% of the nephrons are *cortical nephrons* with short nephron loops, whereas the remaining represent *juxtamedullary nephrons,* with long nephron loops extending deeper into the renal medulla. An elaborate network of blood vessels surrounds the entire nephron. *Glomerular filtration, tubular reabsorption,* and *tubular secretion* represent three processes resulting in urine as the final product.

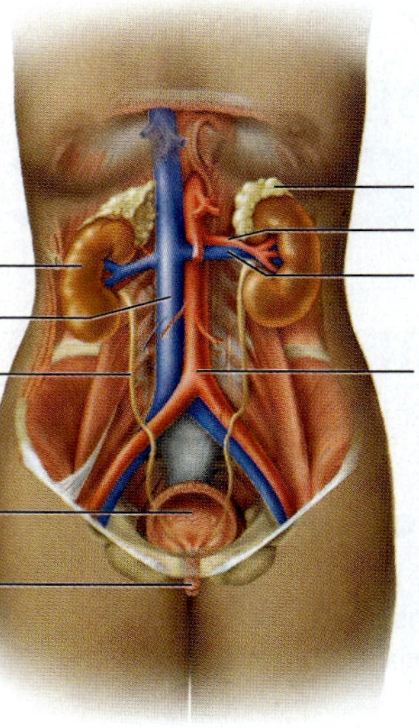

FIGURE 57.1 The urinary system and associated structures. APR

PROCEDURE A: Kidney APR

1. Study figures 57.1 and 57.2 and the list of structures and descriptions.
2. Observe the human torso model and the kidney model. Locate the following:

   kidneys—paired retroperitoneal organs

   ureters—paired tubular organs about 25 cm long that transport urine from kidney to urinary bladder

   urinary bladder—single muscular storage organ for urine

   urethra—conveys urine from urinary bladder to external urethral orifice

   renal sinus—hollow chamber of kidney

   renal pelvis—funnel-shaped sac at superior end of ureter; receives urine from major calyces
   - major calyces (sing. calyx)—2–3 major converging branches into renal pelvis; receive urine from minor calyces
   - minor calyces (sing. calyx)—small subdivisions of major calyces; receive urine from papillary ducts within renal papillae

   renal medulla—deep region of kidney
   - renal pyramids—6–10 conical regions containing blood capillaries, renal tubules, and collecting ducts; composes most of renal medulla
   - renal papillae—projections at ends of renal pyramids with openings into minor calyces

   renal cortex—superficial region of kidney

   renal columns—extensions of renal cortical tissue between renal pyramids

   nephrons—functional units of kidneys
   - cortical nephrons—85% of nephrons close to surface with short nephron loops
   - juxtamedullary nephrons—15% of nephrons close to medulla with long nephron loops; produces concentrated urine to conserve water within the body
3. Obtain a pig or sheep kidney along with a dissecting tray, dissecting instruments, disposable gloves, and protective eyewear and follow these steps:
   a. Rinse the kidney with water to remove as much of the preserving fluid as possible and place it in a dissecting tray.
   b. Carefully remove any adipose tissue from the surface of the specimen.

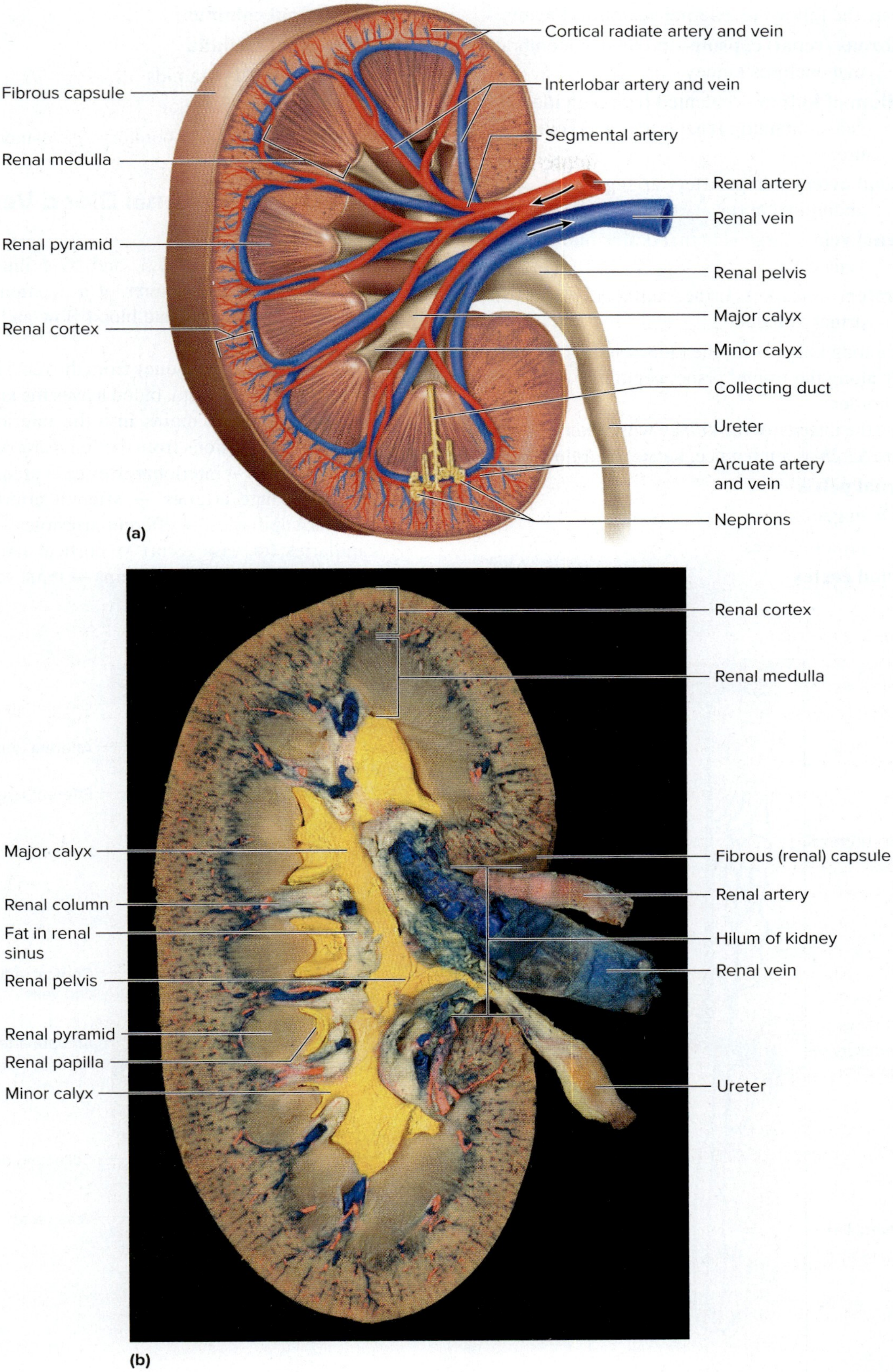

FIGURE 57.2 Frontal section of a kidney: (a) illustration; (b) pig kidney that has a triple injection of latex (*red* in the renal artery, *blue* in the renal vein, and *yellow* in the ureter and renal pelvis). ©J & J Photography APR

c. Locate the following external anatomical features:

fibrous (renal) capsule—protective membrane that encloses kidney

hilum of kidney—indented region on medial side containing renal artery and vein and ureter

renal artery—large artery arising from abdominal aorta

renal vein—large vein that drains into inferior vena cava

ureter—transports urine from renal pelvis to urinary bladder

d. Use a long knife to cut the kidney in half longitudinally along the frontal plane, beginning on the convex border.

e. Rinse the interior of the kidney with water, and using figure 57.2*b* as a reference, locate the following:

renal pelvis

- major calyces
- minor calyces

renal cortex

renal columns

renal medulla

- renal pyramids
- renal papillae

4. Complete Part A of Laboratory Assessment 57.

PROCEDURE B: Renal Blood Vessels and Nephrons

1. Study figures 57.2*a*, 57.3, and 57.4 illustrating renal blood vessels and structures of a nephron. The arrows within the figures indicate blood flow and tubular fluid flow.

2. Blood flows into the kidney from the renal artery arising off the abdominal aorta; blood leaves the kidney through the renal vein that drains into the inferior vena cava. Vascular subdivisions from the renal artery include segmental arteries → interlobar arteries → arcuate arteries → cortical radiate arteries → afferent arterioles → glomerular capillaries → efferent arterioles → peritubular capillaries (or vasa recta) → cortical radiate veins → arcuate veins → interlobar veins → renal vein. The renal

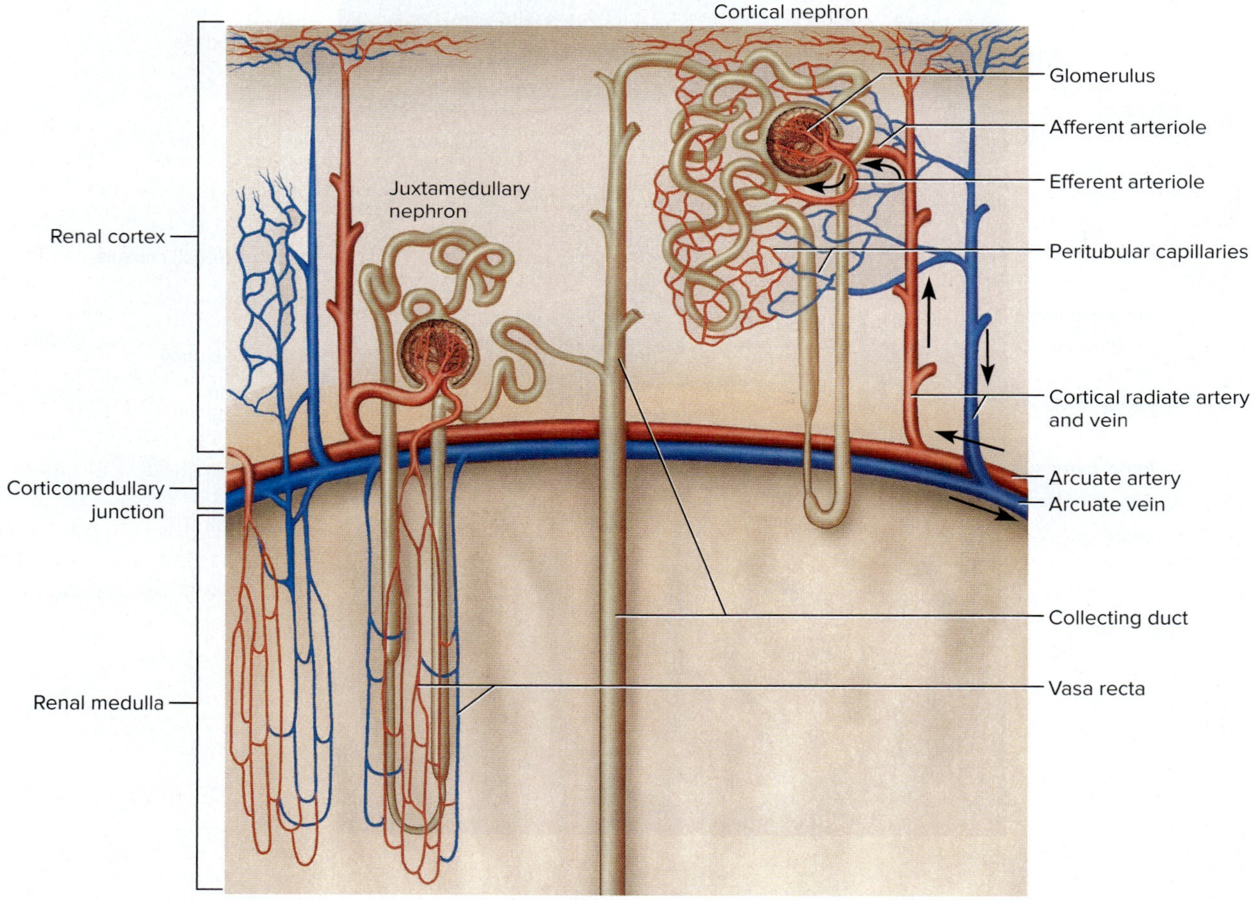

FIGURE 57.3 Renal blood vessels associated with cortical and juxtamedullary nephrons. The arrows indicate the flow of blood. Blood from the renal artery flows through a segmental artery and an interlobar artery to the arcuate artery. Blood from an arcuate vein flows through an interlobar vein to the renal vein. APR

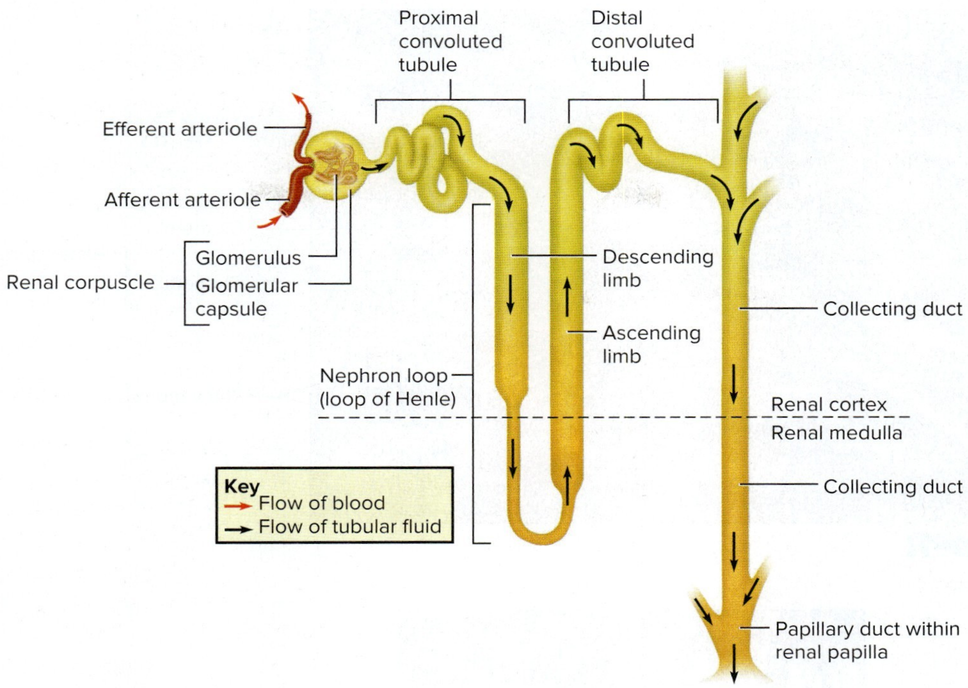

FIGURE 57.4 Structure of a nephron with arrows indicating the flow of tubular fluid. The corticomedullary junction is typical for a cortical nephron.

arteries and veins with comparable names are adjacent to each other within the kidney (figs. 57.2*a* and 57.3). The venous pathway of the kidney does not possess corresponding segmental veins.

3. Obtain a microscope slide of a kidney section and examine it using scanning and then low-power magnification. Locate the *renal capsule,* the *renal cortex* (which appears somewhat granular and may be more darkly stained than the other renal tissues), and the *renal medulla* (fig. 57.5).
4. Examine the renal cortex using high-power magnification. Locate a *renal corpuscle.* These structures appear as isolated circular areas. Identify the *glomerulus,* the capillary cluster inside the corpuscle where filtration of the blood occurs, and the *glomerular (Bowman's) capsule,* that receives the filtrate, appearing as a clear area surrounding the glomerulus. A glomerulus and a glomerular capsule compose a *renal corpuscle.* Also note the numerous sections of renal tubules that occupy the spaces between renal corpuscles (fig. 57.5*a*). The renal cortex contains the proximal and distal convoluted tubules that are involved in reabsorption and secretion.
5. Prepare a labeled sketch of a representative section of the renal cortex in Part B of the laboratory assessment.
6. Examine the renal medulla using high-power magnification. Identify longitudinal views of various collecting ducts and nephron loops. These ducts are lined with simple epithelial cells, which vary in shape from squamous to cuboidal (fig. 57.5*b*).
7. Prepare a labeled sketch of a representative section of the renal medulla in Part B of the laboratory assessment.
8. Complete Part C of the laboratory assessment.

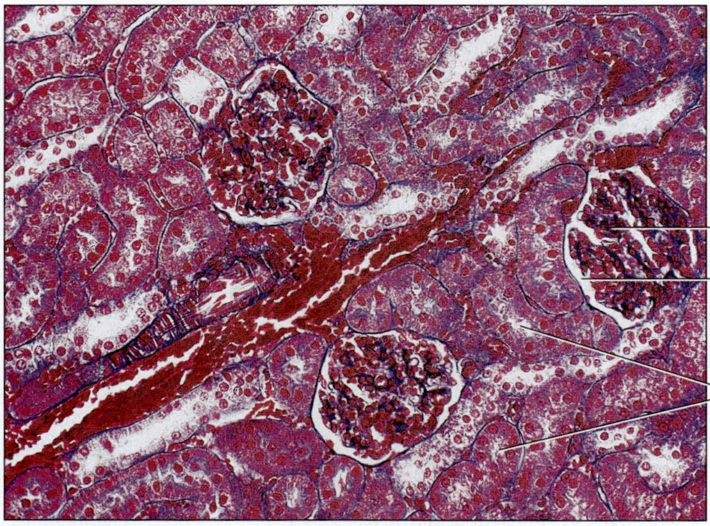

(a) APR

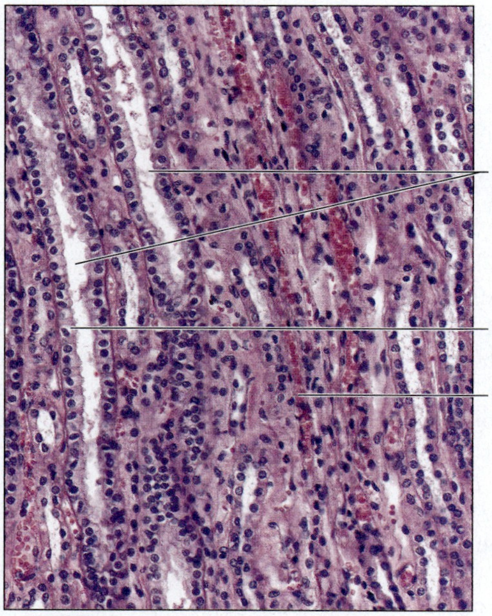

(b) APR

FIGURE 57.5 (a) Micrograph of a section of the renal cortex (220×). (b) Micrograph of a section of the renal medulla (80×). (*Note:* With about a million nephrons in a kidney, there are many different orientations visible in micrograph sections of renal cortex and renal medulla.) (a) Biophoto Associates/Science Source; (b) Al Telser/McGraw Hill

PROCEDURE C: Ureter, Urinary Bladder, and Urethra

The *ureters* are paired tubes about 25 cm (10 inches) long extending from the renal pelvis to the *ureteral openings (ureteric orifices)* of the *urinary bladder.* Peristaltic waves, created from rhythmic contractions of smooth muscle, occur about every 30 seconds, transporting the urine toward the urinary bladder. The urinary bladder possesses mucosal folds called *rugae* that enable it to distend for temporary urine storage. Coarse bundle layers of smooth muscle compose the somewhat broad *detrusor muscle.* The triangular floor of the bladder, the *trigone,* is bordered by the two ureteral openings and the opening into the urethra, which is surrounded by the *internal urethral sphincter.* The internal urethral sphincter is formed by a thickened region of the detrusor muscle (fig. 57.6).

Although the bladder can hold about 600 mL of urine, the desire to urinate from the pressure of a stretched bladder occurs when the bladder has about 150–200 mL of urine. During *micturition (urination)* the powerful detrusor muscle contracts, and the internal urethral sphincter is forced open. The *external urethral sphincter,* at the level of the urogenital diaphragm, is under somatic control. Because the external urethral sphincter is skeletal muscle, there is considerable voluntary control over the voiding of urine.

The *urethra* extends from the exit of the urinary bladder to the *external urethral orifice.* In the female, the urethra is about 4 cm (1.5 inches) long. In the male, the urethra includes passageways for both urine and semen and extends the length of the penis. The male urethra can be divided into three sections from most superior: prostatic urethra, membranous (intermediate) urethra, and spongy urethra (fig. 57.6*b*).

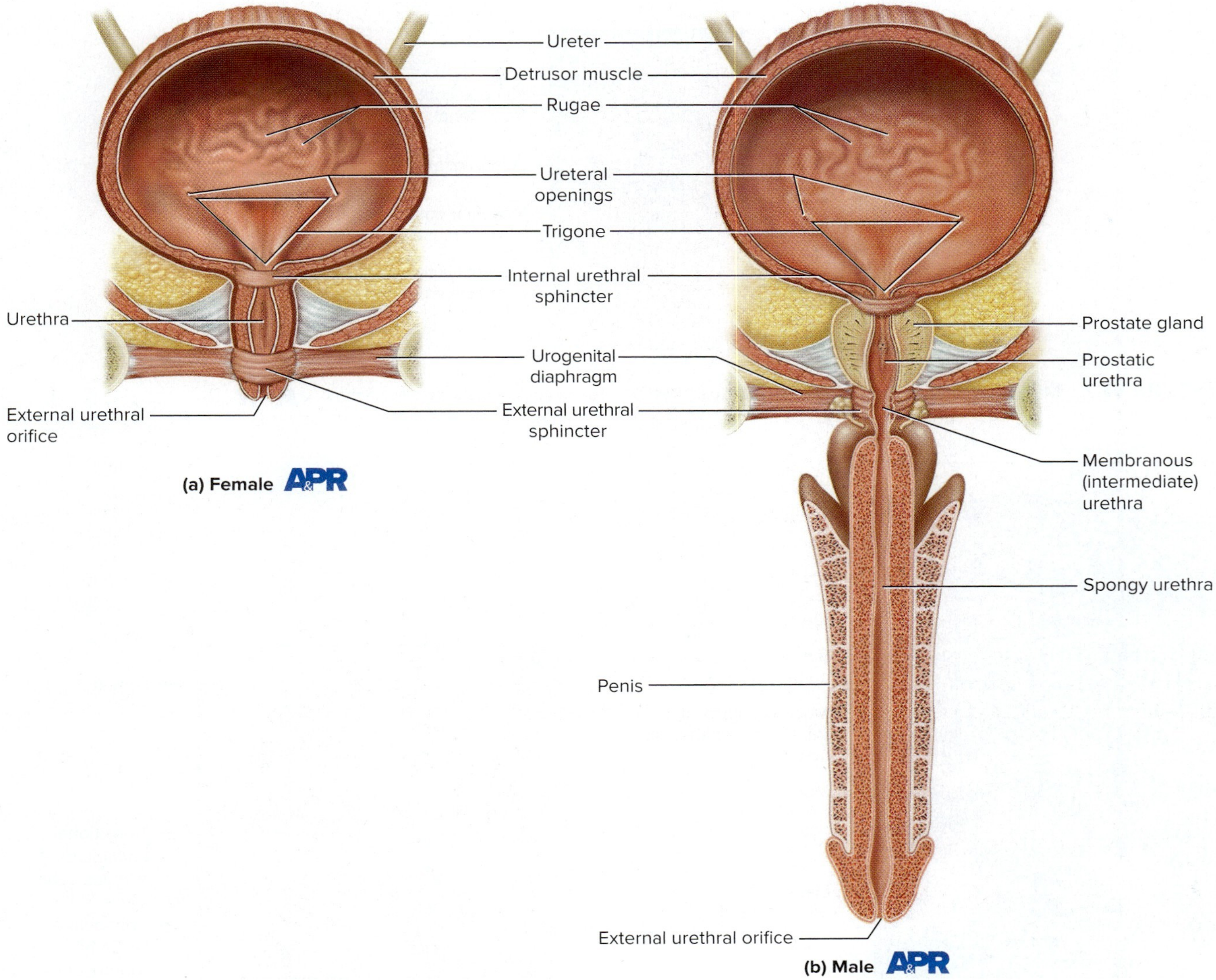

FIGURE 57.6 Ureters and frontal sections of urinary bladder and urethra of (a) female and (b) male.

A total length of about 20 cm (8 inches) would be characteristic for the urethra of a male.

1. Obtain a microscope slide of a cross section of a ureter and examine it using scanning and then low-power magnification. Locate the *mucous coat* (mucosa) next to the lumen. Examine the middle *muscular coat* (muscularis) composed of longitudinal and circular smooth muscle cells responsible for the peristaltic waves that propel urine from the kidneys to the urinary bladder. The outer *adventitia,* composed of connective tissue, secures the ureter in the retroperitoneal position (fig. 57.7).
2. Examine the mucous coat using high-power magnification. The specialized tissue is transitional epithelium, which allows changes in its thickness when unstretched and stretched.
3. Prepare a labeled sketch of a ureter in Part D of the laboratory assessment.
4. Obtain a microscope slide of a segment of the wall of a urinary bladder and examine it using scanning and then low-power magnification. Examine the *mucous coat* (mucosa) next to the lumen and the *submucous coat* (submucosa) composed of connective tissue just beneath the mucous coat. Examine the *muscular coat* (muscularis) composed of bundles of smooth muscle fibers interlaced in many directions. This thick, muscular layer is called the *detrusor muscle* and functions in the elimination of urine. Also note the outer *adventitia* of connective tissue (fig. 57.8).
5. Examine the mucous coat using high-power magnification. The tissue is transitional epithelium, which allows changes in its thickness from unstretched when the bladder is empty (fig. 57.8*b*) to stretched when the bladder distends with urine.

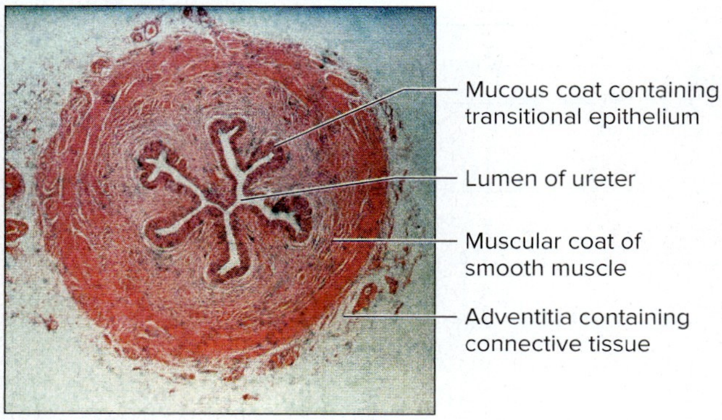

FIGURE 57.7 Micrograph of a cross section of a ureter (40×). Biophoto Associates/Science Source APR

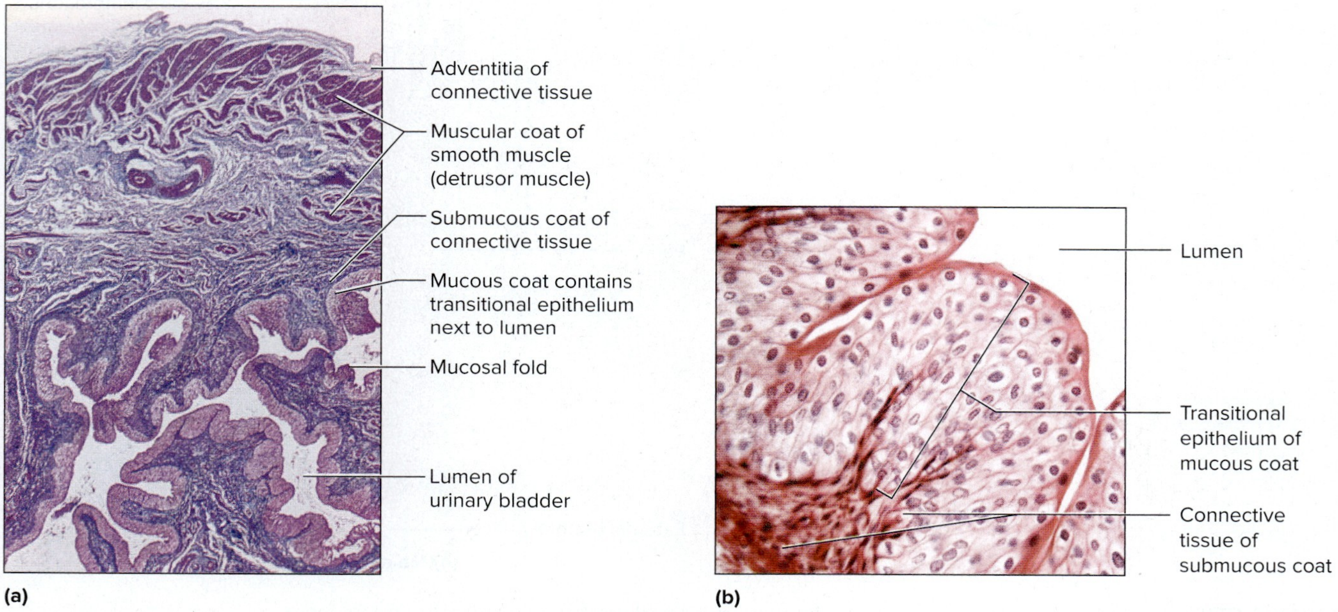

FIGURE 57.8 Micrograph of a segment of (a) the human urinary bladder wall (250×) and (b) the mucous coat and portion of submucosa coat (400×). (*Note:* A serous layer is on the superior surface of the urinary bladder wall instead of the adventitia as the outermost layer.) (a) Biophoto Associates/Science Source; (b) Dennis Strete APR

6. Prepare a labeled sketch of a segment of the urinary bladder wall in Part D of the laboratory assessment.
7. Obtain a microscope slide of a cross section of a urethra and examine it using scanning and then low-power magnification. Locate the muscular coat (muscularis) composed of smooth muscle fibers. Locate groups of mucous glands, called *urethral glands,* in the urethral wall that secrete protective mucus into the lumen of the urethra. Examine the mucous coat (mucosa) using high-power magnification. The specific type of epithelium found in the mucous membrane varies from transitional, to pseudostratified columnar, to stratified squamous epithelium between the urinary bladder and the external urethral orifice. Depending upon where the section of the urethra was taken for the preparation of the microscope slide, the type of epithelial tissue represented could vary (fig. 57.9).
8. Complete Part E of the laboratory assessment.

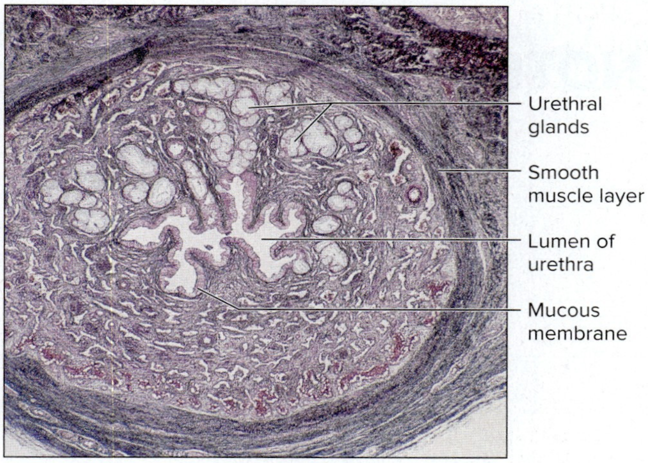

FIGURE 57.9 Cross section through the urethra (10×).
Ed Reschke/Photolibrary/Getty Images

NOTES

LABORATORY
ASSESSMENT 57

Name ______________________

Date ______________________

Section ______________________

The (A) corresponds to the indicated Learning Outcome(s) (O) found at the beginning of the Laboratory Exercise.

Urinary Organs

PART A: Assessments

1. Label the features indicated in figure 57.10.

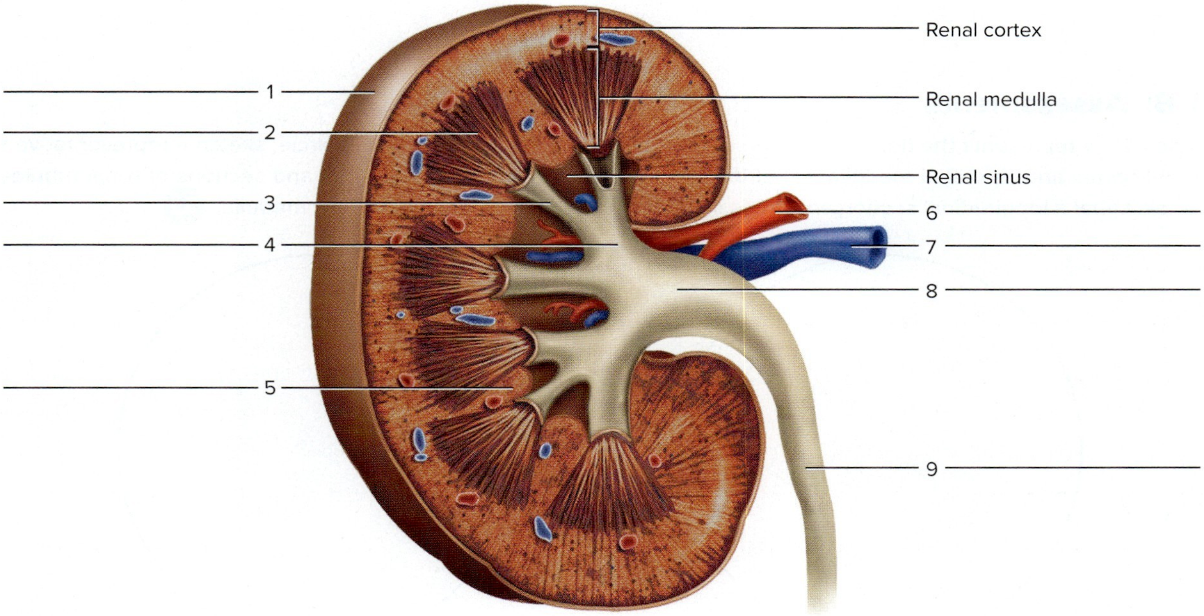

FIGURE 57.10 Label the structures in the frontal section of a kidney. (A1)

2. Match the terms in column A with the descriptions in column B. Place the letter of your choice in the space provided. **A1**

Column A		Column B
a. Calyces	______	**1.** Superficial region around the renal medulla
b. Hilum of kidney	______	**2.** Extensions of renal pelvis containing urine
c. Nephron	______	**3.** Conical mass of tissue within renal medulla
d. Renal column	______	**4.** Projection with tiny openings into a minor calyx
e. Renal cortex	______	**5.** Hollow chamber within kidney
f. Renal papilla	______	**6.** Microscopic functional subunit of kidney
g. Renal pelvis	______	**7.** Located between renal pyramids
h. Renal pyramid	______	**8.** Superior, funnel-shaped sac at end of ureter inside the renal sinus
i. Renal sinus	______	**9.** Medial depression for blood vessels and ureter to enter kidney chamber

PART B: Assessments

Each circle below represents the field of view as seen through the microscope. In each circle, sketch a representative section of the renal cortex and the renal medulla. Label the glomerulus, the glomerular capsule, and sections of renal tubules in the renal cortex. Label a longitudinal section and cross section of a collecting duct in the renal medulla. **A2**

Renal cortex (______×)

Renal medulla (______×)

PART C: Assessments

Complete the following:

1. Distinguish between a renal corpuscle and a renal tubule. **A2** ______________________________

____

____

2. Label the parts of a nephron, some associated blood vessels, and the collecting duct indicated in figure 57.11.

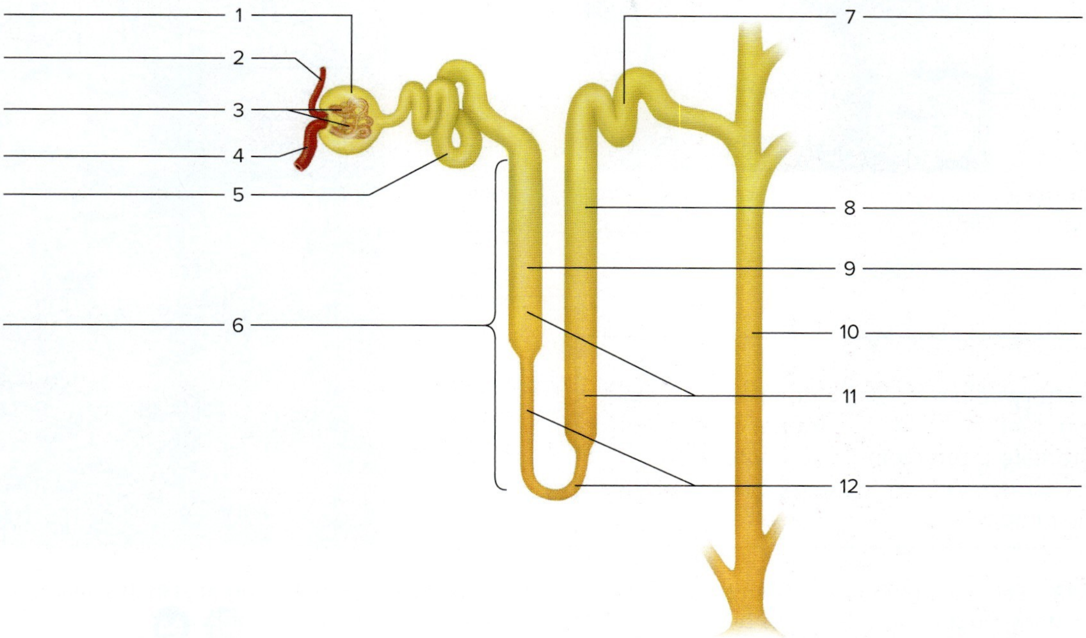

FIGURE 57.11 Label the parts of a nephron, including some associated blood vessels and collecting duct in the illustration. **A1** **A3**

3. Number the following structures to indicate their respective positions in relation to the nephron. Assign the number 1 to the structure nearest the glomerulus. **A3**

______ Ascending limb of nephron loop

______ Collecting duct

______ Descending limb of nephron loop

______ Distal convoluted tubule

______ Glomerular capsule

______ Proximal convoluted tubule

______ Papillary duct in renal papilla

4. Label the features indicated in figure 57.12.

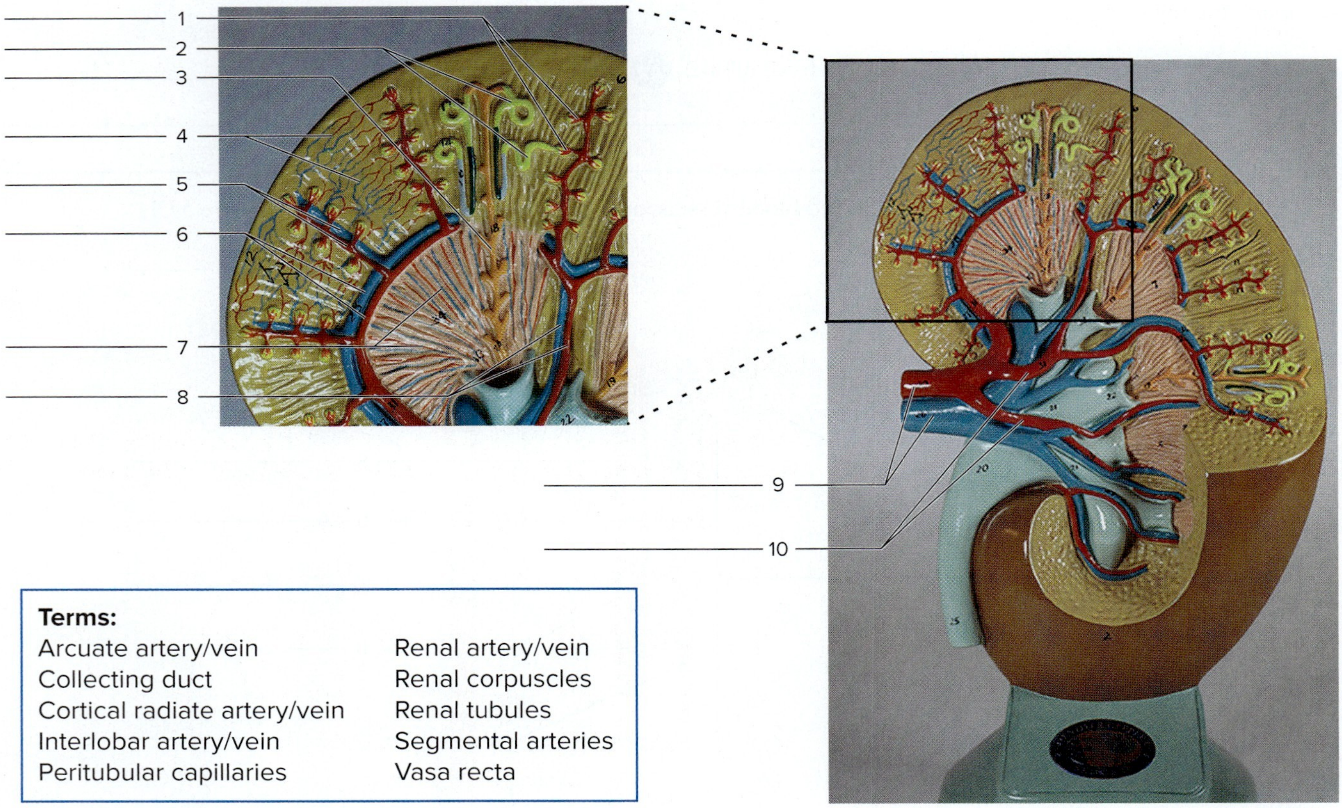

Terms:

Arcuate artery/vein	Renal artery/vein
Collecting duct	Renal corpuscles
Cortical radiate artery/vein	Renal tubules
Interlobar artery/vein	Segmental arteries
Peritubular capillaries	Vasa recta

FIGURE 57.12 Label the blood vessels and general features of the nephron and collecting duct in this image of a frontal section of a kidney model, using the terms provided. ©2017 Denoyer-Geppert Science Company A1 A4

5. Number the following structures to indicate their respective positions in the blood pathway within the kidney. Assign the number 1 to the vessel nearest the abdominal aorta. A4

_______ Afferent arteriole

_______ Arcuate artery

_______ Arcuate vein

_______ Cortical radiate artery

_______ Cortical radiate vein

_______ Efferent arteriole

_______ Glomerulus

_______ Interlobar artery

_______ Interlobar vein

_______ Peritubular capillary (or vasa recta)

_______ Renal artery

_______ Renal vein

_______ Segmental artery

CRITICAL THINKING ASSESSMENT—PART C

Which blood vessel, the afferent arteriole *or* the efferent arteriole, has the larger diameter? A1 A2 A4

______________________________ What is the functional significance of the diameter difference between these two blood vessels?______________________________

____

______________________________.

PART D: Assessments

Each circle below represents the field of view as seen through the microscope. In each circle, sketch a cross section of a ureter and label the three layers and the lumen. Sketch a segment of a urinary bladder and label the four layers and the lumen. A5

Ureter (______×)

Urinary bladder (______×)

PART E: Assessments

1. Number the following structures to indicate respective positions in the pathway of urine flow in a male. Assign the number 1 to the structure nearest the papillary duct in a renal papilla. A6

   ______ External urethral orifice

   ______ Major calyx

   ______ Membranous urethra

   ______ Minor calyx

   ______ Prostatic urethra

   ______ Renal pelvis

   ______ Spongy urethra

   ______ Ureter

   ______ Ureteral opening

   ______ Urinary bladder

2. Label the features indicated in figure 57.13.

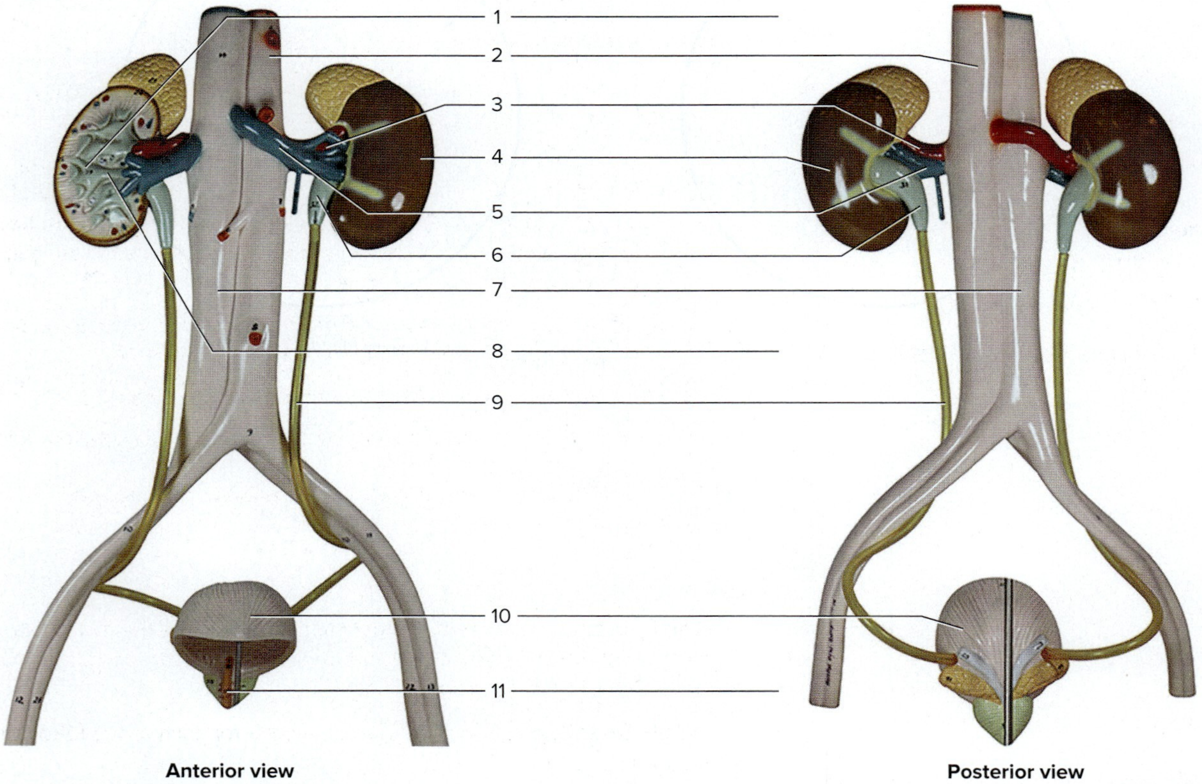

FIGURE 57.13 Label the major organs and associated major blood vessels of the male urinary system in the images of a model. ©2017 Denoyer-Geppert Science Company A1 A6

LABORATORY
EXERCISE 58

Urinalysis

Purpose of the Exercise

To perform observations and tests commonly used to determine the characteristics and composition of urine and to analyze the clinical implications of abnormal results.

MATERIALS NEEDED

Normal and abnormal simulated urine specimens can be substituted for collected urine.
Disposable urine-collecting container
Paper towel
Urinometer cylinder (jar)
Urinometer hydrometer
Laboratory thermometer
Disposable gloves

Reagent strips (individual or combination test strips such as Chemstrip or Multistix) to test for the presence of the following:

Glucose
Protein
Ketones
Specific gravity
pH
Urobilinogen
Nitrites
Leukocytes
Bilirubin
Hemoglobin/occult blood

Compound light microscope
Microscope slide
Coverslip
Centrifuge
Centrifuge tube
Graduated cylinder, 10 mL
Medicine dropper
Sedi-stain

SAFETY

- Review all the safety guidelines in Appendix 1 of your laboratory manual.
- Consider using normal and abnormal simulated urine samples available from various laboratory supply houses or use the spiked urine recipe located in the instructor's manual.
- Use protective eyewear, laboratory coats, and disposable gloves when working with body fluids.
- Work only with your own urine sample.
- Use an appropriate disinfectant to wash the laboratory table before and after the procedures.
- Place glassware in a disinfectant when finished.
- Dispose of contaminated items as directed by your laboratory instructor.
- Wash your hands before leaving the laboratory.

Learning Outcomes

After completing this exercise, you should be able to:

O1 Evaluate the physical characteristics of a urine sample: color, transparency, odor, and specific gravity.

O2 Test, record, and interpret results of the urine sample, using individual and/or combination test strips, for the presence of substances such as: specific gravity, pH, glucose, protein, ketones, bilirubin, urobilinogen, hemoglobin/occult blood, nitrites, and leukocytes.

O3 Analyze the constituents of the urine sample, and provide reasonable clinical implications for abnormal results.

O4 Perform a microscopic study of urine sediment.

O5 Summarize the results of the microscopic sediment observations and tests.

The **O** corresponds to the assessments **A** indicated in the Laboratory Assessment for this Exercise.

Pre-Lab

Carefully read the introductory material and examine the entire lab. Be familiar with normal urine components from lecture or the textbook. Answer the pre-lab questions.

Pre-Lab Questions Select the correct answer for each of the following questions:

1. Urine consists of about _____ percent water and _____ percent solutes.
a. 100; 0 **b.** 95; 5
c. 90; 10 **d.** 50; 50

2. A urinalysis consists of all of the following *except*
a. physical characteristics. **b.** chemical analysis.
c. microscopic examination. **d.** body temperature analysis.

3. The normal color of urine is
a. red to dark brown. **b.** green.
c. pink. **d.** light yellow to amber.

Chapter Opener: ©101cats/Getty Images

4. Microscopic solids are stained with _____ in order to make identifications possible.
   a. reagents
   b. iodine
   c. methylene blue
   d. Sedi-stain
5. Which of the following would be the most typical urinary output in a day?
   a. 0.5 liters
   b. 0.8 liters
   c. 1.2 liters
   d. 2.2 liters
6. Which of the following would represent a normal pH range of urine?
   a. 4.6–8.0
   b. 0–14
   c. 7.35–7.45
   d. 5–10
7. Red blood cells are a normal sediment found in urine.
   a. True
   b. False
8. A urine sample with a high specific gravity may indicate dehydration.
   a. True
   b. False

Urine is the product of three processes of the nephrons within the kidneys: *glomerular filtration, tubular reabsorption,* and *tubular secretion.* As a result of these processes, various waste substances are removed from the blood, and body fluid and electrolyte balance are maintained. Consequently, the composition of urine varies considerably because of differences in dietary intake and physical activity from day to day and person to person. The normal urinary output ranges from 1.0 to 1.8 L per day. Typically, urine consists of 95% water and 5% solutes. The volume of urine produced by the kidneys varies with such factors as fluid intake, environmental temperature, relative humidity, respiratory rate, and body temperature.

An analysis of urine composition and volume often is used to evaluate the functions of the kidneys and other organs. This procedure, called *urinalysis,* is a clinical assessment and a diagnostic tool for certain pathological conditions and general overall health. A urinalysis and a complete blood analysis complement each other for an evaluation of certain diseases, such as diabetes mellitus, and general health.

A urinalysis involves three aspects: *physical characteristics, chemical analysis,* and a *microscopic examination.* Physical characteristics of urine that are noted include volume, color, transparency, and odor. The chemical analysis of solutes in urine addresses urea and other nitrogenous wastes; electrolytes; pigments; and possible glucose, protein, ketones, bilirubin, urobilinogen, and hemoglobin. The specific gravity and pH of urine are greatly influenced by the components and amounts of solutes. An increase in the viscosity of blood due to dehydration, polycythemia, kidney inflammation, or other causes will result in an increase in urine specific gravity. A diet high in protein and whole wheat products lowers urine pH, while bacterial infection may increase urine pH; thus diet and pathological conditions have an effect on urinalysis. An examination of microscopic solids, including cells, casts, and crystals, assists in the diagnosis of injury, various diseases, and urinary infections.

WARNING

While performing the following tests, you should wear disposable gloves and protective eyewear so that skin and eye contact with urine is avoided. Observe all safety procedures listed for this lab. (Normal and abnormal simulated urine specimens can be used instead of real urine for this lab.)

PROCEDURE A: Physical and Chemical Analysis

1. Proceed to the restroom with a clean, disposable container if collected urine is to be used for this lab exercise. The first small volume of urine should not be collected because it contains abnormally high levels of microorganisms from the urethra. Collect a midstream sample of about 50 mL of urine. The best collections are the first specimen in the morning or one taken 3 hours after a meal. Refrigerate samples if they are not used immediately.
2. Place a sample of urine in a clean, transparent container. Describe the *color* of the urine. Normal urine varies from light yellow to amber, depending on the presence of urochromes, end-product pigments produced during the decomposition of hemoglobin.

   Abnormal urine colors include yellow-brown or green, due to elevated concentrations of bile pigments, and red to dark brown, due to the presence of blood. Certain foods, such as beets or carrots, and various drug substances also can cause color changes in urine, but in such cases the colors have no clinical significance. Enter the results of this and the following tests in Part A of Laboratory Assessment 58.
3. Evaluate the *transparency* of the urine sample (judge whether the urine is clear, slightly cloudy, or very cloudy). Normal urine is clear enough to see through. You can read newsprint through slightly cloudy urine; you can no longer read newsprint through cloudy urine. Cloudy urine indicates the presence of various substances, including mucus, bacteria, epithelial cells, fat droplets, and inorganic salts.
4. Gently wave your hand over the urine sample toward your nose to detect the *odor.* Normal urine should have a slight ammonia-like odor due to the nitrogenous wastes. Some vegetables (like asparagus), drugs (like certain vitamins), and diseases (like diabetes mellitus) will influence the odor of urine.
5. Determine the *specific gravity* of the urine sample, which indicates the solute concentration. Specific gravity is the ratio of the weight of something to the weight of an equal volume of pure water. For example, mercury (at 15°C) weighs 13.6 times as much as an equal volume of water; thus, it has a specific gravity of 13.6. Although urine is mostly water, it has substances dissolved in it and is slightly heavier than an equal volume of water. The specific gravity of pure (distilled) water is 1.000, while the specific gravity of urine is higher, varying from 1.003 to 1.035 under normal circumstances. If the specific gravity

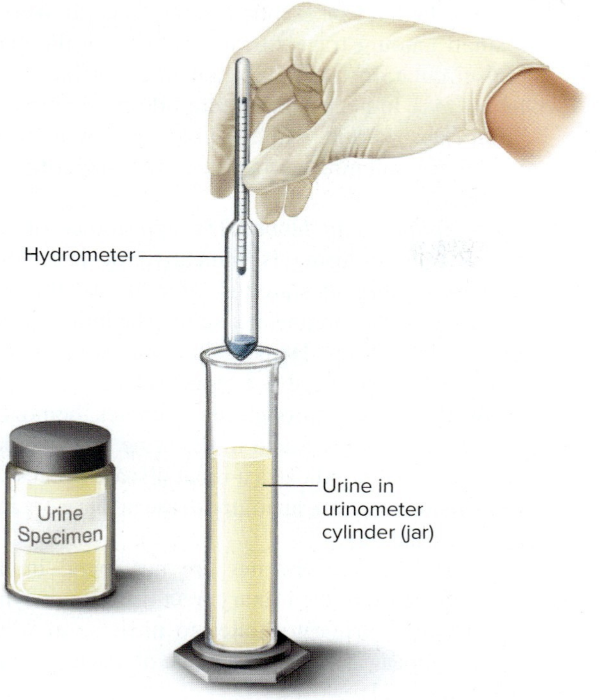

FIGURE 58.1 Float the hydrometer in the urine, making sure that it does not touch the sides or the bottom of the cylinder (jar).

is too low, the urine contains few solutes and represents dilute urine, a likely result of excessive fluid intake or the use of diuretics. A specific gravity above the normal range represents a higher concentration of solutes, likely from a limited fluid intake. Concentrated urine over an extended time increases the risk of the formation of kidney stones (renal calculi). Kidney stones may cause blockage that results in a backup of urine, creating severe pain that may radiate in the lower back (flank region) and/or in the abdominal area. If the stones do not pass on their own, the most common treatment is a noninvasive lithotripsy procedure where high-energy shock (sound) waves are used to break the stones into pieces small enough that they can pass in urine.

To determine the specific gravity of a urine sample, follow these steps:

a. Your instructor may have you use a refractometer to determine specific gravity; however, in this lab procedure a urinometer is used. Pour enough urine into a clean urinometer cylinder to fill it about three-fourths full. Any foam that appears should be removed with a paper towel.

b. Use a laboratory thermometer to measure the temperature of the urine.

c. Gently place the urinometer hydrometer into the urine, *and make sure that the float is not touching the sides or the bottom of the cylinder* (fig. 58.1).

d. Position your eye at the level of the urine surface. Determine which line on the stem of the hydrometer intersects the lowest level of the concave surface (meniscus) of the urine.

e. Liquids tend to contract and become denser as they are cooled, or to expand and become less dense as they are heated, so it may be necessary to make a temperature correction to obtain an accurate specific gravity measurement. To do this, add 0.001 to the hydrometer reading for each 3 degrees of urine temperature above 25°C or subtract 0.001 for each 3 degrees below 25°C. Enter this calculated value in the table in Part A of the laboratory assessment as the test result.

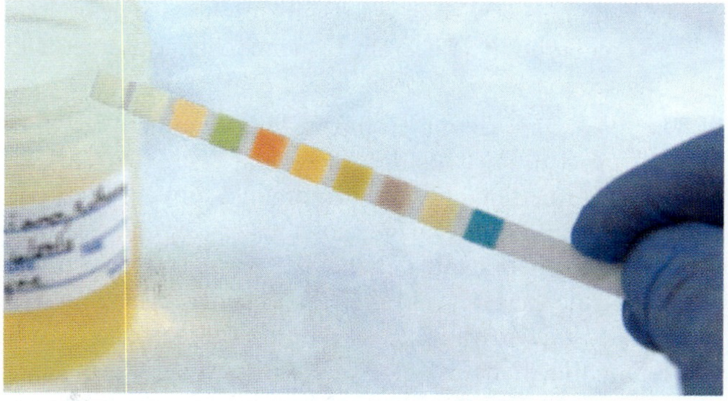

FIGURE 58.2 An example of a reagant test strip dipped into urine to determine a variety of urine components. McGraw Hill

6. Individual or combination reagent strips can be used to perform a variety of urine tests (fig. 58.2). In each case, directions for using the strips are found on the strip container. *Be sure to read them.*

   To perform each test, follow these steps:

   a. Obtain a urine sample and the proper reagent test strip.

   b. Read the directions on the strip container.

   c. Dip the strip in the urine sample as directed on the container.

   d. Remove the strip at an angle and let it touch the inside rim of the urine container to remove any excess liquid.

   e. Wait for the length of time indicated by the directions on the container before you compare the color of the test strip with the standard color scale provided with the container. (Do not touch the strip to the container when comparing color results.) The value or amount represented by the matching color should be used as the test result.

7. Record your urine test results in Part A of the laboratory assessment.
8. If your instructor provides abnormal simulated urine, follow the steps 6a–e to test that urine sample. Then, record your results in Part A of the laboratory assessment.
9. Your urine sample might have been tested using individual test strips, combination test strips, or some from each. Use the following information about any of the tests you conducted to help you interpret the findings

of your sample. (Your test results might have been done and recorded in a different sequence than those listed in this section.)

- *Specific gravity:* Compare your results of the specific gravity using a reagent test strip to that obtained using the urinometer. Given that you used the same urine, your results should be similar. Uncontrolled diabetes mellitus will result in urine that has a higher than normal specific gravity due to the excess solutes (glucose, ketones), while diabetes insipidus will result in urine that has a lower than normal specific gravity due to the diluted solutes.
- *pH test:* The pH of normal urine varies from 4.6 to 8.0, but most commonly, it is near 6.0 (slightly acidic). A vegetarian diet may make the urine pH more alkaline, while a diet high in protein may make urine pH more acidic. Significant daily variations within the broad normal range are a result of dietary excesses of certain foods and liquids consumed.
- *Glucose:* Glucose is normally absent in urine, as it is completely reabsorbed from the renal tubules into circulation. However, glucose may appear in the urine temporarily following a meal high in carbohydrates. Glucose in the urine (called *glucosuria*) is a sign of uncontrolled diabetes mellitus.
- *Protein:* Normally, proteins are not present in urine. However, those of a small molecular size, such as albumins, may appear in trace amounts, particularly after strenuous exercise. Increased amounts of proteins may appear as a result of uncontrolled diabetes mellitus, kidney disease in which the glomeruli are damaged, or high blood pressure (hypertension).
- *Ketones:* Ketones are organic acids that are products of fat catabolism. They normally only appear in urine in very small amounts, if at all. When ketone production exceeds tissue use, *ketoacidosis* results. The most common cause of ketoacidosis is uncontrolled diabetes mellitus—due to insulin deficiency, the cells cannot obtain carbohydrates so they rely on fatty acids for fuels. Fasting or starvation may also cause ketoacidosis.
- *Bilirubin:* Bilirubin, which results from hemoglobin decomposition in the liver, normally is absent in urine. Urochrome, a normal yellow component of urine, is a result of additional breakdown of bilirubin. Elevated bilirubin levels, the result of liver disorders that cause obstruction of the biliary tract and excessive red blood cell destruction (*hemolytic anemia*), make the urine appear dark because of the bilirubin in the urine.
- *Urobilinogen:* Urobilinogen is a colorless breakdown product of bilirubin from the action of intestinal bacteria. Urobilinogen is either reabsorbed into the portal circulation to be re-excreted into the bile by the liver or excreted in the feces; however, a small amount travels to the kidneys in the general circulation to be excreted in the urine. Low levels may indicate inadequate bile production or biliary obstruction, while high levels may indicate hemolytic anemia with infectious hepatitis or cirrhosis.
- *Hemoglobin/occult blood:* The appearance of red blood cells in urine is indicative of pathology because erythrocytes are too large to pass through the filtration membrane; it also may be indicative of menses in females. Hemoglobin resides in the red blood cells, and since red blood cells are not normally able to pass through renal tubules, hemoglobin is not normally found in the urine. Its presence in urine usually indicates a renal disease process, a transfusion reaction, an injury to the urinary organs, or menstrual blood.
- *Nitrites:* Normally, nitrites are not found in the urine. Bacteria in the urinary tract make an enzyme that changes urinary nitrates to nitrites, so when nitrites appear in urine, it is a sign of a urinary tract infection (UTI).
- *Leukocytes:* White blood cells in the urine are indicative of an infection of the urinary tract, urinary bladder, or kidney; normally they are not present.

The clinical significance or conclusive diagnosis of a disease is often difficult to achieve. Most often, several blood and urine characteristics along with other symptoms are assessed collectively in order to make a determination of the abnormality and its potential cause. Any test that is showing abnormal results should be validated by a health care professional.

10. Complete Part A of the laboratory assessment.

PROCEDURE B: Microscopic Sediment Analysis

1. A urinalysis usually includes an analysis of urine sediment—the microscopic solids present in a urine sample. This sediment normally includes mucus, certain crystals, and a variety of cells, such as the epithelial cells that line the urinary tubes and an occasional white blood cell. Some crystals in urine are abnormal; for example, cholesterol crystals may be indicative of renal tubular disease. Other types of solids, such as casts, white blood cells, or red blood cells, may indicate a disease or injury if they are present in excess. (Casts are cylindrical masses of cells or other substances that form in the renal tubules and are flushed out by the flow of urine.) The presence of the parasites trichomonas, yeast, or candida may be indicative of a urinary tract or vaginal infection.

To observe urine sediment, follow these steps:

a. Thoroughly stir or shake your urine sample to suspend the sediment, which tends to settle to the bottom of the container.

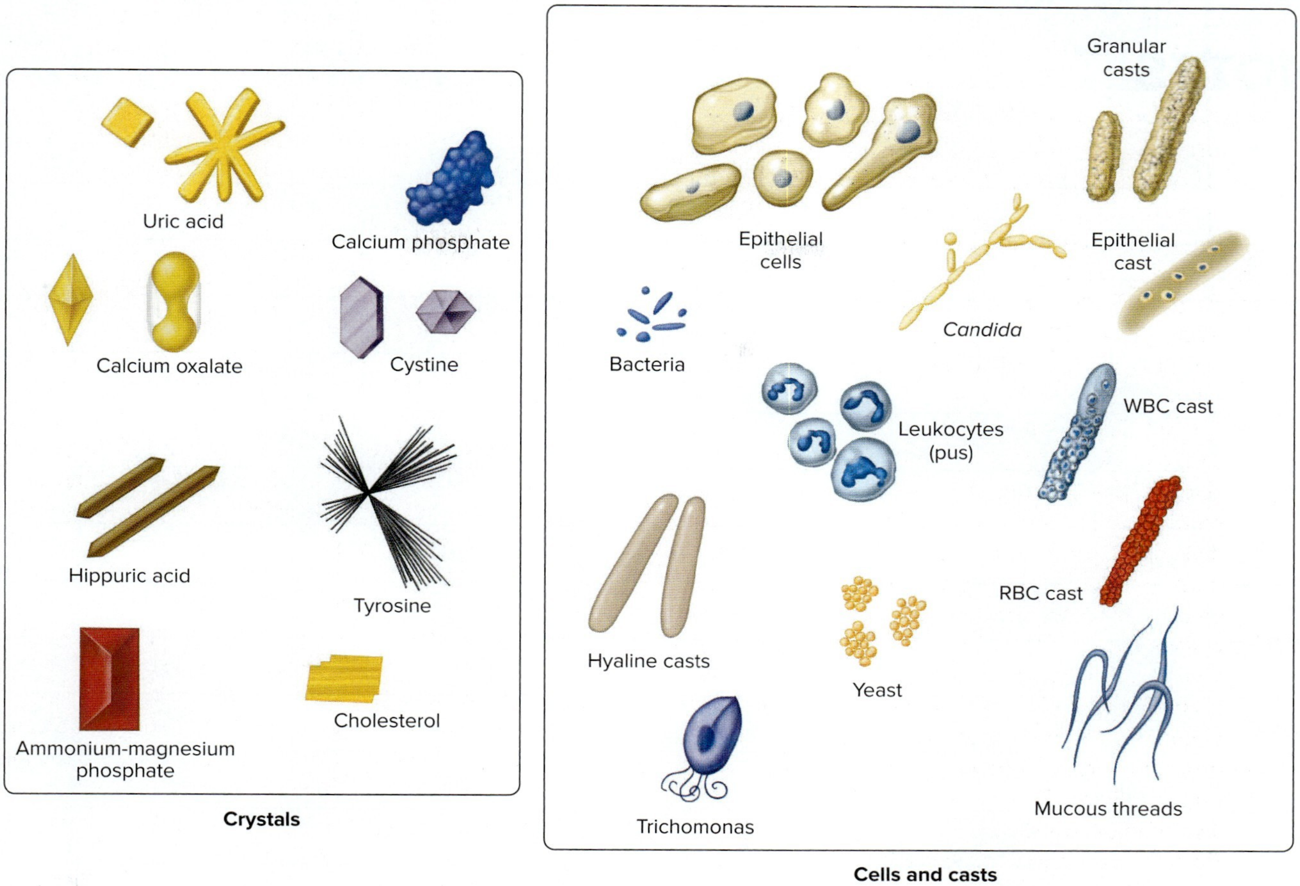

FIGURE 58.3 Types of urine sediment. Healthy individuals lack many of these sediments and possess only occasional to trace amounts of others. (*Note:* Shades of white to purple sediments are most characteristic when using Sedi-stain.)

b. Pour 10 mL of urine into a clean centrifuge tube and centrifuge it for 5 minutes at slow speed (1,500 rpm). Be sure to balance the centrifuge with an even number of tubes filled to the same levels.

c. Carefully decant 9 mL (leave 1 mL) of the liquid from the sediment in the bottom of the centrifuge tube, as directed by your laboratory instructor. Resuspend the 1 mL of sediment.

d. Use a medicine dropper to remove some of the sediment and place it on a clean microscope slide.

e. Add a drop of Sedi-stain to the sample, and add a coverslip.

f. Examine the sediment with scanning, low-power (reduce the light when using low-power), and then high-power magnifications.

g. With the aid of figure 58.3, identify the types of solids present.

2. In Part B of the laboratory assessment, make a sketch of each type of sediment that you observed.

3. Complete Part B of the laboratory assessment.

NOTES

LABORATORY
ASSESSMENT 58

Urinalysis

Name ______________________

Date ______________________

Section ______________________

The (A) corresponds to the indicated Learning Outcome(s) (O) found at the beginning of the Laboratory Exercise.

PART A: Assessments

1. Enter your observations, test results, and evaluations/clinical implications in the following table: A1 A2 A3 A5

Urine Characteristics	Normal Values	Observations and Test Results of Student or Normal Simulated Urine	Observations and Test Results of Abnormal Simulated Urine	Evaluations/ Clinical Implications
Color	Light yellow to amber			
Transparency	Clear			
Odor	Slight ammonia-like			
Specific gravity (corrected for temperature)	1.003–1.035			
pH	4.6–8.0			
Glucose	Negative (0)			
Protein	Negative to trace			
Ketones	Negative (0)			
Bilirubin	Negative (0)			
Urobilinogen	0.1–1.0 mg/dL			
Hemoglobin/occult blood	Negative (0)			
Nitrites	Negative (0)			
Leukocytes	Negative (0)			

2. Summarize the results and clinical implications of the chemical analysis of urine. A5

CRITICAL THINKING ASSESSMENTS—PART A

1. Predict the results of a urinalysis that was conducted after the urine sample sat at room temperature longer than 24 hours. A1 A3

2. Joey complained to his physician that he was constantly thirsty, hungry, and had to urinate frequently. His physician ordered a blood and urine test. The results of his urinalysis showed the following abnormal values:

- pH: 3.5
- glucose: 250 mg/100 mL
- ketones: 10 gm/100 mL
- albumin: 30 mg/100 mL
- specific gravity: 1.045

Given Joey's symptoms and urinalysis results, what do you predict is his diagnosis? Explain and provide reasoning. A3

PART B: Assessments

1. Make a sketch for each type of sediment you observed. Label any from those shown in figure 58.3. A4

2. Summarize the results of the microscopic sediment analysis of urine. A5

____

____

____

____

CRITICAL THINKING ASSESSMENT—PART B

A portion of the glomerular capsule and the fenestrated glomerular capillaries form a delicate filtration membrane that allows water and small molecules through, while excluding other larger-sized molecules and cells. A young patient has been diagnosed with chronic glomerulonephritis, an autoimmune disease, after visiting a doctor because of a noticeable pinkish color to their urine. A1 A2 A3 A5

Evaluate the physical characteristics that would likely be observed in this patient's urine sample.

____

What is causing the abnormal urine color, and what is the name given to this condition?

____

After doing a chem strip test, what other constituent would likely be found in the urinalysis? Explain.

____

____

In the microscopic sediment analysis of this urine sample, what might be observed?

____

NOTES

LABORATORY
EXERCISE 59

Male Reproductive System

Purpose of the Exercise

To examine the microscopic and gross anatomy of the male reproductive organs and to describe their basic physiological functions.

MATERIALS NEEDED

Human torso model
Model of the male reproductive system
Anatomical chart of the male reproductive system
Compound light microscope
Prepared microscope slides of the following:
- Testis section
- Epididymis, cross section
- Ductus deferens, cross section
- Penis, cross section

Learning Outcomes APR

After completing this exercise, you should be able to:

01 Locate and identify the organs of the male reproductive system.
02 Describe the structures and functions of the male reproductive organs.
03 Sketch and label the major features of microscopic sections of the testis, epididymis, ductus deferens, and penis.

The O corresponds to the assessments A indicated in the Laboratory Assessment for this Exercise.

Pre-Lab

Carefully read the introductory material and examine the entire lab. Be familiar with the structures and functions of the male reproductive organs from lecture or the textbook. Answer the pre-lab questions.

Pre-Lab Questions Select the correct answer for each of the following questions:

1. Sperm are produced in the
   a. seminiferous tubules. **b.** interstitial cells.
   c. epididymis. **d.** ductus deferens.
2. The ductus deferens is located within the
   a. testis. **b.** epididymis.
   c. spermatic cord. **d.** penis.
3. The _____ is the common tube for the passage of urine and semen.
   a. ductus deferens **b.** urethra
   c. ureter **d.** ejaculatory duct
4. Which of the following would *not* be visible in a cross (transverse) section of a penis?
   a. urethra **b.** corpus spongiosum
   c. corpora cavernosa **d.** epididymis
5. The interstitial cells produce
   a. sperm. **b.** testosterone.
   c. nutrients. **d.** semen.
6. Semen is composed of sperm and secretions from all the following *except* the
   a. prostate. **b.** seminal vesicle.
   c. bulbourethral gland. **d.** epididymis.
7. The following glands are paired *except* the
   a. prostate. **b.** testis.
   c. seminal vesicle. **d.** bulbourethral.
8. The smooth muscle represents the thickest layer of the wall of the ductus deferens.
   a. True **b.** False
9. The rete testis connects directly to the tail of the epididymis.
   a. True **b.** False

The organs of the male reproductive system are specialized to produce and maintain the male sex cells, to transport these cells together with supporting fluids to the female reproductive tract, and to produce and secrete male sex hormones. These organs include sets of internal and external genitalia.

The *testes* originate within the abdominopelvic cavity, but they descend through an inguinal canal prior to birth to a position in the *scrotum,* which provides a lower temperature necessary for sperm production and storage. Within the testes, numerous *seminiferous tubules* produce *sperm* (*spermatozoa*) by a process called *spermatogenesis (meiosis),* and *interstitial cells* between seminiferous tubules produce testosterone that is released into the bloodstream for transport to target organs. Testosterone influences the sex drive and development of the secondary sex characteristics. The process of spermatogenesis is described in Laboratory Exercise 61.

The sperm produced in the seminiferous tubules enter a tubular network, the *rete testis,* which joins the tightly coiled *epididymis* through *efferent ductules.* Within the head, body, and tail of the epididymis, immature sperm are stored and nourished during their maturation. A muscular *ductus deferens (vas deferens),* located within the *spermatic cord,* passes through the inguinal canal into the pelvic cavity and posterior to the urinary bladder, where it unites with ducts from the *seminal vesicles* to form the ejaculatory ducts. An alkaline fluid containing nutrients and prostaglandins is secreted by the seminal vesicles. The ejaculatory ducts unite with the urethra within the *prostate gland.* Additional secretions are added from the prostate gland and two *bulbourethral glands.* The combination of sperm and the secretions from the accessory glands is *semen.*

The urethra, a common tube to convey semen and urine, passes through the *penis* within the *corpus spongiosum* to the external urethral orifice. The corpus spongiosum plus two *corpora cavernosa* provide vascular spaces that become engorged with blood during an erection. A loose fold of skin, forming a cuff over the glans penis, is called the *prepuce,* and is frequently removed in a procedure called a circumcision shortly after birth.

PROCEDURE A: Male Reproductive Organs APR

1. Study figures 59.1 and 59.2 and the list of structures and descriptions.
2. Observe the human torso model, the model of the male reproductive system, and the anatomical chart of the male reproductive system. Locate the following features:

   testes—paired organs suspended by spermatic cord within scrotum
   - lobules—subdivisions of testes
   - seminiferous tubules—long, tiny coiled tubes; site of spermatogenesis
   - interstitial cells (Leydig cells)—endocrine cells between seminiferous tubules; produce and secrete testosterone
   - rete testis—network of channels from seminiferous tubules to efferent ductules
   - efferent ductules—several ducts from rete testis to epididymis

   inguinal canal—tubelike passageway between abdominopelvic cavity and scrotum; contains the spermatic cord

   spermatic cord—contains ductus deferens, blood vessels, nerves, and lymphatic vessels

   epididymis—site of sperm maturation and storage; tube about 6 m (18 feet) if uncoiled

   ductus deferens (vas deferens)—muscular tube that transports and stores sperm

   ejaculatory duct—tube within prostate gland from ductus deferens to urethra

   seminal vesicles (seminal glands)—paired glands; produce most of seminal fluid that contains fructose and prostaglandins; fructose nourishes sperm and prostaglandins enhance sperm entrance into uterus

   prostate gland—produces and secretes proteins (i.e., prostate-specific antigen) into seminal fluid that help activate sperm; secretes citrate, a nutrient used to make ATP for sperm movement, and PO_4^{3-} that helps buffer acidity of seminal fluid and vagina; single gland

   bulbourethral glands—tiny paired glands; secretions neutralize any residual acidic urine; lubricates penis head prior to intercourse

   scrotum—protective covering of testes and epididymis

   penis—male copulatory organ
   - corpora cavernosa—paired erectile tissue cylinders; engorge with blood during an erection
   - corpus spongiosum—single erectile tissue cylinder; engorges with blood during an erection
   - tunica albuginea—fibrous capsule around testes and corpora erectile tissues
   - glans penis—expanded distal end of penis
   - external urethral orifice (external urethral meatus)—external opening for urine and semen
   - prepuce (foreskin)—fold of skin around glans penis; removed if a surgical circumcision is performed
3. Complete Part A of Laboratory Assessment 59.

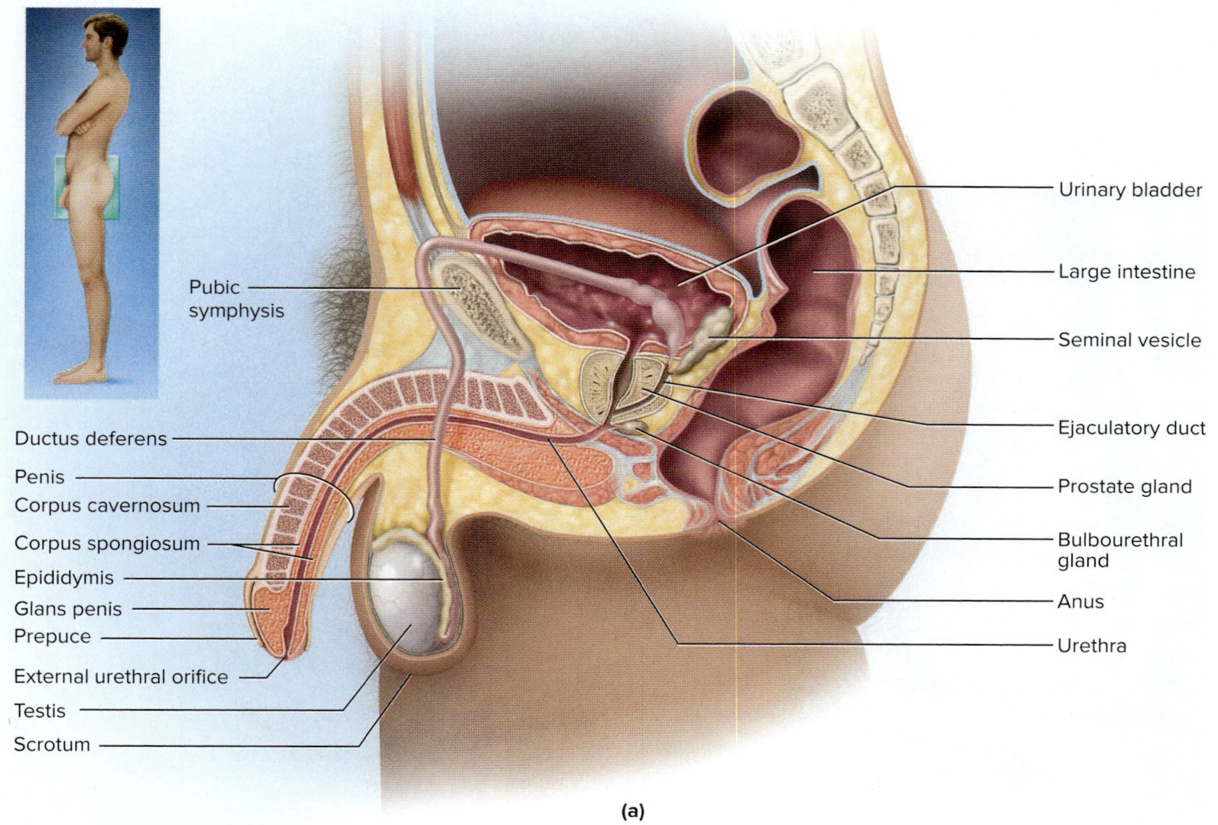

(a)

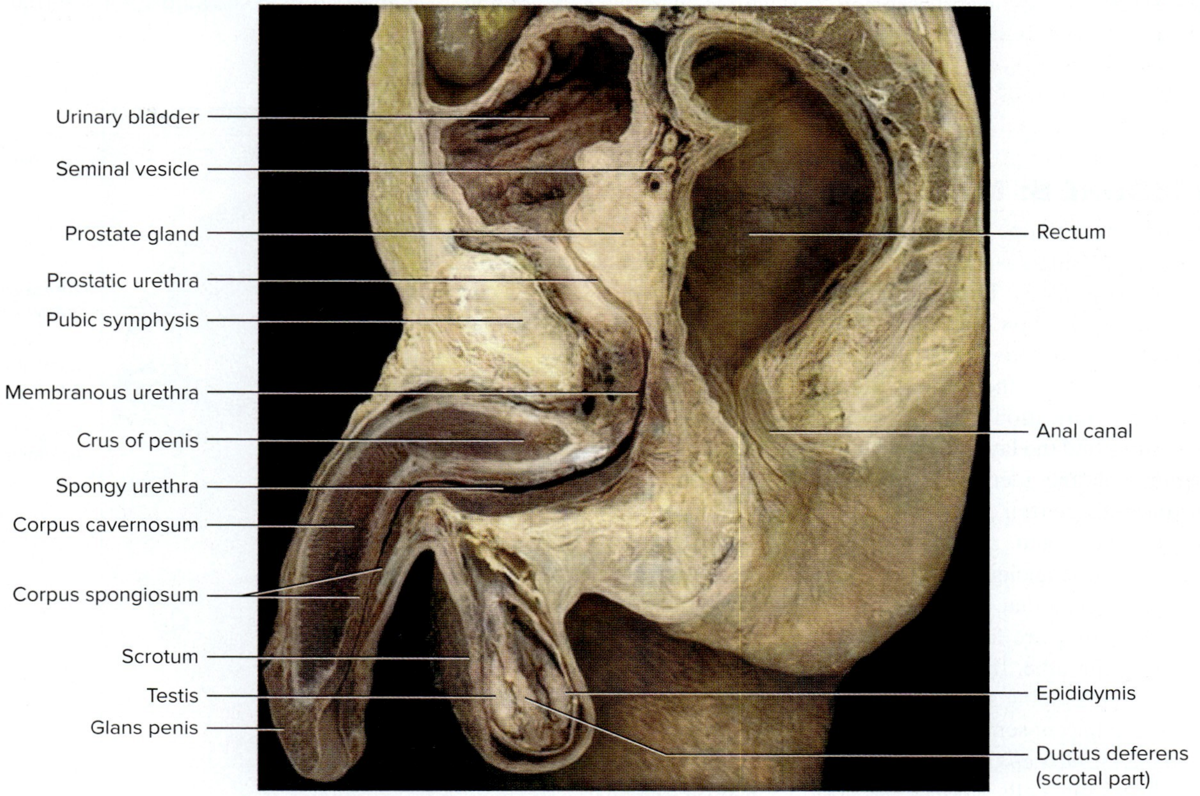

(b)

FIGURE 59.1 The major structures of the male reproductive system in sagittal view: (a) illustration and (b) cadaver.
(b) APR/McGraw Hill APR

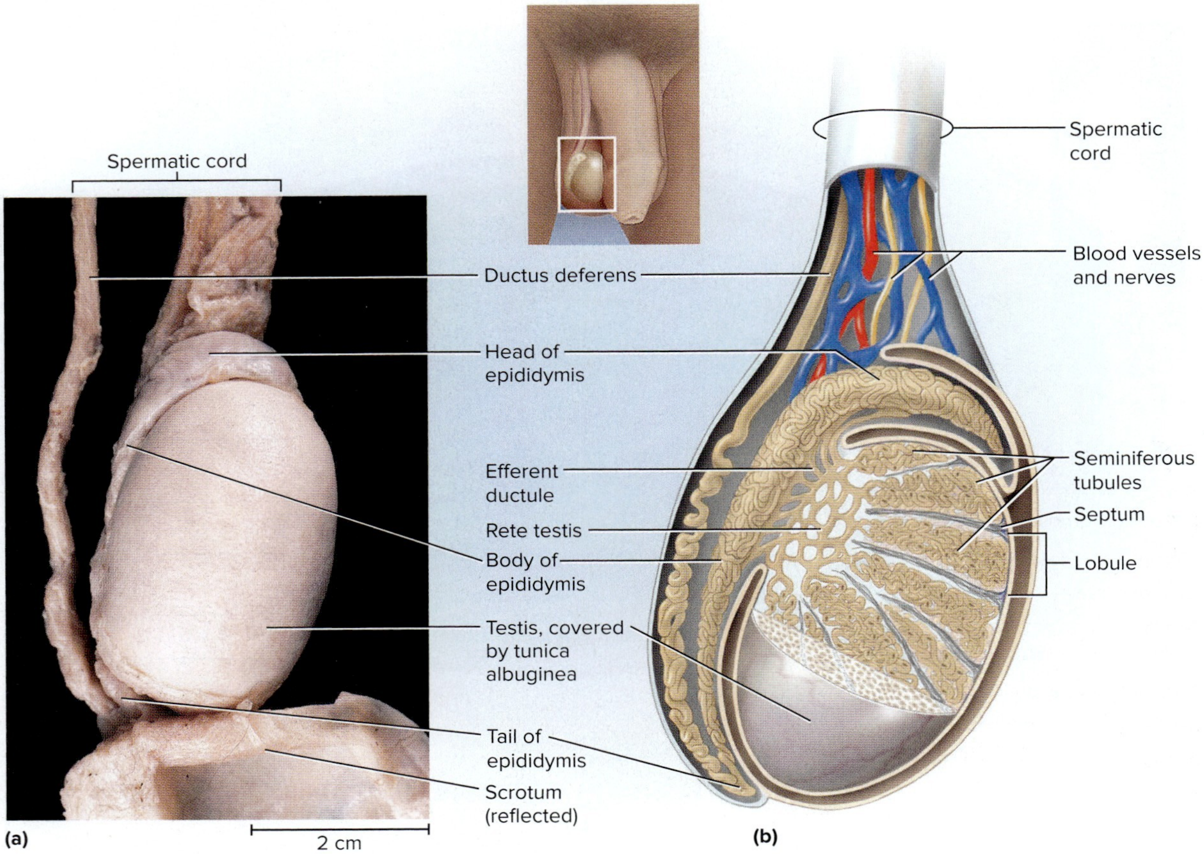

FIGURE 59.2 The testis and associated structures (a) from a cadaver and (b) in an illustration showing some internal structures. (a) Dennis Strete/McGraw Hill APR

PROCEDURE B: Microscopic Anatomy

1. Obtain a microscope slide of a human testis section and examine it using scanning and then low-power magnification (fig. 59.3). Locate the thick *tunica albuginea (fibrous capsule)* on the surface and the numerous sections of *seminiferous tubules* inside.
2. Focus on some of the seminiferous tubules using high-power magnification (fig. 59.4). Locate the *basement membrane* and the layer of *spermatogonia* near the basement membrane. Identify some *sustentacular cells* (supporting cells; Sertoli cells), which have pale, oval-shaped nuclei, and some *spermatogenic cells,* which have smaller, round nuclei. Spermatogonia give rise to spermatogenic cells that are in various stages of spermatogenesis as they are forced toward the lumen. Near the lumen of the tube, find some darkly stained, elongated heads of developing sperm (spermatozoa). In the spaces between adjacent seminiferous tubules, locate some isolated *interstitial cells* (Leydig cells) of the endocrine system. Interstitial cells produce the hormone testosterone, transported by the blood.
3. Prepare a labeled sketch of a representative section of the testis in Part B of the laboratory assessment.

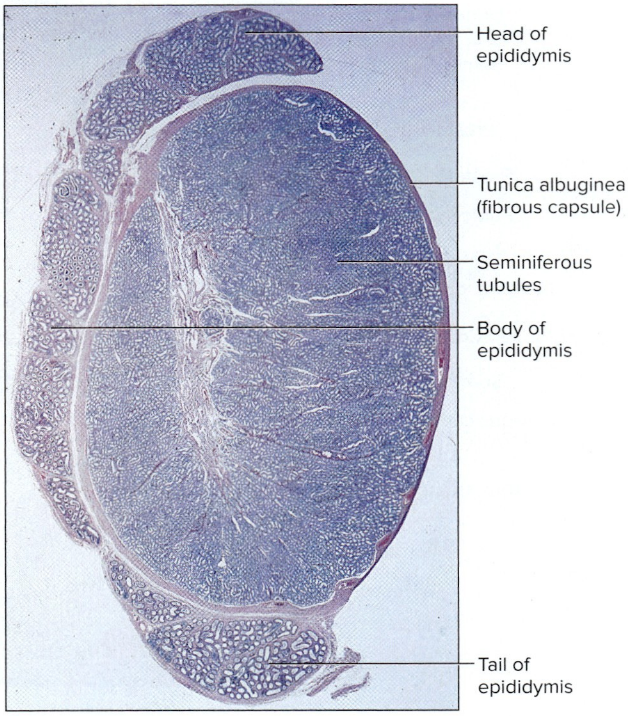

FIGURE 59.3 Micrograph of a human testis (1.7×). Biophoto Associates/Science Source

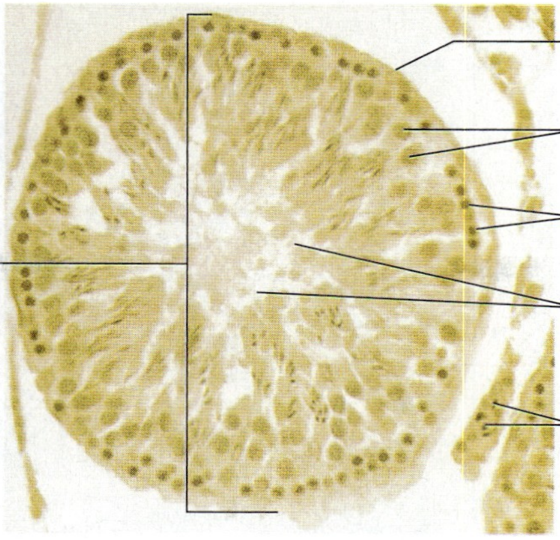

FIGURE 59.4 Micrograph of a seminiferous tubule (250×). Ed Reschke APR

4. Obtain a microscope slide of a cross section of *epididymis,* and examine it using scanning and then low-power magnification (fig. 59.5). Using high-power magnification, note the elongated, *pseudostratified columnar epithelial cells* that compose most of the inner lining. These cells have nonmotile stereocilia (elongated microvilli) on their free surfaces that absorb excess fluid secreted by the testes. Also note the thin layer of smooth muscle and connective tissue surrounding the tube.
5. Prepare a labeled sketch of the epididymis wall in Part B of the laboratory assessment.
6. Obtain a microscope slide of a cross section of *ductus deferens,* and examine it using scanning and then low-power magnification (fig. 59.6*a*). Note the small lumen

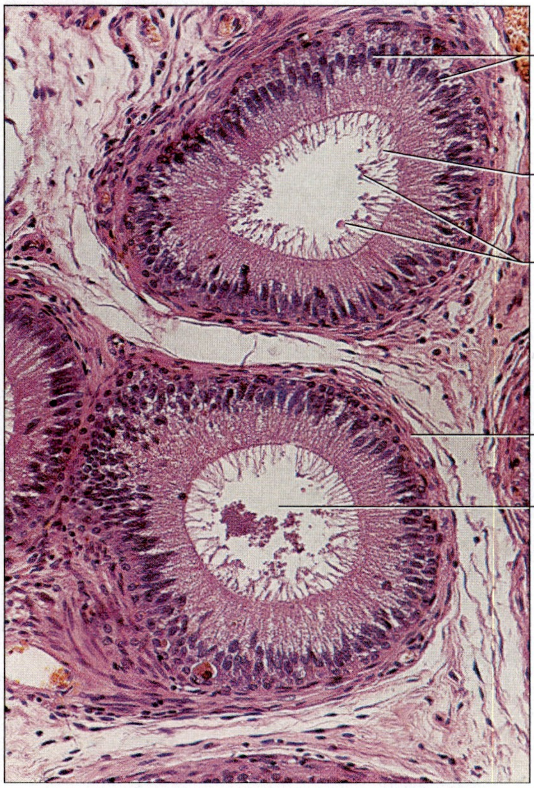

FIGURE 59.5 Micrograph of a cross section of a human epididymis (200×). Ed Reschke

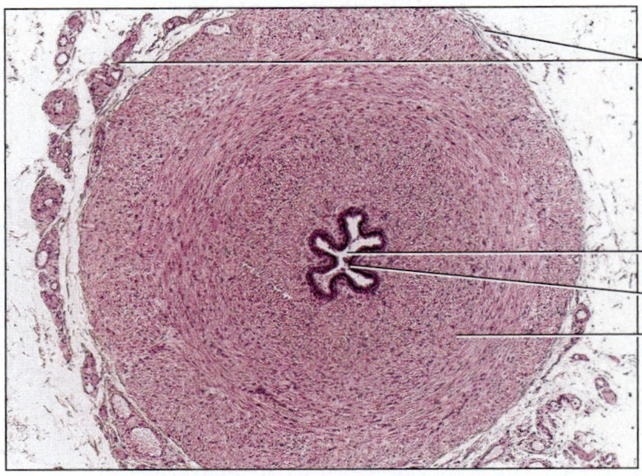

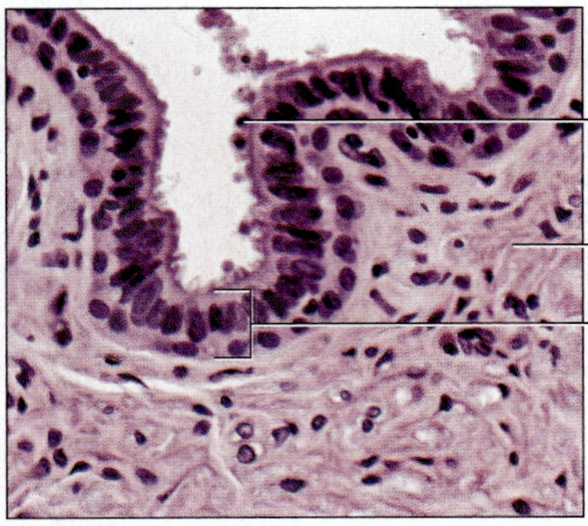

FIGURE 59.6 Micrograph of (a) a cross section of the ductus deferens (40×) and (b) the wall of the ductus deferens (400×). Al Telser/McGraw Hill APR

lined with an epithelium and the thick *smooth muscle* layer composing most of the structure of the duct. Three layers of smooth muscle fibers are distinctive: a middle circular layer between two layers of longitudinal muscle fibers. The peristaltic contractions of the muscular layer transport sperm along the passageway toward the ejaculatory duct and urethra. Sperm storage occurs within the lumen of the ductus deferens and the epididymis. An outer, connective tissue covering of the ductus deferens is called the *adventitia.* Use high-power magnification and examine the epithelium layer composed of pseudostratified columnar epithelial cells (fig. 59.6*b*).

7. Prepare a labeled sketch of a representative section of a ductus deferens in Part B of the laboratory assessment.
8. Study figure 59.7 and the list of structures and descriptions. Obtain a microscope slide of a *penis* cross section, and examine it using scanning and then low-power magnification. Identify the following features:

   corpora cavernosa—two dorsal erectile tissue cylinders; engorge with blood during an erection

   corpus spongiosum—single erectile tissue cylinder around urethra; engorges with blood during an erection

   tunica albuginea—fibrous capsule around corpora cavernosa and corpus spongiosum

   urethra—passageway for urine and semen

   skin—surrounds entire penis
9. Prepare a labeled sketch of a penis cross section in Part B of the laboratory assessment.
10. Complete Part B of the laboratory assessment.

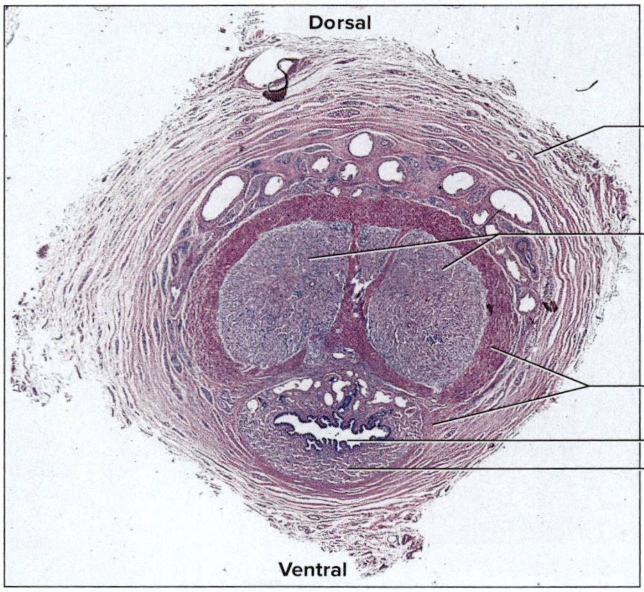

FIGURE 59.7 Micrograph of a cross (transverse) section of the body of the penis (5×). Michael Peres APR

LABORATORY
ASSESSMENT 59

Name ______________________

Date ______________________

Section ______________________

The A corresponds to the indicated Learning Outcome(s) O found at the beginning of the Laboratory Exercise.

Male Reproductive System

PART A: Assessments

1. Label the structures in figures 59.8 and 59.9.

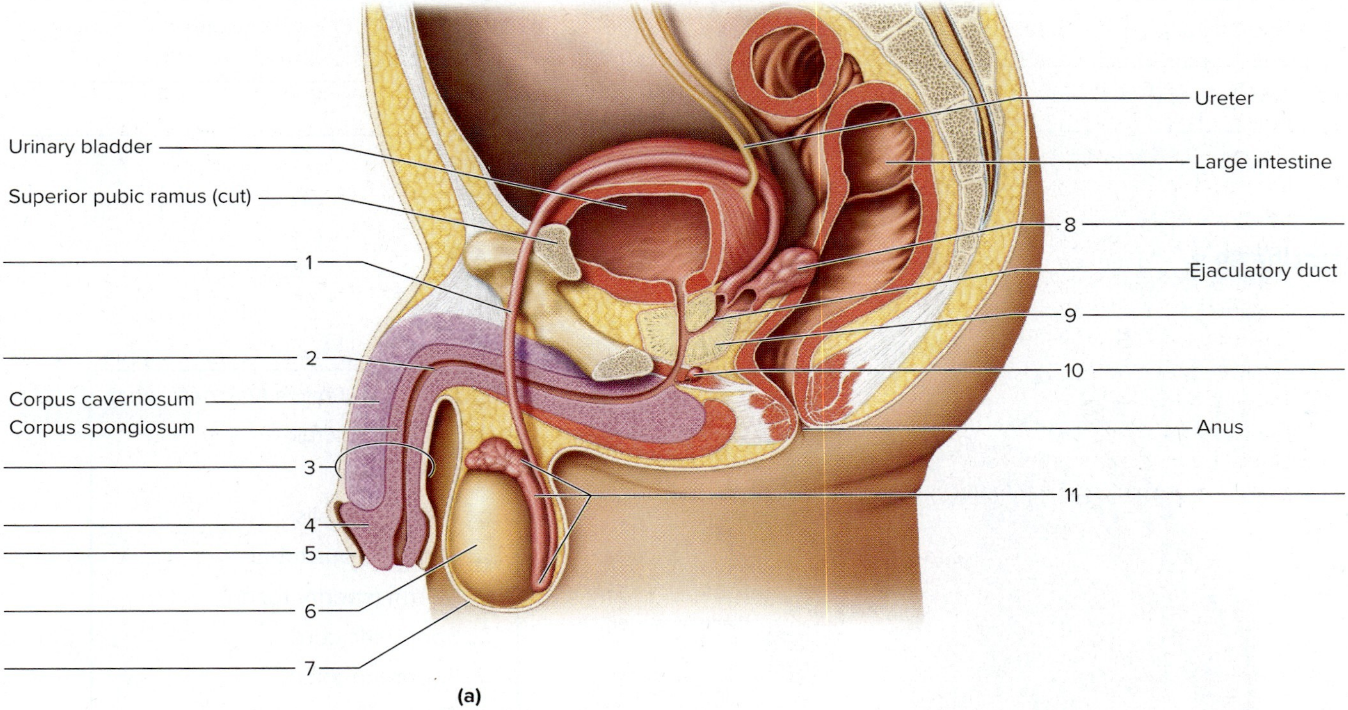

FIGURE 59.8 Label the major structures of the male reproductive system in the sagittal view of (a) an illustration and (b) an image of a model, using the terms provided [for (b) only]. A1

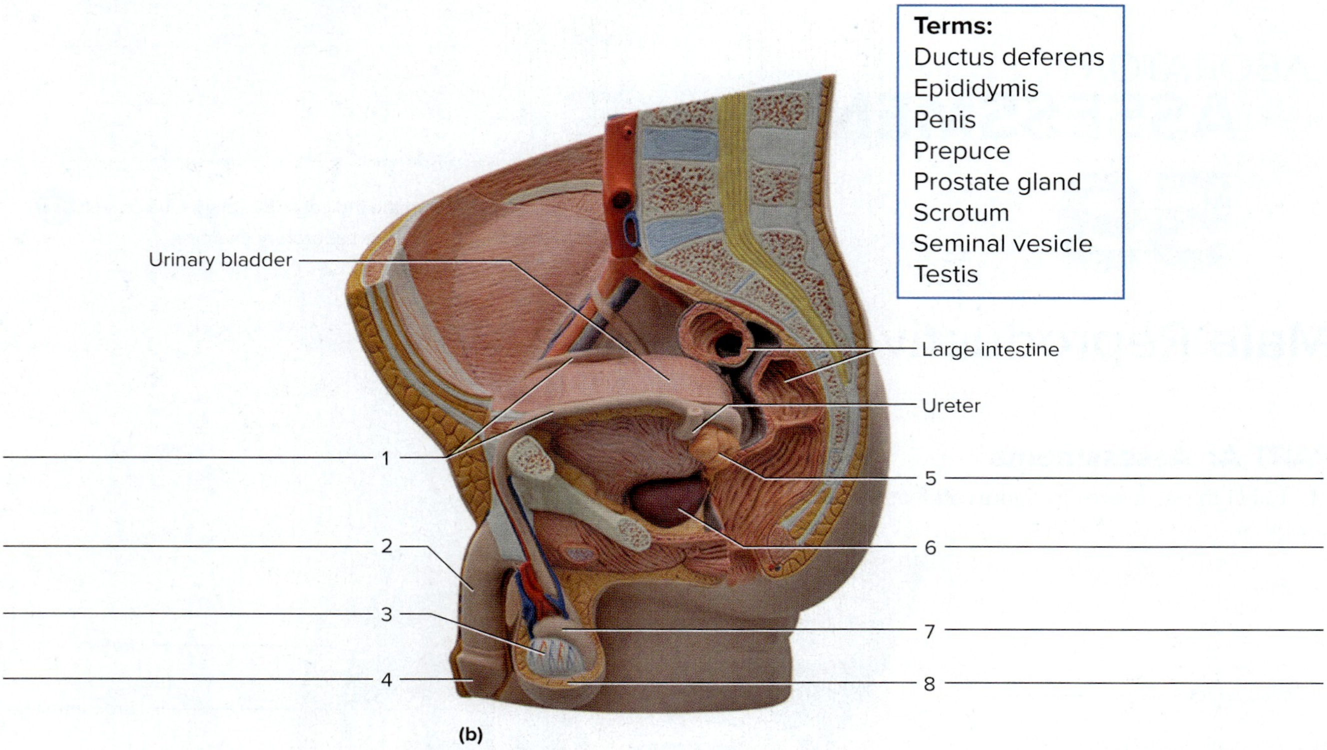

FIGURE 59.8 *Continued.* ©3B Scientific GmbH, Germany, 2017www.3bscientific.com A1

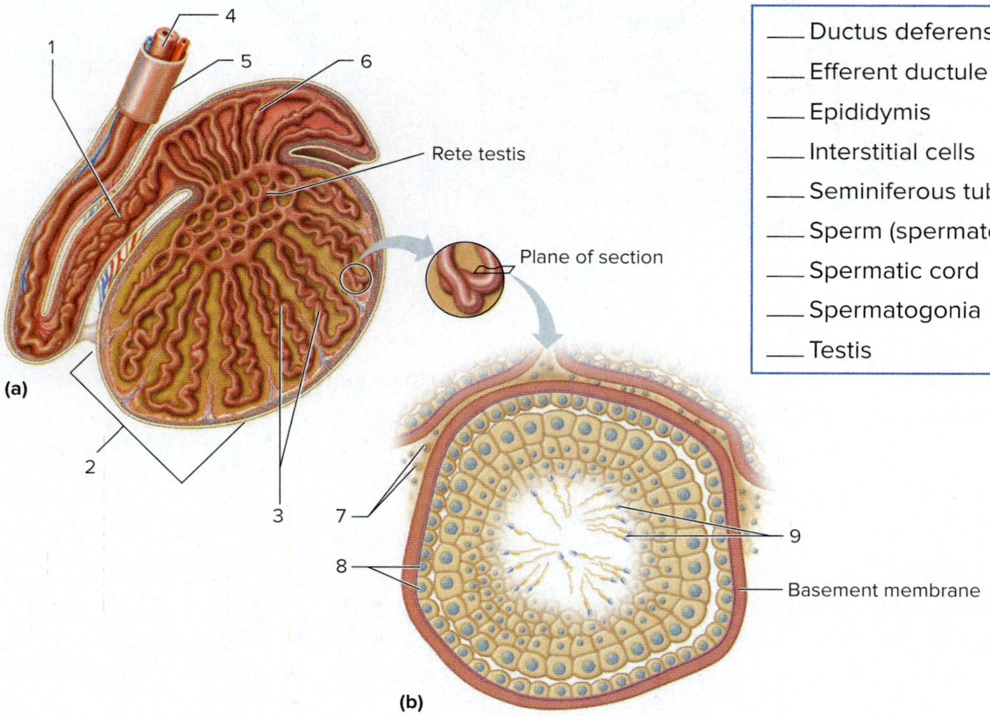

___ Ductus deferens (vas deferens)
___ Efferent ductule
___ Epididymis
___ Interstitial cells
___ Seminiferous tubules
___ Sperm (spermatozoa)
___ Spermatic cord
___ Spermatogonia
___ Testis

FIGURE 59.9 Label the diagram of (a) the sagittal section of a testis and (b) a cross section of a seminiferous tubule by placing the correct numbers in the spaces provided. A1

2. Match the terms in column A with the descriptions in column B. Place the letter of your choice in the space provided. A2

Column A	Column B
a. Bulbourethral glands	______ 1. Process by which sperm are formed
b. Ductus deferens	______ 2. Common tube for urine and semen
c. Ejaculatory duct	______ 3. Sperm duct located within spermatic cord
d. Epididymis	______ 4. Produce male hormones
e. Interstitial cells	______ 5. Location of spermatogenesis
f. Penis	______ 6. Located within prostate gland and can propel semen into urethra
g. Prepuce	______ 7. Contains sperm and secretions from accessory glands
h. Prostate	______ 8. Tiny paired glands that secrete into urethra
i. Semen	______ 9. Maturation and storage site within a coiled tube along testis
j. Seminiferous tubules	______ 10. Surrounds the urethra and secretes a seminal fluid
k. Spermatogenesis	______ 11. Contains three columns of erectile tissue
l. Urethra	______ 12. Removed in a surgical circumcision

CRITICAL THINKING ASSESSMENT—PART A

What is prostatitis? ______________________________ Explain how this may impact the normal function of the urogenital system of the male. A2 ______________________________

____

____

PART B: Assessments

1. Each circle below represents the microscopic field of view. In each circle, sketch a representative area of the organ indicated. Label any of the structures that were observed, and indicate the magnification used for each sketch. A3

Testis (______×)

Epididymis (______×)

Ductus deferens (______×)

Penis (______×)

2. Describe the function of each of the following: A2

   a. Scrotum

   ____

   ____

   b. Spermatogenic cell

   ____

   ____

   c. Interstitial cell (Leydig cell)

   ____

   ____

   d. Epididymis

   ____

   ____

   e. Corpora cavernosa and corpus spongiosum

   ____

   ____

LABORATORY EXERCISE 60

Female Reproductive System

Purpose of the Exercise

To examine the microscopic and gross anatomy of the female reproductive organs and to describe their basic physiological functions.

MATERIALS NEEDED

Human torso model
Model of the female reproductive system
Anatomical chart of the female reproductive system
Compound light microscope
Prepared microscope slides of the following:
- Ovary section with maturing follicles
- Uterine tube, cross section
- Uterine wall section

For Demonstration Activity:

Prepared microscope slides of the following:
- Uterine wall, early proliferative phase
- Uterine wall, secretory phase
- Uterine wall, early menstrual phase

Learning Outcomes APR

After completing this exercise, you should be able to:

01 Locate and identify the organs of the female reproductive system.

02 Describe the structures and functions of the female reproductive organs.

03 Sketch and label the major features of microscopic sections of the ovary, uterine tube, and uterine wall.

The O corresponds to the assessments A indicated in the Laboratory Assessment for this Exercise.

Pre-Lab

Carefully read the introductory material and examine the entire lab. Be familiar with the structures and functions of the female reproductive system from lecture or the textbook. Answer the pre-lab questions.

Pre-Lab Questions Select the correct answer for each of the following questions:

1. Oogenesis originates within the
- **a.** follicle.
- **b.** uterine tube.
- **c.** uterus.
- **d.** corpus albicans.

2. Early cleavage division occurs within the
- **a.** follicle.
- **b.** corpus luteum.
- **c.** uterine tube.
- **d.** uterus.

3. Which of the following is *not* a structure of the uterus?
- **a.** cervix
- **b.** fimbriae
- **c.** endometrium
- **d.** myometrium

4. The _____ are the milk-producing parts of the mammary glands.
- **a.** areolar glands
- **b.** lactiferous ducts
- **c.** adipose cells
- **d.** alveolar glands

5. How many openings are located on a human nipple?
- **a.** 1
- **b.** 2–3
- **c.** 5–10
- **d.** 15–20

6. Which of the following structures occupies the most anterior position of the external genitalia?
- **a.** hymen
- **b.** clitoris
- **c.** external urethral orifice
- **d.** vaginal orifice

7. The broad ligament is the largest ligament involved in holding female reproductive organs in position.
- **a.** True
- **b.** False

8. Primordial follicles are concentrated in the medulla region of an ovary.
- **a.** True
- **b.** False

The organs of the female reproductive system have a large variety of functions. They are specialized to produce and maintain the female sex cells, to transport these cells to the site of fertilization, to provide a favorable environment for a developing offspring, to move the offspring to the outside, and to produce female sex hormones.

The internal female reproductive organs include the *ovaries,* which produce the egg cells (oocytes) and female sex hormones, the uterine tubes, uterus, and vagina. The external genitalia include the labia majora, labia minora, clitoris, and vestibular glands.

The cortex region of an ovary contains numerous follicles in a variety of developmental stages to produce *ova* (*eggs*) by the process called *oogenesis* (*meiosis*). Oogenesis is described in Laboratory Exercise 61. Numerous *primordial follicles,* each containing a single flattened layer of follicular cells around the oocyte, are formed during prenatal development. During the period from puberty to menopause, reproductive cycles occur as individual follicles complete maturation. Each *primary follicle* is surrounded by one layer of cuboidal follicular cells and contains a *primary oocyte* that undergoes oogenesis. Primary follicles develop into *secondary follicles,* each containing two or more layers of cuboidal cells (granulosa cells), and then into *antral (tertiary; vesicular) follicles,* each with pools of follicular fluid that eventually merge into one large cavity called the *antrum.* Finally, a dominant antral follicle enlarges into a *mature follicle* shortly before ovulation of a *secondary oocyte.* The ruptured follicle becomes a *corpus luteum,* which secretes estrogens and progesterone, and eventually degenerates to form a *corpus albicans,* composed of connective tissue.

The *uterine tube* (*oviduct; fallopian tube*) has fingerlike fimbriae that partially envelop the ovary. After ovulation, the secondary oocyte is conveyed through the uterine tube by the current produced by the cilia and peristaltic waves. If the secondary oocyte is fertilized, it completes meiosis, and early cleavage divisions take place as it moves toward the *uterus* during the next several days. The uterus has three regions: *fundus, body,* and *cervix.* There are three layers to its wall: *perimetrium, myometrium,* and *endometrium.* Following implantation of a blastocyst in the endometrium, prenatal development continues in the uterus until birth.

The *vagina* has a mucosal lining, and near the vaginal orifice a membrane (hymen) partially closes the orifice unless broken, often during the first intercourse. This is the tube that sperm enters to fertilize the egg, and the fetus, once mature, exits through the same canal.

The *external genitalia* (*vulva*) are within a diamond-shaped perineum with boundaries established by the pubis, ischial tuberosities, and coccyx. Larger rounded folds of skin, *labia majora,* protect the *labia minora* and other external genitalia. Between the labia minora is a space (*vestibule*) that encloses the urethral and vaginal openings. The *clitoris* at the anterior vulva has structures similar to those of a penis, including a prepuce, glans, and corpora cavernosa.

The *breasts* contain the mammary glands, which are comprised of many lobes, ducts, ligaments, and adipose tissue. At puberty, estrogens stimulate breast development, and at the birth of a baby, various hormones promote mammary gland development and milk production (lactation). Lactiferous ducts terminate in 15–20 openings on the nipple in the center of the pigmented areola.

PROCEDURE A: Female Reproductive Organs APR

1. Study figures 60.1 to 60.4 of the internal reproductive organs and the list of structures and descriptions.
2. Observe the human torso model, the model of the female reproductive system, and the anatomical chart of the female reproductive system. Locate the following features:

ovaries—paired organs that produce ova and sex hormones
- medulla—internal region
- cortex—outer region

ligaments—hold reproductive organs in place
- broad ligament—largest ligament attaches to uterine tube, ovary, and uterus
- suspensory ligament of ovary—holds superior end of ovary to pelvic wall
- ovarian ligament—attaches ovary to uterus
- round ligament of uterus—band within broad ligament from lateral uterus to the connective tissue of external genitalia

uterine tubes (oviducts; fallopian tubes)—paired tubes about 10 cm (4 inches) long; site of fertilization and early cleavage development
- fimbriae—slender extensions on end of infundibulum
- infundibulum—expanded open end near ovary
- ampulla—middle section; typical site of oocyte fertilization
- isthmus—narrow portion near uterus

uterus—houses embryonic and fetal development during pregnancy
- fundus of uterus—wide superior curvature
- body of uterus—large middle section
- cervix of uterus—inferior cylindrical end
  - cervical canal—narrow passage within cervix
  - external os—opening into vagina
- uterine wall—thick and composed of three layers
  - endometrium—inner mucosal layer with simple columnar epithelium and numerous tubular glands
  - myometrium—thick middle layer of smooth muscle; contracts during childbirth
  - perimetrium (serosa; serous coat)—outer layer

rectouterine pouch—inferior recess of peritoneal cavity between rectum and uterus

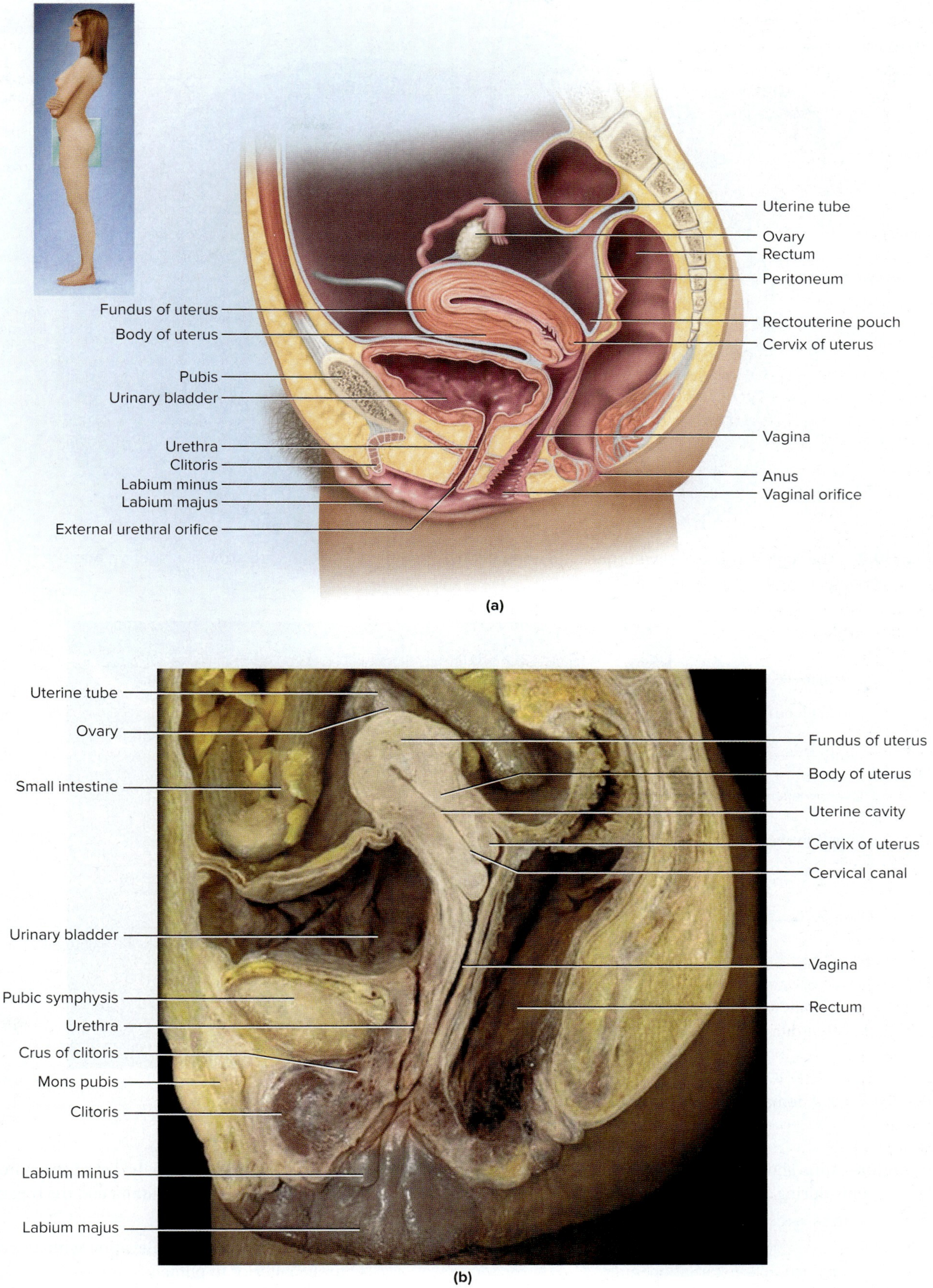

FIGURE 60.1 The structures of the female reproductive system in sagittal view: (a) illustration and (b) cadaver.
(b) APR/McGraw Hill APR

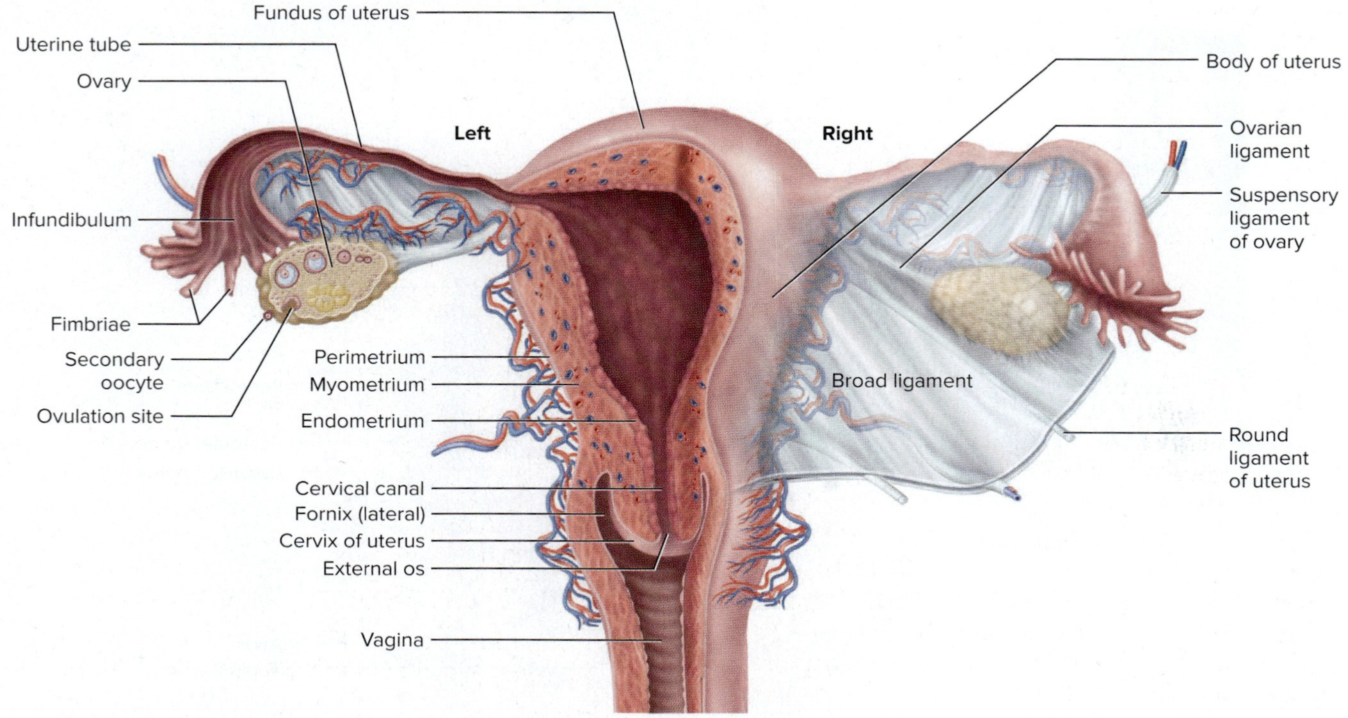

FIGURE 60.2 The female reproductive organs (posterior view). The left-side organs are shown in section and the ligaments are shown on the right side. APR

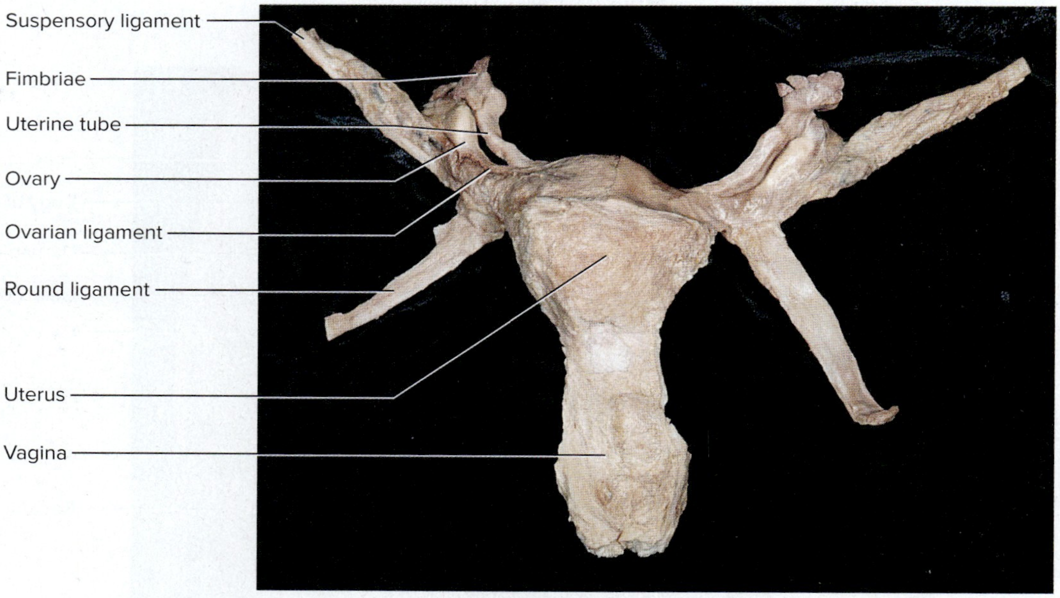

FIGURE 60.3 The female reproductive organs of a cadaver (anterior view). Rebecca Gray/McGraw Hill

vagina—tube about 10 cm (4 inches) long; receives penis during sexual intercourse; birth canal
- fornices—recesses between vaginal wall and cervix
- vaginal orifice—external opening
- hymen—thin fold of mucosa that partially closes vaginal orifice
- mucosal layer—inner vaginal wall of stratified squamous epithelium

3. Study figures 60.4 and 60.5 and the list of structures and descriptions of the external genitalia and the breasts.

external genitalia (vulva)
- mons pubis—rounded elevation with adipose tissue anterior to pubic symphysis
- labia majora—folds of skin with hair that enclose and protect structures between them
- labia minora—lesser hair-free folds between labia majora

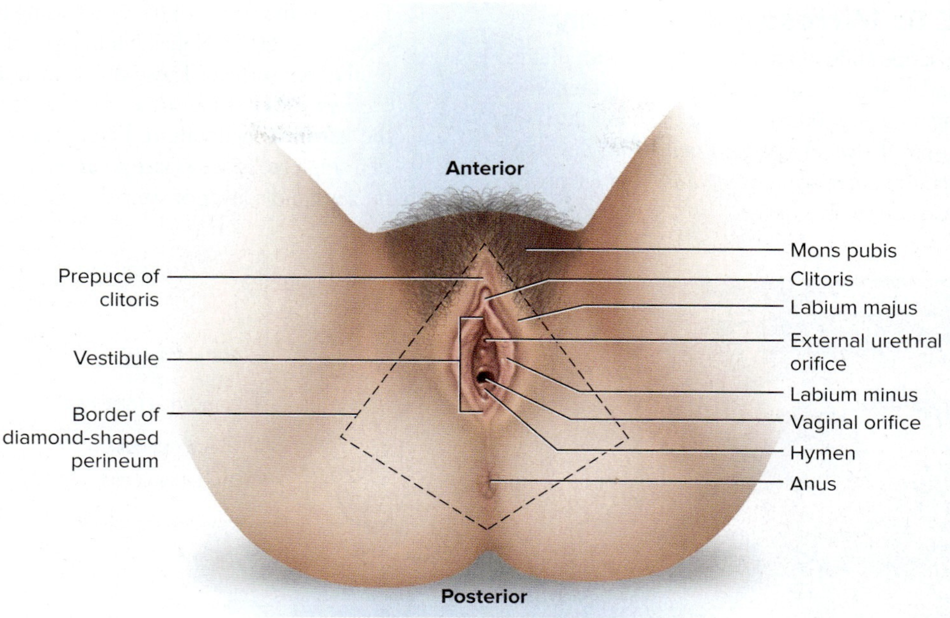

FIGURE 60.4 The structures associated with the female perineum. The external genitalia (vulva) are within the perineum. APR

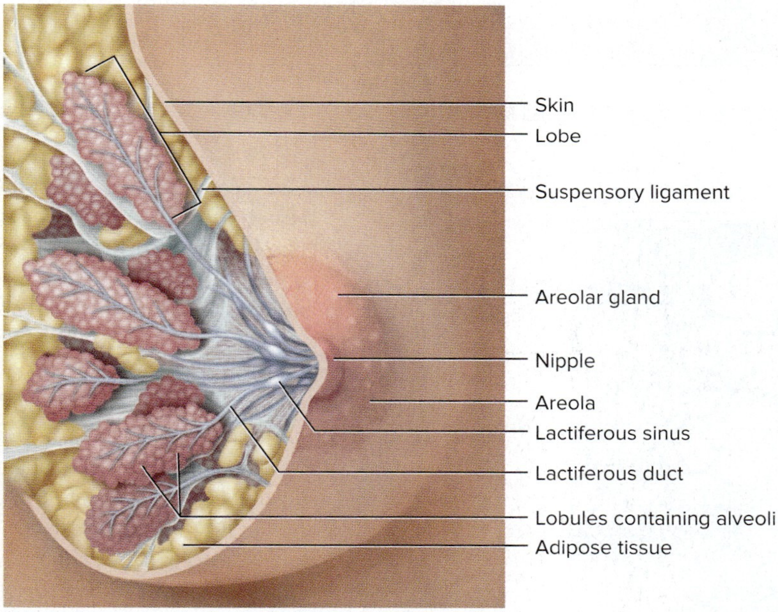

FIGURE 60.5 The structures of a lactating breast (anterior view). APR

- vestibule—region between labia minora
- vestibular glands—located on each side of vagina; secrete a lubricating fluid
- clitoris—small projection at anterior end of vestibule; contains erectile tissue; homologous to penis

breasts—paired elevations containing mammary glands

- nipple—protruding region in center of areola containing openings of the lactiferous ducts
- areola—circular pigmented skin around nipple
- areolar glands—modified glands, intermediary between sweat and true mammary glands, whose secretions help moisten and protect the nipple
- alveolar glands—specialized milk-producing epithelial cells lining hollow cavities (alveoli); develop during pregnancy
- lactiferous ducts—drain the 15–20 lobes of the breast; dilated portion is the lactiferous sinus
- adipose tissue—compose much of internal structure of the breast

4. Complete Part A of Laboratory Assessment 60.

PROCEDURE B: Microscopic Anatomy

1. Obtain a microscope slide of an ovary section with maturing follicles, and examine it using scanning and then low-power magnification (fig. 60.6). Locate the outer layer, or *cortex,* composed of densely packed cells with developing follicles, and the inner layer, or *medulla,* which largely consists of loose connective tissue.
2. Focus on the cortex of the ovary using high-power magnification (fig. 60.7). Note the thin layer of small cuboidal cells on the free surface. These cells constitute the *germinal epithelium.* Also locate some *primordial follicles* just beneath the germinal epithelium. Each primordial follicle consists of a single, relatively large *primary oocyte* with a prominent nucleus and a layer of simple squamous epithelial cells.

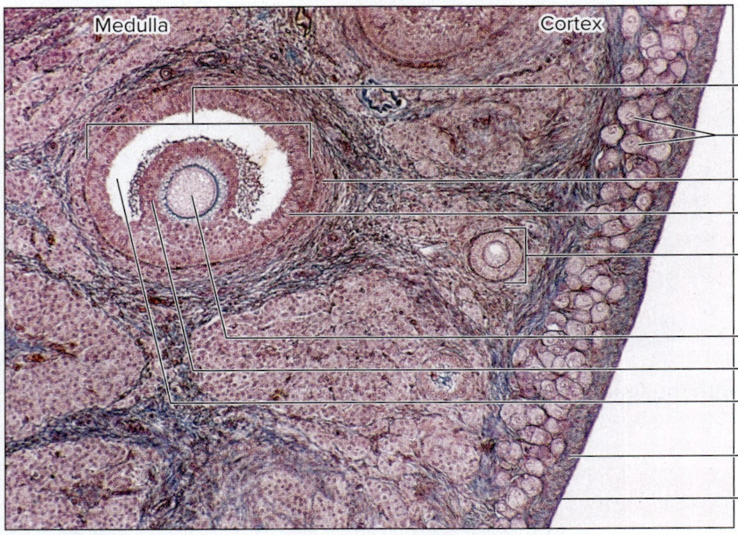

FIGURE 60.6 Micrograph of the ovary (80×). Ed Reschke/Getty Images APR

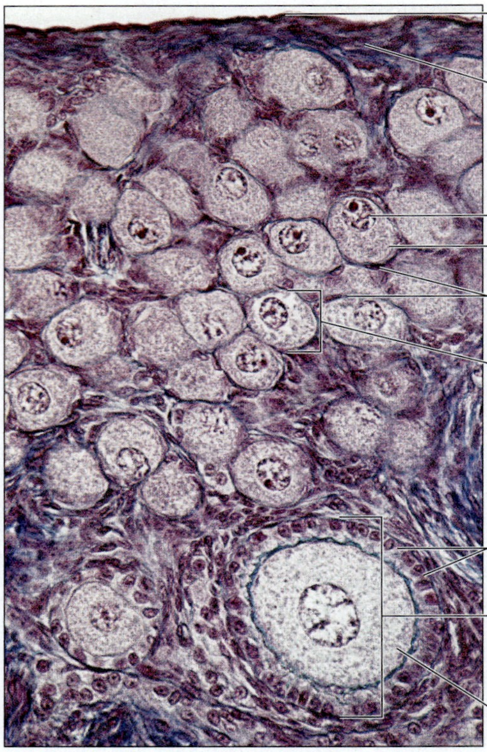

FIGURE 60.7 Micrograph of the ovarian cortex (200×). Ed Reschke/Getty Images APR

3. Use scanning and then low-power magnification to search the ovarian cortex for maturing follicles in various stages of development (fig. 60.6). Locate and compare primordial follicles and *primary follicles.* Primary follicles develop into *secondary follicles,* and then into *antral (tertiary; vesicular) follicles,* and finally a *mature (graafian) follicle* forms just before ovulation. Developing follicles have a covering from a single layer to a double layer of follicular cells. The type of epithelial tissue cells depends upon the stage of development. As follicles develop, a covering of several layers of granulosa cells function to convert androgens to estrogens when stimulated by follicle-stimulating hormone. Surrounding the entire developing follicle is a tough, highly vascular theca folliculi that uses absorbed cholesterol to synthesize the androgens that diffuse into the granulosa cells. The theca is most easily seen when viewing a mature follicle.
4. Prepare labeled sketches in Part B of the laboratory assessment to illustrate the changes that occur in a follicle as it matures.
5. Obtain a microscope slide of a cross section of a uterine tube. Examine it using scanning and then low-power magnification. The shape of the lumen is very irregular because of folds of the mucosa layer.
6. Focus on the inner lining of the uterine tube using high-power magnification (fig. 60.8). The lining is composed of *simple columnar epithelium,* and some of the epithelial cells are ciliated on their free surfaces.

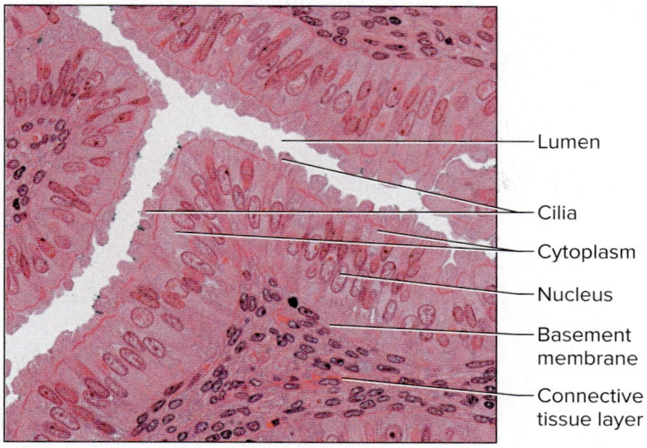

FIGURE 60.8 Micrograph of a cross section of the uterine tube (800×). Michael Peres APR

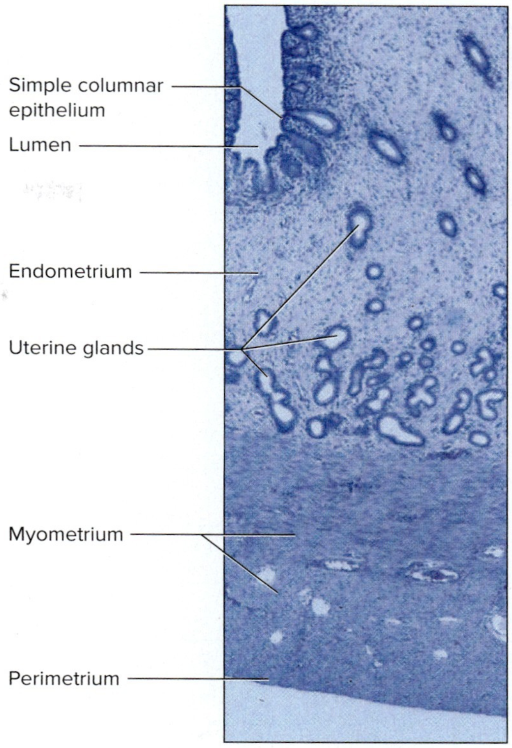

FIGURE 60.9 Micrograph of the uterine wall (10×). Carol D. Jacobson, Ph.D./Dept. of Veterinary Anatomy, Iowa State University/McGraw Hill APR

7. Prepare a labeled sketch of a representative region of the wall of the uterine tube in Part B of the laboratory assessment.
8. Obtain a microscope slide of the uterine wall section (fig. 60.9). Examine it using scanning and then low-power magnification, and locate the following:

   endometrium—inner mucosal layer

   myometrium—middle, thick muscular layer

   perimetrium—outer, thin serosal layer; continuous with broad ligament
9. Prepare a labeled sketch of a representative section of the uterine wall in Part B of the laboratory assessment.
10. Complete Part C of the laboratory assessment.

DEMONSTRATION ACTIVITY

Observe the slides in the demonstration microscopes. Each slide contains a section of uterine mucosa taken during a different phase in the reproductive cycle. In the *early proliferative phase,* note the simple columnar epithelium on the free surface of the mucosa and the many sections of tubular uterine glands in the tissues beneath the epithelium. In the *secretory phase,* the endometrium is thicker, the uterine glands appear more extensive, and they are coiled. In the *early menstrual phase,* the endometrium is thinner because its surface functional layer has been shed, while retaining the basal layer. Also the uterine glands are less apparent, and the spaces between the glands contain many leukocytes. The basal layer of the endometrium renews the functional layer after menstruation ends.

NOTES

Name ______________________

Date ______________________

Section ______________________

The (A) corresponds to the indicated Learning Outcome(s) (O) found at the beginning of the Laboratory Exercise.

LABORATORY ASSESSMENT 60

Female Reproductive System

PART A: Assessments

1. Identify the numbered structures in figures 60.10 and 60.11.

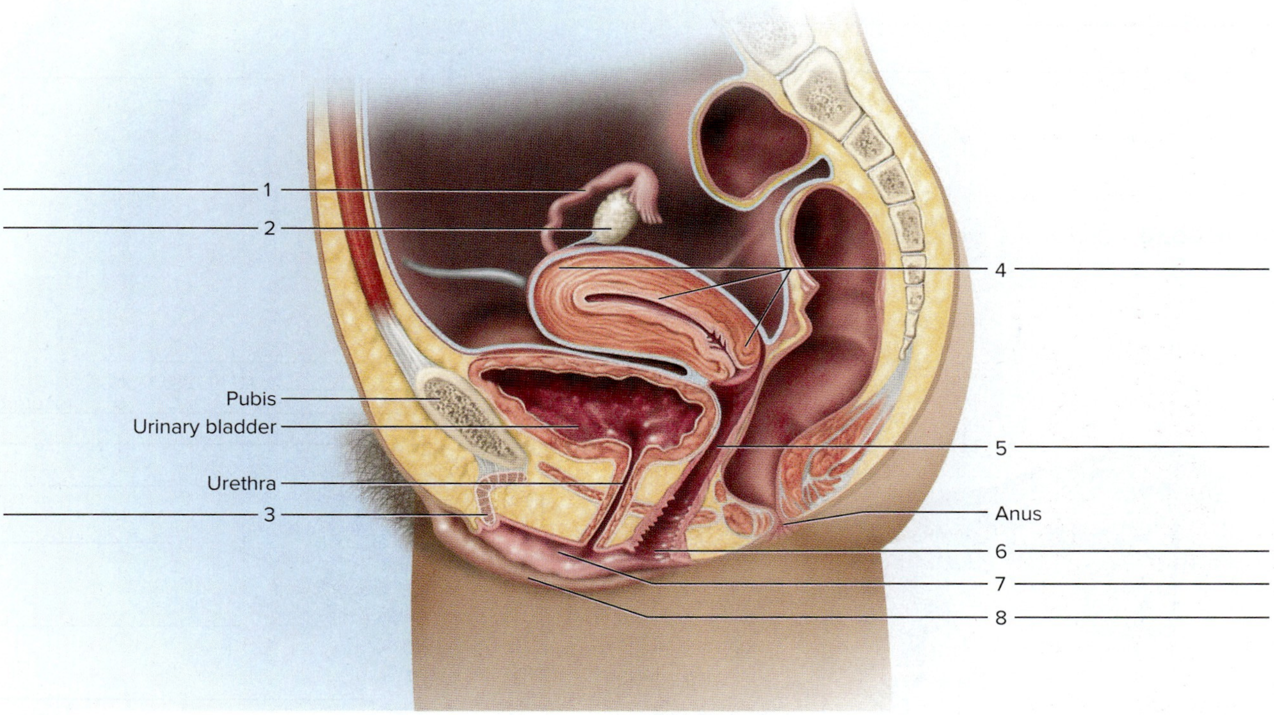

(a)

FIGURE 60.10 Label the major structures of the female reproductive system in the sagittal view of (a) an illustration and (b) an image of a model, using the terms provided [for (b) only]. A1

Terms:

Cervical canal	Myometrium (of uterus)
Clitoris	Ovarian ligament
Fimbriae	Ovary
Labia majus	Round ligament
Labia minus	Uterine tube
Mons pubis	Vagina

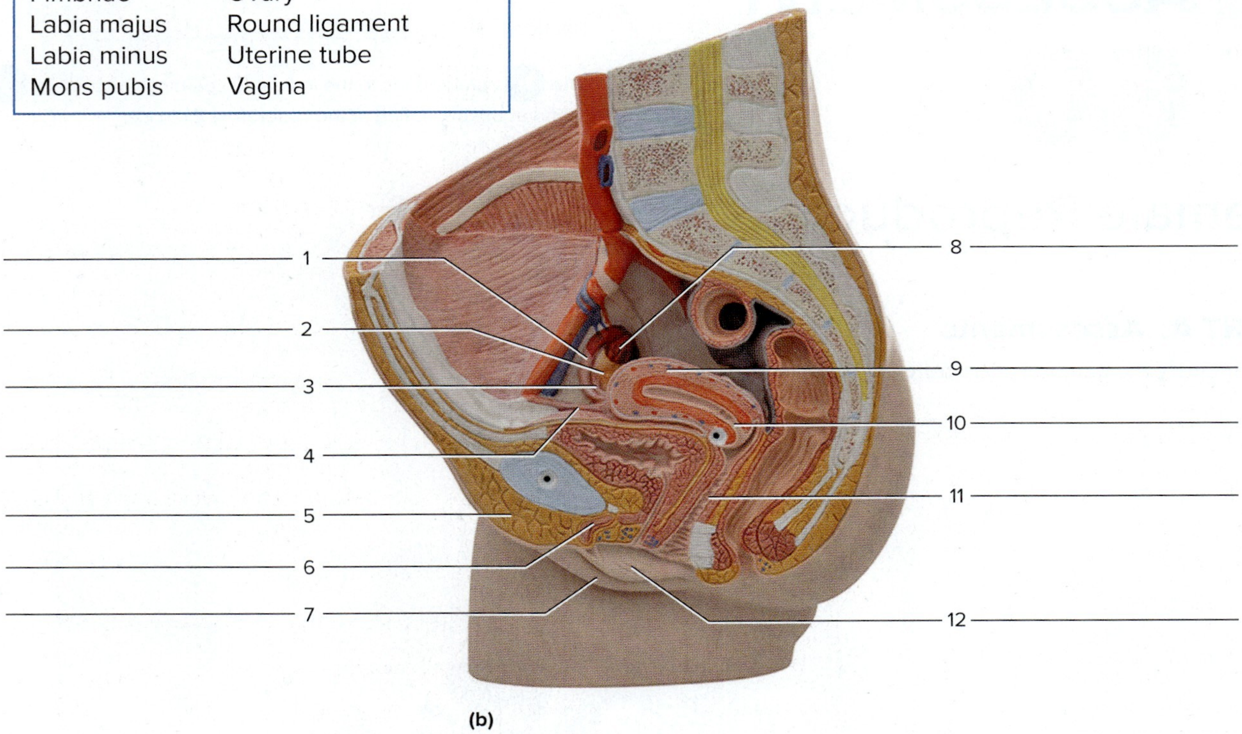

FIGURE 60.10 *Continued.* ©3B Scientific GmbH, Germany, 2017 www.3bscientific.com A1

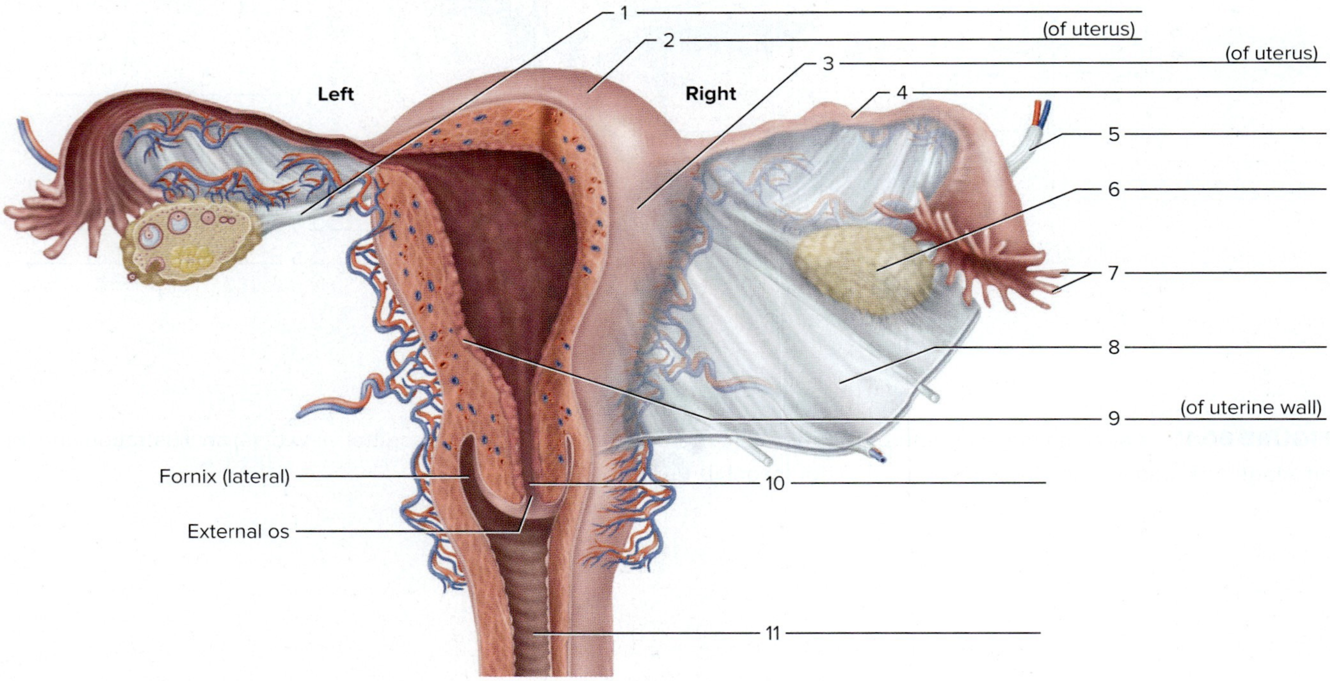

FIGURE 60.11 Label the female reproductive structures in this posterior view. A1

2. Match the terms in column A with the descriptions in column B. Place the letter of your choice in the space provided. A2

Column A	Column B
a. Cervix	______ **1.** Rounded end of uterus near uterine tubes
b. Cilia	______ **2.** Smooth muscle that contracts with force during childbirth
c. Clitoris	______ **3.** Thin membrane that partially closes vaginal orifice
d. Endometrium	______ **4.** Surface region around external genitalia
e. Fimbriae	______ **5.** Secretes lubricating fluid into vagina
f. Fundus	______ **6.** Process by which a secondary oocyte is released from the ovary
g. Hymen	______ **7.** Portion of uterus extending into superior portion of vagina
h. Lactiferous duct	______ **8.** Inner mucosal lining of uterus
i. Myometrium	______ **9.** Transports milk from milk-producing lobes to nipple
j. Ovulation	______ **10.** Help move secondary oocyte through the uterine tube toward uterus
k. Perineum	______ **11.** Contains erectile tissue
l. Suspensory ligament	______ **12.** Fingerlike projections of uterine tube near ovary
m. Vestibular glands	______ **13.** Birth canal during birth process
n. Vagina	______ **14.** Holds ovary to pelvic wall

CRITICAL THINKING ASSESSMENT—PART A

Jill is a 48-year-old mother of three children. After complaining to her physician that she has been feeling a sensation of fullness in the vagina and leakage of urine when coughing and exercising, Jill was diagnosed with an early stage cystocele, also known as a prolapsed bladder. How would this diagnosis result in the symptoms Jill is experiencing, and what might be its cause? A1 A2

____

____

____.

Jill's physician prescribed pelvic floor physical therapy and that Jill be fit for a pessary. What is the purpose of these therapies?

____

____

____.

PART B: Assessments

Each circle below represents the microscopic field of view. In each circle, sketch a representative area of the organ indicated. Label any of the structures observed and indicate the magnification used for each sketch. A3

Primordial or primary follicle (______×)

Secondary or mature follicle (_____×)

Uterine tube (_______×)

Wall of uterus (_____×)

PART C: Assessments

Identify the numbered features in figures 60.12 and 60.13 of a sectioned ovary that represent structures present at some stage during a reproductive cycle.

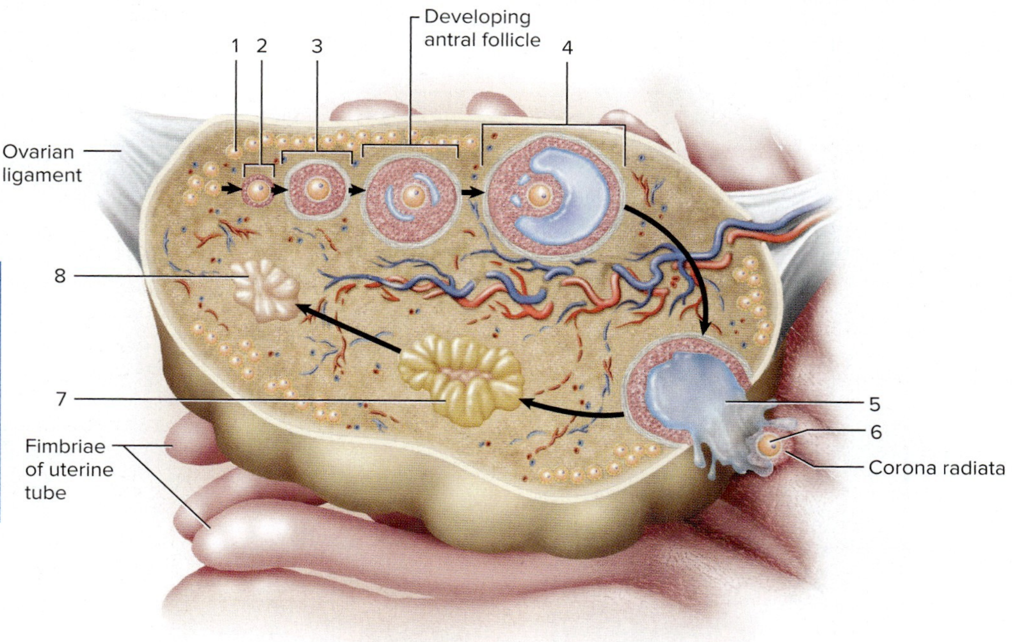

____ Corpus albicans
____ Corpus luteum
____ Mature follicle
____ Ovulation site
____ Primary follicle
____ Primordial follicle
____ Secondary follicle
____ Secondary oocyte

FIGURE 60.12 Label this ovary by placing the correct numbers in the spaces provided. The arrows indicate development over time; follicles do not migrate through an ovary. A1

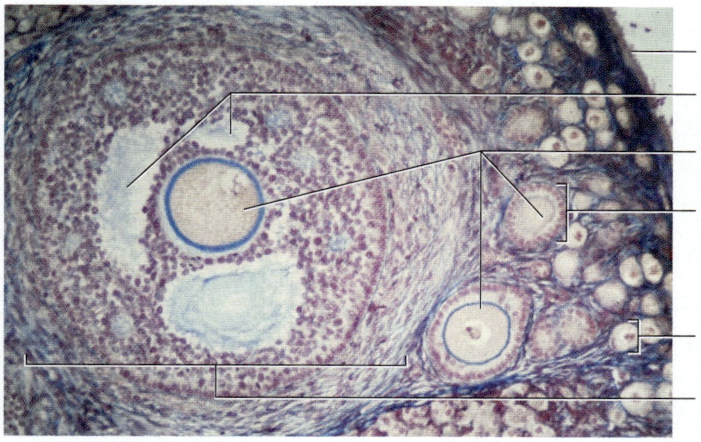

1 ____________
2 ____________
3 ____________
4 ____________
5 ____________
6 ____________

Terms:

Antral follicle	Germinal epithelium	Primary oocytes
Developing antrum	Primary follicle	Primordial follicle

FIGURE 60.13 Label the micrograph of a section of an ovary, using the terms provided (100×). Biophoto Associates/ Science Source A1 A3

NOTES

LABORATORY
EXERCISE 61

Meiosis, Fertilization, and Early Development

Purpose of the Exercise

To describe the process of meiosis and fertilization, to observe sea urchin eggs being fertilized, and to examine various features of the human in early stages of development.

MATERIALS NEEDED

Sea urchin egg suspension*
Sea urchin sperm suspension*
Compound light microscope
Depression microscope slide
Coverslip
Medicine droppers
Models of animal meiosis
Models of stages of human fertilization
Models of human development
Prepared microscope slide of the following:
  Sea urchin embryos (cleavage, blastula, and gastrula stages)

For Learning Extension Activity:
Vaseline
Toothpick

For Demonstration Activity:
Preserved mammalian embryos

*See the Instructor's Manual for a source of materials.

Chapter Opener: ©101cats/Getty Images

Learning Outcomes

After completing this exercise, you should be able to:

O1 Describe meiosis, fertilization, and the early developmental stages of a human.

O2 Identify and sketch the early developmental stages of a sea urchin.

O3 Distinguish the major features of early human development.

The O corresponds to the assessments A indicated in the Laboratory Assessment for this Exercise.

Pre-Lab

Carefully read the introductory material and examine the entire lab. Be familiar with meiosis, fertilization, and early embryonic development from lecture or the textbook. Answer the pre-lab questions.

Pre-Lab Questions Select the correct answer for each of the following questions:

1. Meiosis consists of two divisions in succession and produces _____.
   a. chromosomes **b.** body cells
   c. gametes **d.** blastocysts
2. The ovum and sperm unite and form a diploid cell called the
   a. pronucleus. **b.** blastocyst.
   c. morula. **d.** zygote.
3. The _____ is the pre-embryonic structure that implants in the endometrium.
   a. morula **b.** blastocyst
   c. zygote **d.** two-cell stage
4. Which of the following is *not* an extraembryonic membrane?
   a. umbilical cord **b.** amnion
   c. chorion **d.** yolk sac
5. The placenta is comprised of
   a. only the myometrium. **b.** only the decidua basalis.
   c. only the chorionic villi. **d.** combinations of the decidua basalis and chorionic villi.
6. The embryonic stage of human development exists
   a. from fertilization until implantation.
   b. through the first two weeks of development.
   c. from week three through week eight of development.
   d. from week nine until birth.
7. When the three germ layers have formed, the structure is called a(an) _____.
   a. embryo **b.** blastocyst
   c. morula **d.** fetus

8. The zona pellucida gel layer surrounds the secondary oocyte and early development stages through the morula.
   a. True **b.** False
9. A single completed oogenesis sequence produces four functional ova (eggs).
   a. True **b.** False
10. A single completed spermatogenesis sequence produces four functional sperm.
   a. True **b.** False

Meiosis is a distinctive process of nuclear division in the testes and ovaries that occurs only during *gametogenesis* and produces sex cells called *gametes*. In males the process is more specifically called *spermatogenesis,* and it produces gametes known as *sperm.* In females the process is more specifically called *oogenesis,* and it produces gametes known as *secondary oocytes.* Meiosis is a double division that includes *meiosis I* (*first meiotic division* or *reduction division*) followed by *meiosis II* (*second meiotic division*), and during this entire process, a single parent cell that is diploid ($2n$) with 46 chromosomes is reduced to four cells that are each haploid (n) with 23 chromosomes.

In males the process of spermatogenesis produces four functional haploid sperm from each completed sequence. In females, oogenesis produces a secondary oocyte (sometimes referred to as the ovum or egg) along with a small *polar body* from each completed sequence of meiosis I. Oogenesis begins prior to puberty and continues through menopause. The secondary oocyte completes meiosis II only after sperm penetration occurs, producing a large fertilized ovum and a second tiny polar body that degenerates. The first small polar body from meiosis I will either degenerate or undergo the second meiotic division, resulting in two degenerating polar bodies. Polar bodies degenerate and die because they lack cytoplasmic nutrients.

Ovulation takes place when a secondary oocyte, surrounded by the *corona radiata* (granulosa cells) and by a gel layer called the *zona pellucida,* is expelled from the ovary into the uterine tube. A sperm must make its way between the granulosa cells and then penetrate the zona pellucida in order to enter the secondary oocyte. Upon fertilization, the pronuclei of the ovum and sperm come together and combine their chromosomes, forming a single cell with diploid ($2n$) chromosomes that is called a *zygote.*

Recall that mitosis was described as part of the cell cycle in Laboratory Exercise 7. The events of the cell cycle include interphase, mitosis phases (prophase, metaphase, anaphase, and telophase), and cytokinesis. There are many similarities between the various phases of mitosis and meiosis. However, meiosis differs from mitosis in some significant ways, including:

- Meiosis has two consecutive nuclear divisions: meiosis I followed by meiosis II. Mitosis is a single division.
- Meiosis results in four genetically different daughter cells; mitosis produces two genetically identical daughter cells.
- Meiosis produces daughter cells that are haploid (n); mitosis produces daughter cells that are diploid ($2n$).
- Meiosis occasionally includes exchanges of genetic material between homologous chromosomes, called *crossing-over,* which increase genetic variability. Crossing-over does not occur during mitosis. APR

Human development is divided into three stages: the pre-embryonic, embryonic, and fetal stages. The *pre-embryonic stage* occurs the first two weeks following fertilization. Shortly after fertilization, the zygote undergoes cell division (mitosis and cytokinesis) to form two cells. These two cells become four, then in turn divide into eight, and so forth. A solid cluster of cells forms, the *morula,* which is still surrounded by the zona pellucida. The cells of the morula continue cell cycles, and the hollow *blastocyst* develops. The blastocyst contains an *inner cell mass (embryoblast),* which develops into the embryo, and an outer *trophoblast* layer, which develops into the extraembryonic membranes and a portion of the *placenta.* These early *cleavage* divisions, resulting in an increasing number of cells that are smaller than the zygote, continue for most of the first week of development. The blastocyst implants into the endometrium beginning about 6–7 days after fertilization and is completed by the end of the second week. The early stages of development in some animals, such as sea urchins, are similar to those in humans.

During the *embryonic stage* of development, the blastocyst develops into a *gastrula* forming three primary germ layers: the *ectoderm, endoderm,* and *mesoderm.* These primitive tissues of the embryo give rise to tissues and organs of all major body systems. The extraembryonic membranes include the amnion, yolk sac, allantois, and chorion. These membranes arise during the second week of the pre-embryonic stage and continue to develop throughout the embryonic and fetal stages. The *amnion* extends around the embryo and is filled with amniotic fluid, providing a protective cushion and constant environment. The *yolk sac* forms early blood cells and a portion of the embryonic digestive tube; however, the placental region, not the yolk, provides the nutritive functions. The *allantois* gives rise to the umbilical blood vessels linking the embryo to the placenta, and it eventually forms part of the urinary bladder. The *chorion* extends around the embryo and the other membranes, and it eventually fuses with the amnion to form the amniochorionic membrane. In human development, the embryonic stage begins week three and lasts through the end of the eighth week and is followed by the *fetal stage* until birth. Organ systems mature as the developing fetus increasingly grows.

The *placenta* is comprised of maternal and fetal components. The maternal component is a region of the *endometrium (decidua basalis);* the fetal portion contains an elaborate series of *chorionic villi* formed from the chorion. An examination of the placenta upon birth (the afterbirth) reveals a rough surface of the maternal portion and a smooth surface with the attached umbilical cord of the fetal part.

PROCEDURE A: Meiosis and Fertilization

1. Review the introductory material related to meiosis and study figure 61.1. Examine the models of animal meiosis and place them in the proper sequence of stages of

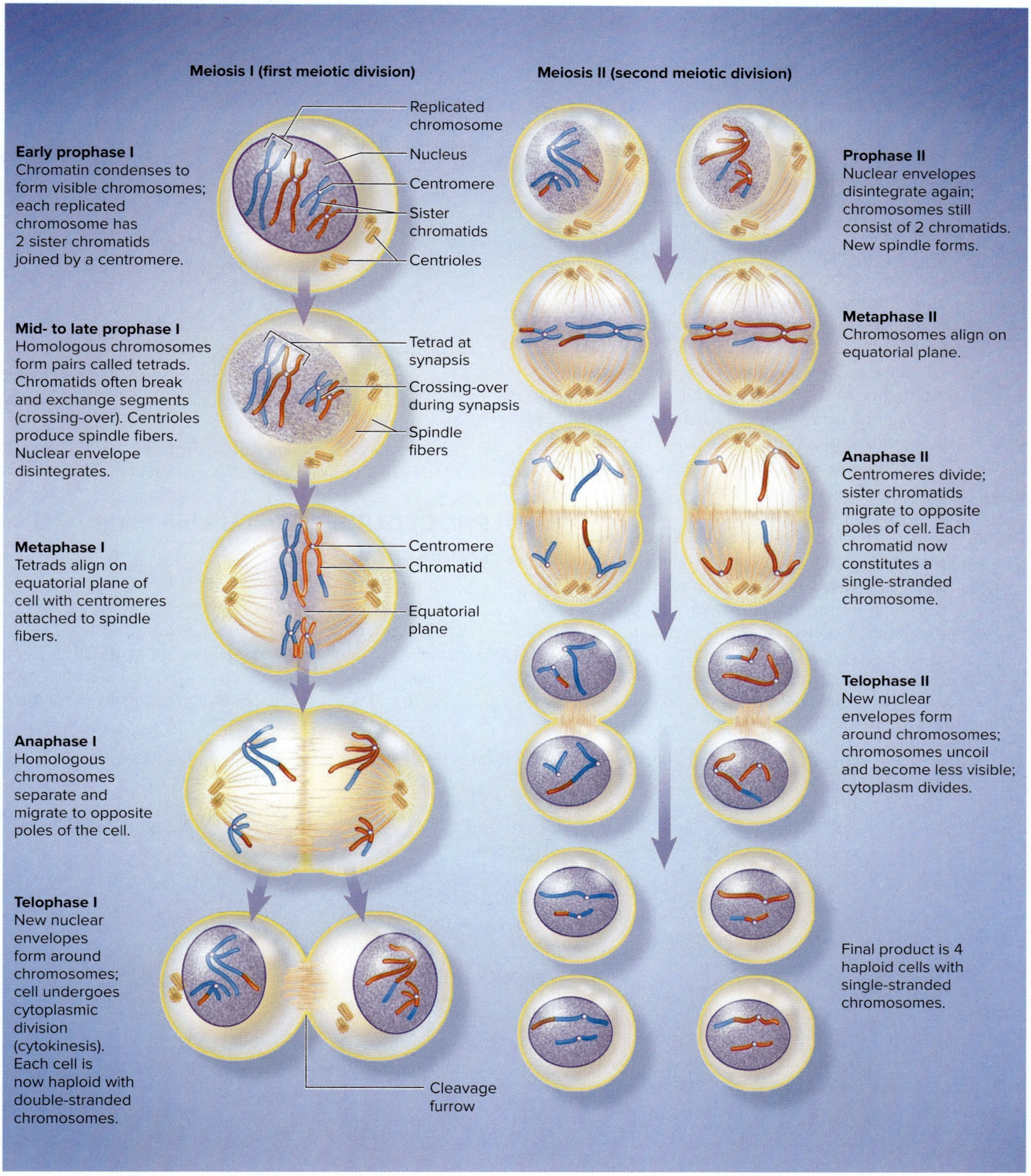

FIGURE 61.1 Meiosis phases illustrated for a cell with a diploid ($2n$) = 4 (2 pairs of homologous chromosomes). Human cells have a diploid ($2n$) = 46 (23 pairs of homologous chromosomes). APR

meiosis. Observe that there are phases during meiosis I and meiosis II that seem similar to some of the events of phases during mitosis. Note that replicated chromosomes are present in prophase I of meiosis, with sister chromatids joined by a common centromere. Also during prophase I, a *tetrad* of pairs of homologous chromosomes forms during a stage called *synapsis*. It is during this tetrad stage that portions of chromatids may be exchanged. This is called *crossing-over,* and it results in greater variations of gene combinations. The two cells produced during meiosis I are haploid as the homologous replicated chromosomes separate. DNA replication does not occur during meiosis II. Sister chromatids separate and migrate to opposite poles during anaphase II. The cells that result from meiosis II are haploid. APR

2. Review the introductory material related to fertilization and pre-embryonic development. Study figure 61.2 and models showing various stages of human fertilization, and figure 61.3 of sea urchin early developmental stages. Answer the questions in Part A of Laboratory Assessment 61.
3. Although it is difficult to observe fertilization in mammals since the process occurs internally, it is possible to view forms of external fertilization. For example, egg and sperm cells can be collected from sea urchins, and the process of fertilization can be observed microscopically. To make this observation, follow these steps:
   - **a.** Place a drop of sea urchin egg-cell suspension in the chamber of a depression slide and add a coverslip.
   - **b.** Examine the egg cells using low-power magnification.
   - **c.** Focus on a single unfertilized egg cell with high-power magnification and sketch the cell in Part A of the laboratory assessment.
   - **d.** Remove the coverslip and add a drop of sea urchin sperm-cell suspension to the depression slide. Replace the coverslip and observe the sperm cells with high-power magnification as they cluster around the egg cells. This attraction is stimulated by gamete secretions.
   - **e.** Observe the egg cells with scanning and then low-power magnification once again and watch for the appearance of *fertilization membranes.* Such a membrane forms as soon as an egg cell is penetrated by a sperm cell; it looks like a clear halo surrounding the egg cell.
   - **f.** Focus on a single fertilized egg cell and sketch it in Part A of the laboratory assessment.

LEARNING EXTENSION ACTIVITY

Use a toothpick to draw a thin line of Vaseline around the chamber of the depression slide containing the fertilized sea urchin egg cells. Place a coverslip over the chamber, and gently press it into the Vaseline to seal the chamber and prevent the liquid inside from evaporating. Keep the slide in a cool place so that the temperature never exceeds 22°C (72°F). Using low-power magnification, examine the slide every 30 minutes, and look for the appearance of two-, four-, and eight-cell stages of developing sea urchin embryos.

PROCEDURE B: Sea Urchin Early Development

1. Obtain a prepared microscope slide of early stages of developing sea urchins. There are various stages of cleavage shown in this slide. Search the slide using scanning and then low-power magnification, and locate two-cell, four-cell, eight-cell, and morula stages (fig. 61.3). Observe that cleavage results in an increase

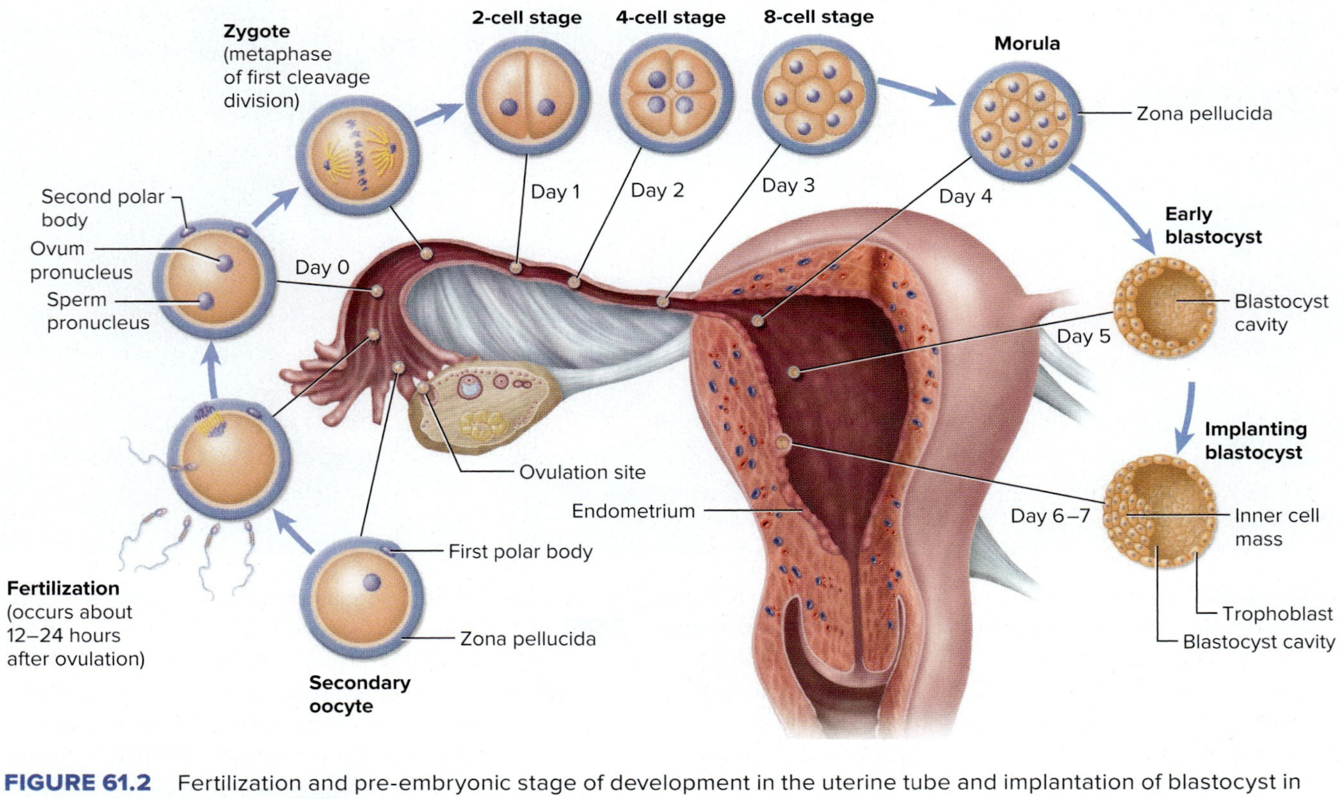

FIGURE 61.2 Fertilization and pre-embryonic stage of development in the uterine tube and implantation of blastocyst in endometrium of uterus. APR

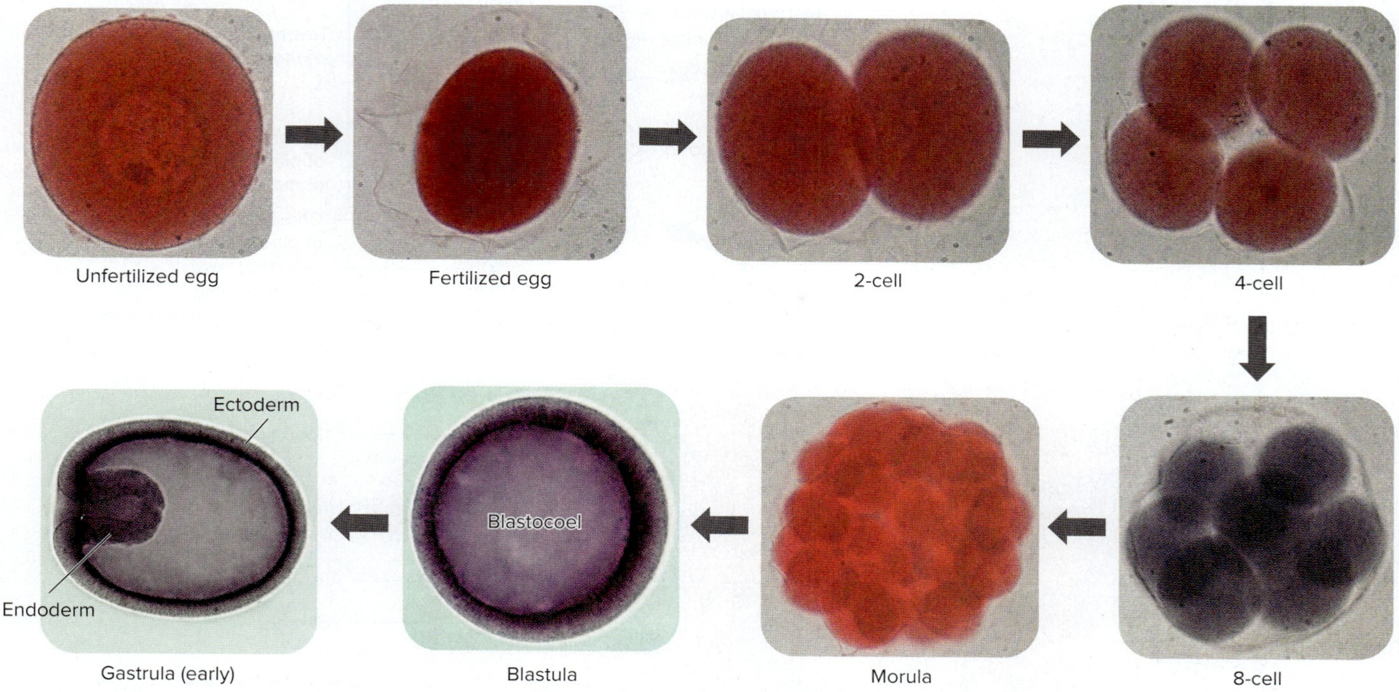

FIGURE 61.3 Sea urchin early stages of development, including an unfertilized egg and fertilized egg (400×). (Unfertilized egg, Fertilized egg, 2-cell, 4-cell, 8-cell, and Morula) Cynthia Prentice-Craver/McGraw Hill; (Blastula and Gastrula) Ed Reschke/Getty Images

of cell numbers; however, the cells get progressively smaller.

2. Using scanning and then low-power magnification, locate the blastula and identify its blastocoel (hollow central cavity). Then, locate the gastrula, and identify the ectoderm (outermost germ layer) and endoderm (innermost germ layer).
3. Prepare a sketch of each stage in Part B of the laboratory assessment.

PROCEDURE C: Human Early Development

1. Study figures 61.4 and 61.5 and the list of structures and descriptions.
2. Observe the models of human development and identify the following features:

blastocyst—attaches and embeds (implants) into the endometrium
- blastocyst cavity (blastocoel)—hollow, fluid-filled part
- inner cell mass (embryoblast)—develops into the actual embryo
- trophoblast—develops into the extraembryonic membranes and part of the placenta

primary germ layers—primitive tissues that develop into all organs
- ectoderm—outer layer; develops into epidermal structures and the nervous system
- endoderm—inner layer; gives rise to linings of digestive, respiratory, and urinary organs
- mesoderm—middle layer; develops into bones, muscles, dermis, and connective tissues

chorion—forms fetal portion of placenta

chorionic villi—branched projections of chorion

amnion—membrane that encloses embryo

amniotic fluid—protective fluid between amnion and embryo

yolk sac—produces early blood cells; forms part of the digestive tract; produces first germ cells (precursor cells to gametes)

allantois—forms part of umbilical cord

umbilical cord—suspends embryo in amniotic fluid; attached to the placenta
- umbilical arteries (2)—transport oxygen-poor blood of fetus to maternal portion of placenta
- umbilical vein (1)—transports oxygen-rich blood from maternal portion of placenta to embryo

placenta—site of exchange of nutrients, gases, and wastes between embryonic and maternal blood

3. Complete Part C of the laboratory assessment.

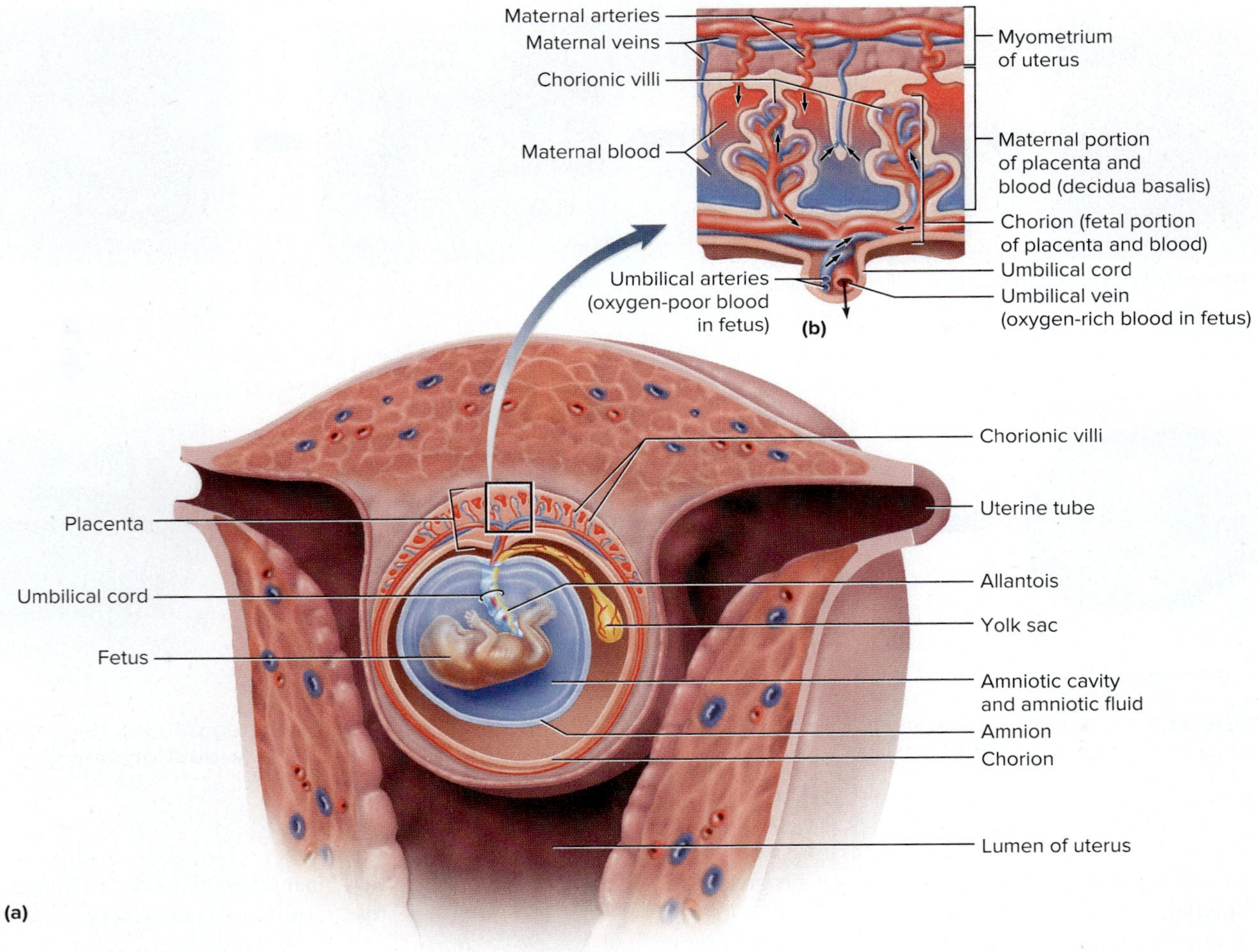

FIGURE 61.4 Structures associated with this twelve-week-old fetus: (a) fetus, membranes, and uterus; (b) fetal and maternal vasculature of placenta.

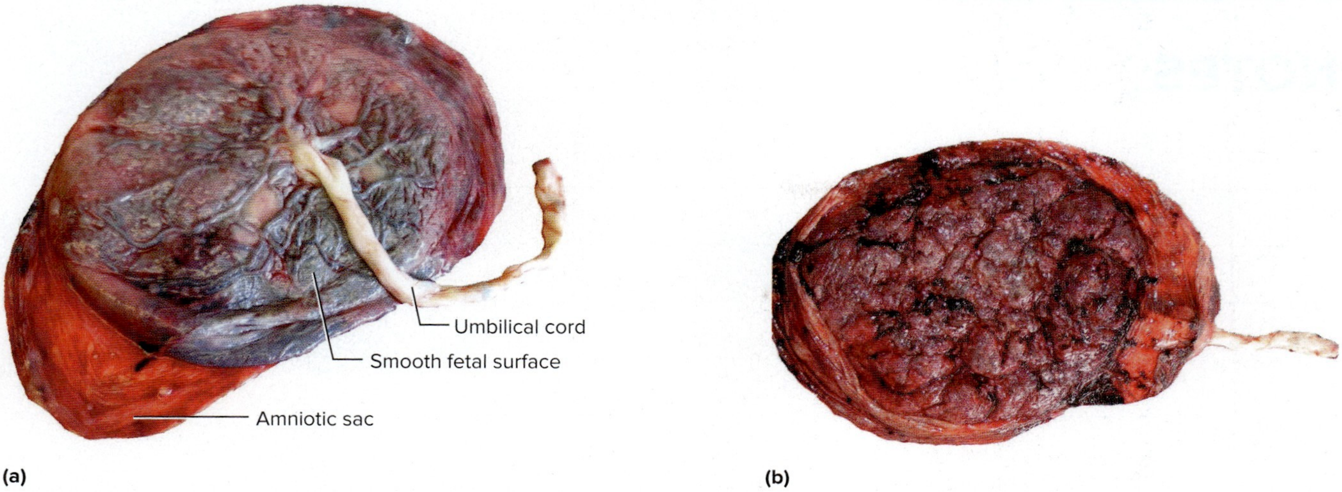

FIGURE 61.5 Human placenta (afterbirth): (a) fetal surface; (b) maternal (uterine) surface. At full term, the placenta is about 18 cm in diameter, 4 cm thick, and weighs about 450 g (one pound). Dr. Kurt Benirschke

DEMONSTRATION ACTIVITY

Observe the preserved mammalian embryos that are on display. In addition to observing the developing external body structures, identify such features as the chorion, chorionic villi, amnion, yolk sac, umbilical cord, and placenta. What special features provide clues to the types of mammals these embryos represent?

____

____

____

NOTES

LABORATORY
ASSESSMENT 61

Name ______________________________

Date ______________________________

Section ______________________________

The A corresponds to the indicated Learning Outcome(s) O found at the beginning of the Laboratory Exercise.

Meiosis, Fertilization, and Early Development

PART A: Assessments

Complete the following statements:

1. Human gametes contain ______________________ chromosomes. A1
2. The diploid number of human chromosomes found in human body cells is ______________________. A1
3. At the end of meiosis, ______________________ haploid daughter cells are produced. A1
4. To create genetic variation, ______________________ occasionally occurs during synapsis with homologous chromosomes that are arranged in a tetrad. A1
5. The cell resulting from fertilization is called a(an) ______________________. A1
6. Tiny cells that form during meiotic divisions of oogenesis and then degenerate are called ______________________. A1
7. Fertilization and the early pre-embryonic stage of development occur within the lumen of the ______________________. A1
8. A solid ball of about sixteen cells is called a(an) ______________________. A1
9. The part of meiosis that is most similar to mitosis is ______________________. A1
10. Meiosis I reduces the chromosome number in half, so it is also called ______________________ division. A1
11. Implantation of the ______________________ into the endometrium occurs around day 6–7 after fertilization. A1
12. The two-cell, four-cell, eight-cell, and morula stages of pre-embryonic development represent the ______________________ phase, where cells increase in number but are smaller than the zygote. A1
13. The primary ______________________ layers are the ectoderm, mesoderm, and endoderm from which all body tissues and organs are derived.
14. The ______________________ stage of development begins week nine and continues to birth.

Each circle below represents the microscopic field of view. In each circle, sketch the following: A2

Single unfertilized sea urchin egg (______×)

Single fertilized sea urchin egg (______×)

PART B: Assessments

Each circle below represents the microscopic field of view. In each circle, sketch the following: A1 A2

Two-cell sea urchin embryo (______×)

Four-cell sea urchin embryo (______×)

Eight-cell sea urchin embryo (______×)

Morula sea urchin embryo (______×)

Blastula of sea urchin (______×)

Gastrula of sea urchin (______×)

PART C: Assessments

1. Match the terms in column A with the descriptions in column B. Place the letter of your choice in the space provided. A3

Column A		Column B
a. Allantois	______	**1.** Gel layer around oocyte and early cleavage stages
b. Amniotic fluid	______	**2.** Hollow ball of cells
c. Blastocyst	______	**3.** Develops into extraembryonic membranes
d. Chorionic villi	______	**4.** Site of exchange of nutrients, gases, and wastes between the embryonic and maternal blood
e. Endometrium	______	**5.** Forms fetal portion of placenta
f. Fetal stage	______	**6.** Forms maternal portion of placenta
g. Gastrula	______	**7.** Forms the umbilical cord blood vessels
h. Inner cell mass	______	**8.** Forms the ectoderm, endoderm, and mesoderm
i. Placenta	______	**9.** Fluid that protects the embryo and fetus from jarring movements and provides a watery environment for development
j. Trophoblast	______	**10.** Part of the blastocyst that develops into the embryo
k. Zona pellucida	______	**11.** Stage of development that begins at the end of week eight

2. Summarize your observations of any similarities noted between sea urchin and human early development stages. A2 A3

____

____

____

____

CRITICAL THINKING ASSESSMENT—PART C

Isotretinoin (Accutane) is an oral medication prescribed to treat severe acne. It is a vitamin A derivative that is a teratogen in humans. What is a teratogen? ____

____

If this drug is taken by a pregnant woman, it can result in malformations to the skull, face, limbs, eyes, heart, and central nervous system of the unborn baby. In what stage of prenatal development (pre-embryonic, embryonic, and fetal) do you think taking this medication would cause the most harm and why? A2 A3

____

____.

NOTES

LABORATORY
EXERCISE 62

Genetics

Purpose of the Exercise

To observe some selected genetic traits, to use pennies and dice to demonstrate laws of probability, and to solve some genetic problems using a Punnett square.

MATERIALS NEEDED

Pennies (or other coins)
Dice
PTC paper
Astigmatism chart
Ichikawa's or Ishihara's color plates for color blindness test

Many examples of human genetics in this laboratory exercise are basic, external features to observe. Some traits are based upon simple Mendelian genetics involving single-gene traits. As our knowledge of genetics continues to develop, traits such as free earlobe may no longer be considered an example of simple Mendelian genetics. Recent evidence involves polygenic inheritance, effects of other genes, or environmental factors as more appropriate explanations. Therefore, the analysis of family genetics is not always feasible or meaningful.

Chapter Opener: ©101cats/Getty Images

Learning Outcomes

After completing this exercise, you should be able to:

O1 Examine and record eleven genotypes and phenotypes of selected genetic traits.

O2 Demonstrate the laws of probability using tossed pennies and dice and interpret the results.

O3 Predict genotypes and phenotypes of complete dominance, incomplete dominance, codominance, and sex-linked problems using Punnett squares.

The O corresponds to the assessments A indicated in the Laboratory Assessment for this Exercise.

Pre-Lab

Carefully read the introductory material and examine the entire lab. Be familiar with genetic terminology and genetic problems from lecture or the textbook. Answer the pre-lab questions.

Pre-Lab Questions Select the correct answer for each of the following questions:

1. The term _____ is used for a person possessing two identical alleles.
   a. homozygous **b.** heterozygous
   c. genotype **d.** phenotype
2. The term _____ is used for the appearance of the person as a result of the way the genes are expressed.
   a. homozygous **b.** heterozygous
   c. genotype **d.** phenotype
3. The _____ is a genetic tool to simulate all possible genetic combinations of offspring.
   a. genome **b.** Punnett square
   c. gamete formation **d.** gene pool
4. The allele is termed _____ if one allele determines the phenotype.
   a. homozygous **b.** heterozygous
   c. dominant **d.** recessive
5. Which of the following represents a dominant human trait?
   a. straight hairline **b.** attached earlobe
   c. dimples **d.** blood type O
6. Certain environmental factors can influence the expression of genes.
   a. True **b.** False
7. An attached earlobe is considered a recessive human trait.
   a. True **b.** False
8. A diploid chromosome number of 46 and a haploid number of 23 are normal for humans.
   a. True **b.** False
9. The possession of freckles is considered a recessive human trait.
   a. True **b.** False

Genetics is the study of inheritance of characteristics from parent to child and the patterns of transmission. The *genes* that transmit this information are coded in segments of DNA in chromosomes that encode specific proteins. Homologous chromosomes possess the same gene at the same location called its *locus.* These genes may exist in variant forms, called *alleles.* For simple patterns of inheritance, one allele is inherited from each parent and these two alleles then determine how a trait is expressed. If a person possesses two identical alleles, the person is said to be *homozygous* for that particular trait. If a person possesses two different alleles, the person is said to be *heterozygous* for that particular trait. The particular combination of these gene variants (alleles) represents the person's *genotype;* the appearance or physical manifestation of the individual that develops as a result of the way the genes are expressed, represents the person's *phenotype.*

If one allele determines the phenotype by masking the expression of the other allele in a heterozygous individual, the allele is termed *dominant.* The allele whose expression is masked is termed *recessive.* A heterozygous person is known as a *carrier* when a recessive gene for a genetic disease "skips" generations and the condition reappears. If the heterozygous condition determines an intermediate phenotype, the inheritance represents *incomplete dominance.*

If more than two allelic forms of a gene exist in the gene pool, it represents an example of *multiple alleles.* Three alleles exist related to the inheritance of the human ABO blood types. However, different alleles are *codominant* if both are expressed in the heterozygous condition.

Some characteristics inherited on the sex chromosomes result in phenotype frequencies that might be more prevalent in males or females. Such characteristics are called *sex-linked (X-linked or Y-linked) characteristics.*

Most inherited traits are controlled by the collaborative effect of more than one gene, called *polygenic inheritance.* These genes may be on the same chromosome or different chromosomes. Examples include eye color and height.

The environment may influence genes, referred to as *multifactorial inheritance.* For example, an individual may inherit genes for tallness, but if they are malnourished as a child, they may not reach full adult height.

As a result of meiosis during the formation of ova (eggs) and sperm, a mother and father each transmit an equal number of chromosomes (the haploid number 23) to form the zygote (diploid number 46). An offspring will receive one allele from each parent. These gametes combine randomly in the formation of each offspring. Hence, the *laws of probability* can be used to predict possible genotypes and phenotypes of offspring. A genetic tool called a *Punnett square* simulates all possible combinations (probabilities) in offspring genotypes and resulting phenotypes.

PROCEDURE A: Human Genotypes and Phenotypes

A complete set of genetic instructions in one human cell constitutes one's *genome.* The human genome contains more than 3 billion base pairs representing approximately 20,000 protein-encoding genes. These instructions represent our genotypes and are expressed as phenotypes sometimes clearly observable on our bodies. Some of these traits are listed in table 62.1 and are discernible in figure 62.1.

A dominant trait might be homozygous or heterozygous; only one capital letter is used along with a blank for the possible second dominant or recessive allele. For a recessive trait, two lowercase letters represent the homozygous recessive genotype for that characteristic. Dominant does not always correlate with the predominance of the allele in the gene pool; dominant means one allele will determine the appearance of the phenotype.

1. **Freckles/no freckles:** The dominant allele (*F*) determines the appearance of freckles. The homozygous recessive condition (*ff*) does not produce freckles (fig. 62.1). Record your results in the table in Part A of the laboratory assessment.

Table 62.1 Examples of Some Common Human Phenotypes

Dominant Traits and Genotypes	Recessive Traits and Genotypes
Freckles (*F*__)	No freckles (*ff*)
Widow's peak (*W*__)	Straight hairline (*ww*)
Dimples (*D*__)	No dimples (*dd*)
Free earlobe (*E*__)	Attached earlobe (*ee*)
Normal skin coloration (*M*__)	Albinism (*mm*)
Astigmatism (*A*__)	Normal vision (*aa*)
Curly hair (*C*__)	Straight hair (*cc*)
PTC taster (*T*__)	Nontaster (*tt*)
Blood type A (I^A __), B (I^B __), or AB ($I^A I^B$)	Blood type O (*ii*)
Normal color vision ($X^C X^C$), ($X^C X^c$), or ($X^C Y$)	Red-green color blindness ($X^c X^c$) or ($X^c Y$)

Dominant Traits | Recessive Traits

(a) Freckles

(b) No freckles

(c) Widow's peak

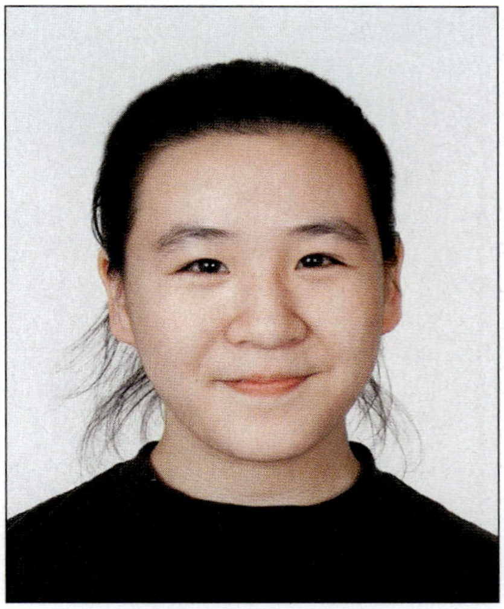

(d) Straight hairline

FIGURE 62.1 Representative genetic traits comparing dominant and recessive phenotypes: (a) freckles; (b) no freckles; (c) widow's peak; (d) straight hairline; (e) dimples; (f) no dimples; (g) free earlobe; (h) attached earlobe.

Dominant Traits

Recessive Traits

(e) Dimples

(f) No dimples

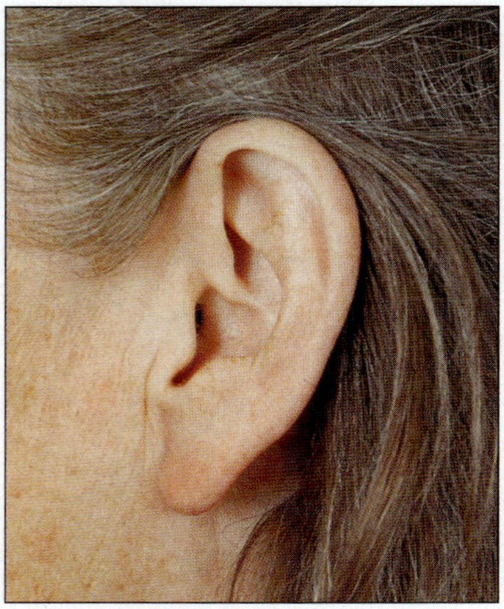

(g) Free earlobe

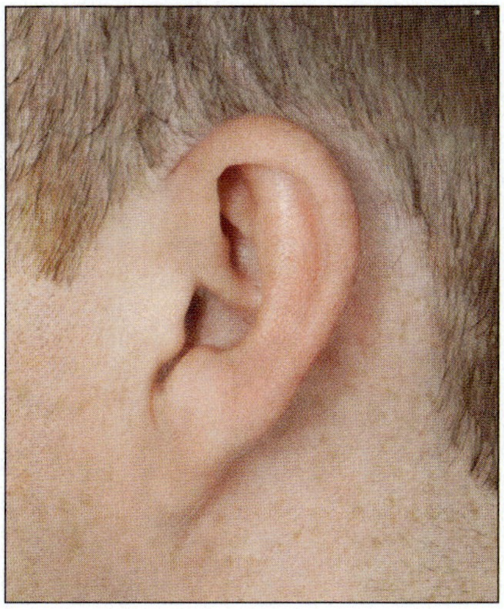

(h) Attached earlobe

FIGURE 62.1 *Continued.*

2. **Widow's peak/straight hairline:** The dominant allele (*W*) determines the appearance of a hairline above the forehead that has a distinct downward point in the center called a widow's peak. The homozygous recessive condition (*ww*) produces a straight hairline (fig. 62.1). A receding hairline would prevent this phenotype determination. Record your results in the table in Part A of the laboratory assessment.
3. **Dimples/no dimples:** The dominant allele (*D*) determines the appearance of a distinct dimple in one or both cheeks upon smiling. The homozygous recessive condition (*dd*) results in the absence of dimples (fig. 62.1). Record your results in the table in Part A of the laboratory assessment.
4. **Free earlobe/attached earlobe:** The dominant allele (*E*) codes for the appearance of an inferior earlobe that hangs freely below the attachment to the head. The homozygous recessive condition (*ee*) determines the earlobe attaching directly to the head at its inferior border (fig. 62.1). Record your results in the table in Part A of the laboratory assessment.
5. **Normal skin coloration/albinism:** The dominant allele (*M*) determines the production of some melanin, producing normal skin coloration. The homozygous recessive condition (*mm*) determines albinism due to the inability to produce or use the enzyme tyrosinase in pigment cells. A person with albinism does not produce melanin in the skin, hair, or the middle tunic (choroid, ciliary body, and iris) of the eye. The absence of melanin in the middle tunic allows the pupil to appear slightly red to nearly black. Remember that the pupil is an opening in the iris filled with transparent aqueous humor. A person with albinism has pale white skin,

flax-white hair, and a pale blue iris. Record your results in the table in Part A of the laboratory report.

6. **Astigmatism/normal vision:** The dominant allele (*A*) results in an abnormal curvature to the cornea or the lens. As a consequence, some portions of the image projected on the retina are sharply focused, and other portions are blurred. The homozygous recessive condition (*aa*) generates normal cornea and lens shapes and normal vision. Use the astigmatism chart and directions to assess this possible astigmatism condition described in Laboratory Exercise 36. Other eye conditions, such as nearsightedness (myopia) and farsightedness (hyperopia), are different genetic traits due to genes at other locations. Record your results in the table in Part A of the laboratory assessment.
7. **Curly hair/straight hair:** The dominant allele (*C*) determines the appearance of curly hair. Curly hair is somewhat flattened in cross section, as the hair follicle of a similar shape served as a mold for the root of the hair during its formation. The homozygous recessive condition (*cc*) produces straight hair. Straight hair is nearly round in cross section from being molded into this shape in the hair follicle. In some populations (European Americans) the heterozygous condition (*Cc*) expresses the intermediate wavy hair phenotype (incomplete dominance). This trait determination assumes no permanents or hair straightening procedures have been performed. Such hair alterations do not change the hair follicle shape, and future hair growth results in original genetic hair conditions. Record your unaltered hair appearance in the table in Part A of the laboratory assessment.
8. **PTC taster/nontaster:** The dominant allele (*T*) determines the ability to experience a bitter sensation when PTC paper is placed on the tongue. About 70% of people possess this dominant gene. The homozygous recessive condition (*tt*) makes a person unable to notice the substance. Place a piece of phenylthiocarbamide (PTC) paper on the upper tongue surface and chew it slightly to see if you notice a bitter sensation from this harmless chemical. The nontaster of the PTC paper does not detect any taste at all from this substance. Record your results in the table in Part A of the laboratory assessment.
9. **Blood type A, B, or AB/blood type O:** Blood type inheritance is an example of multiple-allele inheritance with codominant alleles. There are three alleles (I^A, I^B, and *i*) in the human population affecting RBC membrane structure. These alleles are located on a single pair of homologous chromosomes, so a person possesses either two of the three alleles or two of the same allele. All of the possible combinations of these alleles of genotypes and the resulting phenotypes are depicted in table 62.2. The expression of the blood type AB is a result of both codominant alleles located in the same individual. Possibly you have already determined your blood type in Laboratory Exercise 43 or have it recorded on a blood donor card. (If simulated blood-typing kits were used for Laboratory Exercise 43, those results would not be valid for your genetic factors.) Record your results in the table in Part A of the laboratory assessment.
10. **Sex determination:** A person with sex chromosomes XX displays a female phenotype. A person with sex chromosomes XY displays a male phenotype. Record your results in the table in Part A of the laboratory assessment.
11. **Normal color vision/red-green color blindness:** This condition is a sex-linked (X-linked) characteristic. The alleles for color vision are also on the X chromosome, but absent on the Y chromosome. As a result, a female might possess both alleles (*C* and *c*), one on each of the X chromosomes. The dominant allele (*C*) determines normal color vision; the homozygous recessive condition (*cc*) results in red-green color blindness. However, a male would possess only one of the two alleles for color vision because there is only a single X chromosome in a male. Hence, a male with even a single recessive gene for color blindness possesses the allele for color blindness. Note all the possible genotypes and phenotypes for this condition (table 62.1). Review the color vision test in Laboratory Exercise 36 using the color plate in figure 36.6 and color plates in Ichikawa's or Ishihara's book. Record your results in the table in Part A of the laboratory assessment.
12. Complete Part A of the laboratory assessment.

Table 62.2 Genotypes and Phenotypes (Blood Types)

Genotypes	Phenotypes (Blood Types)
I^AI^A or I^Ai	A
I^BI^B or I^Bi	B
I^AI^B (codominant)	AB
ii	O

PROCEDURE B: Laws of Probability

The laws of probability provide a mathematical way to determine the likelihood of events occurring by chance. This prediction is often expressed as a ratio of the number of results from experimental events to the number of results considered possible. For example, when tossing a coin there is an equal chance of the results displaying heads or tails. Hence, the probability is one-half of obtaining either a heads or a tails (there are two possibilities for each toss). When all of the probabilities of all possible outcomes are considered for the result, they will always add up to 1. To predict the probability of two or more events occurring in succession, multiply the probabilities of each individual event. For example, the probability of tossing a die and displaying a 4 two times in a row is 1/6 × 1/6 = 1/36 (there are six possibilities for each toss). Each toss in a sequence is an *independent event* (chance has no memory). The same laws apply when parents have multiple children (each fertilization is an independent event). Perform the following experiments to demonstrate the laws of probability:

1. Use a single penny (or other coin) and toss it 20 times. Predict the number of heads and tails that would occur from the 20 tosses. Record your prediction and the actual results observed in Part B of the laboratory assessment.
2. Use a single die (*pl.* dice) and toss it 24 times. Predict the number of times a number below 3 (numbers 1 and 2) would occur from the 24 tosses. Record your prediction and the actual results in Part B of the laboratory assessment.
3. Use two pennies and toss them simultaneously 32 times. Predict the number of times two heads, a heads and a tails, and two tails would occur. Record your prediction and the actual results in Part B of the laboratory assessment.
4. Use a pair of dice and toss them simultaneously 32 times. Predict the number of times for both dice coming up with odd numbers, one die an odd and the other an even number, and both dice coming up with even numbers. Record your prediction and the actual results in Part B of the laboratory assessment.
5. Obtain class totals for all of the coins and dice tossed by adding your individual results to a class tally location, as on the blackboard.
6. A Punnett square can be used to visually demonstrate the probable results for two pennies tossed simultaneously. For the purpose of a genetic comparison, an *h* (heads) will represent one "allele" on the coin; a *t* (tails) will represent a different "allele" on the coin.

Coin #1 Possibilities

Coin #2 Possibilities	*h*	*t*
h	*hh*	*ht*
t	*ht*	*tt*

Possible Combinations of Two Tossed Pennies (in boxes)

7. Complete Part B of the laboratory assessment.

PROCEDURE C: Genetic Problems

1. A Punnett square can be constructed to demonstrate a visual display of the predicted offspring from parents with known genotypes. Recall that, in complete dominance, a dominant allele is expressed in the phenotype, as it can mask the other recessive allele on the homologous chromosome pair. During meiosis, the homologous chromosomes with their alleles separate (Mendel's law of segregation) into different gametes. An example of such a cross might be a homozygous dominant mother for dimples (*DD*) who has offspring with a father homozygous recessive (*dd*) for the same trait. The results of such a cross, according to the laws of probability, would be represented by the following Punnett square:

Female Gametes

Male Gametes	*D*	*D*
d	*Dd*	*Dd*
d	*Dd*	*Dd*

Possible Genotypes of Offspring (in boxes)

Results: Genotypes: 100% *Dd* (all heterozygous)

Phenotypes: 100% dimples

In another example, assume that both parents are heterozygous (*Dd*) for dimples. The results of such a cross, according to the laws of probability, would be represented by the following Punnett square:

Female Gametes

Male Gametes	*D*	*d*
D	*DD*	*Dd*
d	*Dd*	*dd*

Possible Genotypes of Offspring (in boxes)

Results: Genotypes: 25% *DD* (homozygous dominant); 50% *Dd* (heterozygous); 25% *dd* (homozygous recessive) (1:2:1 genotypic ratio)

Phenotypes: 75% dimples; 25% no dimples (3:1 phenotypic ratio)

2. Work genetic problems 1 and 2 in Part C of the laboratory assessment.
3. The sickling gene that causes sickle-cell anemia is an example of incomplete dominance. Individuals with homozygous dominant alleles *(SS)* produce normal hemoglobin. Sickle-cell anemia occurs in individuals who have homozygous recessive sickling alleles *(ss)*, producing modified hemoglobin that results in sickle-shaped erythrocytes. These erythrocytes do not hold oxygen well and may get stuck in small blood vessels and block blood flow, causing pain and possibly death. Heterozygous individuals *(Ss)* carry the sickle-cell trait, having received one sickle cell gene from one parent and one normal gene from the other parent. These individuals make normal and modified hemoglobin and are usually asymptomatic unless conditions of low oxygen arise. A Punnett square can be constructed to predict the offspring of parents with known genotypes. In this example, assume the mother's known genotype is Ss and the father's is *SS*. The results of such a cross would be represented by the following Punnett square:

Female Gametes

Male Gametes	*S*	*s*
S	*SS*	*Ss*
S	*SS*	*Ss*

Possible Genotypes of Offspring (in boxes)

Results: Genotypes: 50% *Ss* (heterozygous for sickle-cell trait); 50% *SS* (homozygous dominant) (1:1 genotypic ratio)

Phenotypes: 50% make both normal and abnormal (sickling) hemoglobin; 50% make normal hemoglobin (1:1 phenotypic ratio)

4. Work genetic problem 3 in Part C of the laboratory assessment.
5. The ABO blood type inheritance represents an example of multiple alleles and codominance. Review table 62.2 for the genotypes and phenotypes for the expression of this trait. A Punnett square can be constructed to predict the offspring of parents of known genotypes. In this example, assume the genotype of the mother is I^A I^B, and the father is *ii*. The results of such a cross would be represented by the following Punnett square:

Female Gametes

Male Gametes	I^A	I^B
i	$I^A i$	$I^B i$
i	$I^A i$	$I^B i$

Possible Genotypes of Offspring (in boxes)

Results: Genotypes: 50% I^Ai (heterozygous for A); 50% I^Bi (heterozygous for B) (1:1 genotypic ratio)

Phenotypes: 50% blood type A; 50% blood type B (1:1 phenotypic ratio)

Note: In this cross, all of the children would have blood types unlike either parent.

6. Work genetic problems 4 and 5 in Part C of the laboratory assessment.
7. Review the inheritance of red-green color blindness, an X-linked characteristic, in table 62.1. A Punnett square can be constructed to predict the offspring of parents of known genotypes. In this example, assume the genotype of the mother is heterozygous X^C X^c (normal color vision but a carrier for the color blindness allele), and the father is X^C Y (normal color vision; no allele on the Y chromosome). The results of such a cross would be represented by the following Punnett square:

Female Gametes

Male Gametes	X^C	X^c
X^C	X^CX^C	X^CX^c
Y	X^CY	X^cY

Possible Genotypes of Offspring (in boxes)

Results: Genotypes: 25% X^CX^C; 25% X^CX^c; 25% X^CY; 25% X^cY (1:1:1:1 genotypic ratio)

Phenotypes for sex determination: 50% females; 50% males (1:1 phenotypic ratio)

Phenotypes for color vision: Females 100% normal color vision (however, 50% are heterozygous carriers for color blindness). Males 50% normal; 50% with red-green color blindness (1:1 phenotypic ratio)

Phenotypes for sex determination and color vision combined: 50% normal females; 25% normal males; 25% males with color blindness (2:1:1 phenotypic ratio)

Note: In X-linked inheritance, males with color blindness received the recessive gene from their mothers.

8. Complete Part C of the laboratory assessment.

NOTES

LABORATORY
ASSESSMENT 62

Name ______________________________

Date ______________________________

Section ______________________________

The A corresponds to the indicated Learning Outcome(s) O found at the beginning of the Laboratory Exercise.

Genetics

PART A: Assessments

1. Enter your test results for genotypes and phenotypes in the table. Circle your phenotype and genotype for each of the twelve traits. A1

Trait	Dominant Phenotype	Genotype	Recessive Phenotype	Genotype
Freckles	Freckles	*F*__	No freckles	*ff*
Hairline	Widow's peak	*W*__	Straight	*ww*
Dimples	Dimples	*D*__	No dimples	*dd*
Earlobe	Free	*E*__	Attached	*ee*
Skin coloration	Normal (some melanin)	*M*__	Albinism	*mm*
Vision	Astigmatism	*A*__	Normal	*aa*
Hair shape*	Curly	*C*__	Straight	*cc*
Taste	PTC taster	*T*__	Nontaster for PTC	*tt*
Blood type	A, B, or AB	I^A__; I^B__; or I^AI^B	O	*ii*
Sex		XX or XY		
Color vision	Normal	X^CX__ or X^CY	Red-green color blindness	X^cX^c or X^cY

*In some populations (European Americans) the heterozygous condition (Cc) results in the appearance of wavy hair, which actually represents an example of incomplete dominance for this trait.

2. Choose at least three dominant phenotypes that you circled, and analyze the genotypes for those traits. If it is feasible to observe your biological parents and siblings for any of these traits, are you able to determine if any of your dominant genotypes are homozygous dominant or heterozygous? ________ If so, which ones? ______________________________ Explain the rationale for your response. A1

____

____

____

____

____

____

PART B: Assessments

1. Single penny tossed 20 times and counting heads and tails: A2

Probability (prediction): _____/20 heads _____/20 tails

(Note: Traditionally, probabilities are converted to the lowest fractional representation.)

Actual results: _____ heads _____ tails

Class totals: _____ heads _____ tails

2. Single die tossed 24 times and counting the number of times a number below 3 occurs: A2

Probability: _____/24 number below 3 (numbers 1 and 2)

Actual results: _____ number below 3

Class totals: _____ number below 3 _____ total tosses by class members

3. Two pennies tossed simultaneously 32 times and counting the number of two heads, a heads and a tails, and two tails: A2

Probability: _____/32 of two heads _____/32 of a heads and a tails _____/32 of two tails

Actual results: _____two heads _____heads and tails _____two tails

Class totals: _____two heads _____heads and tails _____two tails

4. Two dice tossed simultaneously 32 times and counting the number of two odd numbers, an odd and an even number, and two even numbers: A2

Probability: _____/32 of two odd numbers _____/32 of an odd and an even number _____/32 of two even numbers

Actual results: _____ two odd numbers _____ an odd and an even number _____ two even numbers

Class totals: _____ two odd numbers _____ an odd and an even number _____ two even numbers

5. Use the example of the two dice tossed 32 times, and construct a Punnett square to represent the possible combinations that could be used to determine the probability (prediction) of odd and even numbers for the resulting tosses. Your construction should be similar to the Punnett square for the two coins tossed that is depicted in Procedure B of the laboratory exercise. A2

6. Complete the following:

a. Are the class totals closer to the predicted probabilities than your results? __________ Explain your response. A2

____

____

b. Does the first toss of the penny or the first toss of the die have any influence on the next toss? __________ Explain your response. A2

____

c. Assume a family has two boys or two girls. What is the probability that the third child will be of the opposite sex? __________ Explain your response. A2

____

d. What is the probability (prediction) that a couple without children will eventually have four children, all girls? __________ Explain your response. A2

____

____

PART C: Assessments

For each of the genetic problems, (a) determine the parents' genotypes, (b) determine the possible gametes for each parent, (c) construct a Punnett square, and (d) record the resulting genotypes and phenotypes as ratios from the cross. Problems 1 and 2 involve examples of complete dominance; problem 3 is an example of incomplete dominance; problems 4 and 5 are examples of codominance; and problem 6 is an example of sex-linked (X-linked) inheritance.

1. Determine the results from a cross of a mother who is heterozygous (*Ee*) for free earlobes with a father who is homozygous recessive (*ee*). A3

2. Determine the results from a cross of a mother and a father who are both heterozygous for freckles. A3

3. Determine the results from a cross of a mother and a father who are both heterozygous and carry the sickle-cell trait. A3

4. Determine the results from a mother who is heterozygous for blood type B and a father who is homozygous dominant for blood type A. A3

5. Determine the results from a mother who is heterozygous for blood type A and a father who is heterozygous for blood type B. A3

6. Color blindness is an example of X-linked inheritance. Hemophilia is another example of X-linked inheritance, also from a recessive allele (*h*). The dominant allele (*H*) determines whether the person possesses normal blood clotting. A person with hemophilia has a permanent tendency for hemorrhaging due to a deficiency of one of the clotting factors (VIII—antihemophilic factor). Determine the offspring from a cross of a mother who is a carrier (heterozygous) for the disease and a father with normal blood coagulation. A3

CRITICAL THINKING ASSESSMENT—PART C

Assume that the genes for hairline and earlobes are on different pairs of homologous chromosomes. Determine the genotypes and phenotypes of the offspring from a cross if both parents are heterozygous for both traits. (1) First determine the genotypes for each parent. (2) Determine the gametes, but remember each gamete has one allele for each trait (gametes are haploid). (3) Construct a Punnett square, with sixteen boxes, that has four different gametes from each parent along the top and the left edges. (This is an application to demonstrate Mendel's law of independent assortment.) (4) List the results of genotypes and phenotypes as ratios. A3

LABORATORY
EXERCISE 63

Cat Dissection: Musculature

Purpose of the Exercise

To examine and identify the skeletal musculature of the cat and to make a comparison with the corresponding musculature of the human.

MATERIALS NEEDED

Preserved cat (double injection)
Cat skeleton
Dissecting tray
Dissecting instruments
Large plastic bag
Identification tag
Disposable gloves
Bone shears
Human torso model
Human upper and lower limb models

SAFETY

- Review all the safety guidelines in Appendix 1 of your laboratory manual.
- Wear disposable gloves and protective eyewear when working on the cat dissection.
- Dispose of tissue remnants and gloves as instructed.
- Wash the dissecting tray and instruments as instructed.
- Wash your laboratory table.
- Wash your hands before leaving the laboratory.

Chapter Opener: ©101cats/Getty Images

Learning Outcomes APR

After completing this exercise, you should be able to:

O1 Locate and identify the major skeletal muscles of the cat.

O2 Name the origins, insertions, and actions of the skeletal muscles designated by the laboratory instructor.

O3 Identify the corresponding skeletal muscles of the human.

The O corresponds to the assessments A indicated in the Laboratory Assessment for this Exercise.

Pre-Lab

Carefully read the introductory material and examine the entire lab. Be familiar with the location, origin, insertion, and action of muscles of the human body from lecture, the textbook, or laboratory exercises.

SKELETAL MUSCLE DISSECTION TECHNIQUES

Dissect: To expose the entire length of the muscle from origin to insertion. Most of the procedures are accomplished with blunt probes used to separate various connective tissues that hold adjacent structures together. It does not mean to remove or to cut into the muscles or other organs.

Transect: To cut through the muscle near its midpoint (belly). The cut is perpendicular to the length of the muscle fibers.

Reflect: To lift a transected muscle aside to expose deeper muscles or other organs.

Although the aim of this exercise is to become more familiar with the human musculature, human cadavers are not always available for dissection. Instead, preserved cats are often used for dissection because they are relatively small and can be purchased from biological suppliers. Because they are mammals, cats have many features in common with humans, including similar skeletal muscles (with similar names).

Despite their similarities, cats use four limbs for locomotion, whereas humans use only two limbs. Because the musculature of each type of organism is adapted to provide for its special needs, comparisons of the muscles of cats and humans may not be precise.

As you continue your dissection of various systems of the cat, many anatomical similarities between cats and humans will be observed. These fundamental similarities are called homologous structures. Although homologous structures have a similar structure and embryological origin, the functions are sometimes different.

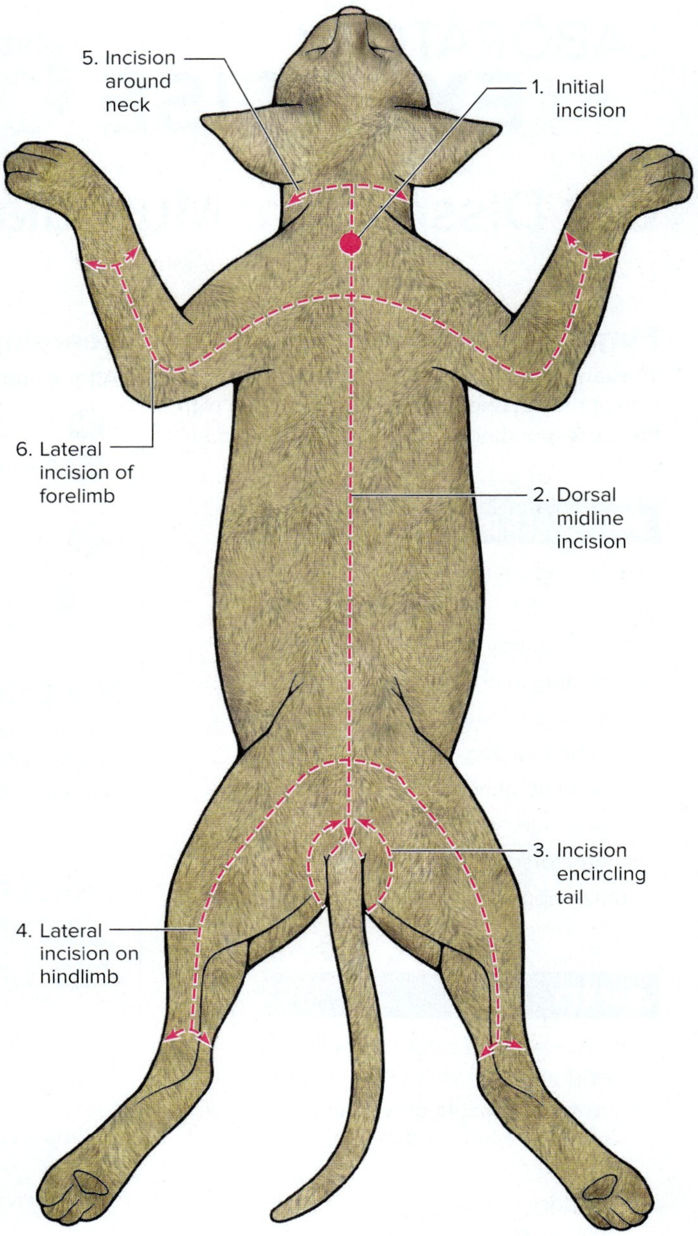

FIGURE 63.1 Incisions to be made for removing the skin of the cat (dorsal view).

PROCEDURE A: Skin Removal

1. Cats usually are preserved with a solution that prevents microorganisms from decomposing the tissues. However, because fumes from this preserving fluid may be annoying and may irritate skin, be sure to work in a well-ventilated room and wear disposable gloves to protect your hands.
2. Obtain a preserved cat, a dissecting tray, a set of dissecting instruments, a large plastic bag, and an identification tag.
3. If the cat is in a storage bag, dispose of any excess preserving fluid in the bag, as directed by the laboratory instructor. However, you may want to save some of the preserving fluid to keep the cat moist until you have completed your work later in the year.
4. If the cat is too wet with preserving fluid, blot the cat dry with paper towels.
5. Remove the skin from the cat. To do this, follow these steps:
   a. Place the cat in the dissecting tray with its ventral surface down.
   b. Use a sharp scalpel to make a short, shallow incision through the skin in the dorsal midline at the base of the cat's neck.
   c. Insert the pointed blade of a dissecting scissors through the opening in the skin and cut along the dorsal midline, following the tips of the vertebral spinous processes to the base of the tail (fig. 63.1).
   d. Use the scissors to make an incision encircling the region of the tail, anus, and genital organs. Do not remove the skin from this area.
   e. Use bone shears or a bone saw to sever the tail at its base and discard it (optional procedure).
   f. From the incision at the base of the tail, cut along the lateral surface of each hindlimb to the paw and encircle the ankle. Also clip off the claws from each paw to prevent them from tearing the plastic storage bag and scratching your skin (optional procedure).
   g. Make an incision from the initial cut at the base of the neck to encircle the neck.
   h. From the incision around the neck, cut along the lateral surface of each forelimb to the paw and encircle the wrist (fig. 63.1).
   i. Grasp the skin on either side of the dorsal midline incision, and carefully pull the skin laterally away from the body. At the same time, use your fingers or a

blunt probe to help remove the skin from the underlying muscles by separating the loose connective tissue (superficial fascia). As you pull the skin away, you may note a thin sheet of muscle tissue attached to it. This muscle is called the *cutaneous maximus,* and it functions to move the cat's skin. Humans lack the cutaneous maximus, but a similar sheet of muscle (platysma) is present in the neck of a human.

j. As you remove the skin, work toward the ventral surface, then work toward the head, and finally work toward the tail. Pull the skin over each limb as if you were removing a glove.

k. If the cat is a female, note the mammary glands on the ventral surface of the thorax and abdomen (fig. 63.2). These glands will adhere to the skin and can be removed from it and discarded.

l. After the skin has been pulled away, carefully remove as much of the remaining connective tissue as possible to expose the underlying skeletal muscles. The muscles should appear light brown and fibrous.

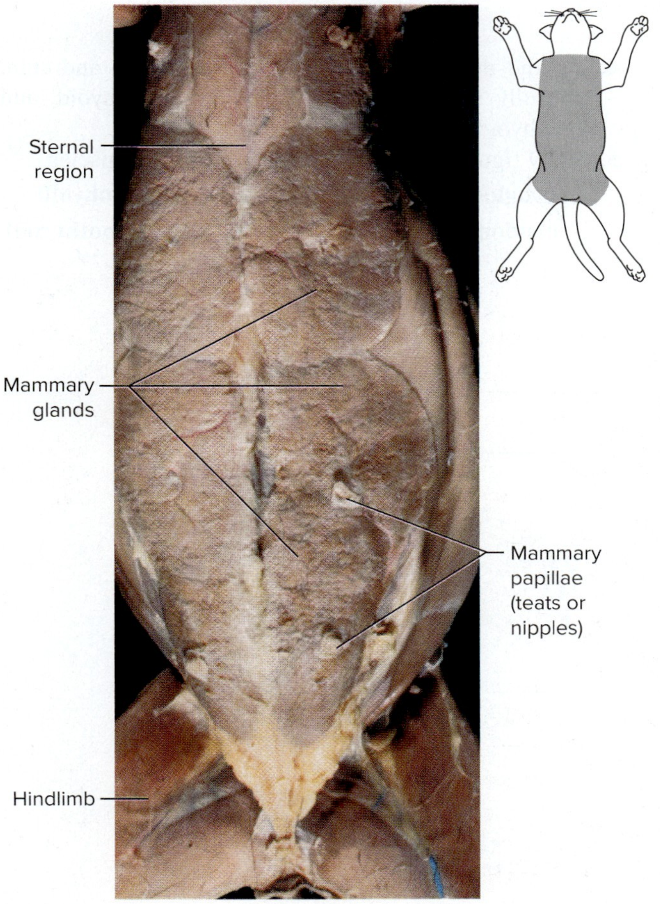

FIGURE 63.2 Mammary glands on the ventral surface of the cat's thorax and abdomen. Dr. Sheril Burton

6. After skinning the cat, follow these steps:
   a. Discard the tissues you have removed as directed by the laboratory instructor.
   b. Wrap the skin around the cat to help keep its body moist, and place it in a plastic storage bag.
   c. Write your name in pencil on an identification tag, and tie the tag to the storage bag so that you can identify your specimen.
7. Observe the recommended safety procedures for the conclusion of a laboratory session.

PROCEDURE B: Skeletal Muscle Dissection

1. The purpose of a skeletal muscle dissection is to separate the individual muscles from any surrounding tissues in order to expose the muscles for observation. To do a muscle dissection, follow these steps:
   a. Use the appropriate figure as a guide and locate the muscle to be dissected in the specimen.
   b. Use a blunt probe or a finger to separate the muscle from the surrounding connective tissue along its natural borders. The muscle should separate smoothly. If the border appears ragged, you probably have torn the muscle fibers.
2. If it is necessary to cut through (transect) a superficial muscle to observe a deeper one, use scissors to transect the muscle about halfway between its origin and insertion. Then, lift aside (reflect) the cut ends, leaving their attachments intact.
3. The following procedures will instruct you to dissect some of the larger and more easily identified muscles of the cat. In each case, the procedure will include the names of the muscles to be dissected, figures illustrating their locations, and tables listing the origins, insertions, and actions of the muscles. Compare the muscle origins and insertions to the cat skeleton (fig. 63.3). (More detailed dissection instructions can be obtained by consulting an additional guide or atlas for cat dissection.) As you dissect each muscle, review the muscle figures and tables in the lab manual and identify any corresponding (homologous) muscles of the human body. Also locate these homologous muscles in the human torso model or models of the human upper and lower limbs.

PROCEDURE C: Muscles of the Head and Neck

1. Place the cat in the dissecting tray with its ventral surface up.
2. Remove the skin and underlying connective tissues from each side of the neck, forward to the chin, and up to each ear.

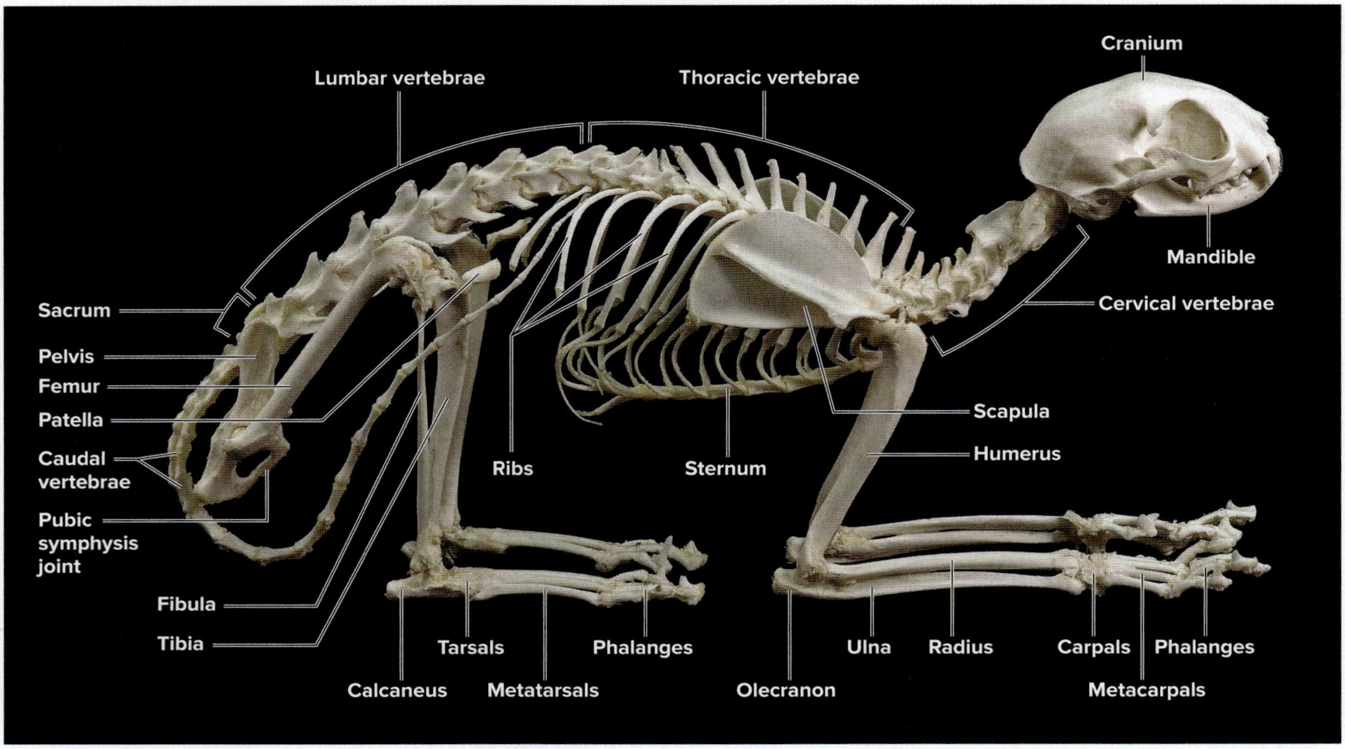

FIGURE 63.3 Skeleton of a cat, lateral view. Christine Eckel/McGraw Hill

3. Study figures 63.4 and 63.5, and then locate and dissect the following muscles:

   sternomastoid **masseter**
   sternohyoid **mylohyoid**
   digastric

4. To find the deep muscles of the cat's neck and chin, carefully remove the sternomastoid, sternohyoid, and mylohyoid muscles from the right side.
5. Study figure 63.6 and locate the following muscles:

   hyoglossus **styloglossus** **geniohyoid**
   cleidomastoid **thyrohyoid** **sternothyroid**

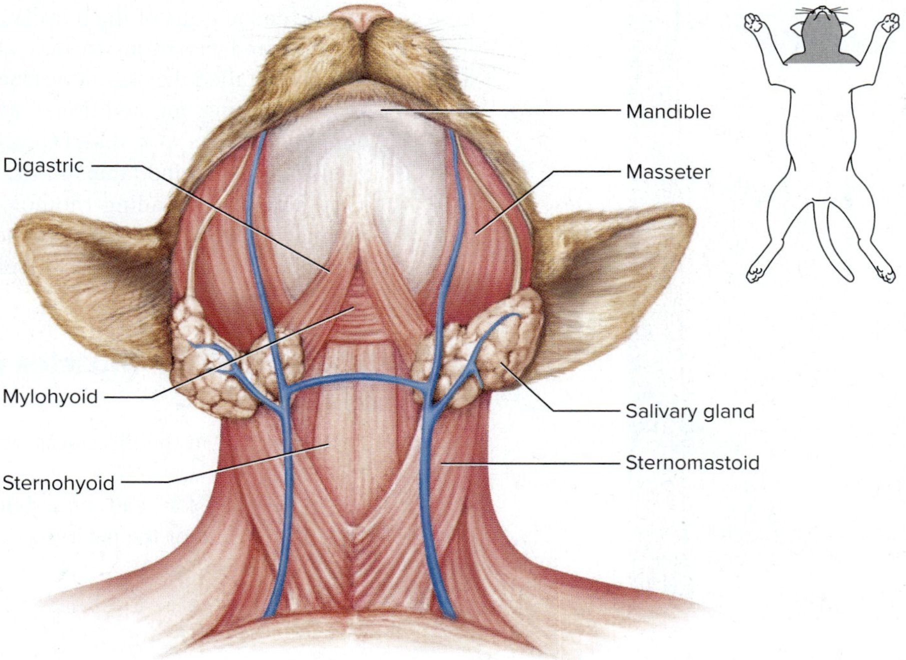

FIGURE 63.4 Muscles on the ventral surface of the cat's head and neck. APR

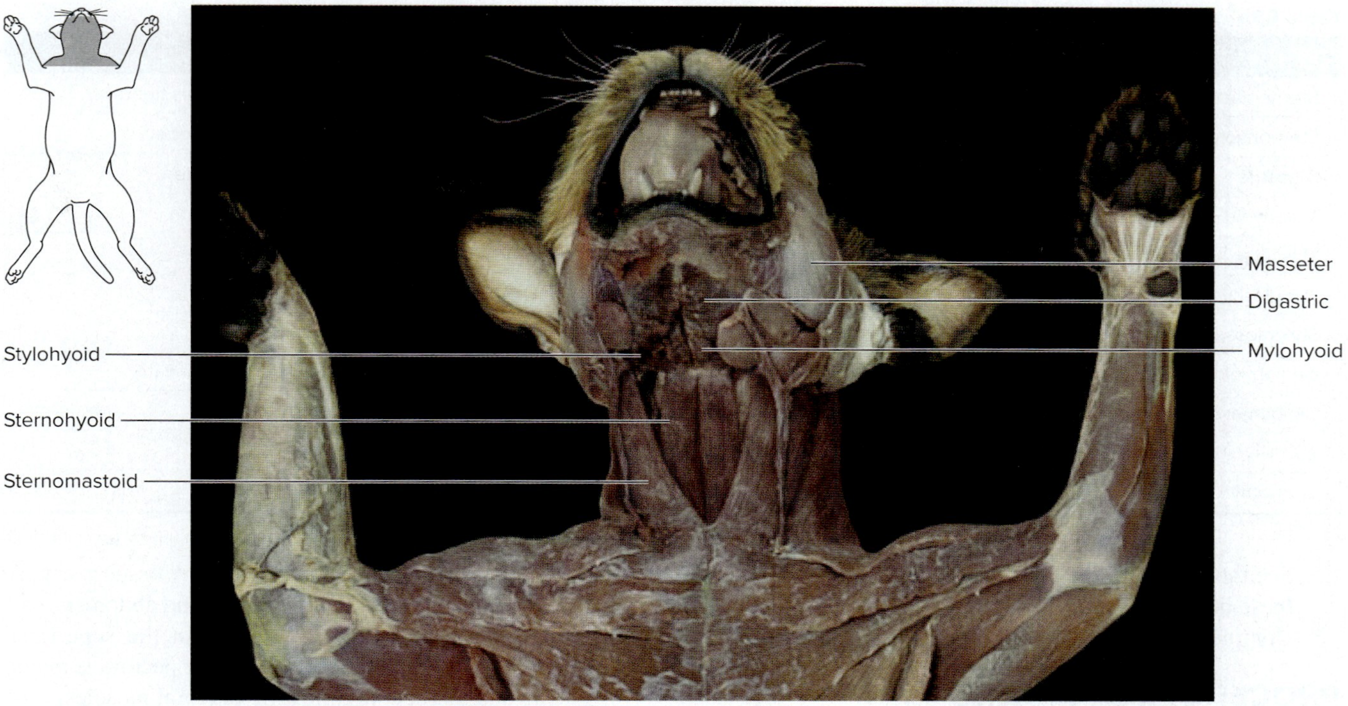

FIGURE 63.5 Deep muscles of the ventral surface of the cat's neck. APR/McGraw Hill

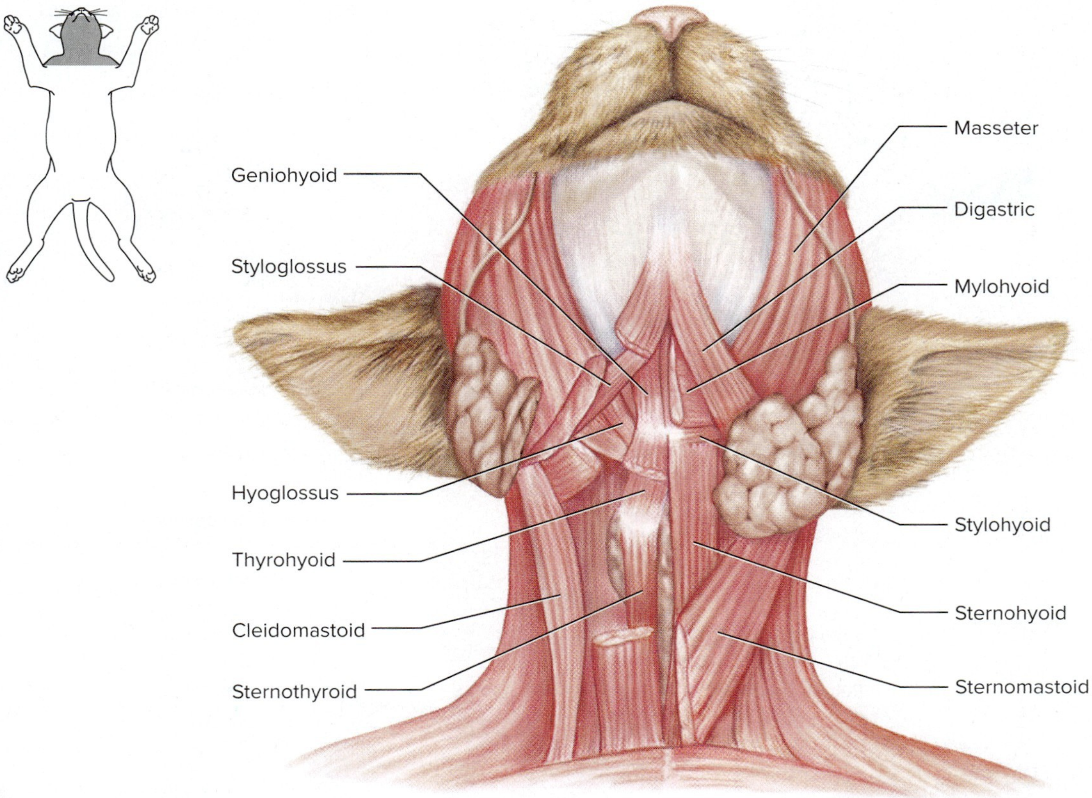

FIGURE 63.6 Deep muscles of the cat's neck and chin regions.

Table 63.1 Muscles of the Head and Neck

Muscle	Origin	Insertion	Action
Sternomastoid	Manubrium of sternum	Mastoid process of temporal bone	Turns and depresses head
Sternohyoid	Costal cartilage	Hyoid bone	Depresses hyoid bone
Digastric	Mastoid process and occipital bone	Ventral border of mandible	Depresses mandible
Mylohyoid	Inner surface of mandible	Hyoid bone	Raises floor of mouth
Masseter	Zygomatic arch	Coronoid fossa of mandible	Elevates mandible
Hyoglossus	Hyoid bone	Tongue	Pulls tongue back
Styloglossus	Styloid process	Tip of tongue	Pulls tongue back
Geniohyoid	Inner surface of mandible	Hyoid bone	Pulls hyoid bone forward
Cleidomastoid	Clavicle	Mastoid process	Turns head to side
Thyrohyoid	Thyroid cartilage	Hyoid bone	Raises larynx
Sternothyroid	Sternum	Thyroid cartilage	Pulls larynx back

6. See table 63.1 for the origins, insertions, and actions of these head and neck muscles.
7. Complete Part A of Laboratory Assessment 63.

PROCEDURE D: Muscles of the Thorax

1. Place the cat in the dissecting tray with its ventral surface up.
2. Remove any remaining fat and connective tissue to expose the muscles in the walls of the thorax and abdomen.
3. Study figures 63.7 and 63.8. Transect the superficial pectoantebrachialis to expose the entire pectoralis major. Locate and dissect the following pectoral muscles:

    pectoantebrachialis—well developed in cats
    pectoralis major
    pectoralis minor
    xiphihumeralis

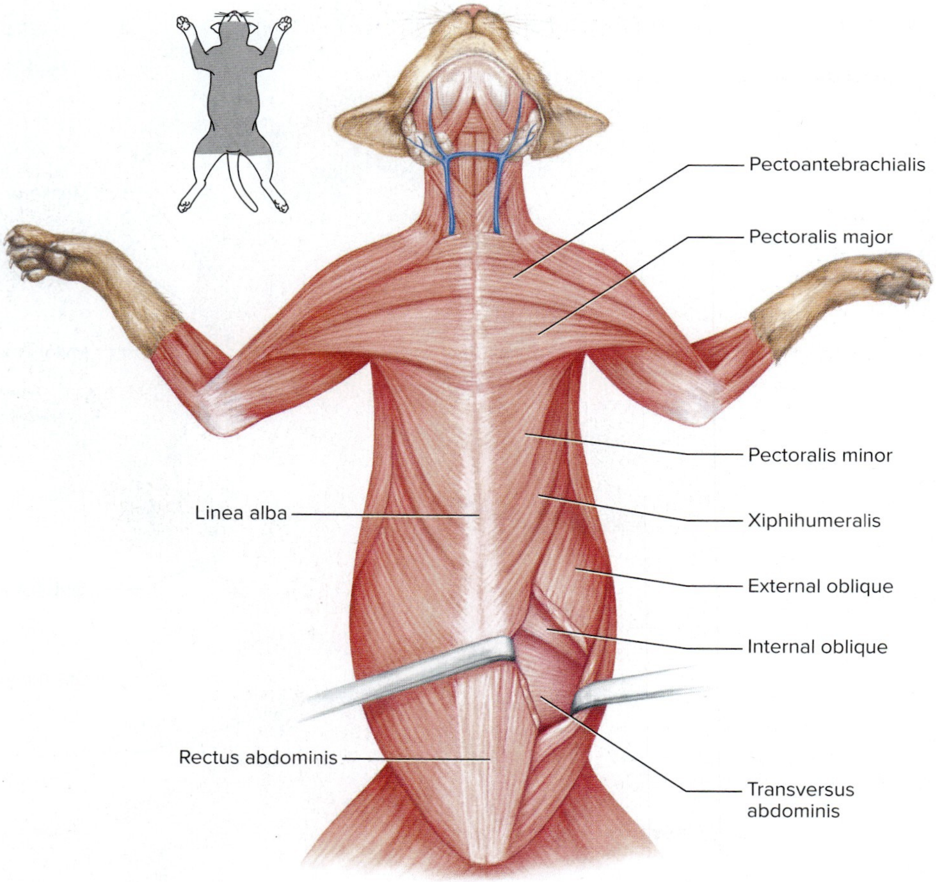

FIGURE 63.7 Muscles of the cat's thorax and abdomen. APR

4. To find the deep thoracic muscles, transect the pectoralis minor, pectoralis major, and pectoantebrachialis muscles, and reflect their cut edges to the sides. Remove the underlying fat and connective tissues to expose the following muscles, as shown in figure 63.9:

scalenus	**levator scapulae**
serratus ventralis	**transversus costarum**

5. See table 63.2 for the origins, insertions, and actions of these muscles.

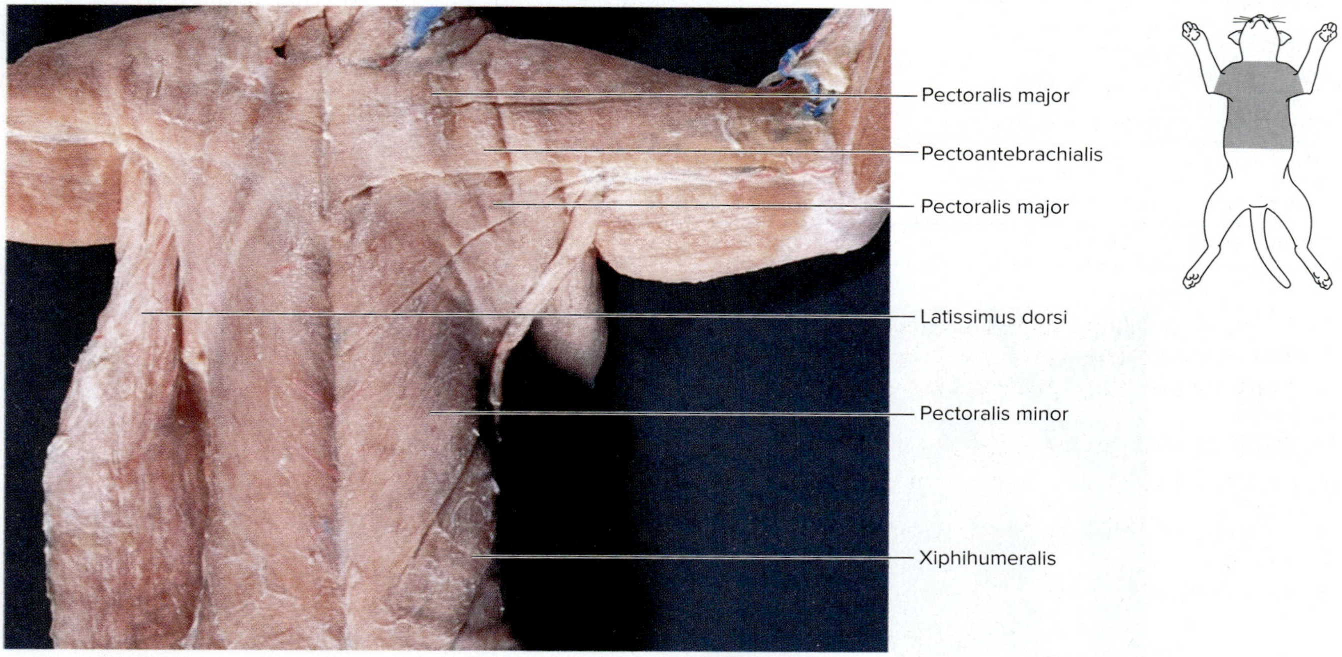

FIGURE 63.8 Superficial muscles of the cat's thorax. Dr. Sheril Burton

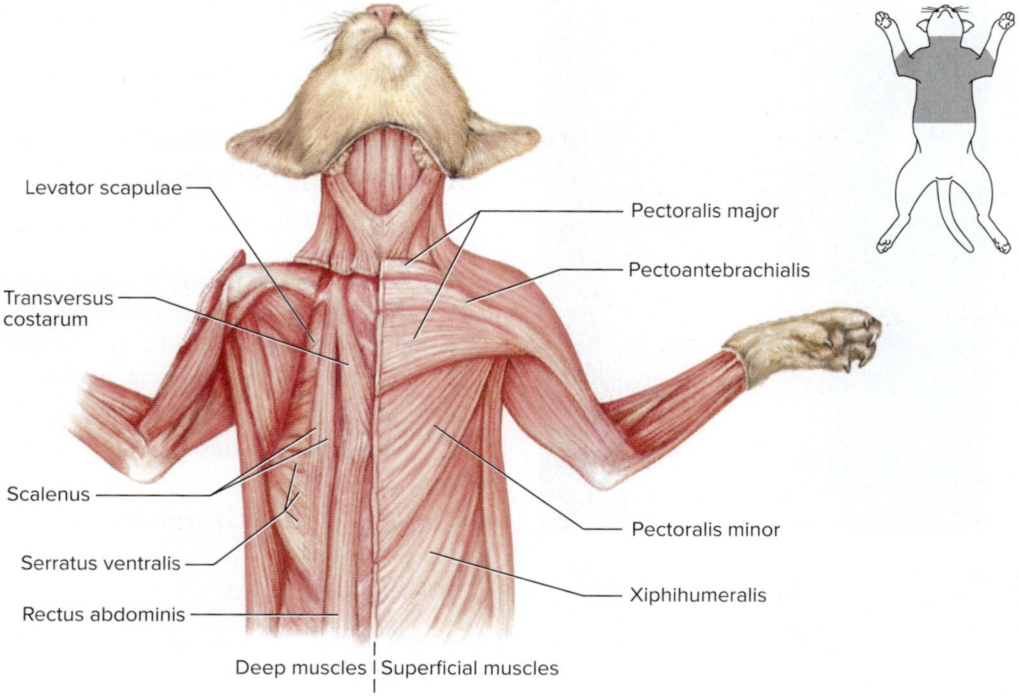

FIGURE 63.9 Superficial and deep muscles of the cat's thorax.

Table 63.2 Muscles of the Thorax

Muscle	Origin	Insertion	Action
Pectoantebrachialis	Manubrium of sternum	Fascia of forelimb	Adducts forelimb
Pectoralis major	Manubrium of sternum and costal cartilages	Greater tubercle of humerus	Adducts and rotates forelimb
Pectoralis minor	Sternum	Proximal end of humerus	Adducts and rotates forelimb
Xiphihumeralis	Xiphoid process of sternum	Proximal end of humerus	Adducts forelimb
Scalenus	Ribs	Cervical vertebrae	Flexes neck
Serratus ventralis	Ribs	Scapula	Pulls scapula ventrally and posteriorly
Levator scapulae	Cervical vertebrae	Scapula	Pulls scapula ventrally and anteriorly
Transversus costarum	Sternum	First rib	Pulls sternum toward head

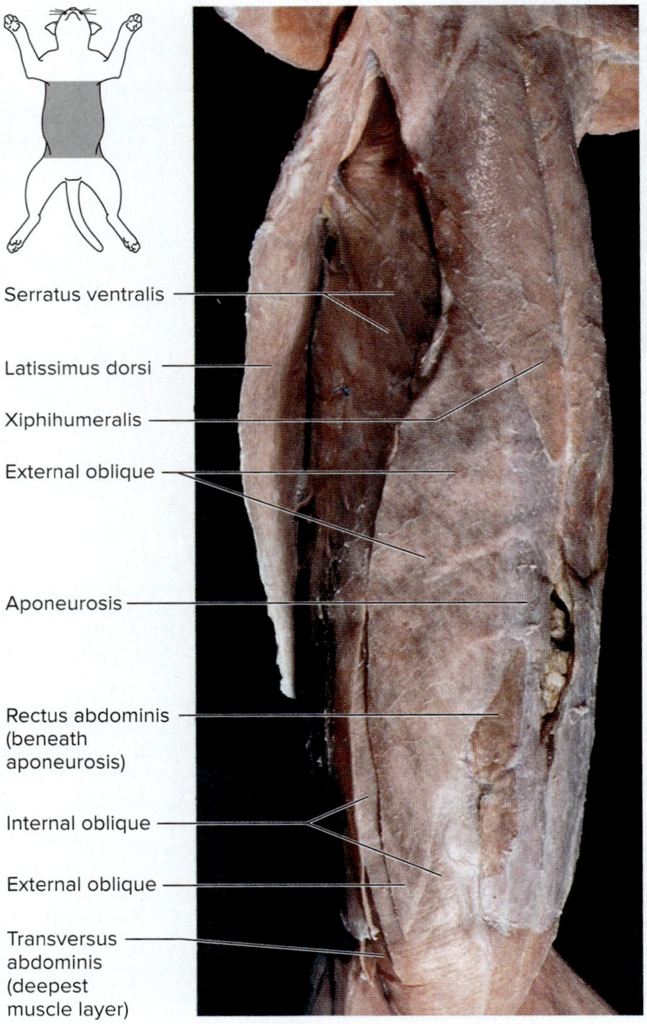

FIGURE 63.10 Muscles of the cat's thorax and abdomen. Some abdominal muscles are partially reflected to show deeper muscles. Dr. Sheril Burton APR

PROCEDURE E: Muscles of the Abdominal Wall

1. Study figures 63.7 and 63.10.
2. Locate the *external oblique* muscle in the abdominal wall.
3. Make a shallow, longitudinal incision through the external oblique. Reflect the cut edge and expose the *internal oblique* muscle beneath (fig. 63.7). The fibers of the internal oblique run at a right angle to those of the external oblique.
4. Make a longitudinal incision through the internal oblique. Reflect the cut edge and expose the *transversus abdominis* muscle.
5. Expose the *rectus abdominis* muscle on one side of the midventral line. This muscle lies beneath an aponeurosis.
6. See table 63.3 for the origins, insertions, and actions of these muscles.
7. Complete Part B of the laboratory assessment.

PROCEDURE F: Muscles of the Shoulder and Back

1. Place the cat in the dissecting tray with its dorsal surface up.
2. Remove any remaining fat and connective tissue to expose the muscles of the shoulder and back.
3. Study figures 63.11 and 63.12, and then locate and dissect the following superficial muscles:

   clavotrapezius
   acromiotrapezius
   spinotrapezius
   levator scapulae ventralis
   clavodeltoid
   acromiodeltoid
   spinodeltoid
   latissimus dorsi

Table 63.3 Muscles of the Abdominal Wall

Muscle	Origin	Insertion	Action
External oblique	Ribs and fascia of back	Linea alba	Compresses abdominal wall
Internal oblique	Fascia of back	Linea alba	Compresses abdominal wall
Transversus abdominis	Lower ribs	Linea alba	Compresses abdominal wall
Rectus abdominis	Pubic symphysis	Sternum and costal cartilages	Compresses abdominal wall and flexes trunk

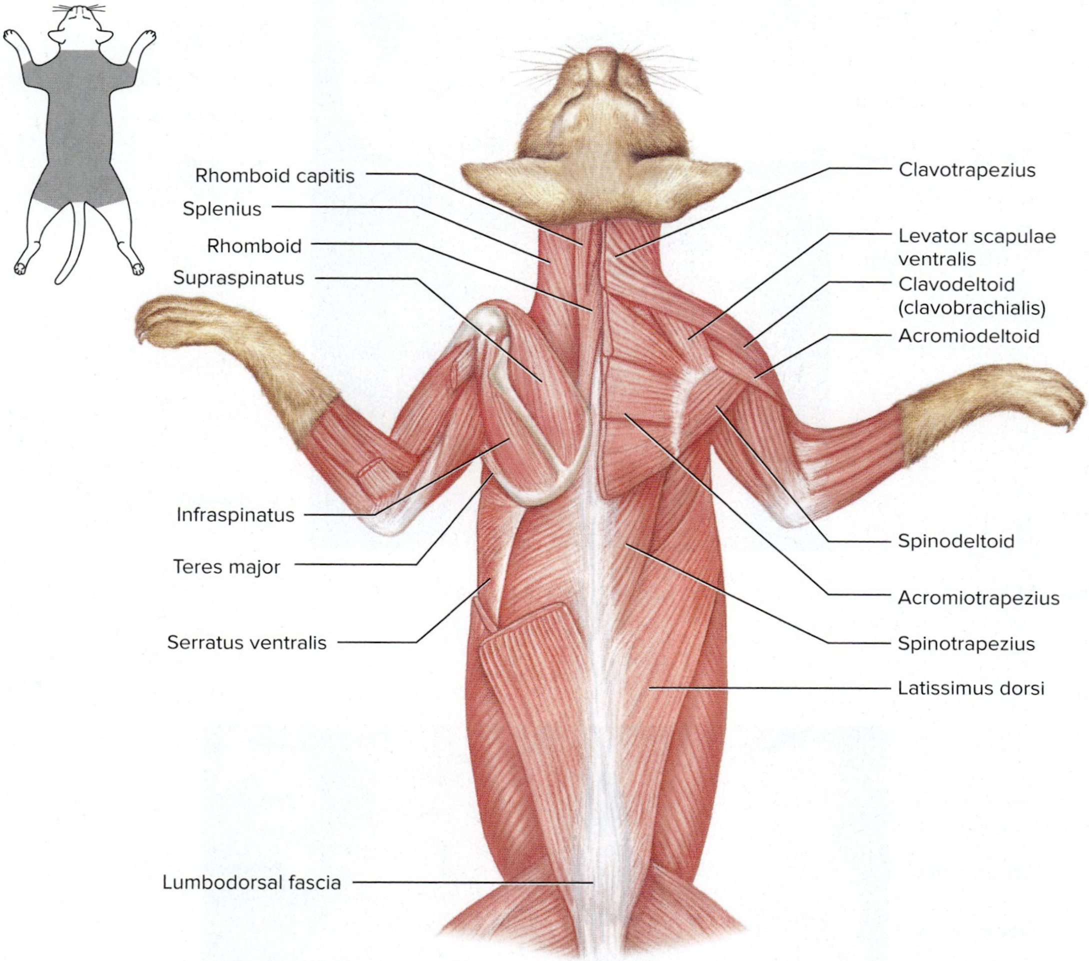

FIGURE 63.11 Superficial muscles of the cat's shoulder and back (*right side*); deep muscles of the cat's shoulder and back (*left side*). APR

Note: The clavodeltoid (clavobrachialis) appears as a continuation of the clavotrapezius.

4. Using scissors, transect the latissimus dorsi and the group of trapezius muscles. Reflect their cut edges and remove any underlying fat and connective tissue. Study figures 63.11 to 63.13, and then locate and dissect the following deep muscles of the shoulder and back:

supraspinatus	**infraspinatus**
teres major	**rhomboid**
rhomboid capitis	**splenius**

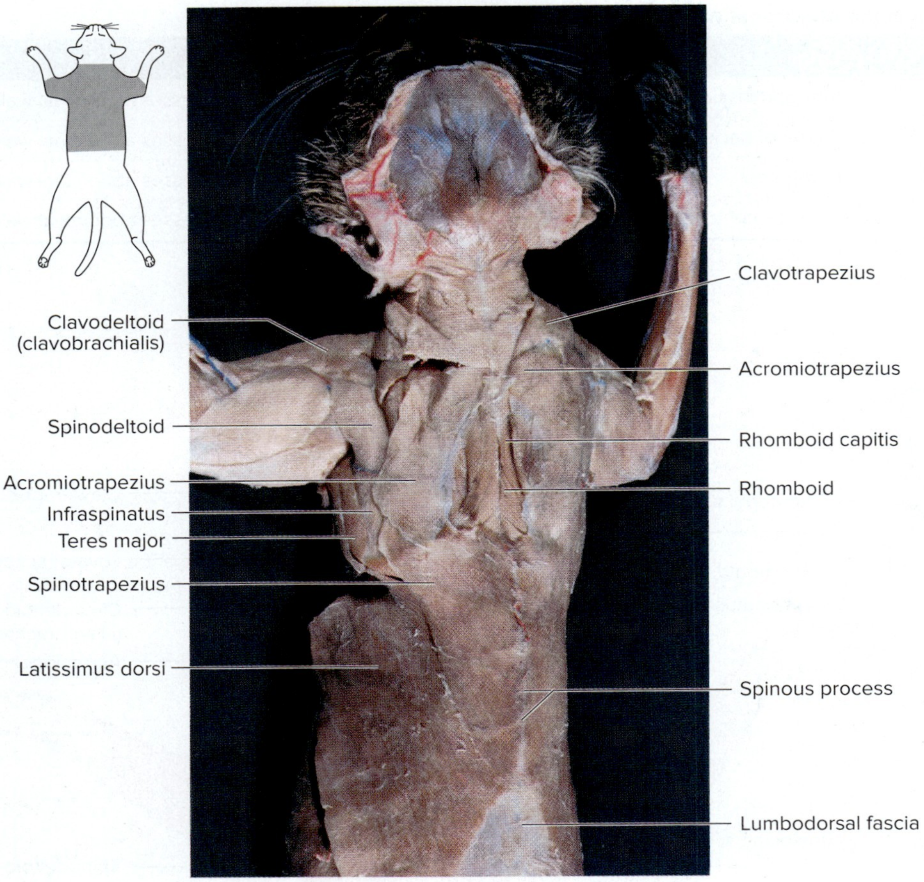

FIGURE 63.12 Muscles of the cat's shoulder and back. Dr. Sheril Burton

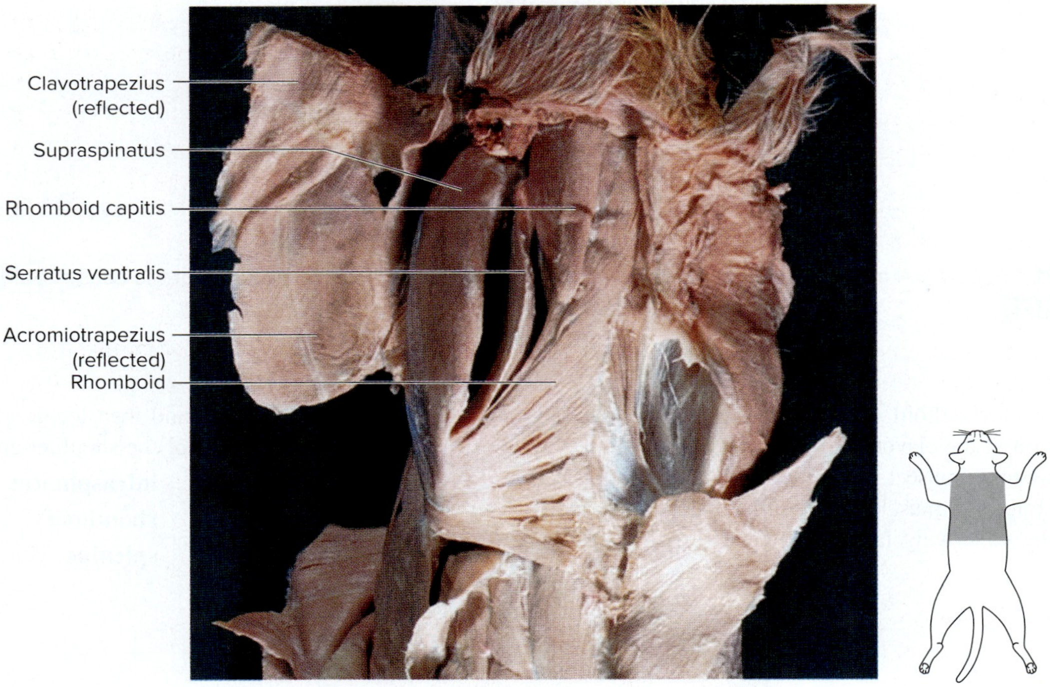

FIGURE 63.13 Deep muscles of the cat's back. The left forelimb is adducted for photograph. Dr. Sheril Burton

Table 63.4 Muscles of the Shoulder and Back

Muscle	Origin	Insertion	Action
Clavotrapezius	Occipital bone of skull and spines of cervical vertebrae	Clavicle	Pulls clavicle upward
Acromiotrapezius	Spines of cervical and thoracic vertebrae	Spine and acromion of scapula	Pulls scapula upward
Spinotrapezius	Spines of thoracic vertebrae	Spine of scapula	Pulls scapula upward and back
Levator scapulae ventralis	Occipital bone of skull	Vertebral border of scapula	Pulls scapula toward head
Clavodeltoid	Clavicle	Ulna near semilunar notch	Flexes forelimb
Acromiodeltoid	Acromion of scapula	Proximal end of humerus	Flexes and rotates forelimb
Spinodeltoid	Spine of scapula	Proximal end of humerus	Flexes and rotates forelimb
Latissimus dorsi	Thoracic and lumbar vertebrae	Proximal end of humerus	Pulls forelimb upward and back
Supraspinatus	Fossa above spine of scapula	Greater tubercle of humerus	Extends forelimb
Infraspinatus	Fossa below spine of scapula	Greater tubercle of humerus	Rotates forelimb
Teres major	Posterior border of scapula	Proximal end of humerus	Rotates forelimb
Rhomboid	Spines of cervical and thoracic vertebrae	Medial border of scapula	Pulls scapula back
Rhomboid capitis	Occipital bone of skull	Medial border of scapula	Pulls scapula forward
Splenius	Fascia of neck	Occipital bone	Turns and raises head

5. See table 63.4 for the origins, insertions, and actions of these muscles in the shoulder and back.
6. Review the locations, origins, insertions, and actions designated by the laboratory instructor without the aid of the figures and tables.
7. Complete Part C of the laboratory assessment.

PROCEDURE G: Muscles of the Forelimb

1. Place the cat in the dissecting tray with its ventral surface up.
2. Remove any remaining fat and connective tissue from a forelimb to expose the muscles.
3. Study figure 63.14*a,* and then locate and dissect the following muscles from the medial surface of the upper forelimb:

    epitrochlearis—well developed in the cat

    triceps brachii

4. Transect the pectoral muscles (pectoantebrachialis, pectoralis major, and pectoralis minor) and reflect their cut edges.
5. Study figures 63.14*b* and 63.15, and then locate and dissect the following muscles:

    coracobrachialis

    biceps brachii

    long head of the triceps brachii

    medial head of the triceps brachii

6. Study figures 63.16*a* and 63.17, which show the lateral surface of the upper forelimb.
7. Locate the lateral head of the triceps muscle, transect it, and reflect its cut edges.
8. Study figure 63.16*b*. Locate and dissect the following deeper muscles:

    anconeus

    brachialis

9. Separate the muscles in the lower forelimb into an *extensor group* (fig. 63.16*a*) and a *flexor group* (fig. 63.14*a*). Generally, the extensors are located on the posterior surface of the lower forelimb, and the flexors are located on the anterior surface. A thick, tough layer of fascia (antebrachial fascia) surrounds the forelimb. Remove this fascia to expose and dissect the following muscles:

    brachioradialis

    extensor carpi radialis

    extensor digitorum communis

    extensor digitorum lateralis

    extensor carpi ulnaris

    flexor carpi ulnaris

    palmaris longus

    flexor carpi radialis

    pronator teres

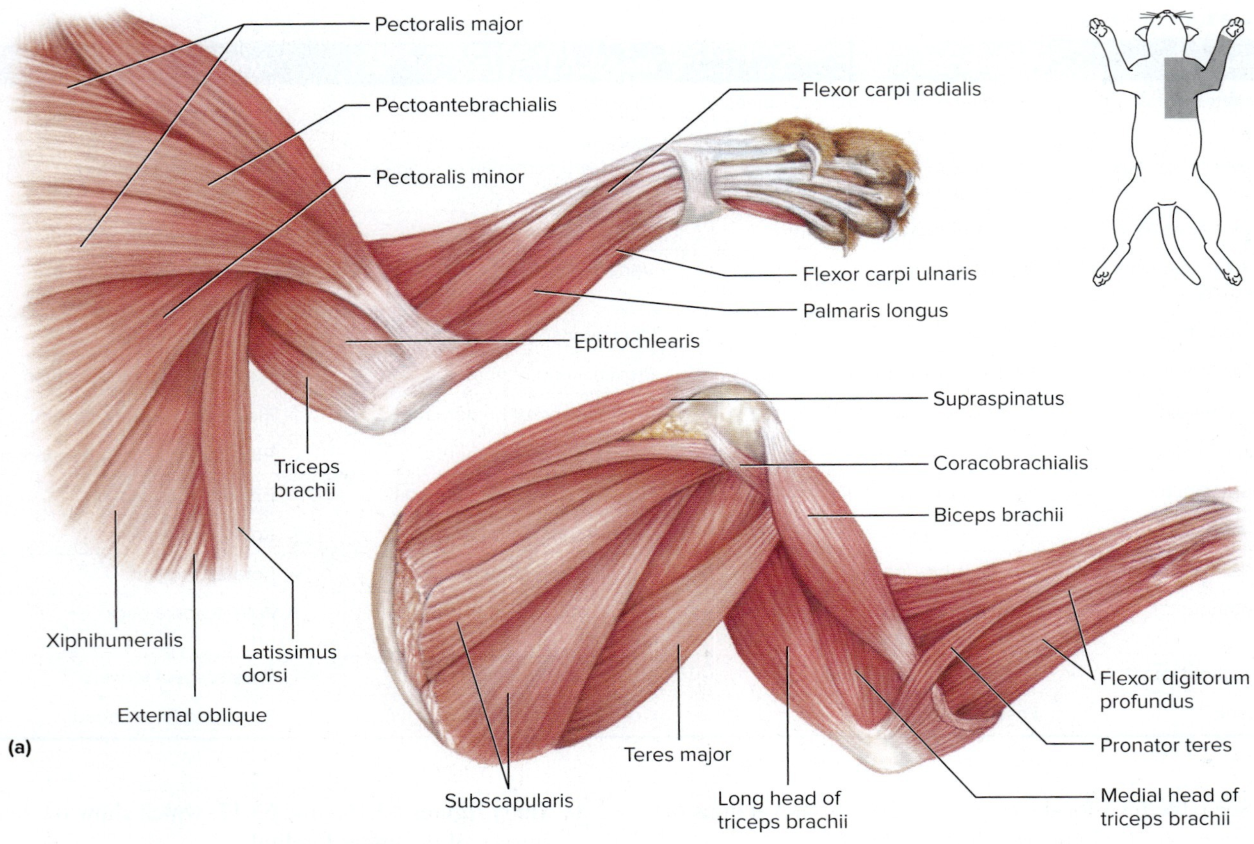

FIGURE 63.14 (a) Superficial muscles of the cat's left medial shoulder and forelimb. (b) Deep muscles of the cat's left medial shoulder and forelimb. APR

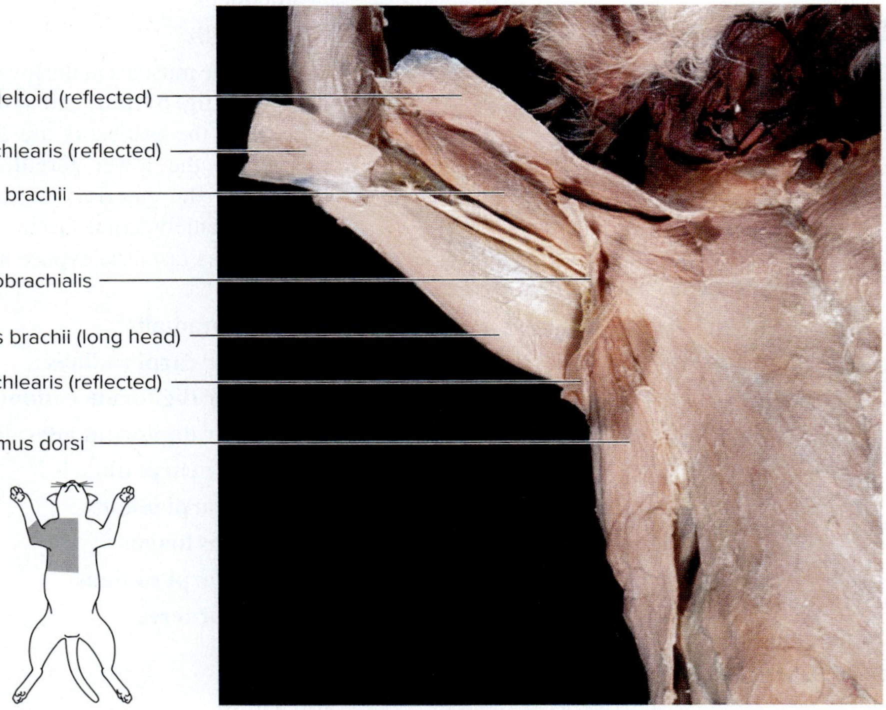

FIGURE 63.15 Deep muscles of the cat's right medial forelimb. Dr. Sheril Burton

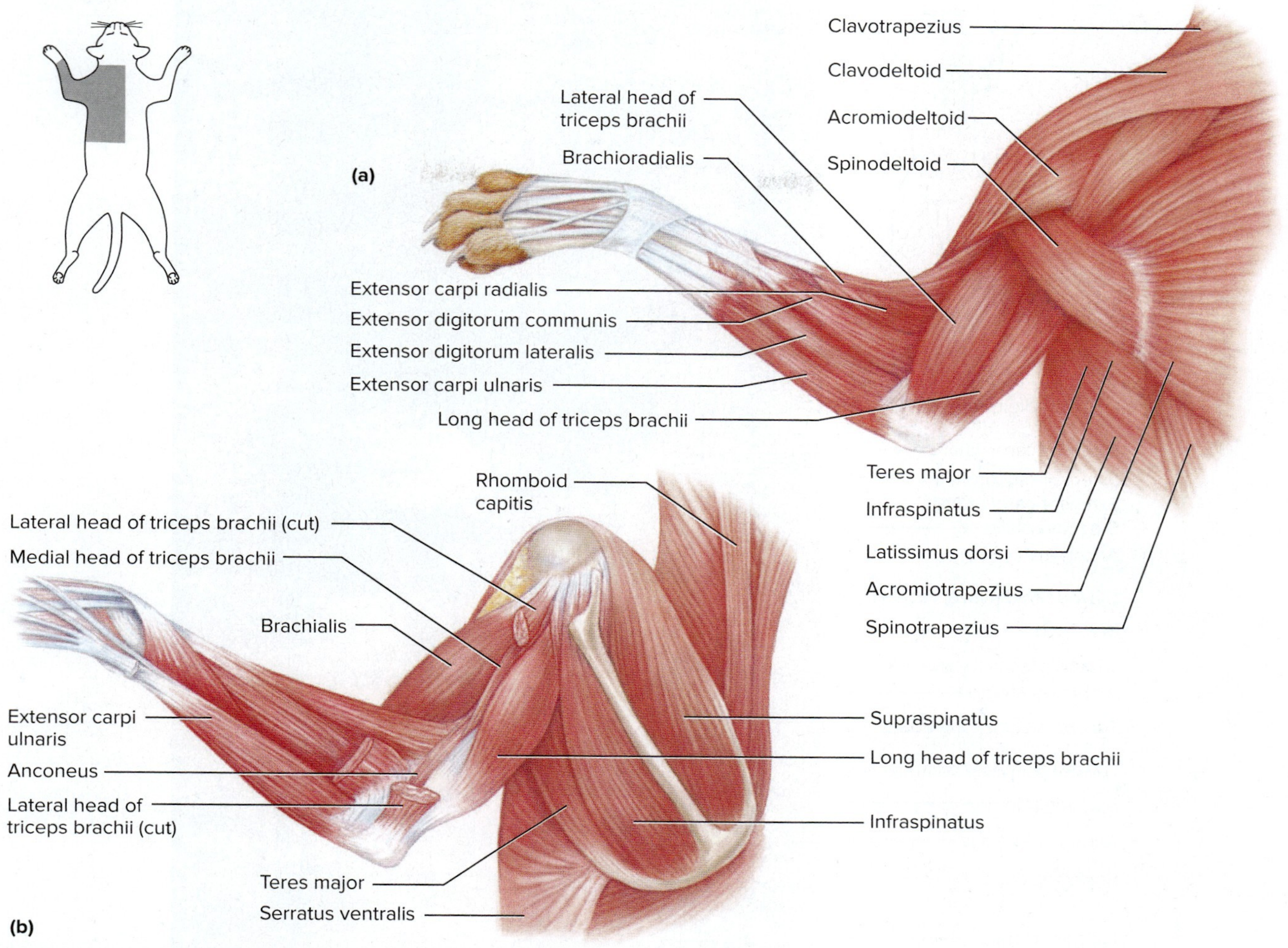

FIGURE 63.16 (a) Superficial muscles of the cat's left lateral shoulder and forelimb. (b) Deep muscles of the cat's left lateral shoulder and forelimb.

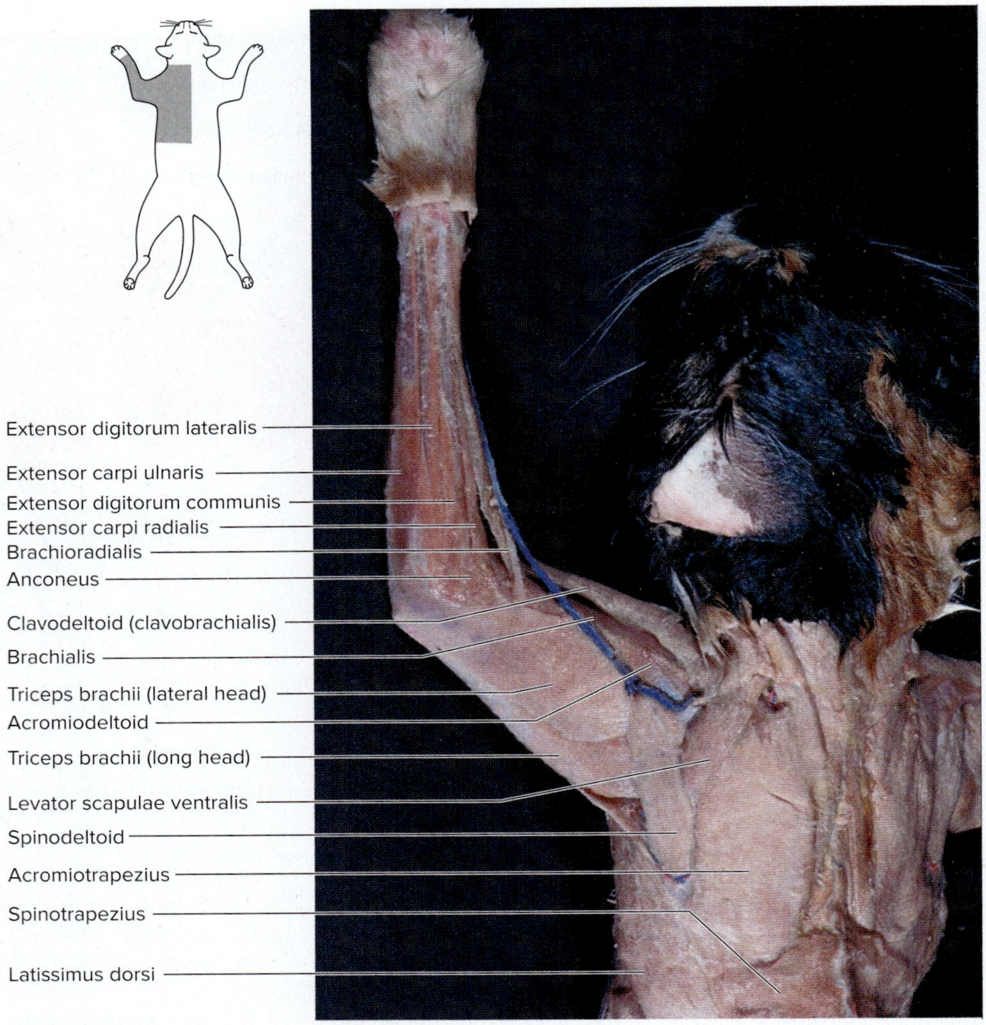

FIGURE 63.17 Superficial muscles of the cat's left lateral shoulder and forelimb, dorsal view. Dr. Sheril Burton

Table 63.5 Muscles of the Forelimb

Muscle	Origin	Insertion	Action
Coracobrachialis	Coracoid process of scapula	Proximal end of humerus	Adducts forelimb
Epitrochlearis	Fascia of latissimus dorsi	Olecranon process of ulna	Extends forelimb
Biceps brachii	Border of glenoid cavity of scapula	Radial tuberosity of radius	Flexes forelimb
Triceps brachii:			
Lateral head	Deltoid tuberosity of humerus	Olecranon process of ulna	Extends forelimb
Long head	Border of glenoid cavity of scapula	Olecranon process of ulna	Extends forelimb
Medial head	Shaft of humerus	Olecranon process of ulna	Extends forelimb
Anconeus	Lateral epicondyle of humerus	Olecranon process of ulna	Extends forelimb
Brachialis	Lateral surface of humerus	Proximal end of ulna	Flexes forelimb
Brachioradialis	Shaft of humerus	Distal end of radius	Supinates lower forelimb
Extensor carpi radialis	Lateral surface of humerus	Second and third metacarpals	Extends wrist
Extensor digitorum communis	Lateral surface of humerus	Digits two to five	Extends digits
Extensor digitorum lateralis	Lateral surface of humerus	Digits two to five	Extends digits
Extensor carpi ulnaris	Lateral epicondyle of humerus and ulna	Fifth metacarpal	Extends wrist
Flexor carpi ulnaris	Medial epicondyle of humerus	Carpals	Flexes wrist
Palmaris longus	Medial epicondyle of humerus	Digits	Flexes digits
Flexor carpi radialis	Medial epicondyle of humerus	Second and third metacarpals	Flexes metacarpals
Pronator teres	Medial epicondyle of humerus	Radius	Pronates lower forelimb
Flexor digitorum profundus	Radius, ulna, and medial epicondyle of humerus	Distal phalanges	Flexes digits

10. Transect the flexor carpi ulnaris, palmaris longus, and flexor carpi radialis to expose the flexor digitorum profundus.
11. See table 63.5 for the origins, insertions, and actions of these muscles of the forelimb.
12. Complete Part D of the laboratory assessment.

PROCEDURE H: Muscles of the Hip and Hindlimb

1. Place the cat in the dissecting tray with its ventral surface up.
2. Remove any remaining fat and connective tissue from the hip and hindlimb to expose the muscles.
3. Study figures 63.18*a* and 63.19, and then locate and dissect the following muscles from the medial surface of the thigh:

   sartorius

   gracilis
4. Using scissors, transect the sartorius and gracilis, and reflect their cut edges to observe the deeper muscles of the thigh.

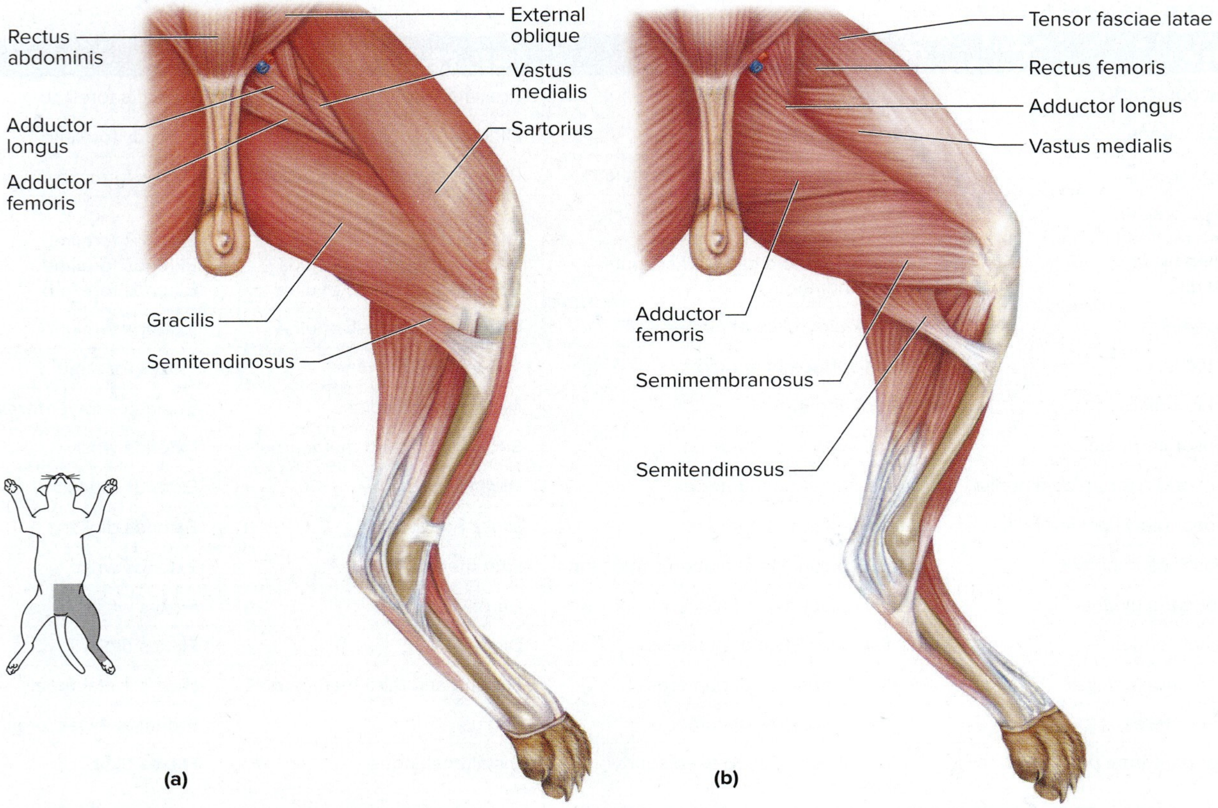

FIGURE 63.18 (a) Superficial muscles of the cat's left medial hindlimb. (b) Deep muscles of the cat's left medial hindlimb. APR

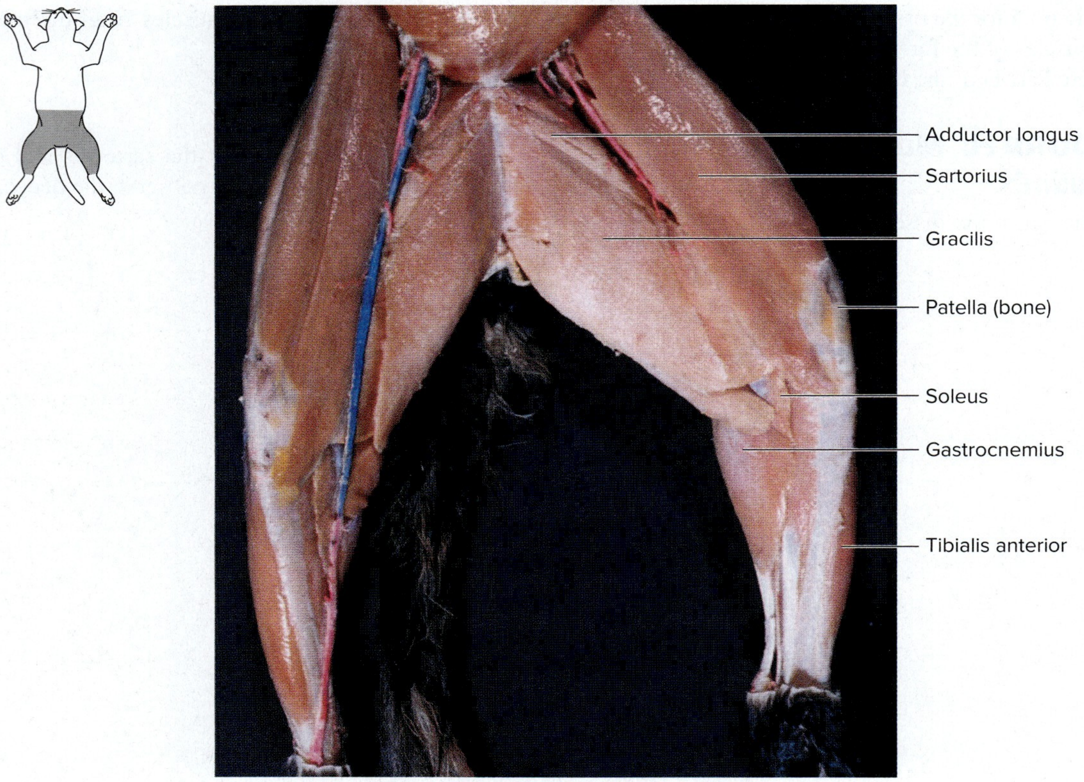

FIGURE 63.19 Superficial muscles of the cat's hindlimb, medial view. Dr. Sheril Burton

5. Study figures 63.18*b* and 63.20, and then locate and dissect the following muscles:
   rectus femoris
   vastus medialis
   adductor longus
   adductor femoris
   semimembranosus
   semitendinosus
   tensor fasciae latae
6. Transect the tensor fasciae latae and rectus femoris muscles and reflect their cut edges. Locate the *vastus intermedius* and *vastus lateralis* muscles beneath (fig. 63.21). (*Note:* In some specimens, the vastus intermedius, vastus lateralis, and vastus medialis muscles are closely united by connective tissue and are difficult to separate.)
7. Study figures 63.22 and 63.23, and then locate and dissect the following muscles from the lateral surface of the hip and thigh:
   biceps femoris
   caudofemoralis
   gluteus maximus—small in cats
8. Using scissors, transect the tensor fasciae latae and biceps femoris, and reflect their cut edges to observe the deeper muscles of the thigh.
9. Locate and dissect the following muscles:
   tenuissimus
   gluteus medius
   (*Note:* The tenuissimus muscle and the sciatic nerve may adhere tightly to the dorsal surface of the biceps femoris muscle. Take care to avoid cutting these structures.)
10. On the lateral surface of the lower hindlimb (figs. 63.22 and 63.24), locate and dissect the following muscles:
   gastrocnemius
   soleus
   tibialis anterior
   fibularis (peroneus) group
11. See table 63.6 for the origins, insertions, and actions of these muscles of the hip and hindlimb.
12. Review the locations, origins, insertions, and actions designated by the laboratory instructor without the aid of the figures and tables.
13. Complete Parts E, F, and G of the laboratory assessment.

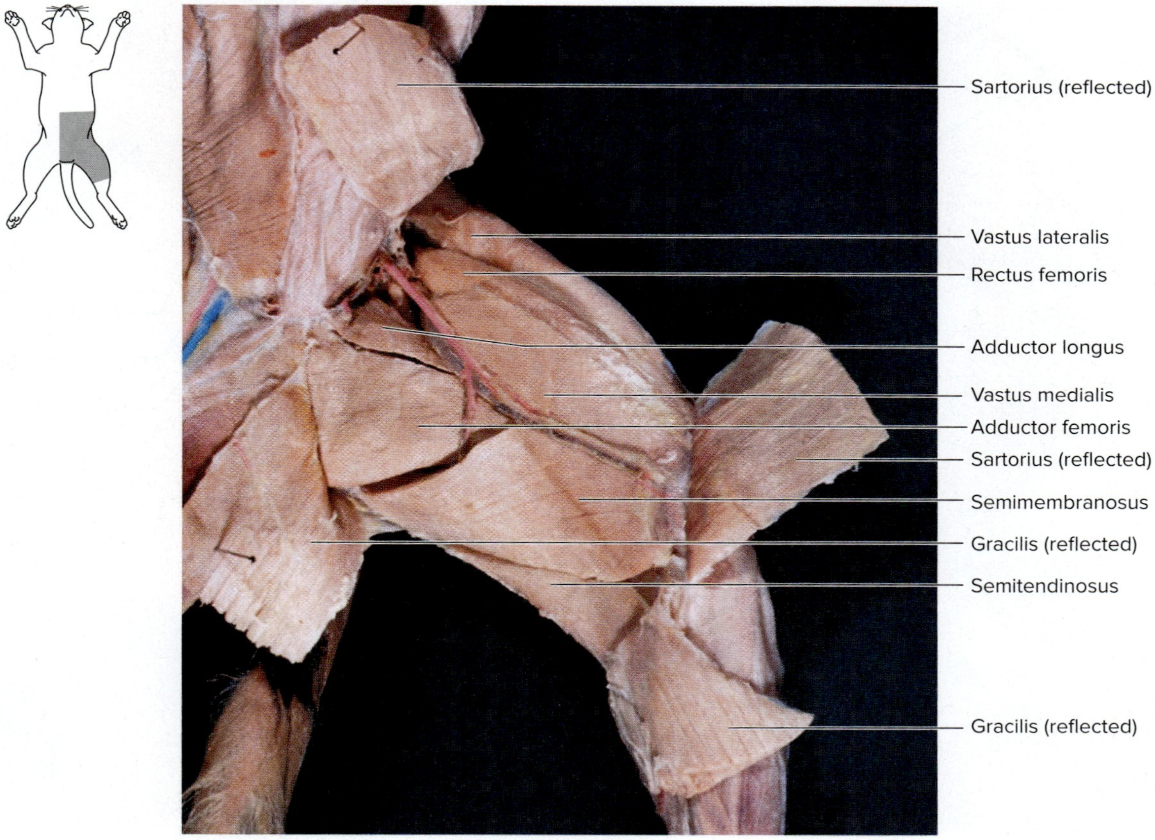

FIGURE 63.20 Deep muscles of the cat's left hindlimb, medial view. Dr. Sheril Burton

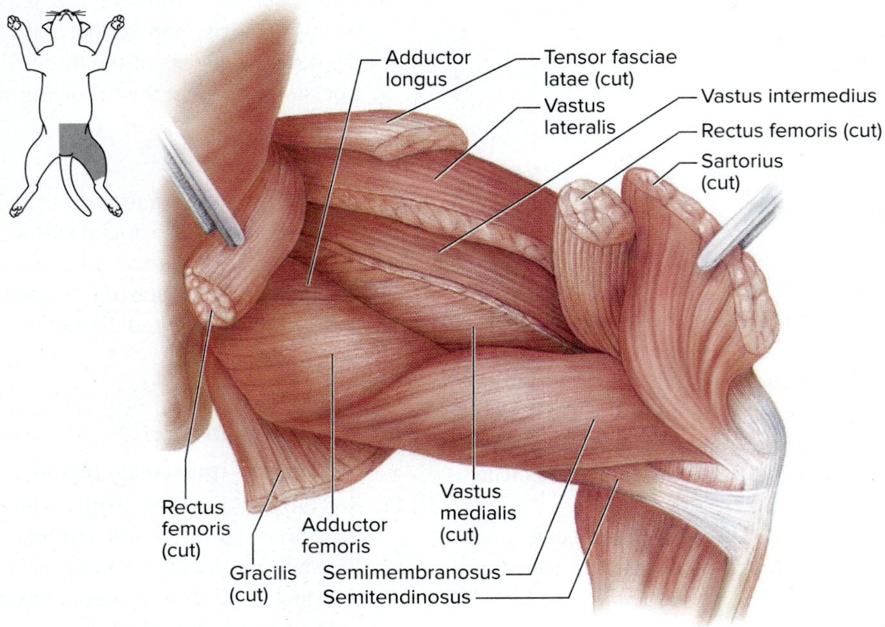

FIGURE 63.21 Other deep muscles of the cat's left thigh, medial view.

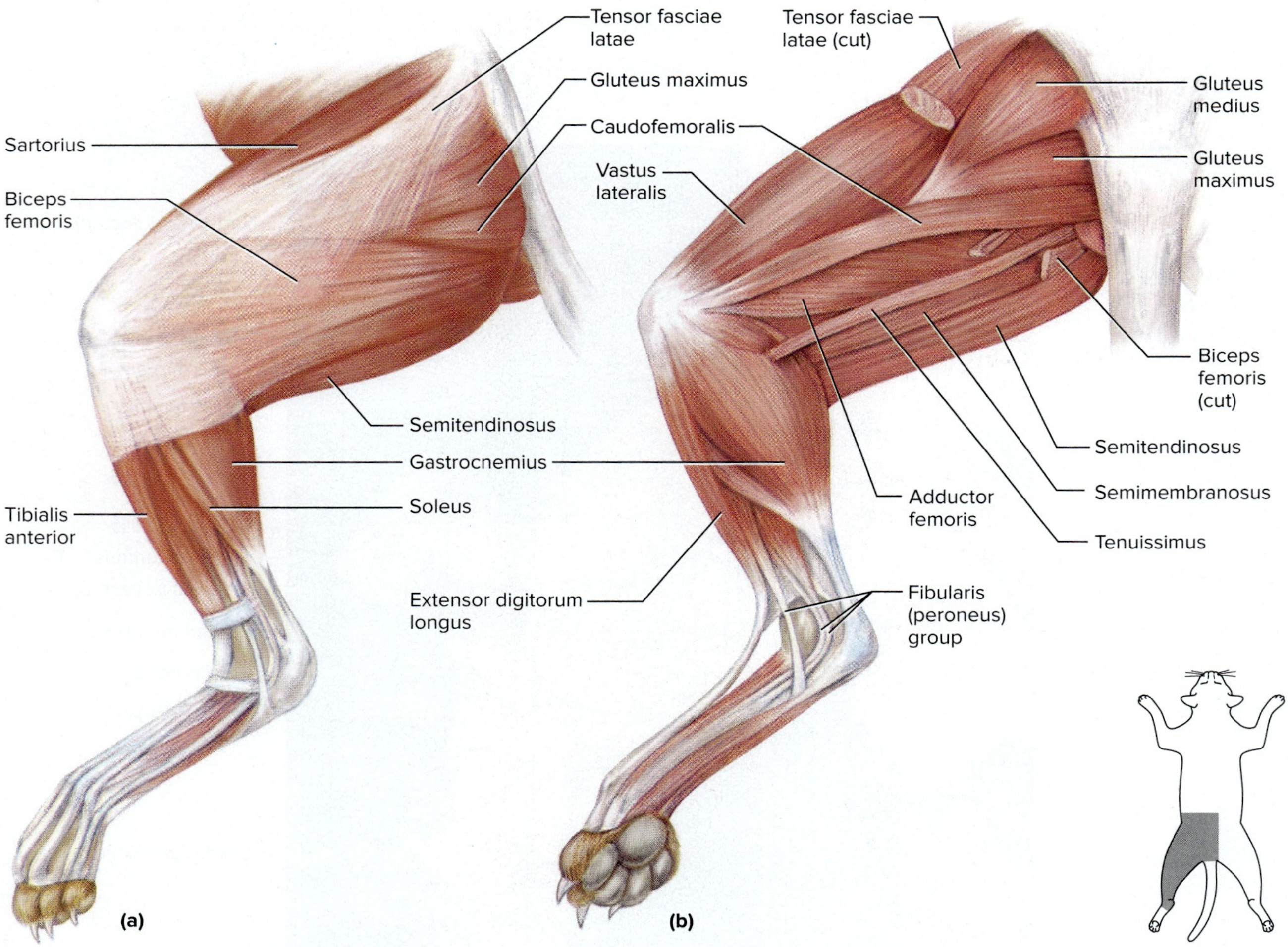

FIGURE 63.22 (a) Superficial muscles of the cat's left lateral hip and hindlimb. (b) Deep muscles of the cat's left lateral hip and hindlimb. APR

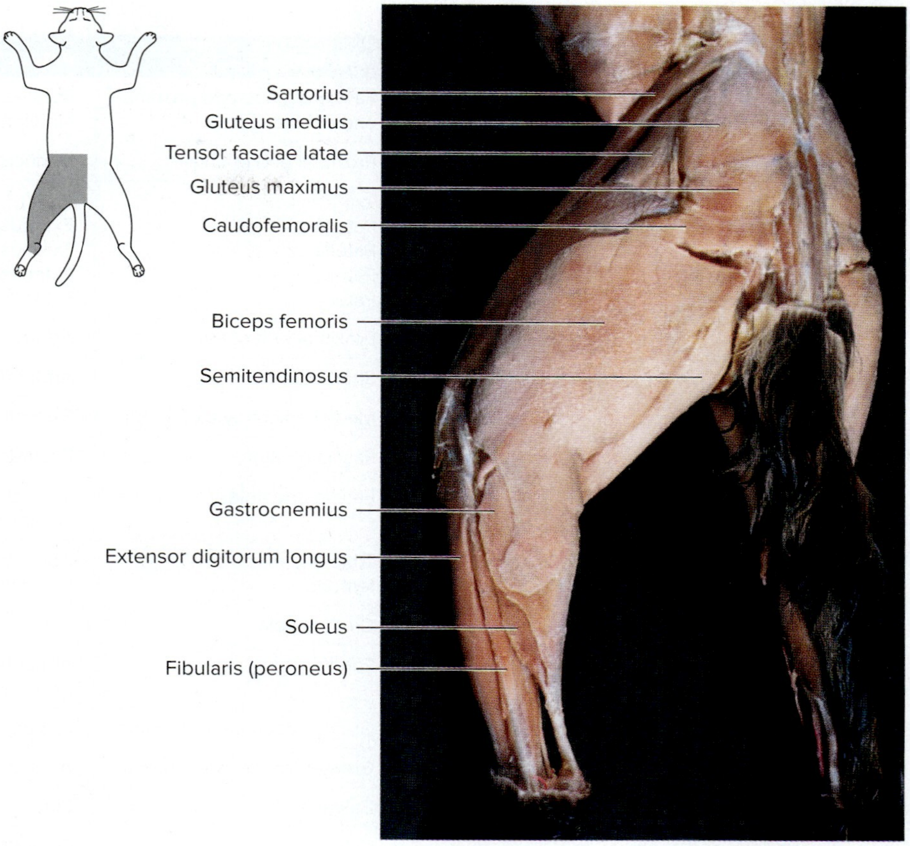

FIGURE 63.23 Muscles of the cat's left hip and hindlimb, lateral view. Dr. Sheril Burton

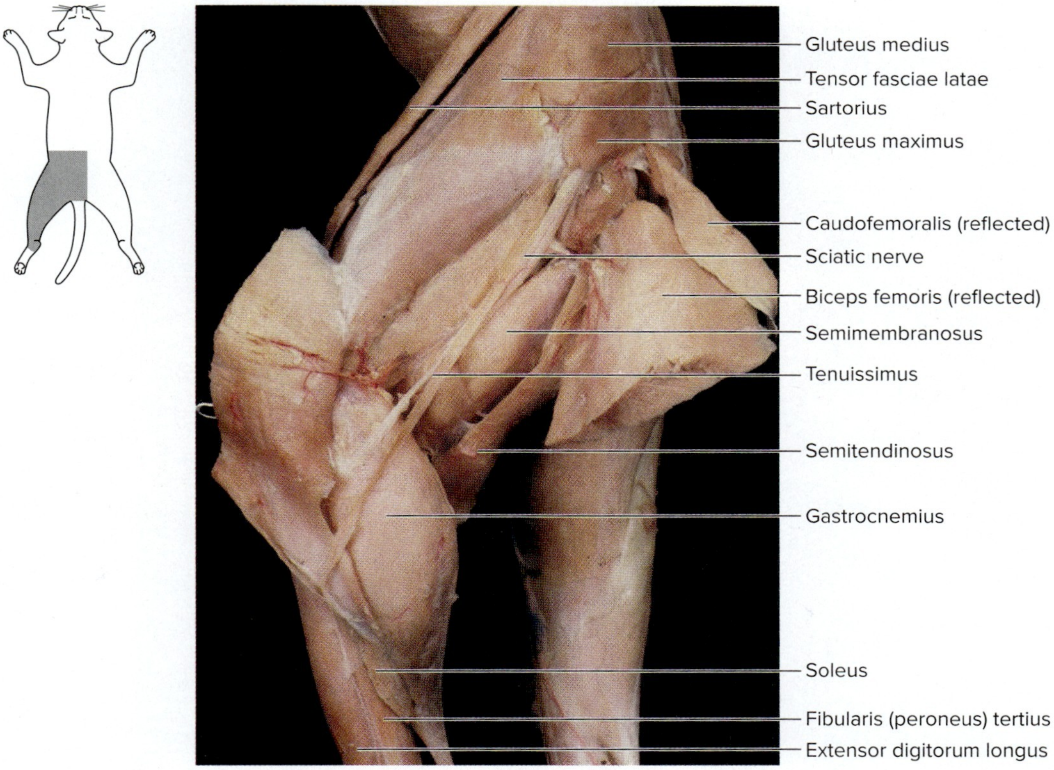

FIGURE 63.24 Deep muscles of the cat's left hindlimb, lateral view. Dr. Sheril Burton

Table 63.6 Muscles of the Hip and Hindlimb

Muscle	Origin	Insertion	Action
Sartorius	Crest of ilium	Fascia of knee and proximal end of tibia	Rotates and extends hindlimb
Gracilis	Pubic symphysis and ischium	Proximal end of tibia	Adducts hindlimb
Quadriceps femoris:			
Vastus lateralis	Shaft of femur and greater trochanter	Patella	Extends hindlimb
Vastus intermedius	Shaft of femur	Patella	Extends hindlimb
Rectus femoris	Ilium	Patella	Extends hindlimb
Vastus medialis	Shaft of femur	Patella	Extends hindlimb
Adductor longus	Pubis	Proximal end of femur	Adducts hindlimb
Adductor femoris	Pubis and ischium	Shaft of femur	Adducts hindlimb
Semimembranosus	Ischial tuberosity	Medial epicondyle of femur	Extends thigh
Tensor fasciae latae	Ilium	Fascia of thigh	Extends thigh
Biceps femoris	Ischial tuberosity	Patella and tibia	Abducts thigh and flexes lower hindlimb
Tenuissimus	Second caudal vertebra	Tibia and fascia of biceps femoris	Abducts thigh and flexes lower hindlimb
Semitendinosus	Ischial tuberosity	Crest of tibia	Flexes lower hindlimb
Caudofemoralis	Caudal vertebrae	Patella	Abducts thigh and extends lower hindlimb
Gluteus maximus	Sacral and caudal vertebrae	Greater trochanter of femur	Abducts thigh
Gluteus medius	Ilium, and sacral and caudal vertebrae	Greater trochanter of femur	Abducts thigh
Gastrocnemius	Lateral and medial epicondyles of femur	Calcaneus	Extends foot
Soleus	Proximal end of fibula	Calcaneus	Extends foot
Tibialis anterior	Proximal end of tibia and fibula	First metatarsal	Flexes foot
Fibularis (peroneus) group	Shaft of fibula	Metatarsals	Flexes foot

Name ______________________________

Date ______________________________

Section ______________________________

The (A) corresponds to the indicated Learning Outcome(s) (O) found at the beginning of the Laboratory Exercise.

LABORATORY ASSESSMENT 63

Cat Dissection: Musculature

PART A: Assessments

Complete the following statements:

1. The ______________________ muscle of the human is homologous to the sternomastoid muscle of the cat. A3
2. The ______________________ muscle elevates the mandible in the human and in the cat. A2 A3
3. Two muscles of the cat that are inserted on the hyoid bone are the ______________________ and the ______________________. A1 A2

PART B: Assessments

Complete the following:

1. Name two muscles that are found in the thoracic wall of the cat but are absent in the human. A1 A3

   ______________________ ______________________

2. Name two muscles that are found in the thoracic wall of the cat and the human. A3

   ______________________ ______________________

3. Name four muscles that are found in the abdominal wall of the cat and the human. Put an asterisk next to the muscle that is the deepest. A3

   ______________________ ______________________

   ______________________ ______________________

PART C: Assessments

Complete the following:

1. Name three muscles of the cat that together correspond to the trapezius muscle in the human. A1 A3

   ______________ ______________ ______________

2. Name three muscles of the cat that together correspond to the deltoid muscle in the human. A1 A3

   ______________ ______________ ______________

3. Name the muscle in the cat and in the human that occupies the fossa above the spine of the scapula. A1 A3

   ____

4. Name the muscle in the cat and in the human that occupies the fossa below the spine of the scapula. A1 A3

   ____

5. Name two muscles found in the cat and in the human that can rotate the forelimb. A2 A3

   ______________________ ______________________

PART D: Assessments

Complete the following:

1. Name two muscles found in the cat and in the human that can flex the forelimb. A2 A3

____________________ ____________________

2. Name a muscle found in the cat but absent in the human that can extend the forelimb. A2 A3

____

3. Name a muscle that has three heads, can extend the forelimb, and is found in the cat and in the human. A2 A3

____

PART E: Assessments

Complete the following:

1. The ____________________ muscle *group* found in the cat and in the human that can extend the hindlimb. A2 A3
2. Name two hindlimb muscles in the cat and in the human that have an insertion in the calcaneus bone. A2 A3

____________________ ____________________

PART F: Assessments

Identify the numbered muscles of the cat in figures 63.25 to 63.28.

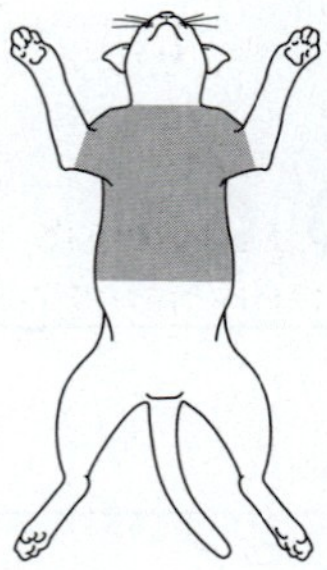

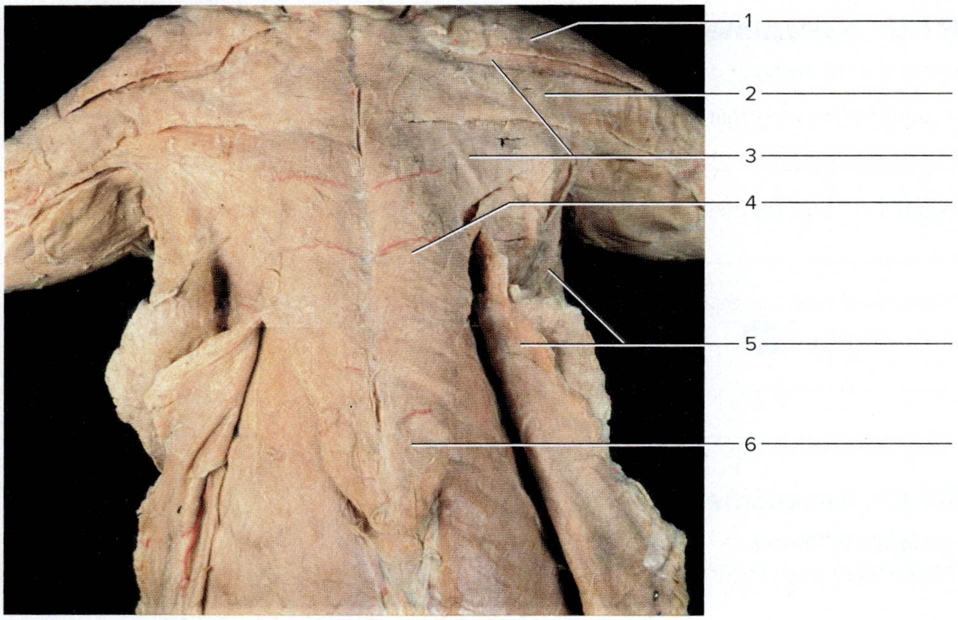

Terms:
Clavodeltoid
Latissimus dorsi
Pectoantebrachialis
Pectoralis major
Pectoralis minor
Xiphihumeralis

FIGURE 63.25 Label the numbered ventral thorax muscles of the dissected cat, using the terms provided. Dennis Strete A1

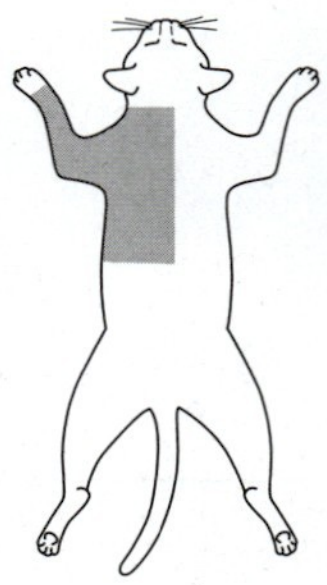

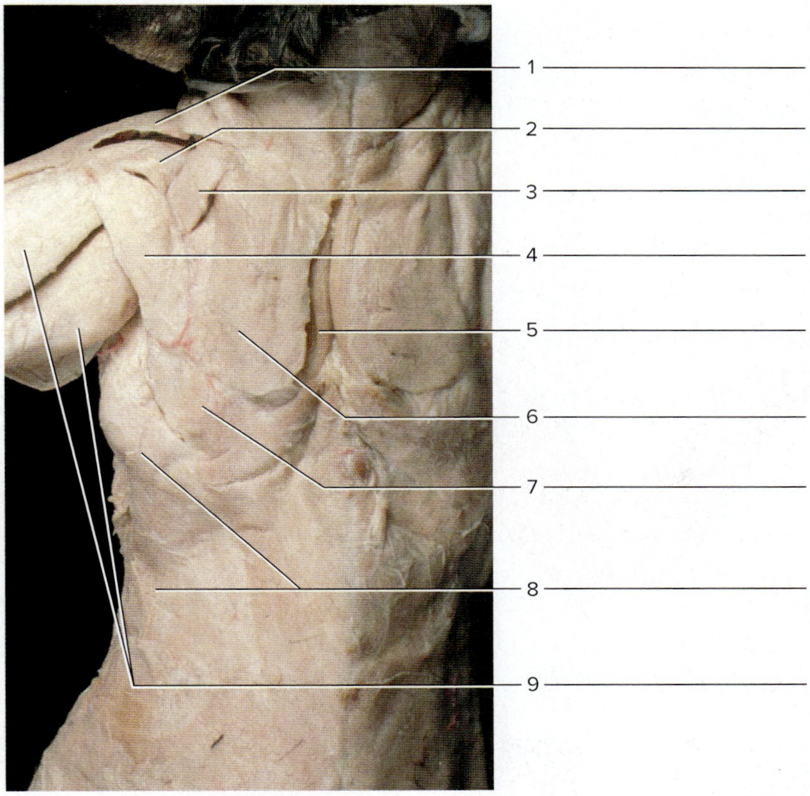

Terms:
Acromiodeltoid
Acromiotrapezius
Clavodeltoid
Latissimus dorsi
Levator scapulae ventralis
Rhomboid
Spinodeltoid
Spinotrapezius
Triceps brachii

(a)

FIGURE 63.26 Label the numbered dorsal muscles of the cat in (a) superficial muscles of a dissected cat, and (b) an illustration of superficial muscles shown on the right side and deep muscles shown on the left side. Use the terms provided for (a) only. Dennis Strete A1

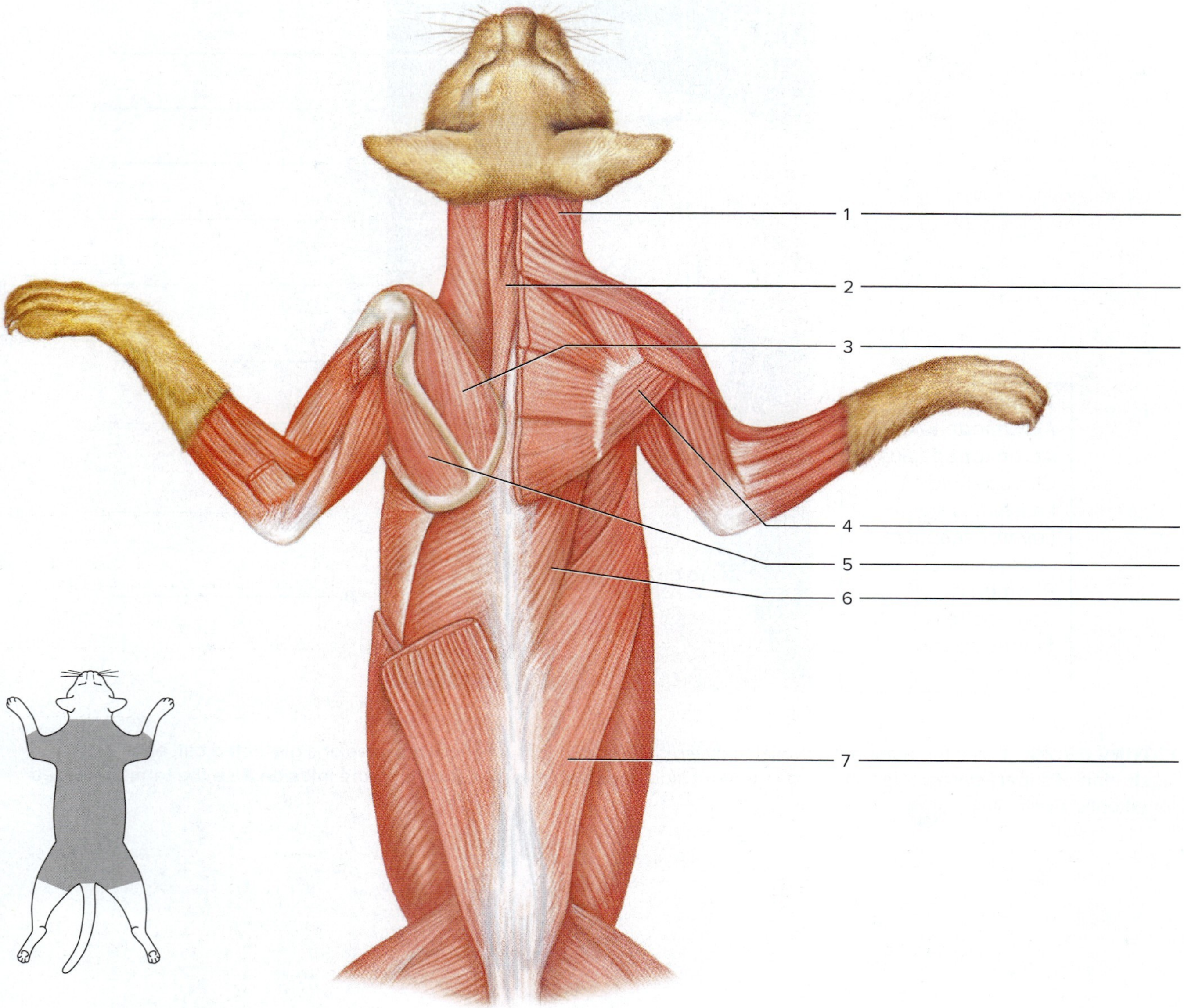

FIGURE 63.26 *Continued.*

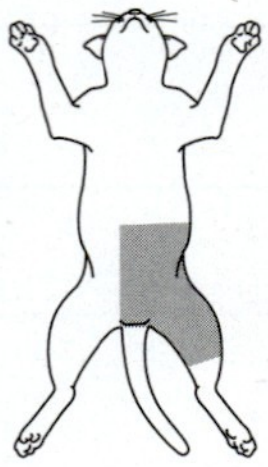

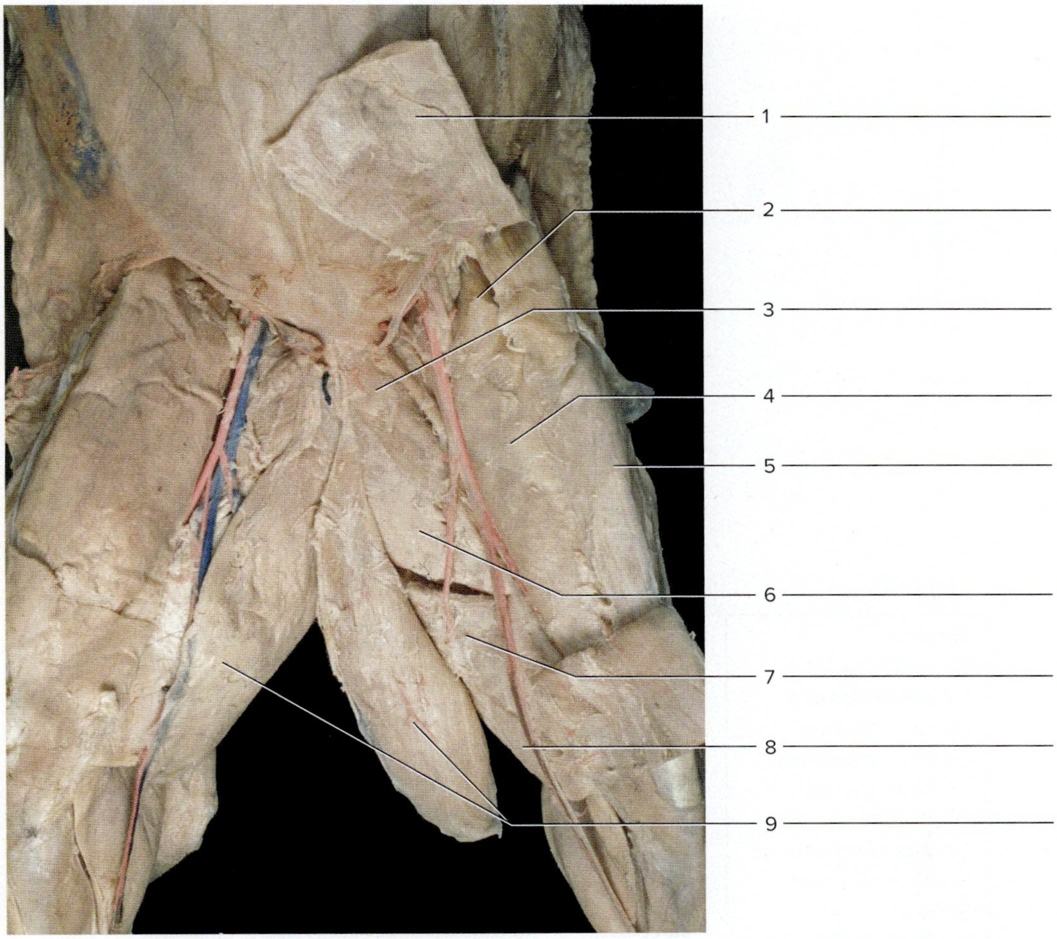

Terms:
Adductor femoris
Adductor longus
Gracilis
Rectus femoris
Sartorius (reflected)
Semimembranosus
Semitendinosus
Vastus medialis
Vastus lateralis

FIGURE 63.27 Label the numbered medial hindlimb muscles of the dissected cat, using the terms provided. Dennis Strete A1

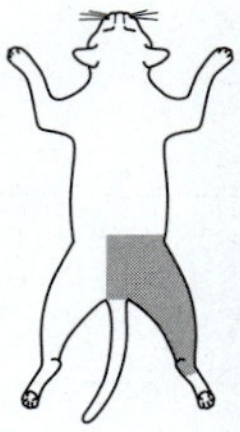

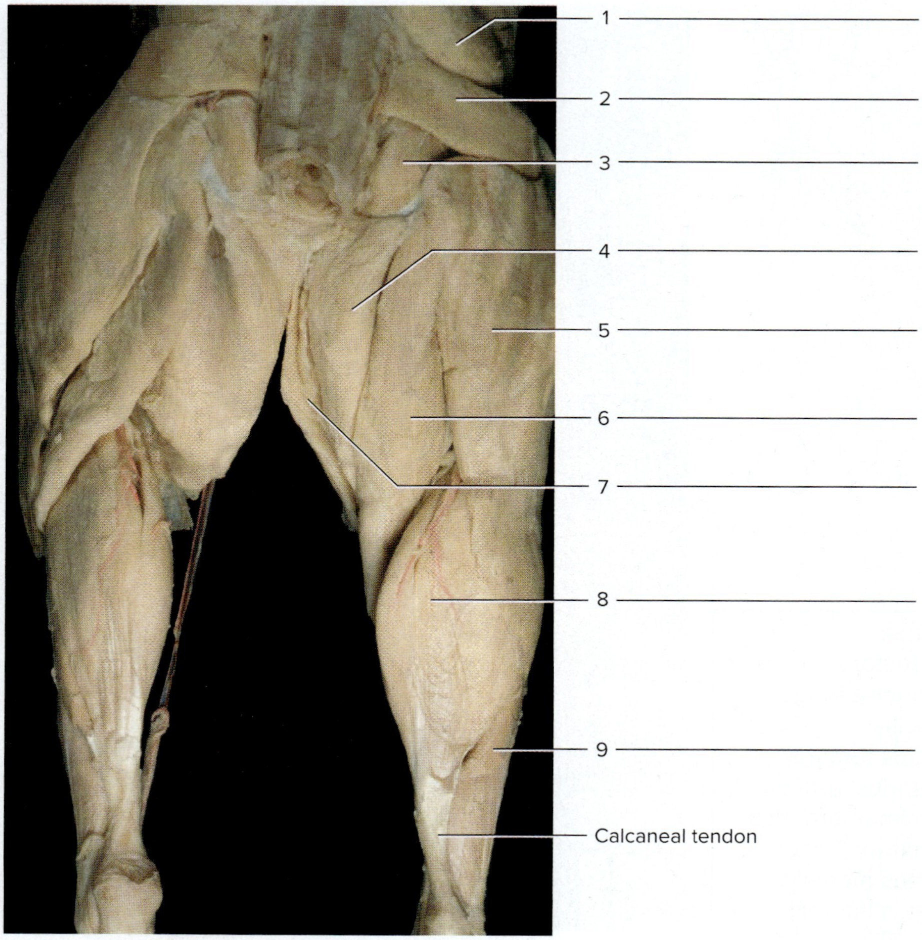

Terms:
Biceps femoris
Caudofemoralis
Gastrocnemius
Gluteus maximus
Gluteus medius
Gracilis
Semimembranosus
Semitendinosus
Soleus

(a)

FIGURE 63.28 Label the numbered muscles of the lateral hip and hindlimb in (a) right superficial muscles of a dissected cat, and in (b) left deep muscles of the cat. Use the terms provided for (a) only. Dennis Strete **A1**

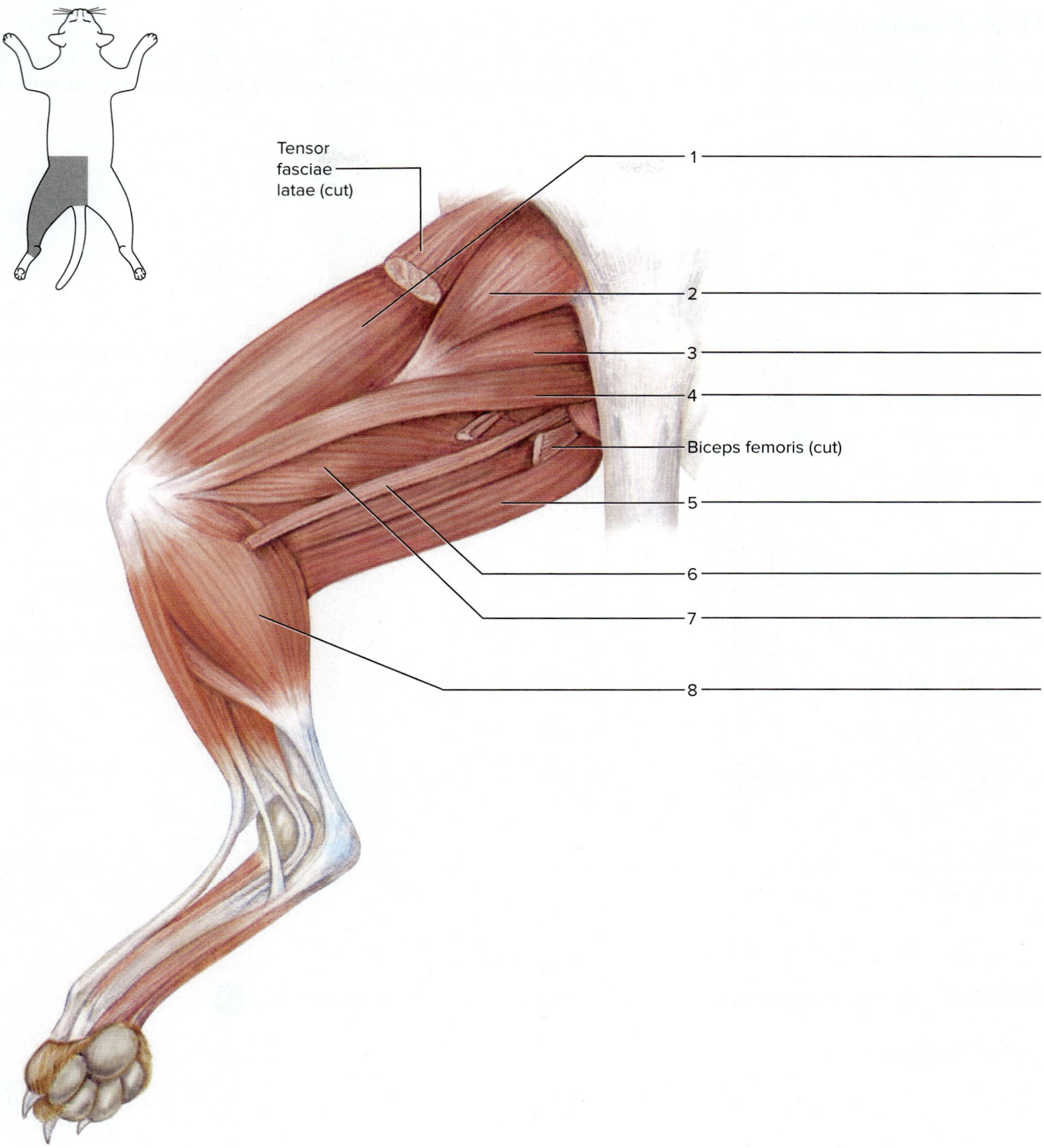

(b)

FIGURE 63.28 *Continued.*

PART G: Assessments

Observe the muscles on the human torso model and the upper and lower limb models. Each of the muscles in figure 63.29 is found both in the cat and in the human. Identify each of the numbered muscles in the figure by placing its name in the space next to its number.

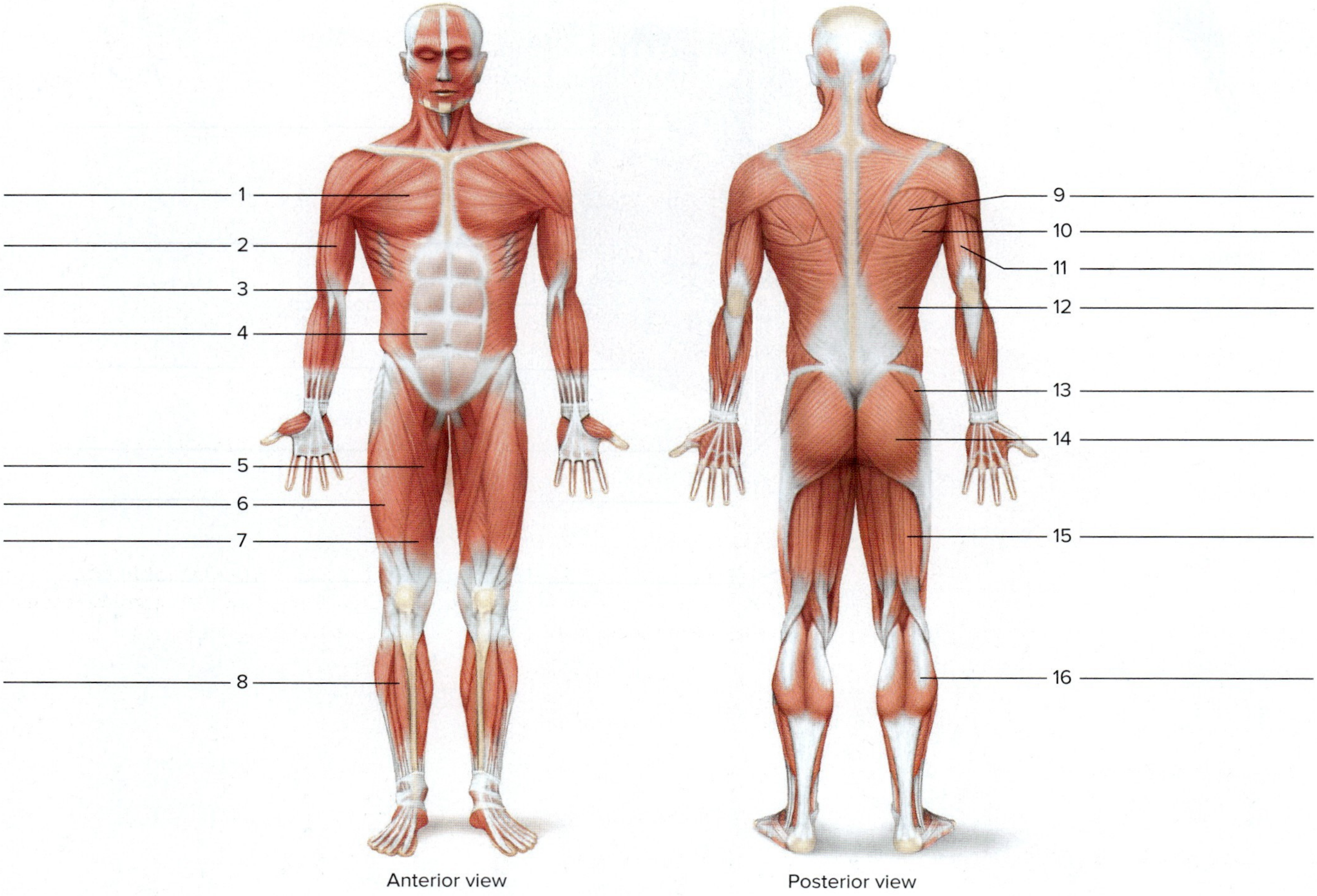

FIGURE 63.29 Identify the numbered muscles that occur in both the cat and the human. A3

LABORATORY
EXERCISE 64

Cat Dissection: Cardiovascular System

Purpose of the Exercise

To examine and identify the major organs of the cardiovascular system of the cat and to compare them with the corresponding organs of the human.

MATERIALS NEEDED

Preserved cat
Dissecting tray
Dissecting instruments
Disposable gloves
Human torso model

SAFETY

- Review all the safety guidelines in Appendix 1 of your laboratory manual.
- Wear disposable gloves and protective eyewear when working on the cat dissection.
- Dispose of tissue remnants and gloves as instructed.
- Wash the dissecting tray and instruments as instructed.
- Wash your laboratory table.
- Wash your hands before leaving the laboratory.

Learning Outcomes APR

After completing this exercise, you should be able to:

O1 Locate and identify the major organs of the cardiovascular system of the cat.

O2 Compare the features of the cardiovascular system of the cat with those of the human.

O3 Identify the corresponding cardiovascular organs of the human.

The O corresponds to the assessments A indicated in the Laboratory Assessment for this Exercise.

Pre-Lab

Carefully read the introductory material and examine the entire lab.
Be familiar with the structures and functions of the human cardiovascular system from lecture, the textbook, or laboratory exercises.

Chapter Opener: ©101cats/Getty Images

In this laboratory exercise, you will dissect the major organs of the cardiovascular system of the cat. As before, while you are examining the organs of the cat, compare them with the corresponding organs of the human torso model.

If the cardiovascular system of the cat has been injected, the systemic arteries will be filled with red latex, and the systemic veins will be filled with blue latex. This will make it easier for you to trace the vessels as you dissect them.

PROCEDURE A: The Arterial System

1. Place the preserved cat in the dissecting tray with its ventral surface up.
2. Open the thoracic cavity and expose its contents. To do this, follow these steps:
   a. Make a longitudinal incision passing anteriorly from the diaphragm along one side of the sternum. Continue the incision through the neck muscles to the mandible. Try to avoid damaging the internal organs as you cut.
   b. Make a lateral cut on each side along the anterior surface of the diaphragm and cut the diaphragm loose from the thoracic wall.
   c. Spread the sides of the thoracic wall outward and use a scalpel to make a longitudinal cut along each side of the inner wall of the rib cage to weaken the ribs. Continue to spread the thoracic wall laterally to break the ribs so that the flaps of the wall will remain open (figs. 64.1 and 64.2).
3. Note the location of the heart and the large blood vessels associated with it. Slit the outer *parietal pericardium* that surrounds the heart by cutting with scissors along the midventral line. Note how this membrane is connected to the *visceral pericardium* that is attached to the surface of the heart. Locate the *pericardial cavity,* the space between the two layers of the pericardium.

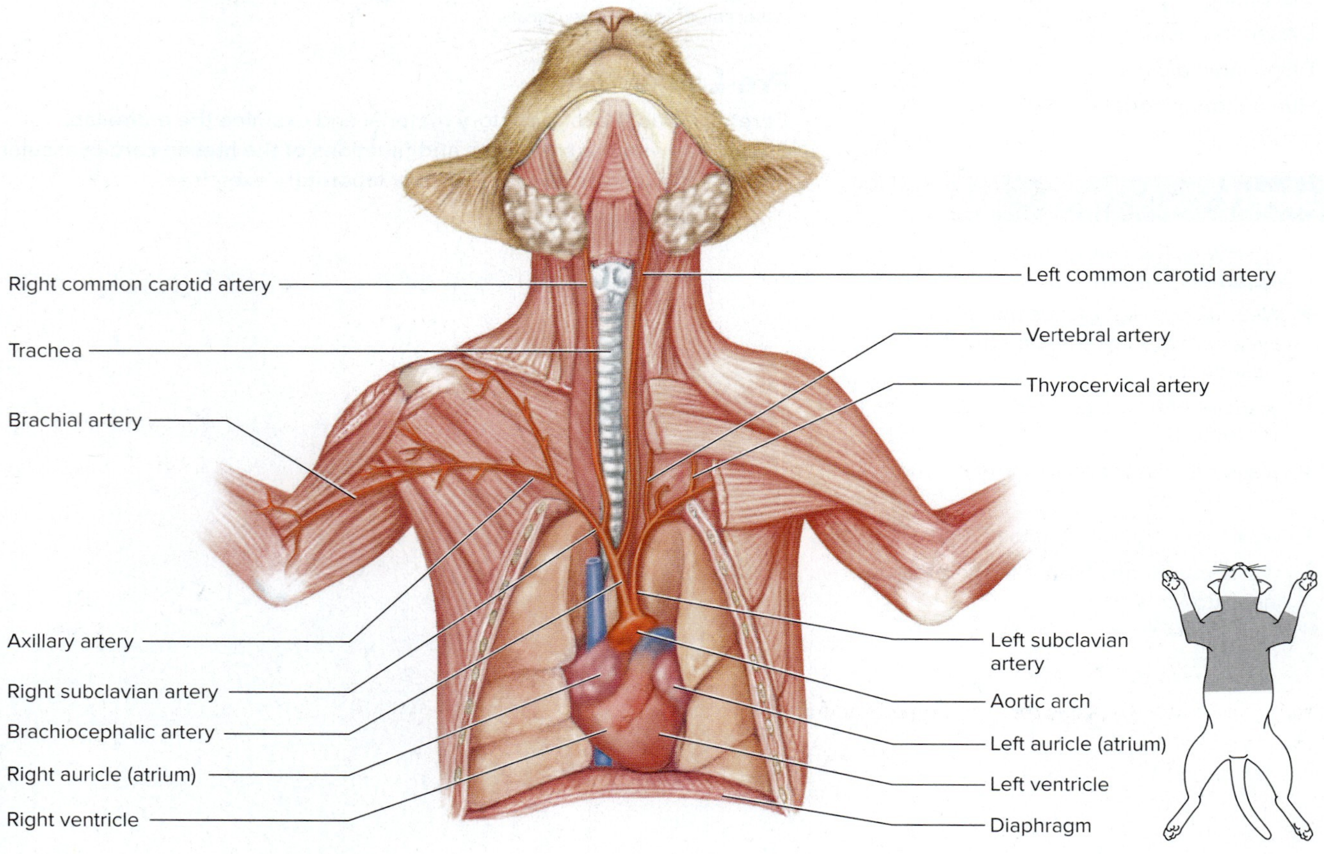

FIGURE 64.1 Arteries of the cat's thorax, neck, and forelimb, ventral view. APR

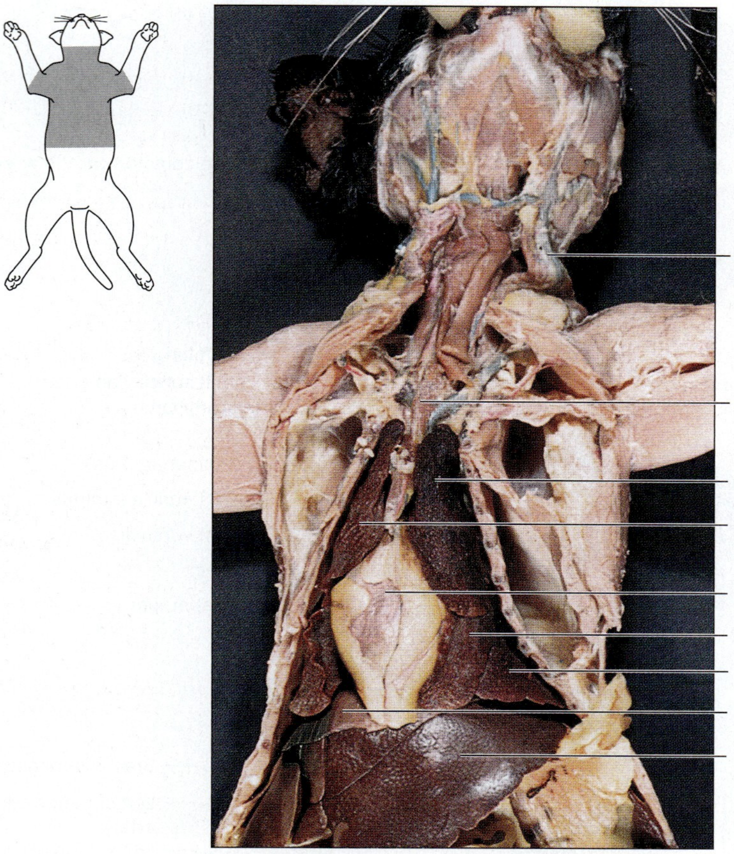

FIGURE 64.2 Features of the cat's neck and thoracic cavity, ventral view. Dr. Sheril Burton

4. Examine the heart (fig. 64.3*a*). The arrangement of blood vessels coming off the aortic arch is different in the cat than in the human body (fig. 64.3*b*). Locate the following:

   right atrium
   left atrium
   right ventricle
   left ventricle
   pulmonary trunk
   aorta
   coronary arteries

5. Use a scalpel to open the heart chambers by making a cut along the frontal plane from its apex to its base. Remove any remaining latex from the chambers. Examine the valves between the chambers, and note the relative thicknesses of the chamber walls (fig. 64.4). Do not remove the heart, as it is needed for future examination of the relationship between major blood vessels.
6. Using figures 64.1, 64.4, and 64.5 as guides, locate and dissect the following arteries of the thorax and neck:

   aortic arch
   brachiocephalic artery—on right only; first branch of the aortic arch
   right subclavian artery
   left subclavian artery
   right common carotid artery
   left common carotid artery

7. Trace the right subclavian artery into the forelimb, and locate the following arteries:

   vertebral artery
   thyrocervical artery
   subscapular artery
   axillary artery
   brachial artery

8. Open the abdominal cavity. To do this, follow these steps:
   a. Use scissors to make an incision through the body wall along the midline from the pubic symphysis to the diaphragm.
   b. Make a lateral incision through the body wall along either side of the inferior border of the diaphragm and along the bases of the thighs.
   c. Reflect the flaps created in the body wall as you would open a book and expose the contents of the abdominal cavity.

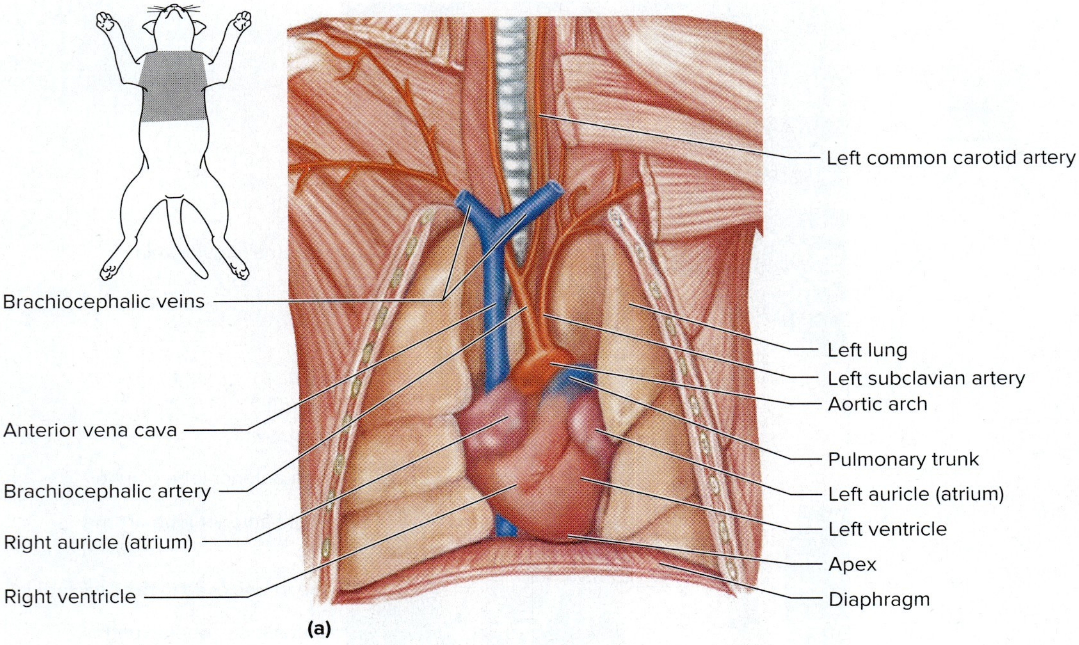

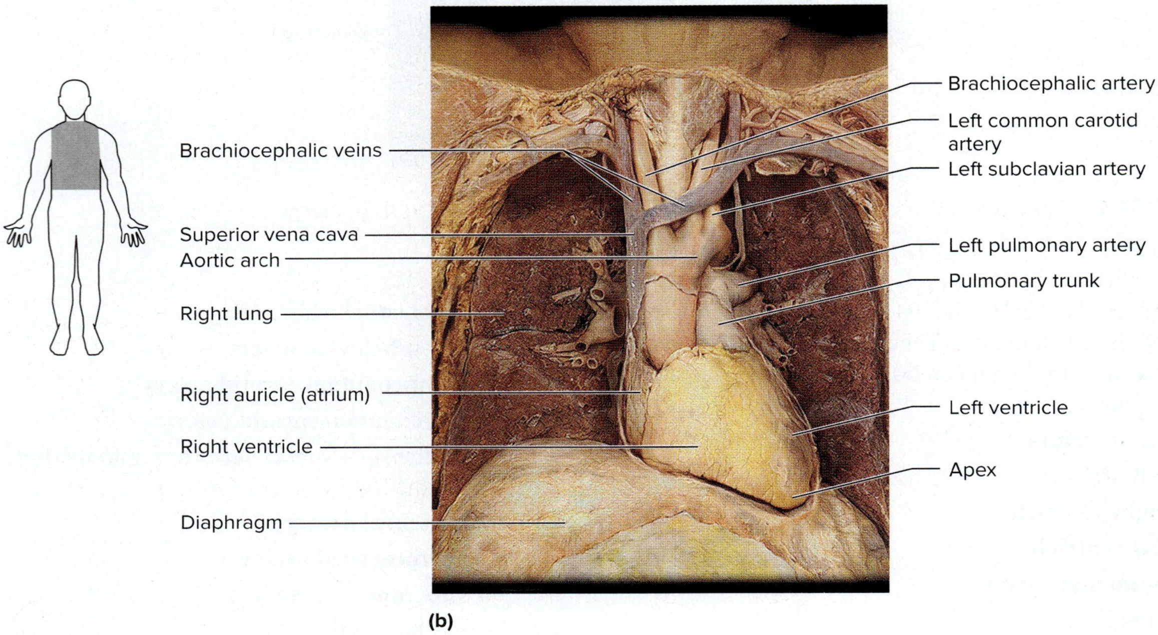

FIGURE 64.3 (a) Heart and associated arteries and veins of the cat, ventral view. (b) Heart and associated arteries and veins of a cadaver, anterior view. (b) APR/McGraw Hill APR

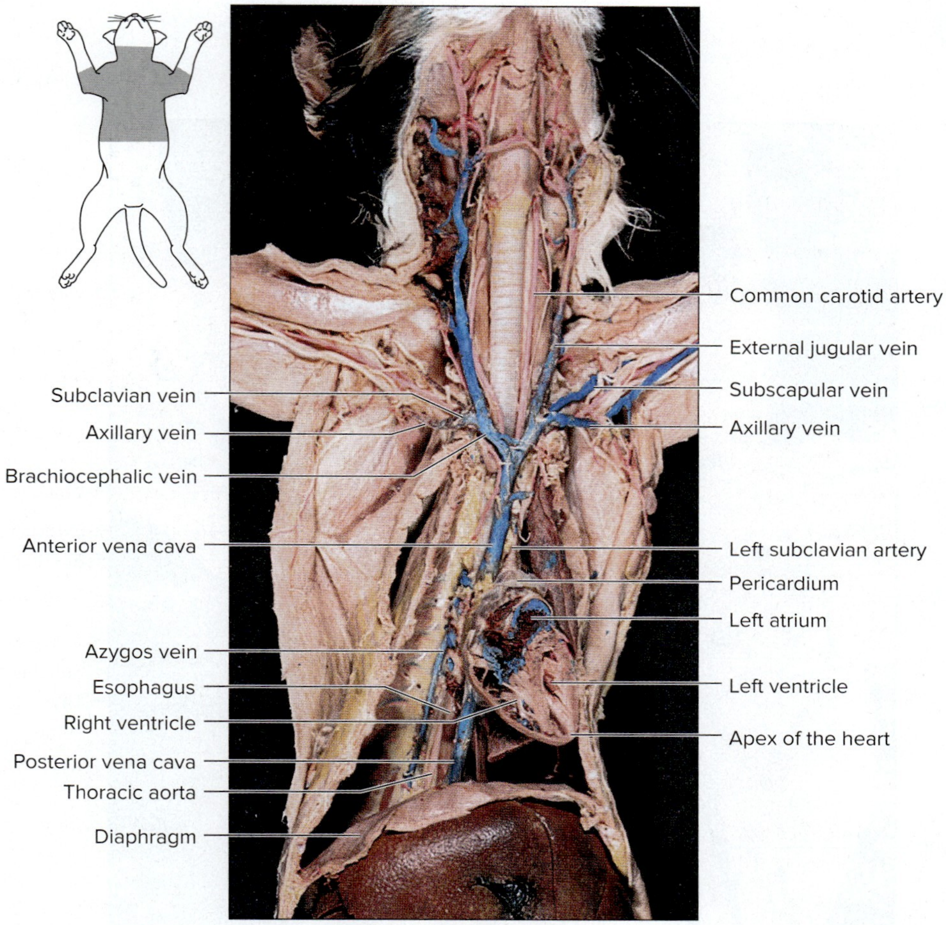

FIGURE 64.4 Blood vessels of the cat's neck and thorax, ventral view. Dr. Sheril Burton

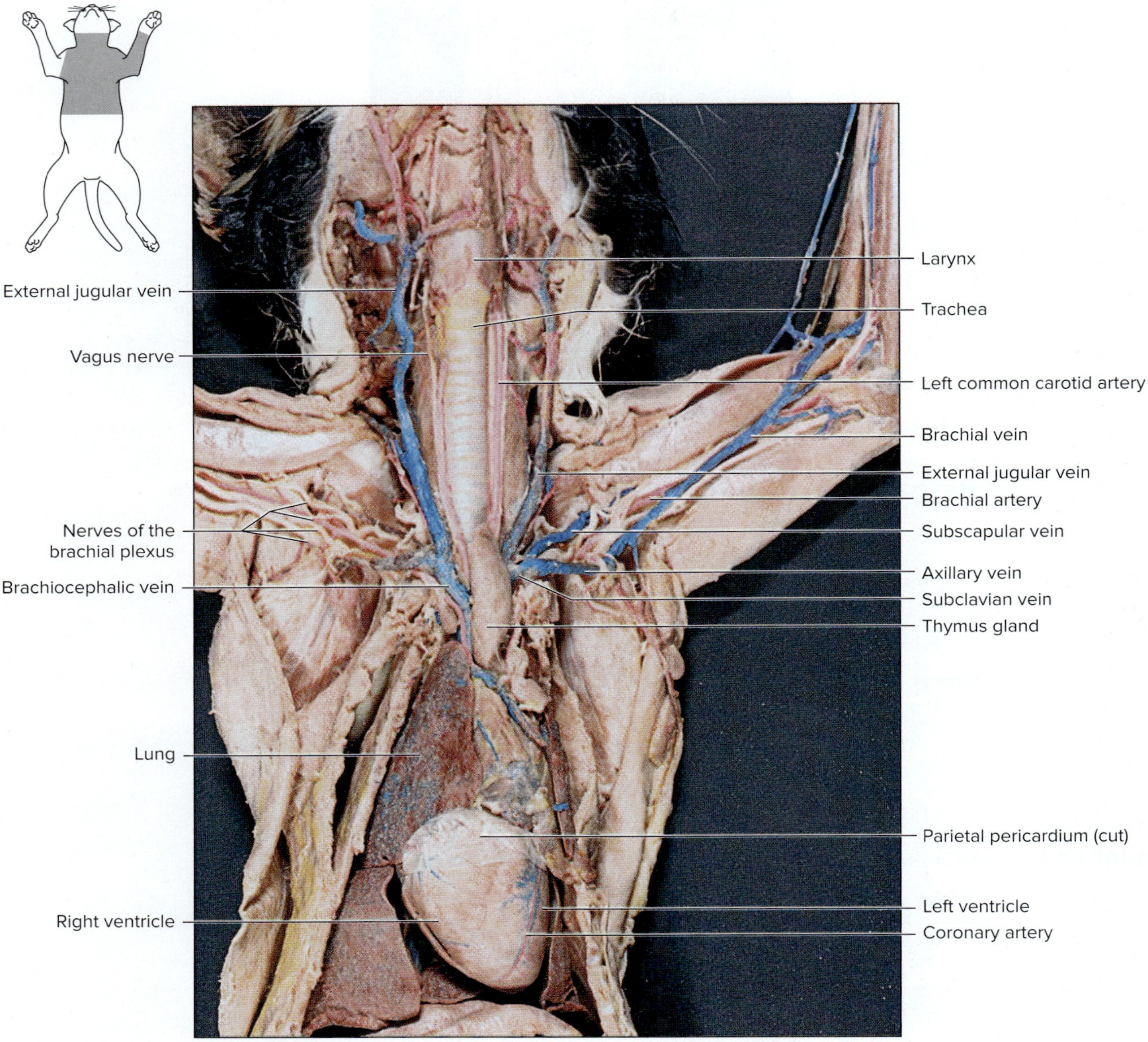

FIGURE 64.5 Blood vessels of the cat's upper forelimb and neck, ventral view. Dr. Sheril Burton APR

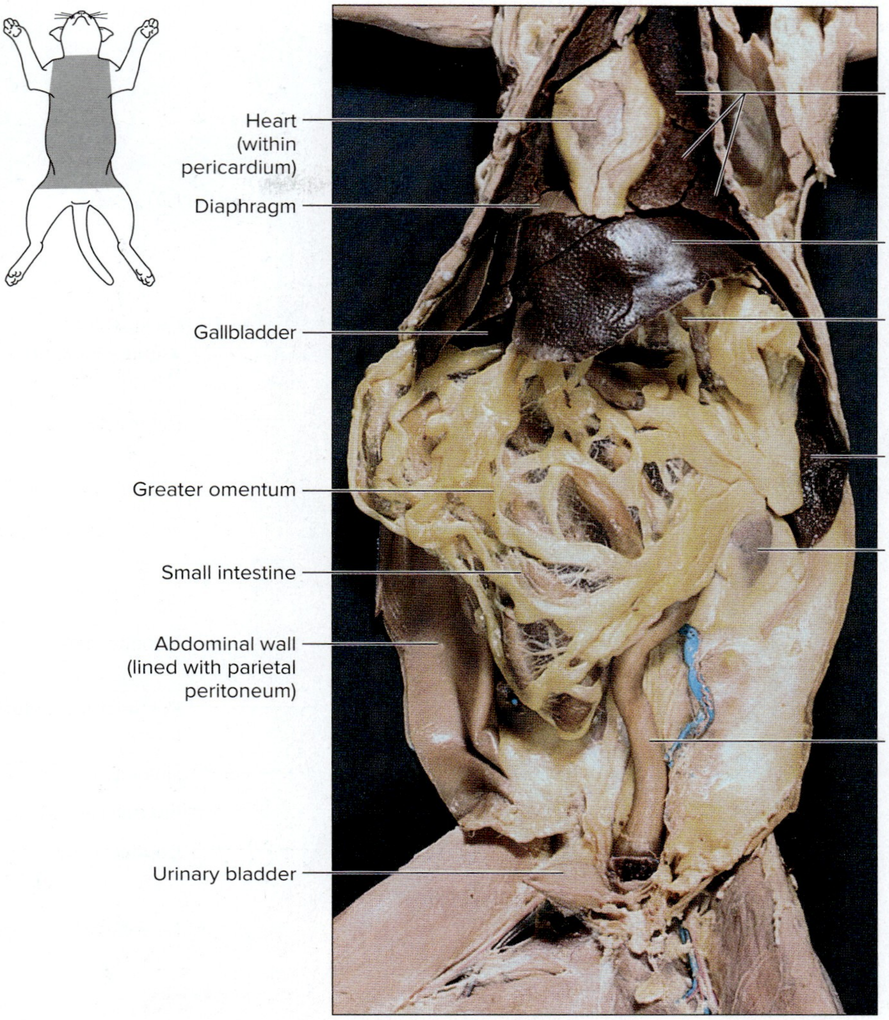

FIGURE 64.6 Organs of the cat's thoracic and abdominal cavities, ventral view. Dr. Sheril Burton

d. Note the *parietal peritoneum* that forms the inner lining of the abdominal wall. Also note the *greater omentum,* a structure composed of a double layer of peritoneum that hangs from the border of the stomach and covers the lower abdominal organs like a fatty apron (fig. 64.6). Reflect the abdominal organs to one side to expose various blood vessels in the abdominopelvic cavity region.

9. As you expose and dissect blood vessels, try not to destroy other visceral organs needed for future studies. Using figures 64.7 to 64.9 as guides, locate and dissect the following arteries of the abdomen:

abdominal aorta—unpaired
celiac artery—unpaired
hepatic artery—unpaired
gastric artery—unpaired
splenic artery—unpaired
anterior mesenteric artery—corresponds to superior mesenteric artery; unpaired
renal arteries—paired
adrenolumbar arteries—paired
gonadal arteries—paired
posterior mesenteric artery—corresponds to inferior mesenteric artery; unpaired
iliolumbar arteries—paired
external iliac arteries—paired
internal iliac arteries —paired
caudal artery—unpaired

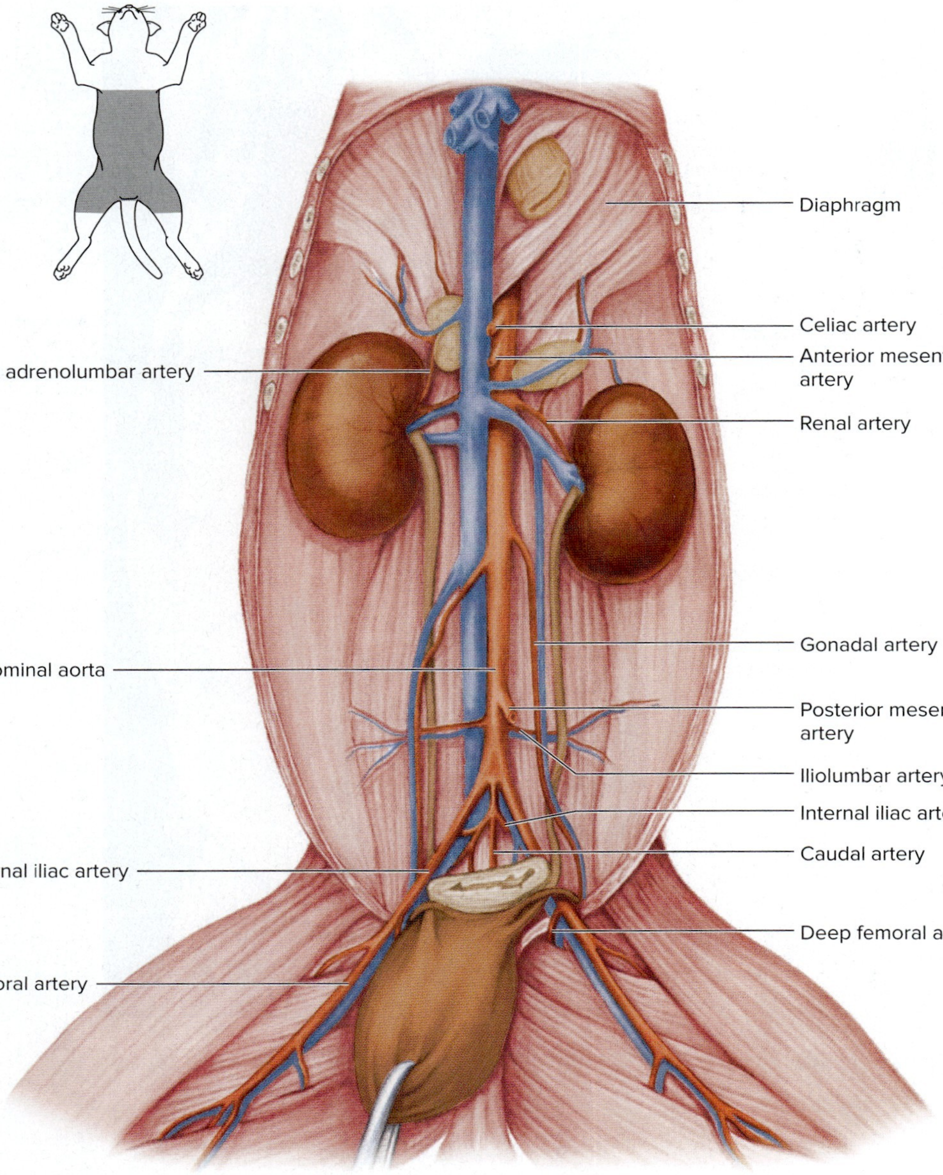

FIGURE 64.7 Arteries of the cat's abdomen and hindlimb, ventral view. APR

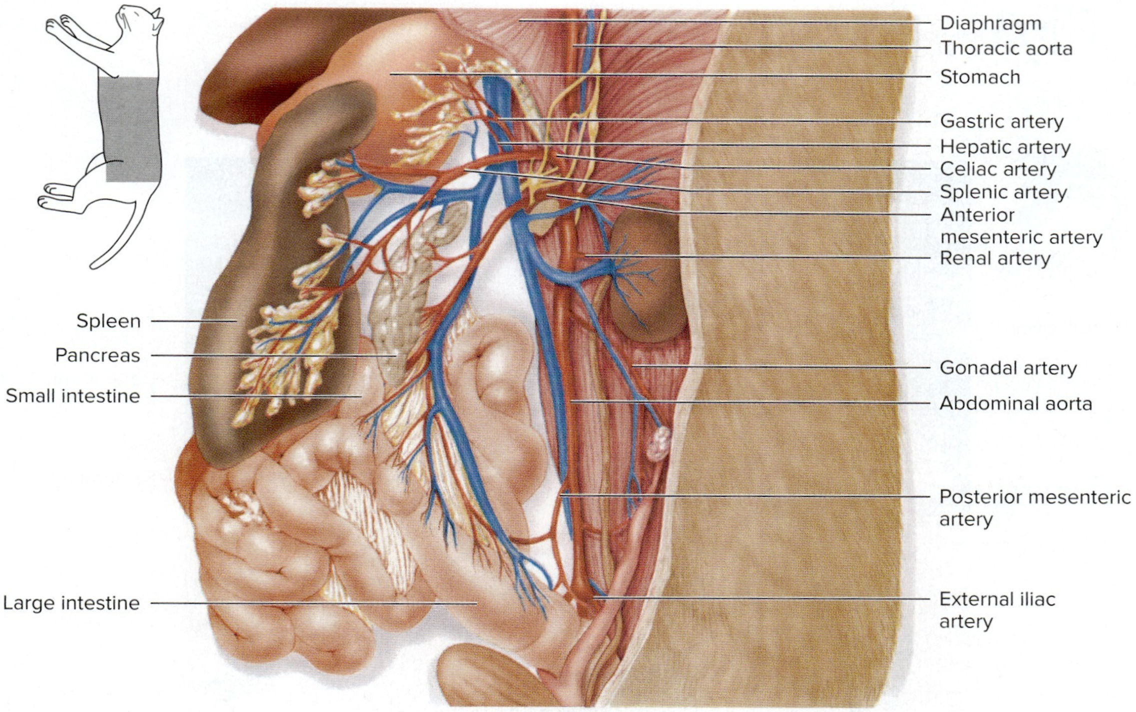

FIGURE 64.8 Arteries of the cat's abdomen, left lateral view.

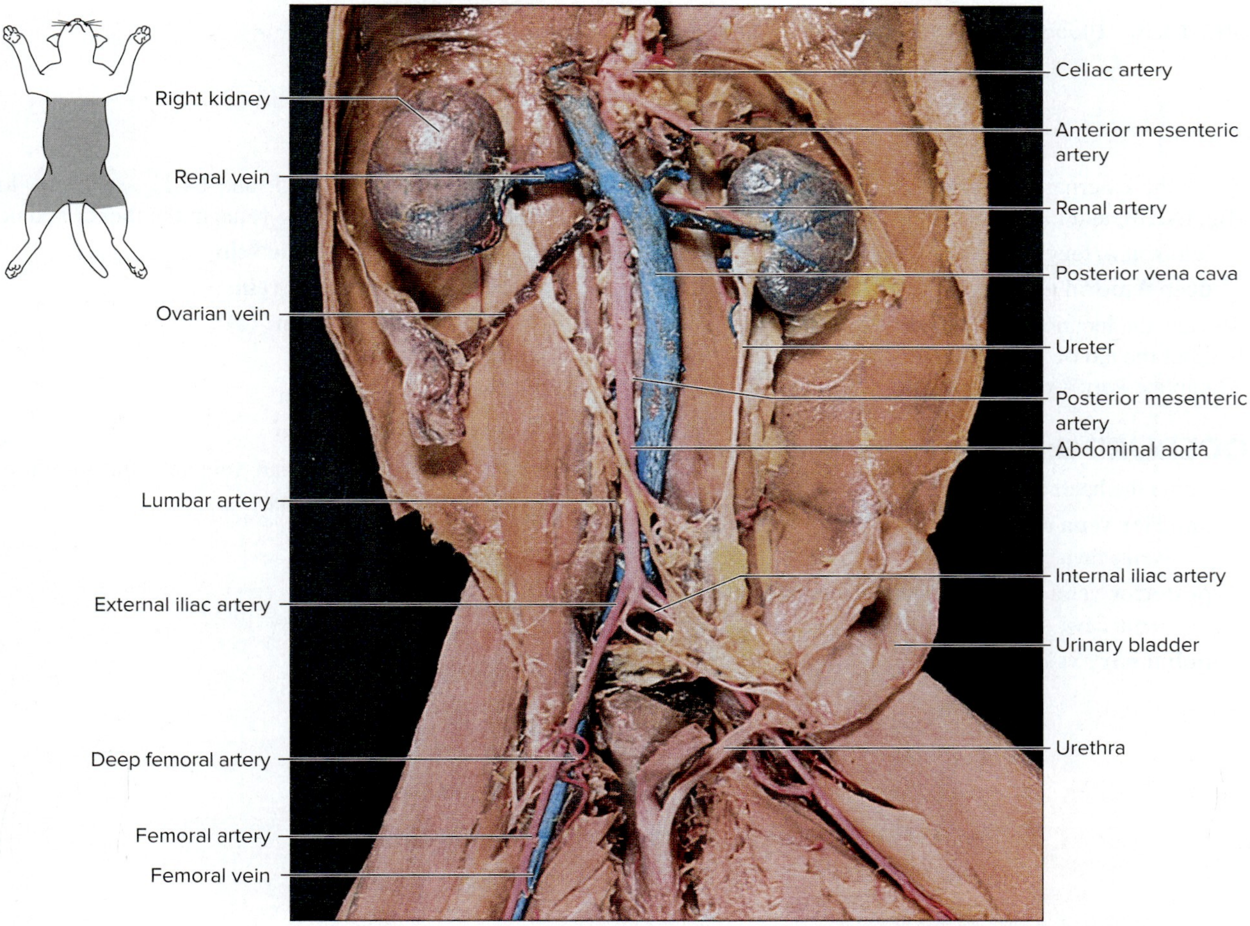

FIGURE 64.9 Blood vessels of the cat's abdominal cavity, with the digestive organs removed, ventral view. Dr. Sheril Burton

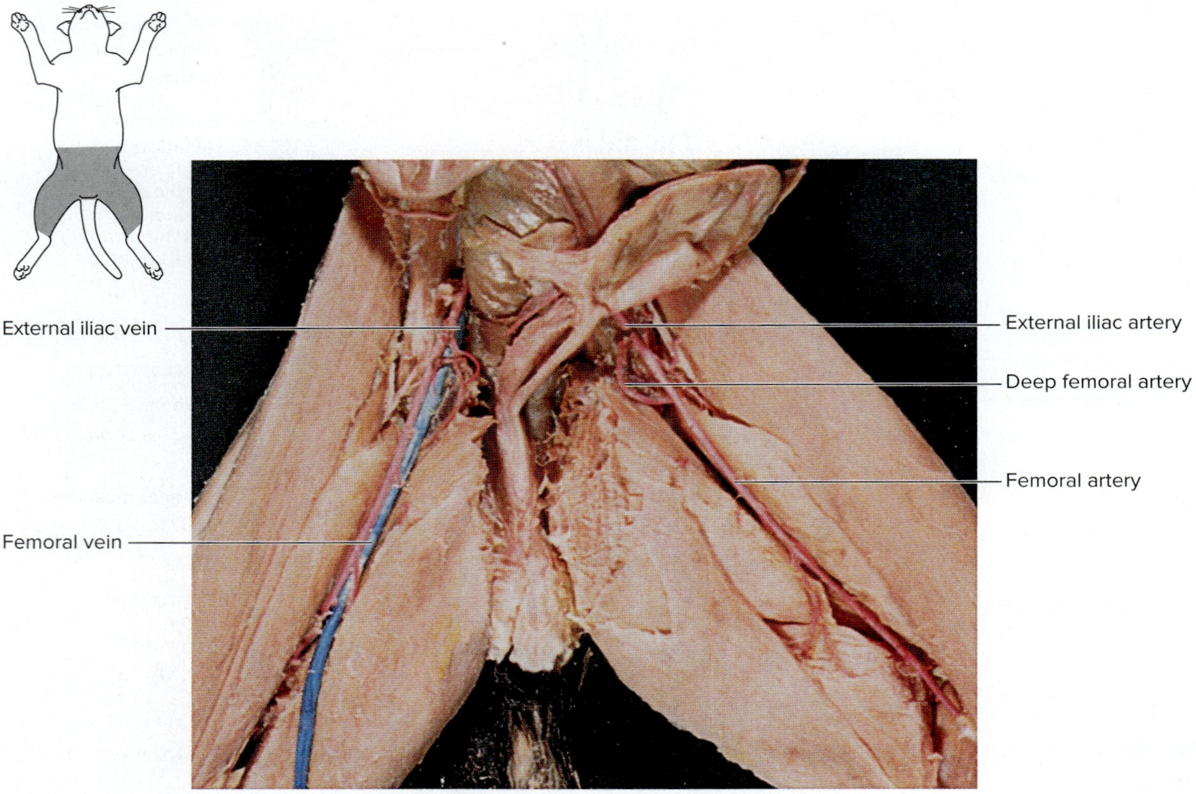

FIGURE 64.10 Blood vessels of the cat's upper hindlimb, ventral view. Dr. Sheril Burton

10. Trace the external iliac artery into the right hindlimb (fig. 64.10), and locate the following:
    - **femoral artery**
    - **deep femoral artery**
11. Review the locations of the heart structures and arteries without the aid of figures.
12. Complete Part A of Laboratory Assessment 64.

PROCEDURE B: The Venous System

1. Examine the heart again and locate the following veins:
    - **anterior vena cava**—corresponds to superior vena cava
    - **posterior vena cava**—corresponds to inferior vena cava
    - **pulmonary veins**
2. Using figures 64.4, 64.5, and 64.11 as guides, locate and dissect the following veins in the thorax and neck:
    - **right brachiocephalic vein**
    - **left brachiocephalic vein**
    - **right subclavian vein**
    - **left subclavian vein**
    - **internal jugular vein**
    - **external jugular vein**
3. Trace the right subclavian vein into the forelimb and locate the following veins:
    - **axillary vein**
    - **subscapular vein**
    - **brachial vein**

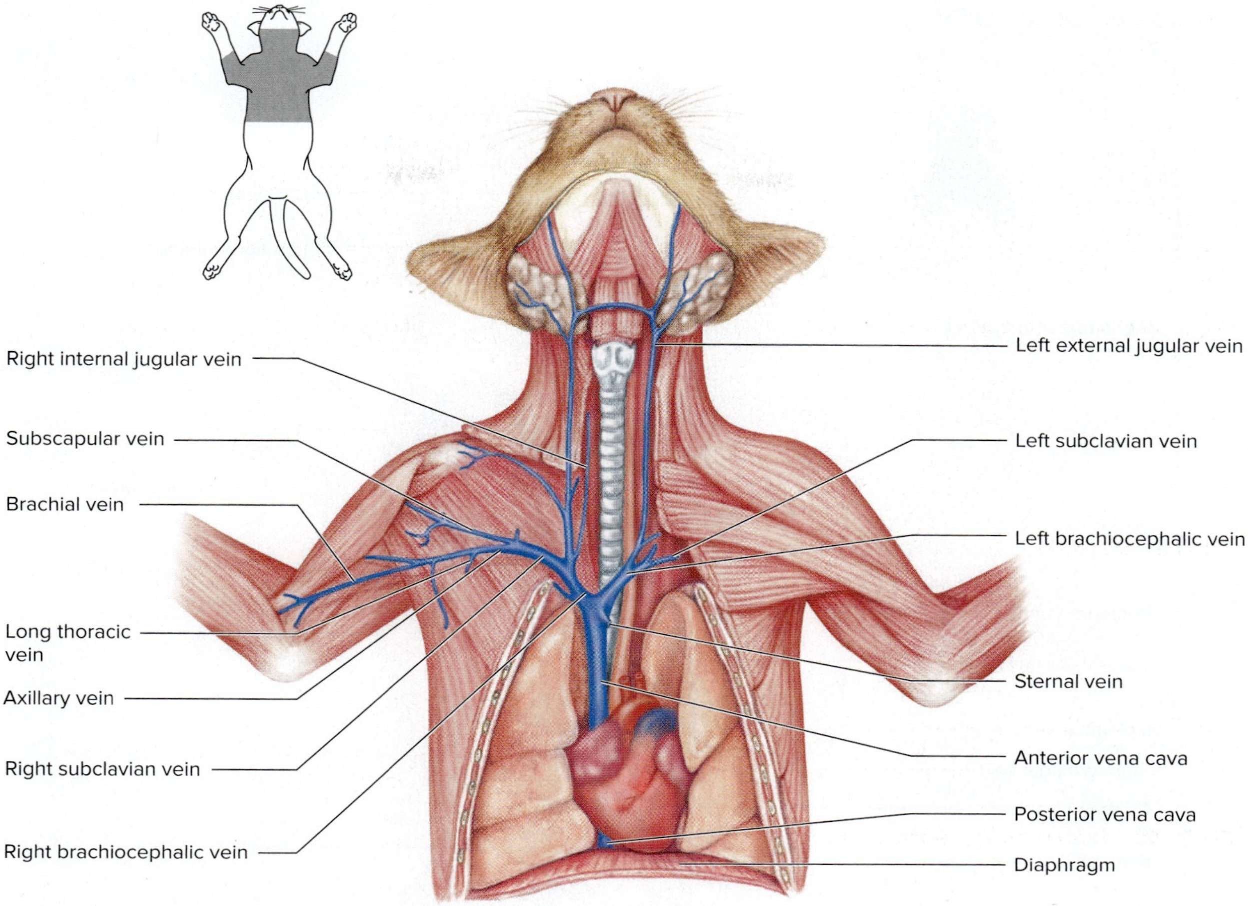

FIGURE 64.11 Veins of the cat's thorax, neck, and forelimb, ventral view. APR

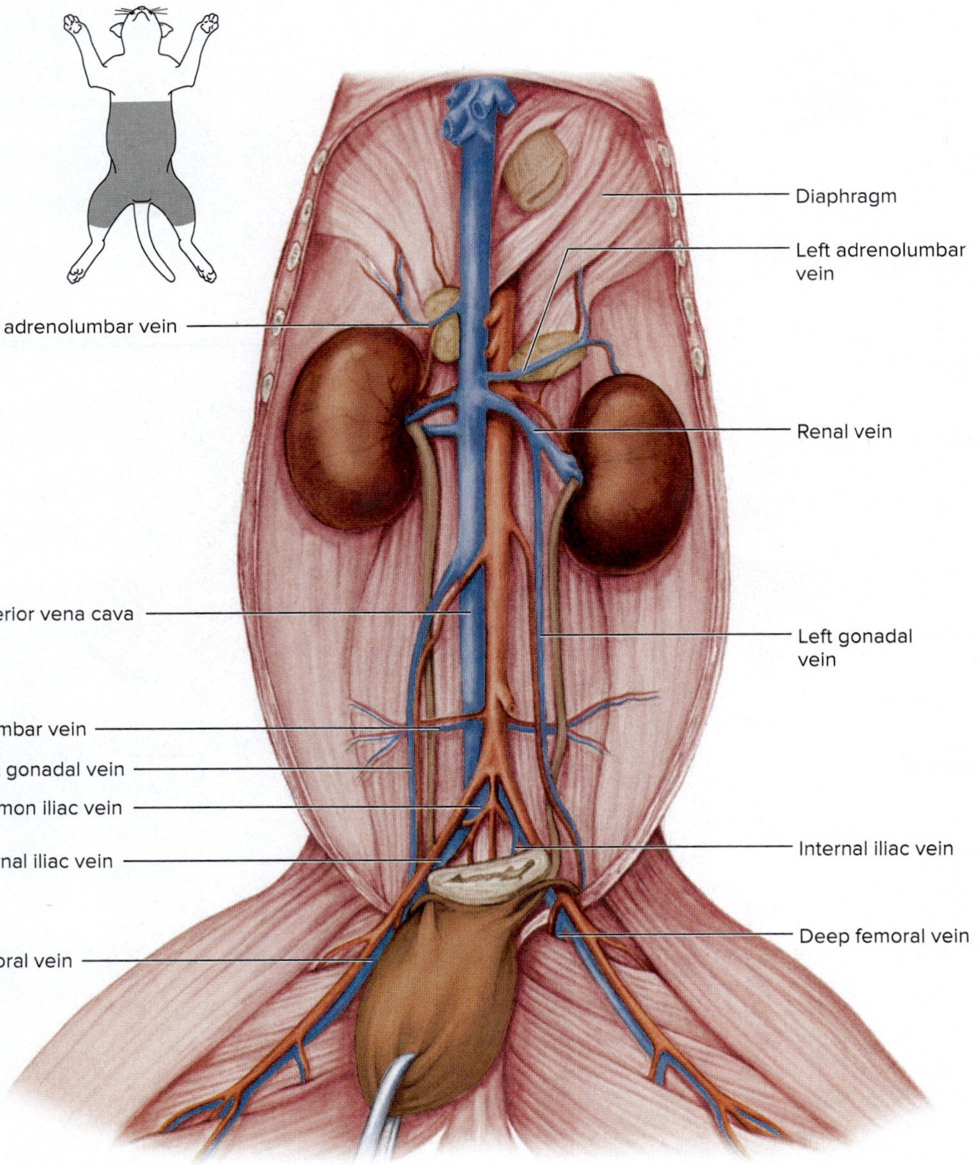

FIGURE 64.12 Veins of the cat's abdomen and hindlimb, ventral view. APR

4. Using figures 64.9, 64.12, and 64.13 as guides, locate and dissect the following veins in the abdomen:
   - **posterior vena cava**
   - **adrenolumbar vein**
   - **anterior mesenteric vein**—corresponds to superior mesenteric vein
   - **posterior mesenteric vein**—corresponds to inferior mesenteric vein
   - **hepatic portal vein**
   - **splenic vein**
   - **renal vein**
   - **gonadal vein**
   - **iliolumbar vein**
   - **common iliac vein**
   - **internal iliac vein**
   - **external iliac vein**
5. Trace the external iliac vein into the hindlimb (fig. 64.10) and locate the following veins:
   - **femoral vein**
   - **deep femoral vein**
6. Review the locations of the veins without the aid of figures.
7. Complete Parts B, C, and D of the laboratory assessment.

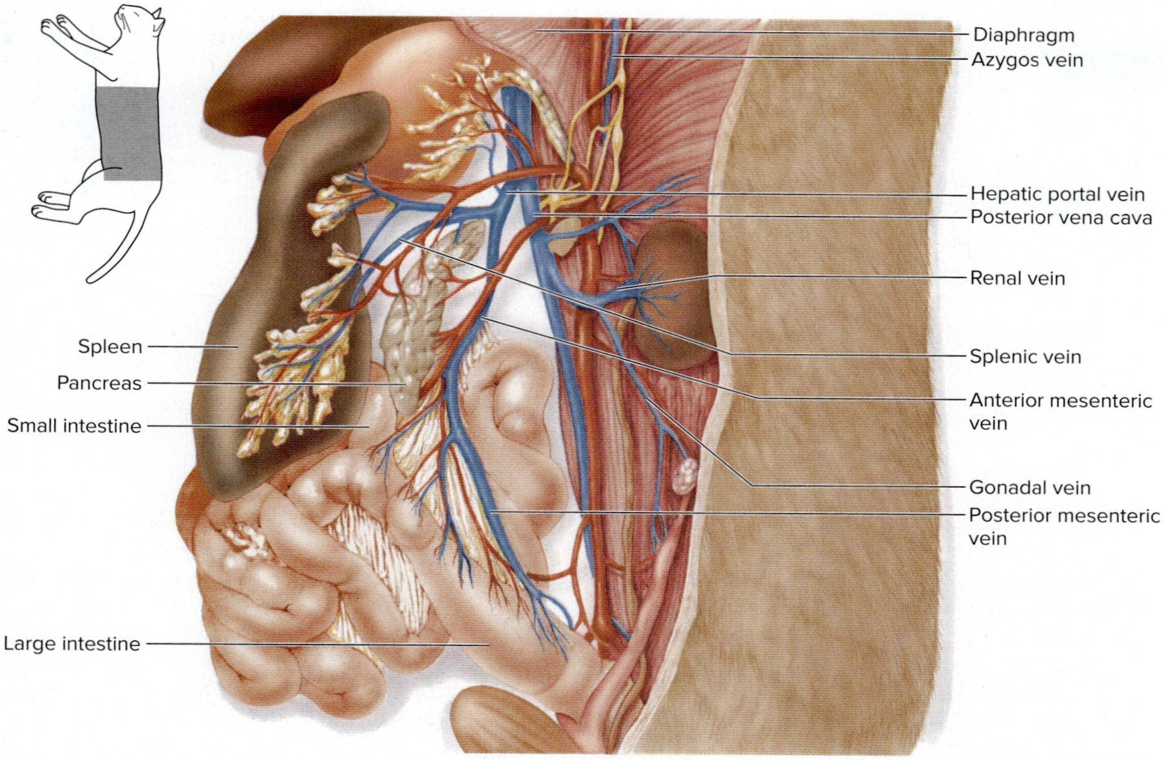

FIGURE 64.13 Veins of the cat's abdomen, left lateral view.

NOTES

LABORATORY
ASSESSMENT 64

Name ______________________________

Date ______________________________

Section ______________________________

The **A** corresponds to the indicated Learning Outcome(s) **O** found at the beginning of the Laboratory Exercise.

Cat Dissection: Cardiovascular System

PART A: Assessments

Complete the following:

1. Describe the location and position of the heart of the cat and the surrounding attachments of the parietal pericardium. **A1**

____

____

2. Describe the relative thicknesses of the walls of the heart chambers of the cat. **A1**

____

____

3. Explain how the wall thicknesses are related to the functions of the chambers. **A1**

____

____

4. Compare the origins of the common carotid arteries of the cat with those of the human. **A2**

____

____

5. Compare the origins of the iliac arteries of the cat with those of the human. **A2**

____

____

PART B: Assessments

Complete the following:

1. Compare the origins of the brachiocephalic veins of the cat with those of the human. **A2**

____

____

2. Compare the relative sizes of the external and internal jugular veins of the cat with those of the human. **A2**

____

____

____

3. List twelve veins that cats and humans have in common. **A2** **A3**

PART C: Assessments

Identify the numbered cardiovascular features of the cat in figures 64.14 to 64.17.

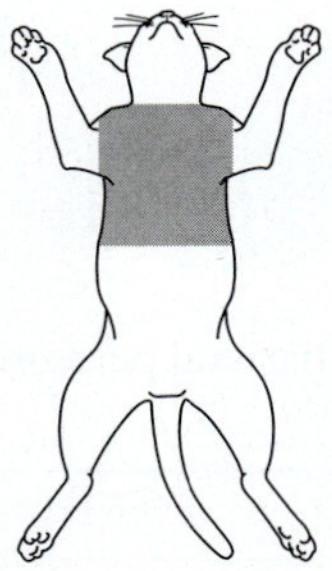

Terms:
Anterior vena cava
Aortic arch
Brachiocephalic artery
Left common carotid artery
Left subclavian artery
Left ventricle
Parietal pericardium
Posterior vena cava
Pulmonary trunk
Right ventricle
Thoracic aorta

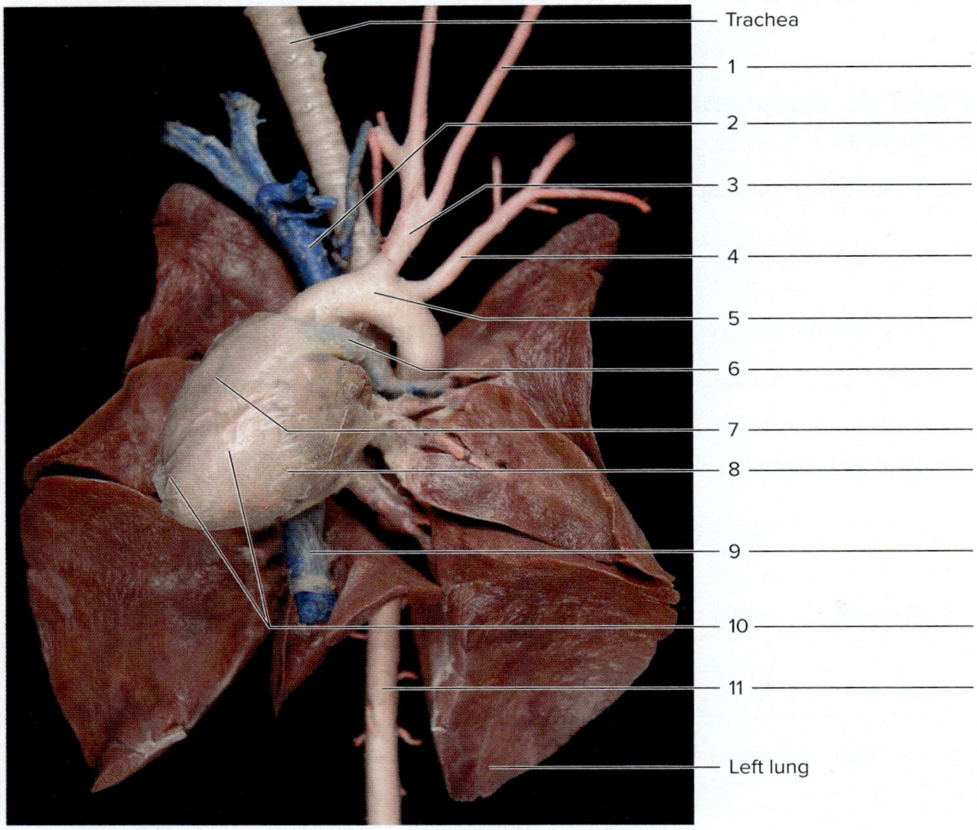

(a)

FIGURE 64.14 Label the numbered cardiovascular structures on the ventral views of the cat in (a) a dissected thoracic cavity, and (b) an illustration. Use terms provided for (a) only. Dennis Strete **A1**

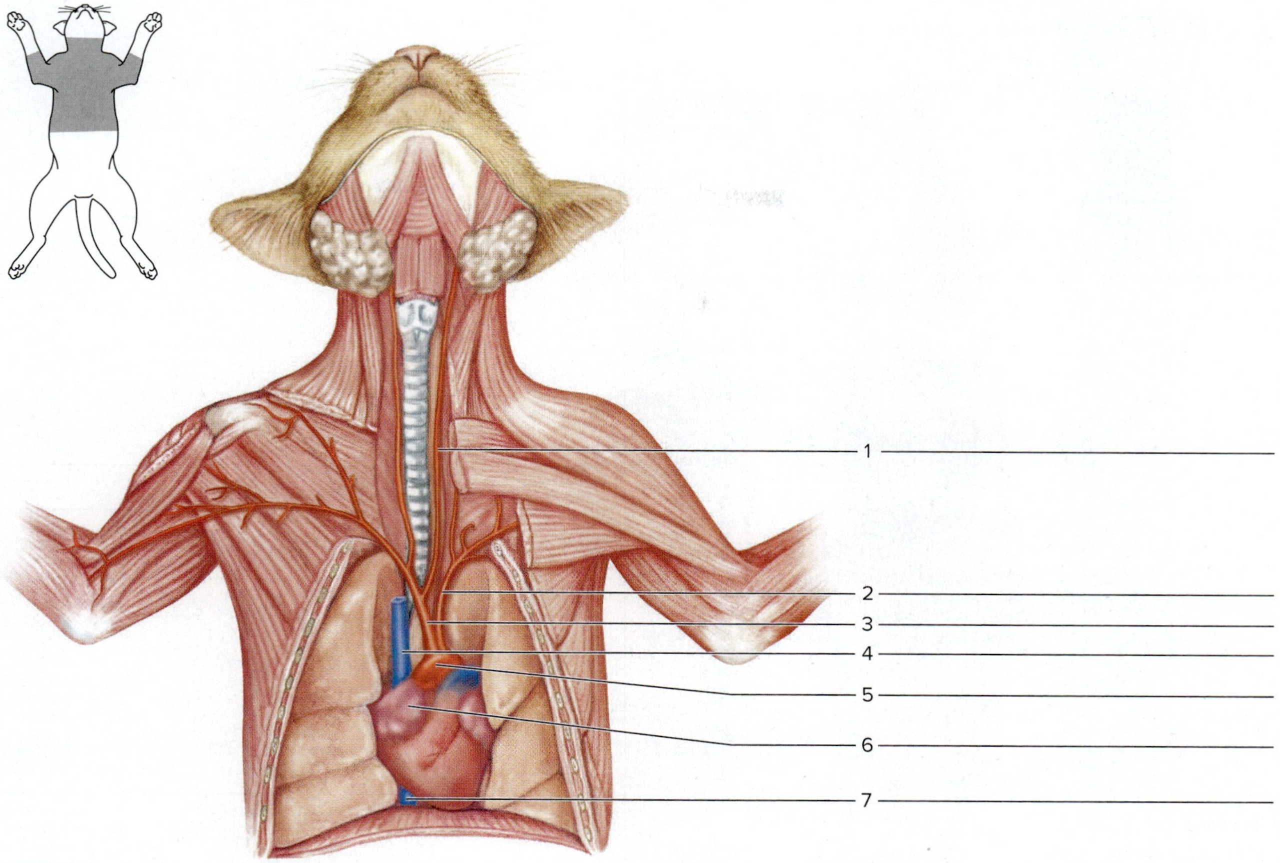

(b)

FIGURE 64.14 *Continued.*

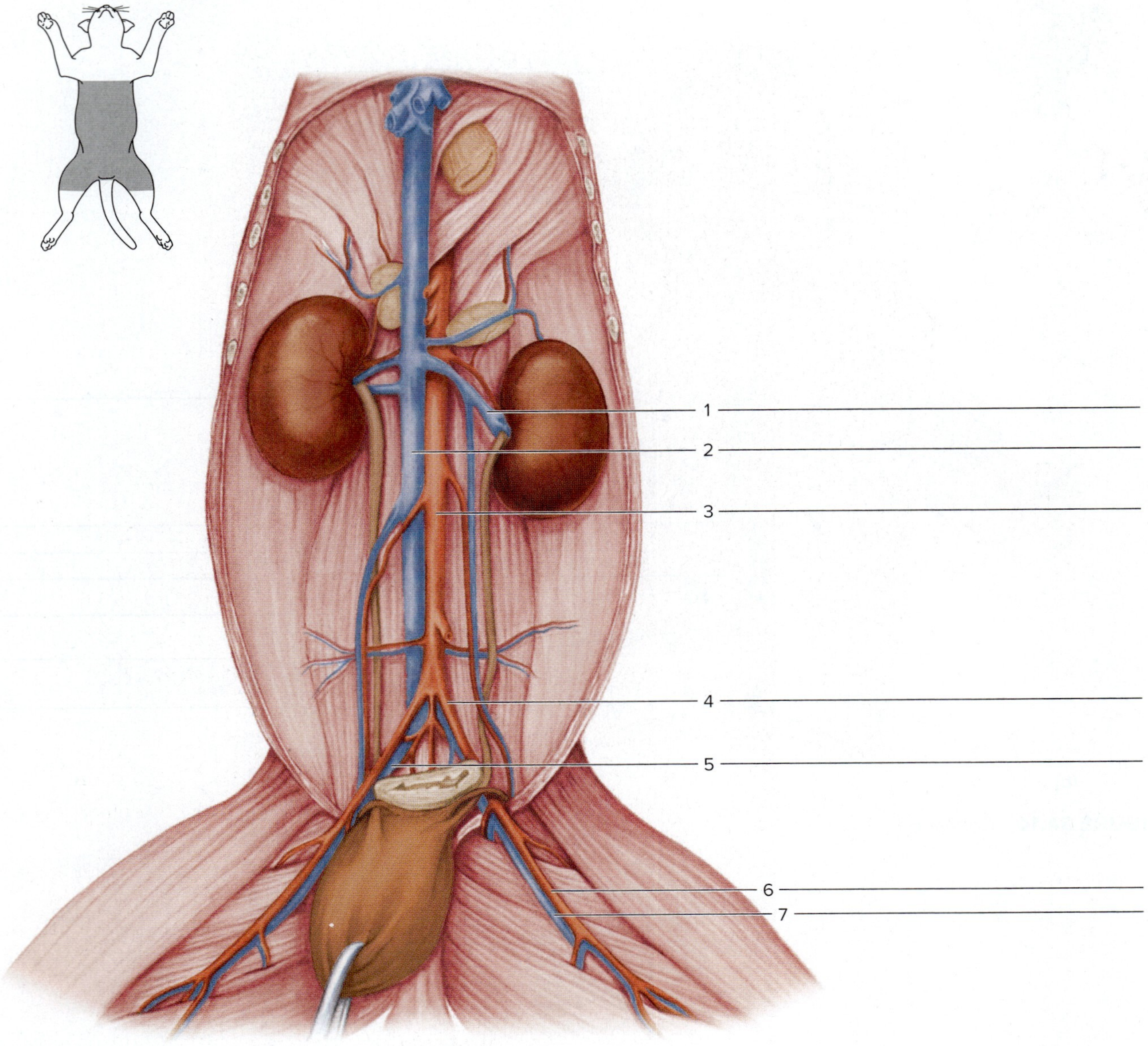

FIGURE 64.15 Label the numbered cardiovascular structures of the cat, ventral view. **A1**

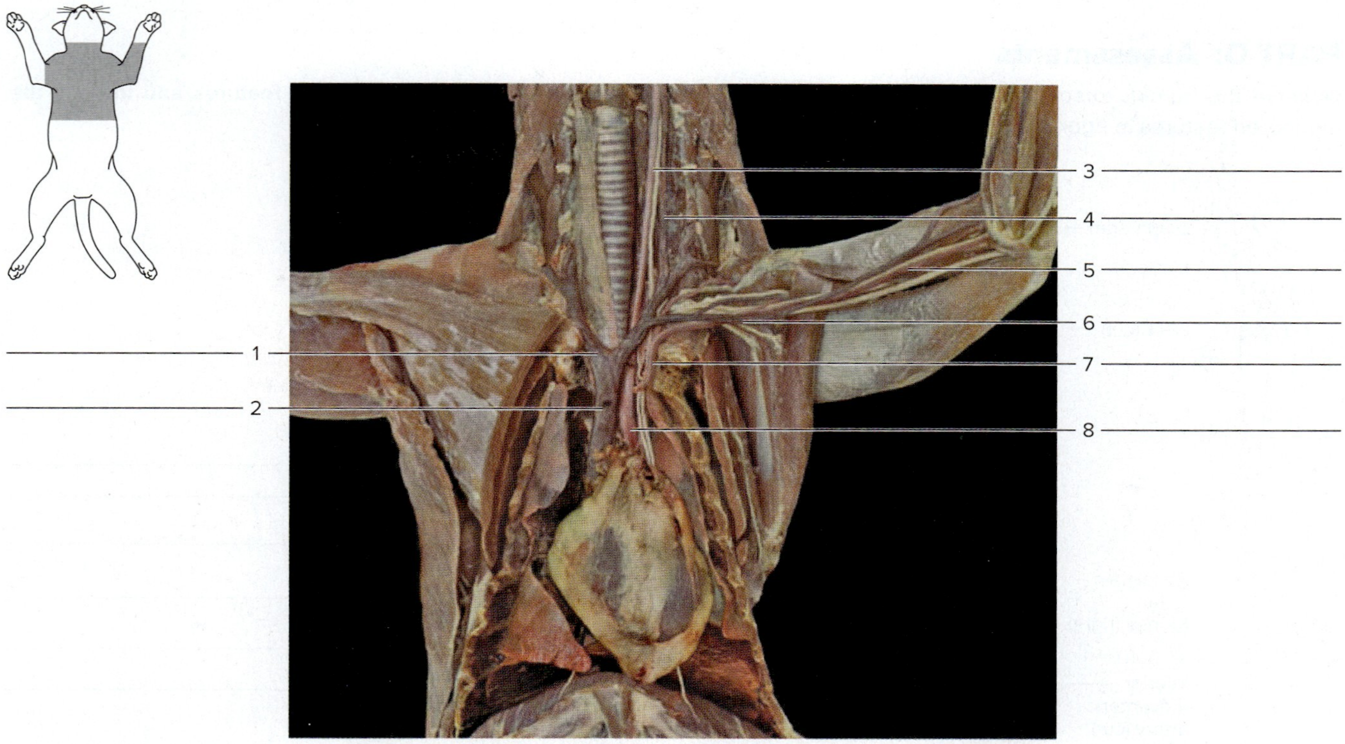

FIGURE 64.16 Label the numbered cardiovascular blood vessels of the cat, ventral view. APR/McGraw Hill A1

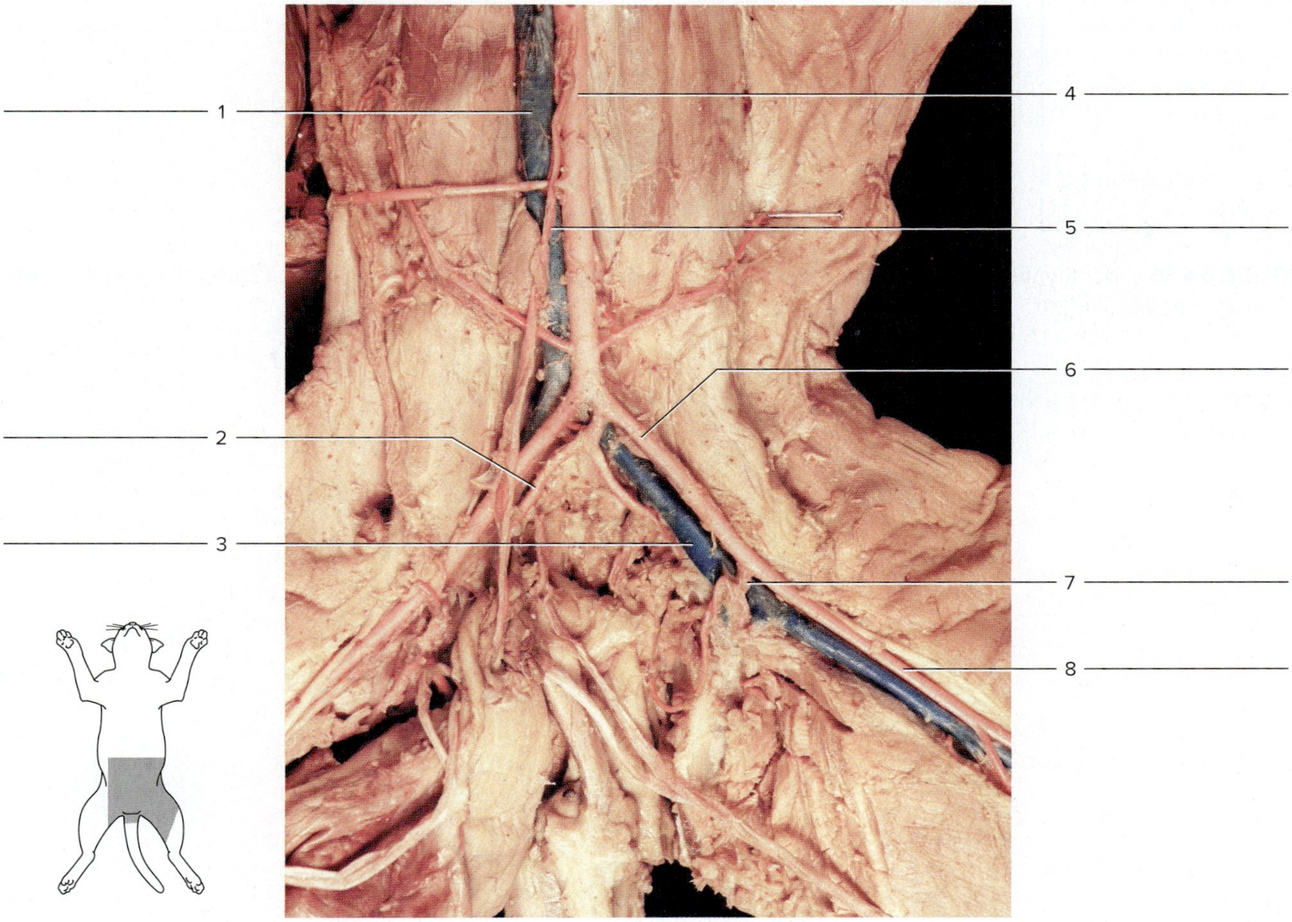

FIGURE 64.17 Label the numbered cardiovascular structures of the cat, ventral view. Dennis Strete A1

PART D: Assessments

Observe the human torso model and figures 64.7, 64.9, and 64.12 of a cat. Locate the labeled features and identify the numbered features in figure 64.18 of a cadaver that were also identified in the cat dissection.

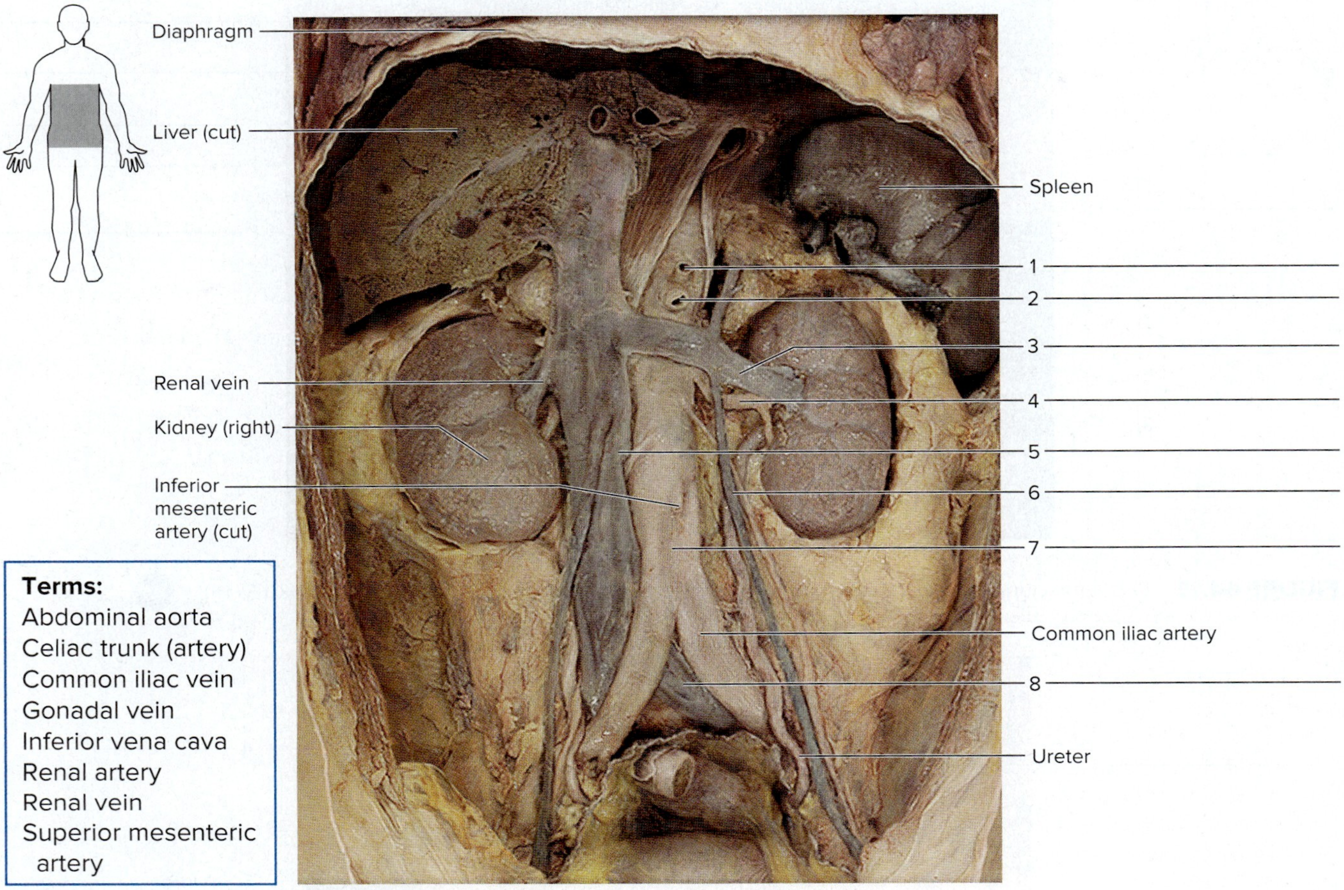

Terms:
Abdominal aorta
Celiac trunk (artery)
Common iliac vein
Gonadal vein
Inferior vena cava
Renal artery
Renal vein
Superior mesenteric artery

FIGURE 64.18 Identify the arteries and veins indicated on this anterior view of the abdomen of a cadaver, using the terms provided. APR/McGraw Hill A2 A3

LABORATORY

EXERCISE 65

Cat Dissection: Respiratory System

Purpose of the Exercise

To examine and identify the major respiratory organs of the cat and to compare and associate them with the organs of the human.

MATERIALS NEEDED

Preserved cat
Dissecting tray
Dissecting instruments
Bone cutter
Disposable gloves
Human torso model

SAFETY

- Review all the safety guidelines in Appendix 1 of your laboratory manual.
- Wear disposable gloves and protective eyewear when working on the cat dissection.
- Dispose of tissue remnants and gloves as instructed.
- Wash the dissecting tray and instruments as instructed.
- Wash your laboratory table.
- Wash your hands before leaving the laboratory.

Learning Outcomes APR

After completing this exercise, you should be able to:

O1 Locate and identify the major respiratory organs of the cat.

O2 Compare the respiratory system of the human with that of the cat.

O3 Identify the corresponding organs in the human torso model.

The O corresponds to the assessments A indicated in the Laboratory Assessment for this Exercise.

Pre-Lab

Carefully read the introductory material and examine the entire lab. Be familiar with the structures and functions of the human respiratory system from lecture, the textbook, or laboratory exercises.

In this laboratory exercise, you will dissect the major respiratory organs of a cat. As you observe these structures in the cat, compare them with those of the human torso model.

PROCEDURE: Respiratory System Dissection

1. Place the preserved cat in a dissecting tray with its ventral surface up.
2. Examine the *external nares* (*nostrils*) and *nasal septum.*
3. Open the oral cavity. To do this, follow these steps:
   a. Use scissors to cut through the soft tissues at the angle of the mouth.
   b. When you reach the bone of the jaw, use a bone cutter to cut through the bone, thus freeing the mandible.
4. Open the oral cavity wide to expose the *hard palate* and the *soft palate* (fig. 65.1). Cut through the tissues of the soft palate and observe the *nasopharynx* above it. Locate the small openings of the *auditory tubes* in the lateral walls of the nasopharynx. Insert a probe into an opening of an auditory tube.

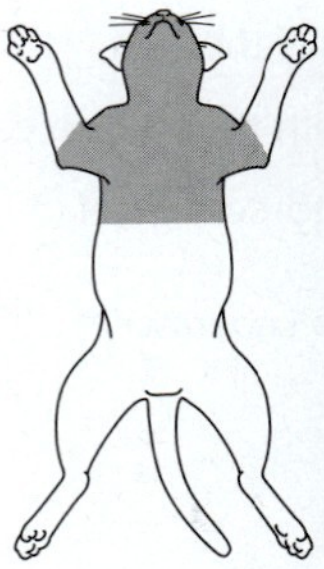

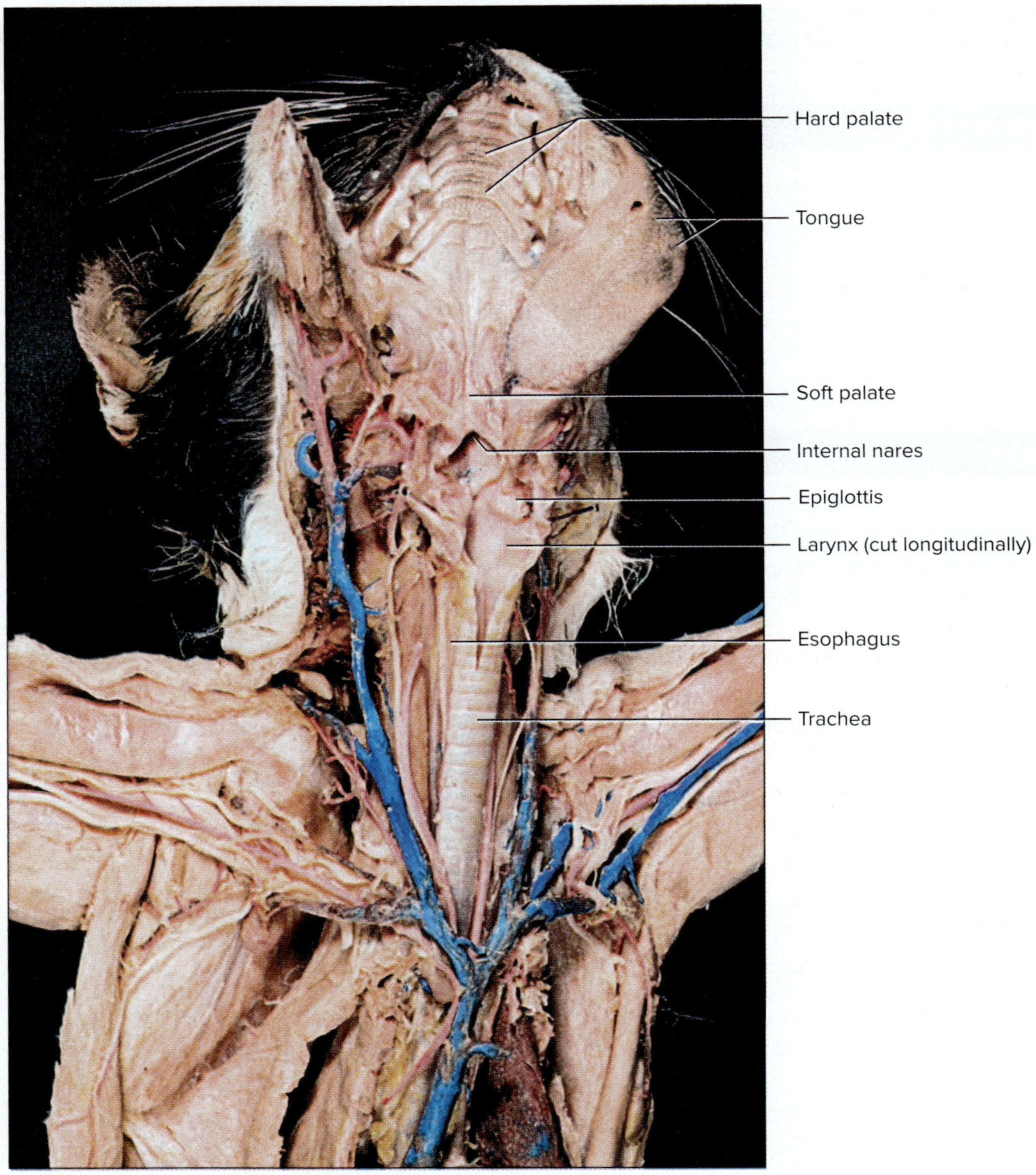

FIGURE 65.1 Features of the cat's neck and oral cavity, ventral view. Courtesy of Dr. Sheril Burton APR

5. Pull the tongue forward, and locate the *epiglottis* near its base. Also identify the *glottis,* an opening between the *vocal folds* (*vocal cords*) into the larynx. Locate the *esophagus,* which is dorsal to the larynx (figs. 65.1 and 65.2).
6. Open the thoracic cavity and expose its contents. See figures 65.1 to 65.3.
7. Dissect the *trachea* in the neck and expose the *larynx* at the anterior end near the base of the tongue. Note the *tracheal rings* and locate the lobes of the *thyroid gland* on each side of the trachea, just posterior to the larynx. Also note the *thymus gland,* a rather diffuse mass of glandular tissue extending along the ventral surface of the trachea into the thorax (figs. 65.1–65.3).
8. Examine the larynx, and identify the *thyroid cartilage* and *cricoid cartilage* in its wall (fig. 65.4*a*). Make a longitudinal incision through the ventral wall of the larynx and locate the vocal folds inside. Compare the structures of the larynx of the cat with those of a human (fig. 65.4*b*).

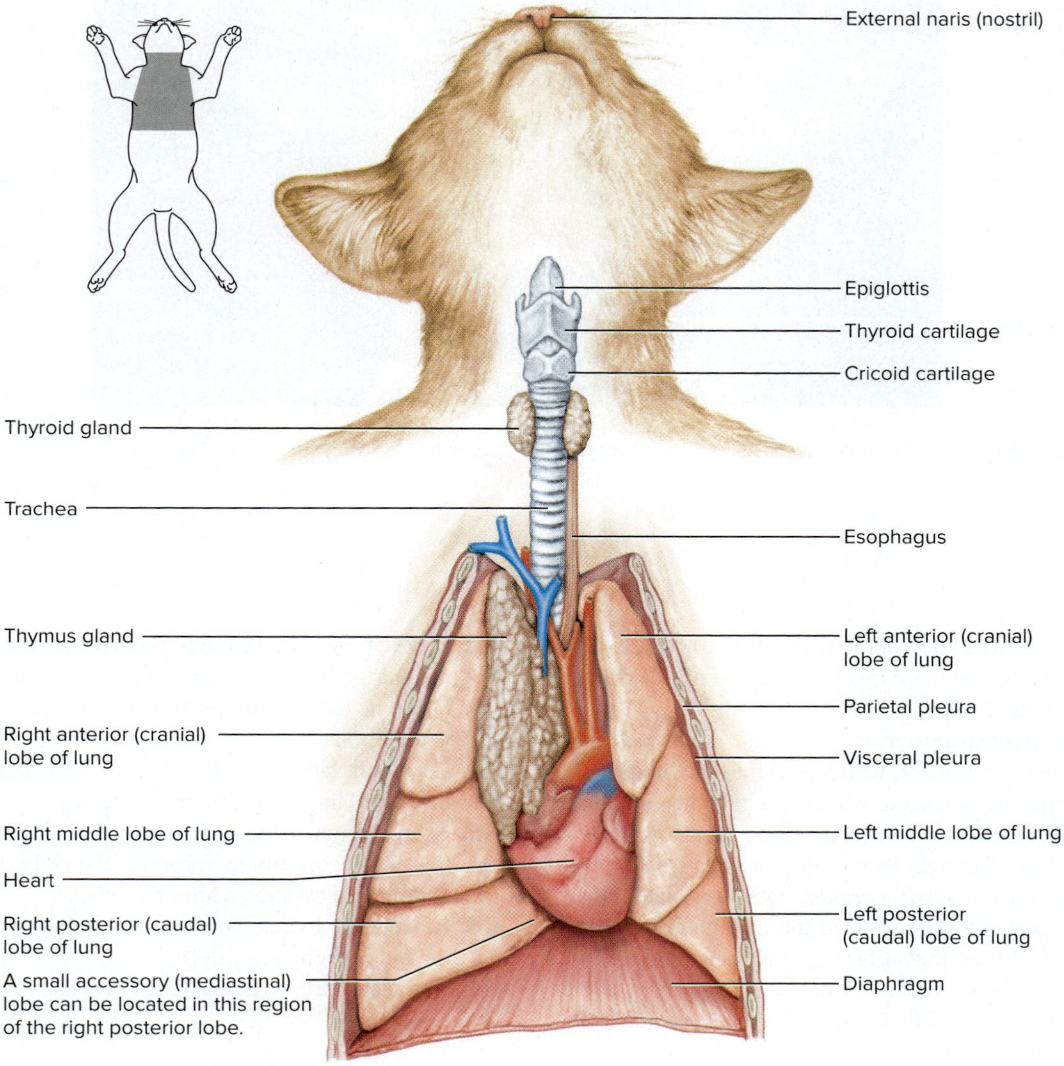

FIGURE 65.2 Ventral view of the cat's thoracic cavity. APR

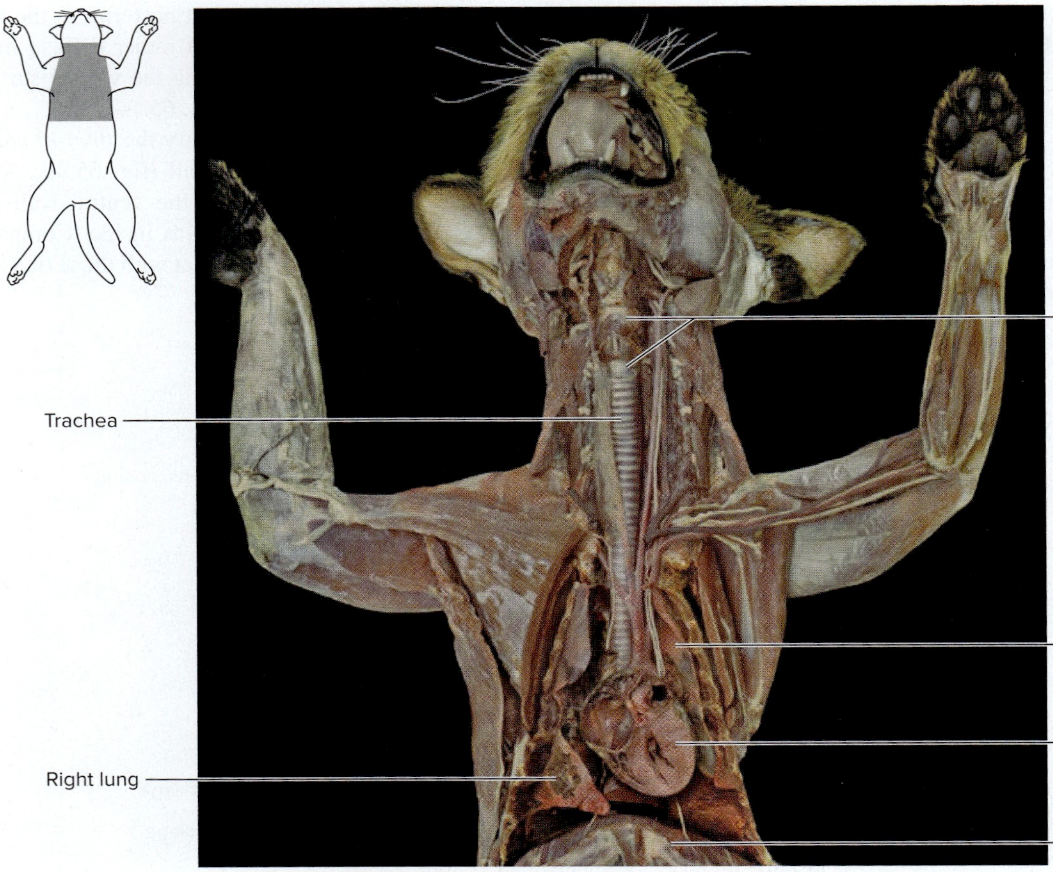

FIGURE 65.3 Structures of the lower respiratory tract of the cat, ventral view. APR/McGraw Hill

9. Observe a sagittal view of the respiratory tract of the cat (fig. 65.5), and compare this to ventral views of the cat (figs. 65.1–65.3).
10. Trace the trachea posteriorly to where it divides into *main (primary) bronchi,* which pass into the lungs.
11. Examine the *lungs,* each of which is subdivided into three lobes—an *anterior* (*cranial*), a *middle,* and a *posterior (caudal) lobe.* The right lung of the cat has a small fourth *accessory* (*mediastinal; cardiac*) lobe associated with the right posterior lobe. Notice the thin membrane, the *visceral pleura,* on the surface of each lung. Also notice the parietal pleura, which forms the inner lining of the thoracic wall, and locate the spaces of the *pleural cavities* (figs. 65.2 and 65.3).
12. Make an incision through a lobe of a lung and examine its interior. Note the branches of the smaller air passages.
13. Examine the *diaphragm* and note that its central portion is tendinous.
14. Locate the *phrenic nerve.* This nerve appears as a white thread passing along the side of the heart to the diaphragm.
15. Examine the organs located in the *mediastinum.* Note that the mediastinum separates the right and left lungs and pleural cavities within the thorax.
16. Review the locations of the respiratory organs without the aid of figures.
17. Complete Parts A, B, and C of Laboratory Assessment 65.

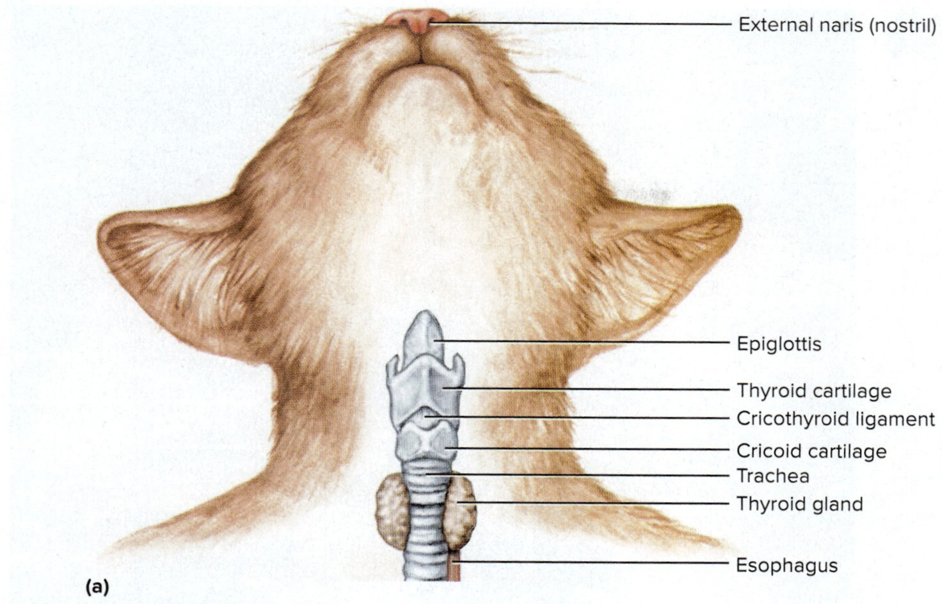

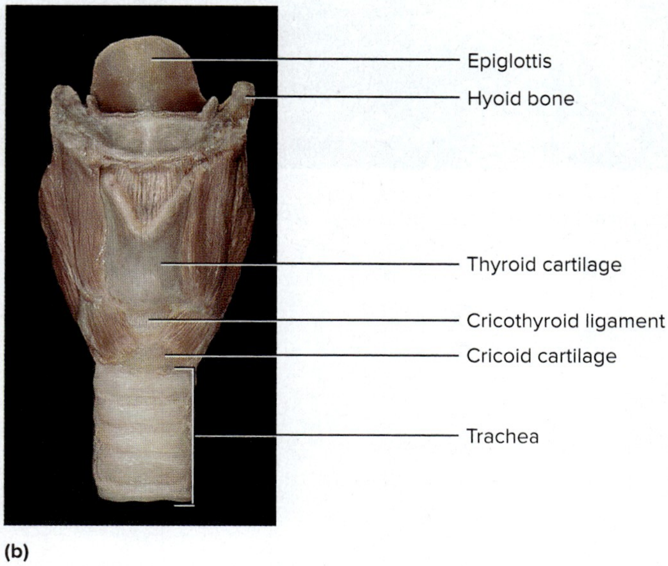

FIGURE 65.4 (a) Larynx and associated structures of the cat, ventral view. (b) Larynx structures of a cadaver, anterior view. (b) APR/McGraw Hill

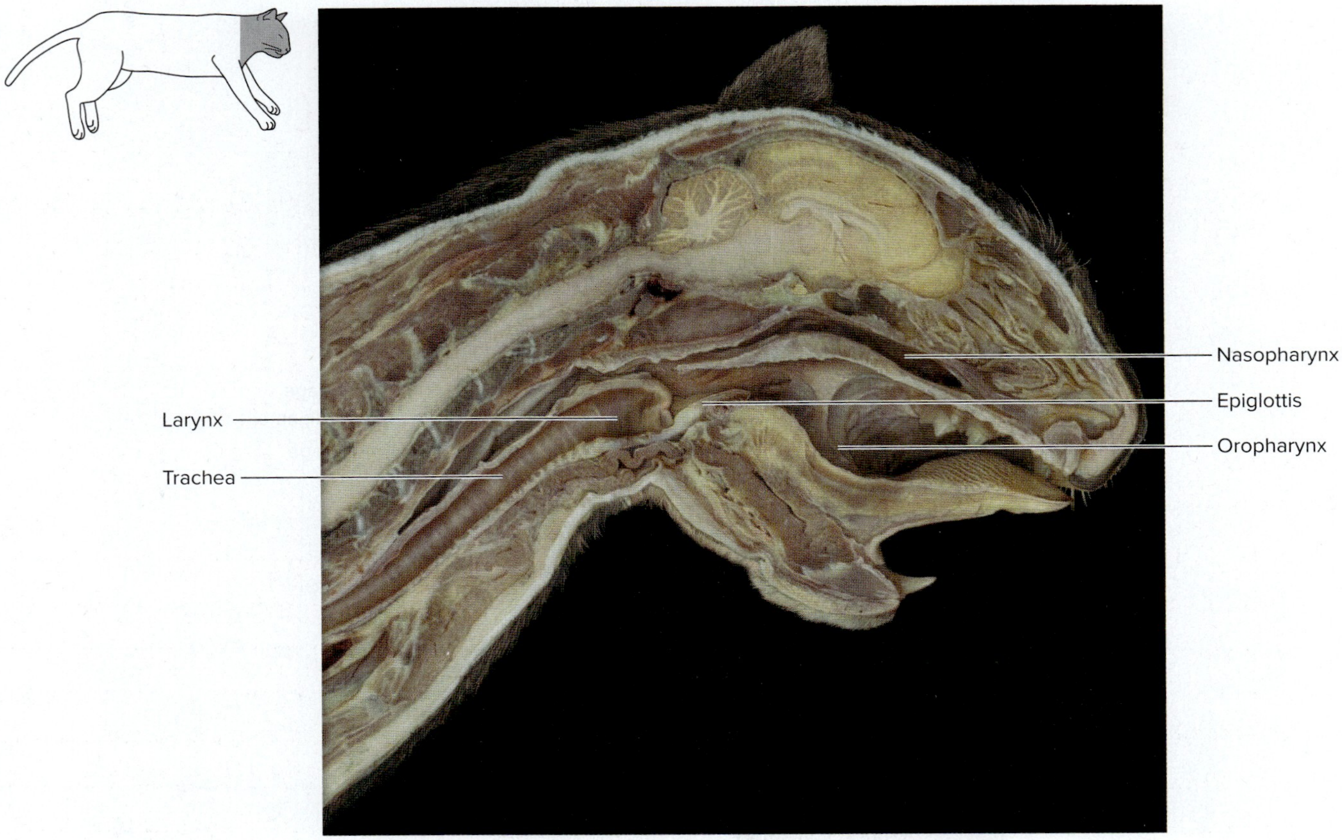

FIGURE 65.5 Upper respiratory tract of the cat, sagittal view. APR/McGraw Hill APR

LABORATORY
ASSESSMENT 65

Name ____________________

Date ____________________

Section ____________________

The (A) corresponds to the indicated Learning Outcome(s) (O) found at the beginning of the Laboratory Exercise.

Cat Dissection: Respiratory System

PART A: Assessments

Complete the following:

1. What is the purpose of the auditory tubes opening into the nasopharynx? (A1) ____________________

2. Distinguish between the glottis and the epiglottis. (A1) ____________________

3. Are the tracheal rings of the cat complete or incomplete? (A1) ____________________ How does this feature compare with that of the human? (A2) ____________________

4. How does the structure of the main (primary) bronchi compare with that of the trachea? (A1) ____________________

5. Compare the number of lobes in the human lungs with the number of lobes in the cat. (A2) ____________________

6. Describe the attachments of the diaphragm. (A1) ____________________

7. What major structures are located within the mediastinum of the cat? (A1) ____________________

   How does this compare to the human mediastinum? (A2) ____________________

PART B: Assessments

Identify the numbered respiratory organs in figure 65.6.

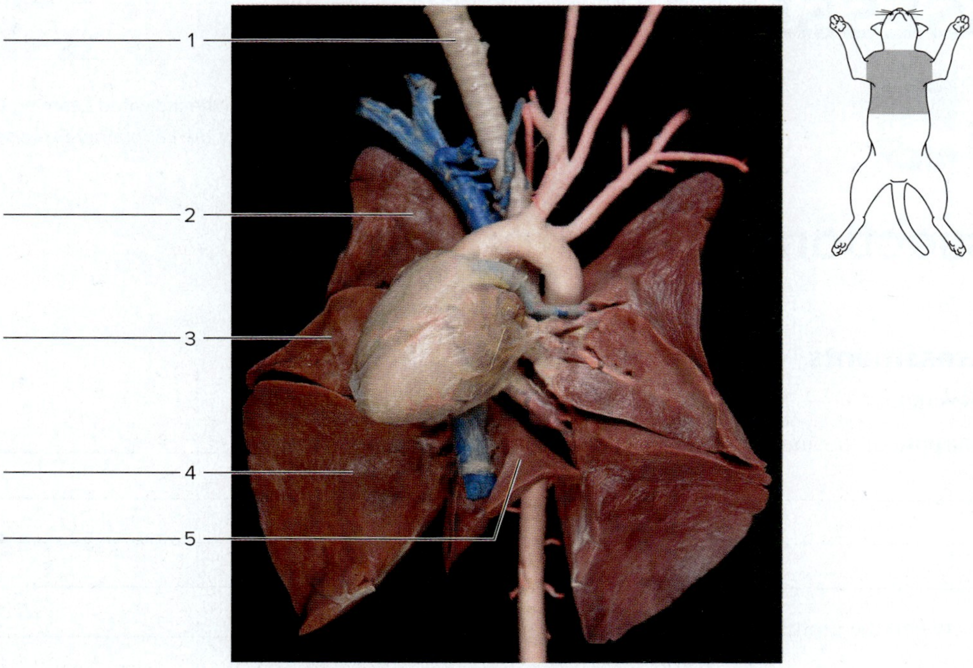

FIGURE 65.6 Label the numbered lobes of the right lung and other respiratory structure shown in the image of a dissected thorax of the cat, ventral view. Dennis Strete A1

PART C: Assessments

Observe the human torso model and figures 65.4*b* and 65.7 of a cadaver. Locate the labeled features and identify the numbered features in figure 65.7 that were also identified in the cat dissection.

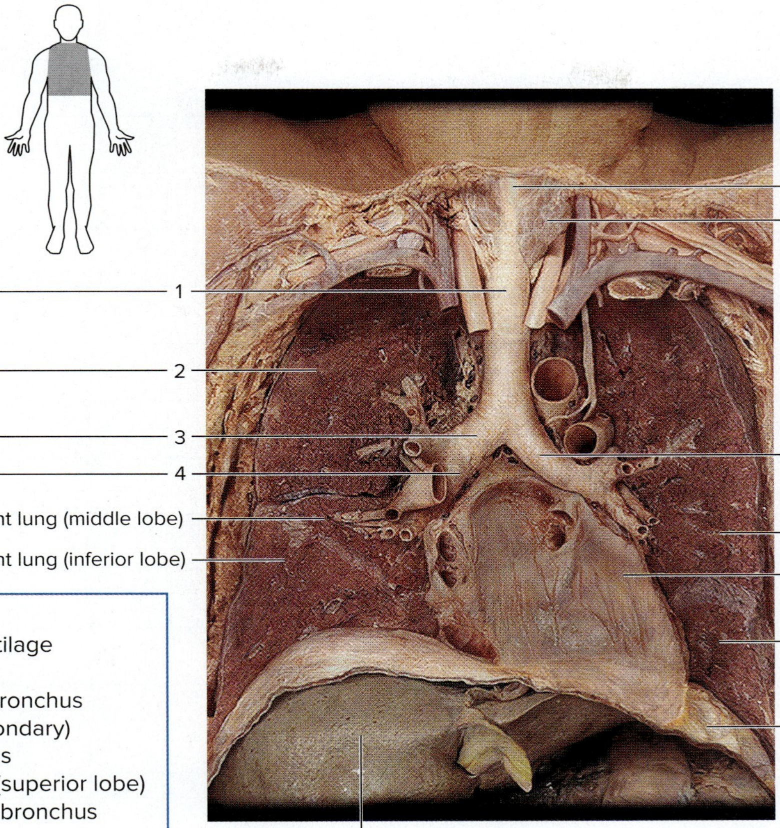

Terms:
Cricoid cartilage
Diaphragm
Left main bronchus
Lobar (secondary) bronchus
Right lung (superior lobe)
Right main bronchus
Trachea

FIGURE 65.7 Identify the respiratory features indicated on this anterior view of the thorax, using the terms provided. APR/McGraw Hill APR A2 A3

NOTES

LABORATORY
EXERCISE 66

Cat Dissection: Digestive System

Purpose of the Exercise

To examine and identify the major digestive organs of the cat and to compare these organs with those of the human.

MATERIALS NEEDED

Preserved cat
Dissecting tray
Dissecting instruments
Bone cutter
Disposable gloves
Hand lens
Human torso model

SAFETY

- Review all the safety guidelines in Appendix 1 of your laboratory manual.
- Wear disposable gloves and protective eyewear when working on the cat dissection.
- Dispose of tissue remnants and gloves as instructed.
- Wash the dissecting tray and instruments as instructed.
- Wash your laboratory table.
- Wash your hands before leaving the laboratory.

Learning Outcomes APR

After completing this exercise, you should be able to:

O1 Locate and identify the major digestive organs of the cat.
O2 Compare the digestive system of the cat with that of the human.
O3 Identify the corresponding digestive organs in the human.

The O corresponds to the assessments A indicated in the Laboratory Assessment for this Exercise.

Pre-Lab

Carefully read the introductory material and examine the entire lab. Be familiar with the structures and functions of the human digestive system from lecture, the textbook, or laboratory exercises.

In this laboratory exercise, you will dissect the major digestive organs of the cat. As you observe these organs, compare them with those of the human by observing the parts of the human torso model; however, keep in mind that the cat is a carnivore (flesh-eating mammal) and that the organs of its digestive system are adapted to capturing, holding, eating, and digesting the bodies of other animals. In contrast, humans are omnivores (eat both plant and animal substances). Because humans are adapted to eating and digesting a greater variety of foods, comparisons of the cat and the human digestive organs may not be precise.

PROCEDURE: Digestive System Dissection

1. Place the preserved cat in the dissecting tray on its left side.
2. Locate the major salivary glands on one side of the head (fig. 66.1*a*). To do this, follow these steps:
   a. Clear away any remaining fascia and other connective tissue from the region below the ear and near the joint of the mandible.
   b. Identify the *parotid gland,* a relatively large mass of glandular tissue just below the ear. Note the parotid duct (Stensen's duct) that passes over the surface of the masseter muscle and opens into the oral cavity.

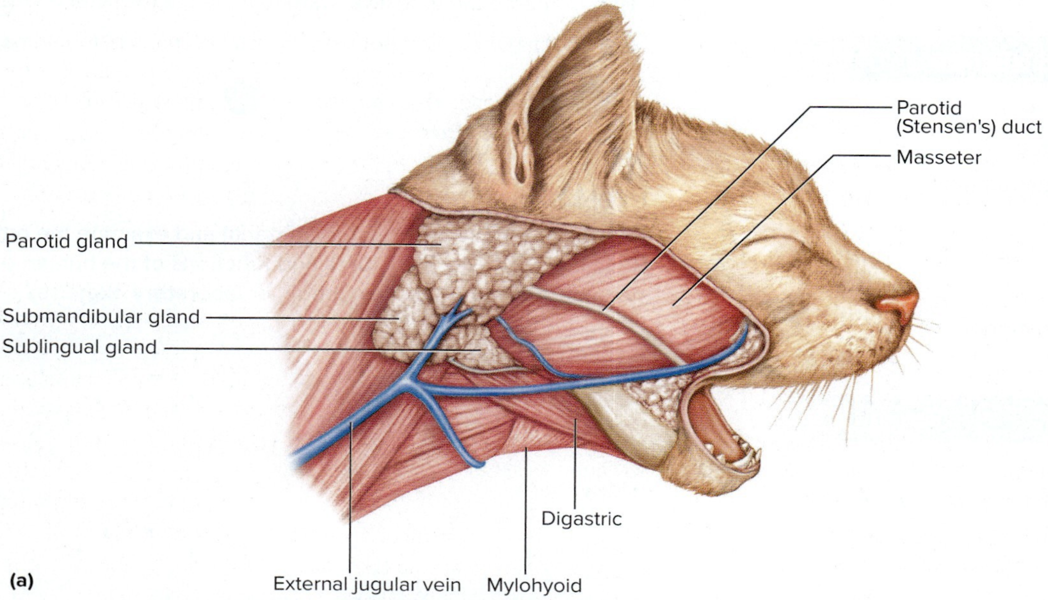

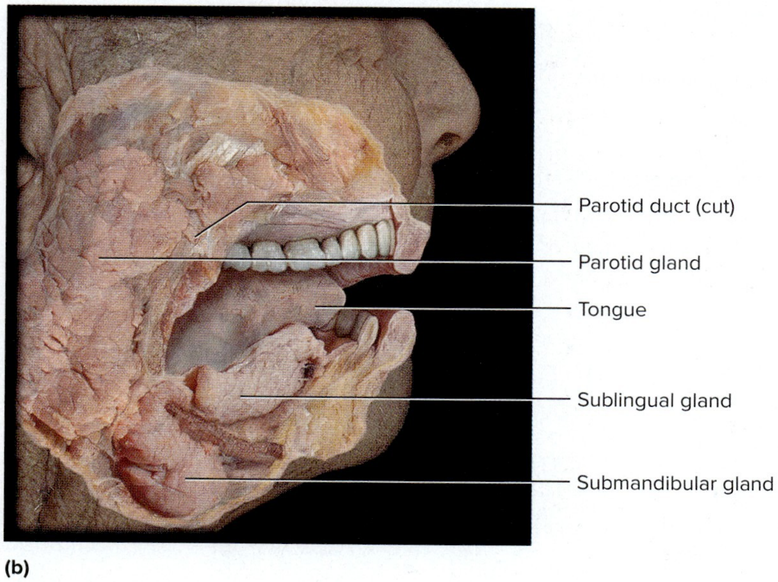

FIGURE 66.1 Major salivary glands of (a) a cat, and (b) a cadaver; both are lateral views. (b) APR/McGraw Hill APR

c. Look for the *submandibular gland* just below the parotid gland, near the angle of the jaw.
d. Locate the *sublingual gland* that is adjacent and medial to the submandibular gland.
e. Compare the location of the salivary glands of the cat with those of a human (fig. 66.1*b*).

3. Open the oral cavity. To do this, follow these steps:
   a. Use scissors to cut through the soft tissues at the angle of the mouth if not previously accomplished.
   b. When you reach the bone of the jaw, use a bone cutter to cut through the bone, thus freeing the mandible.
   c. Open the mouth wide, and locate the following features:

      cheek

      lip

      vestibule

      palate
      - hard palate—has transverse ridges (rugae) to help hold food
      - soft palate—lacks a uvula

      palatine tonsils—small, rounded masses of glandular tissue in the lateral wall of the soft palate

      tongue
      - frenulum
      - papillae—examine with a hand lens to observe spiny projections used to clean fur

4. Examine the teeth of the upper jaw. The adult cat has six incisors, two canines, eight premolars, and two molars. The lower jaw teeth are similar, except four premolars are present.
5. Complete Part A of Laboratory Assessment 66.
6. Examine organs in the abdominal cavity with the cat positioned with its ventral surface up. Review the normal position and function of the greater omentum (fig. 66.2).
7. Examine the *liver,* which is located just beneath the diaphragm and is attached to the central portion of the diaphragm by the *falciform ligament.* Also, locate the *spleen,* which is posterior to the liver on the left side (fig. 66.2). The liver has five lobes—a right medial, right lateral (which is subdivided into two parts by a deep cleft), left medial, left lateral, and caudate lobe. The caudate lobe is the smallest lobe and is located in the median line, where it projects into the curvature of the stomach. The caudate lobe is covered by a sheet of mesentery, the *lesser omentum,* that connects the liver to the stomach (figs. 66.3 and 66.4). Find the greenish *gallbladder* on the inferior surface of the liver on the right side. Also note the *cystic duct,* by which the gallbladder is attached to the *bile duct,* and the *hepatic duct,* which originates in the liver and attaches to the cystic duct. Trace the bile duct to its connection with the duodenum.
8. Locate the *stomach* in the left side of the abdominal cavity. At its anterior end, note the union with the *esophagus,* which passes through the diaphragm. Identify the *cardia, fundus, body,* and *pyloric part* (listed from entrance to exit) of the stomach (figs. 66.3 and 66.4). Use scissors to make an incision along the convex border of the stomach from the cardia region to the pyloric part. The lining of the stomach has numerous *gastric folds* (*rugae*). Examine the *pyloric sphincter,* which creates a constriction between the stomach and small intestine.
9. Locate the *pancreas.* It appears as a grayish, two-lobed, elongated mass of glandular tissue. One lobe lies dorsal to the stomach and extends across to the duodenum. The other lobe is enclosed by the *mesentery,* which supports the duodenum (figs. 66.3 and 66.4).
10. Trace the *small intestine,* beginning at the pyloric sphincter. The first portion, the *duodenum,* travels posteriorly for several centimeters. Then it loops back around a lobe of the pancreas. The proximal half of the remaining portion of the small intestine is the *jejunum,* and the distal half is the *ileum* (figs. 66.3 and 66.4). Open the duodenum and note the velvety appearance of the villi. Note how the mesentery supports the small intestine from the dorsal body wall. The small intestine terminates on the right side where it joins the large intestine.
11. Locate the *large intestine* and identify the *cecum, ascending colon, transverse colon,* and *descending colon.* Also locate the *rectum,* which extends through the pelvic cavity to the *anus* (figs. 66.3 and 66.4). Make an incision at the junction between the ileum and cecum and look for the *ileocecal sphincter.* An appendix is absent in cats.
12. Review the locations of the digestive organs without the aid of figures.
13. Complete Parts B, C, and D of the laboratory assessment.

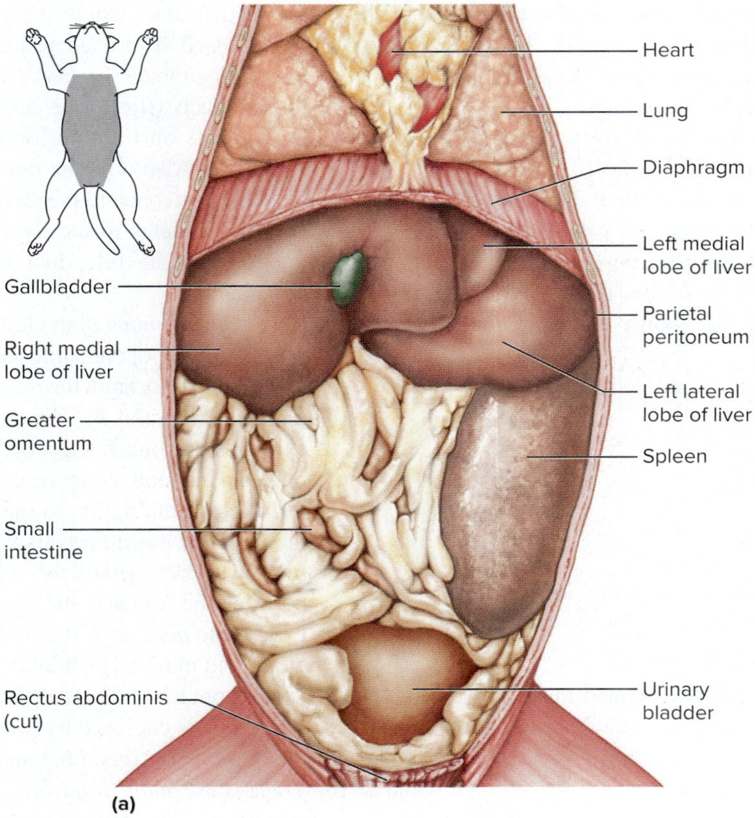

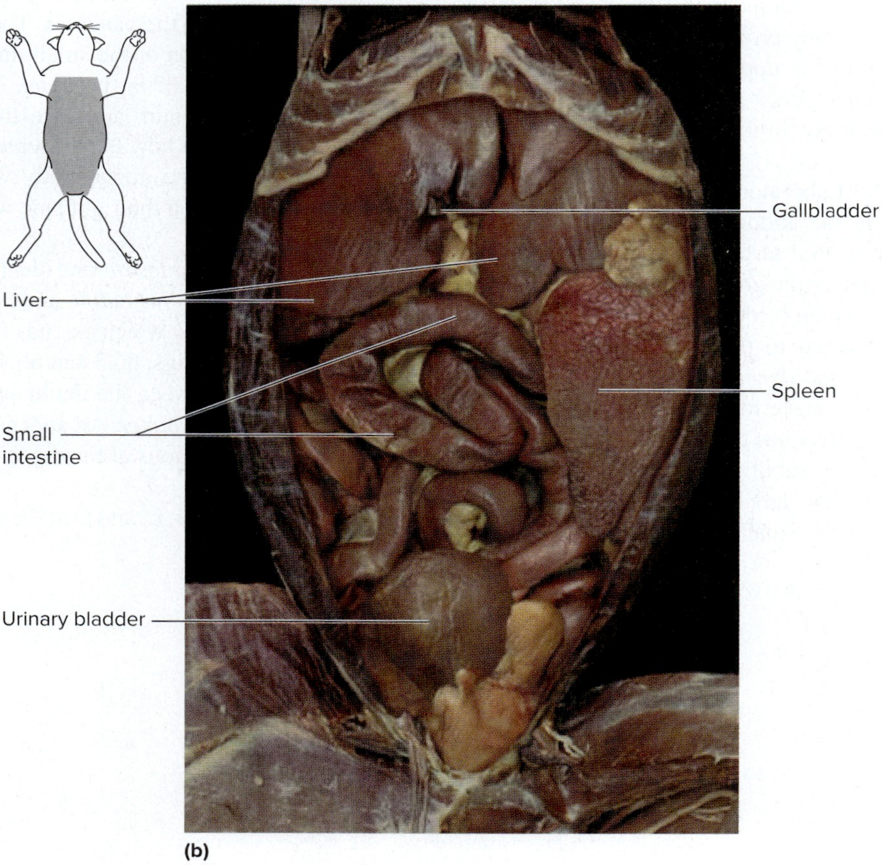

FIGURE 66.2 Ventral view of abdominal organs of the cat: (a) illustration; (b) dissected specimen. (b) APR/McGraw Hill APR

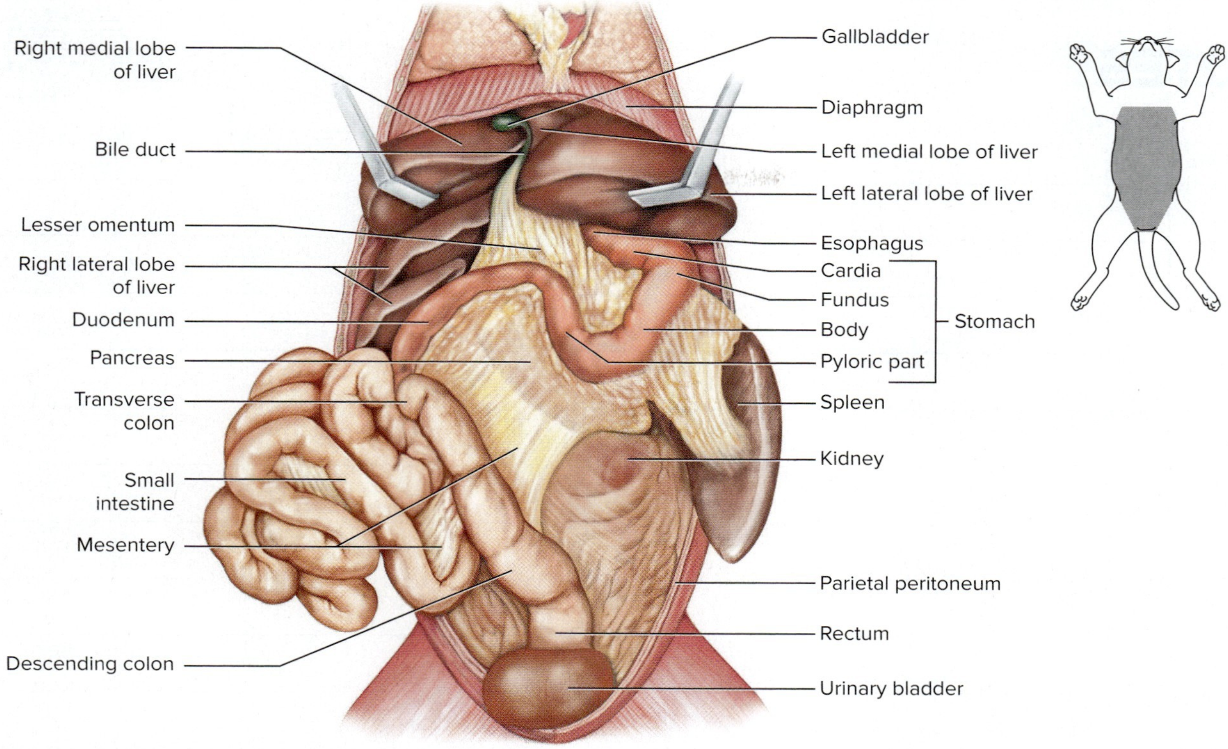

FIGURE 66.3 Ventral view of the cat's abdominal organs with intestines reflected to the right side.

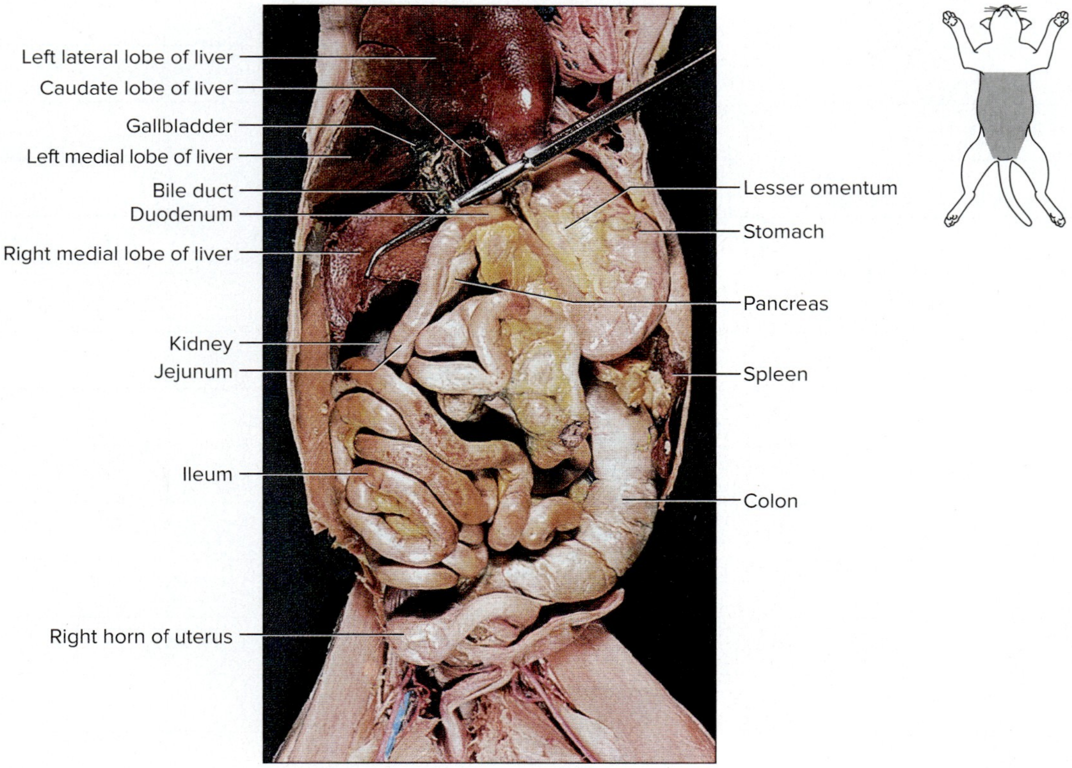

FIGURE 66.4 Ventral view of the abdominal organs of the cat with the greater omentum removed. The left liver lobes have been reflected to the right side. Dr. Sheril Burton APR

NOTES

LABORATORY
ASSESSMENT 66

Name ______________________________

Date ______________________________

Section ______________________________

The (A) corresponds to the indicated Learning Outcome(s) (O) found at the beginning of the Laboratory Exercise.

Cat Dissection: Digestive System

PART A: Assessments

Complete the following:

1. Compare the locations of the major salivary glands of the human with those of the cat. A2 ______________________________

2. Compare the types and numbers of teeth present in the cat's upper and lower jaws with those of the human. A2

3. Describe how the cat's teeth are adapted to a special diet. A1 ______________________________

4. What part of the human soft palate is lacking in the cat? A2 ______________________________

5. What do you think is the function of the transverse ridges (rugae) in the hard palate of the cat? A1 ______________________________

6. How do the papillae on the surface of the human tongue compare with those of the cat? A2

PART B: Assessments

Complete the following:

1. Describe how the peritoneum and mesenteries are associated with the organs in the abdominal cavity. A1 ______________________________

2. Describe the inner lining of the stomach. A1 ______________________________

3. Compare the structure of the human liver with that of the cat. A2 ______________________________

4. Compare the structure and location of the human pancreas with those of the cat. A2 ______________________________

5. What feature of the human cecum is lacking in the cat? A2 ______________________________

PART C: Assessments

Identify the numbered digestive features of the cat in figure 66.5.

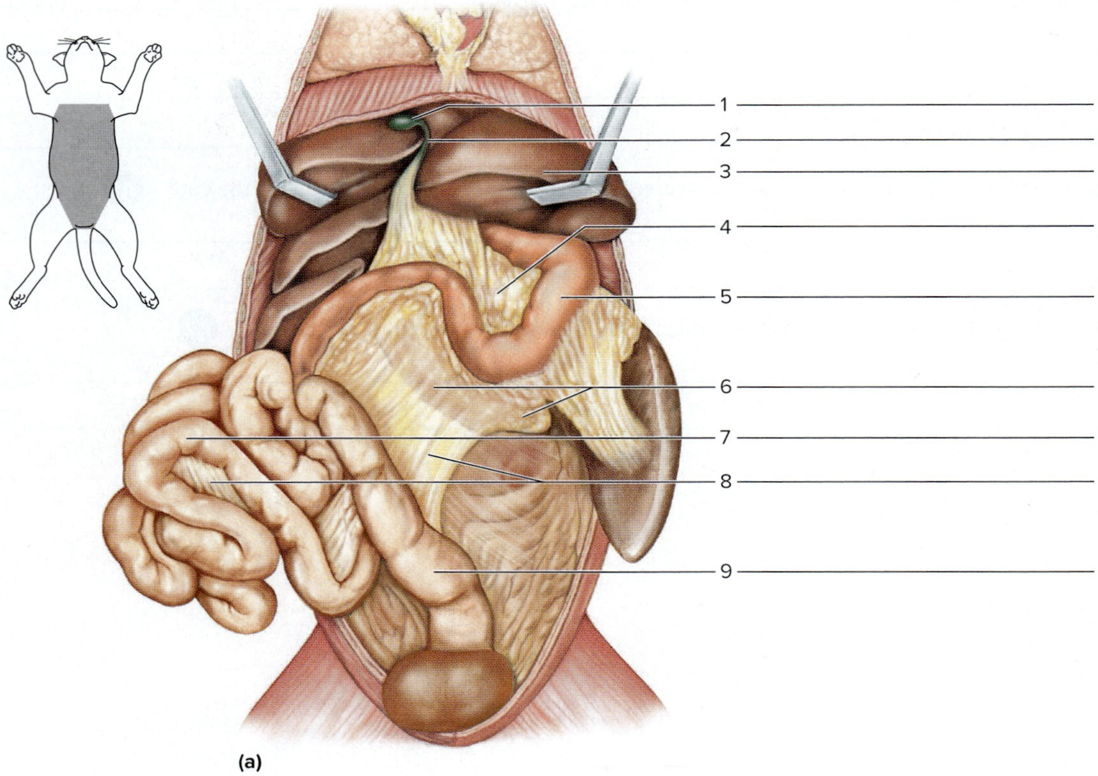

FIGURE 66.5 Label the numbered digestive features on the ventral views of the cat: (a) illustration; (b) dissection. Use terms provided for (b) only. A1

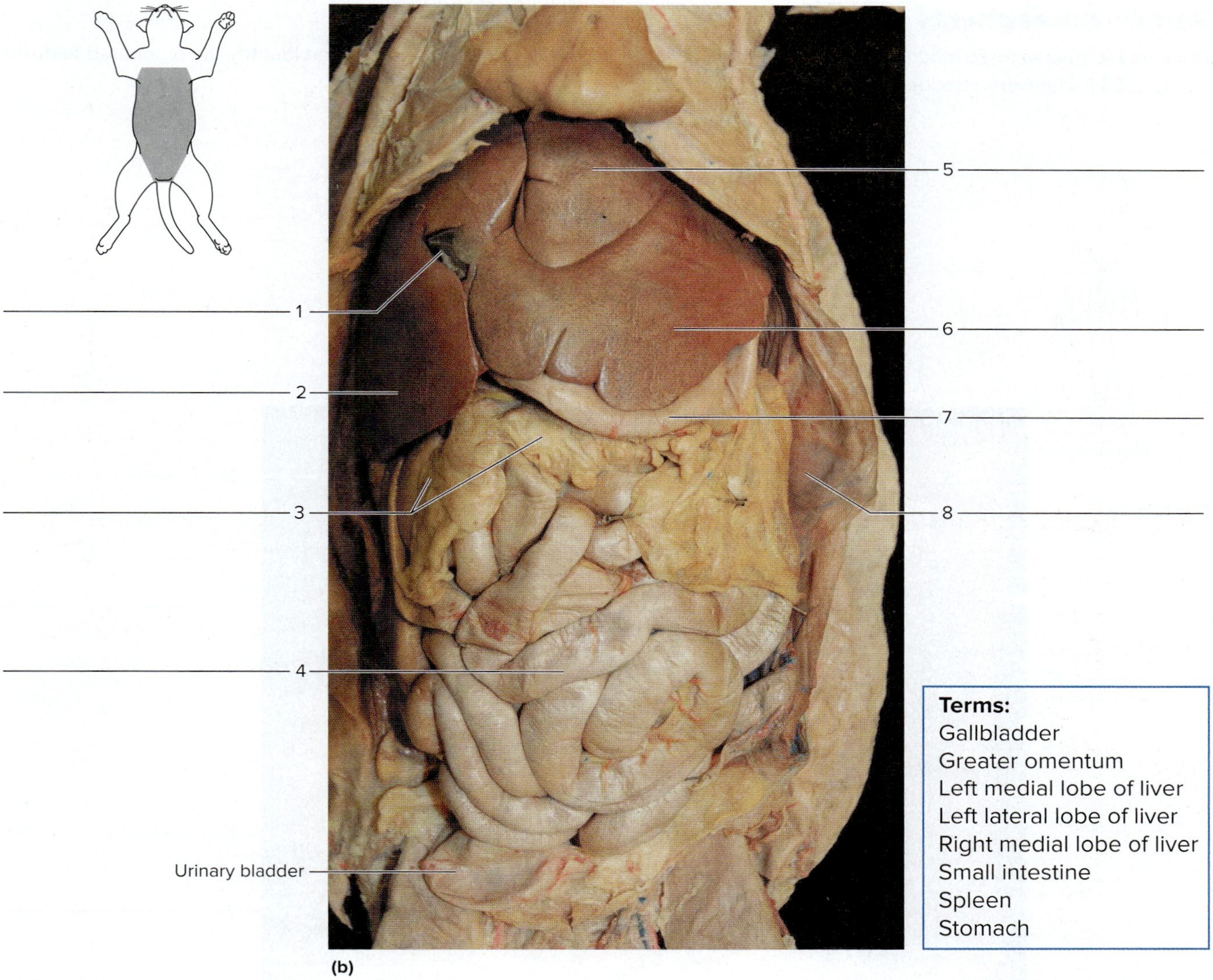

Terms:
Gallbladder
Greater omentum
Left medial lobe of liver
Left lateral lobe of liver
Right medial lobe of liver
Small intestine
Spleen
Stomach

FIGURE 66.5 *Continued.* Dennis Strete

Part D: Assessments

Observe the human torso model and figure 66.6 of a cadaver. Locate the labeled features, and identify the numbered features in figure 66.6 that were also identified in the cat dissection.

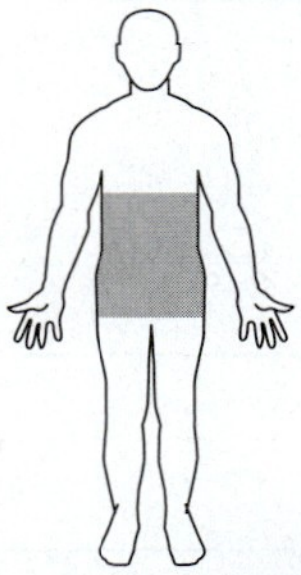

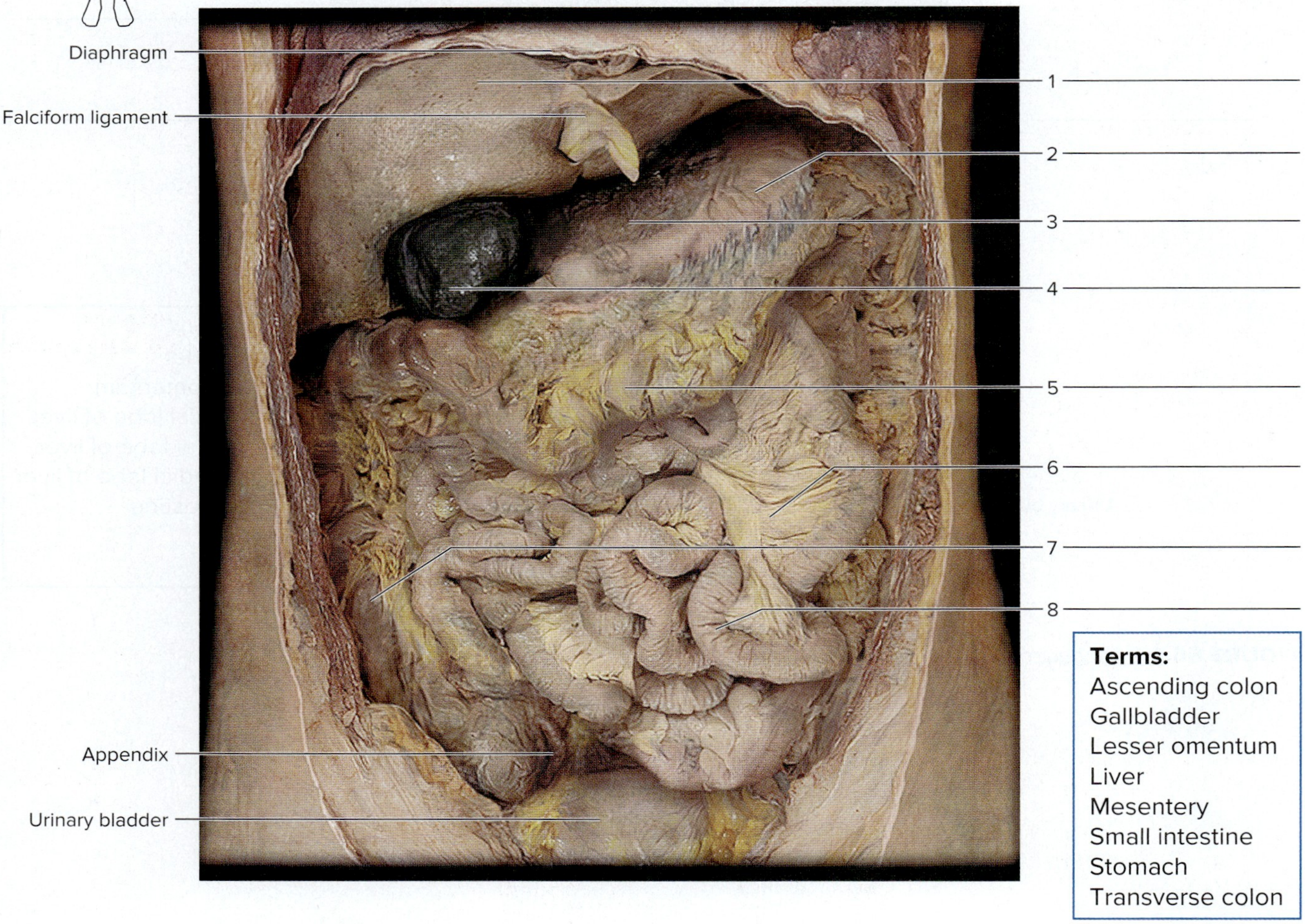

FIGURE 66.6 Identify the digestive features indicated on this anterior view of the abdomen of a cadaver, using the terms provided. APR/McGraw Hill APR A2 A3

LABORATORY
EXERCISE 67

Cat Dissection: Urinary System

Purpose of the Exercise

To examine and identify the urinary organs of the cat and to compare them with those of the human.

MATERIALS NEEDED

Preserved cat
Dissecting tray
Dissecting instruments
Disposable gloves
Human torso model

SAFETY

- Review all the safety guidelines in Appendix 1 of your laboratory manual.
- Wear disposable gloves and protective eyewear when working on the cat dissection.
- Dispose of tissue remnants and gloves as instructed.
- Wash the dissecting tray and instruments as instructed.
- Wash your laboratory table.
- Wash your hands before leaving the laboratory.

Learning Outcomes APR

After completing this exercise, you should be able to:

O1 Locate and identify the urinary organs of the cat.

O2 Compare the urinary organs of the cat with those of the human.

O3 Identify the corresponding organs in the human torso model and a cadaver.

The O corresponds to the assessments A indicated in the Laboratory Assessment for this Exercise.

Pre-Lab

Carefully read the introductory material and examine the entire lab. Be familiar with the structures and functions of the human urinary system from lecture, the textbook, or laboratory exercises.

Chapter Opener: ©101cats/Getty Images

In this laboratory exercise, you will dissect the urinary organs of the cat. As you observe these structures, compare them with the corresponding human organs by observing the parts of the human torso model.

PROCEDURE: Urinary System Dissection

1. Place the preserved cat in a dissecting tray with its ventral surface up.
2. Open the abdominal cavity and remove the liver, stomach, and spleen (figs. 67.1 and 67.2).
3. Push the intestines to one side and locate the *kidneys* in the dorsal abdominal wall on either side of the vertebral column. The kidneys are located dorsal to the *parietal peritoneum* (retroperitoneal).
4. Carefully remove the parietal peritoneum and the adipose tissue surrounding the kidneys. Locate the following, using figures 67.1 to 67.3 as guides:

    ureters
    renal arteries
    renal veins

5. Locate the *adrenal glands,* which lie medially and anteriorly to the kidneys and are surrounded by connective tissues. The adrenal glands are attached to blood vessels and are separated by the abdominal aorta and posterior vena cava.
6. Expose the ureters by cleaning away the connective tissues along their lengths. They enter the *urinary bladder* on the dorsal surface.
7. Examine the urinary bladder, attached to the abdominal wall by folds of peritoneum that form ligaments. Locate the *medial ligament* on the ventral surface of the bladder that connects it with the linea alba and the *two lateral ligaments* on the dorsal surface.
8. Remove the connective tissue from around the urinary bladder. Use a sharp scalpel to open the bladder, and examine its interior. Locate the openings of the ureters and the urethra on the inside.
9. Expose the *urethra* at the posterior of the urinary bladder (figs. 67.2–67.4). The urethra terminates with the vagina in a urogenital sinus in the female cat; the urethra enters the penis in the male cat.

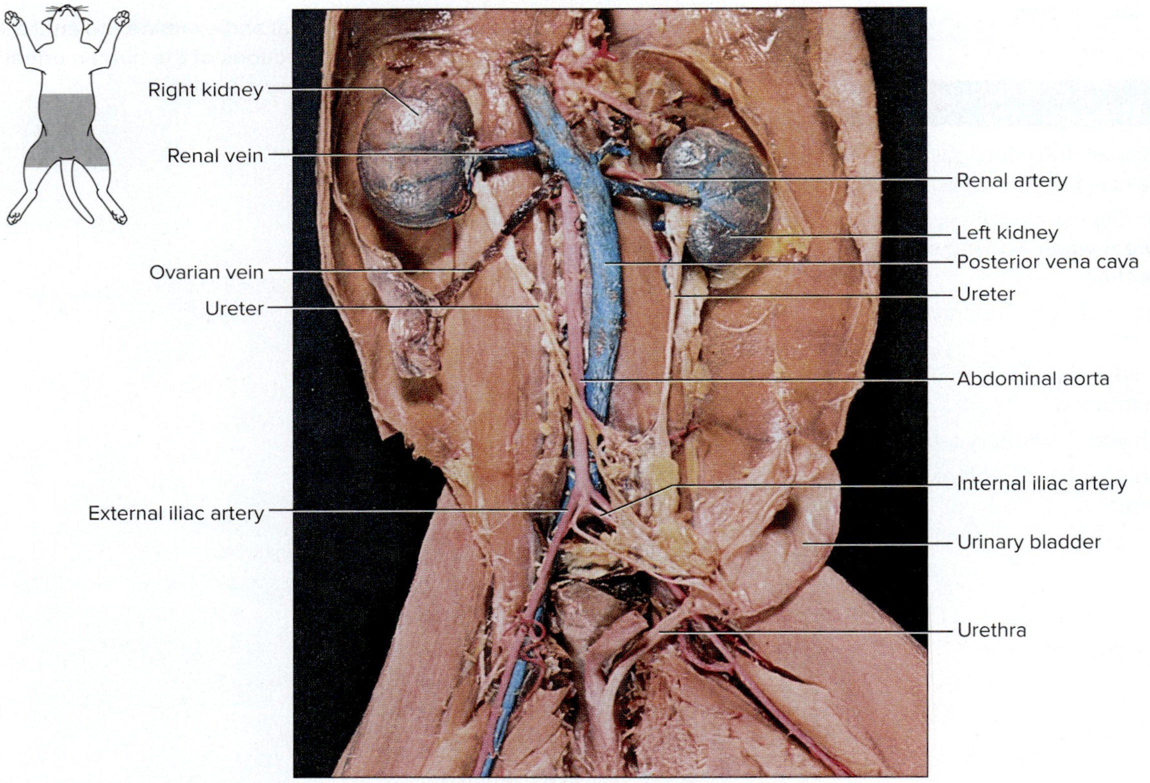

FIGURE 67.1 Urinary system of the female cat's abdominal cavity, with most of the digestive and reproductive organs removed, ventral view. Dr. Sheril Burton APR

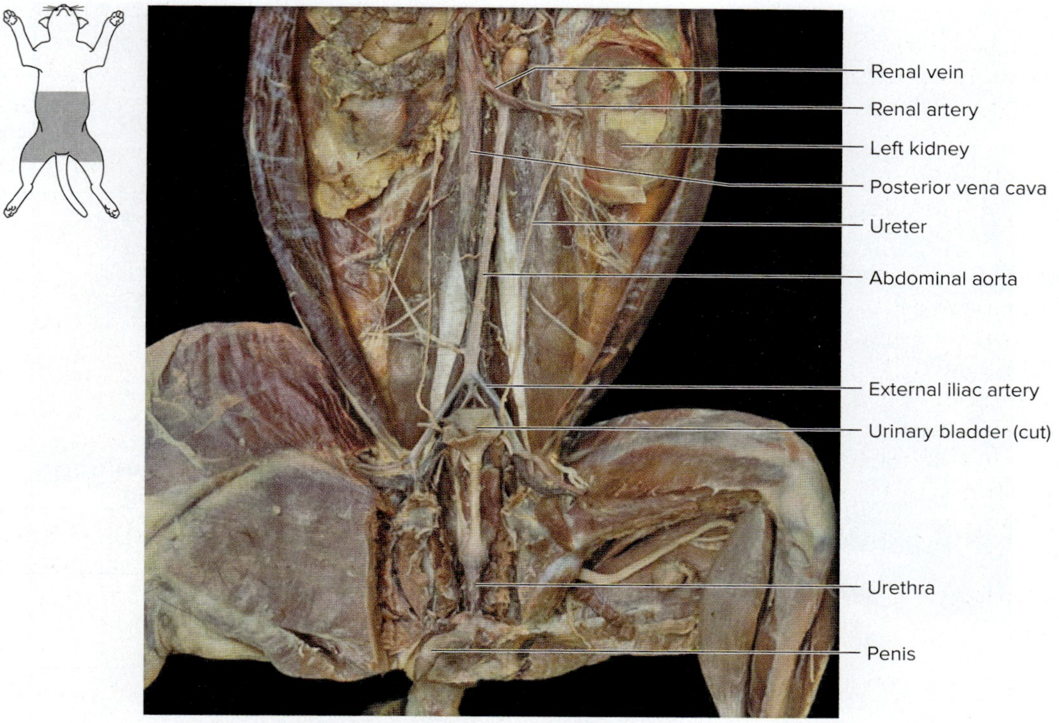

FIGURE 67.2 Urinary system of the male cat's abdominal cavity, with most of the digestive and reproductive organs removed, ventral view. APR/McGraw Hill

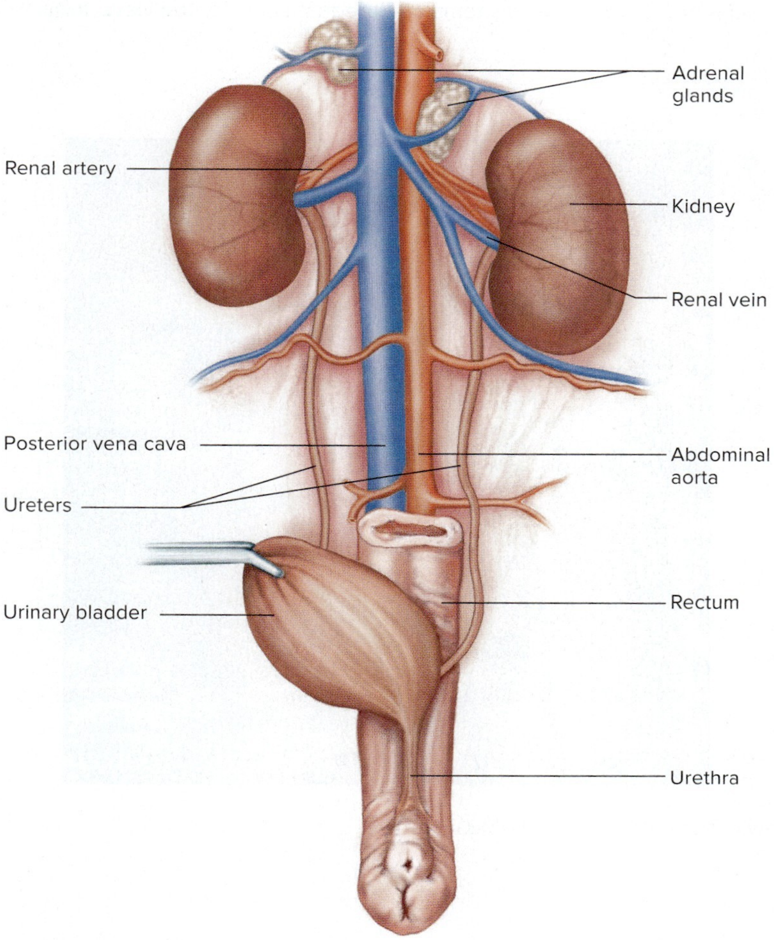

FIGURE 67.3 The urinary organs of a female cat, ventral view.

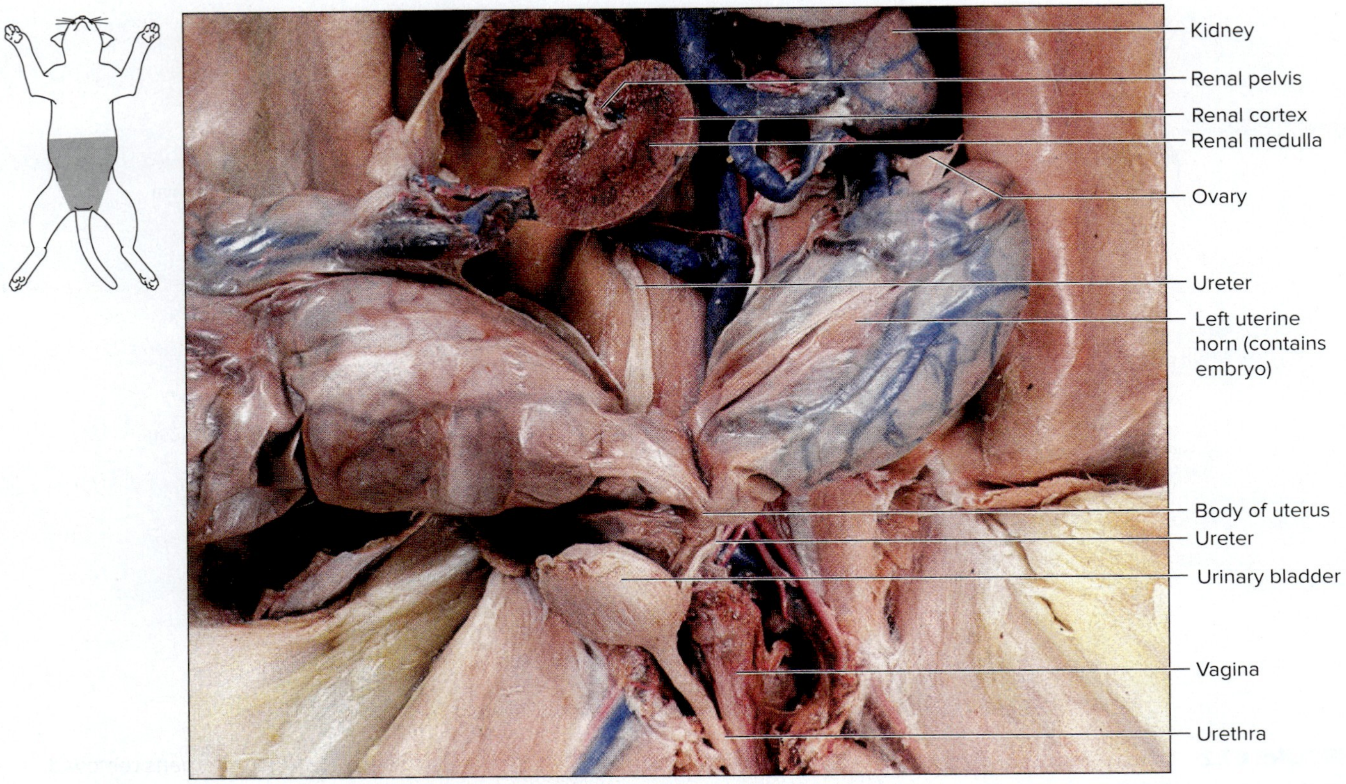

FIGURE 67.4 Female urinary and reproductive systems of a pregnant cat, ventral view. Ralph W. Stevens III, Ph.D/McGraw Hill

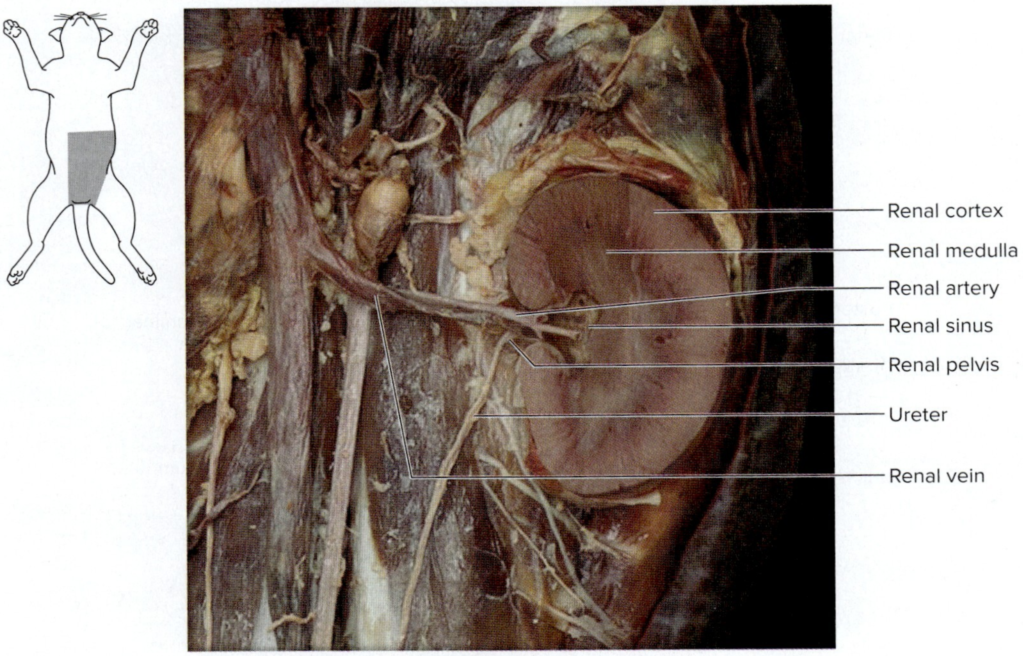

FIGURE 67.5 Ventral view of the cat kidney. APR/McGraw Hill

10. Remove one kidney and section it longitudinally (figs. 67.4–67.6*a*). Identify the following features:

fibrous capsule—encases a kidney

renal cortex—superficial, lighter region

renal medulla—deeper, darker region with striations

renal pyramid—single in cat

renal papilla—single in cat; opens into renal pelvis

renal sinus—hollow chamber of kidney

renal pelvis—funnel-shaped sac at kidney end of ureter

hilum of kidney—indented region containing renal blood vessels and ureter

11. Compare the structures of the kidney of the cat with those of a human (fig. 67.6*b*).
12. Review the locations of the urinary organs without the aid of figures.
13. Discard the organs and tissues that were removed from the cat, as directed by the laboratory instructor.
14. Complete Parts A and B of Laboratory Assessment 67.

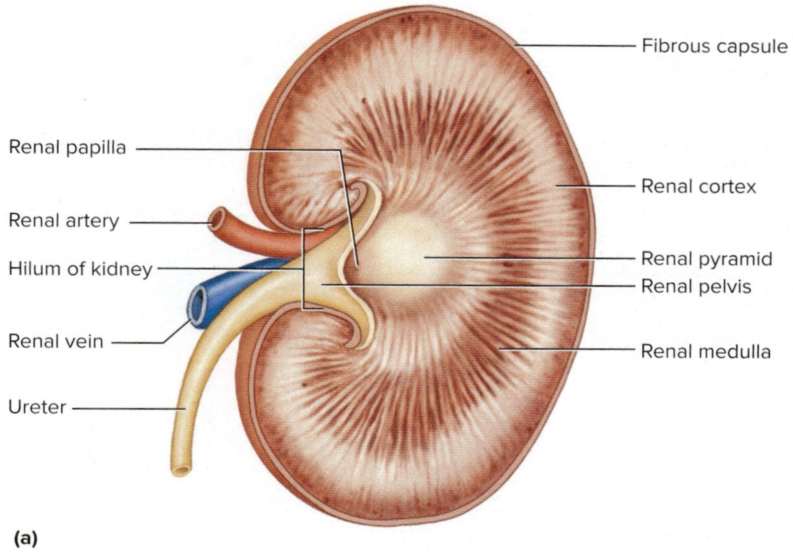

(a)

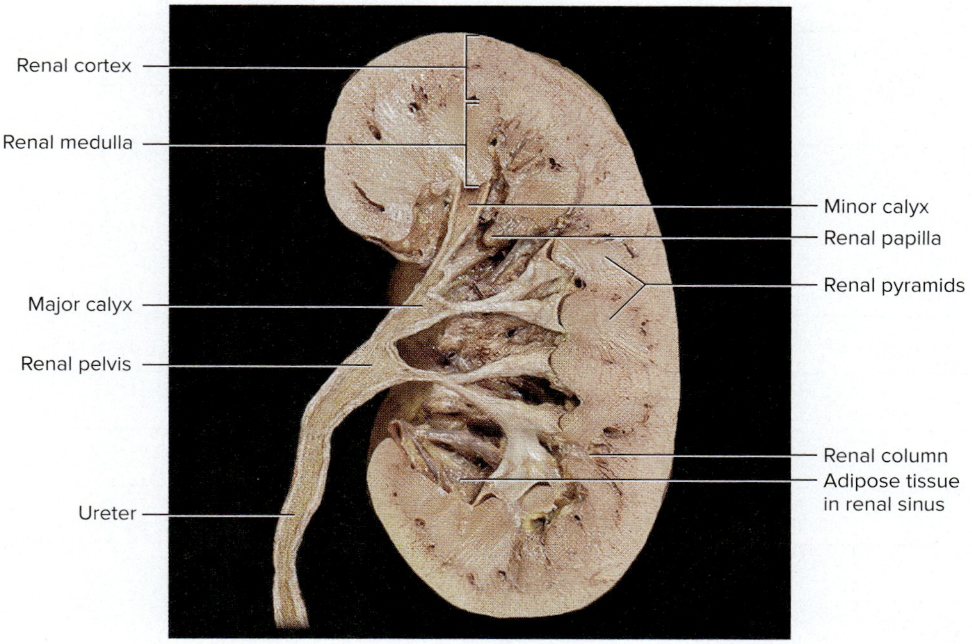

(b)

FIGURE 67.6 Longitudinal section of a kidney in the (a) cat, and (b) cadaver. (b) ©J & J Photography APR

NOTES

LABORATORY
ASSESSMENT 67

Name ______________________________

Date ______________________________

Section ______________________________

The (A) corresponds to the indicated Learning Outcome(s) (O) found at the beginning of the Laboratory Exercise.

Cat Dissection: Urinary System

PART A: Assessments

Complete the following:

1. Compare the positions of the kidneys in the cat with those in the human. A2 ______________________________

2. Compare the locations of the adrenal glands in the cat with those in the human. A2 ______________________________

3. What structures of the cat urinary system are retroperitoneal? A1 ______________________________

4. Describe the wall of the urinary bladder of the cat. A1 ______________________________

5. Compare the renal pyramids and renal papillae of the human kidney with those of the cat. A2 ______________________________

PART B: Assessments

1. Identify the numbered urinary organs in figure 67.7.

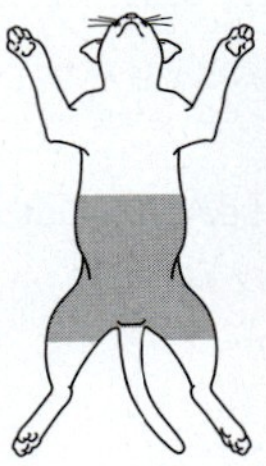

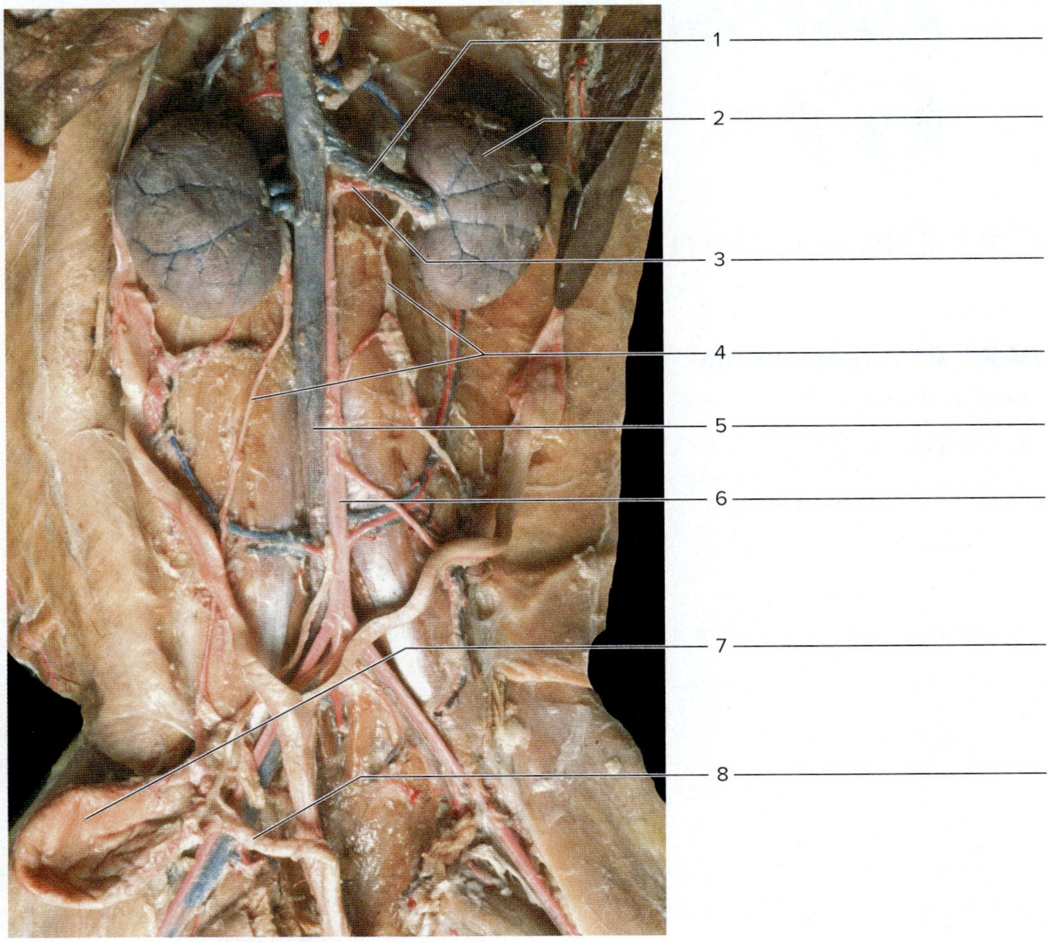

Terms:
Abdominal aorta
Kidney
Posterior vena cava
Renal artery
Renal vein
Ureters
Urethra
Urinary bladder

FIGURE 67.7 Identify the urinary features including associated major blood vessels on a ventral view of a dissected cat, using the terms provided. Dennis Strete **A1**

2. Observe the human torso model and figures 67.6*b* and 67.8 of a cadaver. Locate the labeled features and identify the numbered features in figure 67.8 that were also identified in the cat dissection.

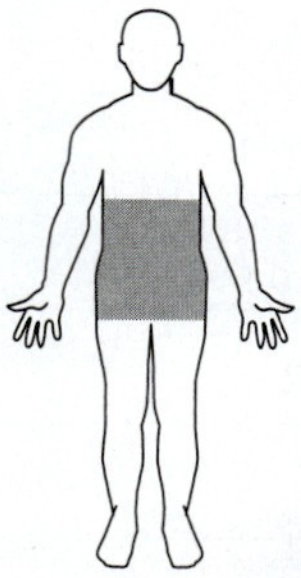

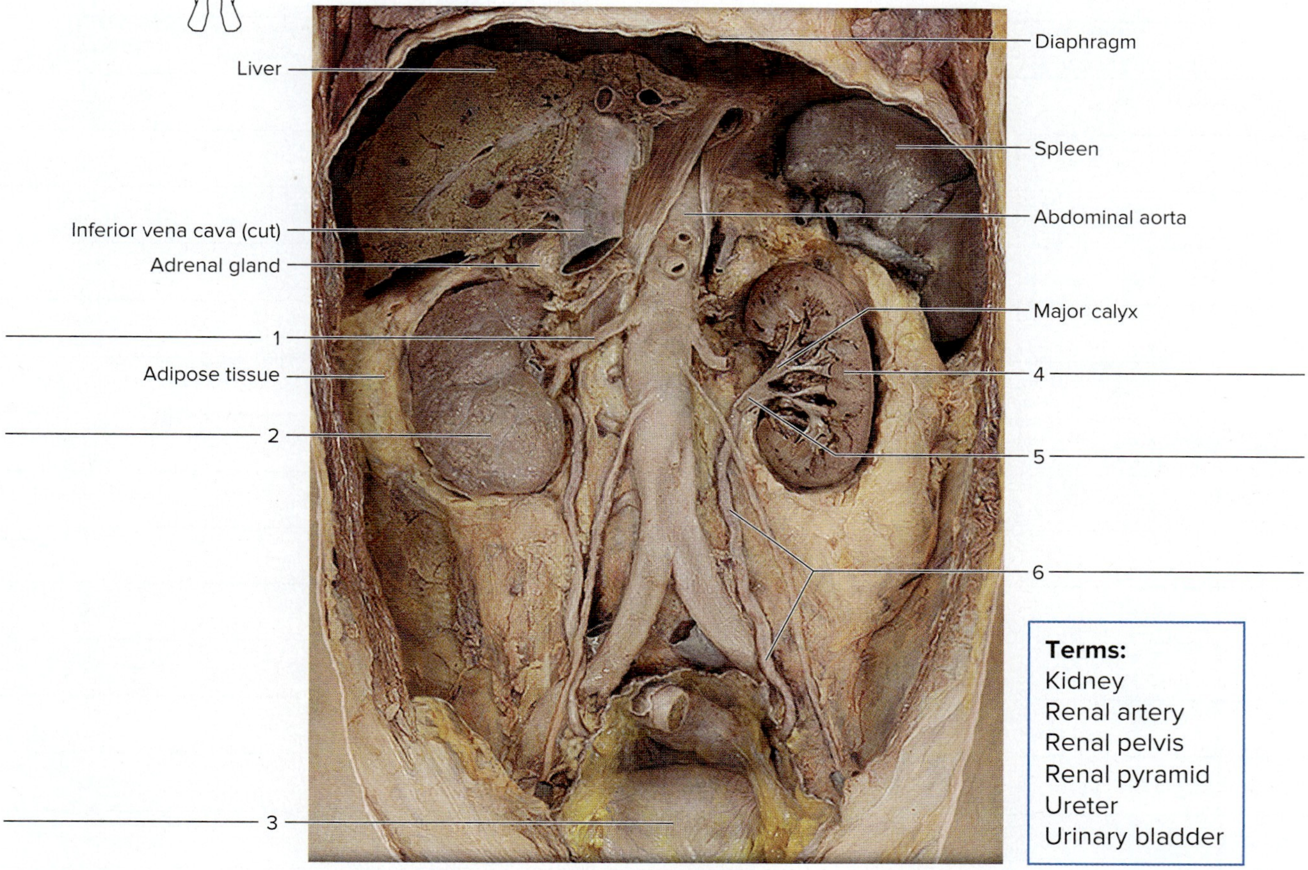

Terms:
Kidney
Renal artery
Renal pelvis
Renal pyramid
Ureter
Urinary bladder

FIGURE 67.8 Identify the urinary features indicated on this anterior view of the abdomen of a cadaver, using the terms provided. APR/McGraw Hill APR A2 A3

NOTES

LABORATORY
EXERCISE 68

Cat Dissection: Reproductive Systems

Purpose of the Exercise

To examine and identify the reproductive organs of the cat and to compare them with the corresponding organs of the human.

MATERIALS NEEDED

Preserved cat
Dissecting tray
Dissecting instruments
Bone cutters
Disposable gloves
Models of human reproductive systems

SAFETY

- Review all the safety guidelines in Appendix 1 of your laboratory manual.
- Wear disposable gloves and protective eyewear when working on the cat dissection.
- Dispose of tissue remnants and gloves as instructed.
- Wash the dissecting tray and instruments as instructed.
- Wash your laboratory table.
- Wash your hands before leaving the laboratory.

Learning Outcomes APR

After completing this exercise, you should be able to:

O1 Locate and identify the reproductive organs of a cat.

O2 Compare the reproductive organs of the cat with those of the human.

O3 Identify the corresponding organs in models of the human reproductive systems and a cadaver.

The O corresponds to the assessments A indicated in the Laboratory Assessment for this Exercise.

Pre-Lab

Carefully read the introductory material and examine the entire lab. Be familiar with the structures and functions of the human reproductive system from lecture, the textbook, or laboratory exercises.

Chapter Opener: ©101cats/Getty Images

In this laboratory exercise, you will dissect the reproductive system of the cat. If you have a female cat, begin with Procedure A. If you have a male cat, begin with Procedure B. After completing the dissection, exchange cats with someone who has dissected one of the opposite sex, and examine its reproductive organs.

As you observe the cat reproductive organs, compare them with the corresponding human organs by examining the models of the human reproductive systems.

PROCEDURE A: Female Reproductive System

1. Place the preserved female cat in a dissecting tray with its ventral surface up and open its abdominal cavity.
2. Locate the small, oval *ovaries* just posterior to the kidneys (figs. 68.1 and 68.2*a* and *b*).
3. Examine an ovary, suspended from the dorsal body wall by a fold of peritoneum called the *mesovarium.* Another attachment, the *ovarian ligament,* connects the ovary to the tubular *uterine horn* (fig. 68.1).
4. Near the anterior end of the ovary, locate the funnel-shaped *infundibulum,* which is at the end of a very small *uterine tube* (*oviduct*). Note the tiny projections, or *fimbriae,* that create a fringe around the edge of the infundibulum. Trace the uterine tube around the ovary to its nearby connection with the uterine horn near the attachment of the ovarian ligament.
5. Examine the uterine horn. It is suspended from the body wall by a fold of peritoneum, the *broad ligament,* that is continuous with the mesovarium of the ovary. Also note the fibrous *round ligament,* which extends from the uterine horn laterally and posteriorly to the body wall (figs. 68.1 and 68.2*a* and *b*).
6. To observe the remaining organs of the reproductive system more easily, remove the left hindlimb by severing it near the hip with bone cutters. Then use the bone cutters to remove the left anterior portion of the pelvic girdle. Also remove the necessary muscles and connective tissue to expose the structures shown in figure 68.3.

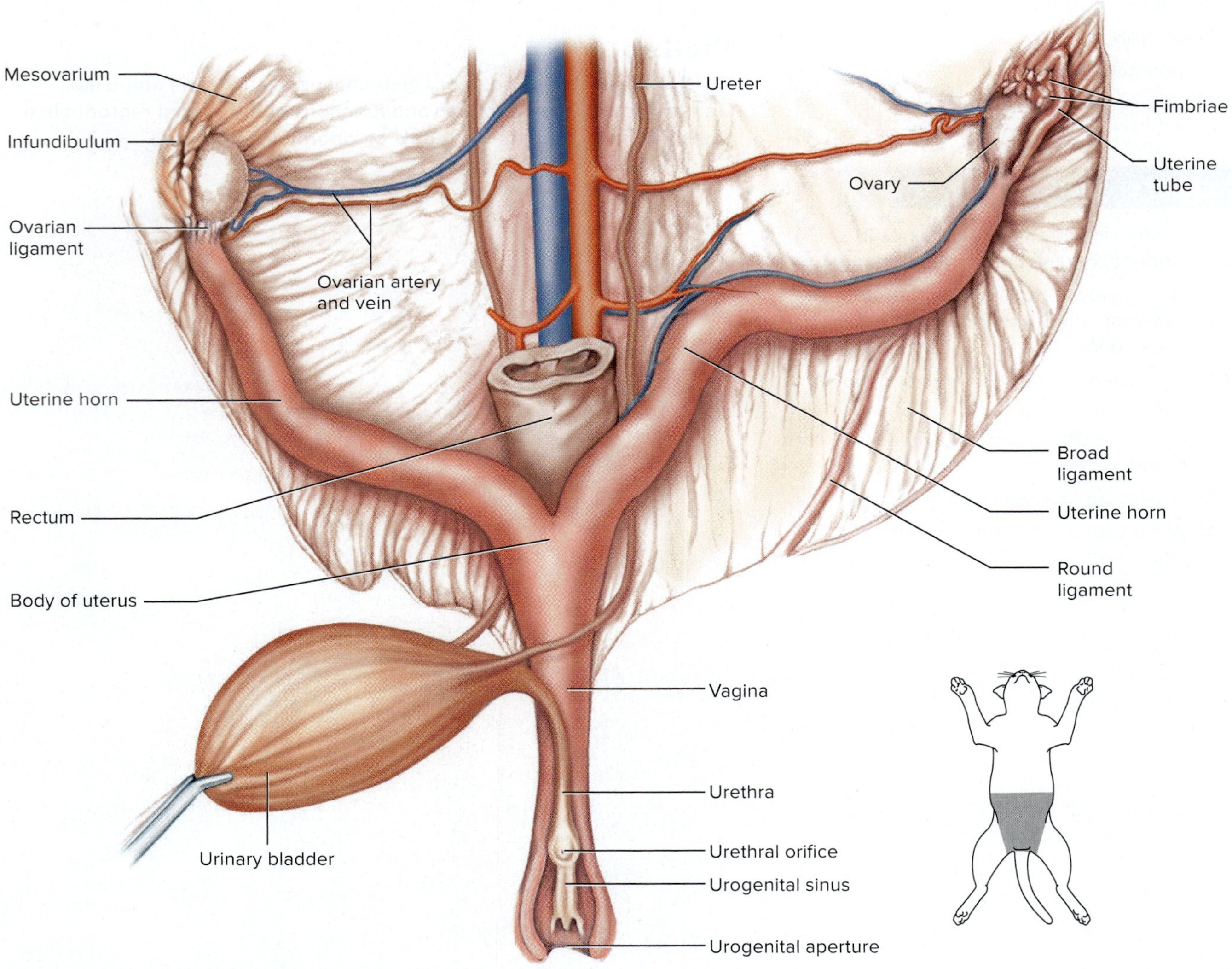

FIGURE 68.1 Ventral view of the female cat's reproductive system.

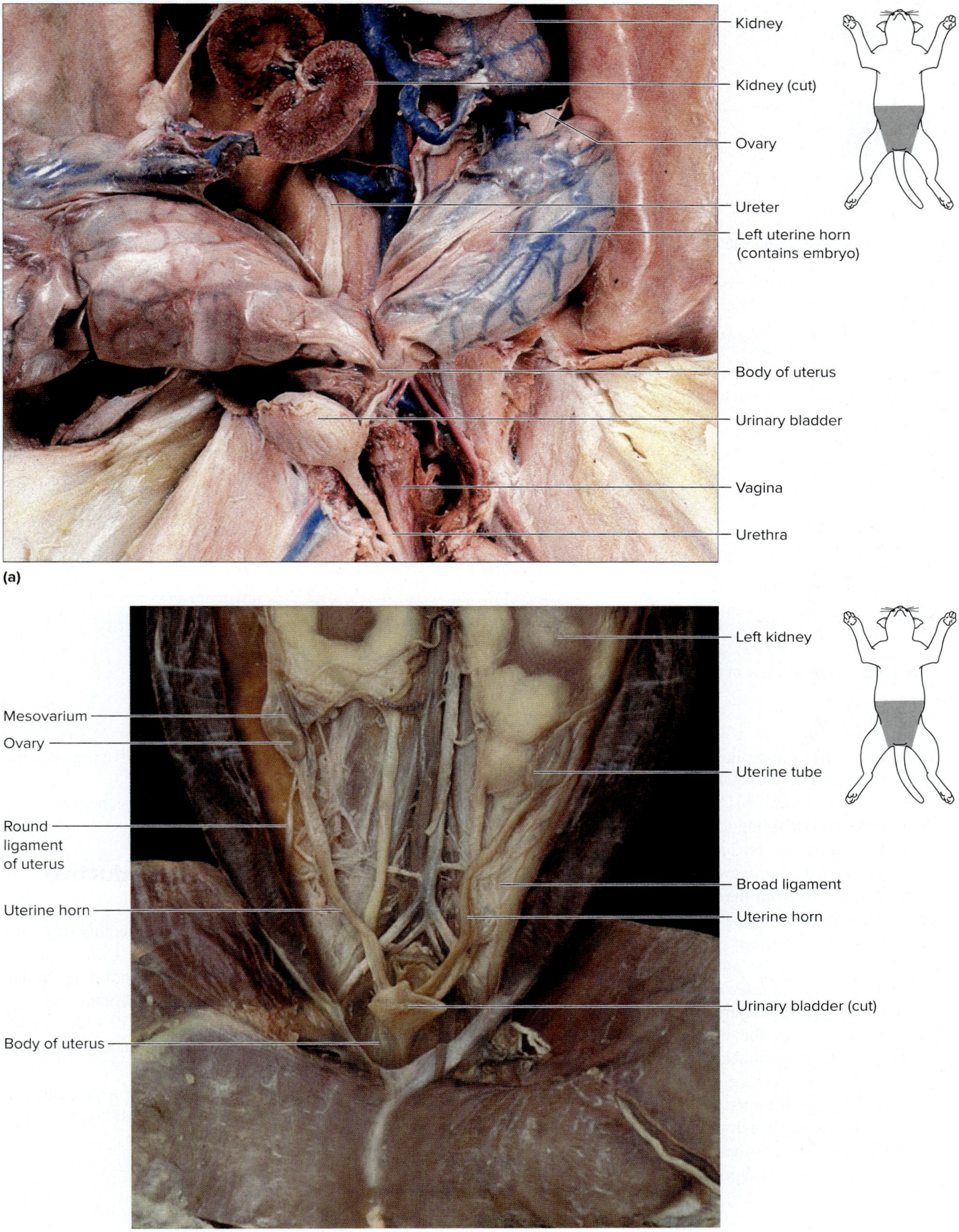

FIGURE 68.2 Ventral view of (a) the female urinary and reproductive systems of a pregnant cat, and (b) the female cat pelvis. (a) Ralph W. Stevens III, Ph.D/McGraw Hill; (b) APR/McGraw Hill

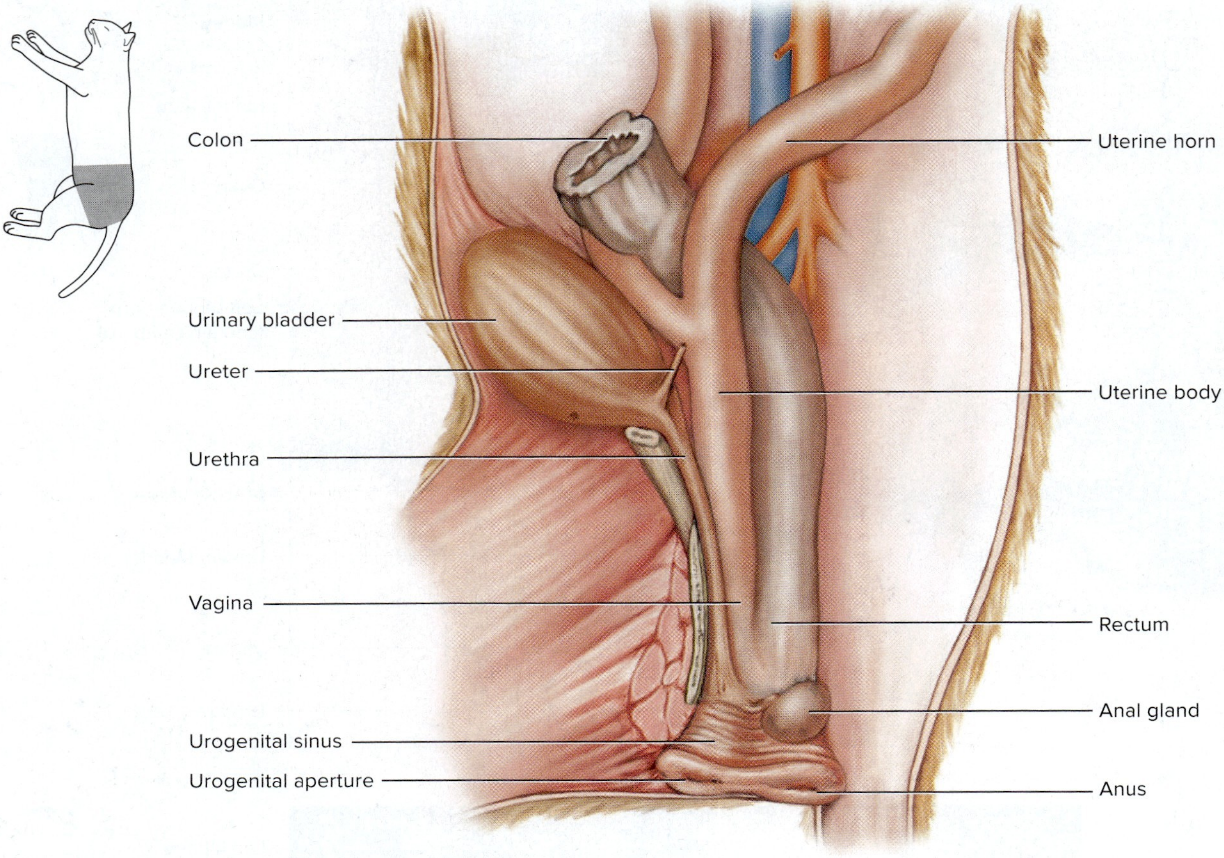

FIGURE 68.3 Lateral view of the female cat's reproductive system.

7. Trace the relatively long uterine horns posteriorly. They unite to form the *uterine body,* which is located between the urethra and rectum. This Y-shaped uterus allows ample space for several offspring to develop at the same time in the uterine horns (fig. 68.2*a* and *b*). The uterine body is continuous with the *vagina,* which leads to the outside.
8. Trace the *urethra* from the urinary bladder posteriorly. The urethra and the vagina open into a common chamber called the *urogenital sinus.* The opening of this chamber, ventral to the anus, is called the *urogenital aperture.* Locate the *labia majora,* which are folds of skin on either side of the urogenital aperture.
9. Use scissors to open the vagina along its lateral wall, beginning at the urogenital aperture and continuing to the body of the uterus. Note the *urethral orifice* in the ventral wall of the urogenital sinus and locate the small, rounded *cervix* of the uterus, which projects into the vagina at its deep end. The *clitoris* is located in the ventral wall of the urogenital sinus near its opening to the outside.
10. Review the locations of the reproductive organs without the aid of figures.
11. Complete Part A of Laboratory Assessment 68.

PROCEDURE B: Male Reproductive System

1. Place the preserved male cat in a dissecting tray with its ventral surface up.
2. Locate the *scrotum* between the hindlimbs. Make an incision along the midline of the scrotum and expose the *testes* inside (fig. 68.4*a* and *b*). The testes are separated by a septum, and each testis is enclosed in a sheath of connective tissue.
3. Remove the sheath surrounding the testes and locate the convoluted *epididymis* on the dorsal surface of each testis (fig. 68.5).

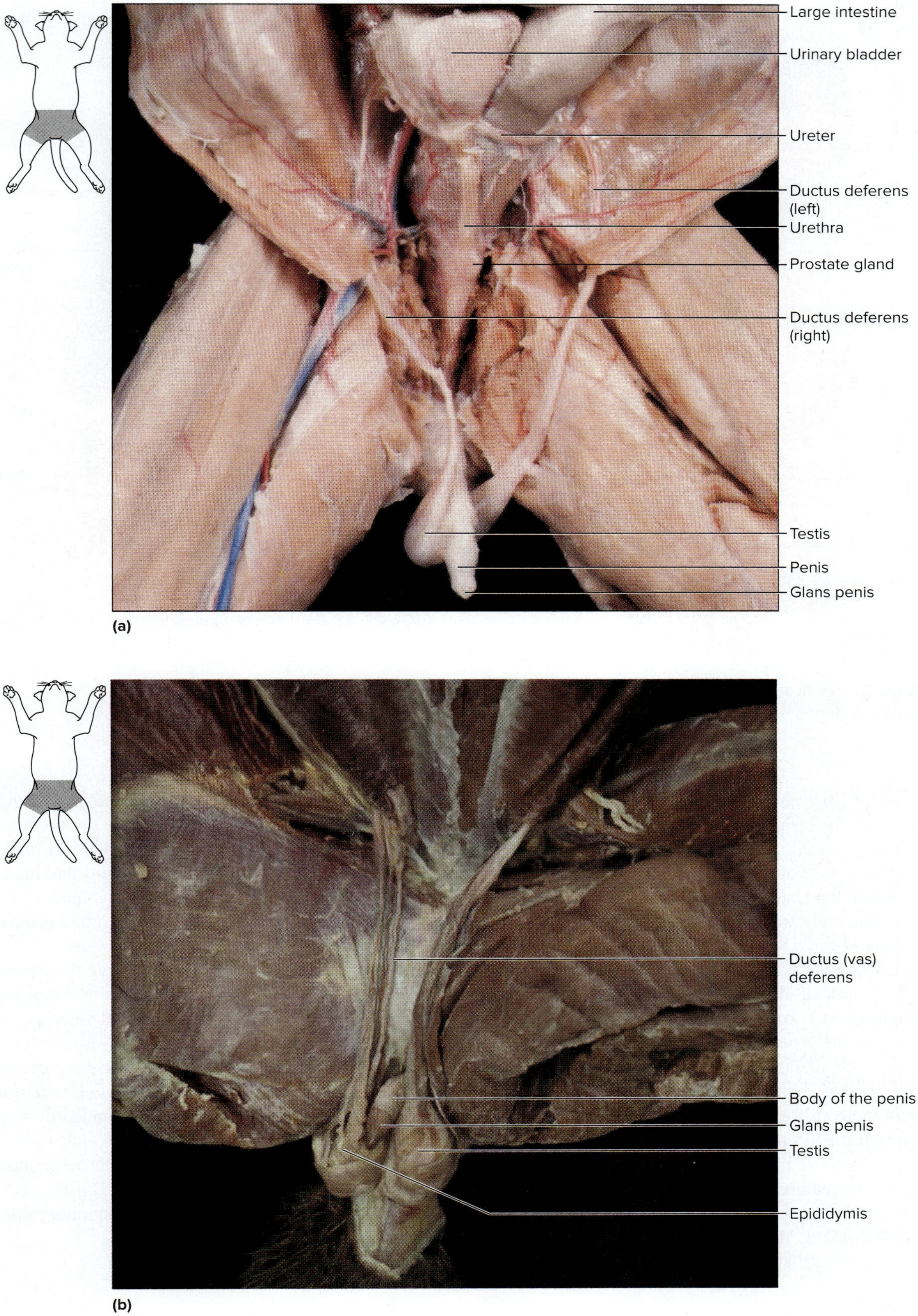

FIGURE 68.4 Ventral view of (a) the male reproductive system of the cat, and (b) the male cat pelvis. (a) Ralph W. Stevens III, Ph.D/McGraw Hill; (b) APR/McGraw Hill APR

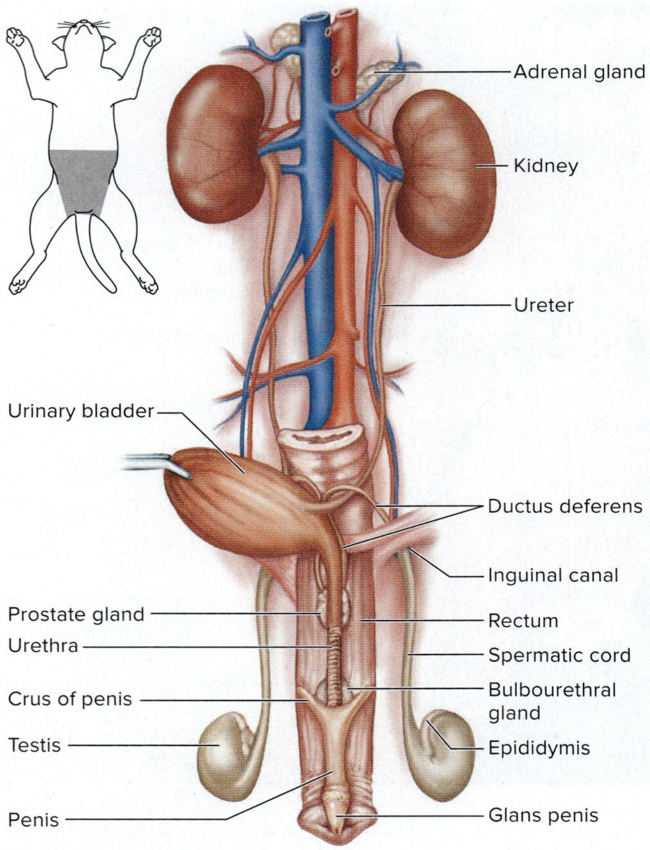

FIGURE 68.5 Ventral view of the male cat's urinary and reproductive systems.

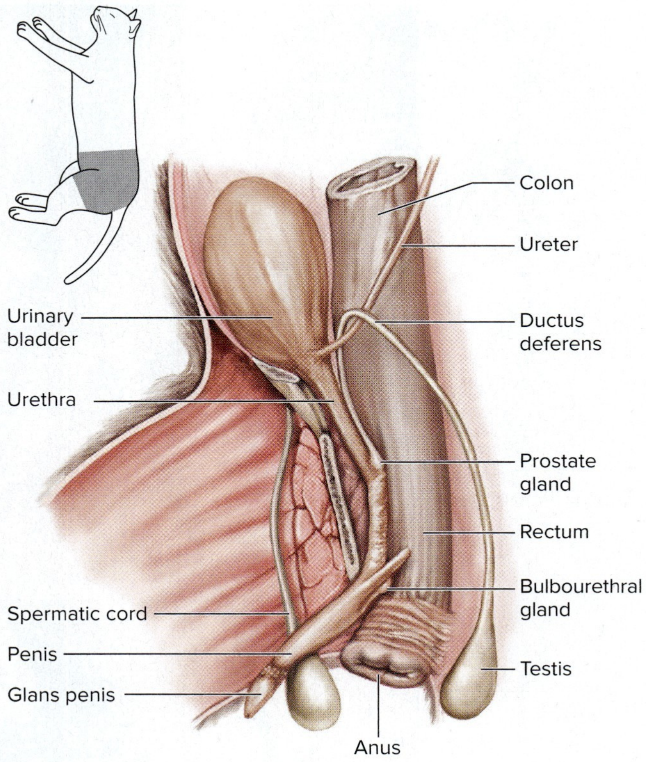

FIGURE 68.6 Lateral view of the male cat's reproductive system.

4. Locate the *spermatic cord* on the right side, leading away from the testis. This spermatic cord contains the *ductus (vas) deferens,* which is continuous with the epididymis, as well as with the nerves and blood vessels that supply the testis on that side. Trace the spermatic cord to the body wall, where its contents pass through the *inguinal canal* and enter the pelvic cavity (fig. 68.5).
5. Locate the *penis,* and identify the *prepuce,* which forms a sheath around the entire penis. Make an incision through the skin of the prepuce and expose the *glans penis.* The glans penis has minute spines on its surface that assist copulation.
6. To observe the remaining structures of the reproductive system more easily, remove the left hindlimb by severing it near the hip with bone cutters. Then use the bone cutters to remove the left anterior portion of the pelvic girdle. Also remove the necessary muscles and connective tissues to expose the organs shown in figure 68.6.
7. Trace the ductus deferens from the inguinal canal to the penis. Note that the ductus deferens loops over the ureter within the pelvic cavity and passes downward behind the urinary bladder to join the urethra. Locate the *prostate gland,* which appears as an enlargement at the junction of the ductus deferens and urethra. The prostate gland is relatively small compared to that in humans and positioned along the urethra some distance from the urinary bladder.
8. Trace the urethra to the penis. Locate the *bulbourethral glands,* which form small swellings on either side at the proximal end of the penis. Seminal vesicles are absent in cats.
9. Use a sharp scalpel to cut a transverse section of the penis. Identify the *corpora cavernosa,* each of which contains many blood spaces surrounded by a sheath of connective tissue.
10. Review the locations of the reproductive organs without the aid of figures.
11. Complete Parts B, C, and D of the laboratory assessment.

Name ______________________

Date ______________________

Section ______________________

The (A) corresponds to the indicated Learning Outcome(s) (O) found at the beginning of the Laboratory Exercise.

LABORATORY ASSESSMENT 68

Cat Dissection: Reproductive Systems

PART A: Assessments

Complete the following:

1. Compare the relative lengths and paths of the uterine tubes (oviducts) of the cat and the human. A1 ______________________

2. How do the shape and structure of the uterus of the cat compare with that of the human? A2 ______________________

3. Assess the function of the uterine horns of the cat. A1 ______________________

4. Compare the relationship of the urethra and the vagina in the cat and in the human. A2 ______________________

PART B: Assessments

Complete the following:

1. Compare the glans penis in the cat and in the human. A2 ______________________

2. How do the location and the relative size of the prostate gland of the cat compare with that of the human? A2

3. What glands associated with the human ductus deferens are missing in the cat? A2 ______________________

4. Compare the prepuce in the cat and in the human. A2 ______________________

PART C: Assessments

Identify the numbered reproductive structures of the female cat in figure 68.7 and the male cat in figure 68.8.

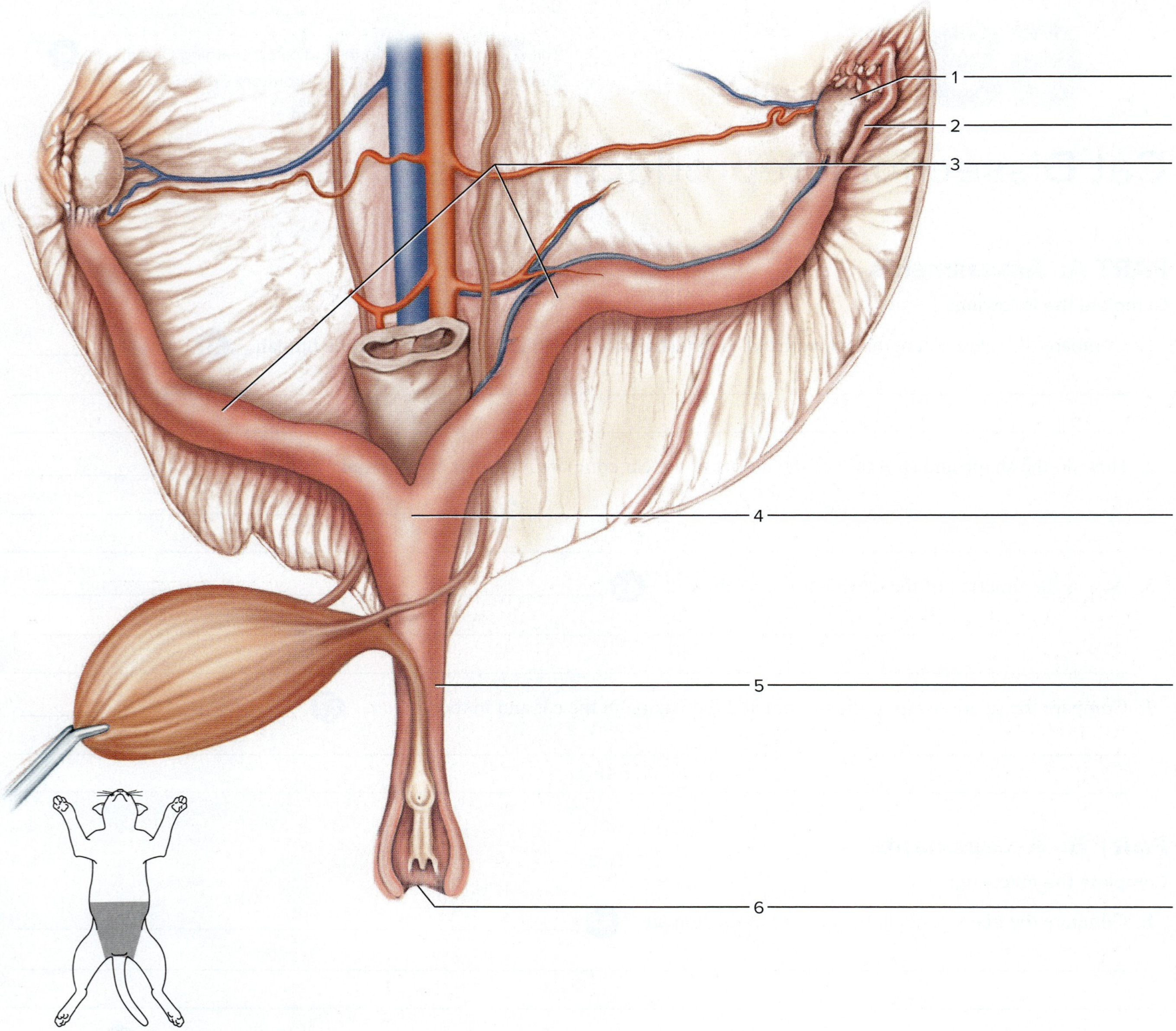

(a)

FIGURE 68.7 Label the numbered reproductive structures on the ventral view of the female cat in (a) an illustration, and (b) an image of a dissected cat. Use the terms provided for (b) only. A1

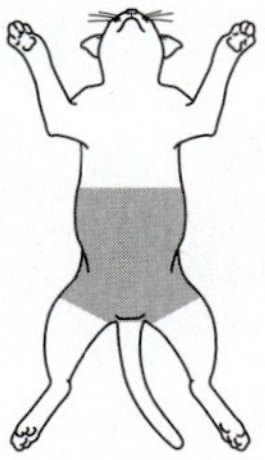

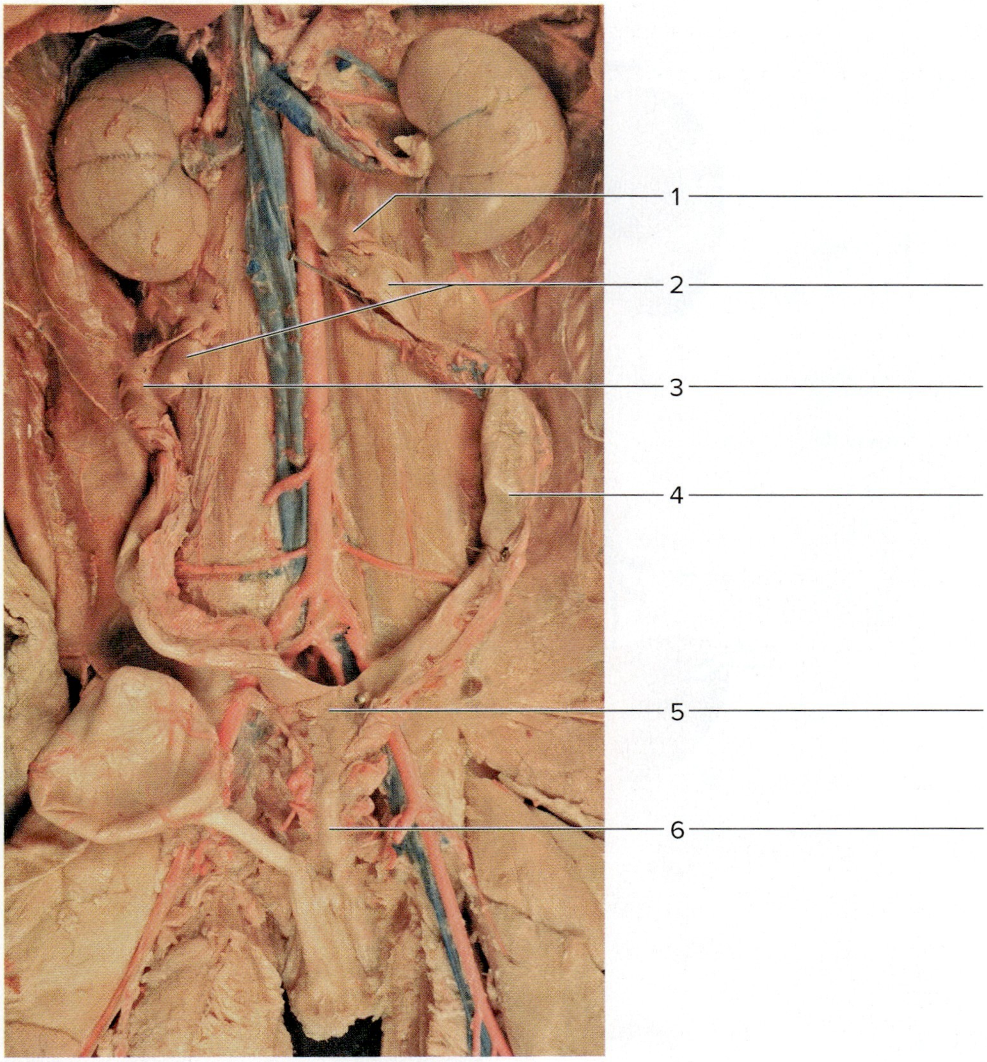

Terms:
Ovarian ligament
Ovaries
Uterine body
Uterine horn
Uterine tube
Vagina

(b)

FIGURE 68.7 *Continued.* Dennis Strete

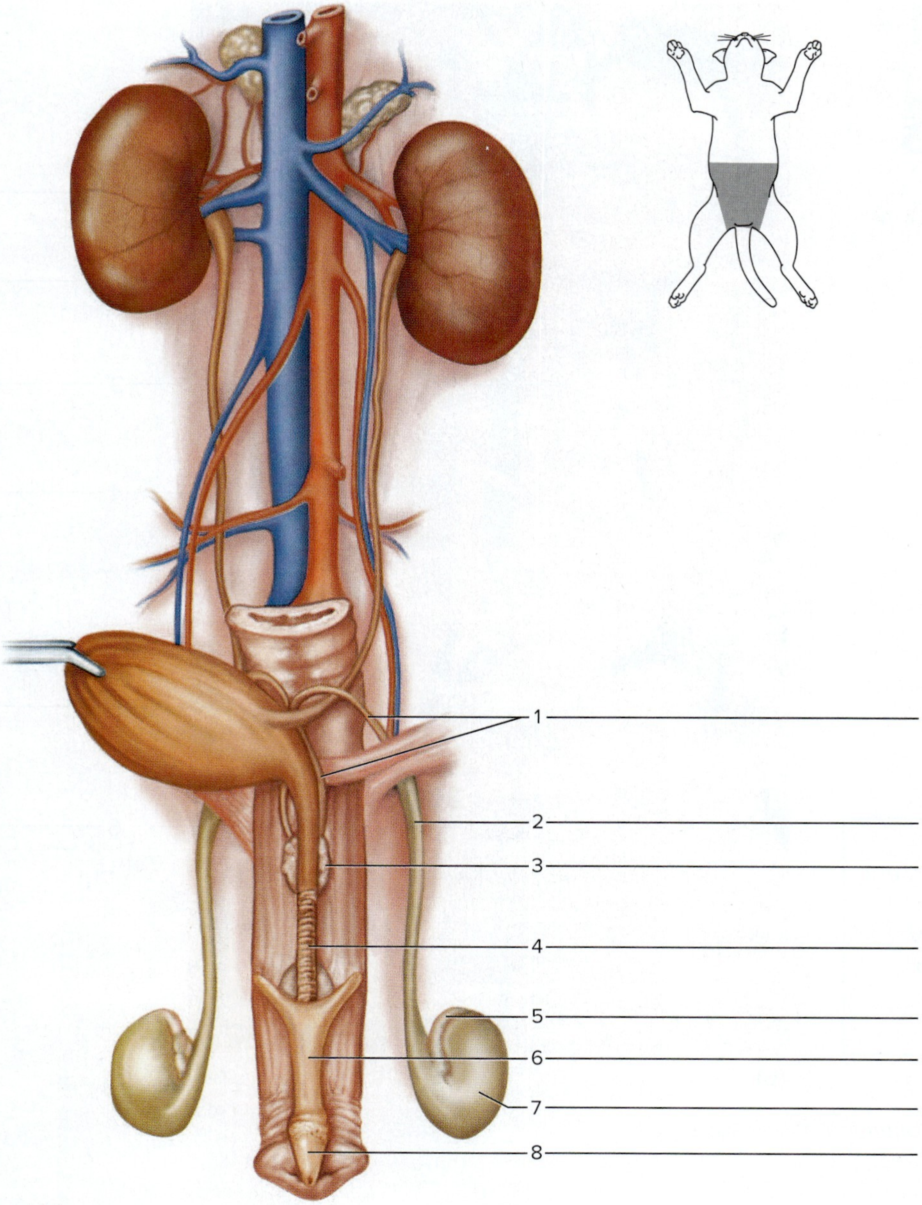

(a)

FIGURE 68.8 Label the numbered reproductive structures on a ventral view of the male cat in (a) an illustration, and (b) an image of a dissected cat. Use the terms provided for (b) only. A1

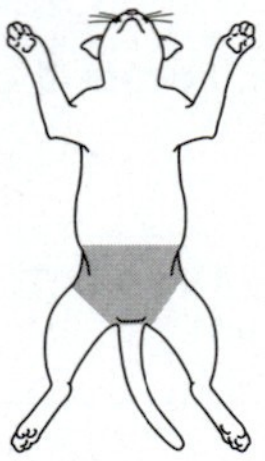

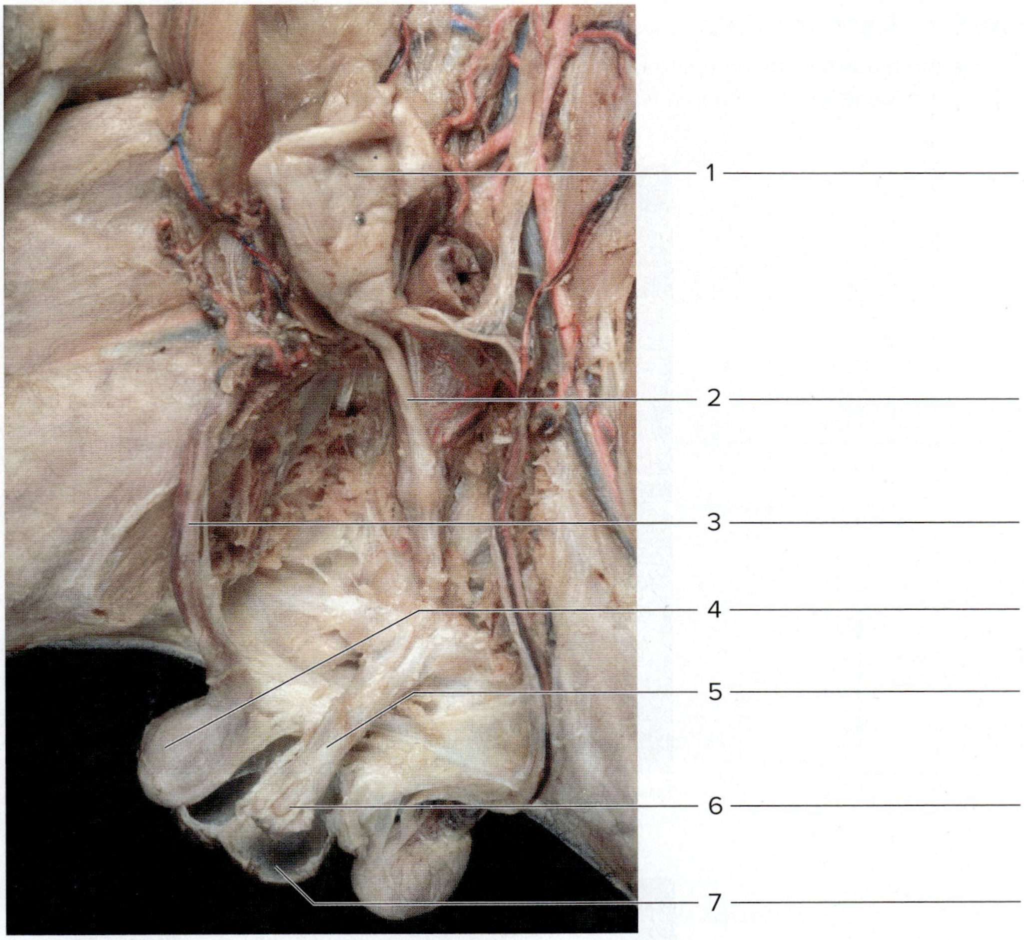

Terms:
Ductus deferens
Glans penis
Penis
Scrotum
Testis
Ureter
Urinary bladder

(b)

FIGURE 68.8 *Continued.* Dennis Strete

Part D: Assessments

Observe the models of the human reproductive system and figure 68.9 of a female and a male cadaver. Locate the labeled features and identify the numbered features in figure 68.9 that were also identified in the cat dissection.

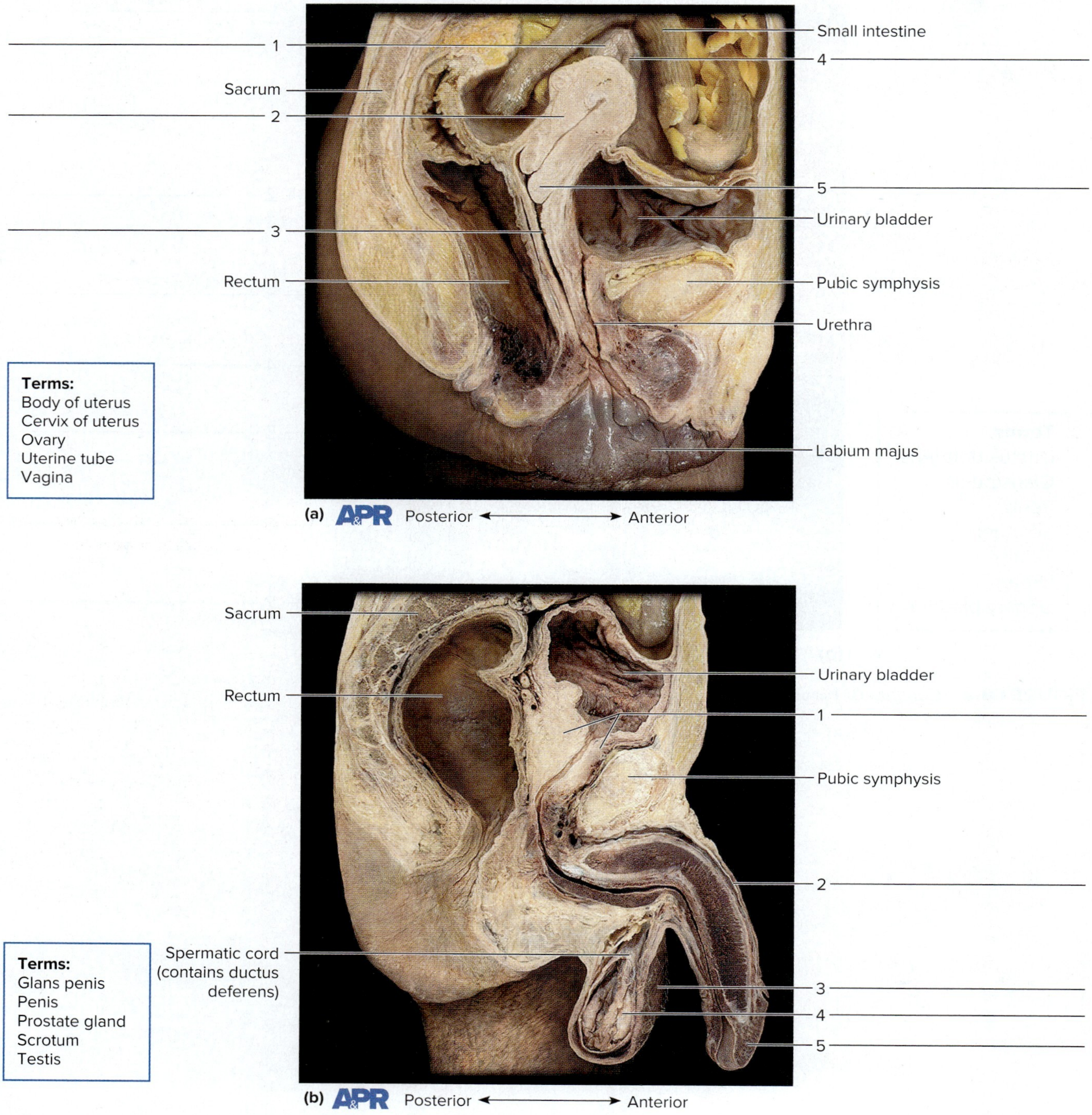

FIGURE 68.9 Sagittal (median; midsagittal) plane of a cadaver: (a) female; (b) male. Identify the reproductive features, using the terms provided. APR/McGraw Hill APR A2 A3

LABORATORY
EXERCISE 69

Fetal Pig Dissection: Musculature

Purpose of the Exercise

To examine and identify the skeletal musculature of the fetal pig and to make a comparison with the corresponding musculature of the human.

MATERIALS NEEDED

Preserved fetal pig (double injection)
Dissecting tray
Dissecting instruments
Twine
Plastic bag
Identification tag
Disposable gloves
Bone shears
Human torso model
Human upper and lower limb models

SAFETY

- Review all the safety guidelines in Appendix 1 of your laboratory manual.
- Wear disposable gloves and protective eyewear when working on the fetal pig dissection.
- Dispose of tissue remnants and gloves as instructed.
- Wash the dissecting tray and instruments as instructed.
- Wash your laboratory table.
- Wash your hands before leaving the laboratory.

Learning Outcomes APR

After completing this exercise, you should be able to:

O1 Locate and identify the major skeletal muscles of the fetal pig.
O2 Name the origins, insertions, and actions of the skeletal muscles designated by the laboratory instructor.
O3 Identify the corresponding skeletal muscles of the human.

The O corresponds to the assessments A indicated in the Laboratory Assessment for this Exercise.

Pre-Lab

Carefully read the introductory material and examine the entire lab. Be familiar with the location, origin, insertion, and action of muscles of the human body from lecture, the textbook, or laboratory exercises.

SKELETAL MUSCLE DISSECTION TECHNIQUES

Dissect: To expose the entire length of the muscle from origin to insertion. Most of the procedures are accomplished with blunt probes used to separate various connective tissues that hold adjacent structures together. It does not mean to remove or to cut into the muscles or other organs.

Transect: To cut through the muscle near its midpoint (belly). The cut is perpendicular to the length of the muscle fibers.

Reflect: To lift a transected muscle aside to expose deeper muscles or other organs.

Although the aim of this exercise is to become more familiar with the human musculature, human cadavers are not always available for dissection. Instead, preserved fetal pigs are often used for dissection because they are relatively small and can be purchased from biological suppliers. Because they are mammals, pigs have many features in common with humans, including similar skeletal muscles (with similar names).

The gestational period of pigs is 112–115 days. At the time of birth, a fetal pig would be about 30 cm (12–13 inches) long measured from nose to base of the tail. A fetal pig that would be approaching 22 cm (9 inches) long would have developed about 100 days. The embryonic and fetal stages of development occur much faster than in humans, for whom the gestational period is closer to forty weeks.

Despite their similarities, pigs use four limbs for locomotion, whereas humans use only two limbs. Because the musculature of each type of organism is adapted to provide for its special needs, comparisons of the muscles of pigs and humans may not be precise.

As you continue your dissection of various systems of the fetal pig, many anatomical similarities between pigs and humans will be observed. These fundamental similarities are called homologous structures. Although homologous structures have a similar structure and embryological origin, the functions are sometimes different.

PROCEDURE A: External Features

1. Fetal pigs usually are preserved with a solution that prevents microorganisms from decomposing the tissues. However, because fumes from this preserving fluid may be annoying and may irritate skin, be sure to work in a well-ventilated room and wear disposable gloves to protect your hands.
2. Obtain a preserved fetal pig, a dissecting tray, a set of dissecting instruments, a large plastic bag, twine, and an identification tag.
3. If the fetal pig is in a storage bag, dispose of any excess preserving fluid in the bag, as directed by the laboratory instructor. However, you may want to save some of the preserving fluid to keep the fetal pig moist until you have completed your work later in the year.
4. Rinse the pig off before dissection begins. If the fetal pig is too wet, blot the pig dry with paper towels.
5. If your fetal pig has an incision in the neck, it marks the location where colored latex (red into arteries and blue into veins) was injected into the blood vessels.
6. Study figures 69.1 to 69.3 and locate the following external anatomical features:

   auricle
   tongue
   external nares
   rostrum (snout)
   eyelid
   shoulder
   arm
   elbow
   forearm
   wrist
   digits
   hip
   thigh
   knee
   leg
   ankle
   umbilical cord
   tail
   mammary papillae
   anus

7. Study the external features of a male and a female pig (fig. 69.2) by comparing your pig to others in the laboratory. Identify the following:

   male
   - urogenital orifice
   - scrotum
   - mammary papillae

   female
   - genital papilla
   - urogenital orifice
   - mammary papillae

PROCEDURE B: Skin Removal

1. Remove the skin from the fetal pig. To do this, follow these steps:
   a. Place the fetal pig in the dissecting tray with its dorsal surface down.
   b. Use a sharp scalpel to make short, shallow incisions through the skin as illustrated in figure 69.3. The incisions are different for the male and female in the genital regions.
   c. Your incisions should encircle the neck, wrists, ankles, and urogenital regions. More skin is left attached on the male pig in the urogenital region.
   d. Use a blunt probe to separate the skin from the muscles. Observe the loose connective tissue (superficial fascia) that binds the skin to muscle. As you pull the skin away, you may note a thin sheet of skeletal muscle attached to it. These are *cutaneous muscles,* and they function to move the skin to get rid of any irritants. Humans lack cutaneous muscles, but a similar sheet of muscle (platysma) is present in the neck of a human.
   e. As you remove the skin, work toward the dorsal surface, then work toward the head, and finally work toward the tail. Pull the skin over each limb as if you were removing a glove.

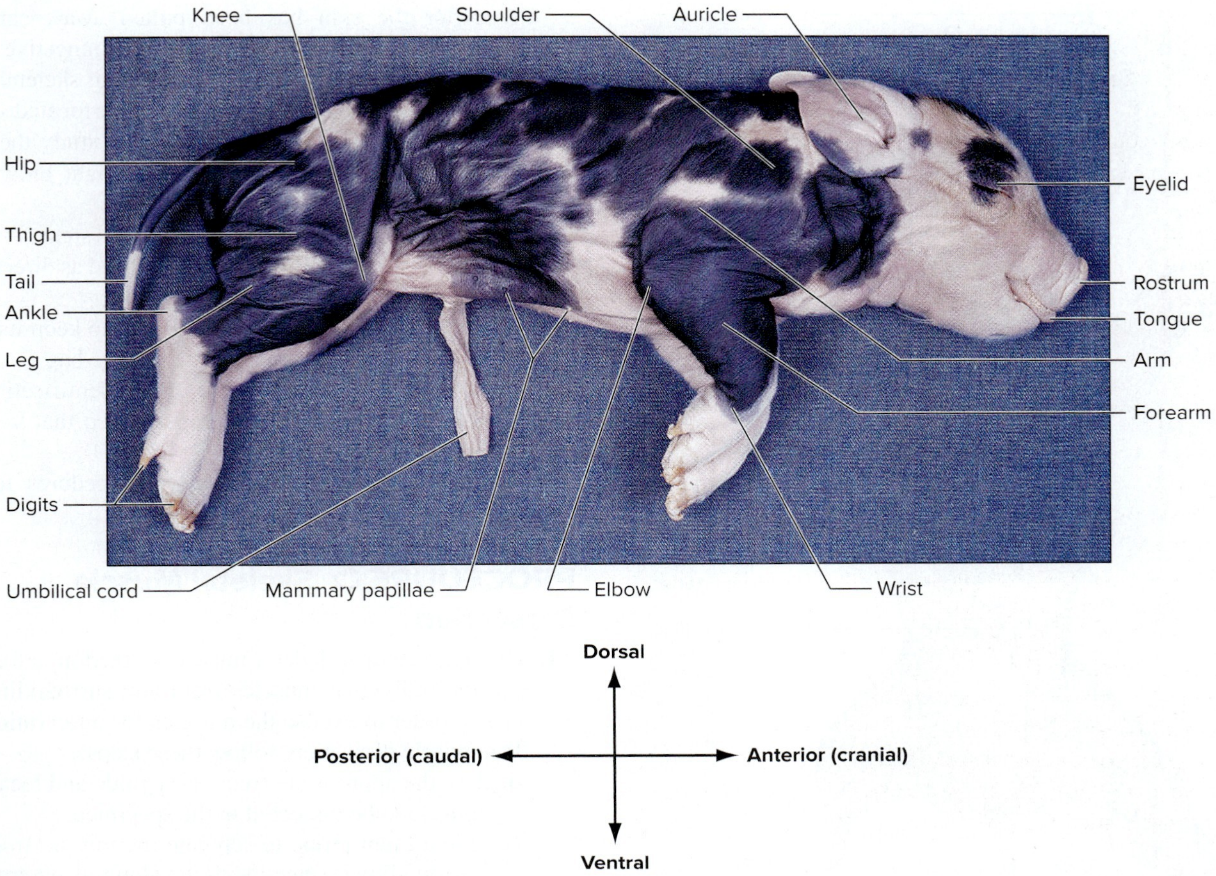

FIGURE 69.1 External features of a fetal pig, lateral view. ©J & J Photography

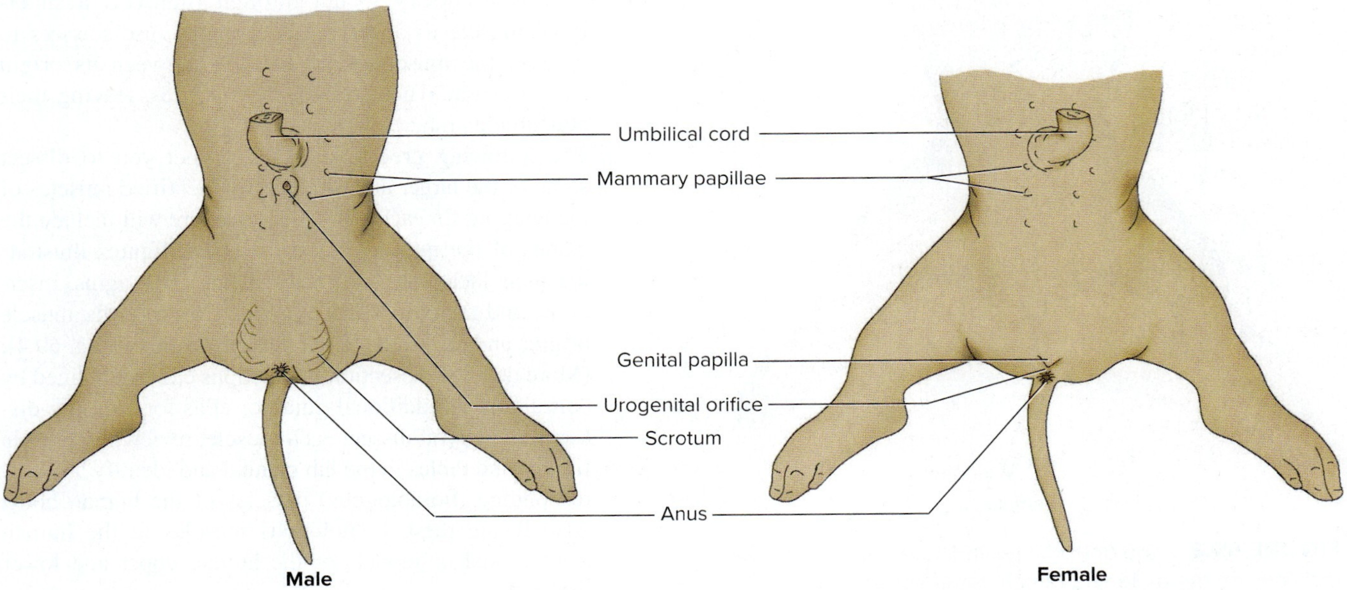

FIGURE 69.2 External features of a male and a female fetal pig, ventral view.

f. After the skin has been pulled away, carefully remove as much of the remaining connective tissue as possible to expose the underlying skeletal muscles. This task only needs to be performed on the right side and ventrally so you can study the muscles. The muscles should appear light brown and fibrous.

2. After skinning the fetal pig, follow these steps:
   a. Discard the tissues you have removed as directed by the laboratory instructor.
   b. Wrap the skin around the pig to help keep its body moist and place it in a plastic storage bag.
   c. Write your name in pencil on an identification tag and tie the tag to the storage bag so that you can identify your specimen.
3. Observe the recommended safety procedures for the conclusion of a laboratory session.

PROCEDURE C: Skeletal Muscle Dissection

1. The purpose of a skeletal muscle dissection is to separate the individual muscles from any surrounding tissues in order to expose the muscles for observation. To do a muscle dissection, follow these steps:
   a. Use the appropriate figure as a guide and locate the muscle to be dissected in the specimen.
   b. Use a blunt probe to separate the muscle from the surrounding connective tissue along its natural borders. The muscle should separate smoothly. If the border appears ragged, you probably have torn the muscle fibers. Because the fetal pig's muscles are incompletely developed, they are easily damaged and hard to distinguish, especially in small to medium sizes.
2. If it is necessary to cut through (transect) a superficial muscle to observe a deeper one, use scissors to transect the muscle about halfway between its origin and insertion. Then, reflect the cut ends, leaving their attachments intact.
3. The following procedures will instruct you to dissect some of the larger and more easily identified muscles of the fetal pig. In each case, the procedure will include the names of the muscles to be dissected, figures illustrating their locations, and tables listing the origins, insertions, and actions of the muscles. Compare the muscle origins and insertions to the fetal pig skeleton (fig. 69.4). (More detailed dissection instructions can be obtained by consulting an additional guide or atlas for fetal pig dissection.) As you dissect each muscle, review the muscle figures and tables in the lab manual and identify any corresponding (homologous) muscles of the human body. Also locate these homologous muscles in the human torso model or models of the human upper and lower limbs.

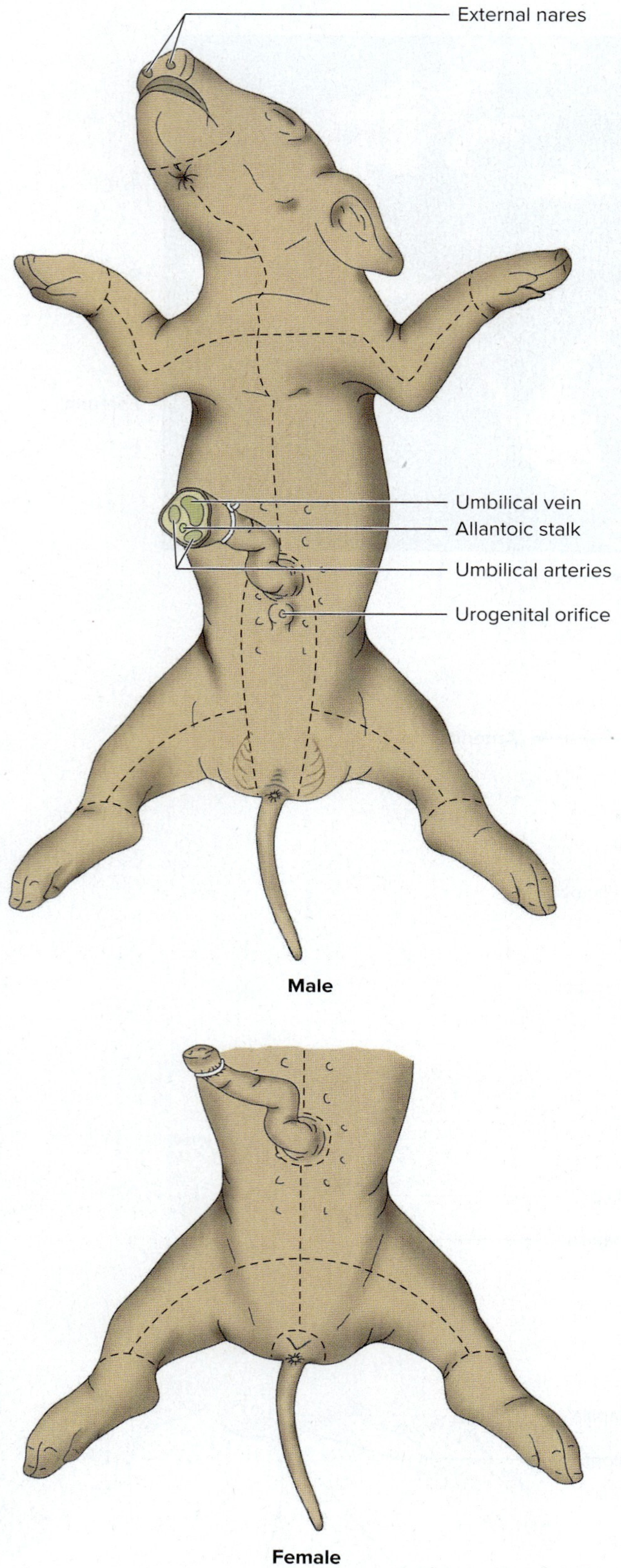

FIGURE 69.3 The dotted lines indicate the shallow skin incisions on the male and female fetal pigs to view the underlying muscles.

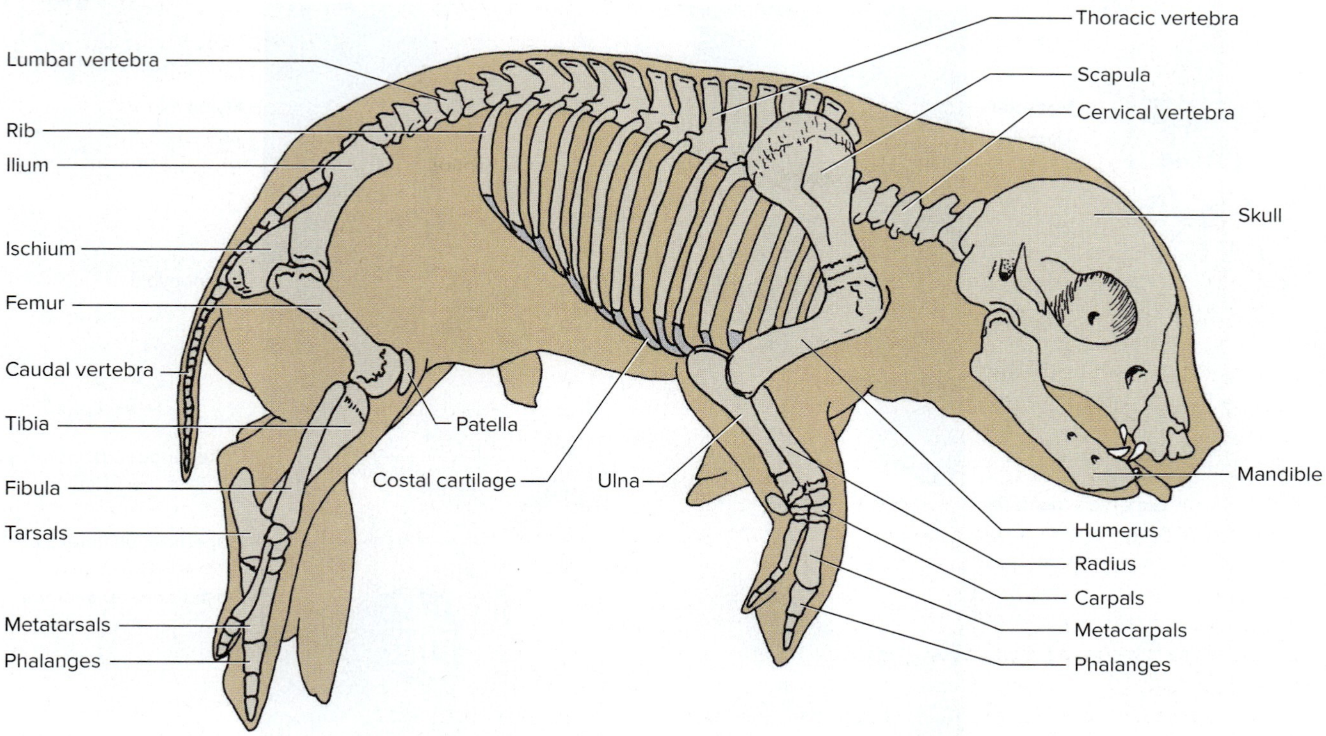

FIGURE 69.4 Skeleton of a fetal pig, lateral view.

PROCEDURE D: Muscles of the Head and Neck

1. Place the fetal pig in the dissecting tray with its ventral surface up.
2. Remove the skin and underlying connective tissues from each side of the neck, forward to the chin, and up to each ear.
3. Study figure 69.5, and then locate and dissect the following muscles:

   sternomastoid
   sternohyoid
   digastric
   mylohyoid
   masseter

4. To find the deep *sternothyroid* muscle in the neck, transect and reflect the sternohyoid muscle from the right side.
5. See table 69.1 for the origins, insertions, and actions of these head and neck muscles.
6. Complete Part A of Laboratory Assessment 69.

PROCEDURE E: Muscles of the Thorax

1. Place the fetal pig in the dissecting tray with its ventral surface up. Spread and secure all four limbs with twine.
2. Remove any remaining fat and connective tissue to expose the muscles in the walls of the thorax and abdomen.

Table 69.1 Muscles of the Head and Neck

Muscle	Origin	Insertion	Action
Sternomastoid	Sternum	Mastoid process of temporal bone	Turns and depresses head
Sternohyoid	Costal cartilage	Hyoid bone	Depresses hyoid bone
Digastric	Mastoid process and occipital bone	Mandible	Depresses mandible
Mylohyoid	Mandible	Hyoid bone	Raises floor of mouth
Masseter	Zygomatic arch	Mandible	Elevates mandible
Sternothyroid	Sternum	Thyroid cartilage	Pulls larynx back

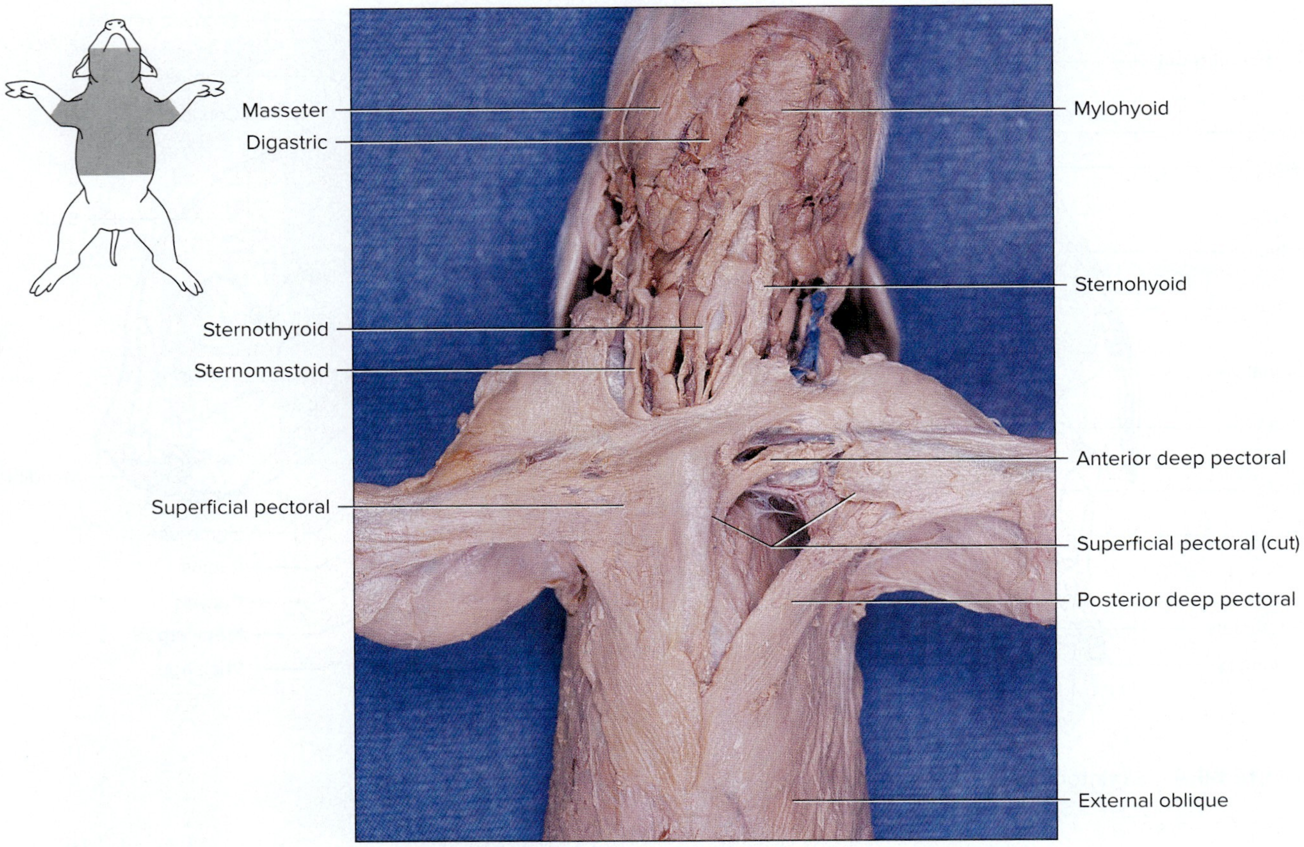

FIGURE 69.5 Muscles of the head, neck, thorax, and abdominal wall, ventral view. ©J & J Photography APR

3. Study figure 69.5. Locate and transect the superficial pectoral to expose the anterior deep pectoral. Locate and dissect the following pectoral muscles:

   superficial pectoral (pectoralis superficialis)—homologous to pectoralis major

   posterior deep pectoral (pectoralis profundus)—homologous to pectoralis minor

   anterior deep pectoral
4. To find the deep *serratus ventralis* thoracic muscle, transect the pectoral muscles and reflect their cut edges to the sides. The fingerlike origins on the ribs can be located by pulling the forelimb away from the thorax.
5. See table 69.2 for the origins, insertions, and actions of these muscles.

PROCEDURE F: Muscles of the Abdominal Wall

1. Study figure 69.5.
2. Locate the *external oblique* muscle in the abdominal wall.
3. Make a shallow, longitudinal incision through the external oblique. Reflect the cut edge and expose the *internal oblique* muscle beneath. The fibers of the internal oblique run at a right angle to those of the external oblique.
4. Make a longitudinal incision through the internal oblique. Reflect the cut edge and expose the *transversus abdominis* muscle.
5. Expose the *rectus abdominis* muscle on one side of the midventral line. This muscle lies beneath an aponeurosis.

Table 69.2 Muscles of the Thorax

Muscle	Origin	Insertion	Action
Superficial pectoral	Sternum	Humerus	Adducts arm
Posterior deep pectoral	Sternum and costal cartilages	Humerus	Adducts arm
Anterior deep pectoral	Sternum and costal cartilages	Scapula	Pulls scapula toward midline of body
Serratus ventralis	Ribs and cervical vertebrae	Vertebral border of scapula	Pulls scapula posteriorly and transfers weight from trunk to pectoral girdle

Table 69.3 Muscles of the Abdominal Wall

Muscle	Origin	Insertion	Action
External oblique	Ribs and fascia of back	Linea alba	Compresses abdominal wall
Internal oblique	Fascia of back	Linea alba	Compresses abdominal wall
Transversus abdominis	Lower ribs and fascia of back	Linea alba	Compresses abdominal wall
Rectus abdominis	Pubic	Sternum and costal cartilages	Compresses abdominal wall and flexes trunk

6. See table 69.3 for the origins, insertions, and actions of these muscles.
7. Complete Part B of the laboratory assessment.

PROCEDURE G: Muscles of the Shoulder and Back

1. Place the fetal pig in the dissecting tray onto its lateral surface.
2. Remove any remaining fat and connective tissue to expose the muscles of the shoulder and back.
3. Study figure 69.6, and then locate and dissect the following superficial muscles:
    brachiocephalic
    acromiotrapezius
    spinotrapezius
    deltoid
    latissimus dorsi

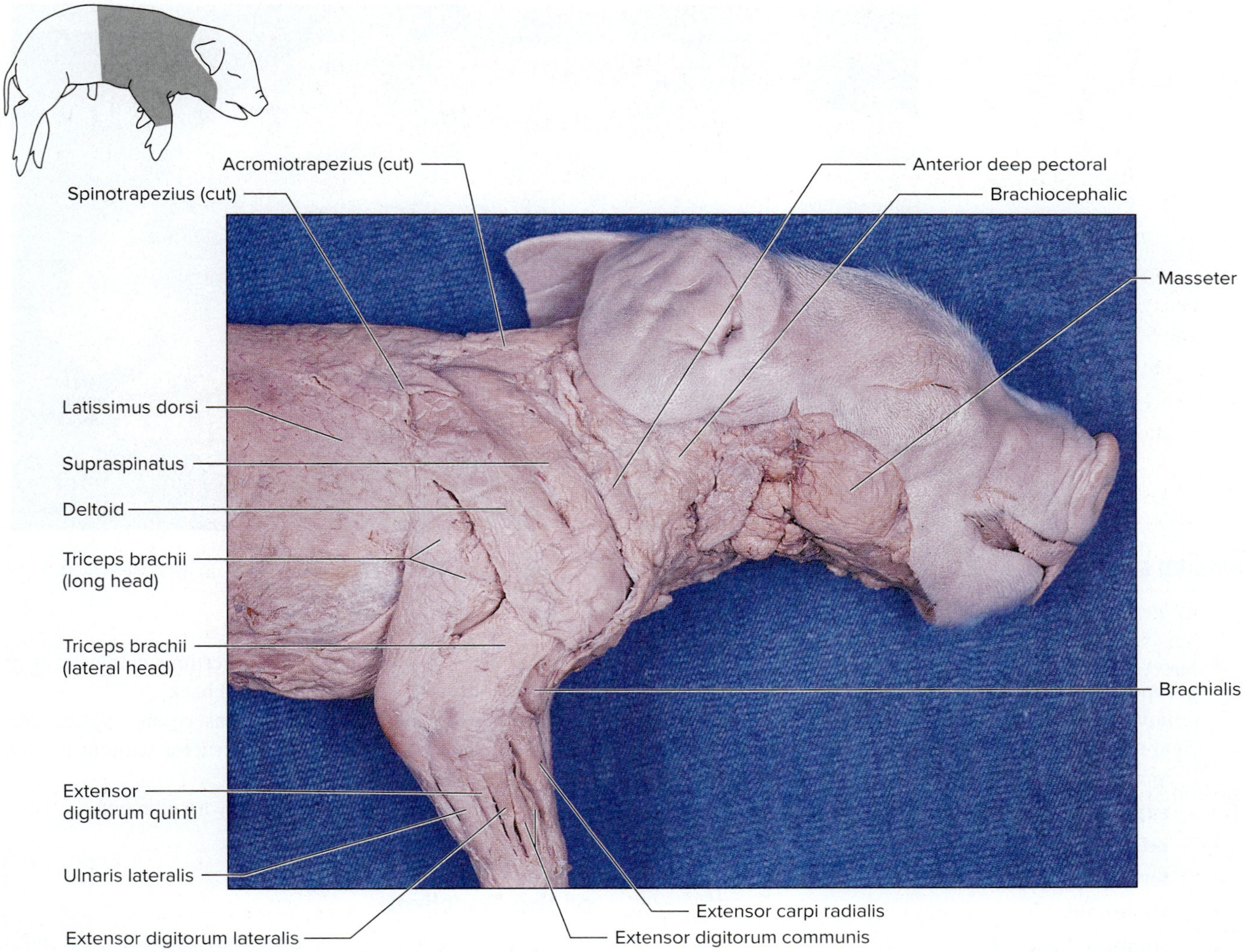

FIGURE 69.6 Muscles of the shoulder, arm, and forearm, lateral view. ©J & J Photography APR

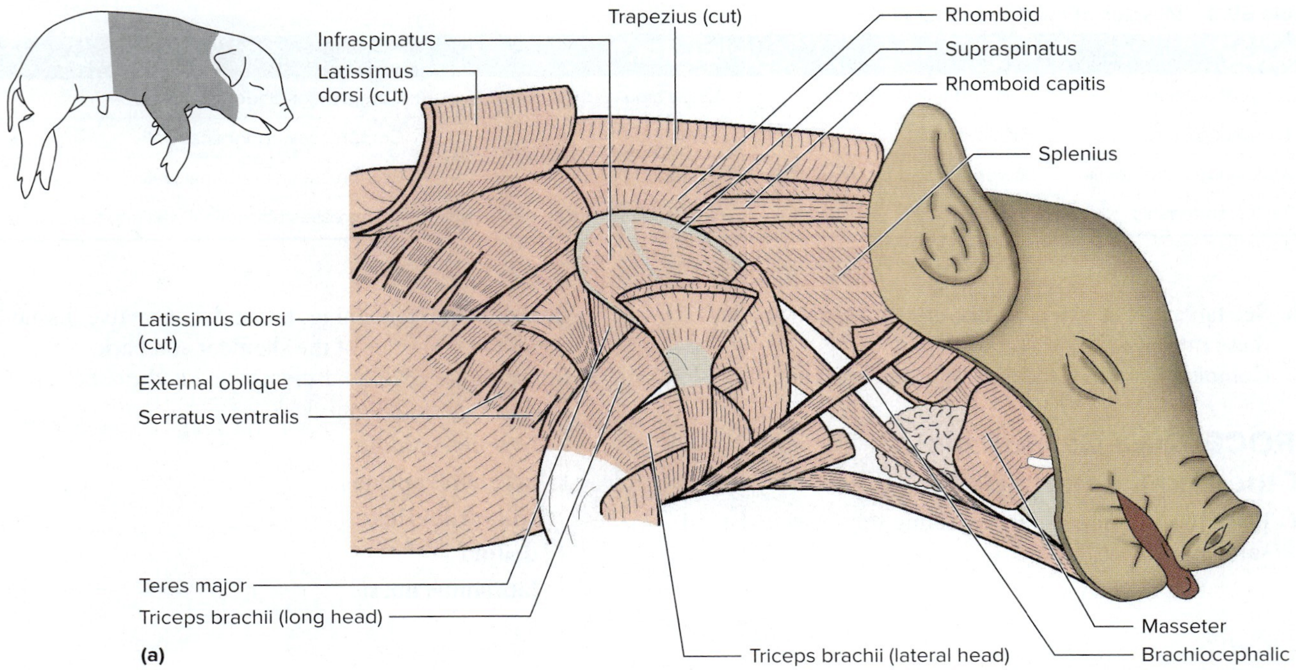

(a)

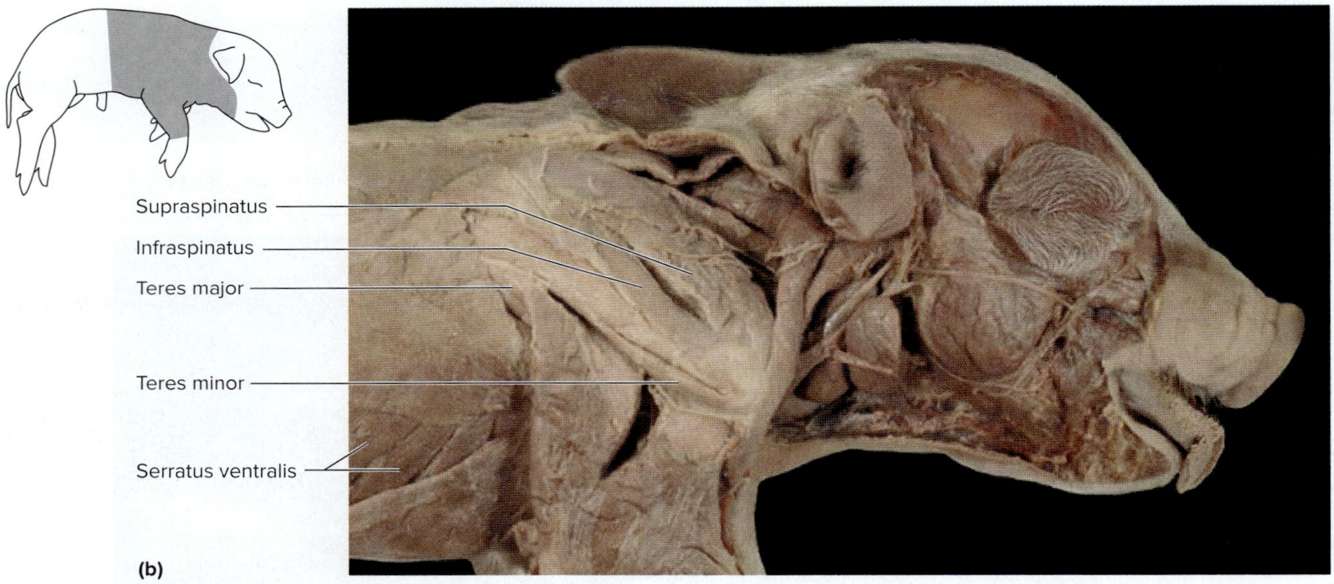

(b)

FIGURE 69.7 Deep muscles of the shoulder and arm, lateral view: (a) illustration; (b) dissection image. (b) APR/McGraw Hill

4. Using scissors, transect the latissimus dorsi and the group of trapezius muscles. Reflect their cut edges and remove any underlying fat and connective tissue. Study figure 69.7, and then locate and dissect the following deep muscles of the shoulder and back:

   supraspinatus
   infraspinatus
   teres major
   rhomboid
   rhomboid capitis
   splenius

5. See table 69.4 for the origins, insertions, and actions of these muscles in the shoulder and back.
6. Review the locations, origins, insertions, and actions designated by the laboratory instructor without the aid of the figures and tables.
7. Complete Part C of the laboratory assessment.

PROCEDURE H: Muscles of the Forelimb

1. Place the fetal pig in the dissecting tray with its ventral surface up.

Table 69.4 Muscles of the Shoulder and Back

Muscle	Origin	Insertion	Action
Acromiotrapezius	Spines of cervical and thoracic vertebrae	Spine and acromion process of scapula	Pulls scapula upward
Spinotrapezius	Spines of thoracic vertebrae	Spine of scapula	Pulls scapula upward and back
Brachiocephalic (clavotrapezius and clavobrachialis)	Occipital bone and mastoid process	Distal end of humerus	Flexes forelimb
Deltoid	Spine of scapula	Proximal end of humerus	Flexes forelimb
Latissimus dorsi	Thoracic and lumbar vertebrae	Proximal end of humerus	Pulls forelimb upward and back
Supraspinatus	Fossa above spine of scapula	Proximal end of humerus	Extends forelimb
Infraspinatus	Fossa below spine of scapula	Proximal end of humerus	Rotates and abducts forelimb
Teres major	Posterior border of scapula	Proximal end of humerus	Extends and adducts forelimb
Rhomboid	Spines of cervical and thoracic vertebrae	Medial border of scapula	Pulls scapula upward and forward
Rhomboid capitis	Occipital bone	Medial border of scapula	Pulls scapula forward
Splenius	Fascia of neck	Occipital bone	Raises head

2. Remove any remaining fat and connective tissue from a forelimb to expose the muscles.
3. Transect the pectoral muscles and reflect their cut edges.
4. Study figure 69.8, and then locate and dissect the following muscles from the medial surface of the arm and forearm of the forelimb:

medial arm muscles

- biceps brachii
- triceps brachii
  - long head
  - medial head

medial forearm muscles

- extensor carpi radialis
- flexor carpi radialis
- flexor digitorum profundus
- flexor digitorum superficialis
- flexor carpi ulnaris

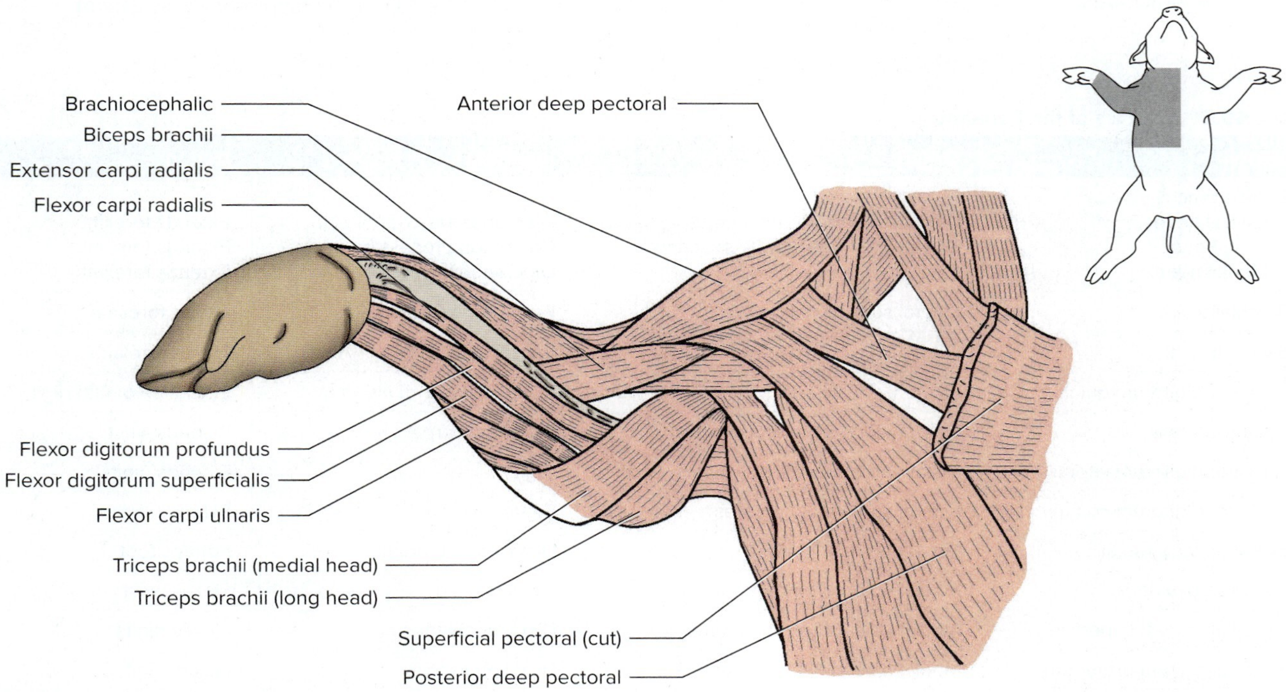

FIGURE 69.8 Medial muscles of the right forelimb, ventral view.

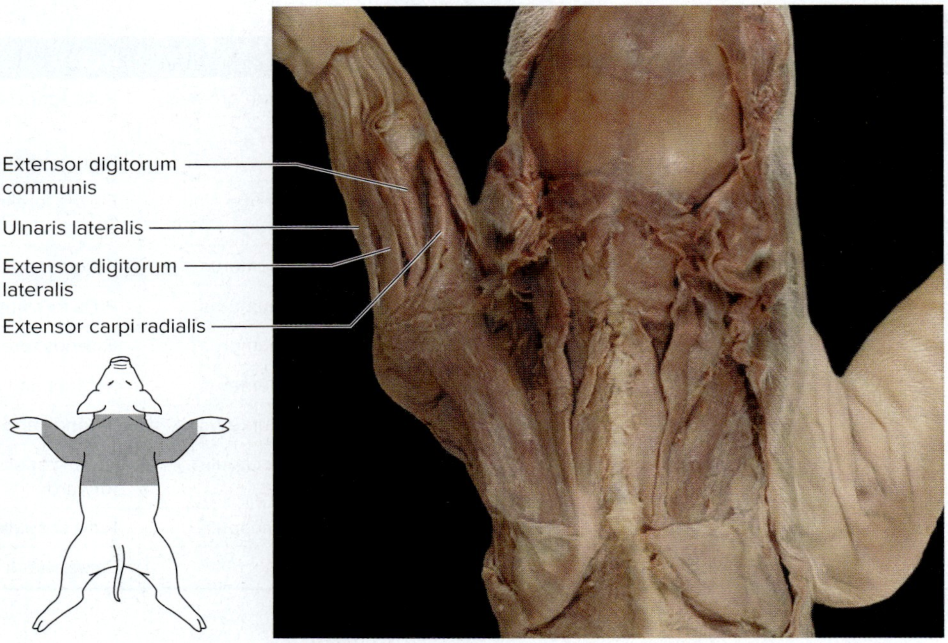

FIGURE 69.9 Lateral muscles of the left forelimb, dorsal view. APR/McGraw Hill APR

5. Place your pig onto its lateral surface. Study figures 69.6 and 69.9, which shows the lateral surface of the forelimb. Locate and dissect the following muscles on the lateral surface of the arm and forearm of the forelimb:

   lateral arm muscles
   - triceps brachii
     - long head
     - lateral head
   - brachialis

   lateral forearm muscles
   - extensor carpi radialis
   - extensor digitorum communis
   - extensor digitorum lateralis
   - extensor digitorum quinti
   - ulnaris lateralis
6. See table 69.5 for the origins, insertions, and actions of these muscles of the forelimb.
7. Complete Part D of the laboratory assessment.

Table 69.5 Muscles of the Forelimb

Muscle	Origin	Insertion	Action
Triceps brachii:			
Lateral head	Deltoid tuberosity of humerus	Olecranon process of ulna	Extends forelimb
Long head	Border of glenoid cavity of scapula	Olecranon process of ulna	Extends forelimb
Medial head	Shaft of humerus	Olecranon process of ulna	Extends forelimb
Brachialis	Lateral surface of humerus	Proximal end of ulna	Flexes forelimb
Biceps brachii	Scapula	Radius and ulna	Flexes forelimb
Extensor digitorum quinti	Humerus	Dorsal surface of fifth digit	Extension of fifth digit
Ulnaris lateralis	Lateral epicondyle of humerus	Fifth metacarpal	Extends wrist
Extensor digitorum lateralis	Distal end of humerus	Digits	Extends digits
Extensor digitorum communis	Lateral surface of humerus	Digits	Extends digits
Extensor carpi radialis	Distal end of humerus	Distal end of radius	Rotates foot
Flexor carpi ulnaris	Medial epicondyle of humerus	Carpals	Flexes wrist
Flexor digitorum superficialis	Medial epicondyle of humerus	Distal phalanges	Flexes digits
Flexor digitorum profundus	Medial epicondyle of humerus	Distal phalanges	Flexes digits
Flexor carpi radialis	Medial epicondyle of humerus	Metacarpals	Flexes metacarpals

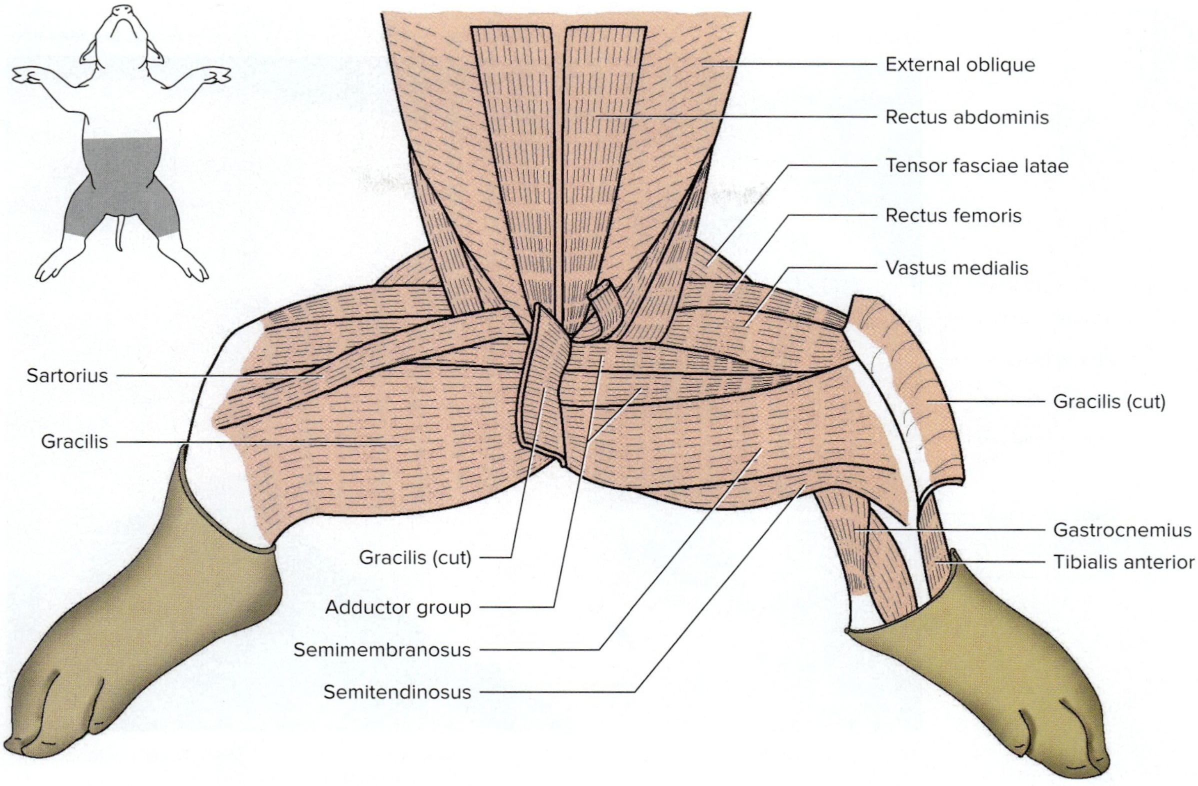

FIGURE 69.10 Medial superficial and deep muscles of the hindlimb, ventral view. APR

PROCEDURE I: Muscles of the Hip and Hindlimb

1. Place the fetal pig in the dissecting tray with its ventral surface up.
2. Remove any remaining fat and connective tissue from the hip and hindlimb to expose the muscles.
3. Study figure 69.10 and then locate and dissect the following muscles from the medial surface of the thigh:

   sartorius

   gracilis
4. Using scissors, transect the sartorius and gracilis and reflect their cut edges to observe the deeper muscles of the thigh.
5. Study figure 69.10 and then locate and dissect the following muscles:

   tensor fasciae latae

   rectus femoris

   vastus medialis

   adductor group

   semimembranosus

   semitendinosus
6. Transect the tensor fasciae latae and rectus femoris muscles and reflect their cut edges. Locate the *vastus intermedius* and *vastus lateralis* muscles beneath. (*Note:* In some specimens, the vastus intermedius, vastus lateralis, and vastus medialis muscles are closely united by connective tissue and are difficult to separate.)
7. Study figure 69.11, and then locate and dissect the following muscles from the lateral surface of the hip and thigh:

   biceps femoris

   gluteus superficialis (gluteus maximus)

   gluteus medius
8. Using scissors, transect the tensor fasciae latae and biceps femoris, and reflect their cut edges to observe the deeper muscles of the thigh.

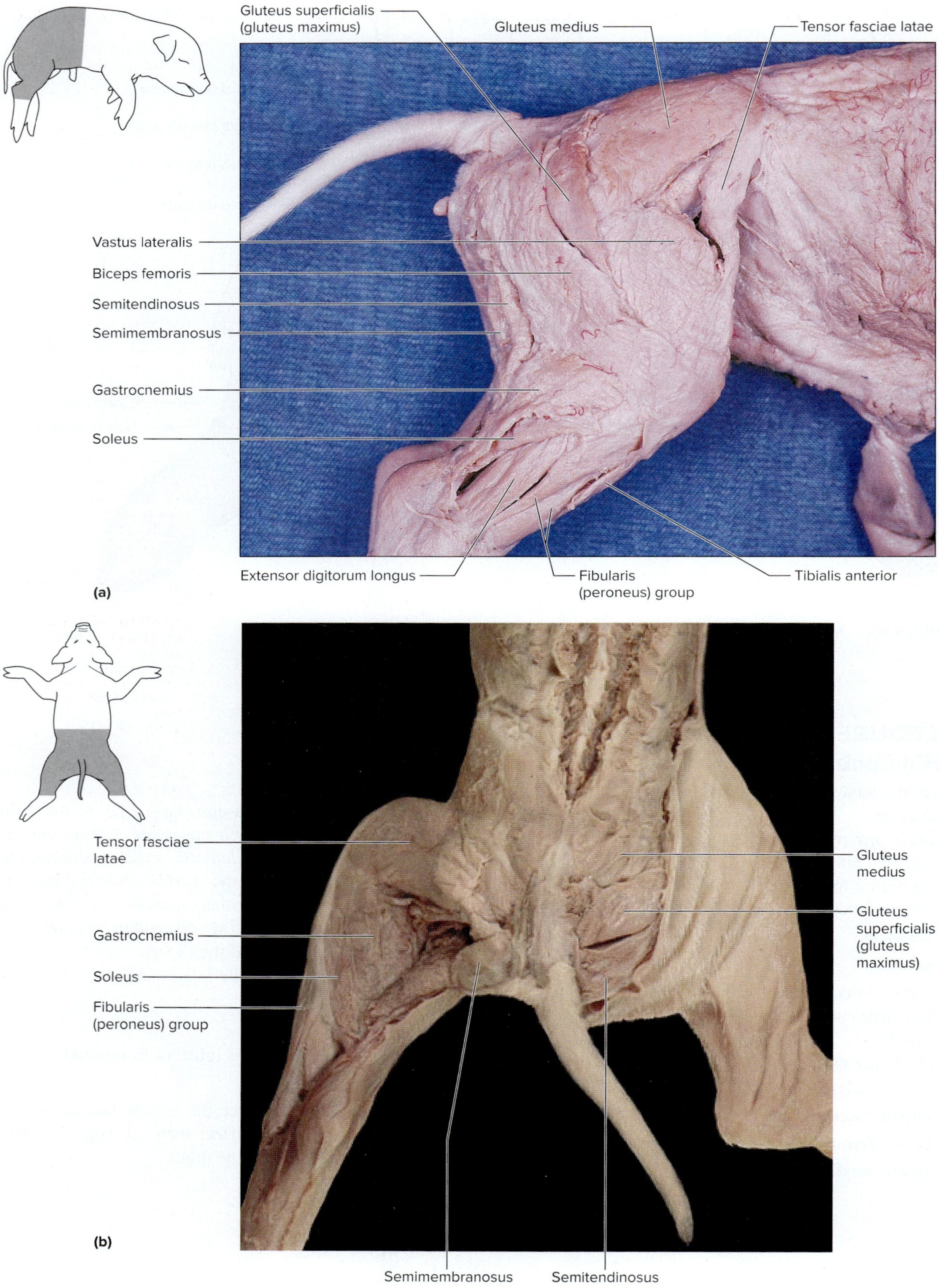

FIGURE 69.11 Muscles of the hip, thigh, and leg: (a) lateral view; (b) dorsal view. (a) ©J & J Photography; (b) APR/McGraw Hill APR

9. On the lateral surface of the leg (see fig. 69.11), locate and dissect the following muscles:
   gastrocnemius
   soleus
   extensor digitorum longus
   tibialis anterior
   fibularis (peroneus) group
10. See table 69.6 for the origins, insertions, and actions of these muscles of the hip and hindlimb.
11. Review the locations, origins, insertions, and actions designated by the laboratory instructor without the aid of the figures and tables.
12. Complete Parts E and F of the laboratory assessment.

Table 69.6 Muscles of the Hip and Hindlimb

Muscle	Origin	Insertion	Action
Tensor fasciae latae	Iliac crest	Fascia lata of thigh	Tighten fascia lata
Gluteus superficialis	Sacral and caudal vertebrae	Greater trochanter of femur	Abducts thigh
Gluteus medius	Ilium	Greater trochanter of femur	Abducts thigh
Quadriceps femoris:			
Vastus lateralis	Shaft of femur and greater trochanter	Patella	Extends hindlimb
Vastus intermedius	Shaft of femur	Patella	Extends hindlimb
Rectus femoris	Ilium	Patella	Extends hindlimb
Vastus medialis	Shaft of femur	Patella	Extends hindlimb
Hamstring muscles:			
Biceps femoris	Ischium	Tibia	Abducts thigh and flexes lower hindlimb
Semimembranosus	Ischium	Tibia	Flexes lower hindlimb
Semitendinosus	Ischium	Tibia	Flexes lower hindlimb
Sartorius	Crest of ilium	Proximal end of tibia	Adducts thigh
Gracilis	Pubis	Proximal end of tibia	Adducts hindlimb
Adductor group	Pubis	Femur	Adducts hindlimb
Gastrocnemius	Lateral and medial epicondyles of femur	Calcaneus	Extends foot
Soleus	Proximal end of fibula	Calcaneus	Extends foot
Tibialis anterior	Proximal end of tibia	Second metatarsal	Flexes foot
Fibularis (peroneus) group	Shaft of tibia and fibula	Metatarsals	Flexes foot
Extensor digitorum longus	Proximal end of tibia and fibula	Digits	Extends digits

NOTES

LABORATORY ASSESSMENT 69

Name ______________________

Date ______________________

Section ______________________

The (A) corresponds to the indicated Learning Outcome(s) (O) found at the beginning of the Laboratory Exercise.

Fetal Pig Dissection: Musculature

PART A: Assessments

Complete the following statements:

1. The ______________ muscle of the human is homologous to the sternomastoid muscle of the pig. A3
2. The ______________ muscle elevates the mandible in the human and in the pig. A2 A3
3. Two muscles of the pig that are inserted on the hyoid bone are the ______________ and the ______________. A2

PART B: Assessments

Complete the following:

1. Name two pectoral muscles that are found in the thoracic wall of the human. A3

   ______________ ______________

2. Name three pectoral muscles that are found in the thoracic wall of the pig. A1

   ______________ ______________ ______________

3. Name four muscles that are found in the abdominal wall of the pig and the human. A1 A3

   ______________ ______________

   ______________ ______________

PART C: Assessments

Complete the following:

1. Name three muscles of the pig that together correspond to the trapezius muscle in the human. A3

   ______________ ______________ ______________

2. Name the muscle in the pig and in the human that occupies the fossa above the spine of the scapula. A3

   ____

3. Name the muscle in the pig and in the human that occupies the fossa below the spine of the scapula. A3

   ____

PART D: Assessments

Complete the following:

Name two muscles found in the pig and in the human that can flex the forelimb. A2 A3

______________ ______________

Name a muscle that has three heads, can extend the forelimb, and is found in the pig and in the human. A2 A3

____

PART E: Assessments

Identify the numbered muscles of the pig in figures 69.12 to 69.16.

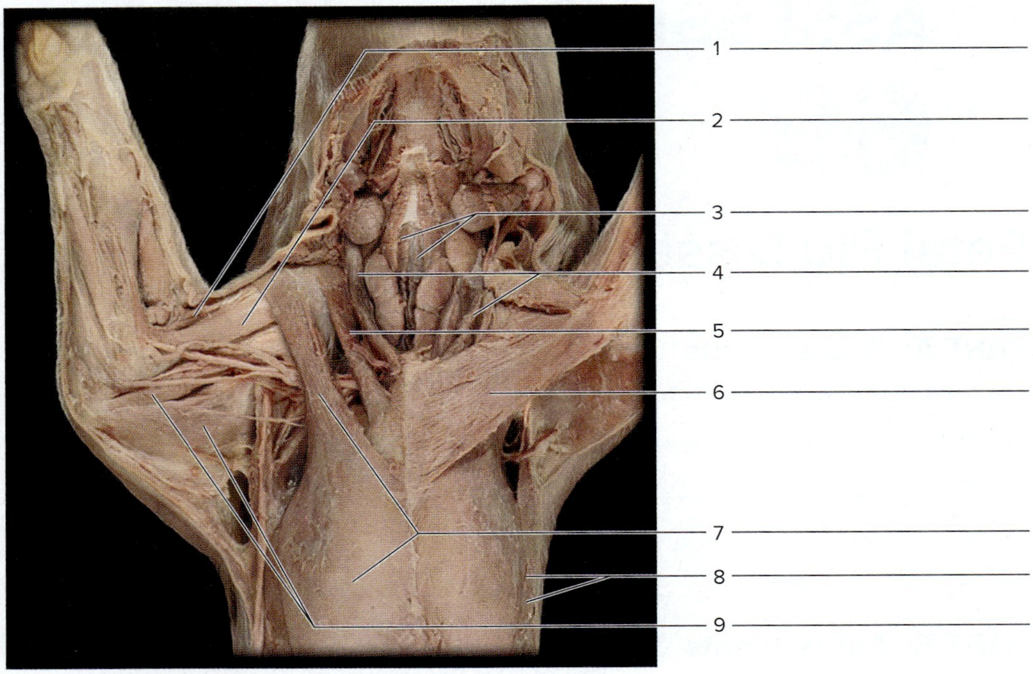

Terms:
Anterior deep pectoral
Biceps brachii
Brachialis
Posterior deep pectoral
Serratus ventralis
Sternomastoid
Sternothyroid
Superficial pectoral
Triceps brachii

FIGURE 69.12 Label the numbered head, neck, thorax, and medial arm muscles of the pig in this ventral view.
APR/McGraw Hill A1

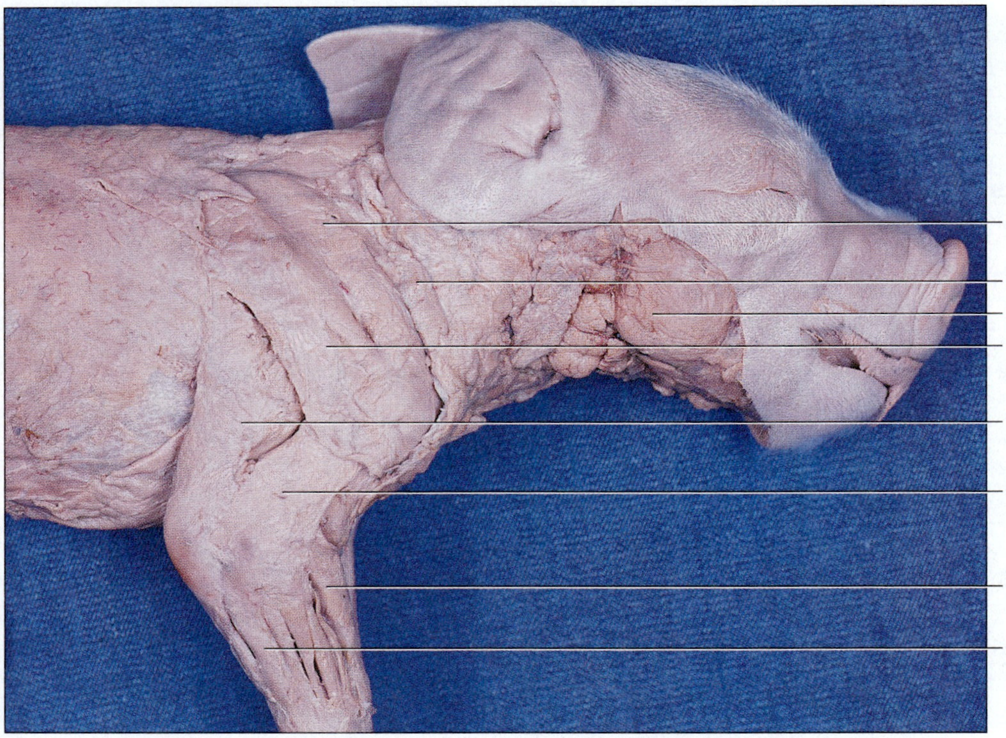

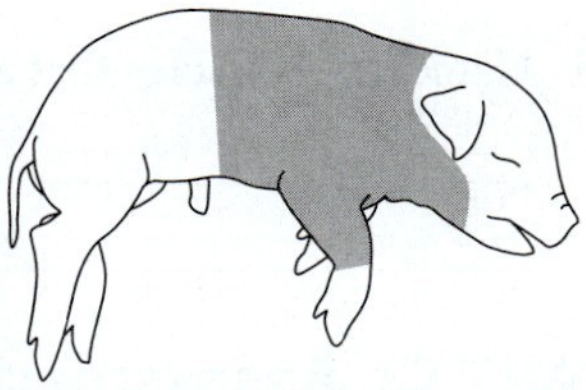

FIGURE 69.13 Label the numbered muscles of the pig in this anterior, lateral view. ©J & J Photography A1

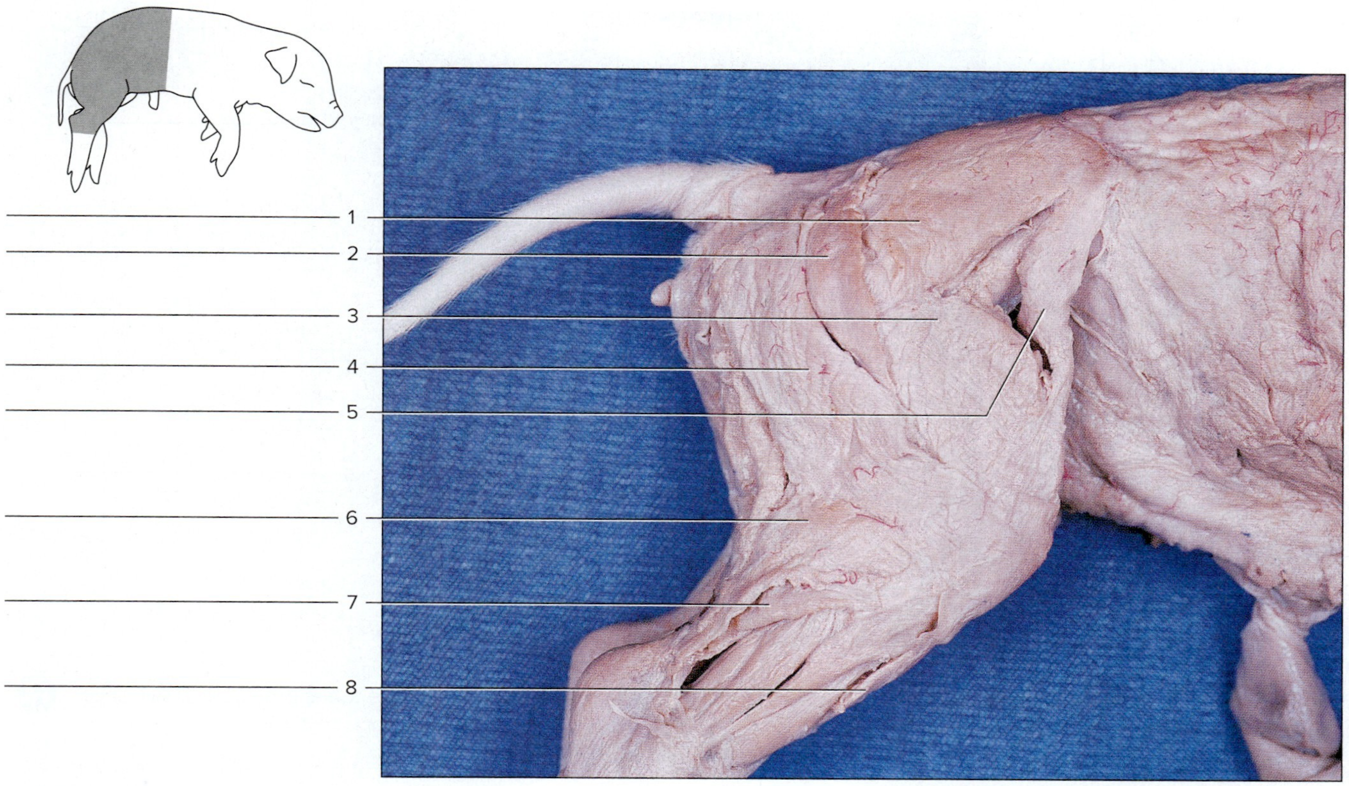

FIGURE 69.14 Label the numbered muscles of the pig in this posterior, lateral view. ©J & J Photography A1

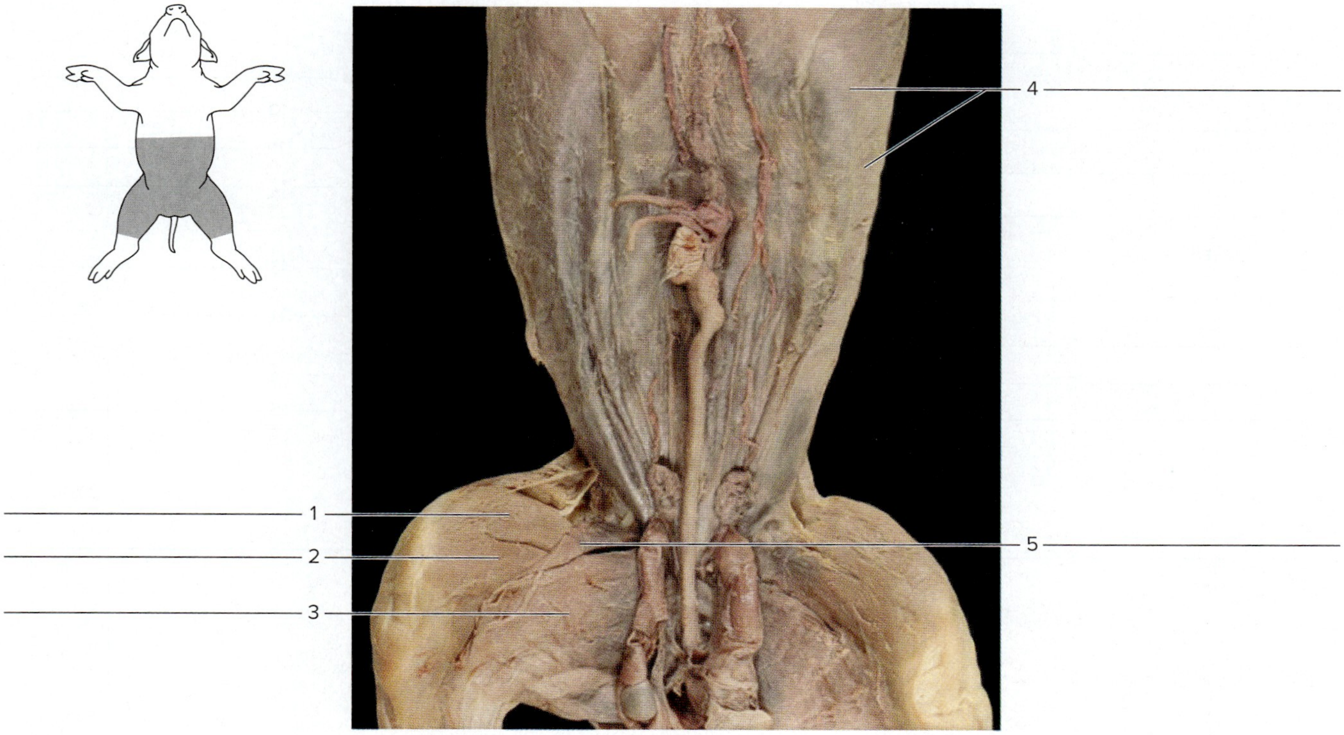

FIGURE 69.15 Label the numbered muscles of the pig in this ventral view. APR/McGraw Hill A1

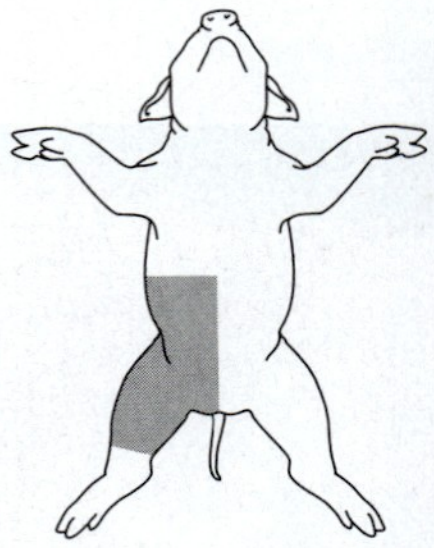

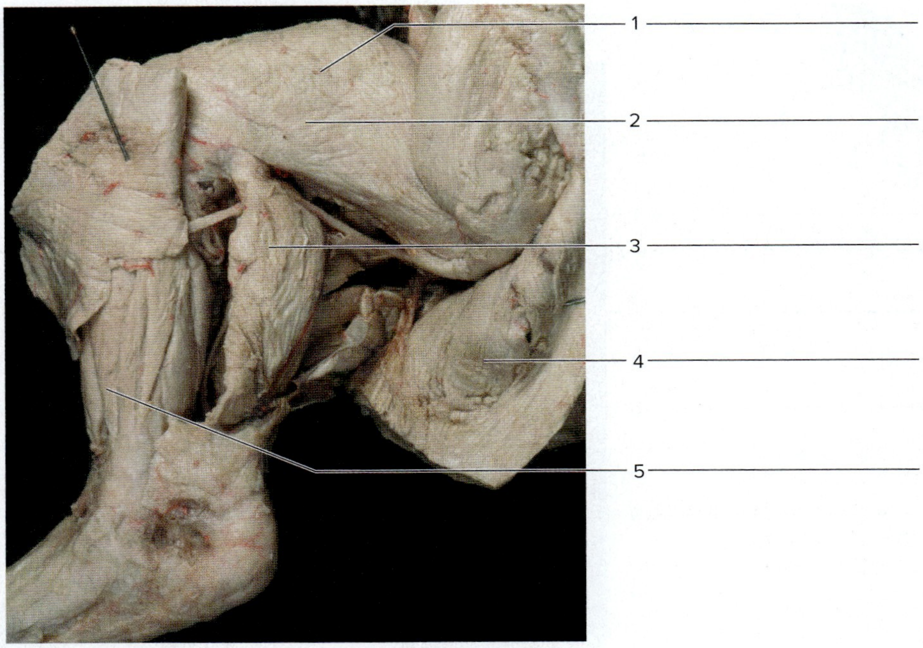

Terms:
Gastrocnemius
Gracilis (cut and reflected)
Rectus femoris
Tibialis anterior
Vastus medialis

FIGURE 69.16 Label the numbered hindlimb muscles of the pig in this ventral view, using the terms provided. Dennis Strete A1

PART F: Assessments

Observe the muscles on the human torso model and the upper and lower limb models. Each of the muscles in figure 69.17 is found both in the pig and in the human. Identify each of the numbered muscles in the figure by placing its name in the space next to its number.

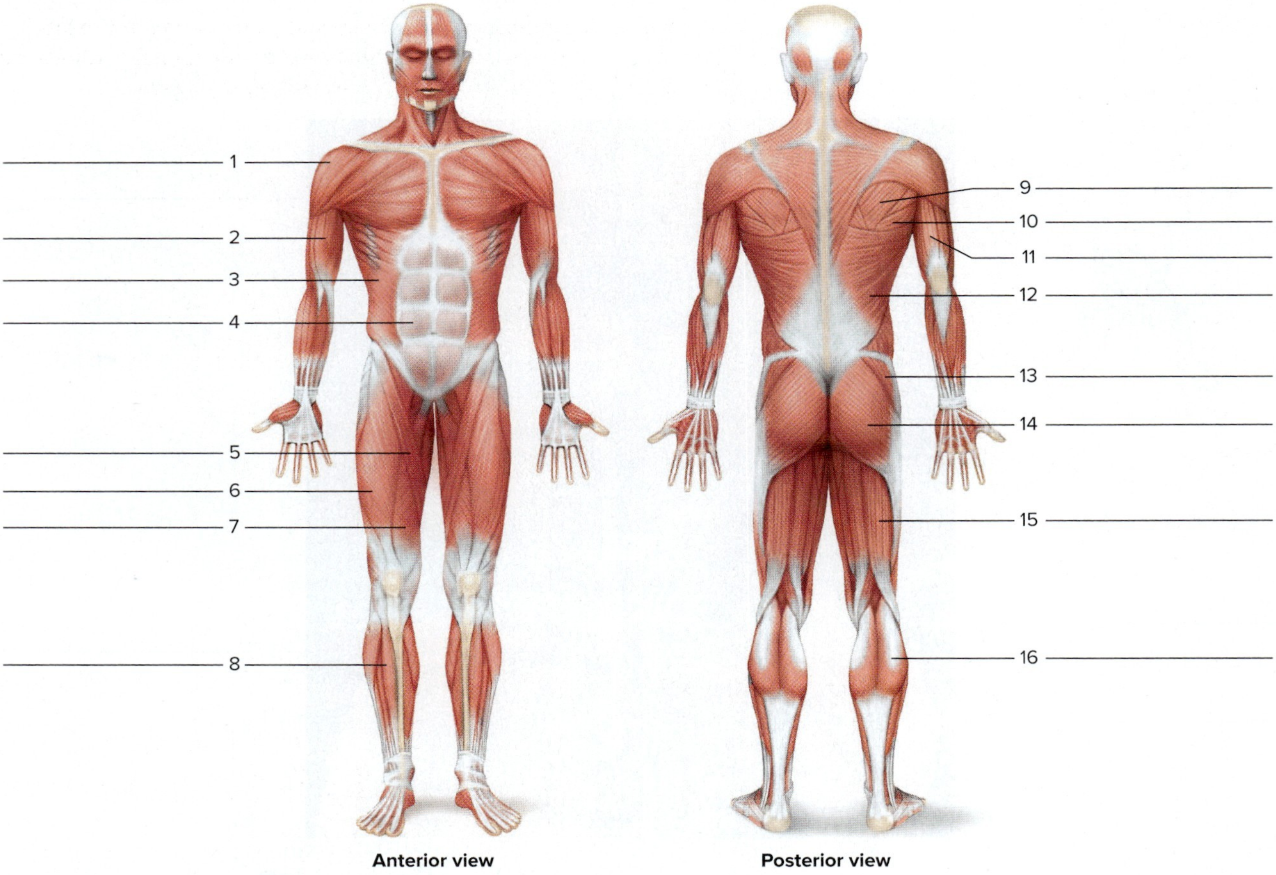

FIGURE 69.17 Identify the numbered muscles that occur in both the pig and the human. A3

LABORATORY
EXERCISE 70

Fetal Pig Dissection: Cardiovascular System

Purpose of the Exercise

To examine and identify the major organs of the cardiovascular system of the fetal pig and to compare them with the corresponding organs of the human.

MATERIALS NEEDED

Preserved fetal pig
Dissecting tray
Dissecting instruments
Twine
Disposable gloves
Human torso model

SAFETY

- Review all the safety guidelines in Appendix 1 of your laboratory manual.
- Wear disposable gloves and protective eyewear when working on the fetal pig dissection.
- Dispose of tissue remnants and gloves as instructed.
- Wash the dissecting tray and instruments as instructed.
- Wash your laboratory table.
- Wash your hands before leaving the laboratory.

Learning Outcomes APR

After completing this exercise, you should be able to:

O1 Locate and identify the major organs of the cardiovascular system of the fetal pig.

O2 Locate and identify fetal circulatory features of the fetal pig and contrast fetal and adult circulation.

O3 Compare the features and organs of the cardiovascular system of the fetal pig with those of the human.

The O corresponds to the assessments A indicated in the Laboratory Assessment for this Exercise.

Pre-Lab

Carefully read the introductory material and examine the entire lab. Be familiar with the structures and functions of the human cardiovascular system from lecture, the textbook, or laboratory exercises.

In this laboratory exercise, you will dissect the major organs of the cardiovascular system of the fetal pig. As before, while you are examining the organs of the fetal pig, compare them with the corresponding organs of the human torso model.

If the cardiovascular system of the fetal pig has been injected, the systemic arteries will be filled with red latex (large arteries may not appear very red through a thick wall), and the systemic veins will be filled with blue latex. This will make it easier for you to trace the vessels as you dissect them.

PROCEDURE A: The Arterial System

1. Place the fetal pig in the dissecting tray with its ventral surface up. Spread and secure all four limbs with twine.
2. Open the thoracic cavity and expose its contents. To do this, follow these steps:
   a. Make a longitudinal incision passing anteriorly from the diaphragm along one side of the sternum. Continue the incision through the neck muscles to the mandible. Try to avoid damaging the internal organs as you cut. Make a lateral cut on each side of the sternum into each arm (fig. 70.1).
   b. Make a lateral cut on each side along the anterior surface of the diaphragm and cut the diaphragm loose from the thoracic wall.
   c. Spread the sides of the thoracic wall outward, and use a scalpel to make a longitudinal cut along each side of the inner wall of the rib cage to weaken the ribs. Continue to spread the thoracic wall laterally to break the ribs so that the flaps of the wall will remain open. It is often necessary to rinse the body cavities to remove any coagulated blood or excess latex that has leaked out during injection.
3. Note the location of the heart and the large blood vessels associated with it. Slit the outer *parietal pericardium* that surrounds the heart by cutting with scissors along the midventral line. Note how this membrane is connected to the *visceral pericardium* that is attached to the surface of the heart. Locate the *pericardial cavity,* the space between the two layers of the pericardium.
4. Remove the thymus gland to expose blood vessels anterior to the heart. Examine the heart (figs. 70.2 and 70.3). The arrangement of blood vessels coming off the aortic arch is different in the pig than in a human body. Locate the following:

   right atrium
   left atrium
   right ventricle
   left ventricle
   pulmonary trunk
   aorta
   coronary arteries

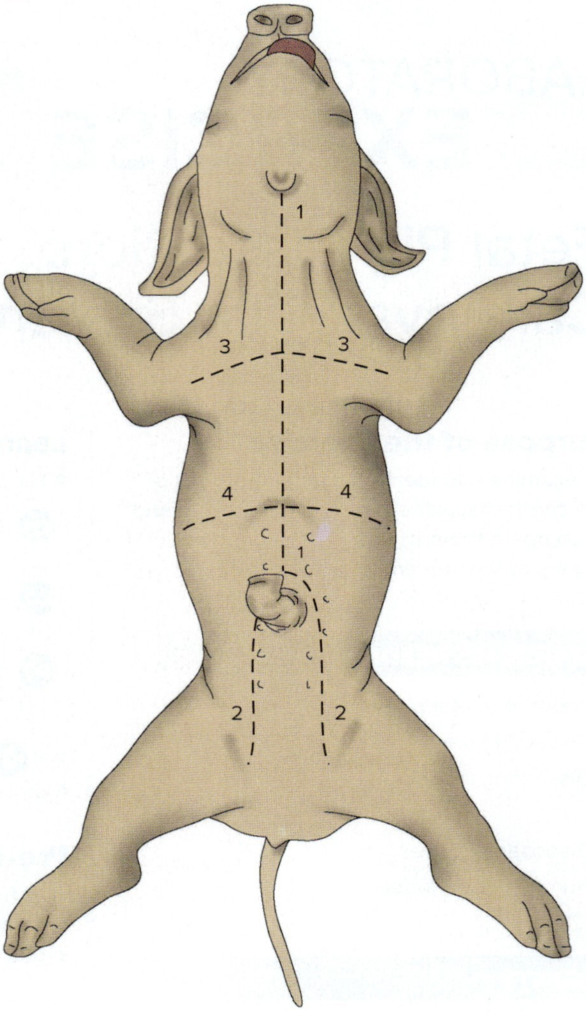

FIGURE 70.1 Incision lines indicate the locations for opening the ventral body cavity.

5. Trace the pulmonary trunk and locate the short vessel, the *ductus arteriosus,* that is connected to the aorta (fig. 70.2*a*). This connection, found in fetal circulation, shunts blood from the pulmonary artery to the aorta as a partial bypass of nonfunctional, developing fetal lungs. This shunt becomes occluded after birth (forming the *ligamentum arteriosum*), allowing a fully functional pulmonary circuit (fig. 70.2*b*).
6. Use a scalpel to open the heart chambers by making a cut along the frontal plane from its apex to its base. Remove any remaining latex from the chambers. Examine the valves between the chambers and note the relative thicknesses of the chamber walls. Do not remove the heart, as it is needed for future study of the relationship of major blood vessels.

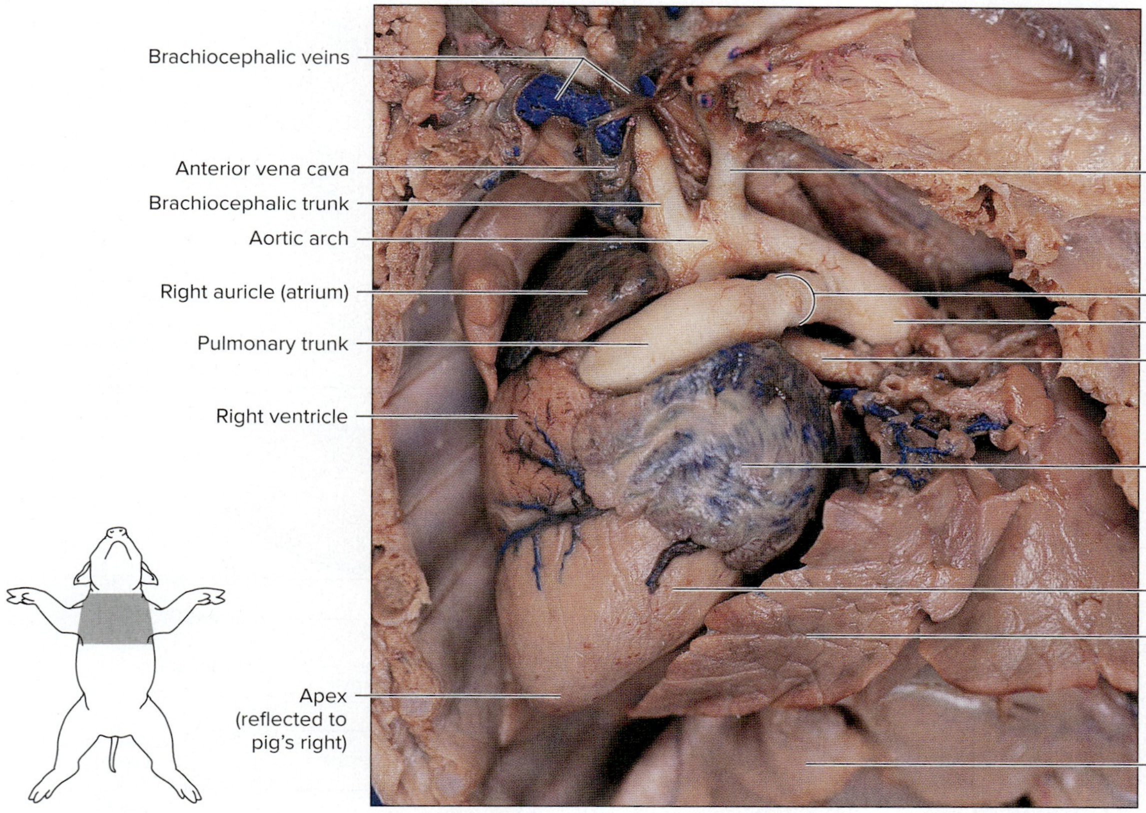

(a)

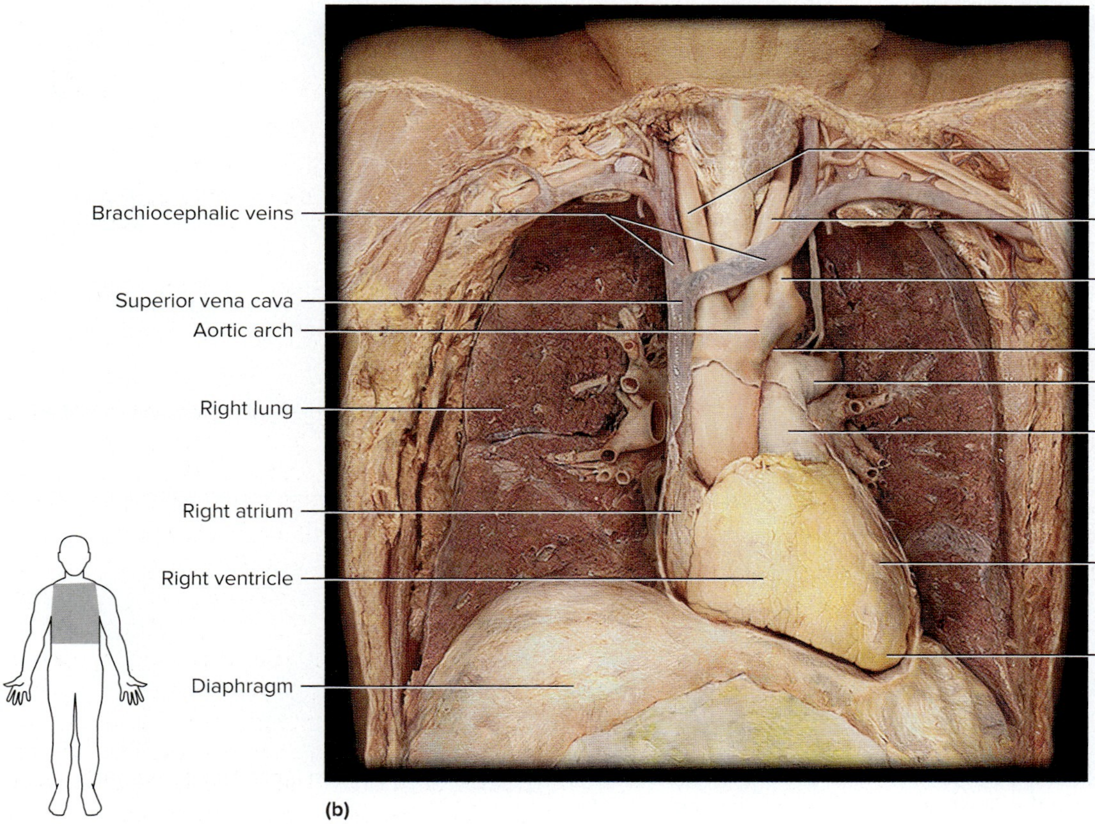

(b)

FIGURE 70.2 (a) Heart and associated arteries and veins of pig's fetal circulation, ventral view. The anterior portion of the left lung has been removed. The apex is reflected to the pig's right to clearly show the arteries. (b) Heart and associated arteries and veins in the thorax of an adult cadaver, anterior view. (a) ©J & J Photography; (b) APR/McGraw Hill APR

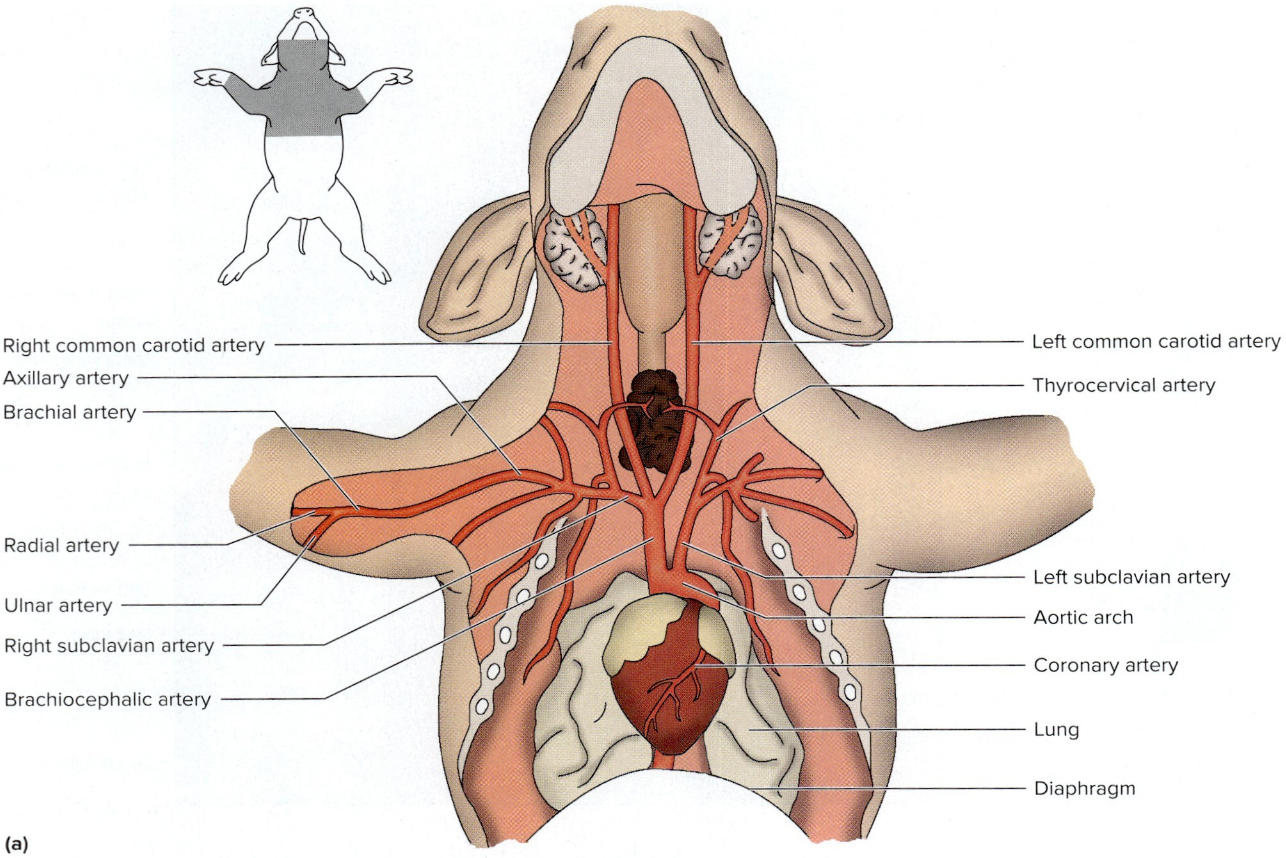

(a)

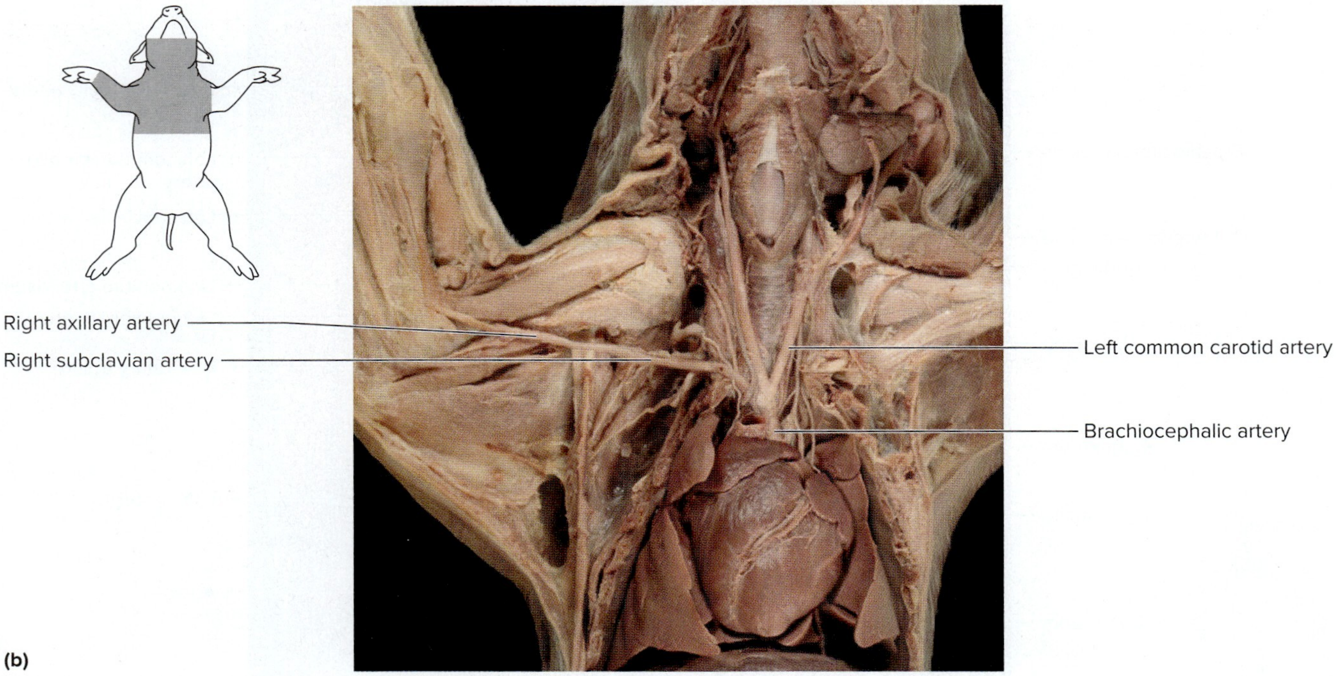

(b)

FIGURE 70.3 Arteries of the trunk, neck, and forelimb of a fetal pig, ventral view: (a) illustration (*note:* the pulmonary arteries are not included in this illustration); (b) dissection image. (b) APR/McGraw Hill

7. Using figure 70.3 as a guide, locate and dissect the following arteries of the thorax and neck:

   aortic arch

   brachiocephalic trunk (artery)—first branch of the aortic arch

   right subclavian artery

   left subclavian artery

   right common carotid artery

   left common carotid artery

8. Trace the right subclavian artery into the forelimb and locate the following arteries:

   axillary artery

   brachial artery

   radial artery

   ulnar artery

9. Open the abdominal cavity. To do this, follow these steps:
   a. Use scissors to make two longitudinal incisions from the pubic bones and continuing around the umbilical cord to meet at the midline. Extend the incision anteriorly to the diaphragm.
   b. Make a lateral incision through the body wall along either side of the inferior border of the diaphragm and along the bases of the thighs.
   c. Reflect the flaps created in the body wall as you would open a book and expose the contents of the abdominal cavity. Sever the *umbilical vein* that extends from the umbilical cord to the liver. Reflect the umbilical cord along with the midventral strip of tissues to expose the two *umbilical arteries* parallel to a central urinary bladder. The umbilical arteries, although injected with red latex, actually transport oxygen-poor fetal blood to the placenta. Flush the abdominal cavity of any coagulated blood or latex that leaked out during injection.
   d. Note the *parietal peritoneum* that forms the inner lining of the abdominal wall. Also note the *visceral peritoneum* that adheres to the surface of organs within the abdominal cavity.

10. As you expose and dissect blood vessels, try not to destroy other visceral organs needed for future studies. At times it will be necessary to displace some abdominal organs to locate some deep blood vessels.

    Using figure 70.4 as a guide, locate and dissect the following arteries of the abdomen:

    abdominal aorta—unpaired

    celiac trunk—unpaired

    anterior mesenteric artery—corresponds to superior mesenteric artery; unpaired

    renal arteries—paired

    posterior mesenteric artery—corresponds to inferior mesenteric artery; unpaired

    external iliac arteries—paired

    internal iliac arteries—paired

    umbilical arteries—paired

11. Trace the external iliac artery into the left hindlimb, and locate the following:

    femoral artery

    deep artery of thigh (deep femoral artery)

12. Review the locations of the heart structures and arteries without the aid of figures.
13. Complete Part A of Laboratory Assessment 70.

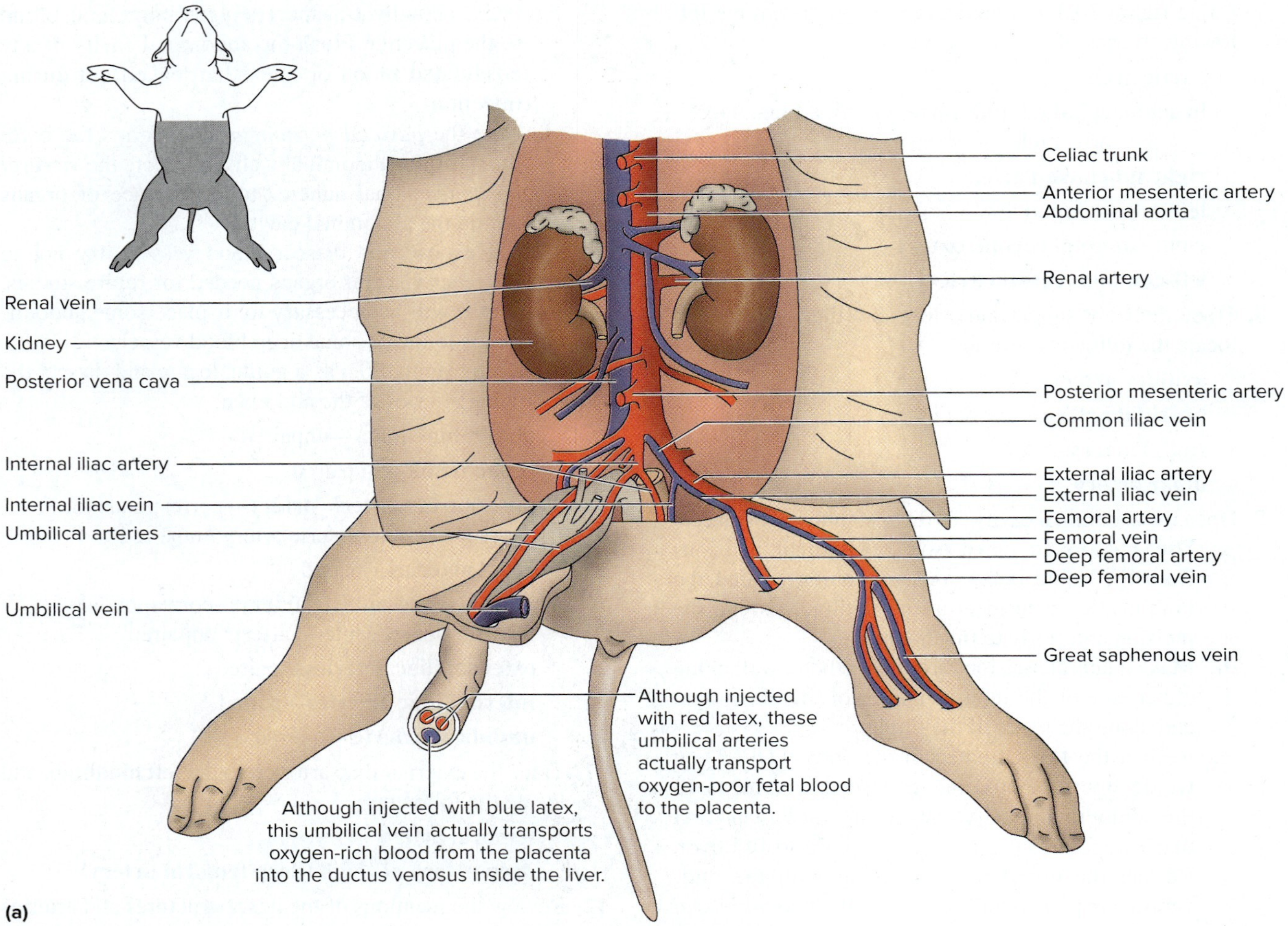

(a)

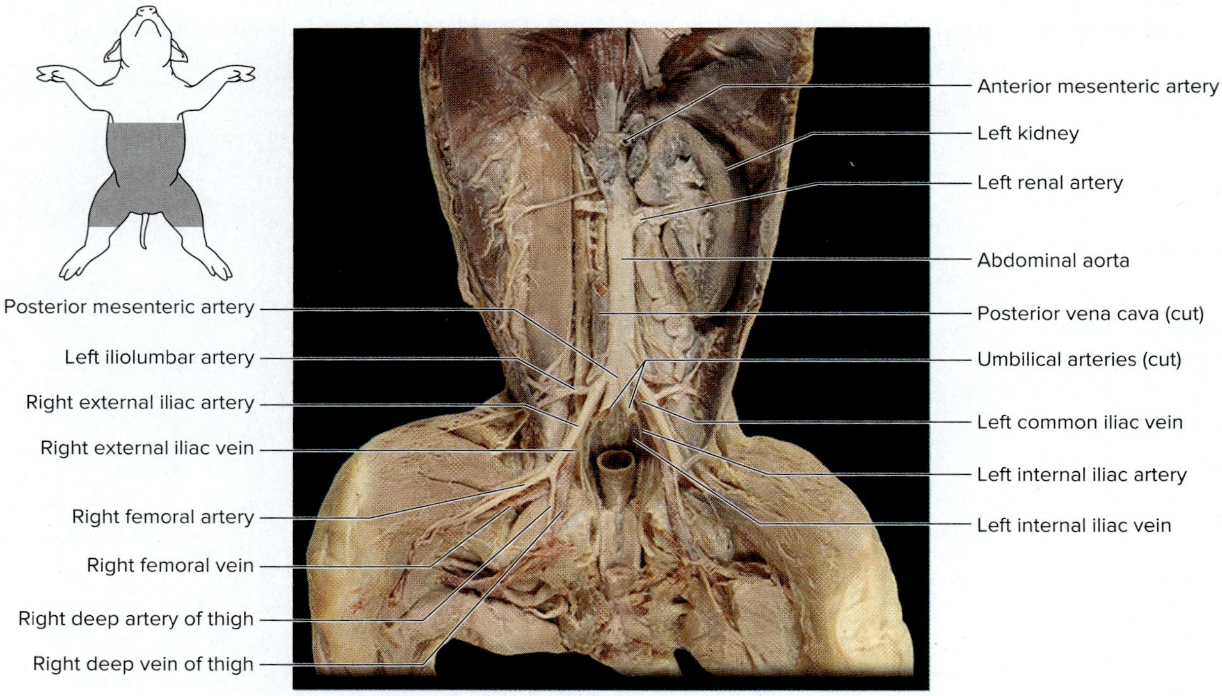

(b)

FIGURE 70.4 Arteries and veins of the abdominal cavity of a fetal pig, ventral view: (a) illustration; (b) dissection image.
(b) APR/McGraw Hill

PROCEDURE B: The Venous System

1. Examine the heart again and locate the following veins:

   anterior vena cava—corresponds to superior vena cava

   posterior vena cava—corresponds to inferior vena cava

2. Using figure 70.5 as a guide, locate and dissect the following veins in the thorax and neck:

   right brachiocephalic vein
   left brachiocephalic vein
   right subclavian vein
   left subclavian vein
   internal jugular vein
   external jugular vein

3. Trace the right subclavian vein into the forelimb and locate the following veins:

   axillary vein
   brachial vein

4. Using figure 70.4 as a guide, locate and dissect the following veins in the abdomen:

   posterior vena cava
   renal veins
   common iliac vein
   internal iliac vein
   external iliac vein

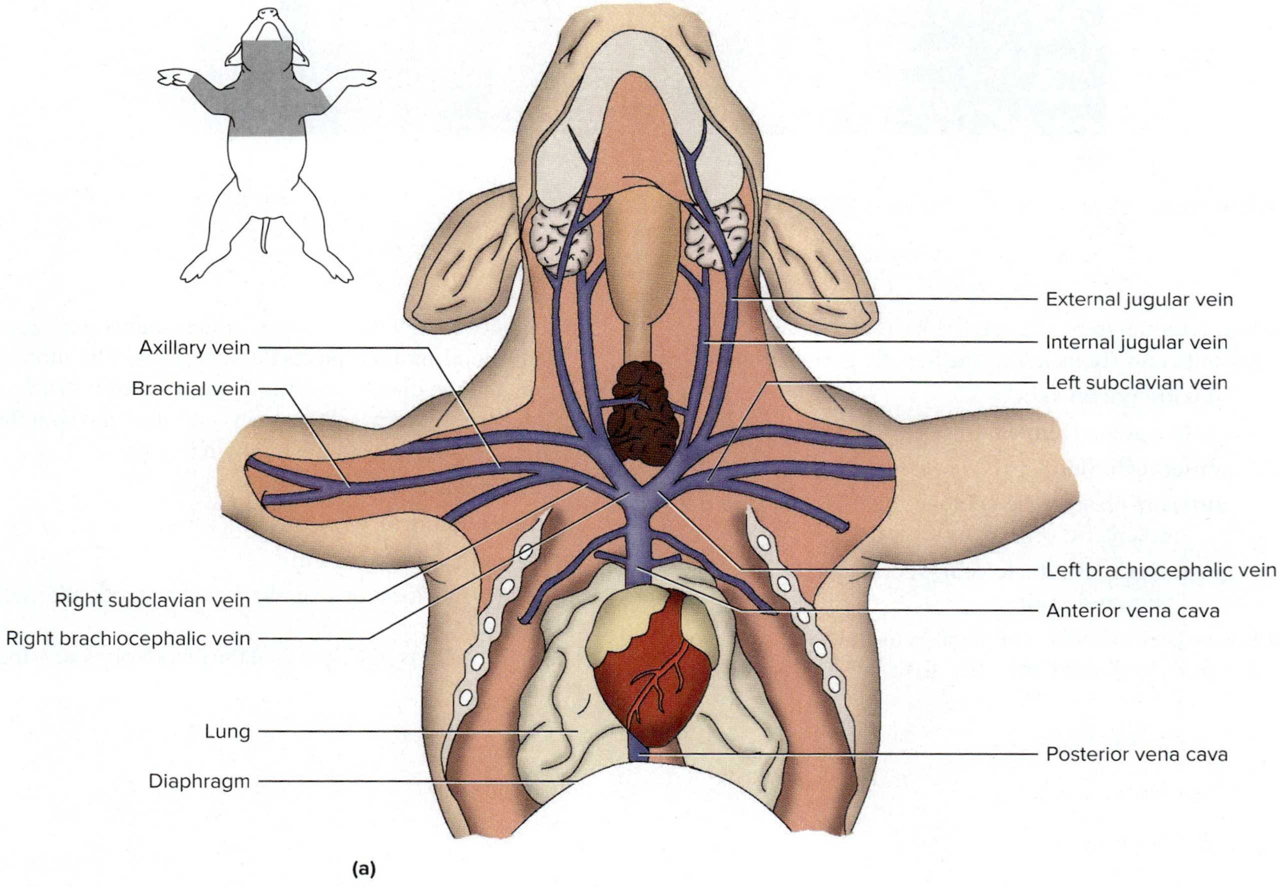

FIGURE 70.5 Veins of the trunk, neck, and forelimb of a fetal pig, ventral view: (a) illustration; (b) dissection image. APR

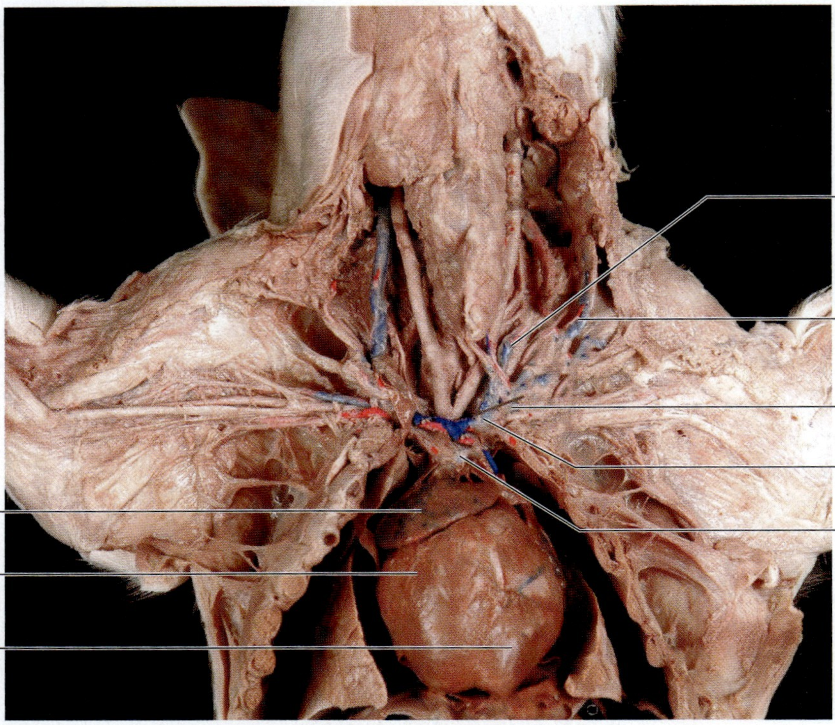

(b)

FIGURE 70.5 *Continued.* Dennis Strete APR

5. Using figure 70.6 as a guide, locate and dissect the following veins (branches of the hepatic portal system):
   - **hepatic portal vein**
   - **gastrosplenic vein**
   - **mesenteric vein**
   - **anterior mesenteric vein**—corresponds to superior mesenteric vein
   - **posterior mesenteric vein**—corresponds to inferior mesenteric vein
6. Locate and trace the umbilical vein (that was cut) from the umbilical cord into the liver. Inside the liver it becomes the *ductus venosus,* which shunts oxygen-rich fetal blood into the posterior vena cava. The umbilical vein and the ductus venosus degenerate after birth.
7. Trace the external iliac vein into the left hindlimb (fig. 70.4) and locate the following veins:
   - **femoral vein**
   - **deep femoral vein**
   - **great saphenous vein**
8. Review the locations of the veins without the aid of figures.
9. Complete Parts B, C, and D of the laboratory assessment.

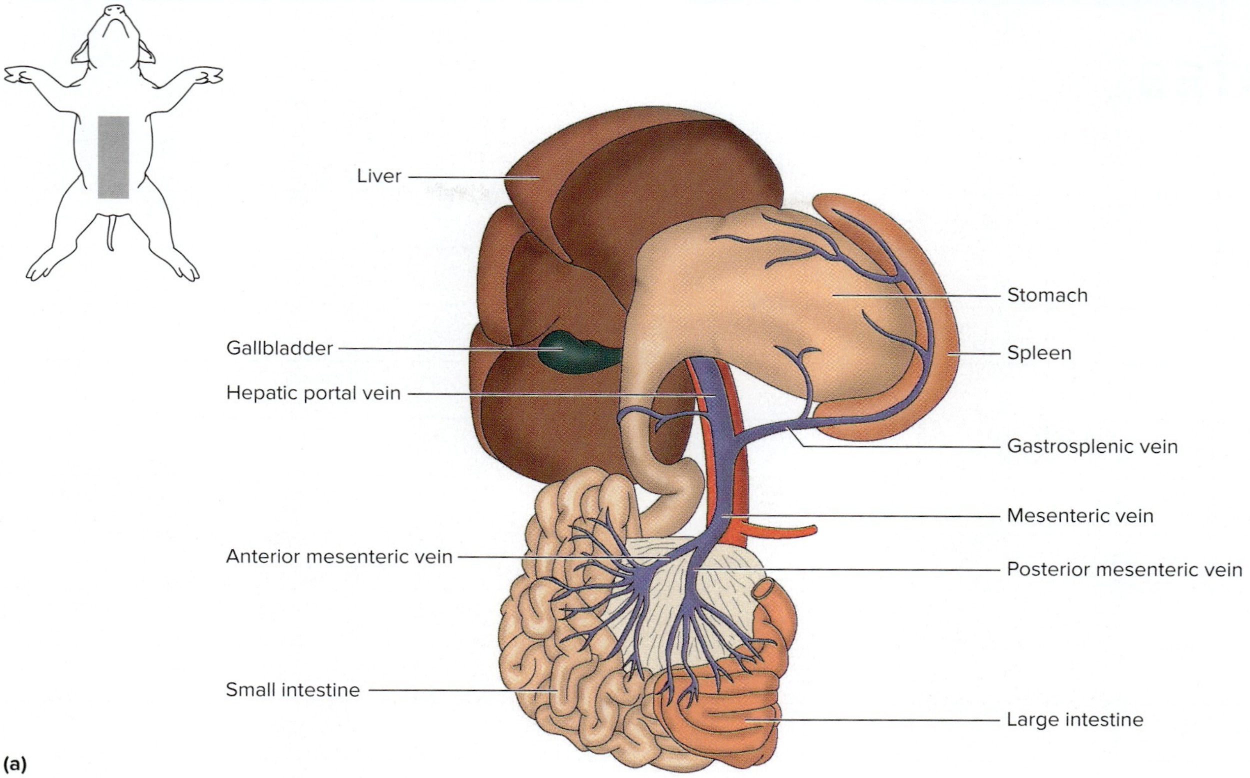

(a)

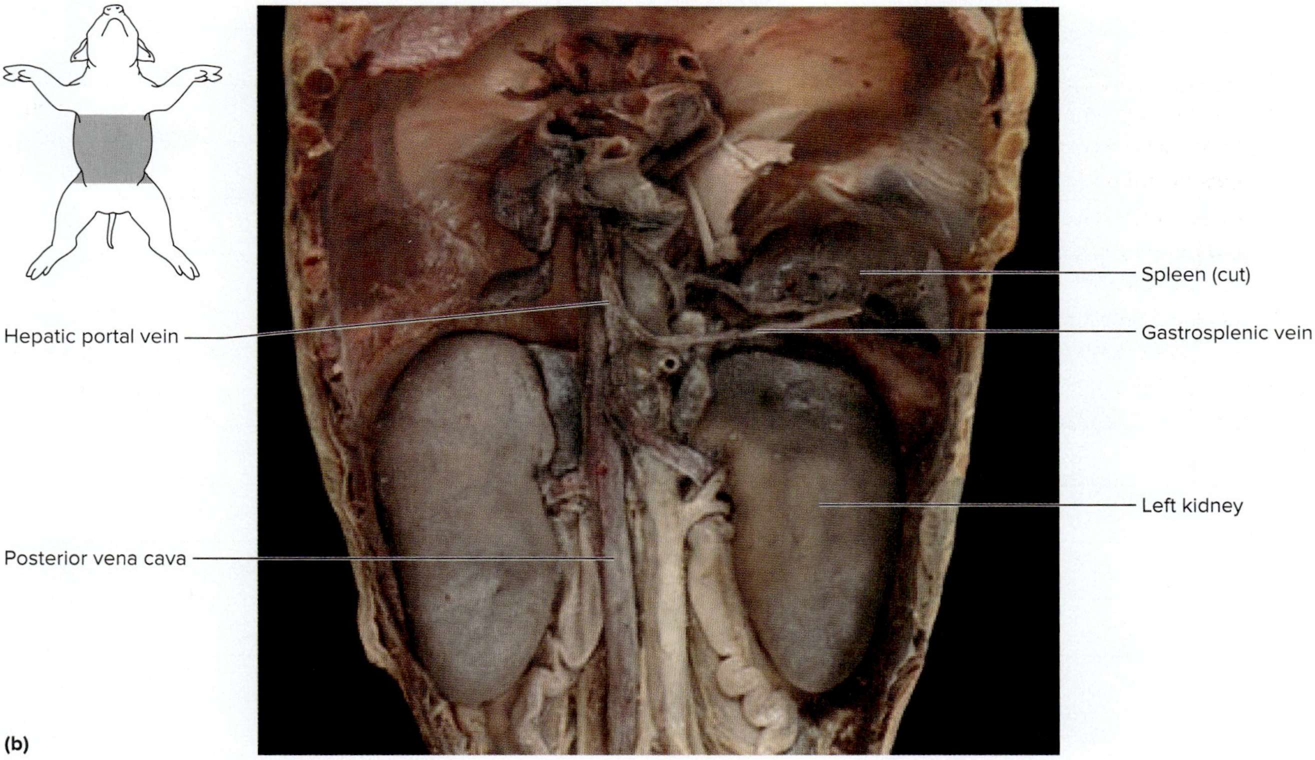

(b)

FIGURE 70.6 Hepatic portal vein and its tributaries of a fetal pig, ventral view: (a) illustration (the liver and stomach are retracted to obtain a clear view); (b) dissection image (liver, gallbladder, stomach, pancreas, and intestines removed).
(b) APR/McGraw Hill

NOTES

Name ______________________

Date ______________________

Section ______________________

The (A) corresponds to the indicated Learning Outcome(s) (O) found at the beginning of the Laboratory Exercise.

LABORATORY ASSESSMENT 70

Fetal Pig Dissection: Cardiovascular System

PART A: Assessments

Complete the following:

1. Describe the relative thicknesses of the walls of the heart chambers of the fetal pig. (A1) ______________________

2. Explain how the wall thicknesses are related to the functions of the chambers. (A1) ______________________

3. Compare the origins of the common carotid arteries of the fetal pig with those of the human. (A3) ______________________

4. Compare the origins of the external and internal iliac arteries of the fetal pig with those of the human. (A3) ______________________

PART B: Assessments

Complete the following:

1. Compare the relative sizes of the external and internal jugular veins of the fetal pig with those of the human. (A3)

2. List twelve veins that pigs and humans have in common. (A3) ______________________

PART C: Assessments

Identify the numbered cardiovascular features of the fetal pig in figures 70.7 and 70.8.

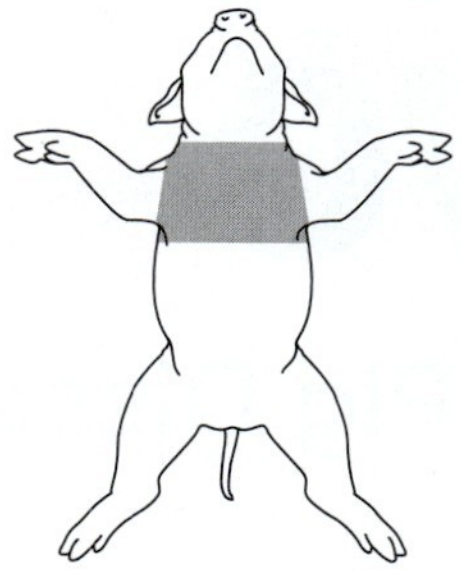

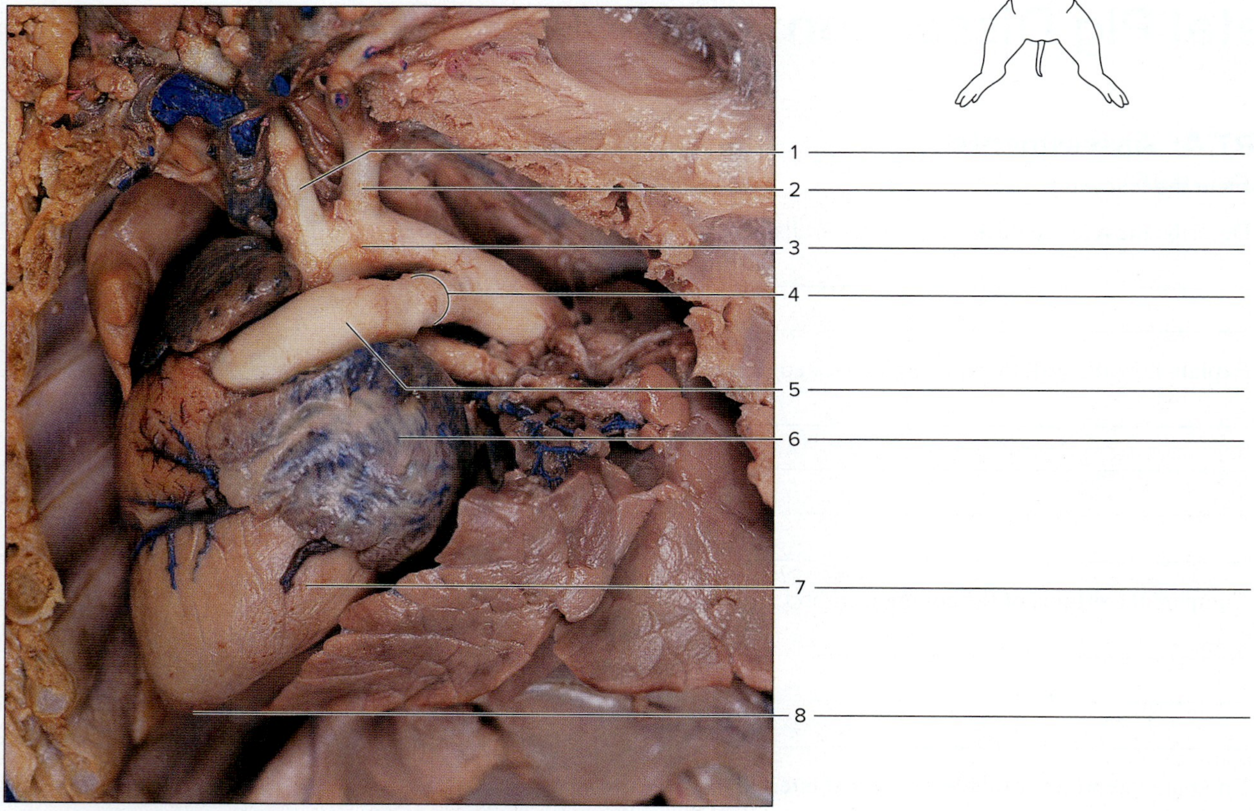

FIGURE 70.7 Label the numbered cardiovascular features of the fetal pig in the thoracic cavity. ©J & J Photography A1 A2

CRITICAL THINKING ASSESSMENT—PART C

Explain the oxygen-rich blood in the umbilical veins of a human and a pig fetus. A2

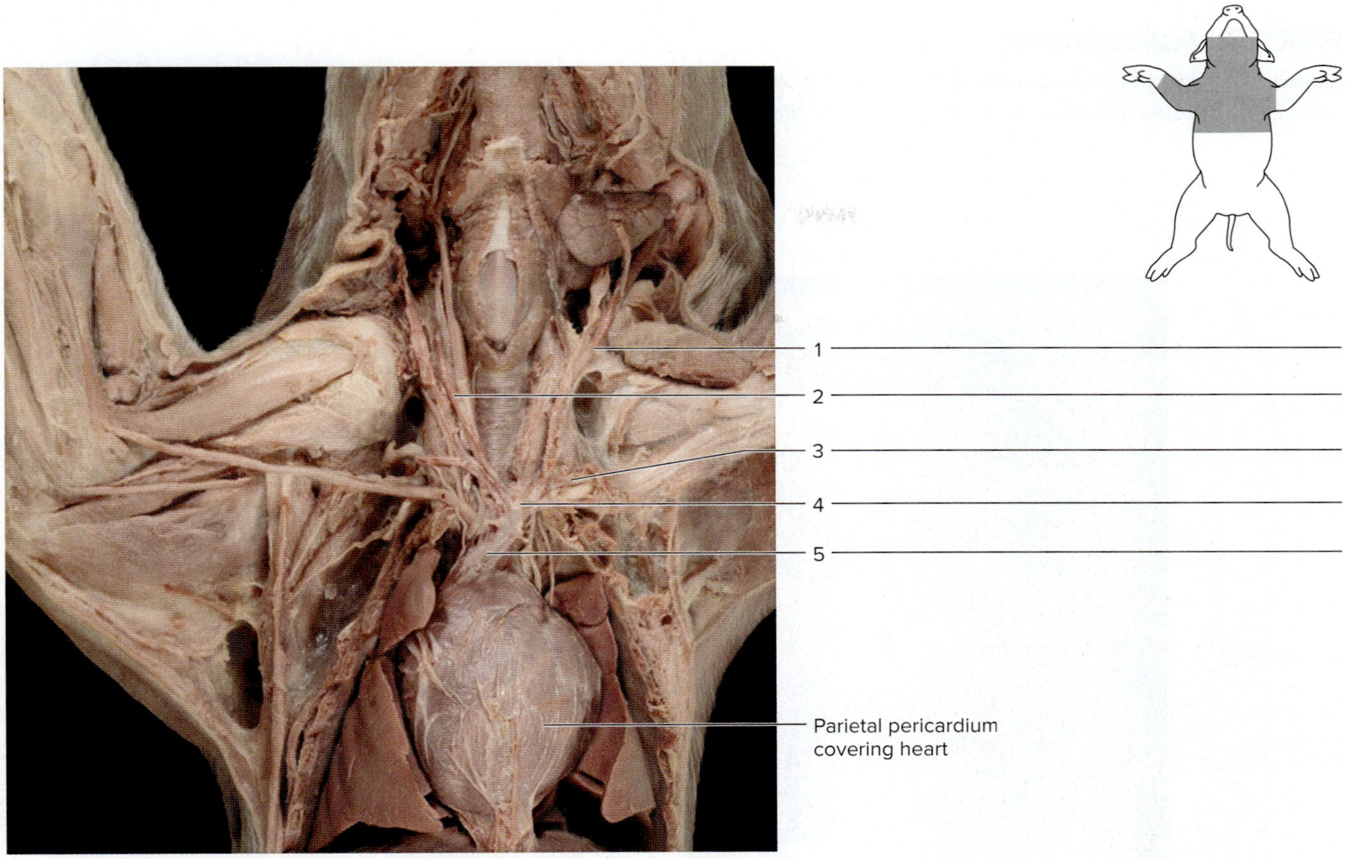

(a)

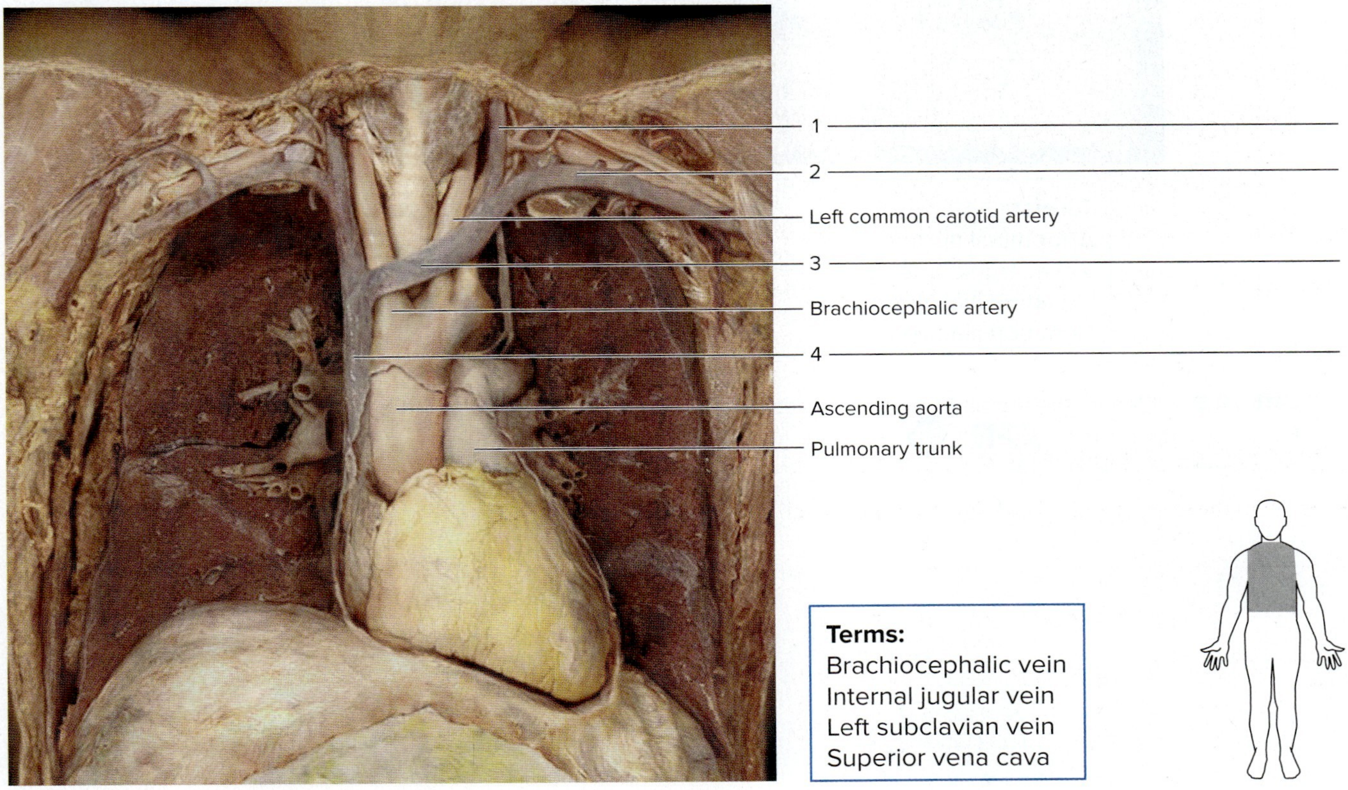

(b)

FIGURE 70.8 Identify the veins on the ventral (anterior) views of the thorax of the (a) fetal pig, and (b) human (cadaver). Use the terms provided for (b) only. APR/McGraw Hill APR A3

PART D: Assessments

Observe the human torso model and figures 70.2*b*, 70.9, and 70.10*b* of a cadaver. Locate the labeled features and identify the numbered features that were also identified in the fetal pig dissection.

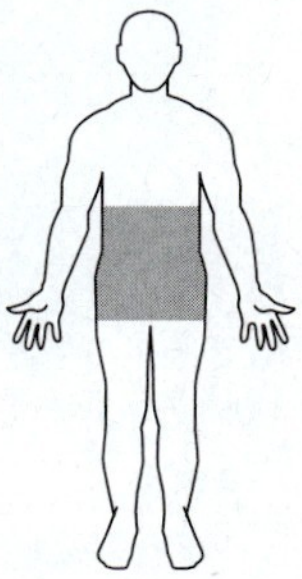

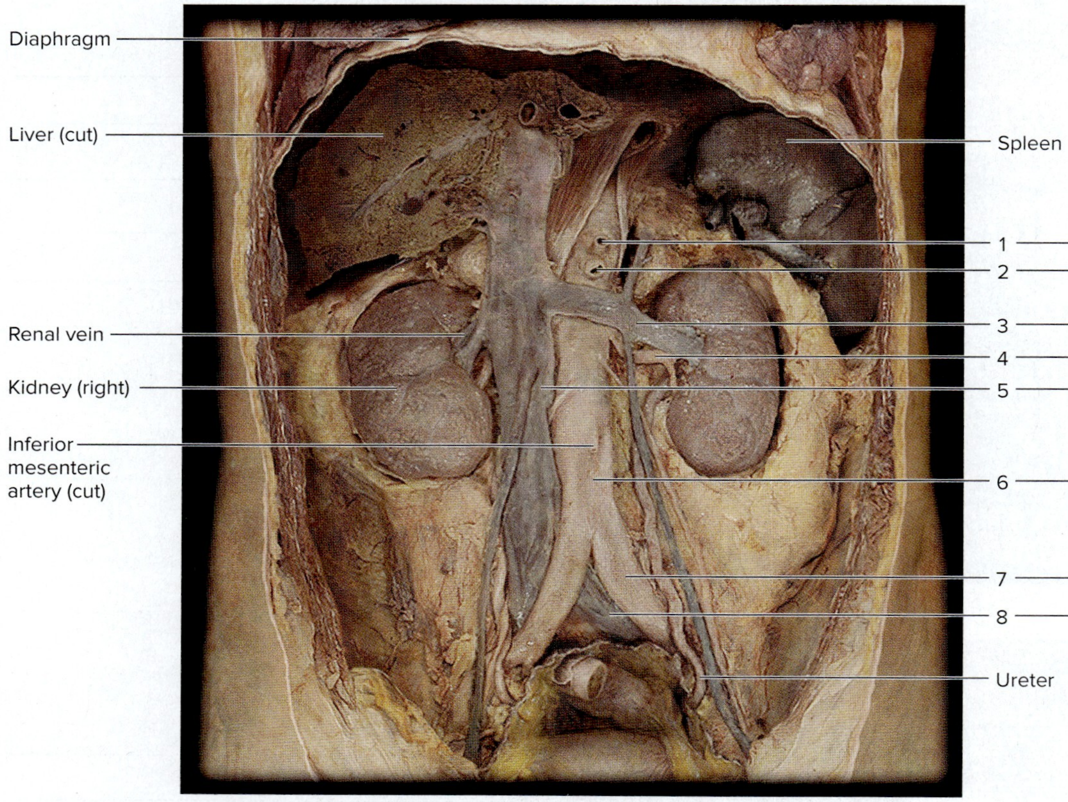

Terms:

Abdominal aorta	Inferior vena cava
Celiac trunk (artery)	Renal artery
Common iliac artery	Renal vein
Common iliac vein	Superior mesenteric artery

FIGURE 70.9 Identify the arteries and veins indicated on this anterior view of the abdomen of a cadaver, using the terms provided. APR/McGraw Hill APR A3

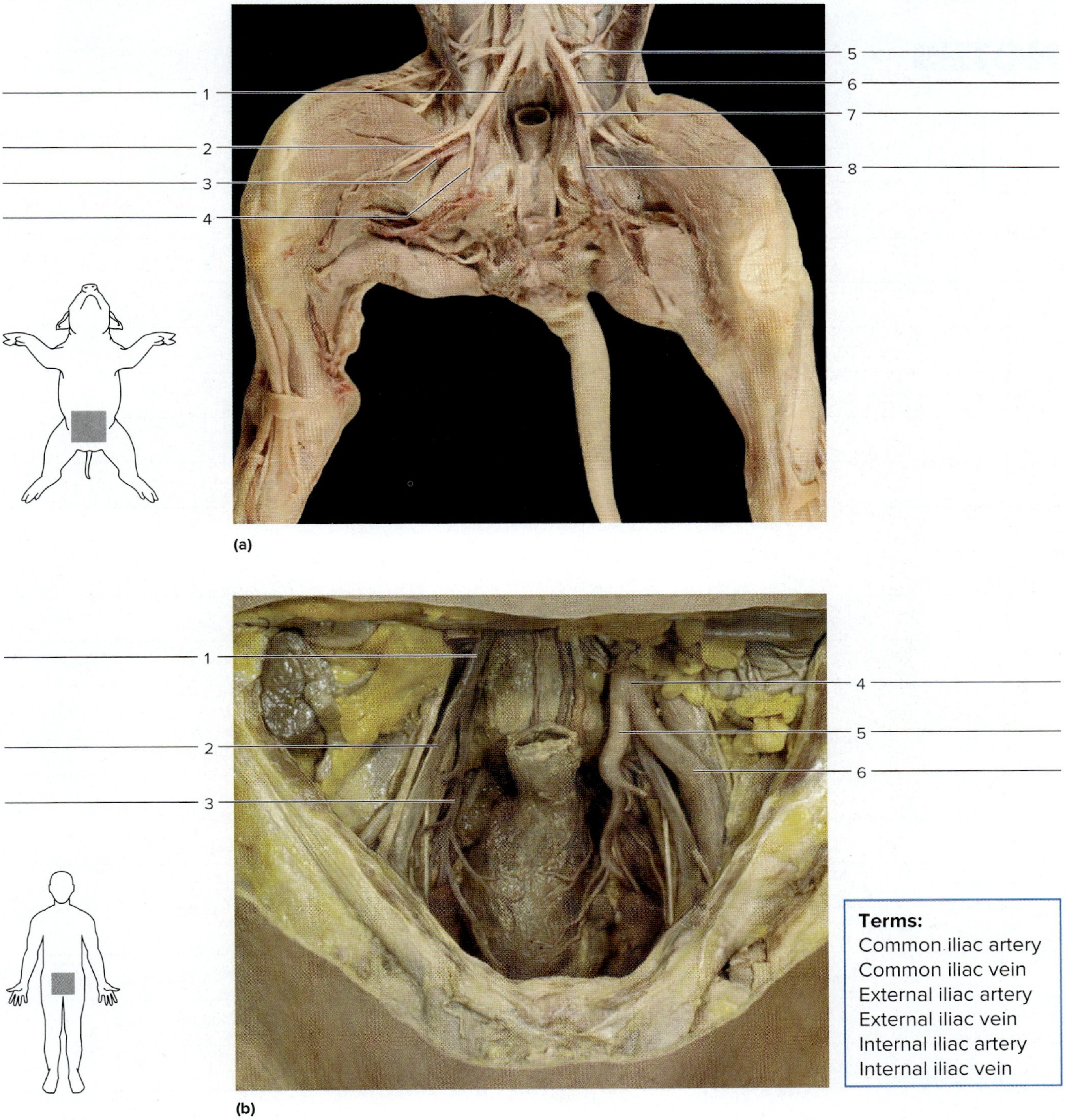

FIGURE 70.10 Identify the arteries and veins on the ventral (anterior) views of the pelvis of the (a) fetal pig, and (b) human (cadaver). Use the terms provided for (b) only. APR/McGraw Hill APR A3

NOTES

LABORATORY
EXERCISE 71

Fetal Pig Dissection: Respiratory System

Purpose of the Exercise

To examine and identify the major respiratory organs of the fetal pig and to compare these organs with those of the human.

MATERIALS NEEDED

Preserved fetal pig
Dissecting tray
Dissecting instruments
Bone cutter
Beaker of water
Disposable gloves
Human torso model

SAFETY

- Review all the safety guidelines in Appendix 1 of your laboratory manual.
- Wear disposable gloves and protective eyewear when working on the fetal pig dissection.
- Dispose of tissue remnants and gloves as instructed.
- Wash the dissecting tray and instruments as instructed.
- Wash your laboratory table.
- Wash your hands before leaving the laboratory.

Learning Outcomes APR

After completing this exercise, you should be able to:

O1 Locate and identify the major respiratory organs of the fetal pig.
O2 Compare the respiratory system of the human with that of the fetal pig.
O3 Identify the corresponding organs in the human torso model.

The O corresponds to the assessments A indicated in the Laboratory Assessment for this Exercise.

Pre-Lab

Carefully read the introductory material and examine the entire lab. Be familiar with the structures and functions of the human respiratory system from lecture, the textbook, or laboratory exercises.

In this laboratory exercise, you will dissect the major respiratory organs of a preserved fetal pig. As you observe these structures in the fetal pig, compare them with those of the human torso model.

PROCEDURE: Respiratory System Dissection

1. Place the fetal pig in a dissecting tray with its ventral surface up.
2. Examine the *external nares (nostrils)* located on the flat *rostrum (snout)*.
3. Open the oral cavity. To do this, follow these steps:
    a. Use scissors to cut through the soft tissues at the angle of the mouth.
    b. When you reach the bone of the jaw, use a bone cutter to cut through the bone, thus freeing the mandible.
4. Open the oral cavity wide to expose the *hard palate* and the *soft palate* (fig. 71.1). Cut through the tissues of the soft palate and observe the *nasopharynx* above it (fig 71.2). Locate the small openings of the *auditory tubes* in the lateral walls of the nasopharynx. Insert a probe into an opening of an auditory tube. Examine the *oropharynx* located near the base of the tongue.
5. Pull the tongue posteriorly and locate the *epiglottis* near its base (fig. 71.2). Also identify the *glottis,* an opening between the *vocal folds (vocal cords)* into the larynx. Locate the *esophagus,* which is dorsal to the larynx.
6. Open the thoracic cavity and expose its contents. Examine the organs located in the *mediastinum.* The mediastinum separates the right and left lungs and pleural cavities within the thorax.
7. Dissect the *trachea* in the neck and expose the *larynx* at the anterior end near the base of the tongue. Note the *tracheal rings* and locate the *thyroid gland* on the trachea, just posterior to the larynx. Locate a small *parathyroid gland* attached to the dorsal surface of the thyroid gland. Also note the *thymus gland,* extending along each side of the trachea into the thorax (fig. 71.3).
8. Examine the larynx by removing any attached muscles and identify the *thyroid cartilage* and *cricoid cartilage* in its wall (fig. 71.4*a*). Make a longitudinal incision through the ventral wall of the larynx and locate the vocal folds that appear as whitish folds inside. Compare the structures of the larynx of the fetal pig with those of the human (fig. 71.4*b*).

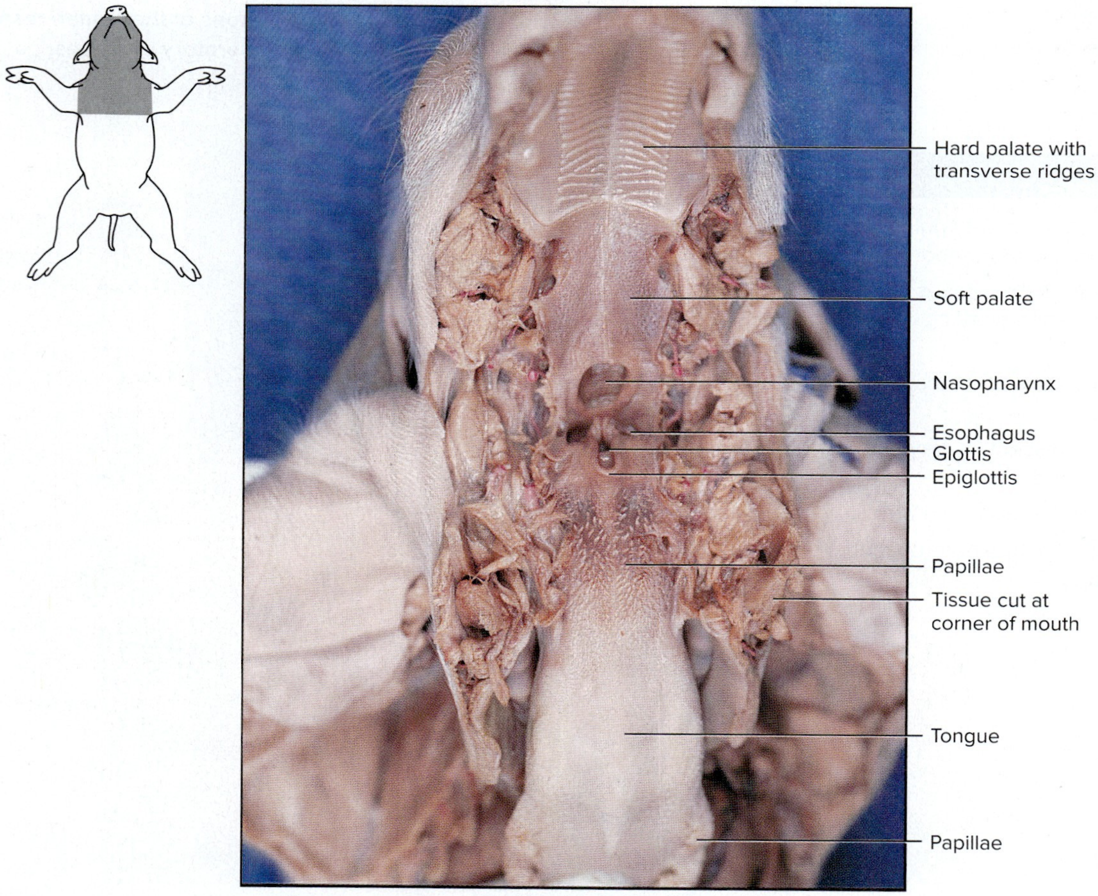

FIGURE 71.1 Oral cavity of the fetal pig with tongue pulled posteriorly, ventral view. ©J & J Photography

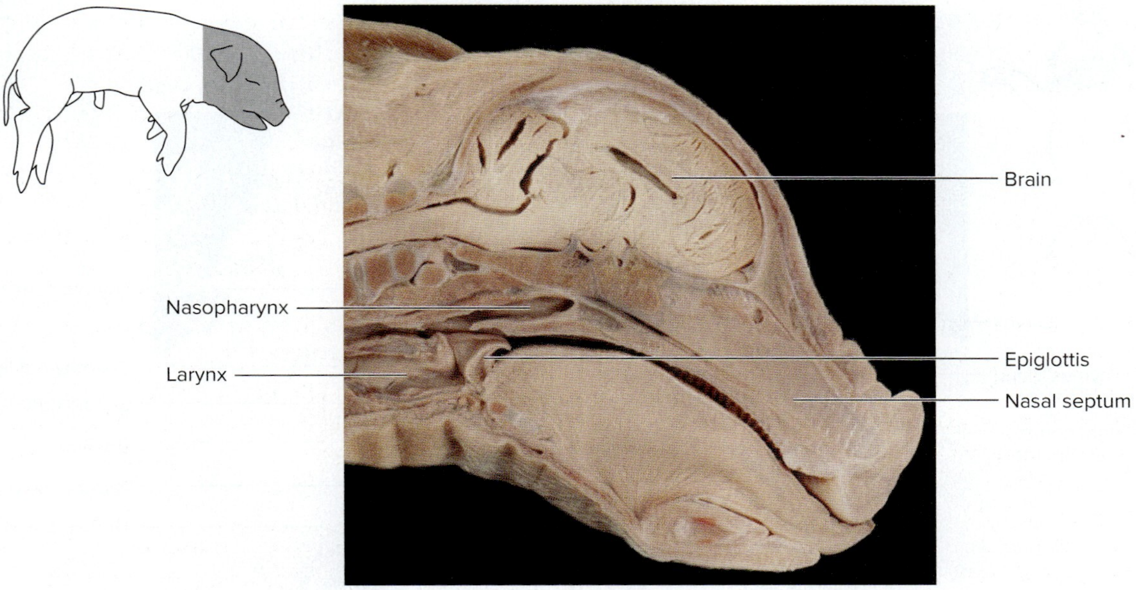

FIGURE 71.2 Sagittal view of upper respiratory structures of the pig. APR/McGraw Hill APR

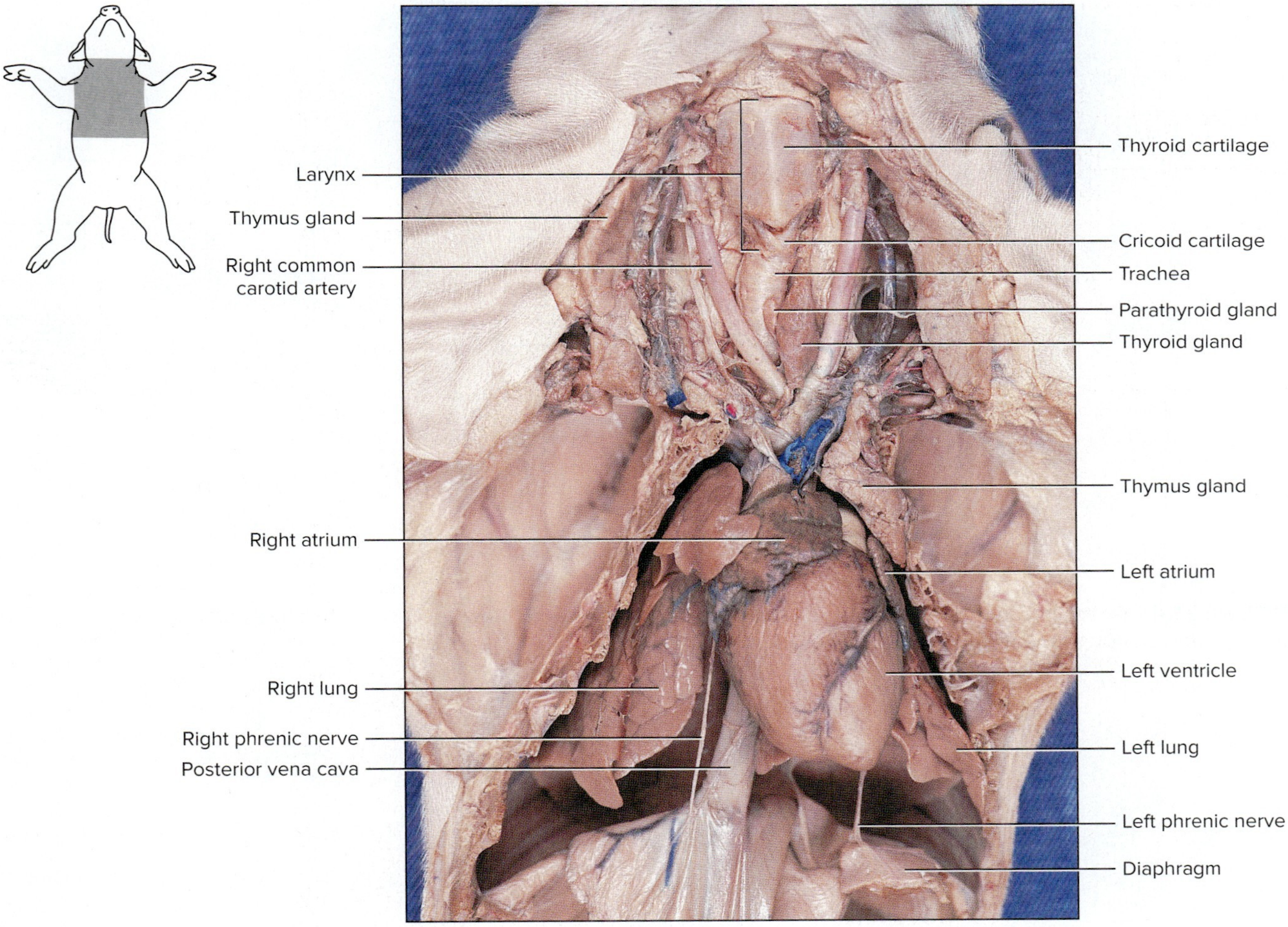

FIGURE 71.3 Respiratory organs and glands of the fetal pig, ventral view. ©J & J Photography APR

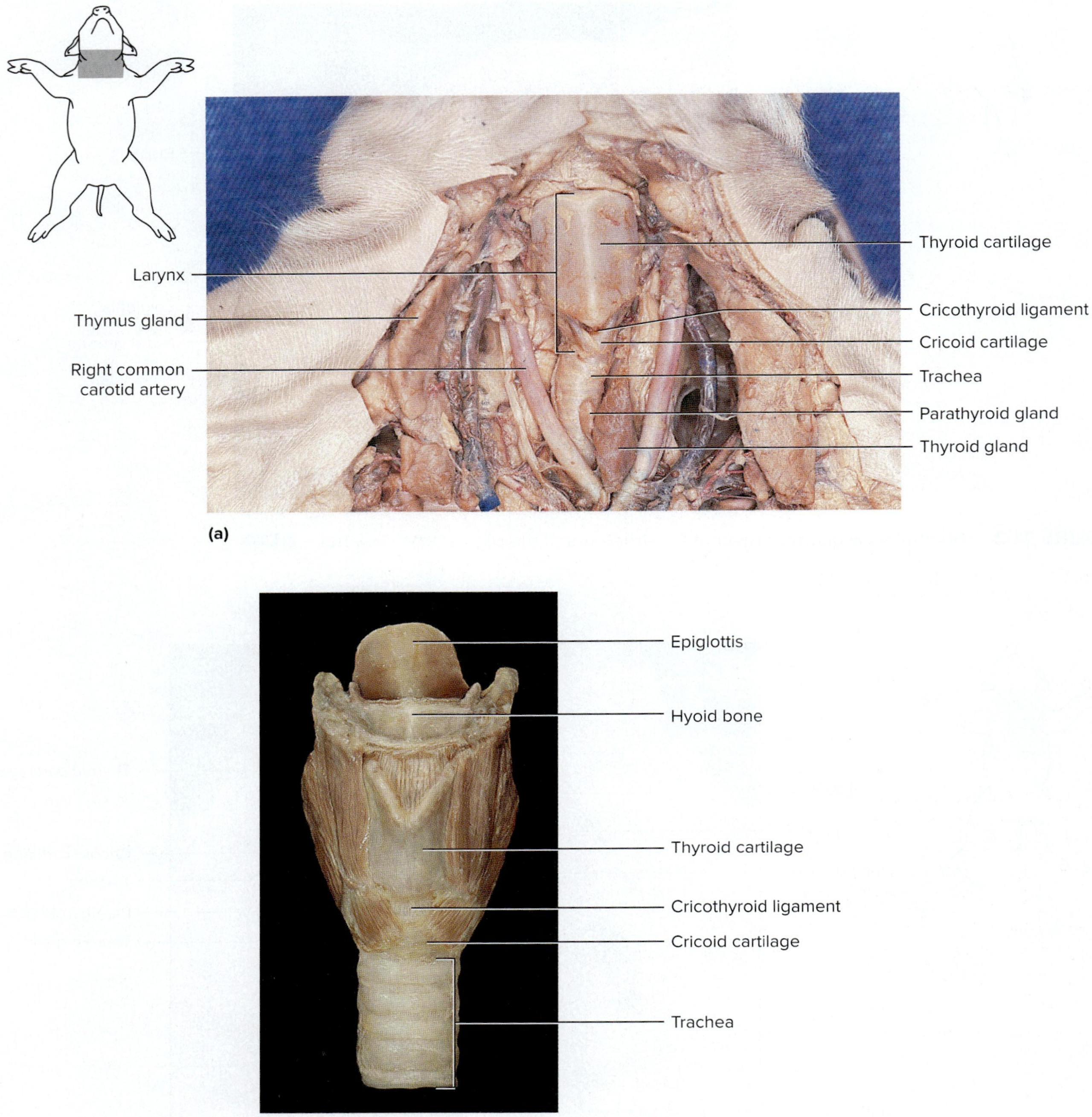

FIGURE 71.4 (a) Larynx and associated structures of the fetal pig, ventral view. (b) Larynx structures of a cadaver, anterior view. (a) ©J & J Photography; (b) APR/McGraw Hill APR

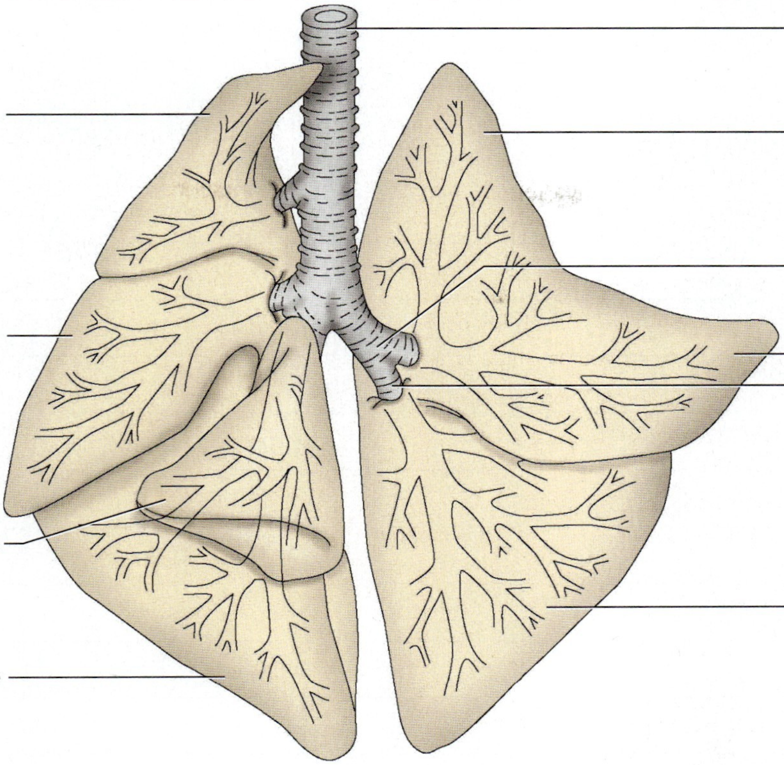

FIGURE 71.5 Lungs and respiratory tubes of the fetal pig, ventral view. APR

9. Remove the heart and trace the trachea posteriorly to where it divides into *bronchi,* which pass into the lungs (fig. 71.5).
10. Examine the *lungs,* each of which is subdivided into lobes—an *anterior* (*cranial*), a *middle,* and a *posterior* (*caudal*) *lobe.* The right lung has an additional deep *accessory lobe.* Notice the thin membrane, the *visceral pleura,* on the surface of each lung. Also notice the *parietal pleura,* which forms the inner lining of the thoracic wall, and locate the spaces of the *pleural cavities.*
11. Make an incision through a lobe of a lung and examine its interior. Note the branches of the smaller air passages. You can see the branches of the bronchial tree if you follow a main (primary) bronchus into the lung and gently scrape away lung tissue with a scalpel.
12. Examine the *diaphragm* and locate the *phrenic nerve.* This nerve appears as a white thread passing along the side of the heart to the diaphragm (fig. 71.3).
13. Remove a small piece of the lung and notice the solid texture of the tissue. Place the piece of lung into a beaker of water. Does the fetal lung tissue float or sink?
14. Review the locations of the respiratory organs without the aid of figures.
15. Complete Parts A, B, and C of Laboratory Assessment 71.

CRITICAL THINKING ACTIVITY

Under what conditions would a piece of lung tissue sink?

___

___

___

Under what conditions would it float?

___

___

___

NOTES

LABORATORY
ASSESSMENT 71

Name ______________________________

Date ______________________________

Section ______________________________

The (A) corresponds to the indicated Learning Outcome(s) (O) found at the beginning of the Laboratory Exercise.

Fetal Pig Dissection: Respiratory System

PART A: Assessments

Complete the following:

1. What is the purpose of the auditory tubes opening into the nasopharynx? Explain why this may lead to plugged ears when you have a cold. (A1) ______________________________

____

____

____

____

2. Distinguish between the glottis and the epiglottis. (A1) ______________________________

____

____

____

3. Are the tracheal rings of the fetal pig complete or incomplete circles? (A1) ______________ How does this feature compare with that of the human? (A2) ______________________________

____

____

4. Compare the number of lobes in the human lungs with the number of lobes in the fetal pig lungs. (A2) ______________

____

____

____

____

5. Describe the attachments of the diaphragm. (A1) ______________________________

____

____

____

6. What major structures are located within the mediastinum? (A1) ______________________________

____

How does this compare to the human mediastinum? (A2) ______________________________

____

____

PART B: Assessments

Identify the numbered respiratory features of the fetal pig in figure 71.6.

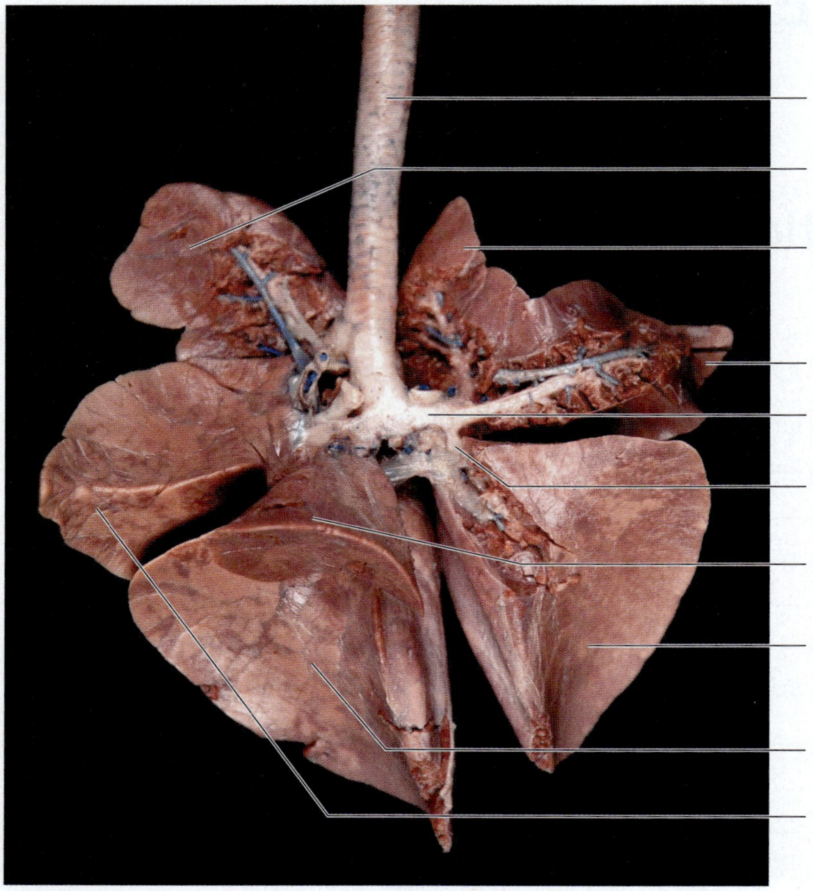

1 ______________________

2 ______________________

3 ______________________

4 ______________________

5 ______________________

6 ______________________

7 ______________________

8 ______________________

9 ______________________

10 ______________________

FIGURE 71.6 Identify the lungs and respiratory tubes of the fetal pig, ventral view. Dennis Strete A1

PART C: Assessments

Observe the human torso model and figures 71.4*b* and 71.7 of a cadaver. Locate the labeled features and identify the numbered features in figure 71.7 that were also identified in the fetal pig dissection.

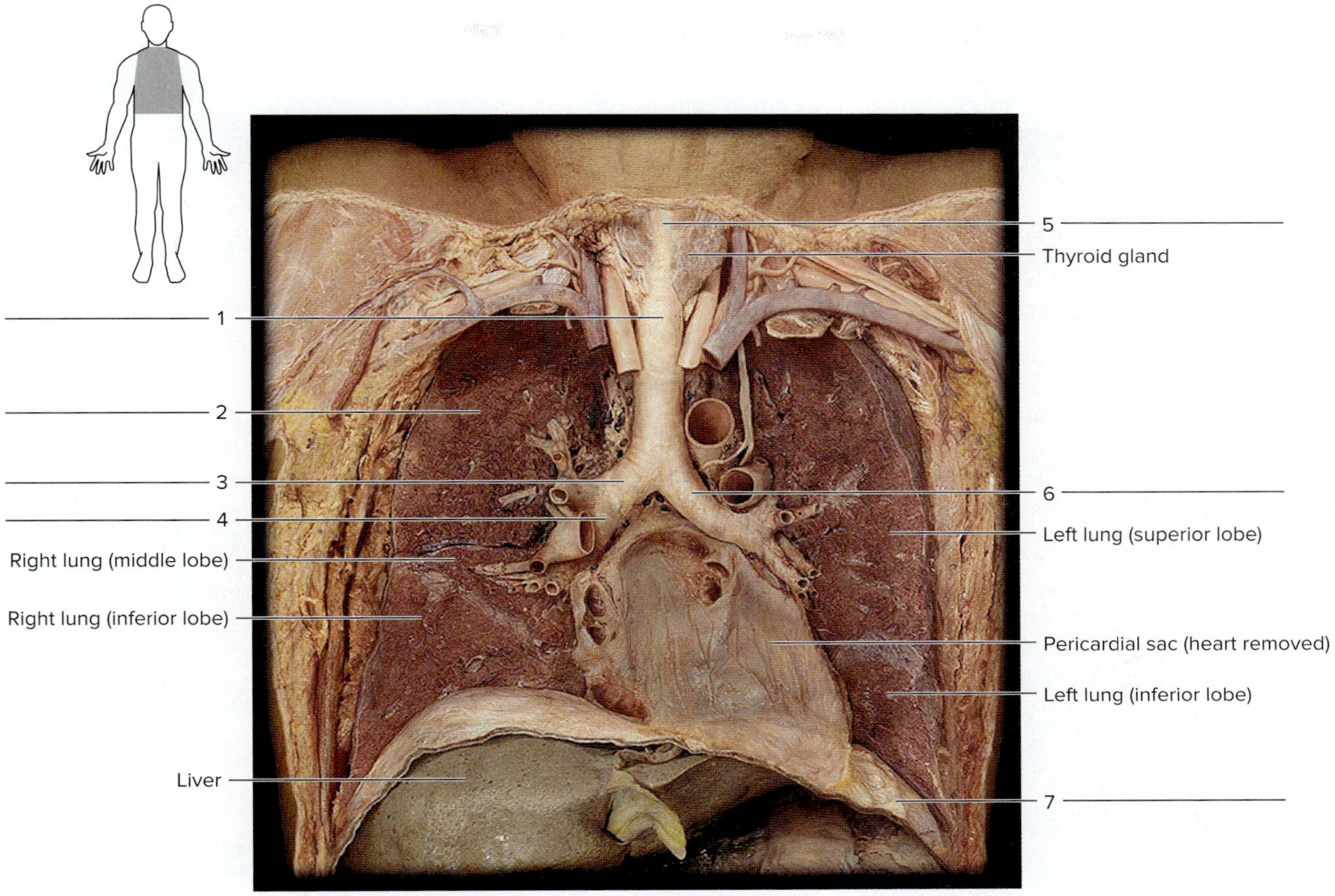

Terms:

- Cricoid cartilage
- Diaphragm
- Left main bronchus
- Lobar (secondary) bronchus
- Right lung (superior lobe)
- Right main bronchus
- Trachea

FIGURE 71.7 Identify the respiratory features indicated on this anterior view of the thorax, using the terms provided.
APR/McGraw Hill A2 A3 APR

NOTES

LABORATORY
EXERCISE 72

Fetal Pig Dissection: Digestive System

Purpose of the Exercise

To examine and identify the major digestive organs of the fetal pig and to compare these organs with those of the human.

MATERIALS NEEDED

Preserved fetal pig
Dissecting tray
Dissecting instruments
Bone cutter
Disposable gloves
Hand lens
Human torso model

SAFETY

- Review all the safety guidelines in Appendix 1 of your laboratory manual.
- Wear disposable gloves and protective eyewear when working on the fetal pig dissection.
- Dispose of tissue remnants and gloves as instructed.
- Wash the dissecting tray and instruments as instructed.
- Wash your laboratory table.
- Wash your hands before leaving the laboratory.

Learning Outcomes APR

After completing this exercise, you should be able to:

01 Locate and identify the major digestive organs of the fetal pig.
02 Compare the digestive system of the fetal pig with that of the human.
03 Identify the corresponding organs in the human torso model.

The O corresponds to the assessments A indicated in the Laboratory Assessment for this Exercise.

Pre-Lab

Carefully read the introductory material and examine the entire lab. Be familiar with the structures and functions of the human digestive system from lecture, the textbook, or laboratory exercises.

In this laboratory exercise, you will dissect the major digestive organs of a fetal pig. As you observe these organs, compare them with those of the human by observing the parts of the human torso model. Because humans are adapted to eating and digesting a greater variety of foods, comparisons of the pig and the human digestive organs may not be precise.

PROCEDURE: Digestive System Dissection

1. Place the fetal pig in the dissecting tray on its left side.
2. Locate the major salivary glands on one side of the head (fig. 72.1*a*). To do this, follow these steps:
   a. Clear away any remaining fascia and other connective tissue from the region below the ear and near the joint of the mandible.

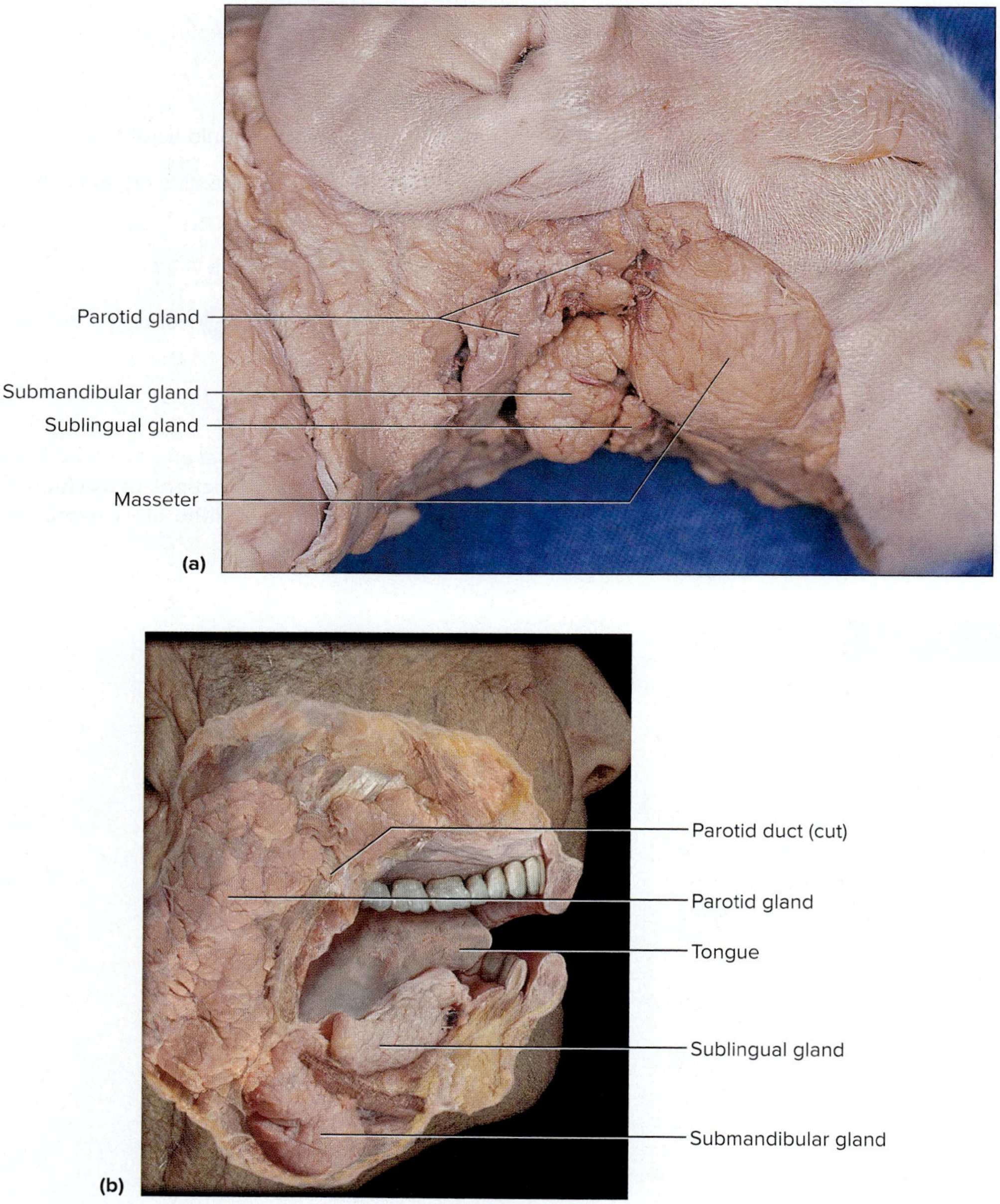

FIGURE 72.1 Salivary glands: (a) the fetal pig, lateral view; (b) the cadaver, lateral view. (a) ©J & J Photography; (b) APR/McGraw Hill APR

b. Identify the *parotid gland,* a relatively large, triangular mass of glandular tissue just below the ear. Although this gland covers a large area, it is poorly developed in the fetal stage of development.
c. Look for the compact *submandibular gland* just below the parotid gland, near the angle of the jaw.
d. Locate the small *sublingual gland* that is adjacent, anterior, and medial to the submandibular gland.
e. Compare the locations of the salivary glands of the fetal pig with those of a human (fig. 72.1*b*).

3. Open the oral cavity (fig. 72.2). To do this, follow these steps:
   a. Use scissors to cut through the soft tissues at the angle of the mouth if not previously done.
   b. When you reach the bone of the jaw, use a bone cutter to cut through the bone, thus freeing the mandible (fig. 72.2*a*).

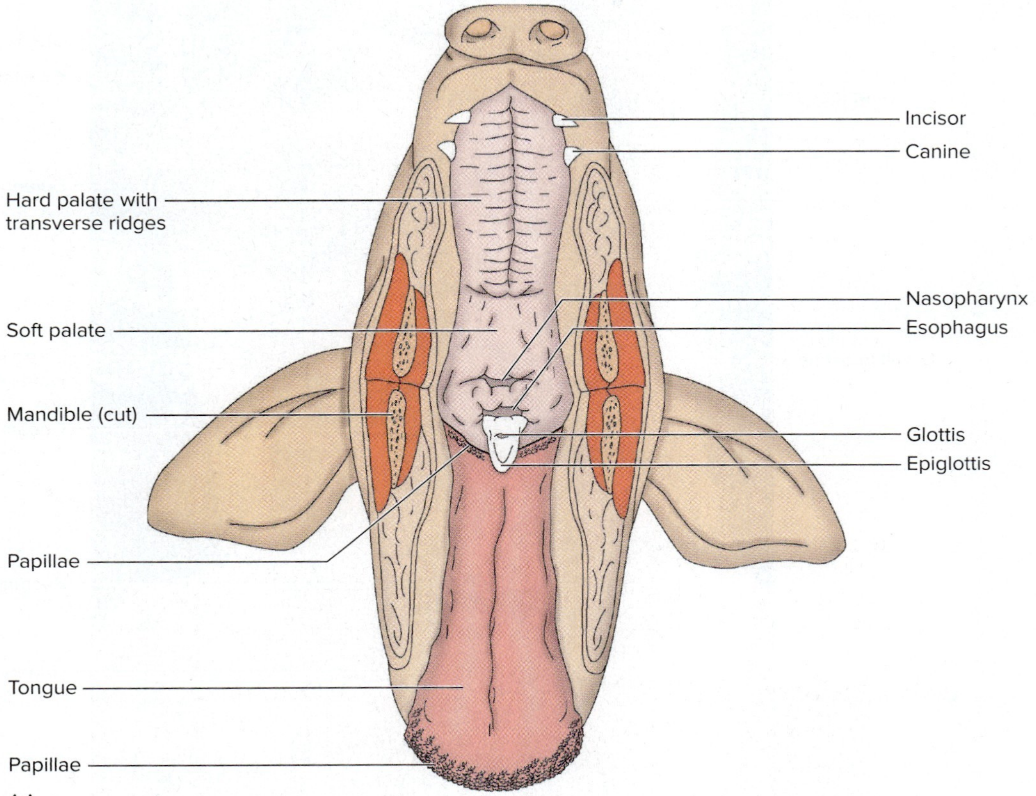

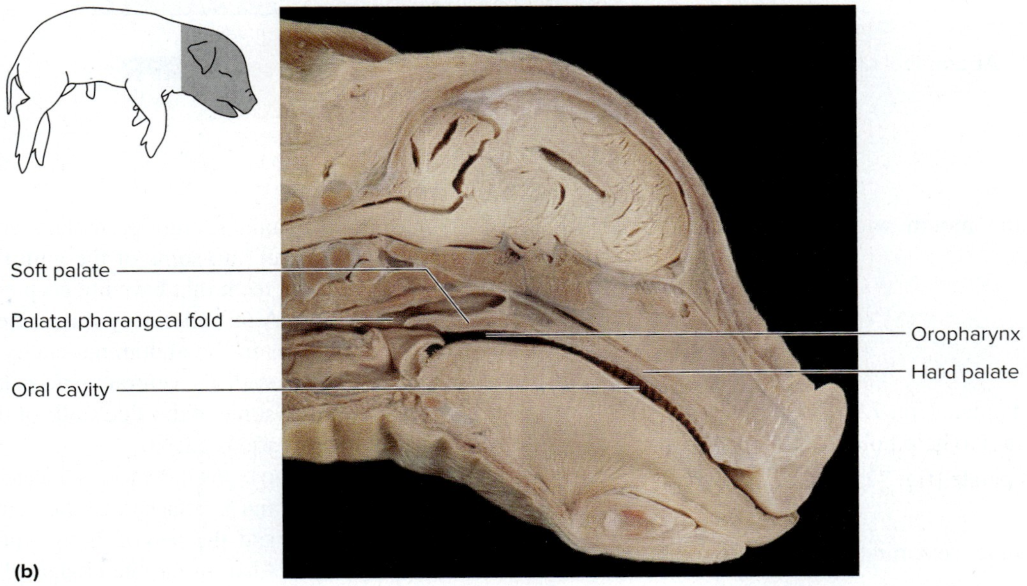

FIGURE 72.2 Oral cavity of the fetal pig: (a) illustration with lower jaw and tongue retracted, ventral view; (b) dissection photograph, sagittal view. (b) APR/McGraw Hill

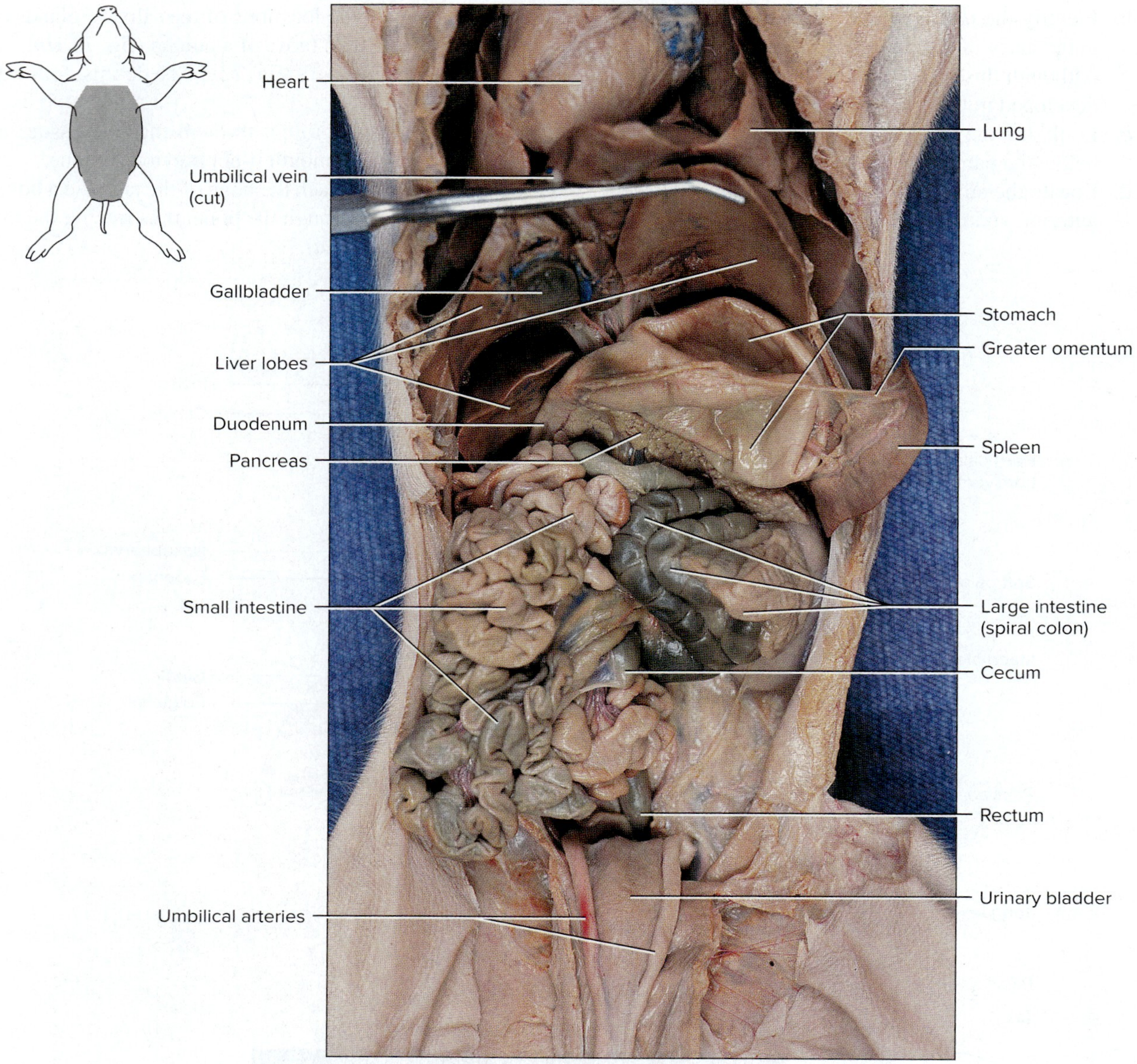

FIGURE 72.3 Abdominal digestive organs of the fetal pig, ventral view. ©J & J Photography APR

c. Open the mouth wide and locate the following features:

cheek

lip

palate

- hard palate (fig. 72.2)—has transverse ridges (rugae) to help hold food
- soft palate (fig. 72.2)—lacks a uvula

tongue

- papillae—examine with a hand lens

4. Examine any erupted teeth of the maxilla. The primary (deciduous) teeth of a young pig would include six incisors, two canines, eight premolars, and zero molars on each jaw. Cut into some of the gum tissue to locate any developing teeth that have not erupted.
5. Complete Part A of Laboratory Assessment 72.
6. Examine organs in the abdominal cavity with the fetal pig positioned with its ventral surface up. You might wish to remove some of the side walls of the body cavity to make observations easier.
7. Examine the large *liver,* which is located just beneath the diaphragm and is attached to the central portion of the diaphragm and the ventral body wall by the *falciform ligament.* Also, locate the elongated *spleen,* which is lateral and ventral to the stomach on the left side (fig. 72.3). Locate the five lobes of the liver. A *greater*

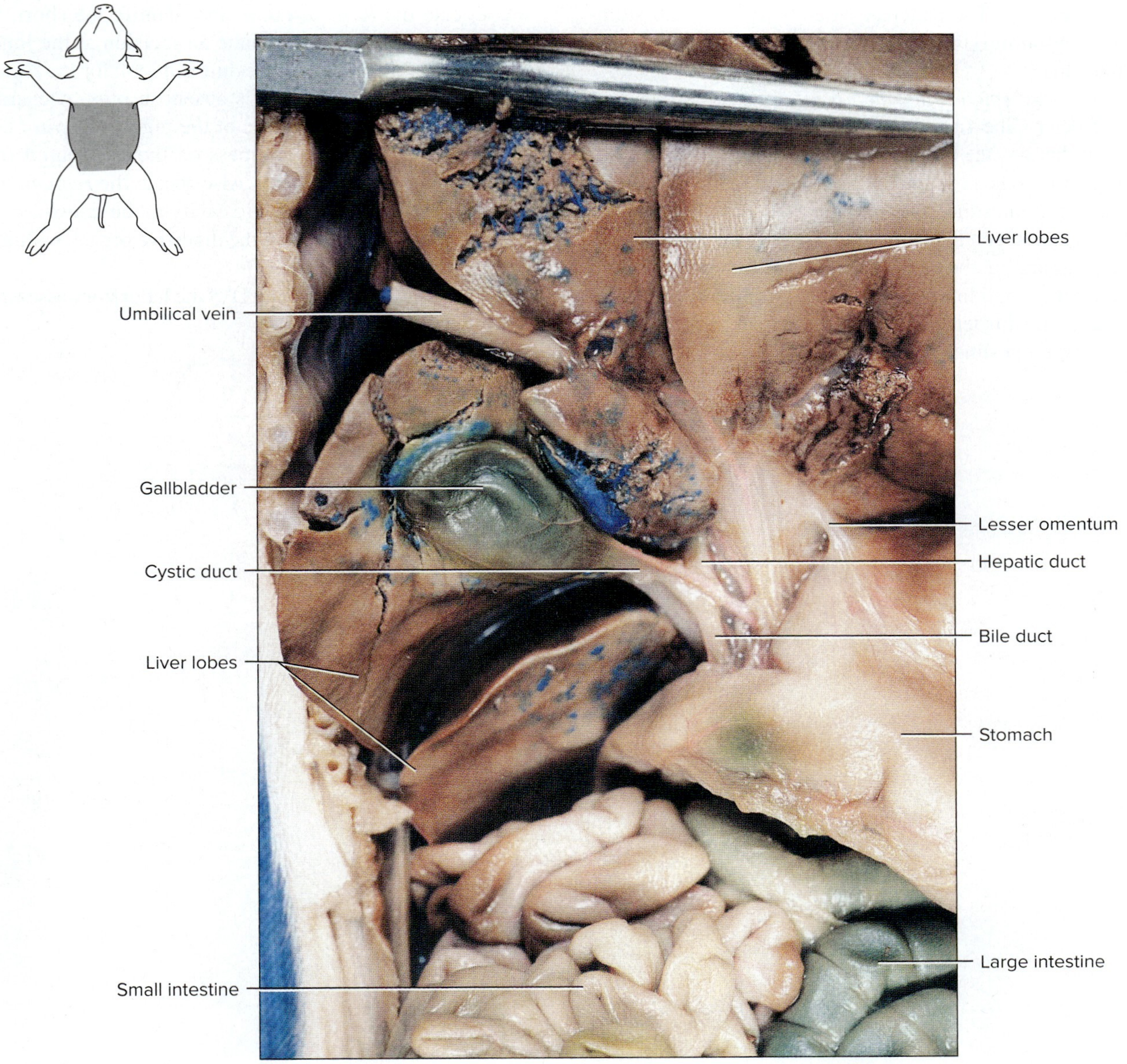

FIGURE 72.4 Digestive organs associated with the gallbladder of a fetal pig, ventral view. The liver is retracted for this view. ©J & J Photography APR

omentum extends from the spleen to the stomach, and the *lesser omentum* connects the liver to the stomach. Lift the liver to find the greenish to nearly colorless *gallbladder* embedded in the underside of the liver on the right side. Also note the *cystic duct,* by which the gallbladder is attached to the *bile duct,* and the *hepatic duct,* which originates in the liver and attaches to the cystic duct. Trace the bile duct to its connection with the duodenum (fig. 72.4).

8. Locate the *stomach* in the left side of the abdominal cavity. At its anterior end, note the union with the *esophagus,* which passes through the diaphragm. Identify the *cardia, fundus, body,* and *pyloric part* (listed from entrance to exit) of the stomach. Use scissors to make an incision along the convex border of the stomach from the cardia region to the pyloric part. The greenish substance found in the stomach and the rest of the digestive tract is called *meconium.* Meconium found in a fetal digestive tract is a combination of sloughed-off epithelial cells, amniotic fluid residues that were swallowed, and bile-stained mucus. It will be the first substance of bowel movements after birth. The lining of the stomach has numerous *gastric folds* (*rugae*). Examine the *pyloric sphincter,* which creates a constriction between the stomach and small intestine.
9. Locate the *pancreas* by lifting the stomach and separating some thin peritoneal membrane over the surface of

the pancreas. The pancreas extends from the left stomach region into the loop of the duodenum of the small intestine.

10. Trace the *small intestine,* beginning at the pyloric sphincter. The first portion, the *duodenum,* is a short loop that has the bile duct and the pancreatic duct leading into it. The proximal half of the remaining portion of the small intestine is the *jejunum,* and the distal half is the *ileum.* Open the small intestine and note the velvety appearance of the villi. Note how the mesentery supports the small intestine from the dorsal body wall. The small intestine terminates on the left side, where it joins the large intestine.

11. Locate the *large intestine* and identify the short, blind sac called the *cecum.* Make an incision at the junction between the ileum and cecum and look for the *ileocecal sphincter.* An appendix is absent in pigs. A characteristic of the large intestine of the pig is the *spiral colon.* This is a tightly coiled mass on the left ventral region of the abdominal cavity. Also locate the *rectum,* which extends through the pelvic cavity to the *anus.*

12. Review the locations of the digestive organs without the aid of figures.

13. Complete Parts B, C, and D of the laboratory assessment.

LABORATORY
ASSESSMENT 72

Name ______________________________

Date ______________________________

Section ______________________________

The (A) corresponds to the indicated Learning Outcome(s) (O) found at the beginning of the Laboratory Exercise.

Fetal Pig Dissection: Digestive System

PART A: Assessments

Complete the following:

1. Compare the locations of the major salivary glands of the human with those of the fetal pig. A2 ______________________________

____

____

2. Compare the types and numbers of primary (deciduous) teeth present in the pig's maxilla with those of the human. A2

____

____

____

3. In what ways do the pig's teeth seem to be adapted to its diet? A1 ______________________________

____

____

4. What part of the human soft palate is lacking in the pig? A2 ______________________________

5. What do you think is the function of the transverse ridges in the hard palate of the pig? A1 ______________________________

____

____

6. Describe the papillae on the surface of the tongue of the pig. A1 ______________________________

____

____

PART B: Assessments

Complete the following:

1. Describe how the peritoneum and mesenteries are associated with the organs in the abdominal cavity. A1 ___

2. Describe the inner lining of the stomach. A1 ___

3. Compare the structure of the human liver with that of the fetal pig. A3 ___

4. Compare the structure and location of the human large intestine with those of the fetal pig. A3 ___

5. What feature of the human cecum is lacking in the fetal pig? A3 ___

PART C: Assessments

Identify the numbered digestive features of the fetal pig in figures 72.5 and 72.6.

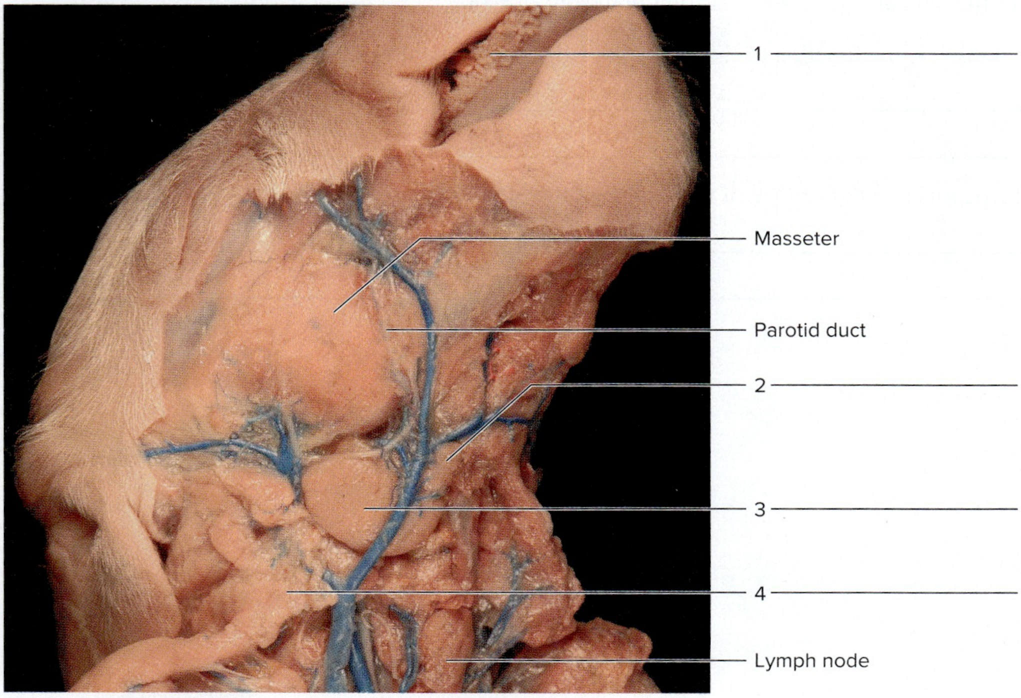

FIGURE 72.5 Identify the salivary glands and certain oral cavity features of the fetal pig, lateral view. Dennis Strete A1

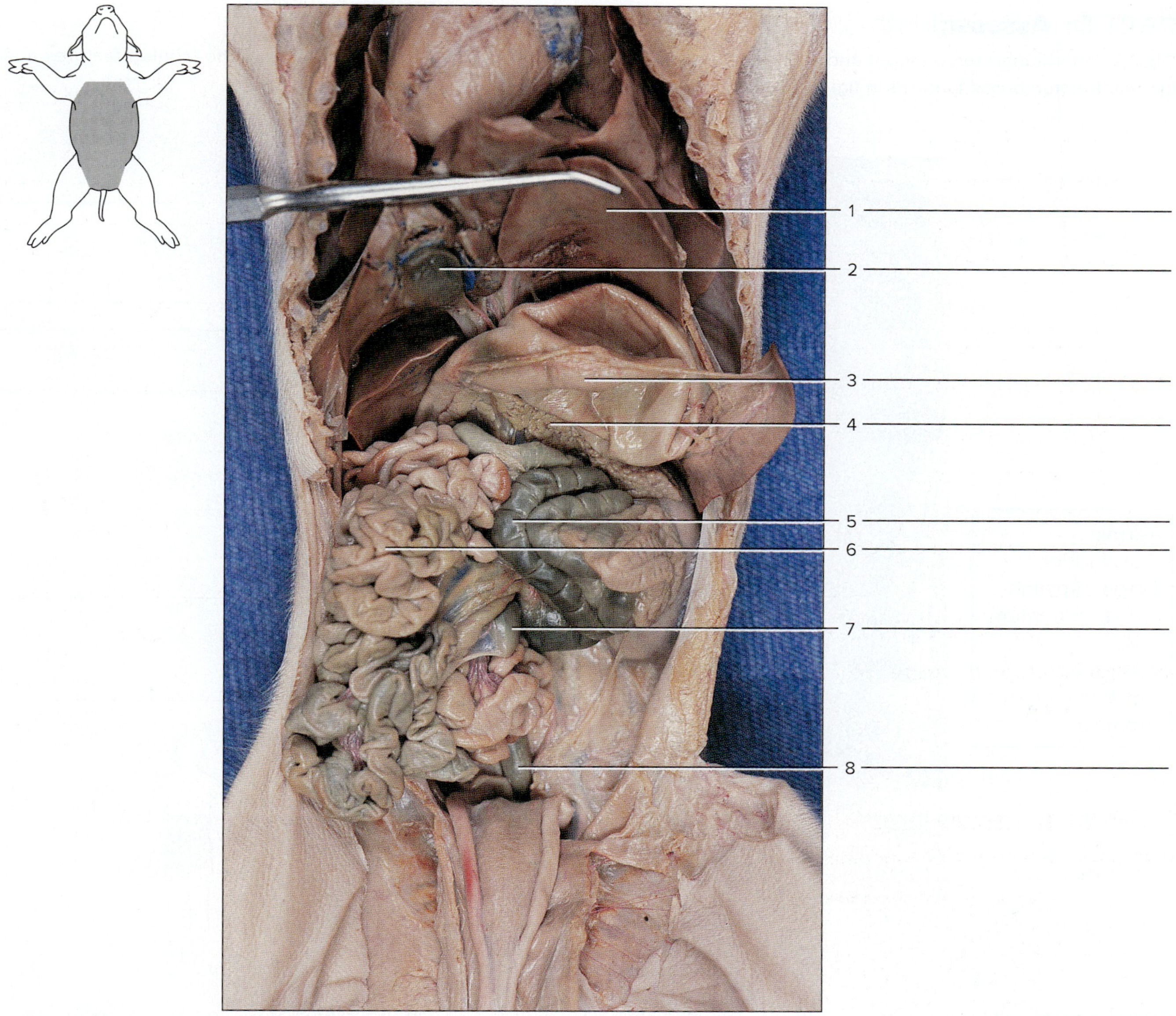

FIGURE 72.6 Label the numbered digestive features of the fetal pig, ventral view. ©J & J Photography **A1**

PART D: Assessments

Observe the human torso model and figure 72.3 of a fetal pig and figure 72.7 of a cadaver. Locate the labeled features and identify the numbered features in figure 72.7 that were also identified in the fetal pig dissection.

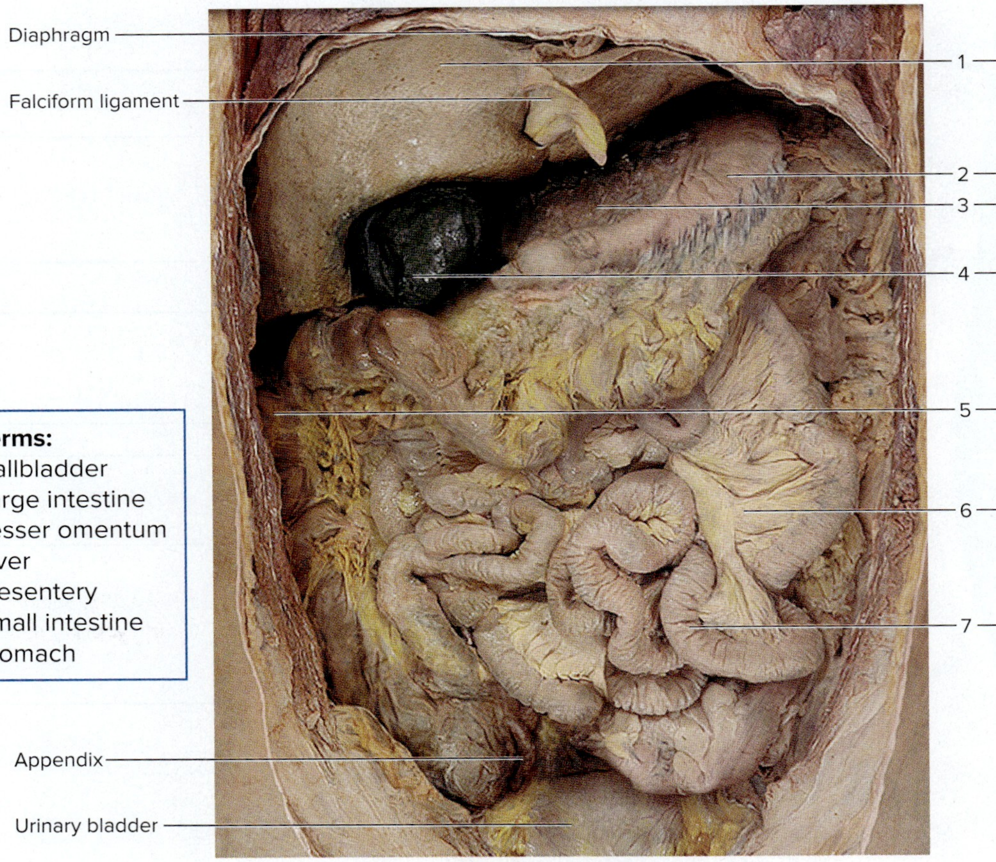

Terms:
Gallbladder
Large intestine
Lesser omentum
Liver
Mesentery
Small intestine
Stomach

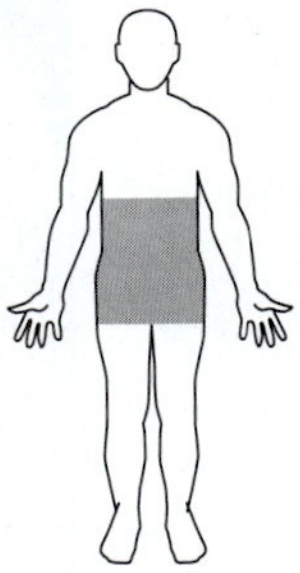

FIGURE 72.7 Identify the digestive features indicated on this anterior view of the abdomen of a cadaver, using the terms provided. APR/McGraw Hill APR A2 A3

LABORATORY
EXERCISE 73

Fetal Pig Dissection: Urinary System

Purpose of the Exercise

To examine and identify the urinary organs of the fetal pig and to compare them with those of the human.

MATERIALS NEEDED

Preserved fetal pig
Dissecting tray
Dissecting instruments
Disposable gloves
Human torso model

SAFETY

- Review all the safety guidelines in Appendix 1 of your laboratory manual.
- Wear disposable gloves and protective eyewear when working on the fetal pig dissection.
- Dispose of tissue remnants and gloves as instructed.
- Wash the dissecting tray and instruments as instructed.
- Wash your laboratory table.
- Wash your hands before leaving the laboratory.

Learning Outcomes APR

After completing this exercise, you should be able to:

O1 Locate and identify the urinary organs of the fetal pig.
O2 Compare the urinary organs of the fetal pig with those of the human.
O3 Identify the corresponding organs in the human torso model and a cadaver.

The O corresponds to the assessments A indicated in the Laboratory Assessment for this Exercise.

Pre-Lab

Carefully read the introductory material and examine the entire lab.
Be familiar with the structures and functions of the human urinary system from lecture, the textbook, or laboratory exercises.

In this laboratory exercise, you will dissect the urinary organs of the fetal pig. As you observe these structures, compare them with the corresponding human organs by observing the parts of the human torso model.

PROCEDURE: Urinary System Dissection

1. Place the fetal pig in a dissecting tray with its ventral surface up.
2. Open the abdominal cavity and remove the liver, stomach, spleen, pancreas, small intestine, and large intestine.
3. Locate the *kidneys* in the dorsal abdominal wall on either side of the vertebral column. The kidneys are located dorsal to the *parietal peritoneum* (retroperitoneal).
4. Carefully remove the parietal peritoneum surrounding the kidneys. Locate the following, using figure 73.1 as a guide:
   ureters
   renal arteries
   renal veins
5. Locate the bandlike *adrenal glands,* which lie medially and anteriorly to the kidneys.
6. Expose the ureters by cleaning away the connective tissues along their lengths. They enter the *fetal urinary bladder* (*allantoic bladder*) on the dorsal surface.
7. Examine the fetal urinary bladder, an elongated, muscular, collapsed sac between two umbilical arteries. Trace the allantoic bladder into the umbilical cord where it

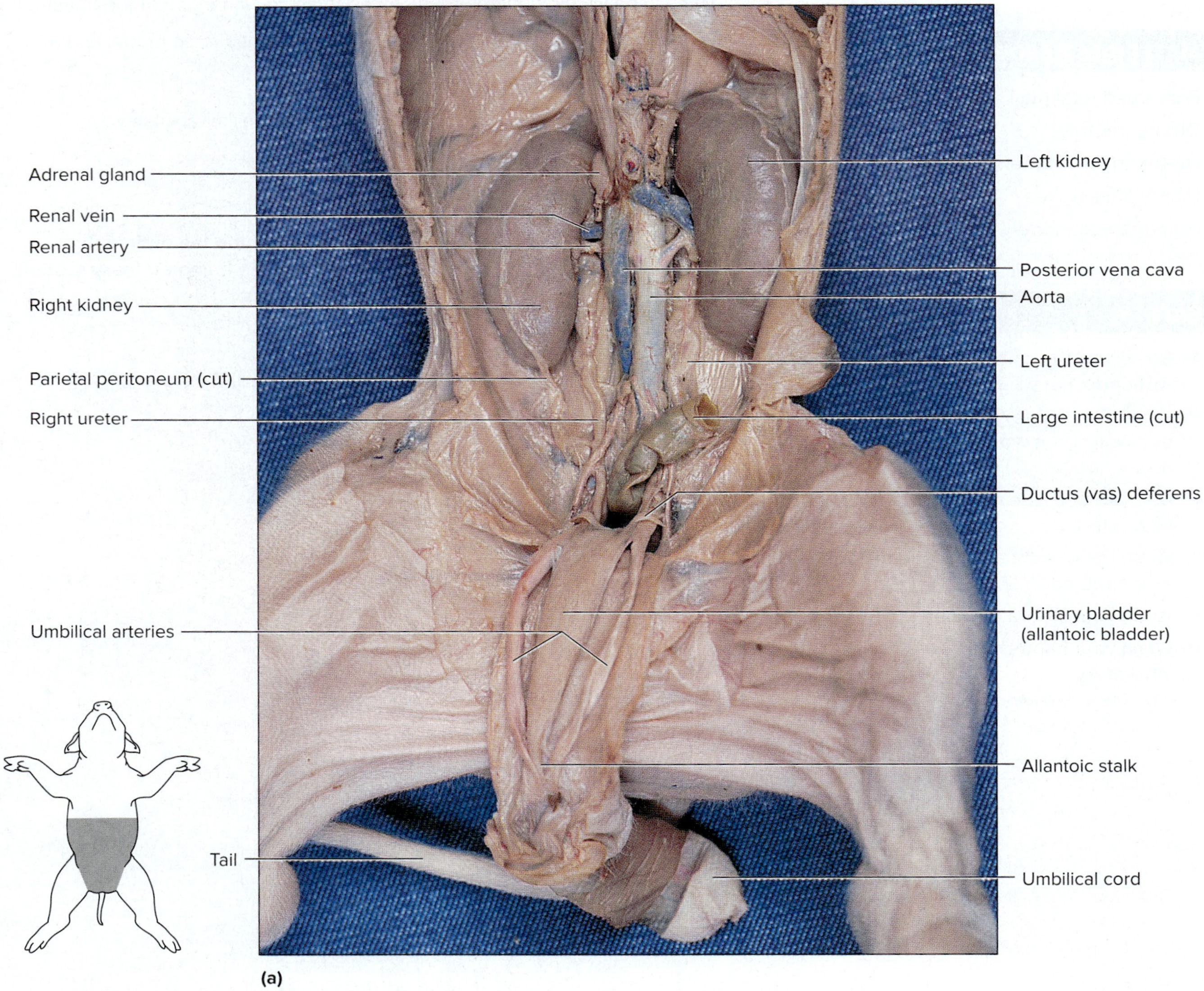

FIGURE 73.1 Urinary system of the male fetal pig with most of the digestive organs removed, ventral view: (a) with umbilical cord structures reflected; (b) with umbilical structures removed. ©J & J Photography APR

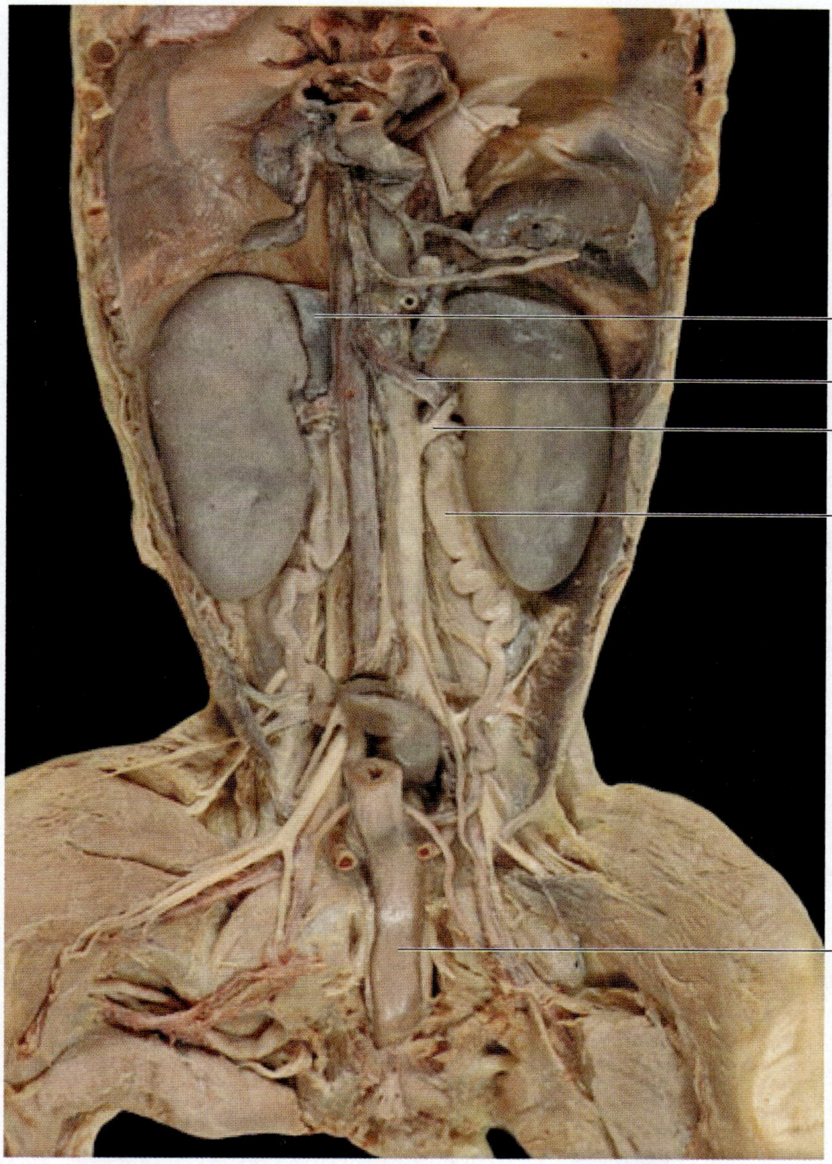

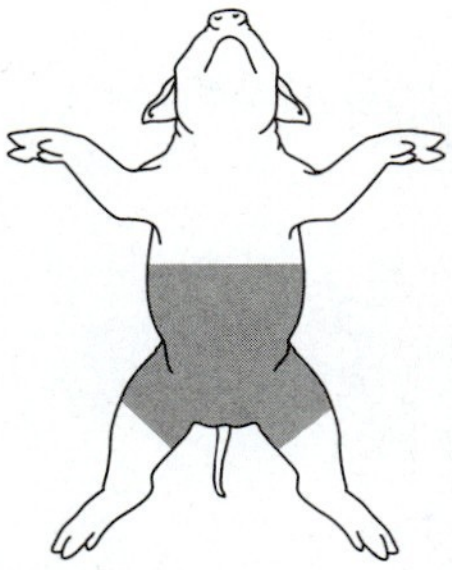

(b)

FIGURE 73.1 *Continued.* APR/McGraw Hill

becomes the *allantoic stalk*. The fetal urinary bladder becomes a functional urinary bladder after birth when the umbilical cord deteriorates and wastes no longer eliminate through the allantoic stalk.

8. Use a sharp scalpel to open the bladder and examine its interior. Locate the openings of the ureters and the urethra on the inside.
9. Expose the *urethra* at the posterior of the urinary bladder. The urethra extends through the *penis* in the male. In the female, the urethra enters the *urogenital sinus* a short distance from the urogenital orifice.
10. Remove one kidney and section it longitudinally. Use figure 73.2*a* as a reference for a kidney of a pig. Identify the following features:

    fibrous capsule
    renal cortex
    renal medulla
    renal pyramid
    minor calyx
    major calyx
    renal pelvis
    renal sinus
    hilum of kidney

11. Compare the structures of the kidney of the mature pig with those of a human (fig. 73.2*b*).
12. Review the locations of the urinary organs without the aid of figures.
13. Discard the organs and tissues that were removed from the fetal pig, as directed by the laboratory instructor.
14. Complete Parts A, B, and C of Laboratory Assessment 73.

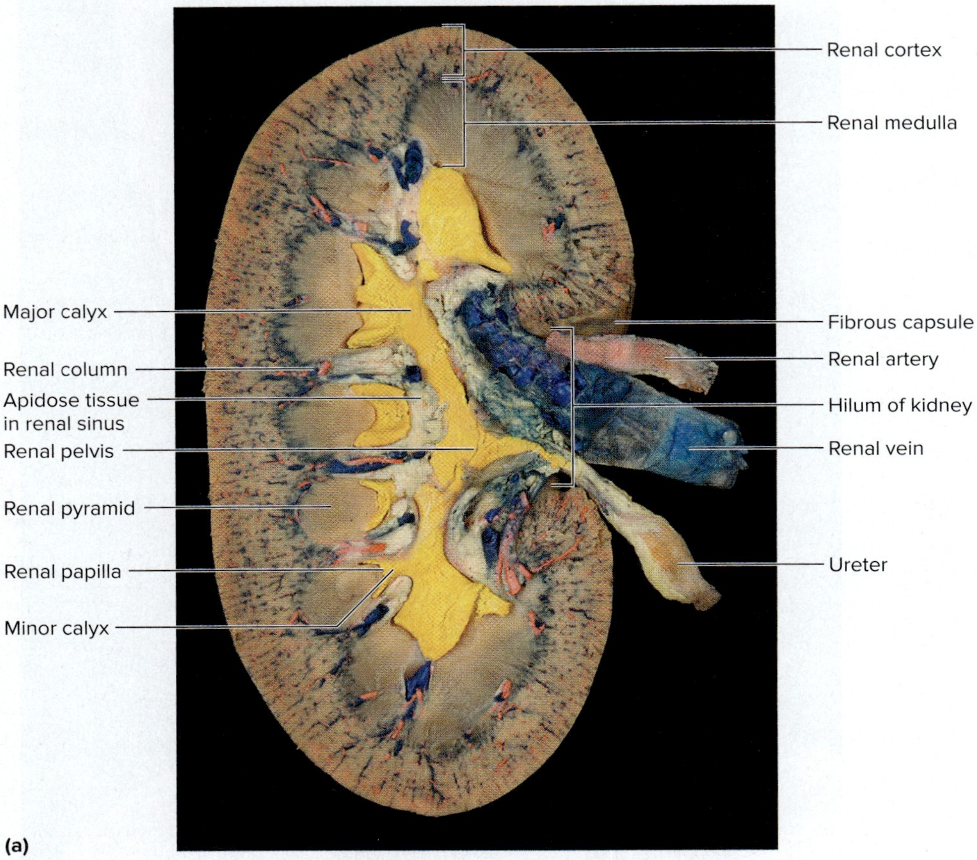

(a)

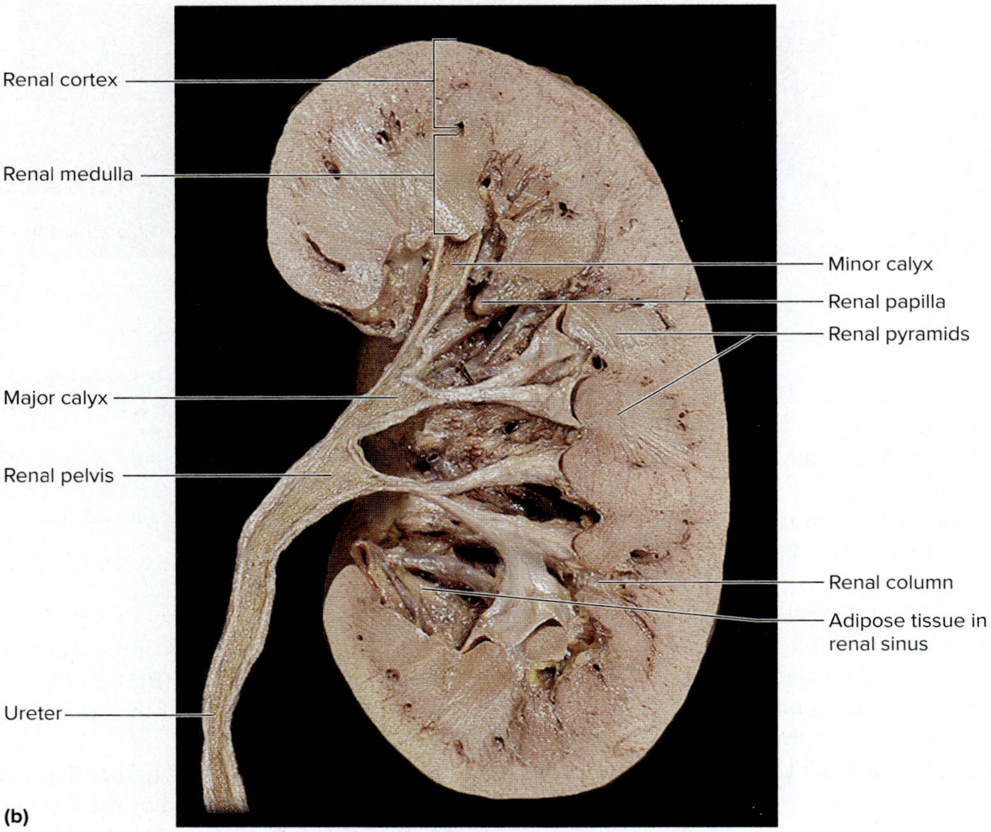

(b)

FIGURE 73.2 Longitudinal section of a kidney: (a) kidney from a mature pig that has a triple injection of latex; (b) cadaver.
 APR

LABORATORY
ASSESSMENT 73

Name ______________________________

Date ______________________________

Section ______________________________

The A corresponds to the indicated Learning Outcome(s) O found at the beginning of the Laboratory Exercise.

Fetal Pig Dissection: Urinary System

PART A: Assessments

Complete the following:

1. Compare the positions of the kidneys in the fetal pig with those in the human. A2 ______________________________

2. Compare the locations of the adrenal glands in the fetal pig with those in the human. A2 ______________________________

3. Trace the elimination of any urinary wastes from the allantoic bladder in a fetus. A1 ______________________________

4. Describe the wall of the urinary bladder of the fetal pig. Include the thickness, texture, and inner surface in your answer. A1

5. Compare the termination of the urethra of the female pig with that of a human female. A2 ______________________________

PART B: Assessments

Identify the numbered urinary features of the fetal pig in figure 73.3.

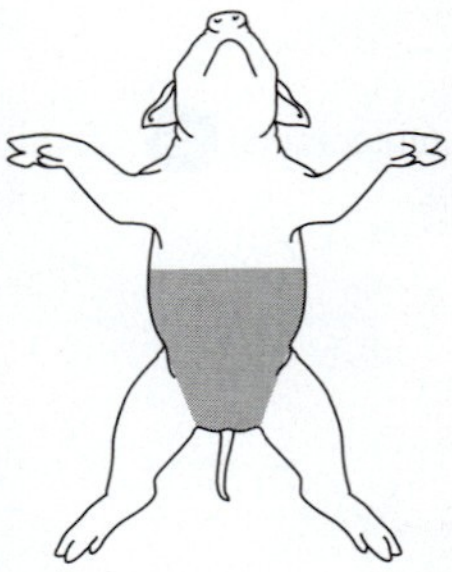

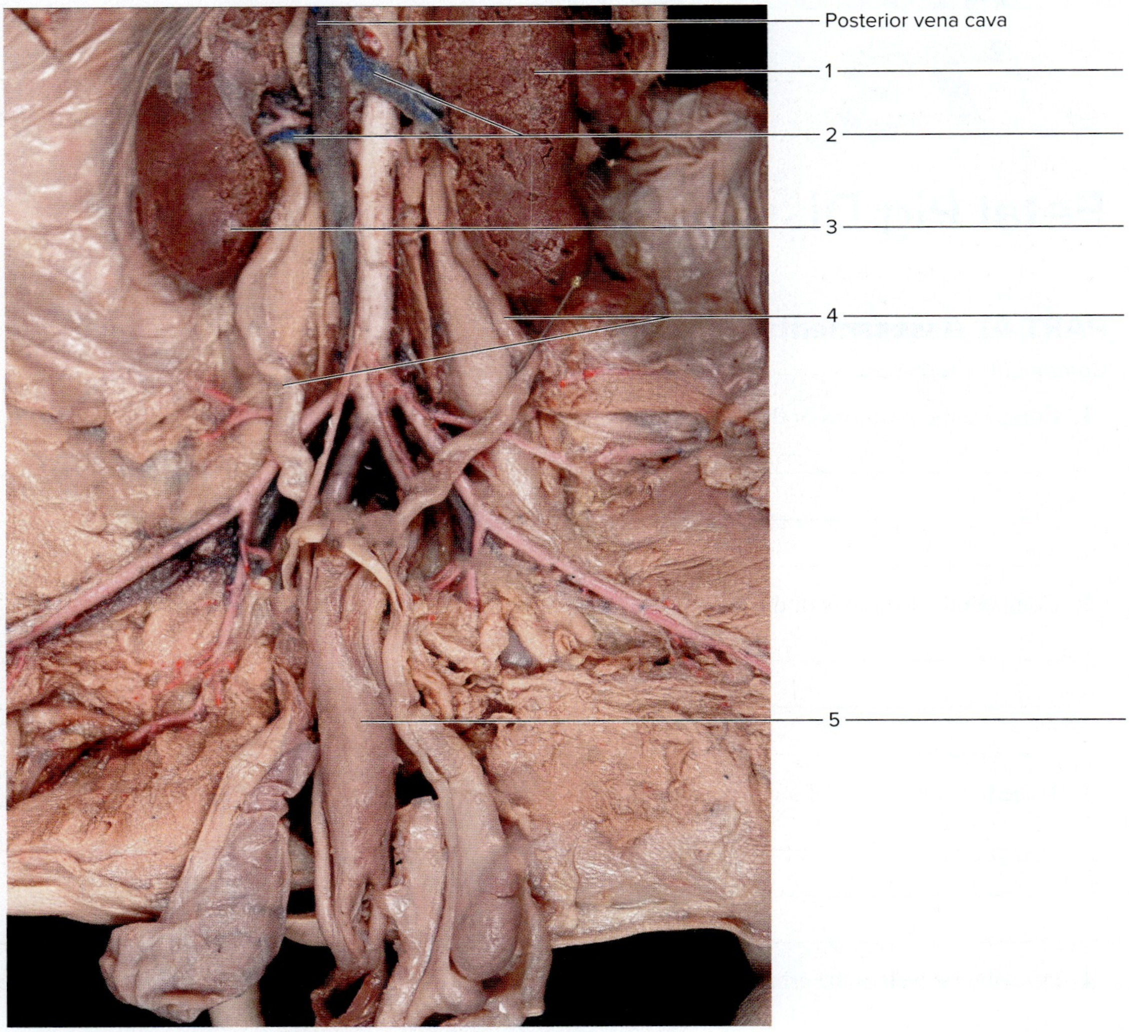

FIGURE 73.3 Identify the urinary features of the fetal pig, ventral view. Dennis Strete A1

PART C: Assessments

Observe the human torso model and figures 73.2*b* and 73.4 of a cadaver. Locate the labeled features and identify the numbered features in figure 73.4 that were also identified in the fetal pig dissection.

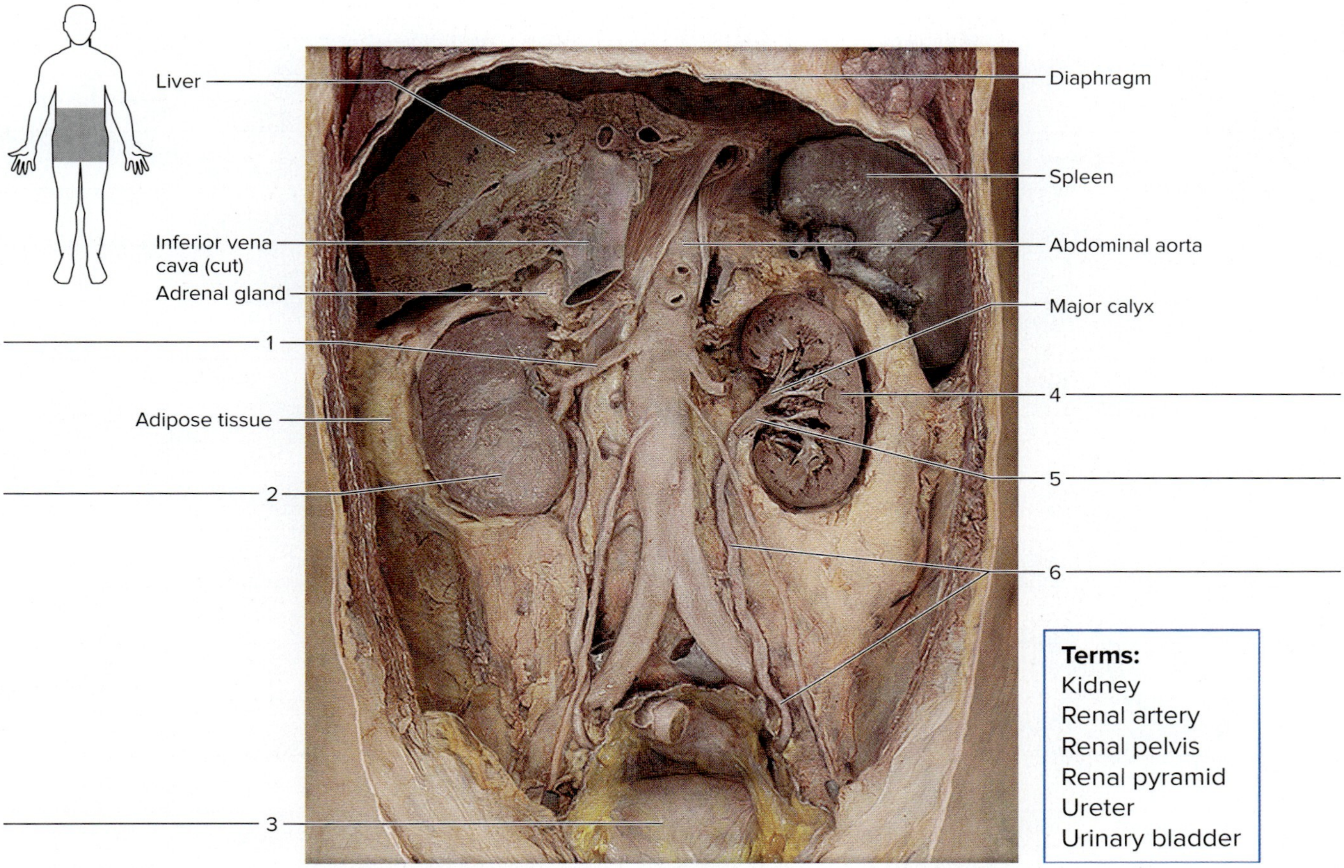

Terms:
Kidney
Renal artery
Renal pelvis
Renal pyramid
Ureter
Urinary bladder

FIGURE 73.4 Identify the urinary features indicated on this anterior view of the abdomen of a cadaver, using the terms provided. APR/McGraw Hill APR A2 A3

NOTES

LABORATORY
EXERCISE 74

Fetal Pig Dissection: Reproductive Systems

Purpose of the Exercise

To examine and identify the reproductive organs of the fetal pig and to compare them with the corresponding organs of the human.

MATERIALS NEEDED

- Preserved fetal pig
- Dissecting tray
- Dissecting instruments
- Magnifying lens
- Disposable gloves
- Models of human reproductive systems

SAFETY

- Review all the safety guidelines in Appendix 1 of your laboratory manual.
- Wear disposable gloves and protective eyewear when working on the fetal pig dissection.
- Dispose of tissue remnants and gloves as instructed.
- Wash the dissecting tray and instruments as instructed.
- Wash your laboratory table.
- Wash your hands before leaving the laboratory.

Learning Outcomes APR

After completing this exercise, you should be able to:

O1 Locate and identify the reproductive organs of a fetal pig.

O2 Compare the reproductive organs of the fetal pig with those of the human.

O3 Identify the corresponding organs in models of the human reproductive systems and a cadaver.

The O corresponds to the assessments A indicated in the Laboratory Assessment for this Exercise.

Pre-Lab

Carefully read the introductory material and examine the entire lab. Be familiar with the structures and functions of the human reproductive system from lecture, the textbook, or laboratory exercises.

In this laboratory exercise, you will dissect the reproductive system of the fetal pig. If you have a female fetal pig, begin with Procedure A. If you have a male fetal pig, begin with Procedure B. After completing the dissection, exchange fetal pigs with someone who has dissected one of the opposite sex and examine its reproductive organs.

As you observe the fetal pig reproductive organs, compare them with the corresponding human organs by examining the models of the human reproductive systems.

PROCEDURE A: Female Reproductive System

1. Place the female fetal pig in a dissecting tray with its ventral surface up and open its abdominal cavity.
2. Locate the small, oval *ovaries* just posterior to the kidneys (fig. 74.1).
3. Near the anterior end of the ovary, locate the funnel-shaped *infundibulum,* which is at the end of the small, coiled *uterine tube (oviduct).* Note the tiny projections, or *fimbriae,* that create a fringe around the edge of the infundibulum. Trace the uterine tube around the ovary to its nearby connection with the *uterine horn.* These structures can best be located with a magnifying lens in fetal pigs.
4. Examine the uterine horn. It is suspended from the body wall by a mesentery, the *broad ligament.* Also note the fibrous *round ligament,* which extends from the uterine horn laterally and posteriorly to the body wall.
5. Use scissors and cut carefully through the midline of the pelvis to observe the remaining organs of the reproductive system.
6. Trace the relatively long uterine horns posteriorly. They unite to form the *uterine body,* which is located between the urethra and rectum. The Y-shaped uterus in a pig

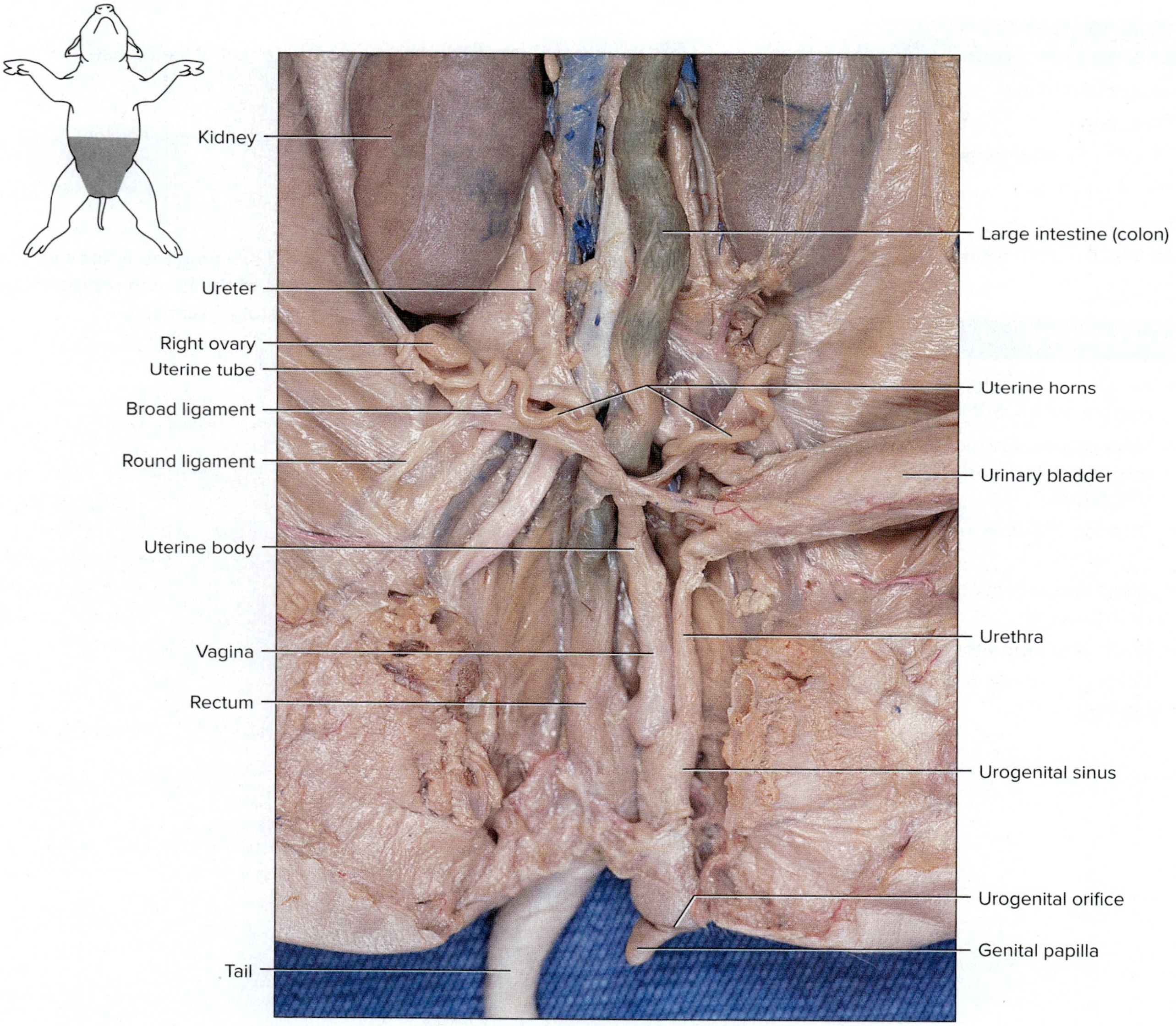

(a)

FIGURE 74.1 (a) and (b) Reproductive system of the female fetal pig, ventral view. ©J & J Photography APR

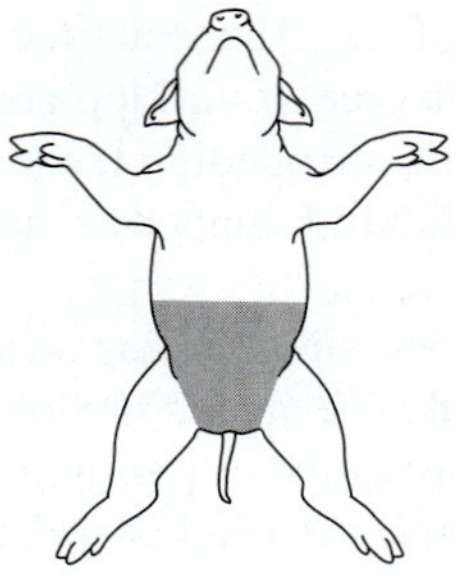

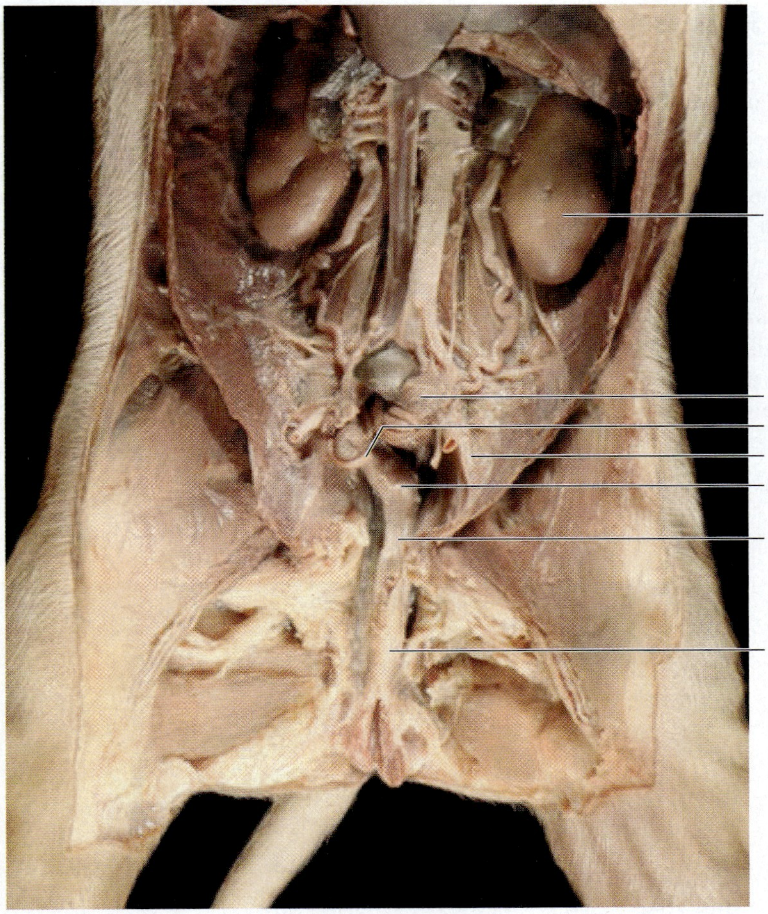

(b)

FIGURE 74.1 *Continued.* APR/McGraw Hill

allows the large uterus to contain a litter of pigs. The uterine body is continuous with the *vagina,* which leads to the outside.

7. Trace the *urethra* from the urinary bladder posteriorly. The urethra and the vagina open into a common chamber called the *urogenital sinus.* The opening of this chamber, ventral to the anus, is called the *urogenital orifice.* Locate the *genital papilla,* a small hood over the urogenital orifice.
8. Use scissors to open the vagina along its lateral wall, beginning at the urogenital orifice and continuing to the body of the uterus. Note the *urethral orifice* in the ventral wall of the urogenital sinus and locate the small, rounded *cervix* of the uterus, which projects into the vagina at its deep end. The *clitoris* is located in the ventral wall of the urogenital sinus near its opening to the outside.
9. Review the locations of the reproductive organs without the aid of figures.
10. Complete Part A of Laboratory Assessment 74.

PROCEDURE B: Male Reproductive System

1. Place the male fetal pig in a dissecting tray with its ventral surface up.
2. Locate the *scrotum,* which appears as an external pouch just ventral to the anus. During fetal development, the *testes* migrate from posterior to the kidneys through the *inguinal canals* into the scrotum by means of a *gubernaculum.* The gubernaculum, a cord of tissue that extends from the *epididymis* to the scrotum, grows more slowly than other tissues and "pulls" each testis into the scrotum. Make an incision on the left side of the scrotum and continue the incision anteriorly to open one of the inguinal canals to locate the *spermatic cord* (fig. 74.2).
3. Locate a testis within the inguinal sac. The exact location will vary depending on the degree of development of the fetal pig. Remove the sheath surrounding the testis and locate the convoluted epididymis cupped around the lateral surface of each testis.
4. Locate the spermatic cord on the left side, leading away from the testis. This cord contains the *ductus deferens* (*vas deferens*), which is continuous with the epididymis, as well as with the nerves and blood vessels that supply the testis on that side. Trace the spermatic cord to the body wall, where its contents pass through the inguinal canal and enter the pelvic cavity (fig. 74.2).
5. Locate the *penis* and identify the *prepuce,* which forms a sheath around the entire penis. Make an incision through the skin of the prepuce and expose the shaft of the penis.

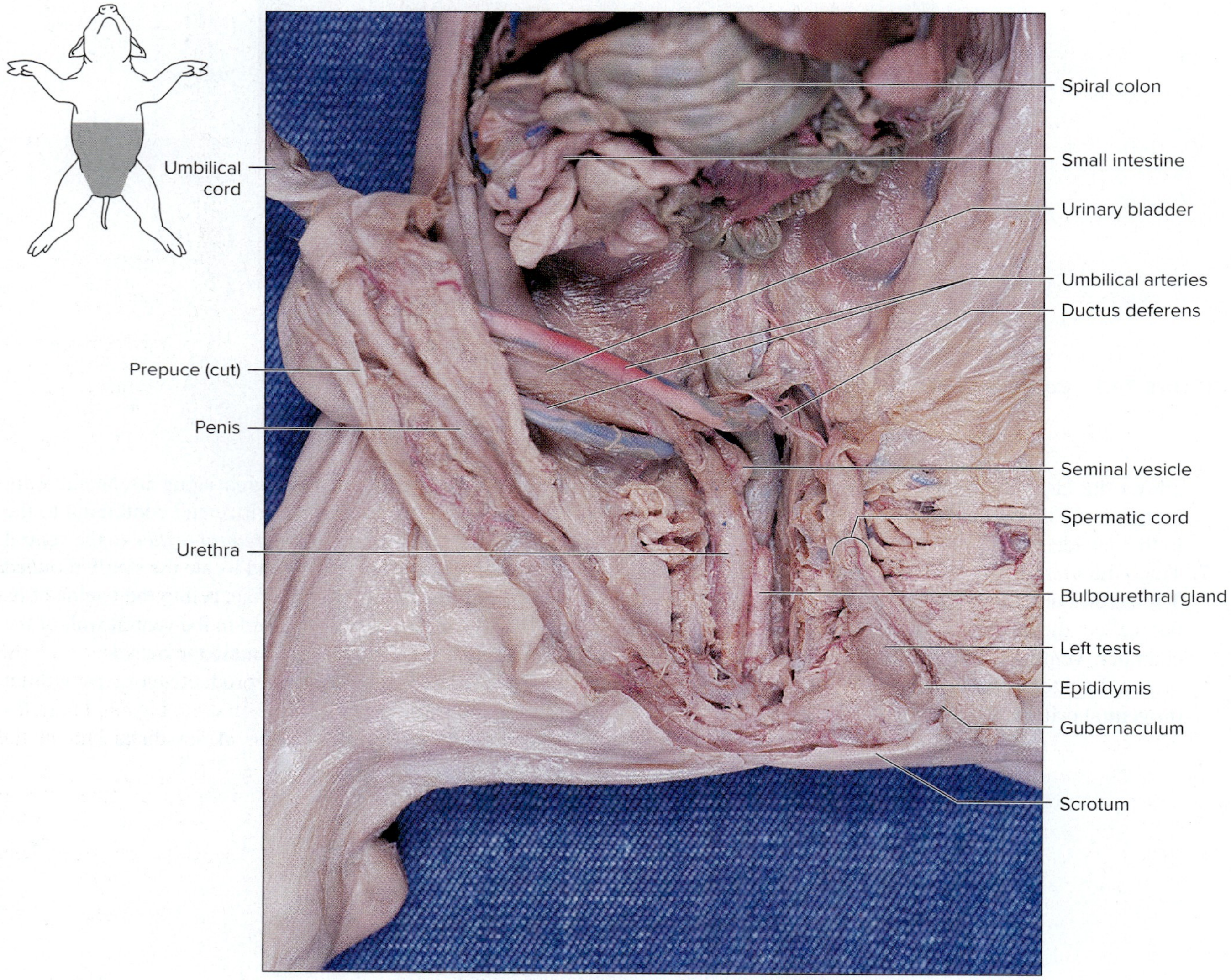

FIGURE 74.2 (a) and (b) Reproductive system of the male fetal pig, ventral view. ©J & J Photography APR

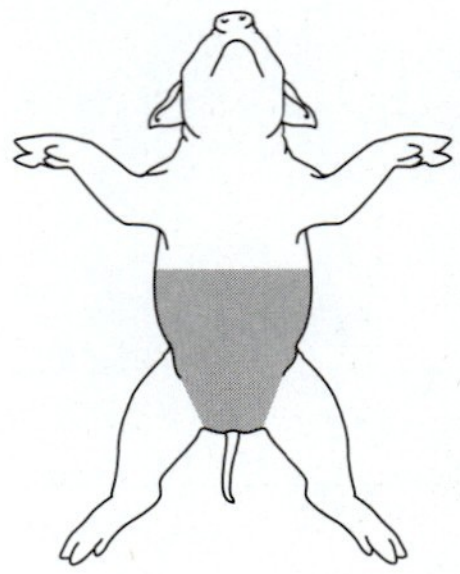

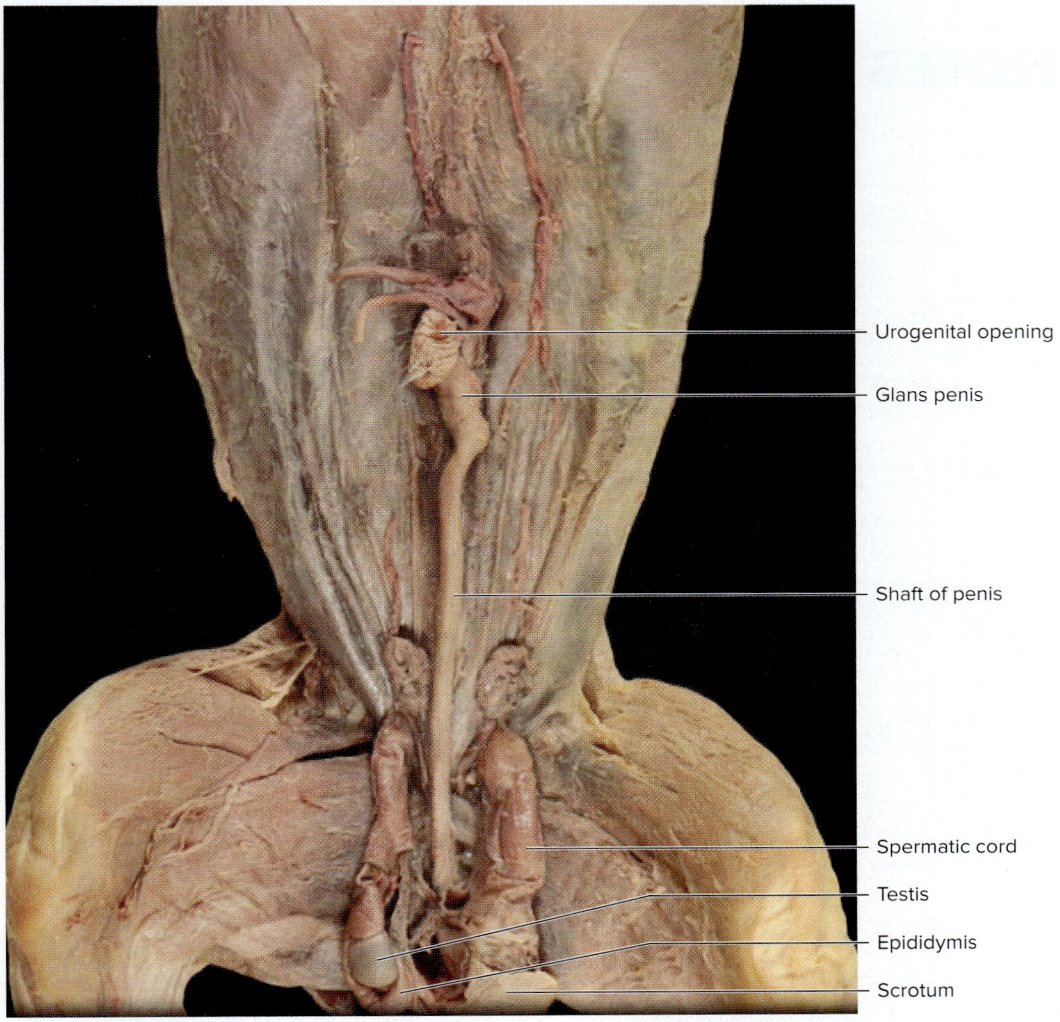

(b)

FIGURE 74.2 *Continued.* APR/McGraw Hill APR

6. Use scissors and cut carefully through the midline of the pelvis to observe the remaining organs of the reproductive system. Trace the ductus deferens from the inguinal canal to the penis. The ductus deferens loops over the ureter within the pelvic cavity and passes downward behind the urinary bladder to join the urethra. Locate the small, paired *seminal vesicles,* which appear as enlargements near the junction of the ductus deferens and urethra. A careful dissection between the seminal vesicles might expose the *prostate gland* near the junction of the urinary bladder and the urethra. (It is difficult to locate in the fetal pig.)
7. Trace the urethra to the penis. Locate the *bulbourethral glands,* which form elongated swellings along the urethra, on either side at the proximal end of the penis.
8. Use a sharp scalpel to cut a transverse section of the penis. Identify the urethra, which serves as a common pathway for urine and semen in male pigs and humans. Locate the *urogenital orifice* at the distal end of the penis.
9. Review the locations of the reproductive organs without the aid of figures.
10. Complete Parts B, C, and D of the laboratory assessment.

NOTES

LABORATORY
ASSESSMENT 74

Name ______________________________

Date ______________________________

Section ______________________________

The A corresponds to the indicated Learning Outcome(s) O found at the beginning of the Laboratory Exercise.

Fetal Pig Dissection: Reproductive Systems

PART A: Assessments

Complete the following:

1. Compare the relative lengths and paths of the uterine tubes (oviducts) of the fetal pig and the human. A2 ______________________________

2. How do the shape and structure of the uterus of the fetal pig compare with that of the human? A2 ______________________________

3. Assess the function of the uterine horns of the pig. A1 ______________________________

4. Compare the relationship of the urethra and the vagina in the fetal pig and in the human. A2 ______________________________

PART B: Assessments

Complete the following:

1. Explain the location of the testes in the fetal pig. A1 ______________________________

2. How do the location and the relative size of the bulbourethral glands of the fetal pig compare with those of the human? A2

3. Compare the location and size of the prepuce in the fetal pig and in the human. A2 ______________________________

____

____

____

PART C: Assessments

Identify the numbered reproductive structures of the female fetal pig in figure 74.3 and the male fetal pig in figure 74.4.

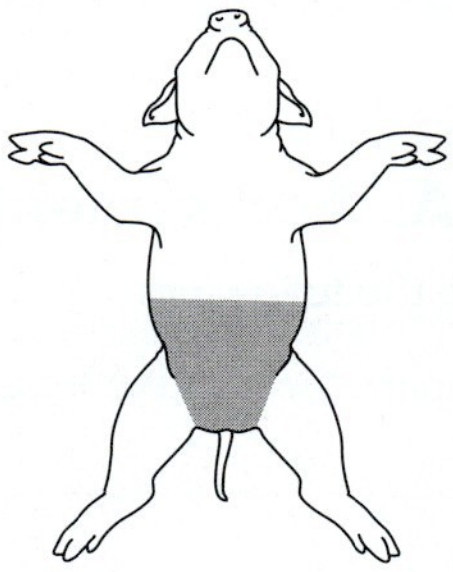

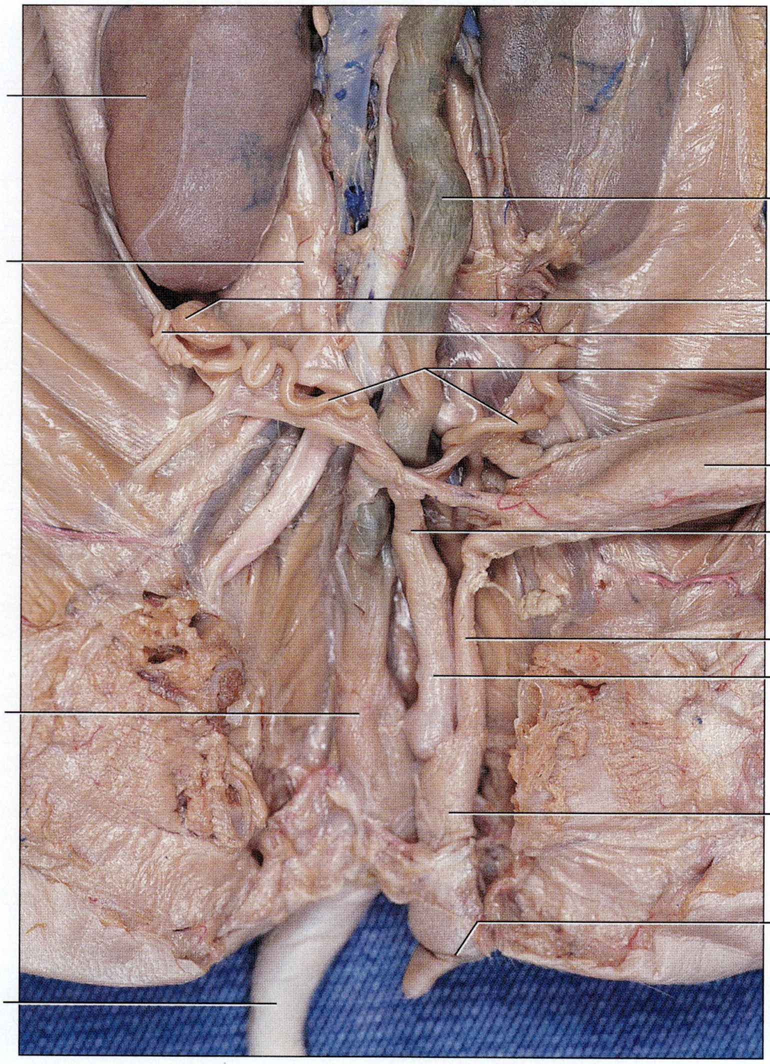

FIGURE 74.3 Label the numbered reproductive structures of the female fetal pig, ventral view. ©J & J Photography A1

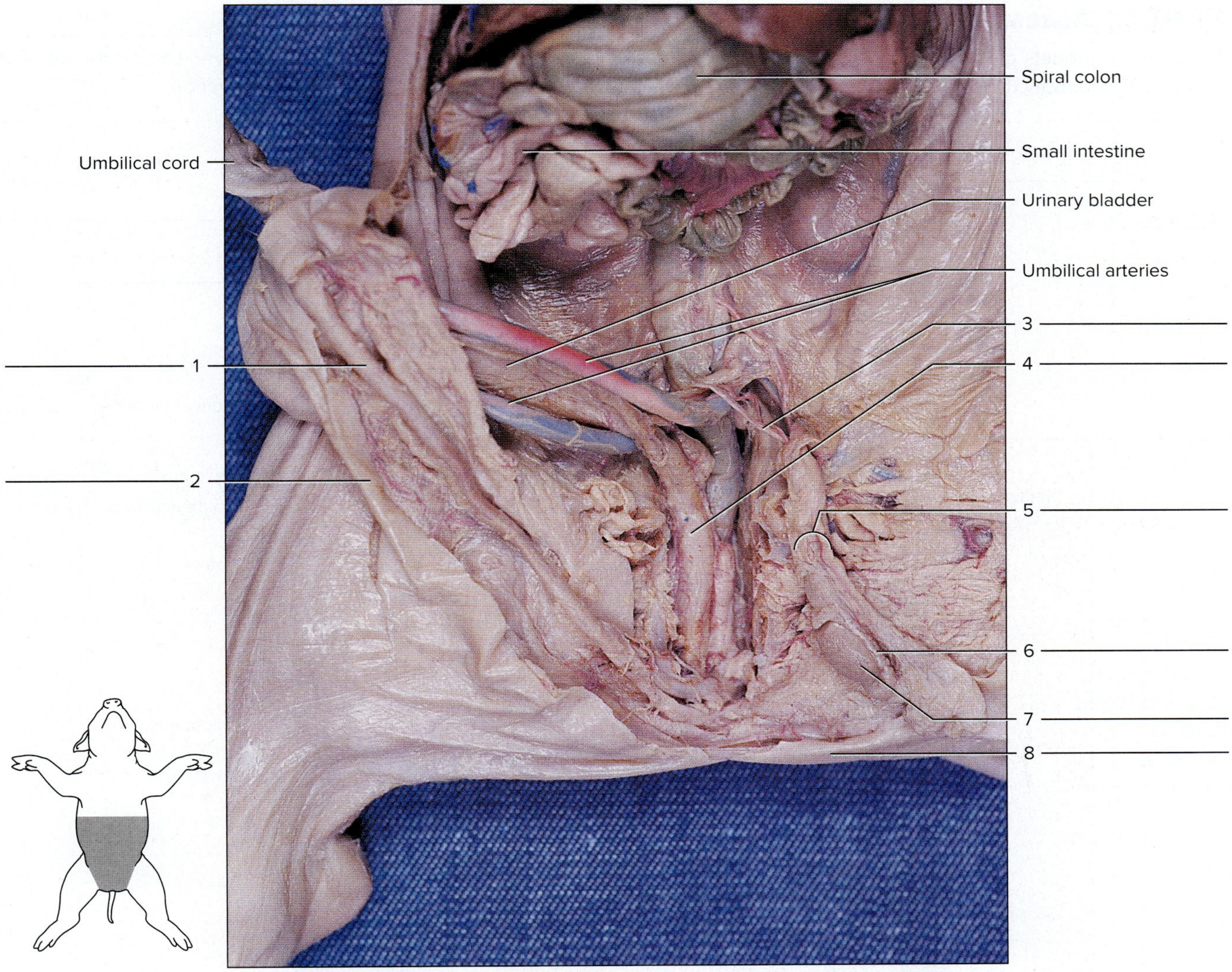

FIGURE 74.4 Label the numbered reproductive structures of the male fetal pig, ventral view. ©J & J Photography A1

PART D: Assessments

Observe the models of the human reproductive system and figure 74.5 of a female and a male cadaver. Locate the labeled features and identify the numbered features in figure 74.5 that were also identified in the fetal pig dissection.

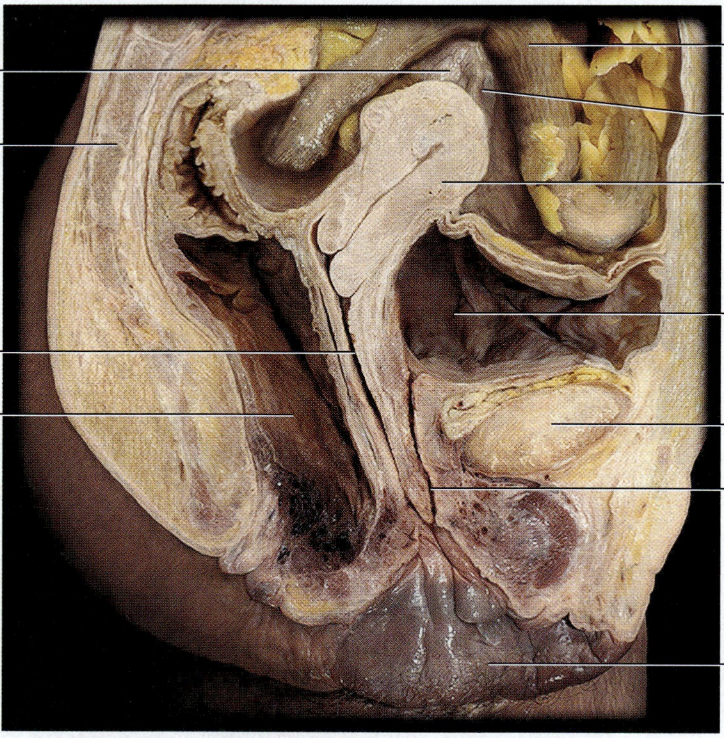

Terms:
Ovary
Uterine tube
Uterus
Vagina

(a) APR Posterior ⟷ Anterior

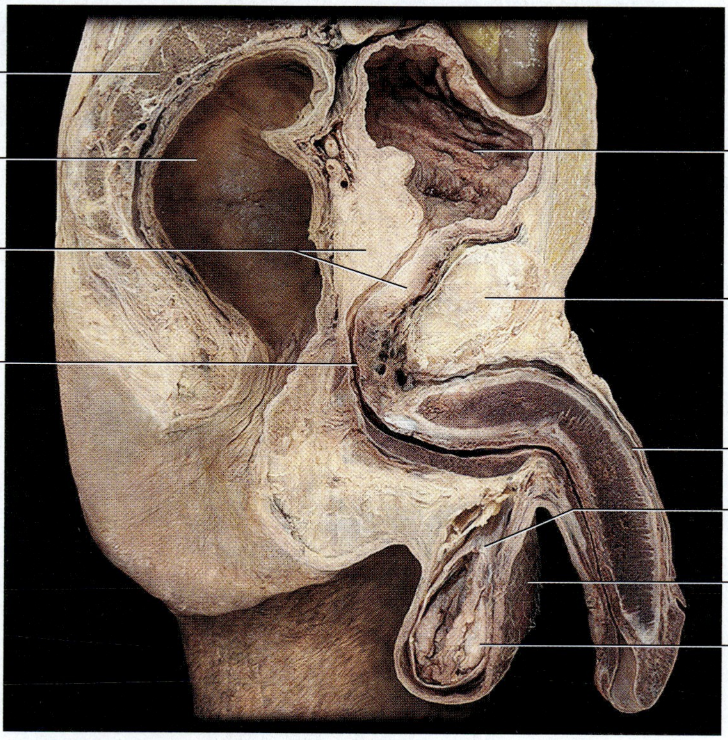

Terms:
Penis
Prostate gland
Scrotum
Testis
Urethra

(b) APR Posterior ⟷ Anterior

FIGURE 74.5 Sagittal (median; midsagittal) plane of a cadaver: (a) female; (b) male. Identify the reproductive features, using the terms provided. APR/McGraw Hill APR A2 A3

APPENDIX 1

Name________________________________

Date________________ Class________________

Laboratory Safety Guidelines

Carefully review the following safety guidelines before starting a lab exercise. Safety guideline reminders are also included in the appropriate lab exercise sections. Avoid tardiness to laboratory sessions, because specific directions and precautions are often given at the beginning of the period. Your instructor should update you about safety procedures, as your school might have safety regulations or modifications in addition to those listed here.

1. *Become familiar with all room exits and the location and operating procedures of all safety equipment [first aid kit, Material Safety Data Sheets (MSDS), eyewash station, safety shower, fire extinguisher, and fire blanket].*
2. *Read all laboratory exercises prior to starting the procedures. Do not work alone in the laboratory.*
3. *Do not smoke, chew, eat, drink, apply cosmetics or lip balm, handle contact lenses (regular eyeglasses should be worn), or work with open wounds in the laboratory. Consider all materials you use as potentially hazardous.*
4. *Use protective eyewear and laboratory coats during labs when handling dangerous chemicals or when heating dangerous materials. Volatile materials should never be heated using an open flame. When heating materials in a test tube, never point the test tube in the direction of a person or leave heated materials unattended.*
5. *Use a commercially prepared disinfectant or a 10% bleach solution to clean laboratory work surfaces before and after any procedures when using animals, body fluids, or dissection specimens.*
6. *Wear enclosed shoes and disposable gloves when working with dangerous chemicals, body fluids, or dissection specimens.*
7. *If body fluids are being studied, work only with your own. Special precautions to prevent contact with body fluids may include wearing disposable gloves, gowns, aprons, or masks, as directed by your instructor.*
8. *Special precautions during dissections may include wearing laboratory coats and protective eyewear. Always point cutting blades away from yourself and laboratory partners.*
9. *Restrain all loose clothing, long hair, and dangling jewelry during laboratory procedures. Keep all unnecessary materials away from the work area to reduce clutter and the possibility of an accident.*
10. *If you have special needs, are taking medications, experience allergic reactions, are pregnant, or are uncomfortable with the procedures, inform your instructor.*
11. *Use only a mechanical pipetting device (never your mouth).*
12. *Immediately report to your instructor any accidents, spills, or damaged equipment.*
13. *Use only disposable lancets and needles, and never attempt to bend, cut, or recap them when finished. Place sharp items in a puncture-resistant container that is marked "Biohazard Container."*
14. *Dispose of chemicals, waste material, body fluids, and dissection specimens according to appropriate directions. Any reusable glassware or utensils that have been contaminated with body fluids should be placed in a disinfectant (10% bleach solution) and later autoclaved.*
15. *Thoroughly wash your hands with soap and warm water immediately after removing disposable gloves and before leaving the laboratory.*
16. *Practice any modified or additional safety guidelines required by your instructor.*

APPENDIX 2

Preparation of Solutions

Amylase solution, 0.5%

Place 0.5 g of bacterial amylase in a graduated cylinder or volumetric flask. Add distilled water to the 100 mL level. Stir until dissolved. The amylase should be free of sugar for best results; a low-maltose solution of amylase yields good results. (Store amylase powder in a freezer until mixing this solution.)

Benedict's solution

Prepared solution is available from various suppliers.

Caffeine, 0.2%

Place 0.2 g of caffeine in a graduated cylinder or volumetric flask. Add distilled water to the 100 mL level. Stir until dissolved.

Calcium chloride, 2.0%

Place 2.0 g of calcium chloride in a graduated cylinder or volumetric flask. Add distilled water to the 100 mL level. Stir until dissolved.

Calcium hydroxide solution (limewater)

Add an excess of calcium hydroxide to 1 L of distilled water. Stopper the bottle and shake thoroughly. Allow the solution to stand for 24 hours. Pour the supernatant fluid through a filter. Store the clear filtrate in a stoppered container.

Epsom salt solution, 0.1%

Place 0.5 g of Epsom salt in a graduated cylinder or volumetric flask. Add distilled water to the 500 mL level. Stir until dissolved.

Glucose solutions

1. *1.0% solution.* Place 1 g of glucose in a graduated cylinder or volumetric flask. Add distilled water to the 100 mL level. Stir until dissolved.
2. *10% solution.* Place 10 g of glucose in a graduated cylinder or volumetric flask. Add distilled water to the 100 mL level. Stir until dissolved.

Iodine-potassium-iodide (IKI solution)

Add 20 g of potassium iodide to 1 L of distilled water, and stir until dissolved. Then add 4.0 g of iodine, and stir again until dissolved. Solution should be stored in a dark, stoppered bottle.

Methylene blue

Dissolve 0.3 g of methylene blue powder in 30 mL of 95% ethyl alcohol. In a separate container, dissolve 0.01 g of potassium hydroxide in 100 mL of distilled water. Mix the two solutions. (Prepared solution is available from various suppliers.)

Monosodium glutamate (MSG) solution, 1%

Place 1 g of monosodium glutamate in a graduated cylinder or volumetric flask. Add distilled water to the 100 mL level. Stir until dissolved.

Physiological saline solution

Place 0.9 g of sodium chloride in a graduated cylinder or volumetric flask. Add distilled water to the 100 mL level. Stir until dissolved.

Potassium chloride, 5%

Place 5.0 g of potassium chloride in a graduated cylinder or volumetric flask. Add distilled water to the 100 mL level. Stir until dissolved.

Quinine sulfate, 0.5%

Place 0.5 g of quinine sulfate in a graduated cylinder or volumetric flask. Add distilled water to the 100 mL level. Stir until dissolved.

Ringer's solution (frog)

Dissolve the following salts in 1 L of distilled water:

6.50 g sodium chloride
0.20 g sodium bicarbonate
0.14 g potassium chloride
0.12 g calcium chloride

Sodium chloride solutions

1. *0.9% solution.* Place 0.9 g of sodium chloride in a graduated cylinder or volumetric flask. Add distilled water to the 100 mL level. Stir until dissolved.
2. *1.0% solution.* Place 1.0 g of sodium chloride in a graduated cylinder or volumetric flask. Add distilled water to the 100 mL level. Stir until dissolved.
3. *3.0% solution.* Place 3.0 g of sodium chloride in a graduated cylinder or volumetric flask. Add distilled water to the 100 mL level. Stir until dissolved.
4. *5.0% solution.* Place 5.0 g of sodium chloride in a graduated cylinder or volumetric flask. Add distilled water to the 100 mL level. Stir until dissolved.

Spiked urine solution

Place one drop yellow food coloring into a beaker of 250 mL distilled water. While stirring, add 3 g $NaNO_2$, 3 g glucose, 2 mL acetone, 2 drops blood, and 2 g egg albumin. Filter solution (not by suction) into a separate beaker. Test solution with a reagent test strip to show positive results on nitrite, protein, glucose, and blood. Add concentrated HCl if necessary to adjust to pH 6. Transfer solution to a sealed bottle, label the bottle, and refrigerate.

Starch solutions

1. *0.5% solution.* Add 5 g of cornstarch to 1 L of distilled water. Heat until the mixture boils. Cool the liquid, and pour it through a filter. Store the filtrate in a refrigerator.
2. *1.0% solution.* Add 10 g of cornstarch to 1 L of distilled water. Heat until the mixture boils. Cool the liquid, and pour it through a filter. Store the filtrate in a refrigerator.
3. *10% solution.* Add 100 g of cornstarch to 1 L of distilled water. Heat until the mixture boils. Cool the liquid, and pour it through a filter. Store the filtrate in a refrigerator.

Sucrose, 5% solution

Place 5.0 g of sucrose in a graduated cylinder or volumetric flask. Add distilled water to the 100 mL level. Stir until dissolved.

Wright's stain

Prepared solution is available from various suppliers.

APPENDIX 3

Assessment of Laboratory Assessments

Many assessment models can be used for laboratory reports. A rubric, which can be used for performance assessments, contains a description of the elements (requirements or criteria) of success to various degrees. The term rubric originated from *rubrica terra,* which is Latin for the application of red earth to indicate anything of importance. A rubric used for assessment contains elements for judging student performance, with points awarded for varying degrees of success in meeting the objectives. The content and the quality level necessary to attain certain points are indicated in the rubric. It is effective if the assessment tool is shared with the students before the laboratory exercise is performed.

Following are two sample rubrics that can easily be modified to meet the needs of a specific course. Some of the elements for these sample rubrics may not be necessary for every laboratory exercise. The generalized rubric needs to contain the possible assessment points that correspond to objectives for a specific course. The point value for each element may vary. The specific rubric example contains performance levels for laboratory assessments. The elements and the point values can easily be altered to meet the value placed on laboratory reports for a specific course.

ASSESSMENT: Generalized Laboratory Report Rubric

Element	Assessment Points Possible	Assessment Points Earned
1. Figures are completely and accurately labeled.		
2. Sketches are accurate, contain proper labels, and are of sufficient detail.		
3. Colored pencils were used extensively to differentiate structures on illustrations.		
4. Matching and fill-in-the-blank answers are completed and accurate.		
5. Short-answer/discussion questions contain complete, thorough, and accurate answers. Some elaboration is evident for some answers.		
6. Data collected are complete, are accurately displayed, and contain a valid explanation.		

TOTAL POINTS: POSSIBLE ________________ EARNED ________________

ASSESSMENT: Specific Laboratory Report Rubric

Element	Excellent Performance (4 points)	Proficient Performance (3 points)	Marginal Performance (2 points)	Novice Performance (1 point)	Points Earned
Figure labels	Labels completed with ≥90% accuracy.	Labels completed with 80%–89% accuracy.	Labels completed with 70%–79% accuracy.	Labels <70% accurate.	
Sketches	Accurate use of scale, details illustrated, and all structures labeled accurately.	Minor errors in sketches. Missing or inaccurate labels on one or more structures.	Sketch not realistic. Missing or inaccurate labels on two or more structures.	Several missing or inaccurate labels.	
Matching and fill-in-the-blanks	All completed and accurate.	One or two errors or omissions.	Three or four errors or omissions.	Five or more errors or omissions.	
Short-answer and discussion questions	Answers complete, valid, and contain some elaboration. No misinterpretations noted.	Answers generally complete and valid. Only minor inaccuracies noted. Minimal elaboration.	Marginal answers to the questions and contains inaccurate information.	Many answers incorrect or fail to address the topic. May be misinterpretations.	
Data collection and analysis	Data complete and displayed with a valid interpretation.	Only minor data missing or a slight misinterpretation.	Some omissions. Not displayed or interpreted accurately	Data incomplete or show serious misinterpretations	

TOTAL POINTS EARNED ______________

APPENDIX 4

Table of Correlations:

- **Laboratory Exercises, Virtual Labs, and Physiology Interactive Lab Simulations (Ph.I.L.S.)**
- **Followed by Virtual Lab Lessons and Ph.I.L.S. Lessons**

**Note: Virtual Lab Lessons are provided in alphabetical order grouped by simulation topic.*

LABORATORY EXERCISE	VIRTUAL LAB	Ph.I.L.S.
Fundamentals of Human Anatomy and Physiology		
1 Scientific Method and Measurements	• Scientific Method – Applying the Scientific Method: Pillbug Preference	
2 Body Organization, Membranes, and Terminology		
3 Chemistry of Life	• Chemical Composition of Cells – Test for Starch • Chemical Composition of Cells – Test for Fat • Chemical Composition of Cells – Test for Proteins	
4 Care and Use of the Microscope	• Microscopy – Operation of Brightfield Microscopy	
Cells		
5 Cell Structure and Function	• Microscopy – Animal Cells (Human Epidermal Cells)	
6 Movements Through Membranes	• Diffusion – Diffusion Across a Selectively Permeable Membrane • Diffusion – Effect of Concentration on the Rate of Diffusion in a Semisolid • Osmosis – Movement of Water Across a Selectively Permeable Membrane • Osmosis – Tonicity in Red Blood Cells	• Ph.I.L.S. Lesson 1: Osmosis and Diffusion: Varying Extracellular Concentration
7 Cell Cycle	• Cell Division – Examining Mitosis	
Tissues		
8 Epithelial Tissues	• Microscopy – Epithelial Tissue Histology	
9 Connective Tissues	• Microscopy – ConnectiveTissue Histology	
10 Muscle and Nervous Tissues	• Microscopy – Muscle Tissue Histology • Microscopy – Nervous Tissue Histology	
Integumentary System		
11 Integumentary system		
Skeletal System		
12 Bone Structure and Classification		
13 Organization of the Skeleton		
14 Skull		
15 Vertebral Column and Thoracic Cage		
16 Pectoral Girdle and Upper Limb		
17 Pelvic Girdle and Lower Limb		
18 Fetal Skeleton		
19 Joint Structure and Movements		

LABORATORY EXERCISE	VIRTUAL LAB	Ph.I.L.S.
Muscular System		
20 Skeletal Muscle Structure and Function	• Skeletal Muscle – Electrical Stimulation	• Ph.I.L.S. Lesson 5: Skeletal Muscle Function: Stimulus-Dependent Force Generation
21 Electromyography: BIOPAC Exercise	• Electromyography: Motor Unit Recruitment • Electromyography: Time to Fatigue	• Ph.I.L.S. Lesson 8: Skeletal Muscle Function: Principles of Summation and Tetanus
22 Muscles of the Head and Neck		
23 Muscles of the Chest, Shoulder, and Upper Limb	• Skeletal Muscle – Shoulder and Elbow Movement Exercise	
24 Muscles of the Vertebral Column, Abdominal Wall, and Pelvic Floor		
25 Muscles of the Hip and Lower Limb		
Surface Anatomy		
26 Surface Anatomy		
Nervous System		
27 Nervous Tissue and Nerves		• Ph.I.L.S. Lesson 10: Resting Potentials: Resting Potential and External [K^+] • Ph.I.L.S. Lesson 11: Resting Potentials: Resting Potential and External [Na^+] • Ph.I.L.S. Lesson 12: Action Potentials: The Compound Action Potential
28 Meninges, Spinal Cord, and Spinal Nerves		
29 Reflex Arc and Somatic Reflexes	• Nervous System – Demonstrate Monosynaptic Reflexes	
30 Brain and Cranial Nerves		
31A Reaction Time: BIOPAC Exercise		
31B Electroencephalography: BIOPAC Exercise		
32 Dissection of the Sheep Brain		
General and Special Senses		
33 General Senses		
34 Smell and Taste		
35 Eye Structure	• Eye and Vision – Eye Dissection	
36 Visual Tests and Demonstrations	• Eye and Vision – Visual Acuity Test • Eye and Vision – Astigmatism Test • Eye and Vision – Accommodation of the Lens • Eye and Vision – Color Vision Test • Eye and Vision – Blind Spot Demonstration • Eye and Vision – Pupillary Reflex Test • Eye and Vision – Convergence Reflex Test	
37 Ear and Hearing		
38 Ear and Equilibrium		
Endocrine System		
39 Endocrine Structure and Function	• Endocrine System – Influence of Thyroid Hormone on Temperature Regulation	• Ph.I.L.S. Lesson 19: Endocrine Function: Thyroid Gland and Metabolic Rate
40 Diabetic Physiology	• Endocrine System – Effects of Blood Glucose Level	
Cardiovascular System		
41 Blood Cells	• Blood – Differential White Blood Cell Count	
42 Blood Testing	• Blood – Hematocrit • Blood – Hemoglobin Content	• Ph.I.L.S. Lesson 34: Blood: pH & Hb-Oxygen Binding
43 Blood Typing	• Blood – Blood Typing	
44 Heart Structure		
45 Cardiac Cycle	• Cardiovascular Physiology – Heart Auscultation	• Ph.I.L.S. Lesson 21: Heart Function: Thermal and Chemical Effects
46 Electrocardiography: BIOPAC Exercise	• Cardiovascular Physiology – Electrocardiography	• Ph.I.L.S. Lesson 26: ECG and Heart Function: The Meaning of Heart Sounds • Ph.I.L.S. Lesson 40: Respiration: Deep Breathing and Cardiac Function
47 Blood Vessel Structure, Arteries, and Veins		
48 Pulse Rate and Blood Pressure	• Cardiovascular Physiology – Pulse Rate • Cardiovascular Physiology – Blood Pressure	

LABORATORY EXERCISE	VIRTUAL LAB	Ph.I.L.S.
Lymphatic System		
49 Lymphatic System		
Respiratory System		
50 Respiratory Organs **51** Breathing and Respiratory Volumes **52** Spirometry: BIOPAC Exercise **53** Control of Breathing	• Respiratory System – Mechanism of Breathing • Respiratory System – Pulmonary Function Tests	• Ph.I.L.S. Lesson 38: Respiration: Alternating Airway Volume
Digestive System		
54 Digestive Organs **55** Action of a Digestive Enzyme **56** Metabolism	• pH Balance – Antacids as Buffers • Digestive System – Enzymes and Digestion	• Ph.I.L.S. Lesson 2: Metabolism: Size and Basal Metabolic Rate
Urinary System		
57 Urinary Organs **58** Urinalysis	• Urinary System - Urinalysis	
Reproductive Systems and Development		
59 Male Reproductive System **60** Female Reproductive System **61** Meiosis, Fertilization, and Early Development **62** Genetics	• Cell Division – Examining Meiosis • Human Genetics – Chromosomal Inheritance During Meiosis • Human Genetics – Genetic Inheritance	
Cat Dissection Exercises		
63 Cat Dissection: Musculature **64** Cat Dissection: Cardiovascular System **65** Cat Dissection: Respiratory System **66** Cat Dissection: Digestive System **67** Cat Dissection: Urinary System **68** Cat Dissection: Reproduction Systems		
Fetal Pig Dissection Exercises		
69 Fetal Pig Dissection: Musculature **70** Fetal Pig Dissection: Cardiovascular System **71** Fetal Pig Dissection: Respiratory System **72** Fetal Pig Dissection: Digestive System **73** Fetal Pig Dissection: Urinary System **74** Fetal Pig Dissection: Reproduction Systems		
Supplemental Laboratory Exercises*		
S-1 Skeletal Muscle Contraction **S-2** Nerve Impulse Stimulation **S-3** Factors Affecting the Cardiac Cycle		• Ph.I.L.S. Lesson 8: Skeletal Muscle Function: Principles of Summation and Tetanus • Ph.I.L.S. Lesson 12: Action Potentials: The Compound Action Potential • Ph.I.L.S. Lesson 21: Frog Heart Function: Thermal and Chemical Effects

*These exercises and their associated Ph.I.L.S. lessons are available in the eBook via Connect Anatomy & Physiology and also online for instructor distribution; see Instructor Resources via Connect Library tab. These same Ph.I.L.S. lessons are also available in the lab manual.

Virtual Labs Applying the Scientific Method: Pillbug Preference

The purpose of this laboratory simulation is to practice using the steps of the scientific method. Students will design and analyze a controlled experiment using pillbugs. A hypothesis will be developed, a strategy will be established to test the hypothesis, experimental and control setups will be determined and conducted, and a conclusion will be drawn.

As you follow these steps, answer the corresponding questions:

1. Click the link to this *Virtual* Lab assignment that was set up by your instructor.
2. Read the Key Concepts.
   a. What are the four main steps in the scientific method process?

      (1) ____________________ (3) ____________________

      (2) ____________________ (4) ____________________

   b. What is the difference between a control group and a test (experimental) group?

      ____

      ____

   c. What step in the scientific method provides a discussion to confirm or reject the original hypothesis?

      ____

   d. If a hypothesis is confirmed, what do you think might be the next steps?

      ____

   e. If a hypothesis is rejected, what do you think might be the next steps?

      ____

3. Read the Overview and watch the short animation.
4. Read Before you begin to receive background information about pillbugs.
5. Continue to the Laboratory Simulation.
6. Complete Phase 1: Hypothesis & Strategy. Each step needs to be completed before moving on to the next step. (Click the bottom tabs to review the Methods, Reset the experiment, add observations in My Notes, record Lab Data, or Show Labels.)
   a. What is your hypothesis for this experiment? ______________________________

      ____

   b. What is your strategy? ______________________________

      ____

7. Complete Phase 2: Experiment.
   a. What is the variable being tested in the experimental group?

      ____

8. Complete Phase 3: Graph and analyze data. Identify the x-axis and y-axis, select line graph from the graph type drop down menu. Click to "CREATE" and "SAVE" the graph. Answer the following questions:

   a. Where was the greatest difference in the number of pillbugs at the end of the 12 minutes?

   ____

   b. Was this greatest difference found in the experimental setup or the control setup or in a combination of both? ____

9. Complete Phase 4: Conclusion.

   a. Did the experiment results confirm or reject your original hypothesis? Explain.

   ____

   ____

10. Complete Phase 5: Save Lab Data. Look for the "Congratulations" message that the finished lab has been automatically submitted. Read the options to save the lab data.

CRITICAL THINKING ASSESSMENT – *Virtual* Lab

Develop a different hypothesis that you could test using pillbugs.

____

____

Share your strategy for testing your hypothesis. ____

____

____

What is your experimental group setup? ____

____

What is your control group setup? ____

____

Blood – Blood Typing

Virtual Labs

The purpose of this laboratory simulation is to determine and evaluate the ABO and Rh blood types of the samples.

As you follow these steps, answer the corresponding questions:

1. Click the link to this *Virtual* Lab assignment that was set up by your instructor.
2. Read the Key Concepts.
   - **a.** Where are antigens that determine ABO and Rh blood types located?

     ____
   - **b.** If present, where are antibodies for blood typing located? ____________________
   - **c.** Complete the table:

Blood type	Antigen	Antibodies	Can donate to	Can receive from
A				
B				
AB				
O				
Rh positive				
Rh negative				

3. Read the Overview.
   - **a.** Why do you think it is critical to know a person's blood type prior to a blood transfusion?

     ____
4. Read Before you begin and make sure you are familiar with the terms.
   - **a.** What type of organic compound is an antigen? ____________________
   - **b.** What do antibodies do? ____________________
   - **c.** When agglutinins bind to agglutinogens, the agglutinogens clump together. What is the clumping of red blood cells in blood typing called? ____________________
   - **d.** What is the fluid containing antibodies called? ____________________
   - **e.** What does "anti-A" mean? ____________________
   - **f.** What does "anti-B" mean? ____________________
   - **g.** What does "anti-D" mean? ____________________
   - **h.** The blood samples that ____________________ determine the blood type.
5. Continue to the Laboratory Simulation.
6. Complete Phase 1: Blood lab equipment. Each step needs to be completed before moving on to the next step. (Click the bottom tabs to review the Methods, Reset the experiment, add observations in My Notes, record Lab Data, or Show Labels.)
7. Complete Phase 2: Blood Sample 1.
   - **a.** What would happen if the same toothpick were used on each blood sample?

     ____
   - **b.** What is the blood type (ABO and Rh) of Sample 1? ____________________

8. Complete Phase 3: Blood Sample 2.

   a. What is the blood type (ABO and Rh) of Sample 2? ____________________

9. Complete Phase 4: Blood Sample 3.

   a. What is the blood type (ABO and Rh) of Sample 3? ____________________

10. Complete Phase 5: Determine transfusion compatibility.

    a. Can Rh– individuals receive blood from Rh+ individuals? Why or why not?

    ____

    ____

    b. The results of this chart show that blood type __________ is a universal recipient.

11. Complete Phase 6: Save Lab Data. Look for the "Congratulations" message that the finished lab has been automatically submitted. Read the options to save the lab data.

CRITICAL THINKING ASSESSMENT – *Virtual* Lab

An individual has blood type AB+. What blood type(s) can this individual donate to?

____

An individual has blood type O+. What blood type(s) can this individual donate to?

____

An individual has blood type B–. What blood type(s) can this individual donate to?

____

What should be removed from the blood of blood type A– before donating it to an individual with AB+ blood?

____

Blood – Differential White Blood Cell Count

Virtual Labs

The purpose of this laboratory simulation is to use a brightfield microscope to conduct a differential white blood cell count (DIFF) to distinguish the types and amount of each type of white blood cells in a prepared slide of whole blood.

As you follow these steps, answer the corresponding questions:

1. Click the link to this *Virtual* Lab assignment that was set up by your instructor.

2. Read the Key Concepts.

a. What is another name for a white blood cell (WBC)? ______________________________

b. What does a DIFF measure? ______________________________

c. Why might a DIFF be done? ______________________________

____

d. The most abundant of the granulocytes are ______________________________

e. The most abundant of the agranulocytes are ______________________________

3. Read the Overview.

a. What magnification will be used to distinguish the WBCs? ______________________________

4. Read Before you begin.

a. How should the microscope slide be moved through one field of view at a time?

____

b. If you count 10 lymphocytes, what is the percentage of lymphocytes in your DIFF? ______________ Is this a normal count for lymphocytes? ____________

5. Continue to the Laboratory Simulation.

6. Complete Phase 1: WBC identification. Each step needs to be completed before moving on to the next step. (Click the bottom tabs to review the Methods, Reset the experiment, add observations in My Notes, record Lab Data, or Show Labels.)

a. Which WBC has bright red granules? ______________________________

b. Which WBC has a bean-shaped nucleus? ______________________________

7. Complete Phase 2: Lab equipment.

8. Complete Phase 3: Preparation of blood smear.

a. What stain was used on the blood smear slide? ______________________________

9. Complete Phase 4: Microscope differential count.

a. What is the most common WBC? ______________________________

b. Which WBC has a multilobed nucleus and pink granules? ______________________________

c. Which WBC is typically the largest in size? ______________________________

10. Complete Phase 5: Lab wrap-up.

a. Which type of leukocyte (WBC) is the least numerous agranulocyte in the blood? ______________________________

b. Which type of leukocyte (WBC) is the least numerous granulocyte in the blood? ______________________________

11. Complete Phase 6: Save Lab Data. Look for the "Congratulations" message that the finished lab has been automatically submitted. Read the options to save the lab data.

CRITICAL THINKING ASSESSMENT – *Virtual* Lab

A cancer patient being treated using chemotherapy has a DIFF that shows a lower than normal number of neutrophils. What is this called? ______________________________

Why might chemotherapy cause this condition? ______________________________

____

What is a concern in having this condition? ______________________________

____

This patient is given Neulasta®, a drug that stimulates myeloid growth factors. What is the target and effect of this drug? ______________________________

____

Virtual Labs — Blood – Hematocrit

The purpose of this laboratory simulation is to measure and evaluate the hematocrit (HCT) of blood samples from three athletes to determine if any of them have blood doped.

As you follow these steps, answer the corresponding questions:

1. Click the link to this *Virtual* Lab assignment that was set up by your instructor.

2. Read the Key Concepts.

a. Define *hematocrit (HCT).* ______________________________

b. What is the normal HCT range for adult males? ______________ Females? ______________

c. What might cause an abnormal elevation in HCT? ______________________________

____

d. What is "blood doping"? ______________________________

3. Read the Overview.

a. A blood sample that is doped is the ______________ control.

b. A blood sample that is not doped is the ______________ control.

4. Read Before you begin.

a. What do you think the centrifuge does to the blood sample? ______________________________

____

5. Continue to the Laboratory Simulation.

6. Complete Phase 1: Identify lab equipment. Each step needs to be completed before moving on to the next step. (Click the bottom tabs to review the Methods, Reset the experiment, add observations in My Notes, record Lab Data, or Show Labels.)

a. Notice that the capillary tubes are heparinized. What do you think this means?

____

7. Complete Phase 2: Positive control sample.

a. Is the clay end of the capillary tube facing the <u>*outside*</u> or <u>*center*</u> of the centrifuge? ______________

8. Complete Phase 3: Negative control sample.

a. Why is the negative control sample capillary tube placed directly across from the positive control sample capillary tube?

____

9. Complete Phase 4: Sample A.

10. Complete Phase 5: Sample B.

11. Complete Phase 6: Sample C.

12. Complete Phase 7: Balance and run the centrifuge.

a. Based upon your lab data, which sample shows that the athlete has blood doped? ______________
Explain your reasoning. ______________________________

____

13. Complete Phase 8: Lab wrap-up.

a. What is the relationship between hematocrit and oxygen-carrying capacity of the blood?

____

14. Complete Phase 9: Apply what you have learned.

a. What blood sample represented a person likely to have been blood doping?

____

b. There is a blood sample that had a higher hematocrit than the negative control but was not likely a person to have been blood doping? Why? ____

____

15. Complete Phase 10: Save Lab Data. Look for the "Congratulations" message that the finished lab has been automatically submitted. Read the options to save the lab data.

CRITICAL THINKING ASSESSMENT – *Virtual* Lab

Injection of erythropoietin (EPO) is a blood doping method that an athlete may use to give a competitive edge; therefore, EPO is used as a performance-enhancing drug. What does EPO do? ____

____

Provide a convincing argument as to the risks associated with taking this drug and why the athlete should refrain from blood doping. ____

____

____

____

Blood – Hemoglobin Content

Virtual Labs

The purpose of this laboratory simulation is to measure the hemoglobin (Hb) content of blood samples from three athletes to determine if any of them have blood doped.

As you follow these steps, answer the corresponding questions:

1. Click the link to this *Virtual* Lab assignment that was set up by your instructor.
2. Read the Key Concepts.
   a. Define *hemoglobin (Hb)*. ______
   b. About how much of a red blood cell consists of hemoglobin? ______
   c. What is the normal Hb range for adult males? ______ Females? ______
   d. "Blood doping" will *raise* or *lower* hemoglobin content. ______
   e. If hemoglobin content increases, so does blood ______.
3. Read the Overview.
   a. A blood sample that is doped is the ______ control.
   b. A blood sample that is not doped is the ______ control.
4. Read Before you begin.
   a. What does a hemoglobinometer do? ______
   ______
   b. Hb is measured in g Hb/100 mL or g Hb/dL. How many milliliters (mL) are in 1 deciliter (dL)? ______
5. Continue to the Laboratory Simulation.
6. Complete Phase 1: Identify lab equipment. Each step needs to be completed before moving on to the next step. (Click the bottom tabs to review the Methods, Reset the experiment, add observations in My Notes, record Lab Data, or Show Labels.)
   a. What does the word *hemolysis* mean? ______
   ______
7. Complete Phase 2: Positive control sample.
   a. Is the hemoglobin (Hb) content *below* or *above* or *within* normal range? ______
8. Complete Phase 3: Negative control sample.
   a. Is the hemoglobin (Hb) content *below* or *above* or *within* normal range? ______
9. Complete Phase 4: Sample A.
   a. Is the hemoglobin (Hb) content *below* or *above* or *within* normal range? ______
10. Complete Phase 5: Sample B.
    a. Is the hemoglobin (Hb) content *below* or *above* or *within* normal range? ______
11. Complete Phase 6: Sample C.
    a. Is the hemoglobin (Hb) content *below* or *above* or *within* normal range? ______
12. Complete Phase 7: Lab wrap-up.
    a. Which athlete likely blood doped? ______

13. Complete Phase 9: Apply what you have learned.

a. What characteristic of blood does Hb content measure? ____________________

b. What other blood test can also be used to measure this characteristic? Explain.

____

14. Complete Phase 10: Save Lab Data. Look for the "Congratulations" message that the finished lab has been automatically submitted. Read the options to save the lab data.

CRITICAL THINKING ***ASSESSMENT*** – *Virtual* Lab

A prevalent nutritional deficiency is iron deficiency. How would this affect hemoglobin content and the relative size of red blood cells (mean corpuscular volume, MCV)? Explain.

____

____

Some of the symptoms an individual with iron deficiency might experience is general weakness, lack of energy, and shortness of breath. Explain why. ____________________

____

____

Cardiovascular Physiology – Blood Pressure

The purpose of this laboratory simulation is to measure and examine blood pressure at different body positions and after exercise.

As you follow these steps, answer the corresponding questions:

1. Click the link to this *Virtual* Lab assignment that was set up by your instructor.
2. Read the Key Concepts.
   a. Define *blood pressure.* ______________________________
   b. Explain what is meant by "blood moves down a pressure gradient." ______________________________
   ____
   c. What are the three primary factors that influence blood pressure? ______________________________
   ____
   d. An individual's blood pressure reading is 110/70. What is the systolic pressure? ____________ The diastolic pressure? ____________
3. Read the Overview.
   a. Define *tissue perfusion.* ______________________________
4. Read Before you begin.
   a. What two pieces of equipment are needed to manually take blood pressure?
   ____
   b. Where is the sphygmomanometer placed? ______________________________
   c. Where is the diaphragm of the stethoscope placed? ______________________________
   d. What is chronically elevated blood pressure called? ______________________________
   e. What is chronically low normal blood pressure called? ______________________________
5. Continue to the Laboratory Simulation.
6. Complete Phase 1: Preparing to take pressure at brachial artery. Each step needs to be completed before moving on to the next step. (Click the bottom tabs to review the Methods, Reset the experiment, add observations in My Notes, record Lab Data, or Show Labels.)
   a. What does the systolic pressure measure? ______________________________
   b. What does the diastolic pressure measure? ______________________________
7. Complete Phase 2: Measure blood pressure after sitting for 5 minutes.
   a. When do Korotkoff sounds stop as the cuff is inflated? ______________________________
   b. Upon deflation, Korotkoff sounds are first heard when pressure in the vessel is <u>above</u> or <u>below</u> the pressure in the blood pressure cuff. ____________________ This is recorded as ____________________ pressure.
   c. Upon deflation, Korotkoff sounds are no longer heard when pressure in the vessel is <u>above</u> or <u>below</u> the pressure in the blood pressure cuff. ____________________ This is recorded as ____________________ pressure.
8. Complete Phase 3: Measure blood pressure after lying down for 5 minutes.
   a. How does the blood pressure after lying down for 5 minutes compare to after sitting for 5 minutes? ______________________________
   ____

9. Complete Phase 4: Measure blood pressure after standing for 5 minutes.

a. When standing, what force pulls blood to the lowest body location? ______________________________

10. Complete Phase 5: Measure blood pressure immediately after exercising.

a. What change in blood pressure was most significant? ______________________________

11. Complete Phase 6: Measure blood pressure immediately 5 minutes after exercising.

12. Complete Phase 7: Apply what you have learned.

a. Explain the results in blood pressure with exercise compared to 5 minutes after exercise. ______________________________

____

____

13. Complete Phase 8: Save Lab Data. Look for the "Congratulations" message that the finished lab has been automatically submitted. Read the options to save the lab data.

CRITICAL THINKING ASSESSMENT – *Virtual* Lab

A student in laboratory classroom is manually taking your blood pressure and tells you that your blood pressure is 123/75. Why might you question the accuracy of this reading?

____

____

Cardiovascular Physiology – Electrocardiography

Virtual Labs

The purpose of this laboratory simulation is to use the electrocardiography (ECG) to assess cardiac function.

As you follow these steps, answer the corresponding questions:

1. Click the link to this *Virtual* Lab assignment that was set up by your instructor.
2. Read the Key Concepts.
   - **a.** What is the heart's physiological pacemaker? ____________________

     Why? ____________________
   - **b.** Where along the conduction system is there a critical delay? ____________________

     Why? ____________________
   - **c.** What part of the conductions system is in the interventricular septum? ____________________
   - **d.** What part of the conduction system stimulates individual groups of ventricular myocardial cells?

     ____
   - **e.** What is an *electrocardiogram (ECG)?* ____________________

     ____
3. Read the Overview.
   - **a.** Why is an ECG used in clinical practice? ____________________

     ____
4. Read Before you begin.
   - **a.** The electrodes are attached to the ____________________ to form the three pairs ____________________ of the Einthoven's triangle.
   - **b.** During a heartbeat, what is the wave on an ECG depicted from the following:

     Depolarization of the atria at SA node = ____________________

     Depolarization of the ventricles = ____________________

     Repolarization of the ventricles = ____________________
   - **c.** Define a *segment.* ____________________
   - **d.** Define an *interval.* ____________________
   - **e.** Define a *complex.* ____________________
   - **f.** The heart rate can be calculated by measuring the ____________________ interval.
5. Continue to the Laboratory Simulation.
6. Complete Phase 1: Components of an ECG. Each step needs to be completed before moving on to the next step. (Click the bottom tabs to review the Methods, Reset the experiment, add observations in My Notes, record Lab Data, or Show Labels.)
7. Complete Phase 2: Prepare ECG.
8. Complete Phase 3: Measure ECG – sitting at rest.
   - **a.** The ____________________ interval decreases as heart rate ____________________.

9. Complete Phase 4: Measure ECG – sitting after exercise.

a. Compare the heart rate at rest vs. after exercise. ______________________________

____

b. After the R-R interval, what waveform had the second most significant change from rest to after exercise? __________

10. Complete Phase 5: Apply what you have learned.

a. When heart rate increases, what is the general change in duration of the waves, intervals, and segments? __________

11. Complete Phase 6: Save Lab Data. Look for the "Congratulations" message that the finished lab has been automatically submitted. Read the options to save the lab data.

CRITICAL THINKING **ASSESSMENT – *Virtual* Lab**

Predict the effect of the following on the R-R interval by placing an **X** in the column:

	Watching a scary movie	Beta blocker medication	High temperature and humidity	Underactive thyroid gland (hypothyroidism)	Sleep	Fever
Increases						
Decreases						

Explain your reasoning. ______________________________

____

____

Cardiovascular Physiology – Heart Auscultation

Virtual Labs

The purpose of this laboratory simulation is to identify the auscultation location for each of the four heart valves and to assess cardiac function through auscultation of the heart with a stethoscope.

As you follow these steps, answer the corresponding questions:

1. Click the link to this *Virtual* Lab assignment that was set up by your instructor.

2. Read the Key Concepts.

a. What causes the heart's valves to open and close? ______________________________

b. What is the main purpose of the heart's valves? ______________________________

c. What are the two sounds heard in a single heartbeat? ______________________________

d. The closing of the tricuspid and mitral valves produces the "__________" sound.

This sound occurs at the start of ventricular *diastole* or *systole*. ______________________________

This sound is associated with the ______________________________ on an ECG.

e. The closing of the aortic and pulmonary semilunar valves produces the "__________" sound.

This sound occurs at the start of ventricular *diastole* or *systole*. ______________________________

This sound is heard toward the end of the ______________________________ on an ECG.

3. Read the Overview.

a. How is auscultation commonly used in clinical practice? ______________________________

____

4. Read Before you begin.

a. Which heart sound is normally louder and longer? ______________________________

b. Where is the first heart sound best heard? ______________________________

c. Where is the second heart sound best heard? ______________________________

d. Consider using the mnemonic device "A PeT Monkey" to help you memorize the locations of the valves for auscultation, starting on the patient's right side going left and then inferior. Identify the valves (in order) of this memory aid:

____

5. Continue to the Laboratory Simulation.

6. Complete Phase 1: Locate specific auscultation areas. Each step needs to be completed before moving on to the next step. (Click the bottom tabs to review the Methods, Reset the experiment, add observations in My Notes, record Lab Data, or Show Labels.)

a. Which valve is being auscultated from the fifth intercostal space on the left side in the mid-clavicular line? ______________________________

7. Complete Phase 2: Recognize heart sounds.

a. Contraction of the ventricles *decreases* or *increases* pressure in the ventricles and the AV valves close.

____

b. Relaxation of the ventricles *decreases* or *increases* pressure in the ventricles and the semilunar valves close.

____

8. Complete Phase 3: Recognize abnormal heart sounds.

a. Which heart sound is affected by mitral valve prolapse? ______________________________

How is that heart sound affected? ______________________________

9. Complete Phase 4: Apply what you have learned.

a. If you hear an abnormal second heart sound, during which phase of the cardiac cycle is the abnormal sound occurring?

____

b. If you hear an abnormal first heart sound, which valves could be affected?

____

10. Complete Phase 7: Save Lab Data. Look for the "Congratulations" message that the finished lab has been automatically submitted. Read the options to save the lab data.

CRITICAL THINKING ASSESSMENT – *Virtual* Lab

Predict which of the following conditions produce an abnormal heart sound by placing an **X** in the column:

	Bradycardia	Aortic valve stenosis	Hypertrophy of ventricular myocardium	Tricuspid valve regurgitation
Abnormal heart sound				

Explain your reasoning. ______________________________

____

____

Cardiovascular Physiology – Pulse Rate

The purpose of this laboratory simulation is to examine the pulse rate and intensity in different arteries, body positions and activity levels.

As you follow these steps, answer the corresponding questions:

1. Click the link to this *Virtual* Lab assignment that was set up by your instructor.
2. Read the Key Concepts.
   a. Explain why pulse is felt during ventricular systole and not during ventricular diastole. ______
   b. If pulse rate is 70 beats per minute, what is the heart rate? ______
   c. How would watching a scary movie influence an individual's pulse rate? ______
3. Read the Overview.
   a. Explain how heart rate can be used to evaluate an individual's health. ______
   b. Systole of which ventricle (*left* or *right*) generates the pulse that is measured in the different systemic arteries? ______
4. Read Before you begin and make sure you are familiar with the terms.
   a. Why is it important to know the different locations a pulse can be palpated in a human body? ______
   b. Which artery can be palpated in the cervical region? ______
   c. Which artery can be palpated in the antecubital region? ______
   d. Which artery can be palpated in the popliteal region? ______
   e. Which artery can be palpated in the inguinal region? ______
   f. Which artery can be palpated on the thumb side of the wrist? ______
   g. Which artery can be palpated on the dorsum (top) of foot? ______
   h. The posterior tibial artery is palpated *medially* or *laterally*? ______
5. Continue to the Laboratory Simulation.
6. Complete Phase 1: Locate arterial pulse sites. Each step needs to be completed before moving on to the next step. (Click the bottom tabs to review the Methods, Reset the experiment, add observations in My Notes, record Lab Data, or Show Labels.)
7. Complete Phase 2: Hypothesis.
   a. What is your hypothesis regarding pulse rate and the artery's proximity to the heart? ______
   b. What is your hypothesis regarding the relative intensity of pulse pressure and the artery's proximity to the heart? ______
8. Complete Phase 3: Measure pulse rate at different locations.
   a. Does the location of the artery affect pulse rate? ______

9. Complete Phase 4: Hypothesis.

a. What is your hypothesis regarding physical activity and pulse rate?

___

10. Complete Phase 5: Measure pulse rate after lying down.

a. Did the pulse rate *increase* or *decrease* after reclining? ___

11. Complete Phase 6: Measure pulse rate after standing up.

a. Did the pulse rate *increase* or *decrease* after standing? ___

Why? ___

12. Complete Phase 7: Measure pulse rate after exercise.

a. Did the pulse rate *increase* or *decrease* after exercise? ___

Why? ___

b. Did the pulse rate *increase* or *decrease* 5 minutes after exercise? ___

Why? ___

13. Complete Phase 8: Lab wrap-up.

a. Based upon the simulation results, do you accept or reject your original hypotheses, and why? ___

___

___

14. Complete Phase 9: Save Lab Data. Look for the "Congratulations" message that the finished lab has been automatically submitted. Read the options to save the lab data.

CRITICAL THINKING **ASSESSMENT** – *Virtual* Lab

A deficiency in red blood cells or hemoglobin concentration results in anemia, which is one of many conditions that can cause *tachycardia*, a rapid resting pulse. Explain the heart's response to this condition. ___

___

___

Cell Division – Examining Meiosis

Virtual Labs

The purpose of this laboratory simulation is to understand and recognize the different phases of meiosis.

As you follow these steps, answer the corresponding questions:

1. Click the link to this *Virtual* Lab assignment that was set up by your instructor.
2. Read the Key Concepts.
   a. What cells of the body undergo meiosis? ____________________
   b. How many daughter cells are produced by meiosis? ____________________
   c. How are the daughter cells of meiosis different than those produced by mitosis? ____________________
   ____
   d. Meiosis involves how many divisions of the genetic material? ____________________
   e. When is the genetic material replicated in meiosis? ____________________
   f. What two processes in meiosis create genetic variability? ____________________
   ____
   g. Why is it important that each gamete has half the number of chromosomes as body cells? ____________________
   ____
3. Read the Overview.
4. Read Before you begin, make sure you are familiar with the terms.
   a. At what point in meiosis do gametes contain twenty-three chromosomes? ____________________
   b. Where does meiosis take place in females? ____________________
   c. Where does meiosis take place in males? ____________________
5. Continue to the Laboratory Simulation.
6. Complete Phase 1: Phases of meiosis I on the whiteboard. Each step needs to be completed before moving on to the next step. (Click the bottom tabs to review the Methods, Reset the experiment, add observations in My Notes, record Lab Data, or Show Labels.)
   a. How many pairs of homologous chromosomes are shown in prophase I on the whiteboard? ____________________
   b. In what phase of meiosis I does crossing over occur? ____________________
7. Complete Phase 2: Phases of meiosis I in micrographs.
8. Complete Phase 3: Phases of meiosis II on the whiteboard.
   a. How many daughter cells are produced at the end of meiosis II? ____________________
9. Complete Phase 4: Phases of meiosis II in micrographs.
10. Complete Phase 5: Ovary cells under the microscope.
11. Complete Phase 6: Labeling follicles in ovary.
    a. What cell can be found in an ovarian follicle? ____________________
12. Complete Phase 7: Testis cells under the microscope.
13. Complete Phase 8: Labeling cells and structures in testis.
    a. What cells do germ-line cells of the testis ultimately produce? ____________________

14. Complete Phase 9: Apply what you have learned.

a. What are the functions of meiosis? ____

____

b. Complete the chart by placing an **X** in the correct column:

	Mitosis	Meiosis
Daughter cells are genetically identical		
Produces body cells		
Produces gametes (oocyte/sperm)		
Replaces dead and damaged cells		
Daughter cells contain half the number of chromosomes as the parent cell		
Daughter cells are each genetically unique		

15. Complete Phase 10: Save Lab Data. Look for the "Congratulations" message that the finished lab has been automatically submitted. Read the options to save the lab data.

CRITICAL THINKING ***ASSESSMENT*** ***– Virtual*** **Lab**

A cell with half the number of chromosomes is a haploid (*n*) cell while a cell with the complete set of chromosomes is a diploid (2*n*) cell. In humans, a haploid cell has __________ chromosomes, and a diploid cell has __________ chromosomes. At the end of meiosis I and at the end of meiosis II, the cells are haploid. How do these haploid cells differ in each division? ____

____

____

Virtual Labs Cell Division – Examining Mitosis

The purpose of this laboratory simulation is to understand and recognize the different phases of mitosis.

As you follow these steps, answer the corresponding questions:

1. Click the link to this *Virtual* Lab assignment that was set up by your instructor.
2. Read the Key Concepts.
   a. The phase of a cell's cycle that occurs prior to mitosis is ______________________.
   What events occur during this phase? ______________________
   ____
   b. What are the four phases of mitosis in order of occurrence? ______________________
   ____
3. Read the Overview.
4. Read Before you begin, make sure you are familiar with the terms.
   a. A photograph taken through a microscope is a(an) ______________________.
   b. Define *mitosis*. ______________________
   c. How is a *chromosome* different from *sister chromatids*? ______________________
   ____
   d. What organelles are found in the centrosome of animal cells? ______________________
   What do these organelles help form? ______________________
5. Continue to the Laboratory Simulation.
6. Complete Phase 1: Phases of mitosis on the whiteboard. Each step needs to be completed before moving on to the next step. (Click the bottom tabs to review the Methods, Reset the experiment, add observations in My Notes, record Lab Data, or Show Labels.)
   a. In what phase of mitosis have the sister chromatids detached from the centromere and migrated to opposite poles of the cell? ______________________
7. Complete Phase 2: Phases of mitosis in micrographs.
8. Complete Phase 3: Mitotic cells under the microscope.
   a. The ____________ power objective should be used first to get the slide in focus using the coarse adjustment knob.
   b. Which power objective has the smallest field of view? ______________________
9. Complete Phase 4: Identifying mitotic cells in onion root tip cells.
   a. In which phase of mitosis are the chromosomes aligned along the equator of the cell? ______________________
   b. In which phase of mitosis does the nuclear envelope and nucleolus form at each pole and the chromosomes decondense?
   ____
   c. In which phase of mitosis does the nuclear envelope break up and the chromosomes condense and easily appear?
   ____

10. Complete Phase 5: Apply what you have learned.

a. What are the functions of mitosis for most cells of the body? ______________________________

____

____

b. Certain cells such as neurons are amitotic. What does this mean? ______________________________

____

11. Complete Phase 6: Save Lab Data. Look for the "Congratulations" message that the finished lab has been automatically submitted. Read the options to save the lab data.

CRITICAL THINKING ASSESSMENT – *Virtual* Lab

Erythrocytes (red blood cells; RBCs) have a life span in blood circulation of about 120 days. These cells lack most organelles, including a nucleus. What does this tell you about a RBC's ability to carry out mitosis? Explain.

____

____

In general, how are new RBCs produced? ______________________________

____

Virtual Labs Chemical Composition of Cells – Test for Fat

The purpose of this laboratory simulation is to determine the presence of a lipid by observing evaporation on paper.

As you follow these steps, answer the corresponding questions:

1. Click the link to this *Virtual* Lab assignment that was set up by your instructor.
2. Read the Key Concepts.
   a. Identify how lipids differ from water. ______________________________
   ____
   ____
3. Read the Overview.
4. Read Before you begin and make sure you are familiar with the terms.
   a. In the Interpreting the Paper Test for Fats and Oils image, describe a positive test for a lipid.
   ____
5. Continue to the Laboratory Simulation.
6. Complete Phase 1: Test for fats. Each step needs to be completed before moving on to the next step. (Click the bottom tabs to review the Methods, Reset the experiment, add observations in My Notes, record Lab Data, or Show Labels.)
7. Complete Phase 2: Interpret your results.
   a. Which paper looks like a glass of ice water had been removed from it? ______________________________
8. Complete Phase 3: Apply what you have learned.
   a. Based on your observations, do you think that Caesar salad dressing would have a *negative* or *positive* result for lipids? ______________ Explain your answer. ______________________________
   ____
9. Complete Phase 4: Save Lab Data. Look for the "Congratulations" message that the finished lab has been automatically submitted. Read the options to save the lab data.

CRITICAL THINKING ASSESSMENT – Virtual Lab

Predict the test for lipid results for the following foods using the table below:

	Gatorade	Mocha latte	Mayonnaise
Appearance of stain			

Explain your reasoning. ______________________________
____
____

CRITICAL THINKING ASSESSMENT – *Virtual* Lab

Adequate dietary fat intake is important to a person's overall health. For example, monounsaturated and polyunsaturated fats are beneficial because they can lower bad cholesterol (LDL) in the blood. Consider where fat is found in the human body, and then, make a case to your friend who wants to lose weight by going on a no-fat diet. ____________________

____

____

____

Virtual Labs Chemical Composition of Cells – Test for Proteins

The purpose of this laboratory simulation is to practice using the Biuret test to determine the presence and concentration of protein on various solutions.

As you follow these steps, answer the corresponding questions:

1. Click the link to this *Virtual* Lab assignment that was set up by your instructor.

2. Read the Key Concepts.

a. What simple molecules are monomers that compose the proteins? ______________________ How many types of these monomers are there in biological systems? ______________

b. Brainstorm a list of proteins that are found in the human body. ______________________________________

____

____

3. Read the Overview.

4. Read Before you begin and make sure you are familiar with the terms.

a. What color is the Biuret reagent? ______________

b. What color would a solution be if a negative result for Biuret test occurs? ______________ A positive result? ______________

5. Continue to the Laboratory Simulation.

6. Complete Phase 1: Solution preparation. Each step needs to be completed before moving on to the next step. (Click the bottom tabs to review the Methods, Reset the experiment, add observations in My Notes, record Lab Data, or Show Labels.)

a. Which test tube is the negative control preparation? ______________________________________

7. Complete Phase 2: Add Biuret reagent to test for protein.

a. Before adding the Biuret reagent, which test tube(s) do you think will show a positive test for protein?______________

Explain your answer. ____

b. After adding the Biuret reagent, were you correct in your thinking? ______________________________________

8. Complete Phase 3: Interpret your results.

a. Which test tube(s) contained amino acids? ______________

9. Complete Phase 4: Apply what you have learned.

a. If the juice from red meat were tested for protein, what do you think the color of the meat solution will be after adding Biuret reagent? ______________ Explain your answer. ______________________________________

____

10. Complete Phase 5: Save Lab Data. Look for the "Congratulations" message that the finished lab has been automatically submitted. Read the options to save the lab data.

CRITICAL THINKING ASSESSMENT – *Virtual* Lab

Predict the test for protein results for the following substances using the table below:

	Egg whites	Tonic water	Hemoglobin	Lemonade
Color				

Explain your reasoning. ____

____

____

Chemical Composition of Cells – Test for Starch

Virtual Labs

The purpose of this laboratory simulation is to practice using positive and negative controls to determine if starch is present in potatoes and onions.

As you follow these steps, answer the corresponding questions:

1. Click the link to this *Virtual* Lab assignment that was set up by your instructor.
2. Read the Key Concepts.
   a. What simple molecules compose the polysaccharide starch? _______________
   b. What color does iodine solution turn when it contacts starch? _______________
3. Read the Overview.
4. Read Before you begin and make sure you are familiar with the terms.
   a. In the Iodine Test for Starch image, which numbered test tube(s) represents a negative test for starch? _______________
   b. In the Iodine Test for Starch image, which numbered test tube(s) represents a positive test for starch? _______________
   c. In the Iodine Test for Starch image, which numbered test tube contains the largest amount of starch? _______________
5. Continue to the Laboratory Simulation.
6. Complete Phase 1: Positive and negative control preparation. Each step needs to be completed before moving on to the next step. (Click the bottom tabs to review the Methods, Reset the experiment, add observations in My Notes, record Lab Data, or Show Labels.)
   a. Which setup(s) is a positive control preparation? _______________
   b. Which setup(s) is a negative control preparation? _______________
7. Complete Phase 2: Potato solution preparation.
   a. Which numbered test tube (1, 2, and/or 3) do you think is most similar to the potato solution (test tube 4)? _______________ Explain your answer. _______________

   ____
   b. Do you think the potato solution contains starch? _______________
8. Complete Phase 3: Onion solution preparation.
   a. Do you think the onion solution contains starch? _______________
9. Complete Phase 4: Test for starch using iodine.
   a. What is the indicator solution being used to test for the presence of starch? _______________
   b. Which test tube (1, 2, 3, 4, or 5) was used to rule out that water in the potato and onion solutions cause a positive result? _______________
   c. Are the colors of the test tubes representative of what you would have hypothesized? _______________ Explain. _______________

   ____
10. Complete Phase 5: Lab wrap-up.
    a. Which food, potato or onion, contains more starch? _______________
11. Complete Phase 6: Apply what you have learned.
    a. If an apple is tested for the presence of starch, what do you think the color of the apple solution will be after adding iodine? _______________ Explain. _______________

    ____

12. Complete Phase 7: Save Lab Data. Look for the "Congratulations" message that the finished lab has been automatically submitted. Read the options to save the lab data.

CRITICAL THINKING ***ASSESSMENT – Virtual*** **Lab**

Predict the test for starch results for the following foods using the table below:

	Corn	Steak	Saltine cracker	Watermelon
Color				
Contains starch (Yes or No)				

Explain your reasoning. ____

____

____

Diffusion – Diffusion Across a Selectively Permeable Membrane

Virtual Labs

The purpose of this laboratory simulation is to examine the diffusion of iodine, starch, and glucose across an artificial membrane and explore factors that affect the rate of diffusion.

As you follow these steps, answer the corresponding questions:

1. Click the link to this *Virtual* Lab assignment that was set up by your instructor.

2. Read the Key Concepts.

a. Define *diffusion*. ______________________________

____

3. Read the Overview.

4. Read Before you begin and make sure you are familiar with the terms.

a. Iodine is an *indicator* for the presence of starch. What do you think this means? ______________________________

____

b. What is a selectively permeable membrane that is part of a cell? ______________________________

5. Continue to the Laboratory Simulation.

6. Complete Phase 1: Hypothesis & Strategy. Each step needs to be completed before moving on to the next step. (Click the bottom tabs to review the Methods, Reset the experiment, add observations in My Notes, record Lab Data, or Show Labels.)

a. What is your hypothesis? ______________________________

____

b. Why should iodine be placed on a separate side of the dialysis tubing from starch?

____

7. Complete Phase 2: Diffusion across a semipermeable membrane.

a. After 10 minutes, which substance did *not* diffuse? ______________________________

How do you know? ______________________________

Why did this substance not diffuse? ______________________________

8. Complete Phase 3: Conclusion.

a. Based upon your experimental results, do you accept or reject your original hypothesis, and why? ______________________________

____

9. Complete Phase 4: Save Lab Data. Look for the "Congratulations" message that the finished lab has been automatically submitted. Read the options to save the lab data.

CRITICAL THINKING ASSESSMENT – *Virtual* Lab

In addition to molecular size, there are other factors that influence the rate of diffusion of iodine, glucose, and/or starch. Choose one factor and develop a hypothesis and strategy you would use to test.

What is your hypothesis for this experiment? ______________________________

____

____

Share your strategy for testing your hypothesis. ______________________________

____

____

____

Diffusion – Effect of Concentration on the Rate of Diffusion in a Semisolid

The purpose of this laboratory simulation is to examine the diffusion of potassium permanganate crystals in an agar gel over time.

As you follow these steps, answer the corresponding questions:

1. Click the link to this *Virtual* Lab assignment that was set up by your instructor.
2. Read the Key Concepts.
   a. Define *diffusion*. ______________________________
   ____
   b. What factors may influence the rate of diffusion? ______________________________
   ____
3. Read the Overview.
4. Read Before you begin and make sure you are familiar with the terms.
   a. What is the solute being used in this experiment? ______________________________
   b. What is the media being used in this experiment? ______________________________
5. Continue to the Laboratory Simulation.
6. Complete Phase 1: Hypothesis & Strategy. Each step needs to be completed before moving on to the next step. (Click the bottom tabs to review the Methods, Reset the experiment, add observations in My Notes, record Lab Data, or Show Labels.)
   a. What is your hypothesis? ______________________________
   ____
   ____
7. Complete Phase 2: Diffusion rate.
   a. Based upon your results, the semisolid media with __________ crystals of potassium permanganate had the greatest rate of diffusion.
   b. The rate of diffusion <u>*increased*</u> or <u>*decreased*</u> or <u>*stayed the same*</u> from the first measurement at ten minutes until the last measurement at thirty minutes for each dish. ______________________________
   Explain why. ______________________________
   ____
8. Complete Phase 3: Graph and analyze data.
   a. What is the variable for the y-axis? ______________________________
   b. What is the variable for the x-axis? ______________________________
   c. Interpret the results of the line graph. ______________________________
   ____
9. Complete Phase 4: Conclusion.
   a. Based upon your experimental results, do you accept or reject your original hypothesis, and why? __________
   ____

10. Complete Phase 5: Save Lab Data. Look for the "Congratulations" message that the finished lab has been automatically submitted. Read the options to save the lab data.

CRITICAL THINKING **ASSESSMENT –** ***Virtual*** **Lab**

What do you do in your daily life that exemplifies the results of this lab simulation and the effect of concentration on the rate of diffusion? Explain.

___

___

___

Virtual Labs

Digestive System – Enzymes and Digestion

The purpose of this laboratory simulation is to investigate how the enzyme amylase depends on pH for its activity.

As you follow these steps, answer the corresponding questions:

1. Click the link to this *Virtual* Lab assignment that was set up by your instructor.
2. Read the Key Concepts.
   a. Amylase is an enzyme found in secretions of the ________________ glands and the ________________.
   b. What is the substance (substrate) that amylase acts on? ________________
      What is the result of the chemical reaction? ________________
3. Read the Overview.
   a. What are the control setups? ________________
   b. A negative result for the Benedict test produces what color? ________________
   c. Which test tube (1–3) in the image contains the least amount of simple sugars? ________________
4. Read Before you begin and make sure you are familiar with the terms.
   a. What monosaccharide and disaccharide are considered simple sugars? ________________
   b. Starch and glycogen are both ________________.
   c. If no simple sugars are present, what color will the solution appear after adding the Benedict's reagent? ________________
5. Continue to the Laboratory Simulation.
6. Complete Phase 1: Amylase and starch at pH 7. Each step needs to be completed before moving on to the next step. (Click the bottom tabs to review the Methods, Reset the experiment, add observations in My Notes, record Lab Data, or Show Labels.)
   a. What is normal human body temperature? ____________ °C
   b. What does the Benedict reaction require to cause a color change? ________________
   c. Does test tube 1 have a _positive_ or _negative_ result? ________________
7. Complete Phase 2: Amylase and starch at pH 2.
   a. Does test tube 2 have a _positive_ or _negative_ result? ________________
8. Complete Phase 3: Maltose and water at pH 7.
   a. Why is this control solution important? ________________
   b. Is the result seen in test tube 3 what you would expect to see? Why or why not?
      ____
9. Complete Phase 4: Starch and water at pH 7.
   a. Why is this control solution important? ________________
   b. Is the result seen in test tube 4 what you would expect to see? Why or why not?
      ____
10. Complete Phase 5: Apply what you have learned.
    a. Which experimental set up test tube number showed the most amount of simple sugars present? ________________
    b. What do the results of this simulation tell you about amylase activity in the stomach versus in the small intestine?
       ____

11. Complete Phase 6: Save Lab Data. Look for the "Congratulations" message that the finished lab has been automatically submitted. Read the options to save the lab data.

CRITICAL THINKING ***ASSESSMENT – Virtual*** **Lab**

A grandma tells her grandchild to chew food twenty times before swallowing. Considering this lab simulation, can you think of any benefits to doing what grandma says?

____

____

____

Virtual Labs Electromyography – Motor Unit Recruitment

The purpose of this laboratory simulation is to investigate how changes in force influence motor unit activity by recording and reading an electromyogram (EMG).

As you follow these steps, answer the corresponding questions:

1. Click the link to this *Virtual* Lab assignment that was set up by your instructor.
2. Read the Key Concepts.
   - **a.** Define a *motor unit.* ______________________________

     ____
   - **b.** An ______________ measures the electrical activity of active motor units in muscle.
3. Read the Overview.
4. Read Before you begin.
   - **a.** The number of action potentials per second is ______________, measured in hertz (Hz).
   - **b.** A duration measurement of 200 ms would be ______________ seconds (s).

     Determine the number of Hz: ______________________________
5. Continue to the Laboratory Simulation.
6. Complete Phase 1: Hypothesis & Strategy. Each step needs to be completed before moving on to the next step. (Click the bottom tabs to review the Methods, Reset the experiment, add observations in My Notes, record Lab Data, or Show Labels.)
   - **a.** As grip strength is increased, what is being measured? ______________________________
   - **b.** Given the different grip strengths (25%, 50%, 75%, and 100%), what is your hypothesis about the relationship between the number of active motor units and force? ______________________________

     ____
7. Complete Phase 2: Observe EMG at 25% grip strength.
   - **a.** What is the number of active motor units at this grip strength? ______________________________
8. Complete Phase 3: Observe EMG at 50% grip strength.
   - **a.** What is the number of active motor units at this grip strength? ______________________________
9. Complete Phase 4: Observe EMG at 75% grip strength.
   - **a.** What is the number of active motor units at this grip strength? ______________________________
10. Complete Phase 5: Observe EMG at 100% grip strength.
    - **a.** What is the number of active motor units at this grip strength? ______________________________
11. Complete Phase 6: Conclusion.
    - **a.** Do you accept or reject your original hypothesis, and why? ______________________________

      ____

      ____

12. Complete Phase 7: Save Lab Data. Look for the "Congratulations" message that the finished lab has been automatically submitted. Read the options to save the lab data.

CRITICAL THINKING ASSESSMENT – *Virtual* Lab

The size of a motor unit differs in muscles of the human body. A muscle with large motor units would have *more* or *less* (circle one) muscle fibers innervated by a somatic motor neuron compared to a muscle with small motor units. Would *small* or *large* motor units enable more precise control of a muscle? _______________ Would *small* or *large* motor units enable gross control of a muscle? _________________

Put an **X** beneath the muscles of the body shown below that would have large motor units:

Gluteus maximus	Extensor digitorum	Orbicularis oculi	Gastrocnemius	Vastus lateralis

Explain your reasoning. ____

____

____

Virtual Labs

Electromyography – Time to Fatigue

The purpose of this laboratory simulation is to use an electromyogram (EMG) as you examine muscle fatigue while squeezing with 100% grip strength.

As you follow these steps, answer the corresponding questions:

1. Click the link to this *Virtual* Lab assignment that was set up by your instructor.
2. Read the Key Concepts.
   - **a.** Define *muscle fatigue.* ______________________________

     ____
   - **b.** An EMG will pick up decreased ____________ activity as contraction strength decreases.
3. Read the Overview.
   - **a.** How is the time to fatigue defined? ______________________________
   - **b.** As ATP and calcium decrease, a fully contracted muscle will ______________________________.
4. Read Before you begin.
5. Continue to the Laboratory Simulation.
6. Complete Phase 1: Hypothesis & Strategy. Each step needs to be completed before moving on to the next step. (Click the bottom tabs to review the Methods, Reset the experiment, add observations in My Notes, record Lab Data, or Show Labels.)
   - **a.** Why is it important to maintain maximum grip strength while conducting this experiment? ____________

     ____
7. Complete Phase 2: Time to fatigue.
   - **a.** What was the time to fatigue in this experiment? ______________________________
8. Complete Phase 3: Graph and analyze data.
   - **a.** What independent variable did you use for the x-axis? ______________________________
   - **b.** What dependent variable did you use for the y-axis? ______________________________
   - **c.** Predict what the force and number of active motor units would be if the elapsed time was 120 seconds. Explain. ______________________________

     ____
9. Complete Phase 4: Conclusion
   - **a.** Is there a direct or indirect relationship between force (kg) and number of active motor units over time (s)? Explain. ______________________________

     ____
10. Complete Phase 5: Save Lab Data. Look for the "Congratulations" message that the finished lab has been automatically submitted. Read the options to save the lab data.

CRITICAL THINKING **ASSESSMENT – *Virtual* Lab**

A friend tells you that a person is just not mentally tough enough if they can't continue a physically demanding task, such as running a marathon, to completion. Make an argument to the contrary. ____________________

____

____

____

Virtual Labs — Endocrine System – Effects of Blood Glucose Level

The purpose of this laboratory simulation is to examine the effects of insulin and glucose by observing the level of activity and alertness of fish.

As you follow these steps, answer the corresponding questions:

1. Click the link to this *Virtual* Lab assignment that was set up by your instructor.
2. Read the Key Concepts.

   a. What role does insulin play in blood glucose homeostasis? ______________________________

   ____

   b. What types of body cells are especially influenced by insulin? ______________________________
3. Read the Overview.

   a. Why is fish being used in this lab simulation? ______________________________
4. Read Before you begin.

   a. What are the signs of a hyperglycemic fish? ______________________________

   b. What are the signs of a hypoglycemic fish? ______________________________

   c. What are the signs of a normoglycemic fish? ______________________________
5. Continue to the Laboratory Simulation.
6. Complete Phase 1: Observation of fish swimming behavior. Each step needs to be completed before moving on to the next step. (Click the bottom tabs to review the Methods, Reset the experiment, add observations in My Notes, record Lab Data, or Show Labels.)
7. Complete Phase 2: Effect of insulin on fish behavior.

   a. What caused the observed behavior in the fish at 1 minute? ______________________________

   ____

   b. Was this fish in insulin shock? ______________________________
8. Complete Phase 3: Effect of glucose on fish behavior.

   a. What caused the observed behavior in the fish at 1 minute? ______________________________

   ____
9. Complete Phase 4: Apply what you have learned.

   a. Why would this improve the diabetic patient's condition in this situation? ______________________________

   ____
10. Complete Phase 5: Save Lab Data. Look for the "Congratulations" message that the finished lab has been automatically submitted. Read the options to save the lab data.

CRITICAL THINKING ***ASSESSMENT – Virtual*** **Lab**

Discuss factors in an insulin-dependent diabetic's daily life that could make the control of blood glucose a balancing act.

Endocrine System – Influence of Thyroid Hormone on Temperature Regulation

Virtual Labs

The purpose of this laboratory simulation is to examine the effect that thyroid hormone has on body temperature regulation and oxygen consumption by administering a drug—propylthiouracil (PTU)—that inhibits thyroid hormone production.

As you follow these steps, answer the corresponding questions:

1. Click the link to this *Virtual* Lab assignment that was set up by your instructor.
2. Read the Key Concepts.
   - **a.** What is shivering? ____________________

     When does this response occur?____________________
   - **b.** Why does shivering cause a greater consumption of oxygen? ____________________

     ____
   - **c.** What does thyroid hormone do in response to low body temperature? ____________________
3. Read the Overview.
   - **a.** Why is it important to have a controlled experiment? ____________________

     ____
   - **b.** What will be measured in the control and experimental groups in this lab simulation? ____________________
4. Read Before you begin.
   - **a.** What is the relationship between thyroid hormone and body temperature? ____________________
   - **b.** What does propylthiouracil (PTU) do? ____________________
   - **c.** What does a respirometer measure? ____________________
   - **d.** Why is soda lime used in the respirometer? ____________________
5. Continue to the Laboratory Simulation.
6. Complete Phase 1: Hypothesis & Strategy. Each step needs to be completed before moving on to the next step. (Click the bottom tabs to review the Methods, Reset the experiment, add observations in My Notes, record Lab Data, or Show Labels.)
   - **a.** Will inhibition of thyroid hormone *raise* or *lower* metabolic rate? ____________________
   - **b.** What is your hypothesis for this lab simulation? ____________________

     ____
   - **c.** Does the weight of the animal need to be taken into consideration when comparing oxygen consumption? __________
7. Complete Phase 2: Weigh mice.
8. Complete Phase 3: Measure O_2 consumption rate for control mouse at 10°C.
9. Complete Phase 4: Measure O_2 consumption rate for control mouse at 14°C.
10. Complete Phase 5: Measure O_2 consumption rate for control mouse at 18°C.
11. Complete Phase 6: Measure O_2 consumption rate for control mouse at 22°C.
    - **a.** What pattern are you observing in your data? ____________________

      ____

12. Complete Phase 7: Measure O_2 consumption rate for experimental mouse at 10°C.

13. Complete Phase 8: Measure O_2 consumption rate for experimental mouse at 14°C.

14. Complete Phase 9: Measure O_2 consumption rate for experimental mouse at 18°C.

15. Complete Phase 10: Measure O_2 consumption rate for experimental mouse at 22°C.

    a. What pattern are you observing in your data? ______________________________

    ____

16. Complete Phase 11: Graph and analyze data.

    a. What is the independent variable did you use for the x-axis? ______________________________

    b. What is the dependent variable did you use for the y-axis? ______________________________

17. Complete Phase 12: Conclusion.

    a. Do you accept or reject your original hypothesis, and why? ______________________________

    ____

    ____

    b. Why does oxygen consumption increase at lower body temperature? ______________________________

    ____

    c. Which mouse, the <u>*control*</u> or <u>*experimental*</u>, had less thyroid hormone? ______________________________

    d. Which mouse, the <u>*control*</u> or <u>*experimental*</u>, consumed more oxygen? ______________________________

    Why? ______________________________

18. Complete Phase 13: Save Lab Data. Look for the "Congratulations" message that the finished lab has been automatically submitted. Read the options to save the lab data.

CRITICAL THINKING ASSESSMENT – *Virtual* Lab

If an individual produces a higher than normal amount of thyroid hormone, known as hyperthyroidism, what would be the effect on metabolic rate? ______________________________

What will happen to body temperature? ______________________________

How will the body respond to help regulate temperature? ______________________________

____

Virtual Labs — Eye and Vision – Accommodation of the Lens

The purpose of this laboratory simulation is to understand lens accommodation and to test the lens accommodation reflex by examining the near point of focus.

As you follow these steps, answer the corresponding questions:

1. Click the link to this *Virtual* Lab assignment that was set up by your instructor.

2. Read the Key Concepts.

a. Define *accommodation of the lens.* ______________________________

____

b. What specific area of the retina is the image being focused as a result of this lens accommodation reflex, and why?

____

____

c. What structure of the eye is responsible for adjusting the shape of the lens during accommodation? ______________

____

d. What division of the autonomic nervous system controls accommodation of the lens? ______________

____

3. Read the Overview.

a. What is the *near point of focus*? ______________________________

b. Which cranial nerve is being tested when doing the accommodation test? ______________

4. Read Before you begin.

a. The lens accommodation for distant vision is a ______________ shaped lens.

b. The lens accommodation for near vision is a ______________ shaped lens.

c. As a person ages, the ______________________ moves farther away.

Why? ______________________________

5. Continue to the Laboratory Simulation.

6. Complete Phase 1: Observe lens accommodation reflex in Patient 1. Each step needs to be completed before moving on to the next step. (Click the bottom tabs to review the Methods, Reset the experiment, add observations in My Notes, record Lab Data, or Show Labels.)

7. Complete Phase 2: Observe lens accommodation reflex in Patient 2.

8. Complete Phase 3: Apply what you have learned.

a. The near point of focus is *closer* or *farther away* in a younger person compared to an older person. ______________

9. Complete Phase 4: Save Lab Data. Look for the "Congratulations" message that the finished lab has been automatically submitted. Read the options to save the lab data.

CRITICAL THINKING ***ASSESSMENT – Virtual*** **Lab**

While dining at a restaurant, you notice your long-time friend who used to have exceptional vision is now reading the menu while holding it at arm's length. Gently tell your friend what you are observing, the likely reason behind it, and what you suggest be done about it.

____

____

____

____

Eye and Vision – Astigmatism Test

Virtual Labs

The purpose of this laboratory simulation is to examine and evaluate patients who have astigmatism, a part of a routine eye exam.

As you follow these steps, answer the corresponding questions:

1. Click the link to this *Virtual* Lab assignment that was set up by your instructor.
2. Read the Key Concepts.
   - **a.** Define *astigmatism*. ______________________________
   - **b.** Would eyeglasses to correct for hyperopia (farsightedness) have *concave* or *convex* lenses? ______________
   - **c.** Would eyeglasses to correct for myopia (nearsightedness) have *concave* or *convex* lenses? ______________
   - **d.** What does the lens power determine? ______________________________
   - **e.** What does having the correct angle ensure? ______________________________
   - **f.** What is the purpose of an astigmatism chart? ______________________________
3. Read the Overview.
   - **a.** For each patient in this simulation, what will be done in the eye examination (in order)? ______________

     ____
4. Read Before you begin and be familiar with how to read the astigmatism chart.
   - **a.** Where do lines on the astigmatism chart shown look the sharpest? ______________________________
   - **b.** If the sharpest lines on the astigmatism chart were instead along the 10 o'clock direction, it would correspond to _________ degrees.
5. Continue to the Laboratory Simulation.
6. Complete Phase 1: Determine vision rating and prescription for Patient 1. Each step needs to be completed before moving on to the next step. (Click the bottom tabs to review the Methods, Reset the experiment, add observations in My Notes, record Lab Data, or Show Labels.)
   - **a.** The lens power for this patient is correcting for *nearsightedness* or *farsightedness*. ______________
7. Complete Phase 2: Determine vision rating and prescription for Patient 2.
   - **a.** The lens power for this patient is correcting for *nearsightedness* or *farsightedness*. ______________
8. Complete Phase 3: Determine vision rating and prescription for Patient 3.
   - **a.** The lens power for this patient is correcting for *nearsightedness* or *farsightedness*. ______________
9. Complete Phase 4: Apply what you have learned.
   - **a.** In an eye exam, what procedural error would make a patient's astigmatism difficult to correct? ______________

     ____
10. Complete Phase 5: Save Lab Data. Look for the "Congratulations" message that the finished lab has been automatically submitted. Read the options to save the lab data.

CRITICAL THINKING ***ASSESSMENT*** *– Virtual* Lab

Astigmatism that is not treated can cause an individual to experience headaches and have trouble seeing at night. Why do you think this is so? ____

____

____

Eye and Vision – Blind Spot Demonstration

Virtual Labs

The purpose of this laboratory simulation is to determine your blind spot and understand why it exists.

As you follow these steps, answer the corresponding questions:

1. Click the link to this *Virtual* Lab assignment that was set up by your instructor.
2. Read the Key Concepts.
   a. Where is the optic disc located? ______________________________
   b. Where is the blind spot located? ______________________________
   c. Why does the optic disc create a blind spot? ______________________________
   d. If the left eye is injured, the blind spot of the ____________ eye will be affected.
3. Read the Overview.
   a. The ______________________ will be visualized when the test chart symbol disappears from your vision.
4. Read Before you begin.
5. Continue to the Laboratory Simulation.
6. Complete Phase 1: Visualize patient's blind spot. Each step needs to be completed before moving on to the next step. (Click the bottom tabs to review the Methods, Reset the experiment, add observations in My Notes, record Lab Data, or Show Labels.)
7. Complete Phase 2: Apply what you have learned.
   a. What evidence during this blind spot test shows that the brain filled in missing visual information based upon the surrounding area? ______________________________

      ____
   b. If the right eye is injured, the __________________________ visual field of the ____________ eye will be most negatively impacted.
8. Complete Phase 3: Save Lab Data. Look for the "Congratulations" message that the finished lab has been automatically submitted. Read the options to save the lab data.

CRITICAL THINKING ASSESSMENT – *Virtual* Lab

If a person has an eye injury and consequently wears a patch over the eye, or has a glass eye, what daily activities should be proceeded with caution or not done at all? Explain.

____

____

____

Virtual Labs

Eye and Vision – Color Vision Test

The purpose of this laboratory simulation is to use and interpret the results of the Ishihara color test to determine the presence and degree of color vision deficiencies.

As you follow these steps, answer the corresponding questions:

1. Click the link to this *Virtual* Lab assignment that was set up by your instructor.
2. Read the Key Concepts.
   a. Which type of photoreceptors are responsible for color vision? ______________________________
   b. How many types of cone photoreceptors are found in the retina? ______________ What are the different wavelengths they detect light? ______________________________
3. Read the Overview.
4. Read Before you begin and be familiar with the Ishihara color test.
   a. A person with ____________________ can read the number on each card.
5. Continue to the Laboratory Simulation.
6. Complete Phase 1: Determine color deficiency and color blindness. Each step needs to be completed before moving on to the next step. (Click the bottom tabs to review the Methods, Reset the experiment, add observations in My Notes, record Lab Data, or Show Labels.)
   a. Is this patient color-blind? ______________________________
7. Complete Phase 2: Interpret your patient's results.
   a. Provide reasoning for your interpretation of your patient's results.
   ____
   ____
8. Complete Phase 3: Apply what you have learned.
   a. What structure is malfunctioning in a person who is color-blind? ______________________________
9. Complete Phase 4: Save Lab Data. Look for the "Congratulations" message that the finished lab has been automatically submitted. Read the options to save the lab data.

CRITICAL THINKING ASSESSMENT – *Virtual* Lab

Most defects in color vision involve the red and green cone pigments. The genes that code for these pigments lie next to each other on the X chromosome. A defective gene can lead to red or green color blindness. What circumstance would need to occur for a female to be born color-blind? ______________________________

____

____

Virtual Labs

Eye and Vision – Convergence Reflex Test

The purpose of this laboratory simulation is to observe the two eyes working together to view and track a close object.

As you follow these steps, answer the corresponding questions:

1. Click the link to this *Virtual* Lab assignment that was set up by your instructor.
2. Read the Key Concepts.
   - **a.** What is the mechanism covered in this lab simulation for focusing light on the retina when viewing objects closer than 20 feet away? ____________________
   - **b.** On which part of the retina are the eyes working to focus light? ____________________

     Why? ____________________
   - **c.** In general, what muscles move the two eyes? ____________________

     What cranial nerves are involved? ____________________
   - **d.** What does the convergence reflex test measure? ____________________

     ____
3. Read the Overview.
4. Read Before you begin.
   - **a.** Define *convergence.* ____________________
5. Continue to the Laboratory Simulation.
6. Complete Phase 1: Demonstrate convergence reflex. Each step needs to be completed before moving on to the next step. (Click the bottom tabs to review the Methods, Reset the experiment, add observations in My Notes, record Lab Data, or Show Labels.)
   - **a.** Did both eyes move in the same direction to track the object? ____________________
7. Complete Phase 2: Apply what you have learned.
   - **a.** What extrinsic eye muscle may be affected? ____________________
   - **b.** Imagine doing this exam with a patient and getting no inferior and lateral movement of the right eye. Which nerve may be functioning improperly? ____________________

     Which extrinsic eye muscle may be affected? ____________________
8. Complete Phase 3: Save Lab Data. Look for the "Congratulations" message that the finished lab has been automatically submitted. Read the options to save the lab data.

CRITICAL THINKING ***ASSESSMENT*** *– Virtual* Lab

A parent notices their child closing one eye to read homework so that words don't float around on the page. The child is often complaining of blurred vision, headaches, and eye soreness. Upon seeing a doctor, what do you think a routine eye exam might show? What can this child do to help treat this condition?

____

____

____

Virtual Labs

Eye and Vision – Eye Dissection

The purpose of this laboratory simulation is to dissect and examine a cow's eye and make comparisons to a human eye.

As you follow these steps, answer the corresponding questions:

1. Click the link to this *Virtual* Lab assignment that was set up by your instructor.
2. Read the Key Concepts.
   a. Where is the tapetum lucidum located? ____________________

   What is its function? ____________________
3. Read the Overview.
   a. What is another term used for a "coronal" cut? ____________________
4. Read Before you begin and be familiar with the terms.
   a. What tunic is the tapetum lucidum a part of? ____________________

   b. Upon dissection, when the vitreous humor seeps out, what tunic might become detached? ____________________

   c. Given that the iris controls the size of the pupil, what primary tissue is it composed of? ____________________

   d. What holds the lens in place? ____________________

   What does this structure secrete? ____________________

   e. What is the function of fat (adipose tissue) around the eye? ____________________
5. Continue to the Laboratory Simulation.
6. Complete Phase 1: Identify tools for an eye dissection. Each step needs to be completed before moving on to the next step. (Click the bottom tabs to review the Methods, Reset the experiment, add observations in My Notes, record Lab Data, or Show Labels.)
   a. Which tool will be used most often to make cuts of the cow eye? ____________________
7. Complete Phase 2: Anterior eye anatomy.
   a. Which structure controls the amount of light entering the eye? ____________________

   b. What is the opening that allows light through? ____________________

   c. What convex transparent structure is continuous with the sclera and bulges anteriorly, allowing light to enter the eye? ____________________

   d. What structure focuses incoming light on the retina? ____________________
8. Complete Phase 3: Posterior eye anatomy.
   a. What structure makes up the "white of the eye"? ____________________

   b. What structure is highly vascular and darkly pigmented? ____________________

   c. What structure produces a blind spot in the field of vision because it does not have any photoreceptors? ____________________

   d. What gelatinous structure holds the retina in place, helps maintain inner pressure of the eye, and transmits light? ____________________
9. Complete Phase 4: Eye dissection. Follow the steps to complete the dissection.
   a. Where does the lens specifically focus the light on the retina? ____________________

   b. What is the function of the vitreous humor? Explain. ____________________

   ____

10. Complete Phase 5: Lab wrap-up.

a. The tapetum lucidum is an iridescent covering of the ________________ that enables vision in dim light.

b. Why is the retina detached in this image? ____

11. Complete Phase 6: Save Lab Data. Look for the "Congratulations" message that the finished lab has been automatically submitted. Read the options to save the lab data.

CRITICAL THINKING ASSESSMENT – Virtual Lab

What characteristic of the cornea makes it suitable for transplant with low risk of rejection? Explain why. ________

____

____

Virtual Labs — Eye and Vision – Pupillary Reflex Test

The purpose of this laboratory simulation is to observe how the pupils respond to varying levels of light exposure and as an object is moved closer to the eyes.

As you follow these steps, answer the corresponding questions:

1. Click the link to this *Virtual* Lab assignment that was set up by your instructor.
2. Read the Key Concepts.
   a. The parasympathetic reflex of the ANS when bright light enters the eye is pupil ______________.
   b. The sympathetic reflex of the ANS when dim light enters the eye is pupil ______________.
   c. What is the mechanism covered in this lab simulation for focusing light on the retina when viewing objects closer than 20 feet away? ______________

      How does this mechanism work? ______________

      ____
3. Read the Overview.
   a. The function of what cranial nerve is being tested by the pupillary light reflex test? ______________
4. Read Before you begin.
   a. Normally, the pupils respond *differently* or *the same*. ______________
5. Continue to the Laboratory Simulation.
6. Complete Phase 1: Observe pupillary light reflex. Each step needs to be completed before moving on to the next step. (Click the bottom tabs to review the Methods, Reset the experiment, add observations in My Notes, record Lab Data, or Show Labels.)
   a. What was the response of both eyes to the light? ______________
   b. How is the pupillary light reflex protective? ______________

      ____
7. Complete Phase 2: Observe pupil constriction in response to viewing a close object.
   a. How did the pupils respond as the penlight moved farther away? ______________
8. Complete Phase 3: Apply what you have learned.
   a. Why do the pupils constrict as objects move closer? ______________

      ____
9. Complete Phase 4: Save Lab Data. Look for the "Congratulations" message that the finished lab has been automatically submitted. Read the options to save the lab data.

CRITICAL THINKING ASSESSMENT – *Virtual* Lab

Why is the pupillary light reflex test done on an athlete who is thought to have a concussion? ______________

____

____

Virtual Labs

Eye and Vision – Visual Acuity Test

The purpose of this laboratory simulation is to use and interpret the results of the Snellen eye chart that is commonly used in an eye exam.

As you follow these steps, answer the corresponding questions:

1. Click the link to this *Virtual* Lab assignment that was set up by your instructor.

2. Read the Key Concepts.

a. Define *visual acuity.* ______________________________

____

b. When light is focused on the retina, it is called ______________, or normal vision.

c. When light is focused in front of the retina, it is called ______________, which is ________sightedness.

d. When light is focused behind the retina, it is called ______________, which is ________sightedness.

3. Read the Overview.

a. Describe what a 20/50 vision rating means. ______________________________

____

Should this person be driving without corrective lenses? Explain why. ______________________________

____

b. Why is 20/10 on the Snellen chart considered excellent vision? ______________________________

____

4. Read Before you begin and be familiar with the terms and steps used to correct vision.

a. Would a positive or negative diopter be used in reading glasses? Explain why. ______________________________

____

b. Concave lenses bend light (more or less) __________ to correct ______________________________.

c. Convex lenses bend light (more or less) __________ to correct ______________________________.

5. Continue to the Laboratory Simulation.

6. Complete Phase 1: Determine vision rating and prescription for Patient 1. Each step needs to be completed before moving on to the next step. (Click the bottom tabs to review the Methods, Reset the experiment, add observations in My Notes, record Lab Data, or Show Labels.)

a. This patient has emmetropia or myopia or hyperopia. ______________________________

7. Complete Phase 2: Determine vision rating and prescription for Patient 2.

a. This patient has emmetropia or myopia or hyperopia. ______________________________

8. Complete Phase 3: Determine vision rating and prescription for Patient 3.

a. This patient has emmetropia or myopia or hyperopia. ______________________________

9. Complete Phase 4: Apply what you have learned.

a. The shape of the corrective lenses in reading glasses would be ______________________________.

10. Complete Phase 5: Save Lab Data. Look for the "Congratulations" message that the finished lab has been automatically submitted. Read the options to save the lab data.

CRITICAL THINKING ASSESSMENT – *Virtual* Lab

Why is it common for an individual to need reading glasses when they reach middle age? Explain.

___

___

___

Human Genetics – Chromosomal Inheritance During Meiosis

Virtual Labs

The purpose of this laboratory simulation is to demonstrate understanding of meiosis and its phases and examine nondisjunction.

As you follow these steps, answer the corresponding questions:

1. Click the link to this *Virtual* Lab assignment that was set up by your instructor.

2. Read the Key Concepts.

a. The cell division that produces gametes is ______________________.

b. What is *spermatogenesis?* ______________________

c. What is *oogenesis?* ______________________

d. What does oogenesis require to complete its process? ______________________

e. What are two results of meiosis? ______________________

f. Define *nondisjunction.* ______________________

3. Read the Overview.

a. During which specific two phases of meiosis might nondisjunction occur? ______________________

4. Read Before you begin and make sure you are familiar with the terms.

a. How many main phases does meiosis consist of? ______________________

b. How many identical sister chromatids does a replicated chromosome have? ______________________

c. How many pairs of autosomes are there in humans? ______________________

d. How many pairs of sex chromosomes are there in humans? ______________________

e. How many chromosomes does a diploid cell have in humans? ______________________

f. How many chromosomes does a haploid cell have in humans? ______________________

5. Continue to the Laboratory Simulation.

6. Complete Phase 1: Normal meiosis I. Each step needs to be completed before moving on to the next step. (Click the bottom tabs to review the Methods, Reset the experiment, add observations in My Notes, record Lab Data, or Show Labels.)

a. Are the cells at the end of meiosis I *diploid* or *haploid*? ______________________

b. What happens to the homologous chromosomes of a homologous pair during anaphase I? ______________________

7. Complete Phase 2: Normal meiosis II.

a. Are the cells at the end of meiosis II *diploid* or *haploid*? ______________________

b. How many cells are produced by meiosis? ______________________

8. Complete Phase 3: Nondisjunction during meiosis I.

a. What is the end product of meiosis I with nondisjunction? ______________________

____

9. Complete Phase 4: Completion of meiosis with nondisjunction during meiosis I.

a. What is the final product of meiosis II with nondisjunction occurring in meiosis I?

____

10. Complete Phase 5: Meiosis I before nondisjunction during meiosis II.

11. Complete Phase 6: Nondisjunction during meiosis II.

a. How many daughter cells contain the normal number of chromosomes at the end of meiosis II with nondisjunction occurring in meiosis II? ____________________

12. Complete Phase 7: Apply what you have learned.

a. Homologous chromosomes are separated in ____________________.

b. Sister chromatids are separated in ____________________.

13. Complete Phase 6: Save Lab Data. Look for the "Congratulations" message that the finished lab has been automatically submitted. Read the options to save the lab data.

CRITICAL THINKING **ASSESSMENT – *Virtual* Lab**

Trisomy 21 is the most common type of Down syndrome, a condition of nondisjunction. How many copies of chromosome 21 do these individuals have in every cell of their body? ____________ What increases a mother's risk of having a baby with trisomy 21?

____

Individuals afflicted with trisomy 21 have common physical features and a range of developmental challenges. How might schools help students with trisomy 21 achieve academic success in regular classes? ____________________

____

____

Virtual Labs

Human Genetics – Genetic Inheritance

The purpose of this laboratory simulation is to learn how to read pedigrees and differentiate between autosomal and X-linked inheritance patterns.

As you follow these steps, answer the corresponding questions:

1. Click the link to this *Virtual* Lab assignment that was set up by your instructor.

2. Read the Key Concepts.

a. What are autosomal traits? ______________________________

b. What are X-linked traits? ______________________________

c. Why can X-linked traits be concealed as recessive in females but not in males? ______________________________

____

3. Read the Overview.

4. Read Before you begin and make sure you are familiar with the terms.

a. Given these alleles—*AA, Aa, aa*—which allele is homozygous dominant? __________ Which allele is homozygous recessive? __________ Which allele is heterozygous dominant? __________

b. Is normal skin pigmentation and albinism a *genotype* or *phenotype*? ______________________________

5. Continue to the Laboratory Simulation.

6. Complete Phase 1: Identifying phenotypes from photos. Each step needs to be completed before moving on to the next step. (Click the bottom tabs to review the Methods, Reset the experiment, add observations in My Notes, record Lab Data, or Show Labels.)

a. What phenotypes do the grandparents have in common? ______________________________

7. Complete Phase 2: Dominant and recessive traits.

a. What are the dominant alleles? ______________________________

b. What are the recessive alleles? ______________________________

8. Complete Phase 3: Genotype of hairline.

a. Define *genotype*. ______________________________

b. What is the eldest male's genotype for hairline? ______________________________

c. How did you come to this conclusion? ______________________________

9. Complete Phase 4: Genotype of eye color.

a. What is the youngest female's genotype for eye color? ______________________________

b. How did you come to this conclusion? ______________________________

10. Complete Phase 5: Genotype of grandparents' earlobes.

a. What are the grandparent's genotypes for earlobes? ______________________________

b. How did you come to this conclusion? ______________________________

11. Complete Phase 6: Color blindness.

a. Color blindness is X-linked. What does this mean? ____

____

b. Is color blindness more common in *males* or *females*? ____

Why? ____

12. Complete Phase 7: Apply what you have learned about X-linked traits.

13. Complete Phase 8: Apply what you have learned about autosomal traits.

14. Complete Phase 6: Save Lab Data. Look for the "Congratulations" message that the finished lab has been automatically submitted. Read the options to save the lab data.

CRITICAL THINKING ASSESSMENT – *Virtual* Lab

Hemophilia A is an X-linked trait. Assume a father does not have hemophilia (X^HY) and a mother is heterozygous (X^HX^h), a carrier of the hemophilia gene.

What are the genotypes of the possible male offspring? ____

What are the genotypes of the possible female offspring? ____

What is the % chance a male offspring could have hemophilia? ____

What is the % chance a female offspring could have hemophilia? ____

What is the % chance a female offspring would be a carrier of the hemophilia gene? ____

If the father did have hemophilia, how would this change the % chance a female offspring could have hemophilia?

____

Microscopy – Animal Cells (Human Epidermal Cells)

Virtual Labs

The purpose of this laboratory simulation is to review steps in preparing a wet mount of a human cheek smear and to use brightfield microscopy to observe human epidermal cells of the cheek smear.

As you follow these steps, answer the corresponding questions:

1. Click the link to this *Virtual* Lab assignment that was set up by your instructor.

2. Read the Key Concepts.

a. For what purposes do microscopes serve? ______________________________

____

b. Are the cells found in humans *prokaryotic* or *eukaryotic*? ______________________________

3. Read the Overview.

a. What personal safety equipment should be used when preparing a wet mount of your cheek cells? ______________

____

4. Read Before you begin, make sure you are familiar with the terms.

a. What objective lens, *4X* or *10X* or *40X*, should be used to start focusing on a slide? ______________

b. Identify the main cell component located between the nucleus and cell membrane. ______________

c. Identify the organelle that houses DNA. ______________________________

d. Identify the organelle where most ATP is synthesized. ______________________________

e. Changing the objective lens from 10X to 40X will *increase* or *decrease* the field of view. ______________

5. Continue to the Laboratory Simulation.

6. Complete Phase 1: Label features of an animal cell. Each step needs to be completed before moving on to the next step. (Click the bottom tabs to review the Methods, Reset the experiment, add observations in My Notes, record Lab Data, or Show Labels.)

a. Which endoplasmic reticulum is covered with ribosomes? ______________________________

7. Complete Phase 2: Focus microscope on human epidermal cells.

a. If a student cannot locate a specimen on the slide at high power objective, what do you suggest the student do? ______________________________

b. If a student cannot get in focus the specimen on the slide at high power objective, what do you suggest the student do? ______________________________

c. What cell structures can you observe with the 40X objective in place? ______________________________

____

d. The ______________ is the boundary between the extracellular and intracellular cellular environment.

8. Complete Phase 3: Identifying human epidermal cell structures.

9. Complete Phase 4: Apply what you have learned.

a. Are you able to identify the feature using the brightfield microscope that is present in animal cells, but absent in plant cells? Explain. ______

b. Predict how the human epidermal cell might appear under the microscope if no stain was added. ______

______

c. Predict how the human epidermal cell might appear under the microscope if too much stain was added. ______

______

10. Complete Phase 5: Lab wrap-up.

a. Where do you think the used toothpick should be placed after preparing the smear on the slide? ______

______

11. Complete Phase 6: Save Lab Data. Look for the "Congratulations" message that the finished lab has been automatically submitted. Read the options to save the lab data.

CRITICAL THINKING ASSESSMENT – *Virtual* Lab

The cells of the cheek that you observed are epidermal cells. What characteristics and functions do these cells have that make them suitable for the location from which they were scraped? ______

______.

When scraping the inside of the cheek, there is no blood loss. Why? ______

______

Microscopy – Connective Tissue Histology

Virtual Labs

The purpose of this laboratory simulation is to observe and identify the different connective tissue types using a brightfield microscope.

As you follow these steps, answer the corresponding questions:

1. Click the link to this *Virtual* Lab assignment that was set up by your instructor.

2. Read the Key Concepts.

a. What general characteristics do connective tissues typically possess? ______________________

____

b. Identify three tissues that are categorized as loose connective tissue proper. ______________________

____

c. Identify three tissues that are categorized as dense connective tissue proper. ______________________

____

d. Identify two general tissues that are categorized as supporting connective tissue. ______________________

e. Identify two tissues that are categorized as fluid connective tissue. ______________________

3. Read the Overview.

a. Identify three structures in connective tissues that can be identified by microscopic observation. ______________________

____

Which two of these three structures make up the extracellular matrix? ______________________

4. Read Before you begin, make sure you are familiar with the terms.

a. What total magnification will result in the smallest field of view? ______________________

b. What organic compound are connective tissue fibers generally made of? ______________________

c. Name the three fiber types. ______________________

5. Continue to the Laboratory Simulation. (*Note:* As you are observing each connective tissue, use the "My Notes" to provide a detailed description.)

6. Complete Phase 1: Identify common connective tissue structures in areolar connective tissue. Each step needs to be completed before moving on to the next step. (Click the bottom tabs to review the Methods, Reset the experiment, add observations in My Notes, record Lab Data, or Show Labels.)

7. Complete Phase 2: Focus microscope on areolar connective tissue.

8. Complete Phase 3: Focus microscope on reticular connective tissue.

a. What connective tissue, *areolar* or *reticular*, makes up the framework of the spleen, lymph nodes, and red bone marrow?

____

9. Complete Phase 4: Focus microscope on adipose connective tissue.

a. What is stored within the cell (adipocyte) of this tissue? ______________________

10. Complete Phase 5: Focus microscope on dense irregular connective tissue.

a. What color are the collagen fibers stained in this micrograph? ______________________

11. Complete Phase 6: Focus microscope on dense regular connective tissue.

a. Compare the arrangement of collagen fibers in dense irregular and dense regular connective tissues. ______________________

____

12. Complete Phase 7: Focus microscope on elastic connective tissue.

   a. What is the function of the elastic fibers in this tissue? ____________________

   ____

13. Complete Phase 8: Connective tissue proper identification.

   a. Which connective tissue is best suited for attachments between bones (ligaments) and between muscles and bones (tendons)? ____________________

   b. Which tissue is best suited to stretch under high blood pressure and recoil to push blood out? ____________________

14. Complete Phase 9: Focus microscope on hyaline cartilage connective tissue.

   a. Cartilage cells are also referred to as ____________________.

   b. Where is hyaline cartilage located in the body? ____________________

15. Complete Phase 10: Focus microscope on fibrocartilage connective tissue.

   a. How would you distinguish between fibrocartilage and hyaline cartilage connective tissues? ____________________

   ____

   b. Where is fibrocartilage located in the body? ____________________

16. Complete Phase 11: Focus microscope on elastic cartilage connective tissue.

   a. Where is elastic cartilage located in the body? ____________________

   b. How would you distinguish elastic cartilage from the other two cartilage connective tissues? ____________________

   ____

17. Complete Phase 12: Focus microscope on compact bone (connective) tissue.

   a. What is the role of the lacunae? ____________________

18. Complete Phase 13: Focus microscope on blood connective tissue.

   a. What is the name given to the extracellular matrix of blood? ____________________

   b. Which cell type is most abundant in blood? ____________________

19. Complete Phase 14: Supporting and fluid connective tissue identification.

   a. Which cartilage connective tissue has a glassy appearance to its extracellular matrix? ____________________

   b. Which connective tissue has a solid extracellular matrix? ____________________

20. Complete Phase 15: Classify connective tissues.

21. Complete Phase 16: Apply what you have learned.

   a. Where are protein fibers of connective tissues found? ____________________

   b. Are cells of connective tissues *close together* or *spread apart* from one another? ____________________

22. Complete Phase 13: Save Lab Data. Look for the "Congratulations" message that the finished lab has been automatically submitted. Read the options to save the lab data.

CRITICAL THINKING ASSESSMENT – *Virtual* Lab

Vitamin C plays an important role in maintaining a healthy amount of collagen in the body. It plays an essential role in synthesizing collagen, as it is a cofactor of enzymes that promote proper folding to establish the stable shape of collagen. Consider the connective tissues that you have observed in this lab simulation and the collagen that they are composed. Predict the effects of severe vitamin C deficiency in the human body. ____________________

____

____

Virtual Labs

Microscopy – Epithelial Tissue Histology

The purpose of this laboratory simulation is to observe and identify the different epithelial tissue types using a brightfield microscope.

As you follow these steps, answer the corresponding questions:

1. Click the link to this *Virtual* Lab assignment that was set up by your instructor.
2. Read the Key Concepts.
   - **a.** What are the broad categories of tissues? ______________________
   - **b.** What general characteristics do all epithelial tissues possess? ______________________

     ____
   - **c.** Multi-layered, flattened epithelial cells are called ______________________.
   - **d.** Single-layered, column-like epithelial cells are called ______________________.
3. Read the Overview.
4. Read Before you begin, make sure you are familiar with the terms.
   - **a.** What objective lens, 4X, 10X, or 40X, should be used to start focusing on a slide? ______________ And, what is its total magnification? ______________
   - **b.** What objective lens magnification will result in the greatest field of view? ______________________
   - **c.** In order to move the slide to find the best view, you will move the ______________________.
   - **d.** The basement membrane is located between epithelial tissue and underlying ______________________ tissue.
5. Continue to the Laboratory Simulation. (*Note:* As you are observing each epithelial tissue, use the "My Notes" to provide a detailed description.)
6. Complete Phase 1: Classify epithelial cell shapes. Each step needs to be completed before moving on to the next step. (Click the bottom tabs to review the Methods, Reset the experiment, add observations in My Notes, record Lab Data, or Show Labels.)
7. Complete Phase 2: Identify structures found in types of simple epithelial tissue.
   - **a.** Which structure is closest to the lumen of this tissue? ______________________
8. Complete Phase 3: Focus microscope on simple squamous epithelial tissue.
9. Complete Phase 4: Focus microscope on simple cuboidal epithelial tissue.
10. Complete Phase 5: Focus microscope on simple columnar epithelial tissue.
11. Complete Phase 6: Focus microscope on pseudostratified columnar epithelial tissue.
12. Complete Phase 7: Identify stratified squamous epithelial tissue structures.
    - **a.** What general function do you think this tissue serves? ______________________
13. Complete Phase 8: Focus microscope on stratified squamous epithelial tissue.
14. Complete Phase 9: Focus microscope on transitional epithelial tissue.
    - **a.** Differentiate between stratified squamous and transitional epithelial tissues. ______________________

      ____
    - **b.** Predict how transitional epithelial tissue will appear when stretched, as with a full urinary bladder. ______________________

      ____

15. Complete Phase 10: Sample epithelial tissue identification.

a. Which tissue identified would be most effective for the exchange of gases, and why? ______________________________

____

b. Which tissue identified would be most effective at moving particles away from the lungs, and why? ______________

____

16. Complete Phase 11: Classify epithelial tissues by number of layers.

17. Complete Phase 12: Apply what you have learned.

a. What are two ways in which epithelial tissues are classified? ______________________________

____

b. A group of similar cells and surrounding material that perform a common function defines a ____________________.

c. Identify two epithelial tissues identified in this lab simulation that contain goblet cells. ____________________

____

d. What epithelial tissue identified in this lab simulation would be best suited in areas of high diffusion?

____

e. In your observations, are there any blood vessels in epithelial tissues? ____________________

How do these epithelial cells receive nourishment? ______________________________

____

18. Complete Phase 13: Save Lab Data. Look for the "Congratulations" message that the finished lab has been automatically submitted. Read the options to save the lab data.

CRITICAL THINKING ASSESSMENT – *Virtual* Lab

Imagine if the epithelial tissue of the large airways of the respiratory tract changed from ciliated pseudostratified columnar to stratified squamous epithelium, how would this affect the function of this area, and why? ____________

____

____

Imagine if the epithelial tissue of the small intestine changed from simple columnar to simple squamous epithelium, how would this affect the function of this area, and why? ______________________________

____

____

Virtual Labs — Microscopy – Muscle Tissue Histology

The purpose of this laboratory simulation is to observe and identify the different muscle tissue types using a brightfield microscope.

As you follow these steps, answer the corresponding questions:

1. Click the link to this *Virtual* Lab assignment that was set up by your instructor.

2. Read the Key Concepts.

a. What are the broad categories of tissues? ______________________

b. What general characteristics do all muscles possess? ______________________

____

c. Name the muscle type that has no striations. ______________________

d. Name the muscle type that is under voluntary control. ______________________

e. Name the muscle type that has intercalated discs. ______________________

3. Read the Overview.

a. Predict the type of muscle shown in the micrograph. ______________________

4. Read Before you begin, make sure you are familiar with the terms.

a. What objective lens, *4X* or *10X* or *40X*, should be used to start focusing on a slide? ______________________

b. In order to move the slide to find the best view, you will move the ______________________.

c. The amount of the slide you are able to see through the eyepiece is the ______________________.

d. What is the total magnification when using the 10X objective lens? ______________________

5. Continue to the Laboratory Simulation.

6. Complete Phase 1: Identify muscle tissue features. Each step needs to be completed before moving on to the next step. (Click the bottom tabs to review the Methods, Reset the experiment, add observations in My Notes, record Lab Data, or Show Labels.)

a. What are the "stripes" located perpendicular to the length of the muscle cell? ______________________

7. Complete Phase 2: Focus microscope on skeletal muscle tissue.

a. Describe your observation of skeletal muscle tissue. ______________________

____

8. Complete Phase 3: Focus microscope on smooth muscle tissue.

a. Describe your observation of smooth muscle tissue. ______________________

____

9. Complete Phase 4: Focus microscope on cardiac muscle tissue.

a. Describe your observation of cardiac muscle tissue. ______________________

____

b. Distinguish striations from intercalated discs. ______________________

____

10. Complete Phase 5: Muscle tissue identification.

11. Complete Phase 6: Apply what you have learned.

a. Which muscle type contains fusiform-shaped cells? ______________________________

b. Which muscle type contains multinucleated cells? ______________________________

c. Which muscle type is only found in the heart? ______________________________

d. Which muscle type is found in the deltoid and trapezius? ______________________________

e. Which muscle type is found in the urinary bladder? ______________________________

12. Complete Phase 7: Save Lab Data. Look for the "Congratulations" message that the finished lab has been automatically submitted. Read the options to save the lab data.

CRITICAL THINKING *ASSESSMENT – Virtual* Lab

Imagine if cardiac muscle were under voluntary control. What do you see as being advantages and disadvantages if this were the case? ______________________________

____

____

____

CRITICAL THINKING *ASSESSMENT – Virtual* Lab

When you lift weights, your muscles get bigger in size (hypertrophy) and become stronger. Which muscle type does this refer to? ______________________________ What organic compound contributes to this hypertrophy? ______________________________

What characteristic of this muscle type enables it to produce these organic compounds, and why?

____

____

____

Virtual Labs Microscopy – Nervous Tissue Histology

The purpose of this laboratory simulation is to observe and identify multipolar neurons within nervous tissue using a brightfield microscope.

As you follow these steps, answer the corresponding questions:

1. Click the link to this *Virtual* Lab assignment that was set up by your instructor.

2. Read the Key Concepts.

a. Distinguish the two main cell types found in nervous tissue. ____________________

____

b. Where is nervous tissue located? ____________________

c. Which basic structure of a neuron takes information it receives and transmits it to the cell body? ____________________

d. What are the possible effector organs that respond to signals transmitted by an axon? ____________________

3. Read the Overview.

a. Structurally, motor neurons and interneurons are ____________________ neurons.

4. Read Before you begin, make sure you are familiar with the terms.

a. What objective lens, *4X* or *10X* or *40X*, should be used to start focusing on a slide? ____________________

5. Continue to the Laboratory Simulation.

6. Complete Phase 1: Focus microscope on nervous tissue. Each step needs to be completed before moving on to the next step. (Click the bottom tabs to review the Methods, Reset the experiment, add observations in My Notes, record Lab Data, or Show Labels.)

a. Describe your observation of nervous tissue. ____________________

____

7. Complete Phase 2: Nervous tissue identification.

a. What organelle(s) are observable in the cell body of the neuron in the micrograph? ____________________

8. Complete Phase 3: Apply what you have learned.

a. Glial cells are *more* or *less* numerous than neurons. ____________________

b. Which would be a single cell process in multipolar neurons, *axon* or *dendrite*? ____________________

9. Complete Phase 7: Save Lab Data. Look for the "Congratulations" message that the finished lab has been automatically submitted. Read the options to save the lab data.

CRITICAL THINKING ***ASSESSMENT*** – *Virtual* Lab

Ribosomes are lacking in axons and axon terminals (synaptic knobs). Lysosomes are found only in the cell body. Given this information and what you have observed in this lab simulation, answer the following questions:

Where are neurotransmitters and most other proteins produced in a neuron? ______________________________

____

Why do microtubules in the axon move damaged membranes and organelles toward the cell body (*retrograde transport*)? ______________________________

____

Microscopy – Operation of a Brightfield Microscope

Virtual Labs

The purpose of this laboratory simulation is to practice using a brightfield microscope as you learn the terminology and function of its parts.

As you follow these steps, answer the corresponding questions:

1. Click the link to this *Virtual* Lab assignment that was set up by your instructor.
2. Read the Key Concepts.
   - **a.** What do brightfield microscopes use in combination to view a specimen? ______________________
3. Read the Overview.
4. Read Before you begin and make sure you are familiar with the terms.
   - **a.** Which objective lens has the smallest field of view? ______________________
   - **b.** What is the function of the condenser? ______________________
   - **c.** The ______________________, also called the iris diaphragm lever, is located on the condensor lens and allows you to move the aperature to decrease or increase the amount of light to the slide.
   - **d.** To decrease contrast, the aperature should be moved to *increase* or *decrease* the light to the slide. __________
5. Continue to the Laboratory Simulation.
6. Complete Phase 1: Power, brightness, and ocular lenses video overview. Each step needs to be completed before moving on to the next step. (Click the bottom tabs to review the Methods, Reset the experiment, add observations in My Notes, record Lab Data, or Show Labels.)
   - **a.** What part of the microscope holds the slide into place? ______________________
7. Complete Phase 2: Power, brightness, and ocular lenses.
   - **a.** What factor affecting image quality can be altered by light intensity? ______________________
   - **b.** Adjusting the ocular lenses to match the spacing of your eyes will help you see a ______________________ image.
   - **c.** What is the magnification of the eyepiece (ocular lenses)? ______________________
8. Complete Phase 3: Focus and stage adjustment video overview.
   - **a.** Adjusting the length between the specimen and objective lens is achieved by moving the ______________________ up or down to get the specimen into focus.

     What parts of the microscope accomplish this task? ______________________
9. Complete Phase 4: Focus the microscope.
   - **a.** Which focus adjustment knob is used first, the *coarse* or *fine*? ______________________
   - **b.** Under high power magnification, the coarse adjustment knob should not be used. Why? ______________________

     ____
10. Complete Phase 5: Adjust microscope stage.
    - **a.** The lower mechanical stage control knob is used to move the stage ______________________.
    - **b.** The upper mechanical stage control knob is used to move the stage ______________________.
11. Complete Phase 6: Magnification video overview.
    - **a.** How can a region of the microscope slide be studied in more detail? ______________________

      ____

12. Complete Phase 7: Adjust magnification.

a. Does the distance between the objective lens and the slide *increase* or *decrease* as the specimen gets into focus using the coarse focus knob? ______________________

b. What focus adjustment knob is used to sharpen the image? ______________________

c. What is the total magnification of a specimen using the 10X objective? ______________________

13. Complete Phase 8: Lab Conclusion.

14. Complete Phase 9: Save Lab Data. Look for the "Congratulations" message that the finished lab has been automatically submitted. Read the options to save the lab data.

CRITICAL THINKING ASSESSMENT – *Virtual* Lab

What can be done when preparing a wet mount slide of epithelial cells to improve contrast when viewing under the microscope? ______________________

____

When observing cells, such as liver cells, using a brightfield microscope, what cell parts can be observed when using a brightfield microscope? (Place an **X** in the column if the cell part can be identified.)

Nucleus	Mitochondria	Nucleolus	Endoplasmic reticulum	Cytoplasm	Cell membrane

CRITICAL THINKING ASSESSMENT – *Virtual* Lab

The brightfield microscope's objective lenses are parfocal. What does this mean, and why is this important? ______________________

____

____

Virtual Labs

Nervous System – Demonstrate Monosynaptic Reflexes

The purpose of this laboratory simulation is to learn about the characteristics and components of a reflex and to demonstrate several monosynaptic reflexes.

As you follow these steps, answer the corresponding questions:

1. Click the link to this *Virtual* Lab assignment that was set up by your instructor.
2. Read the Key Concepts.
   - a. Define a *reflex*. ______________________________

     ____
   - b. Answer the following related to components of a reflex (in order of occurrence):
     - i. Detects the stimulus: ______________________________
     - ii. Conducts signal toward the spinal cord: ______________________________
     - iii. Found in polysynaptic reflexes and located within the spinal cord: ______________________________
     - iv. Conducts signal away from spinal cord to effector: ______________________________
     - v. Executes response (muscles or glands): ______________________________
   - c. *True* or *False*: The central nervous system can override a stretch reflex. ______________________________
3. Read the Overview.
   - a. What is the *integration center* shown in the illustration of a simple reflex arc? ______________________________
4. Read Before you begin.
   - a. Complete the following chart by placing an X in the box that corresponds to the part of the reflex hammer used to test the reflex:

	Biceps brachii reflex	Triceps brachii reflex	Patellar reflex	Calcaneal reflex	Plantar reflex
Broad flat side					
Pointed side					
Tapered metallic tip					

   - b. Why is a monosynaptic reflex referred to as a *stretch reflex*? ______________________________

     ____
5. Continue to the Laboratory Simulation.
6. Complete Phase 1: Biceps reflex – inside elbow. Each step needs to be completed before moving on to the next step. (Click the bottom tabs to review the Methods, Reset the experiment, add observations in My Notes, record Lab Data, or Show Labels.)
   - a. Striking the biceps tendon caused what movement? ______________________________.
7. Complete Phase 2: Triceps reflex – outside elbow.
   - a. Striking the triceps tendon caused what movement? ______________________________.
8. Complete Phase 3: Patellar reflex – below knee.
   - a. Striking the patellar ligament caused what movement? ______________________________.

9. Complete Phase 4: Calcaneal reflex – above heel.

a. Striking the calcaneal tendon caused what movement? ____.

10. Complete Phase 5: Plantar reflex – sole of foot.

a. Testing the plantar reflex caused what movement? ____.

11. Complete Phase 6: Save Lab Data. Look for the "Congratulations" message that the finished lab has been automatically submitted. Read the options to save the lab data.

CRITICAL THINKING ASSESSMENT – *Virtual* Lab

The responses of monosynaptic reflexes are stereotypical. If a reflex response is absent or hyperactive, where in the reflex arc might the problem be and what are the possible causes? ____

____

____

____

Osmosis – Movement of Water Across a Selectively Permeable Membrane

Virtual Labs

The purpose of this laboratory simulation is to examine osmosis by measuring the volume of water moved by osmosis in different solutions.

As you follow these steps, answer the corresponding questions:

1. Click the link to this *Virtual* Lab assignment that was set up by your instructor.
2. Read the Key Concepts.
   a. Water moves toward or away from the side of the membrane with a higher nonpermeable solute concentration. __________
3. Read the Overview.
4. Read Before you begin and make sure you are familiar with the terms.
   a. Define *osmosis*. __________
5. Continue to the Laboratory Simulation.
   a. Given the different solutions (30% corn syrup, 20% corn syrup, 10% corn syrup, and distilled water), what is your hypothesis as to which solution will have the largest change in the volume of water by osmosis, and why? __________
6. Complete Phase 1: 10% Corn syrup. Each step needs to be completed before moving on to the next step. (Click the bottom tabs to review the Methods, Reset the experiment, add observations in My Notes, record Lab Data, or Show Labels.)
   a. Which part of the experimental set up contains hypotonic solution, the beaker or thistle tube? __________
   b. Which part of the experimental set up contains hypertonic solution, the beaker or thistle tube? __________
7. Complete Phase 2: 20% Corn syrup.
   a. How do the results after 10 minutes of the 20% corn syrup compare to the 10% corn syrup? __________
8. Complete Phase 3: 30% Corn syrup.
   a. How do the results after 10 minutes of the 30% corn syrup compare to the 20% corn syrup? __________
   b. Which percentage (%) corn syrup tested in this simulation had the greater osmotic pressure? Explain. __________
9. Complete Phase 4: Apply what you have learned.
   a. The lower the concentrations of corn syrup, the __________ water is drawn up into the thistle tube.
   b. Do you accept or reject your original hypothesis, and why? __________
10. Complete Phase 5: Save Lab Data. Look for the "Congratulations" message that the finished lab has been automatically submitted. Read the options to save the lab data.

CRITICAL THINKING ***ASSESSMENT*** **– *Virtual* Lab**

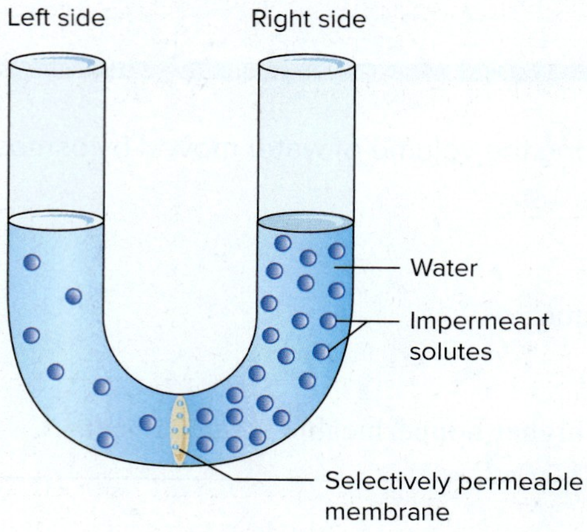

Part one:

Using the illustration on the left and the concept of osmosis, explain where water will move to reach equilibrium. ____________________

____

____

After equilibrium is reached, the osmolarity on the right side of the membrane is *higher* or *lower* or *equal to* the left side. ____________________

Part two:

Apply this simulation to the human body: If the right side is representing extracellular fluid (ECF), and the left side is representing the inside of a cell, and the selectively permeable membrane is representing the cell membrane, what will happen to the cell in this environment? ____________________

____

____

Is this implicative of a person who is *well hydrated* or *dehydrated*? Explain. ____________________

____

____

Osmosis – Tonicity in Red Blood Cells

The purpose of this laboratory simulation is to examine changes in cell volume when red blood cells are placed in different solutions and looking at the effects of tonicity on these cells.

As you follow these steps, answer the corresponding questions:

1. Click the link to this *Virtual* Lab assignment that was set up by your instructor.
2. Read the Key Concepts.
   a. Which osmotic environment is normally found in the body? ______
3. Read the Overview.
   a. Describe the normal red blood cells shown in the image. ______
4. Read Before you begin and make sure you are familiar with the terms.
   a. In what conditions do cell membranes stay intact? ______
   The solution appears *opaque* or *translucent* under these conditions. ______
   b. Define *tonicity*. ______
5. Continue to the Laboratory Simulation.
6. Complete Phase 1: Hypothesis & Strategy. Each step needs to be completed before moving on to the next step. (Click the bottom tabs to review the Methods, Reset the experiment, add observations in My Notes, record Lab Data, or Show Labels.)
   a. Given the different solutions (10% NaCl, 0.9% NaCl, and distilled water), what is your hypothesis as to which solution will result in red blood cells swelling and bursting?
   ______
   b. Predict what a red blood cell would look like in an isotonic solution when examining under the microscope. ______
7. Complete Phase 2: 10% NaCl solution.
   a. A 10% NaCl solution is *hypertonic* or *isotonic* or *hypotonic*. ______
   b. What happened to the red blood cells in this solution? ______
   Why? ______
8. Complete Phase 3: 0.9% NaCl solution.
   a. A 0.9% NaCl solution is *hypertonic* or *isotonic* or *hypotonic*. ______
   b. What can you infer about the solute concentration in this solution compared to the red blood cells? ______
9. Complete Phase 4: Distilled water.
   a. How is the transparency of whole blood in this solution different compared to the previous two solutions, and why?
   ______

10. Complete Phase 5: Conclusion.

a. Do you accept or reject your original hypothesis, and why? ______________________________

____

____

11. Complete Phase 6: Save Lab Data. Look for the "Congratulations" message that the finished lab has been automatically submitted. Read the options to save the lab data.

CRITICAL THINKING ASSESSMENT – Virtual Lab

Explain why it is not a good idea to drink ocean water to replenish thirst.

____

____

____

Virtual Labs

pH Balance – Antacids as Buffers

The purpose of this laboratory simulation is to conduct an experiment to determine how effective different types of antacids are as buffers.

As you follow these steps, answer the corresponding questions:

1. Click the link to this *Virtual* Lab assignment that was set up by your instructor.
2. Read the Key Concepts.
   a. A solution with more hydrogen ions would have a _higher_ or _lower_ pH than a solution with less hydrogen ions. ______
   b. The normal pH range of the blood is 7.35 to 7.45 which would make blood slightly _acidic_ or _neutral_ or _alkaline_. ______
   c. How is the stomach wall protected from its low pH? ______
   d. Most antacids are _strong_ or _weak_ bases. ______
   e. Define *buffer*. ______
3. Read the Overview.
4. Read Before you begin, make sure you are familiar with the terms.
   a. Phenol red is a pH indicator. What does this mean? ______

   ______
5. Continue to the Laboratory Simulation.
6. Complete Phase 1: Hypothesis & Strategy. Each step needs to be completed before moving on to the next step. (Click the bottom tabs to review the Methods, Reset the experiment, add observations in My Notes, record Lab Data, or Show Labels.)
   a. What is your hypothesis as to the effectiveness of antacids on buffering capacity?

   ______

   ______
7. Complete Phase 2: Testing a negative control substance.
   a. What is a *negative control*? ______
   b. What is the negative control in this experiment? ______
8. Complete Phase 3: Determining buffering capacity of calcium carbonate and magnesium hydroxide antacid.
9. Complete Phase 4: Determining buffering capacity of alumina/magnesium trisilicate antacid.
   a. What is HCl simulating in this experiment? ______
10. Complete Phase 5: Determining buffering capacity of famotidine antacid.
11. Complete Phase 6: Determining buffering capacity of sodium bicarbonate citric acid antacid.
12. Complete Phase 7: Determining buffering capacity of chewable calcium carbonate antacid.
13. Complete Phase 8: Graph and analyze data.
    a. Which antacid with phenol red required the most drops of HCl to turn yellow? ______
    The fewest drops? ______

14. Complete Phase 9: Lab summary.

a. Do you accept or reject your original hypothesis, and why? ______________________________

____

b. Based on your experimental results, which antacid is the least effective buffer?

____

c. The most effective antacid requires the *least* or *most* drops of acid to change the color of the solution. ____________________ Why? ______________________________

____

15. Complete Phase 10: Save Lab Data. Look for the "Congratulations" message that the finished lab has been automatically submitted. Read the options to save the lab data.

CRITICAL THINKING ***ASSESSMENT – Virtual*** **Lab**

Certain over-the-counter medications are commonly taken to help relieve symptoms of heartburn. Prilosec® (Omeprazole) is a proton-pump inhibitor. How would taking this medication help? ______________________________

____

Tums® is an over-the-counter antacid that gives quick relief for heartburn. What are Tums® tablets composed of? ________________ Is there a potential concern for an individual who consumes a half bottle of these tablets daily?

____

____

Virtual Labs

Respiratory System – Mechanism of Breathing

The purpose of this laboratory simulation is to examine volume and pressure changes that occur during respiration using a bell jar model.

As you follow these steps, answer the corresponding questions:

1. Click the link to this *Virtual* Lab assignment that was set up by your instructor.
2. Read the Key Concepts.
   - **a.** Where is the pleural cavity located? ______________________________
   - **b.** Air moves ______________ its pressure gradient in order to flow into and out of the lungs.
   - **c.** Define *Boyle's law*. ______________________________
   - **d.** When the diaphragm contracts, it moves *up* or *down*. ____________ Does this *increase* or *decrease* the volume in the thoracic cage? ____________ Does this *increase* or *decrease* the pressure in the thoracic cage? ____________ Will this result in *inspiration* or *expiration* or *neither*? ______________________________
3. Read the Overview.
4. Read Before you begin and make sure you are familiar with the terms.
   - **a.** Predict the effect an increase in the volume of the thoracic cage will have on intra-alveolar pressure. ______________ ______________________________
5. Continue to the Laboratory Simulation.
6. Complete Phase 1: Label model. Each step needs to be completed before moving on to the next step. (Click the bottom tabs to review the Methods, Reset the experiment, add observations in My Notes, record Lab Data, or Show Labels.)
7. Complete Phase 2: Demonstrate breathing mechanism.
   - **a.** Because the lungs are pressed against the internal thoracic cage, lung volume *increases* or *decreases* when the thoracic cage expands. ______________________________
8. Complete Phase 3: Apply what you have learned.
   - **a.** What is intrapleural pressure? ______________________________
   - **b.** Intrapleural pressure is always *higher (positive)* or *lower (negative)* than the intra-alveolar pressure and atmospheric pressure. ______________________________
   - **c.** Why would the lungs not inflate if atmospheric pressure and intrapleural pressure were equal? ______________ ______________________________
9. Complete Phase 4: Save Lab Data. Look for the "Congratulations" message that the finished lab has been automatically submitted. Read the options to save the lab data.

CRITICAL THINKING ASSESSMENT – *Virtual* Lab

Boyle's law is particularly important in SCUBA divers as they descend and ascend vast depths of water. Place an **X** in the boxes that apply to Boyle's law as the diver descends and ascends:

	Descending	Ascending
Pressure on lungs increases		
Pressure on lungs decreases		
Volume of air inside lungs increases		
Volume of air inside lungs decreases		

CRITICAL THINKING ASSESSMENT – *Virtual* Lab

A broken rib may cause a pneumothorax as it pokes into the intrapleural space. What is a pneumothorax? ____________

____

This may lead to a 'collapsed lung.' Why? ____________________________________

____

____

Respiratory System – Pulmonary Function Tests

Virtual Labs

The purpose of this laboratory simulation is to perform a spirometry to conduct pulmonary function tests on patients to determine and diagnose pulmonary diseases.

As you follow these steps, answer the corresponding questions:

1. Click the link to this *Virtual* Lab assignment that was set up by your instructor.

2. Read the Key Concepts.

a. What does a spirometer measure? ______________________________

3. Read the Overview.

a. What reference material will be used in this lab simulation? ______________________________

____

4. Read Before you begin and make sure you are familiar with the terms.

a. Which lung disorder, *obstructive* or *restrictive*, results in less oxygen taken in upon inhalation? ______________

b. Is asthma a(an) *obstructive* or *restrictive* lung disorder? ______________________________

c. TV + IRV + ERV = ______________________________

d. The percentage of the FVC expired in 1 second is the ______________________________.

e. What is used to differentiate obstructive and restrictive lung diseases to confirm diagnosis? Explain. ______________

____

5. Continue to the Laboratory Simulation.

6. Complete Phase 1: Examine patient (provide name): ______________________________.
Each step needs to be completed before moving on to the next step. (Click the bottom tabs to review the Methods, Reset the experiment, add observations in My Notes, record Lab Data, or Show Labels.)

a. Did your final diagnosis match your preliminary diagnosis? ______________________________

7. Complete Phase 2: Examine patient (provide name): ______________________________.

a. Does this patient suffer from a disease? If so, which one? ______________________________

8. Complete Phase 3: Examine patient (provide name): ______________________________.

a. Does this patient suffer from a disease? If so, which one? ______________________________

9. Complete Phase 4: Examine patient (provide name): ______________________________.

a. How do lung values differ slightly in an individual who is not fully grown compared to an individual who is fully grown?

____

b. Does this patient suffer from a disease? If so, which one? ______________________________

c. Which diseases on the diagnosis chart are chronic obstructive pulmonary diseases? ______________________________

____

What do two of these COPDs have in common relative to patient histories? ______________________________

10. Complete Phase 5: Apply what you have learned.

a. A patient with pulmonary fibrosis due to radiation damage also has chronic bronchitis from years of smoking. Which of these is a restrictive lung disorder? ______________ Obstructive lung disorder? ______________

Which one of these two diseases would the flow tests (FEV1 and FEF25-75) be lower than normal values, and why?

____

11. Complete Phase 4: Save Lab Data. Look for the "Congratulations" message that the finished lab has been automatically submitted. Read the options to save the lab data.

CRITICAL THINKING ***ASSESSMENT*** – ***Virtual*** **Lab**

What is pleural effusion? ____

What might cause pleural effusion? ______________________________________

Would pleural effusion be an *obstructive* or *restrictive* lung disorder? ___________________

Explain your reasoning. ____

____

____

Virtual Labs

Skeletal Muscle – Electrical Stimulation

The purpose of this laboratory simulation is to investigate the relationship between the strength of a stimulus and the force generated by the stimulated frog's leg muscle.

As you follow these steps, answer the corresponding questions:

1. Click the link to this *Virtual* Lab assignment that was set up by your instructor.
2. Read the Key Concepts.
   a. What needs to be achieved before a skeletal muscle is able to contract? ______________________
3. Read the Overview.
   a. What is being used in this lab to simulate nerve impulses? ______________________
4. Read Before you begin and watch the short video.
   a. What is used to stimulate the frog muscle? ______________________
   b. What is used to measure the generated force? ______________________
   c. What is used to change the stimulus strength and to measure the transducer's signal? ______________________
5. Continue to the Laboratory Simulation.
6. Complete Phase 1: Set-up experiment. Each step needs to be completed before moving on to the next step. (Click the bottom tabs to review the Methods, Reset the experiment, add observations in My Notes, record Lab Data, or Show Labels.)
7. Complete Phase 2: Data collection.
   a. A muscle will increase its response from threshold stimulus up to ______________________ stimulus, at which point the muscle will not increase its generated force (g).
8. Complete Phase 3: Graph and analyze data.
   a. What independent variable did you use for the x-axis? ______________________
   b. What dependent variable did you use for the y-axis? ______________________
   c. At what output voltage (V) was threshold stimulus achieved? ______________________
   d. At what output voltage (V) was maximal stimulus achieved? ______________________
9. Complete Phase 4: Apply what you have learned.
   a. As output voltage (V) increases, ______________________ muscle fibers are being stimulated.
10. Complete Phase 5: Save Lab Data. Look for the "Congratulations" message that the finished lab has been automatically submitted. Read the options to save the lab data.

CRITICAL THINKING ASSESSMENT – *Virtual* Lab

Do you think the data collected would be the same if a different muscle in the frog were used? Explain. ______________________

____

____

Virtual Labs

Skeletal Muscle – Shoulder and Elbow Movement Exercise

The purpose of this laboratory simulation is to examine certain movements of the shoulder and elbow joints to determine agonist and antagonist muscles involved in these movements.

As you follow these steps, answer the corresponding questions:

1. Click the link to this *Virtual* Lab assignment that was set up by your instructor.
2. Read the Key Concepts.
   - **a.** How do agonist and antagonist muscles differ? ______________________________

     ____
   - **b.** What does a muscle need to do in order to cause movement at a specific joint? ______________________________
3. Read the Overview.
   - **a.** Decreasing the angle at a joint is the movement of ______________________________.
   - **b.** Increasing the angle at a joint is the movement of ______________________________.
   - **c.** Moving away from the midline is the movement of ______________________________.
   - **d.** Moving toward the midline is the movement of ______________________________.
4. Continue to the Laboratory Simulation.
5. Complete Phase 1: Forearm flexion at the elbow joint. Each step needs to be completed before moving on to the next step. (Click the bottom tabs to review the Methods, Reset the experiment, add observations in My Notes, record Lab Data, or Show Labels.)
   - **a.** The agonist is the ____________________ muscle, and it is carrying out the movement of ____________________.
   - **b.** What is the antagonist muscle? ____________________
6. Complete Phase 2: Forearm extension at the elbow joint.
   - **a.** The agonist is the ____________________ muscle, and it is carrying out the movement of ____________________.
   - **b.** What is the antagonist muscle? ____________________
7. Complete Phase 3: Arm abduction at the shoulder joint.
   - **a.** The agonist is the ____________________ muscle, and it is carrying out the movement of ____________________.
   - **b.** What is the antagonist muscle? ____________________
8. Complete Phase 4: Arm adduction at the shoulder joint.
   - **a.** The agonist is the ____________________ muscle, and it is carrying out the movement of ____________________.
   - **b.** What is the antagonist muscle? ____________________
9. Complete Phase 5: Save Lab Data. Look for the "Congratulations" message that the finished lab has been automatically submitted. Read the options to save the lab data.

CRITICAL THINKING ASSESSMENT – *Virtual* Lab

When kicking a soccer ball, what are the agonist and antagonist muscle groups in the thigh and the movements they carry out at the knee joint? ______________________________

____

____

____

Urinary System – Urinalysis

The purpose of this laboratory simulation is to conduct an analysis of a three samples of urine using the dipstick method.

As you follow these steps, answer the corresponding questions:

1. Click the link to this *Virtual* Lab assignment that was set up by your instructor.
2. Read the Key Concepts.
   - **a.** The formation of ________________ is critical to proper health because it maintains normal ________________ composition and removes waste products.
   - **b.** About how much blood is filtered by the kidneys per minute? ________________
   - **c.** What is the functional unit of the kidney? ________________
   - **d.** What force drives the filtration of blood from the glomerulus into the glomerular capsule? ________________
   - **e.** Approximately what percent of tubular fluid is reabsorbed? ________________
   - **f.** What does it mean to be reabsorbed? ________________
3. Read the Overview.
   - **a.** What might abnormal urine composition indicate? ________________
4. Read Before you begin, make sure you are familiar with the terms.
   - **a.** Why are erythrocytes not normally found in urine? ________________
   - **b.** Is glucose normally found in urine? ________________
   - **c.** What are the nitrogenous wastes found in urine?________________
   - **d.** High levels of ________________ would be found in urine when there is excessive fatty acid metabolism.
5. Continue to the Laboratory Simulation.
6. Complete Phase 1: Urinalysis of urine Sample 1. Each step needs to be completed before moving on to the next step. (Click the bottom tabs to review the Methods, Reset the experiment, add observations in My Notes, record Lab Data, or Show Labels.)
   - **a.** A patient who is dehydrated would have <u>*very light yellow*</u> or <u>*light yellow*</u> or <u>*darker yellow*</u> urine. ________________
7. Complete Phase 2: Urinalysis of urine Sample 2.
   - **a.** Cloudy urine may indicate ________________.
8. Complete Phase 3: Urinalysis of urine Sample 3.
   - **a.** Does this patient's urine show any abnormal results? ________________
9. Complete Phase 4: Apply what you have learned.
   - **a.** What two solutes are *not* normally found in urine? ________________
   - **b.** What two cell types are *not* normally found in urine? ________________
10. Complete Phase 5: Analysis.
    - **a.** Sample 1 urinalysis results showed possible ________________.

      Why? ________________
    - **b.** Sample 2 urinalysis results showed possible ________________.

      Why? ________________

11. Complete Phase 6: Save Lab Data. Look for the "Congratulations" message that the finished lab has been automatically submitted. Read the options to save the lab data.

CRITICAL THINKING ***ASSESSMENT*** *– Virtual* Lab

What blood test could be done and what would it show for each abnormal urinalysis in this lab simulation to help confirm diagnosis? Explain. ____

____

____

____

LESSON 1

Osmosis and Diffusion

Varying Extracellular Concentration

The purpose of this exercise is to demonstrate the principle of osmosis by observing and analyzing how the tonicity of a solution—isotonic, hypotonic, and hypertonic—affects red blood cells (erythrocytes). A spectrophotometer is used to measure the color of different blood solutions. A solution consists of a solvent, which is usually water, and a solute, which are the molecules dissolved in the solvent. Molecules tend to move from areas of higher concentration to areas of lower concentration, which allows them to become evenly distributed throughout the solution. This is called diffusion. The concentration of solutes inside a cell is usually different from the concentration of solutes outside the cell, so a concentration gradient is established between the two sides of the cell membrane. Water will move across a selectively permeable membrane from the solution with a lower concentration of nonpenetrating solutes to the solution with the higher concentration of nonpenetrating solutes. This is called osmosis.

In this lesson you will examine the tonicity of sodium chloride solutions in relation to the intracellular fluid of red blood cells. You will measure the color of ten different blood samples each with different concentrations of sodium chloride using a spectrophotometer set at 510 nm. Normally, this wavelength of light is reflected by the cell membranes. If the cells are placed in a hypotonic solution, they will burst, and less light will be reflected. This will cause the transmittance value from the spectrophotometer to increase. If the cells are placed in a hypertonic solution, they will shrink, and more light will be reflected, causing the transmittance value to decrease.

Before you begin, familiarize yourself with the following concepts (use your textbook as a reference):

- Osmolarity and osmotic pressure
- Gradient
- Mechanisms of transport across a membrane
- Distinguish between isotonic, hypotonic, and hypertonic solutions
- Crenation

State your hypothesis regarding the transmittance values of the blood samples in different concentrations of sodium chloride.

____

____

____

____

1. Open Lesson 1: Osmosis and Diffusion: Varying Extracellular Concentration.
2. Read the Objectives. Click "Next."
3. Read the Introduction. Click the terms that contain links to illustrations or animations as you read. Click "Next."
4. Take the Pre-Lab Quiz. Click "Next."
5. Figure P1.1 shows the opening screen for the laboratory exercise for this lesson. Follow the instructions

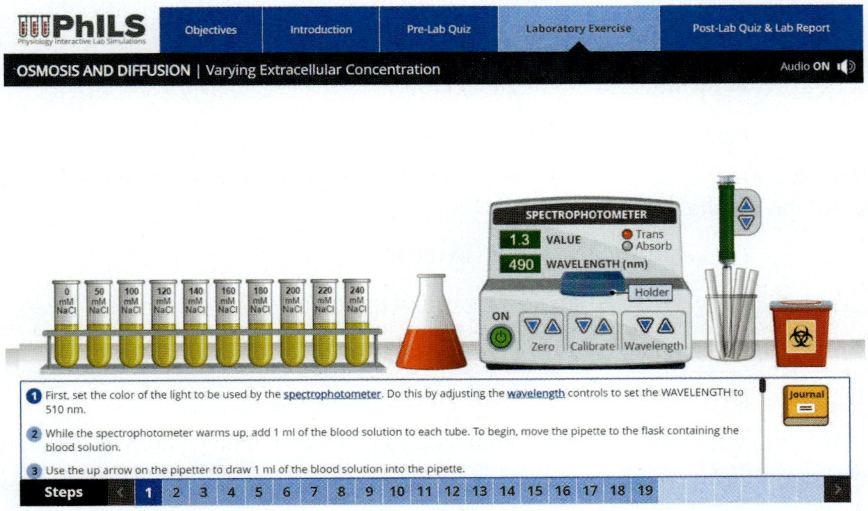

FIGURE P1.1 Opening screen for the laboratory exercise on Osmosis and Diffusion: Varying Extracellular Concentration.
McGraw Hill

step-by-step in the Laboratory Exercise and answer the following questions as you proceed.

6. Transfer the recorded data from your Ph.I.L.S. Journal onto table P1.1. Then, use the table to record your plot points on the graph of the NaCl concentration (x-axis) and percent transmittance (y-axis).
7. On your graph, indicate the areas that correspond to the red blood cells in each of the following solutions: (A) isotonic solution; (B) hypertonic solution; and (C) hypotonic solution.
8. Make note of any pertinent observations here:

____

____

____

9. Normal physiological saline (0.9%) is isotonic. What concentration (mM) would be equivalent to normal physiological saline? ________________________
10. Describe the net movement of water (into the cell, out of the cell, no net movement) when cells are placed in a(n):

    a. isotonic solution ________________________

    b. hypotonic solution ________________________

    c. hypertonic solution ________________________

11. Click the Post-Lab Quiz & Lab Report in the upper menu. Complete the Post-Lab Quiz.
12. Read the Lab Summary. Then, click "Lab Results."
13. Print your Lab Report by clicking "Print Lab" or "Save PDF" to turn in to your instructor.

TABLE P1.1 NaCl Concentration and Percent Transmittance.

[NaCl]	Trans

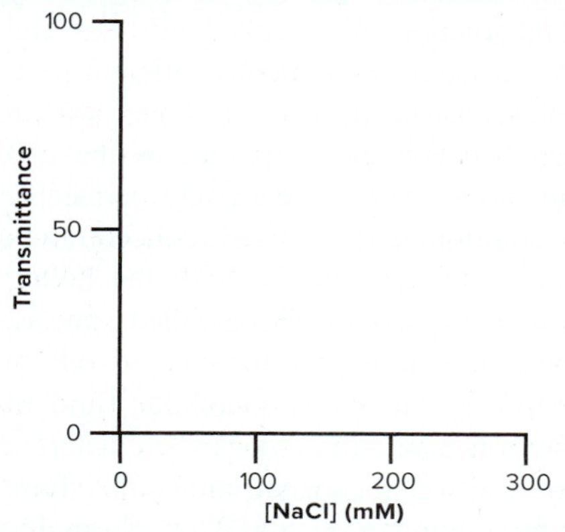

CRITICAL THINKING ASSESSMENT—Ph.I.L.S. LESSON 1

Choose one of the words in parentheses for each blank to accurately complete the sentences.

A patient is given an IV (intravenous) of deionized water (water with no solutes). When this large volume of fluid is administered, plasma becomes

____________________ (*hypotonic* or *hypertonic*) to the red blood cells.

____________________ Water moves from the (*plasma* or *red blood cells*)

____________________ to the (*plasma* or *red blood cells*),

____________________ causing the red blood cells to (*crenate* or *swell*).

A person drinks ocean salt water (3.5% salt). If the person drinks a large volume of the salt water (the salt water is absorbed into the blood), the plasma becomes

____________________ (*hypotonic* or *hypertonic*) to the red blood cells.

____________________ Water moves from the (*plasma* or *red blood cells*)

____________________ to the (*plasma* or *red blood cells*),

____________________ causing the red blood cells to (*crenate* or *swell*).

What do you predict the effect will be if a patient is administered an IV containing a normal physiological saline solution (0.9% sodium chloride)? ____

____

LESSON 2

Metabolism

Size and Basal Metabolic Rate

The purpose of this exercise is to demonstrate how the size of an animal influences the metabolic rate of its body cells. **Metabolism** is the sum total of all the chemical reactions that occur in the body. Cells utilize fuel from the foods you eat and the oxygen you inhale to synthesize ATP. Carbon dioxide, water, and energy in the form of ATP and heat are produced in this reaction. The **basal metabolic rate (BMR)** of an animal is the amount of energy expended in a given amount of time while in a resting state. Warm-blooded animals rely on the production of heat to keep their body temperature within normal limits (homeostasis). Heat normally escapes the body surface toward the outside environment; therefore, the animal's cells must maintain a metabolic rate that produces enough heat to make up for the loss. Because heat moves through the surface of the body, the size of an animal, particularly its *surface area–to–volume ratio,* is a factor that influences BMR. Smaller animals have a greater surface area–to–volume ratio than larger animals, so they lose heat more quickly and their cells have to work harder to make up for this heat loss. To determine the BMR of an animal, the amount of fuel (such as glucose) consumed *could* be measured, or the amount of carbon dioxide (CO_2) produced *could* be measured, or the amount of heat lost *could* be measured, but in this lab, the rate of oxygen (O_2) consumption *will* be indirectly measured. In this laboratory experiment, a mouse will be weighed, placed in a chamber that contains CO_2 absorbent, and observed to see how quickly the mouse consumes oxygen by measuring the movement of a soap bubble along a tube every 15 seconds for up to 2 minutes. There will be four mice of different sizes used in this experiment, and a minimum of two trials should be conducted for each mouse. The average oxygen consumed per minute and the average oxygen consumed per hour per gram of body weight for each mouse will be recorded.

Before you begin, familiarize yourself with the following concepts (use your textbook as a reference):

- Summary reaction of glucose catabolism: $C_6H_{12}O_6 + 6O_2 \rightarrow 6CO_2 + 6H_2O + \text{Energy}$
- Function of adenosine triphosphate (ATP)
- Surface area–to–volume ratio of cells
- Factors that affect metabolic rate (thyroid hormone, exercise, age, and size)

State your hypothesis regarding the size of an animal and the amount of oxygen consumed per hour per gram of body weight.

____

____

____

____

____

1. Open Lesson 2: Metabolism: Size and Basal Metabolic Rate.
2. Read the Objectives. Click "Next."
3. Read the Introduction. Click the terms that contain links to illustrations or animations as you read. Click "Next."
4. Read the Wet Lab. Click the terms that contain links to illustrations or animations as you read. Click "Next."
5. Take the Pre-Lab Quiz. Click "Next."
6. Figure P2.1 shows the opening screen for the laboratory exercise for this lesson. Follow the instructions step-by-step in the Laboratory Exercise and answer the following questions as you proceed.
7. Transfer the recorded data from your Ph.I.L.S. Journal of the individual trials for each mouse onto table P2.1.
8. Which trial for all mice in the journal data shows the highest O_2 consumed (mL) per minute? __________ To what do you attribute this higher oxygen consumption in this particular trial?

   ____

   ____

   ____

   ____

   ____

9. Transfer the recorded data from your Ph.I.L.S. Journal of the final graph onto table P2.2.

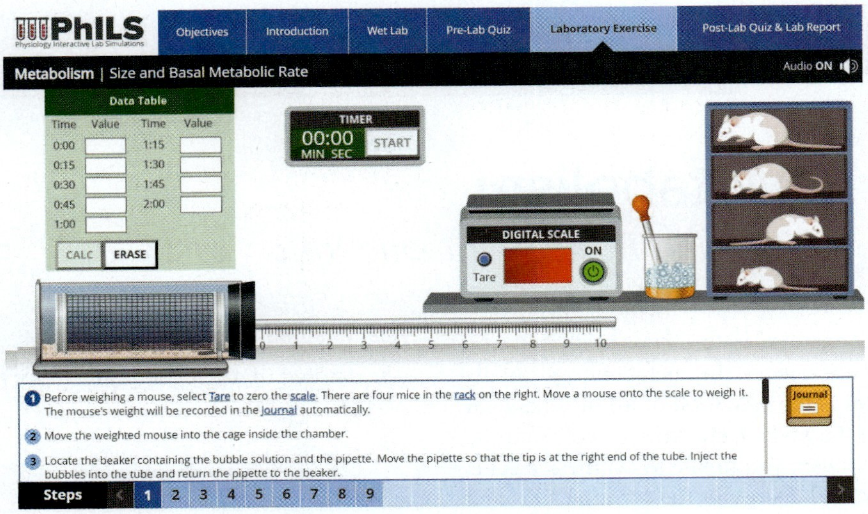

FIGURE P2.1 Opening screen for the laboratory exercise on Metabolism: Size and Basal Metabolic Rate. McGraw Hill

TABLE P2.1 Journal Data of Ph.I.L.S. Lesson 2: Metabolism: Size and Basal Metabolic Rate

Mouse	Weight (g)	Trial 1 (O_2 mL/min)	Trial 2 (O_2 mL/min)	Trial 3 (O_2 mL/min)	Trial 4 (O_2 mL/min)	Trial 5 (O_2 mL/min)
1						
2						
3						
4						

TABLE P2.2 Final Graph Results of Ph.I.L.S. Lesson 2: Metabolism: Size and Basal Metabolic Rate

Mouse	Weight (g)	Avg. O_2/min	Avg. O_2/h/g
1			
2			
3			
4			

10. Using a different color pencil for each final graph, plot the data you recorded in table P2.1 on the axes in (a) below and graph the linear regression. Then, plot the data you recorded in table P2.2 on the axes in (b) below and graph the linear regression.

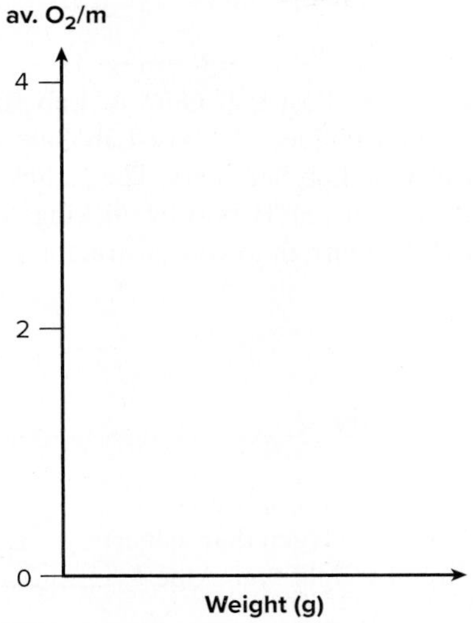

(a)

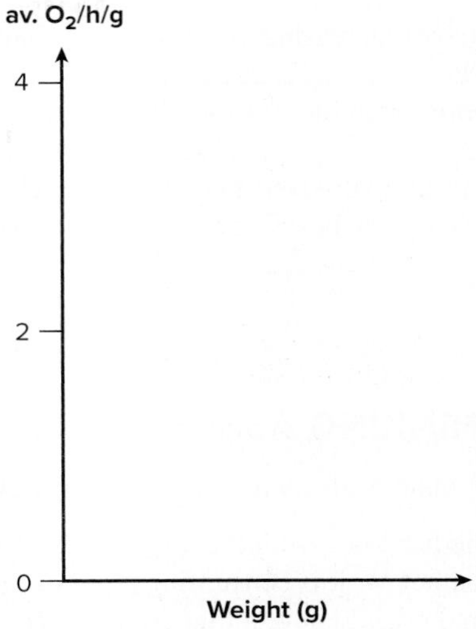

(b)

11. Note any pertinent observations here:

____

____

____

____

____

12. Examine the final graph in (a). On the x-axis is weight (g) and on the y-axis is average oxygen consumed per minute (av. O_2/m). As the weight of the mouse decreases, does the average oxygen consumed per minute *decrease* or *increase*? ______________ Is the relationship between weight and oxygen consumed *direct* or *inverse*?

____

13. Examine the final graph in (b). Weight (g) is on the x-axis and average oxygen consumed per hour per gram (of body weight) (av. O_2/h/g) is on the y-axis. As the weight of the mouse decreases, does the average oxygen consumed per hour per gram *decrease* or *increase*? ______________ Is the relationship between weight and oxygen consumed per gram of body weight *direct* or *inverse*? ______________

14. Based on the results of the final graph in (b) *and* the fact that oxygen consumed is an indirect measure of metabolic rate, as the weight of the mouse decreases, does the metabolic rate *decrease* or *increase*? ______________ Is the relationship between weight and metabolic rate *direct* or *inverse*? ______________

15. Surface Area–to–Volume Ratio: As a (spherical) structure increases in radius, the surface area increases at a rate of r^2 and the volume increases at a rate of r^3. If the radius of the sphere is 2, the calculated surface area $2^2 = 2 \times 2 = 4$ and the calculated volume $2^3 = 2 \times 2 \times 2 = 8$, and the ratio of surface area to volume (SA:V) is 4:8, reduced to 1:2 (0.5). If r is 4, calculate the surface area, volume, and surface area–to–volume ratio. Place your answers in the following box.

$r = 4$
Surface area = ______________ Volume = ______________
Surface area–to–volume (SA:V) ratio = ______________

16. As the (spherical) structure becomes larger, is the *ratio* becoming *larger* or *smaller*? (*Hint:* Compare the surface area–to–volume ratio between two spheres—one with a radius of 2 and the other with a radius of 4.) ______________ Is the volume increasing at a faster rate than the surface area? ______________

17. Suppose the surface area–to–volume ratio is applied to the human body, where volume is the size (weight) of the body and surface area is the area of skin surface. As a person increases in size:

   a. What happens to the ratio of surface area–to–volume?

   ____

   b. Is volume (size) increasing at a faster rate than surface area (skin)? ______________

   c. Is this a more favorable ratio for keeping the body warm? ______________

   d. Compared to a small-sized person, will each individual cell have to be as metabolically active as in a large-sized person to keep the body warm? ______________ Why? ______________

   ____

   e. In a large-sized person, is the metabolic rate of individual cells faster or slower than in individual cells in a small-sized person? ______________

   ____

18. Click the Post-Lab Quiz & Lab Report in the upper menu. Complete the Post-Lab Quiz.

19. Read the Lab Summary. Then, click "Lab Results."

20. Print your Lab Report by clicking "Print Lab" or "Save PDF" to turn in to your instructor.

CRITICAL THINKING ASSESSMENT—Ph.I.L.S. LESSON 2

1. Walt, a 6-foot male, walks with a friend Albert, who is 5-foot-9.

   a. Predict whether Walt, with the larger body, is consuming *more* or *less* oxygen than Albert. ______________

   b. Predict whether Walt is consuming *more* or *less* oxygen per gram of weight than Albert. ______________

   c. Which person would have a higher metabolic rate per gram of weight? ______________
   Why? ______________

   ____

   ____

2. For each of the following factors, indicate whether it would *increase* or *decrease* a person's metabolic rate.

   a. Hyperthyroidism would ______________ metabolic rate.

   b. Hypothyroidism would ______________ metabolic rate.

   c. Exercise would ______________ metabolic rate.

   d. Being sedentary (a "couch potato") would ______________ metabolic rate.

   e. Being an 80-year old as opposed to being a 45-year old would ______________ metabolic rate.

   f. Experiencing stress would ______________ metabolic rate.

LESSON 5

Skeletal Muscle Function

Stimulus-Dependent Force Generation

The purpose of this exercise is to demonstrate how the twitch force generated in skeletal muscle is dependent on stimulus intensity. In this experiment, the gastrocnemius (calf) muscle is removed from a frog and attached to a recording device. As an electrical stimulus is applied at a range of voltages to the muscle, a chart of the timing and force (strength) of contraction is measured and graphed. If no contraction is seen when an electrically weak stimulus is applied, the stimulus is considered **subthreshold.** As the muscle is stimulated with an incremental increase in voltage, the first perceivable contraction means that **threshold** has been reached. Threshold is the minimum voltage required to generate an action potential in the muscle fiber and produce a contraction. In this Ph.I.L.S. lab, the stimulus (voltage) applied to the gastrocnemius muscle mimics the action potential generated by a motor neuron. You will change the voltage from 0 to 1.6 V, while observing how increasing the stimulus intensity relates to an increase in skeletal muscle twitch force to a point when maximum contraction is achieved. Any voltage increase after maximum contraction will produce no further increase in force.

A stimulus at threshold or higher will trigger a skeletal **muscle twitch,** a phenomenon that does not occur *in vivo* because a single twitch by itself would not do any beneficial work. Normally, a succession of action potentials generated by the motor neuron would be delivered to the skeletal muscle. Because the gastrocnemius muscle in this experiment is being held at a fixed length, rather than changing its length as it contracts, the type of contraction is **isometric,** rather than isotonic. Studying an isometric skeletal muscle twitch *in vitro* provides some basic understanding of muscle physiology, allowing us to observe the distinct phases of a single contraction: latent period, contraction phase, and relaxation phase. The *latent period* is a delay between the onset of the stimulus and the onset of the contraction. During this time, the motor neuron action potential leads to the exposure of active sites on the actin filaments, creating an internal tension on the elastic components of the muscle. During the *contraction phase,* an external tension is created as the muscle contracts (shortens). The *relaxation phase* is seen when the level of intracellular Ca^{+2} falls and the crossbridge breaks, releasing the myosin from the actin.

A **motor unit** includes a somatic motor neuron and all of the skeletal muscle fibers that it activates (fig. P5.1). There are multiple motor units in a single muscle. The nervous system responds to the demands on a muscle by stimulating more motor units within that muscle to help it contract as a load increases; this is called *motor unit recruitment.* A single muscle contains only a limited number of motor units, so when all of the motor units are recruited, *maximum stimulation* has been

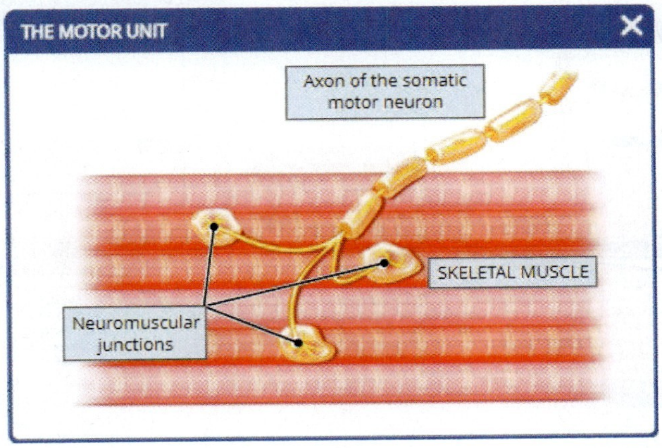

FIGURE P5.1 Motor unit. McGraw Hill APR

reached. Suppose you are trying to move a heavy object. If all the motor units in your muscles are recruited and the force you are generating is not moving the load, then no matter how much willpower you have, you will not be able to generate more force.

Before you begin, familiarize yourself with the following concepts (use your textbook as a reference):

- Motor unit
- Neuromuscular junction
- Skeletal muscle twitch and its parts (latent period, contraction period, and relaxation period)
- Difference between subthreshold, threshold, and maximal stimulus.
- Isometric contraction and isotonic contraction

State your hypothesis regarding how the isometric muscle twitch force of contraction is influenced by the changes in stimulus voltage.

____

____

____

____

____

1. Open Lesson 5: Skeletal Muscle Function: Stimulus Dependent Force Generation.
2. Read Objectives. Click "Next."
3. Read the Introduction. Click the terms that contain links to illustrations or animations as you read. Click "Next."
4. Take the Pre-Lab Quiz. Click "Next."
5. Figure P5.2 shows the opening screen for the laboratory exercise for this lesson. Follow the instructions step-by-step in the Laboratory Exercise and answer the following questions as you proceed.
6. Transfer the recorded data from your Ph.I.L.S. Journal onto table 5.1 and complete the final graph of the muscle tension (g) at each increment of applied volts (V). (Compare to fig. P5.3 graph.)
7. Note any pertinent observations here:

____

____

____

____

8. What was the threshold voltage for stimulation of the frog's gastrocnemius muscle?

____

9. What voltage produced maximum contraction?

____

10. A stimulus voltage from about 0.3 to 1.0 V showed a progressive increase in muscle tension. Explain what is occurring in the muscle during this time. ___________

____

____

____

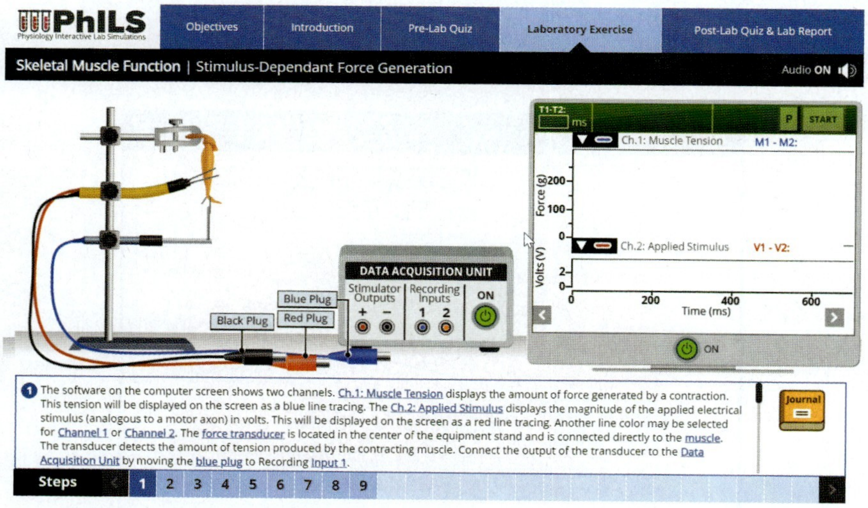

FIGURE P5.2 Opening screen for the laboratory exercise on Skeletal Muscle Function: Stimulus-Dependent Force Generation.
McGraw Hill

TABLE P5.1 Journal Data and Graph of Ph.I.L.S Lesson 5: Skeletal Muscle Function: Stimulus-Dependent Force Generation

Volts (V)	Muscle Tension (g)
0.0	
0.1	
0.2	
0.3	
0.4	
0.5	
0.6	
0.7	
0.8	
0.9	
1.0	
1.1	
1.2	
1.3	
1.4	
1.5	
1.6	

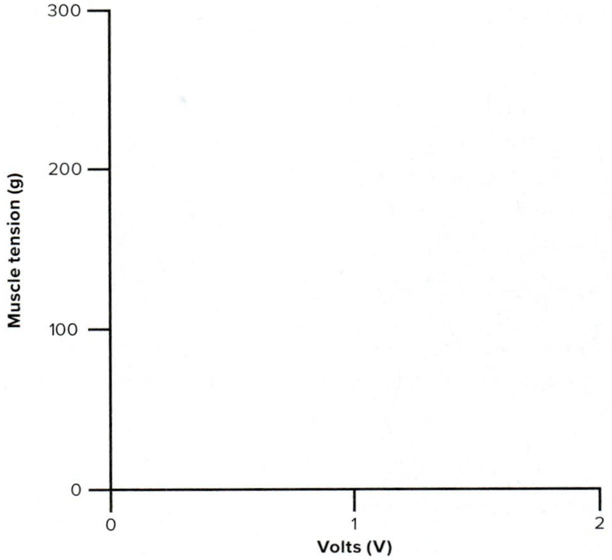

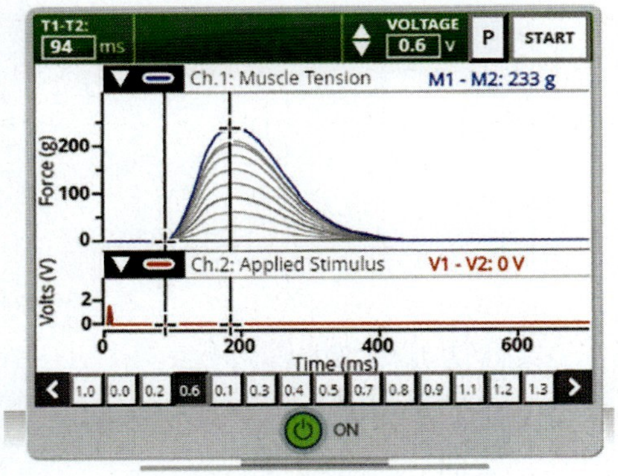

FIGURE P5.3 Example: Line tracings showing the amount of tension produced by applying shocks of different voltages to the exposed frog gastrocnemius. McGraw Hill

11. In the space below, create a graph of a muscle twitch at maximum contraction and label the y-axis, the x-axis, and the following: latent period, contraction phase, and relaxation phase.

12. Click the Post-Lab Quiz & Lab Report in the upper menu. Complete the Post-Lab Quiz.
13. Read the Lab Summary. Then, click "Lab Results."
14. Print your Lab Report by clicking "Print Lab" or "Save PDF" to turn in to your instructor.

LESSON 8

Skeletal Muscle Function:

Principles of Summation and Tetanus

The purpose of this exercise is to explore the mechanics of skeletal muscle contraction by recognizing and analyzing results of a line tracing as electrical stimuli is applied to muscle at different stimulation frequencies. Students will also determine the relationship of calcium release from the sarcoplasmic reticulum to the mechanical events of contraction and relaxation.

A skeletal muscle *twitch* results from the release of neurotransmitter from a somatic motor neuron. The neurotransmitter causes depolarization of the motor endplate and generates an action potential along the skeletal muscle membrane (sarcolemma). The depolarization travels down the transverse tubules to the dihydropyridine receptor, which stimulates the release of calcium ions from the sarcoplasmic reticulum (SR). The calcium ions bind to troponin C and initiate cross bridge cycling. When the stimulus is terminated, the calcium is quickly pumped back into the SR and the skeletal muscle will relax. In this exercise, you will deliver electrical stimuli to mimic the firing of motor neurons.

Voltage will be adjusted in this simulation to initiate individual twitches of the frog muscle. Once individual twitches are being recorded, you will increase the frequency of the stimulation by decreasing the time interval between twitches. An increase in frequency of the electrical stimulation mimics the functioning of the nervous system. Action potentials are all or none and unchanging, so to stimulate the muscle to a greater or lesser degree, the nervous system changes the frequency, or time between, action potentials. As the frequency of action potentials increases (stimuli are closer together), the muscle does not fully relax and calcium concentrations build in the sarcoplasm. With greater concentrations of calcium, a stronger contraction is produced. This phenomenon is known as *wave summation,* in that the waves of stimuli are added together to cause an increase in muscle tension (force). As you continue to increase the frequency, muscle will generate greater tension with shorter and less complete rest periods. This type of contraction is known as *incomplete tetanus.* If the stimuli are increased to the point that the muscle does not relax and there no rest period, the recording will be a flat line. This steady muscular contraction is *complete tetanus* (table P8.1).

State your hypothesis regarding the effect of increased frequency of stimulation on the force produced by a muscle.

____

____

____

____

TABLE P8.1 Observing Muscle Response

Muscle Response	Observation on Computer Screen
Wave summation	Trace moves off the baseline (lower gray line)
Incomplete tetanus	Trace becomes higher than in a single twitch (above upper gray line)
Complete tetanus	Trace is higher than a single twitch (above upper gray line) and becomes smooth

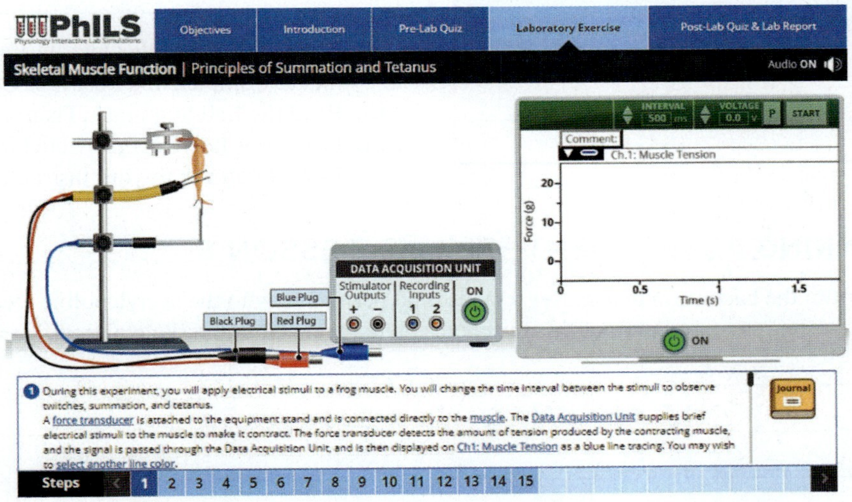

FIGURE P8.1 Opening screen for the laboratory exercise on Skeletal Muscle Function: Principles of Summation and Tetanus.
McGraw Hill

Before you begin, familiarize yourself with the following concepts (use your textbook as a reference):

- Action potential
- Single twitch, summation, incomplete tetanus, and complete tetanus
- Skeletal muscle excitation, excitation-contraction coupling, contraction, and relaxation

1. Open Lesson 8: Skeletal Muscle Function: Principles of Summation and Tetanus.
2. Read Objectives. Click "Next."
3. Read the Introduction. Click the terms that contain links to illustrations or animations as you read. Click "Next."
4. Take the Pre-Lab Quiz. Click "Next."
5. Figure P8.1 shows the opening screen for the laboratory exercise for this lesson. Follow the instructions step-by-step in the Laboratory Exercise and answer the following questions as you proceed.
6. Record the data from your Ph.I.L.S. Journal of the time intervals (frequencies) in milliseconds where you first saw wave summation, incomplete tetanus, and complete tetanus in table P8.2. Then, complete the questions that follow.
7. Make note of any pertinent observations here: ____________________

____

8. Based on your observations and recorded data, what is the relationship between muscle tension and stimulus frequency? ____________________

____

____

TABLE P8.2 Journal Results of Muscle Response Time Intervals

Muscle Response	Time Interval (ms)
Summation	
Incomplete tetanus	
Complete tetanus	

9. How do the results of this laboratory simulation compare to your hypothesis? ____________________

____

10. Complete (fused) tetanus is observed in the laboratory or in a laboratory simulation like this one, but it is not seen in the human body. Why do you think this is so? ____________________

____

11. Calcium is released from the sarcoplasmic reticulum through voltage-gated calcium channels to initiate muscle contraction. When the muscle is no longer stimulated, calcium pumps move the calcium from the sarcoplasm back into the sarcoplasmic reticulum. Complete the following:

    a. During muscle *contraction,* is the net movement of calcium <u>into</u> or <u>out of</u> the sarcoplasm reticulum?.

    ____

    b. During muscle *relaxation,* is the net movement of calcium <u>into</u> or <u>out of</u> the sarcoplasmic reticulum?

    ____

    c. During incomplete tetanus, the amount of tension produced by the muscle is greater than during a single twitch. What does this tell you about the relative amount of calcium in the muscle sarcoplasm during incomplete tetanus? ____________________

    ____

    ____

    d. Describe the net movement of calcium (<u>into</u> or <u>out of</u> the sarcoplasmic reticulum) during complete tetanus. ____________________

    e. Explain why the rate of muscle relaxation would be slower after tetanus than after a single twitch. ____________________

    ____

    ____

12. Click the Post-Lab Quiz & Lab Report in the upper menu. Complete the Post-Lab Quiz.
13. Read the Lab Summary. Then, click "Lab Results."
14. Print your Lab Report by clicking "Print Lab" or "Save PDF" to turn in to your instructor.

CRITICAL THINKING ASSESSMENT—Ph.I.L.S. LESSON 8

The toxins released from the bacterium that causes tetanus bind permanently to the acetylcholine receptors of the skeletal muscle, resulting in continuous stimulation of the muscle. Explain the result to the skeletal muscle of the body.

____

____

____

____

LESSON 10

Resting Potentials:

Resting Potential and External [K^+]

The purpose of this exercise is to examine the effect that different levels of [KCl] (potassium chloride concentration) in saline solution (Ringer's solution) have on the resting membrane potential (RMP) of muscle fibers of a crayfish. This lab simulation is similar to altering the [K^+] in normal extracellular fluid. The potentials will be compared between a probe in the solution and a probe within a muscle cell to determine the RMP of the muscle cells. This lab simulation is intended to be completed before Lesson 11: Resting Potentials: Resting Potential and External [Na^+] in an effort to help in the understanding of the role that Na^+ and K^+ play in the RMP and the effect that adjusting extracellular concentrations of these ions has on the membrane potential. Muscle cells, like neurons, are highly excitable which means that a change in the concentration of Na^+ or K^+ across the membrane will alter their gradients and will therefore change the RMP, moving closer to or further away from threshold.

The cell membrane is selectively permeable, meaning that some molecules are allowed through while others are not. Embedded in the cell membrane are proteins that facilitate the movement of charged molecules or ions. These membrane proteins include leaky K^+ channels and leaky Na^+ channels as well as Na^+/K^+ ATPase pumps. The distribution of ions (cations and anions) across the cell membrane creates a charge difference, the RMP, where there is a negative charge along the inner membrane and a positive charge along the outer membrane.

At rest, sodium (Na^+) ions are more concentrated in the extracellular fluid (ECF) while potassium (K^+) ions are more concentrated in the intracellular fluid (ICF). Certain membrane proteins (leaky channels) allow Na^+ to diffuse into the cell and K^+ to diffuse out of the cell down their chemical gradients. Negatively charged proteins pull K^+ into the cell down an electrical gradient, an opposing force to the chemical gradient of K^+. The Na^+/K^+ ATPase pumps transport these ions against their concentration gradients, and in doing so, keep a higher concentration of K^+ inside the cell (and Na^+ outside the cell) which maintains their chemical gradients. With each ATP molecule consumed, 3 Na^+ are pumped out and 2 K^+ are pumped in. The cell membrane is more permeable to K^+, so although the negatively charged proteins do pull some Na^+ into the cell, it is far less than K^+.

Before you begin, familiarize yourself with the following concepts (use your textbook as a reference):

- Na^+/K^+ ATPase
- Normal Na^+ and K^+ concentration distribution in ECF and ICF
- Cations and anions
- Resting membrane potential
- Threshold
- Depolarization, repolarization, and hyperpolarization

State your hypothesis regarding the concentration of KCl in Ringer's solution and the resulting RMP of the muscle fiber.

____

____

____

1. Open Lesson 10: Resting Potentials: Resting Potential and External [K^+]
2. Read the Objectives. Click "Next."
3. Read the Introduction. Click the terms that contain links to illustrations or animations as you read. Click "Next."
4. Take the Pre-Lab Quiz. Click "Next."
5. Figure P10.1 shows the opening screen for the laboratory exercise for this lesson. Follow the instructions step-by-step in the Laboratory Exercise and answer the following questions as you proceed.
6. What is the blue line deflection on the computer screen measuring? ______________________________

   ____
7. What is used as the control in this lab simulation?

   ____

   What is the purpose of this control? ______________

   ____
8. Transfer the recorded data from your Ph.I.L.S. Journal of the amount of potassium in the saline (mM) onto table P10.1. Use the logarithmic calculations (CALC) to put the average membrane potential (av. Em) at the various log [K^+] (mEq/L) onto table P10.2.

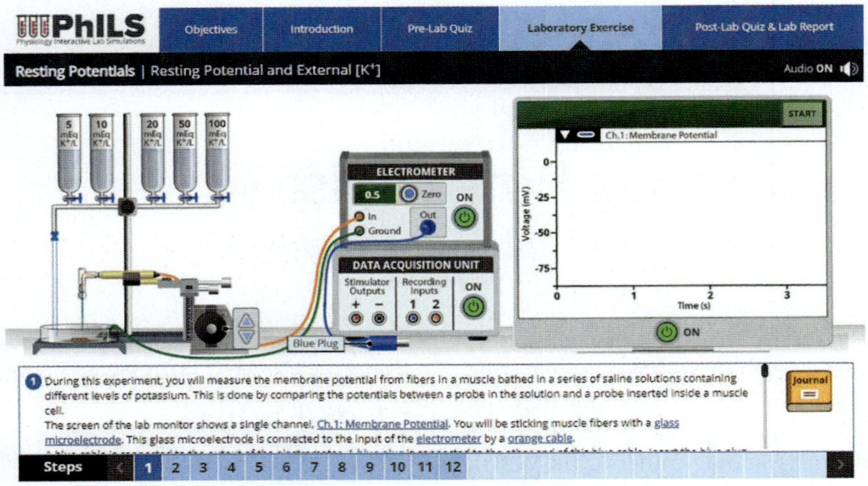

FIGURE P10.1 Opening screen for the laboratory exercise on Resting Potentials: Resting Potential and External [K^+].
McGraw Hill

TABLE P10.1 Journal Data for Ph.I.L.S. Lesson 10: Resting Potentials: Resting Potential and External [K^+]

Tube	5 mEq/L K^+	10 mEq/L K^+	20 mEq/L K^+	50 mEq/L K^+	100 mEq/L K^+
1					
2					
3					
4					
5					

TABLE P10.2 Log Data for Ph.I.L.S. Lesson 10: Resting Potentials: Resting Potential and External [K^+]

[K^+]	Av. Em (mV)
5	
10	
20	
50	
100	

9. Graph the data you recorded in table P10.2 on the axes.

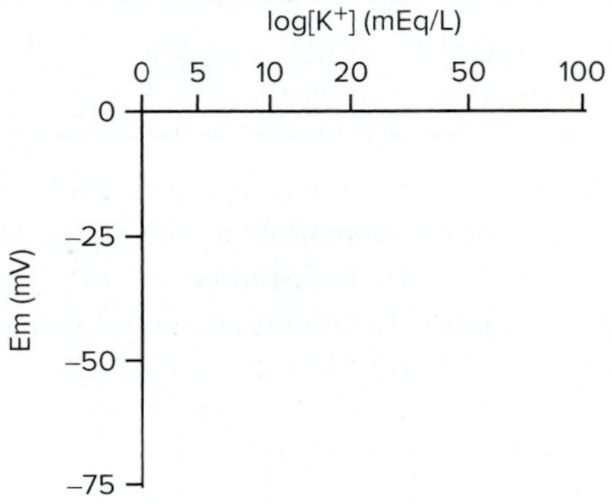

10. Make note of any pertinent observations here:

____

____

11. As the [K^+] increased in the extracellular fluid in this lab simulation, the membrane potential became ________________ (*more* or *less*) negative?

12. When a membrane potential becomes more negative, it is ________________ (*depolarized* or *hyperpolarized*)? When a membrane becomes less negative, it is ________________ (*depolarized* or *hyperpolarized*)?

13. As the [K^+] increased in the extracellular fluid in this lab simulation, the membrane potential moved ________________ (*closer to* or *further from*) action potential threshold?

14. As the [K^+] increased in the extracellular fluid in this lab simulation, an action potential within the muscle fibers of the crayfish was ________________ (*more* or *less*) likely to occur?

15. If the action potential threshold of a muscle fiber is −50 mV, based on your experimental lab simulation results, what is the likelihood that a crayfish muscle bathed in 10 mEq K^+/L contract? Explain. __________

____

____

16. Click the Post-Lab Quiz & Lab Report in the upper menu. Complete the Post-Lab Quiz.
17. Read the Lab Summary. Then, click "Lab Results."
18. Print your Lab Report by clicking "Print Lab" or "Save PDF" to turn in to your instructor.

CRITICAL THINKING ASSESSMENT—Ph.I.L.S. LESSON 10

A crush injury of a body limb results in damaged muscle cells. Would this result in *hypokalemia or hyperkalemia*?
______________________________ Why? ______________________________

____

Predict the effect this may have on cardiac muscle. ______________________________

____

____

LESSON 11

Resting Potentials:

Resting Potential and External [Na^+]

The purpose of this exercise is to examine the effect that different levels of [NaCl] (sodium chloride concentration) in saline solution (Ringer's solution) have on the resting membrane potential (RMP) of muscle fibers of a crayfish. This lab simulation is similar to altering the [Na^+] in normal extracellular fluid. The potentials will be compared between a probe in the solution and a probe within a muscle cell to determine the RMP of the muscle cells. This lab simulation is intended to be completed after Lesson 10: Resting Potentials: Resting Potential and External [K^+] as a continuation in the understanding of the role that Na^+ and K^+ play in the RMP and the effect that adjusting extracellular concentrations of these ions has on the membrane potential. Muscle cells, like neurons, are highly excitable which means that a change in the concentration of Na^+ or K^+ across the membrane will alter their gradients and will therefore change the RMP, moving closer to or further away from threshold.

The cell membrane is selectively permeable, meaning that some molecules are allowed through while others are not. Embedded in the cell membrane are proteins that facilitate the movement of charged molecules or ions. These membrane proteins include leaky K^+ channels and leaky Na^+ channels as well as Na^+/K^+ ATPase pumps. The distribution of ions (cations and anions) across the cell membrane creates a charge difference, the RMP, where there is a negative charge along the inner membrane and a positive charge along the outer membrane.

At rest, sodium (Na^+) ions are more concentrated in the extracellular fluid (ECF) while potassium (K^+) ions are more concentrated in the intracellular fluid (ICF). The Na^+ concentration inside the cell is low. Certain membrane proteins (leaky channels) allow Na^+ to diffuse into the cell and K^+ to diffuse out of the cell down their chemical gradients. There are few Na^+ channels in the membrane that are open, so the permeability of the membrane to Na^+ is limited. Negatively charged proteins in the cell pull some Na^+ into the cell down its electrical gradient, the same direction that Na^+ moves down its chemical gradient. With each ATP molecule consumed, 3 Na^+ are pumped out of (and 2 K^+ are pumped into) the cell by the Na^+/K^+ ATPase pump, which helps keep Na^+ levels very low inside the cell compared to outside the cell and, thus, maintains the chemical gradient of Na^+. The movement of Na^+ creates the driving force in transporting many other molecules and ions through the cell membrane, so maintaining this chemical gradient is important.

Before you begin, familiarize yourself with the following concepts (use your textbook as a reference):

- Na^+/K^+ ATPase
- Normal Na^+ and K^+ concentration distribution in ECF and ICF
- Cations and anions
- Resting membrane potential
- Threshold
- Depolarization, repolarization, and hyperpolarization

State your hypothesis regarding the concentration of NaCl in Ringer's solution and the resulting RMP of the muscle fiber.

____

____

____

1. Open Lesson 11: Resting Potentials: Resting Potential and External [Na^+]
2. Read the Objectives. Click "Next."
3. Read the Introduction. Click the terms that contain links to illustrations or animations as you read. Click "Next."
4. Take the Pre-Lab Quiz. Click "Next."
5. Figure P11.1 shows the opening screen for the laboratory exercise for this lesson. Follow the instructions step-by-step in the Laboratory Exercise and answer the following questions as you proceed.
6. What is the blue line deflection on the computer screen measuring? ________________________

   ____

7. Transfer the recorded data from your Ph.I.L.S. Journal of the amount of sodium in the saline (mM) onto table P11.1. Use the logarithmic calculations (CALC) to put the average membrane potential (av. RMP) at the various log [Na^+] (mEq/L) onto table P11.2.

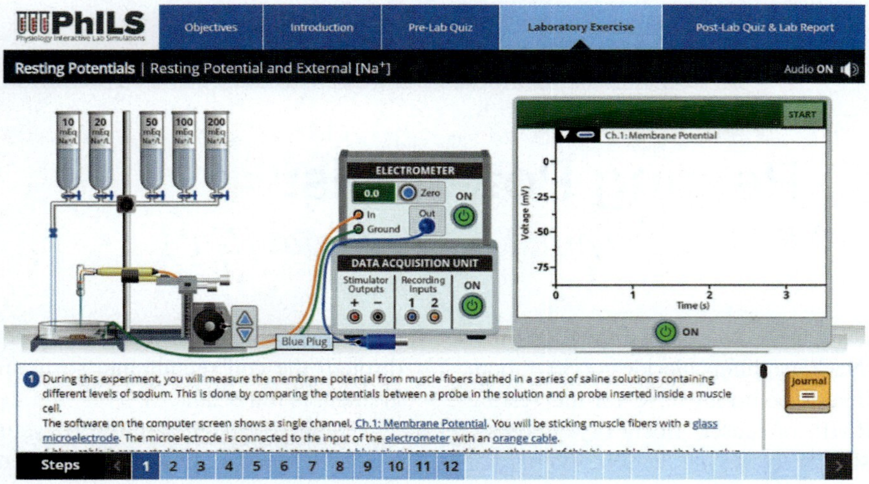

FIGURE P11.1 Opening screen for the laboratory exercise on Resting Potentials: Resting Potential and External [Na+].
McGraw Hill

TABLE P11.1 Journal Data for Ph.I.L.S. Lesson 11: Resting Potentials: Resting Potential and External [Na^+]

Tube	10 mEq/L Na^+	20 mEq/L Na^+	50 mEq/L Na^+	100 mEq/L Na^+	200 mEq/L Na^+
1					
2					
3					
4					
5					

TABLE P11.2 Log Data for Ph.I.L.S. Lesson 11: Resting Potentials: Resting Potential and External [Na^+]

[Na^+]	Av. RMP (mV)
10	
20	
50	
100	
200	

8. Graph the data you recorded in table P11.2 on the axes.

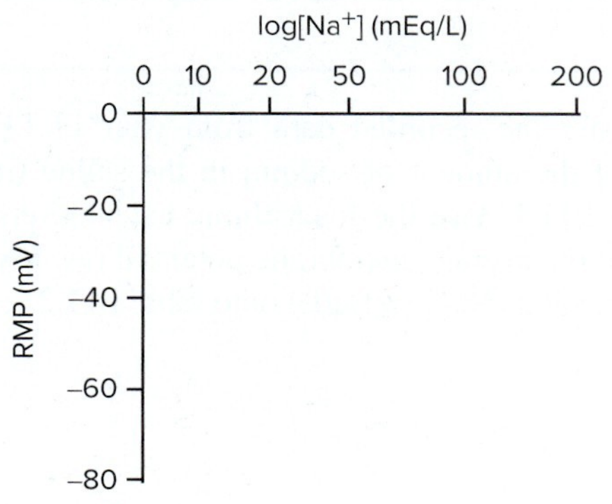

9. Make note of any pertinent observations here:

____

____

10. The RMP of the solution with normal saline was ________________ (*more* or *less*) negative than the rest of the experimental results.

11. As the [Na^+] decreased in the extracellular fluid in this lab simulation, the membrane potential became ________________ (*more* or *less*) negative?

12. When a membrane potential becomes more negative, it is ________________ (*depolarized* or *hyperpolarized*)? When a membrane becomes less negative, it is ________________ (*depolarized* or *hyperpolarized*)?

13. As the [Na^+] decreased in the extracellular fluid in this lab simulation, the membrane potential moved ________________ (*closer to* or *further from*) action potential threshold?

14. As the [Na^+] decreased in the extracellular fluid in this lab simulation, an action potential within the muscle fibers of the crayfish were ________________ (*more* or *less*) likely to occur?

15. If the action potential threshold of a muscle fiber is −50 mV, based on your experimental lab simulation

results, would the sodium concentration (mEq Na^+/L) need to ________________ (*increase* or *decrease*) for the muscle to contract?

16. The cation ________________ (*Na^+* or *K^+*) has the greatest effect on the membrane potential. Therefore, small fluctuations in extracellular [Na^+] have a ________________ (*minimal* or *large*) effect on the membrane potential of most cells of the body.

17. Click the Post-Lab Quiz & Lab Report in the upper menu. Complete the Post-Lab Quiz.
18. Read the Lab Summary. Then, click "Lab Results."
19. Print your Lab Report by clicking "Print Lab" or "Save PDF" to turn in to your instructor.

CRITICAL THINKING ASSESSMENT—Ph.I.L.S. LESSON 11

Patients have died in clinical settings after mistakenly being injected with intravenous (IV) potassium chloride (KCl) instead of 0.9% saline. What is the danger for these patients?

____

____

____

PhILS Simulations

LESSON 12

Action Potentials:

The Compound Action Potential

The purpose of this exercise is to demonstrate the response of nerve fibers (axons) of nerve cells (neurons) to an applied electrical shock using varying degrees of intensity (voltage). The nerve that will be used in this experiment is an excised sciatic nerve of a frog. Like all nerves, the sciatic nerve is composed of bundles of axons that are surrounded by connective tissue. If an electrical shock (stimulus) is applied and the axons do not respond, threshold has not been met, called *subthreshold*. The minimum stimulus needed to elicit an action potential is called *threshold*, and this will be identified by varying the intensity (voltage) of electrical shock to the nerve. In an excitable membrane, such as nerves and muscles, when threshold is met, voltage-regulated sodium channels open allowing sodium (Na^+) to rush into the cells causing *depolarization* and producing an action potential. Sodium channels quickly close and voltageregulated potassium channels open to return the membrane back to its resting potential as potassium (K^+) moves out of the cells, called *repolarization*. This rapid rise and fall of the membrane potential creates the *action potential*, or *impulse*. When axons within a bundle produce action potentials at the same time, they will proceed down the length of the axon at the same conduction velocity, producing a *compound action potential (CAP)*. At low intensity where threshold is just met, a few axons elicit a small CAP. With increased intensity of applied electrical shock to the nerve, more and more axons will be stimulated to elicit action potentials, raising the CAP, and amplifying the nerve response; this is called *recruitment*. When all axons have been stimulated, the amplitude will be at its peak, or *maximum combined action potential*. By adjusting the number of action potentials elicited by axons within a motor nerve, for example, the central nervous system has the ability to control the strength of skeletal muscle contraction innervated by that nerve.

Before you begin, familiarize yourself with the following concepts (use your textbook as a reference):

- Graded (local) potential
- Threshold
- Action potential
- Recruitment

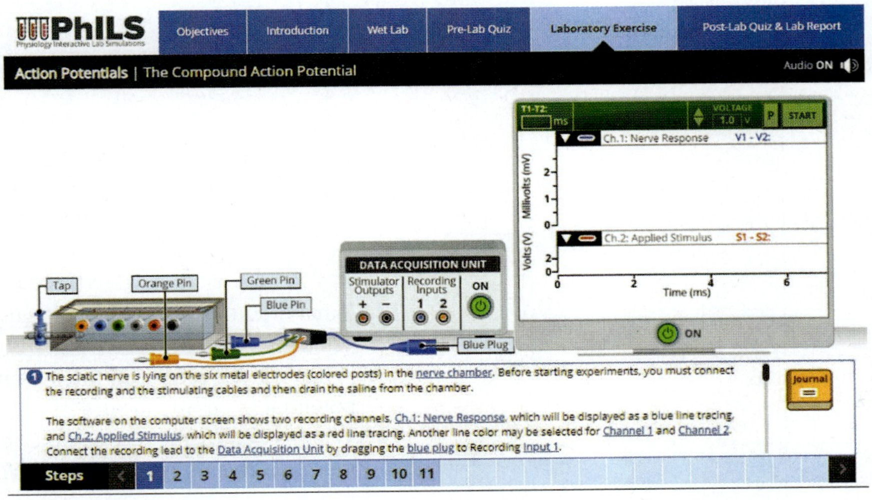

FIGURE P12.1 Opening screen for the laboratory exercise on Action Potentials: The Compound Action Potential.
McGraw Hill

State your hypothesis regarding the intensity of applied electrical stimulation and the compound action potential produced.

____

____

____

____

1. Open Lesson 12: Action Potentials: The Compound Action Potential.
2. Objectives. Click "Next."
3. Read the Introduction. Click the terms that contain links to illustrations or animations as you read. Click "Next."
4. Read the Wet Lab. Click the terms that contain links to illustrations or animations as you read. Click "Next."
5. Take the Pre-Lab Quiz. Click "Next."
6. Figure P12.1 shows the opening screen for the laboratory exercise for this lesson. Follow the instructions step-by-step in the Laboratory Exercise and answer the following questions as you proceed.
7. Define a *compound action potential*. ______________

____

____

8. Transfer the recorded data from your Ph.I.L.S. Journal of the CAP amplitude at various shock voltages onto table P12.1.

TABLE P12.1 Journal Data for Ph.I.L.S. Lesson 12: Action Potentials: The Compound Action Potential

Volts	Amp
0.0	
0.1	
0.2	
0.3	
0.4	
0.5	
0.6	
0.7	
0.8	
0.9	
1.0	
1.1	
1.2	
1.3	
1.4	
1.5	
1.6	

9. Graph the journal data you recorded in table P12.1 below.

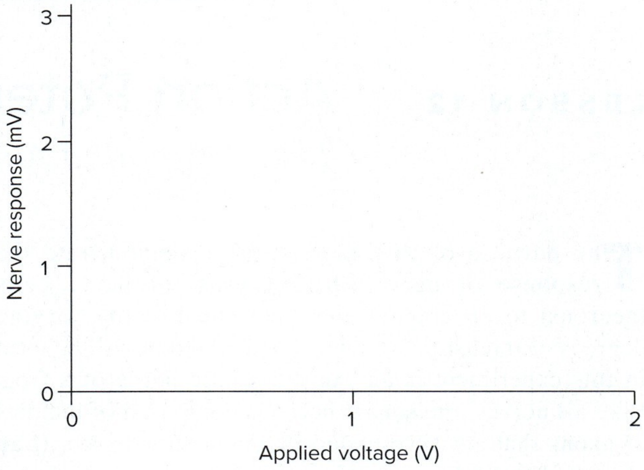

10. Make note of any pertinent observations here:

____

____

____

11. Using your data, graph, and figure P12.2 as reference, graph a compound action potential.

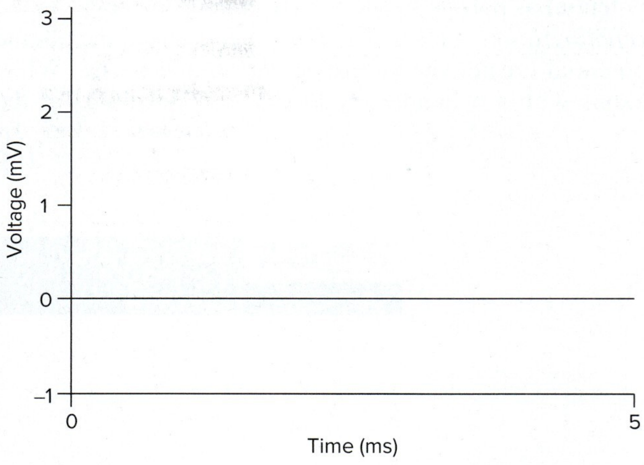

12. What was the threshold voltage for stimulation of the sciatic nerve? ______________
13. What voltage produced the maximum compound action potential? ______________
14. What is occurring between the threshold and the maximum compound action potential? ______________

____

____

____

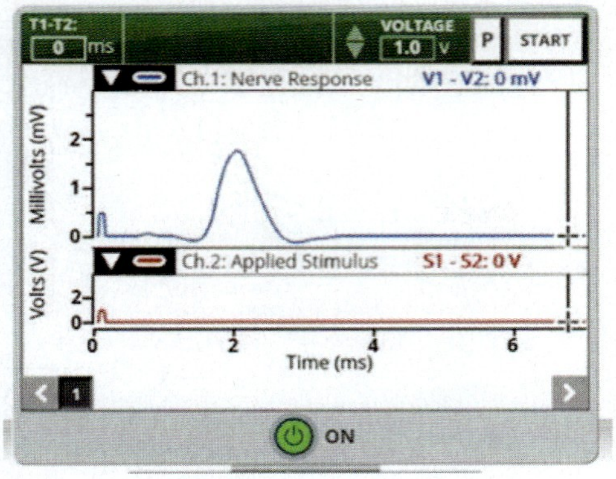

FIGURE P12.2 An example of a graph to show the effect of stimulus voltage on the amplitude of the evoked compound action potential (CAP) recorded from a frog sciatic nerve. McGraw Hill

15. Explain the relationship between increasing voltage stimulation and the recruitment of nerve fibers (axons).

____

____

____

____

16. Click the Post-Lab Quiz & Lab Report in the upper menu. Complete the Post-Lab Quiz.
17. Read the Lab Summary. Then, click "Lab Results."
18. Print the Lab Report by clicking "Print Lab" or "Save PDF" to turn in to your instructor.

CRITICAL THINKING ASSESSMENT—Ph.I.L.S. LESSON 12

Multiple sclerosis is an autoimmune disorder that involves the destruction of the myelin sheath by the immune system. As a result, the nerve cells affected are no longer able to send impulses. What happens to the maximum compound action potential of the affected nerves? Explain. ____

____

____

____

____

PhILS Simulations

LESSON 19

Endocrine Function

Thyroid Gland and Metabolic Rate

The purpose of this exercise is to determine the role that the thyroid gland has on metabolic rate. The endocrine system maintains homeostasis by secreting hormones that affect the actions of cells around the body. Metabolic reactions are a prime target of endocrine hormones, as demonstrated by the actions of thyroid hormone [triiodothyronine (T3) and tetraiodothyronine (T4; thyroxine)] produced by the thyroid gland.

Thyroid hormone increases metabolic activity in nearly all the cells of the body. When body temperature decreases, thyroid hormone secretion increases. The increase in metabolic rate can be inferred from oxygen consumption. In this simulation, you will measure the rate of oxygen consumption, over a range of temperatures, of a normal mouse and a mouse that has been given propylthiouracil (PTU) to reduce the production of T3 and T4 by the thyroid gland. You will create a graph that clearly depicts the effect of temperature on metabolic rate as controlled by the presence or absence of thyroid hormone secretion.

Before you begin, familiarize yourself with the following concepts (use your textbook as a reference):

- Thyroid gland
- T3 and T4
- Thyroglobulin
- Thyroid-releasing hormone (TRH)
- Thyroid-stimulating hormone (TSH)
- Metabolic rate and oxygen consumption
- Metabolic rate and temperature regulation

State your hypothesis regarding the relationship between thyroid hormone secretion from the thyroid gland and metabolic rate.

____

____

____

____

1. Open Lesson 19: Endocrine Function: Thyroid Gland and Metabolic Rate.
2. Read Objectives. Click "Next."
3. Read the Introduction. Click the terms that contain links to illustrations or animations as you read. Click "Next."
4. Read the Wet Lab. Click the terms that contain links to illustrations or animations as you read. Click "Next."
5. Take the Pre-Lab Quiz. Click "Next."
6. Figure P19.1 shows the opening screen for the laboratory exercise for this lesson. Follow the instructions step-by-step in the Laboratory Exercise and answer the following questions as you proceed.
7. Transfer the recorded data from your Ph.I.L.S. Journal of the weight (g) and temperature (°C) for normal and PTU onto table P19.1.

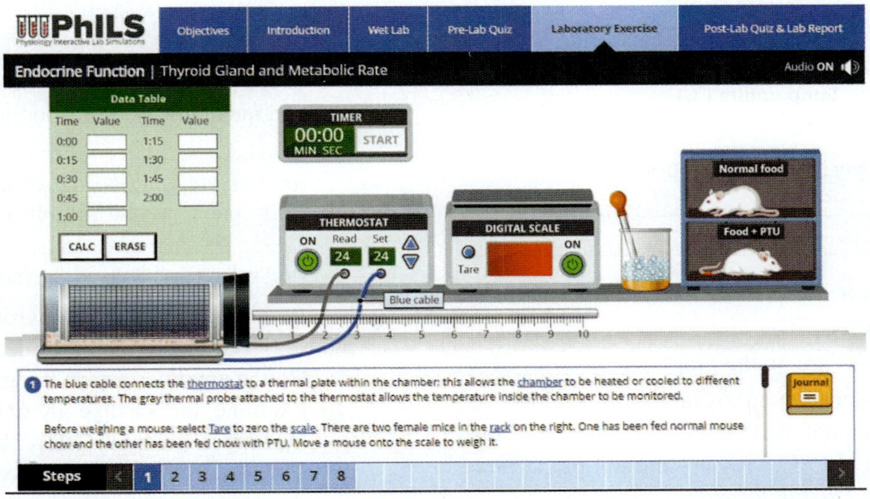

FIGURE P19.1 Opening screen for the laboratory exercise on Endocrine Function: Thyroid Gland and Metabolic Rate.
McGraw Hill

TABLE P19.1 Journal Data of Ph.I.L.S. Lesson 19: Endocrine Function: Thyroid Gland and Metabolic Rate

OXYGEN/M		
Temp	**Normal**	**PTU**
°C	g	g
24		
22		
20		
18		
16		
14		
12		
10		
8		

8. Using the recorded data, complete the final graph of the linear regression of oxygen consumption (O_2/m) and temperature (°C) and compare your graph to fig. P19.2. Use a darker colored pencil for the "NORM" and a lighter colored pencil for the "PTU" graph points. Then, complete the questions that follow.

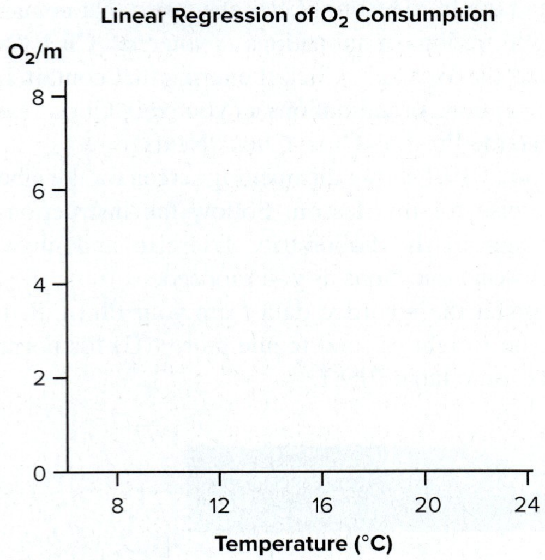

9. Make note of any pertinent observations here:

____

____

10. Using the graph for the "NORM" mouse, answer the following questions:

a. How many lines are on the graph? __________

b. Examine the change in oxygen consumption as the environmental temperature decreases from 24°C to 18°C. Is oxygen consumption <u>increasing</u>, <u>decreasing</u>, or <u>staying the same</u>?

____

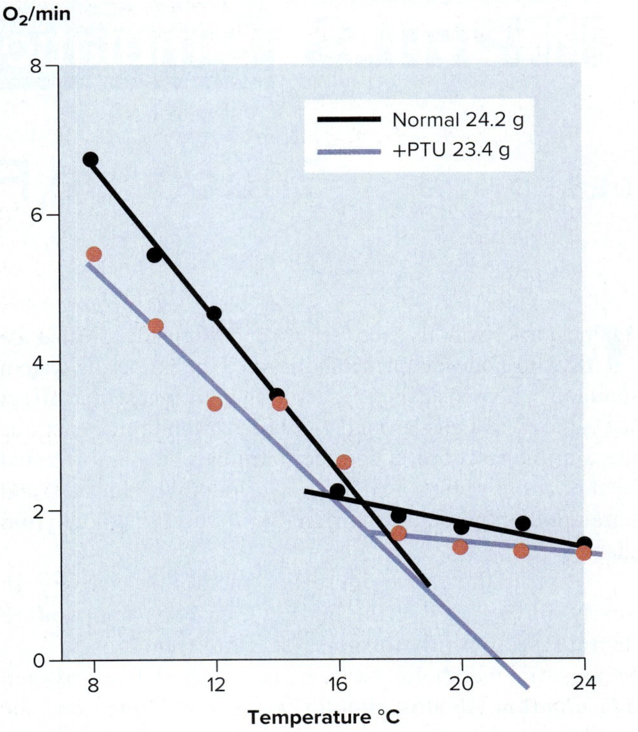

FIGURE P19.2 An example of a graph that shows the effect of thyroid function and temperature on oxygen consumption in mice.

c. Given the relationship of oxygen consumption and metabolic rate, metabolic rate is

____

(<u>increasing</u>, <u>decreasing</u>, or <u>staying the same</u>).

d. This part of the graph can be explained by

____

(<u>the release of thyroid hormone</u> or <u>shivering</u>).

e. Examine the change in oxygen consumption as the environmental temperature decreases from 18°C to 8°C. Is oxygen consumption <u>increasing</u>, <u>decreasing</u>, or <u>staying the same</u>?

____

f. Given the relationship of oxygen consumption and metabolic rate, metabolic rate is

____

(<u>increasing</u>, <u>decreasing</u>, or <u>staying the same</u>).

g. This part of the graph can be explained by

____

(<u>the release of thyroid hormone</u> or <u>shivering</u>).

11. Using the graph for the thyroid hormone–deficient mouse (PTU treated), answer the following questions:

a. How many lines are on the graph? __________

b. Examine the change in oxygen consumption as the environmental temperature decreases from 24°C to 18°C. Is oxygen consumption <u>increasing</u>, <u>decreasing</u>, or <u>staying the same</u>?

____

c. Given the relationship of oxygen consumption and metabolic rate, metabolic rate is

____

(*increasing*, *decreasing*, or *staying the same*).

d. This part of the graph can be explained by

____

(*the impaired release of thyroid hormone* or *shivering*).

e. Examine the change in oxygen consumption as the environmental temperature decreases from 18°C to 8°C. Is oxygen consumption *increasing*, *decreasing*, or *staying the same*?

____

f. Given the relationship of oxygen consumption and metabolic rate, metabolic rate is

____

(*increasing*, *decreasing*, or *staying the same*).

g. This part of the graph can be explained by

____

(*the impaired release of thyroid hormone* or *shivering*).

12. Observe the graphs of BOTH the "NORM" and the "PTU" animals to answer the following questions:

a. For both animals, explain the relationship of temperature and oxygen consumption.

____

____

____

____

____

When is the greatest amount of oxygen consumed—at *cooler* or *warmer* temperatures?

____

b. Compare the different colored lines between 18°C and 24°C. Which is steeper? ______________

Which animal has a greater oxygen consumption?

____

Which animal has a higher metabolic rate?

____

c. The point of intersection of the two lines of the "NORM" (results from the mouse with normal functioning thyroid gland) represents the point at which the animal begins to shiver. At what temperature does this occur? ______________

d. The point of intersection of the two lines of the "PTU" (results from the mouse with malfunctioning thyroid gland) represents the point at which the animal begins to shiver. At what temperature does this occur? ______________

e. Which animals begin to shiver first? ______________

Explain. ______________________

____

____

____

____

13. Click the Post-Lab Quiz & Lab Report in the upper menu. Complete the Post-Lab Quiz.
14. Read the Lab Summary. Then, click "Lab Results."
15. Print your Lab Report by clicking "Print Lab" or "Save PDF" to turn in to your instructor.

CRITICAL THINKING ASSESSMENT—Ph.I.L.S. LESSON 19

Would you predict your thyroid hormone level to *increasing*, *decreasing*, or *staying the same* in the winter? __________ Predict the results (*increasing*, *decreasing*, or *staying the same*) of the following variables for a person with a hyperthyroid condition (overactive thyroid) and a person with a hypothyroid condition (underactive thyroid):

Variable	Change Predicted with Hyperthyroidism	Change Predicted with Hypothyroidism
Thyroid hormone level		
Metabolic rate		
Respiratory rate		
Body temperature		
Body weight		

Who would be more likely to shiver when exposed to decreases in environmental temperature—the person with hyperthyroidism or the person with hypothyroidism? Explain. ______________________

____

____

PhILS Simulations

LESSON 21

Frog Heart Function:

Thermal and Chemical Effects

The purpose of this exercise is to demonstrate the effects of cold Ringer's solution, adrenaline, and acetylcholine on an exposed frog's heart. A frog's heart is similar to the human heart except that it has only one ventricle. A virtual recording instrument will be used to monitor the movement of the frog's heart by recording deflections of a line tracing that represents strength of a heart beat (stroke volume) and heart rate. You will be measuring amplitude of the line deflections and the time between deflections to determine heart beats per minute. Stroke volume and heart rate can be altered by the autonomic nervous system's two divisions: the sympathetic and parasympathetic divisions. The sympathetic division releases adrenaline (epinephrine) and noradrenaline (norepinephrine) to increase heart rate and stroke volume. The parasympathetic division releases acetylcholine to decrease heart rate. Both divisions and their secretions help maintain blood pressure.

Before you begin, familiarize yourself with the following concepts (use your textbook as a reference):

- Cardiac conduction system
- Cardiac muscle contraction
- Action potential
- Stroke volume

State your hypothesis regarding the changes you expect to see in the frog's heart after cold Ringer's solution, adrenaline, and acetylcholine are administered.

____

____

____

____

1. Open Lesson 21: Frog Heart Function: Thermal and Chemical Effects.
2. Objectives. Click "Next."
3. Read the Introduction. Click the terms that contain links to illustrations or animations as you read. Click "Next."
4. Take the Pre-Lab Quiz. Click "Next."
5. Figure P21.1 shows the opening screen for the laboratory exercise for this lesson. Follow the instructions step-by-step in the Laboratory Exercise and answer the following questions as you proceed.
6. Figure P21.2 shows an example of a recording of contractions from an exposed frog heart. Transfer the recorded data from your Ph.I.L.S. Journal for regular, cold, adrenaline, and acetylcholine Ringer's solution (saline) effects on amplitude, interval, frequency, and cardiac output onto table P21.1.

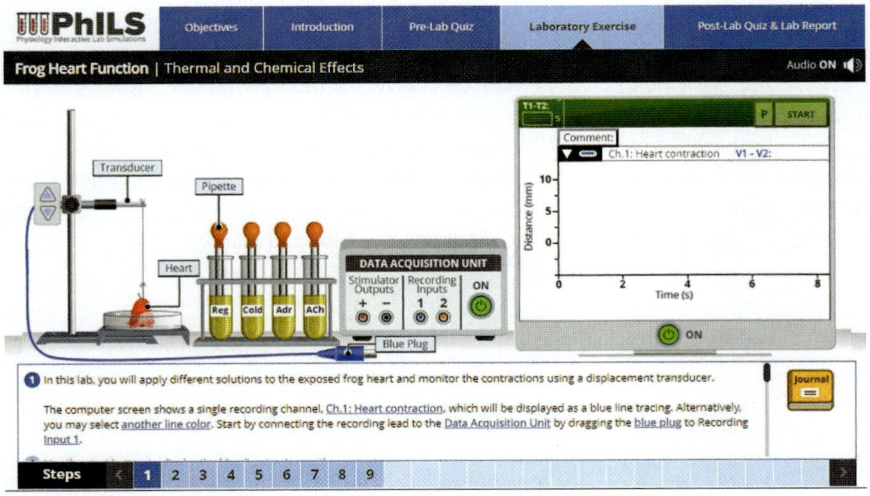

FIGURE P21.1 Opening screen for the laboratory exercise on Frog Heart Function: Thermal and Chemical Effects.
McGraw Hill

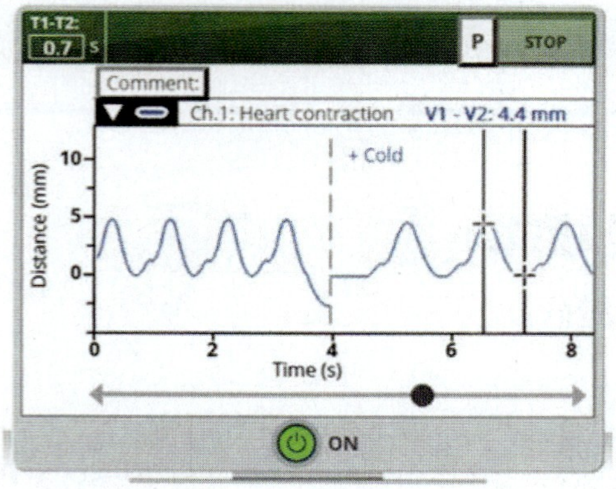

FIGURE P21.2 An example of a recording of contractions (upper trace) from an exposed frog heart. The response in normal Ringer's solution is shown to the left of the vertical dashed line, and the response after the addition of cold saline is shown to the right of the vertical dashed line.
McGraw Hill

7. Make note of any pertinent observations here:

____

____

8. As the temperature decreases, the frequency (heart rate) ____________ (*increases* or *decreases*) and cardiac output ____________ (*increases* or *decreases*). As a result, temperature is ____________ (*directly* or *inversely*) related to heart rate and cardiac output.

TABLE P21.1 Journal Data for Ph.I.L.S. Lesson 21: Heart Function: Thermal and Chemical Effects

Saline	Amplitude (mm)	Interval (seconds)	Frequency (bpm)	Card. Out. (amp x freq)
Regular:				
Cold:				
Adr:				
ACh:				

9. If you added *warm* Ringer's solution, what prediction would you make about heart rate and cardiac output?

____

10. With the addition of adrenaline (epinephrine) to the heart, amplitude ____________ (*increases* or *decreases*), frequency ____________ (*increases* or *decreases*), and cardiac output ____________ (*increases* or *decreases*).
11. With the addition of acetylcholine to the heart, frequency ____________ (*increases* or *decreases*) and cardiac output ____________ (*increases* or *decreases*). Is there a significant change in amplitude? ____________
12. Click the Post-Lab Quiz & Lab Report in the upper menu. Complete the Post-Lab Quiz.
13. Read the Lab Summary. Then, click "Lab Results."
14. Print your Lab Report by clicking "Print Lab" or "Save PDF" to turn in to your instructor.

CRITICAL THINKING ASSESSMENT—Ph.I.L.S. LESSON 21

Body Temperature and Cardiac Output: If a patient enters the hospital with hypothermia (low body temperature), you would predict what change (*increase* or *decrease*) in heart rate? ____________________ In cardiac output? ____________________ If a child has the flu and is "running a fever," you would predict what change (*increase* or *decrease*) in heart rate? ____________________ In cardiac output? ____________________

Autonomic Nervous System and Cardiac Output: When you are excited, exercising, or experiencing an emergency, the sympathetic nervous system is activated. If stimulation of the sympathetic nervous system releases adrenaline (epinephrine), what effect (*increase* or *decrease*) will this have on the amount of blood pumped from the heart per stroke (stroke volume)? ____________________ On heart rate? ____________________ On cardiac output? ____________________ Is there a significant change in blood pressure? ____________________ When you are resting, the parasympathetic nervous system is activated. If stimulation of the parasympathetic nervous system releases acetylcholine, what effect (*increase* or *decrease*) does this have on heart rate? ____________________ On cardiac output? ____________________ Is there a significant change in blood pressure? ____________________

LESSON 26

ECG and Heart Function

The Meaning of Heart Sounds

The purpose of this exercise is to understand the meaning of heart sounds and their relation to the electrical activity of the heart. Heart sounds are very diagnostic of cardiac function. The sounds are produced when valves close following the initiation of ventricular systole and the beginning of ventricular diastole. The first heart sound (lubb) occurs with the closure of the atrioventricular (AV) valves, and the second sound (dupp) occurs with the closure of the semilunar valves. The closure of the valves makes it possible for the blood to flow in only one direction in response to the contraction and relaxation of the ventricles. The electrical activity, measured by an electrocardiogram (ECG), of the heart stimulates the mechanical activity; therefore, the ECG directly relates to mechanical events of the cardiac cycle.

In this simulation, you will demonstrate how the sounds made by the heart are matched to the waveforms of the ECG. Table 45.1 (in lab exercise 45) provides the physiological significance of each wave, segment, and interval of the ECG.

Before you begin, familiarize yourself with the following concepts (use your textbook as a reference):

- Atrioventricular valve function and location(s)
- Semilunar valve function and location(s)
- P, QRS, and T waves of the ECG
- Electrical stimulation of mechanical activity in the heart

State your hypothesis regarding the relationship between heart sounds and the waveforms of a typical electrocardiogram.

____

____

____

____

1. Open Lesson 26: ECG and Heart Function: The Meaning of Heart Sounds.
2. Read Objectives. Click "Next."
3. Read the Introduction. Click the terms that contain links to illustrations or animations as you read. Click "Next."
4. Take the Pre-Lab Quiz. Click "Next."
5. Figure P26.1 shows the opening screen for the laboratory exercise for this lesson. Follow the instructions step-by-step in the Laboratory Exercise and answer the following questions as you proceed. (*Hint:* Remember, there are two sounds, but they will be very

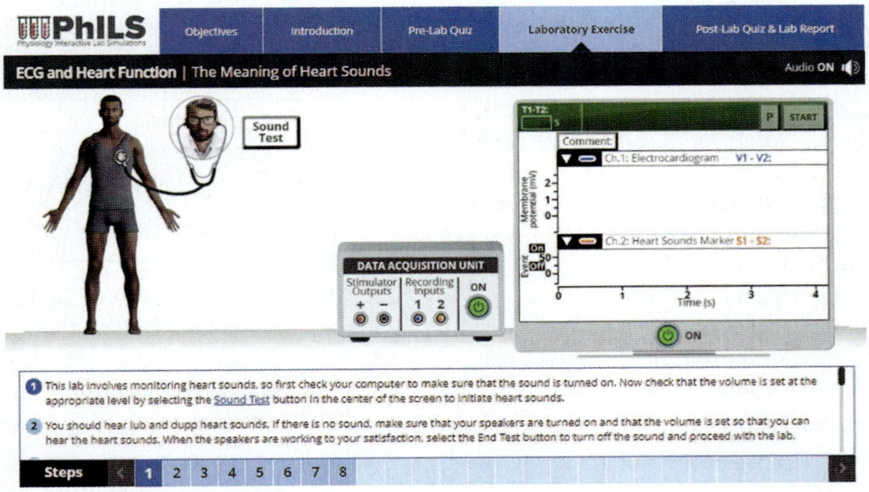

FIGURE P26.1 Opening screen for the laboratory exercise on ECG and Heart Function: The Meaning of Heart Sounds.
McGraw Hill

close together. When you depress the button to record the heart sounds, make sure to record both sounds.)

6. Print a copy or print screen the final line tracing graph you recorded from the virtual computer screen.

7. Using the line tracing graph, diagram the ECG recording from the lab simulation using the graph paper provided. Label the waveforms of the ECG and indicate the heart sounds.

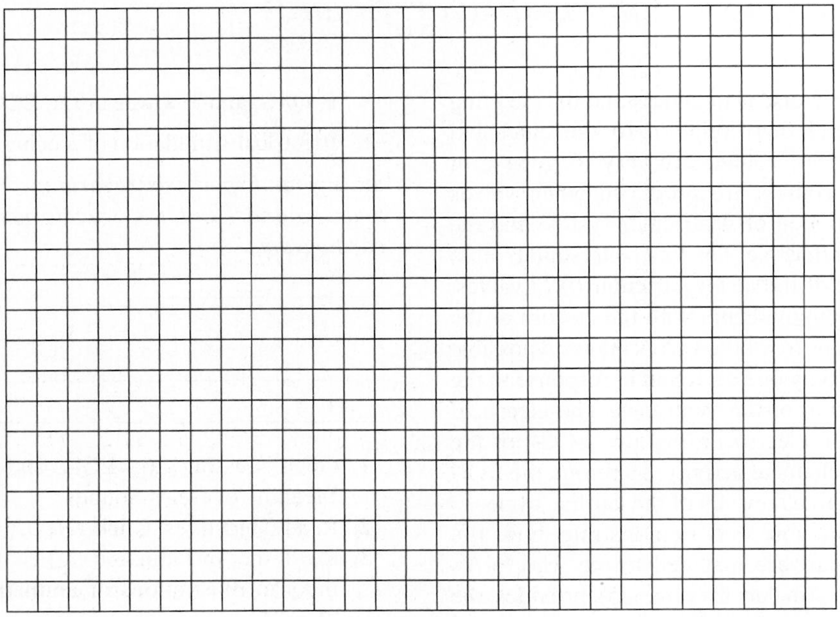

8. Make note of any pertinent observations here:

____

____

9. Answer the following questions regarding the line tracing graph you recorded and diagrammed:

a. The QRS wave is correlated with the ____________ ______________________________ (*first heart sound "lubb"* or *second heart sound "dupp"*).

b. The T wave is correlated with the ____________ ______________________________ (*first heart sound "lubb"* or *second heart sound "dupp"*).

10. Complete the following:

a. Atrial depolarization, causing the ____________ wave of the ECG, triggers contraction of the atria, and blood is pumped from the atria into the ventricles. The position of the valves (AV valves open and semilunar valves closed) remains unchanged.

b. Ventricular depolarization, causing the ____________ wave of the ECG, triggers contraction of the ventricles. The ____________ (*semilunar* or *AV*) valves close, producing the ____________ (*first* or *second*) heart sound. When pressure of blood in the ventricles exceeds pressure in the arteries (aorta and pulmonary trunk), the (*semilunar* or *AV*) valves open and blood is ejected.

c. Ventricular repolarization, causing the ____________ wave of the ECG, occurs as the ventricles are relaxing. As pressure of blood in the ventricles decreases, and to prevent the backflow of blood from the arteries into the ventricles, the ____________ (*semilunar* or *AV*) valves close, producing the ____________ (*first* or *second*) heart sound.

d. At what point do the AV valves reopen? ____________

____

____

11. Click the Post-Lab Quiz & Lab Report in the upper menu. Complete the Post-Lab Quiz.

12. Read the Lab Summary. Then, click "Lab Results."

13. Print your Lab Report by clicking "Print Lab" or "Save PDF" to turn in to your instructor.

LESSON 34

Blood

pH and Hb-Oxygen Binding

The purpose of this exercise is to understand the correlation between pH of the blood and the ability of oxygen to bind to hemoglobin. Transported throughout the body by red blood cells (RBCs), oxygen is essential to all tissues for the production of ATP. The RBCs contain the protein hemoglobin (Hb) that binds oxygen for transport to the tissues. Factors such as partial pressure, temperature, pH, and levels of carbon dioxide can affect the ability of Hb to bind oxygen. In this exercise, you will analyze the effect of various pH conditions on Hb binding.

Hemoglobin samples were prepared for this lab by inducing hemolysis of the RBCs and centrifuging the samples to remove the membranes. Supernatants were then collected and diluted with phosphate buffers of pH 6.8, 7.4, and 8.0. Using a spectrophotometer, you will measure the transmittance of light (620 nm) through the samples. The more oxygen that is bound to Hb (reduced, oxyhemoglobin), the more light will be absorbed and the less light will pass through the sample. The less oxygen that is bound to Hb (oxidized, deoxyhemoglobin), the lighter in color the blood will be and the less light will be absorbed and the more light will pass through the sample.

Before you begin, familiarize yourself with the following concepts (use your textbook as a reference):

- Deoxyhemoglobin and oxyhemoglobin
- Oxygen dissociation curve
- pH effect on Hb binding

State your hypothesis regarding the effect of pH on hemoglobin's ability to bind oxygen.

____

____

____

____

1. Open Lesson 34: Blood: pH and Hb–Oxygen Binding.
2. Read Objectives. Click "Next."
3. Read the Introduction. Click the terms that contain links to illustrations or animations as you read. Click "Next."
4. Take the Pre-Lab Quiz. Click "Next."
5. Figure P34.1 shows the opening screen for the laboratory exercise for this lesson. Follow the instructions step-by-step in the Laboratory Exercise and answer the following questions as you proceed.
6. Transfer the recorded data from your Ph.I.L.S. Journal of the three different pH solutions and the PO_2 on table P34.1.
7. Use the recorded data to complete the final graph of the partial pressure of oxygen and the percent transmittance of light (color) using the graph paper. Use a different color pencil for each of the three pH solutions (6.8, 7.4, and 8.0) graph points and lines. (Compare your graph to fig. P34.2). Then, complete the questions that follow.

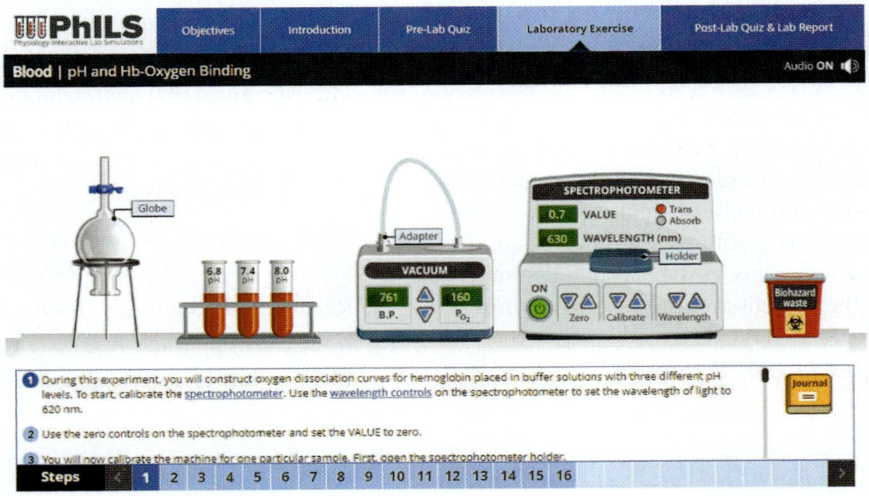

FIGURE P34.1 Opening screen for the laboratory exercise on Blood: pH and Hb-Oxygen Binding. McGraw Hill

TABLE P34.1 Journal Data of Ph.I.L.S. Lesson 34: Blood: pH and Hb-Oxygen Binding

PO_2	pH 6.8	pH 7.4	pH 8.0
160			
140			
120			
100			
80			
60			
40			
20			
0			

Final Graph

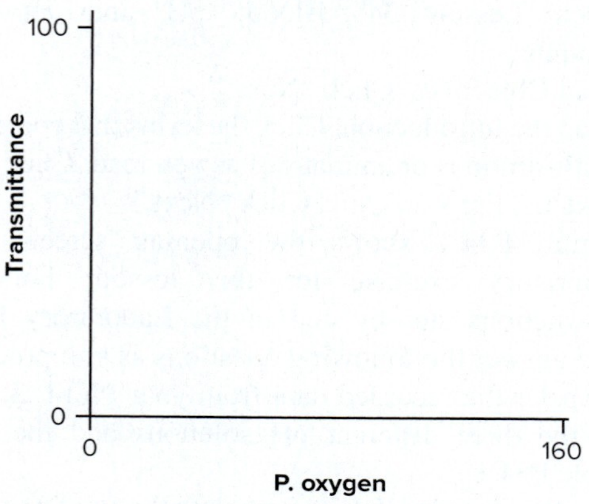

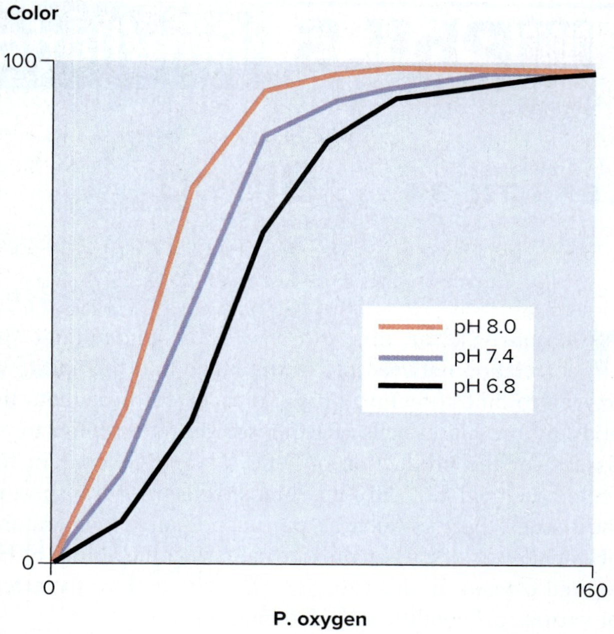

FIGURE P34.2 Example: Graph to show the effect of pH on the color (transmittance) of sheep hemoglobin set at different partial pressures of oxygen. Lower color values indicate that less oxygen is bound to hemoglobin.

8. Make note of any pertinent observations here:

____

____

____

9. Examine the graph you produced. As the partial pressure of oxygen increases, the percent transmittance of light ______________________.

10. There is a direct relationship between percent transmittance of light and percent saturation of hemoglobin. So, as the percent transmittance of light increases, the percent saturation of hemoglobin ______________ (*increases* or *decreases*). Thus, as the partial pressure of oxygen increases, the percent saturation of hemoglobin ______________________ (*increases*, *decreases*, or *stays the same*).

11. Compare the effect of pH on percent transmittance of light by examining the three dissociation curves produced at a pH of 6.8, 7.4, and 8.0 and answer the following questions:

a. When comparing the curve produced at 6.8 to the curve produced at 7.4, the curve produced when the pH is 6.8 would be described as a ______________ (*"shift right"* or *"shift left"*) and at any given partial pressure of oxygen, the percent transmittance of light is ______________ (*higher* or *lower*) and thus, the percent saturation of hemoglobin would be ______________ (*higher* or *lower*).

b. At a given PO_2, hemoglobin has the highest affinity for oxygen (ability to bind oxygen) at a pH of ______________ and the lowest affinity for oxygen at a pH of ______________.

12. Click the Post-Lab Quiz & Lab Report in the upper menu. Complete the Post-Lab Quiz.

13. Read the Lab Summary. Then, click "Lab Results."

14. Print your Lab Report by clicking "Print Lab" or "Save PDF" to turn in to your instructor.

CRITICAL THINKING ASSESSMENT—Ph.I.L.S. LESSON 34

Hydrogen ion (H^+) levels increase within the body from increased production of lactic acid (during anaerobic cellular respiration) and from increased production of carbon dioxide ($CO_2 + H_2O \longleftrightarrow H^+ + HCO_3^-$). The increased amounts of H^+ bind with the amino acids that compose the protein of hemoglobin, slightly altering hemoglobin's shape. This change in shape interferes with the ability of iron within the hemoglobin molecule to bind the oxygen (decreased affinity or attraction). (This is referred to as the *Bohr effect.*)

If the tissues of the body are metabolically active, ______________ (*more* or *less*) H^+ are produced and the pH is ______________ (*decreased* or *increased*). As a result of the Bohr effect, when hemoglobin reaches the systemic capillaries, the hemoglobin will bind the oxygen with ______________ (*greater* or *lesser*) affinity; ______________ (*more* or *less*) oxygen is released to the cells; the saturation of hemoglobin will be ______________ (*greater* or *lesser*), and hemoglobin returns to the lungs with ______________ (*more* or *less*) oxygen than would occur at a normal pH.

LESSON 38

Respiration

Altering Airway Volume

The purpose of this exercise is to demonstrate how changing the *tidal volume* (the amount of air that goes in or out of the lungs during a breathing cycle) will affect the alveolar ventilation (the amount of air that enters the alveoli for gas exchange).

During respiration, air moves in and out of the lungs via the airways. At the end of each breathing cycle there is a certain volume of air left in the airways that never makes it to the alveoli, and the portion of the airways that holds this air (which is expired unchanged) is called the *anatomical dead space*. This volume of air cannot be used for gas exchange. The anatomical dead space is usually kept constant. Any change in tidal volume will cause a change in the amount of air that enters the alveoli.

In this laboratory experiment, you will measure the tidal volume of a virtual volunteer using a spirometer. You will measure the volunteer's tidal volume during normal breathing, and then a measurement will be taken while the volunteer breathes through a plastic tube. Breathing through the plastic tube will increase the volunteer's anatomical dead space.

Before you begin, familiarize yourself with the following concepts (use your textbook as a reference):

- Lung volumes/capacities
- Function of a spirometer
- Breathing mechanism

State your hypothesis concerning what would happen to the anatomical dead space if it were increased. Include effects on tidal volume and the amount of air available for gas exchange.

____

____

____

____

1. Open Lesson 38: Respiration: Altering Airway Volume.
2. Read Objectives. Click "Next."
3. Read the Introduction. Click the terms that contain links to illustrations or animations as you read. Click "Next."
4. Take the Pre-Lab Quiz. Click "Next."
5. Figure P38.1 shows the opening screen for the laboratory exercise for this lesson. Follow the instructions

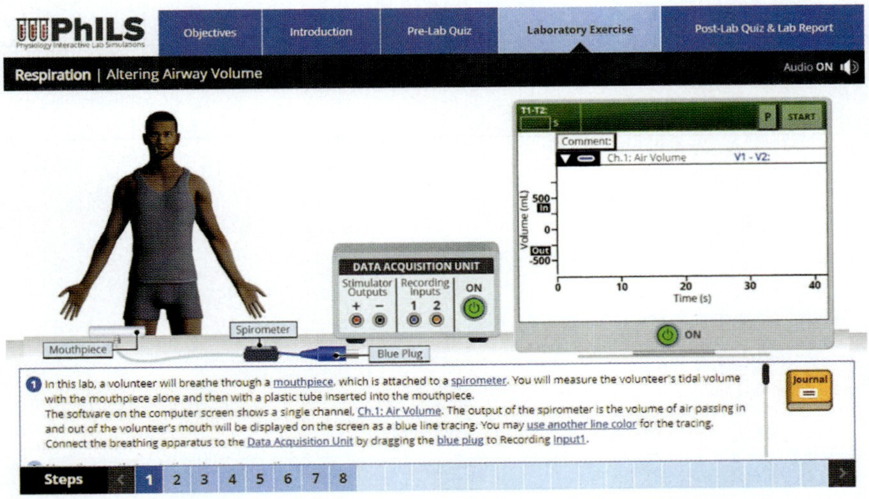

FIGURE P38.1 Opening screen for the laboratory exercise on Respiration: Altering Airway Volume. McGraw Hill

TABLE P38.1 Journal Data of Ph.I.L.S. Lesson 38: Respiration: Altering Airway Volume

Trial	Without Tube (mL)	With Tube (mL)
Trial # 1		
Trial # 2		
Trial # 3		
Average (mean)		

TABLE P38.2 Effect of Breathing with a Plastic Tube on Certain Volume of Air Variables.

Volume of Air	Change (increase, decrease, or remain the same)
Anatomic dead space	
Tidal volume	
Alveolar ventilation	

step-by-step in the Laboratory Exercise and answer the following questions as you proceed.

6. Transfer the recorded data from your Ph.I.L.S. journal onto table 38.1 to compare the tidal volumes of the breathing apparatus alone and the breathing apparatus with a plastic tube.

7. Make note of any pertinent observations here:

____

____

____

8. Which setup had the larger mean tidal volume?

____

9. To maintain normal alveolar ventilation, if the anatomic dead space is increased, the tidal volume will ________________ (increase, decrease, or remain the same).

10. In the setup with the tube attached to the breathing apparatus, did additional air reach the alveoli? ________________

11. Complete table 38.2 for the effect on each variable if a person breathes with the plastic tube.

12. Click the Post-Lab Quiz & Lab Report in the upper menu. Complete the Post-Lab Quiz.

13. Read the Lab Summary. Then, click "Lab Results."

14. Print your Lab Report by clicking "Print Lab" or "Save PDF" to turn in to your instructor.

CRITICAL THINKING ASSESSMENT—Ph.I.L.S. LESSON 38

If the bronchioles are dilated (when the sympathetic nervous system is activated, for example), the anatomic dead space ________________________ (increase, decrease, or remain the same), and tidal volume will ________________________ (increase, decrease, or remain the same). In patients with emphysema, the anatomic dead space increases (because there is a loss of elasticity of lung tissue and the thoracic cage cannot effectively be "pulled back in"). Predict the changes in breathing of a patient with emphysema as they try to compensate for the increase in anatomic dead space. ____

____

____

LESSON 40

Respiration

Deep Breathing and Cardiac Function

The purpose of this exercise is to observe how heart rate varies during the breathing cycle. You will compare how the volume of air changes in the thoracic cavity during slow, deep breathing with changes in stroke volume and heart rate. In a normal breathing cycle, inhalation increases the volume of the thoracic cavity and decreases pressure. Exhalation decreases the volume of the thoracic cavity and increases pressure.

The heart provides sufficient pressure to push blood through the arteries and capillaries to the veins of the body. As the blood travels through the circulatory system there is a gradual decline in pressure that results in low blood pressure in the veins, especially the veins in the lower extremities. The body corrects this problem of low blood pressure by using a couple of mechanisms. One mechanism is the contraction of leg muscles, which compresses the veins and pushes the blood up toward the hips. A second mechanism takes effect when the inspiratory muscles [sternocleidomastoid, external intercostals, pectoralis minor, internal intercostals (intercartilaginous part), diaphragm] are contracted, producing a negative pressure inside the thoracic cavity. This negative pressure helps draw venous blood into the heart from the lower extremities. This also causes an increase in heart rate. When the expiratory muscles [internal intercostals (interosseous part), rectus abdominis, external oblique] are contracted, positive pressure is created inside the thoracic cavity. This decreases the venous blood flow to the heart and decreases the heart rate.

In this laboratory experiment, you will measure the air volume breathed in and out and the pulse and heart rate of a virtual volunteer using a breathing apparatus and a finger pulse unit. You will observe two complete breathing cycles to acquire your data with which to measure thoracic cavity volume, heart rate, and stroke volume changes.

Before you begin, familiarize yourself with the following concepts (use your textbook as a reference):

- Blood pressure
- Stroke volume
- Pulse rate
- The breathing cycle

State your hypothesis regarding the relationship between the breathing cycle and heart rate, including changes in thoracic cavity volume, stroke volume, and heart rate.

____

____

____

____

1. Open Lesson 40: Respiration: Deep Breathing and Cardiac Function.
2. Objectives. Click "Next."
3. Read the Introduction. Click the terms that contain links to illustrations or animations as you read. Click "Next."
4. Take the Pre-Lab Quiz. Click "Next."
5. Figure P40.1 shows the opening screen for the laboratory exercise for this lesson. Follow the instructions step-by-step in the Laboratory Exercise and answer the following questions as you proceed.
6. Print a copy or print screen the final graph you recorded from the virtual computer screen (fig. P40.2).
7. Make note of any pertinent observations here:

____

____

____

8. During inspiration, pressure inside the thoracic cavity ______________________ (increases, decreases or stays the same) and pressure inside the alveoli (intrapulmonary pressure) ______________________ (increases, decreases or stays the same). As a result, atmospheric pressure is ______________________ (higher, lower or the same) than intrapulmonary pressure and air enters the lungs.
9. The pressure in the thoracic cage during inspiration creates a gradient that allows blood to move into the ______________________ (thoracic or abdominal) veins from the ______________________ (thoracic or abdominal) veins.

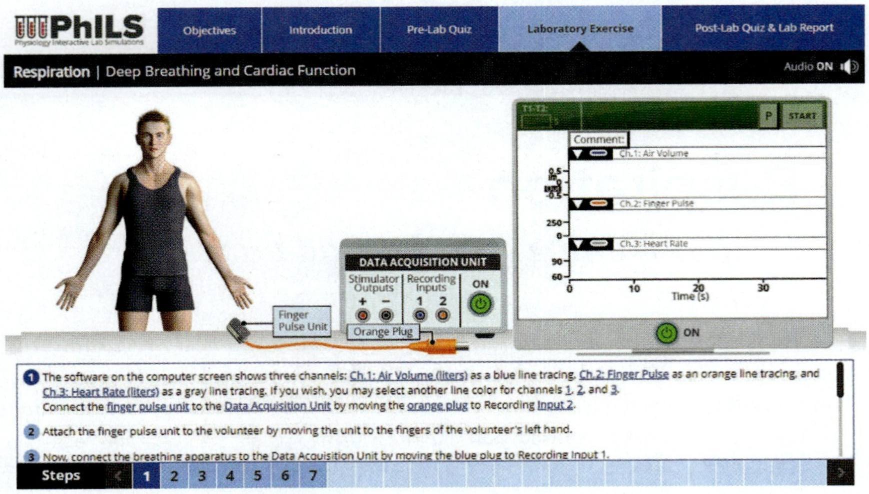

FIGURE P40.1 Opening screen for the laboratory exercise on Respiration: Deep Breathing and Cardiac Function. McGraw Hill

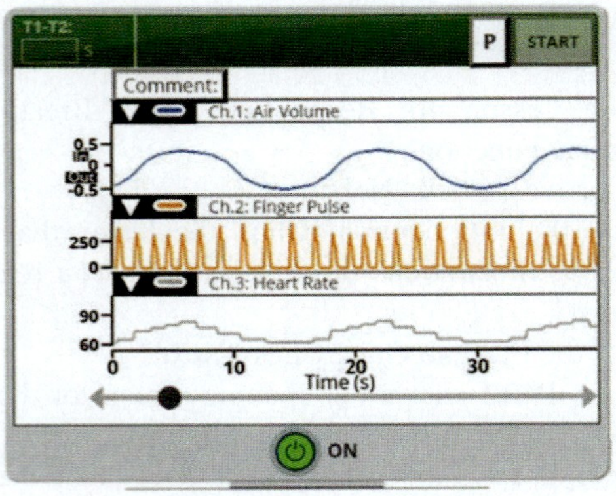

FIGURE P40.2 An example of a recording to show the relationship between slow breathing (upper trace) and cardiac function; finger pulse is on the middle tracing, and heart rate is on the lower. McGraw Hill

10. During inspiration, the breathing curve is a(n) ________ (*upward* or *downward*) deflection, and during expiration the breathing curve is a(n) ____________________ (*upward* or *downward*) deflection.

11. The heart rate is determined by the ________________ (*frequency* or *amplitude*) of the finger pulses.

12. The stroke volume is determined by the ____________ (*frequency* or *amplitude*) of the finger pulses.

13. During inspiration:

   a. The heart rate ____________________________ (*increases*, *decreases*, or *stays the same*).

   b. The stroke volume __________________________ (*increases*, *decreases*, or *stays the same*).

14. During expiration, pressure inside the thoracic cavity ________________________ (*increases*, *decreases*, or *stays the same*) and pressure inside the alveoli (intrapulmonary pressure) ____________________ (*increases*, *decreases*, or *stays the same*). As a result, atmospheric pressure is ______________________ (*higher*, *lower*, or *the same*) than intrapulmonary pressure and air exits the lungs.

15. The intrapulmonary pressure during expiration ________________________ (*increases*, *decreases*, or *stays the same*) the blood flow into the thoracic veins.

16. During expiration:

   a. The heart rate ____________________________ (*increases*, *decreases*, or *stays the same*).

   b. The stroke volume __________________________ (*increases*, *decreases*, or *stays the same*).

17. Cardiac output (CO) is determined by heart rate (HR) and stroke volume (SV): HR x SV = CO. To maintain a consistent cardiac output at rest, as heart rate increases, stroke volume ____________________ (*increases* or *decreases*), and as heart rate decreases, stroke volume ______________________(*increases* or *decreases*). Did you observe this relationship on the graph? ____________________ Is there a *direct* or *inverse* relationship between heart rate and stroke volume when maintaining cardiac output at a set value? ____________________

18. Click the Post-Lab Quiz & Lab Report in the upper menu. Complete the Post-Lab Quiz.

19. Read the Lab Summary. Then, click "Lab Results."

20. Print your Lab Report by clicking "Print Lab" or "Save PDF" to turn in to your instructor.

CRITICAL THINKING ASSESSMENT—Ph.I.L.S. LESSON 40

Explain the relationship between inspiration, pressure changes in the thoracic cavity, venous return (movement of blood in veins), heart rate, stroke volume, and cardiac output at rest.

INDEX

Note: Page number followed by *f* and *t* indicates figure and table respectively. Page numbers for entries from Appendix start with alpha A.

A

C

D

E

F

G

M

S

V

W

X

Y

Z